6e

Basic Mathematics
with *Early Integers*

Alan S. Tussy
Citrus College

Diane R. Koenig
Rock Valley College

⁂ CENGAGE

Australia • Brazil • Mexico • Singapore • United Kingdom • United States

To my daughters, Ashley, Brianna, and Carly, whom I love very much and who make me extremely proud.

—DRK

***Basic Mathematics with Early Integers*, Sixth Edition**

Alan S. Tussy, Diane R. Koenig

Product Manager: Frank Snyder

Content Developers: Samantha Gomez, Alison Duncan

Product Assistant: Abigail DeVeuve

Marketing Manager: Pamela Polk

Content Project Manager: Rebecca Charles

Manufacturing Planner: Becky Cross

IP Analyst: Ashley Maynard

IP Project Manager: Reba Frederics

Senior Art Director: Vernon Boes

Text and Cover Designer: Terri Wright

Cover Image: Kidsada Manchinda/ Alamy Stock Photo

Production Service/Compositor: Beth Asselin, SPi Global

For product information and technology assistance, contact us at
Cengage Customer & Sales Support, 1-800-354-9706.

For permission to use material from this text or product, submit all requests online at **www.cengage.com/permissions**. Further permissions questions can be e-mailed to **permissionrequest@cengage.com**.

Library of Congress Control Number: 2017959959

Student Edition ISBN: 978-1-337-61840-3

Loose-leaf Edition ISBN: 978-1-337-61833-5

Cengage
20 Channel Center Street
Boston, MA 02210
USA

Cengage is a leading provider of customized learning solutions with employees residing in nearly 40 different countries and sales in more than 125 countries around the world. Find your local representative at **www.cengage.com**.

Cengage products are represented in Canada by Nelson Education, Ltd.

To learn more about Cengage platforms and services, visit **www.cengage.com**.

Purchase any of our products at your local college store or at our preferred online store **www.cengagebrain.com**.

Printed in the United States of America
Print Number: 01   Print Year: 2018

CONTENTS

CHAPTER **3**

iStock.com/JoseGirarte

Fractions and Mixed Numbers   199

CHAPTER **4**

Phovoir/Shutterstock.com

Decimals   309

CHAPTER **5**

wong yu liang/Shutterstock.com

Ratio, Proportion, and Measurement   409

CHAPTER 9   An Introduction to Geometry   707

Dmitry Kalinovsky/Shutterstock.com

PREFACE

We are excited to present the Sixth Edition of *Basic Mathematics with Early Integers*, and are confident that the revision process has produced an even stronger experience for students and teachers. A new instructional feature, Look Alikes, has been added to the Study Sets. The objective of Look Alikes is to improve students' problem-recognition skills. Furthermore, all data in the real-life application problems have been updated, and some new, vocationally focused problems have been added.

To strengthen the online experience, the quantity and types of problems available in Enhanced Web Assign® have been expanded. The digital package that accompanies this text introduces a new customized learning tool, MindTap reader.

We want to thank all of you across the country who provided suggestions and input about the previous edition. Your insights have proven invaluable. Throughout the revision process, our fundamental belief has remained the same: Mathematics is a language in its own right. As always, the prime objective of this textbook (and its accompanying ancillaries) is to teach students how to read, write, speak, and think using the language of mathematics.

About the Authors

ALAN S. TUSSY

Alan Tussy grew up in West Covina, California, and in the early 1970's attended the University of Redlands where he majored in mathematics. After taking a fifth year of education courses and completing student teaching requirements, he was hired by the Arcadia Unified School District to teach junior high school math. He later transferred to Arcadia High School where he taught AP Calculus, served as department chair, and coached varsity football. During that time, Alan took evening classes at California State University, Los Angeles, and eventually earned a master's degree in applied mathematics. The master's degree led to a weeknight adjunct assignment at Citrus College in Glendora, California. He was soon hired full-time at Citrus, where he has taught mathematics for 25 years. He has taught up and down the community college curriculum from Arithmetic Fundamentals to Differential Equations, paying special attention to developmental math courses. Alan has written nine math books— a paperback series and a hardcover series. He is a creative and visionary teacher who maintains a keen focus on his students' greatest challenges. He is an extraordinary author dedicated to his students' success. His latest (and very rewarding) endeavor is mentoring and counseling a large number of adjunct math instructors at his school.

DIANE R. KOENIG

A nationally recognized educator and author, Diane Koenig actively shaped several textbooks, ancillaries, and series. Since 1982 when she helped develop the Gustafson/ Frisk series to her work on the Tussy/Koenig/Gustafson series, Diane's writing continues to reflect the expertise she gains from working with students in her Mathematics courses. Throughout her work, she integrates research-based strategies in Mathematics education. She earned a Bachelor of Science degree in Secondary Math Education from Illinois State University in 1980, and began her career at Rock Valley College in 1981, when she became the Math Supervisor for a newly formed Personalized Learning Center.

Earning her Master's Degree in Applied Mathematics from Northern Illinois University in 1984, Diane enjoys the distinction of being the first woman to become a full-time faculty member in the Mathematics department for Rock Valley College. In addition to being awarded AMATYC's Excellence in Teaching Award in 2015, she was chosen as the Rock Valley College Faculty of the Year by her peers in 2005, and the next year she was awarded the NISOD Teaching Excellence Award and the Illinois Mathematics Association of Community Colleges Award for Teaching Excellence. In addition to her teaching, she has been an active member of the Illinois Mathematics Association of Community Colleges (IMACC), serving on the board of directors, on a state-level task force rewriting the course outlines for the Developmental Mathematics courses, and as the Association's newsletter editor.

New to This Edition

EXAMPLE 1

The authors present a comprehensive revision to a classic text that updates the real-world data appearing in worked examples and Study Sets.

APPLICATIONS

New application problems highlighting a variety of occupations and vocations have been added to the Study Sets. These problems provide students with practical context for the mathematical topics they are studying.

LOOK ALIKES

A new *Look Alike* feature builds student problem-recognition skills. *Look Alike* problems require students to distinguish between problems that at first glance appear similar, but in actuality call for different strategies to solve them.

THE LANGUAGE OF ALGEBRA

Success Tip
Caution!

Additional displays, diagrams, and explanations have been added to assist students who are visual learners. A new page design places the *Language of Algebra, Success Tip*, and *Caution* features in the margin for easier reading.

Ancillaries

For the Student	For the Instructor
Online Student Solutions Manual (ISBN: 978-1-337-61582-2) The Online Student Solutions Manual provides worked-out solutions to all of the odd-numbered exercises in the text.	**Online Complete Solutions Manual** (ISBN: 978-1-337-61583-9) The Online Complete Solutions Manual provides worked-out solutions to all of the problems in the text.
	Instructor's Companion Website Everything you need for your course in one place! Access and download a helpful Instructor Manual that paces the chapters, provides a how-to approach and additional opportunities for in-class practice and homework. In addition, you can find the online appendix, PowerPoint presentations, and more on the companion site. This collection of book-specific lecture and class tools is available online via www.cengage.com/login.
⚡ **WEBASSIGN** From Cengage www.webassign.com/cengage (Printed Access Card ISBN: 978-1-337-61831-1, Online Access Code ISBN: 978-1-337-61589-1) Prepare for class with confidence using WebAssign from Cengage for Tussy, *Basic Mathematics with Early Integers,* 6e. This online learning platform, which includes an interactive ebook, fuels practice, so you truly absorb what you learn—and are better prepared come test time. Videos and tutorials walk you through concepts and deliver instant feedback and grading, so you always know where you stand in class. Focus your study time and get extra practice where you need it most. Study smarter with WebAssign! Ask your instructor today how you can get access to WebAssign, or learn about self-study options at www.webassign.com.	⚡ **WEBASSIGN** From Cengage www.webassign.com/cengage (Printed Access Card ISBN: 978-1-337-61831-1, Online Access Code ISBN: 978-1-337-61830-4) WebAssign from Cengage for *Basic Mathematics with Early Integers,* 6e is a fully customizable online solution, including an interactive ebook, for STEM disciplines that empowers you to help your students learn, not just do homework. Insightful tools save you time and highlight exactly where your students are struggling. Decide when and what type of help students can access while working on assignments—and incentivize independent work so help features aren't abused. Meanwhile, your students get an engaging experience, instant feedback, and better outcomes. A total win-win! To try a sample assignment, learn about LMS integration, or connect with our digital course support visit www.webassign.com/cengage.

Acknowledgments

Authoring a textbook is a tremendous undertaking. A revision of this scale would not have been possible without the thoughtful feedback and support from our fellow colleagues. Your contributions to this edition have shaped this revision in countless ways.

We would also like to express our thanks to the editorial, marketing, and production staff of Cengage—Frank Snyder, Rebecca Charles, Samantha Gomez, Alison Duncan, Pamela Polk, and Abigail DeVeuve—for helping us to craft this new edition. Thanks also to Vernon Boes for his work on the design and art program. In addition, our gratitude goes to Beth Asselin and the entire SPi Global team for their copyediting and proofreading expertise.

We want to express our gratitude to those who helped with this project: Brenda Keller, Rhoda Oden, Steve Odrich, Mary Lou Wogan, Paul McCombs, Maria H. Andersen, Sheila Pisa, Laurie McManus, Alexander Lee, Ed Kavanaugh, Karl Hunsicker, Cathy Gong, Dave Ryba, Terry Damron, Marion Hammond, Lin Humphrey, Doug Keebaugh, Robin Carter, Tanja Rinkel, Jeff Cleveland, Jo Morrison, Sheila White, Jim McClain, Paul Swatzel, Matt Stevenson, Carole Carney, Joyce Low, Rob Everest, David Casey, Heddy Paek, Ralph Tippins, Mo Trad, Eagle Zhuang, Chris Scott, Victoria Dominguez, Esme Medrano, Sam M. Ditzion, Lisa Brown, Elaine Tucker, Laura McInerney, Solomon Willis, Tracy Nehnevaji, Philomena Sefranek, and the Citrus College library staff (including Barbara Rugeley) for their help with this project. Your encouragement, suggestions, and insight have been invaluable to us.

Alan S. Tussy
Diane R. Koenig

1 Whole Numbers

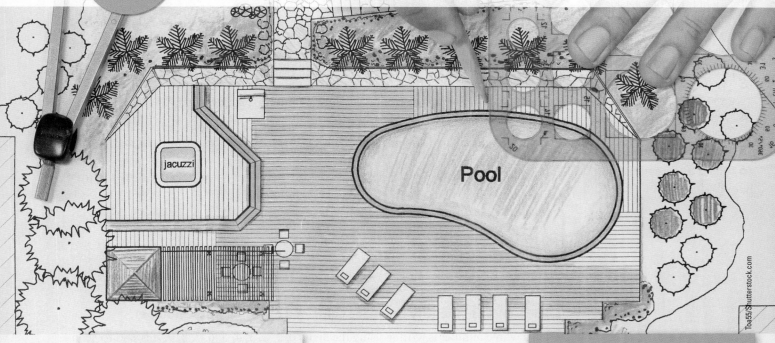

Toa55/Shutterstock.com

from **Campus to Careers**

Landscape Designer

Landscape designers make outdoor places more beautiful and useful. They work on all types of projects. Some focus on yards and parks, others on land around buildings and highways. The training of a landscape designer should include botany classes to learn about plants; art classes to learn about color, line, and form; and mathematics classes to learn how to take measurements and keep business records.

In **Problem 108** of **Study Set 1.5**, you will see how a landscape designer uses division to determine the number of pine trees that are needed to form a windscreen for a flower garden. In **Problem 57** on **Study Set 1.6**, you will see how a landscape designer uses addition and multiplication of whole numbers to calculate the cost of landscaping a yard. And in **Problem 116** of **Study Set 1.9**, the rule for the order of operations is used to calculate the available planting area.

JOB TITLE:
Landscape designer

EDUCATION:
A bachelor's degree in landscape design. Most states require a license.

JOB OUTLOOK:
Expected to grow 5% from 2014 to 2024

ANNUAL EARNINGS:
The average annual salary is $68,600.

FOR MORE INFORMATION
www.thelandlovers.org/
career_LandscapeDesign.asp

OBJECTIVES

1 Identify the place value of a digit in a whole number.

2 Write whole numbers in words and in standard form.

3 Write a whole number in expanded form.

4 Compare whole numbers using inequality symbols.

5 Round whole numbers.

6 Read tables and graphs involving whole numbers.

SECTION 1.1   An Introduction to the Whole Numbers

The **whole numbers** are 0, 1, 2, 3, 4, 5, 6, 7, 8, 9, 10, 11, 12, and so on. They are used to answer questions such as How many?, How fast?, and How far?

- Michael Phelps earned 23 Olympic Gold Medals during his swimming career.
- The average American adult reads at a rate of 250 to 300 words per minute.
- The driving distance from New York City to Los Angeles is 2,786 miles.

The *set of whole numbers* is written using **braces { }**, as shown below. The three dots indicate that the list continues forever—there is no largest whole number. The smallest whole number is 0.

The Set of Whole Numbers

{ 0, 1, 2, 3, 4, 5, 6, 7, 8, 9, 10, 11, 12, . . . }

OBJECTIVE 1 **Identify the place value of a digit in a whole number.**

When a whole number is written using the **digits** 0, 1, 2, 3, 4, 5, 6, 7, 8, 9, it is said to be in **standard form** (also called **standard notation**). The position of a digit in a whole number determines its **place value**. In the number 419, the 9 is in the *ones column*, the 1 is in the *tens column*, and the 4 is in the *hundreds column*.

Tens column
Hundreds column ⌐ ↓ ⌐Ones column
↓ ↓ ↓
4 1 9

To make large whole numbers easier to read, we use commas to separate their digits into groups of three, called **periods**. Each period has a name, such as *ones, thousands, millions, billions,* and *trillions.* The following **place-value chart** shows the place value of each digit in the number 3,302,677,258,000, which is read as:

Three trillion, three hundred two billion, six hundred seventy-seven million, two hundred fifty-eight thousand

In 2015, the federal government collected $3,302,677,258,000 in taxes.

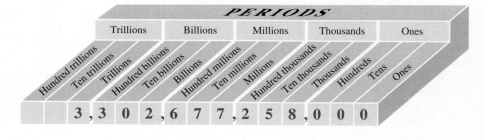

Each of the 2's in 3,302,677,258,000 has a different place value because of its position. The place value of the red 2 is 2 *billions*. The place value of the blue 2 is 2 *hundred thousands*.

LANGUAGE OF MATHEMATICS

As we move to the left in the chart, the place value of each column is 10 times the column directly to its right. This is why we call our number system the *base-10 number system*.

EXAMPLE 1   **Airports.**   Hartsfield-Jackson Atlanta International Airport is the busiest airport in the United States, handling 101,491,106 passengers in 2015. (Source: Airports Council International–North America)
a. What is the place value of the digit 4?
b. Which digit tells the number of millions?

Strategy We will begin in the ones column of 101,491,106. Then, moving to the left, we will name each column (ones, tens, hundreds, and so on) until we reach the digit 4.

WHY It's easier to remember the names of the columns if you begin with the smallest place value and move to the columns that have larger place values.

Solution

a. 101,491,106   Say, "Ones, tens, hundreds, thousands, ten thousands, hundred thousands" as you move from column to column.

    4 hundred thousands is the place value of the digit 4.

b. 101,491,106

    The digit 1 is in the millions column.

Self Check 1

Cell Phones.   In 2015, there were 377,921,241 cellular telephone subscriber connections in the United States. (Source: CTIA The Wireless Association)
a. What is the place value of the digit 3?
b. Which digit tells the number of hundred thousands?

Now Try ➲ **Problem 23**

LANGUAGE OF MATHEMATICS

Each of the worked examples in this textbook includes a *Strategy* and *Why* explanation. A **strategy** is a plan of action to follow to solve the given problem.

OBJECTIVE 2 **Write whole numbers in words and in standard form.**

Since we use whole numbers so often in our daily lives, it is important to be able to read and write them.

Reading and Writing Whole Numbers

To write a whole number in words, start from the left. Write the number in each period followed by the name of the period (except for the *ones period*, which is not used). Use commas to separate the periods.
    To read a whole number out loud, follow the same procedure. The commas are read as slight pauses.

LANGUAGE OF MATHEMATICS

The word **and** should not be said when reading a whole number. It should only be used when reading a mixed number such as 5½ (five *and* one-half) or a decimal such as 3.9 (three *and* nine-tenths).

EXAMPLE 2   Write each number in words:
a. 63   **b.** 499   **c.** 89,015   **d.** 6,070,534

Strategy For the larger numbers in parts c and d, we will name the periods from right to left to find the *greatest* period.

WHY To write a whole number in words, we must give the name of each period (except for the ones period). Finding the largest period helps to start the process.

Solution
a. 63 is written: *sixty-three*.   Use a hyphen to write whole numbers from 21 to 99 in words (except for 30, 40, 50, 60, 70, 80, and 90).

b. 499 is written: *four hundred ninety-nine*.

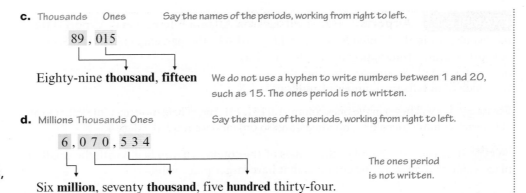

c. Thousands   Ones   Say the names of the periods, working from right to left.

$$89 , 015$$

Eighty-nine **thousand**, **fifteen**   We do not use a hyphen to write numbers between 1 and 20, such as 15. The ones period is not written.

d. Millions Thousands Ones   Say the names of the periods, working from right to left.

$$6 , 0\ 7\ 0 , 5\ 3\ 4$$

The ones period is not written.

Six **million**, seventy **thousand**, five **hundred** thirty-four.

Self Check 2

Write each number in words:
a. 42
b. 798
c. 97,053
d. 23,000,017

Now Try ⮕ Problems 31, 33, and 35

Caution! Two numbers, 40 and 90, are often misspelled: write forty (not fourty) and ninety (not ninty).

EXAMPLE 3   Write each number in standard form:
a. *Twelve thousand, four hundred seventy-two*
b. *Seven hundred one million, thirty-six thousand, six*
c. *Forty-three million, sixty-eight*

Strategy We will locate the commas in the written-word form of each number.

WHY When a whole number is written in words, commas are used to separate periods.

Solution
a. Twelve thousand , four hundred seventy-two

12,472

b. Seven hundred one million , thirty-six thousand , six

701,036,006

c. Forty-three million , sixty-eight   The written-word form does not mention the thousands period.

43,000,068   If a period is not named, three zeros hold its place.

Self Check 3

Write each number in standard form:
a. *Two hundred three thousand, fifty-two*
b. *Nine hundred forty-six million, four hundred sixteen thousand, twenty-two*
c. *Three million, five hundred seventy-nine*

Now Try ⮕ Problems 39 and 45

OBJECTIVE 3 Write a whole number in expanded form.

Success Tip Four-digit whole numbers are sometimes written without a comma. For example, we may write 3,911 or 3911 to represent three thousand, nine hundred eleven.

In the number 6,352, the digit 6 is in the thousands column, 3 is in the hundreds column, 5 is in the tens column, and 2 is in the ones (or units) column. The meaning of 6,352 becomes clear when we write it in **expanded form** (also called **expanded notation**).

$$6{,}352 = 6 \text{ thousands} + 3 \text{ hundreds} + 5 \text{ tens} + 2 \text{ ones}$$

or

$$6{,}352 = \quad 6{,}000 \quad + \quad 300 \quad + \quad 50 + 2$$

EXAMPLE 4   Write each number in expanded form:
a. 85,427   **b.** 1,251,609

Strategy Working from left to right, we will give the place value of each digit and combine them with + symbols.

WHY The term *expanded form* means to write the number as an addition of the place values of each of its digits.

Solution

a. The expanded form of 85,427 is:

8 ten thousands + **5** thousands + **4** hundreds + **2** tens + **7** ones

which can be written as:

80,000   +   5,000   +   400   +   20   +   7

b. The expanded form of 1,251,609 is:

1   **2** hundred   **5** ten   **1**   **6**   **0**   **9**
million + thousands + thousands + thousand + hundreds + tens + ones

Since 0 tens is zero, the expanded form can also be written as:

1   **2** hundred   **5** ten   **1**   **6**   **9**
million + thousands + thousands + thousand + hundreds + ones

which can be written as:

1,000,000 +   2,00,000   +   50,000   +   1,000   +   600   +   9

Self Check 4
Write 708,413 in expanded form.

Now Try ➲ **Problems 49, 53, and 57**

OBJECTIVE 4 Compare whole numbers using inequality symbols.

Whole numbers can be shown by drawing points on a **number line**. Like a ruler, a number line is straight and has uniform markings. To construct a number line, we begin on the left with a point on the line representing the number 0. This point is called the **origin**. We then move to the right, drawing equally spaced marks and labeling them with whole numbers that increase in value. The arrowhead at the right indicates that the number line continues forever.

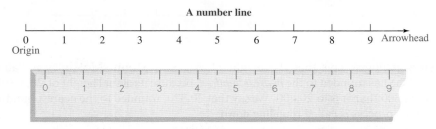

Using a process known as **graphing**, we can represent a single number or a set of numbers on a number line. **The graph of a number** is the point on the number line that corresponds to that number. *To graph a number* means to locate its position on the number line and highlight it with a heavy dot. The graphs of 5 and 8 are shown on the number line below.

As we move to the right on the number line, the numbers increase in value. Because 8 lies to the right of 5, we say that 8 is greater than 5. The **inequality symbol** > ("is greater than") can be used to write this fact:

8 > 5   *Read as "8 is greater than 5."*

Since 8 > 5, it is also true that 5 < 8. We read this as "5 is less than 8."

Success Tip To tell the difference between these two inequality symbols, remember that they always point to the smaller of the two numbers involved.

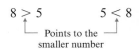

8 > 5   5 < 8

Points to the smaller number

> **Inequality Symbols**
>
> $>$ means *is greater than*
> $<$ means *is less than*

EXAMPLE 5   Place an $<$ or an $>$ symbol in the box to make a true statement: **a.** 3 ☐ 7   **b.** 18 ☐ 16

Strategy To pick the correct inequality symbol to place between a pair of numbers, we need to determine the position of each number on the number line.

WHY For any two numbers on a number line, the number to the *left* is the smaller number and the number to the *right* is the larger number.

Solution

a. Since 3 is to the left of 7 on the number line, we have $3 < 7$.

b. Since 18 is to the right of 16 on the number line, we have $18 > 16$.

OBJECTIVE 5   **Round whole numbers.**

When we don't need exact results, we often round numbers. For example, when a teacher with 36 students orders 40 textbooks, he has rounded the actual number to the *nearest ten*, because 36 is closer to 40 than it is to 30. We say 36, rounded to the nearest 10, is 40. This process is called **rounding up**.

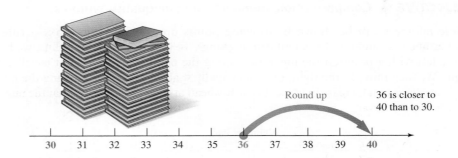

When a geologist says that the height of Alaska's Mount McKinley is "about 20,300 feet," she has rounded to the *nearest hundred*, because its actual height of 20,320 feet is closer to 20,300 than it is to 20,400. We say that 20,320, rounded to the nearest hundred, is 20,300. This process is called **rounding down**.

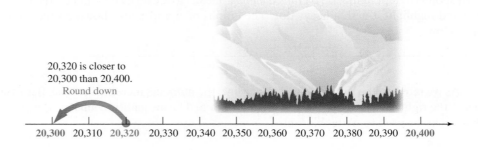

To round a whole number, we follow an established set of rules. To round a number to the nearest ten, for example, we locate the **rounding digit** in the tens column. If the **test digit** to the right of that column (the digit in the ones column) is 5 or greater, we *round up* by increasing the tens digit by 1 and replacing the test digit with 0. If the test digit is less than 5, we *round down* by leaving the tens digit unchanged and replacing the test digit with 0.

EXAMPLE 6   Round each number to the nearest ten: **a.** 3,761   **b.** 12,087

Strategy We will find the digit in the tens column and the digit in the ones column.

WHY To round to the nearest ten, the digit in the tens column is the rounding digit and the digit in the ones column is the test digit.

Solution

a. We find the rounding digit in the tens column, which is 6. Then we look at the test digit to the right of 6, which is the 1 in the ones column. Since $1 < 5$, we round down by leaving the 6 unchanged and replacing the test digit with 0.

   ┌Rounding digit: tens column   ┌Keep the rounding digit: Do not add 1.
  3,761   3,761
    └Test digit: 1 is less than 5.   └Replace with 0.

Thus, 3,761 rounded to the nearest ten is 3,760.

b. We find the rounding digit in the tens column, which is 8. Then we look at the test digit to the right of 8, which is the 7 in the ones column. Because 7 is 5 or greater, we round up by adding 1 to 8 and replacing the test digit with 0.

   ┌Rounding digit: tens column   ┌Add 1.
  12,087   12,087
    └Test digit: 7 is 5 or greater.   └Replace with 0.

Thus, 12,087 rounded to the nearest ten is 12,090.

Now Try ➔ Problem 63

A similar method is used to round numbers to the nearest hundred, the nearest thousand, the nearest ten thousand, and so on.

Rounding a Whole Number

1. To round a number to a certain place value, locate the **rounding digit** in that place.
2. Look at the **test digit**, which is directly to the right of the rounding digit.
3. If the test digit is 5 or greater, round up by adding 1 to the rounding digit and replace all of the digits to its right with 0.

 If the test digit is less than 5, replace it and all of the digits to its right with 0.

EXAMPLE 7   Round each number to the nearest hundred:
a. 18,349   **b.** 7,960

Strategy We will find the rounding digit in the hundreds column and the test digit in the tens column.

LANGUAGE OF MATHEMATICS

When we **round** a whole number, we are finding an approximation of the number. An approximation is close to, but not the same as, the exact value.

Self Check 6

Round each number to the nearest ten:
a. 35,642
b. 9,756

WHY   To round to the nearest hundred, the digit in the hundreds column is the rounding digit and the digit in the tens column is the test digit.

Solution

a. First, we find the rounding digit in the hundreds column, which is 3. Then we look at the test digit 4 to the right of 3 in the tens column. Because $4 < 5$, we round down and leave the 3 in the hundreds column. We then replace the two rightmost digits with 0's.

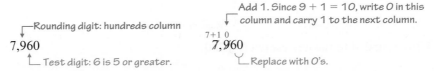

Thus, 18,349 rounded to the nearest hundred is 18,300.

b. First, we find the rounding digit in the hundreds column, which is 9. Then we look at the test digit 6 to the right of 9. Because 6 is 5 or greater, we round up and increase 9 in the hundreds column by 1. Since the 9 in the hundreds column represents 900, increasing 9 by 1 represents increasing 900 to 1,000. Thus, we replace the 9 with a 0 and add 1 to the 7 in the thousands column. Finally, we replace the two rightmost digits with 0's.

Add 1. Since $9 + 1 = 10$, write 0 in this column and carry 1 to the next column.

Rounding digit: hundreds column
7,960
Test digit: 6 is 5 or greater.

7+1 0
7,960
Replace with 0's.

Thus, 7,960 rounded to the nearest hundred is 8,000.

Caution! To round a number, use *only* the test digit directly to the right of the rounding digit to determine whether to round up or round down.

Self Check 7

Round 365,283 to the nearest hundred.

Now Try ➡ Problems 69 and 71

EXAMPLE 8   **U.S. cities.**   Anchorage is the largest city in Alaska. Round the 2017 population of Anchorage shown in the sign to:
a. the nearest thousand
b. the nearest ten thousand

Anchorage
CITY LIMIT
Pop. 299,037   Elev. 102
The city of lights and flowers.

Strategy   In each case, we will identify the rounding digit and the test digit.

WHY   We need to know the value of the test digit to determine whether we round the population up or down.

Solution

a. The rounding digit in the thousands column is 9. Since the test digit 0 is less than 5, we round down.

Rounding digit ┐ ┌ Test digit
299,037

To the nearest thousand, Anchorage's population in 2017 was 299,000.

b. The rounding digit in the ten thousands column is 9. Since the test digit 9 is 5 or greater, we round up by writing 0 in the ten thousands column and carrying a 1 to the hundred thousands column.

Rounding digit ┐ ┌ Test digit
299,037

To the nearest ten thousand, Anchorage's population in 2017 was 300,000.

Self Check 8

U.S. cities.   In 2015, San Antonio, Texas, with a population of 1,469,845, was the nation's 7th largest city. Round the population of San Antonio to:
a. the nearest thousand
b. the nearest million

Now Try ➡ Problems 75 and 79

OBJECTIVE 6 Read tables and graphs involving whole numbers.

The following table is an example of the use of whole numbers. It shows the number of women members of the U.S. House of Representatives for the years 2005 through 2015.

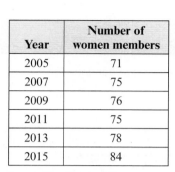

Year	Number of women members
2005	71
2007	75
2009	76
2011	75
2013	78
2015	84

Source: Data from http://www.cawp .rutgers.edu/women-us-congress-2015

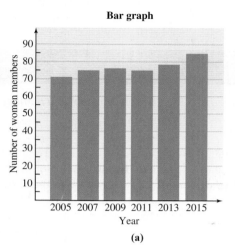

Bar graph

(a)

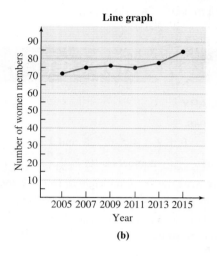

Line graph

(b)

In figure (a), the information in the table is presented in a **bar graph**. The *horizontal* scale is labeled "Year," and units of 2 years are used. The *vertical* scale is labeled "Number of women members," and units of 10 are used. The bar directly over each year extends to a height that shows the number of women members of the House of Representatives that year.

Another way to present the information in the table is with a **line graph**. Instead of using a bar to represent the number of women members, we use a dot drawn at the correct height. After drawing data points for 2005, 2007, 2009, 2011, 2013, and 2015, we connect the points to create the line graph in figure (b).

Think it Through ● RE-ENTRY STUDENTS

"A re-entry student is considered one who is the age of 25 or older, or those students that have had a break in their academic work for 5 years or more. Nationally, this group of students is growing at an astounding rate."
—Student Life and Leadership Department, University Union, Cal Poly University, San Luis Obispo

Some common concerns expressed by adult students considering returning to school are listed below in Column I. Match each concern to an encouraging reply in Column II.

Column I

1. I'm too old to learn.
2. I don't have the time.
3. I didn't do well in school the first time around. I don't think a college would accept me.
4. I'm afraid I won't fit in.
5. I don't have the money to pay for college.

Column II

a. Many students qualify for some type of financial aid.
b. Taking even a single class puts you one step closer to your educational goal.
c. There's no evidence that older students can't learn as well as younger ones.
d. More than 41% of the students in college are older than 25.
e. Typically, community colleges and career schools have an open admissions policy.

Source: Adapted from *Common Concerns for Adult Students*, Minnesota Higher Education Services Office

Answers to Self Checks

1. a. 3 hundred millions   **b.** 9   **2. a.** forty-two   **b.** seven hundred ninety-eight   **c.** ninety-seven thousand, fifty-three   **d.** twenty-three million, seventeen   **3. a.** 203,052   **b.** 946,416,022   **c.** 3,000,579
4. 700,000 + 8,000 + 400 + 10 + 3   **5. a.** >   **b.** <   **6. a.** 35,640   **b.** 9,760   **7.** 365,300
8. a. 1,470,000   **b.** 1,000,000

SECTION 1.1 STUDY SET

VOCABULARY

Fill in the blanks.

1. The numbers 0, 1, 2, 3, 4, 5, 6, 7, 8, and 9 are the _____.

2. The set of _____ numbers is {0, 1, 2, 3, 4, 5, ...}.

3. When we write five thousand eighty-nine as 5,089, we are writing the number in _____ form.

4. To make large whole numbers easier to read, we use commas to separate their digits into groups of three, called _____.

5. When 297 is written as 200 + 90 + 7, we are writing 297 in _____ form.

6. Using a process called *graphing*, we can represent whole numbers as points on a _____ line.

7. The symbols > and < are _____ symbols.

8. If we _____ 627 to the nearest ten, we get 630.

CONCEPTS

9. Copy the following place-value chart. Then enter the whole number 1,342,587,200,946 and fill in the place value names and the periods.

10. **a.** Insert commas in the proper positions for the following whole number written in standard form: 5467010

    b. Insert commas in the proper positions for the following whole number written in words:
    seventy-two million four hundred twelve thousand six hundred thirty-five

11. Write each number in words.
    a. 40   **b.** 90
    c. 68   **d.** 15

12. Write each number in standard form.
    a. 8 ten thousands + 1 thousand + 6 hundreds + 9 tens + 2 ones

    b. 900,000 + 60,000 + 5,000 + 300 + 40 + 7

Graph the following numbers on a number line.

13. 1, 3, 5, 7
    0 1 2 3 4 5 6 7 8 9 10

14. 0, 2, 4, 6, 8
    0 1 2 3 4 5 6 7 8 9 10

15. 2, 4, 5, 8
    0 1 2 3 4 5 6 7 8 9 10

16. 2, 3, 5, 7, 9
    0 1 2 3 4 5 6 7 8 9 10

17. the whole numbers less than 6
    0 1 2 3 4 5 6 7 8 9 10

18. the whole numbers less than 9
    0 1 2 3 4 5 6 7 8 9 10

19. the whole numbers between 2 and 8
    0 1 2 3 4 5 6 7 8 9 10

20. the whole numbers between 0 and 6
    0 1 2 3 4 5 6 7 8 9 10

NOTATION

Fill in the blanks.

21. The symbols { }, called _____, are used when writing a set.

22. The symbol > means ___ _____ _____, and the symbol < means ___ ____ ____.

GUIDED PRACTICE

Find the place values. See Example 1.

23. Consider the number 57,634.
    a. What is the place value of the digit 3?
    b. What digit is in the thousands column?
    c. What is the place value of the digit 6?
    d. What digit is in the ten thousands column?

24. Consider the number 128,940.
    a. What is the place value of the digit 8?
    b. What digit is in the hundreds column?
    c. What is the place value of the digit 2?
    d. What digit is in the hundred thousands column?

25. World hunger. On the website Freerice.com, sponsors donate grains of rice to feed the hungry. From 2007 through 2017, there have been 96,128,453,798 grains of rice donated.

   a. What is the place value of the digit 2?
   b. What digit is in the billions place?
   c. What is the place value of the digit 3?
   d. What digit is in the ten billions place?

26. YouTube views. According to the counter on YouTube, as of January 31, 2017, the video GANGNAM STYLE by PSY has been viewed 2,739,387,518 times.

   a. What is the place value of the digit 5?
   b. What digit is in the ten thousands place?
   c. What is the place value of the digit 2?
   d. What digit is in the hundred millions place?

Write each number in words. **See Example 2.**

27. 93   **28.** 48
29. 732   **30.** 259
31. 154,302   **32.** 615,019
33. 14,432,500   **34.** 104,052,005
35. 970,031,500,104   **36.** 5,800,010,700
37. 82,000,415   **38.** 51,000,201,078

Write each number in standard form. **See Example 3.**

39. Three thousand, seven hundred thirty-seven
40. Fifteen thousand, four hundred ninety-two
41. Nine hundred thirty
42. Six hundred forty
43. Seven thousand, twenty-one
44. Four thousand, five hundred
45. Twenty-six million, four hundred thirty-two
46. Ninety-two billion, eighteen thousand, three hundred ninety-nine

Write each number in expanded form. **See Example 4.**

47. 245   **48.** 518
49. 3,609   **50.** 3,961
51. 72,533   **52.** 73,009
53. 104,401   **54.** 570,003
55. 8,403,613   **56.** 3,519,807
57. 26,000,156   **58.** 48,000,061

Place an < or an > symbol in the box to make a true statement. **See Example 5.**

59. a. 11 ☐ 8   **b.** 29 ☐ 54
60. a. 410 ☐ 609   **b.** 3,206 ☐ 3,231

61. a. 12,321 ☐ 12,209   **b.** 23,223 ☐ 23,231
62. a. 178,989 ☐ 178,898   **b.** 850,234 ☐ 850,342

Round to the nearest ten. **See Example 6.**

63. 98,154   **64.** 26,742
65. 512,967   **66.** 621,116

Round to the nearest hundred. **See Example 7.**

67. 8,352   **68.** 1,845
69. 32,439   **70.** 73,931
71. 65,981   **72.** 5,346,975
73. 2,580,952   **74.** 3,428,961

Round each number to the nearest thousand and then to the nearest ten thousand. **See Example 8.**

75. 52,867   **76.** 85,432
77. 76,804   **78.** 34,209
79. 816,492   **80.** 535,600
81. 296,500   **82.** 498,903

TRY IT YOURSELF

83. Round 79,593 to the nearest …
   a. ten   **b.** hundred
   c. thousand   **d.** ten thousand
84. Round 5,925,830 to the nearest …
   a. thousand   **b.** ten thousand
   c. hundred thousand   **d.** million
85. Round $419,161 to the nearest …
   a. $10   **b.** $100
   c. $1,000   **d.** $10,000
86. Round 5,436,483 ft to the nearest …
   a. 10 ft   **b.** 100 ft
   c. 1,000 ft   **d.** 10,000 ft

Write each number in standard notation.

87. 4 ten thousands + 2 tens + 5 ones
88. 7 millions + 7 tens + 7 ones
89. 200,000 + 2,000 + 30 + 6
90. 7,000,000,000 + 300 + 50
91. Twenty-seven thousand, five hundred ninety-eight
92. Seven million, four hundred fifty-two thousand, eight hundred sixty
93. Ten million, seven hundred thousand, five hundred six
94. Eighty-six thousand, four hundred twelve

LOOK ALIKES

Write each number in standard notation.

95. a. One trillion, six hundred million
   b. One billion, six hundred thousand
   c. One million, six hundred

96. **a.** Ninety-nine billion, ninety-nine

b. Eighty-eight million, eighty-eight

c. Seventy-seven thousand, seventy-seven

97. **a.** 9 billion

b. 9,000 million

98. **a.** 1 billion + 1 million + 1 thousand + 1 one

b. 1,000,000,000 + 1,000,000 + 1,000 + 1

APPLICATIONS

99. Game shows. On *The Price Is Right* television show, the winning contestant is the person who comes closest to (without going over) the price of the item up for bid. Which contestant shown below will win if they are bidding on a bedroom set that has a suggested retail price of $4,745?

100. Presidents. The following list shows the ten youngest U.S. presidents and their ages (in years/days) when they took office. Construct a two-column table that presents the data in order, beginning with the youngest president.

J. Polk 49 yr/122 days	U. Grant 46 yr/236 days
G. Cleveland 47 yr/351 days	J. Kennedy 43 yr/236 days
W. Clinton 46 yr/154 days	F. Pierce 48 yr/101 days
M. Filmore 50 yr/184 days	B. Obama 47 yr/169 days
J. Garfield 49 yr/105 days	T. Roosevelt 42 yr/322 days

101. Renters. The number of renter households in the United States increased for all income brackets from 2005 to 2015. Use the graph in the next column to answer the following questions.

a. Which income bracket had the greatest number of renter households?

b. Which income bracket had the least number of renter households?

c. In 2015, for the income bracket $25,000−$49,999, estimate the number of renter households rounded to the nearest million.

d. In 2005, for the income bracket $100,000 and over, estimate the number of renter households rounded to the nearest million.

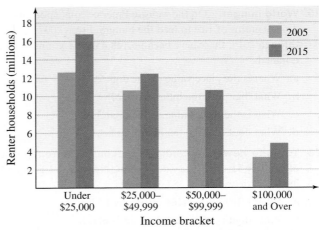

Source: JCHS tabulations of U.S. Census Bureau, Current Population Surveys

102. Sports. The graph shows the maximum recorded ball speeds for five sports.

a. Which sport had the fastest recorded maximum ball speed? Estimate the speed.

b. Which sport had the slowest maximum recorded ball speed? Estimate the speed.

c. Which sport had the second fastest maximum recorded ball speed? Estimate the speed.

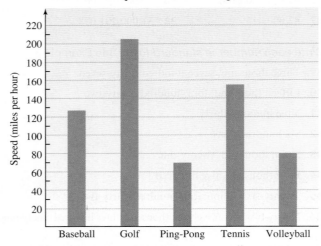

103. Coffee. Complete the bar graph and line graph on the next page using the data in the table.

Starbucks Locations

Year	Number
2000	3,501
2002	5,886
2004	8,569
2006	12,440
2008	16,680
2010	16,858
2012	17,651
2014	21,366

Source: Starbucks Company

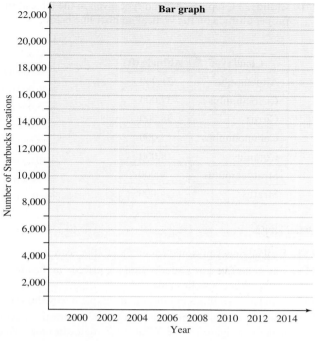

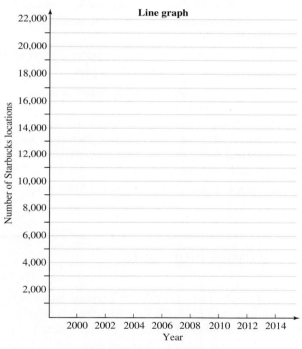

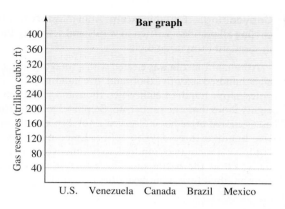

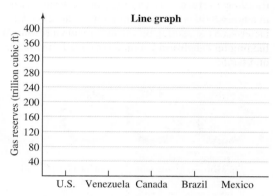

104. Energy reserves. Complete the bar graph and line graph in the next column using the data in the table.

Natural Gas Reserves, 2015 Estimates (in Trillion Cubic Feet)

United States	369
Venezuela	198
Canada	70
Brazil	15
Mexico	11

Source: BP Statistical Review of World Energy, 2015

105. Checking accounts. Complete each check by writing the amount in words on the proper line.

a.

```
DON SMITH                                        7155
1234 MILL STREET
HILLDALE, CA               DATE  March 9, 2017

Payable to   Davis Chevrolet        $  15,601.00

_____ DOLLARS
FIRST FEDERAL BANK
195 JEFFS STREET
HILLDALE, CA

Memo _____        Don Smith
```

b.

```
JUAN DECITO                                      4251
24 ARBOR LANE
ARGENTO, CA                DATE  Aug. 12, 2017

Payable to   DR. ANDERSON           $  3,433.00

_____ DOLLARS
FIRST FEDERAL BANK
195 JEFFS STREET
HILLDALE, CA

Memo _____        Juan Decito
```

106. Announcements. One style used when printing formal invitations and announcements is to write all numbers in words. Use this style to write each of the following phrases.

a. This diploma awarded this 27th day of June, 2017.

b. The suggested contribution for the fundraiser is $850 a plate, or an entire table may be purchased for $5,250.

107. Copyediting. Edit this excerpt from a history text by circling all numbers written in words and rewriting them in standard form using digits.

Abraham Lincoln was elected with a total of one million, eight hundred sixty-five thousand, five hundred ninety-three votes—four hundred eighty-two thousand, eight hundred eighty more than the runner-up, Stephen Douglas. He was assassinated after having served a total of one thousand, five hundred three days in office. Lincoln's Gettysburg Address, a mere two hundred sixty-nine words long, was delivered at the battle site where forty-three thousand, four hundred forty-nine casualties occurred.

108. Reading meters. The amount of electricity used in a household is measured in kilowatt-hours (kwh). Determine the reading on the meter shown below. (When the pointer is between two numbers, read the *smaller* number.)

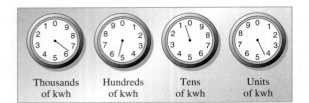

| Thousands of kwh | Hundreds of kwh | Tens of kwh | Units of kwh |

109. Speed of light. The speed of light is 983,571,072 feet per second.

   a. In what place-value column is the 5?

   b. Round the speed of light to the nearest ten million. Give your answer in standard notation and in expanded notation.

   c. Round the speed of light to the nearest hundred million. Give your answer in standard notation and in written-word form.

110. Clouds. Graph each cloud type given in the table at the proper altitude on the vertical number line below.

Cloud type	Altitude (ft)
Altocumulus	21,000
Cirrocumulus	37,000
Cirrus	38,000
Cumulonimbus	15,000
Cumulus	8,000
Stratocumulus	9,000
Stratus	4,000

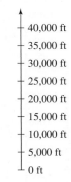

- 40,000 ft
- 35,000 ft
- 30,000 ft
- 25,000 ft
- 20,000 ft
- 15,000 ft
- 10,000 ft
- 5,000 ft
- 0 ft

WRITING

111. Explain how you would round 687 to the nearest ten.

112. The houses in a new subdivision are priced "in the low 130's." What does this mean?

113. A million is a thousand thousands. Explain why this is so.

114. Many television infomercials offer the viewer creative ways to make a six-figure income. What is a six-figure income? What is the smallest and what is the largest six-figure income?

115. What whole number is associated with each of the following words?

| duo | decade | zilch | a grand | four score |
| dozen | trio | century | a pair | nil |

116. Explain what is wrong by reading 20,003 as *twenty thousand and three.*

117. The words *two*, *to*, and *too* sound the same but are spelled differently and have different meanings. (Such words are called **homonyms.**) Write a sentence that contains all three words.

118. Write each statement in words.

   a. $2,016 < 2,106$ **b.** $7,080,008 > 7,008,800$

OBJECTIVES

1 Add whole numbers.

2 Use properties of addition to add whole numbers.

3 Estimate sums of whole numbers.

4 Solve application problems by adding whole numbers.

5 Find the perimeter of a rectangle and a square.

6 Use a calculator to add whole numbers (optional).

SECTION 1.2 Adding Whole Numbers

Everyone uses addition of whole numbers. For example, to prepare an annual budget, an accountant adds separate line item costs. To determine the number of yearbooks to order, a principal adds the number of students in each grade level. A flight attendant adds the number of people in the first-class and economy sections to find the total number of passengers on an airplane.

OBJECTIVE 1 Add whole numbers.

To add whole numbers, think of combining sets of similar objects. For example, if a set of 4 stars is combined with a set of 5 stars, the result is a set of 9 stars.

| A set of 4 stars | A set of 5 stars | A set of 9 stars |

★★★★ ★★★★★   ★★★★★★★★★

We combine these two sets   to get this set.

We can write this addition problem in **horizontal** or **vertical form** using an **addition symbol +**, which is read as "plus." The numbers that are being added are called **addends**, and the answer is called the **sum** or **total**.

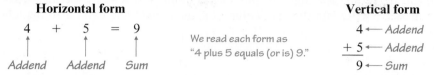

Horizontal form

4 + 5 = 9

Addend   Addend   Sum

We read each form as "4 plus 5 equals (or is) 9."

Vertical form

4 ← *Addend*
+ 5 ← *Addend*
9 ← *Sum*

To add whole numbers that are less than 10, we rely on our understanding of basic addition facts. For example,

2 + 3 = 5,   6 + 4 = 10,   and   9 + 7 = 16

If you need to review the basic addition facts, they can be found in Appendix 1 at the back of the book.

To add whole numbers that are greater than 10, we can use vertical form by stacking them with their corresponding place values lined up. Then we simply add the digits in each corresponding column.

EXAMPLE 1   Add: 421 + 123 + 245

Strategy We will write the addition in vertical form with the ones digits in a column, the tens digits in a column, and the hundreds digits in a column. Then we will add the digits, column by column, working from right to left.

WHY Like money, where pennies are only added to pennies, dimes are only added to dimes, and dollars are only added to dollars, we can only add digits with the same place value: ones to ones, tens to tens, hundreds to hundreds.

Solution
We start at the right and add the ones digits, then the tens digits, and finally the hundreds digits and write each sum below the horizontal bar.

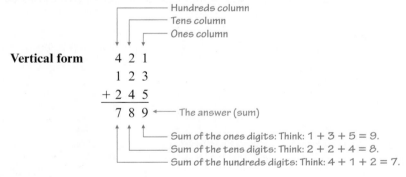

Hundreds column
Tens column
Ones column

Vertical form
```
  4 2 1
  1 2 3
+ 2 4 5
  7 8 9  ← The answer (sum)
```

Sum of the ones digits: Think: 1 + 3 + 5 = 9.
Sum of the tens digits: Think: 2 + 2 + 4 = 8.
Sum of the hundreds digits: Think: 4 + 1 + 2 = 7.

The sum is 789.

Self Check 1

Add: 131 + 232 + 221 + 312

Now Try ➲ Problems 21 and 27

If an addition of the digits in any place-value column produces a sum that is greater than 9, we must **carry**.

EXAMPLE 2   Add: 27 + 18

Strategy We will write the addition in vertical form and add the digits, column by column, working from right to left. We must watch for sums in any place-value column that are greater than 9.

WHY If the sum of the digits in any column is more than 9, we must carry.

Solution
To help you understand the process, each step of this addition is explained separately. Your solution need only look like the *last* step.

We begin by adding the digits in the ones column: $7 + 8 = 15$. Because $15 = 1$ ten $+ 5$ ones, we write 5 in the ones column of the answer and carry 1 to the tens column.

$$
\begin{array}{r}
\overset{1}{2}\,7 \\
+\ 1\,8 \\
\hline
5
\end{array}
$$

Add the digits in the ones column: $7 + 8 = 15$. Carry 1 to the tens column.

Then we add the digits in the tens column.

$$
\begin{array}{r}
\overset{1}{2}\,7 \\
+\ 1\,8 \\
\hline
4\,5
\end{array}
$$

Add the digits in the tens column: $1 + 2 + 1 = 4$. Place the result of 4 in the tens column of the answer.

Your solution should look like this:
$$
\begin{array}{r}
\overset{1}{2}7 \\
+\ 18 \\
\hline
45
\end{array}
$$

The sum is 45.

Self Check 2

Add: $35 + 47$

Now Try ➡ Problems 29 and 33

EXAMPLE 3   Add: $9,835 + 692 + 7,275$

Strategy We will write the numbers in vertical form so that corresponding place-value columns are lined up. Then we will add the digits in each column, watching for any sums that are greater than 9.

WHY If the sum of the digits in any column is more than 9, we must carry.

Solution
We write the addition in vertical form so that the corresponding digits are lined up. Each step of this addition is explained separately. Your solution need only look like the *last* step.

$$
\begin{array}{r}
9,8\,3\,\overset{1}{5} \\
6\,9\,2 \\
+\ 7,2\,7\,5 \\
\hline
2
\end{array}
$$

Add the digits in the ones column: $5 + 2 + 5 = 12$. Write 2 in the ones column of the answer and carry 1 to the tens column.

$$
\begin{array}{r}
9,\overset{2}{8}\,\overset{1}{3}\,5 \\
6\,9\,2 \\
+\ 7,2\,7\,5 \\
\hline
0\,2
\end{array}
$$

Add the digits in the tens column: $1 + 3 + 9 + 7 = 20$. Write 0 in the tens column of the answer and carry 2 to the hundreds column.

$$
\begin{array}{r}
\overset{1}{9},\overset{2}{8}\,\overset{1}{3}\,5 \\
6\,9\,2 \\
+\ 7,2\,7\,5 \\
\hline
8\,0\,2
\end{array}
$$

Add the digits in the hundreds column: $2 + 8 + 6 + 2 = 18$. Write 8 in the hundreds column of the answer and carry 1 to the thousands column.

$$
\begin{array}{r}
\overset{1}{9},\overset{2}{8}\,\overset{1}{3}\,5 \\
6\,9\,2 \\
+\ 7,2\,7\,5 \\
\hline
17,8\,0\,2
\end{array}
$$

Add the digits in the thousands column: $1 + 9 + 7 = 17$.
Write 7 in the thousands column of the answer.
Write 1 in the ten thousands column.

Your solution should look like this:
$$
\begin{array}{r}
\overset{1\ 21}{9,835} \\
692 \\
+\ 7,275 \\
\hline
17,802
\end{array}
$$

The sum is 17,802.

Self Check 3

Add: $675 + 1,497 + 1,527$

Now Try ➡ Problems 37 and 41

Success Tip In Example 3, the digits in each place-value column were added from *top to bottom*. To check the answer, we can instead add from *bottom to top*. Adding down or adding up should give the same result. If it does not, an error has been made and you should re-add. You will learn why the two results should be the same in Objective 2, which follows.

First add
top to
bottom

$$
\begin{array}{r}
17{,}802 \\
9{,}835 \\
692 \\
+\ 7{,}275 \\
\hline
17{,}802
\end{array}
$$

To check,
add
bottom
to top

OBJECTIVE 2 **Use properties of addition to add whole numbers.**

Have you ever noticed that two whole numbers can be added in either order because the result is the same? For example,

$$2 + 8 = 10 \qquad \text{and} \qquad 8 + 2 = 10$$

This example illustrates the **commutative property of addition.**

Commutative Property of Addition

The order in which whole numbers are added does not change their sum.
  For example,

  $$6 + 5 = 5 + 6$$

LANGUAGE OF MATHEMATICS

Commutative is a form of the word *commute*, meaning to go back and forth. *Commuter* trains take people to and from work.

To find the sum of three whole numbers, we add two of them and then add the sum to the third number. In the following examples, we add $3 + 4 + 7$ in two ways. We will use the grouping symbols ( ), called **parentheses**, to show this. It is standard practice to perform the operations within the parentheses first. The steps of the solutions are written in horizontal form.

LANGUAGE OF MATHEMATICS

We read $(3 + 4) + 7$ as "The *quantity* of 3 plus 4," pause slightly, and then say "plus 7." We read $3 + (4 + 7)$ as, "3 plus the *quantity* of 4 plus 7." The word *quantity* alerts the reader to the parentheses that are used as grouping symbols.

Method 1: Group 3 and 4

$$(3 + 4) + 7 = 7 + 7$$
$$= 14$$

Because of the parentheses, add 3 and 4 first to get 7. Then add 7 and 7 to get 14.

Method 2: Group 4 and 7

$$3 + (4 + 7) = 3 + 11$$
$$= 14$$

Because of the parentheses, add 4 and 7 first to get 11. Then add 3 and 11 to get 14.

—— Same result ——

Either way, the answer is 14. This example illustrates that changing the grouping when adding numbers doesn't affect the result. This property is called the **associative property of addition**.

Associative Property of Addition

The way in which whole numbers are grouped does not change their sum.
  For example,

  $$(2 + 5) + 4 = 2 + (5 + 4)$$

LANGUAGE OF MATHEMATICS

Associative is a form of the word *associate*, meaning to join a group. The WNBA (Women's National Basketball *Association*) is a group of 12 professional basketball teams.

Sometimes, an application of the associative property can simplify a calculation.

EXAMPLE 4   Find the sum: $98 + (2 + 17)$

Strategy We will use the associative property to group 2 with 98.

WHY It is helpful to regroup because 98 and 2 are a pair of numbers that are easily added.

Solution
We will write the steps of the solution in horizontal form.

$$98 + (2 + 17) = (98 + 2) + 17 \qquad \text{Use the associative property of addition to regroup the addends.}$$
$$= 100 + 17 \qquad \text{Do the addition within the parentheses first.}$$
$$= 117$$

Self Check 4

Find the sum: $(139 + 25) + 75$

Now Try ➲ **Problems 45 and 49**

Whenever we add 0 to a whole number, the number is unchanged. This property is called the **addition property of 0**.

Addition Property of 0

The sum of any whole number and 0 is that whole number.
    For example,

$$3 + 0 = 3, \qquad 5 + 0 = 5, \qquad \text{and} \qquad 0 + 9 = 9$$

EXAMPLE 5   Add: **a.** $3 + 5 + 17 + 2 + 3$ **b.** $\begin{array}{r} 201 \\ 867 \\ + \ \ 49 \end{array}$

Strategy We will look for groups of two (or three numbers) whose sum is 10 or 20 or 30, and so on.

WHY This method is easier than adding unrelated numbers, and it reduces the chances of a mistake.

Solution
Together, the commutative and associative properties of addition enable us to use any order or grouping to add whole numbers.

a. We will write the steps of the solution in horizontal form.

$$3 + 5 + 17 + 2 + 3 = 20 + 10 \qquad \text{Think: } 3 + 17 = 20 \text{ and } 5 + 2 + 3 = 10.$$
$$= 30$$

b. Each step of the addition is explained separately. Your solution should look like the last step.

$$\begin{array}{r} \overset{1}{} \\ 2\ 0\ \mathbf{1} \\ 8\ 6\ \mathbf{7} \\ + \quad 4\ \mathbf{9} \\ \hline 7 \end{array}$$
Add the bold numbers in the ones column first.
Think: $(9 + 1) + 7 = 10 + 7 = 17$.
Write the 7 and carry the 1.

$$\begin{array}{r} \overset{1}{}\ \overset{1}{} \\ 2\ \mathbf{0}\ 1 \\ 8\ \mathbf{6}\ 7 \\ + \quad \mathbf{4}\ 9 \\ \hline 1\ 7 \end{array}$$
Add the bold numbers in the tens column.
Think: $(6 + 4) + 1 = 10 + 1 = 11$.
Write the 1 and carry the 1.

```
  1  1
  2  0  1     Add the bold numbers in the hundreds column.
  8  6  7     Think: (2 + 8) + 1 = 10 + 1 = 11.
+     4  9
───────────
  1,1  1  7
```

The sum is 1,117.

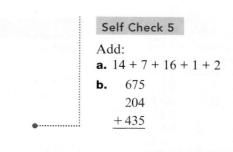

OBJECTIVE 3　Estimate sums of whole numbers.

Estimation is used to find an approximate answer to a problem. Estimates are helpful in two ways. First, they serve as an accuracy check that can find errors. If an answer does not seem reasonable when compared to the estimate, the original problem should be reworked. Second, some situations call for only an approximate answer rather than the exact answer.

There are several ways to estimate, but the objective is the same: Simplify the numbers in the problem so that the calculations can be made easily and quickly. One popular method of estimation is called **front-end rounding**.

Success Tip Estimates can be greater than or less than the exact answer. It depends on how often rounding up and rounding down occurs in the estimation.

EXAMPLE 6　Use front-end rounding to estimate the sum:
3,714 + 2,489 + 781 + 5,500 + 303

Strategy We will use front-end rounding to approximate each addend. Then we will find the sum of the approximations.

WHY Front-end rounding produces addends containing many 0's. Such numbers are easier to add.

Solution

Each of the addends is rounded to its *largest place value* so that all but its first digit is zero. Then we add the approximations using vertical form.

```
  3,714  ⟶     4,000     Round to the nearest thousand.
  2,489  ⟶     2,000     Round to the nearest thousand.
    781  ⟶       800     Round to the nearest hundred.
  5,500  ⟶     6,000     Round to the nearest thousand.
+   303  ⟶  +    300     Round to the nearest hundred.
              ──────
              13,100
```

The estimate is 13,100.

If we calculate 3,714 + 2,489 + 781 + 5,500 + 303, the sum is exactly 12,787. Note that the estimate is close: It's just 313 more than 12,787. This illustrates the tradeoff when using estimation: The calculations are easier to perform and take less time, but the answers are not exact.

OBJECTIVE 4　Solve application problems by adding whole numbers.

Since application problems are almost always written in words, the ability to understand what you read is very important.

LANGUAGE OF MATHEMATICS

Here are some key words and phrases that are often used to indicate **addition**:

gain	total
increase	combined
up	in all
forward	in the future
rise	altogether
more than	extra

Self Check 7

Airline accidents.

The numbers of accidents involving U.S. airlines for the years 2008 through 2015 are listed in the table below. Find the total number of accidents for those years.

Year	Accidents
2008	20
2009	26
2010	28
2011	29
2012	26
2013	19
2014	28
2015	27

Source: National Transportation Safety Board

Now Try ➡ **Problem 104**

EXAMPLE 7

Sharks. The graph on the right shows the number of shark attacks worldwide for the years 2008 through 2015. Find the total number of shark attacks for those years.

Strategy We will carefully read the problem looking for a key word or phrase.

WHY Key words and phrases indicate which arithmetic operation(s) should be used to solve the problem.

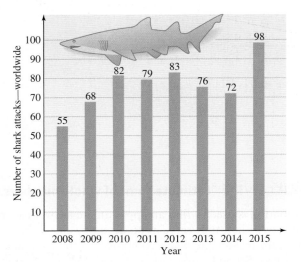

Source: Florida Museum of Natural History, Ichthyology Department

Solution

In the second sentence of the problem, the key word *total* indicates that we should add the number of shark attacks for the years 2008 through 2015. We can use vertical form to find the sum.

$$
\begin{array}{r}
\overset{6\,4}{55} \\
68 \\
82 \\
79 \\
83 \\
76 \\
72 \\
+\ 98 \\
\hline
613
\end{array}
$$

Add the digits, one column at a time, working from right to left. To simplify the calculations, we can look for groups of two or three numbers in each column whose sum is 10.

The total number of shark attacks worldwide for the years 2008 through 2015 was 613.

LANGUAGE OF MATHEMATICS

To solve application problems, we must often **translate** the words of the problem to numbers and symbols. To translate means to change from one form to another, as in translating from Spanish to English.

EXAMPLE 8

iStock.com/kjekol

Endangered wolves. In 1989, there were 1,814 gray wolves in the western Great Lakes region of Minnesota, Wisconsin, and Michigan. By 2015, the number had increased by 1,792. Find the number of gray wolves in 2015 in that region. (Source: U.S. Fish and Wildlife Service)

Strategy We will carefully read the problem looking for key words or phrases.

WHY Key words and phrases indicate which arithmetic operation(s) should be used to solve the problem.

Solution

The phrase *increased by* indicates addition. With that in mind, we translate the words of the problem to numbers and symbols.

| The number of gray wolves in 2015 | is equal to | the number of gray wolves in 1989 | increased by | 1,792. |

| The number of gray wolves in 2015 | = | 1,814 | + | 1,792 |

Use vertical form to perform the addition:

$$\begin{array}{r} \overset{1\ \ 1}{1,814} \\ + 1,792 \\ \hline 3,606 \end{array}$$

In 2015, the number of gray wolves in the western Great Lakes region was 3,606.

Self Check 8

Magazines. In 2011, the monthly circulation of *Game Informer* magazine was 5,954,884 copies. By 2016, the circulation had increased by 398,191 copies per month. What was the monthly circulation of *Game Informer* magazine in 2016? (Source: *The World Almanac and Book of Facts*, 2012, 2017)

Now Try ➲ **Problem 96**

OBJECTIVE 5 Find the perimeter of a rectangle and a square.

Figure (a) below is an example of a four-sided figure called a **rectangle**. Either of the longer sides of a rectangle is called its **length** and either of the shorter sides is called its **width**. Together, the length and width are called the **dimensions** of the rectangle. For any rectangle, opposite sides have the same measure.

When all four of the sides of a rectangle are the same length, we call the rectangle a **square**. An example of a square is shown in figure (b).

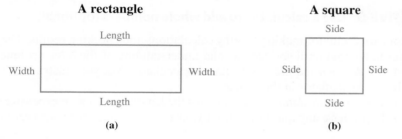

A rectangle

Length
Width   Width
Length

(a)

A square

Side
Side   Side
Side

(b)

LANGUAGE OF MATHEMATICS

When you hear the word ***perimeter***, think of the distance around the "rim" of a flat figure.

The distance around a rectangle or a square is called its **perimeter**. To find the perimeter of a rectangle, we add the lengths of its four sides.

The perimeter of a rectangle = length + length + width + width

To find the perimeter of a square, we add the lengths of its four sides.

The perimeter of a square = side + side + side + side

EXAMPLE 9   Money.   Find the perimeter of the dollar bill shown below.

Width = 65 mm   mm stands for millimeters

Length = 156 mm

Strategy We will add two lengths and two widths of the dollar bill.

WHY A dollar bill is rectangular-shaped, and this is how the perimeter of a rectangle is found.

Solution

We translate the words of the problem to numbers and symbols.

The perimeter of the dollar bill	is equal to	the length of the dollar bill	plus	the length of the dollar bill	plus	the width of the dollar bill	plus	the width of the dollar bill.
The perimeter of the dollar bill	=	156	+	156	+	65	+	65

Use vertical form to perform the addition:

$$
\begin{array}{r}
\overset{2\,2}{156} \\
156 \\
65 \\
+\ \ 65 \\
\hline
442
\end{array}
$$

The perimeter of the dollar bill is 442 mm.

To see whether this result is reasonable, we estimate the answer. Because the rectangle is about 160 mm by 70 mm, its perimeter is approximately $160 + 160 + 70 + 70$, or 460 mm. An answer of 442 mm is reasonable.

Self Check 9

Board games. A Monopoly game board is a square with sides 19 inches long. Find the perimeter of the board.

Now Try ⟳ Problems 65 and 67

OBJECTIVE 6 **Use a calculator to add whole numbers (optional).**

Calculators are useful for making lengthy calculations and checking results. They should not, however, be used until you have a solid understanding of the basic arithmetic facts. This textbook *does not* require you to have a calculator. Ask your instructor if you are allowed to use a calculator in the course.

The *Using Your Calculator* feature explains the keystrokes for an inexpensive scientific calculator. If you have any questions about your specific model, see your user's manual.

Using Your Calculator ▶   The Addition Key: Vehicle Production

In 2015, the top five producers of motor vehicles in the world were Toyota: 10,083,831; Volkswagen: 9,872,424; Hyundai: 7,988,479; General Motors: 7,485,587; and Ford: 6,396,369 (Source: OICA, 2015). We can find the total number of motor vehicles produced by these companies using the addition key ⊞ on a calculator.

10083831 ⊞ 9872424 ⊞ 7988479 ⊞ 7485587 ⊞ 6396369 ⊟ | 41826690 |

On some calculator models, the ⎡Enter⎤ key is pressed instead of the ⊟ for the result to be displayed.
The total number of vehicles produced in 2015 by the top five automakers was 41,826,690.

Answers to Self Checks

1. 896   **2.** 82   **3.** 3,699   **4.** 239   **5. a.** 40   **b.** 1,314   **6.** 16,600   **7.** The total number of accidents for 2008–2015 was 203.   **8.** The monthly circulation in 2016 was 6,353,075.   **9.** The perimeter of the Monopoly board is 76 in.

SECTION 1.2 STUDY SET

VOCABULARY

Fill in the blanks.

1. In the addition problem shown below, label each *addend* and the *sum*.

$$10 \quad + \quad 15 \quad = \quad 25$$

2. When using the vertical form to add whole numbers, if the addition of the digits in any one column produces a sum greater than 9, we must _____.

3. The _____ property of addition states that the order in which whole numbers are added does not change their sum.

4. The _____ property of addition states that the way in which whole numbers are grouped does not change their sum.

5. To see whether the result of an addition is reasonable, we can round the addends and _____ the sum.

6. The words *rise*, *gain*, *total*, and *increase* are often used to indicate the operation of _____.

7. The figure below on the left is an example of a _____. The figure on the right is an example of a _____.

8. Label the *length* and the *width* of the rectangle below. Together, the length and width of a rectangle are called its _____.

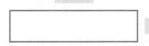

9. When all the sides of a rectangle are the same length, we call the rectangle a _____.

10. The distance around a rectangle is called its _____.

CONCEPTS

11. Which property of addition is shown?

   a. $3 + 4 = 4 + 3$

   b. $(3 + 4) + 5 = 3 + (4 + 5)$

   c. $(36 + 58) + 32 = 36 + (58 + 32)$

   d. $319 + 507 = 507 + 319$

Fill in the blanks.

12. a. Use the commutative property of addition to complete the following:

   $$19 + 33 =$$

b. Use the associative property of addition to complete the following:

   $$3 + (97 + 16) =$$

13. Fill in the blank: Any number added to ▢ stays the same.

14. Fill in the blanks. Use estimation by front-end rounding to determine if the sum shown below (14,735) seems reasonable.

$$
\begin{array}{r}
5,877 \rightarrow \\
402 \rightarrow \\
+8,456 \rightarrow + \\
\hline
14,735 \rightarrow
\end{array}
$$

NOTATION

Fill in the blanks.

15. The addition symbol + is read as " ____."

16. The symbols ( ) are called _____. It is standard practice to perform the operations within them ____.

Write each of the following addition facts in words.

17. $33 + 12 = 45$

18. $28 + 22 = 50$

Complete each step to find the sum.

19. $(36 + 11) + 5 = \boxed{} + 5$

   $= \boxed{}$

20. $12 + (15 + 2) = 12 + \boxed{}$

   $= \boxed{}$

GUIDED PRACTICE

Add. See Example 1.

21. $25 + 13$

22. $47 + 12$

23. $\begin{array}{r} 406 \\ +283 \\ \hline \end{array}$

24. $\begin{array}{r} 213 \\ +751 \\ \hline \end{array}$

25. $21 + 31 + 24$

26. $33 + 43 + 12$

27. $603 + 152 + 121$

28. $462 + 115 + 220$

Add. See Example 2.

29. $19 + 16$

30. $27 + 18$

31. $45 + 47$

32. $37 + 26$

33. $52 + 18$

34. $59 + 31$

35. $\begin{array}{r} 28 \\ +47 \\ \hline \end{array}$

36. $\begin{array}{r} 35 \\ +49 \\ \hline \end{array}$

Add. See Example 3.

37. $156 + 305$

38. $647 + 138$

39. $4,301 + 789 + 3,847$

40. $5,576 + 649 + 1,922$

41. 9,758 + 586 + 7,799

42. 9,339 + 471 + 6,883

43.
346
217
568
+679

44.
290
859
345
+226

Apply the associative property of addition to find the sum. See Example 4.

45. (9 + 3) + 7

46. (7 + 9) + 1

47. (13 + 8) + 12

48. (19 + 7) + 13

49. 94 + (6 + 37)

50. 92 + (8 + 88)

51. 125 + (75 + 41)

52. 240 + (60 + 93)

Use the commutative and associative properties of addition to find the sum. See Example 5.

53. 4 + 8 + 16 + 1 + 1

54. 2 + 1 + 28 + 3 + 6

55. 23 + 5 + 7 + 15 + 10

56. 31 + 6 + 9 + 14 + 20

57.
624
905
+ 86

58.
495
76
+835

59. 457 + 97 + 653

60. 562 + 99 + 848

Use front-end rounding to estimate the sum. See Example 6.

61. 686 + 789 + 12,233 + 24,500 + 5,768

62. 404 + 389 + 11,802 + 36,902 + 7,777

63. 567,897 + 23,943 + 309,900 + 99,113

64. 822,365 + 15,444 + 302,417 + 99,010

Find the perimeter of each rectangle or square. See Example 9.

65.
32 feet (ft)

12 ft

66.
127 meters (m)

91 m

67. 17 inches (in.)

17 in.

68. 5 yards (yd)

5 yd

69. 94 mi (miles)

94 mi

70. 56 ft (feet)

56 ft

71.
87 cm (centimeters)
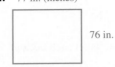
6 cm

72. 77 in. (inches)
76 in.

TRY IT YOURSELF

Add.

73.
8,539
+7,368

74.
5,799
+6,879

75. 51,246 + 578 + 37 + 4,599

76. 4,689 + 73,422 + 26 + 433

77. (45 + 16) + 4

78. 7 + (63 + 23)

79.
632
+347

80.
423
+570

81. 16,427 increased by 13,573

82. 13,567 more than 18,788

83.
76
+ 45

84.
87
+ 56

85. 3,156 + 1,578 + 6,578

86. 2,379 + 4,779 + 2,339

87. 12 + 1 + 8 + 4 + 9 + 16

88. 7 + 15 + 13 + 9 + 5 + 11

LOOK ALIKES

Find the answer to the problem in part a. The answer to part b should then be obvious.

89. a. 299 increased by 99

b. 99 increased by 299

90. a. 3,068 more than 368

b. 368 more than 3,068

91. a.
747
+ 252

b.
252
+ 747

92. a. (913 + 87) + 688

b. 913 + (87 + 688)

APPLICATIONS

93. Dimensions of a house. Find the length of the house shown in the blueprint.

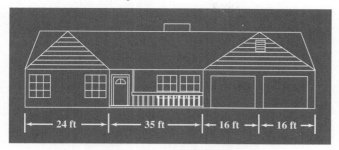

94. Rockets. A Saturn V rocket was used to launch the crew of *Apollo 11* to the moon. The first stage of the rocket was 138 feet tall, the second stage was 98 feet tall, and the third stage was 46 feet tall. Atop the third stage sat the 54-foot-tall lunar module and a 28-foot-tall escape tower. What was the total height of the spacecraft?

95. Fast food. Find the total number of calories in the following lunch from McDonald's: Big Mac (540 calories), French fries (230 calories), Fruit 'n Yogurt Parfait (150 calories), medium Coca-Cola Classic (170 calories).

96. Chief Executive Officer. In 2015, Robert A. Iger, CEO of Walt Disney Company, had a base salary of $2,548,077. He also earned an additional $42,365,536 in Stock and Option Awards, Incentive Plan compensation, pension, and other compensation. Find Robert Iger's total compensation in 2015. (Source: aflcio.org)

97. Websites. The number of persons age 15 or older who visited the Apple iTunes website at least once in June 2016 was 243,547,000. The number that visited the Alibaba website during that month was 64,511,000 greater than the iTunes site. How many visitors did the Alibaba site have? (Source: *The World Almanac and Book of Facts, 2017*)

98. Ice cream. Baskin-Robbins is the world's largest chain of ice cream specialty shops. In 2016, there were 2,524 stores in the United States, and 5,198 stores in 50 other countries of the world. Find the total number of Baskin-Robbins stores in 2016. (Source: entrepreneur.com)

99. Bridge safety. The results of a 2017 report of the condition of U.S. highway bridges is shown below. Each bridge was classified as either *safe*, *in need of repair*, or *should be replaced*. Complete the table.

Number of safe bridges	Number of bridges that need repair	Number of outdated bridges that should be replaced	Total number of bridges
464,859	61,365	84,525	

Source: Bureau of Transportation Statistics

100. Imports. The table below shows the number of new and used passenger cars imported into the United States from various countries in 2015. Find the total number of cars the United States imported from these countries.

Country	Number of passenger cars
Canada	1,969,466
France	28,024
Germany	639,838
Italy	132,316
Japan	1,609,597
Mexico	1,438,840
South Korea	1,065,971
Sweden	37,789
United Kingdom	134,367

Source: Economic Indicators Division, U.S. Census Bureau

101. Weddings. The average wedding costs for 2015 are listed in the table below. Find the total cost of a wedding.

Clothing/hair/makeup	$1,947
Ceremony/music/flowers	$5,358
Photography/video	$4,442
Favors/gifts/invitations	$712
Jewelry	$5,871
Reception	$12,540
Honeymoon	$3,882

Source: jennlanedesign.com, prettypracticalbride.com

102. Budgets. A department head in a company prepared an annual budget with the line items shown. Find the projected number of dollars to be spent.

Line item	Amount
Equipment	$17,242
Utilities	$5,443
Travel	$2,775
Supplies	$10,553
Development	$3,225
Maintenance	$1,075

103. Candy. The graph below shows U.S. candy sales in 2016 during four holiday periods. Find the sum of these seasonal candy sales.

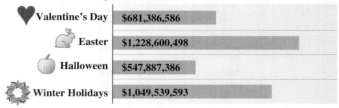

Source: Nielsen AOD

104. Airline safety. The following graph shows the U.S. passenger airlines accident report for the years 2004–2015. How many accidents were there in this 12-year time span?

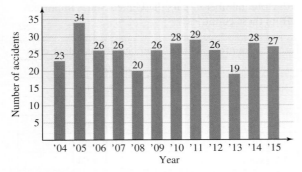

Source: National Transportation Safety Board

105. Flags. To decorate a city flag, yellow fringe is to be sewn around its outside edges, as shown. The fringe is sold by the inch. How many inches of fringe must be purchased to complete the project?

SAN ANTONIO
TEXAS
34 in.
64 in.

106. Decorating. A child's bedroom is rectangular in shape with dimensions 15 feet by 11 feet. How many feet of wallpaper border are needed to wrap around the entire room?

107. Floor mats. Estimate the amount of plastic trim used around the floor mat shown below.

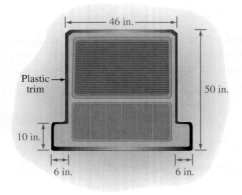

46 in.
Plastic trim
50 in.
10 in.
6 in.   6 in.

108. Fences. A square piece of land measuring 209 feet on all four sides is approximately one *acre*. How many feet of chain-link fencing are needed to enclose a piece of land this size?

109. Traffic accidents. Police used an entire roll of yellow "DO NOT CROSS" barricade tape to seal off a rectangular region around an automobile accident, as shown below. The width of the rectangle was 50 feet and the length was 25 feet more than that. How long was the roll of yellow tape?

POLICE LINE   DO NOT CROSS

110. Arc welding. A "bead" of welding is placed around the outside edge of a square steel plate with sides 37 inches long. How long is the entire weld "bead"?

WRITING

111. Explain why the operation of addition is commutative.

112. Explain why the operation of addition is associative.

113. In this section, it is said that estimation is a *tradeoff*. Give one benefit and one drawback of estimation.

114. A student added three whole numbers top to bottom and then bottom to top, as shown below. What does the result in red indicate? What should the student do next?

$$\begin{array}{r} 1{,}689 \\ 496 \\ 315 \\ +\ \ 788 \\ \hline 1{,}599 \end{array}$$

REVIEW

115. Write each number in expanded notation.

   a. 3,125   **b.** 60,037

116. Round 6,354,784 to the nearest …

   a. ten   **b.** hundred

   c. ten thousand   **d.** hundred thousand

SECTION 1.3   Subtracting Whole Numbers

Everyone uses subtraction of whole numbers. For example, to find the sale price of an item, a store clerk subtracts the discount from the regular price. To measure climate change, a scientist subtracts the high and low temperatures. A trucker subtracts odometer readings to calculate the number of miles driven on a trip.

OBJECTIVES

1 Subtract whole numbers.

2 Subtract whole numbers with borrowing.

3 Check subtractions using addition.

4 Estimate differences of whole numbers.

5 Solve application problems by subtracting whole numbers.

6 Evaluate expressions involving addition and subtraction.

OBJECTIVE 1   Subtract whole numbers.

To subtract two whole numbers, think of taking away objects from a set. For example, if we start with a set of 9 stars and take away a set of 4 stars, a set of 5 stars is left.

We can write this subtraction problem in **horizontal** or **vertical form** using a **subtraction symbol** −, which is read as "minus." We call the number from which another number is subtracted the **minuend**. The number being subtracted is called the **subtrahend**, and the answer is called the **difference**.

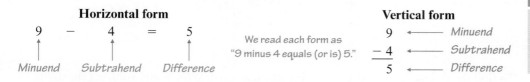

LANGUAGE OF MATHEMATICS

The prefix **sub** means *below*, as in *sub*marine or *sub*way. Notice that in vertical form, the *sub*trahend is written below the minuend.

To subtract two whole numbers that are less than 10, we rely on our understanding of basic subtraction facts. For example,

$$6 - 3 = 3, \qquad 7 - 2 = 5, \qquad \text{and} \qquad 9 - 8 = 1$$

To subtract two whole numbers that are greater than 10, we can use vertical form by stacking them with their corresponding place values lined up. Then we simply subtract the digits in each corresponding column.

EXAMPLE 1   Subtract: 59 − 27

Strategy We will write the subtraction in vertical form with the ones digits in a column and the tens digits in a column. Then we will subtract the digits in each column, working from right to left.

WHY Like money, where pennies are only subtracted from pennies and dimes are only subtracted from dimes, we can only subtract digits with the same place value—ones from ones and tens from tens.

Solution

We start at the right and subtract the ones digits and then the tens digits, and write each difference below the horizontal bar.

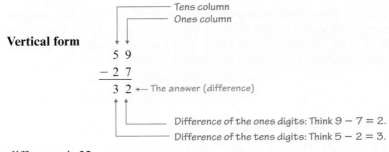

Vertical form

The difference is 32.

Self Check 1

Subtract: 68 − 31

Now Try Problems 15 and 21

EXAMPLE 2 Subtract 235 from 6,496.

Strategy We will translate the sentence to mathematical symbols and then perform the subtraction. We must be careful when translating the instruction to subtract one number *from* another number.

WHY The order of the numbers in the sentence must be reversed when we translate to symbols.

Solution

Since 235 is the number to be subtracted, it is the subtrahend.

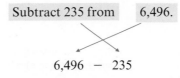

To find the difference, we write the subtraction in vertical form and subtract the digits in each column, working from right to left.

$$\begin{array}{r} 6{,}496 \\ -235 \\ \hline 6{,}261 \end{array}$$

└ Bring down the 6 in the thousands column.

When 235 is subtracted from 6,496, the difference is 6,261.

Caution! When subtracting two numbers, it is very important that we write them in the correct order, because subtraction is *not* commutative. For instance, in Example 2, if we had incorrectly translated "*Subtract 235 from 6,496*" as 235 − 6,496, we see that the difference is not 6,261. In fact, the difference is not even a whole number.

Self Check 2

Subtract: 817 from 1,958.

Now Try Problem 23

OBJECTIVE 2 Subtract whole numbers with borrowing.

If the subtraction of the digits in any place-value column requires that we subtract a larger digit from a smaller digit, we must **borrow** or **regroup**.

EXAMPLE 3 Subtract: 32
                     -15

Strategy As we prepare to subtract in each column, we will compare the digit in the subtrahend (bottom number) to the digit directly above it in the minuend (top number).

WHY If a digit in the subtrahend is greater than the digit directly above it in the minuend, we must borrow (regroup) to subtract in that column.

Solution

To help you understand the process, each step of this subtraction is explained separately. Your solution need only look like the *last* step.

     We write the subtraction in vertical form to line up the tens digits and line up the ones digits.

$$\begin{array}{r} 32 \\ -15 \end{array}$$

Since 5 in the ones column of **15** is greater than 2 in the ones column of **32**, we cannot immediately subtract in that column because $2 - 5$ is *not* a whole number. To subtract in the ones column, we must regroup by borrowing 1 ten from 3 in the tens column. In this regrouping process, we use the fact that 1 ten = 10 ones.

$$\begin{array}{r} {\overset{2}{\cancel{3}}}\,{\overset{12}{2}} \\ -15 \\ \hline 7 \end{array}$$
Borrow 1 ten from 3 in the tens column and change the 3 to 2. Add the borrowed 10 to the digit 2 in the ones column of the minuend to get 12. This step is called regrouping. Then subtract in the ones column: $12 - 5 = 7$.

$$\begin{array}{r} {\overset{2}{\cancel{3}}}\,{\overset{12}{2}} \\ -15 \\ \hline 17 \end{array}$$
Subtract in the tens column: $2 - 1 = 1$.

Your solution should look like this:
$$\begin{array}{r} {\overset{2\ 12}{3\,2}} \\ -15 \\ \hline 17 \end{array}$$

The difference is 17.

Self Check 3

Subtract: 83
             -36

Now Try ➥ **Problem 27**

     Some subtractions require borrowing from two (or more) place-value columns.

EXAMPLE 4 Subtract: $9{,}927 - 568$

Strategy We will write the subtraction in vertical form and subtract as usual. In each column, we must watch for a digit in the subtrahend that is greater than the digit directly above it in the minuend.

WHY If a digit in the subtrahend is greater than the digit above it in the minuend, we need to borrow (regroup) to subtract in that column.

Solution

We write the subtraction in vertical form so that the corresponding digits are lined up. Each step of this subtraction is explained separately. Your solution should look like the *last* step.

$$\begin{array}{r} 9{,}927 \\ -\phantom{9{,}}568 \end{array}$$

Since 8 in the ones column of **568** is greater than 7 in the ones column of **9,927**, we cannot immediately subtract. To subtract in that column, we must regroup by borrowing 1 ten from 2 in the tens column. In this process, we use the fact that 1 ten = 10 ones.

$$\begin{array}{r} {\overset{1\ 17}{9,92\rlap{/}7}} \\ -\ \ 568 \\ \hline 9 \end{array}$$ Borrow 1 ten from 2 in the tens column and change the 2 to 1. Add the borrowed 10 to the digit 7 in the ones column of the minuend to get 17. Then subtract in the ones column: $17 - 8 = 9$.

Since 6 in the tens column of 568 is greater than 1 in the tens column directly above it, we cannot immediately subtract. To subtract in that column, we must regroup by borrowing 1 hundred from 9 in the hundreds column. In this process, we use the fact that 1 hundred = 10 tens.

$$\begin{array}{r} {\overset{\ \ 11}{\overset{8\ \rlap{/}1\ 17}{9,92\rlap{/}7}}} \\ -\ \ 568 \\ \hline 59 \end{array}$$ Borrow 1 hundred from 9 in the hundreds column and change the 9 to 8. Add the borrowed 10 to the digit 1 in the tens column of the minuend to get 11. Then subtract in the tens column: $11 - 6 = 5$.

Complete the solution by subtracting in the hundreds column ($8 - 5 = 3$) and bringing down the 9 in the thousands column.

$$\begin{array}{r} {\overset{\ \ 11}{\overset{8\ \rlap{/}1\ 17}{9,92\rlap{/}7}}} \\ -\ \ 568 \\ \hline 9,359 \end{array}$$

> Your solution should look like this:
> $$\begin{array}{r} {\overset{\ \ 11}{\overset{8\ 1\ 17}{9,92\rlap{/}7}}} \\ -\ \ 568 \\ \hline 9,359 \end{array}$$

The difference is 9,359.

The borrowing process is more difficult when the minuend contains one or more zeros.

Self Check 4

Subtract: $6,734 - 356$

Now Try ➡ **Problem 33**

EXAMPLE 5 Subtract: $42,403 - 1,675$

Strategy We will write the subtraction in vertical form. To subtract in the ones column, we will borrow from the hundreds column of the minuend 42,403.

WHY Since the digit in the tens column of 42,403 is 0, it is not possible to borrow from that column.

Solution
We write the subtraction in vertical form so that the corresponding digits are lined up. Each step of this subtraction is explained separately. Your solution should look like the *last* step.

$$\begin{array}{r} 42,403 \\ -\ 1,675 \end{array}$$

Since 5 in the ones column of 1,675 is greater than 3 in the ones column of 42,403, we cannot immediately subtract. It is not possible to borrow from the digit 0 in the tens column of 42,403. We can, however, borrow from the hundreds column to regroup in the tens column, as shown below. In this process, we use the fact that 1 hundred = 10 tens.

$$\begin{array}{r} {\overset{3\ \ 10}{42,4\rlap{/}0\,3}} \\ -\ 1,675 \end{array}$$ Borrow 1 hundred from 4 in the hundreds column and change the 4 to 3. Add the borrowed 10 to the digit 0 in the tens column of the minuend to get 10.

Now we can borrow from the 10 in the tens column to subtract in the ones column.

$$\begin{array}{r} {\overset{\ \ 9}{\overset{3\ \rlap{/}{10}\ 13}{42,4\rlap{/}0\,3}}} \\ -\ 1,675 \\ \hline 8 \end{array}$$ Borrow 1 ten from 10 in the tens column and change the 10 to 9. Add the borrowed 10 to the digit 3 in the ones column of the minuend to get 13. Then subtract in the ones column: $13 - 5 = 8$.

Next, we perform the subtraction in the tens column: $9 - 7 = 2$.

$$
\begin{array}{r}
{\scriptstyle 9} \\
{\scriptstyle 3\ \cancel{10}\ 13} \\
4\,2\,,\cancel{4}\,\cancel{0}\,\cancel{3} \\
-\ \ \ \ 1\,,6\,7\,5 \\
\hline
2\,8
\end{array}
$$

To subtract in the hundreds column, we borrow from the 2 in the thousands column. In this process, we use the fact that 1 thousand $=$ 10 hundreds.

$$
\begin{array}{r}
{\scriptstyle 13\ \ 9} \\
{\scriptstyle 1\ \ \ 3\ \cancel{10}\ 13} \\
4\,2\,,\cancel{4}\,\cancel{0}\,\cancel{3} \\
-\ \ \ \ 1\,,6\,7\,5 \\
\hline
7\,2\,8
\end{array}
$$

Borrow 1 thousand from 2 in the thousands column and change the 2 to 1.
Add the borrowed 10 to the digit 3 in the hundreds column of the minuend to get 13. Then subtract in the hundreds column: $13 - 6 = 7$.

Complete the solution by subtracting in the thousands column ($1 - 1 = 0$) and bringing down the 4 in the ten thousands column.

$$
\begin{array}{r}
{\scriptstyle 13\ \ 9} \\
{\scriptstyle \cancel{1}\ \ \ 3\ \cancel{10}\ 13} \\
4\,\cancel{2}\,,\cancel{4}\,\cancel{0}\,\cancel{3} \\
-\ \ \ \ 1\,,6\,7\,5 \\
\hline
4\,0\,,7\,2\,8
\end{array}
$$

The difference is 40,728.

> Your solution should look like this:
> $$
> \begin{array}{r}
> {\scriptstyle 13\ \ 9} \\
> {\scriptstyle 1\ \ \ 3\ \cancel{10}\ 13} \\
> 4\,2\,,\cancel{4}\,\cancel{0}\,\cancel{3} \\
> -\ \ \ \ 1\,,6\,7\,5 \\
> \hline
> 4\,0\,,7\,2\,8
> \end{array}
> $$

Self Check 5

Subtract: $65,304 - 1,445$

Now Try ➲ Problem 35

OBJECTIVE 3   Check subtractions using addition.

Every subtraction has a **related addition statement**. For example,

$9 - 4 = 5$	because	$5 + 4 = 9$
$25 - 15 = 10$	because	$10 + 15 = 25$
$100 - 1 = 99$	because	$99 + 1 = 100$

These examples illustrate how we can check subtractions. If a subtraction is done correctly, *the sum of the difference and the subtrahend will always equal the minuend*:

$$\text{Difference} + \text{subtrahend} = \text{minuend}$$

EXAMPLE 6   Check the following subtraction using addition:

$$
\begin{array}{r}
3\,,6\,8\,2 \\
-\ 1\,,9\,5\,4 \\
\hline
1\,,7\,2\,8
\end{array}
$$

Strategy We will add the difference (1,728) and the subtrahend (1,954) and compare that result to the minuend (3,682).

WHY If the sum of the difference and the subtrahend gives the minuend, the subtraction checks.

Solution

The subtraction to check   *Its related addition statement*

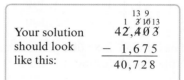

$$
\begin{array}{rl}
3\,,6\,8\,2 & \text{difference} \\
-\ 1\,,9\,5\,4 & +\ \text{subtrahend} \\
\hline
1\,,7\,2\,8 & \text{minuend}
\end{array}
$$

$$
\begin{array}{r}
{\scriptstyle 1\ \ 1} \\
1\,,7\,2\,8 \\
+\ 1\,,9\,5\,4 \\
\hline
3\,,6\,8\,2
\end{array}
$$

Since the sum of the difference and the subtrahend is the minuend, the subtraction is correct.

Self Check 6

Check the following subtraction using addition:

$$
\begin{array}{r}
9\,,7\,8\,4 \\
-\ 4\,,7\,9\,2 \\
\hline
4\,,8\,9\,2
\end{array}
$$

Now Try ➲ Problem 39

OBJECTIVE 4 **Estimate differences of whole numbers.**

Estimation is used to find an approximate answer to a problem.

EXAMPLE 7   Estimate the difference: 89,070 − 5,431

Strategy We will use front-end rounding to approximate the 89,070 and 5,431. Then we will find the difference of the approximations.

WHY Front-end rounding produces whole numbers containing many 0's. Such numbers are easier to subtract.

Solution
Both the minuend and the subtrahend are rounded to their *largest place value* so that all but their first digit is zero. Then we subtract the approximations using vertical form.

$$
\begin{array}{rcl}
89{,}070 & \rightarrow & 90{,}000 \quad \text{Round to the nearest ten thousand.} \\
- \ \ 5{,}431 & \rightarrow & - \ \ 5{,}000 \quad \text{Round to the nearest thousand.} \\
\hline
 & & 85{,}000
\end{array}
$$

The estimate is 85,000. If we calculate 89,070 − 5,431, the difference is exactly 83,639. Note that the estimate is close: It's only 1,361 more than 83,639.

Self Check 7

Estimate the difference: 64,259 − 7,604

Now Try ⟶ Problem 43

OBJECTIVE 5 **Solve application problems by subtracting whole numbers.**

To answer questions about *how much more* or *how many more*, we use subtraction.

EXAMPLE 8   **Horses.** Big Jake, the world's largest horse, weighs 2,600 pounds. Thumbelina, the world's smallest horse, weighs 57 pounds. How much more does Big Jake weigh than Thumbelina? (Source: *Guinness Book of World Records*, 2013)

Strategy We will carefully read the problem, looking for a key word or phrase.

WHY Key words and phrases indicate which arithmetic operation(s) should be used to solve the problem.

Solution
In the second sentence of the problem, the phrase *How much more* indicates that we should subtract the weights of the horses. We translate the words of the problem to numbers and symbols.

The number of pounds more that Big Jake weighs	is equal to	the weight of Big Jake	minus	the weight of Thumbelina.

The number of pounds more that Big Jake weighs	=	2,600	−	57

Use vertical form to perform the subtraction:

$$
\begin{array}{r}
\overset{9}{}\overset{5\ \cancel{10}\ 10}{\cancel{2{,}600}} \\
-\ \ \ \ 57 \\
\hline
2{,}543
\end{array}
$$

Self Check 8

Elephants. An average male African elephant weighs 13,000 pounds. An average male Asian elephant weighs 11,900 pounds. How much more does an African elephant weigh than an Asian elephant?

Now Try ⟶ Problem 87

Big Jake weighs 2,543 pounds more than Thumbelina.

AP Images/Carrie Antlfinger

Brad Barket/Getty Images

EXAMPLE 9   **Radio stations.**   In 2005, there were 763 oldies radio stations in the United States. By 2016, there were 412 fewer. How many oldies radio stations were there in 2016? (Source: *The World Almanac and Book of Facts*, 2017)

Strategy We will carefully read the problem, looking for a key word or phrase.

WHY Key words and phrases indicate which arithmetic operation(s) should be used to solve the problem.

Solution

The key phrase 412 *fewer* indicates subtraction. We translate the words of the problem to numbers and symbols.

The number of oldies radio stations in 2016	is	412	fewer than	the number of oldies radio stations in 2005.

$$\text{The number of oldies radio stations in 2016} = 763 - 412$$

Use vertical form to perform the subtraction:

$$\begin{array}{r} 763 \\ -412 \\ \hline 351 \end{array}$$

In 2016, there were 351 oldies radio stations in the United States.

LANGUAGE OF MATHEMATICS

Here are some more key words and phrases that often indicate *subtraction*:

loss	decrease
down	backward
fell	less than
fewer	reduce
remove	debit
in the past	remains
declined	take away

Self Check 9

Healthy diets. When Dwayne "The Rock" Johnson began to help Nina Gibson lose weight, she weighed 280 pounds. With diet and exercise, she eventually dropped 115 pounds. What was her weight then?
(Source: today.com)

Now Try Problem 99

Using Your Calculator ▶   The Subtraction Key: High School Sports

In the 2015–2016 school year, the number of boys who participated in high school sports was 4,544,574 and the number of girls was 3,324,326. (Source: National Federation of State High School Associations) We can use the subtraction key $\boxed{-}$ on a calculator to determine how many more boys than girls participated in high school sports that year.

4544574 $\boxed{-}$ 3324326 $\boxed{=}$ $\boxed{1220248}$

On some calculator models, the $\boxed{\text{ENTER}}$ key is pressed instead of $\boxed{=}$ for the result to be displayed.

In the 2015–2016 school year, 1,220,248 more boys than girls participated in high school sports.

OBJECTIVE 6 Evaluate expressions involving addition and subtraction.

In arithmetic, numbers are combined with the operations of addition, subtraction, multiplication, and division to create **expressions**. For example,

$$15 + 6, \qquad 873 - 99, \qquad 6{,}512 \times 24, \qquad \text{and} \qquad 42 \div 7$$

are expressions.

Expressions can contain more than one operation. That is the case for the expression $27 - 16 + 5$, which contains addition *and* subtraction. To **evaluate** (find the value of) expressions written in horizontal form that involve addition and subtraction, we perform the operations as they occur *from left to right.*

Caution! When making the calculation in Example 10, we must perform the subtraction first because it occurs first when reading left to right. If the addition is done first, we get the incorrect answer 6.

$$27 - 16 + 5 = 27 - 21$$
$$= 6$$

EXAMPLE 10 Evaluate: $27 - 16 + 5$

Strategy We will perform the subtraction first and add 5 to that result.

WHY The operations of addition and subtraction must be performed as they occur from left to right.

Solution
We will write the steps of the solution in horizontal form.

$$27 - 16 + 5 = 11 + 5 \quad \text{Working left to right, do the subtraction first: } 27 - 16 = 11.$$
$$= 16 \quad \text{Now do the addition.}$$

Answers to Self Checks

1. 37   **2.** 1,141   **3.** 47   **4.** 6,378   **5.** 63,859   **6.** The subtraction is incorrect.   **7.** 52,000
8. An African elephant weighs 1,100 lb more than an Asian elephant.   **9.** After the dieting and exercise program, Nina weighed 165 lb.   **10.** 54

SECTION 1.3 **STUDY SET**

VOCABULARY

Fill in the blanks.

1. In the subtraction problem shown below, label the *minuend*, the *subtrahend*, and the *difference*.

$$25 \quad - \quad 10 \quad = \quad 15$$

2. If the subtraction of the digits in any place-value column requires that we subtract a larger digit from a smaller digit, we must _____ or *regroup*.

3. The words *fall, lose, reduce,* and *decrease* often indicate the operation of _____.

4. Every subtraction has a _____ addition statement. For example,

$$7 - 2 = 5 \text{ because } 5 + 2 = 7$$

5. To see whether the result of a subtraction is reasonable, we can round the minuend and subtrahend and _____ the difference.

6. To *evaluate* an expression such as $58 - 33 + 9$ means to find its _____.

CONCEPTS

Fill in the blanks.

7. The subtraction $7 - 3 = 4$ is related to the addition statement ▢ + ▢ = ▢.

8. The operation of _____ can be used to check the result of a subtraction: If a subtraction is done correctly, *the _____ of the difference and the subtrahend will always equal the minuend.*

9. To *evaluate* (find the value of) an expression that contains both addition and subtraction, we perform the operations as they occur from ____ to ____.

10. To answer questions about *how much more* or *how many more*, we can use _____.

NOTATION

11. Fill in the blank: The subtraction symbol − is read as "_____."

12. Write the following subtraction fact in words:
$$28 - 22 = 6$$

13. Which expression is the correct translation of the sentence: *Subtract 30 from 83.*
$$83 - 30 \quad \text{or} \quad 30 - 83$$

14. Fill in the blanks to complete each step:
$$36 - 11 + 5 = \boxed{} + 5$$
$$= \boxed{}$$

GUIDED PRACTICE

Subtract. See Example 1.

15. $37 - 14$

16. $42 - 31$

17. $\begin{array}{r} 89 \\ -28 \\ \hline \end{array}$

18. $\begin{array}{r} 95 \\ -32 \\ \hline \end{array}$

19. $596 - 372$

20. $869 - 425$

21. $\begin{array}{r} 674 \\ -371 \\ \hline \end{array}$

22. $\begin{array}{r} 257 \\ -155 \\ \hline \end{array}$

Subtract. See Example 2.

23. 347 from 7,989

24. 283 from 9,799

25. 405 from 2,967

26. 304 from 1,736

Subtract. See Example 3.

27. $\begin{array}{r} 53 \\ -17 \\ \hline \end{array}$

28. $\begin{array}{r} 42 \\ -19 \\ \hline \end{array}$

29. $\begin{array}{r} 96 \\ -48 \\ \hline \end{array}$

30. $\begin{array}{r} 94 \\ -37 \\ \hline \end{array}$

Subtract. **See Example 4.**

31. 8,746 − 289

32. 7,531 − 276

33.   6,961
      −   478

34.   4,823
      −   667

Subtract. **See Example 5.**

35. 54,506 − 2,829

36. 69,403 − 4,635

37.   48,402
      −   3,958

38.   39,506
      −   1,729

Check each subtraction using addition. **See Example 6.**

39.   298
      −175
      ───
       123

40.   469
      −237
      ───
       132

41.   4,539
      −3,275
      ─────
       1,364

42.   2,698
      −1,569
      ─────
       1,129

Estimate each difference. **See Example 7.**

43. 67,219 − 4,076

44. 45,333 − 3,410

45. 83,872 − 27,281

46. 74,009 − 37,405

Evaluate each expression. **See Example 10.**

47. 35 − 12 + 6

48. 47 − 23 + 4

49. 56 − 31 + 12

50. 89 − 47 + 6

51. 574 + 47 − 13

52. 863 + 39 − 11

53. 966 + 143 − 61

54. 659 + 235 − 62

TRY IT YOURSELF

Perform the operations.

55. 416 − 357

56. 787 − 696

57.   3,430
      −   529

58.   2,470
      −   863

59. Subtract 199 from 301.

60. Subtract 78 from 2,047.

61.   367
      −347

62.   224
      −122

63. 633 − 598 + 30

64. 600 − 497 + 60

65. 420 − 390

66. 330 − 270

67. 20,007 − 78

68. 70,006 − 48

69. 852 − 695 + 40

70. 397 − 348 + 65

71.   17,246
      −   6,789

72.   34,510
      −   27,593

73.   15,700
      − 15,397

74.   35,600
      − 34,799

75. Subtract 1,249 from 50,009.

76. Subtract 2,198 from 20,020.

77. 120 + 30 − 40

78. 600 + 99 − 54

79.   167,305
      −   23,746

80.   393,001
      −   35,002

81. 29,307 − 10,008

82. 40,012 − 19,045

LOOK ALIKES

Perform each operation.

83. a. 654 subtract 127

   b. 52 subtracted from 321

84. a. 871 subtract 192

   b. 192 subtracted from 413

85. a.   302
          +175

   b.   302
          −175

86. a.   2,135
          +   999

   b.   2,135
          −   999

APPLICATIONS

87. World records.   The world's largest pumpkin weighed in at 2,623 pounds, and the world's largest watermelon weighed in at 351 pounds. How much more did the pumpkin weigh? (Source: ibtimes.com, guinnessworldrecords.com)

88. Trucks.   The Nissan Titan King Cab XE weighs 5,230 pounds and the Honda Ridgeline RTL weighs 4,553 pounds. How much more does the Nissan Titan weigh?

89. Farmer's markets.   See the graph below. How many more farmer's markets were there in 2014 compared to 2010?

90. Farmer's markets.   See the graph below. Between what two years was there the greatest increase in the number of farmer's markets in the U.S.? What was the increase?

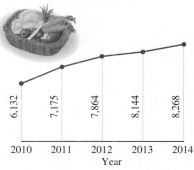

**Number of Farmer's Markets
in the U.S.**

| 2010 | 2011 | 2012 | 2013 | 2014 |
| 6,132 | 7,175 | 7,864 | 8,144 | 8,268 |

Year

Source: USDA-AMS-marketing services division

91. Mileage.   Find the distance (in miles) that a trucker drove on a trip from San Diego to Houston using the odometer readings shown below.

Truck odometer reading leaving San Diego   Truck odometer reading arriving in Houston

92. Diets.   Use the bathroom scale readings shown below to find the number of pounds that a dieter lost.

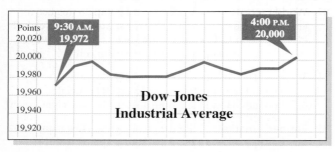

January   October

93. Renting limos.   A group of high school students paid a limousine company $510 to take them to and from their prom. If that included a gratuity (tip) of $85, how much did it cost them to rent the limo?

94. Magazines.   In 2016, *Reader's Digest* had a circulation of 2,662,066. By what amount did this exceed *TV Guide's* circulation of 1,571,537? (Source: *The World Almanac and Book of Facts*, 2017)

95. The stock market.   How many points did the Dow Jones Industrial Average gain on the day described by the graph?

96. Transplants.   See the graph below. Find the decrease in the number of patients waiting for a liver transplant from:

   a. 2006 to 2008   **b.** 2012 to 2014

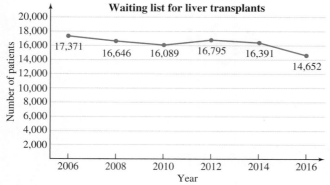

Source: Organ Procurement and Transplant Network, United Network for Organ Sharing

97. Jewelry.   Gold melts at about 1,947°F. The melting point of silver is 183°F lower. What is the melting point of silver?

98. Energy costs.   The electricity cost to run a 10-year-old refrigerator for 1 year is $133. A new energy-saving refrigerator costs $85 less to run for 1 year. What is the electricity cost to run the new refrigerator for 1 year?

99. Zip codes.   As of 2017, the state of Florida has 1,118 fewer zip codes than California. If California has 2,590 zip codes, how many does Florida have?

100. Reading blueprints.   Find the length of the motor on the machine shown in the blueprint.

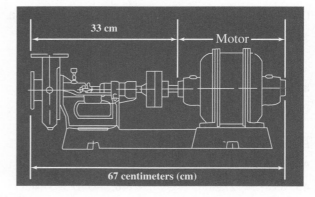

101. Banking.   A savings account contained $1,370. After a withdrawal of $197 and a deposit of $340, how much was left in the account?

102. Physical exams.   A blood test found a man's "bad" cholesterol level to be 205. With a change of eating habits, he lowered it by 27 points in 6 months. One year later, however, the level had risen by 9 points. What was his cholesterol level then?

Refer to the teachers' salary schedule shown below. To use this table, note that a fourth-year teacher (Step 4) in Column 2 makes $52,209 per year.

103. a. What is the salary of a teacher on Step 2/Column 2?

   b. How much more will that teacher make next year when she gains 1 year of teaching experience and moves down to Step 3 in that column?

104. a. What is the salary of a teacher on Step 4/Column 1?

   b. How much more will that teacher make next year when he gains 1 year of teaching experience and takes enough coursework to move over to Column 2?

Teachers' Salary Schedule ABC Unified School District

Years teaching	Column 1	Column 2	Column 3
Step 1	$46,785	$48,243	$49,701
Step 2	$48,107	$49,565	$51,023
Step 3	$49,429	$50,887	$52,345
Step 4	$50,751	$52,209	$53,667
Step 5	$52,073	$53,531	$54,989

WRITING

105. Explain why the operation of subtraction is not commutative.

106. List five words or phrases that indicate subtraction.

107. Explain how addition can be used to check subtraction.

108. The borrowing process is more difficult when the minuend contains one or more zeros. Give an example and explain why.

REVIEW

109. Round 5,370,645 to the indicated place value.
  a. Nearest ten
  b. Nearest ten thousand
  c. Nearest hundred thousand

110. Write 72,001,015
  a. in words
  b. in expanded notation

Find the perimeter of the square and the rectangle.

111.

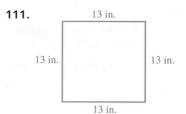

112.

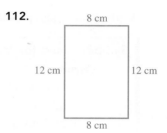

Add.

113.
```
   345
 4,672
+  513
```

114.
```
   813
 7,487
+  654
```

SECTION 1.4 Multiplying Whole Numbers

Everyone uses multiplication of whole numbers. For example, to double a recipe, a cook multiplies the amount of each ingredient by two. To determine the floor space of a dining room, a carpeting salesperson multiplies its length by its width. An accountant multiplies the number of hours worked by the hourly pay rate to calculate the weekly earnings of employees.

OBJECTIVES

1 Multiply whole numbers by one-digit numbers.

2 Multiply whole numbers that end with zeros.

3 Multiply whole numbers by two- (or more) digit numbers.

4 Use properties of multiplication to multiply whole numbers.

5 Estimate products of whole numbers.

6 Solve application problems by multiplying whole numbers.

7 Find the area of a rectangle.

OBJECTIVE 1 Multiply whole numbers by one-digit numbers.

In the following display, there are 4 rows, and each of the rows has 5 stars.

We can find the total number of stars in the display by adding: 5 + 5 + 5 + 5 = 20.

This problem can also be solved using a simpler process called **multiplication**. Multiplication is repeated addition, and it is written using a **multiplication symbol ×**, which is read as "times." Instead of *adding* four 5's to get 20, we can *multiply* 4 and 5 to get 20.

Repeated addition   **Multiplication**

$$5 + 5 + 5 + 5 \quad = \quad 4 \times 5 = 20 \qquad \text{Read as "4 times 5 equals (or is) 20."}$$

We can write multiplication problems in **horizontal** or **vertical form**. The numbers that are being multiplied are called **factors**, and the answer is called the **product**.

Horizontal form
$$4 \times 5 = 20$$
Factor Factor Product

Vertical form
$$\begin{array}{r} 5 \leftarrow \text{Factor} \\ \times\ 4 \leftarrow \text{Factor} \\ \hline 20 \leftarrow \text{Product} \end{array}$$

A **raised dot** · and **parentheses ( )** are also used to write multiplication in horizontal form.

> **Symbols Used for Multiplication**
>
> | Symbol | | Example |
> |---|---|---|
> | $\times$ | times symbol | 4×5 |
> | $\cdot$ | raised dot | $4 \cdot 5$ |
> | ( ) | parentheses | $(4)(5)$ or $4(5)$ or $(4)5$ |

If you need to review the basic multiplication facts, they can be found in Appendix 1 at the back of the book.

To multiply whole numbers that are less than 10, we rely on our understanding of basic multiplication facts. For example,

$$2 \cdot 3 = 6, \qquad 8(4) = 32, \qquad \text{and} \qquad 9 \times 7 = 63$$

To multiply larger whole numbers, we can use vertical form by stacking them with their corresponding place values lined up. Then we make repeated use of basic multiplication facts.

Caution! In this book, we seldom use the $\times$ symbol because it can be confused with the letter x.

EXAMPLE 1   Multiply: 8×47

Strategy We will write the multiplication in vertical form. Then, working right to left, we will multiply each digit of 47 by 8 and carry, if necessary.

WHY This process is simpler than treating the problem as repeated addition and adding eight 47's.

Solution
To help you understand the process, each step of this multiplication is explained separately. Your solution need only look like the *last* step.

$$\begin{array}{r} \text{Tens column} \\ \text{Ones column} \end{array}$$

Vertical form
$$\begin{array}{r} 4\ 7 \\ \times\ \ 8 \end{array}$$

We begin by multiplying 7 by 8.

$$\begin{array}{r} {}^{5}\ \\ 4\ 7 \\ \times\ \ 8 \\ \hline 6 \end{array}$$
Multiply 7 by 8. The product is 56. Write 6 in the ones column of the answer, and carry 5 to the tens column.

$$\begin{array}{r} {}^{5}\ \\ 4\ 7 \\ \times\ \ 8 \\ \hline 3\ 7\ 6 \end{array}$$
Multiply 4 by 8. The product is 32. To the 32, add the carried 5 to get 37. Write 7 in the tens column and the 3 in the hundreds column of the answer.

The product is 376.

> Your solution should look like this:
> $$\begin{array}{r} {}^{5}\ \\ 47 \\ \times\ \ 8 \\ \hline 376 \end{array}$$

Self Check 1

Multiply: 6×54

Now Try ➡ **Problem 19**

OBJECTIVE 2 **Multiply whole numbers that end with zeros.**

An interesting pattern develops when a whole number is multiplied by 10, 100, 1,000, and so on. Consider the following multiplications involving 8:

$8 \cdot 10 = 80$ There is one zero in 10. The product is 8 with one 0 attached.

$8 \cdot 100 = 800$ There are two zeros in 100. The product is 8 with two 0's attached.

$8 \cdot 1,000 = 8,000$ There are three zeros in 1,000. The product is 8 with three 0's attached.

$8 \cdot 10,000 = 80,000$ There are four zeros in 10,000. The product is 8 with four 0's attached.

These examples illustrate the following rule.

Multiplying a Whole Number by 10, 100, 1,000, and So On

To find the product of a whole number and 10, 100, 1,000, and so on, attach the number of zeros in that number to the right of the whole number.

EXAMPLE 2   Multiply: **a.** $6 \times 1,000$ **b.** $45 \cdot 100$ **c.** $912(10,000)$

Strategy For each multiplication, we will identify the factor that ends in zeros and count the number of zeros that it contains.

WHY Each product can then be found by attaching that number of zeros to the other factor.

Solution

a. $6 \times 1,000 = 6,000$ Since 1,000 has three zeros, attach three 0's after 6.

b. $45 \cdot 100 = 4,500$ Since 100 has two zeros, attach two 0's after 45.

c. $912(10,000) = 9,120,000$ Since 10,000 has four zeros, attach four 0's after 912.

Self Check 2

Multiply:
a. $9 \times 1,000$
b. $25 \cdot 100$
c. $875(1,000)$

Now Try ➡ Problems 23 and 25

We can use an approach similar to that of Example 2 for multiplication involving any whole numbers that end in zeros. For example, to find $67 \cdot 2,000$, we have

$67 \cdot 2,000 = 67 \cdot 2 \cdot 1,000$ Write 2,000 as $2 \cdot 1,000$.

$\qquad\qquad = 134 \cdot 1,000$ Working left to right, multiply 67 and 2 to get 134.

$\qquad\qquad = 134,000$ Since 1,000 has three zeros, attach three 0's after 134.

This example suggests that to find $67 \cdot 2,000$ we simply multiply 67 and 2 and attach three zeros to that product. This method can be extended to find products of two factors that *both* end in zeros.

EXAMPLE 3   Multiply: **a.** 14 · 300   **b.** 3,500 · 50,000

Strategy We will multiply the nonzero leading digits of each factor. To that product, we will attach the sum of the number of trailing zeros in the factors.

WHY This method is faster than the standard vertical form multiplication of factors that contain many zeros.

Solution

a. The factor **300** has two trailing zeros.

$14 \cdot 300 = 4,200$ *Attach two 0's after 42.*

Multiply 14 and 3 to get 42.

b. The factors **3,500** and **50,000** have a total of six trailing zeros.

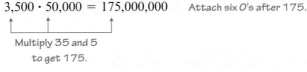

$3,500 \cdot 50,000 = 175,000,000$ *Attach six 0's after 175.*

Multiply 35 and 5 to get 175.

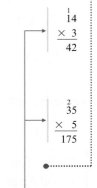

Success Tip Calculations that you cannot perform in your head should be shown outside the steps of your solution.

Self Check 3

Multiply:

a. 15 · 900

b. 3,100 · 7,000

Now Try ➲ **Problems 29 and 33**

OBJECTIVE 3 **Multiply whole numbers by two- (or more) digit numbers.**

EXAMPLE 4   Multiply: 23 · 436

Strategy We will write the multiplication in vertical form. Then we will multiply 436 by 3 and by 20, and add those products.

WHY Since $23 = 3 + 20$, we can multiply 436 by 3 and by 20, and add those products.

Solution

Each step of this multiplication is explained separately. Your solution need only look like the *last* step.

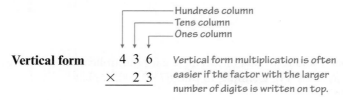

Vertical form

$$\begin{array}{r} 4\ 3\ 6 \\ \times\ \ 2\ 3 \\ \hline \end{array}$$

Vertical form multiplication is often easier if the factor with the larger number of digits is written on top.

We begin by multiplying 436 by 3.

$$\begin{array}{r} 4\ 3\overset{1}{6} \\ \times\ \ 2\ 3 \\ \hline 8 \end{array}$$

Multiply 6 by 3. The product is 18. Write 8 in the ones column and carry 1 to the tens column.

$$\begin{array}{r} 4\overset{1}{3}\overset{1}{6} \\ \times\ \ 2\ 3 \\ \hline 0\ 8 \end{array}$$

Multiply 3 by 3. The product is 9. To the 9, add the carried 1 to get 10. Write the 0 in the tens column and carry the 1 to the hundreds column.

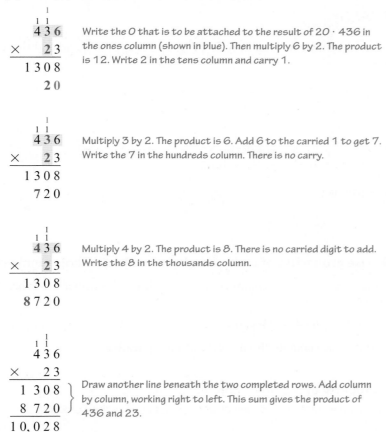

$$\begin{array}{r} \overset{1\ 1}{4\ 3\ 6} \\ \times\ \ \ 2\ 3 \\ \hline 1\ 3\ 0\ 8 \end{array}$$

Multiply 4 by 3. The product is 12. Add the 12 to the carried 1 to get 13. Write 13.

We continue by multiplying 436 by 2 tens, or 20. If we think of 20 as 2 · 10, then we simply multiply 436 by 2 and attach one zero to the result.

$$\begin{array}{r} \overset{1}{\overset{1\ 1}{4\ 3\ 6}} \\ \times\ \ \ 2\ 3 \\ \hline 1\ 3\ 0\ 8 \\ 2\ 0 \end{array}$$

Write the 0 that is to be attached to the result of 20 · 436 in the ones column (shown in blue). Then multiply 6 by 2. The product is 12. Write 2 in the tens column and carry 1.

$$\begin{array}{r} \overset{1}{\overset{1\ 1}{4\ 3\ 6}} \\ \times\ \ \ 2\ 3 \\ \hline 1\ 3\ 0\ 8 \\ 7\ 2\ 0 \end{array}$$

Multiply 3 by 2. The product is 6. Add 6 to the carried 1 to get 7. Write the 7 in the hundreds column. There is no carry.

$$\begin{array}{r} \overset{1\ 1}{4\ 3\ 6} \\ \times\ \ \ 2\ 3 \\ \hline 1\ 3\ 0\ 8 \\ 8\ 7\ 2\ 0 \end{array}$$

Multiply 4 by 2. The product is 8. There is no carried digit to add. Write the 8 in the thousands column.

$$\begin{array}{r} \overset{1\ 1}{4\ 3\ 6} \\ \times\ \ \ 2\ 3 \\ \hline 1\ 3\ 0\ 8 \\ 8\ 7\ 2\ 0 \\ \hline 1\ 0\text{,}0\ 2\ 8 \end{array}$$

Draw another line beneath the two completed rows. Add column by column, working right to left. This sum gives the product of 436 and 23.

The product is 10,028.

Self Check 4

Multiply: 36 · 334

Now Try Problem 37

When a factor in a multiplication contains one or more zeros, we must be careful to enter the correct number of zeros when writing the partial products.

EXAMPLE 5   Multiply: **a.** 406 · 253   **b.** 3,009(2,007)

Strategy We will think of 406 as 6 + 400 and 3,009 as 9 + 3,000.

WHY Thinking of the multipliers (406 and 3,009) in this way is helpful when determining the correct number of zeros to enter in the partial products.

Solution
We will use vertical form to perform each multiplication.

a. Since $406 = 6 + 400$, we will multiply 253 by 6 and by 400, and add those partial products.

$$
\begin{array}{r}
253 \\
\times\ 406 \\
\hline
1\ 518 \quad \leftarrow 6 \cdot 253 \\
101\ 200 \quad \leftarrow 400 \cdot 253.\ \text{Think of 400 as } 4 \cdot 100 \text{ and simply multiply} \\
\hline
102{,}718 \quad\ \ 253 \text{ by 4 and attach two zeros (shown in blue) to the result.}
\end{array}
$$

The product is 102,718.

b. Since $3{,}009 = 9 + 3{,}000$, we will multiply 2,007 by 9 and by 3,000, and add those partial products.

Self Check 5

Multiply:

a. 706(351)

b. 4,004(2,008)

Now Try ⟶ Problem 41

$$
\begin{array}{r}
2{,}007 \\
\times\ 3{,}009 \\
\hline
18\ 063 \quad \leftarrow 9 \cdot 2{,}007 \\
6\ 021\ 000 \quad \leftarrow 3{,}000 \cdot 2{,}007.\ \text{Think of 3,000 as } 3 \cdot 1{,}000 \text{ and simply multiply} \\
\hline
6{,}039{,}063 \quad\ \ 2{,}007 \text{ by 3 and attach three zeros (shown in blue) to the result.}
\end{array}
$$

The product is 6,039,063.

OBJECTIVE 4 **Use properties of multiplication to multiply whole numbers.**

Have you ever noticed that two whole numbers can be multiplied in either order because the result is the same? For example,

$$4 \cdot 6 = 24 \qquad \text{and} \qquad 6 \cdot 4 = 24$$

This example illustrates the **commutative property of multiplication**.

Commutative Property of Multiplication

The order in which whole numbers are multiplied does not change their product.

For example:

$$7 \cdot 5 = 5 \cdot 7$$

Whenever we multiply a whole number by 0, the product is 0. For example,

$$0 \cdot 5 = 0, \qquad 0 \cdot 8 = 0, \qquad \text{and} \qquad 9 \cdot 0 = 0$$

Whenever we multiply a whole number by 1, the number remains the same. For example,

$$3 \cdot 1 = 3, \qquad 7 \cdot 1 = 7, \qquad \text{and} \qquad 1 \cdot 9 = 9$$

These examples illustrate the multiplication properties of 0 and 1.

Success Tip If one (or more) of the factors in a multiplication is 0, the product will be 0. For example,

$$16(27)(0) = 0$$

and

$$109 \cdot 53 \cdot 0 \cdot 2 = 0$$

Multiplication Properties of 0 and 1

The product of any whole number and 0 is 0.
The product of any whole number and 1 is that whole number.

To multiply three numbers, we first multiply two of them and then multiply that result by the third number. In the following examples, we multiply $3 \cdot 2 \cdot 4$ in two ways. The parentheses show us which multiplication to perform first. The steps of the solutions are written in horizontal form.

Method 1: Group 3 · 2

$(3 \cdot 2) \cdot 4 = 6 \cdot 4$ Multiply 3 and 2 to get 6.

$\qquad = 24$ Multiply 6 and 4 to get 24.

Method 2: Group 2 · 4

$3 \cdot (2 \cdot 4) = 3 \cdot 8$ Then multiply 2 and 4 to get 8.

$\qquad = 24$ Then multiply 3 and 8 to get 24.

Same result

Either way, the answer is 24. This example illustrates that changing the grouping when multiplying numbers doesn't affect the result. This property is called the **associative property of multiplication**.

Associative Property of Multiplication

The way in which whole numbers are grouped does not change their product.

For example:

$$(2 \cdot 3) \cdot 5 = 2 \cdot (3 \cdot 5)$$

Sometimes, an application of the associative property can simplify a calculation.

EXAMPLE 6   Find the product: $(17 \cdot 50) \cdot 2$

Strategy We will use the associative property to group 50 with 2.

WHY It is helpful to regroup because 50 and 2 are a pair of numbers that are easily multiplied.

Solution
We will write the solution in horizontal form.

$(17 \cdot 50) \cdot 2 = 17 \cdot (50 \cdot 2)$ Use the associative property of multiplication to regroup the factors.

$\qquad = 17 \cdot 100$ Do the multiplication within the parentheses first.

$\qquad = 1,700$ Since 100 has two zeros, attach two 0's after 17.

Self Check 6

Find the product: $(23 \cdot 25) \cdot 4$

Now Try ➡ **Problem 45**

OBJECTIVE 5 Estimate products of whole numbers.

Estimation is used to find an approximate answer to a problem.

EXAMPLE 7 Estimate the product: $59 \cdot 334$

Strategy We will use front-end rounding to approximate the factors 59 and 334. Then we will find the product of the approximations.

WHY Front-end rounding produces whole numbers containing many 0's. Such numbers are easier to multiply.

Solution
Both of the factors are rounded to their *largest place value* so that all but their first digit is zero.

┌─ Round to the nearest ten. ─┐
$$59 \cdot 334 \qquad\qquad 60 \cdot 300$$
└─ Round to the nearest hundred. ─┘

To find the product of the approximations, $60 \cdot 300$, we simply multiply 6 by 3, to get 18, and attach 3 zeros. Thus, the estimate is 18,**000**.

If we calculate $59 \cdot 334$, the product is exactly 19,706. Note that the estimate is close: It's only 1,706 less than 19,706.

Self Check 7

Estimate the product: $74 \cdot 488$

Now Try Problem 51

OBJECTIVE 6 **Solve application problems by multiplying whole numbers.**

Application problems that involve repeated addition are often more easily solved using multiplication.

EXAMPLE 8 **Daily pay.** In October 2016, the average U.S. manufacturing worker made $26 per hour. At that rate, how much money was earned in an 8-hour workday? (Source: Bureau of Labor Statistics)

Strategy To find the amount earned in an 8-hour workday, we will multiply the hourly rate of $26 by 8.

WHY For each of the 8 hours, the average manufacturing worker earned $26. The amount earned for the day is the sum of eight 26's: $26 + 26 + 26 + 26 + 26 + 26 + 26 + 26$. This repeated addition can be calculated more simply by multiplication.

Solution
We translate the words of the problem to numbers and symbols.

The amount earned in an 8-hr workday	is equal to	the rate per hour	times	8 hours.

$$\text{The amount earned in an 8-hr workday} \quad = \quad 26 \quad \cdot \quad 8$$

Use vertical form to perform the multiplication:

$$\begin{array}{r} \overset{4}{2}6 \\ \times\ \ 8 \\ \hline 208 \end{array}$$

In October 2016, the average U.S. manufacturing worker earned $208 in an 8-hour workday.

Self Check 8

Daily pay. In 2016, the average U.S. construction worker made $28 per hour. At that rate, how much money was earned in an 8-hour workday? (Source: Bureau of Labor Statistics)

Now Try Problem 90

We can use multiplication to count objects arranged in patterns of neatly arranged rows and columns called **rectangular arrays**.

EXAMPLE 9   **Pixels.**   Refer to the illustration at the right. Small squares of color, called *pixels*, create the digital images seen on computer screens. If a 14-inch screen has 640 pixels from side to side and 480 pixels from top to bottom, how many pixels are displayed on the screen?

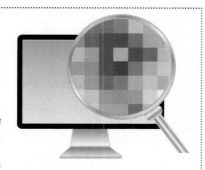

Strategy We will multiply 640 by 480 to determine the number of pixels that are displayed on the screen.

WHY The pixels form a rectangular array with 640 pixels in each row and 480 in each column. Multiplication can be used to count objects in a rectangular array.

Solution

We translate the words of the problem to numbers and symbols.

The number of pixels on the screen	is equal to	the number of pixels in a row	times	the number of pixels in a column.
The number of pixels on the screen	=	640	·	480

To find the product of 640 and 480, we use vertical form to multiply 64 and 48 and attach two zeros to that result.

$$\begin{array}{r} 48 \\ \times\ 64 \\ \hline 192 \\ 2\ 880 \\ \hline 3{,}072 \end{array}$$

Since the product of 64 and 48 is 3,072, the product of 640 and 480 is 307,200. The screen displays 307,200 pixels.

Self Check 9

Pixels. If a 17-inch computer screen has 1,024 pixels from side to side and 768 from top to bottom, how many pixels are displayed on the screen?

Now Try ➔ Problem 97

EXAMPLE 10   **Weight lifting.**   In 1983, Stefan Topurov of Bulgaria was the first man to lift three times his body weight over his head. If he weighed 132 pounds at the time, how much weight did he lift over his head?

Strategy To find how much weight he lifted over his head, we will multiply his body weight by 3.

WHY We can use multiplication to determine the result when a quantity increases in size by 2 *times*, 3 *times*, 4 *times*, and so on.

Solution

We translate the words of the problem to numbers and symbols.

The amount he lifted over his head	was	3	times	his body weight.

Self Check 10

Insects.   Leaf cutter ants can carry pieces of leaves that weigh 30 times their body weight. How much can an ant lift if it weighs 25 milligrams?

Now Try ➲ **Problem 103**

The amount he lifted over his head $= 3 \cdot 132$

Use vertical form to perform the multiplication:

$$\begin{array}{r} 132 \\ \times \quad 3 \\ \hline 396 \end{array}$$

Stefan Topurov lifted 396 pounds over his head.

Using Your Calculator ▶   **The Multiplication Key: Seconds in a Year**

There are 60 seonds in 1 minute, 60 minutes in 1 hour, 24 hours in 1 day, and 365 days in 1 year. We can find the number of seconds in 1 year using the multiplication key $\boxed{\times}$ on a calculator.

60 $\boxed{\times}$ 60 $\boxed{\times}$ 24 $\boxed{\times}$ 364 $\boxed{=}$ $\boxed{31536000}$

On some calculator models, the $\boxed{\text{ENTER}}$ key is pressed instead of the $\boxed{=}$ for the result to be displayed.

There are 31,536,000 seconds in 1 year.

OBJECTIVE 7   **Find the area of a rectangle.**

One important application of multiplication is finding the area of a rectangle. The **area of a rectangle** is the measure of the amount of surface it encloses. Area is measured in square units, such as square inches (written in.²) or square centimeters (written cm²), as shown below.

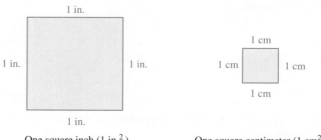

One square inch (1 in.²)   One square centimeter (1 cm²)

The rectangle in the figure below has a length of 5 centimeters and a width of 3 centimeters. Since each small square region covers an area of one square centimeter, each small square region measures 1 cm². The small square regions form a rectangular pattern, with 3 rows of 5 squares.

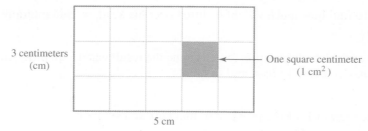

Because there are 5 · 3, or 15, small square regions, the area of the rectangle is 15 cm². This suggests that the area of any rectangle is the product of its length and its width.

$$\boxed{\text{Area of a rectangle}} = \text{length} \cdot \text{width}$$

By using the letter A to represent the area of the rectangle, the letter l to represent the length of the rectangle, and the letter w to represent its width, we can write this **formula** in simpler form. Letters (or symbols), such as A, l, and w, that are used to represent numbers are called **variables**.

Area of a Rectangle

The area, A, of a rectangle is the product of the rectangle's length, l, and its width, w.

   Area = length · width   or $A = l \cdot w$

The formula can be written more simply without the raised dot as $A = lw$.

Caution! Remember that the perimeter of a rectangle is the distance around it and is measured in units such as inches, feet, and miles. The area of a rectangle is the amount of surface it encloses and is measured in square units such as in.², ft², and mi².

EXAMPLE 11 **Gift wrapping.** When completely unrolled, a long sheet of gift wrapping paper has the dimensions shown below. How many square feet of gift wrap are on the roll?

3 ft

12 ft

Strategy We will substitute 12 for the length and 3 for the width in the formula for the area of a rectangle.

WHY To find the number of square feet of paper, we need to find the area of the rectangle shown in the figure.

Solution

$A = lw$ This is the formula for the area of a rectangle.

$A = 12 \cdot 3$ Replace the length l with 12 and the width w with 3.

$A = 36$ Do the multiplication.

There are 36 square feet of wrapping paper on the roll. This can be written in more compact form as 36 ft².

Self Check 11

Advertising. The rectangular posters used on small billboards in the New York subway are 59 inches wide by 45 inches tall. Find the area of a subway poster.

Now Try ➲ Problems 53 and 55

Answers to Self Checks

1. 324   **2. a.** 9,000   **b.** 2,500   **c.** 875,000   **3. a.** 13,500   **b.** 21,700,000   **4.** 12,024   **5. a.** 247,806   **b.** 8,040,032   **6.** 2,300   **7.** 35,000   **8.** $224   **9.** 786,432   **10.** 750 milligrams   **11.** 2,655 in.²

SECTION 1.4   STUDY SET

VOCABULARY

Fill in the blanks.

1. In the multiplication problem shown below, label each *factor* and the *product*.

   5   ·   10   =   50
   ↑   ↑   ↑

2. Multiplication is _____ addition.

3. The _____ property of multiplication states that the order in which whole numbers are multiplied does not change their product. The _____ property of multiplication states that the way in which whole numbers are grouped does not change their product.

4. Letters that are used to represent numbers are called _____.

5. If a square measures 1 inch on each side, its area is 1 _____ inch.

6. The ____ of a rectangle is a measure of the amount of surface it encloses.

CONCEPTS

7. **a.** Write the repeated addition $8 + 8 + 8 + 8$ as a multiplication.

   b. Write the multiplication $7 \cdot 15$ as a repeated addition.

8. **a.** Fill in the blank: A rectangular _____ of red squares is shown below.

   b. Write a multiplication statement that will give the number of red squares shown below.

9. **a.** How many zeros do you attach to the right of 25 to find $25 \cdot 1,000$?

   b. How many zeros do you attach to the right of 8 to find $400 \cdot 2,000$?

10. **a.** Using the numbers 5 and 9, write a statement that illustrates the commutative property of multiplication.

    b. Using the numbers 2, 3, and 4, write a statement that illustrates the associative property of multiplication.

11. Determine whether the concept of *perimeter* or that of *area* should be applied to find each of the following.

    a. The amount of floor space to carpet

    b. The number of inches of lace needed to trim the sides of a handkerchief

    c. The amount of clear glass to be tinted

    d. The number of feet of fencing needed to enclose a playground

12. Perform each multiplication.

    a. $1 \cdot 25$ **b.** $62(1)$

    c. $10 \cdot 0$ **d.** $0(4)$

NOTATION

13. Write three symbols that are used for multiplication.

14. What does ft^2 mean?

15. Write the formula for the area of a rectangle using variables.

16. Which numbers in the work shown below are called partial products?

$$
\begin{array}{r}
86 \\
\times\ 23 \\
\hline
258 \\
1720 \\
\hline
1,978
\end{array}
$$

GUIDED PRACTICE

Multiply. See Example 1.

17. 15×7 18. 19×9

19. 34×8 20. 37×6

Perform each multiplication without using pencil and paper or a calculator. See Example 2.

21. $37 \cdot 100$ 22. $63 \cdot 1,000$

23. 75×10 24. $88 \times 10,000$

25. $107(10,000)$ 26. $323(100)$

27. $512(1,000)$ 28. $673(10)$

Multiply. See Example 3.

29. $68 \cdot 40$ 30. $83 \cdot 30$

31. $56 \cdot 200$ 32. $222 \cdot 500$

33. $130(3,000)$ 34. $630(7,000)$

35. $2,700(40,000)$ 36. $5,100(80,000)$

Multiply. See Example 4.

37. $73 \cdot 128$ 38. $54 \cdot 173$

39. $64(287)$ 40. $72(461)$

Multiply. See Example 5.

41. $602 \cdot 679$ 42. $504 \cdot 729$

43. $3,002(5,619)$ 44. $2,003(1,376)$

Apply the associative property of multiplication to find the product. See Example 6.

45. $(18 \cdot 20) \cdot 5$ 46. $(29 \cdot 2) \cdot 50$

47. $250 \cdot (4 \cdot 135)$ 48. $250 \cdot (4 \cdot 289)$

Estimate each product. See Example 7.

49. 86 · 249

50. 56 · 631

51. 215 · 1,908

52. 434 · 3,789

Find the area of each rectangle or square. See Example 11.

53.

6 in.

14 in.

54.

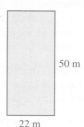

50 m

22 m

55.

12 in.

12 in.

56.

20 cm

20 cm

TRY IT YOURSELF

Multiply.

57.   213
      × 7

58.   863
      × 9

59. 34,474 · 2

60. 54,912 · 4

61.   99
      × 77

62.   73
      × 59

63. 44(55)(0)

64. 81 · 679 · 0 · 5

65. 53 · 30

66. 20 · 78

67.   754
      × 59

68.   846
      × 79

69. (2,978)(3,004)

70. (2,003)(5,003)

71.   916
      × 409

72.   889
      × 507

73. 25 · (4 · 99)

74. (41 · 5) · 20

75. 4,800 × 500

76. 6,400 × 700

77.   2,779
      × 128

78.   3,596
      × 136

79. 370 · 450

80. 280 · 340

LOOK ALIKES

Perform each operation.

81. a.   16
         + 9

b.   16
       − 9

c.   16
       × 9

82. a.   23
         + 18

b.   23
       − 18

c.   23
       × 18

83. a. 405 + 57

b. 405 − 57

c. 405 · 57

84. a. 3,118 + 600

b. 3,118 − 600

c. 3,118 · 600

APPLICATIONS

85. Breakfast cereal.   A cereal maker advertises "Two cups of raisins in every box." Find the number of cups of raisins in a case of 36 boxes of cereal.

86. Snacks.   A candy warehouse sells large four-pound bags of chocolate candies. There are approximately 180 candies per pound. How many candies are there in one bag?

87. Nutrition.   There are 17 grams of fat in one Krispy Kreme chocolate-iced, custard-filled donut. How many grams of fat are there in one dozen of those donuts?

88. Juice.   It takes 13 oranges to make one can of orange juice. Find the number of oranges used to make a case of 24 cans.

89. Birds.   How many times do a hummingbird's wings beat each minute?

65 wingbeats
per second

90. Legal fees.   The average hourly rate for consumer law attorneys in New York with 20 years' experience is $413. If such an attorney bills her client for 15 hours of legal work, what is the fee?

91. Changing units.   There are 12 inches in 1 foot and 5,280 feet in 1 mile. How many inches are there in a mile?

92. Fuel economy.   Mileage figures for a 2016 Ford Mustang GT convertible are shown in the table.

Fuel tank capacity	16 gal
Fuel economy (miles per gallon)	14 city/23 hwy

a. For city driving, how far can it travel on a tank of gas?

b. For highway driving, how far can it travel on a tank of gas?

93. Word count. Generally, the number of words on a page for a published novel is 250. What would be the expected word count for the 308-page children's novel *Harry Potter and the Philosopher's Stone*?

94. Rentals. Mia owns an apartment building with 18 units. Each unit generates a monthly income of $450. Find her total monthly income.

95. Congressional pay. The annual salary of a U.S. House of Representatives member is $174,000. What does it cost per year to pay the salaries of all 36 members of the House from Texas?

96. Crude oil. In 2015, the United States used about 19,389,000 barrels of crude oil per day. One barrel contains 42 gallons of crude oil. How many gallons of crude oil did the United States use in one day?

97. Word processing. A student used the *Insert Table* options shown when typing a report. How many entries will the table hold?

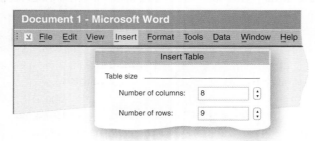

98. Boxing. How many feet of padded rope are needed to make a square boxing ring, 24 feet on each side?

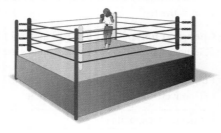

99. Room capacity. A college lecture hall has 17 rows of 33 seats each. A sign on the wall reads, "Occupancy by more than 570 persons is prohibited." If all of the seats are taken, and there is one instructor in the room, is the college breaking the rule?

100. Elevators. There are 14 people in an elevator with a capacity of 2,000 pounds. If the average weight of a person in the elevator is 150 pounds, is the elevator overloaded?

101. Koalas. In one 24-hour period, a koala sleeps 3 times as many hours as it is awake. If it is awake for 6 hours, how many hours does it sleep?

102. Frogs. Bullfrogs can jump as far as 10 times their body length. How far could an 8-inch-long bullfrog jump?

103. Computer hacking. According to a report from Technative, from 2015 to 2016, the number of crypto-ransomware attacks against the corporate sector has grown *six-fold*. If there were 27,000 attacks in 2015, how many were there in 2016? (Source: technative.io)

104. Energy savings. An ENERGY STAR light bulb lasts 8 times as long as a standard 60-watt light bulb. If a standard bulb normally lasts 11 months, how long will an ENERGY STAR bulb last?

105. Prescriptions. How many tablets should a pharmacist put in the container shown in the illustration?

106. Heartbeats. A normal pulse rate for a healthy adult, while resting, can range from 60 to 100 beats per minute.

a. How many beats is that in one day at the lower end of the range?

b. How many beats is that in one day at the upper end of the range?

107. Wrapping presents. When completely unrolled, a long sheet of wrapping paper has the dimensions shown. How many square feet of gift wrap are on the roll?

3 ft

18 ft

108. Poster boards. A rectangular-shaped poster board has dimensions of 24 inches by 36 inches. Find its area.

109. Wyoming. The state of Wyoming is approximately rectangular-shaped, with dimensions 360 miles long and 270 miles wide. Find its perimeter and its area.

110. Comparing rooms. Which has the greater area, a rectangular room that is 14 feet by 17 feet or a square room that is 16 feet on each side? Which has the greater perimeter?

WRITING

111. Explain the difference between 1 foot and 1 square foot.

112. When two numbers are multiplied, the result is 0. What conclusion can be drawn about the numbers?

REVIEW

113. Find the sum of 10,357, 9,809, and 476.

114. Discounts. A radio, originally priced at $367, has been marked down to $179. By how many dollars was the radio discounted?

OBJECTIVES

1 Write the related multiplication statement for a division.

2 Use properties of division to divide whole numbers.

3 Perform long division (no remainder).

4 Perform long division (with a remainder).

5 Use tests for divisibility.

6 Divide whole numbers that end with zeros.

7 Estimate quotients of whole numbers.

8 Solve application problems by dividing whole numbers.

SECTION 1.5 Dividing Whole Numbers

Everyone uses division of whole numbers. For example, to find how many 6-ounce servings a chef can get from a 48-ounce roast, he divides 48 by 6. To split a $36,000 inheritance equally, a brother and sister divide the amount by 2. A professor divides the 35 students in her class into groups of 5 for discussion.

OBJECTIVE 1 Write the related multiplication statement for a division.

To divide whole numbers, think of separating a quantity into equal-sized groups. For example, if we start with a set of 12 stars and divide them into groups of 4 stars, we will obtain 3 groups.

A set of 12 stars.

There are 3 groups of 4 stars.

We can write this division problem using a **division symbol** $\div$, a **long division symbol** $\overline{)}$, or a **fraction bar** $-$. We call the number being divided the **dividend**, and the number that we are dividing by is called the **divisor**. The answer is called the **quotient**.

Division symbol	**Long division symbol**	**Fraction bar**
	Quotient	Dividend Quotient
$12 \div 4 = 3$	$\dfrac{3}{4)\overline{12}}$	$\dfrac{12}{4} = 3$
Dividend Divisor Quotient	Divisor Dividend	Divisor

We read each form as "12 divided by 4 equals (or is) 3."

Recall from Section 1.4 that multiplication is repeated addition. Likewise, division is repeated subtraction. To divide 12 by 4, we ask, "How many 4's can be subtracted from 12?"

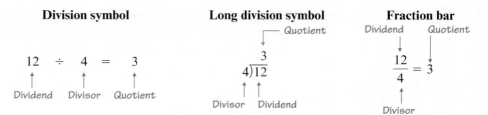

$$\begin{array}{r} 12 \\ -\ 4 \end{array} \Big\}\ \text{Subtract 4 one time.}$$

$$\begin{array}{r} 8 \\ -\ 4 \end{array} \Big\}\ \text{Subtract 4 a second time.}$$

$$\begin{array}{r} 4 \\ -\ 4 \end{array} \Big\}\ \text{Subtract 4 a third time.}$$

$$0$$

Since exactly three 4's can be subtracted from 12 to get 0, we know that $12 \div 4 = 3$.

Another way to answer a division problem is to think in terms of multiplication. For example, the division $12 \div 4$ asks the question, "What must I multiply 4 by to get 12?" Since the answer is 3, we know that

$$12 \div \mathbf{4} = \mathbf{3} \quad \text{because} \quad \mathbf{3} \cdot \mathbf{4} = 12$$

We call $3 \cdot 4 = 12$ the **related multiplication statement** for the division $12 \div 4 = 3$. In general, to write the related multiplication statement for a division, we use:

$$\text{Quotient} \cdot \text{divisor} = \text{dividend}$$

EXAMPLE 1 Write the related multiplication statement for each division.

a. $10 \div 5 = 2$ **b.** $6\overline{)24}$ (quotient 4) **c.** $\dfrac{21}{3} = 7$

Strategy We will identify the quotient, the divisor, and the dividend in each division statement.

WHY A related multiplication statement has the following form: Quotient · divisor = dividend.

Solution

a. $10 \div 5 = 2$ because $2 \cdot 5 = 10$.

Dividend, Quotient, Divisor labeled

b. $6\overline{)24}$ (quotient 4) because $4 \cdot 6 = 24$.   4 is the quotient, 6 is the divisor, and 24 is the dividend.

c. $\dfrac{21}{3} = 7$ because $7 \cdot 3 = 21$.   7 is the quotient, 3 is the divisor, and 21 is the dividend.

Self Check 1

Write the related multiplication statement for each division.
a. $8 \div 2 = 4$

b. $7\overline{)56}$ (quotient 8)

c. $\dfrac{36}{4} = 9$

Now Try Problems 19 and 23

OBJECTIVE 2 **Use properties of division to divide whole numbers.**

Recall from Section 1.4 that *the product of any whole number and 1 is that whole number.* We can use that fact to establish two important properties of division. Consider the following examples where a whole number is divided by 1:

$8 \div 1 = \mathbf{8}$ because $\mathbf{8} \cdot 1 = 8$.

$1\overline{)4}$ (quotient 4) because $\mathbf{4} \cdot 1 = 4$.

$\dfrac{20}{1} = \mathbf{20}$ because $\mathbf{20} \cdot 1 = 20$.

These examples illustrate that *any whole number divided by 1 is equal to the number itself.* Consider the following examples where a whole number is divided by itself:

$6 \div 6 = \mathbf{1}$ because $\mathbf{1} \cdot 6 = 6$.

$9\overline{)9}$ (quotient 1) because $\mathbf{1} \cdot 9 = 9$.

$\dfrac{35}{35} = \mathbf{1}$ because $\mathbf{1} \cdot 35 = 35$.

These examples illustrate that *any nonzero whole number divided by itself is equal to 1.*

Properties of Division

Any whole number divided by 1 is equal to that number.

For example, $\frac{14}{1} = 14$.

Any nonzero whole number divided by itself is equal to 1.

For example, $\frac{14}{14} = 1$.

Recall from Section 1.4 that *the product of any whole number and 0 is 0*. We can use that fact to establish another property of division. Consider the following examples where 0 is divided by a whole number:

$0 \div 2 = 0$ because $\mathbf{0 \cdot 2} = 0$.

$7\overline{)0}\,\,^{\mathbf{0}}$ because $\mathbf{0 \cdot 7} = 0$.

$\frac{0}{42} = \mathbf{0}$ because $\mathbf{0 \cdot 42} = 0$.

These examples illustrate that *0 divided by any nonzero whole number is equal to 0*.

We cannot divide a whole number by 0. To illustrate why, we will attempt to find the quotient when 2 is divided by 0 using the related multiplication statement shown below.

Division statement	*Related multiplication statement*
$\dfrac{2}{0} = ?$	$? \cdot 0 = 2$
	↑
	There is no number that gives
	2 when multiplied by 0.

Since $\frac{2}{0}$ does not have a quotient, we say that division of 2 by 0 is *undefined*. Our observations about division of 0 and division by 0 are listed below.

Division with Zero

1. Zero divided by any nonzero number is equal to 0. For example, $\frac{0}{17} = 0$.
2. Division by 0 is undefined. For example, $\frac{17}{0}$ is undefined.

OBJECTIVE 3 **Perform long division (no remainder).**

A process called **long division** can be used to divide larger whole numbers.

EXAMPLE 2   Divide using long division: $2{,}514 \div 6$. Check the result.

Strategy We will write the problem in long division form and follow a four-step process: **estimate**, **multiply**, **subtract**, and **bring down**.

WHY The repeated subtraction process would take too long to perform, and the related multiplication statement ($? \cdot 6 = 2{,}514$) is too difficult to solve.

Solution
To help you understand the process, each step of this division is explained separately. Your solution need only look like the *last* step.

We write the problem in the form $6\overline{)2514}$. The quotient will appear above the long division symbol. Since 6 will not divide 2,

$$6\overline{)2514}$$

we divide 25 by 6.

$$\begin{array}{r} 4 \\ 6\overline{)2514} \end{array}$$

Ask: "How many times will 6 divide 25?" We estimate that $25 \div 6$ is about 4, and we write the 4 in the hundreds column above the long division symbol.

Next, we multiply 4 and 6, and subtract their product, 24, from 25, to get 1.

$$\begin{array}{r} 4 \\ 6\overline{)2514} \\ -24 \\ \hline 1 \end{array}$$

Now we bring down the next digit in the dividend, the 1, and again estimate, multiply, and subtract.

$$\begin{array}{r} 41 \\ 6\overline{)2514} \\ -24\downarrow \\ \hline 11 \\ -6 \\ \hline 5 \end{array}$$

Ask: "How many times will 6 divide 11?" We estimate that $11 \div 6$ is about 1, and we write the 1 in the tens column above the long division symbol. Multiply 1 and 6, and subtract their product, 6, from 11, to get 5.

To complete the process, we bring down the last digit in the dividend, the 4, and estimate , multiply , and subtract one final time.

$$\begin{array}{r} 419 \\ 6\overline{)2514} \\ -24 \\ \hline 11 \\ -6 \\ \hline 54 \\ -54 \\ \hline 0 \end{array}$$

Ask: "How many times will 6 divide 54?" We estimate that $54 \div 6$ is 9, and we write the 9 in the ones column above the long division symbol. Multiply 9 and 6, and subtract their product, 54, from 54, to get 0.

Your solution should look like this:

$$\begin{array}{r} 419 \\ 6\overline{)2514} \\ -24 \\ \hline 11 \\ -6 \\ \hline 54 \\ -54 \\ \hline 0 \end{array}$$

> **LANGUAGE OF MATHEMATICS**
>
> In Example 2, the long division process ended with a 0. In such cases, we say that the divisor divides the dividend **exactly**.

To check the result, we see if the product of the quotient and the divisor equals the dividend.

Self Check 2

Divide using long division: $2,968 \div 4$. Check the result.

$$\begin{array}{r} {\scriptstyle 1\ 5} \\ 419 \leftarrow \text{Quotient} \\ \times \quad 6 \leftarrow \text{Divisor} \\ \hline 2,514 \leftarrow \text{Dividend} \end{array} \qquad 6\overline{)2514}$$

Now Try Problem 31

The check confirms that $2,514 \div 6 = 419$.

We can see how the long division process works if we write the names of the place-value columns above the quotient. The solution for Example 2 is shown in more detail on the next page.

$$
\begin{array}{r}
\overset{\text{Hundreds Tens Ones}}{} \\
6\overline{)2514} \\
\underline{-2400} \\
114 \\
\underline{-60} \\
54 \\
\underline{-54} \\
0
\end{array}
$$

Here, we are really subtracting $400 \cdot 6$, which is 2,400, from 2,514. That is why the 4 is written in the hundreds column of the quotient.

Here, we are really subtracting $10 \cdot 6$, which is 60, from 114. That is why the 1 is written in the tens column of the quotient.

Here, we are subtracting $9 \cdot 6$, which is 54, from 54. That is why the 9 is written in the ones column of the quotient.

The extra zeros (shown in the steps highlighted in red and blue) are often omitted.

We can use long division to perform divisions when the divisor has more than one digit. The estimation step is often made easier if we approximate the divisor.

EXAMPLE 3　　Divide using long division:　$48\overline{)33{,}888}$

Strategy　We will follow a four-step process: **estimate**, **multiply**, **subtract**, and **bring down**.

WHY　This is how long division is performed.

Solution
To help you understand the process, each step of this division is explained separately. Your solution need only look like the *last* step.

Since 48 will not divide 3, nor will it divide 33, we divide 338 by 48.

$$
\begin{array}{r}
6 \\
48\overline{)33888}
\end{array}
$$

Ask: "How many times will 48 divide 338?" Since 48 is almost 50, we can estimate the answer to that question by thinking $33 \cdot 5$ is about 6, and we write the 6 in the hundreds column of the quotient.

$$
\begin{array}{r}
6 \\
48\overline{)33888} \\
\underline{-288} \\
50
\end{array}
$$

Multiply 6 and 48, and subtract their product, 288, from 338 to get 50. Since 50 is greater than the divisor, 48, the estimate of 6 for the hundreds column of the quotient is *too small*. We will erase the 6 and increase the estimate of the quotient by 1 and try again.

$$
\begin{array}{r}
7 \\
48\overline{)33888} \\
\underline{-336} \\
2
\end{array}
$$

Change the estimate from 6 to 7 in the hundreds column of the quotient. Multiply 7 and 48, and subtract their product, 336, from 338 to get 2. Since 2 is less than the divisor, we can proceed with the long division.

$$
\begin{array}{r}
70 \\
48\overline{)33888} \\
\underline{-336\downarrow} \\
28 \\
\underline{-0} \\
28
\end{array}
$$

Bring down the 8 from the tens column of the dividend. Ask: "How many times will 48 divide 28?" Since 28 cannot be divided by 48, write a 0 in the tens column of the quotient. Multiply 0 and 48, and subtract their product, 0, from 28 to get 28.

$$
\begin{array}{r}
705 \\
48\overline{)33888} \\
-336 \\
\hline
28 \\
-0 \\
\hline
288 \\
-240 \\
\hline
48
\end{array}
$$

Bring down the 8 from the ones column of the dividend. Ask: "How many times will 48 divide 288?" We can estimate the answer to that question by thinking 28 ÷ 5 is about 5, and we write the 5 in the ones column of the quotient. Multiply 5 and 48, and subtract their product, 240, from 288 to get 48. Since 48 is equal to the divisor, the estimate of 5 for the ones column of the quotient is *too small*. We will erase the 5 and increase the estimate of the quotient by 1 and try again.

Caution! If a difference at any time in the long division process is greater than or equal to the divisor, the estimate made at that point should be increased by 1, and you should try again.

$$
\begin{array}{r}
706 \\
48\overline{)33888} \\
-336 \\
\hline
28 \\
-0 \\
\hline
288 \\
-288 \\
\hline
0
\end{array}
$$

Change the estimate from 5 to 6 in the ones column of the quotient. Multiply 6 and 48, and subtract their product, 288, from 288 to get 0. Your solution should look like this.

Self Check 3

Divide using long division:

$57\overline{)45{,}885}$

Now Try ➲ Problem 35

The quotient is 706. Check the result using multiplication.

OBJECTIVE 4 Perform long division (with a remainder).

Sometimes it is not possible to separate a group of objects into a whole number of equal-sized groups. For example, if we start with a set of 14 stars and divide them into groups of 4 stars, we will have 3 groups of 4 stars and 2 stars left over. We call the leftover part the **remainder**.

A set of 12 stars.

There are 3 groups of 4 stars.

In the next long division example, there is a remainder. To check such a problem, we add the remainder to the product of the quotient and divisor. The result should equal the dividend.

(Quotient · divisor) + remainder = dividend   Recall that the operation within the parentheses must be performed first.

EXAMPLE 4   Divide: $23\overline{)832}$. Check the result.

Strategy We will follow a four-step process: **estimate**, **multiply**, **subtract**, and **bring down**.

WHY This is how long division is performed.

Solution

Since 23 will not divide 8, we divide 83 by 23.

$$23\overline{)832}\quad\overset{4}{}$$

Ask: "How many times will 23 divide 83?" Since 23 is about 20, we can estimate the answer to that question by thinking $8 \div 2$ is 4, and we write the 4 in the tens column of the quotient.

$$\begin{array}{r}4\\23\overline{)832}\\-92\end{array}$$

Multiply 4 and 23, and write their product, 92, under the 83. Because 92 is greater than 83, the estimate of 4 for the tens column of the quotient is *too large*. We will erase the 4 and decrease the estimate of the quotient by 1 and try again.

$$\begin{array}{r}3\\23\overline{)832}\\-69\\\hline14\end{array}$$

Change the estimate from 4 to 3 in the tens column of the quotient. Multiply 3 and 23, and subtract their product, 69, from 83, to get 14.

$$\begin{array}{r}3\\23\overline{)832}\\-69\downarrow\\\hline142\end{array}$$

Bring down the 2 from the ones column of the dividend.

$$\begin{array}{r}37\\23\overline{)832}\\-69\\\hline142\\-161\end{array}$$

Ask: "How many times will 23 divide 142?" We can estimate the answer to that question by thinking $14 \div 2$ is 7, and we write the 7 in the ones column of the quotient. Multiply 7 and 23, and write their product, 161, under 142. Because 161 is greater than 142, the estimate of 7 for the ones column of the quotient is *too large*. We will erase the 7 and decrease the estimate of the quotient by 1 and try again.

$$\begin{array}{r}36\\23\overline{)832}\\-69\\\hline142\\-138\\\hline4\end{array}\leftarrow\text{The remainder}$$

Change the estimate from 7 to 6 in the ones column of the quotient. Multiply 6 and 23, and subtract their product, 138, from 142, to get 4.

The quotient is 36, and the remainder is 4. We can write this result as 36 R 4.

To check the result, we multiply the divisor by the quotient and then add the remainder. The result should be the dividend.

Check:

$$\underset{\downarrow}{\text{Quotient}}\quad\underset{\downarrow}{\text{Divisor}}\quad\underset{\downarrow}{\text{Remainder}}$$

$$(36\ \cdot\ 23)\ +\ 4\ = 828 + 4$$
$$= 832 \leftarrow\text{Dividend}$$

Since 832 is the dividend, the answer 36 R 4 is correct.

Self Check 4

Divide: $34\overline{)792}$. Check the result.

Now Try ➡ **Problem 39**

EXAMPLE 5   Divide: $\dfrac{13{,}011}{518}$

Strategy We will write the problem in long-division form and follow a four-step process: **estimate**, **multiply**, **subtract**, and **bring down**.

WHY This is how long division is performed.

Solution

We write the division in the form: $518\overline{)13011}$. Since 518 will not divide 1, nor 13, nor 130, we divide 1,301 by 518.

$$\begin{array}{r} 2 \\ 518\,\overline{)\,13011} \\ -1036 \\ \hline 265 \end{array}$$

Ask: "How many times will 518 divide 1,301?" Since 518 is about 500, we can estimate the answer to that question by thinking $13 \div 5$ is about 2, and we write the 2 in the tens column of the quotient. Multiply 2 and 518, and subtract their product, 1,036, from 1,301, to get 265.

$$\begin{array}{r} 25 \\ 518\,\overline{)\,13011} \\ -1036\downarrow \\ \hline 2651 \\ -2590 \\ \hline 61 \end{array}$$

Bring down the 1 from the ones column of the dividend. Ask: "How many times will 518 divide 2,651?" We can estimate the answer to that question by thinking $26 \div 5$ is about 5, and we write the 5 in the ones column of the quotient. Multiply 5 and 518, and subtract their product, 2,590, from 2,651, to get a remainder of 61.

Self Check 5

Divide: $\dfrac{28{,}992}{629}$

Now Try ➲ **Problem 43**

The result is 25 R 61. To check, verify that $(25 \cdot 518) + 61$ is 13,011.

OBJECTIVE 5 **Use tests for divisibility.**

We have seen that some divisions end with a 0 remainder and others do not. The word *divisible* is used to describe such situations.

Divisibility

One number is **divisible** by another if, when dividing them, we get a remainder of 0.

Since $27 \div 3 = 9$, with a 0 remainder, we say that *27 is divisible by 3*. Since $27 \div 5 = 5$ R 2, we say that *27 is not divisible by 5*.

There are tests to help us decide whether one number is divisible by another.

Tests for Divisibility

A number is divisible by
- 2 if its last digit is divisible by 2.
- 3 if the sum of its digits is divisible by 3.
- 4 if the number formed by its last two digits is divisible by 4.
- 5 if its last digit is 0 or 5.
- 6 if it is divisible by 2 and 3.
- 9 if the sum of its digits is divisible by 9.
- 10 if its last digit is 0.

There are tests for divisibility by a number other than 2, 3, 4, 5, 6, 9, or 10, but they are more complicated. See problems 109 and 110 of Study Set 1.5 for some examples.

EXAMPLE 6   Is 534,840 divisible by:

a. 2   **b.** 3   **c.** 4   **d.** 5   **e.** 6   **f.** 9   **g.** 10

Strategy We will look at the last digit, the last two digits, and the sum of the digits of each number.

WHY The divisibility rules call for these types of examination.

Solution

a. 534,840 is divisible by 2, because its last digit **0** is divisible by 2.

b. 534,840 is divisible by 3, because the sum of its digits is divisible by 3.

$$5 + 3 + 4 + 8 + 4 + 0 = 24 \quad \text{and} \quad 24 \div 3 = 8$$

c. 534,840 is divisible by 4, because the number formed by its last two digits is divisible by 4.

$$40 \div 4 = 10$$

d. 534,840 is divisible by 5, because its last digit is 0 or 5.

e. 534,840 is divisible by 6, because it is divisible by 2 and 3. (See parts a and b.)

f. 534,840 is not divisible by 9, because the sum of its digits is not divisible by 9. There is a remainder.

$$24 \div 9 = 2 \text{ R } 6$$

g. 534,840 is divisible by 10, because its last digit is 0.

Self Check 6

Is 73,311,435 divisible by:

a. 2   **b.** 3   **c.** 5
d. 6   **e.** 9   **f.** 10

Now Try ➲ **Problems 49 and 53**

OBJECTIVE 6 **Divide whole numbers that end with zeros.**

There is a shortcut for dividing a dividend by a divisor when both end with zeros. We simply *remove the ending zeros in the divisor and remove the same number of ending zeros in the dividend.*

EXAMPLE 7   Divide: **a.** $80 \div 10$ **b.** $47,000 \div 100$ **c.** $350\overline{)9,800}$

Strategy We will look for ending zeros in each divisor.

WHY If a divisor has ending zeros, we can simplify the division by removing the same number of ending zeros in the divisor and dividend.

Solution

There is one zero in the divisor.
↓
a. $80 \div 10 = 8 \div 1 = 8$

Remove one zero from the dividend and the divisor, and divide.

There are two zeros in the divisor.
↓
b. $47,000 \div 100 = 470 \div 1 = 470$

Remove two zeros from the dividend and the divisor, and divide.

c. To find

$$350\overline{)9,800}$$

we can drop *one zero* from the divisor and the dividend and perform the division $35\overline{)980}$.

$$
\begin{array}{r}
28 \\
35\overline{)980} \\
-70 \\
\hline
280 \\
-280 \\
\hline
0
\end{array}
$$

Thus, $9{,}800 \div 350$ is 28.

Self Check 7

Divide:

a. $50 \div 10$

b. $62{,}000 \div 100$

c. $12{,}000 \div 1{,}500$

Now Try ➡ **Problems 55 and 57**

OBJECTIVE 7 Estimate quotients of whole numbers.

To estimate quotients, we use a method that approximates both the dividend and the divisor so that they divide easily. There is one rule of thumb for this method: If possible, round both numbers up or both numbers down.

EXAMPLE 8   Estimate the quotient: $170{,}715 \div 57$

Strategy We will round the dividend and the divisor up and find $180{,}000 \div 60$.

WHY The division can be made easier if the dividend and the divisor end with zeros. Also, 6 divides 18 exactly.

Solution

The dividend is approximately

$$170{,}715 \div 57 \qquad 180{,}000 \div 60 = 3{,}000$$

To divide, drop one zero from 180,000 and from 60 and find $18{,}000 \div 6$.

The divisor is approximately

The estimate is 3,000.

If we calculate $170{,}715 \div 57$, the quotient is exactly 2,995. Note that the estimate is close: It's just 5 more than 2,995.

Self Check 8

Estimate the quotient: $33{,}642 \div 42$

Now Try ➡ **Problem 59**

OBJECTIVE 8 Solve application problems by dividing whole numbers.

Application problems that involve forming equal-sized groups can be solved by division.

EXAMPLE 9   **Managing a soup kitchen.**   A soup kitchen plans to feed 1,990 people. Because of space limitations, only 144 people can be served at one time. How many group seatings will be necessary to feed everyone? How many will be served at the last seating?

Strategy We will divide 1,990 by 144.

WHY Separating 1,990 people into equal-sized groups of 144 indicates division.

Solution

We translate the words of the problem to numbers and symbols.

The number of group seatings	is equal to	the number of people to be fed	divided by	the number of people at each seating.
The number of group seatings	=	1,990	÷	144

Use long division to find 1,990 ÷ 144.

```
        13
144) 1,990
    −144
     550
    −432
     118
```

The quotient is 13, and the remainder is 118. This indicates that 14 group seatings are needed: 13 full-capacity seatings and one partial seating to serve the remaining 118 people.

Self Check 9

Movie tickets. On a Saturday, 3,924 movie tickets were purchased at an IMAX theater. Each showing of the movie was sold out, except for the last. If the theater seats 346 people, how many times was the movie shown on Saturday? How many people were at the last showing?

Now Try ➡ **Problem 95**

EXAMPLE 10 **Timeshares.** Every year, the 73 part-owners of a timeshare resort condominium get use of it for an equal number of days. How many days does each part-owner get to stay at the condo? (Use a 365-day year.)

Strategy We will divide 365 by 73.

WHY Since the part-owners get use of the condo for an equal number of days, the phrase *"How many days does each"* indicates division.

Solution

We translate the words of the problem to numbers and symbols.

The number of days each part-owner gets to stay at the condo	is equal to	the number of days in a year	divided by	the number of part-owners.
The number of days each part-owner gets to stay at the condo	=	365	÷	73

Use long division to find 365 ÷ 73.

```
       5
73) 365
   −365
      0
```

Each part-owner gets to stay at the condo for 5 days during the year.

LANGUAGE OF MATHEMATICS

Here are some key words and phrases that are often used to indicate *division*:

how much extra (remainder)
how many left (remainder)
how many does each
distributed equally
shared equally
split equally
goes into
among
per

Self Check 10

Touring. A rock band will take a 275-day world tour and spend the same number of days in each of 25 cities. How long will they stay in each city?

Now Try ➡ **Problem 101**

Using Your Calculator ▶   The Division Key

A beverage company production run of 604,800 bottles of mountain spring water will be shipped to stores on pallets that hold 1,728 bottles each. We can find the number of full pallets to be shipped using the division key $\div$ on a calculator.

604800 $\div$ 1728 $=$ $\boxed{350}$

On some calculator models, the $\boxed{\text{ENTER}}$ key is pressed instead of $\boxed{=}$ for the result to be displayed.

The beverage company will ship 350 full pallets of bottled water.

Answers to Self Checks

1. **a.** $4 \cdot 2 = 8$ **b.** $8 \cdot 7 = 56$ **c.** $9 \cdot 4 = 36$ **2.** 742; $742 \cdot 4 = 2{,}968$ **3.** 805
4. 23 R 10; $(23 \cdot 34); + 10 = 792$ **5.** 46 R 58   **6. a.** no   **b.** yes   **c.** yes   **d.** no   **e.** yes   **f.** no
7. **a.** 5   **b.** 620   **c.** 8   **8.** 800   **9.** 12 showings, 118   **10.** 11 days

---

SECTION **1.5** **STUDY SET**

VOCABULARY

Fill in the blanks.

1. In the three division problems shown below, label the *dividend*, *divisor*, and the *quotient*.

$$12 \div 4 = 3$$
↑   ↑   ↑

$$\begin{array}{r} 3 \\ 4\overline{)12} \end{array} \qquad \dfrac{12}{4} = 3$$

2. We call $5 \cdot 8 = 40$ the related _____ statement for the division $40 \div 8 = 5$.

3. The problem $6\overline{)246}$ is written in _____ division form.

4. If a division is not exact, the leftover part is called the _____.

5. One number is _____ by another number if, when we divide them, the remainder is 0.

6. Phrases such as split *equally* and *how many does each* indicate the operation of _____.

CONCEPTS

7. **a.** Divide the set of objects below into groups of 3. How many groups of 3 are there?

• • • • • • • • • • • • • • • • • • • •

b. Divide the set of objects below into groups of 4. How many groups of 4 are there? How many objects are left over?

* *

8. Tell whether each statement is true or false.
   a. Any whole number divided by 1 is equal to that number.
   b. Any nonzero whole number divided by itself is equal to 1.
   c. Zero divided by any nonzero number is undefined.
   d. Division of a number by 0 is equal to 0.

Fill in the blanks.

9. Divide, if possible.

   a. $\dfrac{25}{25} = \rule{1em}{1em}$ **b.** $\dfrac{6}{1} = \rule{1em}{1em}$

   c. $\dfrac{100}{0}$ is _____   **d.** $\dfrac{0}{12} = \rule{1em}{1em}$

10. To perform long division, we follow a four-step process: _____, _____, _____, and _____ _____.

11. Find the *first* digit of each quotient.

   a. $5\overline{)1147}$ **b.** $9\overline{)587}$

   c. $23\overline{)7501}$ **d.** $16\overline{)892}$

12. a. Quotient · divisor = _____

  b. (Quotient · divisor) + _____ = dividend

13. To check whether the division $9\overline{)333}$ is correct, we use multiplication:

$$\begin{array}{r} 37 \\ \hline \\ \times\ \ 9 \\ \hline \end{array}$$

14. a. A number is divisible by ▨ if its last digit is divisible by 2.

  b. A number is divisible by 3 if the _____ of its digits is divisible by 3.

  c. A number is divisible by 4 if the number formed by its last _____ digits is divisible by 4.

15. a. A number is divisible by 5 if its last digit is ▨ or ▨.

  b. A number is divisible by 6 if it is divisible by ▨ and ▨.

  c. A number is divisible by 9 if the _____ of its digits is divisible by 3.

  d. A number is divisible by ▨ if its last digit is 0.

16. We can simplify the division 43,800 ÷ 200 by removing two _____ from the dividend and the divisor.

NOTATION

17. Write three symbols that can be used for division.

18. In a division, 35 R 4 means "a quotient of 35 and a _____ of 4."

GUIDED PRACTICE

Fill in the blanks. **See Example 1.**

19. $9\overline{)45}^{\,5}$ because ▨ · ▨ = ▨ .

20. $\dfrac{54}{6} = 9$ because ▨ · ▨ = ▨ .

21. 44 ÷ 11 = 4 because ▨ · ▨ = ▨ .

22. 120 ÷ 12 = 10 because ▨ · ▨ = ▨ .

Write the related multiplication statement for each division. **See Example 1.**

23. 21 ÷ 3 = 7

24. 32 ÷ 4 = 8

25. $\dfrac{72}{12} = 6$

26. $15\overline{)75}^{\,5}$

Divide using long division. Check the result. **See Example 2.**

27. 96 ÷ 6

28. 72 ÷ 4

29. $\dfrac{87}{3}$

30. $\dfrac{98}{7}$

31. 2,275 ÷ 7

32. 1,728 ÷ 8

33. $9\overline{)1,962}$

34. $5\overline{)1,635}$

Divide using long division. Check the result. **See Example 3.**

35. $62\overline{)31,248}$

36. $71\overline{)28,613}$

37. $37\overline{)22,274}$

38. $28\overline{)19,712}$

Divide using long division. Check the result. **See Example 4.**

39. $24\overline{)951}$

40. $33\overline{)943}$

41. 999 ÷ 46

42. 979 ÷ 49

Divide using long division. Check the result. **See Example 5.**

43. $\dfrac{24,714}{524}$

44. $\dfrac{29,773}{531}$

45. $178\overline{)3,514}$

46. $164\overline{)2,929}$

If the given number is divisible by 2, 3, 4, 5, 6, 9, or 10, enter a checkmark √ in the box. **See Example 6.**

	Divisible by →	2	3	4	5	6	9	10
47.	2,940							
48.	5,850							
49.	43,785							
50.	72,954							
51.	181,223							
52.	379,157							
53.	9,499,200							
54.	6,653,100							

Use a division shortcut to find each quotient. **See Example 7.**

55. 700 ÷ 10

56. 900 ÷ 10

57. $450\overline{)9,900}$

58. $260\overline{)9,100}$

Estimate each quotient. **See Example 8.**

59. 353,922 ÷ 38

60. 237,621 ÷ 55

61. 46,080 ÷ 933

62. 81,097 ÷ 419

TRY IT YOURSELF

Divide.

63. $\dfrac{25,950}{6}$

64. $\dfrac{23,541}{7}$

65. 54 ÷ 9

66. 72 ÷ 8

67. 273 ÷ 31

68. 295 ÷ 35

69. $\dfrac{64,000}{400}$

70. $\dfrac{125,000}{5,000}$

71. 745 divided by 7

72. 931 divided by 9

73. $29\overline{)14,761}$

74. $27\overline{)10,989}$

75. 539,000 ÷ 175

76. 749,250 ÷ 185

77. 75 ÷ 15

78. 96 ÷ 16

79. $212\overline{)5,087}$

80. $214\overline{)5,777}$

81. $42\overline{)1,273}$

82. $83\overline{)3,363}$

83. $89,000 \div 1,000$

84. $930,000 \div 1,000$

85. $\dfrac{57}{8}$

86. $\dfrac{82}{9}$

LOOK ALIKES

Find the answer to the problem in part a. The answers to parts b and c should then be obvious.

87. a. $368,000 \div 10$

    b. $368,000 \div 100$

    c. $368,000 \div 1,000$

88. a. $475,000 \div 1,000$

    b. $475,000 \div 100$

    c. $475,000 \div 10$

89. a. $12\overline{)607}$

    b. $12\overline{)608}$

    c. $12\overline{)606}$

90. a. $23\overline{)453}$

    b. $23\overline{)452}$

    c. $23\overline{)454}$

APPLICATIONS

91. Ticket sales. A movie theater makes a $4 profit on each ticket sold. How many tickets must be sold to make a profit of $2,500?

92. Running. Brian runs 7 miles each day. In how many days will Brian run 371 miles?

93. Dump trucks. A 15-cubic-yard dump truck must haul 405 cubic yards of dirt to a construction site. How many trips must the truck make?

94. Stocking shelves. After receiving a delivery of 288 bags of potato chips, a store clerk stocked each shelf of an empty display with 36 bags. How many shelves of the display did he stock with potato chips?

95. Lunch time. A fifth grade teacher received 50 half-pint cartons of milk to distribute evenly to his class of 23 students. How many cartons did each child get? How many cartons were left over?

96. Bubble wrap. A furniture manufacturer uses an 11-foot-long strip of bubble wrap to protect a lamp when it is boxed and shipped to a customer. How many lamps can be packaged in this way from a 200-foot-long roll of bubble wrap? How many feet will be left on the roll?

97. Gardening. A metal can holds 640 fluid ounces of gasoline. How many times can the 68-ounce tank of a lawnmower be filled from the can? How many ounces of gasoline will be left in the can?

98. Beverages. A plastic container holds 896 ounces of punch. How many 6-ounce cups of punch can be served from the container? How many ounces will be left over?

99. Lift systems. If the bus shown in the next column weighs 58,000 pounds, how much weight is on each jack?

100. Lottery winners. In 2008, a group of 22 postal workers, who had been buying Pennsylvania Lotto tickets for years, won a $10,282,800 jackpot. If they split the prize evenly, how much money did each person win?

101. Textbook sales. A store received $25,200 on the sale of 240 algebra textbooks. What was the cost of each book?

102. Draining pools. A 950,000-gallon pool is emptied in 20 hours. How many gallons of water are drained each hour?

103. Mileage. A tour bus has a range of 700 miles on one tank (140 gallons) of gasoline. How far does the bus travel on one gallon of gas?

104. Waterfalls. Every 45 seconds, about 33,750,000 gallons of water flow over Niagara Falls. How many gallons flow over the falls in one second? (Source: water.epa.gov)

105. Ordering snacks. How many *dozen* doughnuts must be ordered for a meeting if 156 people are expected to attend and each person will be served one doughnut?

106. Time. A *millennium* is a period of time equal to one thousand years. How many decades are in a millennium?

107. Volleyball. A total of 216 girls are going to play in a city volleyball league. How many girls should be put on each team if the following requirements must be met?

- All the teams are to have the same number of players.
- A reasonable number of players on a team is 7 to 10.
- For scheduling purposes, there must be an even number of teams (2, 4, 6, 8, and so on).

---

from Campus to Careers

108. A landscape designer intends to plant pine trees 12 feet apart to form a windscreen along one side of a flower garden, as shown below. How many trees are needed if the length of the flower garden is 744 feet?

Toa55/Shutterstock.com

12 ft   12 ft

109. Entry-level jobs. The typical starting salaries for 2016 college graduates majoring in health sciences, business, and social sciences are shown below. Complete the last column of the table.

College major	Yearly salary	Monthly salary
Health Sciences	$48,708	
Business	$52,236	
Social Sciences	$46,584	

Source: National Association of Colleges and Employers

110. Population. To find the **population density** of a state, divide its population by its land area (in square miles). The result is the number of people per square mile. Use the data in the table to approximate the population density for each state.

State	2016 Population*	Land area* (square miles)
Arizona	6,840,000	114,000
Oklahoma	3,933,000	69,000
Rhode Island	1,058,000	1,000
South Carolina	4,950,000	30,000

Source: Worldpopulationreview.com   *approximation

WRITING

111. Explain how $24 \div 6$ can be calculated by repeated subtraction.

112. Explain why division of 0 is possible, but division by 0 is impossible.

113. Divisibility test for 7. Use the following rule to show that 308 is divisible by 7. Show each of the steps of your solution in writing.

Subtract twice the units digit from the number formed by the remaining digits. If that result is divisible by 7, then the original number is divisible by 7.

114. Divisibility test for 11. Use the following rule to show that 1,848 is divisible by 11. Show each of the steps of your solution in writing.

Start with the digit in the one's place. From it, subtract the digit in the ten's place. To that result, add the digit in the hundred's place. From that result, subtract the digit in the thousand's place, and so on. If the final result is a number divisible by 11, the original number is divisible by 11.

REVIEW LOOK ALIKES

Perform each operation.

115. a. $272 + 4$ **b.** $272 - 4$
     c. $272 \cdot 4$ **d.** $272 \div 4$

116. a. $430 + 55$ **b.** $430 - 55$
     c. $430 \cdot 55$ **d.** $430 \div 55$

117. a. $1,104 + 46$ **b.** $1,104 - 46$
     c. $1,104 \cdot 46$ **d.** $1,104 \div 46$

118. a. $3,024 + 378$ **b.** $3,024 - 378$
     c. $3,024 \cdot 378$ **d.** $3,024 \div 378$

OBJECTIVES

1 Apply the steps of a problem-solving strategy.

2 Solve problems requiring more than one operation.

3 Recognize unimportant information in application problems.

SECTION 1.6 Problem Solving

The operations of addition, subtraction, multiplication, and division are powerful tools that can be used to solve a wide variety of real-world problems.

OBJECTIVE 1 Apply the steps of a problem-solving strategy.

To become a good problem solver, you need a plan to follow, such as the following five-step strategy.

Strategy for Problem Solving

1. **Analyze the problem** by reading it carefully. What information is given? What are you asked to find? What vocabulary is given? Often, a diagram or table will help you visualize the facts of the problem.
2. **Form a plan** by translating the words of the problem into numbers and symbols.
3. **Perform the calculations.**
4. **State the conclusion** clearly. Be sure to include the units (such as feet, seconds, or pounds) in your answer.
5. **Check the result.** An estimate is often helpful to see whether an answer is reasonable.

To solve application problems, which are usually given in words, we *translate* those words to numbers and mathematical symbols. The following table is a review of some of the key words, phrases, and concepts that were introduced in Sections 1.2 through 1.5.

Addition	Subtraction	Multiplication	Division	Equals
more than	how much more	double	distributed equally	same value
increase	less than	twice	shared equally	results in
gained	decrease	triple	split equally	are
rise	loss	of	per	is
total	fall	times	among	was
in all	fewer	at this rate	goes into	yields
forward	reduce	repeated addition	equal-sized groups	amounts to
altogether	decline	rectangular array	how many does each	the same as

EXAMPLE 1   **Table settings.**   One place setting like that shown on the right costs $94. What is the total cost to purchase these place settings for a restaurant that seats 115 people?

Analyze   At this stage, it is helpful to list the given facts and what you are to find.

- One place setting costs $94.   *Given*
- 115 place settings will be purchased.   *Given*
- What is the total cost to purchase
  115 place settings?   *Find*

Form   The key word *total* suggests addition. In this case, the total cost to purchase the place settings is the sum of one hundred fifteen 94's. This repeated addition can be calculated more simply by multiplication.

We begin with a **verbal model** (shown in blue) that describes the situation in words. Then we translate the words to numbers and symbols.

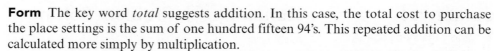

The total cost of the purchase	is equal to	the number of place settings purchased	times	the cost of one place setting.

| The total cost of the purchase | = | 115 | · | $94 |

Calculate   Use vertical form to perform the multiplication:

```
    115
 ×   94
    460
 10 350
 10,810
```

State   It will cost $10,810 to purchase 115 place settings.

Check   We can estimate to check the result. If we use $100 to approximate the cost of one place setting, then the cost of 115 place settings is about 115 · $100 or $11,500. Since the estimate, $11,500, and the result, $10,810, are close, the result seems reasonable.

EXAMPLE 2   **Counting calories.**   A glass of nonfat milk has 63 fewer calories than a glass of whole milk. If a glass of whole milk has 146 calories, how many calories are there in a glass of nonfat milk?

Analyze

- A glass of nonfat milk has 63 fewer calories than a glass of whole milk.   *Given*
- A glass of whole milk has 146 calories.   *Given*
- How many calories are there in a glass of nonfat milk?   *Find*

Form   The word *fewer* indicates subtraction.

A glass of nonfat milk	has	63 fewer calories than	a glass of whole milk.

$$\text{A glass of nonfat milk} = 146 - 63$$

This is the verbal model.

Calculate   Use vertical form to perform the subtraction:

$$\begin{array}{r} 146 \\ -\ 63 \\ \hline 83 \end{array}$$

State   A glass of nonfat milk has 83 calories.

Check   We can use addition to check.

$$\begin{array}{r} 83 \\ +\ 63 \\ \hline 146 \end{array}$$ *Difference + subtrahend = minuend. The result checks.*

Caution!   We must be careful when translating subtraction because order is important. Since the 146 calories in a glass of whole milk is to be made 63 calories fewer, we *reverse* those numbers as we translate from English words to math symbols.

Self Check 2

Lowfat milk.   A glass of lowfat milk has 56 fewer calories than a glass of whole milk. If a glass of whole milk has 146 calories, how many calories are there in a glass of lowfat milk?

Now Try   Problem 19

A diagram is often helpful when analyzing the problem.

EXAMPLE 3   **Tunneling.**   A tunnel boring machine can drill through solid rock at a rate of 33 feet per day. How many days will it take the machine to tunnel through 7,920 feet of solid rock?

AP Images/Manish Swarup

Analyze

- The tunneling machine drills through 33 feet of solid rock per day. *Given*
- The machine has to tunnel through 7,920 feet of solid rock. *Given*
- How many days will it take the machine to tunnel that far? *Find*

In the diagram below, we see that the daily tunneling separates a distance of 7,920 feet into equal-sized lengths of 33 feet. That indicates division.

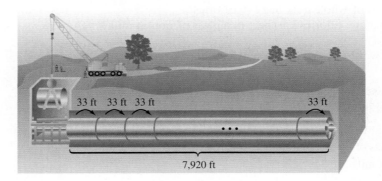

Form We translate the words of the verbal model to numbers and symbols.

The number of days it takes to drill the tunnel	is equal to	the length of the tunnel	divided by	the distance that the machine drills each day.
The number of days it takes to drill the tunnel	=	7,920	÷	33

Calculate Use long division to find 7,920 ÷ 33.

$$
\begin{array}{r}
240 \\
33\overline{)7{,}920} \\
-6\,6 \\
\hline
1\,32 \\
-1\,32 \\
\hline
00 \\
-00 \\
\hline
0
\end{array}
$$

Self Check 3

Oil wells. An offshore oil drilling rig can drill through the ocean floor at a rate of 17 feet per hour. How many hours will it take the machine to drill 578 feet to reach a pocket of crude oil?

Now Try ➡ Problem 21

State It will take the tunneling machine 240 days to drill 7,920 feet through solid rock.

Check We can check using multiplication.

$$
\begin{array}{r}
240 \\
\times\ 33 \\
\hline
720 \\
7200 \\
\hline
7920
\end{array}
$$

Quotient · divisor = dividend. The result checks.

Sometimes it is helpful to organize the given facts of a problem in a table.

EXAMPLE 4   **Orchestras.**   An orchestra consists of a 19-piece woodwind section, a 23-piece brass section, a 54-piece string section, and a two-person percussion section. In all, how many musicians make up the orchestra?

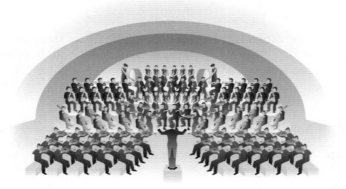

Analyze We can use a table to organize the facts of the problem.

Section	Number of musicians	
Woodwind	19	
Brass	23	*Given*
String	54	
Percussion	2	

Form In the last sentence of the problem, the phrase *in all* indicates addition. We translate the words of the verbal model to numbers and symbols.

The total number of musicians in the orchestra	is equal to	the number in the woodwind section	plus	the number in the brass section	plus	the number in the string section	plus	the number in the percussion section.

The total number of musicians in the orchestra	=	19	+	23	+	54	+	2

Calculate We use vertical form to perform the addition:

$$
\begin{array}{r}
\overset{1}{1}9 \\
23 \\
54 \\
+\ 2 \\
\hline
98
\end{array}
$$

State There are 98 musicians in the orchestra.

Self Check 4

Anatomy. A human skeleton consists of 29 bones in the skull; 26 bones in the spine; 25 bones in the ribs and breastbone; 64 bones in the shoulders, arms, and hands; and 62 bones in the pelvis, legs, and feet. In all, how many bones make up the human skeleton?

Now Try ➲ Problem 23

Check To check the addition, we will add upward.

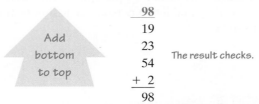

$$\begin{array}{r} \underline{98} \\ 19 \\ 23 \\ 54 \\ + \ 2 \\ \hline 98 \end{array}$$

The result checks.

We could also use estimation to check the result. If we front-end round each addend, we get $20 + 20 + 50 + 2 = 92$. Since the answer, 98, and the estimate, 92, are close, the result seems reasonable.

OBJECTIVE 2 Solve problems requiring more than one operation.

Sometimes more than one operation is needed to solve a problem.

EXAMPLE 5   **Bottled water.**   How many 6-ounce servings are there in a 5-gallon bottle of water? (Hint: There are 128 fluid ounces in 1 gallon.)

Analyze The diagram shown below is helpful in understanding the problem.

- Since each of the 5 gallons of water is 128 ounces, the total number of ounces is the sum of five 128's. This *repeated addition* can be calculated using multiplication.
- Since *equal-sized* servings of water come from the bottle, this suggests division.
- Therefore, to solve this problem, we need to perform two operations: multiplication and division.

128 ounces ⟶
128 ounces ⟶
128 ounces ⟶
128 ounces ⟶
128 ounces ⟶

6 ounces

Form (Step 1) There are 640 ounces of water in the 5-gallon bottle. To find the number of ounces of water in the 5-gallon bottle, we multiply:

$$\begin{array}{r} {\scriptstyle 1\ 4} \\ 128 \\ \times \quad 5 \\ \hline 640 \end{array}$$

Form (Step 2) Now we use the answer from Step 1 to find the number of 6-ounce servings.

The number of servings of water	is equal to	the number of ounces of water in the bottle	divided by	the number of ounces in one serving.
The number of servings of water	=	640	÷	6

Calculate Use long division to find $640 \div 6$.

$$
\begin{array}{r}
106 \\
6\overline{)640} \\
\underline{-6} \\
4 \\
\underline{-0} \\
40 \\
\underline{-36} \\
4 \leftarrow \text{The remainder}
\end{array}
$$

State In a 5-gallon bottle of water, there are 106 6-ounce servings, with 4 ounces of water left over.

Check To check the multiplication, use estimation. To check the division, use the relationship: (Quotient · divisor) + remainder = dividend.

Self Check 5

Bottled water. How many 8-ounce servings are there in a 3-gallon bottle of water? (*Hint:* There are 128 fluid ounces in 1 gallon.)

Now Try ➲ **Problem 25**

OBJECTIVE 3 **Recognize unimportant information in application problems.**

EXAMPLE 6   **Public transportation.**   Forty-seven people were riding on a bus on Route 66. It arrived at the 7th Street stop at 5:30 PM, where 11 people paid the $1.50 fare to board after 16 riders had exited. As the driver pulled away from the stop at 5:32 PM, how many riders were on the bus?

Analyze If we are to find the number of riders on the bus, then the route, the stop, the times, and the fare are not important. It is helpful to cross out that information.

⁴⁷
(Forty-seven) people were riding on a bus on ~~Route 66~~. It arrived at the ~~7th Street~~ stop at ~~5:30 PM~~, where 11 people paid the ~~$1.50 fare~~ to board after 16 riders had exited. As the driver pulled away from that stop at ~~5:32 PM~~, how many riders were on the bus?

If we carefully reread the problem, we see that the phrase *to board* indicates addition and the word *exited* indicates subtraction.

Form We translate the words of the verbal model to numbers and symbols.

The number of riders on the bus after the stop	is equal to	the number of riders on the bus before the stop	plus	the number of riders that boarded	minus	the number of riders that exited.

The number of riders on the bus after the stop	=	47	+	11	−	16

Calculate We will solve the problem in horizontal form. Recall from Section 1.3 that the operations of addition and subtraction must be performed as they occur, from left to right.

$47 + 11 - 16 = 58 - 16$ Working left to right, do the addition first: $47 + 11 = 58$.

$ = 42$ Now do the subtraction.

$$
\begin{array}{rr}
47 & 58 \\
+ 11 & - 16 \\
\hline
58 & 42
\end{array}
$$

Caution! As you read a problem, it is easy to miss numbers that are written in words. It is helpful to circle those words and write the corresponding number above.

Self Check 6

Bus service. Thirty-four people were riding on bus number 481. At 11:45 AM, it arrived at the 103rd Street stop where 6 people got off and 18 people paid the 75¢ fare to board. As the driver pulled away from the stop at 11:47 AM, how many riders were on the bus?

Now Try ➲ **Problem 27**

State There were 42 riders on the bus after the 7th Street stop.

Check The addition can be checked with estimation. To check the subtraction, use: Difference + subtrahend = minuend.

Answers to Self Checks

1. It will cost $11,390 to purchase 85 sets of bed linens.   **2.** There are 90 calories in a glass of lowfat milk.
3. It will take the drilling rig 34 hours to drill 578 feet.   **4.** There are 206 bones in the human skeleton.
5. In a 3-gallon bottle of water, there are 48 8-ounce servings.   **6.** There were 46 riders when the bus left the 103rd Street stop.

SECTION 1.6 STUDY SET

VOCABULARY

Fill in the blanks.

1. A _____ is a careful plan or method.
2. To solve application problems, which are usually given in words, we _____ those words into numbers and mathematical symbols.

Tell whether addition, subtraction, multiplication, or division is indicated by each of the following words and phrases.

3. reduced   4. equal-size groups
5. triple   6. fall
7. gained   8. repeated addition
9. rectangular array   10. in all
11. how many does each   12. rise

CONCEPTS

13. Write the following steps of the problem-solving strategy in the correct order:

    State, Check, Analyze, Form, Calculate

14. A 12-ounce Mountain Dew has 54 milligrams of caffeine. Fill in the blanks to translate the following statement to numbers and symbols.

The number of milligrams of caffeine in a 12-ounce Dr Pepper	is	13 fewer than	the number of milligrams of caffeine in a 12-ounce Mountain Dew.

The number of milligrams of caffeine in a 12-ounce Dr Pepper	=	☐	−	☐

15. Multiply 15 and 8. Then divide that result by 3.
16. Subtract 27 from 100. Then multiply that result by 6.

GUIDED PRACTICE

Solve the following problems. **See Example 1.**

17. **Trucking.** An automobile transport is loaded with 9 new Chevrolet Malibu sedans, each valued at $21,605. What is the total value of the cars carried by the transport?

18. **Gold medals.** The United States team won 46 gold medals at the 2016 Summer Olympic Games in London. At that time, the actual value of a gold medal was estimated to be about $564. What was the total value of the gold medals won by the United States? (Source: forbes.com)

Solve the following problems. **See Example 2.**

19. **TV history.** There were 57 fewer episodes of *I Love Lucy* made than episodes of *Friends*. If there are 236 episodes of *Friends*, how many episodes of *I Love Lucy* are there?

20. **Pets.** In 2015, the number of American households owning a cat was estimated to be 13,100,000 fewer than the number of households owning a dog. If 60,200,000 households owned a dog, how many owned a cat? (Source: avma.org)

Solve the following problems. **See Example 3.**

21. **Chocolate.** A study found that 7 grams of dark chocolate per day is the ideal amount to protect against the risk of a heart attack. How many daily servings are there in a bar of dark chocolate weighing 98 grams? (Source: ScienceDaily.com)

22. **Traveling.** A tourism website claims travelers can see Europe for $95 a day. If a tourist saved $2,185 for a vacation, how many days can he spend in Europe?

Solve the following problems. Use a table to organize the facts of the problem. **See Example 4.**

23. Theater. The play *Romeo and Juliet* by William Shakespeare has five acts. The first act has 5 scenes. The second act has 6 scenes. The third and fourth acts each have 5 scenes, and the last act has 3 scenes. In all, how many scenes are there in the play?

24. Statehood. From 1800 to 1850, 15 states joined the Union. From 1851 to 1900, an additional 14 states entered. Three states joined from 1901 to 1950. Since then, Alaska and Hawaii are the only others to enter the Union. In all, how many states have joined the Union since 1800?

Solve the following problems. Use a diagram to show the facts of the problem. **See Example 5.**

25. Baking. A baker uses 3-ounce pieces of dough to make dinner rolls. How many dinner rolls can he make from 5 pounds of dough? (*Hint:* There are 16 ounces in one pound.)

26. Door mats. There are 7 square yards of carpeting left on a roll. How many 4-square-foot mats can be made from the roll? (*Hint:* There are 9 square feet in one square yard.)

Solve the following problems. **See Example 6.**

27. Laptops. A folder named "Vacation" on a student's Sony Vaio contained 81 HD video files. To free up 15 gigabytes of storage space, he deleted 26 of the files from that folder. Then, 48 hours later, he added 13 new HD video files (2 gigabytes) into it. How many HD video files are now in the student's "Vacation" folder?

28. iPhones. A student had 135 photos saved on her 16-gigabyte iPhone. She deleted 27 of them (40 megabytes) to free up some storage space. Over the next 7 days, she took 19 new photos (28 megabytes) and saved all of them. How many photos are now saved on her phone?

ICP-UK/Alamy

TRY IT YOURSELF

29. Forests. Canada has 1,806,425 fewer square miles of forest than Russia. The United States has 142,758 fewer square miles of forest than Canada. If Russia has 3,146,466 square miles of forest (the most of any country in the world), how many square miles does the United States have? (Source: mapsofworld.com)

30. Apps. In 2016, the most popular smartphone apps in the United States were Facebook, Facebook Messenger, and YouTube, in that order. Facebook Messenger had about 16,300,000 fewer monthly users than Facebook. YouTube had about 16,000,000 fewer monthly users than Facebook Messenger. If Facebook had 146,000,000 monthly users, how many users did YouTube have per month? (Source: cnet.com)

31. Batman. As of 2017, the worldwide box office revenue for the following Batman films were *The Dark Knight Rises* (2012): $1,085 million, *The Dark Knight* (2008): $1,003 million, *Batman v Superman: Dawn of Justice* (2016): $873 million, *Batman* (1989); $411 million, *Batman Forever* (1995): $337 million, *Batman Begins* (2005): $373 million, and *The LEGO Batman Movie* (2017): $294 million. What is the total box office revenue for the films? (Source: www.boxofficemojo.com)

32. Soap operas. The total number of viewers for each of the top four TV soap operas for the week of February 27 – March 3, 2017, were *The Young and the Restless*: 4,429,000, *The Bold and the Beautiful*: 3,621,000, *General Hospital*: 2,606,000, and *Days of Our Lives*: 2,134,000. What was the total number of viewers of these programs for that week? (Source: soapoperanetwork.com)

33. Med school. In 2017, exactly 1,471 fewer women than men applied to U.S. medical schools. If 27,250 men applied, what was the *total* number of people who applied to U.S. medical schools in 2017? (Source: AAMC)

34. Aerobics. A 30-minute high-impact aerobic workout burns 302 calories. A 30-minute low-impact workout burns 64 fewer calories. How many calories are burned during the 30-minute low-impact workout?

35. Travel. How much money will a family of six save on airfare if they take advantage of the offer shown in the advertisement?

Discount Airfare
Roundtrip per person
Los Angeles/Orlando
WAS: $~~593~~
NOW! $516

36. Discount lodging. A hotel is offering rooms that normally go for $129 per night for only $99 a night. How many dollars would a traveler save if he stays in such a room for 5 nights?

37. Painting. One gallon of latex paint covers 350 square feet. How many gallons are needed if the total area of walls and ceilings to be painted is 9,800 square feet, and if two coats must be applied?

38. Asphalt. One bucket of asphalt sealcoat covers 420 square feet. How many buckets are needed if a 5,040-square-foot playground is to be sealed with two coats?

39. iPods. The iPod Touch shown has 32 gigabytes (GB) of storage space. From the information in the screen, determine how many gigabytes of storage space are used and how many are available.

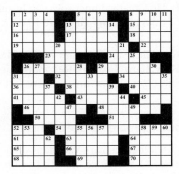

40. Multiple births. Refer to the table shown for U.S. multiple births.

   a. Find the total number of *children* born in a twin, triplet, or quadruplet birth for the year 2010.

   b. Find the total number of *children* born in a twin, triplet, or quadruplet birth for the year 2015.

   c. In which year were more children born in these ways? How many more?

U.S. Multiple Births

Year	Number of sets of twins	Number of sets of triplets	Number of sets of quadruplets
2010	132,562	5,503	313
2015	133,155	3,871	228

Source: cdc.gov and multiples.about.com

41. Trees. The height of the tallest known tree (a California coastal redwood) is 379 feet. Some scientists believe the tallest a tree can grow is 47 feet more than this because it is difficult for water to be raised from the ground any more than that to support further growth. What do the scientists believe to be the maximum height that a tree can reach? (Source: BBC News)

Zack Frank/Shutterstock.com

42. Coffee. A *grande* size cup (16 ounces) of Starbucks' signature medium roast coffee has about twelve times as much caffeine as the same size cup of Starbucks' decaf coffee. If a grande of decaf has 25 milligrams of caffeine, how much caffeine does a grande of regular coffee contain?

43. Time. There are 60 minutes in an hour, 24 hours in a day, and 7 days in a week. How many minutes are there in a week?

44. Length. There are 12 inches in a foot, 3 feet in a yard, and 1,760 yards in a mile. How many inches are in a mile?

45. Fireplaces. A contractor ordered 12 pallets of fireplace brick. Each pallet holds 516 bricks. If it takes 430 bricks to build a fireplace, how many fireplaces can be built from this order? How many bricks will be left over?

46. Roofing. A roofer ordered 108 squares of shingles. (A *square* covers 100 square feet of roof.) In a new development, the houses have 2,800-square-foot roofs. How many houses can be completely roofed with this order?

47. Crossword puzzles. A crossword puzzle is made up of 15 rows and 15 columns of small squares. Forty-six of the squares are blacked out. When completed, how many squares in the crossword puzzle will contain letters?

48. Chess. A chessboard consists of 8 rows, with 8 squares in each row. Each of the two players has 16 chess pieces to place on the board, one per square. At the start of the game, how many squares on the board do not have chess pieces on them?

49. Credit cards. The balance on 10/23/17 on Visa account number 623415 was $1,989. If purchases of $125 and $296 were charged to the card on 10/24/17, a payment of $1,680 was credited on 10/31/17, and no other charges or payments were made, what is the new balance on 11/1/17?

50. Arizona. The average high temperature in Phoenix in January is 60°F. By May, it rises by 28°F, by July it rises another 11°F, and by December it falls 40°F. What is the average temperature in Phoenix in December? (Source: weather.com)

51. Running. Matt Savage has run at least 5 miles every day since September 1, 1979—Including January 3, 1997, the day he got married, and every day on the cruise ship during the honeymoon. The total distance he has run is approximately 9 times around the Earth.

If one trip around the Earth is 7,926 miles, how far has Matt Savage run over the years? (Source: nydailynews.com)

52. How much is a trillion? If a trillion one-dollar bills were laid end to end, they would reach to the moon and back about 202 times. If the distance from the Earth to the moon is 238,900 miles, how far does a trillion one-dollar bills stretch?

53. Blu-rays. A shopper purchased six Blu-ray discs: *Sing* ($16), *Fantastic Beasts and Where to Find Them* ($27), *Doctor Strange* ($19), *Moana* ($22), *The Secret Life of Pets* ($18), and *Trolls* ($24). There was $11 sales tax. If she paid for the DVDs with $20 bills, how many bills were needed? How much did she receive back in change?

54. Redecorating. An interior decorator purchased a painting for $95, a sofa for $225, a chair for $275, and an end table for $155. The tax was $60 and delivery was $75. If she paid for the furniture with $50 bills, how many bills were needed? How much did she receive back in change?

55. Women's basketball. On February 1, 2006, Epiphanny Prince, of New York, broke a national prep record that was held by Cheryl Miller. Prince made 50 two-point baskets, 4 three-point baskets, and 1 free throw. How many points did she score in the game?

56. Collecting trash. After a parade, city workers cleaned the street and filled 8 medium-size (22-gallon) trash bags and 16 large-size (30-gallon) trash bags. How many gallons of trash did the city workers pick up?

from **Campus to Careers**

57. A 27-foot-long by 19-foot-wide rectangular garden is one feature of a landscape design for a community park. A concrete walkway is to run through the garden and will occupy 125 square feet of space. How many square feet are left for planting in the garden?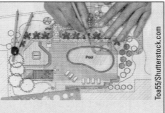

58. Mattresses. A queen-size mattress measures 60 inches by 80 inches, and a full-size mattress measures 54 inches by 75 inches. How much more sleeping surface (area) is there on a queen-size mattress?

59. Drug testing. During a drug trial last year, researchers gave two dozen mice identical doses of a medication. A total of 840 grams of the medication was used. This year, they will perform the same trial with 28 mice. If the medication costs 5 cents per gram, and they plan to give the same-size doses as last time, how much should the researchers expect to spend on the medication this year?

60. Persian rugs. A home decorator had gold-colored tassles sewn on the perimeter of an expensive Persian rug so that it wouldn't fray. (See the illustration below.) If it cost $3 per foot to have the tassles sewn on the rug, and the total cost of the project was $102, how wide is the rug?

14 ft

WRITING

61. Write an application problem that would have the following solution. Use the phrase *less than* in the problem.

25,500
− 6,200
19,300

62. Write an application problem that would have the following solution. Use the word *increase* in the problem.

49,656
+ 22,103
71,759

63. Write an application problem that would have the following solution. Use the phrase *how much does each* in the problem.

410,000
6)2,460,000

64. Write an application problem that would have the following solution. Use the word *twice* in the problem.

55
× 2
110

REVIEW

65. Check the following addition by adding upward. Is the sum correct?

3,714
2,489
781
5,500
+ 303
12,987

66. Check the following subtraction using addition. Is the difference correct?

42,403
− 1,675
40,728

67. Check the following multiplication using estimation. Does the product seem reasonable?

73
× 59
6,407

68. Check the following division using multiplication. Is the quotient correct?

407
27)10,989

OBJECTIVES

1 Factor whole numbers.

2 Identify even and odd whole numbers, prime numbers, and composite numbers.

3 Find prime factorizations using a factor tree.

4 Find prime factorizations using a division ladder.

5 Use exponential notation.

6 Evaluate exponential expressions.

SECTION **1.7**   Prime Factors and Exponents

In this section, we will discuss how to express whole numbers in factored form. The procedures used to find the factored form of a whole number involve multiplication and division.

OBJECTIVE 1 Factor whole numbers.

The statement $3 \cdot 2 = 6$ has two parts: the numbers that are being multiplied and the answer. The numbers that are being multiplied are called *factors*, and the answer is the *product*. We say that 3 and 2 are factors of 6.

Factors
Numbers that are multiplied together are called **factors**.

EXAMPLE 1   Find the factors of 12.

Strategy We will find all the pairs of whole numbers whose product is 12.

WHY Each of the numbers in those pairs is a factor of 12.

Solution
The pairs of whole numbers whose product is 12 are:

$$1 \cdot 12 = 12, \qquad 2 \cdot 6 = 12, \qquad \text{and} \qquad 3 \cdot 4 = 12$$

In order, from least to greatest, the factors of 12 are 1, 2, 3, 4, 6, and 12.

Self Check 1

Find the factors of 20.

Now Try ➡ **Problems 21 and 27**

In Example 1, once we determine the pair 1 and 12 are factors of 12, any remaining factors must be *between* 1 and 12. Once we determine that the pair 2 and 6 are factors of 12, any remaining factors must be *between* 2 and 6. Once we determine that the pair 3 and 4 are factors of 12, any remaining factors of 12 must be *between* 3 and 4. Since there are no whole numbers between 3 and 4, we know that all the possible factors of 12 have been found.

In Example 1, we found that **1, 2, 3, 4, 6,** and **12** are the factors of 12. Notice that each of the factors divides 12 exactly, leaving a remainder of 0.

$$\frac{12}{1} = 12 \qquad \frac{12}{2} = 6 \qquad \frac{12}{3} = 4 \qquad \frac{12}{4} = 3 \qquad \frac{12}{6} = 2 \qquad \frac{12}{12} = 1$$

In general, if a whole number is a factor of a given number, it also divides the given number exactly.

When we say that 3 is a factor of 6, we are using the word *factor* as a noun. The word *factor* is also used as a verb.

Factoring a Whole Number
To **factor** a whole number means to express it as the product of other whole numbers.

EXAMPLE 2   Factor 40 using   **a.** two factors   **b.** three factors

Strategy We will find a pair of whole numbers whose product is 40 and three whole numbers whose product is 40.

WHY To *factor* a number means to express it as the product of two (or more) numbers.

Solution
a. To factor 40 using two factors, there are several possibilities.

$$40 = 1 \cdot 40, \qquad 40 = 2 \cdot 20, \qquad 40 = 4 \cdot 10, \qquad \text{and} \qquad 40 = 5 \cdot 8$$

b. To factor 40 using three factors, there are several possibilities. Two of them are:

$$40 = 5 \cdot 4 \cdot 2 \qquad \text{and} \qquad 40 = 2 \cdot 2 \cdot 10$$

Self Check 2

Factor 18 using **a.** two factors **b.** three factors

Now Try ⟳ Problems 39 and 45

EXAMPLE 3   Find the factors of 17.

Strategy We will find all the pairs of whole numbers whose product is 17.

WHY Each of the numbers in those pairs is a factor of 17.

Solution
The only pair of whole numbers whose product is 17 is:

$$1 \cdot 17 = 17$$

Therefore, the only factors of 17 are 1 and 17.

Self Check 3

Find the factors of 23.

Now Try ⟳ Problem 49

OBJECTIVE 2 **Identify even and odd whole numbers, prime numbers, and composite numbers.**

A whole number is either *even* or *odd*.

> **Even and Odd Whole Numbers**
>
> If a whole number is divisible by 2, it is called an **even** number.
>
> If a whole number is not divisible by 2, it is called an **odd** number.
>
> The even whole numbers are the numbers:
>
>   0, 2, 4, 6, 8, 10, 12, 14, 16, 18, . . .
>
> The odd whole numbers are the numbers:
>
>   1, 3, 5, 7, 9, 11, 13, 15, 17, 19, . . .

LANGUAGE OF MATHEMATICS

The word *infinitely* is a form of the word *infinite*, meaning *unlimited*.

The three dots at the end of each list shown above indicate that there are infinitely many even and infinitely many odd whole numbers.

In Example 3, we saw that the only factors of 17 are 1 and 17. Numbers that have only two factors, 1 and the number itself, are called **prime numbers**.

Prime Numbers

A **prime number** is a whole number greater than 1 that has only 1 and itself as factors.

The prime numbers are the numbers:

2, 3, 5, 7, 11, 13, 17, 19, 23, 29, 31, 37, 41, 43, 47, 53, 59, 61, 67, 71, 73, 79, 83, 89, 97, 101, . . .

There are infinitely many prime numbers.

Note that the only even prime number is 2. Any other even whole number is divisible by 2, and thus has 2 as a factor, in addition to 1 and itself. Also note that not all odd whole numbers are prime numbers. For example, since 15 has factors of 1, 3, 5, and 15, it is not a prime number.

The set of whole numbers contains many prime numbers. It also contains many numbers that are not prime.

Caution! The numbers 0 and 1 are neither prime nor composite because neither is a whole number greater than 1.

Composite Numbers

The **composite numbers** are whole numbers greater than 1 that are *not* prime.

The composite numbers are the numbers:

   4, 6, 8, 9, 10, 12, 14, 15, 16, 18, . . .

There are infinitely many composite numbers.

EXAMPLE 4   **a.** Is 37 a prime number?   **b.** Is 45 a prime number?

Strategy We will determine whether the given number has only 1 and itself as factors.

WHY If that is the case, it is a prime number.

Self Check 4

a. Is 39 a prime number?
b. Is 57 a prime number?

Now Try ➫ Problems 53 and 57

Solution

a. Since 37 is a whole number greater than 1 and its only factors are 1 and 37, it is prime. Since 37 is not divisible by 2, we say it is an odd prime number.

b. The factors of 45 are 1, 3, 5, 9, 15, and 45. Since it has factors other than 1 and 45, 45 is *not* prime. It is an odd composite number.

OBJECTIVE 3 **Find prime factorizations using a factor tree.**

Every composite number can be formed by multiplying a specific combination of prime numbers. The process of finding that combination is called **prime factorization**.

Prime Factorization

To find the **prime factorization** of a whole number means to write it as the product of only prime numbers.

One method for finding the prime factorization of a number is called a **factor tree**. The factor trees shown on the next page are used to find the prime factorization of 90 in two ways.

1. Factor 90 as 9 · 10.

2. Neither 9 nor 10 is prime, so we factor each of them.

3. The process is complete when only prime numbers appear at the bottom of all branches.

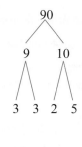

1. Factor 90 as 6 · 15.

2. Neither 6 nor 15 is prime, so we factor each of them.

3. The process is complete when only prime numbers appear at the bottom of all branches.

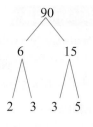

Caution! Remember that there is a difference between the *factors* and the *prime factors* of a number. For example:

The factors of 15 are 1, 3, 5, and 15.

The prime factors of 15 are 3 and 5.

Either way, the prime factorization of 90 contains one factor of 2, two factors of 3, and one factor of 5. Writing the factors in order, from least to greatest, the **prime-factored form** of 90 is 2 · 3 · 3 · 5. It is true that no other combination of prime factors will produce 90. This example illustrates an important fact about composite numbers.

Fundamental Theorem of Arithmetic

Any composite number has exactly one set of prime factors.

EXAMPLE 5   Use a factor tree to find the prime factorization of 210.

Strategy We will factor each number that we encounter as a product of two whole numbers (other than 1 and itself) until *all the factors involved are prime.*

WHY The prime factorization of a whole number contains only prime numbers.

Solution

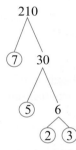

Factor 210 as 7 · 30. (The resulting prime factorization will be the same no matter which two factors of 210 you begin with.) Since 7 is prime, circle it. That branch of the tree is completed.

Since 30 is not prime, factor it as 5 · 6. (The resulting prime factorization will be the same no matter which two factors of 30 you use.) Since 5 is prime, circle it. That branch of the tree is completed.

Since 6 is not prime, factor it as 2 · 3. Since 2 and 3 are prime, circle them. All the branches of the tree are now completed.

The prime factorization of 210 is 7 · 5 · 2 · 3. Writing the prime factors in order, from least to greatest, we have 210 = 2 · 3 · 5 · 7.

Check: Multiply the prime factors. The product should be 210.

$2 · 3 · 5 · 7 = 6 · 5 · 7$ Write the multiplication in horizontal form. Working left to right, multiply 2 and 3.

$= 30 · 7$ Working left to right, multiply 6 and 5.

$= 210$ Multiply 30 and 7. The result checks.

Self Check 5

Use a factor tree to find the prime factorization of 126.

Now Try ➲ Problems 61 and 71

OBJECTIVE 4 Find prime factorizations using a division ladder.

We can also find the prime factorization of a whole number using an inverted division process called a **division ladder**. It is called that because of the vertical "steps" that it produces.

Caution! In Example 6, it would be incorrect to begin the division process with

$$4\,\overline{|280}$$
$$70$$

because 4 is not a prime number.

EXAMPLE 6 Use a division ladder to find the prime factorization of 280.

Strategy We will perform repeated divisions by prime numbers until the final quotient is itself a prime number.

WHY If a prime number is a factor of 280, it will divide 280 exactly.

Solution
It is helpful to begin with the *smallest prime*, 2, as the first trial divisor. Then, if necessary, try the primes 3, 5, 7, 11, 13, … in that order.

Step 1: The prime number 2 divides 280 exactly.

The result is 140, which is not prime. Continue the division process.

$$2\,\overline{|280}$$
$$140$$

Step 2: Since 140 is even, divide by 2 again.

The result is 70, which is not prime. Continue the division process.

$$2\,\overline{|280}$$
$$2\,\overline{|140}$$
$$70$$

Step 3: Since 70 is even, divide by 2 a third time. The result is 35, which is not prime. Continue the division process.

$$2\,\overline{|280}$$
$$2\,\overline{|140}$$
$$2\,\overline{|70}$$
$$35$$

Step 4: Since neither the prime number 2 nor the next greatest prime number 3 divide 35 exactly, we try 5. The result is 7, which is prime. We are done.

The prime factorization of 280 appears in the left column of the division ladder: $2 \cdot 2 \cdot 2 \cdot 5 \cdot 7$. Check this result using multiplication.

$$2\,\overline{|280}$$
$$2\,\overline{|140}$$
$$2\,\overline{|70}$$
$$5\,\overline{|35}$$
$$7 \leftarrow \text{Prime}$$

OBJECTIVE 5 Use exponential notation.

In Example 6, we saw that the prime factorization of 280 is $2 \cdot 2 \cdot 2 \cdot 5 \cdot 7$. Because this factorization has three factors of 2, we call 2 a *repeated factor*. We can use **exponential notation** to write $2 \cdot 2 \cdot 2$ in a more compact form.

Exponent and Base

An **exponent** is used to indicate repeated multiplication. It tells how many times the **base** is used as a factor.

The exponent is 3.

$$2 \cdot 2 \cdot 2 = 2^3 \qquad \text{Read } 2^3 \text{ as "2 to the third power" or "2 cubed."}$$

Repeated factors   The base is 2.

The prime factorization of 280 can be written using exponents: $2 \cdot 2 \cdot 2 \cdot 5 \cdot 7 = 2^3 \cdot 5 \cdot 7$.

In the **exponential expression** 2^3, the number 2 is the base and 3 is the exponent. The expression itself is called a **power of 2**.

EXAMPLE 7   Write each product using exponents:
a. $5 \cdot 5 \cdot 5 \cdot 5$ **b.** $7 \cdot 7 \cdot 11$ **c.** $2(2)(2)(2)(3)(3)(3)$

Strategy We will determine the number of repeated factors in each expression.

WHY An exponent can be used to represent repeated multiplication.

Solution
a. The factor 5 is repeated 4 times. We can represent this repeated multiplication with an exponential expression having a base of 5 and an exponent of 4:

$$5 \cdot 5 \cdot 5 \cdot 5 = 5^4$$

b. $7 \cdot 7 \cdot 11 = 7^2 \cdot 11$ 7 is used as a factor 2 times.

c. $2(2)(2)(2)(3)(3)(3) = 2^4(3^3)$ 2 is used as a factor 4 times, and 3 is used as a factor 3 times.

Self Check 7

Write each product using exponents:

a. $3 \cdot 3 \cdot 7$

b. $5(5)(7)(7)$

c. $2 \cdot 2 \cdot 2 \cdot 3 \cdot 3 \cdot 5$

Now Try Problems 77 and 81

OBJECTIVE 6 **Evaluate exponential expressions.**

We can use the definition of exponent to **evaluate** (find the value of) exponential expressions.

EXAMPLE 8   Evaluate each expression:
a. 7^2 **b.** 2^5 **c.** 10^4 **d.** 6^1

Strategy We will rewrite each exponential expression as a product of repeated factors and then perform the multiplication. This requires that we identify the base and the exponent.

WHY The exponent tells the number of times the base is to be written as a factor.

Solution
We can write the steps of the solutions in horizontal form.

a. $7^2 = 7 \cdot 7$ Read 7^2 as "7 to the second power" or "7 squared." The base is 7, and the exponent is 2. Write the base as a factor 2 times.

$= 49$ Multiply.

b. $2^5 = 2 \cdot 2 \cdot 2 \cdot 2 \cdot 2$ Read 2^5 as "2 to the 5th power." The base is 2, and the exponent is 5. Write the base as a factor 5 times.

$= 4 \cdot 2 \cdot 2 \cdot 2$ Multiply, working left to right.

$= 8 \cdot 2 \cdot 2$

$= 16 \cdot 2$

$= 32$

c. $10^4 = 10 \cdot 10 \cdot 10 \cdot 10$ Read 10^4 as "10 to the 4th power." The base is 10, and the exponent is 4. Write the base as a factor 4 times.

$= 100 \cdot 10 \cdot 10$ Multiply, working left to right.

$= 1,000 \cdot 10$

$= 10,000$

d. $6^1 = 6$ Read 6^1 as "6 to the first power." Write the base 6 once.

Caution! Note that 2^5 means $2 \cdot 2 \cdot 2 \cdot 2 \cdot 2$. It does not mean $2 \cdot 5$. That is, $2^5 = 32$ and $2 \cdot 5 = 10$.

Self Check 8

Evaluate each expression:

a. 9^2 **b.** 6^3

c. 3^4 **d.** 12^1

Now Try Problem 89

EXAMPLE 9 The prime factorization of a number is $2^3 \cdot 3^4 \cdot 5$. What is the number?

Strategy To find the number, we will evaluate each exponential expression and then do the multiplication.

WHY The exponential expressions must be evaluated first.

Solution
We can write the steps of the solutions in horizontal form.

$$2^3 \cdot 3^4 \cdot 5 = 8 \cdot 81 \cdot 5 \qquad \text{Evaluate the exponential expressions: } 2^3 = 8$$
$$\text{and } 3^4 = 81.$$

$$= 648 \cdot 5 \qquad \text{Multiply, working left to right.}$$

$$= 3{,}240 \qquad \text{Multiply.}$$

$2^3 \cdot 3^4 \cdot 5$ is the prime factorization of 3,240.

$$\begin{array}{r} 81 \\ \times\ 8 \\ \hline 648 \end{array}$$

$$\begin{array}{r} \overset{2\ 4}{648} \\ \times\ 5 \\ \hline 3{,}240 \end{array}$$

Self Check 9

The prime factorization of a number is $2 \cdot 3^3 \cdot 5^2$. What is the number?

Now Try ➡ Problems 93 and 97

Success Tip Calculations that you cannot perform in your head should be shown outside the steps of your solution.

Using Your Calculator ▶ The Exponential Key

At the end of 1 hour, a culture contains two bacteria. Suppose the number of bacteria doubles every hour thereafter. Use exponents to determine how many bacteria the culture will contain after 24 hours.

We can use a table to help model the situation. From the table, we see a pattern developing: The number of bacteria in the culture after 24 hours will be 2^{24}.

Time	Number of bacteria
1 hr	$2 = 2^1$
2 hr	$4 = 2^2$
3 hr	$8 = 2^3$
4 hr	$16 = 2^4$
24 hr	$? = 2^{24}$

We can evaluate this exponential expression using the exponential key $\boxed{y^x}$ on a scientific calculator ($\boxed{x^y}$ on some models).

$$2 \;\boxed{y^x}\; 24 \;\boxed{=} \qquad\qquad \boxed{16777216}$$

On a graphing calculator, we use the carat key $\boxed{\wedge}$ to raise a number to a power.

$$2 \;\boxed{\wedge}\; 24 \;\boxed{\text{ENTER}} \qquad\qquad \boxed{16777216}$$

Since $2^{24} = 16{,}777{,}216$, there will be 16,777,216 bacteria after 24 hours.

Answers to Self Checks

1. 1, 2, 4, 5, 10, and 20 **2. a.** $1 \cdot 18$, $2 \cdot 9$, or $3 \cdot 6$ **b.** Two possibilities are $2 \cdot 3 \cdot 3$ and $1 \cdot 2 \cdot 9$ **3.** 1 and 23
4. a. no **b.** no **5.** $2 \cdot 3 \cdot 3 \cdot 7$ **6.** $2 \cdot 2 \cdot 3 \cdot 3 \cdot 3$ **7. a.** $3^2 \cdot 7$ **b.** $5^2(7^2)$ **c.** $2^3 \cdot 3^2 \cdot 5$ **8. a.** 81
b. 216 **c.** 81 **d.** 12 **9.** 1,350

SECTION **1.7** STUDY SET

VOCABULARY

Fill in the blanks.

1. Numbers that are multiplied together are called _____.

2. To _____ a whole number means to express it as the product of other whole numbers.

3. A _____ number is a whole number greater than 1 that has only 1 and itself as factors.

4. Whole numbers greater than 1 that are not prime numbers are called _____ numbers.

5. To prime factor a number means to write it as a product of only _____ numbers.

6. An exponent is used to represent _____ multiplication. It tells how many times the _____ is used as a factor.

7. In the exponential expression 6^4, the number 6 is the _____ and 4 is the _____.

8. We can read 5^2 as "5 to the second power" or as "5 _____." We can read 7^3 as "7 to the third power" or as "7 _____."

CONCEPTS

9. Fill in the blanks to find the pairs of whole numbers whose product is 45.

   $1 \cdot \boxed{} = 45 \qquad 3 \cdot \boxed{} = 45 \qquad 5 \cdot \boxed{} = 45$

   The factors of 45, in order from least to greatest, are: ▇, ▇, ▇, ▇, ▇, ▇

10. Fill in the blanks to find the pairs of whole numbers whose product is 28.

    $1 \cdot \boxed{} = 28 \qquad 2 \cdot \boxed{} = 28 \qquad 4 \cdot \boxed{} = 28$

    The factors of 28, in order from least to greatest, are: ▇, ▇, ▇, ▇, ▇, ▇

11. If 4 is a factor of a whole number, will 4 divide the number exactly?

12. Suppose a number is divisible by 10. Is 10 a factor of the number?

13. a. Fill in the blanks: If a whole number is divisible by 2, it is an _____ number. If it is not divisible by 2, it is an _____ number.

    b. List the first 10 even whole numbers.

    c. List the first 10 odd whole numbers.

14. a. List the first 10 prime numbers.

    b. List the first 10 composite numbers.

15. Fill in the blanks to prime factor 150 using a factor tree.

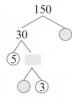

   The prime factorization of 150 is ▇ · ▇ · ▇ · ▇.

16. Which of the whole numbers, 1, 2, 3, 4, 5, 6, 7, 8, 9, and 10, could be at the top of this factor tree?

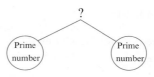

17. Fill in the blanks to prime factor 150 using a division ladder.

    $$\begin{array}{r} \boxed{} \,\lfloor 150 \\ 3\,\lfloor 75 \\ 5\,\lfloor \boxed{} \\ \hline 5 \end{array}$$

    The prime factorization of 150 is ▇ · ▇ · ▇ · ▇.

18. a. When using the division ladder method to find the prime factorization of a number, what is the first divisor to try?

    b. If 2 does not divide the given number exactly, what other divisors should be tried?

NOTATION

19. For each exponential expression, what is the base and the exponent?
    a. 7^6 b. 15^1

20. Consider the expression $2 \cdot 2 \cdot 2 \cdot 3 \cdot 3$.
    a. How many repeated factors of 2 are there?
    b. How many repeated factors of 3 are there?

GUIDED PRACTICE

Find the factors of each whole number. List them from least to greatest. See Example 1.

21. 10	**22.** 6
23. 40	**24.** 75
25. 18	**26.** 32
27. 44	**28.** 65
29. 77	**30.** 81
31. 100	**32.** 441

Factor each whole number using two factors. Do not use the factor 1 in your answer. See Example 2.

33. 8	**34.** 9
35. 27	**36.** 35
37. 49	**38.** 25
39. 20	**40.** 16

Factor each whole number using three factors. Do not use the factor 1 in your answer. See Example 2.

41. 30

42. 28

43. 63

44. 50

45. 54

46. 56

47. 60

48. 64

Find the factors of each whole number. See Example 3.

49. 11

50. 29

51. 37

52. 41

Determine whether each of the following numbers is a prime number. See Example 4.

53. 17

54. 59

55. 99

56. 27

57. 51

58. 91

59. 43

60. 83

Find the prime factorization of each number. Use exponents in your answer, when it is helpful. See Examples 5 and 6.

61. 30

62. 20

63. 39

64. 105

65. 99

66. 400

67. 162

68. 98

69. 64

70. 243

71. 147

72. 140

73. 220

74. 385

75. 102

76. 114

Write each product using exponents. See Example 7.

77. $2 \cdot 2 \cdot 2 \cdot 2 \cdot 2$

78. $3 \cdot 3 \cdot 3 \cdot 3 \cdot 3 \cdot 3$

79. $5 \cdot 5 \cdot 5 \cdot 5$

80. $9 \cdot 9 \cdot 9$

81. $4(4)(8)(8)(8)$

82. $12(12)(12)(16)$

83. $7 \cdot 7 \cdot 7 \cdot 9 \cdot 9 \cdot 7 \cdot 7 \cdot 7 \cdot 7$

84. $6 \cdot 6 \cdot 6 \cdot 5 \cdot 5 \cdot 6 \cdot 6 \cdot 6$

LOOK ALIKES

Evaluate each exponential expression. See Example 8.

85. a. 3^4 **b.** 4^3

86. a. 5^3 **b.** 3^5

87. a. 2^5 **b.** 5^2

88. a. 4^5 **b.** 5^4

89. a. 7^3 **b.** 3^7

90. a. 8^2 **b.** 2^8

91. a. 9^1 **b.** 1^9

92. a. 20^1 **b.** 1^{20}

The prime factorization of a number is given. What is the number? See Example 9.

93. $2 \cdot 3 \cdot 3 \cdot 5$

94. $2 \cdot 2 \cdot 2 \cdot 7$

95. $7 \cdot 11^2$

96. $2 \cdot 3^4$

97. $3^2 \cdot 5^2$

98. $3^3 \cdot 5^3$

99. $2^3 \cdot 3^3 \cdot 13$

100. $2^3 \cdot 3^2 \cdot 11$

APPLICATIONS

101. Perfect number. A whole number is called a **perfect number** when the sum of its factors that are less than the number equals the number. For example, 6 is a perfect number because $1 + 2 + 3 = 6$. Find the factors of 28. Then use addition to show that 28 is also a perfect number.

102. Cryptography. Information is often transmitted in code. Many codes involve writing products of large primes because they are difficult to factor. To see how difficult, try finding two prime factors of 7,663. (*Hint:* Both primes are greater than 70.)

103. Light. The illustration shows that the light energy that passes through the first unit of area, 1 yard away from the bulb, spreads out as it travels away from the source. How much area does that light energy cover 2 yards, 3 yards, and 4 yards from the bulb? Express each answer using exponents.

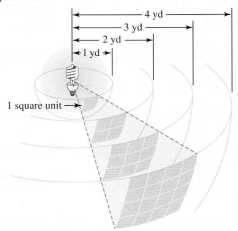

104. Cell division. After 1 hour, a cell has divided to form another cell. In another hour, these two cells have divided so that four cells exist. In another hour, these four cells divide so that eight exist.

   a. How many cells exist at the end of the fourth hour?

   b. The number of cells that exist after each division can be found using an exponential expression. What is the base?

   c. Find the number of cells after 12 hours.

WRITING

105. Explain how to check a prime factorization.

106. Explain the difference between the *factors* of a number and the *prime factors* of a number. Give an example.

107. Find 1^2, 1^3, and 1^4. From the results, what can be said about any power of 1?

108. Use the phrase *infinitely many* in a sentence.

REVIEW

109. Marching bands. When a university band lines up in 8 rows of 15 musicians, there are 5 musicians left over. How many band members are there?

110. U.S. college costs. For the 2016–2017 school year the average yearly cost for tuition and fees at a private four-year college was $33,480. The average yearly cost for tuition and fees at a public four-year college for in-state students was $9,650. At these rates, how much less are the tuition costs and fees at a public college over four years? (Source: The College Board)

SECTION 1.8 **The Least Common Multiple and the Greatest Common Factor**

As a child, you probably learned how to count by 2's and 5's and 10's. Counting in that way is an example of an important concept in mathematics called *multiples*.

OBJECTIVES

1 Find the LCM by listing multiples.

2 Find the LCM using prime factorization.

3 Find the GCF by listing factors.

4 Find the GCF using prime factorization.

OBJECTIVE 1 **Find the LCM by listing multiples.**

The **multiples** of a number are the products of that number and 1, 2, 3, 4, 5, and so on.

EXAMPLE 1   Find the first eight multiples of 6.

Strategy We will multiply 6 by 1, 2, 3, 4, 5, 6, 7, and 8.

WHY The *multiples of a number* are the products of that number and 1, 2, 3, 4, 5, and so on.

Solution

To find the multiples, we proceed as follows:

$6 \cdot 1 = 6$ This is the first multiple of 6.

$6 \cdot 2 = 12$

$6 \cdot 3 = 18$

$6 \cdot 4 = 24$

$6 \cdot 5 = 30$

$6 \cdot 6 = 36$

$6 \cdot 7 = 42$

$6 \cdot 8 = 48$ This is the eighth multiple of 6.

The first eight multiples of 6 are 6, 12, 18, 24, 30, 36, 42, and 48.

Self Check 1

Find the first eight multiples of 9.

Now Try Problems 17 and 89

The first eight multiples of 3 and the first eight multiples of 4 are shown below. The numbers highlighted in red are *common multiples* of 3 and 4.

$3 \cdot 1 = 3$	$4 \cdot 1 = 4$
$3 \cdot 2 = 6$	$4 \cdot 2 = 8$
$3 \cdot 3 = 9$	$4 \cdot 3 = \mathbf{12}$
$3 \cdot 4 = \mathbf{12}$	$4 \cdot 4 = 16$
$3 \cdot 5 = 15$	$4 \cdot 5 = 20$
$3 \cdot 6 = 18$	$4 \cdot 6 = \mathbf{24}$
$3 \cdot 7 = 21$	$4 \cdot 7 = 28$
$3 \cdot 8 = \mathbf{24}$	$4 \cdot 8 = 32$

If we extend each list, it soon becomes apparent that 3 and 4 have infinitely many common multiples.

The common multiples of 3 and 4 are: **12, 24, 36, 48, 60, 72,** ...

Because 12 is the smallest number that is a multiple of both 3 and 4, it is called the **least common multiple (LCM)** of 3 and 4. We can write this in compact form as:

LCM (3, 4) = 12   Read as "The least common multiple of 3 and 4 is 12."

The Least Common Multiple (LCM)

The **least common multiple** of two whole numbers is the smallest common multiple of the numbers.

We have seen that the LCM of 3 and 4 is 12. It is important to note that 12 is divisible by both 3 and 4.

$$\frac{12}{3} = 4 \qquad \text{and} \qquad \frac{12}{4} = 3$$

This observation illustrates an important relationship between divisibility and the least common multiple.

The Least Common Multiple (LCM)

The **least common multiple** of two whole numbers is the smallest whole number that is divisible by both of those numbers.

When finding the LCM of two numbers, writing both lists of multiples can be tiresome. From the previous definition of LCM, it follows that we need only list the multiples of the larger number. The LCM is simply *the first multiple of the larger number that is divisible by the smaller number.* For example, to find the LCM of 3 and 4, we observe that

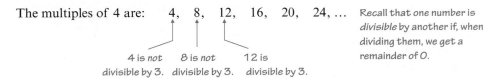

The multiples of 4 are: 4,   8,   12,   16,   20,   24, ...   Recall that one number is *divisible* by another if, when dividing them, we get a remainder of 0.

4 is *not* divisible by 3.   8 is *not* divisible by 3.   12 is divisible by 3.

Since 12 is the first multiple of 4 that is divisible by 3, the LCM of 3 and 4 is 12. As expected, this is the same result that we obtained using the two-list method.

Finding the LCM by Listing the Multiples of the Largest Number

To find the least common multiple of two (or more) whole numbers:

1. Write multiples of the largest number by multiplying it by 1, 2, 3, 4, 5, and so on.
2. Continue this process until you find the first multiple of the larger number that is divisible by each of the smaller numbers. That multiple is their LCM.

EXAMPLE 2   Find the LCM of 6 and 8.

Strategy We will write the multiples of the larger number, 8, until we find one that is divisible by the smaller number, 6.

WHY The LCM of 6 and 8 is the smallest multiple of 8 that is divisible by 6.

Solution

The 1st multiple of 8: $8 \cdot 1 = 8$ ← *8 is not divisible by 6. (When we divide, we get a remainder of 2.) Since 8 is not divisible by 6, find the next multiple.*

The 2nd multiple of 8: $8 \cdot 2 = 16$ ← *16 is not divisible by 6. Find the next multiple.*

The 3rd multiple of 8: $8 \cdot 3 = 24$ ← *24 is divisible by 6. This is the LCM.*

The first multiple of 8 that is divisible by 6 is 24. Thus,

LCM (6, 8) = 24   *Read as "The least common multiple of 6 and 8 is 24."*

Self Check 2

Find the LCM of 8 and 10.

Now Try ⟳ **Problem 25**

We can extend this method to find the LCM of three whole numbers.

EXAMPLE 3   Find the LCM of 2, 3, and 10.

Strategy We will write the multiples of the largest number, 10, until we find one that is divisible by both of the smaller numbers, 2 and 3.

WHY The LCM of 2, 3, and 10 is the smallest multiple of 10 that is divisible by 2 and 3.

Solution

The 1st multiple of 10: $10 \cdot 1 = 10$ ← *10 is divisible by 2, but not by 3. Find the next multiple.*

The 2nd multiple of 10: $10 \cdot 2 = 20$ ← *20 is divisible by 2, but not by 3. Find the next multiple.*

The 3rd multiple of 10: $10 \cdot 3 = 30$ ← *30 is divisible by 2 and by 3. It is the LCM.*

The first multiple of 10 that is divisible by 2 and 3 is 30. Thus,

LCM (2, 3, 10) = 30   *Read as "The least common multiple of 2, 3, and 10 is 30."*

Self Check 3

Find the LCM of 3, 4, and 8.

Now Try ⟳ **Problem 35**

OBJECTIVE 2 **Find the LCM using prime factorization.**

Another method for finding the LCM of two (or more) whole numbers uses prime factorization. This method is especially helpful when working with larger numbers. As an example, we will find the LCM of 36 and 54. First, we find their prime factorizations:

$36 = 2 \cdot 2 \cdot 3 \cdot 3$ *Factor trees (or division ladders) can be used to find the prime factorizations.*

$54 = 2 \cdot 3 \cdot 3 \cdot 3$

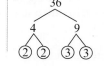

The LCM of 36 and 54 must be divisible by 36 and 54. If the LCM is divisible by 36, it must have the prime factors of 36, which are $2 \cdot 2 \cdot 3 \cdot 3$. If the LCM is divisible by 54, it must have the prime factors of 54, which are $2 \cdot 3 \cdot 3 \cdot 3$. The smallest number that meets both requirements is

These are the prime factors of 36.

$2 \cdot 2 \cdot 3 \cdot 3 \cdot 3$

These are the prime factors of 54.

To find the LCM, we perform the indicated multiplication:

$$\text{LCM } (36, 54) = 2 \cdot 2 \cdot 3 \cdot 3 \cdot 3 = 108$$

Caution! The LCM $(36, 54)$ is not the product of the prime factorization of 36 and the prime factorization of 54. That gives an incorrect answer of 1,944.

$$\text{LCM } (36, 54) = 2 \cdot 2 \cdot 3 \cdot 3 \cdot 2 \cdot 3 \cdot 3 \cdot 3 = 1,944$$

The LCM should contain all the prime factors of 36 and all the prime factors of 54, but the prime factors that 36 and 54 have in common are not repeated.

The prime factorizations of 36 and 54 contain the numbers 2 and 3.

$$36 = 2 \cdot 2 \cdot 3 \cdot 3 \qquad 54 = 2 \cdot 3 \cdot 3 \cdot 3$$

We see that

■ The greatest number of times the factor 2 appears in any one of the prime factorizations is twice and the LCM of 36 and 54 has 2 as a factor twice.

■ The greatest number of times the factor 3 appears in any one of the prime factorizations is three times and the LCM of 36 and 54 has 3 as a factor three times.

These observations suggest a procedure to use to find the LCM of two (or more) numbers using prime factorization.

Finding the LCM Using Prime Factorization

To find the least common multiple of two (or more) whole numbers:

1. Prime factor each number.
2. The LCM is a product of prime factors, where each factor is used the greatest number of times it appears in any one factorization.

EXAMPLE 4   Find the LCM of 24 and 60.

Strategy We will begin by finding the prime factorizations of 24 and 60.

WHY To find the LCM, we need to determine the greatest number of times each prime factor appears in any one factorization.

Solution
Step 1: Prime factor 24 and 60.

$24 = 2 \cdot 2 \cdot 2 \cdot 3$ Division ladders (or factor trees) can be used to find the prime factorizations.

$60 = 2 \cdot 2 \cdot 3 \cdot 5$

2	24		2	60
2	12		2	30
2	6		3	15
	3			5

Step 2: The prime factorizations of 24 and 60 contain the prime factors 2, 3, and 5. To find the LCM, we use each of these factors the greatest number of times it appears in any one factorization.

- We will use the factor 2 three times because 2 appears three times in the factorization of 24. Circle 2 · 2 · 2, as shown below.
- We will use the factor 3 once because it appears one time in the factorization of 24 and one time in the factorization of 60. When the number of times a factor appears are equal, circle either one, but not both, as shown below.
- We will use the factor 5 once because it appears one time in the factorization of 60. Circle the 5, as shown below.

$$24 = (2 \cdot 2 \cdot 2) \cdot (3)$$

$$60 = 2 \cdot 2 \cdot 3 \cdot (5)$$

Since there are no other prime factors in either prime factorization, we have

Use 2 three times.
Use 3 one time.
Use 5 one time.

$$\text{LCM } (24, 60) = 2 \cdot 2 \cdot 2 \cdot 3 \cdot 5 = 120$$

Note that 120 is the smallest number that is divisible by both 24 and 60:

$$\frac{120}{24} = 5 \quad \text{and} \quad \frac{120}{60} = 2$$

Self Check 4

Find the LCM of 18 and 32.

Now Try ➲ **Problem 37**

In Example 4, we can express the prime factorizations of 24 and 60 using exponents. To determine the greatest number of times each factor appears in any one factorization, we circle the factor with the greatest exponent.

$$24 = (2^3) \cdot (3^1)$$ *The greatest exponent on the factor 2 is 3.*
                              The greatest exponent on the factor 3 is 1.

$$60 = 2^2 \cdot 3^1 \cdot (5^1)$$ *The greatest exponent on the factor 5 is 1.*

The LCM of 24 and 60 is

$$2^3 \cdot 3^1 \cdot 5^1 = 8 \cdot 3 \cdot 5 = 120$$ *Evaluate: $2^3 = 8$.*

EXAMPLE 5   Find the LCM of 28, 42, and 45.

Strategy We will begin by finding the prime factorizations of 28, 42, and 45.

WHY To find the LCM, we need to determine the greatest number of times each prime factor appears in any one factorization.

Solution

Step 1: Prime factor 28, 42, and 45.

$28 = (2 \cdot 2) \cdot 7$ *This can be written as $(2^2) \cdot 7^1$.*
$42 = 2 \cdot 3 \cdot (7)$ *This can be written as $2^1 \cdot 3^1 \cdot (7^1)$.*
$45 = (3 \cdot 3) \cdot (5)$ *This can be written as $(3^2) \cdot (5)$.*

Step 2: The prime factorizations of 28, 42, and 45 contain the prime factors 2, 3, 5, and 7. To find the LCM (28, 42, 45), we use each of these factors the greatest number of times it appears in any one factorization.

- We will use the factor 2 two times because 2 appears two times in the factorization of 28. Circle 2 · 2, as shown on previous page.
- We will use the factor 3 twice because it appears two times in the factorization of 45. Circle 3 · 3, as shown on previous page.
- We will use the factor 5 once because it appears one time in the factorization of 45. Circle the 5, as shown on previous page.
- We will use the factor 7 once because it appears one time in the factorization of 28 and one time in the factorization of 42. You may circle either 7, but only circle one of them.

Since there are no other prime factors in either prime factorization, we have

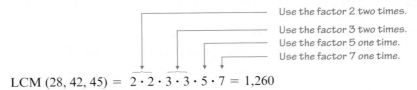

Use the factor 2 two times.
Use the factor 3 two times.
Use the factor 5 one time.
Use the factor 7 one time.

$$\text{LCM } (28, 42, 45) = 2 \cdot 2 \cdot 3 \cdot 3 \cdot 5 \cdot 7 = 1{,}260$$

If we use exponents, we have

$$\text{LCM } (28, 42, 45) = 2^2 \cdot 3^2 \cdot 5 \cdot 7 = 1{,}260$$

Either way, we have found that the LCM (28, 42, 45) = 1,260. Note that 1,260 is the smallest number that is divisible by 28, 42, and 45:

$$\frac{1{,}260}{28} = 45 \qquad \frac{1{,}260}{42} = 30 \qquad \frac{1{,}260}{45} = 28$$

Self Check 5

Find the LCM of 45, 60, and 75.

Now Try Problem 45

EXAMPLE 6   **Patient recovery.**
Two patients recovering from heart surgery exercise daily by jogging around a track. One patient can complete a lap in 4 minutes. The other can complete a lap in 6 minutes. If they begin at the same time and at the same place on the track, in how many minutes will they arrive together at the starting point of their workout?

Tom Wang/Shutterstock.com

Strategy We will find the LCM of 4 and 6.

WHY Since one patient reaches the starting point of the workout every 4 minutes, and the other is there every 6 minutes, we want to find the least common multiple of those numbers. At that time, they will both be at the starting point of the workout.

Solution
To find the LCM, we prime factor 4 and 6, and circle each prime factor the greatest number of times it appears in any one factorization.

$4 = \boxed{2 \cdot 2}$ Use the factor 2 two times, because 2 appears two times in the factorization of 4.

$6 = 2 \cdot \boxed{3}$ Use the factor 3 once, because it appears one time in the factorization of 6.

Since there are no other prime factors in either prime factorization, we have

$$\text{LCM } (4, 6) = 2 \cdot 2 \cdot 3 = 12$$

The patients will arrive together at the starting point 12 minutes after beginning their workout.

Self Check 6

Aquariums. A pet store owner changes the water in a fish aquarium every 45 days, and he changes the pump filter every 20 days. If the water and filter are changed on the same day, in how many days will they be changed again together?

Now Try Problem 91

OBJECTIVE 3 **Find the GCF by listing factors.**

We have seen that two whole numbers can have common multiples. They can also have *common factors*. To explore this concept, let's find the factors of 26 and 39 and see what factors they have in common.

To find the factors of 26, we find all the pairs of whole numbers whose product is 26. There are two possibilities:

$$1 \cdot 26 = 26 \qquad 2 \cdot 13 = 26$$

Each of the numbers in the pairs is a factor of 26. From least to greatest, the factors of 26 are 1, 2, 13, and 26.

To find the factors of 39, we find all the pairs of whole numbers whose product is 39. There are two possibilities:

$$1 \cdot 39 = 39 \qquad 3 \cdot 13 = 39$$

Each of the numbers in the pairs is a factor of 39. From least to greatest, the factors of 39 are 1, 3, 13, and 39. As shown below, the *common factors* of 26 and 39 are 1 and 13.

1, 2, 13, 26   These are the factors of 26.
1, 3, 13, 39   These are the factors of 39.

Because 13 is the largest number that is a factor of both 26 and 39, it is called the **greatest common factor (GCF)** of 26 and 39. We can write this in compact form as:

GCF (26, 39) = 13   Read as "The greatest common factor of 26 and 39 is 13."

The Greatest Common Factor (GCF)

The **greatest common factor** of two whole numbers is the largest common factor of the numbers.

EXAMPLE 7   Find the GCF of 18 and 45.

Strategy We will find the factors of 18 and 45.

WHY Then we can identify the largest factor that 18 and 45 have in common.

Solution
To find the factors of 18, we find all the pairs of whole numbers whose product is 18. There are three possibilities:

$$1 \cdot 18 = 18 \qquad 2 \cdot 9 = 18 \qquad 3 \cdot 6 = 18$$

To find the factors of 45, we find all the pairs of whole numbers whose product is 45. There are three possibilities:

$$1 \cdot 45 = 45 \qquad 3 \cdot 15 = 45 \qquad 5 \cdot 9 = 45$$

The factors of 18 and 45 are listed below. Their common factors are circled.

Factors of 18:   1,   2,   3,   6,   9,   18
Factors of 45:   1,   3,   5,   9,   15,   45

The common factors of 18 and 45 are 1, 3, and 9. Since 9 is their largest common factor,

GCF (18, 45) = 9   Read as "The greatest common factor of 18 and 45 is 9."

Self Check 7

Find the GCF of 30 and 42.

Now Try ➲ **Problem 49**

In Example 7, we found that the GCF of 18 and 45 is 9. Note that 9 is the greatest number that divides 18 and 45.

$$\frac{18}{9} = 2 \qquad \frac{45}{9} = 5$$

In general, the greatest common factor of two (or more) numbers is the largest number that divides them exactly. For this reason, the greatest common factor is also known as the **greatest common divisor (GCD)** and we can write GCD (18, 45) = 9.

OBJECTIVE 4 Find the GCF using prime factorization.

We can find the GCF of two (or more) numbers by listing the factors of each number. However, this method can be lengthy. Another way to find the GCF uses the prime factorization of each number.

Finding the GCF Using Prime Factorization

To find the greatest common factor of two (or more) whole numbers:

1. Prime factor each number.
2. Identify the common prime factors.
3. The GCF is a product of all the common prime factors found in Step 2. If there are no common prime factors, the GCF is 1.

EXAMPLE 8   Find the GCF of 48 and 72.

Strategy We will begin by finding the prime factorizations of 48 and 72.

WHY Then we can identify any prime factors that they have in common.

Solution
Step 1: Prime factor 48 and 72.

$$48 = 2 \cdot 2 \cdot 2 \cdot 2 \cdot 3$$
$$72 = 2 \cdot 2 \cdot 2 \cdot 3 \cdot 3$$

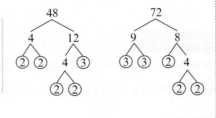

Step 2: The circling shows that 48 and 72 have four common prime factors: Three common factors of 2 and one common factor of 3.

Self Check 8

Find the GCF of 36 and 60.

Now Try Problem 57

Step 3: The GCF is the product of the circled prime factors.

$$\text{GCF}\ (48, 72) = 2 \cdot 2 \cdot 2 \cdot 3 = 24$$

EXAMPLE 9   Find the GCF of 8 and 15.

Strategy We will begin by finding the prime factorizations of 8 and 15.

WHY Then we can identify any prime factors that they have in common.

Solution
The prime factorizations of 8 and 15 are shown below.

$$8 = 2 \cdot 2 \cdot 2$$
$$15 = 3 \cdot 5$$

Self Check 9

Find the GCF of 8 and 25.

Since there are no common factors, the GCF of 8 and 15 is 1. Thus,

Now Try Problem 61

$$\text{GCF}\ (8, 15) = 1 \qquad \textit{Read as "The greatest common factor of 8 and 15 is 1."}$$

EXAMPLE 10   Find the GCF of 20, 60, and 140.

Strategy We will begin by finding the prime factorizations of 20, 60, and 140.

WHY Then we can identify any prime factors that they have in common.

Solution
The prime factorizations of 20, 60, and 140 are shown below.

$$
\begin{aligned}
20 &= 2 \cdot 2 \cdot 5 \\
60 &= 2 \cdot 2 \cdot 3 \cdot 5 \\
140 &= 2 \cdot 2 \cdot 5 \cdot 7
\end{aligned}
$$

The circling above shows that 20, 60, and 140 have three common factors: two common factors of 2 and one common factor of 5. The GCF is the product of the circled prime factors.

GCF (20, 60, 140) = $2 \cdot 2 \cdot 5 = 20$ *Read as "The greatest common factor of 20, 60, and 140 is 20."*

Note that 20 is the greatest number that divides 20, 60, and 140 exactly.

$$\frac{20}{20} = 1 \qquad \frac{60}{20} = 3 \qquad \frac{140}{20} = 7$$

Self Check 10

Find the GCF of 45, 60, and 75.

Now Try ➔ Problem 67

EXAMPLE 11   **Bouquets.**   A florist wants to use 12 white tulips, 30 pink tulips, and 42 purple tulips to make as many identical arrangements as possible. Each bouquet is to have the same number of each color tulip.

a. What is the greatest number of arrangements that she can make?
b. How many of each type of tulip can she use in each bouquet?

Strategy We will find the GCF of 12, 30, and 42.

WHY Since an equal number of tulips of each color will be used to create the identical arrangements, division is indicated. The greatest common factor of three numbers is the largest number that divides them exactly.

Solution
a. To find the GCF, we prime factor 12, 30, and 42 and circle the prime factors that they have in common.

$$
\begin{aligned}
12 &= 2 \cdot 2 \cdot 3 \\
30 &= 2 \cdot 3 \cdot 5 \\
42 &= 2 \cdot 3 \cdot 7
\end{aligned}
$$

The GCF is the product of the circled numbers.

GCF (12, 30, 42) = $2 \cdot 3 = 6$

The florist can make 6 identical arrangements from the tulips.

b. To find the number of white, pink, and purple tulips in each of the 6 arrangements, we divide the number of tulips of each color by 6.

White tulips: Pink tulips: Purple tulips:

$$\frac{12}{6} = 2 \qquad\quad \frac{30}{6} = 5 \qquad\quad \frac{42}{6} = 7$$

Each of the 6 identical arrangements will contain 2 white tulips, 5 pink tulips, and 7 purple tulips.

Self Check 11

School supplies. A bookstore manager wants to use some leftover items (36 markers, 54 pencils, and 108 pens) to make identical gift packs to donate to an elementary school.

a. What is the greatest number of gift packs that can be made?
b. How many of each type of item will be in each gift pack?

Now Try ➔ Problem 97

Answers to Self Checks

1. 9, 18, 27, 36, 45, 54, 63, 72   **2.** 40   **3.** 24   **4.** 288   **5.** 900   **6.** 180 days   **7.** 6   **8.** 12
9. 1   **10.** 15   **11. a.** 18 gift packs **b.** 2 markers, 3 pencils, 6 pens

SECTION 1.8   STUDY SET

VOCABULARY

Fill in the blanks.

1. The _____ of a number are the products of that number and 1, 2, 3, 4, 5, and so on.

2. Because 12 is the smallest number that is a multiple of both 3 and 4, it is the _____ _____ _____ of 3 and 4.

3. One number is _____ by another if, when dividing them, we get a remainder of 0.

4. Because 6 is the largest number that is a factor of both 18 and 24, it is the _____ _____ _____ of 18 and 24.

CONCEPTS

5. a. The LCM of 4 and 6 is 12. What is the smallest whole number divisible by 4 and 6?

   b. Fill in the blank: In general, the LCM of two whole numbers is the _____ whole number that is divisible by both numbers.

6. a. What are the common multiples of 2 and 3 that appear in the list of multiples shown below?

   b. What is the LCM of 2 and 3?

Multiples of 2	Multiples of 3
$2 \cdot 1 = 2$	$3 \cdot 1 = 3$
$2 \cdot 2 = 4$	$3 \cdot 2 = 6$
$2 \cdot 3 = 6$	$3 \cdot 3 = 9$
$2 \cdot 4 = 8$	$3 \cdot 4 = 12$
$2 \cdot 5 = 10$	$3 \cdot 5 = 15$
$2 \cdot 6 = 12$	$3 \cdot 6 = 18$

7. a. The first six multiples of 5 are 5, 10, 15, 20, 25, and 30. What is the first multiple of 5 that is divisible by 4?

   b. What is the LCM of 4 and 5?

8. Fill in the blanks to complete the prime factorization of 24.

9. The prime factorizations of 36 and 90 are:

   $36 = 2 \cdot 2 \cdot 3 \cdot 3$

   $90 = 2 \cdot 3 \cdot 3 \cdot 5$

What is the greatest number of times
   a. 2 appears in any one factorization?
   b. 3 appears in any one factorization?
   c. 5 appears in any one factorization?
   d. Fill in the blanks to find the LCM of 36 and 90:

   LCM = ▯ · ▯ · ▯ · ▯ = ▯

10. The prime factorizations of 14, 70, and 140 are:

   $14 = 2 \cdot 7$
   $70 = 2 \cdot 5 \cdot 7$
   $140 = 2 \cdot 2 \cdot 5 \cdot 7$

What is the greatest number of times
   a. 2 appears in any one factorization?
   b. 5 appears in any one factorization?
   c. 7 appears in any one factorization?
   d. Fill in the blanks to find the LCM of 14, 70, and 140:

   LCM = ▯ · ▯ · ▯ = ▯

11. The prime factorizations of 12 and 54 are:

   $12 = 2^2 \cdot 3^1$
   $54 = 2^1 \cdot 3^3$

What is the greatest number of times
   a. 2 appears in any one factorization?
   b. 3 appears in any one factorization?
   c. Fill in the blanks to find the LCM of 12 and 54:

   LCM = $2^▯ \cdot 3^▯$ = ▯

12. The factors of 18 and 45 are shown below.

   Factors of 18: 1, 2, 3, 6, 9, 18

   Factors of 45: 1, 3, 5, 9, 15, 45

   a. Circle the common factors of 18 and 45.
   b. What is the GCF of 18 and 45?

13. The prime factorizations of 60 and 90 are:

   $60 = 2 \cdot 2 \cdot 3 \cdot 5$
   $90 = 2 \cdot 3 \cdot 3 \cdot 5$

   a. Circle the common prime factors of 60 and 90.
   b. What is the GCF of 60 and 90?

14. The prime factorizations of 36, 84, and 132 are:

   $36 = 2 \cdot 2 \cdot 3 \cdot 3$
   $84 = 2 \cdot 2 \cdot 3 \cdot 7$
   $132 = 2 \cdot 2 \cdot 3 \cdot 11$

   a. Circle the common factors of 36, 84, and 132.
   b. What is the GCF of 36, 84, and 132?

NOTATION

Fill in the blanks.

15. a. The abbreviation for the greatest common factor is

_____.

b. The abbreviation for the least common multiple is

_____.

16. a. We read LCM (2, 15) = 30 as "The _____
_____ multiple ____ 2 and 15 ____ 30."

b. We read GCF (18, 24) = 6 as "The _____
_____ factor ____ 18 and 24 ____ 6."

GUIDED PRACTICE

Find the first eight multiples of each number. See Example 1.

17. 4   **18.** 2

19. 11   **20.** 10

21. 8   **22.** 9

23. 20   **24.** 30

Find the LCM of the given numbers. See Example 2.

25. 3, 5   **26.** 6, 9

27. 8, 12   **28.** 10, 25

29. 5, 11   **30.** 7, 11

31. 4, 7   **32.** 5, 8

Find the LCM of the given numbers. See Example 3.

33. 3, 4, 6   **34.** 2, 3, 8

35. 2, 3, 10   **36.** 3, 6, 15

Find the LCM of the given numbers. See Example 4.

37. 16, 20   **38.** 14, 21

39. 20, 50   **40.** 21, 27

41. 35, 45   **42.** 36, 48

43. 100, 120   **44.** 120, 180

Find the LCM of the given numbers. See Example 5.

45. 6, 24, 36 **46.** 6, 10, 18

47. 5, 12, 15 **48.** 8, 12, 16

Find the GCF of the given numbers. See Example 7.

49. 4, 6   **50.** 6, 15

51. 9, 12   **52.** 10, 12

Find the GCF of the given numbers. See Example 8.

53. 22, 33   **54.** 14, 21

55. 15, 30   **56.** 15, 75

57. 18, 96   **58.** 30, 48

59. 28, 42   **60.** 63, 84

Find the GCF of the given numbers. See Example 9.

61. 16, 51   **62.** 27, 64

63. 81, 125   **64.** 57, 125

Find the GCF of the given numbers. See Example 10.

65. 12, 68, 92   **66.** 24, 36, 40

67. 72, 108, 144   **68.** 81, 108, 162

TRY IT YOURSELF

Find the LCM and the GCF of the given numbers.

69. 100, 120   **70.** 120, 180

71. 14, 140   **72.** 15, 300

73. 66, 198, 242 **74.** 52, 78, 130

75. 8, 9, 49   **76.** 9, 16, 25

77. 120, 125   **78.** 98, 102

79. 34, 68, 102   **80.** 26, 39, 65

81. 46, 69   **82.** 38, 57

83. 50, 81   **84.** 65, 81

LOOK ALIKES

Find the GCF for the numbers given in part a. Then use that result to determine the GCF for the numbers in part b.

85. a. 6, 8   **b.** 60, 80

86. a. 3, 9   **b.** 30, 90

87. a. 4, 6   **b.** 2, 4, 6

88. a. 10, 15   **b.** 5, 10, 15

APPLICATIONS

89. Oil changes. Ford has officially extended the oil change interval for 2008 and newer cars to every 7,500 miles. (It used to be every 5,000 miles.) Complete the table below that shows Ford's new recommended oil change mileages.

1st oil change	2nd oil change	3rd oil change	4th oil change	5th oil change	6th oil change
7,500 mi					

90. ATMs. An ATM offers the customer cash withdrawal choices in multiples of $20. The minimum withdrawal is $20 and the maximum is $200. List the dollar amounts of cash that can be withdrawn from the ATM.

91. Nursing. A nurse is instructed to check a patient's blood pressure every 45 minutes, and another is instructed to take the same patient's temperature every 60 minutes. If both nurses are in the patient's room now, how long will it be until the nurses are together in the room once again?

92. Biorhythms. Some scientists believe that there are natural rhythms of the body, called *biorhythms*, that affect our physical, emotional, and mental cycles. Our physical biorhythm cycle lasts 23 days, the emotional biorhythm cycle lasts 28 days, and our mental biorhythm cycle lasts 33 days. Each biorhythm cycle has a high, low, and critical zone. If your three cycles are together one day, all at their lowest point, in how many more days will they be together again, all at their lowest point?

93. Picnics. A package of hot dogs usually contains 10 hot dogs, and a package of buns usually contains 12 buns. How many packages of hot dogs and buns should a person buy to be sure that there are equal numbers of each?

94. Working couples. A husband works for 6 straight days and then has a day off. His wife works for 7 straight days and then has a day off. If the husband and wife are both off from work on the same day, in how many days will they both be off from work again?

95. Dance floors. A dance floor is to be made from rectangular sheets of plywood that are 6 feet by 8 feet. What is the minimum number of full sheets of plywood that are needed to make a square dance floor, as shown below?

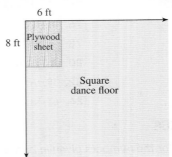

96. Bowls of soup. Each of the bowls shown below holds an exact number of *full ladles* of soup.

   a. If there is no spillage, what is the greatest-size ladle (in ounces) that a chef can use to fill all three bowls?

   b. How many ladles will it take to fill each bowl?

12 ounces   21 ounces   18 ounces

97. Art classes. Students in a painting class must pay an art supplies fee. On the first day of class, the instructor collected a total of $28 in fees from several students. On the second day, she collected $21 more from some other students, and on the third day, she collected another $63 from some more students.

   a. What is the most the art supplies fee could cost per student?

   b. Use your answer in part a to determine how many students paid the art supplies fee each day.

98. Shipping. A toy manufacturer needs to ship 135 brown teddy bears, 105 black teddy bears, and 30 white teddy bears. They can pack only one type of teddy bear in each box, and they must pack the same number of teddy bears in each box. What is the greatest number of teddy bears they can pack in each box?

WRITING

99. Explain how to find the LCM of 8 and 28 using prime factorization.

100. Explain how to find the GCF of 8 and 28 using prime factorization.

101. The prime factorization of 12 is $2 \cdot 2 \cdot 3$, and the prime factorization of 15 is $3 \cdot 5$. Explain why the LCM of 12 and 15 is *not* $2 \cdot 2 \cdot 3 \cdot 3 \cdot 5$.

102. How can you tell by looking at the prime factorizations of two whole numbers that their GCF is 1?

REVIEW

Perform each operation.

103. $9,999 + 1,111$ **104.** $10,000 - 7,989$

105. $305 \cdot 50$ **106.** $2,100 \div 105$

OBJECTIVES

1 Use the order of operations rule.

2 Evaluate expressions containing grouping symbols.

3 Find the mean (average) of a set of values.

SECTION 1.9 Order of Operations

Recall that numbers are combined with the operations of addition, subtraction, multiplication, and division to create **expressions**. We often have to **evaluate** (find the value of) expressions that involve more than one operation. In this section, we introduce an order of operations rule to follow in such cases.

OBJECTIVE 1 **Use the order of operations rule.**

Suppose you are asked to contact a friend if you see a Rolex watch for sale while you are traveling in Europe. While in Switzerland, you find the watch and send the following text message, shown on the left. The next day, you get the response shown on the right from your friend.

You sent this message.   You get this response.

Something is wrong. The first part of the response (No price too high!) says to buy the watch at any price. The second part (No! Price too high.) says not to buy it because it's too expensive. The placement of the exclamation point makes us read the two parts of the response differently, resulting in different meanings. When reading a mathematical statement, the same kind of confusion is possible. For example, consider the expression

$$2 + 3 \cdot 6$$

We can evaluate this expression in two ways. We can add first, and then multiply. Or we can multiply first, and then add. However, the results are different.

$$2 + 3 \cdot 6 = 5 \cdot 6 \quad \text{Add 2 and 3 first.} \quad \bigg| \quad 2 + 3 \cdot 6 = 2 + 18 \quad \text{Multiply 3 and 6 first.}$$
$$= 30 \quad \text{Multiply 5 and 6.} \quad \bigg| \quad = 20 \quad \text{Add 2 and 18.}$$

— Different results —

Every numerical expression has only one correct value. If we don't establish a uniform order of operations, the expression $2 + 3 \cdot 6$ has two different values. To avoid this possibility, always use the following order of operations rule.

Order of Operations

1. Perform all calculations within parentheses and other grouping symbols following the order listed in Steps 2–4 below, working from the innermost pair of grouping symbols to the outermost pair.
2. Evaluate all exponential expressions.
3. Perform all multiplications and divisions as they occur from left to right.
4. Perform all additions and subtractions as they occur from left to right.

When grouping symbols have been removed, repeat Steps 2–4 to complete the calculation.
   If a fraction bar is present, evaluate the expression above the bar (called the **numerator**) and the expression below the bar (called the **denominator**) separately. Then perform the division indicated by the fraction bar, if possible.

It isn't necessary to apply all of these steps in every problem. For example, the expression $2 + 3 \cdot 6$ does not contain any parentheses, and there are no exponential expressions. So we look for multiplications and divisions to perform and proceed as follows:

$$2 + 3 \cdot 6 = 2 + 18 \quad \text{Do the multiplication first.}$$
$$= 20 \quad \text{Do the addition.}$$

EXAMPLE 1   Evaluate: $2 \cdot 4^2 - 8$

Strategy We will scan the expression to determine what operations need to be performed. Then we will perform those operations, one at a time, following the order of operations rule.

WHY Every numerical expression has only one correct value. If we don't follow the correct order of operations, the expression can have more than one value.

Solution
Since the expression does not contain any parentheses, we begin with Step 2 of the order of operations rule: Evaluate all exponential expressions. We will write the steps of the solution in horizontal form.

$$2 \cdot 4^2 - 8 = 2 \cdot 16 - 8 \qquad \textit{Evaluate the exponential expression: } 4^2 = 16.$$

$$= 32 - 8 \qquad \textit{Do the multiplication: } 2 \cdot 16 = 32.$$

$$= 24 \qquad \textit{Do the subtraction.}$$

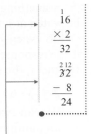

Self Check 1

Evaluate: $4 \cdot 3^3 - 6$

Now Try ➡ Problem 19

Success Tip Calculations that you cannot perform in your head should be shown outside the steps of your solution.

EXAMPLE 2   Evaluate: $80 - 3 \cdot 2 + 16$

Strategy We will perform the multiplication first.

WHY The expression does not contain any parentheses, nor are there any exponents.

Solution
We will write the steps of the solution in horizontal form.

$$80 - 3 \cdot 2 + 16 = 80 - 6 + 16 \qquad \textit{Do the multiplication: } 3 \cdot 2 = 6.$$

$$= 74 + 16 \qquad \textit{Working from left to right, do the subtraction: } 80 - 6 = 74.$$

$$= 90 \qquad \textit{Do the addition.}$$

$$\begin{array}{r} \overset{1}{74} \\ +16 \\ \hline 90 \end{array}$$

Self Check 2

Evaluate: $60 - 2 \cdot 3 + 22$

Now Try ➡ Problem 23

Caution! In Example 2, a common mistake is to forget to work from left to right and *incorrectly* perform the addition before the subtraction. This error produces the wrong answer, 58.

$$80 - 3 \cdot 2 + 16 = 80 - 6 + 16$$
$$= 80 - 22$$
$$= 58$$

Remember to perform additions and subtractions *in the order in which they occur.* The same is true for multiplications and divisions.

EXAMPLE 3   Evaluate: $192 \div 6 - 5(3)2$.

Strategy We will perform the division first.

WHY Although the expression contains parentheses, there are no calculations to perform *within* them. Since there are no exponents, we perform multiplications and divisions as they occur from left to right.

Solution
We will write the steps of the solution in horizontal form.

$$192 \div 6 - 5(3)2 = 32 - 5(3)2 \qquad \textit{Working from left to right, do the division: } 192 \div 6 = 32.$$

$$= 32 - 15(2) \qquad \textit{Working from left to right, do the multiplication: } 5(3) = 15.$$

$$= 32 - 30 \qquad \textit{Complete the multiplication: } 15(2) = 30.$$

$$= 2 \qquad \textit{Do the subtraction.}$$

$$\begin{array}{r} 32 \\ 6{\overline{)192}} \\ -18 \\ \hline 12 \\ -12 \\ \hline 0 \end{array}$$

Self Check 3

Evaluate: $144 \div 9 + 4(2)3$

Now Try ➡ Problem 27

EXAMPLE 4   **Long-distance calls.** The rates that Skype charges for overseas landline calls from the United States are shown below. A newspaper editor in Washington, D.C., made a 60-minute call to Iraq, a 45-minute call to Panama, and a 30-minute call to Vietnam. What was the total cost of the calls?

Landline calls

All rates are per minute.	
Afghanistan	37¢
Iraq	39¢
Haiti	27¢
Panama	11¢
Russia	5¢
Vietnam	18¢
Includes tax	

Solution It is helpful to list the given facts and what you are to find.

- The 60-minute call to Iraq costs 39 cents per minute.   *Given*
- The 45-minute call to Panama costs 11 cents per minute.   *Given*
- The 30-minute call to Vietnam costs 18 cents per minute.   *Given*
- What is the total cost of the calls?   *Find*

Now we translate the words of the problem to numbers and symbols. We can find the cost of each call by multiplying the length of the call (in minutes) by the rate charged per minute (in cents). Since the word *total* indicates addition, we will add to find the total cost of the calls.

The total cost of the calls	is equal to	the cost of the call to Iraq	plus	the cost of the call to Panama	plus	the cost of the call to Vietnam.

$$\text{The total cost of the calls} = 60(39) + 45(11) + 30(18)$$

$$\begin{array}{r} \overset{1}{2,}340 \\ 495 \\ + 540 \\ \hline 3,375 \end{array}$$

To evaluate this expression, we apply the order of operations rule.

$$\begin{aligned} \text{The total cost of the calls} &= 60(39) + 45(11) + 30(18) &&\text{The units are cents.} \\ &= 2,340 + 495 + 540 &&\text{Do the multiplication first.} \\ &= 3,375 &&\text{Do the addition.} \end{aligned}$$

The total cost of the overseas calls is 3,375¢, or $33.75.

Check: We can check the result by finding an estimate using front-end rounding. The total cost of the calls is approximately $60(40¢) + 50(10¢) + 30(20¢) = 2,400¢ + 500¢ + 600¢$ or 3,500¢. The result of 3,375¢ seems reasonable.

Self Check 4

Long-distance calls. A newspaper reporter in Chicago made a 90-minute call to Afghanistan, a 25-minute call to Haiti, and a 55-minute call to Russia. What was the total cost of the calls?

Now Try ➲ **Problem 109**

OBJECTIVE 2   **Evaluate expressions containing grouping symbols.**

Grouping symbols determine the order in which an expression is evaluated. Examples of grouping symbols are parentheses ( ), brackets [ ], braces { }, and the fraction bar —.

EXAMPLE 5   Evaluate each expression: **a.** $12 - 3 + 5$ **b.** $12 - (3 + 5)$

Strategy To evaluate the expression in part a, we will perform the subtraction first. To evaluate the expression in part b, we will perform the addition first.

WHY The similar-looking expression in part b is evaluated in a different order because it contains parentheses. Any operations within parentheses must be performed first.

Solution

a. The expression does not contain any parentheses, nor are there any exponents, nor any multiplication or division. We perform the additions and subtractions as they occur, from left to right.

$$12 - 3 + 5 = 9 + 5 \quad \text{Do the subtraction: } 12 - 3 = 9$$
$$= 14 \quad \text{Do the addition.}$$

b. By the order of operations rule, we must perform the operation within the parentheses first.

$$12 - (3 + 5) = 12 - 8 \quad \text{Do the addition: } 3 + 5 = 8. \text{ Read as "12 minus the quantity of 3 plus 5."}$$

$$= 4 \quad \text{Do the subtraction.}$$

We read the expression $12 - (3 + 5)$ as "12 minus the *quantity* of 3 plus 5." The word quantity alerts the reader to the parentheses that are used as grouping symbols.

Self Check 5

Evaluate each expression:
a. $20 - 7 + 6$
b. $20 - (7 + 6)$

Now Try Problem 33

EXAMPLE 6   Evaluate: $(2 + 6)^3$

Strategy We will perform the operation within the parentheses first.

WHY This is the first step of the order of operations rule.

Solution

$$(2 + 6)^3 = 8^3 \quad \text{Read as "The cube of the quantity of 2 plus 6." Do the addition.}$$
$$= 512 \quad \text{Evaluate the exponential expression: } 8^3 = 8 \cdot 8 \cdot 8 = 512.$$

$$\begin{array}{r} \overset{3}{64} \\ \times\, 8 \\ \hline 512 \end{array}$$

Self Check 6

Evaluate: $(1 + 3)^4$

Now Try Problem 35

EXAMPLE 7   Evaluate: $5 + 2(13 - 5 \cdot 2)$

Strategy We will perform the multiplication within the parentheses first.

WHY When there is more than one operation to perform within parentheses, we follow the order of operations rule. Multiplication is to be performed before subtraction.

Solution
We apply the order of operations rule within the parentheses to evaluate $13 - 5 \cdot 2$.

$$5 + 2(13 - 5 \cdot 2) = 5 + 2(13 - 10) \quad \text{Do the multiplication within the parentheses.}$$

$$= 5 + 2(3) \quad \text{Do the subtraction within the parentheses.}$$
$$= 5 + 6 \quad \text{Do the multiplication: } 2(3) = 6.$$
$$= 11 \quad \text{Do the addition.}$$

Self Check 7

Evaluate: $50 - 4(12 - 5 \cdot 2)$

Now Try Problem 39

Some expressions contain two or more sets of grouping symbols. Since it can be confusing to read an expression such as $16 + 6(4^2 − 3(5 − 2))$, we use a pair of **brackets** in place of the outer pair of parentheses.

$$16 + 6[4^2 − 3(5 − 2)]$$

If an expression contains more than one pair of grouping symbols, we always begin by working within the **innermost pair** and then work to the **outermost pair**.

Innermost parentheses
$$16 + 6[4^2 − 3(5 − 2)]$$
Outermost brackets

LANGUAGE OF MATHEMATICS

Multiplication is indicated when a number is next to a parenthesis or a bracket. For example,

$$16 + 6[4^2 − 3(5 − 2)]$$

Multiplication   Multiplication

EXAMPLE 8   Evaluate: $16 + 6[4^2 − 3(5 − 2)]$

Strategy We will work within the parentheses first and then within the brackets. Within each set of grouping symbols, we will follow the order of operations rule.

WHY By the order of operations, we must work from the *innermost* pair of grouping symbols to the *outermost*.

Solution

$$16 + 6[4^2 − 3(\mathbf{5 − 2})] = 16 + 6[4^2 − 3(\mathbf{3})] \quad \text{Do the subtraction within the parentheses.}$$

$$= 16 + 6[16 − 3(3)] \quad \text{Evaluate the exponential expression: } 4^2 = 16.$$

$$= 16 + 6[16 − 9] \quad \text{Do the multiplication within the brackets.}$$

$$= 16 + 6[7] \quad \text{Do the subtraction within the brackets.}$$

$$= 16 + 42 \quad \text{Do the multiplication: } 6[7] = 42.$$

$$= 58 \quad \text{Do the addition.}$$

Self Check 8

Evaluate:
$130 − 7[2^2 + 3(6 − 2)]$

Now Try ➲ **Problem 43**

Caution! In Example 8, a common mistake is to *incorrectly* add 16 and 6 instead of *correctly* multiplying 6 and 7 first. This error produces a wrong answer, 154.

$$16 + 6[4^2 − 3(5 − 2)] = 16 + 6[4^2 − 3(3)]$$

$$= 16 + 6[16 − 3(3)]$$

$$= 16 + 6[16 − 9]$$

$$= 16 + 6[7]$$

$$= 22[7]$$

$$= 154$$

EXAMPLE 9   Evaluate: $\dfrac{2(13) − 2}{3(2^3)}$

Strategy We will evaluate the expression above and the expression below the fraction bar separately. Then we will do the indicated division, if possible.

WHY Fraction bars are grouping symbols. They group the numerator and the denominator. The expression could be written $[2(13) − 2] ÷ [3(2^3)]$.

Solution

$$\frac{2(13) - 2}{3(2^3)} = \frac{26 - 2}{3(8)}$$ In the numerator, do the multiplication. In the denominator, evaluate the exponential expression within the parentheses.

Self Check 9

Evaluate: $\dfrac{3(14) - 6}{2(3^2)}$

Now Try ➡ **Problem 47**

$$= \frac{24}{24}$$ In the numerator, do the subtraction.
In the denominator, do the multiplication.

$$= 1$$ Do the division indicated by the fraction bar: 24 ÷ 24 = 1.

OBJECTIVE 3 **Find the mean (average) of a set of values.**

The **mean** (sometimes called the **arithmetic mean** or **average**) of a set of numbers is a value around which the values of the numbers are grouped. It gives you an indication of the "center" of the set of numbers. To find the mean of a set of numbers, we must apply the order of operations rule.

> **Finding the Mean**
>
> To find the mean (average) of a set of values, divide the sum of the values by the number of values.

EXAMPLE 10 **NFL offensive linemen.**

The weights of the 2015–2016 Super Bowl champion Denver Broncos starting offensive linemen are shown below. What was their mean (average) weight?

| Left tackle #68 R. Harris 299 lb | Left guard #69 E. Mathis 304 lb | Center #61 M. Paradis 306 lb | Right guard #65 L. Vasquez 330 lb | Right tackle #79 M. Schofield 301 lb |

(Source: nfl.com/New York Giants depth chart)

Strategy We will add 299, 304, 306, 330, and 301 and divide the sum by 5.

WHY To find the mean (average) of a set of values, we divide the sum of the values by the number of values.

Self Check 10

NFL defensive linemen.
The weights of the 2015–2016 Carolina Panthers starting defensive linemen were 315 lb, 325 lb, 295 lb, 310 lb, and 310 lb. What was their mean (average) weight?
(Source: nfl.com)

Now Try ➡ **Problems 51 and 113**

Solution
Since there are 5 weights, divide the sum by 5.

$$\text{Mean} = \frac{299 + 304 + 306 + 330 + 301}{5}$$

$$\text{Mean} = \frac{1,540}{5}$$ In the numerator, do the addition.

$$\text{Mean} = 308$$ Do the indicated division: 1,540 ÷ 5.

In 2015–2016, the mean (average) weight of the starting offensive linemen on the Denver Broncos was 308 pounds.

$$
\begin{array}{r}
{\scriptstyle 1\,2}\\
299\\
304\\
306\\
330\\
+301\\
\hline
1,540
\end{array}
$$

$$
\begin{array}{r}
308\\
5\overline{)1,540}\\
\underline{-1\,5}\\
40\\
\underline{-40}\\
0
\end{array}
$$

Using Your Calculator ▶   Order of Operations and Parentheses

Calculators have the rules for order of operations built in. A left parenthesis key [(] and a right parenthesis key [)] should be used when grouping symbols, including a fraction bar, are needed. For example, to evaluate $\frac{240}{20-5}$, the parentheses keys must be used, as shown below.

240 [÷] [(] 20 [−] 5 [)] [=] | 16 |

On some calculator models, the [ENTER] key is pressed instead of [=] for the result to be displayed.

If the parentheses are not entered, the calculator will find 240 ÷ 20 and then subtract 5 from that result, to produce the wrong answer, 7.

Think it Through   ●   EDUCATION PAYS

"Education does pay. It has a high rate of return for students from all racial/ethnic groups, for men and for women, and for those from all family backgrounds. It also has a high rate of return for society."
—*The College Board, Trends in Higher Education Series*

Attending school requires an investment of time, effort, and sacrifice. Is it all worth it? The graph below shows how average weekly earnings in the U.S. increase as the level of education increases. Begin at the bottom of the graph and work upward. Use the given clues to determine each of the missing weekly earnings amounts.

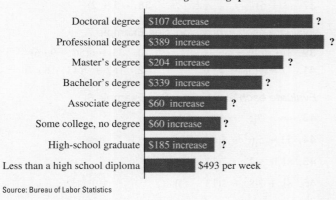

Average earnings per week in 2015

Doctoral degree	$107 decrease ?
Professional degree	$389 increase ?
Master's degree	$204 increase ?
Bachelor's degree	$339 increase ?
Associate degree	$60 increase ?
Some college, no degree	$60 increase ?
High-school graduate	$185 increase ?
Less than a high school diploma	$493 per week

Source: Bureau of Labor Statistics

Answers to Self Checks

1. 102   **2.** 76   **3.** 40   **4.** 4,280¢ = $42.80   **5. a.** 19 **b.** 7   **6.** 256   **7.** 42   **8.** 18   **9.** 2
10. 311 lb

SECTION (1.9) **STUDY SET**

VOCABULARY

Fill in the blanks.

1. Numbers are combined with the operations of addition, subtraction, multiplication, and division to create _____.

2. To *evaluate* the expression 2 + 5 · 4 means to find its _____.

3. The grouping symbols ( ) are called _____, and the symbols [ ] are called _____.

4. The rule for the _____ of operations guarantees that an evaluation of a numerical expression will result in a single answer.

5. In the expression 9 + 6[8 + 6(4 − 1)], the parentheses are the _____ most grouping symbols and the brackets are the _____ most grouping symbols.

6. To find the _____ of a set of values, we add the values and divide by the number of values.

CONCEPTS

7. List the operations in the order in which they should be performed to evaluate each expression. *You do not have to evaluate the expression.*

  a. $5(2)^2 - 1$

  b. $15 + 90 - (2 \cdot 2)^3$

  c. $7 \cdot 4^2$

  d. $(7 \cdot 4)^2$

8. List the operations in the order in which they should be performed to evaluate each expression. *You do not have to evaluate the expression.*

  a. $50 + 8 - 40$

  b. $50 - 40 + 8$

  c. $16 \cdot 2 \div 4$

  d. $16 \div 4 \cdot 2$

9. Consider the expression $\dfrac{5 + 5(7)}{(5 \cdot 20 - 8^2) - 28}$. In the numerator, what operation should be performed first? In the denominator, what operation should be performed first?

10. To find the mean (average) of 15, 33, 45, 12, 6, 19, and 3, we add the values and divide by what number?

NOTATION

11. In the expression $\dfrac{60 - 5 \cdot 2}{5 \cdot 2 + 40}$, what symbol serves as a grouping symbol? What does it group?

12. Use brackets to write $2(12 - (5 + 4))$ in clearer form.

Fill in the blanks.

13. We read the expression $16 - (4 + 9)$ as "16 minus the _____ of 4 plus 9."

14. We read the expression $(8 - 3)^3$ as "The cube of the _____ of 8 minus 3."

Complete the steps to evaluate each expression.

15. $7 \cdot 4 - 5(2)^2 = 7 \cdot 4 - 5(\ )$
$= 28 - \ $
$= \ $

16. $2 + (5 + 6 \cdot 2) = 2 + (5 + \ )$
$= 2 + \ $
$= \ $

17. $[4(2 + 7)] - 4^2 = [4(\ )] - 4^2$
$= \ - 4^2$
$= 36 - \ $
$= \ $

18. $\dfrac{12 + 5 \cdot 3}{3^2 - 2 \cdot 3} = \dfrac{12 + \ }{\ - 6}$
$= \dfrac{\ }{3}$
$= \ $

GUIDED PRACTICE

Evaluate each expression. See Example 1.

19. $3 \cdot 5^2 - 28$

20. $4 \cdot 2^2 - 11$

21. $6 \cdot 3^2 - 41$

22. $5 \cdot 4^2 - 32$

Evaluate each expression. See Example 2.

23. $52 - 6 \cdot 3 + 4$

24. $66 - 8 \cdot 7 + 16$

25. $32 - 9 \cdot 3 + 31$

26. $62 - 5 \cdot 8 + 27$

Evaluate each expression. See Example 3.

27. $192 \div 4 - 4(2)3$

28. $455 \div 7 - 3(4)5$

29. $252 \div 3 - 6(2)6$

30. $264 \div 4 - 7(4)2$

Evaluate each expression. See Example 5.

31. a. $26 - 2 + 9$
  b. $26 - (2 + 9)$

32. a. $37 - 4 + 11$
  b. $37 - (4 + 11)$

33. a. $51 - 16 + 8$
  b. $51 - (16 + 8)$

34. a. $73 - 35 + 9$
  b. $73 - (35 + 9)$

Evaluate each expression. See Example 6.

35. $(4 + 6)^2$

36. $(3 + 4)^2$

37. $(3 + 5)^3$

38. $(5 + 2)^3$

Evaluate each expression. See Example 7.

39. $8 + 4(29 - 5 \cdot 3)$

40. $33 + 6(56 - 9 \cdot 6)$

41. $77 + 9(38 - 4 \cdot 6)$

42. $162 + 7(47 - 6 \cdot 7)$

Evaluate each expression. See Example 8.

43. $46 + 3[5^2 - 4(9 - 5)]$

44. $53 + 5[6^2 - 5(8 - 1)]$

45. $81 + 9[7^2 - 7(11 - 4)]$

46. $81 + 3[8^2 - 7(13 - 5)]$

Evaluate each expression. See Example 9.

47. $\dfrac{2(50) - 4}{2(4^2)}$

48. $\dfrac{4(34) - 1}{5(3^2)}$

49. $\dfrac{25(8) - 8}{6(2^3)}$

50. $\dfrac{6(31) - 26}{4(2^3)}$

Find the mean (average) of each list of numbers. See Example 10.

51. 6, 9, 4, 3, 8

52. 7, 1, 8, 2, 2

53. 3, 5, 9, 1, 7, 5

54. 8, 7, 7, 2, 4, 8

55. 19, 15, 17, 13

56. 11, 14, 12, 11

57. 5, 8, 7, 0, 3, 1

58. 9, 3, 4, 11, 14, 1

TRY IT YOURSELF

Evaluate each expression, if possible.

59. $(8 - 6)^2 + (4 - 3)^2$

60. $(2 + 1)^2 + (3 + 2)^2$

61. $2 \cdot 3^4$

62. $3^3 \cdot 5$

63. $7 + 4 \cdot 5$

64. $6 \cdot 8 - 3$

65. $(7 - 4)^2 + 1$

66. $(9 - 5)^3 + 8$

67. $\dfrac{10 + 5}{52 - 47}$

68. $\dfrac{18 + 12}{61 - 55}$

69. $5 \cdot 10^3 + 2 \cdot 10^2 + 3 \cdot 10^1 + 9$

70. $8 \cdot 10^3 + 0 \cdot 10^2 + 7 \cdot 10^1 + 4$

71. $20 - 10 + 5$

72. $80 - 5 + 4$

73. $25 \div 5 \cdot 5$

74. $6 \div 2 \cdot 3$

75. $150 - 2(2 \cdot 6 - 4)^2$

76. $760 - 2(2 \cdot 3 - 4)^2$

77. $190 - 2[10^2 - (5 + 2^2)] + 45$

78. $161 - 8[6(6) - 6^2] + 2^2(5)$

79. $2 + 3(0)$

80. $5(0) + 8$

81. $\dfrac{(5 - 3)^2 + 2}{4^2 - (8 + 2)}$

82. $\dfrac{(4^3 - 2) + 7}{5(2 + 4) - 7}$

83. $4^2 + 3^2$

84. $12^2 + 5^2$

85. $3 + 2 \cdot 3^4 \cdot 5$

86. $3 \cdot 2^3 \cdot 4 - 12$

87. $60 - \left(6 + \dfrac{40}{2^3}\right)$

88. $7 + \left(5^3 - \dfrac{200}{2}\right)$

89. $\dfrac{(3 + 5)^2 + 2}{2(8 - 5)}$

90. $\dfrac{25 - (2 \cdot 3 - 1)}{2 \cdot 9 - 8}$

91. $(18 - 12)^3 - 5^2$

92. $(9 - 2)^2 - 3^3$

93. $30(1)^2 - 4(2) + 12$

94. $5(1)^3 + (1)^2 + 2(1) - 6$

95. $16^2 - \dfrac{25}{5} + 6(3)4$

96. $15^2 - \dfrac{24}{6} + 8(2)(3)$

97. $\dfrac{3^2 - 2^2}{(3 - 3)^2}$

98. $\dfrac{5^2 + 17}{4 - 2^2}$

99. $3\left(\dfrac{18}{3}\right) - 2(2)$

100. $2\left(\dfrac{12}{3}\right) + 3(5)$

101. $4[50 - (3^3 - 5^2)]$

102. $6[15 + (5 \cdot 2^2)]$

103. $80 - 2[12 - (5 + 4)]$

104. $15 + 5[12 - (2^2 + 4)]$

LOOK ALIKES

Evaluate each expression, if possible.

105. a. $50 \div 5 \div 5$

b. $50 \div (5 \div 5)$

c. $50 \div 5 \cdot 5$

d. $50 \div (5 \cdot 5)$

106. a. $25 - 5^2$

b. $(25 - 5)^2$

107. a. $\dfrac{4 - 2^2}{50 - 32}$

b. $\dfrac{50 - 32}{4 - 2^2}$

108. a. $100 - 53 + 18$

b. $100 - (53 + 18)$

c. $(100 - 53) + 18$

APPLICATIONS

Write an expression to solve each problem and evaluate it.

109. Shopping. At the supermarket, Carlos is buying 3 cases of soda, 4 bags of tortilla chips, and 2 bottles of salsa. Each case of soda costs $7, each bag of chips costs $4, and each bottle of salsa costs $3. Find the total cost of the snacks.

110. Banking. When a customer deposits cash, a teller must complete a currency count on the back of the deposit slip. In the illustration, a teller has written the number of each type of bill to be deposited. What is the total amount of cash being deposited?

Currency count, for financial use only			
24	x 1's		
—	x 2's		
6	x 5's		
10	x 10's		
12	x 20's		
2	x 50's		
1	x 100's		
	TOTAL $		

111. Diving. The scores awarded to a diver by seven judges as well as the degree of difficulty of his dive are shown below. Use the two-step process shown to calculate the diver's overall score.

Step 1 Throw out the lowest score and the highest score.

Step 2 Add the sum of the remaining scores and multiply by the degree of difficulty.

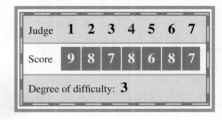

Judge	1	2	3	4	5	6	7
Score	9	8	7	8	6	8	7

Degree of difficulty: **3**

112. Wrapping gifts. How much ribbon is needed to wrap the package shown if 15 inches of ribbon are needed to make the bow?

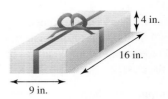

4 in.

16 in.

9 in.

113. Scrabble. Illustration (a) shows part of the game board before and illustration (b) shows it after the words *brick* and *aphid* were played. Determine the scoring for each word. (*Hint:* The number on each tile gives the point value of the letter.)

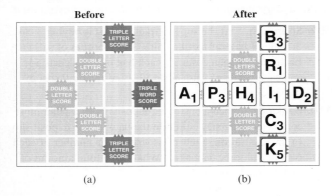

Before   (a)   After   (b)

114. The Gettysburg Address. Here is an excerpt from Abraham Lincoln's Gettysburg Address:

Fourscore and seven years ago, our fathers brought forth on this continent a new nation, conceived in liberty, and dedicated to the proposition that all men are created equal.

Lincoln's comments refer to the year 1776, when the United States declared its independence. If a score is 20 years, in what year did Lincoln deliver the Gettysburg Address?

115. Prime numbers. Show that 87 is the sum of the squares of the first four prime numbers.

from **Campus to Careers**

116. A 27-foot-long by 19-foot-wide rectangular garden is one feature of a landscape design for a community park. A concrete walkway is to run through the garden and will occupy 125 square feet of space. How many square feet are left for planting in the garden?

117. Climate. One December week, the high temperatures in Honolulu, Hawaii, were 75°, 80°, 83°, 80°, 77°, 72°, and 86°. Find the week's mean (average) high temperature.

118. Grades. In a science class, a student had test scores of 94, 85, 81, 77, and 89. He also overslept, missed the final exam, and received a 0 on it. What was his test average (mean) in the class?

119. Energy usage. See the graph in the next column. Find the mean (average) number of therms of natural gas used per month for the year 2017.

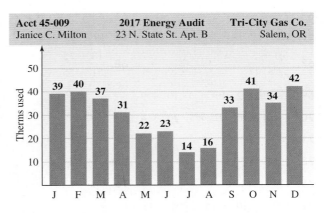

Acct 45-009   **2017 Energy Audit**   Tri-City Gas Co.
Janice C. Milton   23 N. State St. Apt. B   Salem, OR

120. Counting numbers. What is the average (mean) of the first nine counting numbers: 1, 2, 3, 4, 5, 6, 7, 8, and 9?

121. Fast foods. The table below shows the sandwiches Subway advertises on its 6 grams of fat or less menu. What is the mean (average) number of calories for the group of sandwiches?

6-inch subs	Calories
Veggie Delite	230
Turkey Breast	280
Turkey Breast & Ham	295
Ham	290
Roast Beef	290
Subway Club	330
Roasted Chicken Breast	310
Chicken Teriyaki	375

Source: Subway.com/NutritionInfo

122. TV ratings. The table below shows the number of viewers* of the 2016 Major League Baseball World Series between the Chicago Cubs and the Cleveland Indians. How large was the average (mean) audience?

Game 1	Tuesday, Oct. 25	19,368,000
Game 2	Wednesday, Oct. 26	17,395,000
Game 3	Friday, Oct. 28	19,384,000
Game 4	Saturday, Oct. 29	16,704,000
Game 5	Sunday, Oct. 30	23,638,000
Game 6	Tuesday, Nov. 1	23,396,000
Game 7	Wednesday, Nov. 2	40,044,000

* Rounded to the nearest hundred thousand
Source: Nielsen Media Research

123. YouTube. A YouTube video contest is to be part of a kickoff for a new sports drink. The cash prizes to be awarded are shown below.

   a. How many prizes will be awarded?

   b. What is the total amount of money that will be awarded?

   c. What is the average (mean) cash prize?

> **YouTube Video Contest**
>
> **Grand prize: Disney World vacation plus $2,500**
>
> Four 1st place prizes of $500
> Thirty-five 2nd place prizes of $150
> Eighty-five 3rd place prizes of $25

124. Surveys. Some students were asked to rate their college cafeteria food on a scale from 1 to 5. The responses are shown on the tally sheet.

   a. How many students took the survey?

   b. Find the mean (average) rating.

Poor		Fair		Excellent
1	2	3	4	5
IIII	I	TH᷉	I	IIII

WRITING

125. Explain why the order of operations rule is necessary.

126. What does it mean when we say to do all additions and subtractions *as they occur from left to right*? Give an example.

127. Explain the error in the following work:

Evaluate:

$$8 + 2[6 - 3(9 - 8)] = 8 + 2[6 - 3(1)]$$
$$= 8 + 2[6 - 3]$$
$$= 8 + 2(3)$$
$$= 10(3)$$
$$= 30$$

128. Explain the error in the following work:

Evaluate:

$$24 - 4 + 16 = 24 - 20$$
$$= 4$$

REVIEW

Write each number in words.

129. 254,309   **130.** 504,052,040

---

① Summary and Review

SECTION 1.1 ▶	An Introduction to the Whole Numbers

DEFINITIONS AND CONCEPTS	EXAMPLES
The set of **whole numbers** is {0, 1, 2, 3, 4, 5, . . .}. When a whole number is written using the **digits** 0, 1, 2, 3, 4, 5, 6, 7, 8, 9, it is said to be in **standard form**.	Some examples of whole numbers written in standard form are: 2, 16, 530, 7,894, and 3,201,954
The position of a digit in a whole number determines its **place value**. A **place-value chart** shows the place value of each digit in the number. To make large whole numbers easier to read, we use commas to separate their digits into groups of three, called **periods**.	 The place value of the digit 7 is 7 ten millions. The digit 4 tells the number of thousands.

To **write a whole number in words**, start from the left. Write the number in each period followed by the name of the period (except for the *ones period*, which is not used). Use commas to separate the periods. To **read a whole number out loud**, follow the same procedure. The commas are read as slight pauses.	Millions Thousands Ones 2 , 5 6 8 , 0 1 9 Two **million**, five hundred sixty-eight **thousand**, nineteen
To change from the **written-word form of a number to standard form**, look for the commas. Commas are used to separate periods.	Six billion , forty-one million , two hundred eight thousand , thirty-six 6,041,208,036
To write a number in **expanded form** (**expanded notation**) means to write it as an addition of the place values of each of its digits.	The expanded form of 32,159 is: 30,000 + 2,000 + 100 + 50 + 9
Whole numbers can be shown by drawing points on a **number line**.	The graphs of 3 and 7 are shown on the number line below. 0 1 2 3 4 5 6 7 8
Inequality symbols are used to compare whole numbers: $>$ means *is greater than* $<$ means *is less than*	$9 > 8$ and $2,343 > 762$ $1 < 2$ and $9,000 < 12,453$
When we don't need exact results, we often **round** numbers. **Rounding a Whole Number** **1.** To round a number to a certain place value, locate the **rounding digit** in that place. **2.** Look at the **test digit**, which is directly to the right of the rounding digit. **3.** If the test digit is 5 or greater, round up by adding 1 to the rounding digit and replacing all of the digits to its right with 0. If the test digit is less than 5, replace it and all of the digits to its right with 0.	Round 9,842 to the nearest ten. Rounding digit: tens column 9,842 Test digit: Since 2 is less than 5, leave the rounding digit unchanged and replace the test digit with 0. Thus, 9,842 rounded to the nearest ten is 9,840. Round 63,179 to the nearest hundred. Rounding digit: hundreds column 63,179 Test digit: Since 7 is 5 or greater, add 1 to the rounding digit and replace all the digits to its right with 0. Thus, 63,179 rounded to the nearest hundred is 63,200.
Whole numbers are often used in **tables**, **bar graphs**, and **line graphs**.	See page 9 for an example of a table, a bar graph, and a line graph.

REVIEW EXERCISES

Consider the number 41,948,365,720.

1. Which digit is in the ten thousands column?

2. Which digit is in the hundreds column?

3. What is the place value of the digit 1?

4. Which digit tells the number of millions?

5. Write each number in words.

   a. 97,283

   b. 5,444,060,017

6. Write each number in standard form.

   a. Three thousand, two hundred seven

   b. Twenty-three million, two hundred fifty-three thousand, four hundred twelve

   c. 60,000 + 1,000 + 200 + 4

Write each number in expanded form.

7. 570,302

8. 37,309,154

Graph the following numbers on a number line.

9. 0, 2, 8, 10

10. The whole numbers between 3 and 7.

Place an < or an > symbol in the box to make a true statement.

11. 9 ▮ 7   12. 301 ▮ 310

13. Round 2,507,348

    a. to the nearest hundred

    b. to the nearest ten thousand

    c. to the nearest ten

    d. to the nearest million

14. Round 969,501

    a. to the nearest thousand

    b. to the nearest hundred thousand

15. **Construction.**   The following table lists the number of building permits issued in the city of Springsville for the period 2009–2016.

Year	2009	2010	2011	2012	2013	2014	2015	2016
Building permits	12	13	10	7	9	14	6	5

   a. Construct a bar graph of the data.

   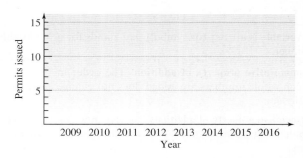

   b. Construct a line graph of the data.

   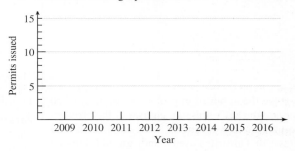

16. **Geography.**   The names and lengths of the five longest rivers in the world are listed below. Write them in order, beginning with the longest.

Amazon (South America)	4,049 mi
Mississippi-Missouri (North America)	3,709 mi
Nile (Africa)	4,160 mi
Ob-Irtysh (Russia)	3,459 mi
Yangtze (China)	3,964 mi

Source: geography.about.com

SECTION 1.2 ▶ Adding Whole Numbers

DEFINITIONS AND CONCEPTS	EXAMPLES
To **add whole numbers**, think of combining sets of similar objects. **Vertical form:** Stack the addends. Add the digits in the ones column, the tens column, the hundreds column, and so on. **Carry** when necessary.	Add: 10,892 + 5,467 + 499 Carrying $\overset{1\ 21}{10,892}$ ← Addend 5,467 ← Addend + 499 ← Addend 16,858 ← Sum To check, add bottom to top
A **variable** is a letter (or symbol) that stands for a number.	Variables: x, a, and y
Commutative property of addition: The order in which whole numbers are added does not change their sum. **Associative property of addition:** The way in which whole numbers are grouped does not change their sum.	$6 + 5 = 5 + 6$ By the commutative property, the sum is the same. $(17 + 5) + 25 = 17 + (5 + 25)$ By the associative property, the sum is the same.
To estimate a sum, use **front-end rounding** to approximate the addends. Then add.	Estimate the sum: $\begin{array}{rcl} 7,219 & \to & 7,000 \\ 592 & \to & 600 \\ +3,425 & \to & +3,000 \\ \hline & & 10,600 \end{array}$ Round to the nearest thousand. Round to the nearest hundred. Round to the nearest thousand. The estimate is 10,600.
To solve the application problems, we must often *translate* the **key words** and **phrases** of the problem to numbers and symbols. Some key words and phrases that are often used to indicate addition are: 	gain	increase	up	forward
rise	more than	total	combined
in all	in the future	extra	altogether		Translate the words to numbers and symbols: **Vacations.** There were 3,882,642 visitors to Yosemite National Park in 2014. The following year, attendance increased by 267,575. How many people visited the park in 2015? The phrase *increased by* indicates addition: The number of visitors to the park in 2015 $=$ 3,882,642 $+$ 267,575
The distance around a rectangle or a square is called its **perimeter**. Perimeter of a rectangle $=$ length + length + width + width Perimeter of a square $=$ side + side + side + side	Find the perimeter of the rectangle shown below. 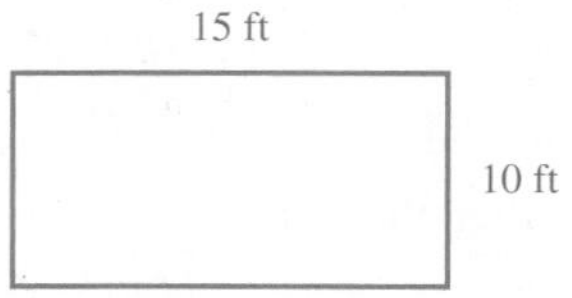 15 ft 10 ft Perimeter = 15 + 15 + 10 + 10 Add the two lengths and the two widths. Perimeter = 50 The perimeter of the rectangle is 50 feet.

REVIEW EXERCISES

Add.

17. 27 + 436

18. 4 + (36 + 19)

19. $\begin{array}{r} 5{,}345 \\ +\ \ 655 \\ \hline \end{array}$

20. 2 + 1 + 38 + 3 + 6

21. 4,447 + 7,478 + 676

22. $\begin{array}{r} 32{,}812 \\ 65{,}034 \\ +54{,}323 \\ \hline \end{array}$

23. Use front-end rounding to estimate the sum.

$$615 + 789 + 14{,}802 + 39{,}902 + 8{,}098$$

24. a. Use the commutative property of addition to complete the following:

$$24 + 61 = \rule{2cm}{0.4pt}$$

   b. Use the associative property of addition to complete the following:

$$9 + (91 + 29) = \rule{2cm}{0.4pt}$$

25. Airports. The nation's three busiest airports in 2016 are listed below. Find the total number of passengers passing through those airports.

Airport	Total passengers
Hartsfield-Jackson Atlanta	101,491,106
Chicago O'Hare	76,949,504
Los Angeles International	74,937,004

Source: Airports Council International–North America

26. Add from bottom to top to check the sum. Is it correct?

$$\begin{array}{r} 1{,}291 \\ 859 \\ 345 \\ +\ \ 226 \\ \hline 1{,}821 \end{array}$$

27. What is the sum of three thousand, seven hundred six and ten thousand, nine hundred fifty-five?

28. What is 451,775 more than 327,891?

29. New homes. In 1974, the average size of a newly built home in the United States was 1,695 square feet. By 2015, the average size had increased by 772 square feet. What was the average size of a new home in 2015? (Source: U.S. Census Bureau)

30. Find the perimeter of the rectangle shown below.

731 ft

642 ft

SECTION 1.3 ▶ **Subtracting Whole Numbers**

DEFINITIONS AND CONCEPTS	EXAMPLES
To **subtract whole numbers**, think of taking away objects from a set. **Vertical form:** Stack the numbers. Subtract the digits in the ones column, the tens column, the hundreds column, and so on. **Borrow** when necessary. To **check:** Difference + subtrahend = minuend	Subtract: 4,957 − 869 Borrowing *Check using addition:* $4{,}9\overset{8\ 4\ 17}{\cancel{9}\cancel{5}\cancel{7}}$ ← Minuend $\overset{1\ 1}{4{,}088}$ $\underline{-\ \ \ 8\ 6\ 9}$ ← Subtrahend $\underline{+\ \ \ 869}$ $4{,}0\ 8\ 8$ ← Difference $4{,}957$
Be careful when translating the instruction to subtract one number *from* another number. The order of the numbers in the sentence must be reversed when we translate to symbols.	Translate the words to numbers and symbols: Subtract 41 from 97. *Since 41 is the number to be subtracted, it is the subtrahend.* 97 − 41
Every subtraction has a **related addition statement**.	10 − 3 = 7 because 7 + 3 = 10

To estimate a difference, use **front-end rounding** to approximate the minuend and subtrahend. Then subtract.	Estimate the difference: $\begin{array}{ll} 59{,}033 \longrightarrow 60{,}000 & \textit{Round to the nearest ten thousand.} \\ -\ 4{,}124 \longrightarrow -\ 4{,}000 & \textit{Round to the nearest thousand.} \\ \hline \phantom{59{,}033 \longrightarrow} 56{,}000 & \end{array}$ The estimate is 56,000.
Some of the **key words** and **phrases** that are often used to indicate subtraction are: 	*loss*	*decrease*	*down*	*backward*
fell	*less than*	*fewer*	*reduce*
remove	*debit*	*in the past*	*remains*
declined			*take away* | To answer questions about *how much more* or *how many more*, we use subtraction.	**Weights of cars.** A Chevy Suburban weighs 5,607 pounds and a Smart car weighs 1,852 pounds. How much heavier is the Suburban? The phrase *how much heavier* indicates subtraction: $\begin{array}{ll} 5{,}607 & \textit{Weight of the Suburban} \\ -1{,}852 & \textit{Weight of the Smart car} \\ \hline 3{,}755 & \end{array}$ The Suburban weighs 3,755 pounds more than the Smart car.
To **evaluate** (find the value of) expressions that involve addition and subtraction written in **horizontal form**, we perform the operations as they occur *from left to right*.	Evaluate: $75 - 23 + 9$ $\begin{array}{ll} \mathbf{75 - 23} + 9 = 52 + 9 & \textit{Working left to right, do the subtraction first.} \\ = 61 & \textit{Now do the addition.} \end{array}$

REVIEW EXERCISES

Subtract.

31. $148 - 87$

32. Subtract 10,218 from 10,435.

33. $750 - 259 + 14$ **34.** $\begin{array}{r} 7{,}800 \\ -5{,}725 \\ \hline \end{array}$

35. Check the subtraction using addition.

$\begin{array}{r} 8{,}017 \\ -6{,}949 \\ \hline 1{,}168 \end{array}$

36. Fill in the blank: $20 - 8 = 12$ because _____ .

37. Estimate the difference: $181{,}232 - 44{,}810$

38. Land area. Use the data in the table below to determine how much larger the land area of Russia is compared to that of Canada.

Country	Land area (square miles)
Russia	6,323,482
Canada	3,551,023

Source: *The World Almanac and Book of Facts*, 2013

39. Banking. A savings account contains $12,975. If the owner makes a withdrawal of $3,800 and later deposits $4,270, what is the new account balance?

40. Sunny days. In the United States, the city of Yuma, Arizona, typically has the most sunny days per year—about 242. The city of Buffalo, New York, typically has 188 days fewer than that. How many sunny days per year does Buffalo have?

SECTION 1.4 ▶ **Multiplying Whole Numbers**

DEFINITIONS AND CONCEPTS	EXAMPLES
Multiplication of whole numbers is repeated addition but with different notation.	Repeated addition: The sum of four 6's Multiplication $6 + 6 + 6 + 6 \ = \ 4 \ \times \ 6 \ = \ 24$
To write multiplication, we use a times symbol $\times$, a raised dot $\cdot$, or parentheses $(\ \ )$.	4×6 $4 \cdot 6$ $4(6)$ or $(4)(6)$ or $(4)6$
Vertical form: Stack the factors. If the bottom factor has more than one digit, multiply in steps to find the partial products. Then add them to find the product.	Multiply: $24 \cdot 163$ $\begin{array}{r} 163 \quad \leftarrow \textit{Factor} \\ \times\ 24 \quad \leftarrow \textit{Factor} \\ \hline 652 \quad \leftarrow \textit{Partial product: } 4 \cdot 163 \\ 3260 \quad \leftarrow \textit{Partial product: } 20 \cdot 163 \\ \hline 3{,}912 \quad \leftarrow \textit{Product} \end{array}$

To find the **product of a whole number and 10, 100, 1,000, and so on**, attach the number of zeros in that number to the right of the whole number. This rule can be extended to multiply any two whole numbers that end in zeros.	Multiply: $8 \cdot 1,000 = 8,000$ Since 1,000 has three zeros, attach three 0's after 8. $43(10,000) = 430,000$ Since 10,000 has four zeros, attach four 0's after 43. $160 \cdot 20,000 = 3,200,000$ 160 and 20,000 have a total of five trailing zeros. Attach five 0's after 32. Multiply 16 and 2 to get 32.
Multiplication Properties of 0 and 1 The product of any whole number and 0 is 0. The product of any whole number and 1 is that whole number.	$0 \cdot 9 = 0$ and $3(0) = 0$ $15 \cdot 1 = 15$ and $1(6) = 6$
Commutative property of multiplication: The order in which whole numbers are multiplied does not change their product. **Associative property of multiplication:** The way in which whole numbers are grouped does not change their product.	$5 \cdot 9 = 9 \cdot 5$ By the commutative property, the product is the same. $(3 \cdot 7) \cdot 10 = 3 \cdot (7 \cdot 10)$ By the associative property, the product is the same.
To **estimate** a product, use **front-end rounding** to approximate the factors. Then multiply.	To estimate the product for $74 \cdot 873$, find $70 \cdot 900$. Round to the nearest ten $74 \cdot 873$ $70 \cdot 900$ Round to the nearest hundred
Application problems that involve **repeated addition** are often more easily solved using multiplication.	**Health care.** A doctor's office is open 210 days a year. Each day the doctor sees 25 patients. How many patients does the doctor see in 1 year? This **repeated addition** can be calculated by multiplication: The number of patients seen each year $= 25 \cdot 210$
We can use multiplication to count objects arranged in rectangular patterns of neatly arranged rows and columns called **rectangular arrays**. Some **key words** that are often used to indicate multiplication are: *double triple twice* *of times tenfold*	**Classrooms.** A large lecture hall has 16 rows of desks, and there are 12 desks in each row. How many desks are in the lecture hall? The **rectangular array** of desks indicates multiplication: The number of desks in the lecture hall $= 16 \cdot 12$
The **area of a rectangle** is the measure of the amount of surface it encloses. Area is measured in square units, such as square inches (written in.^{2}) or square centimeters (written cm^2). Area of a rectangle $=$ length $\cdot$ width or $A = lw$ Letters (or symbols) that are used to represent numbers are called **variables**.	Find the area of the rectangle shown below. 25 in. 4 in. $A = lw$ $A = 25 \cdot 4$ Replace the length l with 25 and the width w with 4. $A = 100$ Multiply. The area of the rectangle is 100 square inches, which can be written in more compact form as 100 in.^{2}

REVIEW EXERCISES

Multiply.

41. 47×9

42. $5 \cdot (7 \cdot 6)$

43. $72 \cdot 10,000$

44. $157 \cdot 59$

45.
$$\begin{array}{r} 5,624 \\ \times\ 281 \\ \hline \end{array}$$

46. $502 \cdot 459$

47. Estimate the product: $6,891 \cdot 438$

48. **a.** Write the repeated addition $7 + 7 + 7 + 7 + 7$ as a multiplication.

   b. Write $2 \cdot t$ in simpler form.

   c. Write $m \cdot n$ in simpler form.

49. Find each product:

   a. $8 \cdot 0$

   b. $7 \cdot 1$

50. What property of multiplication is shown?

   a. $2 \cdot (5 \cdot 7) = (2 \cdot 5) \cdot 7$

   b. $100(50) = 50(100)$

Find the area of the rectangle and the square.

51.

8 cm

4 cm

52.

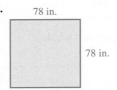

78 in.

78 in.

53. **Sleep.** The National Sleep Foundation recommends that adults get from 7 to 9 hours of sleep each night.

   a. How many hours of sleep is that in one year using the smaller number? (Use a 365-day year.)

   b. How many hours of sleep is that in one year using the larger number?

54. **Graduation.** For a graduation ceremony, the graduates were assembled in a rectangular 22-row and 15-column formation. How many members are in the graduating class?

55. **Paychecks.** Sarah worked 12 hours at $9 per hour, and Santiago worked 14 hours at $8 per hour. Who earned more money?

56. **Shopping.** There are 12 eggs in one dozen and 12 dozen in one gross. How many eggs are in a shipment of 100 gross?

SECTION 1.5 ▶ Dividing Whole Numbers

DEFINITIONS AND CONCEPTS	EXAMPLES
To **divide whole numbers**, think of separating a quantity into equal-sized groups. To write division, we can use a division symbol $\div$, a long division symbol $\overline{)}$, or a fraction bar —.	Dividend Divisor $$8 \div 2 = 4 \qquad 2\overline{)8}\,{}^{4} \qquad \frac{8}{2} = 4$$ Quotient
Another way to answer a division problem is to think in terms of multiplication and write a **related multiplication statement**.	$8 \div 2 = 4$ because $4 \cdot 2 = 8$
A process called **long division** can be used to divide whole numbers. Follow a four-step process: ▪ Estimate ▪ Multiply ▪ Subtract ▪ Bring down	Divide: $8,317 \div 23$ Quotient $$\begin{array}{r} 361 \text{ R } 14 \\ 23\overline{)8,317} \\ -6\ 9 \\ \hline 1\ 41 \\ -1\ 38 \\ \hline 37 \\ -23 \\ \hline 14 \end{array}$$ Dividend Divisor → Remainder

To **check** the result of a division, we multiply the divisor by the quotient and add the remainder. The result should be the dividend.	For the division below, the result checks.

For the division below, the result checks.

Quotient · divisor   remainder

$$(361 \cdot 23) + 14 = 8{,}303 + 14$$
$$= 8{,}317 \leftarrow \text{Dividend}$$

Properties of Division

Any whole number divided by 1 is equal to that number.

$$\frac{4}{1} = 4 \qquad \text{and} \qquad \frac{58}{1} = 58$$

Any nonzero whole number divided by itself is equal to 1.

$$\frac{9}{9} = 1 \qquad \text{and} \qquad \frac{103}{103} = 1$$

Division with Zero

Zero divided by any nonzero number is equal to 0.

$$\frac{0}{7} = 0 \qquad \text{and} \qquad \frac{0}{23} = 0$$

Division by 0 is undefined.

$$\frac{7}{0} \text{ is undefined} \qquad \text{and} \qquad \frac{2{,}190}{0} \text{ is undefined}$$

There are **divisibility tests** to help us decide whether one number is divisible by another. They are listed on page 58.

Is 21,507 divisible by 3?

21,507 is divisible by 3, because the sum of its digits is divisible by 3.

$$2 + 1 + 5 + 0 + 7 = 15 \qquad \text{and} \qquad 15 \div 3 = 5$$

There is a shortcut for **dividing a dividend by a divisor when both end with zeros.** We simply *remove the ending zeros in the divisor and remove the same number of ending zeros in the dividend.*

Divide:

$$64{,}000 \div 1{,}600 = 640 \div 16$$

Remove two zeros from the dividend and the divisor, and divide.

To **estimate quotients**, we use a method that approximates both the dividend and the divisor so that they divide easily.

Estimate the quotient for $154{,}908 \div 51$ by finding $150{,}000 \div 50$.

The dividend is approximately

$$154{,}908 \div 51 \qquad\qquad 150{,}000 \div 50$$

The divisor is approximately

Application problems that involve **forming equal-sized groups** can be solved by division.

Some **key words** and **phrases** that are often used to indicate division:

split equally   *distributed equally*
shared equally   *how many does each*
how many left (remainder)   *per*
how much extra (remainder)   *among*

Braces. An orthodontist offers his patients a plan to pay the $5,400 cost of braces in 36 equal payments. What is the amount of each payment?

The phrase *36 equal payments* indicates division:

$$\boxed{\text{The amount of each payment}} = 5{,}400 \div 36$$

REVIEW EXERCISES

Divide, if possible.

57. $\dfrac{72}{4}$

58. $1{,}443 \div 39$

59. $68\overline{)20{,}876}$

60. $21\overline{)405}$

61. $\dfrac{0}{10}$

62. $\dfrac{165}{0}$

63. $127\overline{)5{,}347}$

64. $1{,}482{,}000 \div 3{,}900$

65. Write the related multiplication statement for $160 \div 4 = 40$

66. Use a check to determine whether the following division is correct.

$$45 \text{ R } 6$$
$$7\overline{)320}$$

67. Is 364,545 divisible by 2, 3, 4, 5, 6, 9, or 10?

68. Estimate the quotient: $210{,}999 \div 53$

69. Treats. If 745 candies are distributed equally among 45 children, how many will each child receive? How many candies will be left over?

70. Purchasing. A county received an $850,000 grant to purchase some new police patrol cars. If a fully equipped patrol car costs $25,000, how many can the county purchase with the grant money?

SECTION 1.6 ▶ Problem Solving

DEFINITIONS AND CONCEPTS	EXAMPLES
To become a good problem solver, you need a plan to follow, such as the following five-step strategy for problem solving: 1. **Analyze the problem** by reading it carefully. What information is given? What are you asked to find? What vocabulary is given? Often, a *diagram* or *table* will help you visualize the facts of the problem. 2. **Form a plan** by translating the words of the problem into numbers and symbols. 3. **Perform the calculations**. 4. **State the conclusion** clearly. Be sure to include the units in your answer. 5. **Check the result.** An estimate is often helpful to see whether an answer is reasonable.	**CEO Pay.** A recent report claimed that in 2015 the top chief executive officers of the largest U.S. companies averaged 335 times more in pay than the average U.S. worker. If the average U.S. worker was paid $37,000 a year, what was the pay of a top CEO? (Source: money.cnn.com) **Analyze** ■ Top CEOs were paid 335 times more than the average worker. *Given* ■ An average worker was paid $37,000 a year. *Given* ■ What was the average pay of a top CEO in 2015? *Find* **Form** Translate the words of the problem to numbers and symbols. The pay of a top CEO in 2015 was equal to 335 times the pay of the average U.S. worker. The pay of a top CEO in 2015 = 335 · 37,000

Calculate Use a shortcut to perform this multiplication.

$$335 \cdot 37000 = 12{,}395{,}000$$

Multiply 335 and 37 to get 12395.

Attach three 0's after 12395.

$$\begin{array}{r} 335 \\ \times\ 37 \\ \hline 2345 \\ 10050 \\ \hline 12395 \end{array}$$

State In 2015 the average pay of a top CEO was $12,395,000.

Check Use front-end rounding to estimate the product: 335 is approximately 300 and 37,000 is approximately 40,000.

$$300 \cdot 40{,}000 = 12{,}000{,}000$$

Since the estimate, $12,000,000, and the result, $12,395,000, are close, the result seems reasonable.

REVIEW EXERCISES

71. Sausage. To make smoked sausage, the sausage is first dried at a temperature of 130°F. Then the temperature is raised 20° to smoke the meat. The temperature is raised another 20° to cook the meat. In the last stage, the temperature is raised another 15°. What is the final temperature in the process?

72. Drive-ins. The largest number of drive-in theaters in the United States was 4,063 in 1958. Since then, the number of drive-ins has decreased by 3,739. How many drive-in theaters are there today? (Source: United Drive-in Theater Owners Association)

73. Weight training. For part of a woman's upper body workout, she does one set of 12 repetitions of 75 pounds on a bench press machine. How many total pounds does she lift in that set?

74. Parking. Parking lot B4 at an amusement park opens at 8:00 AM and closes at 11:00 PM. It costs $5 to park in the lot. If there are 24 rows and each row has 50 parking spaces, how many cars can park in the lot?

75. Production. A manufacturer produces 15,000 light bulbs a day. The bulbs are packaged six to a box. How many boxes of light bulbs are produced each day?

76. Embroidered caps. A digital embroidery machine uses 16 yards of thread to stitch a team logo on the front of a baseball cap. How many hats can be embroidered if the thread comes on a spool of 1,100 yards? How many yards of thread will be left on the spool?

77. Farming. In a shipment of 350 animals, 124 were hogs, 79 were sheep, and the rest were cattle. Find the number of cattle in the shipment.

78. Halloween. A couple bought six bags of mini Snickers bars. Each bag contains 48 pieces of candy. If they plan to give each trick-or-treater three candy bars, to how many children will they be able to give treats?

SECTION 1.7 ▶ Prime Factors and Exponents

DEFINITIONS AND CONCEPTS	EXAMPLES
Numbers that are multiplied together are called **factors**. To **factor** a whole number means to express it as the product of other whole numbers. If a whole number is a factor of a given number, it also *divides the given number exactly.*	The pairs of whole numbers whose product is 6 are: $$1 \cdot 6 = 6 \quad \text{and} \quad 2 \cdot 3 = 6$$ From least to greatest, the factors of 6 are 1, 2, 3, and 6. Each of the factors of 6 divides 6 exactly (no remainder): $$\frac{6}{1} = 6 \qquad \frac{6}{2} = 3 \qquad \frac{6}{3} = 2 \qquad \frac{6}{6} = 1$$
If a whole number is divisible by 2, it is called an **even** number. If a whole number is not divisible by 2, it is called an **odd** number.	Even whole numbers: 0, 2, 4, 6, 8, 10, 12, 14, 16, 18, … Odd whole numbers: 1, 3, 5, 7, 9, 11, 13, 15, 17, 19, …
A **prime number** is a whole number greater than 1 that has only 1 and itself as factors. There are infinitely many prime numbers.	Prime numbers: 2, 3, 5, 7, 11, 13, 17, 19, 23, 29, 31, …
The **composite numbers** are whole numbers greater than 1 that are *not* prime. There are infinitely many composite numbers.	Composite numbers: 4, 6, 8, 9, 10, 12, 14, 15, 16, 18, …
To find the **prime factorization** of a whole number means to write it as the product of only prime numbers. A **factor tree** and a **division ladder** can be used to find prime factorizations.	Use a *factor tree* to find the prime factorization of 30. Factor each number that is encountered as a product of two whole numbers (other than 1 and itself) until all the factors involved are prime. The prime factorization of 30 is $2 \cdot 3 \cdot 5$. Use a *division ladder* to find the prime factorization of 70. Perform repeated divisions by prime numbers until the final quotient is itself a prime number. The prime factorization of 70 is $2 \cdot 5 \cdot 7$.
An **exponent** is used to indicate repeated multiplication. It tells how many times the **base** is used as a factor.	Exponent $$\underbrace{2 \cdot 2 \cdot 2 \cdot 2}_{\text{Repeated factors}} = 2^{4}$$ Base 2^4 is called an exponential expression.
We can use the definition of exponent to **evaluate** (find the value of) exponential expressions.	Evaluate: 7^3 $7^3 = 7 \cdot 7 \cdot 7$ Write the base 7 as a factor 3 times. $\quad\ = 49 \cdot 7$ Multiply, working left to right. $\quad\ = 343$ Multiply. Evaluate: $2^2 \cdot 3^3$ $2^2 \cdot 3^3 = 4 \cdot 27$ Evaluate the exponential expressions first. $\quad\quad\ = 108$ Multiply.

REVIEW EXERCISES

Find all of the factors of each number. List them from least to greatest.

79. 18   **80.** 75

81. Factor 20 using two factors. *Do not use the factor 1 in your answer.*

82. Factor 54 using three factors. *Do not use the factor 1 in your answer.*

Tell whether each number is a prime number, a composite number, or neither.

83. **a.** 31   **b.** 100

    c. 1   **d.** 0

    e. 125   **f.** 47

Tell whether each number is an even or an odd number.

84. **a.** 171   **b.** 214

    c. 0   **d.** 1

Find the prime factorization of each number. Use exponents in your answer, when helpful.

85. 42   **86.** 75

87. 220   **88.** 140

Write each expression using exponents.

89. $6 \cdot 6 \cdot 6 \cdot 6$ **90.** $5(5)(5)(13)(13)$

Evaluate each expression.

91. 5^3 **92.** 11^2

93. $2^4 \cdot 7^2$ **94.** $2^2 \cdot 3^3 \cdot 5^2$

<table>
<tr><td colspan="2">SECTION 1.8 ▶ The Least Common Multiple
and the Greatest Common Factor</td></tr>
</table>

DEFINITIONS AND CONCEPTS	EXAMPLES
The **multiples** of a number are the products of that number and 1, 2, 3, 4, 5, and so on.	Multiples of 2: 2, 4, **6**, 8, 10, **12**, 14, 16, **18**, 20, 22, **24**, … Multiples of 3: 3, **6**, 9, **12**, 15, **18**, 21, **24**, 27, … The common multiples of 2 and 3 are: 6, 12, 18, 24, 30, …
The **least common multiple (LCM)** of two whole numbers is the smallest common multiple of the numbers. The **LCM** of two whole numbers is the *smallest* whole number that is divisible by both of those numbers.	The least common multiple of 2 and 3 is 6, which is written as: LCM (2, 3) = 6. $$\frac{6}{2} = 3 \qquad \text{and} \qquad \frac{6}{3} = 2$$

To **find the LCM** of two (or more) whole numbers by **listing**:
 	
 1. Write multiples of the largest number by multiplying it by 1, 2, 3, 4, 5, and so on.
 2. Continue this process until you find the *first* multiple of the larger number that is divisible by each of the smaller numbers. That multiple is their LCM.
	Find the LCM of 3 and 5.
	
	 Multiples of 5: 5, 10, 15, 20, 25, …
	
	     Not divisible Not divisible Divisible by 3.
	     by 3. by 3.
	
	 Since 15 is the first multiple of 5 that is divisible by 3, the LCM (3, 5) = 15.

To **find the LCM** of two (or more) whole numbers using **prime factorization**:
 	
 1. Prime factor each number.
 2. The LCM is a product of prime factors, where each factor is used the greatest number of times it appears in any one factorization.
	Find the LCM of 6 and 20.
	
	 $6 = 2 \cdot \textcircled{3}$ *The greatest number of times 3 appears is once.*
	 $20 = \boxed{2 \cdot 2} \cdot \textcircled{5}$ *The greatest number of times 2 appears is twice.*
	
	                       The greatest number of times 5 appears is once.
	
	             Use the factor 2 two times.
	                Use the factor 3 one time.
	                  Use the factor 5 one time.
	
	 LCM (6, 20) $= 2 \cdot 2 \cdot 3 \cdot 5 = 60$

The **greatest common factor (GCF)** of two (or more) whole numbers is the largest common factor of the numbers.	The factors of 18: [1], [2], [3], ⟨6, 9, 18 The factors of 30: [1], [2], [3], 5, 6⟩, 10, 15, 30 The common factors of 18 and 30 are 1, 2, 3, and 6. The greatest common factor of 18 and 30 is 6, which is written as: GCF (18, 30) = 6.
The greatest common factor of two (or more) numbers is the *largest* whole number that divides them exactly.	$$\frac{18}{6} = 3 \qquad \text{and} \qquad \frac{30}{6} = 5$$
To **find the GCF** of two (or more) whole numbers using **prime factorization**: 1. Prime factor each number. 2. Identify the common prime factors. 3. The GCF is a product of all the common prime factors found in Step 2. If there are no common prime factors, the GCF is 1.	Find the GCF of 36 and 60. $$36 = \boxed{2} \cdot \boxed{2} \cdot \boxed{3} \cdot 3$$ *36 and 60 have two common factors* *of 2 and one common factor of 3.* $$60 = \boxed{2} \cdot \boxed{2} \cdot \boxed{3} \cdot 5$$ The GCF is the product of the circled prime factors. $$\text{GCF } (36, 60) = 2 \cdot 2 \cdot 3 = 12$$

REVIEW EXERCISES

95. Find the first ten multiples of 9.

96. a. Find the common multiples of 6 and 8 in the lists below.

     Multiples of 6: 6, 12, 18, 24, 30, 36, 42, 48, 54, …

     Multiples of 8: 8, 16, 24, 32, 40, 48, 56, 64, 72, …

    b. Find the common factors of 6 and 8 in the lists below.

       Factors of 6: 1, 2, 3, 6

       Factors of 8: 1, 2, 4, 8

Find the LCM of the given numbers.

97. 4, 6   **98.** 3, 4

99. 9, 15   **100.** 12, 18

101. 18, 21   **102.** 24, 45

103. 4, 14, 20   **104.** 21, 28, 42

Find the GCF of the given numbers.

105. 8, 12   **106.** 9, 12

107. 30, 40   **108.** 30, 45

109. 63, 84   **110.** 112, 196

111. 48, 72, 120   **112.** 88, 132, 176

113. Meetings. The Rotary Club meets every 14 days, and the Kiwanis Club meets every 21 days. If both clubs have a meeting on the same day, in how many more days will they again meet on the same day?

114. Flowers. A florist is making flower arrangements for a 4th of July party. He has 32 red carnations, 24 white carnations, and 16 blue carnations. He wants each arrangement to be identical.

    a. What is the greatest number of arrangements that he can make if every carnation is used?

    b. How many of each type of carnation will be used in each arrangement?

SECTION 1.9 ▶ Order of Operations

DEFINITIONS AND CONCEPTS	EXAMPLES
To **evaluate** (find the value of) expressions that involve more than one operation, use the order of operations rule. **Order of Operations** 1. Perform all calculations within parentheses and other grouping symbols following the order listed in Steps 2–4 below, working from the innermost pair of grouping symbols to the outermost pair. 2. Evaluate all exponential expressions. 3. Perform all multiplications and divisions as they occur from left to right. 4. Perform all additions and subtractions as they occur from left to right. When grouping symbols have been removed, repeat Steps 2–4 to complete the calculation. If a fraction bar is present, evaluate the expression above the bar (called the **numerator**) and the expression below the bar (called the **denominator**) separately. Then perform the division indicated by the fraction bar, if possible.	Evaluate: $10 + 3[2^4 - 3(5 - 2)]$ Work within the *innermost* parentheses first and then within the *outermost* brackets. $10 + 3[2^4 - 3(5 - 2)] = 10 + 3[2^4 - 3(3)]$ Subtract within the parentheses. $= 10 + 3[16 - 3(3)]$ Evaluate the exponential expression within the brackets: $2^4 = 16$. $= 10 + 3[16 - 9]$ Multiply within the brackets. $= 10 + 3[7]$ Subtract within the brackets. $= 10 + 21$ Multiply: $3[7] = 21$. $= 31$ Do the addition. **Caution!** A common error is to incorrectly add 10 and 3 in the fifth step of the solution. $= 10 + 3[7]$ $= 13[7]$ Multiply before adding. $= 91$ Evaluate: $\dfrac{3^3 + 8}{7(15 - 14)}$ Evaluate the expressions above and below the fraction bar separately. $\dfrac{3^3 + 8}{7(15 - 14)} = \dfrac{27 + 8}{7(1)}$ In the numerator, evaluate the exponential expression. In the denominator, subtract. $= \dfrac{35}{7}$ In the numerator, add. In the denominator, multiply. $= 5$ Divide.
The **mean**, or **average**, of a set of numbers is a value around which the values of the numbers are grouped. To **find the mean (average)** of a set of values, divide the sum of the values by the number of values.	Find the mean (average) of the test scores 74, 83, 79, 91, and 73. $\text{Mean} = \dfrac{74 + 83 + 79 + 91 + 73}{5}$ Since there are 5 scores, divide by 5. $\text{Mean} = \dfrac{400}{5}$ Do the addition in the numerator. $\text{Mean} = 80$ Divide. The mean (average) test score is 80.

REVIEW EXERCISES

Evaluate each expression.

115. $3^2 + 12 \cdot 3$

116. $35 - 5 \cdot 3 + 3$

117. $(6 \div 2 \cdot 3)^2 \cdot 3$

118. $(35 - 5 \cdot 3) \div 5$

119. $2^3 \cdot 5 - 4 \div 2 \cdot 4$

120. $8 \cdot (5 - 4 \div 2)^2$

121. $2 + 3\left(\dfrac{100}{10} - 2^2 \cdot 2\right)$

122. $4(4^2 - 5 \cdot 3 + 2) - 4$

123. $\dfrac{4(6) - 6}{2(3^2)}$

124. $\dfrac{6 \cdot 2 + 3 \cdot 7}{5^2 - 2(7)}$

125. $7 + 3[3^3 - 10(4 - 2)]$

126. $5 + 2\left[\left(2^4 - 3 \cdot \dfrac{8}{2}\right) - 2\right]$

Find the arithmetic mean (average) of each set of test scores.

127.

Test	1	2	3	4
Score	80	74	66	88

128.

Test	1	2	3	4	5
Score	73	77	81	0	69

1. a. The set of _____ numbers is {0, 1, 2, 3, 4, 5, ... }.

b. The symbols > and < are _____ symbols.

c. To *evaluate* an expression such as $58 - 33 + 9$ means to find its _____.

d. The _____ of a rectangle is a measure of the amount of surface it encloses.

e. One number is _____ by another number if, when we divide them, the remainder is 0.

f. The grouping symbols ( ) are called _____, and the symbols [ ] are called _____.

g. A _____ number is a whole number greater than 1 that has only 1 and itself as factors.

2. Graph the whole numbers less than 7 on a number line.

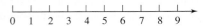

3. Consider the whole number 402,198.

a. What is the place value of the digit 1?

b. What digit is in the ten thousands column?

4. a. Write 7,018,641 in words.

b. Write "one million, three hundred eighty-five thousand, two hundred sixty-six" in standard form.

c. Write 92,561 in expanded form.

5. Place an < or an > symbol in the box to make a true statement.

a. 15 ☐ 10   **b.** 1,247 ☐ 1,427

6. Round 34,759,841 to

a. the nearest million

b. the nearest hundred thousand

c. the nearest thousand

7. The NHL. The table below shows the number of teams in the National Hockey League at various times during its history. Use the data to complete the bar graph in the next column.

Year	1960	1970	1980	1990	2000	2010
Number of teams	6	14	21	21	28	30

Source: www.rauzulusstreet.com

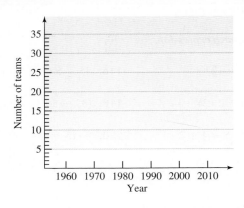

8. Subtract 287 from 535. Show a check of your result.

9. Add: 136,231
82,574
+ 6,359

10. Subtract: 4,521
− 3,579

11. Multiply: 53
× 8

12. Multiply: $74 \cdot 562$

13. Divide: $6\overline{)432}$

14. Divide: $8,379 \div 73$. Show a check of your result.

15. Find the product of 23,000 and 600.

16. Find the quotient of 125,000 and 500.

17. Use front-end rounding to estimate the difference: $49,213 - 7,198$

18. A rectangle is 327 inches wide and 757 inches long. Find its perimeter.

19. Find the area of the square shown.

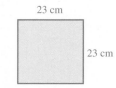

23 cm

23 cm

20. a. Find the factors of 12.

b. Find the first six multiples of 4.

c. Write $5 + 5 + 5 + 5 + 5 + 5 + 5 + 5$ as a multiplication.

21. Find the prime factorization of 1,260.

22. Teeth. Children have one set of primary (baby) teeth used in early development. These 20 teeth are generally replaced by a second set of larger permanent (adult) teeth. Determine the number of adult teeth if there are 12 more of those than baby teeth.

23. Tossing a coin. During World War II, John Kerrich, a prisoner of war, tossed a coin 10,000 times and wrote down the results. If he recorded 5,067 heads, how many tails occurred? (Source: *Figure This!*)

24. P.E. classes. In a physical education class, the students stand in a rectangular formation of 8 rows and 12 columns when the instructor takes attendance. How many students are in the class?

25. Floor space. The men's, women's, and children's departments in a clothing store occupy a total of 12,255 square feet. Find the square footage of each department if they each occupy the same amount of floor space.

26. Mileage. The fuel tank of a Hummer H3 holds 23 gallons of gasoline. How far can a Hummer travel on one tank of gas if it gets 18 miles per gallon on the highway?

27. Inheritance. A father willed his estate, valued at $1,350,000, to his four adult children. Upon his death, the children paid legal expenses of $26,000 and then split the remainder of the inheritance equally among themselves. How much did each one receive?

28. What property is illustrated by each statement?

   a. $18 \cdot (9 \cdot 40) = (18 \cdot 9) \cdot 40$

   b. $23,999 + 1 = 1 + 23,999$

29. Perform each operation, if possible.

   a. $15 \cdot 0$ **b.** $\dfrac{0}{15}$

   c. $\dfrac{8}{8}$ **d.** $\dfrac{8}{0}$

30. Find the LCM of 15 and 18.

31. Find the LCM of 8, 9, and 12.

32. Find the GCF of 30 and 54.

33. Find the GCF of 24, 28, and 36.

34. Stocking shelves. Boxes of rice are being stacked next to boxes of instant mashed potatoes on the same bottom shelf in a supermarket display. The boxes of rice are 8 inches tall, and the boxes of instant potatoes are 10 inches high.

   a. What is the shortest height at which the two stacks will be the same height?

   b. How many boxes of rice and how many boxes of potatoes will be used in each stack?

35. Is 521,340 divisible by 2, 3, 4, 5, 6, 9, or 10?

36. Grades. A student scored 73, 52, 95, and 70 on four exams and received 0 on one missed exam. Find his mean (average) exam score.

Evaluate each expression.

37. $9 + 4 \cdot 5$

38. $3^4 \cdot 10 - 2(6)(4)$

39. $20 + 2[4^2 - 2(6 - 2^2)]$

40. $\dfrac{3^3 - 2(15 - 14)^2}{33 - 9 + 1}$

2 The Integers

StockLite/Shutterstock.com

from Campus to Careers

Personal Financial Adviser

Personal financial advisers help people manage their money and teach them how to make their money grow. They offer advice on how to budget for monthly expenses, as well as how to save for retirement. A bachelor's degree in business, accounting, finance, economics, or statistics provides good preparation for the occupation. Strong communication and problem-solving skills are equally important to achieve success in this field.

In **Problem 94** of **Study Set 2.2**, you will see how a personal financial adviser uses integers to determine whether a duplex rental unit would be a moneymaking investment for a client.

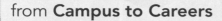

JOB TITLE:
Personal Financial Adviser

EDUCATION:
Must have at least a bachelor's degree. Some states require a certificate or license.

JOB OUTLOOK:
Excellent—Jobs are projected to grow by 30% over the next decade.

ANNUAL EARNINGS:
In 2015, the median annual salary was $89,160.

FOR MORE INFORMATION:
https://money.usnews.com/careers/best-jobs/financial-advisor

CHAPTER OUTLINE

OBJECTIVES

1 Define the set of integers.

2 Graph integers on a number line.

3 Use inequality symbols to compare integers.

4 Find the absolute value of an integer.

5 Find the opposite of an integer.

SECTION **2.1**   An Introduction to the Integers

We have seen that whole numbers can be used to describe many situations that arise in everyday life. However, we cannot use whole numbers to express temperatures below zero, the balance in a checking account that is overdrawn, or how far an object is below sea level. In this section, we will see how negative numbers can be used to describe these three situations as well as many others.

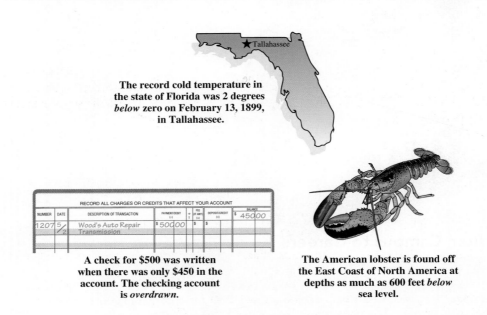

★Tallahassee

The record cold temperature in the state of Florida was 2 degrees *below* zero on February 13, 1899, in Tallahassee.

RECORD ALL CHARGES OR CREDITS THAT AFFECT YOUR ACCOUNT

NUMBER	DATE	DESCRIPTION OF TRANSACTION	PAYMENT/DEBIT (-)	FEE IF ANY (-)	DEPOSIT/CREDIT (+)	BALANCE $ 450 00
1207	5/2	Wood's Auto Repair Transmission	$ 500 00	$	$	

A check for $500 was written when there was only $450 in the account. The checking account is *overdrawn*.

The American lobster is found off the East Coast of North America at depths as much as 600 feet *below* sea level.

OBJECTIVE 1   **Define the set of integers.**

To describe a temperature of 2 degrees above zero, a balance of $50, or 600 feet above sea level, we can use numbers called **positive numbers**. All positive numbers are greater than 0, and we can write them with or without a **positive sign** $+$.

In words	In symbols	Read as
2 degrees above zero	2	two
A balance of $50	50	fifty
600 feet above sea level	600	six hundred

To describe a temperature of 2 degrees below zero, $50 overdrawn, or 600 feet below sea level, we need to use negative numbers. **Negative numbers** are numbers less than 0, and they are written using a **negative sign** $-$.

In words	In symbols	Read as
2 degrees below zero	-2	negative two
$50 overdrawn	-50	negative fifty
600 feet below sea level	-600	negative six hundred

Together, positive and negative numbers are called **signed numbers**.

> **Positive and Negative Numbers**
> **Positive numbers** are greater than 0. **Negative numbers** are less than 0.

Caution! Zero is neither positive nor negative.

The collection of positive whole numbers, the negatives of the whole numbers, and 0 is called the set of **integers** (read as "in-ti-jers").

> **The Set of Integers**
> $\{ \ldots, -5, -4, -3, -2, -1, 0, 1, 2, 3, 4, 5, \ldots \}$

The three dots on the right indicate that the list continues forever—there is no largest integer. The three dots on the left indicate that the list continues forever—there is no smallest integer. The set of **positive integers** is $\{1, 2, 3, 4, 5, \ldots \}$, and the set of **negative integers** is $\{ \ldots, -5, -4, -3, -2, -1 \}$.

The set of integers → $\{ \ldots, -5, -4, -3, -2, -1, \underbrace{0, 1, 2, 3, 4, 5, \ldots}_{\text{The set of whole numbers}} \}$

LANGUAGE OF MATHEMATICS

Since every whole number is an integer, we say that the set of whole numbers is a **subset** of the integers.

OBJECTIVE 2 **Graph integers on a number line.**

In Section 1.1, we introduced the number line. We can use an extension of the number line to learn about negative numbers.

Negative numbers can be represented on a number line by extending the line to the left and drawing an arrowhead. Beginning at the origin (the 0 point), we move to the left, marking equally spaced points as shown below. As we move to the right on the number line, the values of the numbers increase. As we move to the left, the values of the numbers decrease.

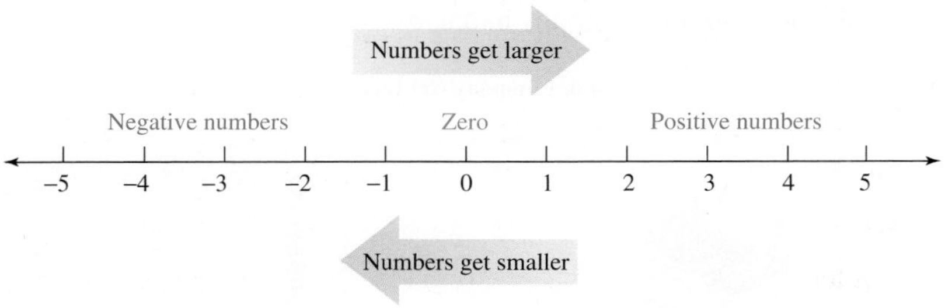

Numbers get larger →

Negative numbers   Zero   Positive numbers

-5 -4 -3 -2 -1 0 1 2 3 4 5

← Numbers get smaller

The thermometer shown on the next page is an example of a vertical number line. It is scaled in degrees and shows a temperature of $-10°$. The time line is an example of a horizontal number line. It is scaled in units of 500 years.

MAYA CIVILIZATION

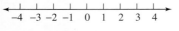

Based on data from *People in Time and Place, Western Hemisphere* (Silver Burdett & Ginn., 1991), p. 129

A vertical number line   **A horizontal number line**

EXAMPLE 1   Graph −3, 2, −1, and 4 on a number line.

Strategy We will locate the position of each integer on the number line and draw a bold dot.

WHY To *graph a number* means to make a drawing that represents the number.

Solution
The position of each negative integer is to the left of 0. The position of each positive integer is to the right of 0.

Self Check 1

Graph −4, −2, 1, and 3 on a number line.

Now Try ➡ **Problem 23**

By extending the number line to include negative numbers, we can represent more situations using bar graphs and line graphs. For example, the following bar graph shows the net income of the Kodak Company for the years 2003 through 2015. Since the net income in 2007 was positive $676 million, the company made a *profit*. Since the net income in 2012 was −$1,379 million, the company had a *loss*.

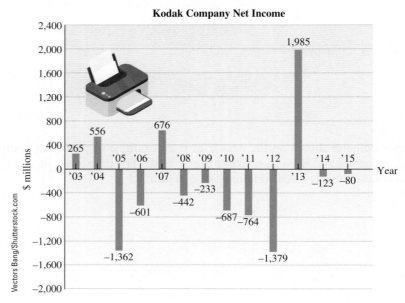

Source: morningstar.com and google.com/finance

Think it Through   •   CREDIT CARD DEBT

"The most dangerous pitfall for many college students is the overuse of credit cards. Many banks do their best to entice new card holders with low or zero-interest cards."
—*Gary Schatsky, certified financial planner*

Which numbers on the credit card statement below are actually debts and, therefore, could be represented using negative numbers?

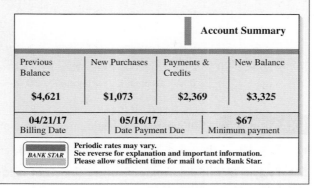

	Account Summary		
Previous Balance	New Purchases	Payments & Credits	New Balance
$4,621	$1,073	$2,369	$3,325

04/21/17 Billing Date	05/16/17 Date Payment Due	$67 Minimum payment

BANK STAR   Periodic rates may vary.
See reverse for explanation and important information.
Please allow sufficient time for mail to reach Bank Star.

OBJECTIVE 3 Use inequality symbols to compare integers.

Recall that the symbol < means "is less than" and that > means "is greater than." The figure below shows the graph of the integers −2 and 1. Since −2 is to the left of 1 on the number line, −2 < 1. Since −2 < 1, it is also true that 1 > −2.

$$-4 \quad -3 \quad -2 \quad -1 \quad 0 \quad 1 \quad 2 \quad 3 \quad 4$$

EXAMPLE 2   Place an < or an > symbol in the box to make a true statement.
a. 4 ☐ −5   **b.** −8 ☐ −7

Strategy To pick the correct inequality symbol to place between the pair of numbers, we will determine the position of each number on the number line.

WHY For any two numbers on a number line, the number to the *left* is the smaller number and the number on the *right* is the larger number.

Solution

a. Since 4 is to the right of −5 on the number line, 4 > −5.

b. Since −8 is to the left of −7 on the number line, −8 < −7.

Self Check 2

Place an < or an > symbol in the box to make a true statement.
a. 6 ☐ −6
b. −11 ☐ −10

Now Try Problems 31 and 35

There are three other commonly used inequality symbols.

Inequality Symbols

≠ means *is not equal to*
≥ means *is greater than or equal to*
≤ means *is less than or equal to*

$-5 \neq -2$ Read as "−5 is not equal to −2."

$-6 \leq 10$ Read as "−6 is less than or equal to 10."
This statement is true because $-6 < 10$.

$12 \leq 12$ Read as "12 is less than or equal to 12."
This statement is true because $12 = 12$.

$-15 \geq -17$ Read as "−15 is greater than or equal to −17."
This statement is true because $-15 > -17$.

$-20 \geq -20$ Read as "−20 is greater than or equal to −20."
This statement is true because $-20 = -20$.

EXAMPLE 3   Tell whether each statement is true or false.

a. $-9 \geq -9$ **b.** $-1 \leq -5$ **c.** $-27 \geq 6$ **d.** $-32 \leq -31$

Strategy We will determine if either the strict inequality or the equality that the symbols $\leq$ and $\geq$ allow is true.

WHY If either is true, then the given statement is true.

Solution

a. $-9 \geq -9$ This statement is true because $-9 = -9$.

b. $-1 \leq -5$ This statement is false because neither $-1 < -5$ nor $-1 = -5$ is true.

c. $-27 \geq 6$ This statement is false because neither $-27 > 6$ nor $-27 = 6$ is true.

d. $-32 \leq -31$ This statement is true because $-32 < -31$.

OBJECTIVE 4 **Find the absolute value of an integer.**

Using a number line, we can see that the numbers 3 and −3 are both a distance of 3 units away from 0, as shown below.

The **absolute value** of a number gives the distance between the number and 0 on the number line. To indicate absolute value, the number is inserted between two vertical bars, called the **absolute value symbol**. For example, we can write $|-3| = 3$. This is read as "The absolute value of negative 3 is 3," and it tells us that the distance between −3 and 0 on the number line is 3 units. From the figure, we also see that $|-3| = 3$.

Caution! Absolute value expresses distance. The absolute value of a number is always positive or 0. It is never negative.

Absolute Value

The absolute value of a number is the distance on the number line between the number and 0.

EXAMPLE 4   Find each absolute value: **a.** $|8|$ **b.** $|-5|$ **c.** $|0|$

Strategy We need to determine the distance that the number within the vertical absolute value bars is from 0 on a number line.

WHY The absolute value of a number is the distance between 0 and the number on a number line.

Solution

a. On the number line, the distance between 8 and 0 is 8. Therefore,

$$|8| = 8$$

b. On the number line, the distance between -5 and 0 is 5. Therefore,

$$|-5| = 8$$

c. On the number line, the distance between 0 and 0 is 0. Therefore,

$$|0| = 0$$

Self Check 4

Find each absolute value:

a. $|-9|$

b. $|4|$

Now Try ➡ Problems 47 and 49

OBJECTIVE 5 **Find the opposite of an integer.**

Opposites

Two numbers that are the same distance from 0 on the number line, but on opposite sides of it, are called **opposites**.

The figure below shows that for each integer on the number line, there is a corresponding integer, called its *opposite*, on the opposite side of 0. For example, we see that 3 and -3 are opposites, as are -5 and 5. Note that 0 is its own opposite.

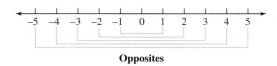

Opposites

To write the opposite of a number, a $-$ symbol is used. For example, the opposite of 5 is -5 (read as "negative 5"). Parentheses are needed to express the opposite of a negative number. The opposite of -5 is written as $-(-5)$. Since 5 and -5 are the same distance from 0, the opposite of -5 is 5. Therefore, $-(-5) = 5$. This illustrates the following rule.

The Opposite of the Opposite Rule

The opposite of the opposite of a number is that number.

Number	Opposite	
57	-57	Read as "negative fifty-seven."
-8	$-(-8) = 8$	Read as "the opposite of negative eight is eight."
0	$-0 = 0$	Read as "the opposite of 0 is 0."

The concept of opposite can also be applied to an absolute value. For example, the opposite of the absolute value of -8 can be written as $-|-8|$. Think of this as a two-step process, where the absolute value symbol serves as a grouping symbol. Find the absolute value first, and then attach a $-$ sign to that result.

First, find the absolute value.

$$-|-8| = -8$$

Read as "the opposite of the absolute value of negative eight is negative eight."

Then attach a $-$ sign.

EXAMPLE 5   Simplify each expression: **a.** $-(-44)$ **b.** $-|11|$ **c.** $-|-225|$

Strategy We will find the opposite of each number.

WHY In each case, the $-$ symbol written outside the grouping symbols means "the opposite of."

Solution

a. $-(-44)$ means the opposite of -44. Since the opposite of -44 is 44, we write

$$-(-44) = 44$$

b. $-|11|$ means the opposite of the absolute value of 11. Since $|11| = 11$, and the opposite of 11 is -11, we write

$$-|11| = -11$$

c. $-|-225|$ means the opposite of the absolute value of -225. Since $|-225| = 225$, and the opposite of 225 is -225, we write

$$-|-225| = -225$$

Self Check 5

Simplify each expression:

a. $-(-1)$

b. $-|4|$

c. $-|-99|$

Now Try ➡ Problems 55, 65, and 67

The $-$ symbol is used to indicate a negative number, the opposite of a number, and the operation of subtraction. The key to reading the $-$ symbol correctly is to examine the context in which it is used.

Reading the $-$ Symbol		
-12	Negative twelve	A $-$ symbol directly in front of a number is read as "negative."
$-(-12)$	The opposite of negative twelve	The first $-$ symbol is read as "the opposite of" and the second as "negative."
$12 - 5$	Twelve minus five	Notice the space used before and after the $-$ symbol. This indicates subtraction and is read as "minus."

Answers to Self Checks

1.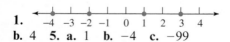

b. 4 **5. a.** 1 **b.** -4 **c.** -99

2. a. $>$ **b.** $<$ **3. a.** false **b.** true **c.** true **d.** false   **4. a.** 9

SECTION 2.1   STUDY SET

VOCABULARY

Fill in the blanks.

1. _____ numbers are greater than 0, and _____ numbers are less than 0.

2. $\{\dots, -5, -4, -3, -2, -1, 0, 1, 2, 3, 4, 5, \dots\}$ is called the set of _____ .

3. To _____ an integer means to locate it on the number line and highlight it with a dot.

4. The symbols $>$ and $<$ are called _____ symbols.

5. The _____ _____ of a number is the distance between the number and 0 on the number line.

6. Two numbers that are the same distance from 0 on the number line, but on opposite sides of it, are called _____ .

CONCEPTS

7. Represent each of these situations using a signed number.

a. $225 overdrawn

b. 10 seconds before liftoff

c. 3 degrees below normal

d. A deficit of $12,000

e. A 1-mile retreat by an army

8. Represent each of these situations using a signed number, and then describe its opposite in words.

   a. A trade surplus of $3 million

   b. A bacteria count 70 more than the standard

   c. A profit of $67

   d. A business $1 million in the "black"

   e. 20 units over their quota

9. Determine what is wrong with each number line.

   a.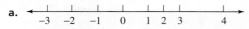

   b.

   c.

   d.

10. a. If a number is less than 0, what type of number must it be?

    b. If a number is greater than 0, what type of number must it be?

11. On the number line, what number is

    a. 3 units to the right of -7?

    b. 4 units to the left of 2?

12. Name two numbers on the number line that are a distance of

    a. 5 away from -3.   b. 4 away from 3.

13. a. Which number is closer to -3 on the number line: 2 or -7?

    b. Which number is farther from 1 on the number line: -5 or 8?

14. Is there a number that is both greater than 10 and less than 10 at the same time?

15. a. Express the fact $-12 < 15$ using an $>$ symbol.

    b. Express the fact $-4 > -5$ using an $<$ symbol.

16. Fill in the blank: The opposite of the _____ of a number is that number.

17. Complete the table by finding the opposite and the absolute value of the given numbers.

Number	Opposite	Absolute value
-25		
39		
0		

18. Is the absolute value of a number always positive?

NOTATION

19. Translate each phrase to mathematical symbols.

    a. The opposite of negative eight

    b. The absolute value of negative eight

    c. Eight minus eight

    d. The opposite of the absolute value of negative eight

20. a. Write the set of integers.

    b. Write the set of positive integers.

    c. Write the set of negative integers.

21. Fill in the blanks.

    a. We read $\geq$ as "is _____ than or _____ to."

    b. We read $\leq$ as "is _____ than or _____ to."

22. Which of the following expressions contains a minus sign?

    $15 - 8$ $-(-15)$ -15

GUIDED PRACTICE

Graph the following numbers on a number line. **See Example 1.**

23. $-3, 4, 3, 0, -1$

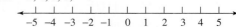

24. $2, -4, 5, 1, -1$

25. The integers that are less than 3 but greater than -5

26. The integers that are less than 4 but greater than -3

27. The opposite of -3, the opposite of 5, and the absolute value of -2

28. The absolute value of 3, the opposite of 3, and the number that is 1 less than -3

29. 2 more than 0, 4 less than 0, 2 more than negative 5, and 5 less than 4

30. 4 less than 0, 1 more than 0, 3 less than -2, and 6 more than -4

Place an $<$ or an $>$ symbol in the box to make a true statement. **See Example 2.**

31. $-5 \quad 5$ 32. $0 \quad -1$

33. $-12 \quad -6$ 34. $-7 \quad -6$

35. $-10 \quad -17$ 36. $-11 \quad -20$

37. $-325 \quad -532$ 38. $-401 \quad -104$

Tell whether each statement is true or false. **See Example 3.**

39. $-15 \leq -14$ 40. $-77 \leq -76$

41. $210 \geq 210$ 42. $37 \geq 37$

43. $-1,255 \geq -1,254$ 44. $-6,546 \geq -6,465$

45. $0 \leq -8$ 46. $-6 \leq -6$

Find each absolute value. **See Example 4.**

47. $|9|$ **48.** $|12|$

49. $|-8|$ **50.** $|-1|$

51. $|-14|$ **52.** $|-85|$

53. $|180|$ **54.** $|371|$

Simplify each expression. **See Example 5.**

55. $-(-11)$ **56.** $-(-1)$

57. $-(-4)$ **58.** $-(-9)$

59. $-(-102)$ **60.** $-(-295)$

61. $-(-561)$ **62.** $-(-703)$

63. $-|20|$ **64.** $-|143|$

65. $-|6|$ **66.** $-|0|$

67. $-|-253|$ **68.** $-|-11|$

69. $-|-0|$ **70.** $-|97|$

TRY IT YOURSELF

Place an < or an > symbol in the box to make a true statement.

71. $|-12|$ ☐ $-(-7)$ **72.** $|-50|$ ☐ $-(-40)$

73. $-|-71|$ ☐ $-|-65|$ **74.** $-|-163|$ ☐ $-(-150)$

75. $-(-343)$ ☐ $-(-161)$ **76.** $-(-999)$ ☐ $-(-998)$

77. $-|-30|$ ☐ $-|-(-8)|$ **78.** $-|-100|$ ☐ $-|-(-88)|$

Write the integers in order, from least to greatest.

79. $82, -52, 52, -22, 12, -12$

80. $49, -9, 19, -39, 89, -49$

Fill in the blanks to continue each pattern.

81. $5, 3, 1, -1,$ ☐ , ☐ , ☐ , …

82. $4, 2, 0, -2,$ ☐ , ☐ , ☐ , …

LOOK ALIKES

Simplify each expression.

83. a. $-(-18)$ **b.** $-|-18|$

84. a. $-|-27|$ **b.** $-(-27)$

85. a. $|-(-44)|$ **b.** $-|-(-44)|$

86. a. $-|-(-106)|$ **b.** $-(-(-106))$

Tell whether each inequality is true or false.

87. a. $38 < 39$ **b.** $38 > 39$

   c. $38 \geq 39$ **d.** $38 \leq 39$

88. a. $75 > 75$ **b.** $75 \leq 75$

   c. $75 \geq 75$ **d.** $75 < 75$

89. a. $0 \geq -(-5)$ **b.** $0 < -(-5)$

   c. $0 > -(-5)$ **d.** $0 \leq -(-5)$

90. a. $-1 < 0$ **b.** $0 < -1$

   c. $0 > -1$ **d.** $-1 > 0$

APPLICATIONS

Use signed numbers to solve each problem.

91. Horse racing. In the 1973 Belmont Stakes, Secretariat won by 31 lengths over second-place finisher Twice a Prince. Some experts call it the greatest performance by a thoroughbred in the history of racing. Express the position of Twice a Prince compared to Secretariat as a signed number. (Source: ezinearticles.com)

Cheryl Ann Quigley/Shutterstock.com

92. NASCAR. In the NASCAR driver standings, negative numbers are used to tell how many points behind the leader a given driver is. Jimmie Johnson was the leading driver in 2016. The other drivers in the top ten, listed alphabetically, were Kurt Busch (-2744), Kyle Busch (-5), Carl Edwards (-33), Chase Elliot (-2755), Denny Hamlin (-2720), Kevin Harvick (-2751), Matt Kenseth (-2710), Kyle Larson (-2752), and Joey Logano (-3). Use this information to rank the drivers in the table below. (Source: NASCAR.com)

Action Sports Photography/Shutterstock.com

2016 NASCAR Final Driver Standings

Rank	Driver	Points behind leader
1	Jimmie Johnson	Leader
2		
3		
4		
5		
6		
7		
8		
9		
10		

Source: NASCAR.com

93. Free fall. A boy launches a water balloon from the top of a building, as shown on the next page. At that instant, his friend starts a stopwatch and keeps track of the time as the balloon sails above the building and then falls to the ground. Use the number line to estimate the position of the balloon at each time listed in the table below.

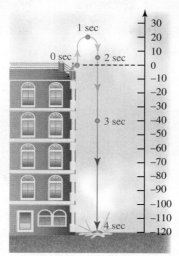

Time	Position of balloon
0 sec	
1 sec	
2 sec	
3 sec	
4 sec	

94. iPhones. You can get a more accurate reading of an iPhone's signal strength by dialing *3001#12345#*. Field test mode is then activated, and the standard signal strength bars (in the upper left corner of the display) are replaced by a negative number. The closer the negative number is to zero, the stronger the signal. Which iPhone shown below is receiving the strongest signal?

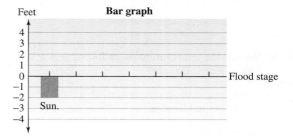

95. Technology. The readout from a testing device is shown below. Use the number line to find the height of each of the peaks and the depth of each of the valleys.

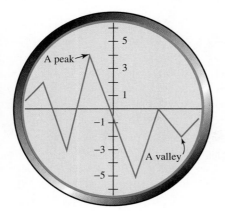

96. Flooding. A week of daily reports listing the height of a river in comparison to flood stage is given in the table. Complete the bar graph shown below.

Flood Stage Report

Sun.	2 ft below
Mon.	3 ft over
Tue.	4 ft over
Wed.	2 ft over
Thu.	1 ft below
Fri.	3 ft below
Sat.	4 ft below

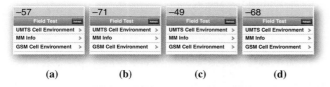

97. Golf. In golf, *par* is the standard number of strokes considered necessary on a given hole. A score of −2 indicates that a golfer used 2 strokes less than par. A score of +2 means 2 more strokes than par were used. In the graph below, each golf ball represents the score of a professional golfer on the 16th hole of a certain course.

a. What score was shot most often on this hole?

b. What was the best score on this hole?

c. Explain why this hole appears to be too easy for a professional golfer.

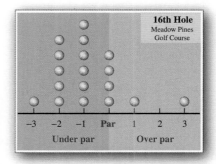

98. Paychecks. Examine the items listed on the following paycheck stub. Then write two columns on your paper—one headed "positive" and the other "negative." List each item under the proper heading.

Tom Dryden	Dec. 17	Christmas bonus	$100
Gross pay	$2,000	**Reductions**	
Overtime	$300	Retirement	$200
Deductions		**Taxes**	
Union dues	$30	Federal withholding	$160
U.S. Bonds	$100	State withholding	$35

99. Weather maps. The illustration shows the predicted Fahrenheit temperatures for a day in mid-January.

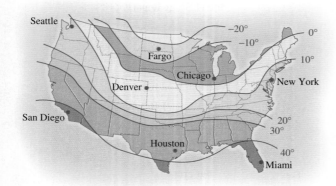

a. What is the temperature range for the region including Fargo, North Dakota?

b. According to the prediction, what is the warmest it should get in Houston?

c. According to this prediction, what is the coldest it should get in Seattle?

100. Internet companies. The graph below shows the net income of Amazon.com for the years 2002–2015. (Source: Morningstar)

a. In what years did Amazon suffer a loss? Estimate each loss.

b. In what year did Amazon first turn a profit? Estimate it.

c. In what year did Amazon have the greatest profit? Estimate it.

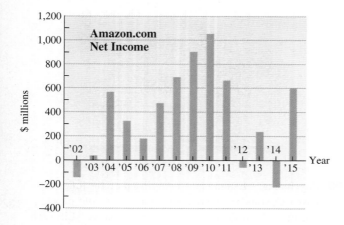

101. History. Number lines can be used to display historical data. Some important world events are shown on the time line in the next column.

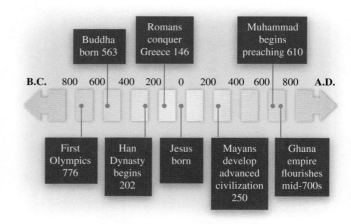

a. What basic unit is used to scale this time line?

b. What can be thought of as positive numbers?

c. What can be thought of as negative numbers?

d. What important event distinguishes the positive from the negative numbers?

102. Astronomy. Astronomers use an inverted vertical number line called the *apparent magnitude scale* to denote the brightness of objects in the sky. The brighter an object appears to an observer on Earth, the more negative is its apparent magnitude. Graph each of the following on the scale to the right.

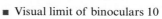

■ Visual limit of binoculars 10

■ Visual limit of large telescope 20

■ Visual limit of naked eye 6

■ Full moon −12

■ Pluto 15

■ Sirius (a bright star) −2

■ Sun −26

■ Venus −4

103. Line graphs. Each thermometer in the illustration gives the daily high temperature in degrees Fahrenheit. Use the data to complete the line graph on the next page.

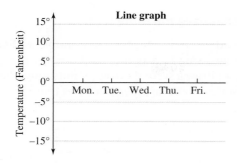

Line graph

104. Gardening. The illustration shows the depths at which the bottoms of various types of flower bulbs should be planted. (The symbol " represents inches.)

   a. At what depth should a tulip bulb be planted?

   b. How much deeper are hyacinth bulbs planted than gladiolus bulbs?

   c. Which bulb must be planted the deepest? How deep?

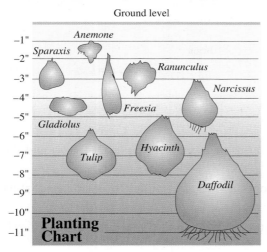

Planting Chart

WRITING

105. Explain the concept of *the opposite of a number*.

106. What real-life situation do you think gave rise to the concept of a negative number?

107. Explain why the absolute value of a number is never negative.

108. Give an example of the use of the number line that you have seen in another course.

109. Diving. Divers use the terms *positive buoyancy*, *neutral buoyancy*, and *negative buoyancy* as shown. What do you think each of these terms means?

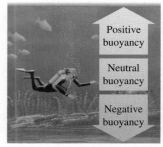

110. Geography. Much of the Netherlands is low-lying, with half of the country below sea level. Explain why it is not under water.

111. Suppose integer *A* is greater than integer *B*. Is the opposite of integer *A* greater than integer *B*? Explain why or why not. Use an example.

112. Explain why -11 is less than -10.

REVIEW

113. Round 23,456 to the nearest hundred.

114. Evaluate: $19 - 2 \cdot 3$

115. Subtract 2,081 from 2,842.

116. Divide 346 by 15.

117. Give the name of the property shown below:
$$(13 \cdot 2) \cdot 5 = 13 \cdot (2 \cdot 5)$$

118. Write *four times five* using three different symbols.

SECTION 2.2   Adding Integers

An amazing change in temperature occurred in 1943 in Spearfish, South Dakota. On January 22, at 7:30 A.M., the temperature was -4 degrees Fahrenheit. Strong warming winds suddenly kicked up and, in just 2 minutes, the temperature rose 49 degrees! To calculate the temperature at 7:32 A.M., we need to add 49 to -4.

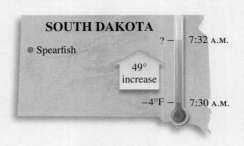

SOUTH DAKOTA
Spearfish
49° increase
? — 7:32 A.M.
$-4°$F — 7:30 A.M.

$$-4 + 49$$

To perform this addition, we must know how to add positive and negative integers. In this section, we develop rules to help us make such calculations.

OBJECTIVES

1 Add two integers that have the same sign.

2 Add two integers that have different signs.

3 Perform several additions to evaluate expressions.

4 Identify opposites (additive inverses) when adding integers.

5 Solve application problems by adding integers.

OBJECTIVE **1** Add two integers that have the same sign.

We can use the number line to explain addition of integers. For example, to find $4 + 3$, we begin at 0 and draw an arrow 4 units long that points to the right. It represents positive 4. From the tip of that arrow, we draw a second arrow, 3 units long, that points to the right. It represents positive 3. Since we end up at 7, it follows that $4 + 3 = 7$.

To check our work, let's think of the problem in terms of money. If you had $4 and earned $3 more, you would have a total of $7.

To find $-4 + (-3)$ on a number line, we begin at 0 and draw an arrow 4 units long that points to the left. It represents -4. From the tip of that arrow, we draw a second arrow, 3 units long, that points to the left. It represents -3. Since we end up at -7, it follows that $-4 + (-3) = -7$.

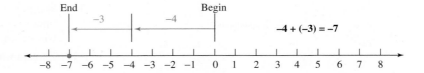

Let's think of this problem in terms of money. If you lost $4 ($-4$) and then lost another $3 ($-3$), overall, you would have lost a total of $7 ($-7$).

Here are some observations about the process of adding two numbers that have the same sign on a number line.

- The arrows representing the integers point in the same direction, and they build upon each other.
- The answer has the same sign as the integers that we added.

These observations illustrate the following rules.

Adding Two Integers That Have the Same (Like) Signs

1. To add two positive integers, add them as usual. The final answer is positive.
2. To add two negative integers, add their absolute values and make the final answer negative.

EXAMPLE 1   Add: **a.** $-3 + (-5)$ **b.** $-26 + (-65)$ **c.** $-456 + (-177)$

Strategy We will use the rule for adding two integers that have the *same sign*.

WHY In each case, we are asked to add two negative integers.

Solution

a. To add two negative integers, we add the absolute values of the integers and make the final answer negative. Since $|-3| = 3$ and $|-5| = 5$, we have

$$-3 + (-5) = -8$$

⤷ Add their absolute values, 3 and 5, to get 8. Then make the final answer negative.

b. Find the absolute values: $|-26| = 26$ and $|-65| = 65$

$$-26 + (-65) = -91$$

⤷ Add their absolute values, 26 and 65, to get 91. Then make the final answer negative.

c. Find the absolute values: $|-456| = 456$ and $|-177| = 177$

$$-456 + (-177) = -633$$

⤷ Add their absolute values, 456 and 177, to get 633. Then make the final answer negative.

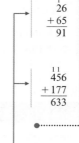

$$\begin{array}{r} 1 \\ 26 \\ +\,65 \\ \hline 91 \end{array}$$

$$\begin{array}{r} 1\,1 \\ 456 \\ +\,177 \\ \hline 633 \end{array}$$

LANGUAGE OF MATHEMATICS

When writing additions that involve integers, write negative integers within parentheses to separate the negative sign − from the plus symbol +.

   $9 + (-4)$ ~~$9 + -4$~~
and
   $-9 + (-4)$ ~~$-9 + -4$~~

Self Check 1

Add:

a. $-7 + (-2)$

b. $-25 + (-48)$

c. $-325 + (-169)$

Now Try ➡ Problems 19, 23, and 27

Success Tip Calculations that you cannot perform in your head should be shown outside the steps of your solution.

OBJECTIVE 2 **Add two integers that have different signs.**

To find $4 + (-3)$ on a number line, we begin at 0 and draw an arrow 4 units long that points to the right. This represents positive 4. From the tip of that arrow, we draw a second arrow, 3 units long, that points to the left. It represents -3. Since we end up at 1, it follows that $4 + (-3) = 1$.

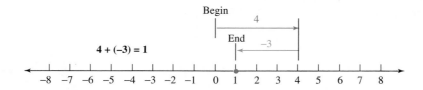

In terms of money, if you won \$4 and then lost \$3 (-3), overall, you would have \$1 left.

To find $-4 + 3$ on a number line, we begin at 0 and draw an arrow 4 units long that points to the left. It represents -4. From the tip of that arrow, we draw a second arrow, 3 units long, that points to the right. It represents positive 3. Since we end up at -1, it follows that $-4 + 3 = -1$.

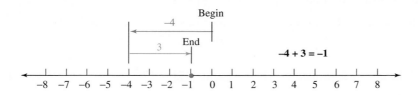

In terms of money, if you lost \$4 (-4) and then won \$3, overall, you have lost \$1 (-1).

Here are some observations about the process of adding two integers that have different signs on a number line.

- The arrows representing the integers point in opposite directions.
- The longer of the two arrows determines the sign of the answer. If the longer arrow represents a positive integer, the sum is positive. If it represents a negative integer, the sum is negative.

These observations suggest the following rules.

Adding Two Integers That Have Different (Unlike) Signs

To add a positive integer and a negative integer, subtract the smaller absolute value from the larger.

1. If the positive integer has the larger absolute value, the final answer is positive.
2. If the negative integer has the larger absolute value, make the final answer negative.

EXAMPLE 2 Add: $5 + (-7)$

Strategy We will use the rule for adding two integers that have different signs.

WHY The addend 5 is positive, and the addend -7 is negative.

Solution
Step 1: To add two integers with different signs, we first subtract the smaller absolute value from the larger absolute value. Since $|5|$, which is 5, is smaller than $|-7|$, which is 7, we begin by subtracting 5 from 7.

$$7 - 5 = 2$$

Step 2: Since the negative number, -7, has the larger absolute value, we attach a negative sign $-$ to the result from step 1. Therefore,

$$5 + (-7) = -2$$

Make the final answer negative.

Self Check 2
Add: $6 + (-9)$

Now Try ➲ Problem 31

EXAMPLE 3 Add: **a.** $8 + (-4)$ **b.** $-41 + 17$ **c.** $-206 + 568$

Strategy We will use the rule for adding two integers that have different signs.

WHY In each case, we are asked to add a positive integer and a negative integer.

Solution
a. Find the absolute values: $|8| = 8$ and $|-4| = 4$

$8 + (-4) = 4$ Subtract the smaller absolute value from the larger: $8 - 4 = 4$. Since the positive number, 8, has the larger absolute value, the final answer is positive.

Caution! Did you notice that the answers to the addition problems in Examples 2 and 3 were found using subtraction? This is the case when the addition involves two integers that have *different signs*.

b. Find the absolute values: $|-41| = 41$ and $|17| = 17$

$-41 + 17 = -24$ Subtract the smaller absolute value from the larger: $41 - 17 = 24$. Since the negative number, -41, has the larger absolute value, make the final answer negative.

$$\begin{array}{r} \overset{3\ 11}{4\rlap{/}1} \\ -\ 17 \\ \hline 24 \end{array}$$

c. Find the absolute values: $|-206| = 206$ and $|568| = 568$

$$-206 + 568 = 362$$

Subtract the smaller absolute value from the larger: $568 - 206 = 362$. Since the positive number, 568, has the larger absolute value, the answer is positive.

$$\begin{array}{r} 568 \\ -206 \\ \hline 362 \end{array}$$

Self Check 3

Add:
a. $7 + (-2)$ **b.** $-53 + 39$
c. $-506 + 888$

Now Try Problems 33, 35, and 39

Think it Through • **CASH FLOW**

"College can be trial by fire—a test of how to cope with pressure, freedom, distractions, and a flood of credit card offers. It's easy to get into a cycle of overspending and unnecessary debt as a student."
— *Planning for College, Wells Fargo Bank*

If your income is less than your expenses, you have a *negative* cash flow. A negative cash flow can be a red flag that you should increase your income and/or reduce your expenses. Which of the following activities can increase income and which can decrease expenses?

- Buy generic or store-brand items.
- Get training and/or more education.
- Use your student ID to get discounts at stores, events, etc.
- Work more hours.
- Turn a hobby or skill into a money-making business.
- Tutor young students.
- Stop expensive habits, like smoking, buying snacks every day, etc.
- Attend free activities and free or discounted days at local attractions.
- Sell rarely used items, like an old CD player.
- Compare the prices of at least three products or at three stores before buying.

Based on the *Building Financial Skills* by National Endowment for Financial Education.

OBJECTIVE 3 **Perform several additions to evaluate expressions.**

To evaluate expressions that contain several additions, we make repeated use of the rules for adding two integers.

EXAMPLE 4 Evaluate: $-3 + 5 + (-12) + 2$

Strategy Since there are no calculations within parentheses, no exponential expressions, and no multiplication or division, we will perform the additions, working from the left to the right.

WHY This is step 4 of the order of operations rule that was introduced in Section 1.9.

Solution

$$\begin{aligned}-3 + 5 + (-12) + 2 &= 2 + (-12) + 2 \\ &= -10 + 2 \\ &= -8\end{aligned}$$

Use the rule for adding two integers that have different signs: $-3 + 5 = 2$.

Use the rule for adding two integers that have different signs: $2 + (-12) = -10$.

Use the rule for adding two integers that have different signs.

Self Check 4

Evaluate:
$-12 + 8 + (-6) + 1$

Now Try Problem 43

The properties of addition that were introduced in Section 1.2, *Adding Whole Numbers*, are also true for integers.

Commutative Property of Addition

The order in which integers are added does not change their sum.

Associative Property of Addition

The way in which integers are grouped does not change their sum.

Another way to evaluate an expression like that in Example 4 is to use these properties to reorder and regroup the integers in a helpful way.

EXAMPLE 5   Use the commutative and/or associative properties of addition to help evaluate the expression: $-3 + 5 + (-12) + 2$

Strategy We will use the commutative and/or associative properties of addition so that we can add the positives and add the negatives separately. Then we will add those results to obtain the final answer.

WHY It is easier to add integers that have the same sign than integers that have different signs. This approach lessens the possibility of an error, because we only have to add integers that have different signs once.

Solution

$$-3 + 5 + (-12) + 2$$

$$= -3 + (-12) + 5 + 2 \qquad \text{Use the commutative property of addition to reorder the integers.}$$

$$\underset{\text{Negatives}}{= [-3 + (-12)]} + \underset{\text{Positives}}{(5 + 2)} \qquad \text{Use the associative property of addition to group the negatives and group the positives.}$$

$$= -15 + 7 \qquad \text{Use the rule for adding two integers that have the same sign twice. Add the negatives within the brackets. Add the positives within the parentheses.}$$

$$= -8 \qquad \text{Use the rule for adding two integers that have different signs. This is the same result as in Example 4.}$$

Self Check 5

Use the commutative and/or associative properties of addition to help evaluate the expression:

$-12 + 8 + (-6) + 1$

Now Try ⟹ **Problem 45**

EXAMPLE 6   Evaluate: $[-21 + (-5)] + (-17 + 6)$

Strategy We will perform the addition within the brackets and the addition within the parentheses first. Then we will add those results.

WHY By the order of operations rule, we must perform the calculations within the grouping symbols first.

Solution Use the rule for adding two integers that have the same sign to do the addition within the brackets and the rule for adding two integers that have different signs to do the addition within the parentheses.

$$[-21 + (-5)] + (-17 + 6) = -26 + (-11) \qquad \text{Add within each pair of grouping symbols.}$$

$$= -37 \qquad \text{Use the rule for adding two integers that have the same sign.}$$

Self Check 6

Evaluate:

$(-6 + 8) + [10 + (-17)]$

Now Try ⟹ **Problem 47**

OBJECTIVE 4 **Identify opposites (additive inverses) when adding integers.**

Recall from Section 1.2 that when 0 is added to a whole number, the whole number remains the same. This is true for any number. For example, $-5 + 0 = -5$ and $0 + (-43) = -43$. Because of this, we call 0 the **additive identity**.

LANGUAGE OF MATHEMATICS

Identity is a form of the word *identical*, meaning the same. You have probably seen *identical* twins.

Addition Property of 0

The sum of any integer and 0 is that integer.
  For example,

$$-3 + 0 = -3, \qquad -19 + 0 = -19, \quad \text{and} \quad 0 + (-76) = -76$$

There is another important fact about the operation of addition and 0. To illustrate it, we use the number line below to add 6 and its opposite, -6. Notice that $6 + (-6) = 0$.

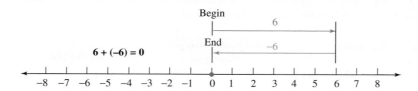

If the sum of two numbers is 0, the numbers are said to be **additive inverses** of each other. Since $6 + (-6) = 0$, we say that 6 and -6 are additive inverses. Likewise, -7 is the additive inverse of 7, and 51 is the additive inverse of -51.

We can now classify a pair of integers such as 6 and -6 in two ways: as opposites or as additive inverses.

Addition Property of Opposites

The sum of a number and its opposite (additive inverse) is 0.
  For example,

$$4 + (-4) = 0, \qquad -53 + 53 = 0, \quad \text{and} \quad 710 + (-710) = 0$$

At certain times, the addition property of opposites can be used to make addition of several integers easier.

EXAMPLE 7   Evaluate: $12 + (-5) + 6 + 5 + (-12)$

Strategy Instead of working from left to right, we will use the commutative and associative properties of addition to add *pairs of opposites*.

 WHY Since the sum of an integer and its opposite is 0, it is helpful to identify such pairs in an addition.

Solution

$$\underbrace{12 + (-5) + 6 + 5 + (-12)}_{\text{opposites}} = 0 + 0 + 6 \qquad \text{Locate pairs of opposites and add them to get 0.}$$

$$= 6 \qquad \text{The sum of any integer and 0 is that integer.}$$

opposites

Self Check 7

Evaluate:
$8 + (-1) + 6 + (-8) + 1$

Now Try ➭ **Problem 51**

OBJECTIVE 5 **Solve application problems by adding integers.**

Since application problems are almost always written in words, the ability to understand what you read is very important. Recall from Chapter 1 that words and phrases such as *gained*, *increased by*, and *rise* indicate addition.

EXAMPLE 8   **Record temperature change.**   At the beginning of this section, we learned that at 7:30 A.M. on January 22, 1943, in Spearfish, South Dakota, the temperature was −4°F. The temperature then rose 49 degrees in just 2 minutes. What was the temperature at 7:32 A.M.?

Strategy We will carefully read the problem looking for a key word or phrase.

 WHY   Key words and phrases indicate what arithmetic operations should be used to solve the problem.

Solution The phrase *rose 49 degrees* indicates addition. With that in mind, we translate the words of the problem to numbers and symbols.

The temperature at 7:32 A.M.	was	the temperature at 7:30 A.M.	plus	49 degrees.
The temperature at 7:32 A.M.	=	−4	+	49

To find the sum, we will use the rule for adding two integers that have different signs. First, we find the absolute values: $|-4| = 4$ and $|49| = 49$.

$$-4 + 49 = 45$$ Subtract the smaller absolute value from the larger absolute value: 49 − 4 = 45. Since the positive number, 49, has the larger absolute value, the final answer is positive.

At 7:32 A.M., the temperature was 45°F.

Self Check 8

Temperature change. On the morning of February 21, 1918, in Granville, North Dakota, the morning low temperature was −33°F. By the afternoon, the temperature had risen a record 83 degrees. What was the afternoon high temperature in Granville? (Source: *Extreme Weather* by Christopher C. Burt)

Now Try ⟳ **Problem 87**

Using Your Calculator ▶   Entering Negative Numbers

Canada is the largest U.S. trading partner. To calculate the 2015 U.S. trade balance with Canada, we add the $281 billion worth of U.S. exports to Canada (considered positive) to the $296 billion worth of U.S. imports from Canada (considered negative). We can use a calculator to perform the addition: 281 + (−296).

We do not have to do anything special to enter a positive number. Negative numbers are entered using either **direct** or **reverse** entry, depending on the type of calculator you have.

To enter −296 using reverse entry, press the change-of-sign key $\boxed{+/-}$ *after entering* 296. To enter −296 using direct entry, press the negative key $\boxed{(-)}$ *before* entering 296. In either case, note that $\boxed{+/-}$ and the $\boxed{(-)}$ keys are different from the subtraction key $\boxed{-}$.

Reverse entry: 281 $\boxed{+}$ 296 $\boxed{+/-}$ $\boxed{=}$

Direct entry: 281 $\boxed{+}$ $\boxed{(-)}$ 296 $\boxed{\text{ENTER}}$ $\boxed{\qquad -15}$

In 2015, the United States had a trade balance of −$15 billion with Canada. Because the result is negative, it is called a trade deficit.

SECTION 2.2 STUDY SET

VOCABULARY

Fill in the blanks.

1. Two negative integers, as well as two positive integers, are said to have the same or _____ signs.

2. A positive integer and a negative integer are said to have different or _____ signs.

3. When 0 is added to a number, the number remains the same. We call 0 the additive _____.

4. Since −5 + 5 = 0, we say that 5 is the additive _____ of −5. We can also say that 5 and −5 are _____.

5. _____ property of addition: The order in which integers are added does not change their sum.

6. _____ property of addition: The way in which integers are grouped does not change their sum.

CONCEPTS

7. **a.** What is the absolute value of 10? What is the absolute value of −12?

   b. Which number has the larger absolute value, 10 or −12?

   c. Using your answers to part a, subtract the smaller absolute value from the larger absolute value. What is the result?

8. **a.** If you lost $6 and then lost $8, overall, what amount of money was lost?

   b. If you lost $6 and then won $8, overall, what amount of money have you won?

Fill in the blanks.

9. To add two integers with unlike signs, _____ their absolute values, the smaller from the larger. Then attach to that result the sign of the number with the _____ absolute value.

10. To add two integers with like signs, add their _____ values and attach their common _____ to the sum.

11. **a.** Is the sum of two positive integers always positive?

    b. Is the sum of two negative integers always negative?

    c. Is the sum of a positive integer and a negative integer always positive?

    d. Is the sum of a positive integer and a negative integer always negative?

12. Complete the table by finding the additive inverse, opposite, and absolute value of the given numbers.

Number	Additive inverse	Opposite	Absolute value
19			
−2			
0			

13. **a.** What is the sum of an integer and its additive inverse?

    b. What is the sum of an integer and its opposite?

14. **a.** What number must be added to −5 to obtain 0?

    b. What number must be added to 8 to obtain 0?

NOTATION

Complete each step to evaluate the expression.

15. $-16 + (-2) + (-1) = \boxed{} + (-1)$

    $= \boxed{}$

16. $-8 + (-2) + 6 = \boxed{} + 6$

    $= \boxed{}$

17. $(-3 + 8) + (-3) = \boxed{} + (-3)$

    $= \boxed{}$

18. $-5 + [2 + (-9)] = -5 + (\boxed{})$

    $= \boxed{}$

GUIDED PRACTICE

Add. See Example 1.

19. $-6 + (-3)$ 20. $-2 + (-3)$

21. $-5 + (-5)$ 22. $-8 + (-8)$

23. $-51 + (-11)$ 24. $-43 + (-12)$

25. $-69 + (-27)$ 26. $-55 + (-36)$

27. $-248 + (-131)$ 28. $-423 + (-164)$

29. $-565 + (-309)$ 30. $-709 + (-187)$

Add. See Examples 2 and 3.

31. $-8 + 5$ 32. $-9 + 3$

33. $7 + (-6)$ 34. $4 + (-2)$

35. $20 + (-42)$ 36. $-18 + 10$

37. $71 + (-23)$ **38.** $75 + (-56)$

39. $479 + (-122)$ **40.** $589 + (-242)$

41. $-339 + 279$ **42.** $-704 + 649$

Evaluate each expression. See Examples 4 and 5.

43. $9 + (-3) + 5 + (-4)$

44. $-3 + 7 + (-4) + 1$

45. $6 + (-4) + (-13) + 7$

46. $59 + (-24) + (-31) + (-7)$

Evaluate each expression. See Example 6.

47. $[-3 + (-4)] + (-5 + 2)$

48. $[9 + (-10)] + (-7 + 9)$

49. $(-1 + 34) + [16 + (-8)]$

50. $(-32 + 13) + [5 + (-14)]$

Evaluate each expression. See Example 7.

51. $23 + (-5) + 3 + 5 + (-23)$

52. $41 + (-1) + 9 + 1 + (-41)$

53. $-10 + (-1) + 10 + (-6) + 1$

54. $-14 + (-30) + 14 + (-9) + 9$

TRY IT YOURSELF

Add.

55. $-2 + 6 + (-1)$ **56.** $4 + (-3) + (-2)$

57. $-7 + 0$ **58.** $0 + (-15)$

59. $24 + (-15)$ **60.** $-4 + 14$

61. $-435 + (-127)$ **62.** $-346 + (-273)$

63. $-7 + 9$ **64.** $-3 + 6$

65. $2 + (-2)$ **66.** $-10 + 10$

67. $-5 + (-9 + 8) + (-2)$

68. $(-9 + 12) + (-4)$

69. $-9 + 1 + (-2) + (-1) + 9$

70. $5 + 4 + (-6) + (-4) + (-5)$

71. $[6 + (-4)] + [8 + (-11)]$

72. $[5 + (-8)] + [9 + (-15)]$

73. $(-4 + 8) + (-11 + 4)$

74. $(-12 + 6) + (-6 + 8)$

75. $-675 + (-456) + 99$

76. $-9,750 + (-780) + 2,345$

77. Find the sum of -6, -7, and -8.

78. Find the sum of -11, -12, and -13.

79. $-2 + [789 + (-9,135)]$

80. $-8 + [2,701 + (-4,089)]$

81. What is 25 more than -45?

82. What is 31 more than -65?

LOOK ALIKES

Add.

83. a. $91 + (-7)$ **b.** $-91 + (-7)$

    c. $-91 + 7$ **d.** $91 + 7$

84. a. $73 + (-8)$ **b.** $-73 + (-8)$

    c. $-73 + 8$ **d.** $73 + 8$

85. a. $-232 + (-141) + 57$

    b. $232 + (-141) + 57$

86. a. What is 63 more than -88?

    b. What is 88 more than -63?

APPLICATIONS

Use signed numbers to solve each problem.

87. Record temperatures.   The lowest recorded temperatures for Michigan and Minnesota are shown below. Use the given information to find the highest recorded temperature for each state.

State	Lowest temperature	Highest temperature
Michigan	Feb. 9, 1934: $-51°F$	July 13, 1936: 163°F warmer than the record low
Minnesota	Feb. 2, 1996: $-60°F$	July 29, 1917: 175°F warmer than the record low

Source: *The World Almanac and Book of Facts*, 2017

88. Elevations.   The lowest point in the United States is Death Valley, California, with an elevation of -282 feet (282 feet below sea level). Mt. McKinley (Alaska) is the highest point in the United States. Its elevation is 20,602 feet higher than Death Valley. What is the elevation of Mt. McKinley? (Source: *The World Almanac and Book of Facts*, 2013)

89. Sunken ships.   Refer to the map below.

a. The German battleship *Bismarck*, one of the most feared warships of World War II, was sunk by the British in 1941. It lies on the ocean floor 15,720 feet below sea level off the west coast of France. Represent that depth using a signed number.

b. In 1912, the luxury liner *Titanic* sank after striking an iceberg. It lies on the North Atlantic ocean floor, 3,220 feet higher than the *Bismarck*. At what depth is the *Titanic* resting?

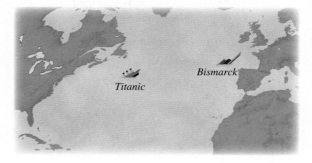

90. **Jogging.** A businessman's lunchtime workout includes jogging up ten stories of stairs in his high-rise office building. He starts the workout on the fourth level below ground in the underground parking garage.

    a. Represent that level using a signed number.

    b. On what story of the building will he finish his workout?

91. **Flooding.** After a heavy rainstorm, a river that had been 9 feet under flood stage rose 11 feet in a 48-hour period.

    a. Represent the level of the river before the storm using a signed number.

    b. Find the height of the river after the storm in comparison to flood stage.

92. **Atoms.** An atom is composed of protons, neutrons, and electrons. A proton has a positive charge (represented by 1), a neutron has no charge, and an electron has a negative charge (-1). Two simple models of atoms are shown below.

    a. How many protons does the atom in figure (a) have? How many electrons?

    b. What is the net charge of the atom in figure (a)?

    c. How many protons does the atom in figure (b) have? How many electrons?

    d. What is the net charge of the atom in figure (b)?

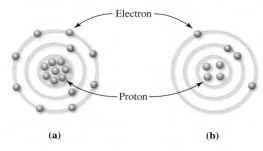

| (a) | (b) |

93. **Chemistry.** The three steps of a chemistry lab experiment are listed here. The experiment begins with a compound that is stored at $-40°F$.

    Step 1 Raise the temperature of the compound $200°$.

    Step 2 Add sulfur and then raise the temperature $10°$.

    Step 3 Add 10 milliliters of water, stir, and raise the temperature $25°$.

    What is the resulting temperature of the mixture after step 3?

---

from Campus to Careers

Personal Financial Adviser

94. Suppose you are a personal financial adviser and your clients are considering purchasing income property. You find a duplex apartment unit that is for sale and learn  that the maintenance costs, utilities, and taxes on it total $900 per month. If the current owner receives monthly rental payments of $450 and $380 from the tenants, does the duplex produce a positive cash flow each month?

---

95. **Health.** Find the point total for the six risk factors (shown with blue headings) on the medical questionnaire below. Then use the table at the bottom of the form (under the red heading) to determine the risk of contracting heart disease for the man whose responses are shown.

Age		Total Cholesterol	
Age	Points	Reading	Points
35	−4	280	3

Cholesterol		Blood Pressure	
HDL	Points	Systolic/Diastolic	Points
62	−3	124/100	3

Diabetic		Smoker	
	Points		Points
Yes	4	Yes	2

10-Year Heart Disease Risk			
Total Points	**Risk**	Total Points	**Risk**
−2 or less	1%	5	4%
−1 to 1	2%	6	6%
2 to 3	3%	7	6%
4	4%	8	7%

Source: National Heart, Lung, and Blood Institute

96. **Political polls.** Six months before a general election, the incumbent senator found himself trailing the challenger by 18 points. To overtake his opponent, the campaign staff decided to use a four-part strategy. Each part of this plan is shown below, with the anticipated point gain.

    Part 1 Intense TV ad blitz: gain 10 points

    Part 2 Ask for union endorsement: gain 2 points

    Part 3 Voter mailing: gain 3 points

    Part 4 Get-out-the-vote campaign: gain 1 point

    With these gains, will the incumbent overtake the challenger on election day?

97. **Military science.** During a battle, an army retreated 1,500 meters, regrouped, and advanced 3,500 meters. The next day, it advanced 1,250 meters. Find the army's net gain.

98. **Department stores.** The graph on the next page shows the annual net income for Sears during the years 2007–2016.

    a. Estimate the company's total net income over this span of ten years in millions of dollars. Use front-end rounding.

    b. About how many billions of dollars is your answer to part a?

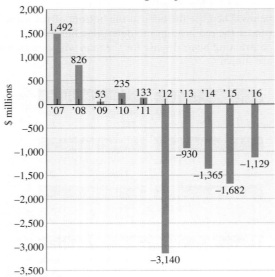

Sears Holdings Corp Net Income

Source: financials.morningstar.com

99. Accounting. On a financial balance sheet, debts (considered negative numbers) are written within parentheses. Assets (considered positive numbers) are written without parentheses. What is the 2017 fund balance for the preschool whose financial records are shown below?

ABC Preschool Balance Sheet, June 2017

Fund	Balance $
Classroom supplies	$5,889
Emergency needs	$927
Holiday program	($2,928)
Insurance	$1,645
Janitorial	($894)
Licensing	$715
Maintenance	($6,321)
BALANCE	?

100. Spreadsheets. Monthly rain totals for four counties are listed in the spreadsheet below. The −1 entered in cell B1 means that the rain total for Suffolk County for a certain month was 1 inch *below* average. We can analyze these data by asking the computer to perform various operations.

Book 1

	File	Edit	View	Insert	Format	Tools	Data	Window	Help

	A	B	C	D	E	F
1	Suffolk	−1	−1	0	+1	+1
2	Marin	0	−2	+1	+1	−1
3	Logan	−1	+1	+2	+1	+1
4	Tipton	−2	−2	+1	−1	−3
5						

a. To ask the computer to add the numbers in cells B1, B2, B3, and B4, we type SUM(B1:B4). Find this sum.

b. Find SUM(F1:F4).

101. U.S. federal budget. The table below shows the receipts (money received) and the outlays (money spent) of the U.S. Government for the fiscal year 2016 by month. (Source: Department of the Treasury)

a. For what months did the government receipts exceed the government outlays?

b. Was there a government surplus or deficit for the fiscal year? How much was it?

c. Write the number in your answer to part b in words.

Fiscal Year 2016		
Month	**Receipts ($ millions)**	**Outlays ($ millions)**
Oct.	211,046	347,604
Nov.	204,968	269,517
Dec.	349,631	364,075
Jan.	313,579	258,416
Feb.	169,147	361,757
March	227,848	335,891
April	438,432	331,977
May	224,604	277,111
June	329,572	323,320
July	209,998	322,817
Aug.	231,327	338,438
Sept.	356,537	323,178
Total	3,266,688	3,854,100

102. U.S. federal budget. The graph below shows the receipts (money received) and the outlays (money spent) of the U.S. Government for the fiscal year 2015.

a. What number should be used to label the deficit bar in the graph?

b. What dollar amount is represented by the deficit bar in the graph?

c. Write your answer to part b in words.

Fiscal Year 2015

Source: Department of the Treasury

WRITING

103. Is the sum of a positive and a negative number always positive? Explain why or why not.

104. How do you explain the fact that when asked to *add* −4 and 8, we must actually *subtract* to obtain the result?

105. Explain why the sum of two negative numbers is a negative number.

106. Write an application problem that will require adding −50 and −60.

107. If the sum of two integers is 0, what can be said about the integers? Give an example.

108. Explain why the expression −6 + −5 is not written correctly. How should it be written?

109. Explain the commutative property of addition in your own words. Give an example using a positive integer and a negative integer.

110. Explain the associative property of addition in your own words. Give an example using two positive integers and a negative integer.

REVIEW

111. **a.** Find the perimeter of the rectangle shown below.

   b. Find the area of the rectangle shown below.

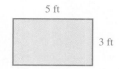

112. What property is illustrated by the statement $5 \cdot 15 = 15 \cdot 5$?

113. Prime factor 250. Use exponents to express the result.

114. Divide: $\dfrac{144}{12}$

SECTION 2.3 Subtracting Integers

In this section, we will discuss a rule that is helpful when subtracting signed numbers.

OBJECTIVES

1 Use the subtraction rule.

2 Evaluate expressions involving subtraction and addition.

3 Solve application problems by subtracting integers.

OBJECTIVE 1 Use the subtraction rule.

The subtraction problem $6 - 4$ can be thought of as taking away 4 from 6. We can use a number line to illustrate this. Beginning at 0, we draw an arrow of length 6 units long that points to the right. It represents positive 6. From the tip of that arrow, we draw a second arrow, 4 units long, that points to the left. It represents taking away 4. Since we end up at 2, it follows that $6 - 4 = 2$.

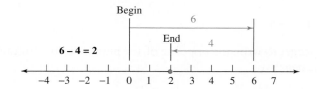

Note that the illustration above also represents the *addition* $6 + (-4) = 2$. We see that

This observation suggests the following rule.

> **Rule for Subtraction**
> To subtract two numbers, add the first number to the opposite (additive inverse) of the number to be subtracted.

Put more simply, this rule says that *subtraction is the same as adding the opposite*.

After rewriting a subtraction as addition of the opposite, we then use one of the rules for the addition of signed numbers discussed in Section 2.2 to find the result.

You won't need to use this rule for every subtraction problem. For example, $6 - 4$ is obviously 2; it does not need to be rewritten as adding the opposite. But for more complicated problems such as $-6 - 4$ or $3 - (-5)$, where the result is not obvious, the subtraction rule will be quite helpful.

EXAMPLE 1   Subtract and check the result:

a. $-6 - 4$ **b.** $3 - (-5)$ **c.** $7 - 23$

Strategy To find each difference, we will apply the rule for subtraction: Add the first integer to the opposite of the integer to be subtracted.

WHY It is easy to make an error when subtracting signed numbers. We will probably be more accurate if we write each subtraction as addition of the opposite.

Solution

a. We read $-6 - 4$ as "negative six *minus* four." Thus, the number to be subtracted is 4. Subtracting 4 is the same as adding its opposite, -4.

Change the subtraction to addition.

$$-6 - 4 \quad = \quad -6 + (-4) = -10 \qquad \text{Use the rule for adding two integers with the same sign.}$$

Change the number being subtracted to its opposite.

To check, we add the *difference*, -10, and the *subtrahend*, 4. We should get the *minuend*, -6.

Check: $-10 + 4 = -6$ The result checks.

Caution! Don't forget to write the opposite of the number to be subtracted within parentheses if it is negative.

$$-6 - 4 = -6 + (-4)$$

b. We read $3 - (-5)$ as "three *minus* negative five." Thus, the number to be subtracted is -5. Subtracting -5 is the same as adding its opposite, 5.

Add ...

$$3 - (-5) \quad = \quad 3 + 5 = 8$$

... the opposite

Check: $8 + (-5) = 3$ The result checks.

c. We read $7 - 23$ as "seven *minus* twenty-three." Thus, the number to be subtracted is 23. Subtracting 23 is the same as adding its opposite, -23.

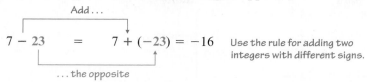

Add...

$$7 - 23 \quad = \quad 7 + (-23) = -16$$

...the opposite

Use the rule for adding two integers with different signs.

Check: $-16 + 23 = 7$ The result checks.

Self Check 1

Subtract and check the result:
a. $-2 - 3$
b. $4 - (-8)$
c. $6 - 85$

Now Try ⟹ Problems 21, 25, and 29

Caution! When applying the subtraction rule, *do not change* the first number.

$$-6 - 4 = -6 + (-4) \qquad 3 - (-5) = 3 + 5$$

EXAMPLE 2   **a.** Subtract -12 from -8.   **b.** Subtract -8 from -12.

Strategy We will translate each phrase to mathematical symbols and then perform the subtraction. We must be careful when translating the instruction to subtract one number *from* another number.

WHY The order of the numbers in each word phrase must be reversed when we translate it to mathematical symbols.

Solution

a. Since -12 is the number to be subtracted, we reverse the order in which -12 and -8 appear in the sentence when translating to symbols.

Subtract -12 from -8

$$-8 - (-12) \quad \text{Write } -12 \text{ within parentheses.}$$

To find this difference, we write the subtraction as addition of the opposite:

Add...

$$-8 - (-12) = -8 + 12 = 4 \quad \begin{array}{l}\text{Use the rule for adding two} \\ \text{integers with different signs.}\end{array}$$

...the opposite

b. Since -8 is the number to be subtracted, we reverse the order in which -8 and -12 appear in the sentence when translating to symbols.

Subtract -8 from -12

$$-12 - (-8) \quad \text{Write } -8 \text{ within parentheses.}$$

To find this difference, we write the subtraction as addition of the opposite:

Add...

$$-12 - (-8) = -12 + 8 = -4 \quad \begin{array}{l}\text{Use the rule for adding two} \\ \text{integers with different signs.}\end{array}$$

...the opposite

Self Check 2

a. Subtract -10 from -7.
b. Subtract -7 from -10.

Now Try ⟹ Problem 33

LANGUAGE OF MATHEMATICS

When we change a number to its opposite, we say we have **changed** (or **reversed**) its sign.

Remember that any subtraction problem can be rewritten as an equivalent addition. We just add the opposite of the number that is to be subtracted. Here are four examples:

- $4 - 8 = 4 + (-8) = -4$
- $4 - (-8) = 4 + 8 = 12$
- $-4 - 8 = -4 + (-8) = -12$
- $-4 - (-8) = -4 + 8 = 4$

Any subtraction can be written as addition of the opposite of the number to be subtracted.

OBJECTIVE 2 Evaluate expressions involving subtraction and addition.

Expressions can involve repeated subtraction or combinations of subtraction and addition. To evaluate them, we use the order of operations rule discussed in Section 1.9.

EXAMPLE 3 Evaluate: $-1 - (-2) - 10$

Strategy This expression involves two subtractions. We will write each subtraction as addition of the opposite and then evaluate the expression using the order of operations rule.

WHY It is easy to make an error when subtracting signed numbers. We will probably be more accurate if we write each subtraction as addition of the opposite.

Solution We apply the rule for subtraction twice and then perform the additions, working from left to right. (We could also add the positives and the negatives separately, and then add those results.)

$$-1 - (-2) - 10 = -1 + 2 + (-10)$$ Add the opposite of -2, which is 2. Add the opposite of 10, which is -10.

$$= 1 + (-10)$$ Work from left to right. Add $-1 + 2$ using the rule for adding integers that have different signs.

$$= -9$$ Use the rule for adding integers that have different signs.

Self Check 3

Evaluate: $-3 - 5 - (-1)$

Now Try Problem 37

EXAMPLE 4 Evaluate: $-80 - (-2 - 24)$

Strategy We will consider the subtraction within the parentheses first and rewrite it as addition of the opposite.

WHY By the order of operations rule, we must perform all calculations within parentheses first.

Solution

$$-80 - (-2 - 24) = -80 - [-2 + (-24)]$$ Add the opposite of 24, which is -24. Since -24 must be written within parentheses, we write $-2 + (-24)$ within brackets.

$$= -80 - (-26)$$ Within the brackets, add -2 and -24. Since only one set of grouping symbols is now needed, we can write the answer, -26, within parentheses.

$$\begin{array}{r} {\scriptstyle 7\,10} \\ 8\!\!\!/0 \\ -26 \\ \hline 54 \end{array}$$

$$= -80 + 26$$ Add the opposite of -26, which is 26.

$$= -54$$ Use the rule for adding integers that have different signs.

Self Check 4

Evaluate: $-72 - (-6 - 51)$

Now Try Problem 49

EXAMPLE 5   Evaluate: $-(-6) + (-18) - 4 - (-51)$

Strategy This expression involves one addition and two subtractions. We will write each subtraction as addition of the opposite and then evaluate the expression.

WHY It is easy to make an error when subtracting signed numbers. We will probably be more accurate if we write each subtraction as addition of the opposite.

Solution We apply the rule for subtraction twice. Then we will add the positives and the negatives separately, and add those results. (By the commutative and associative properties of addition, we can add the integers in any order.)

$$-(-6) + (-18) - 4 - (-51)$$
$$= 6 + (-18) + (-4) + 51$$
Simplify: $-(-6) = 6$. Add the opposite of 4, which is -4, and add the opposite of -51, which is 51.

$$= (6 + 51) + [(-18) + (-4)]$$
Reorder the integers. Then group the positives together and group the negatives together.

$$= 57 + (-22)$$
Add the positives and add the negatives.

$$= 35$$
Use the rule for adding integers that have different signs.

Self Check 5
Evaluate:
$-(-3) + (-16) - 9 - (-28)$

Now Try ➡ Problem 55

OBJECTIVE 3 **Solve application problems by subtracting integers.**

Subtraction finds the *difference* between two numbers. When we find the difference between the maximum value and the minimum value of a collection of measurements, we are finding the **range** of the values.

> Range = maximum value − minimum value

EXAMPLE 6   **The windy city.** The record high temperature for Chicago, Illinois, is 104°F. The record low is −27°F. Find the temperature range for these extremes. (Source: weather.com)

Chicago •
ILLINOIS
Springfield •

Strategy We will subtract the minimum temperature (−27°F) from the maximum temperature (104°F).

WHY The *range* of a collection of data indicates the spread of the data. It is the difference between the maximum and minimum values.

Solution We apply the rule for subtraction and add the opposite of −27.

$$104 - (-27) = 104 + 27$$ 104° is the highest temperature and −27° is the lowest.

$$= 131$$

The temperature range for these extremes is 131°F.

Self Check 6

The gateway city. The record high temperature for St. Louis, Missouri, is 125°F. The record low temperature is −18°F. Find the temperature range for these extremes. (Source: weather.com)

Now Try ➡ Problem 105

Things are constantly changing in our daily lives. The amount of money we have in the bank, the price of gasoline, and our ages are examples. In mathematics, the operation of subtraction is used to measure change. To find the **change** in a quantity, we subtract the earlier value from the later value.

> Change = later value − earlier value

The five-step problem-solving strategy introduced in Section 1.6 can be used to solve more complicated application problems.

EXAMPLE 7	Water management.

On Monday, the water level in a city storage tank was 16 feet above normal. By Friday, the level had fallen to a mark 14 feet below normal. Find the change in the water level from Monday to Friday.

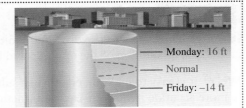

Monday: 16 ft
Normal
Friday: −14 ft

Analyze It is helpful to list the given facts and what you are to find.

- On Monday, the water level was 16 feet above normal. *Given*
- On Friday, the water level was 14 feet below normal. *Given*
- Find the change in the water level. *Find*

Form To find the change in the water level, we *subtract the earlier value from the later value.* The water levels of 16 feet above normal (the earlier value) and 14 feet below normal (the later value) can be represented by 16 and −14.

We translate the verbal model of the problem to numbers and symbols.

The change in the water level	is equal to	the later water level (Friday)	minus	the earlier water level (Monday).
The change in the water level	=	−14	−	16

Self Check 7

Crude oil. On Wednesday, the level of crude oil in a storage tank was 5 feet above standard capacity. Thursday, after a large refining session, the level fell to a mark 76 feet below standard capacity. Find the change in the crude oil level from Wednesday to Thursday.

Now Try ➲ Problem 107

Calculate We can use the rule for subtraction to find the difference.

$$-14 - 16 = -14 + (-16)$$ *Add the opposite of 16, which is −16.*
$$= -30$$ *Use the rule for adding integers with the same sign.*

State The negative result means the water level *fell* 30 feet from Monday to Friday.

Check If we represent the change in water level on a horizontal number line, we see that the water level fell 16 + 14 = 30 units. The result checks.

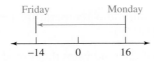

Friday   Monday
−14   0   16

Using Your Calculator ▶   Subtraction With Negative Numbers

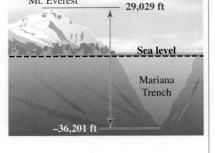

Mt. Everest ——— 29,029 ft

Sea level

Mariana Trench

−36,201 ft

The world's highest peak is Mount Everest in the Himalayas. The greatest ocean depth yet measured lies in the Mariana Trench near the island of Guam in the western Pacific. To find the range between the highest peak and the greatest depth, we must subtract:

$$29,029 - (-36,201)$$

To perform this subtraction on a calculator, we enter the following:

Reverse entry: 29029 $\boxed{-}$ 36201 $\boxed{+/-}$ $\boxed{=}$

Direct entry: 29029 $\boxed{-}$ $\boxed{(-)}$ 36201 $\boxed{\text{ENTER}}$

$\boxed{\text{65230}}$

The range is 65,230 feet between the highest peak and the lowest depth. (We could also write 29,029 − (−36,201) as 29,029 + 36,201 and then use the addition key $\boxed{+}$ to find the answer.)

SECTION 2.3   STUDY SET

VOCABULARY

Fill in the blanks.

1. -8 is the _____ of 8.
2. When we change a number to its opposite, we say we have *changed* (or *reversed*) its _____.
3. To evaluate an expression means to find its _____.
4. The difference between the maximum and the minimum value of a collection of measurements is called the _____ of the values.

CONCEPTS

Fill in the blanks.

5. To subtract two integers, add the first integer to the _____ of the integer to be subtracted.
6. Subtracting is the same as _____ the opposite.
7. Subtracting 3 is the same as adding ___.
8. Subtracting -6 is the same as adding ___.
9. We can find the _____ in a quantity by subtracting the earlier value from the later value.
10. After rewriting a subtraction as addition of the opposite, we then use one of the rules for the _____ of signed numbers discussed in the previous section to find the result.
11. In each case, determine what number is being subtracted.
    a. $-7 - 3$ **b.** $1 - (-12)$
12. Fill in the blanks to rewrite each subtraction as addition of the opposite of the number being subtracted.
    a. $2 - 7 = 2 + $ ▨
    b. $2 - (-7) = 2 + $ ▨
    c. $-2 - 7 = -2 + $ ▨
    d. $-2 - (-7) = -2 + $ ▨
13. Apply the rule for subtraction and fill in the three blanks.

    $3 - (-6) = 3$ ▨ ▨ $= $ ▨

14. Use addition to check this subtraction: $14 - (-2) = 12$. Is the result correct?

NOTATION

15. Write each phrase using symbols.
    a. negative eight minus negative four
    b. negative eight subtracted from negative four

16. Write each phrase in words.
    a. $7 - (-2)$
    b. $-2 - (-7)$

Complete each step to evaluate each expression.

17. $1 - 3 - (-2) = 1 + ($ ▨ $) + 2$
    $= -2 + $ ▨
    $= $ ▨

18. $-6 + 5 - (-5) = -6 + 5 + $ ▨
    $= $ ▨ $+ 5$
    $= $ ▨

19. $(-8 - 2) - (-6) = [-8 + ($ ▨ $)] - (-6)$
    $= $ ▨ $- (-6)$
    $= -10 + $ ▨
    $= $ ▨

20. $-(-5) - (-1 - 4) = $ ▨ $- [-1 + ($ ▨ $)]$
    $= 5 - ($ ▨ $)$
    $= 5 + $ ▨
    $= $ ▨

GUIDED PRACTICE

Subtract. **See Example 1.**

21. $-4 - 3$	**22.** $-4 - 1$
23. $-5 - 5$	**24.** $-7 - 7$
25. $8 - (-1)$	**26.** $3 - (-8)$
27. $11 - (-7)$	**28.** $18 - (-25)$
29. $3 - 21$	**30.** $8 - 32$
31. $15 - 65$	**32.** $12 - 82$

LOOK ALIKES

Perform the indicated operation. **See Example 2.**

33. **a.** Subtract -1 from -11.
    b. Subtract -11 from -1.
34. **a.** Subtract -2 from -19.
    b. Subtract -19 from -2.
35. **a.** Subtract -41 from -16.
    b. Subtract -16 from -41.
36. **a.** Subtract -57 from -15.
    b. Subtract -15 from -57.

Evaluate each expression. **See Example 3.**

37. $-4 - (-4) - 15$ **38.** $-3 - (-3) - 10$

39. $10 - 9 - (-8)$ **40.** $16 - 14 - (-9)$

41. $-1 - (-3) - 4$ **42.** $-2 - 4 - (-1)$

43. $-5 - 8 - (-3)$ **44.** $-6 - 5 - (-1)$

Evaluate each expression. **See Example 4.**

45. $-1 - (-4 - 6)$ **46.** $-7 - (-2 - 14)$

47. $-42 - (-16 - 14)$ **48.** $-45 - (-8 - 32)$

49. $-9 - (6 - 7)$ **50.** $-13 - (6 - 12)$

51. $-8 - (4 - 12)$ **52.** $-9 - (1 - 10)$

Evaluate each expression. **See Example 5.**

53. $-(-5) + (-15) - 6 - (-48)$

54. $-(-2) + (-30) - 3 - (-66)$

55. $-(-3) + (-41) - 7 - (-19)$

56. $-(-1) + (-52) - 4 - (-21)$

57. $-1{,}557 - 890$ **58.** $20{,}007 - (-496)$

59. $-979 - (-44{,}879)$ **60.** $-787 - 1{,}654 - (-232)$

TRY IT YOURSELF

Evaluate each expression.

61. $-5 + 7 - (-1) - 8$ **62.** $-5 + 9 - (-4) - 16$

63. Subtract -3 from 7.   **64.** Subtract 8 from -2.

65. $-2 - (-10)$ **66.** $-6 - (-12)$

67. $0 - (-5)$ **68.** $0 - 8$

69. $(6 - 4) - (1 - 2)$ **70.** $(5 - 3) - (4 - 6)$

71. $-5 - (-4)$ **72.** $-12 - (-7)$

73. $-3 - 3 - 3$ **74.** $-1 - 1 - 1$

75. $-(-9) + (-20) - 14 - (-3)$

76. $-(-8) + (-33) - 7 - (-21)$

77. $[-4 + (-8)] - (-6) + 15$

78. $[-5 + (-4)] - (-2) + 22$

79. Subtract -6 from -10.

80. Subtract -4 from -9.

81. $-3 - (-3)$ **82.** $-5 - (-5)$

83. $-8 - [4 - (-6)]$ **84.** $-1 - [5 - (-2)]$

85. $4 - (-4)$ **86.** $-3 - 3$

87. $(-6 - 5) - 3 + (-11)$ **88.** $(-2 - 1) - 5 + (-19)$

LOOK ALIKES

Evaluate each expression.

89. a. $-40 + (-6)$ **b.** $-40 - (-6)$

90. a. $18 + (-52)$ **b.** $18 - (-52)$

91. a. $-99 + 14$ **b.** $-99 - 14$

92. a. $532 + 847$ **b.** $532 - 847$

APPLICATIONS

Use signed numbers to solve each problem.

93. Submarines. A submarine was traveling 2,000 feet below the ocean's surface when the sonar system warned of a possible collision with another sub. The captain ordered the navigator to dive an additional 200 feet and then level off. Use a signed number to represent the depth of the submarine after the dive.

94. Scuba diving. A diver jumps from his boat into the water and descends to a depth of 50 feet. He pauses to check his equipment and then descends an additional 70 feet. Use a signed number to represent the diver's final depth.

95. Geography. Death Valley, California, is the lowest land point in the United States, at 282 feet below sea level. The lowest land point on Earth is the Dead Sea, which is 1,411 feet below sea level. How much lower is the Dead Sea than Death Valley?

96. History. Two of the greatest Greek mathematicians were Archimedes (287–212 B.C.) and Pythagoras (569–500 B.C.).
   a. Express the year of Archimedes' birth as a negative number.
   b. Express the year of Pythagoras' birth as a negative number.
   c. How many years apart were they born?

97. Amperage. During normal operation, the ammeter on a car reads 5. If the headlights are turned on, they lower the ammeter reading 7 amps. If the radio is turned on, it lowers the reading 6 amps. What number will the ammeter register if they are both turned on?

98. Gin rummy. After a losing round, a card player must deduct the value of each of the cards left in his hand from his previous point total of 21. If face cards are counted as 10 points, what is his new score?

99. Football. A college football team records the outcome of each of its plays during a game on a stat sheet. Find the net gain (or loss) after the third play.

Down	Play	Result
1st	Run	Lost 1 yd
2nd	Pass—sack!	Lost 6 yd
Penalty	Delay of game	Lost 5 yd
3rd	Pass	Gained 8 yd

100. Accounting. Complete the balance sheet below. Then, determine the overall financial condition of the company by subtracting the total debts from the total assets.

Walker Corporation
Balance Sheet 2017

Assets		
Cash	$11	1 0 9
Supplies		7 8 6 2
Land		67 5 4 3
Total assets	$	
Debts		
Accounts payable	$79	0 3 7
Income taxes		20 1 8 1
Total debts	$	

101. Overdraft protection. A student forgot that she had only $15 in her bank account and wrote a check for $25, used an ATM to get $40 cash, and used her debit card to buy $30 worth of groceries. On each of the three transactions, the bank charged her a $20 overdraft protection fee. Find the new account balance.

102. Checking accounts. Michael has $1,303 in his checking account. Can he pay his car insurance premium of $676, his utility bills of $121, and his rent of $750 without having to make another deposit? Explain.

103. Temperature extremes. The highest and lowest temperatures ever recorded in several cities are shown below. List the cities in order, from the largest to smallest range in temperature extremes.

Extreme Temperatures

City	Highest	Lowest
Portland, ME	103	−39
Barrow, AK	79	−56
Kansas City, MO	109	−23
Atlantic City, NJ	106	−11
Norfolk, VA	105	−3

Source: *The World Almanac and Book of Facts*, 2017

104. Eyesight. *Nearsightedness*, the condition in which near objects are clear and far objects are blurry, is measured using negative numbers. *Farsightedness*, the condition in which far objects are clear and near objects are blurry, is measured using positive numbers. Find the range in the measurements shown below. (The symbol +4 is used in optometry and means positive 4.)

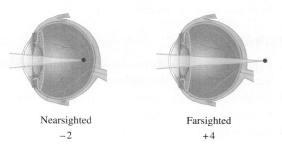

Nearsighted   Farsighted
−2   +4

105. Freeze drying. To make freeze-dried coffee, the coffee beans are roasted at a temperature of 360°F and then the ground coffee bean mixture is frozen at a temperature of −110°F. What is the temperature range of the freeze-drying process?

106. Weather. Rashawn flew from his New York home to Hawaii for a week of vacation. He left blizzard conditions and a temperature of −6°F, and stepped off the airplane into 85°F weather. What temperature change did he experience?

107. Reading programs. In a state reading test given at the start of a school year, an elementary school's performance was 23 points below the county average. The principal immediately began a special tutorial program. At the end of the school year, retesting showed the students to be only 7 points below the average. How did the school's reading score change over the year?

108. Lie detector tests. On one lie detector test, a burglar scored −18, which indicates deception. However, on a second test, he scored −1, which is inconclusive. Find the change in his scores.

109. Music. A computer screenshot of a person listening to a song by Bruno Mars is shown below. The symbol 1:27 appearing on the left means 1 minute and 27 seconds of the song have already played.
   a. What does the symbol −2:19 appearing on the right mean?
   b. How long is the entire song?

110. Termites. One method to kill drywood termites uses liquid nitrogen. It dramatically lowers the temperature in the infested area to −20°F, and since termites require moderate temperatures to survive, the extreme cold kills them. If liquid nitrogen was applied to termite infested wood in a garage, and the temperature in the garage at the time was 83°F, by how many degrees did the nitrogen treatment lower the temperature in the wood?

WRITING

111. Explain what is meant when we say that subtraction is the same as addition of the opposite.

112. Give an example showing that it is possible to subtract something from nothing.

113. Explain how to check the result: $-7 - 4 = -11$

114. Explain why students don't need to change every subtraction they encounter to an addition of the opposite. Give some examples.

REVIEW

115. a. Round 24,085 to the nearest ten.

    b. Round 5,999 to the nearest hundred.

116. List the factors of 20 from least to greatest.

117. It takes 13 oranges to make one can of orange juice. Find the number of oranges used to make 12 cans of orange juice.

118. a. Find the LCM of 15 and 18.

    b. Find the GCF of 15 and 18.

OBJECTIVES

1 Multiply two integers that have different signs.

2 Multiply two integers that have the same sign.

3 Perform several multiplications to evaluate expressions.

4 Evaluate exponential expressions that have negative bases.

5 Solve application problems by multiplying integers.

SECTION **2.4** Multiplying Integers

Multiplication of integers is very much like multiplication of whole numbers. The only difference is that we must determine whether the answer is positive or negative.

When we multiply two nonzero integers, they either have different signs or they have the same sign. This means that there are two possibilities to consider.

OBJECTIVE **1** **Multiply two integers that have different signs.**

To develop a rule for multiplying two integers that have different signs, we will find $4(-3)$, which is the product of a positive integer and negative integer. We say that the signs of the factors are *unlike*. By the definition of multiplication, $4(-3)$ means that we are to add −3 four times.

$$4(-3) = (-3) + (-3) + (-3) + (-3)$$ Write −3 as an addend four times.
$$= -12$$ Use the rule for adding two integers that have the same sign.

The result is negative. As a check, think in terms of money. If you lose $3 four times, you have lost a total of $12, which is written −$12. This example illustrates the following rule.

Multiplying Two Integers That Have Different (Unlike) Signs

To multiply a positive integer and a negative integer, multiply their absolute values. Then make the final answer negative.

EXAMPLE 1 Multiply:

a. $7(-5)$ **b.** $20(-8)$ **c.** $-93 \cdot 16$ **d.** $-34(1,000)$

Strategy We will use the rule for multiplying two integers that have different (unlike) signs.

WHY In each case, we are asked to multiply a positive integer and a negative integer.

Solution

a. Find the absolute values: $|7| = 7$ and $|-5| = 5$.

$$7(-5) = -35$$ Multiply the absolute values, 7 and 5, to get 35.
    Then make the final answer negative.

b. Find the absolute values: $|20| = 20$ and $|-8| = 8$.

$$20(-8) = -160$$ Multiply the absolute values, 20 and 8, to get 160.
    Then make the final answer negative.

c. Find the absolute values: $|-93| = 93$ and $|16| = 16$.

$$-93 \cdot 16 = -1,488$$ Multiply the absolute values, 93 and 16, to get 1,488.
    Then make the final answer negative.

$$\begin{array}{r} 93 \\ \times\ 16 \\ \hline 558 \\ 930 \\ \hline 1,488 \end{array}$$

d. Recall from Section 1.4, to find the product of a whole number and 10, 100, 1,000, and so on, *attach the number of zeros in that number to the right of the whole number.* This rule can be extended to products of integers and 10, 100, 1,000, and so on.

$$-34(1,000) = -34,000$$ Since 1,000 has three zeros, attach three 0's after −34.

Caution! When writing multiplication involving signed numbers, do not write a negative sign − next to a raised dot · (the multiplication symbol). Instead, use parentheses to show the multiplication.

$$6(-2) \quad 6 \cdot {-2}$$

and

$$-6(-2) \quad {-6} \cdot {-2}$$

Self Check 1

Multiply:

a. $2(-6)$

b. $30(-4)$

c. $-75 \cdot 17$

d. $-98(1,000)$

Now Try Problems 21, 25, 29, and 31

OBJECTIVE 2 **Multiply two integers that have the same sign.**

To develop a rule for multiplying two integers that have the same sign, we will first consider $4(3)$, which is the product of two positive integers. We say that the signs of the factors are *like*. By the definition of multiplication, $4(3)$ means that we are to add 3 four times.

$$4(3) = 3 + 3 + 3 + 3$$ Write 3 as an addend four times.

$$= 12$$ The result is 12, which is a positive number.

As expected, the result is positive.

To develop a rule for multiplying two negative integers, consider the following list, where we multiply -4 by factors that decrease by 1. We know how to find the first four products. Graphing those results on a number line is helpful in determining the last three products.

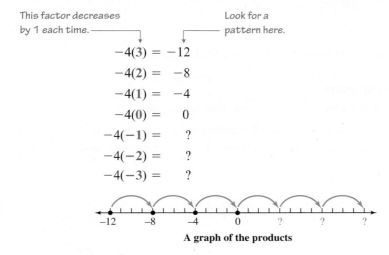

This factor decreases by 1 each time. —

Look for a pattern here. —

$$-4(3) = -12$$
$$-4(2) = -8$$
$$-4(1) = -4$$
$$-4(0) = 0$$
$$-4(-1) = ?$$
$$-4(-2) = ?$$
$$-4(-3) = ?$$

A graph of the products

From the pattern, we see that the product increases by 4 each time. Thus,

$$-4(-1) = 4, \qquad -4(-2) = 8, \qquad \text{and} \qquad -4(-3) = 12$$

These results illustrate that *the product of two negative integers is positive.* As a check, think of it as losing four debts of $3. This is equivalent to gaining $12. Therefore, $-4(-\$3) = \$12$.

We have seen that the product of two positive integers is positive, and the product of two negative integers is also positive. Those results illustrate the following rule.

Multiplying Two Integers That Have the Same (Like) Signs

To multiply two integers that have the same sign, multiply their absolute values. The final answer is positive.

EXAMPLE 2   Multiply:

a. $-5(-9)$ **b.** $-8(-10)$ **c.** $-23(-42)$ **d.** $-2,500(-30,000)$

Strategy We will use the rule for multiplying two integers that have the same (like) signs.

WHY In each case, we are asked to multiply two negative integers.

Solution

a. Find the absolute values: $|-5| = 5$ and $|-9| = 9$.

   $-5(-9) = 45$ Multiply the absolute values, 5 and 9, to get 45.
                      The final answer is positive.

b. Find the absolute values: $|-8| = 8$ and $|-10| = 10$.

   $-8(-10) = 80$ Multiply the absolute values, 8 and 10, to get 80.
                       The final answer is positive.

Self Check 2

Multiply:
a. $-9(-7)$
b. $-12(-2)$
c. $-34(-15)$
d. $-4,100(-20,000)$

Now Try Problems 33, 37, 41, and 43

c. Find the absolute values: $|-23| = 23$ and $|-42| = 42$.

   $-23(-42) = 966$ Multiply the absolute values, 23 and 42, to get 966.
                         The final answer is positive.

$$\begin{array}{r} 42 \\ \times\ 23 \\ \hline 126 \\ 840 \\ \hline 966 \end{array}$$

d. We can extend the method discussed in Section 1.4 for multiplying whole-number factors with trailing zeros to products of integers with trailing zeros.

   $-2,500(-30,000) = 75,000,000$ Attach six 0's after 75.

Multiply -25 and -3 to get 75.

We now summarize the multiplication rules for two integers.

Multiplying Two Integers

To multiply two nonzero integers, multiply their absolute values.

1. The product of two integers that have the same (*like*) signs is positive.

2. The product of two integers that have different (*unlike*) signs is negative.

Using Your Calculator ▶ Multiplication With Negative Numbers

At Thanksgiving time, a large supermarket chain offered customers a free turkey with every grocery purchase of $200 or more. Each turkey cost the store $8, and 10,976 people took advantage of the offer. Since each of the 10,976 turkeys given away represented a loss of $8, (which can be expressed as −$8), the company lost a total of 10,976 (−$8). To perform this multiplication using a calculator, we enter the following:

Reverse entry: 10976 ⊠ 8 +/− =   ┃ −87808 ┃

Direct entry: 10976 ⊠ (−) 8 ENTER   ┃ −87808 ┃

The negative result indicates that with the turkey giveaway promotion, the supermarket chain lost $87,808.

OBJECTIVE 3 **Perform several multiplications to evaluate expressions.**

To evaluate expressions that contain several multiplications, we make repeated use of the rules for multiplying two integers.

EXAMPLE 3 Evaluate each expression:
a. $6(-2)(-7)$ **b.** $-9(8)(-1)$ **c.** $-3(-5)(2)(-4)$

Strategy Since there are no calculations within parentheses and no exponential expressions, we will perform the multiplications, working from the left to the right.

WHY This is step 3 of the order of operations rule that was introduced in Section 1.9.

Solution
a. $6(-2)(-7) = -12(-7)$ Use the rule for multiplying two integers that have different signs: $6(-2) = -12$.

$= 84$ Use the rule for multiplying two integers that have the same sign.

$$\begin{array}{r} {\scriptstyle 1} \\ 12 \\ \times 7 \\ \hline 84 \end{array}$$

b. $-9(8)(-1) = -72(-1)$ Use the rule for multiplying two integers that have different signs: $-9(8) = -72$.

$= 72$ Use the rule for multiplying two integers that have the same sign.

c. $-3(-5)(2)(-4) = 15(2)(-4)$ Use the rule for multiplying two integers that have the same sign: $-3(-5) = 15$.

$= 30(-4)$ Use the rule for multiplying two integers that have the same sign: $15(2) = 30$.

$= -120$ Use the rule for multiplying two integers that have different signs.

Self Check 3
Evaluate each expression:
a. $3(-12)(-2)$
b. $-1(9)(-6)$
c. $-4(-5)(8)(-3)$

Now Try ➲ Problems 45, 47, and 49

The properties of multiplication that were introduced in Section 1.4, *Multiplying Whole Numbers*, are also true for integers.

Properties of Multiplication

Commutative property of multiplication: The order in which integers are multiplied does not change their product.

Associative property of multiplication: The way in which integers are grouped does not change their product.

Multiplication property of 0: The product of any integer and 0 is 0.

Multiplication property of 1: The product of any integer and 1 is that integer.

Another approach to evaluate expressions like those in Example 3 is to use the properties of multiplication to reorder and regroup the factors in a helpful way.

EXAMPLE 4 Use the commutative and/or associative properties of multiplication to evaluate each expression from Example 3 in a different way:

a. $6(-2)(-7)$ **b.** $-9(8)(-1)$ **c.** $-3(-5)(2)(-4)$

Strategy When possible, we will use the commutative and/or associative properties of multiplication to multiply pairs of negative factors.

WHY The product of two negative factors is positive. With this approach, we work with fewer negative numbers, and that lessens the possibility of an error.

Solution

a. $6(-2)(-7) = 6(\mathbf{14})$ — Multiply the last two negative factors to produce a positive product: $-7(-2) = 14$.

$= 84$

$$\begin{array}{r} \overset{2}{14} \\ \times 6 \\ \hline 84 \end{array}$$

b. $-9(8)(-1) = 9(8)$ — Multiply the negative factors to produce a positive product: $-9(-1) = 9$.

$= 72$

c. $-3(-5)(2)(-4) = 15(-8)$ — Multiply the first two negative factors to produce a positive product. Multiply the last two factors.

$= -120$ — Use the rule for multiplying two integers that have different signs.

$$\begin{array}{r} \overset{4}{15} \\ \times 8 \\ \hline 120 \end{array}$$

Self Check 4

Use the commutative and/or associative properties of multiplication to evaluate each expression from Self Check 3 in a different way:

a. $3(-12)(-2)$
b. $-1(9)(-6)$
c. $-4(-5)(8)(-3)$

Now Try Problems 45, 47, and 49

EXAMPLE 5 Evaluate: **a.** $-2(-4)(-5)$ **b.** $-3(-2)(-6)(-5)$

Strategy When possible, we will use the commutative and/or associative properties of multiplication to multiply pairs of negative factors.

WHY The product of two negative factors is positive. With this approach, we work with fewer negative numbers, and that lessens the possibility of an error.

Solution

a. Note that this expression is the product of three (an odd number) negative integers.

$-2(-4)(-5) = 8(-5)$ — Multiply the first two negative factors to produce a positive product.

$= -40$ — The product is negative.

b. Note that this expression is the product of four (an even number) negative integers.

$-3(-2)(-6)(-5) = 6(30)$ — Multiply the first two negative factors and the last two negative factors to produce positive products.

$= 180$ — The product is positive.

Self Check 5

Evaluate each expression:
a. $-1(-2)(-5)$
b. $-2(-7)(-1)(-2)$

Now Try Problems 53 and 57

Example 5, part a, illustrates that a product is negative when there is an odd number of negative factors. Example 5, part b, illustrates that a product is positive when there is an even number of negative factors.

> **Multiplying an Even and an Odd Number of Negative Integers**
> The product of an even number of negative integers is positive.
> The product of an odd number of negative integers is negative.

OBJECTIVE 4 **Evaluate exponential expressions that have negative bases.**

Recall that exponential expressions are used to represent repeated multiplication. For example, 2 to the third power, or 2^3, is a shorthand way of writing $2 \cdot 2 \cdot 2$. In this expression, the *exponent* is 3 and the base is *positive* 2. In the next example, we evaluate exponential expressions with bases that are negative numbers.

EXAMPLE 6   Evaluate each expression: **a.** $(-2)^4$ **b.** $(-5)^3$ **c.** $(-1)^5$

Strategy We will write each exponential expression as a product of repeated factors and then perform the multiplication. This requires that we identify the base and the exponent.

WHY The exponent tells the number of times the base is to be written as a factor.

Solution
a. We read $(-2)^4$ as "negative two raised to the fourth power" or as "the fourth power of negative two." Note that the exponent is even.

$$(-2)^4 = (-2)(-2)(-2)(-2)$$ Write the base, −2, as a factor 4 times.

$$= 4(4)$$ Multiply the first two negative factors and the last two negative factors to produce positive products.

$$= 16$$ The result is positive.

b. We read $(-5)^3$ as "negative five raised to the third power" or as "the third power of negative five," or as "negative five, cubed." Note that the exponent is odd.

$$(-5)^3 = (-5)(-5)(-5)$$ Write the base, −5, as a factor 3 times.

$$= 25(-5)$$ Multiply the first two negative factors to produce a positive product.

$$= -125$$ The result is negative.

$$\begin{array}{r} \overset{2}{25} \\ \times\, 5 \\ \hline 125 \end{array}$$

c. We read $(-1)^5$ as "negative one raised to the fifth power" or as "the fifth power of negative one." Note that the exponent is odd.

$$(-1)^5 = (-1)(-1)(-1)(-1)(-1)$$ Write the base, −1, as a factor 5 times.

$$= 1(1)(-1)$$ Multiply the first and second negative factors and multiply the third and fourth negative factors to produce positive products.

$$= -1$$ The result is negative.

Self Check 6

Evaluate each expression:
a. $(-3)^4$
b. $(-4)^3$
c. $(-1)^7$

Now Try ➲ Problems 61, 65, and 67

In Example 6, part a, −2 was raised to an even power, and the answer was positive. In parts b and c, −5 and −1 were raised to odd powers, and, in each case, the answer was negative. These results suggest a general rule.

> **Even and Odd Powers of a Negative Integer**
> When a negative integer is raised to an even power, the result is positive.
> When a negative integer is raised to an odd power, the result is negative.

Caution! The base of an exponential expression *does not include* the negative sign unless parentheses are used.

-7^3 Positive base: 7

$(-7)^3$ Negative base: -7

Although the exponential expressions $(-3)^2$ and -3^2 look similar, they are not the same. We read $(-3)^2$ as "negative 3 squared" and -3^2 as "the opposite of the square of three." When we evaluate them, it becomes clear that they are not equivalent.

$$(-3)^2 = (-3)(-3)$$ Because of the parentheses, the base is -3. The exponent is 2.
$$= 9$$

$$-3^2 = -(3 \cdot 3)$$ Since there are no parentheses around -3, the base is 3. The exponent is 2.
$$= -9$$

Different results

EXAMPLE 7   Evaluate: -2^2

Strategy We will rewrite the expression as a product of repeated factors, and then perform the multiplication. We must be careful when identifying the base. It is 2, not -2.

WHY Since there are no parentheses around -2, the base is 2.

Solution

$$-2^2 = -(2 \cdot 2)$$ Read as "the opposite of the square of two."
$$= -4$$ Do the multiplication within the parentheses to get 4. Then write the opposite of that result.

Self Check 7

Evaluate: -4^2

Now Try Problem 71

Using Your Calculator ▶   Raising a Negative Number to a Power

We can find powers of negative integers, such as $(-5)^6$, using a calculator. The keystrokes that are used to evaluate such expressions vary from model to model, as shown below. You will need to determine which keystrokes produce the positive result that we would expect when raising a negative number to an even power.

5 [+/−] [y^x] 6 [=]   Some calculators don't require the parentheses to be entered.

[(] 5 [+/−] [)] [y^x] 6 [=]   Other calculators require the parentheses to be entered.

[(] [(−)] 5 [)] [^] 6 [ENTER] | 15625 |

From the calculator display, we see that $(-5)^6 = 15,625$.

OBJECTIVE 5 Solve application problems by multiplying integers.

Problems that involve repeated addition are often more easily solved using multiplication.

EXAMPLE 8   **Oceanography.** Scientists lowered an underwater vessel called a *submersible* into the Pacific Ocean to record the water temperature. The first measurement was made 75 feet below sea level, and more were made every 75 feet until it reached the ocean floor. Find the depth of the submersible when the 25th measurement was made.

Emory Kristof/National Geographic Creative

Analyze

- The first measurement was made 75 feet below sea level.   *Given*
- More measurements were made every 75 feet.   *Given*
- Find the depth of the submersible when it made the 25th measurement.   *Find*

Form   If we use negative numbers to represent the depths at which the measurements were made, then the first was at -75 feet. The depth (in feet) of the submersible when the 25th measurement was made can be found by adding -75 twenty-five times. This repeated addition can be calculated more simply by multiplication.

We translate the words of the problem to numbers and symbols.

The depth of the submersible when it made the 25th measurement	is equal to	the number of measurements made	times	the amount it was lowered each time.
The depth of the submersible when it made the 25th measurement	=	25	·	(−75)

Calculate   To find the product, we use the rule for multiplying two integers that have different signs. First, we find the absolute values: $|25| = 25$ and $|-75| = 75$.

$$25(-75) = -1,875$$

Multiply the absolute values, 25 and 75, to get 1,875. Since the integers have different signs, make the final answer negative.

$$\begin{array}{r} 75 \\ \times\, 25 \\ \hline 375 \\ 1\ 500 \\ \hline 1,875 \end{array}$$

State   The depth of the submersible was 1,875 feet below sea level ($-1,875$ feet) when the 25th temperature measurement was taken.

Check   We can use estimation or simply perform the actual multiplication again to see if the result seems reasonable.

Self Check 8

Gasoline leaks.   To determine how badly a gasoline tank was leaking, inspectors used a drilling process to take soil samples nearby. The first sample was taken 6 feet below ground level, and more were taken every 6 feet after that. The 14th sample was the first one that did not show signs of gasoline. How far below ground level was that?

Now Try ➡ **Problem 105**

Answers to Self Checks

1. a. −12   **b.** −120   **c.** −1,275   **d.** −98,000   **2. a.** 63   **b.** 24   **c.** 510   **d.** 82,000,000   **3. a.** 72   **b.** 54
c. −480   **4. a.** 72   **b.** 54   **c.** −480   **5. a.** −10   **b.** 28   **6. a.** 81   **b.** −64   **c.** −1   **7.** −16
8. 84 ft below ground level (−84 ft)

SECTION 2.4   STUDY SET

VOCABULARY

Fill in the blanks.

1. In the multiplication problem shown below, label each *factor* and the *product*.

$$-5 \quad \cdot \quad 10 \quad = \quad -50$$
↑   ↑   ↑

2. Two negative integers, as well as two positive integers, are said to have the same signs or _____ signs.

3. A positive integer and a negative integer are said to have different signs or _____ signs.

4. _____ property of multiplication: The order in which integers are multiplied does not change their product.

5. _____ property of multiplication: The way in which integers are grouped does not change their product.

6. In the expression $(-3)^5$, the _____ is -3, and 5 is the _____.

CONCEPTS

Fill in the blanks.

7. Multiplication of integers is very much like multiplication of whole numbers. The only difference is that we must determine whether the answer is _____ or _____.

8. When we multiply two nonzero integers, they either have _____ signs or _____ signs.

9. To multiply a positive integer and a negative integer, multiply their absolute values. Then make the final answer _____.

10. To multiply two integers that have the same sign, multiply their absolute values. The final answer is _____.

11. The product of two integers with _____ signs is negative.

12. The product of two integers with _____ signs is positive.

13. The product of any integer and 0 is ▨ .

14. The product of an even number of negative integers is _____, and the product of an odd number of negative integers is _____.

15. Find each absolute value.
    a. $|-3|$ b. $|-12|$

16. If each of the following expressions were evaluated, what would be the *sign* of the result?
    a. $(-5)^{13}$ b. $(-3)^{20}$

NOTATION

17. For each expression, identify the base and the exponent.
    a. -8^4 b. $(-7)^9$

18. Translate to mathematical symbols.
    a. negative three times negative two
    b. negative five squared
    c. the opposite of the square of five

Complete each step to evaluate the expression.

19. $-3(-2)(-4) = (-4)$
    $= $

20. $(-3)^4 = (-3)(-3)(-3)$
    $= (9)$
    $= $

GUIDED PRACTICE

Multiply. See Example 1.

21. $5(-3)$ 22. $4(-6)$

23. $9(-2)$ 24. $5(-7)$

25. $18(-4)$ 26. $17(-8)$

27. $21(-6)$ 28. $39(-3)$

29. $-45 \cdot 37$ 30. $-42 \cdot 24$

31. $-94 \cdot 1,000$ 32. $-76 \cdot 1,000$

Multiply. See Example 2.

33. $(-8)(-7)$ 34. $(-9)(-3)$

35. $-7(-1)$ 36. $-5(-1)$

37. $-3(-52)$ 38. $-4(-73)$

39. $-6(-46)$ 40. $-8(-48)$

41. $-59(-33)$ 42. $-61(-29)$

43. $-60,000(-1,200)$ 44. $-20,000(-3,200)$

Evaluate each expression. See Examples 3 and 4.

45. $6(-3)(-5)$ 46. $9(-3)(-4)$

47. $-5(10)(-3)$ 48. $-8(7)(-2)$

49. $-2(-4)(6)(-8)$ 50. $-3(-5)(2)(-9)$

51. $-8(-3)(7)(-2)$ 52. $-9(-3)(4)(-2)$

Evaluate each expression. See Example 5.

53. $-4(-2)(-6)$ 54. $-4(-6)(-3)$

55. $-3(-9)(-3)$ 56. $-5(-2)(-5)$

57. $-1(-3)(-2)(-6)$ 58. $-1(-4)(-2)(-4)$

59. $-9(-4)(-1)(-4)$ 60. $-6(-3)(-6)(-1)$

Evaluate each expression. See Example 6.

61. $(-3)^3$ 62. $(-6)^3$

63. $(-2)^5$ 64. $(-3)^5$

65. $(-5)^4$ 66. $(-7)^4$

67. $(-1)^8$ 68. $(-1)^{10}$

Evaluate each expression. See Example 7.

69. $(-7)^2$ and -7^2

70. $(-5)^2$ and -5^2

71. $(-12)^2$ and -12^2

72. $(-11)^2$ and -11^2

TRY IT YOURSELF

Evaluate each expression.

73. $6(-5)(2)$ 74. $4(-2)(2)$

75. $-8(0)$ 76. $0(-27)$

77. $(-4)^3$ 78. $(-8)^3$

79. $(-2)10$ 80. $(-3)8$

81. $-2(-3)(3)(-1)$ 82. $5(-2)(3)(-1)$

83. Find the product of -6 and the opposite of 10.

84. Find the product of the opposite of 9 and the opposite of 8.

85. $-6(-4)(-2)$ 86. $-3(-2)(-3)$

87. $-42 \cdot 200,000$ 88. $-56 \cdot 10,000$

89. -5^4 90. -2^4

91. $-12(-12)$ 92. $-5(-5)$

93. $(-1)^6$ 94. $(-1)^5$

95. $(-1)(-2)(-3)(-4)(-5)$

96. $(-10)(-8)(-6)(-4)(-2)$

LOOK ALIKES

Find the answer to the problem in part a. The answers to parts b, c, and d should then be obvious.

97. a. $83(-29)$ **b.** $-83(-29)$

     c. $-83(29)$ **d.** $83(29)$

98. a. $-49(-49)$ **b.** $49(49)$

     c. $-49(49)$ **d.** $49(-49)$

99. a. $-17(-32)(8)$ **b.** $-17(32)(8)$

     c. $17(-32)(-8)$ **d.** $17(-32)(8)$

100. a. $-21(-29)(-3)$ **b.** $-21(29)(3)$

     c. $-21(29)(-3)$ **d.** $21(-29)(3)$

Perform the indicated operations.

101. a. $15 + (-7)$ **b.** $15 - (-7)$

     c. $15(-7)$

102. a. $18(-8)$ **b.** $18 + (-8)$

     c. $18 - (-8)$

103. a. $-41 - (-34)$ **b.** $-41(-34)$

     c. $-41 + (-34)$

104. a. $-62 - 22$ **b.** $(-62)22$

     c. $-62 + 22$

APPLICATIONS

Use signed numbers to solve each problem.

105. Submarines. As part of a training exercise, the captain of a submarine ordered it to descend 250 feet, level off for 5 minutes, and then repeat the process several times. If the sub was on the ocean's surface at the beginning of the exercise, what signed number represents its depth after the 8th dive?

106. Building a pier. *A pile driver* uses a heavy weight to pound tall poles into the ocean floor. If each strike of a pile driver on the top of a pole sends it 6 inches deeper, what signed number represents the depth of the pole after 20 strikes?

Image Source/Getty Images

107. Magnification. A mechanic used an electronic testing device to check the smog emissions of a car. The results of the test are displayed on a screen in the next column.

     a. Find the high and low values for this test as shown on the screen in the next column.

     b. By switching a setting, the picture on the screen can be magnified. What would be the new high and new low if every value were doubled?

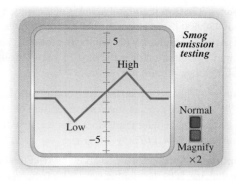

108. Light. Sunlight is a mixture of all colors. When sunlight passes through water, the water absorbs different colors at different rates, as shown.

     a. Use a signed number to represent the depth to which red light penetrates water.

     b. Green light penetrates four times as deep as red light. How deep is this?

     c. Blue light penetrates three times as deep as orange light. How deep is this?

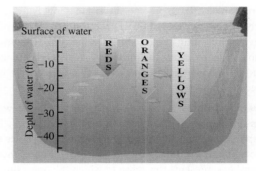

109. Job losses. The recession of 2008 was a major worldwide economic downturn that caused tremendous job losses in the United States. See the bar graph below. Find the number of jobs lost in . . .

     a. September 2008 if it was about six times the number lost in April.

     b. October 2008 if it was about nine times the number lost in May.

     c. November 2008 if it was about seven times the number lost in February.

     d. December 2008 if it was about six times the number lost in March.

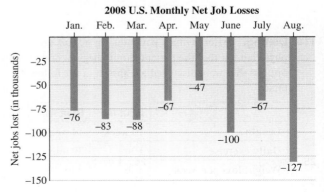

Source: Labor Department

110. Japan. The population of Japan is decreasing by about 470,600 per year because of low birth rates. If this pattern continues, what will be the total decline in Japan's population over the next 25 years? (Source: Washingtonpost.com)

111. Planets. The average surface temperature of Mars is −81°F. Find the average surface temperature of Uranus if it is four times colder than Mars. (Source: *The World Almanac and Book of Facts*, 2017)

112. Crop loss. A farmer, worried about his fruit trees suffering frost damage, calls the weather service for temperature information. He is told that temperatures will be decreasing approximately 5 degrees every hour for the next five hours. What signed number represents the total change in temperature expected over the next five hours?

113. Tax write-off. For each of the last six years, a businesswoman has filed a $200 depreciation allowance on her income tax return for an office computer system. What signed number represents the total amount of depreciation written off over the six-year period?

114. Erosion. A levee protects a town in a low-lying area from flooding. According to geologists, the banks of the levee are eroding at a rate of 2 feet per year. If something isn't done to correct the problem, what signed number indicates how much of the levee will erode during the next decade?

115. Deck supports. After a winter storm, a homeowner has an engineering firm inspect his damaged deck. Their report concludes that the original foundation poles were not sunk deep enough, by a factor of 3. What signed number represents the depth to which the poles should have been sunk?

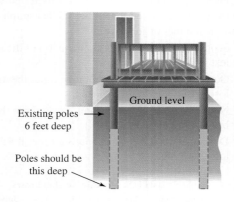

Ground level

Existing poles
6 feet deep

Poles should be
this deep

116. Dieting. After giving a patient a physical exam, a physician felt that the patient should begin a diet. The two options that were discussed are shown in the following table.

	Plan #1	Plan #2
Length	10 weeks	14 weeks
Daily exercise	1 hr	30 min
Weight loss per week	3 lb	2 lb

a. Find the expected weight loss from Plan 1. Express the answer as a signed number.

b. Find the expected weight loss from Plan 2. Express the answer as a signed number.

c. With which plan should the patient expect to lose more weight? Explain why the patient might not choose it.

117. Advertising. On Thursday, March 12, 2016, Rodeo Houston had an all-time record paid attendance: 75,508. Suppose a local radio station gave a sports bag, worth $3, to all paying customers. What signed number expresses the radio station's financial loss from this giveaway?

118. Health care. A health care provider for a company estimates that 75 hours per week are lost by employees suffering from stress-related or preventable illness. In a 52-week year, how many hours are lost? Use a signed number to answer.

119. Explain why the product of a positive number and a negative number is negative, using $5(-3)$ as an example.

120. Explain the multiplication rule for integers that is shown in the pattern of signs below.

$$(-)(-) = +$$
$$(-)(-)(-) = -$$
$$(-)(-)(-)(-) = +$$
$$(-)(-)(-)(-)(-) = -$$
.
.
.

121. When a number is multiplied by −1, the result is the opposite of the original number. Explain why.

122. A student claimed, "A positive and a negative is a negative." What is wrong with this statement?

123. List the first ten prime numbers.

124. Enrollment. The number of students attending a college went from 10,250 to 12,300 in one year. What was the increase in enrollment?

125. Divide: $978 \div 49$

126. What does the symbol < mean?

SECTION **2.5** Dividing Integers

In this section, we will develop rules for division of integers, just as we did earlier for multiplication of integers.

OBJECTIVE **1** Divide two integers.

Recall from Section 1.5 that every division has a related multiplication statement. For example,

$$\frac{6}{3} = 2 \quad \text{because} \quad 2(3) = 6$$

and

$$\frac{20}{5} = 4 \quad \text{because} \quad 4(5) = 20$$

We can use the relationship between multiplication and division to help develop rules for dividing integers. There are four cases to consider.

Case 1: A positive integer divided by a positive integer

From years of experience, we already know that the result is positive. Therefore, *the quotient of two positive integers is positive.*

Case 2: A negative integer divided by a negative integer

As an example, consider the division $\frac{-12}{-2} = ?$. We can find ? by examining the related multiplication statement.

Related multiplication statement

$?(-2) = -12$

This must be *positive 6* if the product is to be *negative 12*.

Division statement

$\frac{-12}{-2} = ?$

So the quotient is *positive 6*.

Therefore, $\frac{-12}{-2} = 6$. This example illustrates that *the quotient of two negative integers is positive.*

Case 3: A positive integer divided by a negative integer

Let's consider $\frac{12}{-2} = ?$. We can find ? by examining the related multiplication statement.

Related multiplication statement

$?(-2) = 12$

This must be -6 if the product is to be *positive 12*.

Division statement

$\frac{12}{-2} = ?$

So the quotient is -6.

Therefore, $\frac{12}{-2} = -6$. This example illustrates that *the quotient of a positive integer and a negative integer is negative.*

Case 4: A negative integer divided by a positive integer

Let's consider $\frac{-12}{2} = ?$. We can find ? by examining the related multiplication statement.

Related multiplication statement	Division statement
$?(2) = -12$	$\dfrac{-12}{2} = ?$
└── This must be -6 if the product is to be -12.	└── So the quotient is -6.

Therefore, $\frac{-12}{2} = -6$. This example illustrates that *the quotient of a negative integer and a positive integer is negative.*

We now summarize the results from the previous examples and note that they are similar to the rules for multiplication.

Dividing Two Integers

To divide two integers, divide their absolute values.

1. The quotient of two integers that have the same (*like*) signs is positive.
2. The quotient of two integers that have different (*unlike*) signs is negative.

EXAMPLE 1   Divide and check the result:

a. $\dfrac{-14}{7}$ **b.** $30 \div (-5)$ **c.** $\dfrac{176}{-11}$ **d.** $-24{,}000 \div 600$

Strategy We will use the rule for dividing two integers that have unlike signs.

WHY Each division involves a positive and a negative integer.

Solution

a. Find the absolute values: $|-14| = 14$ and $|7| = 7$.

$$\dfrac{-14}{7} = -2 \qquad \text{Divide the absolute values, 14 by 7, to get 2.}$$
$$\qquad\qquad\quad\text{Then make the final answer negative.}$$

To check, we multiply the *quotient*, -2, and the *divisor*, 7. We should get the *dividend*, -14.

Check: $-2(7) = -14$ The result checks.

b. Find the absolute values: $|30| = 30$ and $|-5| = 5$.

$$30 \div (-5) = -6 \qquad \text{Divide the absolute values, 30 by 5, to get 6.}$$
$$\qquad\qquad\qquad\quad\text{Then make the final answer negative.}$$

Check: $-6(-5) = 30$ The result checks.

c. Find the absolute values: $|176| = 176$ and $|-11| = 11$.

$$\dfrac{176}{-11} = -16 \qquad \text{Divide the absolute values, 176 by 11, to get 16.}$$
$$\qquad\qquad\qquad\text{Then make the final answer negative.}$$

Check: $-16(-11) = 176$ The result checks.

$$\begin{array}{r} 16 \\ 11\overline{)176} \\ -11 \\ \hline 66 \\ -66 \\ \hline 0 \end{array}$$

d. Recall from Section 1.5 that if a divisor has ending zeros, we can simplify the division by removing the same number of ending zeros in the divisor and dividend.

There are two zeros in the divisor.

$$-24{,}000 \div 600 = -240 \div 6 = -40 \qquad \begin{array}{l}\text{Divide the absolute values, 240 by 6,}\\ \text{to get 40.}\end{array}$$

Remove two zeros from the dividend and the divisor, and divide.

Then make the final answer negative.

Check: $-40(600) = -24{,}000$ Use the original divisor and dividend in the check.

Self Check 1

Divide and check the result:

a. $\dfrac{-45}{5}$

b. $28 \div (-4)$

c. $\dfrac{336}{-14}$

d. $-18{,}000 \div 300$

Now Try Problems 13, 17, 21, and 27

EXAMPLE 2   Divide and check the result:

a. $\dfrac{-12}{-3}$ **b.** $-48 \div (-6)$ **c.** $\dfrac{-315}{-9}$ **d.** $-200 \div (-40)$

Strategy We will use the rule for dividing two integers that have the same (like) signs.

WHY In each case, we are asked to find the quotient of two negative integers.

Solution

a. Find the absolute values: $|-12| = 12$ and $|-3| = 3$.

$$\dfrac{-12}{-3} = 4 \qquad \text{\small Divide the absolute values, 12 by 3, to get 4.}$$
$$\text{\small The final answer is positive.}$$

> *Check:* $4(-3) = -12$ *The result checks.*

b. Find the absolute values: $|-48| = 48$ and $|-6| = 6$.

$$-48 \div (-6) = 8 \qquad \text{\small Divide the absolute values, 48 by 6, to get 8.}$$
$$\text{\small The final answer is positive.}$$

> *Check:* $8(-6) = -48$ *The result checks.*

c. Find the absolute values: $|-315| = 315$ and $|-9| = 9$.

$$\dfrac{-315}{-9} = 35 \qquad \text{\small Divide the absolute values, 315 by 9, to get 35.}$$
$$\text{\small The final answer is positive.}$$

$$\begin{array}{r} 35 \\ 9)\overline{315} \\ -27 \\ \hline 45 \\ -45 \\ \hline 0 \end{array}$$

> *Check:* $35(-9) = -315$ *The result checks.*

d. We can simplify the division by removing the same number of ending zeros in the divisor and dividend.

> There is one zero in the divisor.

$$-200 \div (-40) \;=\; -20 \div (-4) \;=\; 5 \qquad \text{\small Divide the absolute values, 20 by 4,}$$
$$\text{\small to get 5. The final answer is positive.}$$

> Remove one zero from the dividend and the divisor, and divide.

> *Check:* $5(-40) = -200$ *The result checks.*

Self Check 2

Divide and check the result:

a. $\dfrac{-27}{-3}$

b. $-24 \div (-4)$

c. $\dfrac{-301}{-7}$

d. $-400 \div (-20)$

Now Try ➡ Problems 31, 35, 39, and 43

OBJECTIVE 2 Identify *division of 0* and *division by 0.*

To review the concept of division of 0, we consider $\dfrac{0}{-2} = ?$. We can attempt to find ? by examining the related multiplication statement.

Related multiplication statement	**Division statement**

$$(?)(-2) = 0$$
$$\text{\small This must be 0 if the}$$
$$\text{\small product is to be 0.}$$

$$\dfrac{0}{-2} = ?$$
$$\text{\small So the quotient is 0.}$$

Therefore, $\dfrac{0}{-2} = 0$. This example illustrates that *the quotient of 0 divided by any nonzero number is 0.*

To review division by 0, let's consider $\dfrac{-2}{0} = ?$. We can attempt to find ? by examining the related multiplication statement.

Related multiplication statement

$$(?)0 = -2$$

↑ There is no number that gives -2 when multiplied by 0.

Division statement

$$\frac{-2}{0} = ?$$

↑ There is no quotient.

Therefore, $\frac{-2}{0}$ does not have an answer and we say that $\frac{-2}{0}$ is undefined. This example illustrates that *the quotient of any nonzero number divided by 0 is undefined.*

Division With 0

1. If 0 is divided by any nonzero number, the quotient is 0.

2. Division of any nonzero number by 0 is undefined.

Self Check 3

Divide, if possible:

a. $\dfrac{-12}{0}$

b. $0 \div (-6)$

Now Try ➡ Problems 45 and 47

EXAMPLE 3   Divide, if possible: **a.** $\dfrac{-4}{0}$ **b.** $0 \div (-8)$

Strategy In each case, we need to determine if we have division *of* 0 or division *by* 0.

WHY *Division of 0* by a nonzero integer is defined, and the answer is 0. However, *division of a nonzero integer by 0* is undefined; there is no answer.

Solution

a. $\dfrac{-4}{0}$ is undefined.   *This is division by 0.*

b. $0 \div (-8) = 0$ because $0(-8) = 0.$ *This is division of 0.*

OBJECTIVE 3 **Solve application problems by dividing integers.**

Problems that involve forming equal-sized groups can be solved by division.

EXAMPLE 4   **Real estate.** Over the course of a year, a homeowner reduced the price of his house by an equal amount each month because it was not selling. By the end of the year, the price was $11,400 less than at the beginning of the year. By how much was the price of the house reduced each month?

Laura Gangi Pond/Shutterstock.com

Analyze

- The homeowner dropped the price $11,400 in 1 year.   *Given*
- The price was reduced by an equal amount each month.   *Given*
- By how much was the price of the house reduced each month?   *Find*

Form We can express the drop in the price of the house for the year as −$11,400. The phrase *reduced by an equal amount each month* indicates division.

We translate the words of the problem to numbers and symbols.

The amount the price was reduced each month	is equal to	the drop in the price of the house for the year	divided by	the number of months in 1 year.

The amount the price was reduced each month	=	−11,400	÷	12

Calculate To find the quotient, we use the rule for dividing two integers that have different signs. First, we find the absolute values: $|-11,400| = 11,400$ and $|12| = 12$.

$$-11,400 \div 12 = -950$$

Divide the absolute values, 11,400 and 12, to get 950. Then make the final answer negative.

$$
\begin{array}{r}
950 \\
12\overline{)11,400} \\
-10\,8 \\
\hline
60 \\
-60 \\
\hline
00 \\
-00 \\
\hline
0
\end{array}
$$

State The negative result indicates that the price of the house was *reduced* by $950 each month.

Check We can use estimation to check the result. A reduction of $1,000 each month would cause the price to drop $12,000 in 1 year. It seems reasonable that a reduction of $950 each month would cause the price to drop $11,400 in a year.

Self Check 4

Selling boats. The owner of a sailboat reduced the price of the boat by an equal amount each month because there were no interested buyers. After 8 months, and a $960 reduction in price, the boat sold. By how much was the price of the boat reduced each month?

Now Try ➔ **Problem 85**

Using Your Calculator ▶ **Division With Negative Numbers**

According to the Labor Department, the United States lost 630,000 construction jobs during the recession year 2008. Because the jobs were lost, we write this as −630,000. To find the average number of construction jobs lost each month, we divide: $\frac{-630,000}{12}$. We can use a calculator to perform the division.

Reverse entry: 630000 [+/−] [÷] 12 [=]

Direct entry: [(−)] 630000 [÷] 12 [ENTER] | −52500 |

The average number of construction jobs lost each month in 2008 was 52,500.

Answers to Self Checks

1. **a.** −9 **b.** −7 **c.** −24 **d.** −60   2. **a.** 9 **b.** 6 **c.** 43 **d.** 20   3. **a.** undefined **b.** 0
4. The price was reduced by $120 each month.

SECTION 2.5 STUDY SET

VOCABULARY

Fill in the blanks.

1. In the division problems shown below, label the *dividend*, *divisor*, and *quotient*.

$$12 \div (-4) = -3$$

$$\frac{12}{-4} = -3$$

2. The related _____ statement for $\frac{-6}{3} = -2$ is $-2(3) = -6$.

3. $\frac{-3}{0}$ is division _____ 0, and $\frac{0}{-3} = 0$ is division _____ 0.

4. Division of a nonzero integer by 0, such as $\frac{-3}{0}$, is _____ .

CONCEPTS

5. Write the related multiplication statement for each division.

   a. $\frac{-25}{5} = -5$ b. $-36 \div (-6) = 6$ c. $\frac{0}{-15} = 0$

6. Using multiplication, check to determine whether $-720 \div 45 = -12$.

7. Fill in the blanks.
   To divide two integers, divide their absolute values.

   a. The quotient of two integers that have the same (*like*) signs is _____ .

   b. The quotient of two integers that have different (*unlike*) signs is _____ .

8. If a divisor has ending zeros, we can simplify the division by removing the same number of ending zeros in the divisor and dividend. Fill in the blank:
   $-2,400 \div 60 = -240 \div$

9. Fill in the blanks.

   a. If 0 is divided by any nonzero integer, the quotient is ____ .

   b. Division of any nonzero integer by 0 is _____ .

10. What operation can be used to solve problems that involve forming equal-sized groups?

11. Determine whether each statement is always true, sometimes true, or never true.

    a. The product of a positive integer and a negative integer is negative.

    b. The sum of a positive integer and a negative integer is negative.

    c. The quotient of a positive integer and a negative integer is negative.

12. Determine whether each statement is always true, sometimes true, or never true.

    a. The product of two negative integers is positive.

    b. The sum of two negative integers is negative.

    c. The quotient of two negative integers is negative.

GUIDED PRACTICE

Divide and check the result. See Example 1.

13. $\frac{-14}{2}$ 14. $\frac{-10}{5}$

15. $\frac{-20}{5}$ 16. $\frac{-24}{3}$

17. $36 \div (-6)$ 18. $36 \div (-9)$

19. $24 \div (-3)$ 20. $42 \div (-6)$

21. $\frac{264}{-12}$ 22. $\frac{364}{-14}$

23. $\frac{702}{-18}$ 24. $\frac{396}{-12}$

25. $-9,000 \div 300$

26. $-12,000 \div 600$

27. $-250,000 \div 5,000$

28. $-420,000 \div 7,000$

Divide and check the result. See Example 2.

29. $\frac{-8}{-4}$ 30. $\frac{-12}{-4}$

31. $\frac{-45}{-9}$ 32. $\frac{-81}{-9}$

33. $-63 \div (-7)$ 34. $-21 \div (-3)$

35. $-32 \div (-8)$ 36. $-56 \div (-7)$

37. $\frac{-400}{-25}$ 38. $\frac{-490}{-35}$

39. $\frac{-651}{-31}$ 40. $\frac{-736}{-32}$

41. $-800 \div (-20)$ 42. $-800 \div (-40)$

43. $-15,000 \div (-30)$ 44. $-36,000 \div (-60)$

Divide, if possible. **See Example 3.**

45. a. $\dfrac{-3}{0}$ **b.** $\dfrac{0}{-3}$

46. a. $\dfrac{-5}{0}$ **b.** $\dfrac{0}{-5}$

47. a. $\dfrac{0}{-24}$ **b.** $\dfrac{-24}{0}$

48. a. $\dfrac{0}{-32}$ **b.** $\dfrac{-32}{0}$

TRY IT YOURSELF

Divide, if possible.

49. $-36 \div (-12)$ **50.** $-45 \div (-15)$

51. $\dfrac{425}{-25}$ **52.** $\dfrac{462}{-42}$

53. $\dfrac{0}{120}$ **54.** $\dfrac{0}{-11}$

55. Find the quotient of -45 and 9.

56. Find the quotient of -36 and -4.

57. $-2,500 \div 500$ **58.** $-52,000 \div 4,000$

59. $\dfrac{-6}{0}$ **60.** $\dfrac{-8}{0}$

61. $\dfrac{-19}{1}$ **62.** $\dfrac{-9}{1}$

63. $-23 \div (-23)$ **64.** $-11 \div (-11)$

65. $\dfrac{40}{-2}$ **66.** $\dfrac{35}{-7}$

67. $9 \div (-9)$ **68.** $15 \div (-15)$

69. $\dfrac{-10}{-1}$ **70.** $\dfrac{-12}{-1}$

71. $\dfrac{-888}{37}$ **72.** $\dfrac{-456}{24}$

73. $\dfrac{3,000}{-100}$ **74.** $\dfrac{-60,000}{-1,000}$

75. Divide 8 by -2.   **76.** Divide -16 by -8.

77. $\dfrac{-13,550}{25}$ **78.** $\dfrac{3,876}{-19}$

79. $\dfrac{27,778}{-17}$ **80.** $\dfrac{-168,476}{-77}$

LOOK ALIKES

Perform the indicated operation.

81. a. $-18 + 2$ **b.** $-18 - 2$

     c. $-18(2)$ **d.** $\dfrac{-18}{2}$

82. a. $-20(-5)$ **b.** $-20 - (-5)$

     c. $\dfrac{-20}{-5}$ **d.** $-20 + (-5)$

83. a. $-72 - (-8)$ **b.** $-72(-8)$

     c. $\dfrac{-72}{-8}$ **d.** $-72 + (-8)$

84. a. $-75 - 25$ **b.** $-75 + 25$

     c. $-75(25)$ **d.** $\dfrac{-75}{25}$

APPLICATIONS

Use signed numbers to solve each problem.

85. Lowering prices. A furniture store owner reduced the price of an oak table an equal amount each week because it was not selling. After six weeks and a $210 reduction in price, the table was purchased. By how much did the price of the table change each week?

86. Temperature drop. During a five-hour period, the temperature steadily dropped 20°F. By how many degrees did the temperature change each hour?

87. Submarines. In a series of three equal dives, a submarine is programmed to reach a depth of 3,030 feet below the ocean surface. What signed number describes how deep each of the dives will be?

88. Grand Canyon. A mule train is to travel from a stable on the rim of the Grand Canyon to a camp on the canyon floor, approximately 5,500 feet below the rim. If the guide wants the mules to be rested after every 500 feet of descent, how many stops will be made on the trip?

89. Chemistry. During an experiment, a solution was steadily chilled and the times and temperatures were recorded, as shown in the illustration below. What signed number represents the change in temperature each minute?

Beginning of experiment   End of experiment
8:00 A.M.   8:06 A.M.

90. Ocean exploration. The Mariana Trench is the deepest part of the world's oceans. It is located in the North Pacific Ocean near the Philippines and has a maximum depth of 36,201 feet. If a remote-controlled vessel is sent to the bottom of the trench in a series of 11 equal descents, what signed number describes how far the vessel will descend on each dive? (Source: marianatrench.com)

91. Baseball trades. At the midway point of the season, a baseball team finds itself 12 games behind the league leader. Team management decides to trade for a talented hitter, in hopes of making up at least half of the deficit in the standings by the end of the year. What signed number describes where in the league standings management expects to finish at season's end?

92. Budget deficits. A politician proposed a two-year plan for cutting a county's $20-million budget deficit, as shown. If this plan is put into effect, how will the deficit change in two years?

	Plan	Prediction
1st year	Raise taxes, drop failing programs	Will cut deficit in half
2nd year	Search out waste and fraud	Will cut remaining deficit in half

93. Markdowns. The owner of a clothing store decides to reduce the price on a line of jeans that are not selling. She feels she can afford to lose $300 of projected income on these pants. What signed number represents the amount that she can mark down each of the 20 pairs of jeans?

94. Water storage. Over a week's time, engineers at a city water reservoir steadily released enough water to lower the water level 105 feet. What signed number represents the daily change in the water level?

95. The stock market. On Monday, the value of Maria's 255 shares of stock was at an all-time high. By Friday, the value had fallen $4,335. What signed number represents her per-share loss that week?

96. Cutting budgets. In a cost-cutting effort, a company decides to cut $5,840,000 from its annual budget. To do this, all of the company's 160 departments will have their budgets reduced by an equal amount. What signed number represents each department's budget reduction?

WRITING

97. Explain why the quotient of two negative integers is positive.

98. How do the rules for multiplying integers compare with the rules for dividing integers?

99. Use a specific example to explain how multiplication can be used as a check for division.

100. Explain what it means when we say that division by 0 is undefined.

101. Explain the division rules for integers that are shown below using symbols.

$$\frac{+}{+} = + \qquad \frac{-}{-} = + \qquad \frac{-}{+} = - \qquad \frac{+}{-} = -$$

102. Explain the difference between *division of 0* and *division by 0*.

REVIEW

103. Evaluate: $\dfrac{6(32) - 72}{3(2^3)}$

104. Find the prime factorization of 210.

105. The statement $(4 + 8) + 10 = 4 + (8 + 10)$ illustrates what property?

106. Is $17 \geq 17$ a true statement?

107. Does $8 - 2 = 2 - 8$?

108. Sharif has scores of 55, 70, 80, and 75 on four mathematics tests. What is his mean (average) score?

SECTION 2.6   Order of Operations and Estimation

OBJECTIVES

1 Use the order of operations rule.

2 Evaluate expressions containing grouping symbols.

3 Evaluate expressions containing absolute values.

4 Estimate the value of an expression.

In this chapter, we have discussed the rules for adding, subtracting, multiplying, and dividing integers. Now we will use those rules in combination with the order of operations rule from Section 1.9 to evaluate expressions involving more than one operation.

OBJECTIVE 1 Use the order of operations rule.

Recall that every numerical expression has only one correct value. If we don't establish a uniform order of operations, an expression such as $2 + 3 \cdot 6$ can have more than one value. To avoid this possibility, always use the following rule for the order of operations.

Order of Operations

1. Perform all calculations within parentheses and other grouping symbols in the following order listed in Steps 2–4 below, working from the innermost pair of grouping symbols to the outermost pair.

2. Evaluate all the exponential expressions.

3. Perform all multiplications and divisions as they occur from left to right.

4. Perform all additions and subtractions as they occur from left to right.

When grouping symbols have been removed, repeat Steps 2–4 to complete the calculation.

   If a fraction bar is present, evaluate the expression above the bar (called the **numerator**) and the expression below the bar (the **denominator**) separately. Then perform the division indicated by the fraction bar, if possible.

We can use this rule to evaluate expressions involving integers.

EXAMPLE 1   Evaluate: $-4(-3)^2 - (-2)$

Strategy We will scan the expression to determine what operations need to be performed. Then we will perform those operations, one at a time, following the order of operations rule.

  WHY Every numerical expression has only one correct value. If we don't follow the correct order of operations, the expression can have more than one value.

Solution Although the expression contains parentheses, there are no calculations to perform *within* them. We begin with Step 2 of the order of operations rule: Evaluate all exponential expressions.

$$-4(-3)^2 - (-2) = -4(9) - (-2) \quad \text{Evaluate the exponential expression:}$$
$$(-3)^2 + 9.$$

$$= -36 - (-2) \quad \text{Do the multiplication: } -4(9) = -36.$$

$$= -36 + 2 \quad \text{If it is helpful, use the subtraction rule:}$$
$$\text{Add the opposite of } -2, \text{ which is 2.}$$

$$= -34 \quad \text{Do the addition.}$$

Self Check 1

Evaluate: $-5(-2)^2 - (-6)$

Now Try ➲ Problem 13

EXAMPLE 2   Evaluate: $12(3) + (-5)(-3)(-2)$

Strategy We will perform the multiplication first.

  WHY There are no operations to perform within parentheses, nor are there any exponents.

Solution

$$12(3) + (-5)(-3)(-2) = 36 + (-30) \quad \text{Working from left to right,}$$
$$\text{do the multiplications.}$$

$$= 6 \quad \text{Do the addition.}$$

Self Check 2

Evaluate:
$4(9) + (-4)(-3)(-2)$

Now Try ➲ Problem 17

Self Check 3

Evaluate: $45 \div (-5)3$

Now Try Problem 21

| **EXAMPLE 3** | Evaluate: $40 \div (-4)5$ |

Strategy This expression contains the operations of division and multiplication. We will perform the divisions and multiplications as they occur from left to right.

WHY There are no operations to perform within parentheses, nor are there any exponents.

Solution

$$40 \div (-4)5 = -10 \cdot 5 \qquad \textit{Do the division first: } 40 \div (-4) = -10.$$
$$= -50 \qquad \textit{Do the multiplication.}$$

Caution! In Example 3, a common mistake is to forget to work from left to right and incorrectly perform the multiplication first. This produces the wrong answer, -2.

$$40 \div (-4)5 = 40 \div (-20)$$
$$= -2$$

| **EXAMPLE 4** | Evaluate: $-2^2 - (-2)^2$ |

Strategy There are two exponential expressions to evaluate and a subtraction to perform. We will begin with the exponential expressions.

WHY Since there are no operations to perform within parentheses, we begin with Step 2 of the order of operations rule: Evaluate all exponential expressions.

Solution Recall from Section 2.4 that the values of -2^2 and $(-2)^2$ are not the same.

Self Check 4

Evaluate: $-3^2 - (-3)^2$

Now Try Problem 25

$$-2^2 - (-2)^2 = -4 - 4 \qquad \begin{array}{l}\textit{Evaluate the exponential expressions:} \\ -2^2 = -(2 \cdot 2) = -4 \textit{ and } (-2)^2 = -2(-2) = 4.\end{array}$$
$$= -4 + (-4) \qquad \begin{array}{l}\textit{If it is helpful, use the subtraction rule: Add the opposite of 4,} \\ \textit{which is } -4.\end{array}$$
$$= -8 \qquad \textit{Do the addition.}$$

OBJECTIVE 2 **Evaluate expressions containing grouping symbols.**

Recall that **parentheses ( )**, **brackets [ ]**, **absolute value symbols** $| \ |$, and the **fraction bar** — are called **grouping symbols**. When evaluating expressions, we must perform all calculations within parentheses and other grouping symbols first.

| **EXAMPLE 5** | Evaluate: $-15 + 3(-4 + 7 \cdot 2)$ |

Strategy We will begin by evaluating the expression $-4 + 7 \cdot 2$ that is within the parentheses. Since it contains more than one operation, we will use the order of operations rule to evaluate it. We will perform the multiplication first and then the addition.

WHY By the order of operations rule, we must perform all calculations within the parentheses first following the order listed in Steps 2–4 of the rule.

Solution

$$-15 + 3(-4 + 7 \cdot 2) = -15 + 3(-4 + \mathbf{14})$$ *Do the multiplication within the parentheses: 7 · 2 = 14.*

$$= -15 + 3(10)$$ *Do the addition within the parentheses: −4 + 14 = 10.*

$$= -15 + 30$$ *Do the multiplication: 3(10) = 30.*

$$= 15$$ *Do the addition.*

Self Check 5

Evaluate:
$-18 + 6(-7 + 9 \cdot 2)$

Now Try ➲ Problem 29

Expressions can contain two or more pairs of grouping symbols. To evaluate the following expression, we begin within the innermost pair of grouping symbols, the parentheses. Then we work within the outermost pair, the brackets.

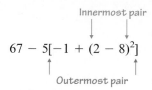

Innermost pair

$$67 - 5[-1 + (2 - 8)^2]$$

Outermost pair

EXAMPLE 6   Evaluate: $67 - 5[-1 + (2 - 8)^2]$

Strategy We will work within the parentheses first and then within the brackets. Within each pair of grouping symbols, we will follow the order of operations rule.

WHY We must work from the *innermost* pair of grouping symbols to the *outermost*.

Solution

$$67 - 5[-1 + (\mathbf{2} - \mathbf{8})^2]$$

$$= 67 - 5[-1 + (\mathbf{-6})^2]$$ *Do the subtraction within the parentheses: 2 − 8 = −6.*

$$= 67 - 5[-1 + 36]$$ *Evaluate the exponential expression within the brackets.*

$$= 67 - 5[35]$$ *Do the addition within the brackets: −1 + 36 = 35.*

$$= 67 - \mathbf{175}$$ *Do the multiplication: 5(35) = 175.*

$$= 67 + (\mathbf{-175})$$ *If it is helpful, use the subtraction rule: Add the opposite of 175, which is −175.*

$$= -108$$ *Do the addition.*

$$\begin{array}{r} \overset{2}{35} \\ \times \ 5 \\ \hline 175 \end{array}$$

$$\begin{array}{r} \overset{6\,15}{17\!\!\!/5} \\ -\ 67 \\ \hline 108 \end{array}$$

Self Check 6

Evaluate:
$81 - 4[-2 + (5 - 9)^2]$

Now Try ➲ Problem 33

Success Tip Any arithmetic steps that you cannot perform in your head should be shown outside of the horizontal steps of your solution.

EXAMPLE 7 Evaluate: $-\left[1 - \left(2^4 + \dfrac{66}{-6}\right)\right]$

Strategy We will work within the parentheses first and then within the brackets. Within each pair of grouping symbols, we will follow the order of operations rule.

WHY We must work from the *innermost* pair of grouping symbols to the *outermost*.

Solution

$-\left[1 - \left(2^4 + \dfrac{66}{-6}\right)\right] = -\left[1 - \left(16 + \dfrac{66}{-6}\right)\right]$ Evaluate the exponential expression within the parentheses: $2^4 = 16$.

$= -\left[1 - \left(16 + (-11)\right)\right]$ Do the division within the parentheses: $66 \div (-6) = -11$.

$= -[1 - 5]$ Do the addition within the parentheses: $16 + (-11) = 5$.

$= -[-4]$ Do the subtraction within the brackets: $1 - 5 = -4$.

$= 4$ The opposite of -4 is 4.

Self Check 7

Evaluate:

$-\left[8 - \left(3^3 + \dfrac{90}{-9}\right)\right]$

Now Try ➡ **Problem 37**

EXAMPLE 8 Evaluate: $\dfrac{-20 + 3(-5)}{21 - (-4)^2}$

Strategy We will evaluate the expression above and the expression below the fraction bar separately. Then we will do the indicated division, if possible.

WHY Fraction bars are grouping symbols that group the numerator and the denominator. The expression could be written $[-20 + 3(-5)] \div [21 - (-4)^2]$.

Solution

$\dfrac{-20 + 3(-5)}{21 - (-4)^2} = \dfrac{-20 + (-15)}{21 - 16}$ In the numerator, do the multiplication: $3(-5) = -15$. In the denominator, evaluate the exponential expression: $(-4)^2 = 16$.

$= \dfrac{-35}{5}$ In the numerator, add: $-20 + (-15) = -35$. In the denominator, subtract: $21 - 16 = 5$.

$= -7$ Do the division indicated by the fraction bar.

Self Check 8

Evaluate: $\dfrac{-9 + 6(-4)}{28 - (-5)^2}$

Now Try ➡ **Problem 41**

OBJECTIVE 3 **Evaluate expressions containing absolute values.**

Earlier in this chapter, we found the absolute values of integers. For example, recall that $|-3| = 3$ and $|10| = 10$. We use the order of operations rule to evaluate more complicated expressions that contain absolute values.

EXAMPLE 9 Evaluate each expression: **a.** $|-4(3)|$ **b.** $|-6 + 1|$

Strategy We will perform the calculation within the absolute value symbols first. Then we will find the absolute value of the result.

WHY Absolute value symbols are grouping symbols, and by the order of operations rule, all calculations within grouping symbols must be performed first.

Solution

a. $|-4(3)| = |-12|$ *Do the multiplication within the absolute value symbol:* $-4(3) = -12.$

    $= 12$ *Find the absolute value of* $-12.$

b. $|-6 + 1| = |-5|$ *Do the addition within the absolute value symbol:* $-6 + 1 = -5.$

    $= 5$ *Find the absolute value of* $-5.$

Self Check 9

Evaluate each expression:

a. $|(-6)(5)|$

b. $|-3 + 96|$

Now Try ➡ **Problem 45**

EXAMPLE 10   Evaluate: $8 - 4|-6 - 2|$

Strategy The absolute value bars are grouping symbols. We will perform the subtraction within them first.

WHY By the order of operations rule, we must perform all calculations within parentheses and other grouping symbols (such as absolute value bars) first.

Solution

$8 - 4|-6 - 2| = 8 - 4|-6 + (-2)|$ *If it is helpful, use the subtraction rule within the absolute value symbol: Add the opposite of 2, which is* $-2.$

$= 8 - 4|-8|$ *Do the addition within the absolute value symbol:* $-6 + (-2) = -8.$

$= 8 - 4(8)$ *Find the absolute value:* $|-8| = 8.$

$= 8 - 32$ *Do the multiplication:* $4(8) = 32.$

$= 8 + (-32)$ *If it is helpful, use the subtraction rule: Add the opposite of 32, which is* $-32.$

$= -24$ *Do the addition.*

$$\begin{array}{r} 2\,1\,2 \\ 3\,2 \\ -\,8 \\ \hline 2\,4 \end{array}$$

LANGUAGE OF MATHEMATICS

Multiplication is indicated when a number is outside and next to an absolute value symbol. For example,

$$8 - 4|-6 - 2|$$

means

$$8 - 4 \cdot |-6 - 2|$$

Self Check 10

Evaluate: $7 - 5|-1 - 6|$

Now Try ➡ **Problem 49**

OBJECTIVE 4 **Estimate the value of an expression.**

Recall that the idea behind estimation is to simplify calculations by using rounded numbers that are close to the actual values in the problem. When an exact answer is not necessary and a quick approximation will do, we can use estimation.

EXAMPLE 11 **The stock market.** The change in the Dow Jones Industrial Average is announced at the end of each trading day to give a general picture of how the stock market is performing. A positive change means a good performance, while a negative change indicates a poor performance. The week of October 13–17, 2008, had some record changes, as shown on the next page. Round each number to the nearest ten and estimate the net gain or loss of points in the Dow that week.

Strategy To estimate the net gain or loss, we will round each number to the nearest ten and *add* the approximations.

Bart Sadowski/Shutterstock.com

+936	-78	-733	+402	-123
Monday	Tuesday	Wednesday	Thursday	Friday
Oct. 13, 2008	Oct. 14, 2008	Oct. 15, 2008	Oct. 16, 2008	Oct. 17, 2008
(largest 1-day		(second-largest		
increase)		1-day decline)		

Source: finance.yahoo.com

Self Check 11

The stock market. For the week of September 12–16, 2016, the Dow Jones Industrial Average performance was as follows, Monday: 243, Tuesday: –258, Wednesday: –53, Thursday: 178, Friday: –89. Round each number to the nearest ten and estimate the net gain or loss of points in the Dow for that week. (Source: finance.yahoo.com)

Now Try ⟿ Problems 53 and 101

WHY The phrase *net gain or loss* refers to what remains after all of the losses and gains have been combined (added).

Solution To nearest ten:

936 rounds to 940   −78 rounds to −80   −733 rounds to −730
402 rounds to 400   −123 rounds to −120

To estimate the net gain or loss for the week, we add the rounded numbers.

$$940 + (-80) + (-730) + 400 + (-120)$$
$$= 1{,}340 + (-930) \quad \text{Add the positives and the negatives separately.}$$
$$= 410 \quad \text{Do the addition.}$$

$$\begin{array}{r} \overset{13}{1,\!340} \\ -930 \\ \hline 410 \end{array}$$

The positive result means there was a net gain that week of approximately 410 points in the Dow.

Answers to Self Checks

1. −14   **2.** 12   **3.** −27   **4.** −18   **5.** 48   **6.** 25   **7.** 9   **8.** −11   **9. a.** 30   **b.** 93
10. −28   **11.** There was a net gain that week of approximately 20 points.

SECTION 2.6 STUDY SET

VOCABULARY

Fill in the blanks.

1. To evaluate expressions that contain more than one operation, we use the _____ of operations rule.

2. Absolute value symbols, parentheses, and brackets are types of _____ symbols.

3. In the expression $-9 + 2[-5 - 6(-3 - 1)]$, the parentheses are the _____ most grouping symbols and the brackets are the _____ most grouping symbols.

4. In situations where an exact answer is not needed, an approximation or _____ is a quick way of obtaining a rough idea of the size of the actual answer.

CONCEPTS

5. Refer to the expressions in the next column. List the operations in the order in which they should be performed to evaluate each expression. *You do not have to evaluate the expression.*

   a. $-5(-2)^2 - 1$

   b. $15 - 3 + (-5 \cdot 2)^3$

   c. $4 + 2(-7 - 3)$

   d. $-2 \cdot 3^2$

6. Consider the expression $\dfrac{5 + 5(7)}{2 + (4 - 8)}$. In the numerator, what operation should be performed first? In the denominator, what operation should be performed first?

NOTATION

7. Give the name of each grouping symbol:
   ( )   [ ] | | —

8. What operation is indicated?

   $$-2 + 9|8 - (-2 + 4)|$$

Complete each step to evaluate the expression.

9. $-8 - 5(-2)^2 = -8 - 5(\boxed{})$

$\qquad\qquad\qquad = -8 - \boxed{}$

$\qquad\qquad\qquad = -8 + (\boxed{})$

$\qquad\qquad\qquad = \boxed{}$

10. $2 + (5 - 6 \cdot 2) = 2 + (5 - \boxed{})$

$\qquad\qquad\qquad = 2 + [5 + (\boxed{})]$

$\qquad\qquad\qquad = 2 + (\boxed{})$

$\qquad\qquad\qquad = \boxed{}$

11. $-9 + 5(-4 \cdot 2 + 7) = -9 + 5(\boxed{} + 7)$

$\qquad\qquad\qquad = -9 + 5(\boxed{})$

$\qquad\qquad\qquad = -9 + (\boxed{})$

$\qquad\qquad\qquad = \boxed{}$

12. $\dfrac{|-9 + (-3)|}{9 - 6} = \dfrac{|\boxed{}|}{3}$

$\qquad\qquad\qquad = \dfrac{\boxed{}}{3}$

$\qquad\qquad\qquad = \boxed{}$

GUIDED PRACTICE

Evaluate each expression. See Example 1.

13. $-2(-3)^2 - (-8)$ **14.** $-6(-2)^2 - (-9)$

15. $-5(-4)^2 - (-18)$ **16.** $-3(-5)^2 - (-24)$

Evaluate each expression. See Example 2.

17. $9(7) + (-6)(-2)(-4)$

18. $9(8) + (-2)(-5)(-7)$

19. $8(6) + (-2)(-9)(-2)$

20. $7(8) + (-3)(-6)(-2)$

Evaluate each expression. See Example 3.

21. $30 \div (-5)2$ **22.** $50 \div (-2)5$

23. $60 \div (-3)4$ **24.** $120 \div (-4)3$

Evaluate each expression. See Example 4.

25. $-6^2 - (-6)^2$ **26.** $-7^2 - (-7)^2$

27. $-10^2 - (-10)^2$ **28.** $-8^2 - (-8)^2$

Evaluate each expression. See Example 5.

29. $-14 + 2(-9 + 6 \cdot 3)$

30. $-18 + 3(-10 + 3 \cdot 7)$

31. $-23 + 3(-15 + 8 \cdot 4)$

32. $-31 + 6(-12 + 5 \cdot 4)$

Evaluate each expression. See Example 6.

33. $77 - 2[-6 + (3 - 9)^2]$

34. $84 - 3[-7 + (5 - 8)^2]$

35. $99 - 4[-9 + (6 - 10)^2]$

36. $67 - 5[-6 + (4 - 7)^2]$

Evaluate each expression. See Example 7.

37. $-\left[4 - \left(3^3 + \dfrac{22}{-11}\right)\right]$

38. $-\left[1 - \left(2^3 + \dfrac{40}{-20}\right)\right]$

39. $-\left[50 - \left(5^3 + \dfrac{50}{-2}\right)\right]$

40. $-\left[12 - \left(2^5 + \dfrac{40}{-4}\right)\right]$

Evaluate each expression. See Example 8.

41. $\dfrac{-24 + 3(-4)}{42 - (-6)^2}$ **42.** $\dfrac{-18 + 6(-2)}{52 - (-7)^2}$

43. $\dfrac{-38 + 11(-2)}{69 - (-8)^2}$ **44.** $\dfrac{-36 + 8(-2)}{85 - (-9)^2}$

Evaluate each expression. See Example 9.

45. a. $|-6(2)|$ **b.** $|-12 + 7|$

46. a. $|-4(9)|$ **b.** $|-15 + 6|$

47. a. $|15(-4)|$ **b.** $|16 + (-30)|$

48. a. $|12(-5)|$ **b.** $|47 + (-70)|$

Evaluate each expression. See Example 10.

49. $16 - 6|-2 - 1|$ **50.** $15 - 6|-3 - 1|$

51. $17 - 2|-6 - 4|$ **52.** $21 - 9|-3 - 1|$

Estimate the value of each expression by rounding each number to the nearest ten. See Example 11.

53. $-379 + (-13) + 287 + (-671)$

54. $-363 + (-781) + 594 + (-42)$

Estimate the value of each expression by rounding each number to the nearest hundred. See Example 11.

55. $-3,887 + (-5,806) + 4,701$

56. $-5,684 + (-2,270) + 3,404 + 2,689$

TRY IT YOURSELF

Evaluate each expression.

57. $(-3)^2 - 4^2$ **58.** $9 - 9 \cdot 4$

59. $3^2 - 4(-2)(-1)$ **60.** $2^3 - 3^3$

61. $|-3 \cdot 4 + (-5)|$ **62.** $|6 - 4(1 - 5)|$

63. $(2 - 5)(5 + 2)$ **64.** $-3(2)^2 4$

65. $6 + \dfrac{25}{-5} + 6 \cdot 3$

66. $-(5 \cdot 7 - 4 \cdot 8)^4$

67. $\dfrac{-6 - 2^3}{-2 - (-4)}$

68. $\dfrac{-6 - 6}{-2 - 2}$

69. $-12 \div (-2)2$

70. $-60(-2) \div 3$

71. $-16 - 4 \div (-2)$

72. $80 - 40 \div 10 + 5$

73. $-|2 \cdot 7 - (-5)^2|$

74. $-|8 \div (-2) - 5|$

75. $-3|4 - 7|$

76. $|-2 + 6 - 5|$

77. $(7 - 5)^2 - (1 - 4)^2$

78. $5^2 - (-9 - 3)$

79. $-1(2^2 - 2 + 1^2)$

80. $(-7 - 4)^2 - (-1)$

81. $\dfrac{-5 - 5}{1^4 + 1^5}$

82. $\dfrac{-7 - (-3)}{2 - 2^2}$

83. $-50 - 2(-3)^3(4)$

84. $8 - 3[5^2 - (7 - 3)^2]$

85. $-6^2 + 6^2$

86. $-9^2 + 9^2$

87. $3\left(\dfrac{-18}{3}\right) - 2(-2)$

88. $2\left(\dfrac{-12}{3}\right) + 3(-5)$

89. $2|1 - 8| \cdot |-8|$

90. $2(5) - 6(|-3|)^2$

91. $\dfrac{2 + 3[5 - (1 - 10)]}{|2(-8 + 2) + 10|}$

92. $\dfrac{11 + (-2 \cdot 2 + 3)}{|15 + (-3 \cdot 4 - 8)|}$

93. $-2 + |6 - 4^2|$

94. $-3 - 4|6 - 7|$

95. $\dfrac{-4(-5) - 2}{3 - 3^2}$

96. $\dfrac{(-6)^2 - 1}{-(2^2 - 3)}$

LOOK ALIKES

Evaluate each expression.

97. a. $(-8 - 3)(-2)$ **b.** $(-8 - 3) - 2$

98. a. $2 \cdot 5^3$ **b.** $(2 \cdot 5)^3$

99. a. $-200 \div 5 \cdot 2$ **b.** $-200 \div (5 \cdot 2)$

100. a. $9 + 2[-2 - (5 + 1)]$ **b.** $(9 + 2)[-2 - (5 + 1)]$

APPLICATIONS

101. The prairie state. Illinois population first began to drop in 2014, when the state lost 11,961 people. That number more than doubled in 2015, with a loss of 28,497 people. The trend continued in 2016 when the state lost 37,508 people. Round each number to the nearest thousand and estimate Illinois' total population loss from 2014 through 2016.

102. Stock market records. Refer to the tables in the next column. Round each of the record Dow Jones point gains and losses to the nearest hundred and then add all ten of them. What is the result?

5 Greatest Dow Jones Daily Point Gains

Rank	Date	Gain
1	10/13/2008	936
2	10/28/2008	889
3	8/26/2015	619
4	11/13/2008	553
5	3/16/2000	499

5 Greatest Dow Jones Daily Point Losses

Rank	Date	Loss
1	9/29/2008	-778
2	10/15/2008	-733
3	9/17/2001	-685
4	12/1/2008	-680
5	10/9/2008	-679

Source: Dow Jones Indexes

103. Testing.   In an effort to discourage her students from guessing on multiple-choice tests, a professor uses the grading scale shown in the table. If unsure of an answer, a student does best to skip the question, because incorrect responses are penalized very heavily. Find the test score of a student who gets 12 correct and 3 wrong and leaves 5 questions blank.

Response	Value
Correct	3
Incorrect	-4
Left blank	-1

104. Spreadsheets.   The table shows the data from a chemistry experiment in spreadsheet form. To obtain a result, the chemist needs to add the values in row 1, double that sum, and then divide that number by the smallest value in column C. What is the final result of these calculations?

	A	B	C	D
1	12	-5	6	-2
2	15	4	5	-4
3	6	4	-2	8

105. Business takeovers.   Six investors are taking over a poorly managed company, but first they must repay the debt that the company built up over the past four quarters. (See graph on the next page.) If the investors plan equal ownership, how much of the company's total debt is each investor responsible for?

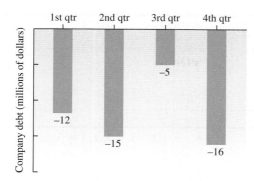

106. Declining enrollment. Find the drop in enrollment for each high school shown in the table below. Express each drop as a negative number. Then find the mean (average) drop in enrollment for these four schools.

High school	2016 enrollment	2017 enrollment	Drop
Lincoln	2,683	2,573	
Camble	2,754	2,662	
Munson	1,948	1,875	
Hardy	2,257	2,192	

107. The federal budget. See the graph below. Suppose you were hired to write a speech for a politician who wanted to highlight the improvement in the federal government's finances during the 1990s. Would it be better for the politician to talk about the mean (average) budget deficit/surplus for the last half of the decade or for the last four years of that decade? Explain your reasoning.

U.S. Budget Deficit/Surplus
($ billions)

Deficit	Year	Surplus
−164	1995	
−107	1996	
−22	1997	
	1998	+70
	1999	+123

108. Scouting reports. The illustration below shows a football coach how successful his opponent was running a "28 pitch" the last time the two teams met. What was the opponent's mean (average) gain with this play?

Play: 28 pitch

Gain 16 yd	Gain 10 yd	Loss 2 yd	No gain
Gain 4 yd	Loss 4 yd	TD Gain 66 yd	Loss 2 yd

109. Estimation. Quickly determine a reasonable estimate of the exact answer in each of the following situations. Use front-end rounding.

a. A scuba diver, swimming at a depth of 34 feet below sea level, spots a sunken ship beneath him. He dives down another 57 feet to reach it. What is the depth of the sunken ship?

b. A dental hygiene company offers a money-back guarantee on its tooth whitener kit. When the kit is returned by a dissatisfied customer, the company loses the $11 it cost to produce it, because it cannot be resold. How much money has the company lost because of this return policy if 56 kits have been mailed back by customers?

c. A tram line makes a 7,891-foot descent from a mountaintop in 18 equal stages. How much does it descend in each stage?

110. Estimation. Quickly determine a reasonable estimate of the exact answer in each of the following situations. Use front-end rounding.

a. A submarine, cruising at a depth of −175 feet, descends another 605 feet. What is the depth of the submarine?

b. A married couple has assets that total $840,756 and debts that total $265,789. What is their net worth?

c. According to pokerlistings.com, the top five online poker losses for 2015 were −$3,554,314, −$1,129,700, −$1,034,396, −$910,779, and −$838,821. Find the total amount lost.

WRITING

111. When evaluating expressions, why is the order of operations rule necessary?

112. In the rules for the order of operations, what does the phrase *as they occur from left to right* mean?

113. Explain the error in each evaluation below.

a. $80 \div (-2)4 = 80 \div (-8)$
$$= -10$$

b. $-1 + 8|4 - 9| = -1 + 8|-5|$
$$= 7|-5|$$
$$= 35$$

114. Describe a situation in daily life in which you use estimation.

REVIEW

115. On the number line, what number is

a. 4 units to the right of −7?

b. 6 units to the left of 2?

116. Is 834,540 divisible by: **a.** 2   **b.** 3   **c.** 4   **d.** 5   **e.** 6   **f.** 9   **g.** 10

117. Elevators. An elevator has a weight capacity of 1,000 pounds. Seven people, with an average weight of 140 pounds, are in it. Is it overloaded?

118. a. Find the LCM of 12 and 44.

b. Find the GCF of 12 and 44.

2 Summary and Review

SECTION 2.1 ▸ An Introduction to the Integers

DEFINITIONS AND CONCEPTS	EXAMPLES
The collection of positive whole numbers, the negatives of the whole numbers, and 0 is called the set of **integers**.	The set of integers: $\{\ldots, -5, -4, -3, -2, -1, 0, 1, 2, 3, 4, 5, \ldots\}$
Positive numbers are greater than 0, and **negative numbers** are less than 0.	The set of positive integers: $\{1, 2, 3, 4, 5, \ldots\}$ The set of negative integers: $\{\ldots, -5, -4, -3, -2, -1\}$
Negative numbers can be represented on a **number line** by extending the line to the left and drawing an arrowhead. As we move to the right on the number line, the values of the numbers increase. As we move to the left, the values of the numbers decrease.	Graph $-1, 6, 0, -4$, and 3 on a number line. Numbers get larger Negative numbers Zero Positive numbers $-6\ \ -5\ \ -4\ \ -3\ \ -2\ \ -1\ \ \ 0\ \ \ 1\ \ \ 2\ \ \ 3\ \ \ 4\ \ \ 5\ \ \ 6$ Numbers get smaller
Inequality symbols: $\neq$ means *is not equal to* $\geq$ means *is greater than or equal to* $\leq$ means *is less than or equal to*	Each of the following statements is true: $5 \neq -3$ Read as "5 is not equal to -3." $4 \geq -6$ Read as "4 is greater than or equal to -6." $-2 \leq -2$ Read as "-2 is less than or equal to -2."
The **absolute value** of a number is the distance on a number line between the number and 0.	Find each absolute value: $\lvert 12 \rvert = 12$ $\lvert -9 \rvert = 9$ $\lvert 0 \rvert = 0$
Two numbers that are the same distance from 0 on the number line, but on opposite sides of it, are called **opposites**.	The opposite of 4 is -4. The opposite of -77 is 77. The opposite of 0 is 0.
The opposite of the opposite rule The opposite of the opposite of a number is that number.	Simplify each expression: $-(-6) = 6$ $-\lvert 8 \rvert = -8$ $-\lvert -26 \rvert = -26$
The $-$ **symbol** is used to indicate a negative number, the opposite of a number, and the operation of subtraction.	-2 *negative 2* $-(-4)$ *the opposite of negative four* $6 - 1$ *six minus one*

REVIEW EXERCISES

1. Write the set of integers.

2. Represent each of the following situations using a signed number.

   a. a deficit of $1,200

   b. 10 seconds before going on the air

3. **Water pressure.** Salt water exerts a pressure of approximately 29 pounds per square inch at a depth of 33 feet. Express the depth using a signed number.

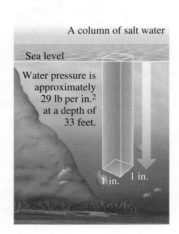

A column of salt water

Sea level

Water pressure is approximately 29 lb per in.^{2} at a depth of 33 feet.

1 in.   1 in.

4. Graph the following integers on a number line.

a. −3, 0, 4, −1

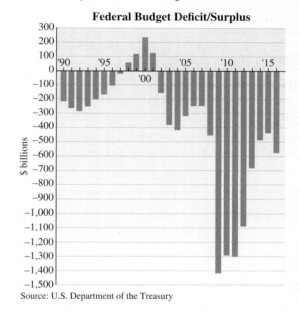

b. the integers greater than −3 but less than 4

5. Place an < or an > symbol in the box to make a true statement.

a. 0 ☐ −7   b. −20 ☐ −19

6. Tell whether each statement is true or false.

a. −17 ≥ −16   b. −56 ≤ −56

7. Find each absolute value.

a. |5| b. |−43| c. |0|

8. a. What is the opposite of 8?

b. What is the opposite of −8?

c. What is the opposite of 0?

9. Simplify each expression.

a. −|12|

b. −(−12)−(−12)

c. −0−0

10. Explain the meaning of each red − symbol.

a. −5   b. −(−5)

c. −(−5)   d. 5 − (−5)

11. Ladies professional golf association. The scores of the top six finishers of the 2015 KIA Classic Tournament and their final scores related to par are listed alphabetically: Cristie Kerr (−20), Hyo Joo Kim (−14), Lydia Ko (−17), Alison Lee (−16), Mirim Lee (−18), and Inbee Park (−15). Complete the table in the next column. Remember, in golf, the lowest score wins.

Position	Player	Score to Par
1		
2		
3		
4		
5		
6		

Source: lpga.com

12. Federal budget. The graph shows the U.S. government's deficit/surplus budget data for the years 1990–2016.

a. When did the first budget surplus occur? Estimate it.

b. In what year was there the largest surplus? Estimate it.

c. In what year was there the greatest deficit? Estimate it.

Federal Budget Deficit/Surplus

Source: U.S. Department of the Treasury

SECTION 2.2 ▶ **Adding Integers**

DEFINITIONS AND CONCEPTS	EXAMPLES
Adding two integers that have the same (like) signs **1.** To add two positive integers, add them as usual. The final answer is positive. **2.** To add two negative integers, add their absolute values and make the final answer negative.	Add: −5 + (−10) Find the absolute values: |−5| = 5 and |−10| = 10. $-5 + (-10) = -15$ *Add their absolute values, 5 and 10, to get 15. Then make the final answer negative.*
Adding two integers that have different (unlike) signs To add a positive integer and a negative integer, subtract the smaller absolute value from the larger. **1.** If the positive integer has the larger absolute value, the final answer is positive. **2.** If the negative integer has the larger absolute value, make the final answer negative.	Add: −7 + 12 Find the absolute values: |−7| = 7 and |12| = 12. $-7 + 12 = 5$ *Subtract the smaller absolute value from the larger: 12 − 7 = 15. Since the positive number, 12, has the larger absolute value, the final answer is positive.* Add: −8 + 3 Find the absolute values: |−8| = 8 and |3| = 3. $-8 + 3 = -5$ *Subtract the smaller absolute value from the larger: 8 − 3 = 5. Since the negative number, −8, has the larger absolute value, make the final answer negative.*

To **evaluate expressions** that contain several additions, we make repeated use of the rules for adding two integers.	Evaluate: $-7 + 1 + (-20) + 1$ Perform the additions working left to right. $$-7 + 1 + (-20) + 1 = -6 + (-20) + 1$$ $$= -26 + 1$$ $$= -25$$
We can use the **commutative** and **associative properties of addition** to *reorder* and *regroup* addends.	Another way to evaluate this expression is to add the negatives and add the positives separately. Then add those results. Negatives Positives $$-7 + 1 + (-20) + 1 = [-7 + (-20)] + (1 + 1)$$ $$= -27 + 2$$ $$= -25$$
Addition property of 0 The sum of any number and 0 is that number.	$-2 + 0 = -2$ and $0 + (-25) = -25$
If the sum of two numbers is 0, the numbers are said to be **additive inverses** of each other.	3 and -3 are *additive inverses* because $3 + (-3) = 0$.
Addition property of opposites The sum of an integer and its opposite (additive inverse) is 0.	$4 + (-4) = 0$ and $-712 + 712 = 0$

At certain times, the **addition property of opposites** can be used to make addition of several integers easier.	Evaluate: $14 + (-9) + 8 + 9 + (-14)$
	Locate pairs of opposites and add them to get 0.
	
	            Opposites
	$$14 + (-9) + 8 + 9 + (-14) = 0 + 0 + 8$$
	        Opposites $= 8$
	                                  The sum of any integer and
	                                  0 is that integer.

REVIEW EXERCISES

Add.

13. $-6 + (-4)$

14. $-3 + (-6)$

15. $-28 + 60$

16. $93 + (-20)$

17. $-8 + 8$

18. $73 + (-73)$

19. $-1 + (-4) + (-3)$

20. $3 + (-2) + (-4)$

21. $[7 + (-9)] + (-4 + 16)$

22. $(-2 + 11) + [(-5) + 4]$

23. $-4 + 0$

24. $0 + (-20)$

25. $-2 + (-1) + (-76) + 1 + 2$

26. $-5 + (-31) + 9 + (-9) + 5$

27. Find the sum of -102, 73, and -345.

28. What is 3,187 more than -59?

29. a. What is the additive inverse of -11?

    b. What is the additive inverse of 4?

    c. Is the sum of two positive integers always positive?

    d. Is the sum of two negative integers always negative?

    e. Is the sum of a positive integer and a negative integer always positive?

    f. Is the sum of a positive integer and a negative integer always negative?

30. Population change. The number of births, deaths, net migration, and net immigration for a large country are shown in the graph below. Determine the net change in the population for 2016.

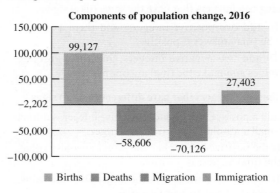

Components of population change, 2016

31. Drought. During a drought, the water level in a reservoir fell to a point 100 feet below normal. After a lot of rain in April, it rose 16 feet, and after even more rain in May, it rose another 18 feet.

   a. Express the water level of the reservoir before the rainy months as a signed number.

   b. What was the water level after the rain?

32. Temperature extremes. The world record for lowest temperature is $-129°F$. It was set on July 21, 1983, in Antarctica. The world record for highest temperature is an amazing 263°F warmer. It was set on July 10, 1913, in Death Valley, California. Find the record high temperature. (Source: *The World Almanac and Book of Facts*, 2017)

SECTION 2.3 ▶ Subtracting Integers

DEFINITIONS AND CONCEPTS	EXAMPLES
The **rule for subtraction** is helpful when subtracting signed numbers. To subtract two integers, add the first integer to the opposite of the integer to be subtracted. *Subtracting is the same as adding the opposite.*	Subtract: $3 - (-5)$ Add … $3 - (-5) = 3 + 5 = 8$ Use the rule for adding two integers with the same sign. …the opposite Check using addition: $8 + (-5) = 3$
After rewriting a subtraction as addition of the opposite, use one of the rules for the addition of signed numbers discussed in Section 2.2 to find the result.	Subtract: $-3 - 5 = -3 + (-5) = -8$ Add the opposite of 5, which is -5. $-4 - (-7) = -4 + 7 = 3$ Add the opposite of -7, which is 7.
Be careful when translating the instruction to subtract one number *from* another number.	Subtract -6 from -9. $-9 - (-6)$ The number to be subtracted is -6.
Expressions can involve repeated subtraction or combinations of subtraction and addition. To evaluate them, we use the **order of operations rule** discussed in Section 1.9.	Evaluate: $-43 - (-6 - 15)$ $-43 - (-6 - 15) = -43 - [-6 + (-15)]$ Within the parentheses, add the opposite of 15, which is -15. $= -43 - [-21]$ Within the brackets, add -6 and -15. $= -43 + 21$ Add the opposite of -21, which is 21. $= -22$ Use the rule for adding integers that have different signs.
When we find the difference between the maximum value and the minimum value of a collection of measurements, we are finding the **range** of the values. Range = maximum value − minimum value	**Geography.** The highest point in the United States is Mt. McKinley at 20,230 feet. The lowest point is -282 feet at Death Valley, California. Find the range between the highest and lowest points. Range = $20{,}230 - (-282)$ Range = $20{,}230 + 282$ Add the opposite of -282, which is 282. Range = $20{,}602$ Do the addition. The range between the highest point and lowest point in the United States is 20,602 feet.
To find the **change** in a quantity, we subtract the *earlier value* from the *later value*. Change = later value − earlier value	**Submarines.** A submarine was traveling at a depth of 165 feet below sea level. The captain ordered it to a new position of only 8 feet below the surface. Find the change in the depth of the submarine. We can represent 165 feet below sea level as -165 feet and 8 feet below the surface as -8 feet. Change of depth = $-8 - (-165)$ Subtract the earlier depth from the later depth. Change of depth = $-8 + 165$ Add the opposite of -165, which is 165. Change of depth = 157 Use the rule for adding integers that have different signs. The change in the depth of the submarine was 157 feet.

33. Fill in the blank: Subtracting an integer is the same as adding the _____ of that integer.

34. Write each phrase using symbols.
  a. negative nine minus negative one
  b. negative ten subtracted from negative six

Subtract.

35. $5 - 8$

36. $-9 - 12$

37. $-4 - (-8)$

38. $-8 - (-2)$

39. $-6 - 106$

40. $-7 - 1$

41. $0 - 37$

42. $0 - (-30)$

Evaluate each expression.

43. $12 - 2 - (-6)$

44. $-16 - 9 - (-1)$

45. $-9 - 7 + 12$

46. $-5 - 6 + 33$

47. $1 - (2 - 7)$

48. $-12 - (6 - 10)$

49. $-70 - [(-6) - 2]$

50. $89 - [(-2) - 12]$

51. $-(-5) + (-28) - 2 - (-100)$

52. a. Subtract 27 from -50.

  b. Subtract -50 from 27.

Use signed numbers to solve each problem.

53. Mining. Some miners discovered a small vein of gold at a depth of 150 feet. This encouraged them to continue their exploration. After descending another 75 feet, they came upon a much larger find. Use a signed number to represent the depth of the second discovery.

54. Record temperatures. The lowest and highest recorded temperatures for Alaska and Virginia are shown. For each state, find the range between the record high and low temperatures.

Alaska	Virginia
Low: $-80°$ Jan. 23, 1971	Low: $-30°$ Jan. 21, 1985
High: $100°$ June 27, 1915	High: $110°$ July 15, 1954

55. Politics. On July 20, 2007, a CNN/Opinion Research poll had Barack Obama trailing Hillary Clinton in the South Carolina Democratic presidential primary race by 16 points. On January 26, 2008, Obama finished 28 points ahead of Clinton in the actual primary. Find the point change in Barack Obama's support.

56. Overdraft fees. A student had a balance of $255 in her checking account. She wrote a check for rent for $300, and when it arrived at the bank, she was charged an overdraft fee of $35. What is the new balance in her account?

SECTION 2.4 ▶ **Multiplying Integers**

DEFINITIONS AND CONCEPTS	EXAMPLES
Multiplying two integers that have different (unlike) signs To multiply a positive integer and a negative integer, multiply their absolute values. Then make the final answer negative.	Multiply: $6(-8)$ Find the absolute values: $\|6\| = 6$ and $\|-8\| = 8$. $6(-8) = -48$ Multiply the absolute values, 6 and 8, to get 48. Then make the final answer negative.
Multiplying two integers that have the same (like) signs To multiply two integers that have the same sign, multiply their absolute values. The final answer is positive.	Multiply: $-2(-7)$ Find the absolute values: $\|-2\| = 2$ and $\|-7\| = 7$. $-2(-7) = 14$ Multiply the absolute values, 2 and 7, to get 14. The final answer is positive.
To **evaluate expressions** that contain several multiplications, we make repeated use of the rules for multiplying two integers. Another approach to evaluate expressions is to use the **commutative** and/or **associative properties of multiplication** to reorder and regroup the factors in a helpful way.	Evaluate $-5(3)(-6)$ in two ways. Perform the multiplications, working left to right. $-5(3)(-6) = -15(-6)$ $= 90$ First, multiply the pair of negative factors. $-5(3)(-6) = 30(3)$ Multiply the negative factors to produce a positive product. $= 90$

Multiplying an even and an odd number of negative integers The product of an even number of negative integers is positive. The product of an odd number of negative integers is negative.	Four negative factors: $-5(-1)(-6)(-2) = 60$ ⟶ positive Five negative factors: $-2(-4)(-3)(-1)(-5) = -120$ ⟶ negative
Even and odd powers of a negative integer When a negative integer is raised to an even power, the result is positive. When a negative integer is raised to an odd power, the result is negative.	Evaluate: $(-3)^4 = (-3)(-3)(-3)(-3)$ The exponent is even. $\qquad = 9(9)$ Multiply pairs of integers. $\qquad = 81$ The answer is positive. Evaluate: $(-2)^3 = (-2)(-2)(-2)$ The exponent is odd. $\qquad = -8$ The answer is negative.
Although the exponential expressions $(-6)^2$ and -6^2 look similar, they are not the same. The bases are different.	Evaluate: $(-6)^2$ and -6^2 Because of the parentheses, the base is -6. The exponent is 2. $\quad\mid\quad$ Since there are no parentheses around -6, the base is 6. The exponent is 2. $(-6)^2 = (-6)(-6)$ $\quad\mid\quad$ $-6^2 = -(6 \cdot 6)$ $\qquad = 36$ $\quad\mid\quad$ $\qquad = -36$
Application problems that involve **repeated addition** are often more easily solved using multiplication.	**Chemistry.** A chemical compound that is normally stored at 0°F had its temperature lowered 8°F each hour for 6 hours. What signed number represents the change in temperature of the compound after 6 hours? $-8 \cdot 6 = -48$ Multiply the change in temperature each hour by the number of hours. The change in temperature of the compound is -48°F.

REVIEW EXERCISES

Multiply.

57. $7(-2)$

58. $(-8)(47)$

59. $-23(-14)$

60. $-5(-5)$

61. $-1 \cdot 25$

62. $(6)(-34)$

63. $-4,000(17,000)$

64. $-100,000(-300)$

65. $(-6)(-2)(-3)$

66. $-4(-3)(-3)$

67. $(-3)(4)(2)(-5)$

68. $(-1)(10)(10)(-1)$

69. Find the product of -15 and the opposite of 30.

70. Find the product of the opposite of 16 and the opposite of 3.

71. Deficits.   A state treasurer's prediction of a tax shortfall was two times worse than the actual deficit of $130 million. The governor's prediction of the same shortfall was even worse—three times the amount of the actual deficit. Complete the labeling of the vertical axis of the graph below to show the two incorrect predictions.

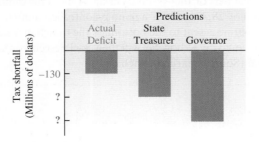

72. Mining.   An elevator is used to lower coal miners from the ground level entrance to various depths in the mine. The elevator stops every 45 vertical feet to let off miners. At what depth do the miners work who get off the elevator at the 12th stop?

Evaluate each expression.

73. $(-5)^3$

74. $(-2)^5$

75. $(-8)^4$

76. $(-4)^4$

77. When $(-17)^9$ is evaluated, will the result be positive or negative?

78. Explain the difference between -9^2 and $(-9)^2$ and then evaluate each expression.

SECTION 2.5 ▶ Dividing Integers

DEFINITIONS AND CONCEPTS	EXAMPLES

Dividing two integers
To divide two integers, divide their absolute values.
 	
1. The quotient of two integers that have the same (*like*) signs is positive.
 	
2. The quotient of two integers that have different (*unlike*) signs is negative.
 	
To **check division** of integers, multiply the *quotient* and the *divisor*. You should get the *dividend*.
	Divide: $\dfrac{-21}{-7}$
	
	Find the absolute values: $\|-21\| = 21$ and $\|-7\| = 7$.
	
	$\dfrac{-21}{-7} = 3$ Divide the absolute values, 21 by 7, to get 3.
	    The final answer is positive.
	
	Check: $3(-7) = -21$ The result checks.
	
	Divide: $-54 \div 9$
	
	Find the absolute values: $\|-54\| = 54$ and $\|9\| = 9$.
	
	$-54 \div 9 = -6$ Divide the absolute values, 54 by 9, to get 6.
	    Then make the final answer negative.
	
	Check: $-6(9) = -54$ The result checks.

Division with 0 If 0 is divided *by* any nonzero number, the quotient is 0. Division *of* any nonzero number by 0 is undefined.	$\dfrac{0}{-8} = 0$ $0 \div (-20) = 0$ $\dfrac{-2}{0}$ is undefined. $-6 \div 0$ is undefined.
Problems that involve forming **equal-sized groups** can be solved by division.	**Used car sales.** The price of a used car was reduced each day by an equal amount because it was not selling. After 7 days , and a \$1,050 reduction in price, the car was finally purchased. By how much was the price of the car reduced each day? $\dfrac{-1,050}{7} = -150$ Divide the change in the price of the car by the number of days the price was reduced. The negative result indicates that the price of the car was *reduced* by \$150 each day.

REVIEW EXERCISES

79. Fill in the blanks: We know that $\dfrac{-15}{5} = -3$ because

  ▢ (▢) = ▢ .

80. Check using multiplication to determine whether $-152 \div (-8) = 18$.

Divide, if possible.

81. $\dfrac{25}{-5}$ **82.** $\dfrac{-14}{7}$

83. $-64 \div (-8)$ **84.** $72 \div (-9)$

85. $\dfrac{-10}{-1}$ **86.** $\dfrac{-673}{-673}$

87. $-150,000 \div 3,000$ **88.** $-24,000 \div (-60)$

89. $\dfrac{-1,058}{-46}$ **90.** $-272 \div 16$

91. $\dfrac{0}{-5}$ **92.** $\dfrac{-4}{0}$

93. Divide -96 by 3.

94. Find the quotient of -125 and -25.

95. Production time.   Because of improved production procedures, the time needed to produce an electronic component dropped by 12 minutes over the past 6 months. If the drop in production time was uniform, how much did it change each month over this period of time?

96. Ocean exploration.   The Puerto Rico Trench is the deepest part of the Atlantic Ocean. It has a maximum depth of 28,374 feet. If a remote-controlled unmanned submarine is sent to the bottom of the trench in a series of 6 equal dives, how far will the vessel descend on each dive? (Source: marianatrench.com)

SECTION 2.6 ▶ **Order of Operations and Estimation**

DEFINITIONS AND CONCEPTS	EXAMPLES
Order of operations	Evaluate: $-3(-5)^2 - (-40)$
1. Perform all calculations within parentheses and other grouping symbols following the order listed in Steps 2–4 below, working from the innermost pair of grouping symbols to the outermost pair.	$-3(-5)^2 - (-40) = -3(25) - (-40)$ *Evaluate the exponential expression.*
2. Evaluate all exponential expressions.	$= -75 - (-40)$ *Do the multiplication.*
3. Perform all multiplications and divisions as they occur from left to right.	$= -75 + 40$ *Use the subtraction rule: Add the opposite of −40.*
4. Perform all additions and subtractions as they occur from left to right.	$= -35$ *Do the addition.*
When grouping symbols have been removed, repeat Steps 2–4 to complete the calculation.	Evaluate: $\dfrac{-6 + 4(-2)}{16 - (-3)^2}$
If a fraction bar is present, evaluate the expression above the bar (called the **numerator**) and the expression below the bar (called the **denominator**) separately. Then perform the division indicated by the fraction bar, if possible.	$\dfrac{-6 + 4(-2)}{16 - (-3)^2} = \dfrac{-6 + (-8)}{16 - 9}$ *In the numerator, do the multiplication. In the denominator, evaluate the exponential expression.*
	$= \dfrac{-14}{7}$ *In the numerator, do the addition. In the denominator, do the subtraction.*
	$= -2$ *Do the division.*
Absolute value symbols are grouping symbols, and by the order of operations rule, all calculations within grouping symbols must be performed first.	Evaluate: $10 - 2\,|-8 + 1|$
	$10 - 2\,|-8 + 1| = 10 - 2\,|-7|$ *Do the addition within the absolute value symbols.*
	$= 10 - 2(7)$ *Find the absolute value of −7.*
	$= 10 - 14$ *Do the multiplication.*
	$= -4$ *Do the subtraction.*
When an exact answer is not necessary and a quick approximation will do, we can use **estimation**.	Estimate the value of $-56 + (-67) + 89 + (-41) + 14$ by rounding each number to the nearest ten.
	$-60 + (-70) + 90 + (-40) + 10$
	$= -170 + 100$ *Add the positives and the negatives separately.*
	$= -70$ *Do the addition.*

REVIEW EXERCISES

Evaluate each expression.

97. $7 - 8 \cdot 6$

98. $7 - (-2)^2 + 1$

99. $65 - 8(9) - (-47)$

100. $-3(-2)^3 - 16$

101. $-2(5)(-4) + \dfrac{|-9|}{3^2}$

102. $-4^2 + (-4)^2$

103. $-12 - (8 - 9)^2$

104. $7|-8| - 2(3)(4)$

105. $-4\left(\dfrac{15}{-3}\right) - 2^3$

106. $-20 + 2(12 - 5 \cdot 2)$

107. $-20 + 2[12 - (-7 + 5)^2]$

108. $8 - 6|-3 \cdot 4 + 5|$

109. $\dfrac{2 \cdot 5 + (-6)}{-3 - 1^5}$

110. $\dfrac{3(-6) - 11 + 1}{4^2 - 3^2}$

111. $-\left[1 - \left(2^3 + \dfrac{100}{-50}\right)\right]$

112. $-\left[45 - \left(5^3 + \dfrac{100}{-4}\right)\right]$

113. Round each number to the nearest hundred to estimate the value of the following expression: $-4{,}471 + 7{,}935 + 2{,}094 + (-3{,}188)$

114. Find the mean (average) of $-8, 4, 7, -11, 2, 0, -6,$ and -4.

1. Fill in the blanks.

   a. $\{\,\ldots,-5,-4,-3,-2,-1,0,1,2,3,4,5,\ldots\,\}$ is called the set of _____.

   b. The symbols $>$ and $<$ are called _____ symbols.

   c. The _____ _____ of a number is the distance between the number and 0 on the number line.

   d. Two numbers that are the same distance from 0 on the number line, but on opposite sides of it, are called _____.

   e. In the expression $(-3)^5$, the _____ is -3 and 5 is the _____.

2. Insert one of the symbols $>$ or $<$ in the blank to make the statement true.

   a. -8 ___ -9 b. -213 ___ 123 c. -5 ___ 0

3. Tell whether each statement is true or false.

   a. $19 \geq 19$ b. $-(-8) = 8$

   c. $-|-2| > |6|$ d. $-7 + 0 = 0$

   e. $-5(0) = 0$

4. **School enrollment.** According to the projections in the table, which high school will face the greatest shortage of classroom seats in the year 2025?

High Schools with Shortage of Classroom Seats by 2025	
Lyons	-669
Tolbert	$-1,630$
Poly	$-2,488$
Cleveland	-350
Samuels	-586
South	$-2,379$
Van Owen	$-1,690$
Twin Park	-462
Heywood	$-1,004$
Hampton	-774

5. Graph the following numbers on a number line: $-3, 4, -1,$ and 3.

6. Add.

   a. $-16 + 14$ b. $-72 + (-73)$

   c. $8 + (-6) + (-9) + 5 + 1$

   d. $(-31 + 12) + [3 + (-16)]$

   e. $-24 + (-3) + 24 + (-5) + 5$

7. Subtract.

   a. $-7 - 6$ b. $-7 - (-6)$

   c. $82 - (-109)$ d. $0 - 15$

   e. $-29 - (-60) - 71$

8. Multiply.

   a. $-10 \cdot 7$ b. $(-4)(-8)$

   c. $-4(2)(-6)$ d. $-9(-3)(-1)(-2)$

   e. $-20,000(1,300)$

9. Write the related multiplication statement for $\dfrac{-20}{-4} = 5$.

10. Divide and check the result.

    a. $\dfrac{-32}{4}$ b. $24 \div (-3)$

    c. $-54 \div (-6)$ d. $\dfrac{408}{-12}$

    e. $-560,000 \div 7,000$

11. a. What is 15 more than -27?

    b. Subtract -19 from -1.

    c. Divide -28 by -7.

    d. Find the product of 10 and the opposite of 8.

12. a. What property is shown: $-3 + 5 = 5 + (-3)$

    b. What property is shown: $-4(-10) = -10(-4)$

    c. Fill in the blank: Subtracting is the same as _____ the opposite.

13. Divide, if possible.

    a. $\dfrac{-21}{0}$ b. $\dfrac{-5}{1}$

    c. $\dfrac{0}{-6}$ d. $\dfrac{-18}{-18}$

14. Evaluate each expression:

    a. $(-4)^2$ b. -4^2

Evaluate each expression.

15. $4 - (-3)^2 - (-6)$ 16. $-18 \div 2 \cdot 3$

17. $-3 + \left(\dfrac{-16}{4}\right) - 3^3$ 18. $94 - 3[-7 + (5 - 8)^2]$

19. $\dfrac{4(-6) - 4^2 + (-2)}{-3 - 4 \cdot 1^5}$ 20. $6(-2 \cdot 6 + 5 \cdot 4)$

21. $21 - 9|-3 - 4 + 2|$

22. $-\left[2 - \left(4^3 + \dfrac{20}{-5}\right)\right]$

23. **Chemistry.** In a lab, the temperature of a fluid was reduced 6°F per hour for 12 hours. What signed number represents the change in temperature?

24. **Gambling.** On the first hand of draw poker, a player won the chips shown on the left. On the second hand, he lost the chips shown on the right. Determine his net gain or loss for the first two hands. The dollar value of each colored poker chip is shown.

25. **Geography.** The lowest point on the African continent is the Qattara Depression in the Sahara Desert, 436 feet below sea level. The lowest point on the North American continent is Death Valley, California, 282 feet below sea level. Find the difference in these elevations.

26. **Trams.** A tram line makes a 5,250-foot descent from a mountaintop to the base of the mountain in 15 equal stages. How much does it descend in each stage?

27. **Card games.** After the first round of a card game, Tommy had a score of 8. When he lost the second round, he had to deduct the value of the cards left in his hand from his first-round score. (See the illustration.) What was his score after two rounds of the game? For scoring, face cards (Kings, Queens, and Jacks) are counted as 10 points and aces as 1 point.

28. **Bank takeovers.** Before three investors can take over a failing bank, they must repay the losses that the bank had over the past three quarters, as shown below. If the investors plan equal ownership, how much of the bank's total losses is each investor responsible for?

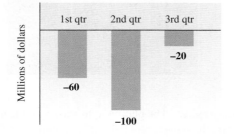

29. **Banking.** After making deposits of $125 and $100, a student's account was still $19 overdrawn. What was her account balance before the deposits?

30. **Elevators.** The weight of the passengers on board an elevator as it traveled from the first to the second floor was 165 pounds *under capacity*. When the doors opened at the second floor, no one exited, and several people entered. The weight of the passengers in the elevator was then 85 pounds *over* capacity. What was the weight of the people who boarded the elevator on the second floor?

31. **World's coldest ice cream.** Dippin'Dots is an ice cream snack that is produced by flash freezing an ice cream mixture in liquid nitrogen. The temperature at which the multicolored dots are produced is about nine times as cold as the temperature at which they are stored for sale. If Dippin'Dots are produced at $-360°F$, at what temperature are they stored?

32. **Crocs.** If you divide the 2015 net income of Crocs Company (makers of shoes, sandals, and clogs) by 5, the result is approximately the net income for the company in 2016. If the company lost $83,196,000 in 2015, what was its net income in 2016? (Source: wikinvest.com)

1. Consider the number 7,326,549. [Section 1.1]

   a. What is the place value of the digit 7?
   b. Which digit is in the hundred thousands column?
   c. Round to the nearest hundred.
   d. Round to the nearest ten thousand.

2. Bids. A school district received the bids shown in the table for electrical work. If the lowest bidder wins, which company should be awarded the contract?
[Section 1.1]

Citrus Unified School District Bid 02-9899 Cabling and Conduit Installation	
Datatel	$2,189,413
Walton Electric	$2,201,999
Advanced Telecorp	$2,175,081
CRF Cable	$2,174,999
Clark & Sons	$2,175,801

3. Nuclear reactors. The table gives the number of nuclear reactors operating in the United States for selected years. Complete the bar graph using the given data. [Section 1.1]

Year	1984	1989	1994	1999	2004	2009	2014
Plants	87	111	109	104	104	104	99

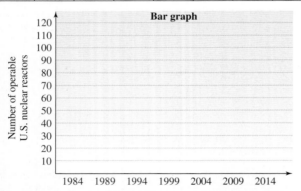

Source: allcountries.org and *The World Almanac and Book of Facts*, 2017

4. Thread count. The thread count of a fabric is the sum of the number of horizontal and vertical threads woven in one square inch of fabric. One square inch of a bed sheet is shown below. Find the thread count. [Section 1.2]

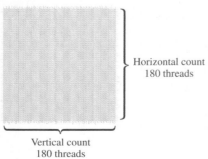

Horizontal count 180 threads

Vertical count 180 threads

Add. [Section 1.2]

5. 1,237 + 68 + 549

6.
$$\begin{array}{r} 8,907 \\ 2,345 \\ 7,899 \\ + 5,237 \end{array}$$

Subtract. [Section 1.3]

7. 6,375 − 2,569

8.
$$\begin{array}{r} 5,369 \\ - 685 \end{array}$$

9.
$$\begin{array}{r} 39,506 \\ - 1,729 \end{array}$$

10. Subtract 304 from 1,736. [Section 1.3]

11. Check the subtraction below using addition. Is it correct? [Section 1.3]

$$\begin{array}{r} 469 \\ - 237 \\ \hline 132 \end{array}$$

12. Shipping furniture. In a shipment of 147 pieces of furniture, 27 pieces were sofas, 55 were leather chairs, and the rest were wooden chairs. Find the number of wooden chairs. [Section 1.3]

Multiply. [Section 1.4]

13. 435 · 27

14.
$$\begin{array}{r} 9,183 \\ \times 602 \end{array}$$

15. 3,100 · 7,000

16. Packaging. There are 3 tennis balls in one can, 24 cans in one case, and 12 cases in one box. How many tennis balls are there in one box? [Section 1.4]

17. Gardening. Find the perimeter and the area of the rectangular garden shown below. [Section 1.4]

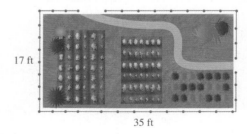

17 ft

35 ft

18. Photography. The photographs shown on the next page are the same except that different numbers of *pixels* (squares of color) are used to display them. The number of pixels in each row and each column of the photographs are given. Find the total number of pixels in each photograph. [Section 1.4]

5 pixels   12 pixels

100 pixels

Divide. [Section 1.5]

19. $\dfrac{701}{8}$

20. $1{,}261 \div 97$

21. $38\overline{)17{,}746}$

22. $350\overline{)9{,}800}$

23. Check the division below using multiplication. Is it correct? [Section 1.5]

$9\overline{)1{,}962} = 218$

24. Gardening. A metal can holds 320 fluid ounces of gasoline. How many times can the 30-ounce tank of a lawnmower be filled from the can? How many ounces of gasoline will be left in the can? [Section 1.5]

25. Baking. A baker uses 4-ounce pieces of bread dough to make dinner rolls. How many dinner rolls can he make from 15 pounds of dough? (*Hint:* There are 16 ounces in one pound.) [Section 1.6]

26. List the factors of 18, from least to greatest. [Section 1.7]

27. Identify each number as a prime number, a composite number, or neither. Then identify it as an even number or an odd number. [Section 1.7]
   a. 17   **b.** 18
   c. 0   **d.** 1

28. Find the prime factorization of 504. Use exponents to express your answer. [Section 1.7]

29. Write the expression $11 \cdot 11 \cdot 11 \cdot 11$ using an exponent. [Section 1.7]

30. Evaluate: $5^2 \cdot 7$ [Section 1.7]

31. Find the LCM of 8 and 12. [Section 1.8]

32. Find the LCM of 3, 6, and 15. [Section 1.8]

33. Find the GCF of 30 and 48. [Section 1.8]

34. Find the GCF of 81, 108, and 162. [Section 1.8]

Evaluate each expression. [Section 1.9]

35. $4 - 2[26 + 2(5 - 3)]$

36. $264 \div 4 - 7(4)2$ **37.** $\dfrac{4^2 - 2 \cdot 3}{2 + (3^2 - 3 \cdot 2)}$

38. Speed checks. A traffic officer used a radar gun and found that the speeds of several cars traveling on Main Street were:

   38 mph, 42 mph, 36 mph, 38 mph, 48 mph, 44 mph

   What was the mean (average) speed of the cars traveling on Main Street? [Section 1.9]

39. Graph the following integers on a number line. [Section 2.1]
   a. $-2, -1, 0, 2$

   ![number line from -3 to 3]

   b. The integers greater than -4 but less than 2

   ![number line from -4 to 4]

40. Find the sum of -11, 20, -13, and 1. [Section 2.2]

Use signed numbers to answer each problem.

41. Lie detector tests. A burglar scored -18 on a polygraph test, a score that indicates deception. However, on a second test, he scored 3, a score that is uncertain. Find the change in the scores. [Section 2.3]

42. Banking. A student has \$48 in his checking account. He then writes a check for \$105 to purchase books. The bank honors the check but charges the student an additional \$22 service fee for being overdrawn. What is the student's new checking account balance? [Section 2.3]

43. Chemistry. The *melting point* of a solid is the temperature range at which it changes state from solid to liquid. The melting point of helium is seven times colder than the melting point of mercury. If the melting point of mercury is $-39°$ Celsius, what is the melting point of helium? (Source: chemicalelements.com) [Section 2.4]

44. Buying a business. When 12 investors decided to buy a bankrupt company, they agreed to assume equal shares of the company's debt of \$660,000. How much debt was each investor responsible for? [Section 2.5]

Evaluate each expression. [Section 2.6]

45. $5 + (-3)(-7)(-2)$

46. $-2[-6(5 \cdot 1^3) - 5]$

47. $\dfrac{10 - (-5)}{1 - 2 \cdot 3}$

48. $\dfrac{3(-6) - 10}{3^2 - 4^2}$

49. $3^4 + 6(-12 + 5 \cdot 4)$

50. $15 - 2|\,-3 - 4|$

51. $2\left(\dfrac{-12}{3}\right) + 3(-5)$

52. $-9^2 + (-9)^2$

53. $-\left|\dfrac{45}{-9} - (-9)\right|$

54. $\dfrac{-4(-5) - 2}{3 - 3^2}$

For Exercises 55 and 56, quickly determine a reasonable estimate of the exact answer. [Section 2.6]

55. Camping. Hikers make a 1,150-foot descent into a canyon in 12 stages. What signed number represents how far they descend in each stage?

56. Recalls. An automobile maker has to recall 19,250 cars because they have a faulty engine mount. If it costs $195 to repair each car, what signed number represents the loss the company will suffer because of the recall?

3 Fractions and Mixed Numbers

iStock.com/JoseGirarte

from **Campus to Careers**

School Guidance Counselor

School guidance counselors plan academic programs and help students choose the best courses to take to achieve their educational goals. Counselors often meet with students to discuss the life skills needed for personal and social growth. To prepare for this career, guidance counselors take classes in an area of mathematics called *statistics*, where they learn how to collect, analyze, explain, and present data.

In **Problem 115** of **Study Set 3.4**, you will see how a counselor must be able to add fractions to better understand a graph that shows students' study habits.

JOB TITLE:
School Guidance Counselor

EDUCATION:
A master's degree is usually required to be licensed as a counselor. However, some schools accept a bachelor's degree with the appropriate counseling courses.

JOB OUTLOOK:
The number of openings will increase by about 8% from 2014 through 2024.

ANNUAL EARNINGS:
The average (median) salary in 2015 was $58,660.

FOR MORE INFORMATION:
www.bls.gov/ooh/
community-and-social-service

OBJECTIVES

1 Identify the numerator and denominator of a fraction.

2 Simplify special fraction forms.

3 Define equivalent fractions.

4 Build equivalent fractions.

5 Simplify fractions.

SECTION **3.1** An Introduction to Fractions

Whole numbers are used to count objects, such as cars, stamps, eggs, and books. When we need to describe a part of a whole, such as one-half of a pie, three-quarters of an hour, or a one-third-pound burger, we can use *fractions*.

One-half of a cherry pie	Three-quarters of an hour	A one-third-pound burger
$\dfrac{1}{2}$	$\dfrac{3}{4}$	$\dfrac{1}{3}$

LANGUAGE OF MATHEMATICS

The word *fraction* comes from the Latin word *fractio* meaning "breaking in pieces."

OBJECTIVE 1 **Identify the numerator and denominator of a fraction.**

A **fraction** describes the number of equal parts of a whole. For example, consider the figure below with 5 of the 6 equal parts colored red. We say that $\frac{5}{6}$ (five-sixths) of the figure is shaded.

In a fraction, the number above the **fraction bar** is called the **numerator**, and the number below is called the **denominator**.

Fraction bar $\longrightarrow$ $\dfrac{5}{6}$ $\begin{array}{l}\longleftarrow \text{numerator} \\ \longleftarrow \text{denominator}\end{array}$

EXAMPLE 1 Identify the numerator and denominator of each fraction:

a. $\dfrac{11}{12}$ **b.** $\dfrac{8}{3}$

Strategy We will find the number above the fraction bar and the number below it.

WHY The number above the fraction bar is the numerator, and the number below is the denominator.

Solution

a. $\dfrac{11}{12}$ $\begin{array}{l}\longleftarrow \text{numerator} \\ \longleftarrow \text{denominator}\end{array}$ **b.** $\dfrac{8}{3}$ $\begin{array}{l}\longleftarrow \text{numerator} \\ \longleftarrow \text{denominator}\end{array}$

Self Check 1

Identify the numerator and denominator of each fraction:

a. $\dfrac{7}{9}$

b. $\dfrac{21}{20}$

Now Try ➡ Problem 21

If the numerator of a fraction is less than its denominator, the fraction is called a **proper fraction**. A proper fraction is less than 1. If the numerator of a fraction is greater than or equal to its denominator, the fraction is called an **improper fraction**. An improper fraction is greater than or equal to 1.

LANGUAGE OF MATHEMATICS

The phrase *improper fraction* is somewhat misleading. In algebra and other mathematics courses, we often use such fractions "properly" to solve many types of problems.

Proper fractions

$$\frac{1}{4}, \quad \frac{2}{3}, \quad \text{and} \quad \frac{98}{99}$$

Improper fractions

$$\frac{7}{2}, \quad \frac{98}{97}, \quad \frac{16}{16}, \quad \text{and} \quad \frac{5}{1}$$

EXAMPLE 2   Write fractions that represent the shaded and unshaded portions of the figure below.

Strategy We will determine the number of equal parts into which the figure is divided. Then we will determine how many of those parts are shaded.

WHY The denominator of a fraction shows the number of equal parts in the whole. The numerator shows how many of those parts are being considered.

Solution
Since the figure is divided into 3 equal parts, the denominator of the fraction is 3. Since 2 of those parts are shaded, the numerator is 2, and we say that

$\frac{2}{3}$ of the figure is shaded.   Write: $\dfrac{\text{number of parts shaded}}{\text{number of equal parts}}$

Since 1 of the 3 equal parts of the figure is not shaded, the numerator is 1, and we say that

$\frac{1}{3}$ of the figure is not shaded.   Write: $\dfrac{\text{number of parts not shaded}}{\text{number of equal parts}}$

Self Check 2

Write fractions that represent the portion of the month that has passed and the portion that remains.

DECEMBER

X̶	X̶	X̶	X̶	X̶	X̶	X̶
X̶	X̶	X̶0	X̶1	12	13	14
15	16	17	18	19	20	21
22	23	24	25	26	27	28
29	30	31				

Now Try ⟶ **Problems 25 and 105**

There are times when a negative fraction is needed to describe a quantity. For example, if an earthquake causes a road to sink seven-eighths of an inch, the amount of downward movement can be represented by $-\frac{7}{8}$. Negative fractions can be written in three ways. The negative sign can appear in the numerator, in the denominator, or in front of the fraction.

$$\frac{-7}{8} = \frac{7}{-8} = -\frac{7}{8} \qquad \qquad \frac{-15}{4} = \frac{15}{-4} = -\frac{15}{4}$$

Notice that the examples above agree with the rule from Chapter 2 for dividing integers with different (unlike) signs: *the quotient of a negative integer and a positive integer is negative.*

OBJECTIVE 2 Simplify special fraction forms.

Recall from Section 1.5 that a fraction bar indicates division. This fact helps us simplify four special fraction forms.

- *Fractions that have the same numerator and denominator:* In this case, we have a number divided by itself. The result is 1 (provided the numerator and denominator are not 0). We call each of the following fractions a **form of 1**.

$$1 = \frac{1}{1} = \frac{2}{2} = \frac{3}{3} = \frac{4}{4} = \frac{5}{5} = \frac{6}{6} = \frac{7}{7} = \frac{8}{8} = \frac{9}{9} = \cdots$$

- *Fractions that have a denominator of 1:* In this case, we have a number divided by 1. The result is simply the numerator.

$$\frac{5}{1} = 5 \qquad \frac{24}{1} = 24 \qquad \frac{-7}{1} = -7$$

- *Fractions that have a numerator of 0:* In this case, we have division of 0. The result is 0 (provided the denominator is not 0).

$$\frac{0}{8} = 0 \qquad \frac{0}{56} = 0 \qquad \frac{0}{-11} = 0$$

- *Fractions that have a denominator of 0:* In this case, we have division by 0. The division is undefined.

$$\frac{7}{0} \text{ is undefined} \qquad \frac{-18}{0} \text{ is undefined}$$

EXAMPLE 3   Simplify, if possible: **a.** $\dfrac{12}{12}$ **b.** $\dfrac{0}{24}$ **c.** $\dfrac{18}{0}$ **d.** $\dfrac{9}{1}$

Strategy To simplify each fraction, we will divide the numerator by the denominator, if possible.

WHY A fraction bar indicates division.

Solution

a. $\dfrac{12}{12} = 1$ *This corresponds to dividing a quantity into 12 equal parts and then considering all 12 of them. We would get 1 whole quantity.*

b. $\dfrac{0}{24} = 0$ *This corresponds to dividing a quantity into 24 equal parts and then considering 0 (none) of them. We would get 0.*

c. $\dfrac{18}{0}$ is undefined   *This corresponds to dividing a quantity into 0 equal parts and then considering 18 of them. That is not possible.*

d. $\dfrac{9}{1} = 9$ *This corresponds to "dividing" a quantity into 1 equal part and then considering 9 of them. We would get 9 of those quantities.*

Self Check 3

Simplify, if possible:

a. $\dfrac{4}{4}$ **b.** $\dfrac{51}{1}$ **c.** $\dfrac{45}{0}$ **d.** $\dfrac{0}{6}$

Now Try ➲ Problem 33

OBJECTIVE 3 Define equivalent fractions.

Fractions can look different but still represent the same part of a whole. To illustrate this, consider the identical rectangular regions on the next page. The first one is divided into 10 equal parts. Since 6 of those parts are red, $\frac{6}{10}$ of the figure is shaded.

The second figure is divided into 5 equal parts. Since 3 of those parts are red, $\frac{3}{5}$ of the figure is shaded. We can conclude that $\frac{6}{10} = \frac{3}{5}$ because $\frac{6}{10}$ and $\frac{3}{5}$ represent the same shaded portion of the figure. We say that $\frac{6}{10}$ and $\frac{3}{5}$ are *equivalent fractions*.

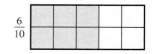

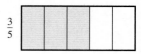

> **Equivalent Fractions**
>
> Two fractions are **equivalent** if they represent the same number. **Equivalent fractions** represent the same portion of a whole.

OBJECTIVE 4   Build equivalent fractions.

Writing a fraction as an equivalent fraction with a *larger* denominator is called **building the fraction**. To build a fraction, we use a familiar property from Chapter 1 that is also true for fractions:

> **Multiplication Property of 1**
>
> The product of any fraction and 1 is that fraction.

We also use the following rule for multiplying fractions. (It will be discussed in greater detail in the next section.)

> **Multiplying Fractions**
>
> To multiply two fractions, multiply the numerators and multiply the denominators.

To build an equivalent fraction for $\frac{1}{2}$ with a denominator of 8, we first ask, "What number times 2 equals 8?" To answer that question we *divide* 8 by 2 to get 4. Since we need to multiply the denominator of $\frac{1}{2}$ by 4 to obtain a denominator of 8, it follows that $\frac{4}{4}$ should be the form of 1 that is used to build an equivalent fraction for $\frac{1}{2}$.

$$\frac{1}{2} = \frac{1}{2} \cdot \boxed{\frac{4}{4}}$$ Multiply $\frac{1}{2}$ by 1 in the form of $\frac{4}{4}$. Note the form of 1 highlighted in red.

$$= \frac{1 \cdot 4}{2 \cdot 4}$$ Use the rule for multiplying two fractions. Multiply the numerators. Multiply the denominators.

$$= \frac{4}{8}$$

We have found that $\frac{4}{8}$ is equivalent to $\frac{1}{2}$. To build an equivalent fraction for $\frac{1}{2}$ with a denominator of 8, we *multiplied by a factor equal to 1* in the form of $\frac{4}{4}$. Multiplying $\frac{1}{2}$ by $\frac{4}{4}$ changes its appearance but does not change its value because we are multiplying it by 1.

LANGUAGE OF MATHEMATICS

Building an equivalent fraction with a larger denominator is also called *expressing a fraction in* **higher terms**.

> **Building Fractions**
>
> To build a fraction, *multiply it by a factor equal to 1* in the form of $\frac{2}{2}, \frac{3}{3}, \frac{4}{4}, \frac{5}{5}$, and so on.

| EXAMPLE 4 | Write $\frac{3}{5}$ as an equivalent fraction with a denominator of 35. |

Strategy We will compare the given denominator to the required denominator and ask, "What number times 5 equals 35?"

WHY The answer to that question helps us determine the form of 1 to use to build an equivalent fraction.

Solution

To answer the question "What number times 5 equals 35?" we *divide* 35 by 5 to get 7. Since we need to multiply the denominator of $\frac{3}{5}$ by 7 to obtain a denominator of 35, it follows that $\frac{7}{7}$ should be the form of 1 that is used to build an equivalent fraction for $\frac{3}{5}$.

$$\frac{3}{5} = \frac{3}{5} \cdot \boxed{\frac{7}{7}} \qquad \text{Multiply } \tfrac{3}{5} \text{ by a form of 1: } \tfrac{7}{7} = 1.$$

$$= \frac{3 \cdot 7}{5 \cdot 7} \qquad \begin{array}{l}\text{Multiply the numerators.}\\\text{Multiply the denominators.}\end{array}$$

$$= \frac{21}{35}$$

Self Check 4

Write $\frac{5}{8}$ as an equivalent fraction with a denominator of 24.

Now Try ➡ **Problem 37**

We have found that $\frac{21}{35}$ is equivalent to $\frac{3}{5}$.

Success Tip To build an equivalent fraction in Example 4, we multiplied $\frac{3}{5}$ by 1 in the form of $\frac{7}{7}$. As a result of that step, the numerator and the denominator of $\frac{3}{5}$ were multiplied by 7:

$$\frac{3 \cdot 7}{5 \cdot 7} \quad \begin{array}{l}\longleftarrow \text{ The numerator is multiplied by 7.}\\\longleftarrow \text{ The denominator is multiplied by 7.}\end{array}$$

This process illustrates the following property of fractions.

The Fundamental Property of Fractions

If the numerator and denominator of a fraction are multiplied by the same nonzero number, the resulting fraction is equivalent to the original fraction.

     Since multiplying the numerator and denominator of a fraction by the same nonzero number produces an equivalent fraction, your instructor may allow you to begin your solution to problems like Example 4 as shown in the Success Tip above.

| EXAMPLE 5 | Write 4 as an equivalent fraction with a denominator of 6. |

Strategy We will express 4 as the fraction $\frac{4}{1}$ and build an equivalent fraction by multiplying it by $\frac{6}{6}$.

WHY Since we need to multiply the denominator of $\frac{4}{1}$ by 6 to obtain a denominator of 6, it follows that $\frac{6}{6}$ should be the form of 1 that is used to build an equivalent fraction for $\frac{4}{1}$.

Solution

$$4 = \frac{4}{1}$$ Write 4 as a fraction: $4 = \frac{4}{1}$.

$$= \frac{4}{1} \cdot \frac{6}{6}$$ Build an equivalent fraction by multiplying $\frac{4}{1}$ by a form of 1: $\frac{6}{6} = 1$.

$$= \frac{4 \cdot 6}{1 \cdot 6}$$ Multiply the numerators.
Multiply the denominators.

$$= \frac{24}{6}$$

Self Check 5

Write 10 as an equivalent fraction with a denominator of 3.

Now Try ➲ **Problem 49**

OBJECTIVE 5 Simplify fractions.

Every fraction can be written in infinitely many equivalent forms. For example, some equivalent forms of $\frac{10}{15}$ are:

$$\frac{2}{3} = \frac{4}{6} = \frac{6}{9} = \frac{8}{12} = \frac{10}{15} = \frac{12}{18} = \frac{14}{21} = \frac{16}{24} = \frac{18}{27} = \frac{20}{30} = \cdots$$

Of all of the equivalent forms in which we can write a fraction, we often need to determine the one that is in *simplest form*.

Simplest Form of a Fraction

A fraction is in **simplest form**, or **lowest terms**, when the numerator and denominator have no common factors other than 1.

EXAMPLE 6   Are the following fractions in simplest form?

a. $\frac{12}{27}$ **b.** $\frac{5}{8}$

Strategy We will determine whether the numerator and denominator have any common factors other than 1.

WHY If the numerator and denominator have no common factors other than 1, the fraction is in simplest form.

Solution

a. The factors of the numerator, 12, are: **1**, 2, **3**, 4, 6, 12
The factors of the denominator, 27, are: **1**, **3**, 9, 27

Since the numerator and denominator have a common factor of 3, the fraction $\frac{12}{27}$ is *not* in simplest form.

b. The factors of the numerator, 5, are: **1**, 5
The factors of the denominator, 8, are: **1**, 2, 4, 8

Since the only common factor of the numerator and denominator is 1, the fraction $\frac{5}{8}$ is in simplest form.

Self Check 6

Are the following fractions in simplest form?

a. $\frac{4}{21}$

b. $\frac{6}{20}$

Now Try ➲ **Problem 61**

To **simplify a fraction**, we write it in simplest form by *removing a factor equal to 1*. For example, to simplify $\frac{10}{15}$, we note that the greatest factor common to the numerator and denominator is 5 and proceed as follows:

$$\frac{10}{15} = \frac{2 \cdot 5}{3 \cdot 5}$$ Factor 10 and 15. Note the form of 1 highlighted in red.

$$= \frac{2}{3} \cdot \frac{5}{5} \qquad \text{Use the rule for multiplying fractions in reverse:}$$
write $\frac{2 \cdot 5}{3 \cdot 5}$ as the product of two fractions, $\frac{2}{3}$ and $\frac{5}{5}$.

$$= \frac{2}{3} \cdot 1 \qquad \text{A number divided by itself is equal to 1: } \frac{5}{5} = 1.$$

$$= \frac{2}{3} \qquad \text{Use the multiplication property of 1: The product}$$
of any fraction and 1 is that fraction.

We have found that the simplified form of $\frac{10}{15}$ is $\frac{2}{3}$. To simplify $\frac{10}{15}$, we *removed a factor equal to 1* in the form of $\frac{5}{5}$. The result, $\frac{2}{3}$, is equivalent to $\frac{10}{15}$.

To streamline the simplifying process, we can replace pairs of factors common to the numerator and denominator with the equivalent fraction $\frac{1}{1}$.

EXAMPLE 7 Simplify each fraction:

a. $\dfrac{6}{10}$ **b.** $\dfrac{7}{21}$

Strategy We will factor the numerator and denominator. Then we will look for any factors common to the numerator and denominator and remove them.

WHY We need to make sure that the numerator and denominator have no common factors other than 1. If that is the case, then the fraction is in *simplest form*.

Solution

a. $\dfrac{6}{10} = \dfrac{2 \cdot 3}{2 \cdot 5}$ To prepare to simplify, factor 6 and 10. Note the form of 1 highlighted in red.

$$= \frac{\overset{1}{\cancel{2}} \cdot 3}{\underset{1}{\cancel{2}} \cdot 5} \qquad \begin{array}{l}\text{Simplify by removing the common factor of 2 from the numerator and}\\ \text{denominator. A slash / and the 1's are used to show that } \frac{2}{2} \text{ is replaced by}\\ \text{the equivalent fraction } \frac{1}{1}. \text{ A factor equal to 1 in the form of } \frac{2}{2} \text{ was removed.}\end{array}$$

$$= \frac{3}{5} \qquad \begin{array}{l}\text{Multiply the remaining factors in the numerator: } 1 \cdot 3 = 3.\\ \text{Multiply the remaining factors in the denominator: } 1 \cdot 5 = 5.\end{array}$$

Since 3 and 5 have no common factors (other than 1), $\dfrac{3}{5}$ is in simplest form.

b. $\dfrac{7}{21} = \dfrac{7}{3 \cdot 7}$ To prepare to simplify, factor 21.

$$= \frac{\overset{1}{\cancel{7}}}{3 \cdot \underset{1}{\cancel{7}}} \qquad \begin{array}{l}\text{Simplify by removing the common factor of 7 from the numerator}\\ \text{and denominator.}\end{array}$$

$$= \frac{1}{3} \qquad \text{Multiply the remaining factors in the denominator: } 1 \cdot 3 = 3.$$

Caution! Don't forget to write the 1's when removing common factors of the numerator and the denominator. Failure to do so can lead to the common mistake shown below.

$$\frac{7}{21} = \frac{\cancel{7}}{3 \cdot \cancel{7}} = \frac{0}{3}$$

Self Check 7

Simplify each fraction:

a. $\dfrac{10}{25}$

b. $\dfrac{3}{9}$

Now Try ➲ Problems 65 and 69

We can easily identify common factors of the numerator and the denominator of a fraction if we write them in prime-factored form.

EXAMPLE 8   Simplify each fraction, if possible: **a.** $\dfrac{90}{105}$ **b.** $\dfrac{25}{27}$

Strategy We begin by prime factoring the numerator, 90, and denominator, 105. Then we look for any factors common to the numerator and denominator and remove them.

WHY When the numerator and/or denominator of a fraction are large numbers, such as 90 and 105, writing their prime factorizations is helpful in identifying any common factors.

Solution

a. $\dfrac{90}{105} = \dfrac{2 \cdot 3 \cdot 3 \cdot 5}{3 \cdot 5 \cdot 7}$

To prepare to simplify, write 90 and 105 in prime-factored form.

$= \dfrac{2 \cdot \overset{1}{\cancel{3}} \cdot 3 \cdot \overset{1}{\cancel{5}}}{\underset{1}{\cancel{3}} \cdot \underset{1}{\cancel{5}} \cdot 7}$

Remove the common factors of 3 and 5 from the numerator and denominator. Slashes and 1's are used to show that $\frac{3}{3}$ and $\frac{5}{5}$ are replaced by the equivalent fraction $\frac{1}{1}$. A factor equal to 1 in the form of $\frac{3 \cdot 5}{3 \cdot 5} = \frac{15}{15}$ was removed.

$= \dfrac{6}{7}$

Multiply the remaining factors in the numerator: $2 \cdot 1 \cdot 3 \cdot 1 = 6$.
Multiply the remaining factors in the denominator: $1 \cdot 1 \cdot 7 = 7$.

Since 6 and 7 have no common factors (other than 1), $\frac{6}{7}$ is in simplest form.

b. $\dfrac{25}{27} = \dfrac{5 \cdot 5}{3 \cdot 3 \cdot 3}$ Write 25 and 27 in prime-factored form.

Since 25 and 27 have no common factors, other than 1, the fraction $\frac{25}{27}$ is in simplest form.

Self Check 8

Simplify each fraction, if possible:

a. $\dfrac{70}{126}$

b. $\dfrac{16}{81}$

Now Try ➜ Problems 73 and 81

EXAMPLE 9   Simplify: $\dfrac{63}{36}$

Strategy We will prime factor the numerator and denominator. Then we will look for any factors common to the numerator and denominator and remove them.

WHY We need to make sure that the numerator and denominator have no common factors other than 1. If that is the case, then the fraction is in *simplest form*.

Solution

$\dfrac{63}{36} = \dfrac{3 \cdot 3 \cdot 7}{2 \cdot 2 \cdot 3 \cdot 3}$ To prepare to simplify, write 63 and 36 in prime-factored form.

$= \dfrac{\overset{1}{\cancel{3}} \cdot \overset{1}{\cancel{3}} \cdot 7}{2 \cdot 2 \cdot \underset{1}{\cancel{3}} \cdot \underset{1}{\cancel{3}}}$ Simplify by removing the common factors of 3 from the numerator and denominator.

$= \dfrac{7}{4}$ Multiply the remaining factors in the numerator: $1 \cdot 1 \cdot 7 = 7$.
Multiply the remaining factors in the denominator: $2 \cdot 2 \cdot 1 \cdot 1 = 4$.

$\begin{array}{r} 3\,\lfloor\underline{63} \\ 3\,\lfloor\underline{21} \\ 7 \end{array}$ $\begin{array}{r} 2\,\lfloor\underline{36} \\ 2\,\lfloor\underline{18} \\ 3\,\lfloor\underline{9} \\ 3 \end{array}$

Success Tip If you recognized that 63 and 36 have a common factor of 9, you may remove that common factor from the numerator and denominator without writing the prime factorizations. However, make sure that the numerator and denominator of the resulting fraction do not have any common factors. If they do, continue to simplify.

$\dfrac{63}{36} = \dfrac{7 \cdot \overset{1}{\cancel{9}}}{4 \cdot \underset{1}{\cancel{9}}} = \dfrac{7}{4}$ Factor 63 as $7 \cdot 9$ and 36 as $4 \cdot 9$, and then remove the common factor of 9 from the numerator and denominator.

Self Check 9

Simplify: $\dfrac{162}{72}$

Now Try ➜ Problem 89

Use the following steps to simplify a fraction.

> **Simplifying Fractions**
>
> To simplify a fraction, *remove factors equal to* 1 of the form $\frac{2}{2}, \frac{3}{3}, \frac{4}{4}, \frac{5}{5}$, and so on, using the following procedure:
>
> 1. Factor (or prime factor) the numerator and denominator to determine their common factors.
> 2. Remove factors equal to 1 by replacing each pair of factors common to the numerator and denominator with the equivalent fraction $\frac{1}{1}$.
> 3. Multiply the remaining factors in the numerator and in the denominator.

Negative fractions are simplified in the same way as positive fractions. Just remember to write a negative sign − in front of each step of the solution. For example, to simplify $-\frac{15}{33}$ we proceed as follows:

$$-\frac{15}{33} = -\frac{\overset{1}{\cancel{3}} \cdot 5}{\underset{1}{\cancel{3}} \cdot 11}$$

$$= -\frac{5}{11}$$

Answers to Self Checks

1. a. numerator: 7; denominator: 9   **b.** numerator: 21; denominator: 20   **2.** $\frac{11}{31}, \frac{20}{31}$ **3. a.** 1   **b.** 51

c. undefined   **d.** 0   **4.** $\frac{15}{24}$ **5.** $\frac{30}{3}$ **6. a.** yes   **b.** no   **7. a.** $\frac{2}{5}$ **b.** $\frac{1}{3}$ **8. a.** $\frac{5}{9}$ **b.** in simplest form

9. $\frac{9}{4}$

VOCABULARY

Fill in the blanks.

1. A _____ describes the number of equal parts of a whole.

2. For the fraction $\frac{7}{8}$, the _____ is 7 and the _____ is 8.

3. If the numerator of a fraction is less than its denominator, the fraction is called a _____ fraction. If the numerator of a fraction is greater than or equal to its denominator, it is called an _____ fraction.

4. Each of the following fractions is a form of ___ .
$$\frac{1}{1} = \frac{2}{2} = \frac{3}{3} = \frac{4}{4} = \frac{5}{5} = \frac{6}{6} = \frac{7}{7} = \frac{8}{8} = \frac{9}{9} = \ldots$$

5. Two fractions are _____ if they represent the same number.

6. _____ fractions, such as $\frac{1}{2}, \frac{2}{4}, \frac{3}{6}$, and $\frac{4}{8}$, represent the same portion of a whole.

7. Writing a fraction as an equivalent fraction with a larger denominator is called _____ the fraction.

8. A fraction is in _____ form, or lowest terms, when the numerator and denominator have no common factors other than 1.

CONCEPTS

9. What concept studied in this section is shown on the right?

10. What concept studied in this section does the following statement illustrate?
$$\frac{1}{2} = \frac{2}{4} = \frac{3}{6} = \frac{4}{8} = \frac{5}{10} = \ldots$$

11. Classify each fraction as a proper fraction or an improper fraction.

    a. $\dfrac{37}{24}$ **b.** $\dfrac{1}{3}$

    c. $\dfrac{71}{100}$ **d.** $\dfrac{9}{9}$

12. Remove the common factors of the numerator and denominator to simplify the fraction:

$$\frac{2 \cdot 3 \cdot 3 \cdot 5}{2 \cdot 3 \cdot 5 \cdot 7}$$

13. What common factor (other than 1) do the numerator and the denominator of the fraction $\frac{10}{15}$ have?

Fill in the blanks.

14. Multiplication property of 1: The product of any fraction and 1 is that _____ .

15. Multiplying fractions: To multiply two fractions, multiply the _____ and multiply the denominators.

16. a. Consider the following work: $\dfrac{2}{3} = \dfrac{2}{3} \cdot \dfrac{4}{4}$

                          $= \dfrac{8}{12}$

    To build an equivalent fraction for $\frac{2}{3}$ with a denominator of 12, _____ it by a factor equal to 1 in the form of ▮ .

    b. Consider the following work: $\dfrac{15}{27} = \dfrac{\overset{1}{\cancel{3}} \cdot 5}{\underset{1}{\cancel{3}} \cdot 9}$

                            $= \dfrac{5}{9}$

    To simplify the fraction $\frac{15}{27}$, _____ a factor equal to 1 of the form ▮ .

NOTATION

17. Write the fraction $\frac{7}{-8}$ in two other ways.

18. Write each integer as a fraction.

    a. 8   **b.** -25

Fill in the blanks to complete each step.

19. Build an equivalent fraction for $\frac{1}{6}$ with a denominator of 18.

$$\frac{1}{6} = \frac{1}{6} \cdot \frac{3}{▮}$$

$$= \frac{▮ \cdot 3}{6 \cdot ▮}$$

$$= \frac{3}{▮}$$

20. Simplify:

$$\frac{18}{24} = \frac{2 \cdot ▮ \cdot 3}{2 \cdot 2 \cdot 2 \cdot ▮}$$

$$= \frac{\overset{1}{\cancel{2}} \cdot 3 \cdot \overset{1}{\cancel{3}}}{2 \cdot 2 \cdot 2 \cdot \underset{1}{\cancel{3}}}$$

$$= \frac{3}{▮}$$

GUIDED PRACTICE

Identify the numerator and denominator of each fraction. See Example 1.

21. $\dfrac{4}{5}$ **22.** $\dfrac{7}{8}$

23. $\dfrac{17}{10}$ **24.** $\dfrac{29}{21}$

Write a fraction to describe what part of the figure is shaded. Write a fraction to describe what part of the figure is not shaded. See Example 2.

25.   **26.**

27.   **28.**

29.   **30.**

31.    **32.**

Simplify, if possible. See Example 3.

33. a. $\dfrac{4}{1}$ **b.** $\dfrac{8}{8}$ **c.** $\dfrac{0}{12}$ **d.** $\dfrac{1}{0}$

34. a. $\dfrac{25}{1}$ **b.** $\dfrac{14}{14}$ **c.** $\dfrac{0}{1}$ **d.** $\dfrac{83}{0}$

35. a. $\dfrac{5}{0}$ **b.** $\dfrac{0}{50}$ **c.** $\dfrac{33}{33}$ **d.** $\dfrac{75}{1}$

36. a. $\dfrac{0}{64}$ **b.** $\dfrac{27}{0}$ **c.** $\dfrac{125}{125}$ **d.** $\dfrac{98}{1}$

Write each fraction as an equivalent fraction with the indicated denominator. See Example 4.

37. $\frac{7}{8}$, denominator 40

38. $\frac{3}{4}$, denominator 24

39. $\frac{4}{9}$, denominator 27

40. $\frac{5}{7}$, denominator 49

41. $\frac{5}{6}$, denominator 54

42. $\frac{2}{3}$, denominator 27

43. $\frac{2}{7}$, denominator 14

44. $\frac{3}{10}$, denominator 50

45. $\frac{1}{2}$, denominator 30

46. $\frac{1}{3}$, denominator 60

47. $\frac{11}{16}$, denominator 32

48. $\frac{9}{10}$, denominator 60

49. $\frac{5}{4}$, denominator 28

50. $\frac{9}{4}$, denominator 44

51. $\frac{16}{15}$, denominator 45

52. $\frac{13}{12}$, denominator 36

Write each whole number as an equivalent fraction with the indicated denominator. See Example 5.

53. 4, denominator 9

54. 4, denominator 3

55. 6, denominator 8

56. 3, denominator 6

57. 3, denominator 5

58. 7, denominator 4

59. 14, denominator 2

60. 10, denominator 9

Are the following fractions in simplest form? See Example 6.

61. a. $\frac{12}{16}$ b. $\frac{3}{25}$

62. a. $\frac{9}{24}$ b. $\frac{7}{36}$

63. a. $\frac{35}{36}$ b. $\frac{18}{21}$

64. a. $\frac{22}{45}$ b. $\frac{21}{56}$

Simplify each fraction, if possible. See Example 7.

65. $\frac{6}{9}$

66. $\frac{15}{20}$

67. $\frac{16}{20}$

68. $\frac{25}{35}$

69. $\frac{5}{15}$

70. $\frac{6}{30}$

71. $\frac{2}{48}$

72. $\frac{2}{42}$

Simplify each fraction, if possible. See Example 8.

73. $\frac{36}{96}$

74. $\frac{48}{120}$

75. $\frac{16}{17}$

76. $\frac{14}{25}$

77. $\frac{60}{108}$

78. $\frac{75}{275}$

79. $\frac{50}{55}$

80. $\frac{22}{88}$

81. $\frac{55}{62}$

82. $\frac{41}{51}$

83. $\frac{180}{210}$

84. $\frac{90}{120}$

Simplify each fraction, if possible. See Example 9.

85. $\frac{306}{234}$

86. $\frac{208}{117}$

87. $\frac{15}{6}$

88. $\frac{24}{16}$

89. $\frac{420}{144}$

90. $\frac{216}{189}$

91. $-\frac{4}{68}$

92. $-\frac{3}{42}$

93. $-\frac{90}{105}$

94. $-\frac{98}{126}$

95. $-\frac{16}{26}$

96. $-\frac{81}{132}$

TRY IT YOURSELF

Tell whether each pair of fractions are equivalent by simplifying each fraction.

97. a. $\frac{2}{14}$ and $\frac{6}{36}$ b. $\frac{4}{30}$ and $\frac{12}{90}$

98. a. $\frac{22}{34}$ and $\frac{33}{51}$ b. $\frac{3}{12}$ and $\frac{4}{24}$

99. Shade $\frac{1}{6}$ of the figure. As a result of the shading, what fraction equivalent to $\frac{1}{6}$ is shown?

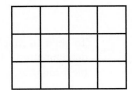

100. Shade the *same portion* of each of the figures below so that a fraction equivalent to $\frac{1}{4}$ is showing. Then, for each figure, give the equivalent fraction that is illustrated.

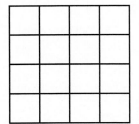 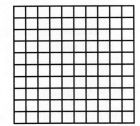

LOOK ALIKES

101. a. Simplify: $\frac{6}{24}$

   b. Write $\frac{1}{4}$ as an equivalent fraction with denominator 24.

102. a. Simplify: $\frac{8}{12}$

   b. Write $\frac{2}{3}$ as an equivalent fraction with denominator 12.

103. a. Simplify: $\frac{21}{30}$

   b. Write $\frac{7}{10}$ as an equivalent fraction with denominator 30.

104. a. Simplify: $\dfrac{30}{35}$

b. Write $\dfrac{6}{7}$ as an equivalent fraction with denominator 35.

APPLICATIONS

105. Dentistry. Refer to the dental chart.

a. How many teeth are shown on the chart?

b. What fraction of this set of teeth have fillings?

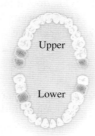

106. Time clocks. For each clock, what fraction of the hour has passed? Write your answers in simplified form. (*Hint:* There are 60 minutes in an hour.)

a.    **b.**

c.    **d.**

107. Rulers. The illustration to the right shows a ruler.

a. How many spaces are there between the numbers 0 and 1?

b. To what fraction is the arrow pointing? Write your answer in simplified form.

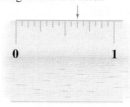

108. Sinkholes. The illustration below shows a side view of a drop in the sidewalk near a sinkhole. Describe the movement of the sidewalk using a signed fraction.

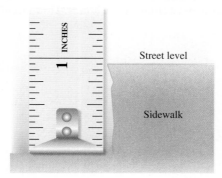

Street level

Sidewalk

109. Political parties. The graph shows the number of Democratic, Republican, and Independent governors of the 50 states, as of January 2017.

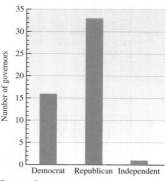

Source: thegreenpapers.com

a. How many Democratic governors are there? How many Republican governors are there? How many Independent governors are there?

b. What fraction of the governors are Democrats?

c. What fraction of the governors are Republicans? Write your answer in simplified form.

d. What fraction of the governors are Independents?

110. Gas tanks. Write fractions to describe the amount of gas left in the tank and the amount of gas that has been used.

111. Selling condos. The model below shows a new condominium development. The condos that have been sold are shaded.

a. How many units are there in the development?

b. What fraction of the units in the development have been sold? What fraction have not been sold? Write your answers in simplified form.

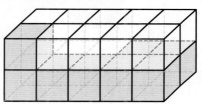

112. Architecture. The floor plan of a movie theater complex is shown on the next page. It has three Screen Rooms, a Lobby (that *includes* the Men's and Women's Restrooms), and an Entrance.

a. What fraction of the complex floor space is used for Screen Room 1?

b. What fraction of the complex floor space is used for the Lobby?

c. What fraction of the complex floor space is used for the Lobby and Entrance (combined)?

d. What fraction of the Lobby floor space is used for the Men's and Women's Restrooms (combined)?

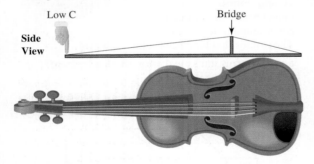

113. Accounting. Businesses often divide a calendar year into four quarters for accounting purposes. On what day would the second quarter begin and on what day would it end?

114. Music. The illustration shows a side view of the finger position needed to produce a length of string (from the *bridge* to the *fingertip*) that gives low C on a violin. To play other notes, fractions of that length are used. Draw arrows to show these finger positions on the string.

a. $\frac{1}{2}$ of the length gives middle C.

b. $\frac{3}{4}$ of the length gives F above low C.

c. $\frac{2}{3}$ of the length gives G.

115. Prime numbers. What fraction of the numbers in the chart below are prime numbers?

1	2	3	4	5	6	7	8	9	10
11	12	13	14	15	16	17	18	19	20
21	22	23	24	25	26	27	28	29	30
31	32	33	34	35	36	37	38	39	40
41	42	43	44	45	46	47	48	49	50
51	52	53	54	55	56	57	58	59	60
61	62	63	64	65	66	67	68	69	70
71	72	73	74	75	76	77	78	79	80
81	82	83	84	85	86	87	88	89	90
91	92	93	94	95	96	97	98	99	100

116. Fundraising. A student has raised $2,400 toward taking a trip that costs $3,600. What fraction of the cost of the trip has been raised so far?

WRITING

117. Explain the concept of equivalent fractions. Give an example.

118. What does it mean for a fraction to be in simplest form? Give an example.

119. Why can't we say that $\frac{2}{5}$ of the figure is shaded?

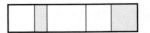

120. Perhaps you have heard the following joke. Explain what is wrong with the customer's thinking.

> *A pizza parlor waitress asks a customer if he wants the pizza cut into four pieces or six pieces or eight pieces. The customer then declares that he wants it cut into either four or six pieces "because I can't eat eight."*

121. What type of problem is shown below? Explain the steps used to find the answer.

a. $\dfrac{1}{2} = \dfrac{1}{2} \cdot \dfrac{4}{4} = \dfrac{4}{8}$

b. $\dfrac{15}{35} = \dfrac{3 \cdot \overset{1}{\cancel{5}}}{\underset{1}{\cancel{5}} \cdot 7} = \dfrac{3}{7}$

122. Explain the difference in the two approaches used to simplify $\frac{20}{28}$. Are the results the same?

$$\dfrac{\overset{1}{\cancel{4}} \cdot 5}{\underset{1}{\cancel{4}} \cdot 7} \quad \text{and} \quad \dfrac{\overset{1}{\cancel{2}} \cdot \overset{1}{\cancel{2}} \cdot 5}{\underset{1}{\cancel{2}} \cdot \underset{1}{\cancel{2}} \cdot 7}$$

REVIEW

123. Paychecks. *Gross pay* is what a worker makes before deductions and *net pay* is what is left after taxes, health benefits, union dues, and other deductions are taken out. Suppose a worker's monthly gross pay is $3,575. If deductions of $235, $782, $148, and $103 are taken out of his check, what is his monthly net pay?

124. Horse racing. One day, a man bet on all eight horse races at Santa Anita Racetrack. He won $168 on the first race and he won $105 on the fourth race. He lost his $50-bets on each of the other races. Overall, did he win or lose money betting on the horses? How much?

SECTION 3.2   Multiplying Fractions

In the next three sections, we discuss how to add, subtract, multiply, and divide fractions. We begin with the operation of multiplication.

OBJECTIVES

1 Multiply fractions.

2 Simplify answers when multiplying fractions.

3 Evaluate exponential expressions that have fractional bases.

4 Solve application problems by multiplying fractions.

5 Find the area of a triangle.

OBJECTIVE 1   Multiply fractions.

To develop a rule for multiplying fractions, let's consider a real-life application.

Suppose $\frac{3}{5}$ of the last page of a school newspaper is devoted to campus sports coverage. To show this, we can divide the page into fifths and shade three of them red.

Sports coverage: $\frac{3}{5}$ of the page

Furthermore, suppose that $\frac{1}{2}$ of the sports coverage is about women's teams. We can show that portion of the page by dividing the already colored region into two halves and shading one of them in purple.

Women's teams coverage: $\frac{1}{2}$ of $\frac{3}{5}$ of the page

To find the fraction represented by the purple-shaded region, the page needs to be divided into equal-sized parts. If we extend the dashed line downward, we see there are 10 equal-sized parts. The purple-shaded parts are 3 out of 10, or $\frac{3}{10}$, of the page. Thus, $\frac{3}{10}$ of the last page of the school newspaper is devoted to women's sports.

Women's teams coverage: $\frac{3}{10}$ of the page

Success Tip In the newspaper example, we found a *part of a part* of a page. Multiplying proper fractions can be thought of in this way. When taking a *part of a part* of something, the result is always smaller than the original part that you began with.

In this example, we have found that

$$\frac{1}{2} \quad \text{of} \quad \frac{3}{5} \quad \text{is} \quad \frac{3}{10}$$

$$\frac{1}{2} \quad \cdot \quad \frac{3}{5} \quad = \quad \frac{3}{10}$$

Since the key word *of* indicates multiplication, and the key word *is* means equals, we can translate this statement to symbols.

Two observations can be made from this result.

■ The numerator of the answer is the product of the numerators of the original fractions.

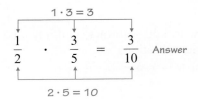

■ The denominator of the answer is the product of the denominators of the original fractions.

These observations illustrate the following rule for multiplying two fractions.

Multiplying Fractions

To multiply two fractions, multiply the numerators and multiply the denominators. Simplify the result, if possible.

EXAMPLE 1   Multiply: **a.** $\dfrac{1}{6} \cdot \dfrac{1}{4}$ **b.** $\dfrac{7}{8} \cdot \dfrac{3}{5}$

Strategy We will multiply the numerators and denominators and make sure that the result is in simplest form.

WHY This is the rule for multiplying two fractions.

Solution

a. $\dfrac{1}{6} \cdot \dfrac{1}{4} = \dfrac{1 \cdot 1}{6 \cdot 4}$ Multiply the numerators.
Multiply the denominators.

$= \dfrac{1}{24}$ Since 1 and 24 have no common factors other than 1, the result is in simplest form.

b. $\dfrac{7}{8} \cdot \dfrac{3}{5} = \dfrac{7 \cdot 3}{8 \cdot 5}$ Multiply the numerators.
Multiply the denominators.

$= \dfrac{21}{40}$ Since 21 and 40 have no common factors other than 1, the result is in simplest form.

Self Check 1

Multiply:

a. $\dfrac{1}{2} \cdot \dfrac{1}{8}$

b. $\dfrac{5}{9} \cdot \dfrac{2}{3}$

Now Try ➡ Problems 17 and 21

The sign rules for multiplying integers also hold for multiplying fractions. When we multiply two fractions with *like* signs, the product is positive. When we multiply two fractions with *unlike* signs, the product is negative.

EXAMPLE 2   Multiply: $-\dfrac{3}{4}\left(\dfrac{1}{8}\right)$

Strategy We will use the rule for multiplying two fractions that have different (unlike) signs.

WHY One fraction is positive and one is negative.

Solution

$$-\frac{3}{4}\left(\frac{1}{8}\right) = -\frac{3 \cdot 1}{4 \cdot 8}$$

Multiply the numerators.
Multiply the denominators.
Since the fractions have unlike signs, make the answer negative.

$$= -\frac{3}{32}$$

Since 3 and 32 have no common factors other than 1, the result is in simplest form.

Self Check 2

Multiply: $\frac{5}{6}\left(-\frac{1}{3}\right)$

Now Try ➡ Problem 25

EXAMPLE 3 Multiply: $\frac{1}{2} \cdot 3$

Strategy We will begin by writing the integer 3 as a fraction.

WHY Then we can use the rule for multiplying two fractions to find the product.

Solution

$$\frac{1}{2} \cdot 3 = \frac{1}{2} \cdot \frac{3}{1}$$ Write 3 as a fraction: $3 = \frac{3}{1}$.

$$= \frac{1 \cdot 3}{2 \cdot 1}$$ Multiply the numerators.
Multiply the denominators.

$$= \frac{3}{2}$$ Since 3 and 2 have no common factors other than 1, the result is in simplest form.

Self Check 3

Multiply: $\frac{1}{3} \cdot 7$

Now Try ➡ Problem 29

OBJECTIVE 2 **Simplify answers when multiplying fractions.**

After multiplying two fractions, we need to simplify the result, if possible. To do that, we can use the procedure discussed in Section 3.1 by removing pairs of common factors of the numerator and denominator.

EXAMPLE 4 Multiply and simplify: $\frac{5}{8} \cdot \frac{4}{5}$

Strategy We will multiply the numerators and denominators and make sure that the result is in simplest form.

WHY This is the rule for multiplying two fractions.

Solution

$$\frac{5}{8} \cdot \frac{4}{5} = \frac{5 \cdot 4}{8 \cdot 5}$$ Multiply the numerators.
Multiply the denominators.

$$= \frac{5 \cdot 2 \cdot 2}{2 \cdot 2 \cdot 2 \cdot 5}$$ To prepare to simplify, write 4 and 8 in prime-factored form.

$$= \frac{\overset{1}{\cancel{5}} \cdot \overset{1}{\cancel{2}} \cdot \overset{1}{\cancel{2}}}{\underset{1}{\cancel{2}} \cdot \underset{1}{\cancel{2}} \cdot 2 \cdot \underset{1}{\cancel{5}}}$$ To simplify, remove the common factors of 2 and 5 from the numerator and denominator.

$$= \frac{1}{2}$$ Multiply the remaining factors in the numerator: $1 \cdot 1 \cdot 1 = 1$.
Multiply the remaining factors in the denominator: $1 \cdot 1 \cdot 2 \cdot 1 = 2$.

Success Tip If you recognized that 4 and 8 have a common factor of 4, you may remove that common factor from the numerator and denominator of the product without writing the prime factorizations. However, make sure that the numerator and denominator of the resulting fraction do not have any common factors. If they do, continue to simplify.

Self Check 4

Multiply and simplify: $\dfrac{11}{25} \cdot \dfrac{10}{11}$

Now Try ➩ **Problem 33**

$$\frac{5}{8} \cdot \frac{4}{5} = \frac{5 \cdot 4}{8 \cdot 5} = \frac{\overset{1}{\cancel{5}} \cdot \overset{1}{\cancel{4}}}{2 \cdot \underset{1}{\cancel{4}} \cdot \underset{1}{\cancel{5}}} = \frac{1}{2}$$

Factor 8 as 2 · 4, and then remove the common factors of 4 and 5 in the numerator and denominator.

The rule for multiplying two fractions can be extended to find the product of three or more fractions.

EXAMPLE 5   Multiply and simplify: $\dfrac{2}{3}\left(-\dfrac{9}{14}\right)\left(-\dfrac{7}{10}\right)$

Strategy We will multiply the numerators and denominators and make sure that the result is in simplest form.

  WHY This is the rule for multiplying three (or more) fractions.

Solution Recall from Section 2.4 that a product is positive when there are an even number of negative factors. Since $\frac{2}{3}\left(-\frac{9}{14}\right)\left(-\frac{7}{10}\right)$ has *two* negative factors, the product is positive.

$$\frac{2}{3}\left(-\frac{9}{14}\right)\left(-\frac{7}{10}\right) = \frac{2}{3}\left(\frac{9}{14}\right)\left(\frac{7}{10}\right)$$

Since the answer is positive, drop both − signs and continue.

Self Check 5

Multiply and simplify:

$$\frac{2}{5}\left(-\frac{15}{22}\right)\left(-\frac{11}{26}\right)$$

Now Try ➩ **Problem 37**

$$= \frac{2 \cdot 9 \cdot 7}{3 \cdot 14 \cdot 10}$$

Multiply the numerators.
Multiply the denominators.

$$= \frac{2 \cdot 3 \cdot 3 \cdot 7}{3 \cdot 2 \cdot 7 \cdot 2 \cdot 5}$$

To prepare to simplify, write 9, 14, and 10 in prime-factored form.

$$= \frac{\overset{1}{\cancel{2}} \cdot \overset{1}{\cancel{3}} \cdot 3 \cdot \overset{1}{\cancel{7}}}{\underset{1}{\cancel{3}} \cdot 2 \cdot \underset{1}{\cancel{7}} \cdot \underset{1}{\cancel{2}} \cdot 5}$$

To simplify, remove the common factors of 2, 3, and 7 from the numerator and denominator.

$$= \frac{3}{10}$$

Multiply the remaining factors in the numerator.
Multiply the remaining factors in the denominator.

Caution! In Example 5, it was very helpful to prime factor and simplify when we did (the third step of the solution). If, instead, you find the product of the numerators and the product of the denominators, the resulting fraction is difficult to simplify because the numerator, 126, and the denominator, 420, are large.

$$\frac{2}{3} \cdot \frac{9}{14} \cdot \frac{7}{10} \quad = \quad \frac{2 \cdot 9 \cdot 7}{3 \cdot 14 \cdot 10} \quad = \quad \frac{126}{420}$$

Factor and simplify at this stage, before multiplying in the numerator and denominator.

Don't multiply in the numerator and denominator and then try to simplify the result. You will get the same answer, but it takes much more work.

OBJECTIVE 3 Evaluate exponential expressions that have fractional bases.

We have evaluated exponential expressions that have whole-number bases and integer bases. If the base of an exponential expression is a fraction, the exponent tells us how many times to write that fraction as a factor. For example,

$$\left(\frac{2}{3}\right)^2 = \frac{2}{3} \cdot \frac{2}{3} = \frac{2 \cdot 2}{3 \cdot 3} = \frac{4}{9}$$ *Since the exponent is 2, write the base, $\frac{2}{3}$, as a factor 2 times.*

EXAMPLE 6   Evaluate each expression: **a.** $\left(\frac{1}{4}\right)^3$ **b.** $\left(-\frac{2}{3}\right)^2$ **c.** $-\left(\frac{2}{3}\right)^2$

Strategy We will write each exponential expression as a product of repeated factors and then perform the multiplication. This requires that we identify the base and the exponent.

WHY The exponent tells the number of times the base is to be written as a factor.

Solution
Recall that exponents are used to represent repeated multiplication.

a. We read $\left(\frac{1}{4}\right)^3$ as "one-fourth raised to the third power," or as "one-fourth, cubed."

$$\left(\frac{1}{4}\right)^3 = \frac{1}{4} \cdot \frac{1}{4} \cdot \frac{1}{4}$$ *Since the exponent is 3, write the base, $\frac{1}{4}$, as a factor 3 times.*

$$= \frac{1 \cdot 1 \cdot 1}{4 \cdot 4 \cdot 4}$$ *Multiply the numerators.*
Multiply the denominators.

$$= \frac{1}{64}$$

b. We read $\left(-\frac{2}{3}\right)^2$ as "negative two-thirds raised to the second power," or as "negative two-thirds, squared."

$$\left(-\frac{2}{3}\right)^2 = \left(-\frac{2}{3}\right)\left(-\frac{2}{3}\right)$$ *Since the exponent is 2, write the base, $-\frac{2}{3}$, as a factor 2 times.*

$$= \frac{2 \cdot 2}{3 \cdot 3}$$ *The product of two fractions with like signs is positive: Drop the − signs. Multiply the numerators. Multiply the denominators.*

$$= \frac{4}{9}$$

c. We read $-\left(\frac{2}{3}\right)^2$ as "the opposite of two-thirds squared." Recall that if the − symbol is not within the parentheses, it is not part of the base.

$$-\left(\frac{2}{3}\right)^2 = -\frac{2}{3} \cdot \frac{2}{3}$$ *Since the exponent is 2, write the base, $\frac{2}{3}$, as a factor 2 times.*

$$= -\frac{2 \cdot 2}{3 \cdot 3}$$ *Multiply the numerators.*
Multiply the denominators.

$$= -\frac{4}{9}$$

Self Check 6

Evaluate each expression:

a. $\left(\frac{2}{5}\right)^3$

b. $\left(-\frac{3}{4}\right)^2$

c. $-\left(\frac{3}{4}\right)^2$

Now Try Problems 41 and 43

OBJECTIVE 4 Solve application problems by multiplying fractions.

The key word *of* often appears in application problems involving fractions. When a fraction is followed by the word *of*, such as $\frac{1}{2}$ *of* or $\frac{3}{4}$ *of*, it indicates that we are to find a part of some quantity using multiplication.

EXAMPLE 7 **How a bill becomes law.** If the President vetoes (refuses to sign) a bill, it takes $\frac{2}{3}$ of those voting in the House of Representatives (and the Senate) to override the veto for it to become law. If all 435 members of the House cast a vote, how many of their votes does it take to override a presidential veto?

Analyze

- It takes $\frac{2}{3}$ *of* those voting to override a veto.   *Given*
- All 435 members of the House cast a vote.   *Given*
- How many votes does it take to override a presidential veto?   *Find*

Form The key phrase $\frac{2}{3}$ *of* suggests that we are to find a part of the 435 possible votes using multiplication.

We translate the words of the problem to numbers and symbols.

The number of votes needed in the House to override a veto	is equal to	$\frac{2}{3}$	of	the number of House members that vote.

$$\begin{array}{ccc} \text{The number of votes needed in the House to override a veto} & = & \frac{2}{3} \quad \cdot \quad 435 \end{array}$$

Calculate To find the product, we will express 435 as a fraction and then use the rule for multiplying two fractions.

$$\frac{2}{3} \cdot 435 = \frac{2}{3} \cdot \frac{435}{1}$$ Write 435 as a fraction: $435 = \frac{435}{1}$.

$$= \frac{2 \cdot 435}{3 \cdot 1}$$ Multiply the numerators. Multiply the denominators.

$$= \frac{2 \cdot 3 \cdot 5 \cdot 29}{3 \cdot 1}$$ To prepare to simplify, write 435 in prime-factored form: $3 \cdot 5 \cdot 29$.

$$= \frac{2 \cdot \overset{1}{\cancel{3}} \cdot 5 \cdot 29}{\underset{1}{\cancel{3}} \cdot 1}$$ Remove the common factor of 3 from the numerator and denominator.

$$= \frac{290}{1}$$ Multiply the remaining factors in the numerator: $2 \cdot 1 \cdot 5 \cdot 29 = 290$. Multiply the remaining factors in the denominator: $1 \cdot 1 = 1$.

$$= 290$$ Any number divided by 1 is equal to that number.

435
3 145
5 29

Self Check 7

How a bill becomes law.
If only 96 senators are present and cast a vote, how many of their votes does it take to override a presidential veto?

Now Try Problems 45 and 87

State It would take 290 votes in the House to override a veto.

Check We can estimate to check the result. We will use 440 to approximate the number of House members voting. Since $\frac{1}{2}$ of 440 is 220, and since $\frac{2}{3}$ is a greater part than $\frac{1}{2}$, we would expect the number of votes needed to be *more than* 220. The result of 290 seems reasonable.

OBJECTIVE 5   Find the area of a triangle.

As the figures below show, a triangle has three sides. The length of the base of the triangle can be represented by the letter *b* and the height by the letter *h*. The height of a triangle is always perpendicular (makes a square corner) to the base. This is shown by using the symbol ⌐.

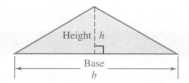 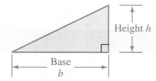

Recall that the area of a figure is the amount of surface that it encloses. The area of a triangle can be found by using the following formula.

Area of a Triangle

The area *A* of a triangle is one-half the product of its base *b* and its height *h*.

$$\text{Area} = \frac{1}{2}(\text{base})(\text{height}) \quad \text{or} \quad A = \frac{1}{2} \cdot b \cdot h$$

The formula can be written without the raised dots as $A = \frac{1}{2}bh$.

EXAMPLE 8   **Geography.**   Approximate the area of the state of Virginia (in square miles) using the triangle shown below.

Strategy We will find the product of $\frac{1}{2}$, 405, and 200.

WHY The formula for the area of a triangle is $A = \frac{1}{2}(\text{base})(\text{height})$.

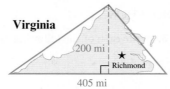

Virginia

Solution

$$A = \frac{1}{2}bh \qquad \text{This is the formula for the area of a triangle.}$$

$$A = \frac{1}{2} \cdot 405 \cdot 200 \qquad \frac{1}{2}bh \text{ means } \frac{1}{2} \cdot b \cdot h. \text{ Substitute 405 for } b \text{ and 200 for } h.$$

$$A = \frac{1}{2} \cdot \frac{405}{1} \cdot \frac{200}{1} \qquad \text{Write 405 and 200 as fractions.}$$

$$A = \frac{1 \cdot 405 \cdot 200}{2 \cdot 1 \cdot 1} \qquad \substack{\text{Multiply the numerators.}\\ \text{Multiply the denominators.}}$$

$$A = \frac{1 \cdot 405 \cdot \overset{1}{2} \cdot 100}{\underset{1}{2} \cdot 1 \cdot 1} \qquad \substack{\text{Factor 200 as } 2 \cdot 100. \text{ Then remove the common}\\ \text{factor of 2 from the numerator and denominator.}}$$

$$A = 40,500 \qquad \text{In the numerator, multiply: } 405 \cdot 100 = 40,500.$$

The area of the state of Virginia is approximately 40,500 square miles. This can be written as 40,500 mi².

Caution! Remember that area is measured in square units, such as in.², ft², and cm². Don't forget to write the units in your answer when finding the area of a figure.

Self Check 8

Find the area of the triangle shown below.

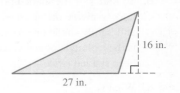

Now Try ➡ Problems 49 and 103

Answers to Self Checks

1. a. $\frac{1}{16}$ **b.** $\frac{10}{27}$ **2.** $-\frac{5}{18}$ **3.** $\frac{7}{3}$ **4.** $\frac{2}{5}$ **5.** $\frac{3}{26}$ **6. a.** $\frac{8}{125}$ **b.** $\frac{9}{16}$ **c.** $-\frac{9}{16}$ **7.** 64 votes **8.** 216 in.²

SECTION 3.2 STUDY SET

VOCABULARY

Fill in the blanks.

1. When a fraction is followed by the word *of*, such as $\frac{1}{3}$ *of*, it indicates that we are to find a part of some quantity using _____.

2. The answer to a multiplication is called the _____.

3. To _____ a fraction, we remove common factors of the numerator and denominator.

4. In the expression $\left(\frac{1}{4}\right)^3$, the _____ is $\frac{1}{4}$ and the _____ is 3.

5. The _____ of a triangle is the amount of surface that it encloses.

6. Label the *base* and the *height* of the triangle shown below.

CONCEPTS

7. Fill in the blanks: To multiply two fractions, multiply the _____ and multiply the _____. Then _____, if possible.

8. Use the following rectangle to find $\frac{1}{3} \cdot \frac{1}{4}$.

   a. Draw three vertical lines that divide the given rectangle into four equal parts and lightly shade one part. What fractional part of the rectangle did you shade?

   b. To find $\frac{1}{3}$ of the shaded portion, draw two horizontal lines to divide the given rectangle into three equal parts and lightly shade one part. Into how many equal parts is the rectangle now divided? How many parts have been shaded twice?

   c. What is $\frac{1}{3} \cdot \frac{1}{4}$?

9. Determine whether each product is positive or negative. *You do not have to find the answer.*

   a. $-\frac{1}{8} \cdot \frac{3}{5}$

   b. $-\frac{7}{16}\left(-\frac{2}{21}\right)$

   c. $-\frac{4}{5}\left(\frac{1}{3}\right)\left(-\frac{1}{8}\right)$

   d. $-\frac{3}{4}\left(-\frac{8}{9}\right)\left(-\frac{1}{2}\right)$

10. Translate each phrase to symbols. *You do not have to find the answer.*

    a. $\frac{7}{10}$ of $\frac{4}{9}$

    b. $\frac{1}{5}$ of 40

Fill in the blanks.

11. Area of a triangle $= \frac{1}{2}($ ▨ $)($ ▨ $)$

    or

    $A =$ ▨

12. Area is measured in _____ units, such as in.² and ft².

NOTATION

13. Write each of the following as a fraction.

    a. 4

    b. –3

    c. x

14. Fill in the blanks: $\left(\frac{1}{2}\right)^2$ represents the repeated multiplication ▨ · ▨.

Fill in the blanks to complete each step.

15. *Multiply and simplify:*

$$\frac{5}{8} \cdot \frac{7}{15} = \frac{5 \cdot \,▨}{8 \cdot \,▨}$$

$$= \frac{5 \cdot 7}{▨ \cdot 2 \cdot 2 \cdot \,▨ \cdot 5}$$

$$= \frac{\overset{1}{\cancel{▨}} \cdot 7}{2 \cdot 2 \cdot 2 \cdot 3 \cdot \underset{1}{\cancel{▨}}}$$

$$= \frac{7}{▨}$$

16. *Multiply and simplify:*

$$\frac{7}{12} \cdot \frac{4}{21} = \frac{7 \cdot 4}{\boxed{} \cdot \boxed{}}$$

$$= \frac{7 \cdot \boxed{} \cdot 4}{\boxed{} \cdot 4 \cdot 3 \cdot \boxed{} \cdot \boxed{}}$$

$$= \frac{\overset{1}{\cancel{7}} \cdot \overset{1}{\cancel{7}} \cdot 4}{3 \cdot 4 \cdot 3 \cdot \underset{1}{\cancel{7}} \cdot \underset{1}{\cancel{7}}}$$

$$= \frac{\boxed{}}{9}$$

GUIDED PRACTICE

Multiply. **See Example 1.**

17. $\dfrac{1}{4} \cdot \dfrac{1}{2}$ **18.** $\dfrac{1}{3} \cdot \dfrac{1}{5}$

19. $\dfrac{1}{9} \cdot \dfrac{1}{5}$ **20.** $\dfrac{1}{2} \cdot \dfrac{1}{8}$

21. $\dfrac{2}{3} \cdot \dfrac{7}{9}$ **22.** $\dfrac{3}{4} \cdot \dfrac{5}{7}$

23. $\dfrac{8}{11} \cdot \dfrac{3}{7}$ **24.** $\dfrac{11}{13} \cdot \dfrac{2}{3}$

Multiply. **See Example 2.**

25. $-\dfrac{4}{5} \cdot \dfrac{1}{3}$ **26.** $-\dfrac{7}{9} \cdot \dfrac{1}{4}$

27. $\dfrac{5}{6}\left(-\dfrac{7}{12}\right)$ **28.** $\dfrac{2}{15}\left(-\dfrac{4}{3}\right)$

Multiply. **See Example 3.**

29. $\dfrac{1}{8} \cdot 9$ **30.** $\dfrac{1}{6} \cdot 11$

31. $\dfrac{1}{2} \cdot 5$ **32.** $\dfrac{1}{2} \cdot 21$

Multiply. Write the product in simplest form. **See Example 4.**

33. $\dfrac{11}{10} \cdot \dfrac{5}{11}$ **34.** $\dfrac{5}{4} \cdot \dfrac{2}{5}$

35. $\dfrac{6}{49} \cdot \dfrac{7}{6}$ **36.** $\dfrac{13}{4} \cdot \dfrac{4}{39}$

Multiply. Write the product in simplest form. **See Example 5.**

37. $\dfrac{3}{4}\left(-\dfrac{8}{35}\right)\left(-\dfrac{7}{12}\right)$ **38.** $\dfrac{9}{10}\left(-\dfrac{4}{15}\right)\left(-\dfrac{5}{18}\right)$

39. $-\dfrac{5}{8}\left(\dfrac{16}{27}\right)\left(-\dfrac{9}{25}\right)$ **40.** $-\dfrac{15}{28}\left(\dfrac{7}{9}\right)\left(-\dfrac{18}{35}\right)$

Evaluate each expression. **See Example 6.**

41. a. $\left(\dfrac{3}{5}\right)^2$ **b.** $\left(-\dfrac{3}{5}\right)^2$

42. a. $\left(\dfrac{4}{9}\right)^2$ **b.** $\left(-\dfrac{4}{9}\right)^2$

43. a. $-\left(-\dfrac{1}{6}\right)^2$ **b.** $\left(-\dfrac{1}{6}\right)^3$

44. a. $-\left(-\dfrac{2}{5}\right)^2$ **b.** $\left(-\dfrac{2}{5}\right)^3$

Find each product. Write your answer in simplest form. **See Example 7.**

45. $\dfrac{3}{4}$ of $\dfrac{5}{8}$ **46.** $\dfrac{4}{5}$ of $\dfrac{3}{7}$

47. $\dfrac{1}{6}$ of 54 **48.** $\dfrac{1}{9}$ of 36

Find the area of each triangle. **See Example 8.**

49.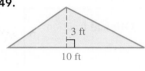

50.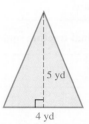

51.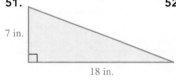

52.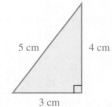

53.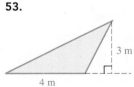

54.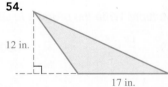

55.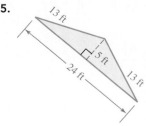

56.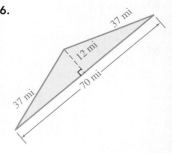

TRY IT YOURSELF

57. Complete the multiplication table of fractions.

$\cdot$	$\frac{1}{2}$	$\frac{1}{3}$	$\frac{1}{4}$	$\frac{1}{5}$	$\frac{1}{6}$
$\frac{1}{2}$					
$\frac{1}{3}$					
$\frac{1}{4}$					
$\frac{1}{5}$					
$\frac{1}{6}$					

58. Complete the table by finding the original fraction, given its square.

Original fraction squared	Original fraction
$\frac{1}{9}$	
$\frac{1}{100}$	
$\frac{4}{25}$	
$\frac{16}{49}$	
$\frac{81}{36}$	
$\frac{9}{121}$	

Multiply. Write the product in simplest form.

59. $-\dfrac{15}{24} \cdot \dfrac{8}{25}$

60. $-\dfrac{20}{21} \cdot \dfrac{7}{16}$

61. $\dfrac{3}{8} \cdot \dfrac{7}{16}$

62. $\dfrac{5}{9} \cdot \dfrac{2}{7}$

63. $\dfrac{1}{2}\left(-\dfrac{1}{3}\right)\left(-\dfrac{1}{4}\right)$

64. $\left(\dfrac{3}{8}\right)\left(-\dfrac{2}{3}\right)\left(-\dfrac{12}{27}\right)$

65. $-\dfrac{5}{6} \cdot 18$

66. $6\left(-\dfrac{2}{3}\right)$

67. $\left(-\dfrac{3}{4}\right)^3$

68. $\left(-\dfrac{2}{5}\right)^3$

69. $\dfrac{3}{4} \cdot \dfrac{4}{3}$

70. $\dfrac{4}{5} \cdot \dfrac{5}{4}$

71. $\dfrac{5}{3}\left(-\dfrac{6}{15}\right)(-4)$

72. $\dfrac{5}{6}\left(-\dfrac{2}{3}\right)(-12)$

73. $-\dfrac{11}{12} \cdot \dfrac{18}{55} \cdot 5$

74. $-\dfrac{24}{5} \cdot \dfrac{7}{12} \cdot \dfrac{1}{14}$

75. $\left(-\dfrac{11}{21}\right)\left(-\dfrac{14}{33}\right)$

76. $\left(-\dfrac{16}{35}\right)\left(-\dfrac{25}{48}\right)$

77. $-\left(-\dfrac{5}{9}\right)^2$

78. $-\left(-\dfrac{5}{6}\right)^2$

79. $\dfrac{7}{10}\left(\dfrac{20}{21}\right)$

80. $\dfrac{7}{6}\left(\dfrac{9}{49}\right)$

81. $\dfrac{3}{4}\left(\dfrac{5}{7}\right)\left(\dfrac{2}{3}\right)\left(\dfrac{7}{3}\right)$

82. $-\dfrac{5}{4}\left(\dfrac{8}{15}\right)\left(\dfrac{2}{3}\right)\left(\dfrac{7}{2}\right)$

83. $-\dfrac{14}{15}\left(-\dfrac{11}{8}\right)$

84. $-\dfrac{15}{16}\left(-\dfrac{8}{3}\right)$

85. $\dfrac{3}{16} \cdot 4 \cdot \dfrac{2}{3}$

86. $5 \cdot \dfrac{7}{5} \cdot \dfrac{3}{14}$

LOOK ALIKES

Evaluate each expression

87. **a.** $\left(\dfrac{1}{2}\right)^2$ **b.** $\left(\dfrac{1}{2}\right)^3$

  c. $\left(\dfrac{1}{2}\right)^4$ **d.** $\left(\dfrac{1}{2}\right)^5$

88. **a.** $\left(-\dfrac{1}{3}\right)^2$ **b.** $\left(-\dfrac{1}{3}\right)^3$

  c. $\left(-\dfrac{1}{3}\right)^4$ **d.** $\left(-\dfrac{1}{3}\right)^5$

In part a, find the product. Then use that result to answer part b. No work is neccessary.

89. **a.** $\dfrac{9}{10}$ of $\dfrac{1}{5}$ **b.** $\dfrac{1}{5}$ of $-\dfrac{9}{10}$

90. **a.** $\dfrac{3}{4}$ of $\dfrac{1}{8}$ **b.** $\dfrac{1}{8}$ of $-\dfrac{3}{4}$

APPLICATIONS

91. Senate rules. A *filibuster* is a method U.S. senators sometimes use to block passage of a bill or appointment. It takes $\frac{3}{5}$ of those voting in the Senate to break a filibuster. If all 100 senators cast a vote, how many of their votes does it take to break a filibuster?

92. Genetics. Gregor Mendel (1822–1884), an Augustinian monk, is credited with developing a model that became the foundation of modern genetics. In his experiments, he crossed purple-flowered plants with white-flowered plants and found that $\frac{3}{4}$ of the offspring plants had purple flowers and $\frac{1}{4}$ of them had white flowers. Refer to the illustration below, which shows a group of offspring plants. According to this concept, when the plants begin to flower, how many will have purple flowers?

93. Bouncing balls. A tennis ball is dropped from a height of 54 inches. Each time it hits the ground, it rebounds one-third of the previous height that it fell. Find the three missing rebound heights in the illustration.

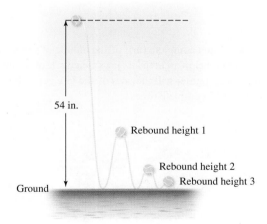

54 in.

Rebound height 1

Rebound height 2

Rebound height 3

Ground

94. Elections. The final election returns for a city bond measure are shown below.

a. Find the total number of votes cast.

b. Find two-thirds of the total number of votes cast.

c. Did the bond measure pass?

MEASURE 1
100% of the precincts reporting

Fire–Police–Paramedics General Obligation Bonds
(Requires two-thirds vote to pass)

YES	No
125,599	62,801

95. Cooking. Use the recipe below, along with the concept of multiplication of fractions, to find how much sugar and how much molasses are needed to make *one dozen* cookies. (*Hint:* this recipe is for *two dozen* cookies.)

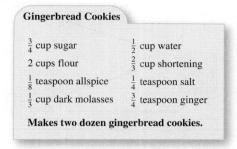

Gingerbread Cookies

$\frac{3}{4}$ cup sugar $\frac{1}{2}$ cup water

2 cups flour $\frac{2}{3}$ cup shortening

$\frac{1}{8}$ teaspoon allspice $\frac{1}{4}$ teaspoon salt

$\frac{1}{3}$ cup dark molasses $\frac{3}{4}$ teaspoon ginger

Makes two dozen gingerbread cookies.

96. The Earth's surface. The surface of Earth covers an area of approximately 196,800,000 square miles. About $\frac{3}{4}$ of that area is covered by water. Find the number of square miles of the surface covered by water.

97. Botany. In an experiment, monthly growth rates of three types of plants doubled when nitrogen was added to the soil. Complete the graph by drawing the improved growth rate bar next to each normal growth rate bar.

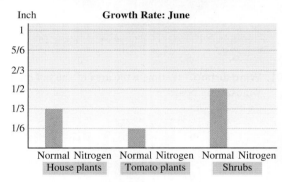

Inch   **Growth Rate: June**

1

5/6

2/3

1/2

1/3

1/6

Normal Nitrogen   Normal Nitrogen   Normal Nitrogen
House plants   Tomato plants   Shrubs

98. Icebergs. About $\frac{9}{10}$ of the volume of an iceberg is below the water line.

a. What fraction of the volume of an iceberg is *above* the water line?

b. Suppose an iceberg has a total volume of 18,700 cubic meters. What is the volume of the part of the iceberg that is above the water line?

99. Kitchen design. Find the area of the *kitchen work triangle* formed by the paths between the refrigerator, the range, and the sink shown below.

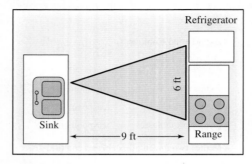

100. Stars and stripes. The illustration shows a folded U.S. flag. When it is placed on a table as part of an exhibit, how much area will it occupy?

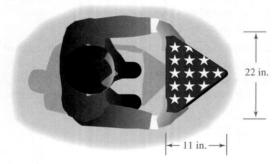

101. Windsurfing. Estimate the area of the sail on the windsurfing board.

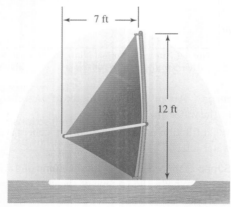

102. Tile design. A design for bathroom tile is shown. Find the amount of area on a tile that is blue.

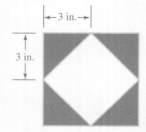

103. Geography. Estimate the area of the state of New Hampshire, using the triangle in the illustration.

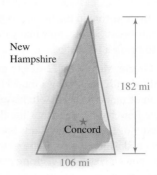

104. Stamps. The best designs in a contest to create a wildlife stamp are shown. To save on paper costs, the postal service has decided to choose the stamp that has the smaller area. Which one did the postal service choose? (*Hint:* use the formula for the area of a rectangle.)

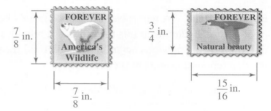

105. Vises. Each complete turn of the handle of the bench vise shown below tightens its jaws exactly $\frac{1}{16}$ of an inch. How much tighter will the jaws of the vise get if the handle is turned 12 complete times?

106. Woodworking. Each time a board is passed through a power sander, the machine removes $\frac{1}{64}$ of an inch of thickness. If a rough pine board is passed through the sander 6 times, by how much will its thickness change?

107. Eyelashes. In a recent study, researchers at the Georgia Institute of Technology found that eyelashes create a "speed bump" that diverts airflow away from the surface of the eyes, keeps them moist, and protects them from incoming particles. However, if the eyelashes are too long, they actually funnel airflow into the eye, causing more evaporation and dryness. The study concluded that the ideal eyelash length is one-third of the width of the eye. If the human eye is about three-quarters of an inch wide, how long are the ideal eyelashes according to the study?

marinafrost/Shutterstock.com

108. Font size.   A *point* is a unit of measure of font height that is widely used in the publishing trade, particularly in design and typesetting. It is defined to be $\frac{1}{72}$ of an inch. Microsoft Word uses point sizes to specify the height of all its fonts. If a capital letter, say E, is printed in a 48 point font, what is the height of the letter in inches? Express the result in simplest form.

E   48 points

WRITING

109. In a word problem, when a fraction is followed by the word *of*, multiplication is usually indicated. Give three real-life examples of this type of use of the word *of*.

110. Can you multiply the number 5 and another number and obtain an answer that is less than 5? Explain why or why not.

111. A majority.   The definition of the word *majority* is as follows: "a number greater than *one-half of* the total." Explain what it means when a teacher says, "A majority of the class voted to postpone the test until Monday." Give an example.

112. What does area measure? Give an example.

113. In the following solution, what step did the student forget to use that caused him to have to work with such large numbers?
Multiply. Simplify the product, if possible.

$$\frac{44}{63} \cdot \frac{27}{55} = \frac{44 \cdot 27}{63 \cdot 55}$$

$$= \frac{1,188}{3,465}$$

114. Is the product of two proper fractions always smaller than either of those fractions? Explain why or why not.

REVIEW

Divide and check each result using multiplication.

115. $\dfrac{-8}{4}$

116. $21 \div (-3)$

117. $-36 \div (-9)$

118. $\dfrac{-200}{-200}$

SECTION 3.3   Dividing Fractions

We will now discuss how to divide fractions. The fraction multiplication skills that you learned in Section 3.2 will also be useful in this section.

OBJECTIVES

1 Find the reciprocal of a fraction.

2 Divide fractions.

3 Solve application problems by dividing fractions.

OBJECTIVE 1   Find the reciprocal of a fraction.

Division with fractions involves working with *reciprocals*. To present the concept of reciprocal, we consider the problem $\frac{7}{8} \cdot \frac{8}{7}$.

$$\frac{7}{8} \cdot \frac{8}{7} = \frac{7 \cdot 8}{8 \cdot 7}$$ Multiply the numerators.
                                                              Multiply the denominators.

$$= \frac{\overset{1}{7} \cdot \overset{1}{8}}{\underset{1}{8} \cdot \underset{1}{7}}$$ To simplify, remove the common factors of 7 and 8 from the numerator and denominator.

$$= \frac{1}{1}$$ Multiply the remaining factors in the numerator.
                     Multiply the remaining factors in the denominator.

$$= 1$$ Any whole number divided by 1 is equal to that number.

The product of $\frac{7}{8}$ and $\frac{8}{7}$ is 1.

Whenever the product of two numbers is 1, we say that those numbers are *reciprocals*. Therefore, $\frac{7}{8}$ and $\frac{8}{7}$ are reciprocals. To find the reciprocal of a fraction, *we invert the numerator and the denominator.*

Caution! Zero does not have a reciprocal because the product of 0 and a number can never be 1.

> **Reciprocals**
>
> Two numbers are called **reciprocals** if their product is 1.

EXAMPLE 1   For each number, find its reciprocal and show that their product is 1:

a. $\frac{2}{3}$ **b.** $-\frac{3}{4}$ **c.** 5

Strategy To find each reciprocal, we will invert the numerator and denominator.

WHY This procedure will produce a new fraction that, when multiplied by the original fraction, gives a result of 1.

Caution! Don't confuse the concepts of the *opposite* of a negative number and the *reciprocal* of a negative number. For example:

The reciprocal of $-\frac{9}{16}$ is $-\frac{16}{9}$.

The opposite of $-\frac{9}{16}$ is $\frac{9}{16}$.

Solution

a. Fraction   Reciprocal

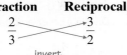

The reciprocal of $\frac{2}{3}$ is $\frac{3}{2}$.

Check: $\frac{2}{3} \cdot \frac{3}{2} = \frac{\overset{1}{2} \cdot \overset{1}{3}}{\underset{1}{3} \cdot \underset{1}{2}} = 1$

b. Fraction   Reciprocal

$-\frac{3}{4} \quad\longrightarrow\quad -\frac{4}{3}$
invert

The reciprocal of $-\frac{3}{4}$ is $-\frac{4}{3}$.

Check: $-\frac{3}{4}\left(-\frac{4}{3}\right) = \frac{\overset{1}{3} \cdot \overset{1}{4}}{\underset{1}{4} \cdot \underset{1}{3}} = 1$ The product of two fractions with like signs is positive.

Self Check 1

For each number, find its reciprocal and show that their product is 1.

a. $\frac{3}{5}$ **b.** $-\frac{5}{6}$ **c.** 8

Now Try Problem 13

c. Since $5 = \frac{5}{1}$, the reciprocal of 5 is $\frac{1}{5}$.

Check: $5 \cdot \frac{1}{5} = \frac{5}{1} \cdot \frac{1}{5} = \frac{\overset{1}{5} \cdot 1}{1 \cdot \underset{1}{5}} = 1$

OBJECTIVE 2 Divide fractions.

To develop a rule for dividing fractions, let's consider a real-life application.

Suppose that the manager of a candy store buys large bars of chocolate and divides each one into four equal parts to sell. How many fourths can be obtained from 5 bars?

We are asking, "How many $\frac{1}{4}$'s are there in 5?" To answer the question, we need to use the operation of division. We can represent this division as $5 \div \frac{1}{4}$.

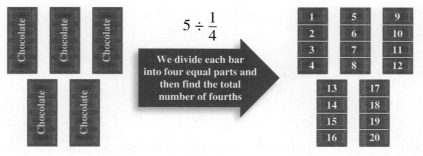

5 bars of chocolate

Total number of fourths = 5 • 4 = 20

There are 20 fourths in the 5 bars of chocolate. Two observations can be made from this result.

- This division problem involves a fraction: $5 \div \frac{1}{4}$.
- Although we were asked to find $5 \div \frac{1}{4}$, we solved the problem using *multiplication* instead of *division*: $5 \cdot 4 = 20$. That is, division by $\frac{1}{4}$ (a fraction) is the same as multiplication by 4 (its reciprocal).

$$5 \div \frac{1}{4} = 5 \cdot 4$$

These observations suggest the following rule for dividing two fractions.

Dividing Fractions

To divide two fractions, multiply the first fraction by the reciprocal of the second fraction. Simplify the result, if possible.

For example, to find $\frac{5}{7} \div \frac{3}{4}$, we multiply $\frac{5}{7}$ by the reciprocal of $\frac{3}{4}$.

Change the division to multiplication.

$$\frac{5}{7} \div \frac{3}{4} \quad = \quad \frac{5}{7} \cdot \frac{4}{3}$$

The reciprocal of $\frac{3}{4}$ is $\frac{4}{3}$.

$$= \frac{5 \cdot 4}{7 \cdot 3} \qquad \text{Multiply the numerators.}$$
$$\text{Multiply the denominators.}$$

$$= \frac{20}{21}$$

Thus, $\frac{5}{7} \div \frac{3}{4} = \frac{20}{21}$. We say that the *quotient* of $\frac{5}{7}$ and $\frac{3}{4}$ is $\frac{20}{21}$.

EXAMPLE 2   Divide: $\frac{1}{3} \div \frac{4}{5}$

Strategy We will multiply the first fraction, $\frac{1}{3}$, by the reciprocal of the second fraction, $\frac{4}{5}$. Then, if possible, we will simplify the result.

WHY This is the rule for dividing two fractions.

Solution

$$\frac{1}{3} \div \frac{4}{5} = \frac{1}{3} \cdot \frac{5}{4}$$ Multiply $\frac{1}{3}$ by the reciprocal of $\frac{4}{5}$, which is $\frac{5}{4}$.

$$= \frac{1 \cdot 5}{3 \cdot 4}$$ Multiply the numerators.
Multiply the denominators.

$$= \frac{5}{12}$$

Since 5 and 12 have no common factors other than 1, the result is in simplest form.

Self Check 2

Divide: $\frac{2}{3} \div \frac{7}{8}$

Now Try ➲ **Problem 17**

EXAMPLE 3 Divide and simplify: $\frac{9}{16} \div \frac{3}{20}$

Strategy We will multiply the first fraction, $\frac{9}{16}$, by the reciprocal of the second fraction, $\frac{3}{20}$. Then, if possible, we will simplify the result.

WHY This is the rule for dividing two fractions.

Solution

$$\frac{9}{16} \div \frac{3}{20} = \frac{9}{16} \cdot \frac{20}{3}$$ Multiply $\frac{9}{16}$ by the reciprocal of $\frac{3}{20}$, which is $\frac{20}{3}$.

$$= \frac{9 \cdot 20}{16 \cdot 3}$$ Multiply the numerators.
Multiply the denominators.

$$= \frac{\overset{1}{\cancel{3}} \cdot 3 \cdot \overset{1}{\cancel{4}} \cdot 5}{\underset{1}{\cancel{4}} \cdot 4 \cdot \underset{1}{\cancel{3}}}$$ To simplify, factor 9 as 3 · 3, factor 20 as 4 · 5, and factor 16 as 4 · 4. Then remove the common factors of 3 and 4 from the numerator and denominator.

$$= \frac{15}{4}$$ Multiply the remaining factors in the numerator: 1 · 3 · 1 · 5 = 15.
Multiply the remaining factors in the denominator: 1 · 4 · 1 = 4.

Self Check 3

Divide and simplify: $\frac{4}{5} \div \frac{8}{25}$

Now Try ➲ **Problem 21**

EXAMPLE 4 Divide and simplify: $120 \div \frac{10}{7}$

Strategy We will write 120 as a fraction and then multiply the first fraction by the reciprocal of the second fraction.

WHY This is the rule for dividing two fractions.

Solution

$$120 \div \frac{10}{7} = \frac{120}{1} \div \frac{10}{7}$$ Write 120 as a fraction: $120 = \frac{120}{1}$.

$$= \frac{120}{1} \cdot \frac{7}{10}$$ Multiply $\frac{120}{1}$ by the reciprocal of $\frac{10}{7}$, which is $\frac{7}{10}$.

$$= \frac{120 \cdot 7}{1 \cdot 10}$$ Multiply the numerators.
Multiply the denominators.

$$= \frac{\overset{1}{\cancel{10}} \cdot 12 \cdot 7}{1 \cdot \underset{1}{\cancel{10}}}$$ To simplify, factor 120 as 10 · 12, then remove the common factor of 10 from the numerator and denominator.

$$= \frac{84}{1}$$ Multiply the remaining factors in the numerator: 1 · 12 · 7 = 84.
Multiply the remaining factors in the denominator: 1 · 1 = 1.

$$= 84$$ Any whole number divided by 1 is the same number.

Self Check 4

Divide and simplify:

$80 \div \frac{20}{11}$

Now Try ➲ **Problem 27**

Because of the relationship between multiplication and division, the sign rules for *dividing* fractions are the same as those for *multiplying* fractions.

EXAMPLE 5   Divide and simplify: $\dfrac{1}{6} \div \left(-\dfrac{1}{18} \right)$

Strategy We will multiply the first fraction, $\frac{1}{6}$, by the reciprocal of the second fraction, $-\frac{1}{18}$. To determine the sign of the result, we will use the rule for multiplying two fractions that have different (unlike) signs.

WHY   One fraction is positive and one is negative.

Solution

$$\frac{1}{6} \div \left(-\frac{1}{18} \right) = \frac{1}{6}\left(-\frac{18}{1} \right)$$ Multiply $\frac{1}{6}$ by the reciprocal of $-\frac{1}{18}$, which is $-\frac{18}{1}$.

$$= -\frac{1 \cdot 18}{6 \cdot 1}$$ Multiply the numerators.
Multiply the denominators.
Since the fractions have unlike signs, make the answer negative.

$$= -\frac{1 \cdot 3 \cdot \overset{1}{\cancel{6}}}{\underset{1}{\cancel{6}} \cdot 1}$$ To simplify, factor 18 as $3 \cdot 6$. Then remove the common factor of 6 from the numerator and denominator.

$$= -\frac{3}{1}$$ Multiply the remaining factors in the numerator.
Multiply the remaining factors in the denominator.

$$= -3$$

Self Check 5

Divide and simplify:

$$\frac{2}{3} \div \left(-\frac{7}{6} \right)$$

Now Try ➧ **Problem 29**

EXAMPLE 6   Divide and simplify: $-\dfrac{21}{36} \div (-3)$

Strategy We will multiply the first fraction, $-\frac{21}{36}$, by the reciprocal of -3. To determine the sign of the result, we will use the rule for multiplying two fractions that have the same (like) signs.

WHY   Both fractions are negative.

Solution

$$-\frac{21}{36} \div (-3) = -\frac{21}{36}\left(-\frac{1}{3} \right)$$ Multiply $-\frac{21}{36}$ by the reciprocal of -3, which is $-\frac{1}{3}$.

$$= \frac{21}{36}\left(\frac{1}{3} \right)$$ Since the product of two negative fractions is positive, drop both $-$ signs and continue.

$$= \frac{21 \cdot 1}{36 \cdot 3}$$ Multiply the numerators.
Multiply the denominators.

$$= \frac{\overset{1}{\cancel{3}} \cdot 7 \cdot 1}{36 \cdot \underset{1}{\cancel{3}}}$$ To simplify, factor 21 as $3 \cdot 7$. Then remove the common factor of 3 from the numerator and denominator.

$$= \frac{7}{36}$$ Multiply the remaining factors in the numerator:
$1 \cdot 7 \cdot 1 = 7$.
Multiply the remaining factors in the denominator:
$36 \cdot 1 = 36$.

Self Check 6

Divide and simplify:

$$-\frac{35}{16} \div (-7)$$

Now Try ➧ **Problem 33**

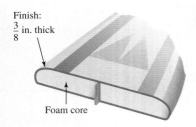

Finish:
$\frac{3}{8}$ in. thick

Foam core

OBJECTIVE 3 Solve application problems by dividing fractions.

Problems that involve forming equal-sized groups can be solved by division.

EXAMPLE 7   **Surfboard designs.**   Most surfboards are made of a foam core covered with several layers of fiberglass to keep them watertight. How many layers are needed to build up a finish $\frac{3}{8}$ of an inch thick if each layer of fiberglass has a thickness of $\frac{1}{16}$ of an inch?

Analyze

- The surfboard is to have a $\frac{3}{8}$-inch-thick fiberglass finish.   *Given*
- Each layer of fiberglass is $\frac{1}{16}$ of an inch thick.   *Given*
- How many layers of fiberglass need to be applied?   *Find*

Form Think of the $\frac{3}{8}$-inch-thick finish separated into an unknown number of equally thick layers of fiberglass. This indicates division.

We translate the words of the problem to numbers and symbols.

The number of layers of fiberglass that are needed	is equal to	the thickness of the finish	divided by	the thickness of 1 layer of fiberglass.
The number of layers of fiberglass that are needed	=	$\frac{3}{8}$	÷	$\frac{1}{16}$

Calculate To find the quotient, we will use the rule for dividing two fractions.

$$\frac{3}{8} \div \frac{1}{16} = \frac{3}{8} \cdot \frac{16}{1}$$ Multiply $\frac{3}{8}$ by the reciprocal of $\frac{1}{16}$, which is $\frac{16}{1}$.

$$= \frac{3 \cdot 16}{8 \cdot 1}$$ Multiply the numerators.
Multiply the denominators.

$$= \frac{3 \cdot 2 \cdot \overset{1}{\cancel{8}}}{\underset{1}{\cancel{8}} \cdot 1}$$ To simplify, factor 16 as 2 · 8. Then remove the common factor of 8 from the numerator and denominator.

$$= \frac{6}{1}$$ Multiply the remaining factors in the numerator: 3 · 2 · 1 = 6.
Multiply the remaining factors in the denominator: 1 · 1 = 1.

$$= 6$$ Any number divided by 1 is the same number.

State The number of layers of fiberglass needed is 6.

Check If 6 layers of fiberglass, each $\frac{1}{16}$ of an inch thick, are used, the finished thickness will be $\frac{6}{16}$ of an inch. If we simplify $\frac{6}{16}$, we see that it is equivalent to the desired finish thickness:

$$\frac{6}{16} = \frac{\overset{1}{\cancel{2}} \cdot 3}{\underset{1}{\cancel{2}} \cdot 8} = \frac{3}{8}$$

The result checks.

Self Check 7

Cooking. A recipe calls for 4 cups of sugar, and the only measuring container you have holds $\frac{1}{3}$ cup. How many $\frac{1}{3}$ cups of sugar would you need to add to follow the recipe?

Now Try Problem 93

SECTION 3.3   STUDY SET

VOCABULARY

Fill in the blanks.

1. The _____ of $\frac{5}{12}$ is $\frac{12}{5}$.

2. To find the reciprocal of a fraction, _____ the numerator and denominator.

3. The answer to a division is called the _____.

4. To simplify $\frac{2 \cdot 2 \cdot 3}{2 \cdot 3 \cdot 5 \cdot 7}$, we _____ common factors of the numerator and denominator.

CONCEPTS

5. Fill in the blanks.

a. To divide two fractions, _____ the first fraction by the _____ of the second fraction.

b.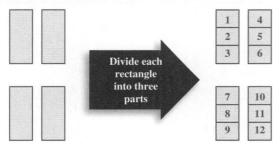

$\dfrac{1}{2} \div \dfrac{2}{3} = \dfrac{1}{2}$ ☐ ☐

6. a. What division problem is illustrated below?

b. What is the answer?

Divide each rectangle into three parts

1	4
2	5
3	6

7	10
8	11
9	12

7. Determine whether each quotient is positive or negative. *You do not have to find the answer.*

a. $-\dfrac{1}{4} \div \dfrac{3}{4}$ **b.** $-\dfrac{7}{8} \div \left(-\dfrac{21}{32}\right)$

8. Complete the table.

Number	Opposite	Reciprocal
$\frac{3}{10}$		
$-\frac{7}{11}$		
6		

9. a. Multiply $\frac{4}{5}$ and its reciprocal. What is the result?

b. Multiply $-\frac{3}{5}$ and its reciprocal. What is the result?

10. a. Find: $15 \div 3$

b. Rewrite $15 \div 3$ as multiplication by the reciprocal of 3, and find the result.

c. Complete this statement: Division by 3 is the same as multiplication by ☐.

NOTATION

Fill in the blanks to complete each step.

11. $\dfrac{4}{9} \div \dfrac{8}{27} = \dfrac{4}{9} \cdot \dfrac{☐}{8}$

$= \dfrac{4 \cdot ☐}{9 \cdot ☐}$

$= \dfrac{4 \cdot 3 \cdot ☐}{9 \cdot ☐ \cdot ☐}$

$= \dfrac{\cancel{4} \cdot 3 \cdot \overset{1}{\cancel{9}}}{\underset{1}{\cancel{4}} \cdot 2 \cdot \underset{1}{\cancel{4}}}$

$= \dfrac{☐}{2}$

12. $\dfrac{25}{31} \div 10 = \dfrac{25}{31} \div \dfrac{10}{☐}$

$= \dfrac{25}{31} \cdot \dfrac{1}{☐}$

$= \dfrac{25 \cdot ☐}{31 \cdot ☐}$

$= \dfrac{5 \cdot ☐ \cdot 1}{31 \cdot 2 \cdot 5}$

$= \dfrac{\overset{1}{\cancel{5}} \cdot 5 \cdot 1}{31 \cdot 2 \cdot \underset{1}{\cancel{5}}}$

$= \dfrac{5}{☐}$

GUIDED PRACTICE

Find the reciprocal of each number. **See Example 1 and Objective 3.**

13. a. $\dfrac{6}{7}$ **b.** $-\dfrac{15}{8}$ **c.** 10

14. a. $\dfrac{2}{9}$ **b.** $-\dfrac{9}{4}$ **c.** 7

15. a. $\dfrac{11}{8}$ **b.** $-\dfrac{1}{14}$ **c.** -63

16. a. $\dfrac{13}{2}$ **b.** $-\dfrac{1}{5}$ **c.** -21

Divide. Simplify each quotient, if possible. **See Example 2.**

17. $\dfrac{1}{8} \div \dfrac{2}{3}$ **18.** $\dfrac{1}{2} \div \dfrac{8}{9}$

19. $\dfrac{2}{23} \div \dfrac{1}{7}$ **20.** $\dfrac{4}{21} \div \dfrac{1}{5}$

Divide. Simplify each quotient, if possible. **See Example 3.**

21. $\dfrac{25}{32} \div \dfrac{5}{28}$ **22.** $\dfrac{4}{25} \div \dfrac{2}{35}$

23. $\dfrac{27}{32} \div \dfrac{9}{8}$ **24.** $\dfrac{16}{27} \div \dfrac{20}{21}$

Divide. Simplify each quotient, if possible. See Example 4.

25. $50 \div \dfrac{10}{9}$ **26.** $60 \div \dfrac{10}{3}$

27. $150 \div \dfrac{15}{32}$ **28.** $170 \div \dfrac{17}{6}$

Divide. Simplify each quotient, if possible. See Example 5.

29. $\dfrac{1}{8} \div \left(-\dfrac{1}{32}\right)$ **30.** $\dfrac{1}{9} \div \left(-\dfrac{1}{27}\right)$

31. $\dfrac{2}{5} \div \left(-\dfrac{4}{35}\right)$ **32.** $\dfrac{4}{9} \div \left(-\dfrac{16}{27}\right)$

Divide. Simplify each quotient, if possible. See Example 6.

33. $-\dfrac{28}{55} \div (-7)$ **34.** $-\dfrac{32}{45} \div (-8)$

35. $-\dfrac{33}{23} \div (-11)$ **36.** $-\dfrac{21}{31} \div (-7)$

TRY IT YOURSELF

Divide. Simplify each quotient, if possible.

37. $120 \div \dfrac{12}{5}$ **38.** $360 \div \dfrac{36}{5}$

39. $\dfrac{1}{2} \div \dfrac{3}{5}$ **40.** $\dfrac{1}{7} \div \dfrac{5}{6}$

41. $\left(-\dfrac{7}{4}\right) \div \left(-\dfrac{21}{8}\right)$ **42.** $\left(-\dfrac{15}{16}\right) \div \left(-\dfrac{5}{8}\right)$

43. $\dfrac{4}{5} \div \dfrac{4}{5}$ **44.** $\dfrac{2}{3} \div \dfrac{2}{3}$

45. Divide $-\dfrac{15}{32}$ by $\dfrac{3}{4}$.   **46.** Divide $-\dfrac{7}{10}$ by $\dfrac{4}{5}$.

47. $3 \div \dfrac{1}{12}$ **48.** $9 \div \dfrac{3}{4}$

49. $-\dfrac{4}{5} \div (-6)$ **50.** $-\dfrac{7}{8} \div (-14)$

51. $\dfrac{15}{16} \div 180$ **52.** $\dfrac{7}{8} \div 210$

53. $-\dfrac{9}{10} \div \dfrac{4}{15}$ **54.** $-\dfrac{3}{4} \div \dfrac{3}{2}$

55. $\dfrac{9}{10} \div \left(-\dfrac{3}{25}\right)$ **56.** $\dfrac{11}{16} \div \left(-\dfrac{9}{16}\right)$

57. $\dfrac{3}{16} \div \dfrac{1}{9}$ **58.** $\dfrac{5}{8} \div \dfrac{2}{9}$

59. $-\dfrac{1}{8} \div 8$ **60.** $-\dfrac{1}{15} \div 15$

The following problems involve multiplication and division. Perform each operation. Simplify the result, if possible.

61. $\dfrac{7}{6}\left(\dfrac{9}{49}\right)$ **62.** $\dfrac{7}{10}\left(\dfrac{20}{21}\right)$

63. $-\dfrac{1}{3} \div \dfrac{4}{5}$ **64.** $-\dfrac{2}{3} \div \left(-\dfrac{3}{2}\right)$

65. $\dfrac{13}{16} \div 2$ **66.** $\dfrac{7}{8} \div 6$

67. $\left(-\dfrac{11}{21}\right)\left(-\dfrac{14}{33}\right)$ **68.** $\left(-\dfrac{16}{35}\right)\left(-\dfrac{25}{48}\right)$

69. $-\dfrac{15}{32} \div \dfrac{5}{64}$ **70.** $-\dfrac{28}{15} \div \dfrac{21}{10}$

71. $11 \cdot \dfrac{1}{6}$ **72.** $9 \cdot \dfrac{1}{8}$

73. $\dfrac{3}{4} \cdot \dfrac{5}{7}$ **74.** $\dfrac{2}{3} \cdot \dfrac{7}{9}$

75. $\dfrac{25}{7} \div \left(-\dfrac{30}{21}\right)$ **76.** $\dfrac{39}{25} \div \left(-\dfrac{13}{10}\right)$

LOOK ALIKES

The following problems involve multiplication and division. Perform each operation. Simplify the result, if possible.

77. a. $\dfrac{3}{4} \cdot \dfrac{5}{7}$ **b.** $\dfrac{3}{4} \div \dfrac{5}{7}$

78. a. $\dfrac{2}{11} \cdot \dfrac{4}{5}$ **b.** $\dfrac{2}{11} \div \dfrac{4}{5}$

79. a. $\dfrac{8}{15} \cdot \dfrac{3}{16}$ **b.** $-\dfrac{8}{15} \div \dfrac{3}{16}$

80. a. $-\dfrac{12}{33} \cdot \dfrac{11}{9}$ **b.** $\dfrac{12}{33} \div \dfrac{11}{9}$

APPLICATIONS

81. Patio furniture. A production process applies several layers of a clear plastic coat to outdoor furniture to help protect it from the weather. If each protective coat is $\frac{3}{32}$-inch thick, how many applications will be needed to build up $\frac{3}{8}$ inch of clear finish?

82. Marathons. Each lap around a stadium track is $\frac{1}{4}$ mile. How many laps would a runner have to complete to get a 26-mile workout?

83. Cooking. A recipe calls for $\frac{3}{4}$ cup of flour, and the only measuring container you have holds $\frac{1}{8}$ cup. How many $\frac{1}{8}$ cups of flour would you need to add to follow the recipe?

84. Lasers. A technician uses a laser to slice thin pieces of aluminum off the end of a rod that is $\frac{7}{8}$-inch long. How many $\frac{1}{64}$-inch-wide slices can be cut from this rod? (Assume that there is no waste in the process.)

85. Underground cables. Refer to the illustration and table below.

   a. How many days will it take to install underground TV cable from the broadcasting station to the new homes using route 1?

   b. How long is route 2?

   c. How many days will it take to install the cable using route 2?

   d. Which route will require the fewer number of days to install the cable?

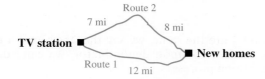

Proposal	Amount of cable installed per day	Comments
Route 1	$\frac{2}{5}$ of a mile	Ground very rocky
Route 2	$\frac{3}{5}$ of a mile	Longer than Route 1

86. Production planning. The materials used to make a pillow are shown. Examine the inventory list to decide how many pillows can be manufactured in one production run with the materials in stock.

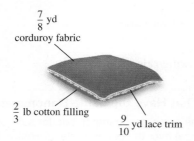

Factory Inventory List

Materials	Amount in stock
Lace trim	135 yd
Corduroy fabric	154 yd
Cotton filling	98 lb

87. Note cards. Ninety 3 × 5 cards are stacked next to a ruler, as shown in the next column.

   a. Into how many parts is 1 inch divided on the ruler?

   b. How thick is the stack of cards?

   c. How thick is one 3 × 5 card?

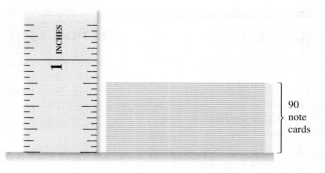

88. Computer printers. The illustration shows how the letter E is formed by a dot matrix printer. What is the height of one dot?

89. Forestry. A set of forestry maps divides the 6,284 acres of an old-growth forest into $\frac{4}{5}$-acre sections. How many sections do the maps contain?

90. Hardware. A hardware chain purchases large amounts of nails and packages them in $\frac{9}{16}$-pound bags for sale. How many of these bags of nails can be obtained from 2,871 pounds of nails?

WRITING

91. Explain how to divide two fractions.

92. Why do you need to know how to multiply fractions to be able to divide fractions?

93. Explain why 0 does not have a reciprocal.

94. What number is its own reciprocal? Explain why this is so.

95. Write an application problem that could be solved by finding $10 \div \frac{1}{5}$.

96. Explain why dividing a fraction by 2 is the same as finding $\frac{1}{2}$ of it. Give an example.

REVIEW

Fill in the blanks.

97. The symbol $<$ means ___ ____ _____.

98. The statement $9 \cdot 8 = 8 \cdot 9$ illustrates the _____ property of multiplication.

99. _____ is neither positive nor negative.

100. The sum of two negative numbers is _____.

101. Find the LCM of 6 and 9.

102. Evaluate each expression.

   a. 3^5 **b.** $(-2)^5$

OBJECTIVES

1 Add and subtract fractions that have the same denominator.

2 Add and subtract fractions that have different denominators.

3 Find the LCD to add and subtract fractions.

4 Identify the greater of two fractions.

5 Solve application problems by adding and subtracting fractions.

SECTION 3.4 Adding and Subtracting Fractions

In mathematics and everyday life, we can only add (or subtract) objects that are similar. For example, we can add dollars to dollars, but we cannot add dollars to oranges. This concept is important when adding or subtracting fractions.

OBJECTIVE 1   Add and subtract fractions that have the same denominator.

Consider the problem $\frac{3}{5} + \frac{1}{5}$. When we write it in words, it is apparent that we are adding similar objects.

three-**fifths**   +   one-**fifth**
└─── Similar objects ───┘

Because the denominators of $\frac{3}{5}$ and $\frac{1}{5}$ are the same, we say that they have a **common denominator**. Since the fractions have a common denominator, we can add them. The following figure explains the addition process.

three-fifths	one-fifth	four-fifths
$\frac{3}{5}$	$+$ $\frac{1}{5}$	$=$ $\frac{4}{5}$

We can make some observations about the addition shown in the figure.

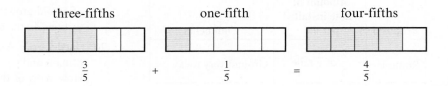

The sum of the numerators is the numerator of the answer.

$$\frac{3}{5} + \frac{1}{5} = \frac{4}{5}$$

The answer is a fraction that has the same denominator as the two fractions that were added.

These observations illustrate the following rule.

Adding and Subtracting Fractions That Have the Same Denominator

To add (or subtract) fractions that have the same denominator, add (or subtract) their numerators and write the sum (or difference) over the common denominator. Simplify the result, if possible.

> **EXAMPLE 1**   Perform each operation and simplify the result, if possible.
>
> **a.** Add: $\dfrac{1}{8} + \dfrac{5}{8}$ **b.** Subtract: $\dfrac{11}{15} - \dfrac{4}{15}$
>
> **Strategy** We will use the rule for adding and subtracting fractions that have *the same* denominator.
>
> **WHY** In part a, the fractions have the same denominator, 8. In part b, the fractions have the same denominator, 15.
>
> **Solution**
>
> **a.** $\dfrac{1}{8} + \dfrac{5}{8} = \dfrac{1+5}{8}$ Add the numerators and write the sum over the common denominator 8.
>
> $\qquad\qquad = \dfrac{6}{8}$ This fraction can be simplified.
>
> $\qquad\qquad = \dfrac{\overset{1}{\cancel{2}} \cdot 3}{\underset{1}{\cancel{2}} \cdot 4}$ To simplify, factor 6 as 2 · 3 and 8 as 2 · 4. Then remove the common factor of 2 from the numerator and denominator.
>
> $\qquad\qquad = \dfrac{3}{4}$ Multiply the remaining factors in the numerator: 1 · 3 = 3.
> Multiply the remaining factors in the denominator: 1 · 4 = 4.
>
> **b.** $\dfrac{11}{15} - \dfrac{4}{15} = \dfrac{11-4}{15}$ Subtract the numerators and write the difference over the common denominator 15.
>
> $\qquad\qquad = \dfrac{7}{15}$
>
> Since 7 and 15 have no common factors other than 1, the result is in simplest form.

Caution! We **do not** add fractions by adding the numerators and adding the denominators!

$$\cancel{\dfrac{1}{8} + \dfrac{5}{8} = \dfrac{1+5}{8+8} = \dfrac{6}{16}}$$

The same caution applies when subtracting fractions.

Self Check 1

Perform each operation and simplify the result, if possible.

a. Add: $\dfrac{5}{12} + \dfrac{1}{12}$

b. Subtract: $\dfrac{8}{9} - \dfrac{1}{9}$

Now Try ➲ Problems 17 and 21

The rule for subtraction from Section 2.3 can be extended to subtraction involving signed fractions:

To subtract two fractions, add the first to the opposite of the fraction to be subtracted.

> **EXAMPLE 2**   Subtract: $-\dfrac{7}{3} - \left(-\dfrac{2}{3}\right)$
>
> **Strategy** To find the difference, we will apply the rule for subtraction.
>
> **WHY** It is easy to make an error when subtracting signed fractions. We will probably be more accurate if we write the subtraction as addition of the opposite.
>
> **Solution**
>
> We read $-\dfrac{7}{3} - \left(-\dfrac{2}{3}\right)$ as "negative seven-thirds *minus* negative two-thirds." Thus, the number to be subtracted is $-\dfrac{2}{3}$. Subtracting $-\dfrac{2}{3}$ is the same as adding its opposite, $\dfrac{2}{3}$.
>
> $$-\dfrac{7}{3} - \left(-\dfrac{2}{3}\right) = -\dfrac{7}{3} + \dfrac{2}{3}$$ Add the opposite of $-\dfrac{2}{3}$, which is $\dfrac{2}{3}$.
>
> $$\qquad\qquad = \dfrac{-7}{3} + \dfrac{2}{3}$$ Write $-\dfrac{7}{3}$ as $\dfrac{-7}{3}$.

$$= \frac{-7 + 2}{3} \qquad \text{Add the numerators and write the sum over the common denominator 3.}$$

Subtract: $-\dfrac{9}{11} - \left(-\dfrac{3}{11}\right)$

Now Try ➭ **Problem 25**

$$= \frac{-5}{3} \qquad \text{Use the rule for adding two integers with different signs: } -7 + 2 = -5.$$

$$= -\frac{5}{3} \qquad \text{Rewrite the result with the } - \text{ sign in front: } \frac{-5}{3} = -\frac{5}{3}. \text{ This fraction is in simplest form.}$$

EXAMPLE 3   Perform the operations and simplify: $\dfrac{18}{25} - \dfrac{2}{25} - \dfrac{1}{25}$

Strategy We will use the rule for subtracting fractions that have *the same* denominator.

WHY All three fractions have the same denominator, 25.

Solution

$$\frac{18}{25} - \frac{2}{25} - \frac{1}{25} = \frac{18 - 2 - 1}{25} \qquad \text{Subtract the numerators and write the difference over the common denominator 25.}$$

Perform the operations and simplify:

$$\frac{2}{9} + \frac{2}{9} + \frac{2}{9}$$

Now Try ➭ **Problem 29**

$$= \frac{15}{25} \qquad \text{This fraction can be simplified.}$$

$$= \frac{3 \cdot \overset{1}{\cancel{5}}}{\underset{1}{\cancel{5}} \cdot 5} \qquad \text{To simplify, factor 15 as } 3 \cdot 5 \text{ and 25 as } 5 \cdot 5. \text{ Then remove the common factor of 5 from the numerator and denominator.}$$

$$= \frac{3}{5} \qquad \text{Multiply the remaining factors in the numerator: } 3 \cdot 1 = 3. \text{ Multiply the remaining factors in the denominator: } 1 \cdot 5 = 5.$$

OBJECTIVE 2 **Add and subtract fractions that have different denominators.**

Now we consider the problem $\frac{3}{5} + \frac{1}{3}$. Since the denominators are different, we cannot add these fractions in their present form.

three-**fifths**   +   one-**third**

└─ Not similar objects ─┘

To add (or subtract) fractions with different denominators, we express them as equivalent fractions that have a common denominator. The smallest common denominator, called the **least** or **lowest common denominator**, is usually the easiest common denominator to use.

Least Common Denominator

The **least common denominator (LCD)** for a set of fractions is the smallest number each denominator will divide exactly (divide with no remainder).

The denominators of $\frac{3}{5}$ and $\frac{1}{3}$ are 5 and 3. The numbers 5 and 3 divide many numbers exactly (30, 45, and 60, to name a few), but the smallest number that they divide exactly is 15. Thus, 15 is the LCD for $\frac{3}{5}$ and $\frac{1}{3}$.

To find $\frac{3}{5} + \frac{1}{3}$, we *build* equivalent fractions that have denominators of 15. (This procedure was introduced in Section 3.1.) Then we use the rule for adding fractions that have the same denominator.

$$\frac{3}{5} + \frac{1}{3} = \frac{3}{5} \cdot \frac{3}{3} + \frac{1}{3} \cdot \frac{5}{5}$$

We need to multiply this denominator by 5 to obtain 15.
It follows that $\frac{5}{5}$ should be the form of 1 used to build $\frac{1}{3}$.

We need to multiply this denominator by 3 to obtain 15.
It follows that $\frac{3}{3}$ should be the form of 1 that is used to build $\frac{3}{5}$.

$$= \frac{9}{15} + \frac{5}{15}$$

Multiply the numerators. Multiply the denominators.
Note that the denominators are now the same.

$$= \frac{9+5}{15}$$

Add the numerators and write the sum over
the common denominator 15.

$$= \frac{14}{15}$$

Since 14 and 15 have no common factors other
than 1, this fraction is in simplest form.

The figure below shows $\frac{3}{5}$ and $\frac{1}{3}$ expressed as equivalent fractions with a denominator of 15. Once the denominators are the same, the fractions are similar objects and can be added easily.

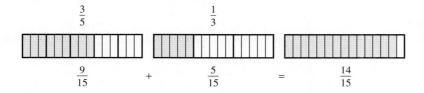

$$\frac{3}{5} \qquad \frac{1}{3}$$

$$\frac{9}{15} \qquad + \qquad \frac{5}{15} \qquad = \qquad \frac{14}{15}$$

We can use the following steps to add or subtract fractions with different denominators.

Adding and Subtracting Fractions That Have Different Denominators

1. Find the LCD.
2. Rewrite each fraction as an equivalent fraction with the LCD as the denominator. To do so, build each fraction using a form of 1 that involves any factors needed to obtain the LCD.
3. Add or subtract the numerators and write the sum or difference over the LCD.
4. Simplify the result, if possible.

EXAMPLE 4   Add: $\frac{1}{7} + \frac{2}{3}$

Strategy We will express each fraction as an equivalent fraction that has the LCD as its denominator. Then we will use the rule for adding fractions that have the same denominator.

WHY To add (or subtract) fractions, the fractions must have *like* denominators.

Solution
Since the smallest number the denominators 7 and 3 divide exactly is 21, the LCD is 21.

$$\frac{1}{7} + \frac{2}{3} = \frac{1}{7} \cdot \frac{3}{3} + \frac{2}{3} \cdot \frac{7}{7}$$

To build $\frac{1}{7}$ and $\frac{2}{3}$ so that their denominators are 21, multiply each by a form of 1.

$$= \frac{3}{21} + \frac{14}{21}$$

Multiply the numerators. Multiply the denominators. The denominators are now the same.

$$= \frac{3 + 14}{21}$$

Add the numerators and write the sum over the common denominator 21.

$$= \frac{17}{21}$$

Since 17 and 21 have no common factors other than 1, this fraction is in simplest form.

Self Check 4

Add: $\frac{1}{2} + \frac{2}{5}$

Now Try Problem 35

EXAMPLE 5   Subtract: $\frac{5}{2} - \frac{7}{3}$

Strategy We will express each fraction as an equivalent fraction that has the LCD as its denominator. Then we will use the rule for subtracting fractions that have the same denominator.

WHY To add (or subtract) fractions, the fractions must have *like* denominators.

Solution
Since the smallest number the denominators 2 and 3 divide exactly is 6, the LCD is 6.

$$\frac{5}{2} - \frac{7}{3} = \frac{5}{2} \cdot \frac{3}{3} - \frac{7}{3} \cdot \frac{2}{2}$$

To build $\frac{5}{2}$ and $\frac{7}{3}$ so that their denominators are 6, multiply each by a form of 1.

$$= \frac{15}{6} - \frac{14}{6}$$

Multiply the numerators. Multiply the denominators. The denominators are now the same.

$$= \frac{15 - 14}{6}$$

Subtract the numerators and write the difference over the common denominator 6.

$$= \frac{1}{6}$$

This fraction is in simplest form.

Self Check 5

Subtract: $\frac{6}{7} - \frac{3}{5}$

Now Try Problem 37

Success Tip In Example 6, did you notice that the denominator 5 is a factor of the denominator 15, and that the LCD is 15? In general, when adding (or subtracting) two fractions with different denominators, *if the smaller denominator is a factor of the larger denominator, the larger denominator is the LCD*.

EXAMPLE 6   Subtract: $\frac{2}{5} - \frac{11}{15}$

Strategy Since the smallest number the denominators 5 and 15 divide exactly is 15, the LCD is 15. We will only need to build an equivalent fraction for $\frac{2}{5}$.

WHY We do not have to build the fraction $\frac{11}{15}$ because it already has a denominator of 15.

Solution

$$\frac{2}{5} - \frac{11}{15} = \frac{2}{5} \cdot \frac{3}{3} - \frac{11}{15}$$

To build $\frac{2}{5}$ so that its denominator is 15, multiply it by a form of 1.

$$= \frac{6}{15} - \frac{11}{15}$$

Multiply the numerators. Multiply the denominators. The denominators are now the same.

$$= \frac{6 - 11}{15}$$

Subtract the numerators and write the difference over the common denominator 15.

$$= -\frac{5}{15}$$

If it is helpful, use the subtraction rule and add the opposite in the numerator: $6 + (-11) = -5$. Write the $-$ sign in front of the fraction.

$$= -\frac{\overset{1}{\cancel{5}}}{3 \cdot \underset{1}{\cancel{5}}}$$

To simplify, factor 15 as $3 \cdot 5$. Then remove the common factor of 5 from the numerator and denominator.

$$= -\frac{1}{3}$$

Multiply the remaining factors in the denominator: $3 \cdot 1 = 3$.

Self Check 6

Subtract: $\dfrac{2}{3} - \dfrac{13}{6}$

Now Try ➭ Problem 41

Caution! You might not have to build each fraction when adding or subtracting fractions with different denominators. For instance, the step in blue shown below is unnecessary when solving Example 6.

$$\frac{2}{5} - \frac{11}{15} = \frac{2}{5} \cdot \frac{3}{3} - \frac{11}{15} \cdot \cancel{\frac{1}{1}}$$

EXAMPLE 7   Add: $-5 + \dfrac{3}{4}$

Strategy We will write -5 as the fraction $\dfrac{-5}{1}$. Then we will follow the steps for adding fractions that have different denominators.

WHY The fractions $\dfrac{-5}{1}$ and $\dfrac{3}{4}$ have different denominators.

Solution

Since the smallest number the denominators 1 and 4 divide exactly is 4, the LCD is 4.

$$-5 + \frac{3}{4} = \frac{-5}{1} + \frac{3}{4}$$

Write -5 as $\frac{-5}{1}$.

$$= \frac{-5}{1} \cdot \frac{4}{4} + \frac{3}{4}$$

To build $\frac{-5}{1}$ so that its denominator is 4, multiply it by a form of 1.

$$= \frac{-20}{4} + \frac{3}{4}$$

Multiply the numerators. Multiply the denominators. The denominators are now the same.

$$= \frac{-20 + 3}{4}$$

Add the numerators and write the sum over the common denominator 4.

$$= \frac{-17}{4}$$

Use the rule for adding two integers with different signs: $-20 + 3 = -17$.

$$= -\frac{17}{4}$$

Write the result with the $-$ sign in front: $\frac{-17}{4} = -\frac{17}{4}$. This fraction is in simplest form.

Self Check 7

Add: $-6 + \dfrac{3}{8}$

Now Try ➭ Problem 45

OBJECTIVE **3** **Find the LCD to add and subtract fractions.**

When we add or subtract fractions that have different denominators, the least common denominator is not always obvious. We can use a concept studied earlier to determine the LCD for more difficult problems that involve larger denominators. To illustrate this, let's find the least common denominator of $\frac{3}{8}$ and $\frac{1}{10}$. (Note, the LCD *is not* 80.)

We have learned that both 8 and 10 must divide the LCD exactly. This divisibility requirement should sound familiar. Recall the following fact from Section 1.8.

The Least Common Multiple (LCM)

The **least common multiple (LCM)** of two whole numbers is the smallest whole number that is divisible by both of those numbers.

Thus, the least common denominator of $\frac{3}{8}$ and $\frac{1}{10}$ is simply the *least common multiple* of 8 and 10.

We can find the LCM of 8 and 10 by listing multiples of the larger number, 10, until we find one that is divisible by the smaller number, 8. (This method is explained in Example 2 of Section 1.8.)

Multiples of 10: 10, 20, 30, **40**, 50, 60, . . .

This is the first multiple of 10 that
is divisible by 8 (no remainder).

Since the LCM of 8 and 10 is 40, it follows that the LCD of $\frac{3}{8}$ and $\frac{1}{10}$ is 40.

We can also find the LCM of 8 and 10 using prime factorization. We begin by prime factoring 8 and 10. (This method is explained in Example 4 of Section 1.8.)

$$8 = \boxed{2 \cdot 2 \cdot 2}$$
$$10 = 2 \cdot \boxed{5}$$

The LCM of 8 and 10 is a product of prime factors, where each factor is used the greatest number of times it appears in any one factorization.

- We will use the factor 2 three times because 2 appears three times in the factorization of 8. Circle 2 · 2 · 2, as shown above.

- We will use the factor 5 once because it appears one time in the factorization of 10. Circle 5 as shown above.

Since there are no other prime factors in either prime factorization, we have

Use 2 three times.
Use 5 one time.

$$\text{LCM } (8, 10) = 2 \cdot 2 \cdot 2 \cdot 5 = 40$$

Finding the LCD

The least common denominator (LCD) of a set of fractions is the least common multiple (LCM) of the denominators of the fractions. Two ways to find the LCM of the denominators are as follows:

- Write the multiples of the largest denominator in increasing order, until one is found that is divisible by the other denominators.

- Prime factor each denominator. The LCM is a product of prime factors, where each factor is used the greatest number of times it appears in any one factorization.

EXAMPLE 8 Add: $\dfrac{7}{15} + \dfrac{3}{10}$

Strategy We begin by expressing each fraction as an equivalent fraction that has the LCD for its denominator. Then we use the rule for adding fractions that have the same denominator.

WHY To add (or subtract) fractions, the fractions must have *like* denominators.

Solution

To find the LCD, we find the prime factorization of both denominators and use each prime factor the *greatest* number of times it appears in any one factorization:

$$\left.\begin{array}{l} 15 = ③\cdot⑤ \\ 10 = ②\cdot 5 \end{array}\right\} \text{LCD} = 2\cdot 3\cdot 5 = 30$$

2 appears once in the factorization of 10.
3 appears once in the factorization of 15.
5 appears once in the factorizations of 15 and 10.

The LCD for $\dfrac{7}{15}$ and $\dfrac{3}{10}$ is 30.

$$\dfrac{7}{15} + \dfrac{3}{10} = \dfrac{7}{15}\cdot\dfrac{2}{2} + \dfrac{3}{10}\cdot\dfrac{3}{3}$$

To build $\frac{7}{15}$ and $\frac{3}{10}$ so that their denominators are 30, multiply each by a form of 1.

$$= \dfrac{14}{30} + \dfrac{9}{30}$$

Multiply the numerators. Multiply the denominators. The denominators are now the same.

$$= \dfrac{14 + 9}{30}$$

Add the numerators and write the sum over the common denominator 30.

$$= \dfrac{23}{30}$$

Since 23 and 30 have no common factors other than 1, this fraction is in simplest form.

Self Check 8

Add: $\dfrac{1}{8} + \dfrac{5}{6}$

Now Try ➡ **Problem 49**

EXAMPLE 9 Subtract and simplify: $\dfrac{13}{28} - \dfrac{1}{21}$

Strategy We begin by expressing each fraction as an equivalent fraction that has the LCD for its denominator. Then we use the rule for subtracting fractions with *like* denominators.

WHY To add (or subtract) fractions, the fractions must have like denominators.

Solution

To find the LCD, we find the prime factorization of both denominators and use each prime factor the *greatest* number of times it appears in any one factorization:

$$\left.\begin{array}{l} 28 = ②\cdot 2\cdot⑦ \\ 21 = ③\cdot 7 \end{array}\right\} \text{LCD} = 2\cdot 2\cdot 3\cdot 7 = 84$$

2 appears twice in the factorization of 28.
3 appears once in the factorization of 21.
7 appears once in the factorizations of 28 and 21.

The LCD for $\frac{13}{28}$ and $\frac{1}{21}$ is 84.

We will compare the prime factorizations of 28, 21, and the prime factorization of the LCD, 84, to determine what forms of 1 to use to build equivalent fractions for $\frac{13}{28}$ and $\frac{1}{21}$ with a denominator of 84.

LCD = $2\cdot 2\ \cdot\ 3\ \cdot\ 7$	LCD = $2\cdot 2\cdot\ 3\cdot 7$
Cover the prime factorization of 28. Since 3 is left uncovered, use $\frac{3}{3}$ to build $\frac{13}{28}$.	Cover the prime factorization of 21. Since $2\cdot 2 = 4$ is left uncovered, use $\frac{4}{4}$ to build $\frac{1}{21}$.

$$\frac{13}{28} - \frac{1}{21} = \frac{13}{28} \cdot \frac{3}{3} - \frac{1}{21} \cdot \frac{4}{4}$$

To build $\frac{13}{28}$ and $\frac{1}{21}$ so that their denominators are 84, multiply each by a form of 1.

$$= \frac{39}{84} - \frac{4}{84}$$

Multiply the numerators. Multiply the denominators. The denominators are now the same.

$$= \frac{39 - 4}{84}$$

Subtract the numerators and write the difference over the common denominator.

$$= \frac{35}{84}$$

This fraction is not in simplest form.

$$= \frac{5 \cdot \overset{1}{\cancel{7}}}{2 \cdot 2 \cdot 3 \cdot \underset{1}{\cancel{7}}}$$

To simplify, factor 35 and 84. Then remove the common factor of 7 from the numerator and denominator.

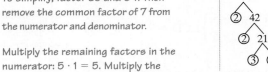

$$= \frac{5}{12}$$

Multiply the remaining factors in the numerator: $5 \cdot 1 = 5$. Multiply the remaining factors in the denominator: $2 \cdot 2 \cdot 3 \cdot 1 = 12$.

Self Check 9

Subtract and simplify:

$$\frac{21}{56} - \frac{9}{40}$$

Now Try ➡ Problem 53

OBJECTIVE 4 Identify the greater of two fractions.

If two fractions have the same denominator, the fraction with the greater numerator is the greater fraction.

For example,

$$\frac{7}{8} > \frac{3}{8} \quad \text{because } 7 > 3 \qquad -\frac{1}{3} > -\frac{2}{3} \quad \text{because } -1 > -2$$

If the denominators of two fractions are different, we need to write the fractions with a common denominator (preferably the LCD) before we can make a comparison.

EXAMPLE 10   Which fraction is larger: $\dfrac{5}{6}$ or $\dfrac{7}{8}$?

Strategy We will express each fraction as an equivalent fraction that has the LCD for its denominator. Then we will compare their numerators.

WHY We cannot compare the fractions as given. They are not similar objects.

five-**sixths**   seven-**eighths**

Solution

Since the smallest number the denominators will divide exactly is 24, the LCD for $\frac{5}{6}$ and $\frac{7}{8}$ is 24.

$$\frac{5}{6} = \frac{5}{6} \cdot \frac{4}{4} \qquad \Bigg| \qquad \frac{7}{8} = \frac{7}{8} \cdot \frac{3}{3}$$

To build $\frac{5}{6}$ and $\frac{7}{8}$ so that their denominators are 24, multiply each by a form of 1.

$$= \frac{20}{24} \qquad \Bigg| \qquad = \frac{21}{24}$$

Multiply the numerators.
Multiply the denominators.

Self Check 10

Which fraction is larger:

$$\frac{7}{12} \quad \text{or} \quad \frac{3}{5}?$$

Now Try ➡ Problem 61

Next, we compare the numerators. Since $21 > 20$, it follows that $\frac{21}{24}$ is greater than $\frac{20}{24}$. Thus, $\frac{7}{8} > \frac{5}{6}$.

OBJECTIVE 5 Solve application problems by adding and subtracting fractions.

EXAMPLE 11   **Television viewing habits.**   Students on a college campus were asked to estimate to the nearest hour how much television they watched each day. The results are given in the **circle graph** (also called a **pie chart**) to the right. For example, the chart tells us that $\frac{1}{4}$ of those responding watched 1 hour per day. What fraction of the student body watches from 0 to 2 hours daily?

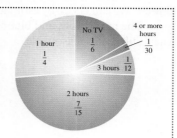

Analyze

- $\frac{1}{6}$ of the student body watches no TV daily.   *Given*

- $\frac{1}{4}$ of the student body watches 1 hour of TV daily.   *Given*

- $\frac{7}{15}$ of the student body watches 2 hours of TV daily.   *Given*

- What fraction of the student body watches 0 to 2 hours of TV daily?   *Find*

Form We translate the words of the problem to numbers and symbols.

| The fraction of the student body that watches from 0 to 2 hours of TV daily | is equal to | the fraction that watches no TV daily | plus | the fraction that watches 1 hour of TV daily | plus | the fraction that watches 2 hours of TV daily. |

$$\begin{array}{ccccccc}\text{The fraction of the student body that watches from 0 to 2 hours of TV daily} & = & \frac{1}{6} & + & \frac{1}{4} & + & \frac{7}{15}\end{array}$$

Calculate We must find the sum of three fractions with different denominators. To find the LCD, we prime factor the denominators and use each prime factor the *greatest* number of times it appears in any one factorization:

$$\left.\begin{array}{l} 6 = 2 \cdot \boxed{3} \\ 4 = \boxed{2 \cdot 2} \\ 15 = 3 \cdot \boxed{5} \end{array}\right\} \text{LCD} = 2 \cdot 2 \cdot 3 \cdot 5 = 60$$

2 appears twice in the factorization of 4.
3 appears once in the factorization of 6 and 15.
5 appears once in the factorization of 15.

The LCD for $\frac{1}{6}$, $\frac{1}{4}$, and $\frac{7}{15}$ is 60.

$$\frac{1}{6} + \frac{1}{4} + \frac{7}{15} = \frac{1}{6} \cdot \frac{10}{10} + \frac{1}{4} \cdot \frac{15}{15} + \frac{7}{15} \cdot \frac{4}{4}$$

Build each fraction so that its denominator is 60.

$$= \frac{10}{60} + \frac{15}{60} + \frac{28}{60}$$

Multiply the numerators. Multiply the denominators. The denominators are now the same.

$$= \frac{10 + 15 + 28}{60}$$

Add the numerators and write the sum over the common denominator 60.

$$= \frac{53}{60}$$

This fraction is in simplest form.

$$\begin{array}{r} \frac{1}{10} \\ 15 \\ + \ 28 \\ \hline 53 \end{array}$$

Self Check 11

Refer to the circle graph in Example 11. Find the fraction of the student body that watches 2 or more hours of television daily.

Now Try ➡ Problems 65 and 115

State The fraction of the student body that watches 0 to 2 hours of TV daily is $\frac{53}{60}$.

Check We can check by estimation. The result, $\frac{53}{60}$, is approximately $\frac{50}{60}$, which simplifies to $\frac{5}{6}$. The red, yellow, and blue shaded areas appear to shade about $\frac{5}{6}$ of the pie chart. The result seems reasonable.

Answers to Self Checks

1. a. $\frac{1}{2}$ **b.** $\frac{7}{9}$ **2.** $-\frac{6}{11}$ **3.** $\frac{2}{3}$ **4.** $\frac{9}{10}$ **5.** $\frac{9}{35}$ **6.** $-\frac{3}{2}$ **7.** $-\frac{45}{8}$ **8.** $\frac{23}{24}$ **9.** $\frac{3}{20}$ **10.** $\frac{3}{5}$ **11.** $\frac{7}{12}$

Think it Through • BUDGETS

"Putting together a budget is crucial if you don't want to spend your way into serious problems. You're also developing a habit that can serve you well throughout your life."
—Liz Pulliam Weston, MSN Money

The circle graph to the right shows a suggested budget for new college graduates as recommended by Springboard, a nonprofit consumer credit counseling service. What fraction of net take-home pay should be spent on housing?

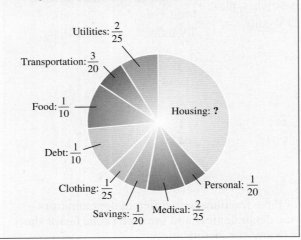

VOCABULARY

Fill in the blanks.

1. Because the denominators of $\frac{3}{8}$ and $\frac{7}{8}$ are the same number, we say that they have a _____ denominator.

2. The _____ common denominator for a set of fractions is the smallest number each denominator will divide exactly (no remainder).

3. Consider the solution below. To _____ an equivalent fraction with a denominator of 18, we multiply $\frac{4}{9}$ by a 1 in the form of [ ].

$$\frac{4}{9} = \frac{4}{9} \cdot \frac{2}{2}$$
$$= \frac{8}{18}$$

4. Consider the solution below. To _____ the fraction $\frac{15}{27}$, we factor 15 and 27, and then remove the common factor of 3 from the _____ and the _____ .

$$\frac{15}{27} = \frac{\overset{1}{\cancel{3}} \cdot 5}{\underset{1}{\cancel{3}} \cdot 3 \cdot 3}$$

$$= \frac{5}{9}$$

CONCEPTS

Fill in the blanks.

5. To add (or subtract) fractions that have the same denominator, add (or subtract) their _____ and write the sum (or difference) over the _____ denominator. _____ the result, if possible.

6. To add (or subtract) fractions that have different denominators, we express each fraction as an equivalent fraction that has the _____ for its denominator. Then we use the rule for adding (subtracting) fractions that have the _____ denominator.

7. When adding (or subtracting) two fractions with different denominators, if the smaller denominator is a factor of the larger denominator, the _____ denominator is the LCD.

8. Write the subtraction as addition of the opposite:

$$-\frac{1}{8} - \left(-\frac{5}{8}\right) = \boxed{} \ \boxed{} \ \boxed{}$$

9. Consider $\frac{3}{4}$. By what form of 1 should we multiply the numerator and denominator to express it as an equivalent fraction with a denominator of 36?

10. The *denominators* of two fractions are given. Find the least common denominator.

    a. 2 and 3   **b.** 3 and 5

    c. 4 and 8   **d.** 6 and 36

11. Consider the following prime factorizations:

$$24 = 2 \cdot 2 \cdot 2 \cdot 3$$
$$90 = 2 \cdot 3 \cdot 3 \cdot 5$$

For any one factorization, what is the greatest number of times

    a. a 5 appears?

    b. a 3 appears?

    c. a 2 appears?

12. The *denominators* of two fractions have their prime-factored forms shown below. Fill in the blanks to find the LCD for the fractions.

$$\left. \begin{array}{l} 20 = 2 \cdot 2 \cdot 5 \\ 30 = 2 \cdot 3 \cdot 5 \end{array} \right\} \text{LCD} = \boxed{} \cdot \boxed{} \cdot \boxed{} \cdot \boxed{} = \boxed{}$$

13. The *denominators* of three fractions have their prime-factored forms shown below. Fill in the blanks to find the LCD for the fractions.

$$\left. \begin{array}{l} 20 = 2 \cdot 2 \cdot 5 \\ 30 = 2 \cdot 3 \cdot 5 \\ 90 = 2 \cdot 3 \cdot 3 \cdot 5 \end{array} \right\} \text{LCD} = \boxed{} \cdot \boxed{} \cdot \boxed{} \cdot \boxed{} = \boxed{}$$

14. Place a $>$ or $<$ symbol in the blank to make a true statement.

    a. $\dfrac{32}{35} \ \boxed{} \ \dfrac{31}{35}$

    b. $-\dfrac{13}{17} \ \boxed{} \ -\dfrac{11}{17}$

NOTATION

Fill in the blanks to complete each step.

15. $\dfrac{2}{5} + \dfrac{1}{7} = \dfrac{2}{5} \cdot \dfrac{\boxed{}}{\boxed{}} + \dfrac{1}{7} \cdot \dfrac{5}{5}$

$$= \frac{\boxed{}}{35} + \frac{5}{\boxed{}}$$

$$= \frac{\boxed{} + \boxed{}}{35}$$

$$= \frac{\boxed{}}{35}$$

16. $\dfrac{7}{8} - \dfrac{2}{3} = \dfrac{7}{21} \cdot \dfrac{3}{\boxed{}} - \dfrac{2}{3} \cdot \dfrac{\boxed{}}{8}$

$$= \frac{21}{\boxed{}} - \frac{16}{\boxed{}}$$

$$= \frac{21 - 16}{\boxed{}}$$

$$= \frac{\boxed{}}{24}$$

GUIDED PRACTICE

Perform each operation and simplify, if possible. See Example 1.

17. $\dfrac{4}{9} + \dfrac{1}{9}$ **18.** $\dfrac{3}{7} + \dfrac{1}{7}$

19. $\dfrac{3}{8} + \dfrac{1}{8}$ **20.** $\dfrac{7}{12} + \dfrac{1}{12}$

21. $\dfrac{11}{15} - \dfrac{7}{15}$ **22.** $\dfrac{10}{21} - \dfrac{5}{21}$

23. $\dfrac{11}{20} - \dfrac{3}{20}$ **24.** $\dfrac{7}{18} - \dfrac{5}{18}$

Subtract and simplify, if possible. See Example 2.

25. $-\dfrac{11}{5} - \left(-\dfrac{8}{5}\right)$

26. $-\dfrac{15}{9} - \left(-\dfrac{11}{9}\right)$

27. $-\dfrac{7}{21} - \left(-\dfrac{2}{21}\right)$

28. $-\dfrac{21}{25} - \left(-\dfrac{9}{25}\right)$

Perform the operations and simplify, if possible. See Example 3.

29. $\dfrac{19}{40} - \dfrac{3}{40} - \dfrac{1}{40}$

30. $\dfrac{11}{24} - \dfrac{1}{24} - \dfrac{7}{24}$

31. $\dfrac{13}{33} + \dfrac{1}{33} + \dfrac{7}{33}$

32. $\dfrac{21}{50} + \dfrac{1}{50} + \dfrac{13}{50}$

Add and simplify, if possible. See Example 4.

33. $\dfrac{1}{3} + \dfrac{1}{7}$

34. $\dfrac{1}{4} + \dfrac{1}{5}$

35. $\dfrac{2}{5} + \dfrac{1}{9}$

36. $\dfrac{2}{7} + \dfrac{1}{2}$

Subtract and simplify, if possible. See Example 5.

37. $\dfrac{4}{5} - \dfrac{3}{4}$

38. $\dfrac{2}{3} - \dfrac{3}{5}$

39. $\dfrac{3}{4} - \dfrac{2}{7}$

40. $\dfrac{6}{7} - \dfrac{2}{3}$

Subtract and simplify, if possible. See Example 6.

41. $\dfrac{11}{12} - \dfrac{2}{3}$

42. $\dfrac{11}{18} - \dfrac{1}{6}$

43. $\dfrac{9}{14} - \dfrac{1}{7}$

44. $\dfrac{13}{15} - \dfrac{2}{3}$

Add and simplify, if possible. See Example 7.

45. $-2 + \dfrac{5}{9}$

46. $-3 + \dfrac{5}{8}$

47. $-3 + \dfrac{9}{4}$

48. $-1 + \dfrac{7}{10}$

Add and simplify, if possible. See Example 8.

49. $\dfrac{1}{6} + \dfrac{5}{8}$

50. $\dfrac{7}{12} + \dfrac{3}{8}$

51. $\dfrac{4}{9} + \dfrac{5}{12}$

52. $\dfrac{1}{9} + \dfrac{5}{6}$

Subtract and simplify, if possible. See Example 9.

53. $\dfrac{9}{10} - \dfrac{3}{14}$

54. $\dfrac{11}{12} - \dfrac{11}{30}$

55. $\dfrac{11}{12} - \dfrac{7}{15}$

56. $\dfrac{7}{15} - \dfrac{5}{12}$

Determine which fraction is larger. See Example 10.

57. $\dfrac{3}{8}$ or $\dfrac{5}{16}$

58. $\dfrac{5}{6}$ or $\dfrac{7}{12}$

59. $\dfrac{4}{5}$ or $\dfrac{2}{3}$

60. $\dfrac{7}{9}$ or $\dfrac{4}{5}$

61. $\dfrac{7}{9}$ or $\dfrac{11}{12}$

62. $\dfrac{3}{8}$ or $\dfrac{5}{12}$

63. $\dfrac{23}{20}$ or $\dfrac{7}{6}$

64. $\dfrac{19}{15}$ or $\dfrac{5}{4}$

Add and simplify, if possible. See Example 11.

65. $\dfrac{1}{6} + \dfrac{5}{18} + \dfrac{2}{9}$

66. $\dfrac{1}{10} + \dfrac{1}{8} + \dfrac{1}{5}$

67. $\dfrac{4}{15} + \dfrac{2}{3} + \dfrac{1}{6}$

68. $\dfrac{1}{2} + \dfrac{3}{5} + \dfrac{3}{20}$

TRY IT YOURSELF

Perform each operation and simplify, if possible.

69. $-\dfrac{1}{12} - \left(-\dfrac{5}{12}\right)$

70. $-\dfrac{1}{16} - \left(-\dfrac{15}{16}\right)$

71. $\dfrac{4}{5} + \dfrac{2}{3}$

72. $\dfrac{1}{4} + \dfrac{2}{3}$

73. $\dfrac{12}{25} - \dfrac{1}{25} - \dfrac{1}{25}$

74. $\dfrac{7}{9} + \dfrac{1}{9} + \dfrac{1}{9}$

75. $-\dfrac{7}{20} - \dfrac{1}{5}$

76. $-\dfrac{5}{8} - \dfrac{1}{3}$

77. $-\dfrac{7}{16} + \dfrac{1}{4}$

78. $-\dfrac{17}{20} + \dfrac{4}{5}$

79. $\dfrac{11}{12} - \dfrac{2}{3}$

80. $\dfrac{2}{3} - \dfrac{1}{6}$

81. $\dfrac{2}{3} + \dfrac{4}{5} + \dfrac{5}{6}$

82. $\dfrac{3}{4} + \dfrac{2}{5} + \dfrac{3}{10}$

83. $\dfrac{9}{20} - \dfrac{1}{30}$

84. $\dfrac{5}{6} - \dfrac{3}{10}$

85. $\dfrac{27}{50} + \dfrac{5}{16}$

86. $\dfrac{49}{50} - \dfrac{15}{16}$

87. $\dfrac{13}{20} - \dfrac{1}{5}$

88. $\dfrac{71}{100} - \dfrac{1}{10}$

89. $\dfrac{37}{103} - \dfrac{17}{103}$

90. $\dfrac{54}{53} - \dfrac{52}{53}$

91. $-\dfrac{3}{4} - 5$

92. $-2 - \dfrac{7}{8}$

93. $\dfrac{4}{27} + \dfrac{1}{6}$ **94.** $\dfrac{8}{9} - \dfrac{7}{12}$

95. $\dfrac{7}{30} - \dfrac{19}{75}$ **96.** $\dfrac{73}{75} - \dfrac{31}{30}$

97. Find the difference of $\dfrac{11}{60}$ and $\dfrac{2}{45}$.

98. Find the sum of $\dfrac{9}{48}$ and $\dfrac{7}{40}$.

99. Subtract $\dfrac{5}{12}$ from $\dfrac{2}{15}$.

100. What is the sum of $\dfrac{11}{24}$ and $\dfrac{7}{36}$ increased by $\dfrac{5}{48}$?

LOOK ALIKES

Perform each operation and simplify, if possible.

101. a. $\dfrac{1}{4} + \dfrac{1}{8}$ **b.** $\dfrac{1}{4} - \dfrac{1}{8}$

   c. $\dfrac{1}{4} \cdot \dfrac{1}{8}$ **d.** $\dfrac{1}{4} \div \dfrac{1}{8}$

102. a. $\dfrac{5}{21} + \dfrac{3}{14}$ **b.** $\dfrac{5}{21} - \dfrac{3}{14}$

   c. $\dfrac{5}{21} \cdot \dfrac{3}{14}$ **d.** $\dfrac{5}{21} \div \dfrac{3}{14}$

103. a. $\dfrac{1}{10} + \dfrac{3}{10}$ **b.** $\dfrac{1}{10} - \dfrac{3}{10}$

   c. $\dfrac{1}{10} \cdot \dfrac{3}{10}$ **d.** $\dfrac{1}{10} \div \dfrac{3}{10}$

104. a. $\dfrac{3}{8} + \dfrac{6}{13}$ **b.** $\dfrac{3}{8} - \dfrac{6}{13}$

   c. $\dfrac{3}{8} \cdot \dfrac{6}{13}$ **d.** $\dfrac{3}{8} \div \dfrac{6}{13}$

APPLICATIONS

105. Botany.   To determine the effects of smog on tree development, a scientist cut down a pine tree and measured the width of the growth rings for two five-year time periods. See illustration in the next column.
   a. What was the total growth over the ten-year period?

   b. What is the difference in growth for the two five-year time periods?

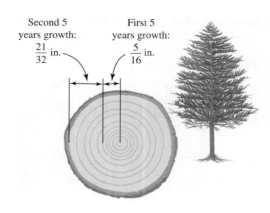

Second 5 years growth: $\dfrac{21}{32}$ in.   First 5 years growth: $\dfrac{5}{16}$ in.

106. Garage door openers.   What is the difference in strength between a $\frac{1}{3}$-hp and a $\frac{1}{2}$-hp garage door opener?

107. Magazine covers.   The page design for the magazine cover shown below includes a blank strip at the top, called a *header*, and a blank strip at the bottom of the page, called a *footer*. How much page length is lost because of the header and footer?

$\dfrac{3}{8}$ in. header

Page length

$\dfrac{5}{16}$ in. footer

108. Delivery trucks.   A truck can safely carry a one-ton load. Should it be used to deliver one-half ton of sand, one-third ton of gravel, and one-fifth ton of cement in one trip to a job site?

109. Dinners.   A family bought two large pizzas for dinner. Some pieces of each pizza were not eaten, as shown.
   a. What fraction of the first pizza was not eaten?

   b. What fraction of the second pizza was not eaten?

   c. What fraction of a pizza was left?

   d. Could the family have been fed with just one pizza?

110. Gasoline barrels. Three identical-sized barrels are shown below. If the contents of two of the barrels are poured into the empty third barrel, what fraction of the third barrel will be filled?

111. Weights and measures. A consumer protection agency determines the accuracy of butcher shop scales by placing a known three-quarter-pound weight on the scale and then comparing that to the scale's readout. According to the illustration, by how much is this scale off? Does it result in undercharging or overcharging customers on their meat purchases?

112. Infinite sums. In advanced mathematics courses, students learn how to find the sum of an infinite (never ending) list of numbers. This problem will give you a brief look at what you can expect to study in the future if you take such a course.

a. Fill in the blanks to complete the list shown below.

$$\frac{1}{2} + \frac{1}{4} + \frac{1}{8} + \frac{1}{16} + \frac{1}{32} + \boxed{} + \boxed{}$$

b. Find the sum of the fractions listed in part a.

c. If the list in part a continued forever, what do you think would be the sum of all the fractions?

113. Hardware. Find the inside diameter of the circular washer shown below.

Outside diameter
$\frac{7}{8}$ in.

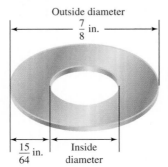

$\frac{15}{64}$ in.   Inside diameter

114. Figure drawing. As an aid in drawing the human body, artists divide the body into three parts. Each part is then expressed as a fraction of the total body height. For example, the torso is $\frac{4}{15}$ of the body height. What fraction of body height is the head?

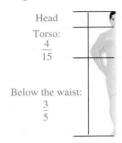

Head

Torso:
$\frac{4}{15}$

Below the waist:
$\frac{3}{5}$

from **Campus to Careers**

School Guidance Counselor

115. Suppose you work as a school guidance counselor at a community college and your department has conducted a survey of the full-time students to learn more about their study habits. As part of a Power Point presentation of the survey results to the school board, you show the following circle graph. At that time, you are asked, "What fraction of the full-time students study 2 hours or more daily?" What would you answer?

More than 2 hr
$\frac{3}{10}$

2 hr
$\frac{2}{5}$

Less than 1 hr
$\frac{1}{10}$

$\frac{1}{5}$

1 hr

116. Health statistics. The circle graph below shows the leading causes of death in the United States for 2014. For example, $\frac{6}{25}$ of all of the deaths that year were caused by heart disease. What fraction of all the deaths were caused by heart disease, cancer, respiratory diseases, or stroke, combined?

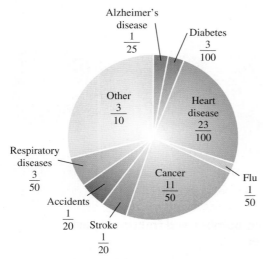

Source: National Center for Health Statistics

117. Musical notes. The notes used in music have fractional values. Their names and the symbols used to represent them are shown in illustration (a). In common time, the values of the notes in each measure must add to 1. Is the measure in illustration (b) complete?

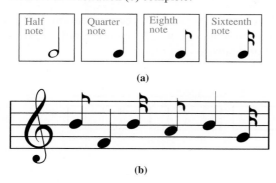

118. Tools. A mechanic likes to hang his wrenches above his tool bench in order of narrowest to widest. What is the proper order of the wrenches in the illustration?

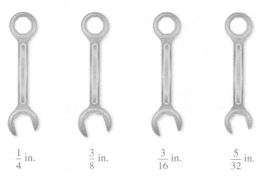

119. Tire tread. A mechanic measured the tire tread depth on each of the tires on a car and recorded them on the form shown below. (The letters LF stand for *left front*, RR stands for *right rear*, and so on.)

   a. Which tire has the most tread?

   b. Which tire has the least tread?

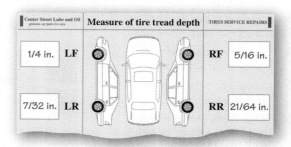

120. Hiking. The illustration below shows the length of each part of a three-part hike. Rank the lengths of the parts from longest to shortest.

WRITING

121. Explain why we cannot add or subtract the fractions $\frac{2}{9}$ and $\frac{2}{5}$ as they are written.

122. To multiply fractions, must they have the same denominators? Explain why or why not. Give an example.

REVIEW

Simplify each expression.

123. $\dfrac{15}{28} \cdot \dfrac{14}{25}$

124. $-\dfrac{10}{3} \cdot \dfrac{12}{35}$

125. $-\dfrac{12}{7} \div \dfrac{36}{35}$

126. $\dfrac{8}{5} \div \dfrac{16}{15}$

OBJECTIVES

1 Identify the whole-number and fractional parts of a mixed number.

2 Write mixed numbers as improper fractions.

3 Write improper fractions as mixed numbers.

4 Graph fractions and mixed numbers on a number line.

5 Multiply and divide mixed numbers.

6 Solve application problems by multiplying and dividing mixed numbers.

SECTION **3.5** Multiplying and Dividing Mixed Numbers

In the next two sections, we show how to add, subtract, multiply, and divide *mixed numbers*. These numbers are widely used in daily life.

The recipe calls for $2\frac{1}{3}$ cups of flour.

(Read as "two and one-third.")

It took $3\frac{3}{4}$ hours to paint the living room.

(Read as "three and three-fourths.")

The entrance to the park is $1\frac{1}{2}$ miles away.

(Read as "one and one-half.")

OBJECTIVE 1 **Identify the whole-number and fractional parts of a mixed number.**

A **mixed number** is the *sum* of a whole number and a proper fraction. For example, $3\frac{3}{4}$ is a mixed number.

$$3\frac{3}{4} \qquad = \qquad 3 \qquad + \qquad \frac{3}{4}$$

Mixed number Whole-number part Fractional part

Caution! Note that $3\frac{3}{4}$ means $3 + \frac{3}{4}$, even though the + symbol is not written. Do not confuse $3\frac{3}{4}$ with $3 \cdot \frac{3}{4}$ or $3\left(\frac{3}{4}\right)$, which indicate the multiplication of 3 by $\frac{3}{4}$.

Mixed numbers can be represented by shaded regions. In the illustration below, each rectangular region outlined in black represents one whole. To represent $3\frac{3}{4}$, we shade 3 *whole* rectangular regions and 3 out of 4 *parts* of another.

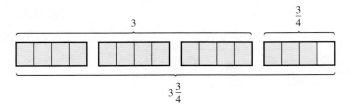

3

$\frac{3}{4}$

$3\frac{3}{4}$

EXAMPLE 1 In the illustration below, each disk represents one whole. Write an improper fraction and a mixed number to represent the shaded portion.

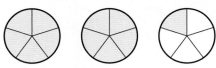

Strategy We will determine the number of equal parts into which a disk is divided. Then we will determine how many of those *parts* are shaded and how many of the *whole* disks are shaded.

WHY To write an improper fraction, we need to find its numerator and its denominator. To write a mixed number, we need to find its whole number part and its fractional part.

Solution

Since each disk is divided into 5 equal parts, the denominator of the improper fraction is 5. Since a total of 11 of those parts are shaded, the numerator is 11, and we say that

$\frac{11}{5}$ is shaded.   Write: $\frac{\text{total number of parts shaded}}{\text{number of equal parts in one disk}}$

Since 2 whole disks are shaded, the whole number part of the mixed number is 2. Since 1 out of 5 of the parts of the last disk is shaded, the fractional part of the mixed number is $\frac{1}{5}$, and we say that

$2\frac{1}{5}$ is shaded.

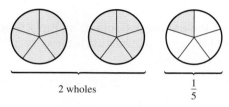

2 wholes $\frac{1}{5}$

Self Check 1

In the illustration below, each oval region represents one whole. Write an improper fraction and a mixed number to represent the shaded portion.

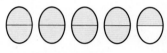

Now Try ➡ **Problem 19**

In this section, we will work with negative as well as positive mixed numbers. For example, the negative mixed number $-3\frac{3}{4}$ could be used to represent $3\frac{3}{4}$ feet below sea level. Think of $-3\frac{3}{4}$ as $-3 - \frac{3}{4}$ or as $-3 + \left(-\frac{3}{4}\right)$.

OBJECTIVE 2 Write mixed numbers as improper fractions.

In Example 1, we saw that the shaded portion of the illustration can be represented by the mixed number $2\frac{1}{5}$ and by the improper fraction $\frac{11}{5}$. To develop a procedure to write any mixed number as an improper fraction, consider the following steps that show how to do this for $2\frac{1}{5}$. The objective is to find how many *fifths* that the mixed number $2\frac{1}{5}$ represents.

$$2\frac{1}{5} = 2 + \frac{1}{5} \qquad \text{Write the mixed number } 2\frac{1}{5} \text{ as a sum.}$$

$$= \frac{2}{1} + \frac{1}{5} \qquad \text{Write 2 as a fraction: } 2 = \frac{2}{1}.$$

$$= \frac{2}{1} \cdot \frac{5}{5} + \frac{1}{5} \qquad \text{To build } \frac{2}{1} \text{ so that its denowminator is 5, multiply it by a form of 1.}$$

$$= \frac{10}{5} + \frac{1}{5} \qquad \begin{array}{l}\text{Multiply the numerators.}\\\text{Multiply the denominators.}\end{array}$$

$$= \frac{11}{5} \qquad \begin{array}{l}\text{Add the numerators and write the sum over}\\\text{the common denominator 5.}\end{array}$$

Thus, $2\frac{1}{5} = \frac{11}{5}$.

We can obtain the same result with far less work. To change $2\frac{1}{5}$ to an improper fraction, we simply multiply 5 by 2 and add 1 to get the numerator, and keep the denominator of 5.

$$2\frac{1}{5} = \frac{5 \cdot 2 + 1}{5} = \frac{10 + 1}{5} = \frac{11}{5}$$

This example illustrates the following procedure.

Writing a Mixed Number as an Improper Fraction

To write a mixed number as an improper fraction:

1. Multiply the denominator of the fraction by the whole-number part.
2. Add the numerator of the fraction to the result from step 1.
3. Write the sum from step 2 over the original denominator.

EXAMPLE 2   Write the mixed number $7\frac{5}{6}$ as an improper fraction.

Strategy We will use the three-step procedure to find the improper fraction.

WHY It's faster than writing $7\frac{5}{6}$ as $7 + \frac{5}{6}$, building to get an LCD, and adding.

Solution
To find the numerator of the improper fraction, multiply 6 by 7, and add 5 to that result. The denominator of the improper fraction is the same as the denominator of the fractional part of the mixed number.

Step 2: Add.

$$7\frac{5}{6} = \frac{6 \cdot 7 + 5}{6} = \frac{42 + 5}{6} = \frac{47}{6}$$

By the order of operations rule, multiply first and then add.

Step 1: Multiply.   Step 3: Use the same denominator.

To write a *negative mixed number* in fractional form, ignore the $-$ sign and use the method shown in Example 2 for a positive mixed number. Once that procedure is completed, write a $-$ sign in front of the result. For example,

$$-6\frac{1}{4} = -\frac{25}{4} \qquad -1\frac{9}{10} = -\frac{19}{10} \qquad -12\frac{3}{8} = -\frac{99}{8}$$

OBJECTIVE 3 **Write improper fractions as mixed numbers.**

To write an improper fraction as a mixed number, we must find two things: the *whole-number part* and the *fractional part* of the mixed number. To develop a procedure to do this, let's consider the improper fraction $\frac{7}{3}$. To find the number of groups of 3 in 7, we can divide 7 by 3. This will find the whole-number part of the mixed number. The remainder is the numerator of the fractional part of the mixed number.

Whole-number part

$$
\begin{array}{r}
2 \\
3\overline{)7} \\
-6 \\
\hline
1
\end{array}
\qquad
2\frac{1}{3}
$$

The remainder is the numerator of the fractional part.

The divisor is the denominator of the fractional part.

This example suggests the following procedure.

Writing an Improper Fraction as a Mixed Number

To write an improper fraction as a mixed number:

1. Divide the numerator by the denominator to obtain the whole-number part.
2. The remainder over the divisor is the fractional part.

EXAMPLE 3   Write each improper fraction as a mixed number or a whole number:

a. $\frac{29}{6}$ **b.** $\frac{40}{16}$ **c.** $\frac{84}{3}$ **d.** $-\frac{9}{5}$

Strategy We will divide the numerator by the denominator and write the remainder over the divisor.

WHY A fraction bar indicates division.

Solution

a. To write $\frac{29}{6}$ as a mixed number, divide 29 by 6:

$$\begin{array}{r} 4 \\ 6\overline{)29} \\ -24 \\ \hline 5 \end{array}$$

← The whole-number part is 4.

← Write the remainder 5 over the divisor 6 to get the fractional part.

Thus, $\dfrac{29}{6} = 4\dfrac{5}{6}$.

b. To write $\frac{40}{16}$ as a mixed number, divide 40 by 16:

$$\begin{array}{r} 2 \\ 16\overline{)40} \\ -32 \\ \hline 8 \end{array}$$

Thus, $\dfrac{40}{16} = 2\dfrac{8}{16} = 2\dfrac{1}{2}$.   *Simplify the fractional part:* $\dfrac{8}{16} = \dfrac{\overset{1}{\cancel{8}}}{2 \cdot \underset{1}{\cancel{8}}} = \dfrac{1}{2}$.

c. For $\frac{84}{3}$, divide 84 by 3:

$$\begin{array}{r} 28 \\ 3\overline{)84} \\ -6 \\ \hline 24 \\ -24 \\ \hline 0 \end{array}$$

Thus, $\dfrac{84}{3} = 28$.

← Since the remainder is 0, the improper fraction represents a whole number.

d. To write $-\frac{9}{5}$ as a mixed number, ignore the − sign, and use the method for the positive improper fraction $\frac{9}{5}$. Once that procedure is completed, write a − sign in front of the result.

$$\begin{array}{r} 1 \\ 5\overline{)9} \\ -5 \\ \hline 4 \end{array}$$

Thus, $-\dfrac{9}{5} = -1\dfrac{4}{5}$.

Self Check 3

Write each improper fraction as a mixed number or a whole number:

a. $\dfrac{31}{7}$ **b.** $\dfrac{50}{26}$

c. $\dfrac{51}{3}$ **d.** $-\dfrac{10}{3}$

Now Try ➡ Problems 31, 35, 39, and 43

OBJECTIVE 4   Graph fractions and mixed numbers on a number line.

In Chapters 1 and 2, we graphed whole numbers and integers on a number line. Fractions and mixed numbers can also be graphed on a number line.

EXAMPLE 4   Graph $-2\frac{3}{4}$, $-1\frac{1}{2}$, $-\frac{1}{8}$, and $\frac{13}{5}$ on a number line.

Strategy We will locate the position of each fraction and mixed number on the number line and draw a bold dot.

WHY To *graph a number* means to make a drawing that represents the number.

Solution

- Since $-2\frac{3}{4} < -2$, the graph of $-2\frac{3}{4}$ is to the left of -2 on the number line.
- The number $-1\frac{1}{2}$ is between -1 and -2.
- The number $-\frac{1}{8}$ is less than 0.
- Expressed as a mixed number, $\frac{13}{5} = 2\frac{3}{5}$.

Self Check 4

Graph $-1\frac{7}{8}$, $-\frac{2}{3}$, $\frac{3}{5}$, and $\frac{9}{4}$ on a number line.

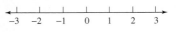

Now Try ➡ Problem 47

OBJECTIVE 5 Multiply and divide mixed numbers.

We will use the same procedures for multiplying and dividing mixed numbers as we used in Sections 3.2 and 3.3 to multiply and divide fractions. However, we must write the mixed numbers as improper fractions before we actually multiply or divide.

> **Multiplying and Dividing Mixed Numbers**
>
> To multiply or divide mixed numbers, first change the mixed numbers to improper fractions. Then perform the multiplication or division of the fractions. Write the result as a mixed number or a whole number in simplest form.

The sign rules for multiplying and dividing integers also hold for multiplying and dividing mixed numbers.

EXAMPLE 5 Multiply and simplify, if possible.

a. $1\frac{3}{4} \cdot 2\frac{1}{3}$ **b.** $5\frac{1}{5} \cdot \left(1\frac{2}{13}\right)$ **c.** $-4\frac{1}{9}(3)$

Strategy We will write the mixed numbers and whole numbers as improper fractions.

WHY Then we can use the rule for multiplying two fractions from Section 3.2.

Solution

a. $1\dfrac{3}{4} \cdot 2\dfrac{1}{3} = \dfrac{7}{4} \cdot \dfrac{7}{3}$ Write $1\frac{3}{4}$ and $2\frac{1}{3}$ as improper fractions.

$= \dfrac{7 \cdot 7}{4 \cdot 3}$ Use the rule for multiplying two fractions. Multiply the numerators and the denominators.

$= \dfrac{49}{12}$ Since there are no common factors to remove, perform the multiplication in the numerator and in the denominator. The result is an improper fraction.

$= 4\dfrac{1}{12}$ Write the improper fraction $\frac{49}{12}$ as a mixed number.

$$\begin{array}{r} 4 \\ 12\overline{)49} \\ -48 \\ \hline 1 \end{array}$$

b. $5\dfrac{1}{5}\left(1\dfrac{2}{13}\right) = \dfrac{26}{5} \cdot \dfrac{15}{13}$ Write $5\frac{1}{5}$ and $1\frac{2}{13}$ as improper fractions.

$= \dfrac{26 \cdot 15}{5 \cdot 13}$ Multiply the numerators. Multiply the denominators.

$$= \frac{2 \cdot 13 \cdot 3 \cdot 5}{5 \cdot 13}$$

To prepare to simplify, factor 26 as 2 · 13 and 15 as 3 · 5.

$$= \frac{2 \cdot \overset{1}{\cancel{13}} \cdot 3 \cdot \overset{1}{\cancel{5}}}{\underset{1}{\cancel{5}} \cdot \underset{1}{\cancel{13}}}$$

Remove the common factors of 13 and 5 from the numerator and denominator.

$$= \frac{6}{1}$$

Multiply the remaining factors in the numerator: 2 · 1 · 3 · 1 = 6.
Multiply the remaining factors in the denominator: 1 · 1 = 1.

$$= 6$$

Any whole number divided by 1 remains the same.

c. $-4\frac{1}{9} \cdot 3 = -\frac{37}{9} \cdot \frac{3}{1}$

Write $-4\frac{1}{9}$ as an improper fraction and write 3 as a fraction.

$$= -\frac{37 \cdot 3}{9 \cdot 1}$$

Multiply the numerators and multiply the denominators.
Since the fractions have unlike signs, make the answer negative.

$$= -\frac{37 \cdot \overset{1}{\cancel{3}}}{\underset{1}{\cancel{3}} \cdot 3 \cdot 1}$$

To simplify, factor 9 as 3 · 3, and then remove the common factor of 3 from the numerator and denominator.

$$= -\frac{37}{3}$$

Multiply the remaining factors in the numerator and in the denominator. The result is an improper fraction.

$$= -12\frac{1}{3}$$

Write the negative improper fraction $-\frac{37}{3}$ as a negative mixed number.

$$\begin{array}{r} 12 \\ 3\overline{)37} \\ -3 \\ \hline 7 \\ -6 \\ \hline 1 \end{array}$$

Self Check 5

Multiply and simplify, if possible.

a. $3\frac{1}{3} \cdot 2\frac{1}{3}$

b. $9\frac{3}{5} \cdot \left(3\frac{3}{4}\right)$

c. $-4\frac{5}{6}(2)$

Now Try ➲ Problems 51, 55, and 57

Success Tip We can use estimation by rounding to check the results when multiplying mixed numbers. If the fractional part of the mixed number is $\frac{1}{2}$ or greater, round up by adding 1 to the whole-number part and dropping the fraction. If the fractional part of the mixed number is less than $\frac{1}{2}$, round down by dropping the fraction and using only the whole-number part. To check the answer $4\frac{1}{12}$ from Example 5a, we proceed as follows:

$$1\frac{3}{4} \cdot 2\frac{1}{3} \approx 2 \cdot 2 = 4$$

Since $\frac{3}{4}$ is greater than $\frac{1}{2}$, round $1\frac{3}{4}$ up to 2.
Since $\frac{1}{3}$ is less than $\frac{1}{2}$, round $2\frac{1}{3}$ down to 2.

Since $4\frac{1}{12}$ is close to 4, it is a reasonable answer.

EXAMPLE 6   Divide and simplify, if possible:

a. $-3\frac{3}{8} \div \left(-2\frac{1}{4}\right)$ **b.** $1\frac{11}{16} \div \frac{3}{4}$

Strategy We will write the mixed numbers as improper fractions.

WHY Then we can use the rule for dividing two fractions from Section 3.3.

Solution

a. $-3\frac{3}{8} \div \left(-2\frac{1}{4}\right) = -\frac{27}{8} \div \left(-\frac{9}{4}\right)$

Write $-3\frac{3}{8}$ and $-2\frac{1}{4}$ as improper fractions.

$$= -\frac{27}{8}\left(-\frac{4}{9}\right)$$

Use the rule for dividing two fractions:
Multiply $-\frac{27}{8}$ by the reciprocal of $-\frac{9}{4}$, which is $-\frac{4}{9}$.

$$= \frac{27}{8}\left(\frac{4}{9}\right)$$

Since the product of two negative fractions is positive, drop both − signs and continue.

$$= \frac{27 \cdot 4}{8 \cdot 9}$$

Multiply the numerators.
Multiply the denominators.

$$= \frac{3 \cdot \overset{1}{\cancel{9}} \cdot \overset{1}{\cancel{4}}}{2 \cdot \underset{1}{\cancel{4}} \cdot \underset{1}{\cancel{9}}}$$ To simplify, factor 27 as $3 \cdot 9$ and 8 as $2 \cdot 4$. Then remove the common factors of 9 and 4 from the numerator and denominator.

$$= \frac{3}{2}$$ Multiply the remaining factors in the numerator: $3 \cdot 1 \cdot 1 = 3$.
Multiply the remaining factors in the denominator: $2 \cdot 1 \cdot 1 = 2$.

$$= 1\frac{1}{2}$$ Write the improper fraction $\frac{3}{2}$ as a mixed number by dividing 3 by 2.

b. $1\dfrac{11}{16} \div \dfrac{3}{4} = \dfrac{27}{16} \div \dfrac{3}{4}$ Write $1\frac{11}{16}$ as an improper fraction.

$$= \frac{27}{16} \cdot \frac{4}{3}$$ Multiply $\frac{27}{16}$ by the reciprocal of $\frac{3}{4}$, which is $\frac{4}{3}$.

$$= \frac{27 \cdot 4}{16 \cdot 3}$$ Multiply the numerators.
Multiply the denominators.

$$= \frac{\overset{1}{\cancel{3}} \cdot 9 \cdot \overset{1}{\cancel{4}}}{\underset{1}{\cancel{4}} \cdot 4 \cdot \underset{1}{\cancel{3}}}$$ To simplify, factor 27 as $3 \cdot 9$ and 16 as $4 \cdot 4$. Then remove the common factors of 3 and 4 from the numerator and denominator.

$$= \frac{9}{4}$$ Multiply the remaining factors in the numerator and in the denominator. The result is an improper fraction.

$$= 2\frac{1}{4}$$ Write the improper fraction $\frac{9}{4}$ as a mixed number by dividing 9 by 4.

Self Check 6

Divide and simplify, if possible:

a. $-3\dfrac{4}{15} \div \left(-2\dfrac{1}{10}\right)$

b. $5\dfrac{3}{5} \div \dfrac{7}{8}$

Now Try ➡ Problems 59 and 65

OBJECTIVE 6 Solve application problems by multiplying and dividing mixed numbers.

EXAMPLE 7 **Toys.** The dimensions of the rectangular-shaped screen of an Etch-a-Sketch are shown in the illustration below. Find the area of the screen.

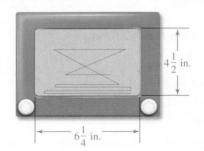

Strategy To find the area, we will multiply $6\frac{1}{4}$ by $4\frac{1}{2}$.

WHY The formula for the area of a rectangle is Area = length · width.

Solution

$A = lw$ This is the formula for the area of a rectangle.

$A = 6\dfrac{1}{4} \cdot 4\dfrac{1}{2}$ Substitute $6\frac{1}{4}$ for l and $4\frac{1}{2}$ for w.

$A = \dfrac{25}{4} \cdot \dfrac{9}{2}$ Write $6\frac{1}{4}$ and $4\frac{1}{2}$ as improper fractions.

$$A = \frac{25 \cdot 9}{4 \cdot 2}$$ Multiply the numerators.
Multiply the denominators.

$$A = \frac{225}{8}$$ Since there are no common factors to remove, perform the multiplication in the numerator and in the denominator. The result is an improper fraction.

$$A = 28\frac{1}{8}$$ Write the improper fraction $\frac{225}{8}$ as a mixed number.

```
       28
   8)225
     −16
      65
     −64
       1
```

The area of the screen of an Etch-a-Sketch is $28\frac{1}{8}$ in.²

EXAMPLE 8   **Government grants.** If $\$12\frac{1}{2}$ million is to be split equally among five cities to fund recreation programs, how much will each city receive?

Analyze

- There is $\$12\frac{1}{2}$ million in grant money.   *Given*
- Five cities will split the money equally.   *Given*
- How much grant money will each city receive?   *Find*

Form The key phrase *split equally* suggests division. We translate the words of the problem to numbers and symbols.

The amount of money that each city will receive (in millions of dollars)	is equal to	the total amount of grant money (in millions of dollars)	divided by	the number of cities receiving money.
The amount of money that each city will receive (in millions of dollars)	=	$12\frac{1}{2}$	÷	5

Calculate To find the quotient, we will express $12\frac{1}{2}$ and 5 as fractions and then use the rule for dividing two fractions.

$$12\frac{1}{2} \div 5 = \frac{25}{2} \div \frac{5}{1}$$ Write $12\frac{1}{2}$ as an improper fraction, and write 5 as a fraction.

$$= \frac{25}{2} \cdot \frac{1}{5}$$ Multiply by the reciprocal of $\frac{5}{1}$, which is $\frac{1}{5}$.

$$= \frac{25 \cdot 1}{2 \cdot 5}$$ Multiply the numerators.
Multiply the denominators.

$$= \frac{\overset{1}{5} \cdot 5 \cdot 1}{2 \cdot \underset{1}{5}}$$ To simplify, factor 25 as $5 \cdot 5$. Then remove the common factor of 5 from the numerator and denominator.

$$= \frac{5}{2}$$ Multiply the remaining factors in the numerator.
Multiply the remaining factors in the denominator.

$$= 2\frac{1}{2}$$ Write the improper fraction $\frac{5}{2}$ as a mixed number by dividing 5 by 2. The units are millions of dollars.

State Each city will receive $\$2\frac{1}{2}$ million in grant money.

Check We can estimate to check the result. If there was $10 million in grant money, each city would receive $\frac{\$10 \text{ million}}{5}$, or $2 million. Since there is actually $\$12\frac{1}{2}$ million in grant money, the answer that each city would receive $\$2\frac{1}{2}$ million seems reasonable.

Answers to Self Checks

1. $\frac{9}{2}$, $4\frac{1}{2}$ **2.** $\frac{27}{8}$ **3. a.** $4\frac{3}{7}$ **b.** $1\frac{12}{13}$ **c.** 17   **d.** $-3\frac{1}{3}$ **4.**

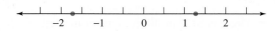

5. a. $7\frac{7}{9}$ **b.** 36   **c.** $-9\frac{2}{3}$ **6. a.** $1\frac{5}{9}$ **b.** $6\frac{2}{5}$ **7.** $26\frac{13}{16}$ in.²   **8.** $3\frac{3}{4}$ min

---

SECTION 3.5   STUDY SET

VOCABULARY

Fill in the blanks.

1. A _____ number, such as $8\frac{4}{5}$, is the sum of a whole number and a proper fraction.

2. In the mixed number $8\frac{4}{5}$, the _____ -number part is 8 and the _____ part is $\frac{4}{5}$.

3. The numerator of an _____ fraction is greater than or equal to its denominator.

4. To _____ a number means to locate its position on a number line and highlight it using a dot.

CONCEPTS

5. What signed mixed number could be used to describe each situation?

   a. A temperature of five and one-third degrees above zero

   b. The depth of a sprinkler pipe that is six and seven-eighths inches below the sidewalk

6. What signed mixed number could be used to describe each situation?

   a. A rain total two and three-tenths of an inch lower than the average

   b. Three and one-half minutes after the liftoff of a rocket

Fill in the blanks.

7. To write a mixed number as an improper fraction:

   1. _____ the denominator of the fraction by the whole-number part.

   2. _____ the numerator of the fraction to the result from step 1.

   3. Write the sum from step 2 over the original _____.

8. To write an improper fraction as a mixed number:

   1. _____ the numerator by the denominator to obtain the whole-number part.

   2. The _____ over the divisor is the fractional part.

9. What fractions have been graphed on the number line?

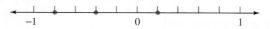

10. What mixed numbers have been graphed on the number line?

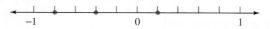

11. Fill in the blank: To multiply or divide mixed numbers, first change the mixed numbers to _____ fractions. Then perform the multiplication or division of the fractions as usual.

12. Simplify the fractional part of each mixed number.

   a. $11\frac{2}{4}$

   b. $1\frac{3}{9}$

   c. $7\frac{15}{27}$

13. Use *estimation* to determine whether the following answer seems reasonable:

$$4\frac{1}{5} \cdot 2\frac{5}{7} = 7\frac{2}{35}$$

14. What is the formula for the

   a. area of a rectangle?

   b. area of a triangle?

NOTATION

15. Fill in the blanks.

   a. We read $5\frac{11}{16}$ as "five _____ eleven- _____."

   b. We read $-4\frac{2}{3}$ as " _____ four and _____ -thirds."

16. Determine the sign of the result. *You do not have to find the answer.*

   a. $1\frac{1}{9}\left(-7\frac{3}{14}\right)$

   b. $-3\frac{4}{15} \div \left(-1\frac{5}{6}\right)$

Fill in the blanks to complete each step.

17. Multiply:

$$5\frac{1}{4} \cdot 1\frac{1}{7} = \frac{21}{\blacksquare} \cdot \frac{\blacksquare}{7}$$

$$= \frac{21 \cdot \blacksquare}{\blacksquare \cdot 7}$$

$$= \frac{3 \cdot \overset{1}{\cancel{7}} \cdot 2 \cdot \overset{1}{\blacksquare}}{\underset{1}{\cancel{\blacksquare}} \cdot \underset{1}{\cancel{7}}}$$

$$= \frac{\blacksquare}{1}$$

$$= \blacksquare$$

18. Divide:

$$-5\frac{5}{6} \div 2\frac{1}{12} = -\frac{\blacksquare}{6} \div \frac{25}{\blacksquare}$$

$$= -\frac{\blacksquare}{6} \cdot \frac{12}{\blacksquare}$$

$$= -\frac{35 \cdot 12}{6 \cdot \blacksquare}$$

$$= -\frac{\overset{1}{\cancel{5}} \cdot \blacksquare \cdot 2 \cdot \overset{1}{\cancel{6}}}{\underset{1}{\cancel{6}} \cdot \underset{1}{\cancel{5}} \cdot \blacksquare}$$

$$= -\frac{\blacksquare}{5}$$

$$= -2\frac{\blacksquare}{5}$$

GUIDED PRACTICE

Each region outlined in black represents one whole. Write an improper fraction and a mixed number to represent the shaded portion. See Example 1.

19.

20.

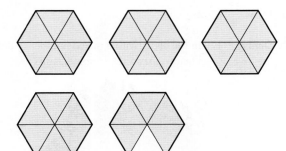

21.

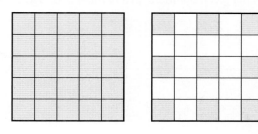

22.

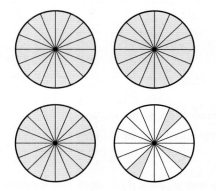

Write each mixed number as an improper fraction. See Example 2.

23. $6\frac{1}{2}$　　　**24.** $8\frac{2}{3}$

25. $20\frac{4}{5}$　　　**26.** $15\frac{3}{8}$

27. $-7\frac{5}{9}$　　　**28.** $-7\frac{1}{12}$

29. $-8\frac{2}{3}$　　　**30.** $-9\frac{3}{4}$

Write each improper fraction as a mixed number or a whole number. Simplify the result, if possible. See Example 3.

31. $\frac{13}{4}$　　　**32.** $\frac{41}{6}$

33. $\frac{28}{5}$　　　**34.** $\frac{28}{3}$

35. $\frac{42}{9}$　　　**36.** $\frac{62}{8}$

37. $\frac{84}{8}$　　　**38.** $\frac{93}{9}$

39. $\frac{52}{13}$　　　**40.** $\frac{80}{16}$

41. $\frac{34}{17}$　　　**42.** $\frac{38}{19}$

43. $-\frac{58}{7}$　　　**44.** $-\frac{33}{7}$

45. $-\frac{20}{6}$　　　**46.** $-\frac{28}{8}$

Graph the given numbers on a number line. See Example 4.

47. $-2\frac{8}{9}, 1\frac{2}{3}, \frac{16}{5}, -\frac{1}{2}$

$$\xleftarrow{\;\;\;\;\;}\underset{-5\;\;-4\;\;-3\;\;-2\;\;-1\;\;\;0\;\;\;1\;\;\;2\;\;\;3\;\;\;4\;\;\;5}{\mid\;\mid}\xrightarrow{\;\;\;\;\;}$$

48. $-\frac{3}{4}, -3\frac{1}{4}, \frac{5}{2}, 4\frac{3}{4}$

$$\xleftarrow{\;\;\;\;\;}\underset{-5\;\;-4\;\;-3\;\;-2\;\;-1\;\;\;0\;\;\;1\;\;\;2\;\;\;3\;\;\;4\;\;\;5}{\mid\;\mid}\xrightarrow{\;\;\;\;\;}$$

49. $3\frac{1}{7}, -\frac{98}{99}, -\frac{10}{3}, \frac{3}{2}$

$$\xleftarrow{\;\;\;\;\;}\underset{-5\;\;-4\;\;-3\;\;-2\;\;-1\;\;\;0\;\;\;1\;\;\;2\;\;\;3\;\;\;4\;\;\;5}{\mid\;\mid}\xrightarrow{\;\;\;\;\;}$$

50. $-2\frac{1}{5}, \frac{4}{5}, -\frac{11}{3}, \frac{17}{4}$

$$\xleftarrow{\;\;\;\;\;}\underset{-5\;\;-4\;\;-3\;\;-2\;\;-1\;\;\;0\;\;\;1\;\;\;2\;\;\;3\;\;\;4\;\;\;5}{\mid\;\mid}\xrightarrow{\;\;\;\;\;}$$

Multiply and simplify, if possible. See Example 5.

51. $3\frac{1}{2} \cdot 2\frac{1}{3}$　　　　**52.** $1\frac{5}{6} \cdot 1\frac{1}{2}$

53. $2\frac{2}{5}\left(3\frac{1}{12}\right)$　　　**54.** $\frac{40}{13}\left(\frac{26}{5}\right)$

55. $6\frac{1}{2} \cdot 1\frac{3}{13}$　　　**56.** $12\frac{3}{5} \cdot 1\frac{3}{7}$

57. $-2\frac{1}{2}(4)$　　　　**58.** $-3\frac{3}{4}(8)$

Divide and simplify, if possible. See Example 6.

59. $-1\frac{13}{15} \div \left(-4\frac{1}{5}\right)$　　**60.** $-2\frac{5}{6} \div \left(-8\frac{1}{2}\right)$

61. $15\frac{1}{3} \div 2\frac{2}{9}$　　　**62.** $6\frac{1}{4} \div 3\frac{3}{4}$

63. $1\frac{3}{4} \div \frac{3}{4}$　　　　**64.** $5\frac{3}{5} \div \frac{9}{10}$

65. $1\frac{7}{24} \div \frac{7}{8}$　　　**66.** $4\frac{1}{2} \div \frac{3}{17}$

TRY IT YOURSELF

Perform each operation and simplify, if possible.

67. $-6 \cdot 2\frac{7}{24}$　　　**68.** $-7 \cdot 1\frac{3}{28}$

69. $-6\frac{3}{5} \div 7\frac{1}{3}$　　　**70.** $-4\frac{1}{4} \div 4\frac{1}{2}$

71. $\left(1\frac{2}{3}\right)^2$　　　　**72.** $\left(3\frac{1}{2}\right)^2$

73. $8 \div 3\frac{1}{5}$　　　　**74.** $15 \div 3\frac{1}{3}$

75. $-20\frac{1}{4} \div \left(-1\frac{11}{16}\right)$　**76.** $-2\frac{7}{10} \div \left(-1\frac{1}{14}\right)$

77. $3\frac{1}{16} \cdot 4\frac{4}{7}$　　　**78.** $5\frac{3}{5} \cdot 1\frac{11}{14}$

79. Find the quotient of $-4\frac{1}{2}$ and $2\frac{1}{4}$.

80. Find the quotient of 25 and $-10\frac{5}{7}$.

81. $2\frac{1}{2}\left(-3\frac{1}{3}\right)$　　　**82.** $\left(-3\frac{1}{4}\right)\left(1\frac{1}{5}\right)$

83. $7 \cdot 1\frac{3}{28}$　　　　**84.** $4\frac{2}{3} \cdot 7$

85. $8 \div 3\frac{1}{5}$　　　　**86.** $4\frac{2}{5} \div 11$

87. Find the product of $1\frac{2}{3}$, 6, and $-\frac{1}{8}$.

88. Find the product of $-\frac{5}{6}$, -8, and $-2\frac{1}{10}$.

89. $\left(-1\frac{1}{3}\right)^3$　　　　**90.** $\left(-1\frac{1}{5}\right)^3$

LOOK ALIKES

Perform each operation and simplify, if possible.

91. a. $2 \cdot 3\frac{3}{4}$　　　**b.** $2 \div 3\frac{3}{4}$

92. a. $5 \cdot 1\frac{1}{15}$　　　**b.** $5 \div 1\frac{1}{15}$

93. a. $-4\frac{1}{2} \cdot 1\frac{2}{9}$　　**b.** $-4\frac{1}{2} \div 1\frac{2}{9}$

94. a. $3\frac{2}{5} \cdot \left(-2\frac{2}{5}\right)$　**b.** $3\frac{2}{5} \div \left(-2\frac{2}{5}\right)$

APPLICATIONS

95. In the illustration below, each barrel represents one whole.

　a. Write a mixed number to represent the shaded portion.

　b. Write an improper fraction to represent the shaded portion.

96. Draw $\frac{17}{8}$ pizzas.

97. Diving. Fill in the blank with a mixed number to describe the dive shown below: forward ▢ somersaults

98. Product labeling. Several mixed numbers appear on the label shown below. Write each mixed number as an improper fraction.

Laundry Basket
1³/₄ Bushel
23¼" L X 18⅞" W X 10½" H

•Easy-grip rim
is reinforced to
handle the biggest loads

99. Reading meters.

a. Use a mixed number to describe the value to which the arrow is currently pointing.

b. If the arrow moves twelve tick marks to the left, to what value will it be pointing?

100. Reading meters.

a. Use a mixed number to describe the value to which the arrow is currently pointing.

b. If the arrow moves up six tick marks, to what value will it be pointing?

101. Online shopping. A mother is ordering a pair of jeans for her daughter from the screen shown below. If the daughter's height is $60\frac{3}{4}$ in. and her waist is $24\frac{1}{2}$ in., on what size and what cut (regular or slim) should the mother point and click?

Girl's jeans- regular cut					
Size	7	8	10	12	14	16
Height	50-52	52-54	54-56	56¼-58½	59-61	61-62
Waist	22¼-22¾	22¾-23¼	23¾-24¼	24¾-25¼	25¾-26¼	26¼-28

Girl's jeans- slim cut					
Size	7	8	10	12	14	16
Height	50-52	52-54	54-56	56½-58½	59-61	61-62
Waist	20¾-21¼	21¼-21¾	22¼-22¾	23¼-23¾	24¼-24¾	25-26½

To order:
Point arrow ➤ to proper size/cut and click

102. Sewing. Use the following table to determine the number of yards of fabric needed …

a. to make a size 16 top if the fabric to be used is 60 inches wide.

b. to make size 18 pants if the fabric to be used is 45 inches wide.

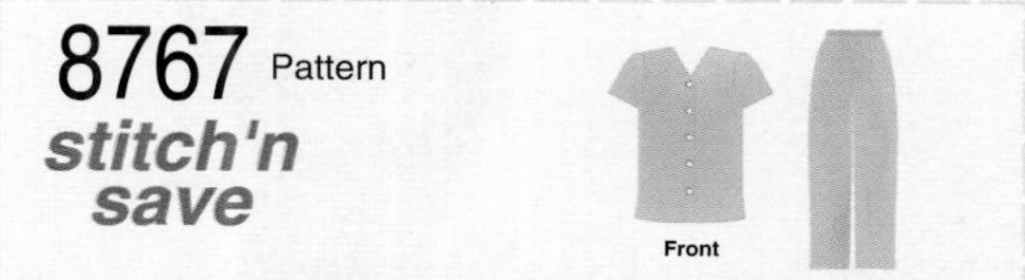

8767 Pattern
stitch'n save

Front

SIZES	8	10	12	14	16	18	20	
Top 45"	2¼	2⅜	2⅜	2⅜	2½	2⅝	2¾	
60"	2	2	2⅛	2⅛	2⅛	2⅛	2⅛	Yds
Pants 45"	2⅝	2⅝	2⅝	2⅝	2⅝	2⅝	2⅝	
60"	1¾	2	2¼	2¼	2¼	2¼	2½	Yds

103. License plates. Find the area of the license plate shown below.

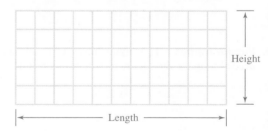

$12\frac{1}{4}$ in.

$6\frac{1}{4}$ in.

WB COUNTY   UTAH   18
123 ABC

104. Graph paper. Mathematicians use specially marked paper, called graph paper, when drawing figures. It is made up of squares that are $\frac{1}{4}$-inch long by $\frac{1}{4}$-inch high.

a. Find the length of the piece of graph paper shown below.

b. Find its height.

c. What is the area of the piece of graph paper?

Height

Length

105. Emergency exits. The following sign marks the emergency exit on a school bus. Find the area of the sign.

$8\frac{1}{4}$ in.

EMERGENCY EXIT

$10\frac{1}{3}$ in.

106. Clothing design. Find the number of square yards of material needed to make the triangular-shaped shawl shown in the illustration.

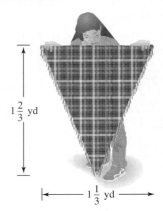

$1\frac{2}{3}$ yd

$1\frac{1}{3}$ yd

107. Calories. A company advertises that its mints contain only $3\frac{1}{5}$ calories a piece. What is the calorie intake if you eat an entire package of 20 mints?

108. Cement mixers. A cement mixer can carry $9\frac{1}{2}$ cubic yards of concrete. If it makes 8 trips to a job site, how much concrete will be delivered to the site?

109. Shopping. In the illustration, what is the cost of buying the fruit in the scale? Give your answer in cents and in dollars.

Oranges
84 cents a pound

110. Picture frames. How many inches of molding is needed to make the square picture frame below?

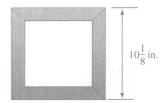

$10\frac{1}{8}$ in.

111. Breakfast cereal. A box of cereal contains about $13\frac{3}{4}$ cups. Refer to the nutrition label shown below and determine the recommended size of one serving.

Nutrition Facts
Serving size : ? cups
Servings per container: 11

112. Breakfast cereal. A box of cereal contains about $14\frac{1}{4}$ cups. Refer to the nutrition label shown below. Determine how many servings there are for children under 4 in one box.

Nutrition Facts
Serving size
Children under 4: $\frac{3}{4}$ cup

Servings per container
Children under 4: ?

113. Catering. How many people can be served $\frac{1}{3}$-pound hamburgers if a caterer purchases 200 pounds of ground beef?

114. Subdivisions. A developer donated to the county 100 of the 1,000 acres of land she owned. She divided the remaining acreage into $1\frac{1}{3}$-acre lots. How many lots were created?

115. Horse racing. The race tracks on which thoroughbred horses run are marked off in $\frac{1}{8}$-mile-long segments called *furlongs*. How many furlongs are there in a $1\frac{1}{16}$-mile race?

116. Fire escapes. Part of the fire escape stairway for one story of an office building is shown below. Each riser is $7\frac{1}{2}$ inches high and each story of the building is 105 inches high.

   a. How many stairs are there in one story of the fire escape stairway?

   b. If the building has 43 stories, how many stairs are there in the entire fire escape stairway?

Step

Step

Step

Fire escape stair case

Riser →

WRITING

117. Explain the difference between $2\frac{3}{4}$ and $2\left(\frac{3}{4}\right)$.

118. Give three examples of how you use mixed numbers in daily life.

REVIEW

Find the LCM of the given numbers.

119. 5, 12, 15   **120.** 8, 12, 16

Find the GCF of the given numbers.

121. 12, 68, 92   **122.** 24, 36, 40

OBJECTIVES

1 Add mixed numbers.

2 Add mixed numbers in vertical form.

3 Subtract mixed numbers.

4 Solve application problems by adding and subtracting mixed numbers.

SECTION 3.6 Adding and Subtracting Mixed Numbers

In this section, we discuss several methods for adding and subtracting mixed numbers.

OBJECTIVE 1 Add mixed numbers.

We can add mixed numbers by writing them as improper fractions. To do so, we follow these steps.

Adding Mixed Numbers: Method 1

1. Write each mixed number as an improper fraction.
2. Write each improper fraction as an equivalent fraction with a denominator that is the LCD.
3. Add the fractions.
4. Write the result as a mixed number, if desired.

Method 1 works well when the whole-number parts of the mixed numbers are small.

EXAMPLE 1 Add: $4\dfrac{1}{6} + 2\dfrac{3}{4}$

Strategy We will write each mixed number as an improper fraction and then use the rule for adding two fractions that have different denominators.

WHY We cannot add the mixed numbers as they are; their fractional parts are not similar objects.

$$4\dfrac{1}{6} + 2\dfrac{3}{4}$$

Four and one-sixth ⟶↑   ↑⟵Two and three-fourths

Solution

$$4\dfrac{1}{6} + 2\dfrac{3}{4} = \dfrac{25}{6} + \dfrac{11}{4} \qquad \text{Write } 4\tfrac{1}{6} \text{ and } \tfrac{3}{4} \text{ as improper fractions.}$$

By inspection, we see that the lowest common denominator is 12.

$$= \dfrac{25}{6} \cdot \dfrac{2}{2} + \dfrac{11}{4} \cdot \dfrac{3}{3} \qquad \begin{array}{l}\text{To build } \tfrac{25}{6} \text{ and } \tfrac{11}{4} \text{ so that their denominators} \\ \text{are 12, multiply each by a form of 1.}\end{array}$$

$$= \dfrac{50}{12} + \dfrac{33}{12} \qquad \begin{array}{l}\text{Multiply the numerators.} \\ \text{Multiply the denominators.}\end{array}$$

$$= \dfrac{83}{12} \qquad \begin{array}{l}\text{Add the numerators and write the sum} \\ \text{over the common denominator 12. The} \\ \text{result is an improper fraction.}\end{array}$$

$$= 6\dfrac{11}{12} \qquad \begin{array}{l}\text{Write the improper fraction } \tfrac{83}{12} \\ \text{as a mixed number.}\end{array}$$

$$\begin{array}{r} 6 \\ 12\overline{)83} \\ -72 \\ \hline 11 \end{array}$$

Self Check 1

Add: $3\dfrac{2}{3} + 1\dfrac{1}{5}$

Now Try ➲ Problem 13

Success Tip We can use estimation by rounding to check the results when adding (or subtracting) mixed numbers. To check the answer $6\frac{11}{12}$ from Example 1, we proceed as follows:

$$4\frac{1}{6} + 2\frac{3}{4} \approx 4 + 3 = 7$$

Since $\frac{1}{6}$ is less than $\frac{1}{2}$, round $4\frac{1}{6}$ down to 4.

Since $\frac{3}{4}$ is greater than $\frac{1}{2}$, round $2\frac{3}{4}$ up to 3.

Since $6\frac{11}{12}$ is close to 7, it is a reasonable answer.

EXAMPLE 2   Add: $-3\frac{1}{8} + 1\frac{1}{2}$

Strategy We will write each mixed number as an improper fraction and then use the rule for adding two fractions that have different denominators.

WHY We cannot add the mixed numbers as they are; their fractional parts are not similar objects.

$$-3\frac{1}{8} + 1\frac{1}{2}$$

Negative three and one-eighth ⟶   ⟵ One and one-half

Solution

$$-3\frac{1}{8} + 1\frac{1}{2} = -\frac{25}{8} + \frac{3}{2}$$ Write $-3\frac{1}{8}$ and $1\frac{1}{2}$ as improper fractions.

Since the smallest number the denominators 8 and 2 divide exactly is 8, the LCD is 8. We will only need to build an equivalent fraction for $\frac{3}{2}$.

$$= -\frac{25}{8} + \frac{3}{2} \cdot \frac{4}{4}$$ To build $\frac{3}{2}$ so that its denominator is 8, multiply it by a form of 1.

$$= -\frac{25}{8} + \frac{12}{8}$$ Multiply the numerators.
Multiply the denominators.

$$= \frac{-25 + 12}{8}$$ Add the numerators and write the sum over the common denominator 8.

$$= \frac{-13}{8}$$ Use the rule for adding integers that have different signs: $-25 + 12 = -13$.

$$= -1\frac{5}{8}$$ Write $\frac{-13}{8}$ as a negative mixed number by dividing 13 by 8.

Self Check 2

Add: $-4\frac{1}{12} + 2\frac{1}{4}$

Now Try ➡ Problem 17

We can also add mixed numbers by adding their whole-number parts and their fractional parts. To do so, we follow these steps.

Adding Mixed Numbers: Method 2

1. Write each mixed number as the sum of a whole number and a fraction.

2. Use the commutative property of addition to write the whole numbers together and the fractions together.

3. Add the whole numbers and the fractions separately.

4. Write the result as a mixed number, if necessary.

Method 2 works well when the whole-number parts of the mixed numbers are large.

EXAMPLE 3   Add: $168\frac{3}{7} + 85\frac{2}{9}$

Strategy We will write each mixed number as the sum of a whole number and a fraction. Then we will add the whole numbers and the fractions separately.

WHY If we change each mixed number to an improper fraction, build equivalent fractions, and add, the resulting numerators will be very large and difficult to work with.

Solution

We will write the solution in *horizontal* form.

$$168\frac{3}{7} + 85\frac{2}{9} = 168 + \frac{3}{7} + 85 + \frac{2}{9}$$ Write each mixed number as the sum of a whole number and a fraction.

$$= 168 + 85 + \frac{3}{7} + \frac{2}{9}$$ Use the commutative property of addition to change the order of the addition so that the whole numbers are together and the fractions are together.

$$= 253 + \frac{3}{7} + \frac{2}{9}$$ Add the whole numbers.

$$\begin{array}{r} \overset{1\,1}{168} \\ +\ 85 \\ \hline 253 \end{array}$$

$$= 253 + \frac{3}{7} \cdot \frac{9}{9} + \frac{2}{9} \cdot \frac{7}{7}$$ Prepare to add the fractions. To build $\frac{3}{7}$ and $\frac{2}{9}$ so that their denominators are 63, multiply each by a form of 1.

$$= 253 + \frac{27}{63} + \frac{14}{63}$$ Multiply the numerators. Multiply the denominators.

$$= 253 + \frac{41}{63}$$ Add the numerators and write the sum over the common denominator 63.

$$\begin{array}{r} \overset{1}{27} \\ +\ 14 \\ \hline 41 \end{array}$$

$$= 253\frac{41}{63}$$ Write the sum as a mixed number.

Self Check 3

Add: $275\frac{1}{6} + 81\frac{3}{5}$

Now Try ➭ **Problem 21**

Caution! If we use method 1 to add the mixed numbers in Example 3, the numbers we encounter are very large. As expected, the result is the same: $253\frac{41}{63}$.

$$168\frac{3}{7} + 85\frac{2}{9} = \frac{1{,}179}{7} + \frac{767}{9}$$ Write $168\frac{3}{7}$ and $85\frac{2}{9}$ as improper fractions.

$$= \frac{1{,}179}{7} \cdot \frac{9}{9} + \frac{767}{9} \cdot \frac{7}{7}$$ The LCD is 63.

$$= \frac{10{,}611}{63} + \frac{5{,}369}{63}$$ Note how large the numerators are.

$$= \frac{15{,}980}{63}$$ Add the numerators and write the sum over the common denominator 63.

$$= 253\frac{41}{63}$$ To write the improper fraction as a mixed number, divide 15,980 by 63.

Generally speaking, the larger the whole-number parts of the mixed numbers, the more difficult it becomes to add those mixed numbers using method 1.

OBJECTIVE 2 Add mixed numbers in vertical form.

We can add mixed numbers quickly when they are written in **vertical form** by working in columns. The strategy is the same as in Example 3: Add whole numbers to whole numbers and fractions to fractions.

EXAMPLE 4   Add: $25\dfrac{3}{4} + 31\dfrac{1}{5}$

Strategy We will perform the addition in *vertical form* with the fractions in a column and the whole numbers lined up in columns. Then we will add the fractional parts and the whole-number parts separately.

WHY It is often easier to add the fractional parts and the whole-number parts of mixed numbers vertically—especially if the whole-number parts contain two or more digits, such as 25 and 31.

Solution

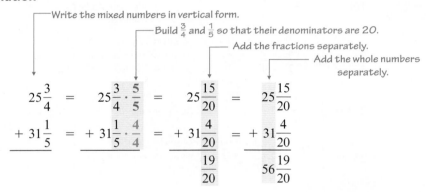

Self Check 4

Add: $71\dfrac{5}{8} + 23\dfrac{1}{3}$

Now Try ➠ Problem 25

The sum is $56\dfrac{19}{20}$.

EXAMPLE 5   Add and simplify, if possible: $75\dfrac{1}{12} + 43\dfrac{1}{4} + 54\dfrac{1}{6}$

Strategy We will write the problem in *vertical form*. We will make sure that the fractional part of the answer is in simplest form.

WHY When adding, subtracting, multiplying, or dividing fractions or mixed numbers, the answer should always be written in simplest form.

Solution

The LCD for $\dfrac{1}{12}, \dfrac{1}{4}$, and $\dfrac{1}{6}$ is 12.

Write the mixed numbers in vertical form.

Build $\dfrac{1}{4}$ and $\dfrac{1}{6}$ so that their denominators are 12.

Add the fractions separately.

Add the whole numbers separately.

$$75\dfrac{1}{12} \;=\; 75\dfrac{1}{12} \;=\; 75\dfrac{1}{12} \;=\; \overset{11}{75}\dfrac{1}{12}$$

$$43\dfrac{1}{4} \;=\; 43\dfrac{1}{4}\cdot\dfrac{3}{3} \;=\; 43\dfrac{3}{12} \;=\; 43\dfrac{3}{12}$$

$$+\;54\dfrac{1}{6} \;=\; +\;54\dfrac{1}{6}\cdot\dfrac{2}{2} \;=\; +\;54\dfrac{2}{12} \;=\; +\;54\dfrac{2}{12}$$

$$\dfrac{6}{12} \qquad\qquad 172\dfrac{6}{12} = 172\dfrac{1}{2}$$

Simplify:

$$\dfrac{6}{12} = \dfrac{\overset{1}{\cancel{6}}}{2\cdot\underset{1}{\cancel{6}}} = \dfrac{1}{2}.$$

Self Check 5

Add and simplify, if possible:

$$68\dfrac{1}{6} + 37\dfrac{5}{18} + 52\dfrac{1}{9}$$

Now Try ➠ Problem 29

The sum is $172\dfrac{1}{2}$.

When we add mixed numbers, sometimes the sum of the fractions is an improper fraction.

EXAMPLE 6   Add: $45\dfrac{2}{3} + 96\dfrac{4}{5}$

Strategy We will write the problem in *vertical form*. We will make sure that the fractional part of the answer is in simplest form.

WHY When adding, subtracting, multiplying, or dividing fractions or mixed numbers, the answer should always be written in simplest form.

Solution

The LCD for $\frac{2}{3}$ and $\frac{4}{5}$ is 15.

— Write the mixed numbers in vertical form.

— Build $\frac{2}{3}$ and $\frac{4}{5}$ so that their denominators are 15.

— Add the fractions separately.

— Add the whole numbers separately.

$$
\begin{array}{lllll}
45\dfrac{2}{3} & = & 45\dfrac{2}{3}\cdot\dfrac{5}{5} & = & 45\dfrac{10}{15} \\[2mm]
+\,96\dfrac{4}{5} & = & +\,96\dfrac{4}{5}\cdot\dfrac{3}{3} & = & +\,96\dfrac{12}{15} \\[1mm]
\hline
& & & & \dfrac{22}{15}
\end{array}
\qquad
\begin{array}{l}
45\dfrac{10}{15} \\[2mm]
+\,96\dfrac{12}{15} \\[1mm]
\hline
141\dfrac{22}{15}
\end{array}
$$

— The fractional part of the answer is greater than 1.

Since we don't want an improper fraction in the answer, we write $\frac{22}{15}$ as a mixed number. Then we *carry* 1 from the fraction column to the whole-number column.

$141\dfrac{22}{15} = 141 + \dfrac{22}{15}$ Write the mixed number as the sum of a whole number and a fraction.

$= 141 + 1\dfrac{7}{15}$ To write the improper fraction as a mixed number, divide 22 by 15.

$= 142\dfrac{7}{15}$ Carry the 1 and add it to 141 to get 142.

$$
\begin{array}{r}
1 \\
15\overline{)22} \\
-15 \\
\hline
7
\end{array}
$$

Self Check 6

Add: $76\dfrac{11}{12} + 49\dfrac{5}{8}$

Now Try ➲ **Problem 33**

OBJECTIVE 3 Subtract mixed numbers.

Subtracting mixed numbers is similar to adding mixed numbers.

EXAMPLE 7   Subtract and simplify, if possible: $16\dfrac{7}{10} - 9\dfrac{8}{15}$

Strategy We will perform the subtraction in *vertical form* with the fractions in a column and the whole numbers lined up in columns. Then we will subtract the fractional parts and the whole-number parts separately.

WHY It is often easier to subtract the fractional parts and the whole-number parts of mixed numbers vertically.

Solution

The LCD for $\frac{7}{10}$ and $\frac{8}{15}$ is 30.

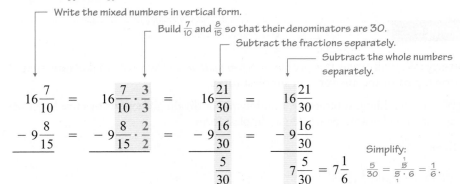

Write the mixed numbers in vertical form.

Build $\frac{7}{10}$ and $\frac{8}{15}$ so that their denominators are 30.

Subtract the fractions separately.

Subtract the whole numbers separately.

$$16\frac{7}{10} = 16\frac{7}{10}\cdot\frac{3}{3} = 16\frac{21}{30} = 16\frac{21}{30}$$

$$-9\frac{8}{15} = -9\frac{8}{15}\cdot\frac{2}{2} = -9\frac{16}{30} = -9\frac{16}{30}$$

$$\frac{5}{30} \qquad 7\frac{5}{30} = 7\frac{1}{6}$$

Simplify:

$\frac{5}{30} = \frac{\overset{1}{\cancel{5}}}{\underset{1}{\cancel{5}}\cdot 6} = \frac{1}{6}.$

The difference is $7\frac{1}{6}$.

Self Check 7

Subtract and simplify, if possible:

$12\frac{9}{20} - 8\frac{1}{30}$

Now Try ➡ Problem 37

Subtraction of mixed numbers (like subtraction of whole numbers) sometimes involves borrowing. When the fraction we are subtracting is greater than the fraction we are subtracting it from, it is necessary to borrow.

EXAMPLE 8 Subtract: $34\frac{1}{8} - 11\frac{2}{3}$

Strategy We will perform the subtraction in *vertical form* with the fractions in a column and the whole numbers lined up in columns. Then we will subtract the fractional parts and the whole-number parts separately.

WHY It is often easier to subtract the fractional parts and the whole-number parts of mixed numbers vertically.

Solution

The LCD for $\frac{1}{8}$ and $\frac{2}{3}$ is 24.

Write the mixed numbers in vertical form.

Build $\frac{1}{8}$ and $\frac{2}{3}$ so that their denominators are 24.

$$34\frac{1}{8} = 34\frac{1}{8}\cdot\frac{3}{3} = 34\frac{3}{24}$$

$$-11\frac{2}{3} = -11\frac{2}{3}\cdot\frac{8}{8} = -11\frac{16}{24}$$

Note that $\frac{16}{24}$ is greater than $\frac{3}{24}$.

Since $\frac{16}{24}$ is greater than $\frac{3}{24}$, borrow 1 (in the form of $\frac{24}{24}$) from 34 and add it to $\frac{3}{24}$ to get $\frac{27}{24}$.

Subtract the fractions separately.

Subtract the whole numbers separately.

$$34\overset{3}{}\frac{3}{24} + \frac{24}{24} = 33\frac{27}{24} = 33\frac{27}{24}$$

$$-11\frac{16}{24} = -11\frac{16}{24} = -11\frac{16}{24}$$

$$\frac{11}{24} \qquad 22\frac{11}{24}$$

Self Check 8

Subtract: $258\frac{3}{4} - 175\frac{15}{16}$

Now Try ➡ Problem 41

The difference is $22\frac{11}{24}$.

Success Tip We can use estimation by rounding to check the results when subtracting mixed numbers. To check the answer $22\frac{11}{24}$ from Example 8, we proceed as follows:

$$34\frac{1}{8} - 11\frac{2}{3} \approx 34 - 12 = 22$$

Since $\frac{1}{8}$ is less than $\frac{1}{2}$, round $34\frac{1}{8}$ down to 34.
Since $\frac{2}{3}$ is greater than $\frac{1}{2}$, round $11\frac{2}{3}$ up to 12.

Since $22\frac{11}{24}$ is close to 22, it is a reasonable answer.

EXAMPLE 9   Subtract: $419 - 53\frac{11}{16}$

Strategy We will write the numbers in vertical form and borrow 1 $\left(\text{in the form of } \frac{16}{16}\right)$ from 419.

WHY In the fraction column, we need to have a fraction from which to subtract $\frac{11}{16}$.

Solution

Write the numbers in vertical form.

Borrow 1 (in the form of $\frac{16}{16}$) from 419.
Then subtract the fractions separately.

Subtract the whole numbers separately. This also requires borrowing.

$$
\begin{array}{rcrcr}
419 & = & 418\dfrac{16}{16} & = & \overset{3\ 11}{\cancel{41}8}\dfrac{16}{16} \\[2ex]
-\ 53\dfrac{11}{16} & = & -\ 53\dfrac{11}{16} & = & -\ 53\dfrac{11}{16} \\[2ex]
\hline
& & 365\dfrac{5}{16} & & 365\dfrac{5}{16}
\end{array}
$$

The difference is $365\frac{5}{16}$.

Self Check 9

Subtract: $2{,}300 - 129\frac{31}{32}$

Now Try ➡ **Problem 45**

OBJECTIVE 4 **Solve application problems by adding and subtracting mixed numbers.**

EXAMPLE 10   **Horse racing.** In order to become the Triple Crown Champion, a thoroughbred horse must win three races: the Kentucky Derby ($1\frac{1}{4}$ miles long), the Preakness Stakes ($1\frac{3}{16}$ miles long), and the Belmont Stakes ($1\frac{1}{2}$ miles long). What is the combined length of the three races of the Triple Crown?

Olga_i/Shutterstock.com

Analyze

- The Kentucky Derby is $1\frac{1}{4}$ miles long.   *Given*
- The Preakness Stakes is $1\frac{3}{16}$ miles long.   *Given*
- The Belmont Stakes is $1\frac{1}{2}$ miles long.   *Given*
- What is the combined length of the three races?   *Find*

Form The key phrase *combined length* indicates addition.
We translate the words of the problem to numbers and symbols.

The combined length of the three races	is equal to	the length of the Kentucky Derby	plus	the length of the Preakness Stakes	plus	the length of the Belmont Stakes.

The combined length of the three races	=	$1\frac{1}{4}$	+	$1\frac{3}{16}$	+	$1\frac{1}{2}$

Calculate To find the sum, we will write the mixed numbers in vertical form. To add in the fraction column, the LCD for $\frac{1}{4}$, $\frac{3}{16}$, and $\frac{1}{2}$ is 16.

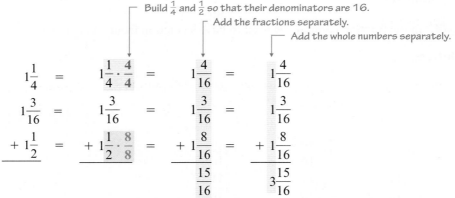

Build $\frac{1}{4}$ and $\frac{1}{2}$ so that their denominators are 16.
Add the fractions separately.
Add the whole numbers separately.

$$1\frac{1}{4} = 1\frac{1}{4} \cdot \frac{4}{4} = 1\frac{4}{16} = 1\frac{4}{16}$$
$$1\frac{3}{16} = 1\frac{3}{16} = 1\frac{3}{16} = 1\frac{3}{16}$$
$$+\,1\frac{1}{2} = +\,1\frac{1}{2} \cdot \frac{8}{8} = +\,1\frac{8}{16} = +\,1\frac{8}{16}$$
$$\frac{15}{16} \qquad 3\frac{15}{16}$$

State The combined length of the three races of the Triple Crown is $3\frac{15}{16}$ miles.

Check We can estimate to check the result. If we round $1\frac{1}{4}$ down to 1, round $1\frac{3}{16}$ down to 1, and round $1\frac{1}{2}$ up to 2, the approximate combined length of the three races is $1 + 1 + 2 = 4$ miles. Since $3\frac{15}{16}$ is close to 4, the result seems reasonable.

Self Check 10

Salads. A three-bean salad calls for one can of green beans ($14\frac{1}{2}$ ounces), one can of garbanzo beans ($10\frac{3}{4}$ ounces), and one can of kidney beans ($15\frac{7}{8}$ ounces). How many ounces of beans are called for in the recipe?

Now Try ➲ **Problem 97**

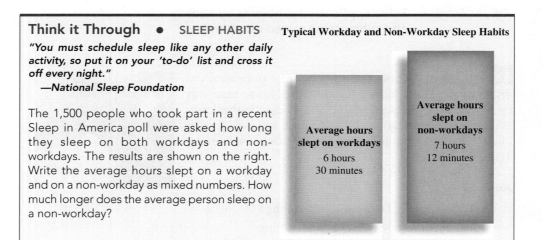

Think it Through   ● SLEEP HABITS

"You must schedule sleep like any other daily activity, so put it on your 'to-do' list and cross it off every night."
—*National Sleep Foundation*

The 1,500 people who took part in a recent Sleep in America poll were asked how long they sleep on both workdays and non-workdays. The results are shown on the right. Write the average hours slept on a workday and on a non-workday as mixed numbers. How much longer does the average person sleep on a non-workday?

Typical Workday and Non-Workday Sleep Habits

Average hours slept on workdays
6 hours
30 minutes

Average hours slept on non-workdays
7 hours
12 minutes

Source: National Sleep Foundation

EXAMPLE 11 **Baking.** How much butter is left in a 10-pound tub after $2\frac{2}{3}$ pounds are used for a wedding cake?

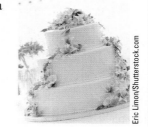

Eric Limon/Shutterstock.com

Analyze

- The tub contained 10 pounds of butter.   *Given*
- $2\frac{2}{3}$ pounds of butter are used for a cake.   *Given*
- How much butter is left in the tub?   *Find*

Form The key phrase *how much butter is left* indicates subtraction.

We translate the words of the problem to numbers and symbols.

The amount of butter left in the tub	is equal to	the amount of butter in one tub	minus	the amount of butter used for the cake.

$$\text{The amount of butter left in the tub} = 10 - 2\frac{2}{3}$$

Calculate To find the difference, we will write the numbers in vertical form and borrow 1 (in the form of $\frac{3}{3}$) from 10.

In the fraction column, we need to have a fraction from which to subtract $\frac{2}{3}$.

Subtract the fractions separately.

Subtract the whole numbers separately.

$$10 = \overset{9}{\cancel{10}}\frac{3}{3} = \overset{9}{\cancel{10}}\frac{3}{3}$$
$$-\,2\frac{2}{3} = -\,2\frac{2}{3} = -\,2\frac{2}{3}$$
$$\frac{1}{3} \qquad 7\frac{1}{3}$$

State There are $7\frac{1}{3}$ pounds of butter left in the tub.

Check We can check using addition. If $2\frac{2}{3}$ pounds of butter were used and $7\frac{1}{3}$ pounds of butter are left in the tub, then the tub originally contained $2\frac{2}{3} + 7\frac{1}{3} = 9\frac{3}{3} = 10$ pounds of butter. The result checks.

Self Check 11

Trucking. The mixing barrel of a cement truck holds 9 cubic yards of concrete. How much concrete is left in the barrel if $6\frac{3}{4}$ cubic yards have already been unloaded?

Now Try ➡ **Problem 103**

Answers to Self Checks

1. $4\frac{13}{15}$ 2. $-1\frac{5}{6}$ 3. $356\frac{23}{30}$ 4. $94\frac{23}{24}$ 5. $157\frac{5}{9}$ 6. $126\frac{13}{24}$ 7. $4\frac{5}{12}$ 8. $82\frac{13}{16}$ 9. $2,170\frac{1}{32}$
10. $41\frac{1}{8}$ oz   11. $2\frac{1}{4}$ yd³

SECTION **3.6** **STUDY SET**

VOCABULARY

Fill in the blanks.

1. A _____ number, such as $1\frac{7}{8}$, contains a whole-number part and a fractional part.

2. We can add (or subtract) mixed numbers quickly when they are written in _____ form by working in columns.

3. To add (or subtract) mixed numbers written in vertical form, we add (or subtract) the _____ separately and the _____ numbers separately.

4. A fraction such as $\frac{11}{8}$, whose numerator is greater than or equal to the denominator, is called an _____ fraction.

5. Consider the following problem:

$$36\frac{5}{7}$$
$$+\,42\frac{4}{7}$$
$$\overline{78\frac{9}{7}} = 78 + 1\frac{2}{7} = 79\frac{2}{7}$$

Since we don't want an improper fraction in the answer, we write $\frac{9}{7}$ as $1\frac{2}{7}$, _____ the 1, and add it to 78 to get 79.

6. Consider the following problem:

$$86\frac{1}{3} = \quad 86\overset{5}{\cancel{}}\frac{1}{3} + \frac{3}{3}$$
$$-\,24\frac{2}{3} = -24\frac{2}{3}$$

To subtract in the fraction column, we _____ 1 from 86 in the form of $\frac{3}{3}$.

CONCEPTS

7. a. For $76\frac{3}{4}$, list the whole-number part and the fractional part.

b. Write $76\frac{3}{4}$ as a sum.

8. Use the commutative property of addition to rewrite the following expression with the whole numbers together and the fractions together. *You do not have to find the answer.*

$$14 + \frac{5}{8} + 53 + \frac{1}{6}$$

9. The *denominators* of two fractions are given. Find the least common denominator.

   a. 3 and 4   **b.** 5 and 6

   c. 6 and 9   **d.** 8 and 12

10. Simplify.

   a. $9\frac{17}{16}$ **b.** $1{,}288\frac{7}{3}$

   c. $16\frac{12}{8}$ **d.** $45\frac{24}{20}$

NOTATION

Fill in the blanks to complete each step.

11.
$$6\frac{3}{5} = \quad 6\frac{3}{5} \cdot \frac{7}{7} = \quad 6\frac{}{35}$$
$$+\,3\frac{2}{7} = \quad +\,3\frac{2}{7} \cdot \frac{}{} = \quad +\,3\frac{10}{}$$
$$\overline{\qquad\qquad\qquad\qquad\qquad\qquad 9\frac{}{}}$$

12.
$$67\frac{3}{8} = \quad 67\frac{3}{8} \cdot \frac{}{} = \quad 67\frac{9}{24} = \quad 67\overset{6}{\cancel{}}\frac{9}{24} + \frac{}{} = \quad 66\frac{}{24}$$
$$-\,23\frac{2}{3} = -23\frac{2}{3} \cdot \frac{8}{8} = -23\frac{}{24} = -23\frac{16}{24} \qquad\qquad = -23\frac{16}{24}$$
$$\overline{\qquad\qquad\qquad\qquad\qquad\qquad\qquad\qquad\qquad \frac{}{24}}$$

GUIDED PRACTICE

Add. **See Example 1.**

13. $1\frac{1}{4} + 2\frac{1}{3}$ **14.** $2\frac{2}{5} + 3\frac{1}{4}$

15. $2\frac{1}{3} + 4\frac{2}{5}$ **16.** $4\frac{1}{3} + 1\frac{1}{7}$

Add. **See Example 2.**

17. $-4\frac{1}{8} + 1\frac{3}{4}$ **18.** $-3\frac{11}{15} + 2\frac{1}{5}$

19. $-6\frac{5}{6} + 3\frac{2}{3}$ **20.** $-6\frac{3}{14} + 1\frac{2}{7}$

Add. **See Example 3.**

21. $334\frac{1}{7} + 42\frac{2}{3}$ **22.** $259\frac{3}{8} + 40\frac{1}{3}$

23. $667\frac{1}{5} + 47\frac{3}{4}$ **24.** $568\frac{1}{6} + 52\frac{3}{4}$

Add. **See Example 4.**

25. $41\frac{2}{9} + 18\frac{2}{5}$ **26.** $60\frac{3}{11} + 24\frac{2}{3}$

27. $89\frac{6}{11} + 43\frac{1}{3}$ **28.** $77\frac{5}{8} + 55\frac{1}{7}$

Add and simplify, if possible. **See Example 5.**

29. $14\frac{1}{4} + 29\frac{1}{20} + 78\frac{3}{5}$ **30.** $11\frac{1}{12} + 59\frac{1}{4} + 82\frac{1}{6}$

31. $106\frac{5}{18} + 22\frac{1}{2} + 19\frac{1}{9}$ **32.** $75\frac{2}{5} + 43\frac{7}{30} + 54\frac{1}{3}$

Add and simplify, if possible. **See Example 6.**

33. $39\frac{5}{8} + 62\frac{11}{12}$ **34.** $53\frac{5}{6} + 47\frac{3}{8}$

35. $82\frac{8}{9} + 46\frac{11}{15}$ **36.** $44\frac{2}{9} + 76\frac{20}{21}$

Subtract and simplify, if possible. **See Example 7.**

37. $19\dfrac{11}{12} - 9\dfrac{2}{3}$ **38.** $32\dfrac{2}{3} - 7\dfrac{1}{6}$

39. $21\dfrac{5}{6} - 8\dfrac{3}{10}$ **40.** $41\dfrac{2}{5} - 6\dfrac{3}{20}$

Subtract. **See Example 8.**

41. $47\dfrac{1}{11} - 15\dfrac{2}{3}$ **42.** $58\dfrac{4}{11} - 15\dfrac{1}{2}$

43. $84\dfrac{5}{8} - 12\dfrac{6}{7}$ **44.** $95\dfrac{4}{7} - 23\dfrac{5}{6}$

Subtract. **See Example 9.**

45. $674 - 94\dfrac{11}{15}$ **46.** $437 - 63\dfrac{6}{23}$

47. $112 - 49\dfrac{9}{32}$ **48.** $221 - 88\dfrac{35}{64}$

TRY IT YOURSELF

Add or subtract and simplify, if possible.

49. $140\dfrac{5}{6} - 129\dfrac{4}{5}$ **50.** $291\dfrac{1}{4} - 289\dfrac{1}{12}$

51. $4\dfrac{1}{6} + 1\dfrac{1}{5}$ **52.** $2\dfrac{2}{5} + 3\dfrac{1}{4}$

53. $5\dfrac{1}{2} + 3\dfrac{4}{5}$ **54.** $6\dfrac{1}{2} + 2\dfrac{2}{3}$

55. $2 + 1\dfrac{7}{8}$ **56.** $3\dfrac{3}{4} + 5$

57. $8\dfrac{7}{9} - 3\dfrac{1}{9}$ **58.** $9\dfrac{9}{10} - 6\dfrac{3}{10}$

59. $140\dfrac{3}{16} - 129\dfrac{3}{4}$ **60.** $442\dfrac{1}{8} - 429\dfrac{2}{3}$

61. $380\dfrac{1}{6} + 17\dfrac{1}{4}$ **62.** $103\dfrac{1}{2} + 210\dfrac{2}{5}$

63. $-2\dfrac{5}{6} + 1\dfrac{3}{8}$ **64.** $-4\dfrac{5}{9} + 2\dfrac{1}{6}$

65. $3\dfrac{1}{4} + 4\dfrac{1}{4}$ **66.** $2\dfrac{1}{8} + 3\dfrac{3}{8}$

67. $-3\dfrac{3}{4} + \left(-1\dfrac{1}{2}\right)$ **68.** $-3\dfrac{2}{3} + \left(-1\dfrac{4}{5}\right)$

69. $7 - \dfrac{2}{3}$ **70.** $6 - \dfrac{1}{8}$

71. $12\dfrac{1}{2} + 5\dfrac{3}{4} + 35\dfrac{1}{6}$ **72.** $31\dfrac{1}{3} + 20\dfrac{2}{5} + 10\dfrac{1}{15}$

73. $16\dfrac{1}{4} - 13\dfrac{3}{4}$ **74.** $40\dfrac{1}{7} - 19\dfrac{6}{7}$

75. $-4\dfrac{5}{8} - 1\dfrac{1}{4}$ **76.** $-2\dfrac{1}{16} - 3\dfrac{7}{8}$

77. $6\dfrac{5}{8} - 3$ **78.** $10\dfrac{1}{2} - 6$

79. $\dfrac{7}{3} + 2$ **80.** $\dfrac{9}{7} + 3$

81. $58\dfrac{7}{8} + 340\dfrac{1}{2} + 61\dfrac{3}{4}$ **82.** $191\dfrac{1}{2} + 233\dfrac{1}{16} + 16\dfrac{5}{8}$

83. $9 - 8\dfrac{3}{4}$ **84.** $11 - 10\dfrac{4}{5}$

For each problem, fill in the blanks of the estimate. Then tell whether the given result seems reasonable.

85.

Exact	Estimate
$167\dfrac{3}{8}$	▨
$+\ 44\dfrac{3}{4}$	$+$ ▨
$202\dfrac{5}{8}$	▨

86.

Exact	Estimate
$293\dfrac{5}{9}$	▨
$+\ 458\dfrac{7}{18}$	$+$ ▨
$751\dfrac{17}{18}$	▨

87.

Exact	Estimate
$616\dfrac{2}{5}$	▨
$-\ 488\dfrac{2}{3}$	$-$ ▨
$127\dfrac{11}{15}$	▨

88.

Exact	Estimate
$207\dfrac{1}{2}$	▨
$-\ 189\dfrac{7}{11}$	$-$ ▨
$27\dfrac{19}{22}$	▨

LOOK ALIKES

Evaluate each expression.

89. a. $5 - \left(\dfrac{3}{7}\right)$ **b.** $5 - \left(-\dfrac{3}{7}\right)$

 c. $5\left(-\dfrac{3}{7}\right)$ **d.** $5 \div \left(-\dfrac{3}{7}\right)$

90. a. $3 - \left(\dfrac{2}{5}\right)$ **b.** $3 - \left(-\dfrac{2}{5}\right)$

 c. $3\left(-\dfrac{2}{5}\right)$ **d.** $3 \div \left(-\dfrac{2}{5}\right)$

91. a. $3\dfrac{2}{7} + 2\dfrac{1}{7}$ **b.** $3\dfrac{2}{7} + \left(-2\dfrac{1}{7}\right)$

 c. $\left(3\dfrac{2}{7}\right)\left(-2\dfrac{1}{7}\right)$ **d.** $\left(3\dfrac{2}{7}\right) \div \left(-2\dfrac{1}{7}\right)$

92. a. $5\dfrac{2}{3} - 2$ **b.** $5\dfrac{2}{3} + 2$

 c. $\left(5\dfrac{2}{3}\right)(-2)$ **d.** $\left(5\dfrac{2}{3}\right) \div (-2)$

APPLICATIONS

93. Air travel. A businesswoman's flight left Los Angeles and in $3\dfrac{3}{4}$ hours she landed in Minneapolis. She then boarded a commuter plane in Minneapolis and arrived at her final destination in $1\dfrac{1}{2}$ hours. Find the total time she spent on the flights.

94. Shipping. A passenger ship and a cargo ship left San Diego harbor at midnight. During the first hour, the passenger ship traveled south at $16\dfrac{1}{2}$ miles per hour, while the cargo ship traveled north at a rate of $5\dfrac{1}{5}$ miles per hour. How far apart were they at 1:00 A.M.?

95. Trail mix. How many cups of trail mix will the recipe shown below make?

Trail Mix		
A healthy snack—great for camping trips	$2\dfrac{3}{4}$ cups peanuts	$\dfrac{1}{3}$ cup coconut
	$\dfrac{1}{2}$ cup sunflower seeds	$2\dfrac{2}{3}$ cups oat flakes
	$\dfrac{2}{3}$ cup raisins	$\dfrac{1}{4}$ cup pretzels

96. Hardware. Refer to the illustration below. How long should the threaded part of the bolt be?

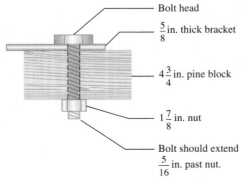

- Bolt head
- $\dfrac{5}{8}$ in. thick bracket
- $4\dfrac{3}{4}$ in. pine block
- $1\dfrac{7}{8}$ in. nut
- Bolt should extend $\dfrac{5}{16}$ in. past nut.

97. Octuplets. On January 26, 2009, at Kaiser Permanente Bellflower Medical Center in California, Nadya Suleman gave birth to eight babies. (The United States' first live octuplets were born in Houston in 1998 to Nkem Chukwu and Iyke Louis Udobi). Find the combined birthweights of the babies from the information shown below. (Source: The Nadya Suleman family website)

No. 1: Noah, male, $2\dfrac{11}{16}$ pounds

No. 2: Maliah, female, $2\dfrac{3}{4}$ pounds

No. 3: Isaiah, male, $3\dfrac{1}{4}$ pounds

No. 4: Nariah, female, $2\dfrac{1}{2}$ pounds

No. 5: Makai, male, $1\dfrac{1}{2}$ pounds

No. 6: Josiah, male, $2\dfrac{3}{4}$ pounds

No. 7: Jeremiah, male, $1\dfrac{15}{16}$ pounds

No. 8: Jonah, male, $2\dfrac{11}{16}$ pounds

98. Septuplets. On November 19, 1997, at Iowa Methodist Medical Center, Bobbie McCaughey gave birth to seven babies. Find the combined birthweights of the babies from the following information. (Source: *Los Angeles Times*, Nov. 20, 1997)

Kenneth Robert $3\dfrac{1}{4}$ lb

Nathanial Roy $2\dfrac{7}{8}$ lb

Kelsey Ann $2\dfrac{5}{16}$ lb

Brandon James $3\dfrac{3}{16}$ lb

Natalie Sue $2\dfrac{5}{8}$ lb

Joel Steven $2\dfrac{15}{16}$ lb

Alexis May $2\dfrac{11}{16}$ lb

99. Historical documents. The Declaration of Independence on display at the National Archives in Washington, D.C., is $24\dfrac{1}{2}$ inches wide by $29\dfrac{3}{4}$ inches high. How many inches of molding would be needed to frame it?

100. Stamp collecting. The Pony Express Stamp, shown below, was issued in 1940. It is a favorite of collectors all over the world. A Postal Service document describes its size in an unusual way:

"The dimensions of the stamp are $\dfrac{84}{100}$ by $1\dfrac{44}{100}$ inches, arranged horizontally."

To display the stamp, a collector wants to frame it with gold braid. How many inches of braid are needed?

CTR Photos/Shutterstock.com

101. Freeway signs. A freeway exit sign is shown. How far apart are the Citrus Ave. and Grand Ave. exits?

102. Basketball. See the graph below. What is the difference in height between the tallest and the shortest of the starting players?

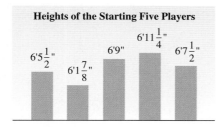

103. Hose repairs. To repair a bad connector, a gardener removes $1\frac{1}{2}$ feet from the end of a 50-foot hose. How long is the hose after the repair?

104. Haircuts. A mother makes her child get a haircut when his hair measures 3 inches in length. His barber uses clippers with attachment #2 that leaves $\frac{3}{8}$-inch of hair. How many inches does the child's hair grow between haircuts?

105. Gasoline. Use the service station sign below to answer the following questions.

   a. What is the difference in price per gallon between the least and most expensive types of gasoline at the self-serve pump?

   b. For each type of gasoline, how much more is the cost per gallon for full service compared to self-serve?

106. Water slides. An amusement park added a new section to a water slide to create a slide $311\frac{5}{12}$ feet long. How long was the slide before the addition?

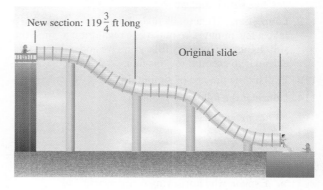

107. Jewelry. A jeweler cut a 7-inch-long silver wire into three pieces. To do this, he aligned a 6-inch-long ruler directly below the wire and made the indicated cuts. Find the length of piece 2 of the wire.

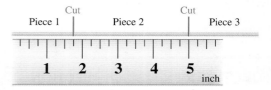

108. Sewing. To make some draperies, an interior decorator needs $12\frac{1}{4}$ yards of material for the den and $8\frac{1}{2}$ yards for the living room. If the material comes only in 21-yard bolts, how much will be left over after completing both sets of draperies?

109. Luggage. On many flights, airlines do not accept luggage whose total dimensions (length + width + height) exceed 62 inches. Estimate the total dimensions of the piece of luggage shown below.

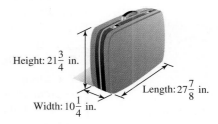

110. Freeway signs. Estimate the distance between the Downtown San Diego off ramp and the Sea World off ramp on the 5 Freeway.

WRITING

111. Of the methods studied to add mixed numbers, which do you like better, and why?

112. Leap year. It actually takes Earth $365\frac{1}{4}$ days, give or take a few minutes, to make one revolution around the sun. Explain why every four years we add a day to the calendar to account for this fact.

113. Explain the process of simplifying $12\frac{7}{5}$.

114. Consider the following problem:

$$108\frac{1}{3}$$
$$-99\frac{2}{3}$$

  a. Explain why borrowing is necessary.

  b. Explain how the borrowing is done.

115. Explain why $18\frac{13}{25}$ rounds to 19.

116. Explain why $18\frac{12}{25}$ rounds to 18.

REVIEW LOOK ALIKES

Perform each operation and simplify, if possible.

117. a. $\frac{1}{2} + \frac{1}{4}$ **b.** $\frac{1}{2} - \frac{1}{4}$

   c. $\frac{1}{2} \cdot \frac{1}{4}$ **d.** $\frac{1}{2} \div \frac{1}{4}$

118. a. $\frac{1}{8} + \frac{5}{6}$ **b.** $\frac{1}{8} - \frac{5}{6}$

   c. $\frac{1}{8} \cdot \frac{5}{6}$ **d.** $\frac{1}{8} \div \frac{5}{6}$

OBJECTIVES

1 Use the order of operations rule.

2 Solve application problems by using the order of operations rule.

3 Evaluate formulas.

4 Simplify complex fractions.

SECTION 3.7 Order of Operations and Complex Fractions

We have seen that the order of operations rule is used to evaluate expressions that contain more than one operation. In Chapter 1, we used it to evaluate expressions involving whole numbers, and in Chapter 2, we used it to evaluate expressions involving integers. We will now use it to evaluate expressions involving fractions and mixed numbers.

OBJECTIVE 1 Use the order of operations rule.

Recall from Section 1.9 that every numerical expression has only one correct value. If we don't establish a uniform order of operations, an expression can have more than one value. To avoid this possibility, always use the following rule for the order of operations.

Order of Operations

1. Perform all calculations within parentheses and other grouping symbols following the order listed in steps 2–4 below, working from the innermost pair of grouping symbols to the outermost pair.

2. Evaluate all exponential expressions.

3. Perform all multiplications and divisions as they occur from left to right.

4. Perform all additions and subtractions as they occur from left to right.

When grouping symbols have been removed, repeat steps 2–4 to complete the calculation.

   If a fraction bar is present, evaluate the expression above the bar (called the **numerator**) and the expression below the bar (called the **denominator**) separately. Then perform the division indicated by the fraction bar, if possible.

EXAMPLE 1   Evaluate: $\dfrac{3}{4} + \dfrac{5}{3}\left(-\dfrac{1}{2}\right)^3$

Strategy We will scan the expression to determine what operations need to be performed. Then we will perform those operations, one at a time, following the order of operations rule.

WHY Every numerical expression has only one correct value. If we don't follow the correct order of operations, the expression can have more than one value.

Solution
Although the expression contains parentheses, there are no calculations to perform *within* them. We will begin with step 2 of the rule: Evaluate all exponential expressions. We will write the steps of the solution in horizontal form.

$$\dfrac{3}{4} + \dfrac{5}{3}\left(-\dfrac{1}{2}\right)^3 = \dfrac{3}{4} + \dfrac{5}{3}\left(-\dfrac{1}{8}\right)$$ Evaluate: $\left(-\frac{1}{2}\right)^3 = \left(-\frac{1}{2}\right)\left(-\frac{1}{2}\right)\left(-\frac{1}{2}\right) = -\frac{1}{8}$.

$$= \dfrac{3}{4} + \left(-\dfrac{5}{24}\right)$$ Multiply: $\frac{5}{3}\left(-\frac{1}{8}\right) = -\frac{5 \cdot 1}{3 \cdot 8} = -\frac{5}{24}$.

$$= \dfrac{3}{4} \cdot \dfrac{6}{6} + \left(-\dfrac{5}{24}\right)$$ Prepare to add the fractions: Their LCD is 24. To build the first fraction so that its denominator is 24, multiply it by a form of 1.

$$= \dfrac{18}{24} + \left(-\dfrac{5}{24}\right)$$ Multiply the numerators: $3 \cdot 6 = 18$. Multiply the denominators: $4 \cdot 6 = 24$.

$$= \dfrac{13}{24}$$ Add the numerators: $18 + (-5) = 13$. Write the sum over the common denominator 24.

Self Check 1

Evaluate: $\dfrac{7}{8} + \dfrac{3}{2}\left(-\dfrac{1}{4}\right)^2$

Now Try ➡ **Problem 15**

If an expression contains grouping symbols, we perform the operations within the grouping symbols first.

EXAMPLE 2   Evaluate: $\left(\dfrac{7}{8} - \dfrac{1}{4}\right) \div \left(-2\dfrac{3}{16}\right)$

Strategy We will perform any operations within parentheses first.

WHY This is the first step of the order of operations rule.

Solution
We will begin by performing the subtraction within the first set of parentheses. The second set of parentheses does not contain an operation to perform.

$$\left(\dfrac{7}{8} - \dfrac{1}{4}\right) \div \left(-2\dfrac{3}{16}\right)$$

$$= \left(\dfrac{7}{8} - \dfrac{1}{4} \cdot \dfrac{2}{2}\right) \div \left(-2\dfrac{3}{16}\right)$$ Within the first set of parentheses, prepare to subtract the fractions: Their LCD is 8. Build $\frac{1}{4}$ so that its denominator is 8.

$$= \left(\dfrac{7}{8} - \dfrac{2}{8}\right) \div \left(-2\dfrac{3}{16}\right)$$ Multiply the numerators: $1 \cdot 2 = 2$. Multiply the denominators: $4 \cdot 2 = 8$.

$$= \dfrac{5}{8} \div \left(-2\dfrac{3}{16}\right)$$ Subtract the numerators: $7 - 2 = 5$. Write the difference over the common denominator 8.

$$= \frac{5}{8} \div \left(-\frac{35}{16}\right)$$ *Write the mixed number as an improper fraction.*

$$= \frac{5}{8}\left(-\frac{16}{35}\right)$$ *Use the rule for division of fractions: Multiply the first fraction by the reciprocal of $-\frac{35}{16}$.*

$$= -\frac{5 \cdot 16}{8 \cdot 35}$$ *Multiply the numerators and multiply the denominators. The product of two fractions with unlike signs is negative.*

Self Check 2

Evaluate:

$$\left(\frac{19}{21} - \frac{2}{3}\right) \div \left(-2\frac{1}{7}\right)$$

Now Try ➡ **Problem 19**

$$= -\frac{\overset{1}{\cancel{5}} \cdot 2 \cdot \overset{1}{\cancel{8}}}{\underset{1}{\cancel{8}} \cdot \underset{1}{\cancel{5}} \cdot 7}$$ *To simplify, factor 16 as 2 · 8 and factor 35 as 5 · 7. Remove the common factors of 5 and 8 from the numerator and denominator.*

$$= -\frac{2}{7}$$ *Multiply the remaining factors in the numerator. Multiply the remaining factors in the denominator.*

EXAMPLE 3   Add $7\frac{1}{3}$ to the difference of $\frac{5}{6}$ and $\frac{1}{4}$.

Strategy We will translate the words of the problem to numbers and symbols. Then we will use the order of operations rule to evaluate the resulting expression.

WHY Since the expression involves two operations, addition and subtraction, we need to perform them in the proper order.

Solution

The key word *difference* indicates subtraction. Since we are to add $7\frac{1}{3}$ to the difference, the difference should be written first within parentheses, followed by the addition.

Add $7\frac{1}{3}$ to   the difference of $\frac{5}{6}$ and $\frac{1}{4}$.

$$\left(\frac{5}{6} - \frac{1}{4}\right) + 7\frac{1}{3}$$ *Translate from words to numbers and mathematical symbols.*

$$\left(\frac{5}{6} - \frac{1}{4}\right) + 7\frac{1}{3} = \left(\frac{5}{6} \cdot \frac{2}{2} - \frac{1}{4} \cdot \frac{3}{3}\right) + 7\frac{1}{3}$$ *Prepare to subtract the fractions within the parentheses. Build the fractions so that their denominators are the LCD 12.*

$$= \left(\frac{10}{12} - \frac{3}{12}\right) + 7\frac{1}{3}$$ *Multiply the numerators. Multiply the denominators.*

$$= \frac{7}{12} + 7\frac{1}{3}$$ *Subtract the numerators: 10 − 3 = 7. Write the difference over the common denominator 12.*

Self Check 3

Add $2\frac{1}{4}$ to the difference of $\frac{7}{8}$ and $\frac{2}{3}$.

Now Try ➡ **Problem 23**

$$= \frac{7}{12} + 7\frac{4}{12}$$ *Prepare to add the fractions. Build $\frac{1}{3}$ so that its denominator is 12: $\frac{1}{3} \cdot \frac{4}{4} = \frac{4}{12}$.*

$$= 7\frac{11}{12}$$ *Add the numerators of the fractions: 7 + 4 = 11. Write the sum over the common denominator 12.*

OBJECTIVE 2 **Solve application problems by using the order of operations rule.**

Sometimes more than one operation is needed to solve a problem.

EXAMPLE 4   **Masonry.**   To build a wall, a mason will use blocks that are $5\frac{3}{4}$ inches high, held together with $\frac{3}{8}$-inch-thick layers of mortar. If the plans call for 8 layers, called *courses*, of blocks, what will be the height of the wall when completed?

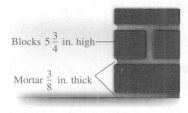

Blocks $5\frac{3}{4}$ in. high

Mortar $\frac{3}{8}$ in. thick

Analyze

- The blocks are $5\frac{3}{4}$ inches high.   *Given*
- A layer of mortar is $\frac{3}{8}$ inch thick.   *Given*
- There are 8 layers (courses) of blocks.   *Given*
- What is the height of the wall when completed?   *Find*

Form To find the height of the wall when it is completed, we could add the heights of 8 blocks and 8 layers of mortar. However, it will be simpler if we find the height of one block and one layer of mortar, and multiply that result by 8.

The height of the wall when completed	is equal to	8	times	(	the height of one block	plus	the thickness of one layer of mortar.	)

$$\text{The height of the wall when completed} \;=\; 8 \left(5\frac{3}{4} \;+\; \frac{3}{8} \right)$$

Calculate To evaluate the expression, we use the order of operations rule.

$$8\left(5\frac{3}{4} + \frac{3}{8} \right) = 8\left(5\frac{6}{8} + \frac{3}{8} \right)$$

Prepare to add the fractions within the parentheses: Their LCD is 8. Build $\frac{3}{4}$ so that its denominator is 8: $\frac{3}{4} \cdot \frac{2}{2} = \frac{6}{8}$.

$$= 8\left(5\frac{9}{8} \right)$$

Add the numerators of the fractions: $6 + 3 = 9$. Write the sum over the common denominator 8.

$$= \frac{8}{1}\left(\frac{49}{8} \right)$$

Prepare to multiply the fractions. Write $5\frac{9}{8}$ as an improper fraction.

$$= \frac{\overset{1}{8} \cdot 49}{1 \cdot \underset{1}{8}}$$

Multiply the numerators and multiply the denominators. To simplify, remove the common factor of 8 from the numerator and denominator.

$$= 49$$

Simplify: $\frac{49}{1} = 49$.

State The completed wall will be 49 inches high.

Check We can estimate to check the result. Since one block and one layer of mortar is about 6 inches high, eight layers of blocks and mortar would be $8 \cdot 6$ inches, or 48 inches high. The result of 49 inches seems reasonable.

Self Check 4

Masonry.   Find the height of a wall if 8 layers (called *courses*) of $7\frac{3}{8}$-inch-high blocks are held together by $\frac{1}{4}$-inch-thick layers of mortar.

Now Try ➲ Problem 81

OBJECTIVE 3 Evaluate formulas.

To evaluate a formula, we replace its letters, called **variables**, with specific numbers and evaluate the right side using the order of operations rule.

EXAMPLE 5   The formula for the area of a trapezoid is $A = \frac{1}{2}h(a + b)$, where A is the area, h is the height, and a and b are the lengths of its bases. Find A when $h = 1\frac{2}{3}$ in., $a = 2\frac{1}{2}$ in., and $b = 5\frac{1}{2}$ in.

Strategy In the formula, we will replace the letter h with $1\frac{2}{3}$, the letter a with $2\frac{1}{2}$, and the letter b with $5\frac{1}{2}$.

WHY Then we can use the order of operations rule to find the value of the expression on the right side of the = symbol.

Solution

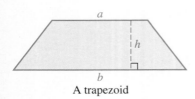

A trapezoid

$$A = \frac{1}{2}h\,(a + b)$$ This is the formula for the area of a trapezoid.

$$A = \frac{1}{2}\left(1\frac{2}{3}\right)\left(2\frac{1}{2} + 5\frac{1}{2}\right)$$ Replace h, a, and b with the given values.

$$A = \frac{1}{2}\left(1\frac{2}{3}\right)(8)$$ Do the addition within the parentheses: $2\frac{1}{2} + 5\frac{1}{2} = 8$.

$$A = \frac{1}{2}\left(\frac{5}{3}\right)\left(\frac{8}{1}\right)$$ To prepare to multiply fractions, write $1\frac{2}{3}$ as an improper fraction and 8 as $\frac{8}{1}$.

$$A = \frac{1 \cdot 5 \cdot 8}{2 \cdot 3 \cdot 1}$$ Multiply the numerators. Multiply the denominators.

$$A = \frac{1 \cdot 5 \cdot \overset{1}{2} \cdot 4}{\underset{1}{2} \cdot 3 \cdot 1}$$ To simplify, factor 8 as $2 \cdot 4$. Then remove the common factor of 2 from the numerator and denominator.

$$A = \frac{20}{3}$$ Multiply the remaining factors in the numerator. Multiply the remaining factors in the denominator.

$$A = 6\frac{2}{3}$$ Write the improper fraction $\frac{20}{3}$ as a mixed number by dividing 20 by 3.

The area of the trapezoid is $6\frac{2}{3}$ in.^{2}.

Self Check 5

The formula for the area of a triangle is $A = \frac{1}{2}bh$. Find the area of a triangle whose base is $12\frac{1}{2}$ meters long and whose height is $15\frac{1}{3}$ meters.

Now Try Problems 27 and 91

OBJECTIVE 4 Simplify complex fractions.

Fractions whose numerators and/or denominators contain fractions are called *complex fractions*. Here is an example of a complex fraction:

A fraction in the numerator ⟶ $\dfrac{\dfrac{3}{4}}{\dfrac{7}{8}}$ ⟵ The main fraction bar

A fraction in the denominator ⟶

Complex Fraction

A **complex fraction** is a fraction whose numerator or denominator, or both, contain one or more fractions or mixed numbers.

Here are more examples of complex fractions:

$$\cfrac{-\dfrac{1}{4}-\dfrac{4}{5}}{2\dfrac{4}{5}} \qquad\begin{array}{l}\longleftarrow\text{ Numerator }\longrightarrow\\ \longleftarrow\text{ Main fraction bar }\longrightarrow\\ \longleftarrow\text{ Denominator }\longrightarrow\end{array}\qquad \cfrac{\dfrac{1}{3}+\dfrac{1}{4}}{\dfrac{1}{3}-\dfrac{1}{4}}$$

To *simplify* a complex fraction means to express it as a fraction in simplified form.

The following method for simplifying complex fractions is based on the fact that the main fraction bar indicates division.

$$\cfrac{\dfrac{1}{4}}{\dfrac{2}{5}} \quad\longleftarrow\quad \begin{array}{l}\textit{The main fraction bar means}\\ \textit{"divide the fraction in the}\\ \textit{numerator by the fraction in}\\ \textit{the denominator."}\end{array}\quad\longrightarrow\quad \dfrac{1}{4}\div\dfrac{2}{5}$$

> ## Simplifying a Complex Fraction
>
> To simplify a complex fraction:
>
> **1.** Add or subtract in the numerator and/or denominator so that the numerator is a single fraction and the denominator is a single fraction.
>
> **2.** Perform the indicated division by multiplying the numerator of the complex fraction by the reciprocal of the denominator.
>
> **3.** Simplify the result, if possible.
>
> After simplifying a complex fraction, the result is often an improper fraction. Check with your instructor to see if he or she will accept an answer in that form or prefers that you express the final result as a mixed number.

EXAMPLE 6

Simplify: $\cfrac{\dfrac{1}{4}}{\dfrac{2}{5}}$

Strategy We will perform the division indicated by the main fraction bar using the rule for dividing fractions from Section 3.3.

WHY We can skip step 1 and immediately divide because the numerator and the denominator of the complex fraction are already single fractions.

Solution

$$\cfrac{\dfrac{1}{4}}{\dfrac{2}{5}} = \dfrac{1}{4}\div\dfrac{2}{5} \qquad \begin{array}{l}\textit{Write the division indicated by the main fraction bar using}\\ \textit{a }\div\textit{ symbol.}\end{array}$$

$$= \dfrac{1}{4}\cdot\dfrac{5}{2} \qquad \begin{array}{l}\textit{Use the rule for dividing fractions: Multiply the first fraction}\\ \textit{by the reciprocal of }\frac{2}{5}\textit{, which is }\frac{5}{2}.\end{array}$$

$$= \dfrac{1\cdot 5}{4\cdot 2} \qquad \begin{array}{l}\textit{Multiply the numerators.}\\ \textit{Multiply the denominators.}\end{array}$$

$$= \dfrac{5}{8}$$

Self Check 6

Simplify: $\cfrac{\dfrac{1}{6}}{\dfrac{3}{8}}$

Now Try ➲ Problem 31

EXAMPLE 7

Simplify: $\dfrac{-\dfrac{1}{4} + \dfrac{2}{5}}{\dfrac{1}{2} - \dfrac{4}{5}}$

Strategy Recall that a fraction bar is a type of grouping symbol. We will work above and below the main fraction bar separately to write $-\frac{1}{4} + \frac{2}{5}$ and $\frac{1}{2} - \frac{4}{5}$ as single fractions.

WHY The numerator and the denominator of the complex fraction must be written as single fractions before dividing.

Solution To write the numerator as a single fraction, we build $-\frac{1}{4}$ and $\frac{2}{5}$ to have an LCD of 20, and then add. To write the denominator as a single fraction, we build $\frac{1}{2}$ and $\frac{4}{5}$ to have an LCD of 10, and subtract.

$$\frac{-\dfrac{1}{4} + \dfrac{2}{5}}{\dfrac{1}{2} - \dfrac{4}{5}} = \frac{-\dfrac{1}{4} \cdot \dfrac{5}{5} + \dfrac{2}{5} \cdot \dfrac{4}{4}}{\dfrac{1}{2} \cdot \dfrac{5}{5} - \dfrac{4}{5} \cdot \dfrac{2}{2}}$$

The LCD for the numerator is 20. Build each fraction so that each has a denominator of 20.

The LCD for the denominator is 10. Build each fraction so that each has a denominator of 10.

$$= \frac{-\dfrac{5}{20} + \dfrac{8}{20}}{\dfrac{5}{10} - \dfrac{8}{10}}$$

Multiply in the numerator.
Multiply in the denominator.

$$= \frac{\dfrac{3}{20}}{-\dfrac{3}{10}}$$

In the numerator of the complex fraction, add the fractions.

In the denominator, subtract the fractions.

$$= \frac{3}{20} \div \left(-\frac{3}{10}\right)$$

Write the division indicated by the main fraction bar using a ÷ symbol.

$$= \frac{3}{20}\left(-\frac{10}{3}\right)$$

Multiply the first fraction by the reciprocal of $-\frac{3}{10}$, which is $-\frac{10}{3}$.

$$= -\frac{3 \cdot 10}{20 \cdot 3}$$

The product of two fractions with unlike signs is negative. Multiply the numerators. Multiply the denominators.

$$= -\frac{\overset{1}{3} \cdot \overset{1}{10}}{2 \cdot \underset{1}{10} \cdot \underset{1}{3}}$$

To simplify, factor 20 as 2 · 10. Then remove the common factors of 3 and 10 from the numerator and denominator.

$$= -\frac{1}{2}$$

Multiply the remaining factors in the numerator.
Multiply the remaining factors in the denominator.

Self Check 7

Simplify: $\dfrac{-\dfrac{5}{8} + \dfrac{1}{3}}{\dfrac{3}{4} - \dfrac{1}{3}}$

Now Try ➡ **Problem 35**

EXAMPLE 8

Simplify: $\dfrac{7 - \dfrac{2}{3}}{4\dfrac{5}{6}}$

Strategy Recall that a fraction bar is a type of grouping symbol. We will work above and below the main fraction bar separately to write $7 - \frac{2}{3}$ as a single fraction and $4\frac{5}{6}$ as an improper fraction.

WHY The numerator and the denominator of the complex fraction must be written as single fractions before dividing.

Solution

$$\frac{7 - \dfrac{2}{3}}{4\dfrac{5}{6}} = \frac{\dfrac{7}{1} \cdot \dfrac{3}{3} - \dfrac{2}{3}}{\dfrac{29}{6}}$$

In the numerator, write 7 as $\frac{7}{1}$. The LCD for the numerator is 3. Build $\frac{7}{1}$ so that it has a denominator of 3.
In the denominator, write $4\frac{5}{6}$ as the improper fraction $\frac{29}{6}$.

$$= \frac{\dfrac{21}{3} - \dfrac{2}{3}}{\dfrac{29}{6}}$$

Multiply in the numerator.

$$= \frac{\dfrac{19}{3}}{\dfrac{29}{6}}$$

In the numerator of the complex fraction, subtract the numerators: $21 - 2 = 19$. Then write the difference over the common denominator 3.

$$= \frac{19}{3} \div \frac{29}{6}$$

Write the division indicated by the main fraction bar using a ÷ symbol.

$$= \frac{19}{3} \cdot \frac{6}{29}$$

Multiply the first fraction by the reciprocal of $\frac{29}{6}$, which is $\frac{6}{29}$.

$$= \frac{19 \cdot 6}{3 \cdot 29}$$

Multiply the numerators.
Multiply the denominators.

$$= \frac{19 \cdot 2 \cdot \overset{1}{\cancel{3}}}{\underset{1}{\cancel{3}} \cdot 29}$$

To simplify, factor 6 as $2 \cdot 3$. Then remove the common factor of 3 from the numerator and denominator.

$$= \frac{38}{29}$$

Multiply the remaining factors in the numerator.
Multiply the remaining factors in the denominator.

Self Check 8

Simplify: $\dfrac{5 - \dfrac{3}{4}}{1\dfrac{7}{8}}$

Now Try Problem 39

Answers to Self Checks

1. $\frac{31}{32}$ **2.** $-\frac{1}{9}$ **3.** $2\frac{11}{24}$ **4.** 61 in. **5.** $95\frac{5}{6}$ m² **6.** $\frac{4}{9}$ **7.** $-\frac{7}{10}$ **8.** $\frac{34}{15}$

SECTION 3.7 STUDY SET

VOCABULARY

Fill in the blanks.

1. We use the order of _____ rule to evaluate expressions that contain more than one operation.

2. To evaluate a formula such as $A = \frac{1}{2}h(a + b)$, we substitute specific numbers for the letters, called _____, in the formula and find the value of the right side.

3. $\dfrac{\dfrac{1}{2}}{\dfrac{3}{4}}$ and $\dfrac{\dfrac{7}{8} + \dfrac{2}{5}}{\dfrac{1}{2} - \dfrac{1}{3}}$ are examples of _____ fractions.

4. In the complex fraction $\dfrac{\dfrac{2}{5} + \dfrac{1}{4}}{\dfrac{2}{5} - \dfrac{1}{4}}$, the _____ is $\frac{2}{5} + \frac{1}{4}$ and the _____ is $\frac{2}{5} - \frac{1}{4}$.

CONCEPTS

5. What operations are involved in this expression?

$$5\left(6\frac{1}{3}\right) + \left(-\frac{1}{4}\right)^3$$

6. a. To evaluate $\frac{7}{8} + \left(\frac{1}{3}\right)\left(\frac{1}{4}\right)$, what operation should be performed first?

   b. To evaluate $\frac{7}{8} + \left(\frac{1}{3} - \frac{1}{4}\right)^2$, what operation should be performed first?

7. Translate the following to numbers and symbols. *You do not have to find the answer.*

   Add $1\frac{2}{15}$ to the difference of $\frac{2}{3}$ and $\frac{1}{10}$.

8. Refer to the trapezoid shown below. Label the length of the upper base $3\frac{1}{2}$ inches, the length of the lower base $5\frac{1}{2}$ inches, and the height $2\frac{2}{3}$ inches.

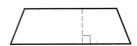

9. What division is represented by this complex fraction?

$$\frac{\frac{2}{3}}{\frac{1}{5}}$$

10. Consider: $\dfrac{\frac{2}{3} - \frac{1}{5}}{\frac{1}{2} + \frac{4}{5}}$

   a. What is the LCD for the fractions in the numerator of this complex fraction?

   b. What is the LCD for the fractions in the denominator of this complex fraction?

11. Write the denominator of the following complex fraction as an improper fraction.

$$\frac{\frac{1}{8} - \frac{3}{16}}{5\frac{3}{4}}$$

12. When this complex fraction is simplified, will the result be positive or negative?

$$\frac{-\frac{2}{3}}{\frac{3}{4}}$$

NOTATION

Fill in the blanks to complete each step.

13. $\dfrac{7}{12} - \dfrac{1}{2} \cdot \dfrac{1}{3} = \dfrac{7}{12} - \dfrac{1 \cdot 1}{2 \cdot }$

$= \dfrac{7}{12} - \dfrac{1}{}$

$= \dfrac{7}{12} - \dfrac{1}{6} \cdot \dfrac{}{}$

$= \dfrac{7}{12} - \dfrac{\blacksquare}{12}$

$= \dfrac{\blacksquare}{12}$

14. $\dfrac{\frac{1}{8}}{\frac{3}{4}} = \dfrac{1}{8} \div \dfrac{}{}$

$= \dfrac{1}{8} \cdot \dfrac{}{}$

$= \dfrac{1 \cdot }{8 \cdot 3}$

$= \dfrac{1 \cdot \overset{1}{\diagup}}{2 \cdot \underset{1}{\diagup} \cdot 3}$

$= \dfrac{1}{}$

GUIDED PRACTICE

Evaluate each expression. See Example 1.

15. $\dfrac{3}{4} + \dfrac{2}{5}\left(-\dfrac{1}{2}\right)^2$

16. $\dfrac{1}{4} + \dfrac{8}{27}\left(-\dfrac{3}{2}\right)^2$

17. $\dfrac{1}{6} + \dfrac{9}{8}\left(-\dfrac{2}{3}\right)^3$

18. $\dfrac{1}{5} + \dfrac{1}{9}\left(-\dfrac{3}{2}\right)^3$

Evaluate each expression. See Example 2.

19. $\left(\dfrac{3}{4} - \dfrac{1}{6}\right) \div \left(-2\dfrac{1}{6}\right)$

20. $\left(\dfrac{7}{8} - \dfrac{3}{7}\right) \div \left(-1\dfrac{3}{7}\right)$

21. $\left(\dfrac{15}{16} - \dfrac{1}{8}\right) \div \left(-9\dfrac{3}{4}\right)$

22. $\left(\dfrac{19}{36} - \dfrac{1}{6}\right) \div \left(-8\dfrac{2}{3}\right)$

Evaluate each expression. See Example 3.

23. Add $5\dfrac{4}{15}$ to the difference of $\dfrac{5}{6}$ and $\dfrac{2}{3}$.

24. Add $8\dfrac{5}{24}$ to the difference of $\dfrac{3}{4}$ and $\dfrac{1}{6}$.

25. Add $2\dfrac{7}{18}$ to the difference of $\dfrac{7}{9}$ and $\dfrac{1}{2}$.

26. Add $1\dfrac{19}{30}$ to the difference of $\dfrac{4}{5}$ and $\dfrac{1}{2}$.

Evaluate the formula $A = \frac{1}{2}h(a + b)$ for the given values. See Example 5.

27. $a = 2\frac{1}{2}, b = 7\frac{1}{2}, h = 5\frac{1}{4}$

28. $a = 4\frac{1}{2}, b = 5\frac{1}{2}, h = 2\frac{1}{8}$

29. $a = 1\frac{1}{4}, b = 6\frac{3}{4}, h = 4\frac{1}{2}$

30. $a = 1\frac{1}{3}, b = 4\frac{2}{3}, h = 2\frac{2}{5}$

Simplify each complex fraction. See Example 6.

31. $\dfrac{\frac{1}{16}}{\frac{2}{5}}$

32. $\dfrac{\frac{2}{11}}{\frac{3}{4}}$

33. $\dfrac{\frac{5}{8}}{\frac{3}{4}}$

34. $\dfrac{\frac{1}{5}}{\frac{8}{15}}$

Simplify each complex fraction. See Example 7.

35. $\dfrac{-\frac{1}{4} + \frac{2}{3}}{\frac{5}{6} + \frac{2}{3}}$

36. $\dfrac{-\frac{1}{2} + \frac{7}{8}}{\frac{3}{4} - \frac{1}{2}}$

37. $\dfrac{\frac{1}{3} - \frac{3}{4}}{\frac{1}{6} + \frac{2}{3}}$

38. $\dfrac{\frac{1}{3} - \frac{3}{4}}{\frac{1}{6} + \frac{1}{3}}$

Simplify each complex fraction. See Example 8.

39. $\dfrac{5 - \frac{5}{6}}{1\frac{1}{12}}$

40. $\dfrac{4 - \frac{3}{4}}{1\frac{7}{8}}$

41. $\dfrac{4 - \frac{7}{8}}{3\frac{1}{4}}$

42. $\dfrac{6 - \frac{2}{7}}{6\frac{2}{3}}$

TRY IT YOURSELF

Evaluate each expression or formula. Simplify each complex fraction.

43. $\frac{7}{8} - \left(\frac{4}{5} + 1\frac{3}{4}\right)$

44. $\left(\frac{5}{4}\right)^2 + \left(\frac{2}{3} - 2\frac{1}{6}\right)$

45. $\dfrac{-\frac{14}{15}}{\frac{7}{10}}$

46. $\dfrac{\frac{5}{27}}{-\frac{5}{9}}$

47. $A = \frac{1}{2}bh$ for $b = 10$ and $h = 7\frac{1}{5}$

48. $V = lwh$ for $l = 12$, $w = 8\frac{1}{2}$, and $h = 3\frac{1}{3}$

49. $\frac{2}{3}\left(-\frac{1}{4}\right) + \frac{1}{2}$

50. $-\frac{7}{8} - \left(\frac{1}{8}\right)\left(\frac{2}{3}\right)$

51. $\frac{1}{2}\left(\frac{1}{8}\right) + \left(-\frac{1}{4}\right)^2$

52. $-\frac{3}{16} - \left(-\frac{1}{2}\right)^3$

53. $\dfrac{\frac{3}{8} + \frac{1}{4}}{\frac{3}{8} - \frac{1}{4}}$

54. $\dfrac{\frac{2}{3} - \frac{5}{2}}{\frac{2}{3} - \frac{3}{2}}$

55. Add $12\frac{11}{12}$ to the difference of $5\frac{1}{6}$ and $3\frac{7}{8}$.

56. Add $18\frac{1}{3}$ to the difference of $11\frac{3}{5}$ and $9\frac{11}{15}$.

57. $\dfrac{5\frac{1}{2}}{-\frac{1}{4} + \frac{3}{4}}$

58. $\dfrac{4\frac{1}{4}}{\frac{2}{3} + \left(-\frac{1}{6}\right)}$

59. $\left|\frac{2}{3} - \frac{9}{10}\right| \div \left(-\frac{1}{5}\right)$

60. $\left|-\frac{3}{16} \div 2\frac{1}{4}\right| + \left(-2\frac{1}{8}\right)$

61. $\dfrac{\frac{1}{5} - \left(-\frac{1}{4}\right)}{\frac{1}{4} + \frac{4}{5}}$

62. $\dfrac{\dfrac{1}{8} - \left(-\dfrac{1}{2}\right)}{\dfrac{1}{4} + \dfrac{3}{8}}$

63. $1\dfrac{3}{5}\left(\dfrac{1}{2}\right)^2\left(\dfrac{3}{4}\right)$

64. $2\dfrac{3}{5}\left(-\dfrac{1}{3}\right)^2\left(\dfrac{1}{2}\right)$

65. $A = lw$ for $l = 5\dfrac{5}{6}$ and $w = 7\dfrac{3}{5}$

66. $P = 2l + 2w$ for $l = \dfrac{7}{8}$ and $w = \dfrac{3}{5}$

67. $\left(2 - \dfrac{1}{2}\right)^2 + \left(2 + \dfrac{1}{2}\right)^2$

68. $\left(\dfrac{9}{20} \div 2\dfrac{2}{5}\right) + \left(\dfrac{3}{4}\right)^2$

69. $\dfrac{-\dfrac{5}{6}}{-1\dfrac{7}{8}}$

70. $\dfrac{-\dfrac{4}{3}}{-2\dfrac{5}{6}}$

71. Subtract $9\dfrac{1}{10}$ from the sum of $7\dfrac{3}{7}$ and $3\dfrac{1}{5}$.

72. Subtract $3\dfrac{2}{3}$ from the sum of $2\dfrac{5}{12}$ and $1\dfrac{5}{8}$.

73. $\dfrac{\dfrac{3}{5} - 2}{\dfrac{2}{5} - 2}$

74. $\dfrac{\dfrac{1}{3} + \dfrac{1}{4}}{\dfrac{1}{3} - \dfrac{1}{4}}$

75. $\left(\dfrac{8}{5} - 1\dfrac{1}{3}\right) - \left(-\dfrac{4}{5} \cdot 10\right)$

76. $\left(1 - \dfrac{3}{4}\right)\left(1 + \dfrac{3}{4}\right)$

LOOK ALIKES

Evaluate each expression.

77. a. $\dfrac{4}{27} + \dfrac{4}{9} \div \dfrac{2}{3}$ **b.** $\left(\dfrac{4}{27} + \dfrac{4}{9}\right) \div \dfrac{2}{3}$

78. a. $-\dfrac{1}{4} - \left(\dfrac{2}{3}\right)^3$ **b.** $\left(\dfrac{1}{4} - \dfrac{2}{3}\right)^3$

Simplify each complex fraction.

79. a. $\dfrac{\dfrac{9}{10}}{\dfrac{3}{25}}$ **b.** $\dfrac{\dfrac{9}{10} + \dfrac{1}{5}}{\dfrac{3}{25}}$

80. a. $\dfrac{\dfrac{7}{8}}{\dfrac{21}{32}}$ **b.** $\dfrac{\dfrac{7}{8} - \dfrac{1}{4}}{\dfrac{21}{32}}$

APPLICATIONS

81. Remodeling a bathroom. A handyman installed 20 rows of grout and tile on a bathroom wall using the pattern shown below. How high above floor level does the tile work reach? (*Hint:* There is no grout line above the last row of tiles.)

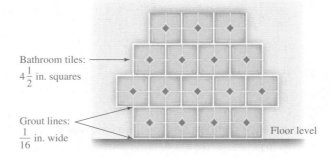

Bathroom tiles: $4\dfrac{1}{2}$ in. squares

Grout lines: $\dfrac{1}{16}$ in. wide

Floor level

82. Plywood. To manufacture a sheet of plywood, several thin layers of wood are glued together, as shown. Then an exterior finish is attached to the top and the bottom, as shown below. How thick is the final product?

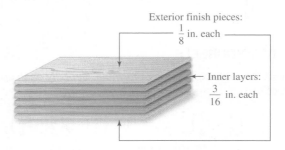

Exterior finish pieces: $\dfrac{1}{8}$ in. each

Inner layers: $\dfrac{3}{16}$ in. each

83. Postage rates. Can the advertising package shown below be mailed for the 1-ounce rate?

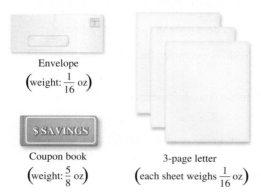

Envelope
$\left(\text{weight: } \frac{1}{16} \text{ oz}\right)$

$SAVINGS

Coupon book
$\left(\text{weight: } \frac{5}{8} \text{ oz}\right)$

3-page letter
$\left(\text{each sheet weighs } \frac{1}{16} \text{ oz}\right)$

84. Physical therapy. After back surgery, a patient followed a walking program shown in the table below to strengthen her muscles. What was the total distance she walked over this three-week period?

Week	Distance per day
#1	$\frac{1}{4}$ mile
#2	$\frac{1}{2}$ mile
#3	$\frac{3}{4}$ mile

85. Reading programs. To improve reading skills, elementary school children read silently at the end of the school day for $\frac{1}{4}$ hour on Mondays and for $\frac{1}{2}$ hour on Fridays. For the month of January, how many total hours did the children read silently in class?

S	M	T	W	T	F	S
	1	2	3	4	5	6
7	8	9	10	11	12	13
14	15	16	17	18	19	20
21	22	23	24	25	26	27
28	29	30	31			

86. Physical fitness. Two people begin their workouts from the same point on a bike path and travel in opposite directions, as shown below. How far apart are they in $1\frac{1}{2}$ hours? Use the table to help organize your work.

	Rate (mph) ·	Time (hr) =	Distance (mi)
Jogger			
Cyclist			

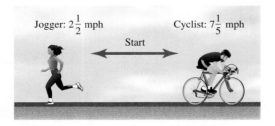

Jogger: $2\frac{1}{2}$ mph   Cyclist: $7\frac{1}{5}$ mph

Start

87. Hiking. A scout troop plans to hike from the campground to Glenn Peak, as shown below. Since the terrain is steep, they plan to stop and rest after every $\frac{2}{3}$ mile. With this plan, how many parts will there be to this hike?

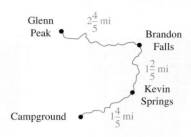

Glenn Peak $2\frac{4}{5}$ mi   Brandon Falls

$1\frac{2}{5}$ mi

Kevin Springs

Campground $1\frac{4}{5}$ mi

88. Deli shops. A sandwich shop sells a $\frac{1}{2}$-pound club sandwich made of turkey and ham. The owner buys the turkey in $1\frac{3}{4}$-pound packages and the ham in $2\frac{1}{2}$-pound packages. If he mixes two packages of turkey and one package of ham together, how many sandwiches can he make from the mixture?

89. Skin creams. Using a formula of $\frac{1}{2}$ ounce of sun block, $\frac{2}{3}$ ounce of moisturizing cream, and $\frac{3}{4}$ ounce of lanolin, a beautician mixes her own brand of skin cream. She packages it in $\frac{1}{4}$-ounce tubes. How many full tubes can be produced using this formula? How much skin cream is left over?

90. Sleep. The graph below compares the amount of sleep a 1-month-old baby got to the $15\frac{1}{2}$-hour daily requirement recommended by Children's Hospital of Orange County, California. For the week, how far below the baseline was the baby's daily average?

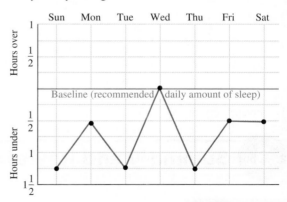

91. Camping. The four sides of a tent are all the same trapezoid-shape. (See the illustration below.) How many square yards of canvas are used to make one of the sides of the tent?

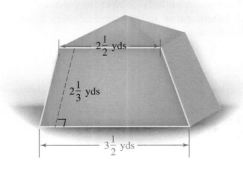

$2\frac{1}{2}$ yds

$2\frac{1}{3}$ yds

$3\frac{1}{2}$ yds

92. Sewing. A seamstress begins with a trapezoid-shaped piece of denim to make the back pocket on a pair of jeans. (See the illustration below.) How many square inches of denim are used to make the pocket?

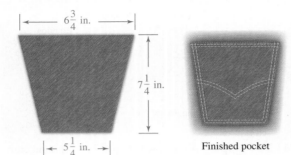

Finished pocket

93. Amusement parks. At the end of a ride at an amusement park, a boat splashes into a pool of water. The time (in seconds) that it takes two pipes to refill the pool is given by

$$\dfrac{1}{\dfrac{1}{10} + \dfrac{1}{15}}$$

Simplify the complex fraction to find the time.

94. Algebra. Complex fractions, like the one shown below, are seen in an algebra class when the topic of *slope of a line* is studied. Simplify this complex fraction and, as is done in algebra, write the answer as an improper fraction.

$$\dfrac{\dfrac{1}{2} - \dfrac{1}{3}}{\dfrac{1}{4} - \dfrac{1}{5}}$$

WRITING

95. Why is an order of operations rule necessary?

96. What does it mean to evaluate a formula?

97. What is a complex fraction?

98. In the complex fraction $\dfrac{\dfrac{3}{8} + \dfrac{1}{4}}{\dfrac{3}{8} - \dfrac{1}{4}}$, the fraction bar serves as a grouping symbol. Explain why this is so.

REVIEW

99. Find the sum: $8 + 19 + 124 + 2{,}097$

100. Subtract 879 from 1,023.

101. Multiply 879 by 23.

102. Divide 1,665 by 45.

103. List the factors of 24.

104. Find the prime factorization of 24.

③ Summary and Review

SECTION 3.1	An Introduction to Fractions

DEFINITIONS AND CONCEPTS	EXAMPLES
A **fraction** describes the number of equal parts of a whole. In a fraction, the number above the **fraction bar** is called the **numerator**, and the number below is called the **denominator**.	Since 3 of 8 equal parts are colored red, $\frac{3}{8}$ (three-eighths) of the figure is shaded. Fraction bar ⟶ $\dfrac{3}{8}$ ⟵ numerator, denominator

If the numerator of a fraction is less than its denominator, the fraction is called a **proper fraction**. If the numerator of a fraction is greater than or equal to its denominator, the fraction is called an **improper fraction**.	Proper fractions: $\dfrac{1}{5}, \dfrac{7}{8}$, and $\dfrac{999}{1,000}$ *Proper fractions are less than 1.* Improper fractions: $\dfrac{3}{2}, \dfrac{41}{16}$, and $\dfrac{15}{15}$ *Improper fractions are greater than or equal to 1.*
There are four **special fraction forms** that involve 0 and 1. Each of these fractions is a **form of 1**: $1 = \dfrac{1}{1} = \dfrac{2}{2} = \dfrac{3}{3} = \dfrac{4}{4} = \dfrac{5}{5} = \dfrac{6}{6} = \dfrac{7}{7} = \dfrac{8}{8} = \dfrac{9}{9} = \cdots$	Simplify each fraction: $\dfrac{0}{8} = 0 \qquad \dfrac{7}{0}$ is undefined $\qquad \dfrac{5}{1} = 5 \qquad \dfrac{20}{20} = 1$
Two fractions are **equivalent** if they represent the same number. **Equivalent fractions** represent the same portion of a whole.	$\dfrac{2}{3}, \dfrac{4}{6}$, and $\dfrac{8}{12}$ are equivalent fractions. They represent the same shaded portion of the figure. $\dfrac{2}{3} \quad = \quad \dfrac{4}{6} \quad = \quad \dfrac{8}{12}$
To **build a fraction**, we multiply it by a factor equal to 1 in the form of $\dfrac{2}{2}, \dfrac{3}{3}, \dfrac{4}{4}, \dfrac{5}{5}$, and so on.	Write $\dfrac{3}{4}$ as an equivalent fraction with a denominator of 36. $\dfrac{3}{4} = \dfrac{3}{4} \cdot \dfrac{9}{9}$ *We must multiply the denominator of $\frac{3}{4}$ by 9 to obtain a denominator of 36. It follows that $\frac{9}{9}$ should be the form of 1 that is used to build $\frac{3}{4}$.* $= \dfrac{3 \cdot 9}{4 \cdot 9}$ *Multiply the numerators.* *Multiply the denominators.* $= \dfrac{27}{36}$ $\dfrac{27}{36}$ is equivalent to $\dfrac{3}{4}$.
A fraction is **in simplest form**, or **lowest terms**, when the numerator and denominator have no common factors other than 1.	Is $\dfrac{6}{14}$ in simplest form? The factors of the numerator, 6, are: **1**, **2**, 3, 6. The factors of the denominator, 14, are: **1**, **2**, 7, 14. Since the numerator and denominator have a common factor of 2, the fraction $\dfrac{6}{14}$ is *not* in simplest form.
To **simplify a fraction**, we write it in simplest form by removing a factor equal to 1: 1. Factor (or prime factor) the numerator and denominator to determine their common factors. 2. Remove factors equal to 1 by replacing each pair of factors common to the numerator and denominator with the equivalent fraction $\dfrac{1}{1}$. 3. Multiply the remaining factors in the numerator and in the denominator.	Simplify: $\dfrac{12}{30}$ $\dfrac{12}{30} = \dfrac{2 \cdot 2 \cdot 3}{2 \cdot 3 \cdot 5}$ *Prime factor 12 and 30.* $= \dfrac{2 \cdot 2 \cdot \overset{1}{\cancel{3}}}{2 \cdot \cancel{3} \cdot 5}$ *Remove the common factors of 2 and 3 from the numerator and denominator.* *Multiply the remaining factors in the numerator:* $1 \cdot 2 \cdot 1 = 2$. $= \dfrac{2}{5}$ *Multiply the remaining factors in the denominator:* $1 \cdot 1 \cdot 5 = 5$. Since 2 and 5 have no common factors other than 1, we say that $\dfrac{2}{5}$ is in simplest form.

REVIEW EXERCISES

1. Identify the numerator and denominator of the fraction $\frac{11}{16}$. Is it a proper or an improper fraction?

2. Write fractions that represent the shaded and unshaded portions of the figure to the right.

3. In the illustration below, why can't we say that $\frac{3}{4}$ of the figure is shaded?

4. Write the fraction $\frac{2}{-3}$ in two other ways.

5. Simplify, if possible:

   a. $\frac{5}{5}$ b. $\frac{0}{10}$

   c. $\frac{18}{1}$ d. $\frac{7}{0}$

6. What concept about fractions is illustrated below?

   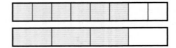

Write each fraction as an equivalent fraction with the indicated denominator.

7. $\frac{2}{3}$, denominator 18   8. $\frac{3}{8}$, denominator 16

9. $\frac{7}{15}$, denominator 45   10. $\frac{13}{12}$, denominator 60

11. Write 5 as an equivalent fraction with denominator 9.

12. Are the following fractions in simplest form?

    a. $\frac{6}{9}$ b. $\frac{10}{81}$

Simplify each fraction, if possible.

13. $\frac{15}{45}$ 14. $\frac{20}{48}$

15. $\frac{66}{108}$ 16. $\frac{117}{208}$

17. $\frac{81}{64}$

18. Tell whether $\frac{8}{12}$ and $\frac{176}{264}$ are equivalent by simplifying each fraction.

19. **Sleep.** If a woman gets seven hours of sleep each night, write a fraction to describe the part of a whole day that she spends sleeping and another to describe the part of a whole day that she is not sleeping.

20. a. What type of problem is shown below? Explain the step shown in red.

    $$\frac{5}{8} = \frac{5}{8} \cdot \frac{2}{2} = \frac{10}{16}$$

    b. What type of problem is shown below? Explain the step shown in red.

    $$\frac{4}{6} = \frac{\overset{1}{2} \cdot 2}{\underset{1}{2} \cdot 3} = \frac{2}{3}$$

SECTION 3.2 ▶ Multiplying Fractions

DEFINITIONS AND CONCEPTS	EXAMPLES

To **multiply two fractions**, multiply the numerators and multiply the denominators. Simplify the result, if possible.	Multiply and simplify, if possible: $\dfrac{4}{5} \cdot \dfrac{2}{3}$
	
	 $\dfrac{4}{5} \cdot \dfrac{2}{3} = \dfrac{4 \cdot 2}{5 \cdot 3}$ Multiply the numerators.
	                Multiply the denominators.
	
	       $= \dfrac{8}{15}$
	
	 Since 8 and 15 have no common factors other than 1, the result is in simplest form.

Multiplying signed fractions The product of two fractions with the same (like) signs is positive. The product of two fractions with different (unlike) signs is negative.	Multiply and simplify, if possible: $-\dfrac{3}{4} \cdot \dfrac{2}{27}$ $-\dfrac{3}{4} \cdot \dfrac{2}{27} = -\dfrac{3 \cdot 2}{4 \cdot 27}$ Multiply the numerators. Multiply the denominators. Since the fractions have unlike signs, make the answer negative. $= -\dfrac{\overset{1}{\cancel{3}} \cdot 2}{2 \cdot 2 \cdot \underset{1}{\cancel{3}} \cdot 3 \cdot 3}$ Prime factor 4 and 27. Then simplify, by removing the common factors of 2 and 3 from the numerator and denominator. $= -\dfrac{1}{18}$ Multiply the remaining factors in the numerator: $1 \cdot 1 = 1$. Multiply the remaining factors in the denominator: $1 \cdot 2 \cdot 1 \cdot 3 \cdot 3 = 18$.
The base of an **exponential expression** can be a positive or a negative fraction. The rule for multiplying two fractions can be extended to find the product of three or more fractions.	Evaluate: $\left(\dfrac{2}{3}\right)^3$ $\left(\dfrac{2}{3}\right)^3 = \dfrac{2}{3} \cdot \dfrac{2}{3} \cdot \dfrac{2}{3}$ Write the base, $\frac{2}{3}$, as a factor 3 times. $= \dfrac{2 \cdot 2 \cdot 2}{3 \cdot 3 \cdot 3}$ Multiply the numerators. Multiply the denominators. $= \dfrac{8}{27}$ This fraction is in simplest form.
When a **fraction is followed by the word _of_**, it indicates that we are to find a part of some quantity using multiplication.	To find $\frac{2}{5}$ of 35, we multiply: $\dfrac{2}{5} \; of \; 35 = \dfrac{2}{5} \cdot 35$ The word _of_ indicates multiplication. $= \dfrac{2}{5} \cdot \dfrac{35}{1}$ Write 35 as a fraction: $35 = \frac{35}{1}$. $= \dfrac{2 \cdot 35}{5 \cdot 1}$ Multiply the numerators. Multiply the denominators. $= \dfrac{2 \cdot \overset{1}{\cancel{5}} \cdot 7}{\underset{1}{\cancel{5}} \cdot 1}$ Prime factor 35. Then simplify by removing the common factor of 5 from the numerator and denominator. $= \dfrac{14}{1}$ Multiply the remaining factors in the numerator and in the denominator. $= 14$ Any number divided by 1 is equal to that number.

The formula for the area of a triangle

Area of a triangle $= \dfrac{1}{2}$ (base)(height)

or

$$A = \dfrac{1}{2}bh$$

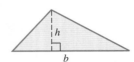

Find the area of the triangle shown on the right.

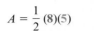

$A = \dfrac{1}{2}$ (**base**)(**height**)

$A = \dfrac{1}{2}$ (8)(5)   Substitute 8 for the base and 5 for the height.

$A = \dfrac{1}{2}\left(\dfrac{8}{1}\right)\left(\dfrac{5}{1}\right)$ Write 8 and 5 as fractions.

$A = \dfrac{1 \cdot 8 \cdot 5}{2 \cdot 1 \cdot 1}$ Multiply the numerators.
Multiply the denominators.

$A = \dfrac{1 \cdot \overset{1}{2} \cdot 2 \cdot 2 \cdot 5}{\underset{1}{2} \cdot 1 \cdot 1}$ Prime factor 8. Then simplify by removing the common factor of 2 from the numerator and denominator.

$A = 20$

The area of the triangle is 20 ft².

REVIEW EXERCISES

21. Fill in the blanks: To multiply two fractions, multiply the _____ and multiply the _____. Then _____, if possible.

22. Translate the following phrase to symbols. *You do not have to find the answer.*

$$\dfrac{5}{6} \text{ of } \dfrac{2}{3}$$

Multiply. Simplify the product, if possible.

23. $\dfrac{1}{2} \cdot \dfrac{1}{3}$

24. $\dfrac{2}{5}\left(-\dfrac{7}{9}\right)$

25. $\dfrac{9}{16} \cdot \dfrac{20}{27}$

26. $-\dfrac{5}{6}\left(-\dfrac{1}{15}\right)\left(-\dfrac{18}{25}\right)$

27. $\dfrac{3}{5} \cdot 7$

28. $-4\left(-\dfrac{9}{16}\right)$

29. $3\left(\dfrac{1}{3}\right)$

30. $-\dfrac{6}{7}\left(-\dfrac{7}{6}\right)$

Evaluate each expression.

31. $-\left(\dfrac{3}{4}\right)^2$

32. $\left(-\dfrac{5}{2}\right)^3$

33. $\left(-\dfrac{2}{5}\right)^3$

34. $\left(\dfrac{2}{3}\right)^2$

35. Drag racing. A top-fuel dragster had to make 8 trial runs on a quarter-mile track before it was ready for competition. Find the total distance it covered on the trial runs.

36. Gravity. Objects on the moon weigh only one-sixth of their weight on Earth. How much will an astronaut weigh on the moon if he weighs 180 pounds on Earth?

37. Find the area of the triangular sign.

38. Find the area of the triangle shown below.

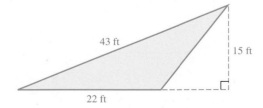

SECTION 3.3 ▶ **Dividing Fractions**

DEFINITIONS AND CONCEPTS	EXAMPLES
One number is the **reciprocal** of another if their product is 1. To find the **reciprocal of a fraction**, invert the numerator and denominator.	The reciprocal of $\frac{4}{5}$ is $\frac{5}{4}$ because $\frac{4}{5} \cdot \frac{5}{4} = 1$. Fraction Reciprocal $\frac{4}{5}$ ⟶ $\frac{5}{4}$ Invert
To **divide two fractions**, multiply the first fraction by the reciprocal of the second fraction. Simplify the result, if possible.	Divide and simplify, if possible: $\dfrac{4}{35} \div \dfrac{2}{21}$ $\dfrac{4}{35} \div \dfrac{2}{21} = \dfrac{4}{35} \cdot \dfrac{21}{2}$ Multiply $\frac{4}{35}$ by the reciprocal of $\frac{2}{21}$, which is $\frac{21}{2}$. $= \dfrac{4 \cdot 21}{35 \cdot 2}$ Multiply the numerators. Multiply the denominators. $= \dfrac{2 \cdot 2 \cdot 3 \cdot 7}{5 \cdot 7 \cdot 2}$ To prepare to simplify, write 4, 21, and 35 in prime-factored form. $= \dfrac{\overset{1}{2} \cdot 2 \cdot 3 \cdot \overset{1}{7}}{5 \cdot \underset{1}{7} \cdot \underset{1}{2}}$ To simplify, remove the common factors of 2 and 7 from the numerator and denominator. $= \dfrac{6}{5}$ Multiply the remaining factors in the numerator: $1 \cdot 2 \cdot 3 \cdot 1 = 6$. Multiply the remaining factors in the denominator: $5 \cdot 1 \cdot 1 = 5$.
The **sign rules for dividing fractions** are the same as those for multiplying fractions.	Divide and simplify, if possible: $\dfrac{9}{16} \div (-3)$ $\dfrac{9}{16} \div (-3) = \dfrac{9}{16} \cdot \left(-\dfrac{1}{3}\right)$ Multiply $\frac{9}{16}$ by the reciprocal of -3, which is $-\frac{1}{3}$. $= -\dfrac{9 \cdot 1}{16 \cdot 3}$ Multiply the numerators. Multiply the denominators. Since the fractions have unlike signs, make the answer negative. $= -\dfrac{\overset{1}{3} \cdot 3 \cdot 1}{16 \cdot \underset{1}{3}}$ To simplify, factor 9 as $3 \cdot 3$. Then remove the common factor of 3 from the numerator and denominator. $= -\dfrac{3}{16}$ Multiply the remaining factors in the numerator: $1 \cdot 3 \cdot 1 = 3$. Multiply the remaining factors in the denominator: $16 \cdot 1 = 16$.

Problems that involve forming **equal-sized groups** can be solved by division.	**Sewing.** How many Halloween costumes, which require $\frac{3}{4}$ yard of material, can be made from 6 yards of material?

Since 6 yards of material is to be separated into an unknown number of equal-sized $\frac{3}{4}$-yard pieces, division is indicated.

$$6 \div \frac{3}{4} = \frac{6}{1} \cdot \frac{4}{3}$$ Write 6 as a fraction: $6 = \frac{6}{1}$.
Multiply $\frac{6}{1}$ by the reciprocal of $\frac{3}{4}$, which is $\frac{4}{3}$.

$$= \frac{6 \cdot 4}{1 \cdot 3}$$ Multiply the numerators.
Multiply the denominators.

$$= \frac{2 \cdot \overset{1}{\cancel{3}} \cdot 4}{1 \cdot \underset{1}{\cancel{3}}}$$ To simplify, factor 6 as $2 \cdot 3$. Then remove the common factor of 3 from the numerator and denominator.

$$= \frac{8}{1}$$ Multiply the remaining factors in the numerator.
Multiply the remaining factors in the denominator.

$$= 8$$ Any number divided by 1 is the same number.

The number of Halloween costumes that can be made from 6 yards of material is 8.

REVIEW EXERCISES

39. Find the reciprocal of each number.

   a. $\frac{1}{8}$

   b. $-\frac{11}{12}$

   c. 5

   d. $\frac{8}{7}$

40. Fill in the blanks: To divide two fractions, _____ the first fraction by the _____ of the second fraction.

Divide. Simplify the quotient, if possible.

41. $\frac{1}{6} \div \frac{11}{25}$

42. $-\frac{7}{32} \div \frac{1}{4}$

43. $-\frac{39}{25} \div \left(-\frac{13}{10}\right)$

44. $54 \div \frac{63}{5}$

45. $-\frac{3}{8} \div \frac{1}{4}$

46. $\frac{4}{5} \div \frac{1}{2}$

47. $\frac{2}{3} \div (-120)$

48. $\frac{7}{15} \div \frac{7}{15}$

49. Making jewelry.   How many $\frac{1}{16}$-ounce silver angel pins can be made from a $\frac{3}{4}$-ounce bar of silver?

50. Sewing.   How many pillow cases, which require $\frac{2}{3}$ yard of material, can be made from 20 yards of cotton cloth?

SECTION 3.4 ▶ Adding and Subtracting Fractions

DEFINITIONS AND CONCEPTS	EXAMPLES
To **add (or subtract) fractions that have the same denominator**, add (or subtract) the numerators and write the sum (or difference) over the common denominator. Simplify the result, if possible.	Add: $\dfrac{3}{16} + \dfrac{5}{16}$

$$\frac{3}{16} + \frac{5}{16} = \frac{3 + 5}{16}$$ Add the numerators and write the sum over the common denominator 16.

$$= \frac{8}{16}$$ The resulting fraction can be simplified.

$$= \frac{\overset{1}{\cancel{8}}}{2 \cdot \underset{1}{\cancel{8}}}$$ To simplify, factor 16 as $2 \cdot 8$. Then remove the common factor of 8 from the numerator and denominator.

$$= \frac{1}{2}$$ Multiply the remaining factors in the denominator: $2 \cdot 1 = 2$.

Adding and subtracting fractions that have different denominators

1. Find the LCD.
2. Rewrite each fraction as an equivalent fraction with the LCD as the denominator. To do so, build each fraction using a form of 1 that involves any factors needed to obtain the LCD.
3. Add or subtract the numerators and write the sum or difference over the LCD.
4. Simplify the result, if possible.

Subtract: $\dfrac{4}{7} - \dfrac{1}{3}$

Since the smallest number the denominators 7 and 3 divide exactly is 21, the LCD is 21.

$$\frac{4}{7} - \frac{1}{3} = \frac{4}{7} \cdot \frac{3}{3} - \frac{1}{3} \cdot \frac{7}{7}$$ To build $\frac{4}{7}$ and $\frac{1}{3}$ so that their denominators are 21, multiply each by a form of 1.

$$= \frac{12}{21} - \frac{7}{21}$$ Multiply the numerators. Multiply the denominators. The denominators are now the same.

$$= \frac{12 - 7}{21}$$ Subtract the numerators and write the difference over the common denominator 21.

$$= \frac{5}{21}$$ This fraction is in simplest form.

The **least common denominator (LCD)** of a set of fractions is the **least common multiple (LCM)** of the denominators of the fractions. Two ways to find the LCM of the denominators are as follows:

- Write the multiples of the largest denominator in increasing order, until one is found that is divisible by the other denominators.
- Prime factor each denominator. The LCM is a product of prime factors, where each factor is used the greatest number of times it appears in any one factorization.

Add and simplify: $\dfrac{9}{20} + \dfrac{7}{15}$

To find the LCD, find the prime factorization of both denominators and use each prime factor the *greatest* number of times it appears in any one factorization:

$$\left.\begin{array}{l} 20 = (2 \cdot 2) \cdot 5 \\ 15 = 3 \cdot 5 \end{array}\right\} \text{LCD} = 2 \cdot 2 \cdot 3 \cdot 5 = 60$$

$$\frac{9}{20} + \frac{7}{15} = \frac{9}{20} \cdot \frac{3}{3} + \frac{7}{15} \cdot \frac{4}{4}$$ To build $\frac{9}{20}$ and $\frac{7}{15}$ so that their denominators are 60, multiply each by a form of 1.

$$= \frac{27}{60} + \frac{28}{60}$$ Multiply the numerators. Multiply the denominators. The denominators are now the same.

$$= \frac{27 + 28}{60}$$ Add the numerators and write the sum over the common denominator 60.

$$= \frac{55}{60}$$ This fraction is not in simplest form.

$$= \frac{\overset{1}{\cancel{5}} \cdot 11}{2 \cdot 2 \cdot 3 \cdot \underset{1}{\cancel{5}}}$$ To simplify, prime factor 55 and 60. Then remove the common factor of 5 from the numerator and denominator.

$$= \frac{11}{12}$$ Multiply the remaining factors in the numerator and in the denominator.

Comparing fractions

If two fractions have the **same denominator**, the fraction with the greater numerator is the greater fraction.

If two fractions have **different denominators**, express each of them as an equivalent fraction that has the LCD for its denominator. Then compare numerators.

Which fraction is larger: $\dfrac{11}{18}$ or $\dfrac{7}{18}$?

$$\frac{11}{18} > \frac{7}{18} \text{ because } 11 > 7$$

Which fraction is larger: $\dfrac{2}{3}$ or $\dfrac{3}{4}$?

Build each fraction to have a denominator that is the LCD, 12.

$$\frac{2}{3} = \frac{2}{3} \cdot \frac{4}{4} = \frac{8}{12} \qquad\qquad \frac{3}{4} = \frac{3}{4} \cdot \frac{3}{3} = \frac{9}{12}$$

Since $9 > 8$, it follows that $\dfrac{9}{12} > \dfrac{8}{12}$ and therefore, $\dfrac{3}{4} > \dfrac{2}{3}$.

REVIEW EXERCISES

Add or subtract and simplify, if possible.

51. $\frac{2}{7} + \frac{3}{7}$

52. $\frac{3}{4} - \frac{1}{4}$

53. $\frac{7}{8} + \frac{3}{8}$

54. $-\frac{3}{5} - \frac{3}{5}$

55. a. Add the fractions represented by the figures below.

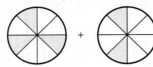

b. Subtract the fractions represented by the figures below.

56. Fill in the blanks. Use the prime factorizations below to find the least common denominator for fractions with denominators of 45 and 30.

$$\left.\begin{array}{l} 45 = 3 \cdot 3 \cdot 5 \\ 30 = 2 \cdot 3 \cdot 5 \end{array}\right\} \text{LCD} = \boxed{} \cdot \boxed{} \cdot \boxed{} \cdot \boxed{} = \boxed{}$$

Add or subtract and simplify, if possible.

57. $\frac{1}{6} + \frac{2}{3}$

58. $-\frac{2}{5} - \frac{3}{8}$

59. $\frac{5}{24} + \frac{3}{16}$

60. $3 - \frac{1}{7}$

61. $-\frac{19}{18} + \frac{5}{12}$

62. $\frac{17}{20} - \frac{4}{15}$

63. $-6 + \frac{13}{6}$

64. $\frac{1}{3} + \frac{1}{4} + \frac{1}{5}$

65. Machine shops.   How much must be milled off the $\frac{3}{4}$-inch-thick steel rod below so that the collar will slip over the end of it?

Steel rod

66. Polls.   A group of adults were asked to rate the transportation system in their community. The results are shown below in a circle graph. What fraction of the group responded by saying either excellent, good, or fair?

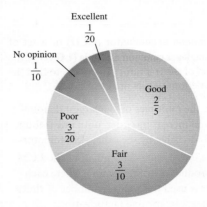

67. Telemarketing.   In the first hour of work, a telemarketer made 2 sales out of 9 telephone calls. In the second hour, she made 3 sales out of 11 calls. During which hour was the rate of sales to calls better?

68. Cameras.   When the shutter of a camera stays open longer than $\frac{1}{125}$ second, any movement of the camera will probably blur the picture. With this in mind, if a photographer is taking a picture of a fast-moving object, should she select a shutter speed of $\frac{1}{60}$ or $\frac{1}{250}$?

SECTION 3.5 ▶ Multiplying and Dividing Mixed Numbers

DEFINITIONS AND CONCEPTS	EXAMPLES
A **mixed number** is the sum of a whole number and a proper fraction.	$$2\frac{3}{4} \quad = \quad 2 \quad + \quad \frac{3}{4}$$ Mixed number Whole-number part Fractional part
There is a relationship between **mixed numbers and improper fractions** that can be seen using shaded regions.	Each disk represents one whole. $$2\frac{3}{4} = \frac{11}{4}$$
To write a mixed number as an improper fraction: 1. Multiply the denominator of the fraction by the whole-number part. 2. Add the numerator of the fraction to the result from Step 1. 3. Write the sum from Step 2 over the original denominator.	Write $3\frac{4}{5}$ as an improper fraction. Step 2: Add $$3\frac{4}{5} = \frac{5 \cdot 3 + 4}{5} = \frac{15 + 4}{5} = \frac{19}{5}$$ Step 1: Multiply Step 3: Use the same denominator Thus, $3\frac{4}{5} = \frac{19}{5}$.
To write an improper fraction as a mixed number: 1. Divide the numerator by the denominator to obtain the whole-number part. 2. The remainder over the divisor is the fractional part.	Write $\frac{47}{6}$ as a mixed number. $$\begin{array}{r} 7 \\ 6\overline{)47} \\ -42 \\ \hline 5 \end{array}$$ ← The whole-number part is 7. ← Write the remainder 5 over the divisor 6 to get the fractional part. Thus, $\frac{47}{6} = 7\frac{5}{6}$.
Fractions and mixed numbers can be **graphed** on a number line.	Graph $-3\frac{1}{3}$, $1\frac{1}{4}$, $\frac{18}{5}$, and $-\frac{7}{8}$ on a number line.

To **multiply mixed numbers**, first change the mixed numbers to improper fractions. Then perform the multiplication of the fractions. Write the result as a mixed number or whole number in simplest form.

Multiply and simplify: $10\frac{1}{2} \cdot 1\frac{1}{6}$

$$10\frac{1}{2} \cdot 1\frac{1}{6} = \frac{21}{2} \cdot \frac{7}{6}$$ Write $10\frac{1}{2}$ and $1\frac{1}{6}$ as improper fractions.

$$= \frac{21 \cdot 7}{2 \cdot 6}$$ Use the rule for multiplying two fractions. Multiply the numerators. Multiply the denominators.

$$= \frac{\overset{1}{\cancel{3}} \cdot 7 \cdot 7}{2 \cdot 2 \cdot \underset{1}{\cancel{3}}}$$ To simplify, factor 21 as $3 \cdot 7$ and 6 as $2 \cdot 3$. Then remove the common factor of 3 from the numerator and denominator.

$$= \frac{49}{4}$$ Multiply the remaining factors in the numerator and in the denominator. The result is an improper fraction.

$$= 12\frac{1}{4}$$ Write the improper fraction $\frac{49}{4}$ as a mixed number.

$$\begin{array}{r} 12 \\ 4\overline{)49} \\ \underline{-4} \\ 09 \\ \underline{-8} \\ 1 \end{array}$$

To **divide mixed numbers,** first change the mixed numbers to improper fractions. Then perform the division of the fractions. Write the result as a mixed number or whole number in simplest form.

Divide and simplify: $5\frac{2}{3} \div \left(-3\frac{7}{9}\right)$

$$5\frac{2}{3} \div \left(-3\frac{7}{9}\right) = \frac{17}{3} \div \left(-\frac{34}{9}\right)$$ Write $5\frac{2}{3}$ and $3\frac{7}{9}$ as improper fractions.

$$= \frac{17}{3}\left(-\frac{9}{34}\right)$$ Multiply $\frac{17}{3}$ by the reciprocal of $-\frac{34}{9}$, which is $-\frac{9}{34}$.

$$= -\frac{17 \cdot 9}{3 \cdot 34}$$ Multiply the numerators. Multiply the denominators. Since the fractions have unlike signs, make the answer negative.

$$= -\frac{\overset{1}{\cancel{17}} \cdot \overset{1}{\cancel{3}} \cdot 3}{\underset{1}{\cancel{3}} \cdot 2 \cdot \underset{1}{\cancel{17}}}$$ To simplify, factor 9 as $3 \cdot 3$ and 34 as $2 \cdot 17$. Then remove the common factors of 3 and 17 from the numerator and denominator.

$$= -\frac{3}{2}$$ Multiply the remaining factors in the numerator and in the denominator. The result is a negative improper fraction.

$$= -1\frac{1}{2}$$ Write the negative improper fraction $-\frac{3}{2}$ as a negative mixed number.

REVIEW EXERCISES

69. In the illustration below, each triangular region outlined in black represents one whole. Write a mixed number and an improper fraction to represent what is shaded.

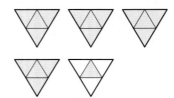

70. Graph $-2\frac{2}{3}$, $\frac{8}{9}$, $-\frac{3}{4}$, and $\frac{59}{24}$ on a number line.

$$\xleftarrow{\hspace{0.5em}|\quad|\quad|\quad|\quad|\quad|\quad|\quad|\quad|\quad|\quad|\hspace{0.5em}}\rightarrow$$
$$-5\ -4\ -3\ -2\ -1\ \ 0\ \ 1\ \ 2\ \ 3\ \ 4\ \ 5$$

Write each improper fraction as a mixed number or a whole number.

71. $\dfrac{16}{5}$

72. $-\dfrac{47}{12}$

73. $\dfrac{51}{3}$

74. $\dfrac{14}{6}$

Write each mixed number as an improper fraction.

75. $9\frac{3}{8}$

76. $-2\frac{1}{5}$

77. $3\frac{11}{14}$

78. $1\frac{99}{100}$

Multiply or divide and simplify, if possible.

79. $1\frac{2}{5} \cdot 1\frac{1}{2}$

80. $-3\frac{1}{2} \div 3\frac{2}{3}$

81. $-6\left(-6\frac{2}{3}\right)$

82. $8 \div 3\frac{1}{5}$

83. $-11\frac{1}{5} \div \left(-\frac{7}{10}\right)$

84. $5\frac{2}{3}\left(-7\frac{1}{5}\right)$

85. $\left(-2\frac{3}{4}\right)^2$

86. $1\frac{5}{16} \cdot 1\frac{7}{9} \cdot 2\frac{2}{3}$

87. Photography. Each leg of a camera tripod can be extended to become $5\frac{1}{2}$ times its original length. If a leg is originally $8\frac{3}{4}$ inches long, how long will it become when it is completely extended?

88. Pet doors. Find the area of the opening provided by the rectangular-shaped pet door shown below.

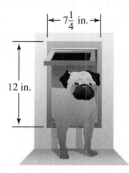

$7\frac{1}{4}$ in.

12 in.

89. Printing. It takes a color copier $2\frac{1}{4}$ minutes to print a movie poster. How many posters can be printed in 90 minutes?

90. Storm damage. A truck can haul $7\frac{1}{2}$ tons of trash in one load. How many loads would it take to haul away $67\frac{1}{2}$ tons from a hurricane cleanup site?

SECTION 3.6 ▶ **Adding and Subtracting Mixed Numbers**

DEFINITIONS AND CONCEPTS	EXAMPLES
To add (or subtract) mixed numbers, we can change each to an improper fraction and use the method of Section 3.4.	Add: $3\frac{1}{2} + 1\frac{3}{5}$

$3\frac{1}{2} + 1\frac{3}{5} = \frac{7}{2} + \frac{8}{5}$ Write $3\frac{1}{2}$ and $1\frac{3}{5}$ as mixed numbers.

$= \frac{7}{2} \cdot \frac{5}{5} + \frac{8}{5} \cdot \frac{2}{2}$ To build $\frac{7}{2}$ and $\frac{8}{5}$ so that their denominators are 10, multiply both by a form of 1.

$= \frac{35}{10} + \frac{16}{10}$ Multiply the numerators. Multiply the denominators.

$= \frac{51}{10}$ Add the numerators and write the sum over the common denominator 10.

$= 5\frac{1}{10}$ To write the improper fraction $\frac{51}{10}$ as a mixed number, divide 51 by 10.

To add (or subtract) mixed numbers, we can also write them in **vertical form** and add (or subtract) the whole-number parts and the fractional parts separately.

Add: $42\dfrac{1}{3} + 89\dfrac{6}{7}$

Build to get the LCD, 21.
Add the fractions.
Add the whole numbers.

$$42\dfrac{1}{3} = \quad 42\dfrac{1}{3}\cdot\dfrac{7}{7} = \quad 42\dfrac{7}{21} = \quad \overset{1}{42}\dfrac{7}{21}$$
$$+\,89\dfrac{6}{7} = +\,89\dfrac{6}{7}\cdot\dfrac{3}{3} = +\,89\dfrac{18}{21} = +\,89\dfrac{18}{21}$$
$$\dfrac{25}{21} \qquad 131\dfrac{25}{21}$$

We don't want an improper fraction in the answer.

When we add mixed numbers, sometimes the sum of the fractions is an improper fraction. If that is the case, write the improper fraction as a mixed number and **carry** its whole-number part to the whole-number column.

Write $\dfrac{25}{21}$ as $1\dfrac{4}{21}$, carry the 1 to the whole-number column, and add it to 131 to get 132:

$$131\dfrac{25}{21} = 131 + 1\dfrac{4}{21} = 132\dfrac{4}{21}$$

Subtraction of mixed numbers in vertical form sometimes involves **borrowing**. When the fraction we are subtracting is greater than the fraction we are subtracting it from, borrowing is necessary.

Subtract: $23\dfrac{1}{4} - 17\dfrac{5}{9}$

Build to get the LCD, 36.
Since $\dfrac{20}{36}$ is greater than $\dfrac{9}{36}$, we must borrow from 28.

$$28\dfrac{1}{4} = \quad 28\dfrac{1}{4}\cdot\dfrac{9}{9} = \quad 28\dfrac{9}{36} = \quad \overset{7}{28}\dfrac{9}{36}+\dfrac{36}{36} = \quad \overset{7}{28}\dfrac{45}{36}$$
$$-\,17\dfrac{5}{9} = -\,17\dfrac{5}{9}\cdot\dfrac{4}{4} = -\,17\dfrac{20}{36} = -\,17\dfrac{20}{36} \qquad = -\,17\dfrac{20}{36}$$
$$10\dfrac{25}{36}$$

REVIEW EXERCISES

Add or subtract and simplify, if possible.

91. $1\dfrac{3}{8} + 2\dfrac{1}{5}$

92. $3\dfrac{1}{2} + 2\dfrac{2}{3}$

93. $2\dfrac{5}{6} - 1\dfrac{3}{4}$

94. $3\dfrac{7}{16} - 2\dfrac{1}{8}$

95. $157\dfrac{11}{30} + 98\dfrac{7}{12}$

96. $6\dfrac{3}{14} + 17\dfrac{7}{10}$

97. $33\dfrac{8}{9} + 49\dfrac{1}{6}$

98. $98\dfrac{11}{20} + 14\dfrac{4}{5}$

99. $50\dfrac{5}{8} - 19\dfrac{1}{6}$

100. $375\dfrac{3}{4} - 59$

101. $23\dfrac{1}{3} - 2\dfrac{5}{6}$

102. $39 - 4\dfrac{5}{8}$

103. Painting supplies. In a project to restore a house, painters used $10\dfrac{3}{4}$ gallons of primer, $21\dfrac{1}{2}$ gallons of latex paint, and $7\dfrac{2}{3}$ gallons of enamel. Find the total number of gallons of paint used.

104. Passports. The required dimensions for a passport photograph are shown below. What is the distance from the subject's eyes to the top of the photograph?

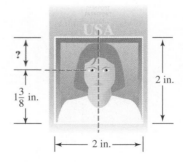

SECTION 3.7 ▶ Order of Operations and Complex Fractions

DEFINITIONS AND CONCEPTS	EXAMPLES
Order of Operations 1. Perform all calculations within parentheses and other grouping symbols following the order listed in Steps 2–4 below, working from the innermost pair of grouping symbols to the outermost pair. 2. Evaluate all exponential expressions. 3. Perform all multiplications and divisions as they occur from left to right. 4. Perform all additions and subtractions as they occur from left to right. When grouping symbols have been removed, repeat Steps 2–4 to complete the calculation. If a fraction bar is present, evaluate the expression above the bar and the expression below the bar separately. Then perform the division indicated by the fraction bar, if possible.	Evaluate: $\left(\dfrac{1}{3}\right)^2 \div \left(\dfrac{3}{4} - \dfrac{1}{3}\right)$

Evaluate: $\left(\dfrac{1}{3}\right)^2 \div \left(\dfrac{3}{4} - \dfrac{1}{3}\right)$

First, we perform the subtraction within the second set of parentheses. (There is no operation to perform within the first set.)

$$\left(\dfrac{1}{3}\right)^2 \div \left(\dfrac{3}{4} - \dfrac{1}{3}\right)$$

$$= \left(\dfrac{1}{3}\right)^2 \div \left(\dfrac{3}{4}\cdot\dfrac{3}{3} - \dfrac{1}{3}\cdot\dfrac{4}{4}\right)$$ Within the parentheses, build each fraction so that its denominator is the LCD 12.

$$= \left(\dfrac{1}{3}\right)^2 \div \left(\dfrac{9}{12} - \dfrac{4}{12}\right)$$ Multiply the numerators. Multiply the denominators.

$$= \left(\dfrac{1}{3}\right)^2 \div \dfrac{5}{12}$$ Subtract the numerators: $9 - 4 = 5$. Write the difference over the common denominator 12.

$$= \dfrac{1}{9} \div \dfrac{5}{12}$$ Evaluate the exponential expression: $\left(\dfrac{1}{3}\right)^2 = \dfrac{1}{3}\cdot\dfrac{1}{3} = \dfrac{1}{9}$.

$$= \dfrac{1}{9} \cdot \dfrac{12}{5}$$ Use the rule for dividing fractions: Multiply the first fraction by the reciprocal of $\dfrac{5}{12}$, which is $\dfrac{12}{5}$.

$$= \dfrac{1\cdot 12}{9\cdot 5}$$ Multiply the numerators. Multiply the denominators.

$$= \dfrac{1\cdot \overset{1}{\cancel{3}}\cdot 4}{\underset{1}{\cancel{3}}\cdot 3\cdot 5}$$ To simplify, factor 12 as $3\cdot 4$ and 9 as $3\cdot 3$. Then remove the common factor of 3 from the numerator and denominator.

$$= \dfrac{4}{15}$$ Multiply the remaining factors in the numerator. Multiply the remaining factors in the denominator.

DEFINITIONS AND CONCEPTS	EXAMPLES
To **evaluate a formula**, we replace its variables (letters) with specific numbers and evaluate the right side using the order of operations rule.	Evaluate: $A = \dfrac{1}{2}h(a + b)$ for $a = 1\dfrac{1}{3}$, $b = 2\dfrac{2}{3}$, and $h = 2\dfrac{4}{5}$.

Evaluate: $A = \dfrac{1}{2}h(a + b)$ for $a = 1\dfrac{1}{3}$, $b = 2\dfrac{2}{3}$, and $h = 2\dfrac{4}{5}$.

$$A = \dfrac{1}{2}h(a + b)$$ This is the given formula.

$$A = \dfrac{1}{2}\left(2\dfrac{4}{5}\right)\left(1\dfrac{1}{3} + 2\dfrac{2}{3}\right)$$ Replace h, a, and b with the given values.

$$A = \dfrac{1}{2}\left(2\dfrac{4}{5}\right)(4)$$ Do the addition within the parentheses.

$$A = \dfrac{1}{2}\left(\dfrac{14}{5}\right)\left(\dfrac{4}{1}\right)$$ To prepare to multiply fractions, write $2\dfrac{4}{5}$ as an improper fraction and 4 as $\dfrac{4}{1}$.

$$A = \dfrac{1\cdot 14\cdot 4}{2\cdot 5\cdot 1}$$ Multiply the numerators. Multiply the denominators.

$$A = \dfrac{1\cdot 14\cdot \overset{1}{\cancel{2}}\cdot 2}{\underset{1}{\cancel{2}}\cdot 5\cdot 1}$$ To simplify, factor 4 as $2\cdot 2$. Then remove the common factor of 2 from the numerator and denominator.

$$A = \dfrac{28}{5}$$ Multiply the remaining factors in the numerator. Multiply the remaining factors in the denominator.

$$A = 5\dfrac{3}{5}$$ Write the improper fraction $\dfrac{28}{5}$ as a mixed number by dividing 28 by 5.

A **complex fraction** is a fraction whose numerator or denominator, or both, contain one or more fractions or mixed numbers.	Complex fractions: $$\dfrac{\dfrac{9}{10}}{\dfrac{27}{5}} \qquad \dfrac{\dfrac{2}{5}-\dfrac{1}{3}}{\dfrac{3}{7}+\dfrac{1}{5}} \qquad \dfrac{-7\dfrac{1}{4}}{2-\dfrac{1}{9}}$$
The method for **simplifying complex fractions** is based on the fact that the main fraction bar indicates division.	Simplify: $\dfrac{\dfrac{9}{10}}{\dfrac{27}{5}}$ $\dfrac{\dfrac{9}{10}}{\dfrac{27}{5}} = \dfrac{9}{10} \div \dfrac{27}{5}$ Write the division indicated by the main fraction bar using a ÷ symbol. $= \dfrac{9}{10} \cdot \dfrac{5}{27}$ Use the rule for dividing fractions: Multiply the first fraction by the reciprocal of $\frac{27}{5}$, which is $\frac{5}{27}$. $= \dfrac{9 \cdot 5}{10 \cdot 27}$ Multiply the numerators. Multiply the denominators. $= \dfrac{\overset{1}{\cancel{9}} \cdot \overset{1}{\cancel{5}}}{2 \cdot \underset{1}{\cancel{5}} \cdot 3 \cdot \underset{1}{\cancel{9}}}$ To simplify, factor 10 as 2 · 5 and 27 as 3 · 9. Then remove the common factors of 9 and 5 from the numerator and denominator. $= \dfrac{1}{6}$ Multiply the remaining factors in the numerator. Multiply the remaining factors in the denominator.
To **simplify a complex fraction**: 1. Add or subtract in the numerator and/or denominator so that the numerator is a single fraction and the denominator is a single fraction. 2. Perform the indicated division by multiplying the numerator of the complex fraction by the reciprocal of the denominator. 3. Simplify the result, if possible.	Simplify: $\dfrac{\dfrac{2}{5}-\dfrac{1}{3}}{\dfrac{3}{7}+\dfrac{1}{5}}$ $\dfrac{\dfrac{2}{5}-\dfrac{1}{3}}{\dfrac{3}{7}+\dfrac{1}{5}} = \dfrac{\dfrac{2}{5}\cdot\dfrac{3}{3}-\dfrac{1}{3}\cdot\dfrac{5}{5}}{\dfrac{3}{7}\cdot\dfrac{5}{5}+\dfrac{1}{5}\cdot\dfrac{7}{7}}$ In the numerator, build each fraction so that each has a denominator of 15. In the denominator, build each fraction so that each has a denominator of 35. $= \dfrac{\dfrac{6}{15}-\dfrac{5}{15}}{\dfrac{15}{35}+\dfrac{7}{35}}$ Multiply the numerators. Multiply the denominators. $= \dfrac{\dfrac{1}{15}}{\dfrac{22}{35}}$ Subtract the numerators and write the difference over the common denominator 15. Add the numerators and write the sum over the common denominator 35. $= \dfrac{1}{15} \div \dfrac{22}{35}$ Write the division indicated by the main fraction bar using a ÷ symbol. $= \dfrac{1}{15} \cdot \dfrac{35}{22}$ Use the rule for dividing fractions: Multiply the first fraction by the reciprocal of $\frac{22}{35}$, which is $\frac{35}{22}$. $= \dfrac{1 \cdot 35}{15 \cdot 22}$ Multiply the numerators. Multiply the denominators. $= \dfrac{1 \cdot \overset{1}{\cancel{5}} \cdot 7}{3 \cdot \underset{1}{\cancel{5}} \cdot 22}$ To simplify, factor 35 as 5 · 7 and 15 as 3 · 5. Then remove the common factor of 5 from the numerator and denominator. $= \dfrac{7}{66}$ Multiply the remaining factors in the numerator. Multiply the remaining factors in the denominator.

REVIEW EXERCISES

Evaluate each expression.

105. $\dfrac{3}{4} + \left(-\dfrac{1}{3}\right)^2\left(\dfrac{5}{4}\right)$

106. $\left(\dfrac{2}{3} \div \dfrac{16}{9}\right) - \left(1\dfrac{2}{3} \cdot \dfrac{1}{15}\right)$

107. $\left(\dfrac{11}{5} - 1\dfrac{2}{3}\right) - \left(-\dfrac{4}{9} \cdot 18\right)$

108. $\left|-\dfrac{9}{16} \div 2\dfrac{1}{4}\right| + \left(-3\dfrac{7}{8}\right)$

Simplify each complex fraction.

109. $\dfrac{\dfrac{3}{5}}{-\dfrac{17}{20}}$

110. $\dfrac{4 - \dfrac{2}{7}}{4\dfrac{1}{7}}$

111. $\dfrac{\dfrac{2}{3} - \dfrac{1}{6}}{-\dfrac{3}{4} - \dfrac{1}{2}}$

112. $\dfrac{5\dfrac{1}{4}}{\dfrac{7}{4} + \left(-\dfrac{1}{3}\right)}$

113. Subtract $4\dfrac{1}{8}$ from the sum of $5\dfrac{1}{5}$ and $1\dfrac{1}{2}$.

114. Add $12\dfrac{11}{16}$ to the difference of $4\dfrac{5}{8}$ and $3\dfrac{1}{4}$.

115. Evaluate the formula $A = \dfrac{1}{2}h(a + b)$ for $a = 1\dfrac{1}{8}$, $b = 4\dfrac{7}{8}$, and $h = 2\dfrac{7}{9}$.

116. Evaluate the formula $P = 2\ell + 2w$ for $\ell = 2\dfrac{1}{3}$ and $w = 3\dfrac{1}{4}$.

117. Dermatology. A dermatologist mixes $1\dfrac{1}{2}$ ounces of cucumber extract, $2\dfrac{2}{3}$ ounces of aloe vera cream, and $\dfrac{3}{4}$ ounce of vegetable glycerin to make his own brand of anti-wrinkle cream. He packages it in $\dfrac{5}{6}$-ounce tubes. How many full tubes can be produced using this formula? How much cream is left over?

118. Guitar design. Find the missing dimension on the vintage 1962 Stratocaster body shown below.

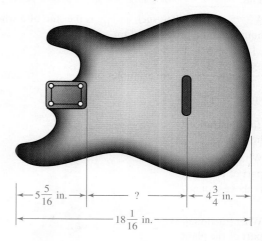

$5\dfrac{5}{16}$ in.   ? $4\dfrac{3}{4}$ in.

$18\dfrac{1}{16}$ in.

1. Fill in the blanks.

a. For the fraction $\frac{6}{7}$, the _____ is 6 and the _____ is 7.

b. Two fractions are _____ if they represent the same number.

c. A fraction is in _____ form when the numerator and denominator have no common factors other than 1.

d. To _____ a fraction, we remove common factors of the numerator and denominator.

e. The _____ of $\frac{4}{5}$ is $\frac{5}{4}$.

f. A _____ number, such as $1\frac{9}{16}$, is the sum of a whole number and a proper fraction.

g. $\dfrac{\frac{1}{8}}{\frac{7}{12}}$ and $\dfrac{\frac{3}{4}+\frac{1}{3}}{\frac{5}{12}-\frac{1}{4}}$ are examples of _____ fractions.

2. See the illustration to the right.

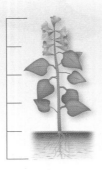

a. What fractional part of the plant is above ground?

b. What fractional part of the plant is below ground?

3. Each region outlined in black represents one whole. Write an improper fraction and a mixed number to represent the shaded portion.

4. Graph $2\frac{4}{5}$, $-\frac{2}{5}$, $-1\frac{1}{7}$, and $\frac{7}{6}$ on a number line.

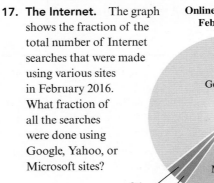

5. Are $\frac{1}{3}$ and $\frac{5}{15}$ equivalent?

6. a. Express $\frac{4}{5}$ as an equivalent fraction with denominator 45.

b. Express $\frac{7}{8}$ as an equivalent fraction with denominator 24.

7. Simplify each fraction, if possible.

a. $\dfrac{0}{15}$

b. $\dfrac{9}{0}$

8. Simplify each fraction.

a. $\dfrac{27}{36}$

b. $\dfrac{72}{180}$

9. Add and simplify, if possible: $\dfrac{3}{16} + \dfrac{7}{16}$

10. Multiply and simplify, if possible: $-\dfrac{3}{4}\left(\dfrac{1}{5}\right)$

11. Divide and simplify, if possible: $\dfrac{4}{3} \div \dfrac{2}{9}$

12. Subtract and simplify, if possible: $\dfrac{11}{12} - \dfrac{11}{30}$

13. Add and simplify, if possible: $-\dfrac{3}{7} + 2$

14. Multiply and simplify, if possible: $\dfrac{9}{10}\left(-\dfrac{4}{15}\right)\left(-\dfrac{25}{18}\right)$

15. Which fraction is larger: $\dfrac{8}{9}$ or $\dfrac{9}{10}$?

16. Coffee drinkers. Two-fifths of 100 adults surveyed said they started their morning with a cup of coffee. Of the 100, how many would this be?

17. The Internet. The graph shows the fraction of the total number of Internet searches that were made using various sites in February 2016. What fraction of all the searches were done using Google, Yahoo, or Microsoft sites?

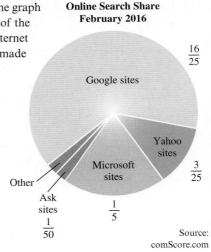

Online Search Share February 2016

Google sites — $\frac{16}{25}$

Yahoo sites — $\frac{3}{25}$

Microsoft sites — $\frac{1}{5}$

Ask sites — $\frac{1}{50}$

Other

Source: comScore.com

18. a. Write $\frac{55}{6}$ as a mixed number.

  b. Write $1\frac{18}{21}$ as an improper fraction.

19. Find the sum of $157\frac{3}{10}$ and $103\frac{13}{15}$. Simplify the result.

20. Subtract and simplify, if possible: $67\frac{1}{4} - 29\frac{5}{6}$

21. Divide and simplify, if possible: $6\frac{1}{4} \div 3\frac{3}{4}$

22. Boxing. Two of the greatest heavyweight boxers of all time are Muhammad Ali and George Foreman. Refer to the "Tale of the Tape" comparison shown below.

  a. Which fighter weighed more? How much more?

  b. Which fighter had the larger waist measurement? How much larger?

  c. Which fighter had the larger forearm measurement? How much larger?

Tale of the Tape		
Muhammad Ali		**George Foreman**
6-3	Height	6-4
210½ lb	Weight	250 lb
82 in.	Reach	79 in.
43 in.	Chest (Normal)	48 in.
45½ in.	Chest (Expanded)	50 in.
34 in.	Waist	39½ in.
12½ in.	Fist	13½ in.
15 in.	Forearm	14¾ in.

Source: The International Boxing Hall of Fame

23. Evaluate the formula $P = 2\ell + 2w$ for $\ell = \frac{1}{3}$ and $w = \frac{1}{9}$.

24. Sports contracts. A basketball player signed a nine-year contract for $13\frac{1}{2}$ million. How much is this per year?

25. Sewing. When cutting material for a $10\frac{1}{2}$-inch-wide placemat, a seamstress allows $\frac{5}{8}$ inch at each end for a hem, as shown below. How wide should the material be cut to make a placemat?

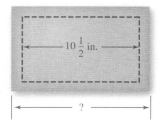

26. Find the perimeter and the area of the triangle shown to the right.

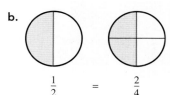

20 in. $22\frac{2}{3}$ in. $10\frac{2}{3}$ in.

27. Nutrition. A box of Tic Tacs contains 40 of the $1\frac{1}{2}$-calorie breath mints. How many calories are there in a box of Tic Tacs?

28. Cooking. How many servings are there in an 8-pound roast, if the suggested serving size is $\frac{2}{3}$ pound?

29. Evaluate:

$$\left(\frac{2}{3} \cdot \frac{5}{16}\right) - \left(-1\frac{3}{5} \div 4\frac{4}{5}\right)$$

30. Evaluate: $\left(\frac{1}{2}\right)^3 \div \left(\frac{3}{4} - \frac{1}{3}\right)$

31. Simplify:

$$\frac{\dfrac{5}{6}}{\dfrac{7}{8}}$$

32. Simplify:

$$\frac{\dfrac{1}{2} + \dfrac{1}{3}}{-\dfrac{1}{6} - \dfrac{1}{3}}$$

33. Explain what is meant when we say, "The product of any number and its reciprocal is 1." Give an example.

34. Explain each mathematical concept that is shown below.

  a. $\dfrac{6}{8} = \dfrac{\overset{1}{\cancel{2}} \cdot 3}{\underset{1}{\cancel{2}} \cdot 4} = \dfrac{3}{4}$

  b.

  (circle divided in half, shaded) $\dfrac{1}{2}$ = $\dfrac{2}{4}$ (circle divided in quarters, shaded)

  c. $\dfrac{3}{5} = \dfrac{3}{5} \cdot \dfrac{4}{4} = \dfrac{12}{20}$

1. Consider the number 5,896,619. [Section 1.1]

   a. What digit is in the millions column?

   b. What is the place value of the digit 8?

   c. Round to the nearest hundred.

   d. Round to the nearest ten thousand.

2. **Banks.** In 2015, the world's largest bank, with total assets of $3,616,390,000,000 was the Deutsche Bank of Germany. In what place-value column is the digit 1? (Source: gfmag.com) [Section 1.1]

3. **Population.** Rank the following counties in order, from greatest to least population. [Section 1.1]

County	2015 Population
Dallas County, TX	2,553,385
Kings County, NY	2,636,735
Miami-Dade County, FL	2,693,117
San Bernardino County, CA	2,128,133
Queens County, NY	2,339,150
Riverside, CA	2,361,026

   (Source: *The World Almanac and Book of Facts*, 2016)

4. **Pool construction.** Refer to the rectangular swimming pool shown below.

   a. Find the perimeter of the pool. [Section 1.2]

   b. Find the area of the pool's surface. [Section 1.4]

   150 ft
   75 ft

5. Add: 7,897 [Section 1.2]
   6,909
   1,812
   +14,378

6. Subtract 3,456 from 20,000. Check the result. [Section 1.3]

7. **Sheets of stickers.** There are twenty rows of twelve gold stars on one sheet of stickers. If a packet contains ten sheets, how many stars are there in one packet? [Section 1.4]

8. Multiply: 5,345 [Section 1.4]
   × 56

9. Divide: $35\overline{)34,685}$. Check the result. [Section 1.5]

10. **Discount Lodging.** A hotel is offering rooms that normally go for $119 per night for only $79 a night. How many dollars will a traveler save if she stays in such a room for 4 nights? [Section 1.6]

11. List factors of 24, from least to greatest. [Section 1.7]

12. Find the prime factorization of 450. [Section 1.7]

13. Find the LCM of 16 and 20. [Section 1.8]

14. Find the GCF of 63 and 84. [Section 1.8]

15. Evaluate: $15 + 5[12 - (2^2 + 4)]$ [Section 1.9]

16. **Real estate.** A homeowner, wishing to sell his house, had it appraised by three different real estate agents. The appraisals were: $158,000, $163,000, and $147,000. He decided to use the mean of the appraisals as the listing price. For what amount was the home listed? [Section 1.9]

17. Write the set of integers. [Section 2.1]

18. Is the statement $-9 \le -8$ true or false? [Section 2.1]

19. Find the sum of -20, 6, and -1. [Section 2.2]

20. Subtract: $-50 - (-60)$ [Section 2.3]

21. **Gold mining.** An elevator lowers gold miners from the ground-level entrance to different depths in the mine. The elevator stops every 25 vertical feet to let off miners. At what depth do the miners work if they get off the elevator at the 8th stop? [Section 2.4]

22. **Temperature drop.** During a five-hour period, the temperature steadily dropped 55°F. By how many degrees did the temperature change each hour? [Section 2.5]

Evaluate each expression. [Section 2.6]

23. $7(8) + (-3)(-6)(-2)$

24. $(-2)^3 - 3^3$

25. $-5 + 3|-4 - (-6)|$

26. $\dfrac{2(3^2 - 4^2)}{-2(3) - 1}$

Simplify each fraction. [Section 3.1]

27. $\dfrac{21}{28}$

28. $\dfrac{40}{16}$

Perform each operation.

29. $\dfrac{6}{5}\left(-\dfrac{2}{3}\right)$ [Section 3.2]

30. $\dfrac{14}{8} \div \dfrac{7}{2}$ [Section 3.3]

31. $\dfrac{2}{3} + \dfrac{3}{4}$ [Section 3.4]

32. $\dfrac{4}{7} - \dfrac{3}{5}$ [Section 3.4]

33. Shaving. Advertisements for an electric shaver claim that men can shave in one-third of the time it takes them using a razor. If a man normally spends 90 seconds shaving using a razor, how long will it take him if he uses the electric shaver? [Section 3.2]

34. Fire hazards. Two terminals in an electrical switch were so close that electricity could jump the gap and start a fire. The illustration below shows a newly designed switch that will keep this from happening. By how much was the distance between the ground terminal and the hot terminal increased? [Section 3.4]

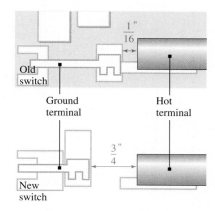

35. Write $\frac{75}{7}$ as a mixed number. [Section 3.5]

36. Write $-6\frac{5}{8}$ as an improper fraction. [Section 3.5]

Perform each operation. Simplify, if possible.

37. $2\frac{2}{5}\left(3\frac{1}{12}\right)$ [Section 3.5]

38. $15\frac{1}{3} \div 2\frac{2}{9}$ [Section 3.5]

39. $4\frac{2}{3} + 5\frac{1}{4}$ [Section 3.6]

40. $14\frac{2}{5} - 8\frac{2}{3}$ [Section 3.6]

41. Lumber. As shown below, 2-by-4's from the lumber yard do not really have dimensions of 2 inches by 4 inches. How wide and how high is the stack of 2-by-4's in the illustration? [Section 3.5]

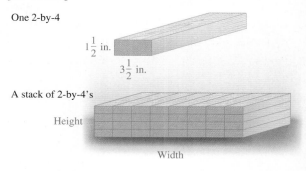

One 2-by-4

$1\frac{1}{2}$ in.

$3\frac{1}{2}$ in.

A stack of 2-by-4's

Height

Width

42. Gas stations. How much gasoline is left in a 500-gallon storage tank if $225\frac{3}{4}$ gallons have been pumped out of it? [Section 3.6]

43. Find the perimeter of the triangle shown below. [Section 3.6]

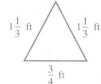

$1\frac{1}{3}$ ft $1\frac{1}{3}$ ft

$\frac{3}{4}$ ft

44. Evaluate: $\left(-\frac{3}{4} \cdot \frac{9}{16}\right) + \left(\frac{1}{2} - \frac{1}{8}\right)$ [Section 3.7]

45. Simplify: $\dfrac{\frac{2}{3}}{\frac{4}{5}}$ [Section 3.7]

46. Simplify: $\dfrac{\frac{3}{7} + \left(-\frac{1}{2}\right)}{1\frac{3}{4}}$ [Section 3.7]

4 Decimals

from **Campus to Careers**

Home Health Aide

Home health aides provide personalized care to the elderly and the disabled in the patient's own home. They help their patients take medicine, eat, dress, and bathe. Home health aides need to have a good number sense. They must accurately take the patient's temperature, pulse, and blood pressure, and monitor the patient's calorie intake and sleeping schedule.

In **Problem 105** of **Study Set 4.2**, you will see how a home health aide uses decimal addition and subtraction to chart a patient's temperature.

JOB TITLE:
Home Health Aide

EDUCATION:
Successful completion of a home health aide training program as required by state law or federal regulation.

JOB OUTLOOK:
Excellent due to rapid employment growth and high replacement needs.

ANNUAL EARNINGS:
The average (median) salary in 2015 was $21,920.

FOR MORE INFORMATION:
www.bls.gov/oes/current/oes311011.htm

CHAPTER OUTLINE

OBJECTIVES

1 Identify the place value of a digit in a decimal number.

2 Write decimals in expanded form.

3 Read decimals and write them in standard form.

4 Compare decimals using inequality symbols.

5 Graph decimals on a number line.

6 Round decimals.

7 Read tables and graphs involving decimals.

SECTION 4.1 An Introduction to Decimals

The place-value system for whole numbers that was introduced in Section 1.1 can be extended to create the **decimal numeration system**. Numbers written using **decimal notation** are often simply called **decimals**. They are used in measurement because it is easy to put them in order and compare them. And as you probably know, our money system is based on decimals.

The decimal 1,537.6 on the odometer represents the distance, in miles, that the car has traveled.

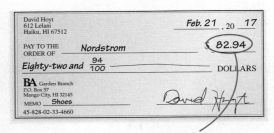

The decimal 82.94 represents the amount of the check, in dollars.

OBJECTIVE 1 **Identify the place value of a digit in a decimal number.**

Like fraction notation, decimal notation is used to represent part of a whole. However, when writing a number in decimal notation, we don't use a fraction bar, nor is a denominator shown. For example, consider the rectangular region below that has 1 of 10 equal parts colored red. We can use the fraction $\frac{1}{10}$ or the decimal 0.1 to describe the amount of the figure that is shaded. Both are read as "one-tenth," and we can write:

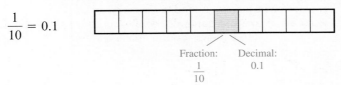

$$\frac{1}{10} = 0.1$$

Fraction: $\frac{1}{10}$

Decimal: 0.1

The square region below has 1 of 100 equal parts colored red. We can use the fraction $\frac{1}{100}$ or the decimal 0.01 to describe the amount of the figure that is shaded. Both are read as "one one-hundredth," and we can write:

$$\frac{1}{100} = 0.01$$

Fraction: $\frac{1}{100}$

Decimal: 0.01

Decimals are written by entering the digits 0, 1, 2, 3, 4, 5, 6, 7, 8, and 9 into place-value columns that are separated by a **decimal point**. The following **place-value chart** shows the names of the place-value columns. Those to the left of the decimal point form the **whole-number part** of the decimal number; they have the familiar names ones, tens, hundreds, and so on. The columns to the right of the decimal point form the **fractional part**. Their place-value names are similar to those in the whole-number part, but they end in "*ths*." Notice that there is no one*ths* place in the chart.

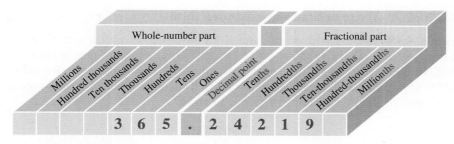

The decimal 365.24219, entered in the place-value chart above, represents the number of days it takes Earth to make one full orbit around the sun. We say that the decimal is written in **standard form** (also called **standard notation**). Each of the 2's in 365.24219 has a different place value because of its position. The place value of the red 2 is two tenths. The place value of the blue 2 is two thousandths.

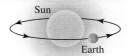

EXAMPLE 1   Consider the decimal number: 2,864.709531

a. What is the place value of the digit 5?

b. Which digit tells the number of millionths?

Strategy We will locate the decimal point in 2,864.709531. Then, moving to the right, we will name each column (tenths, hundredths, and so on) until we reach 5.

WHY It's easier to remember the names of the columns if you begin at the decimal point and move to the right.

Solution

a. 2,864.709531   Say "Tenths, hundredths, thousandths, ten-thousandths" as you
                     move from column to column.

5 ten-thousandths is the place value of the digit 5.

b. 2,864.709531   Say "Tenths, hundredths, thousandths, ten-thousandths,
                     hundred-thousandths, millionths" as you move from column to column.

The digit 1 is in the millionths column.

Self Check 1

Consider the decimal number: 56,081.639724

a. What is the place value of the digit 9?

b. Which digit tells the number of hundred-thousandths?

Now Try Problem 17

We can write a whole number in decimal notation by placing a decimal point immediately to its right and then entering a zero, or zeros, to the right of the decimal point. For example,

$$99 \quad = \quad 99.0 \quad = \quad 99.00 \quad \text{Because } 99 = 99\frac{0}{10} = 99\frac{00}{100}.$$

A whole number   Place a decimal point here and enter
                  a zero, or zeros, to the right of it.

When there is no whole-number part of a decimal, we can show that by entering a zero directly to the left of the decimal point. For example,

$$.83 \quad = \quad 0.83 \quad \text{Because } \frac{83}{100} = 0\frac{83}{100}.$$

No whole-number part   Enter a zero here, if desired.

Caution! We *do not* separate groups of three digits on the right side of the decimal point with commas as we do on the left side. For example, it would be incorrect to write:

2,864.709,531

Negative decimals are used to describe many situations that arise in everyday life, such as temperatures below zero and the balance in a checking account that is overdrawn. For example, the coldest natural temperature ever recorded on Earth was −128.6°F at the Russian Vostok Station in Antarctica on July 21, 1983.

OBJECTIVE 2 **Write decimals in expanded form.**

The decimal 4.458, entered in the place-value chart below, represents the time (in seconds) that it took women's record holder Melanie Troxel to cover a quarter mile in her top-fuel dragster. Notice that the place values of the columns for the whole-number part are 1, 10, 100, 1,000, and so on. We learned in Section 1.1 that the value of each of those columns is 10 times greater than the column directly to its right.

Whole-number part						Fractional part			
Hundred thousands	Ten thousands	Thousands	Hundreds	Tens	Ones	Decimal point · Tenths	Hundredths	Thousandths	Ten-thousandths / Hundred-thousandths
					4	. 4	5	8	
100,000	10,000	1,000	100	10	1	$\frac{1}{10}$	$\frac{1}{100}$	$\frac{1}{1,000}$	$\frac{1}{10,000}$ $\frac{1}{100,000}$

The place values of the columns for the fractional part of a decimal are $\frac{1}{10}, \frac{1}{100}, \frac{1}{1,000}$, and so on. Each of those columns has a value that is $\frac{1}{10}$ of the value of the place directly to its left. For example,

- The value of the tenths column is $\frac{1}{10}$ of the value of the ones column: $1 \cdot \frac{1}{10} = \frac{1}{10}$.
- The value of the hundredths column is $\frac{1}{10}$ of the value of the tenths column: $\frac{1}{10} \cdot \frac{1}{10} = \frac{1}{100}$.
- The value of the thousandths column is $\frac{1}{10}$ of the value of the hundredths column: $\frac{1}{100} \cdot \frac{1}{10} = \frac{1}{1,000}$.

The meaning of the decimal 4.458 becomes clear when we write it in **expanded form** (also called **expanded notation**).

$$4.458 = \mathbf{4}\text{ ones} + \mathbf{4}\text{ tenths} + \mathbf{5}\text{ hundredths} + \mathbf{8}\text{ thousandths}$$

which can be written as:

$$4.458 = 4 + \frac{4}{10} + \frac{5}{100} + \frac{8}{1,000}$$

LANGUAGE OF MATHEMATICS

The word *decimal* comes from the Latin word *decima*, meaning a tenth part.

EXAMPLE 2 Write the decimal number 592.8674 in expanded form.

Strategy Working from left to right, we will give the place value of each digit and combine them with + symbols.

WHY The term *expanded form* means to write the number as an addition of the place values of each of its digits.

Solution The expanded form of 592.8674 is:

$\mathbf{5}$ hundreds + $\mathbf{9}$ tens + $\mathbf{2}$ ones + $\mathbf{8}$ tenths + $\mathbf{6}$ hundredths
    + $\mathbf{7}$ thousandths + $\mathbf{4}$ ten-thousandths

which can be written as

$$500 + 90 + 2 + \frac{8}{10} + \frac{6}{100} + \frac{7}{1,000} + \frac{4}{10,000}$$

Self Check 2

Write the decimal number 1,277.9465 in expanded form.

Now Try ➡ Problems 23 and 27

OBJECTIVE 3 Read decimals and write them in standard form.

To understand how to read a decimal, we will examine the expanded form of 4.458 in more detail. Recall that

$$4.458 = 4 + \frac{4}{10} + \frac{5}{100} + \frac{8}{1,000}$$

To add the fractions, we need to build $\frac{4}{10}$ and $\frac{5}{100}$ so that each has a denominator that is the LCD, 1,000.

$$4.458 = 4 + \frac{4}{10} \cdot \frac{\mathbf{100}}{\mathbf{100}} + \frac{5}{100} \cdot \frac{\mathbf{10}}{\mathbf{10}} + \frac{8}{1,000}$$

$$= 4 + \frac{400}{1,000} + \frac{50}{1,000} + \frac{8}{1,000}$$

$$= 4 + \frac{458}{1,000}$$

$$= 4\frac{458}{1,000}$$

We have found that 4.458 = $4\frac{458}{1,000}$
(Whole-number part / Fractional part)

We read 4.458 as "four and four hundred fifty-eight thousandths" because 4.458 is the same as $4\frac{458}{1,000}$. Notice that the last digit in 4.45**8** is in the thousandths place. This observation suggests the following method for reading decimals.

Reading a Decimal

To read a decimal:

1. Look to the left of the decimal point and say the name of the whole number.

2. The decimal point is read as "and."

3. Say the fractional part of the decimal as a whole number followed by the name of the last place-value column of the digit that is the farthest to the right.

We can use the steps for reading a decimal to write it in words.

EXAMPLE 3   Write each decimal in words and then as a fraction or mixed number. **You do not have to simplify the fraction.**

a. *Sputnik*, the first satellite launched into space, weighed 184.3 pounds.
b. Usain Bolt of Jamaica holds the men's world record in the 100-meter dash: 9.58 seconds.
c. A one-dollar bill is 0.0043 inch thick.
d. Liquid mercury freezes solid at −37.7°F.

Strategy We will identify the whole number to the left of the decimal point, the fractional part to its right, and the name of the place-value column of the digit the farthest to the right.

Self Check 3

Write each decimal in words and then as a fraction or mixed number. **You do not have to simplify the fraction.**

a. The average normal body temperature is generally accepted as 98.6°F.

b. The planet Venus makes one full orbit around the sun every 224.7007 Earth days.

c. One gram is about 0.035274 ounce.

d. Liquid nitrogen freezes solid at −345.748°F.

Now Try Problems 31, 35, and 39

WHY We need to know those three pieces of information to read a decimal or write it in words.

Solution

a.   **184** . **3**   The whole-number part is 184. The fractional part is 3.
The digit the farthest to the right, 3, is in the tenths place.

One hundred eighty-four and three tenths
Written as a mixed number, 184.3 is $184\frac{3}{10}$.

b.   **9** . **58**   The whole-number part is 9. The fractional part is 58.
The digit the farthest to the right, 8, is in the hundredths place.

Nine and fifty-eight hundredths
Written as a mixed number, 9.58 is $9\frac{58}{100}$.

c.   **0** . **0043**   The whole-number part is 0. The fractional part is 43.
The digit the farthest to the right, 3, is in the ten-thousandths place.

Forty-three ten-thousandths   Since the whole-number part is 0, we need not write it nor the word *and*.

Written as a fraction, 0.0043 is $\frac{43}{10,000}$.

d.   **−37** . **7**   This is a negative decimal.

Negative *thirty-seven and seven tenths*.

Written as a negative mixed number, −37.7 is $-37\frac{7}{10}$.

The procedure for reading a decimal can be applied in reverse to convert from written-word form to standard form.

EXAMPLE 4   Write each number in standard form:

a. *One hundred seventy-two and forty-three hundredths*
b. *Eleven and fifty-one thousandths*

Strategy We will locate the word *and* in the written-word form and translate the phrase that appears before it and the phrase that appears after it separately.

WHY The whole-number part of the decimal is described by the phrase that appears before the word *and*. The fractional part of the decimal is described by the phrase that follows the word *and*.

Solution

a.   *One hundred seventy-two*   **and**   *forty-three hundredths*

172.43

This is the hundredths place-value column.

b. Sometimes, when changing from written-word form to standard form, we must insert placeholder 0's in the fractional part of a decimal so that the last digit appears in the proper place-value column.

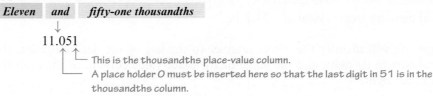

Eleven   *and*   *fifty-one thousandths*

11.051

This is the thousandths place-value column.
A place holder 0 must be inserted here so that the last digit in 51 is in the thousandths column.

Caution! If a placeholder 0 is not written in 11.051, an incorrect answer of 11.51 (eleven and fifty-one *hundredths,* not *thousandths*) results.

Self Check 4

Write each number in standard form:

a. *Eight hundred six and ninety-two hundredths*
b. *Twelve and sixty-seven ten-thousandths*

Now Try Problems 41, 45, and 47

OBJECTIVE 4   Compare decimals using inequality symbols.

To develop a way to compare decimals, let's consider 0.3 and 0.271. Since 0.271 contains more digits, it may appear that 0.271 is greater than 0.3. However, the opposite is true. To show this, we write 0.3 and 0.271 in fraction form:

$$0.3 = \frac{3}{10} \qquad 0.271 = \frac{271}{1,000}$$

Now we build $\frac{3}{10}$ into an equivalent fraction so that it has the same denominator as $\frac{271}{1,000}$.

$$0.3 = \frac{3}{10} \cdot \frac{100}{100} = \frac{300}{1,000}$$

Since $\frac{300}{1,000} > \frac{271}{1,000}$, it follows that $0.3 > 0.271$. This observation suggests a quicker method for comparing decimals.

Comparing Decimals

To compare two decimals:

1. Make sure both numbers have the same number of decimal places to the right of the decimal point. Write any additional zeros necessary to achieve this.
2. Compare the digits of each decimal, column by column, working from left to right.
3. *If the decimals are positive:* When two digits differ, the decimal with the greater digit is the greater number. *If the decimals are negative:* When two digits differ, the decimal with the smaller digit is the greater number.

EXAMPLE 5   Place an $<$ or $>$ symbol in the box to make a true statement:
a. 1.2679 ▮ 1.2658   **b.** 54.9 ▮ 54.929   **c.** -10.419 ▮ -10.45

Strategy We will stack the decimals and then, working from left to right, we will scan their place-value columns looking for a difference in their digits.

WHY We need only look in that column to determine which digit is the greater.

Solution
a. Since both decimals have the same number of places to the right of the decimal point, we can immediately compare the digits, column by column.

$$1.26\mathbf{7}9$$
$$1.26\mathbf{5}8$$

Same digit ⟶
Same digit ⟶
Same digit ⟶ — These digits are different: Since 7 is greater than 5, it follows that the first decimal is greater than the second.

Thus, 1.2679 is greater than 1.2658 and we can write $1.2679 > 1.2658$.

b. We can write two zeros after the 9 in 54.9 so that the decimals have the same number of digits to the right of the decimal point. This makes the comparison easier.

$$54.9\mathbf{0}0$$
$$54.9\mathbf{2}9$$

As we work from left to right, this is the first column in which the digits differ. Since $2 > 0$, it follows that 54.929 is greater than 54.9 (or 54.9 is less than 54.929) and we can write $54.9 < 54.929$.

Success Tip Writing additional zeros *after the last digit to the right of the decimal point does not change the value of the decimal.* Also, deleting additional zeros after the last digit to the right of the decimal point does not change the value of the decimal. For example,

$$54.9 = 54.90 = 54.900$$

Because $54\frac{90}{100}$ and $54\frac{900}{1,000}$ in simplest form are equal to $54\frac{9}{10}$.

These additional zeros do not change the value of the decimal.

c. We are comparing two negative decimals. In this case, when two digits differ, the decimal with the smaller digit is the greater number.

$$-10.4\mathbf{1}9$$
$$-10.4\mathbf{5}0 \quad \text{Write a zero after 5 to help in the comparison.}$$

As we work from left to right, this is the first column in which the digits differ. Since $1 < 5$, it follows that -10.419 is greater than -10.45 and we can write $-10.419 > -10.45$.

OBJECTIVE 5 Graph decimals on a number line.

Decimals can be shown by drawing points on a number line.

EXAMPLE 6 Graph -1.8, -1.23, -0.3, and 1.89 on a number line.

Strategy We will locate the position of each decimal on the number line and draw a bold dot.

WHY To *graph a number* means to make a drawing that represents the number.

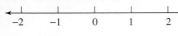
Solution The graph of each negative decimal is to the left of 0 and the graph of each positive decimal is to the right of 0. Since $-1.8 < -1.23$, the graph of -1.8 is to the left of -1.23.

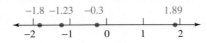

OBJECTIVE 6 Round decimals.

When we don't need exact results, we can approximate decimal numbers by **rounding**. To round the decimal part of a decimal number, we use a method similar to that used to round whole numbers.

Rounding a Decimal

1. To round a decimal to a certain decimal place value, locate the rounding digit in that place.

2. Look at the test digit directly to the right of the rounding digit.

3. If the test digit is 5 or greater, round up by adding 1 to the rounding digit and dropping all the digits to its right. If the test digit is less than 5, round down by keeping the rounding digit and dropping all the digits to its right.

EXAMPLE 7   **Chemistry.** A student in a chemistry class uses a digital balance to weigh a compound in grams. Round the reading shown on the balance to the nearest thousandth of a gram.

Strategy We will identify the digit in the thousandths column and the digit in the ten-thousandths column.

 WHY   To round to the nearest thousandth, the digit in the thousandths column is the rounding digit and the digit in the ten-thousandths column is the test digit.

Solution The rounding digit in the thousandths column is 8. Since the test digit 7 is 5 or greater, we round up.

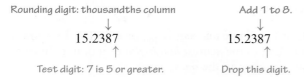

Rounding digit: thousandths column   Add 1 to 8.

$$15.2387 \qquad 15.2387$$

Test digit: 7 is 5 or greater.   Drop this digit.

The reading on the balance is approximately 15.239 grams.

Self Check 7

Round 24.41658 to the nearest ten-thousandth.

Now Try ➲ Problems 65 and 69

EXAMPLE 8   Round each decimal to the indicated place value:
a. -645.1358 to the nearest tenth   **b.** 33.096 to the nearest hundredth

Strategy In each case, we will first identify the rounding digit. Then we will identify the test digit and determine whether it is less than 5 or greater than or equal to 5.

 WHY   If the test digit is less than 5, we round down; if it is greater than or equal to 5, we round up.

Solution

a. Negative decimals are rounded in the same ways as positive decimals. The rounding digit in the tenths column is 1. Since the test digit 3 is less than 5, we round down.

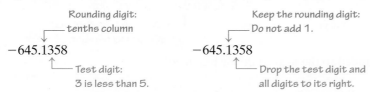

Rounding digit: tenths column   Keep the rounding digit: Do not add 1.

$$-645.1358 \qquad -645.1358$$

Test digit: 3 is less than 5.   Drop the test digit and all digits to its right.

Thus, -645.1358 rounded to the nearest tenth is -645.1.

b. The rounding digit in the hundredths column is 9. Since the test digit 6 is 5 or greater, we round up.

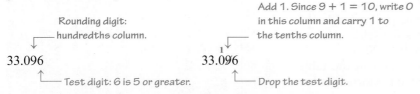

Rounding digit: hundredths column.   Add 1. Since $9 + 1 = 10$, write 0 in this column and carry 1 to the tenths column.

$$33.096 \qquad 33.\overset{1}{0}96$$

Test digit: 6 is 5 or greater.   Drop the test digit.

Thus, 33.096 rounded to the nearest hundredth is 33.10.

Caution! It would be incorrect to drop the 0 in the answer 33.10. If asked to round to a certain place value (in this case, hundredths), that place must have a digit, even if the digit is 0.

Self Check 8

Round each decimal to the indicated place value:

a. -708.522 to the nearest tenth

b. 9.1198 to the nearest thousandth

Now Try ➲ Problems 73 and 77

There are many situations in our daily lives that call for rounding amounts of money. For example, a grocery shopper might round the unit cost of an item to the nearest cent or a taxpayer might round his or her income to the nearest dollar when filling out an income tax return.

EXAMPLE 9

a. Utility bills. A utility company calculates a homeowner's monthly electric bill by multiplying the unit cost of $0.06421 by the number of kilowatt hours used that month. Round the unit cost to the nearest cent.

b. Annual income. A secretary earned $36,500.91 in one year. Round her income to the nearest dollar.

Strategy In part a, we will round the decimal to the nearest hundredth. In part b, we will round the decimal to the ones column.

WHY Since there are 100 cents in a dollar, each cent is $\frac{1}{100}$ of a dollar. To round to the *nearest cent* is the same as rounding to the *nearest hundredth* of a dollar. To round to the *nearest dollar* is the same as rounding to the *ones place*.

Solution

a. The rounding digit in the hundredths column is 6. Since the test digit 4 is less than 5, we round down.

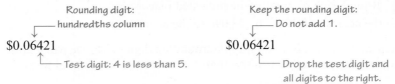

Thus, $0.06421 rounded to the nearest cent is $0.06.

b. The rounding digit in the ones column is 0. Since the test digit 9 is 5 or greater, we round up.

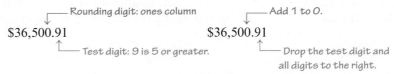

Thus, $36,500.91 rounded to the nearest dollar is $36,501.

Self Check 9

a. Round $0.076601 to the nearest cent

b. Round $24,908.53 to the nearest dollar.

Now Try Problems 85 and 87

OBJECTIVE 7 Read tables and graphs involving decimals.

The table on the left is an example of the use of decimals. It shows the number of pounds of trash generated daily per person in the United States for selected years from 1960 through 2013.

When the data in the table are presented in the form of a **bar graph**, a trend is apparent. The amount of trash generated daily per person increased steadily until the year 2000. Since then, it appears to have dropped slightly.

Year	Pounds
1960	2.68
1970	3.25
1980	3.66
1990	4.57
2000	4.74
2010	4.44
2013	4.40

Source: U.S. Environmental Protection Agency

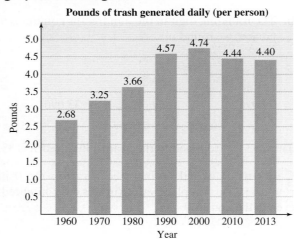

Answers to Self Checks

1. a. 9 thousandths **b.** 2   **2.** $1,000 + 200 + 70 + 7 + \frac{9}{10} + \frac{4}{100} + \frac{6}{1,000} + \frac{5}{10,000}$ **3. a.** ninety-eight and six tenths, $98\frac{6}{10}$ **b.** two hundred twenty-four and seven thousand seven ten-thousandths, $224\frac{7,007}{10,000}$ **c.** thirty-five thousand, two hundred seventy-four millionths, $\frac{35,274}{1,000,000}$ **d.** negative three hundred forty-five and seven hundred forty-eight thousandths, $-345\frac{748}{1,000}$ **4. a.** 806.92 **b.** 12.0067   **5. a.** < **b.** > **c.** <

6.
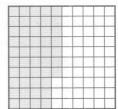

7. 24.4166   **8. a.** −708.5 **b.** 9.120   **9. a.** $0.08 **b.** $24,909

SECTION 4.1 STUDY SET

VOCABULARY

Fill in the blanks.

1. Decimals are written by entering the digits 0, 1, 2, 3, 4, 5, 6, 7, 8, and 9 into place-value columns that are separated by a decimal _____ .

2. The place-value columns to the left of the decimal point form the whole-number part of a decimal number, and the place-value columns to the right of the decimal point form the _____ part.

3. We can show the value represented by each digit of the decimal 98.6213 by using _____ form:

$$98.6213 = 90 + 8 + \frac{6}{10} + \frac{2}{100} + \frac{1}{1,000} + \frac{3}{10,000}$$

4. When we don't need exact results, we can approximate decimal numbers by _____ .

CONCEPTS

5. Write the name of each column in the following place-value chart.

4 , 7 8 9 . 0 2 6 5

6. Write the value of each column in the following place-value chart.

7	2	.	3	1	9	5	8

7. Fill in the blanks.
   a. The value of each place in the whole-number part of a decimal number is ▢ times greater than the column directly to its right.
   b. The value of each place in the fractional part of a decimal number is ▢ of the value of the place directly to its left.

8. Represent each situation using a signed number.
   a. A checking account overdrawn by $33.45
   b. A river 6.25 feet above flood stage
   c. 3.9 degrees below zero
   d. 17.5 seconds after liftoff

9. a. Represent the shaded part of the rectangular region as a fraction and a decimal.

   b. Represent the shaded part of the square region as a fraction and a decimal.

10. Write $400 + 20 + 8 + \frac{9}{10} + \frac{1}{100}$ as a decimal.

11. Fill in the blanks in the following illustration to label the *whole-number part* and the *fractional part.*

$$63.37 \qquad = \qquad 63\frac{37}{100}$$

12. Fill in the blanks.
    a. To round $0.13506 to the *nearest cent*, the rounding digit is ▢ and the test digit is ▢.
    b. To round $1,906.47 to the *nearest dollar*, the rounding digit is ▢ and the test digit is ▢.

NOTATION

Fill in the blanks.

13. The columns to the right of the decimal point in a decimal number form its fractional part. Their place-value names are similar to those in the whole-number part, but they end in the three letters "____."

14. When reading a decimal, such as 2.37, we can read the decimal point as "_____" or as "_____."

15. Write a decimal number that has . . .

6 in the ones column,

1 in the tens column,

0 in the tenths column,

8 in the hundreds column,

2 in the hundredths column,

9 in the thousands column,

4 in the thousandths column,

7 in the ten thousands column, and

5 in the ten-thousandths column.

16. Determine whether each statement is true or false.

   a. $0.9 = 0.90$

   b. $1.260 = 1.206$

   c. $-1.2800 = -1.280$

   d. $0.001 = .0010$

GUIDED PRACTICE

Answer the following questions about place value. See Example 1.

17. Consider the decimal number: 145.926

   a. What is the place value of the digit 9?

   b. Which digit tells the number of thousandths?

   c. Which digit tells the number of tens?

   d. What is the place value of the digit 5?

18. Consider the decimal number: 304.817

   a. What is the place value of the digit 1?

   b. Which digit tells the number of thousandths?

   c. Which digit tells the number of hundreds?

   d. What is the place value of the digit 7?

19. Consider the decimal number: 6.204538

   a. What is the place value of the digit 8?

   b. Which digit tells the number of hundredths?

   c. Which digit tells the number of ten-thousandths?

   d. What is the place value of the digit 6?

20. Consider the decimal number: 4.390762

   a. What is the place value of the digit 6?

   b. Which digit tells the number of thousandths?

   c. Which digit tells the number of ten-thousandths?

   d. What is the place value of the digit 4?

Write each decimal number in expanded form. See Example 2.

21. 37.89

22. 26.93

23. 124.575

24. 231.973

25. 7,498.6468

26. 1,946.7221

27. 6.40941

28. 8.70214

Write each decimal in words and then as a fraction or mixed number. See Example 3.

29. 0.3

30. 0.9

31. 50.41

32. 60.61

33. 19.529

34. 12.841

35. 304.0003

36. 405.0007

37. -0.00137

38. -0.00613

39. $-1,072.499$

40. $-3,076.177$

Write each number in standard form. See Example 4.

41. Six and one hundred eighty-seven thousandths

42. Four and three hundred ninety-two thousandths

43. Ten and fifty-six ten-thousandths

44. Eleven and eighty-six ten-thousandths

45. Negative sixteen and thirty-nine hundredths

46. Negative twenty-seven and forty-four hundredths

47. One hundred four and four millionths

48. Two hundred three and three millionths

Place an $<$ or an $>$ symbol in the box to make a true statement. See Example 5.

49. 2.59 ▨ 2.55

50. 5.17 ▨ 5.14

51. 45.103 ▨ 45.108

52. 13.874 ▨ 13.879

53. 3.28724 ▨ 3.2871

54. 8.91335 ▨ 8.9132

55. 379.67 ▨ 379.6088

56. 446.166 ▨ 446.2

57. -23.45 ▨ -23.1

58. -301.98 ▨ -302.45

59. -0.065 ▨ -0.066

60. -3.99 ▨ -3.9888

Graph each number on a number line. See Example 6.

61. $0.8, -0.7, -3.1, 4.5, -3.9$

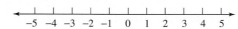

62. $0.6, -0.3, -2.7, 3.5, -2.2$

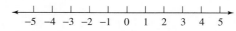

63. $-1.21, -3.29, -4.25, 2.75, -1.84$

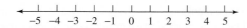

64. $-3.19, -0.27, -3.95, 4.15, -1.66$

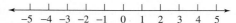

Round each decimal number to the indicated place value. See Example 7.

65. 506.198 nearest tenth

66. 51.451 nearest tenth

67. 33.0832 nearest hundredth

68. 64.0059 nearest hundredth

69. 4.2341 nearest thousandth

70. 8.9114 nearest thousandth

71. 0.36563 nearest ten-thousandth

72. 0.77623 nearest ten-thousandth

Round each decimal number to the indicated place value. See Example 8.

73. −0.137 nearest hundredth

74. −808.0897 nearest hundredth

75. −2.718218 nearest tenth

76. −3,987.8911 nearest tenth

77. 3.14959 nearest thousandth

78. 9.50966 nearest thousandth

79. 1.4142134 nearest millionth

80. 3.9998472 nearest millionth

81. 16.0995 nearest thousandth

82. 67.0998 nearest thousandth

83. 290.303496 nearest hundred-thousandth

84. 970.457297 nearest hundred-thousandth

Round each given dollar amount. See Example 9.

85. $0.284521 nearest cent

86. $0.312906 nearest cent

87. $27,841.52 nearest dollar

88. $44,633.78 nearest dollar

LOOK ALIKES

Round each decimal number to the indicated place value.

89. 8,506.2941

   a. nearest hundred   **b.** nearest hundredth

90. 134,677.01193

   a. nearest thousand   **b.** nearest thousandth

Write each number in standard form.

91. a. One hundred   **b.** One hundredth

   c. One thousand   **d.** One thousandth

   e. One million   **f.** One millionth

   g. One billion   **h.** One billionth

92. a. Seven billionths   **b.** Seven billion

   c. Seven millionths   **d.** Seven million

   e. Seven thousandths   **f.** Seven thousand

   g. Seven hundredths   **h.** Seven hundred

APPLICATIONS

93. Reading meters.   To what decimal is the arrow pointing?

94. Measurement.   Estimate a length of 0.3 inch on the 1-inch-long line segment below.

_____

95. Checking accounts.   Complete the check shown by writing in the amount, using a decimal.

> Ellen Russell
> 455 Santa Clara Ave.
> Parker, CO 25413
>
> April 14, 20 17
>
> PAY TO THE
> ORDER OF ___ *Citicorp* ___ $ ___
>
> *One thousand twenty-five and* $\frac{78}{100}$ ___ DOLLARS
>
> **BA** Downtown Branch
> P.O. Box 2456
> Colorado Springs, CO 23712
> MEMO Mortgage
> 45-828-02-33-4660
>
> *Ellen Russell*

96. Money.   We use a decimal point when working with dollars, but the decimal point is not necessary when working with cents. For each dollar amount in the table, give the equivalent amount expressed as cents.

Dollars	Cents
$0.50	
$0.05	
$0.55	
$5.00	
$0.01	

97. Injections.   A syringe is shown below. Use an arrow to show to what point the syringe should be filled if a 0.38-cc dose of medication is to be given. ("cc" stands for "cubic centimeters.")

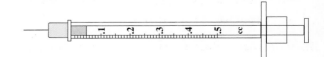

98. Lasers. The laser used in laser vision correction is so precise that each pulse can remove 39 millionths of an inch of tissue in 12 billionths of a second. Write each of these numbers as a decimal.

99. NASCAR. On April 17, 2011, in the Aaron's 499 Sprint Cup Series race at Talladega Superspeedway, Jimmie Johnson beat Clint Bowyer by a mere 0.002 second. This margin of victory tied for the closest finish since NASCAR switched to electronic timing. Write the decimal in words and then as a fraction in simplest form. (Source: deadspin.com)

100. The metric system. The metric system is widely used in science to measure length (meters), weight (grams), and capacity (liters). Round each decimal to the nearest hundredth.
  a. 1 ft is 0.3048 meter.
  b. 1 mi is 1,609.344 meters.
  c. 1 lb is 453.59237 grams.
  d. 1 gal is 3.785306 liters.

101. Utility bills. A portion of a homeowner's electric bill is shown below. Round each decimal dollar amount to the nearest cent.

Billing Period

From: 06/05/17   To: 07/05/17   Meter Number: 10694435

The Gas Company

Next Meter Reading Date on or about Aug 03 2017

Summary of Charges

Customer Charge	30 Days	× $0.16438
Baseline	14 Therms	× $1.01857
Over Baseline	11 Therms	× $1.20091
State Regulatory Fee	25 Therms	× $0.00074
Public Purpose Surcharge	25 Therms	× $0.09910

102. Income tax. A portion of a W-2 tax form is shown below. Round each dollar amount to the nearest dollar.

Form **W-2** Wage and Tax Statement   **2017**

1 Wages, tips, other comp $35,673.79	**2** Fed inc tax withheld $7,134.28
3 Social security wages $38,204.16	**4** SS tax withheld $2,368.65
5 Medicare wages & tips $38,204.16	**6** Medicare tax withheld $550.13
7 Social security tips	**8** Allocated tips

103. The Dewey Decimal System. When stacked on the shelves, the library books shown below are to be in numerical order, least to greatest, *from left to right*. How should the titles be rearranged to be in the proper order?

Crafts — 745.51
Modern art — 745.601
Hobbies — 745.58
Folk dolls — 745.6
Candlemaking — 745.49

104. 2016 Olympics. The top six finishers in the women's individual all-around gymnastics competition at the Rio de Janeiro Olympic Games are shown below in alphabetical order. If the highest score wins, which gymnasts won the gold (1st place), silver (2nd place), and bronze (3rd place) medals?

	Name	**Nation**	**Score**
	Simone Biles	U.S.A.	62.198
	Elsabeth Black	Canada	58.298
	Shang Chunsong	China	58.549
	Aliya Mustafina	Russia	58.665
	Aly Raisman	U.S.A.	60.098
	Wang Yan	China	58.032

105. Tuneups. The six spark plugs from the engine of a Nissan Quest were removed, and the spark plug gap was checked. If vehicle specifications call for the gap to be from 0.031 to 0.035 inch, which of the plugs should be replaced?

Cylinder 1: 0.035 in.
Cylinder 2: 0.029 in.
Cylinder 3: 0.033 in.
Cylinder 4: 0.039 in.
Cylinder 5: 0.031 in.
Cylinder 6: 0.032 in.

Spark plug gap

106. Geology. Geologists classify types of soil according to the grain size of the particles that make up the soil. The four major classifications of soil are shown below. Classify each of the samples (A, B, C, and D) in the table as clay, silt, sand, or granule.

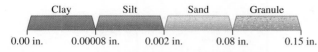

Clay	Silt	Sand	Granule
0.00 in.	0.00008 in.	0.002 in.	0.08 in.	0.15 in.

Sample	Location found	Grain size (in.)	Classification
A	Riverbank	0.009	
B	Pond	0.0007	
C	NE corner	0.095	
D	Dry lake	0.00003	

107. Microscopes. A microscope used in a lab is capable of viewing structures that range in size from 0.1 to as small as 0.0001 centimeter. Which of the structures listed in the table would be visible through this microscope?

Structure	Size (cm)
Bacterium	0.00011
Plant cell	0.015
Virus	0.000017
Animal cell	0.00093
Asbestos fiber	0.0002

108. Acceleration. Refer to the graph below. Find the time it takes each midsized SUV to accelerate from 0 to 60 mph. (Source: consumerreports.org)

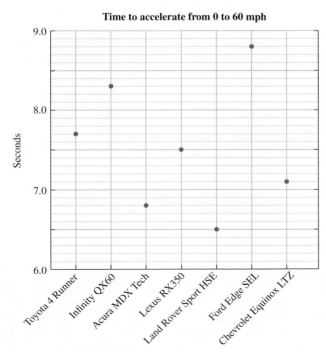

Time to accelerate from 0 to 60 mph

Source: consumerreports.org

109. Digging for clams. The following graph shows the time each day that the lowest tide occurs on a beach in Alaska. The negative numbers indicate the water level will be below the normal low water mark. If a tide of −4.0 feet or less is needed to expose the clam beds, on what days should there be good clamming?

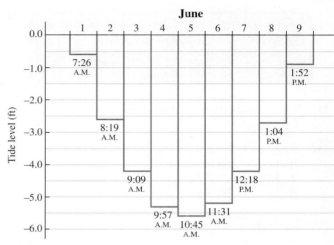

Source: ycharts.com

110. Gasoline prices. Refer to the graph below.
  a. In what month of 2015 was the retail price of a gallon of gasoline the least and in what month was it the greatest? Estimate each price.
  b. In what month of 2016 was the retail price of a gallon of gasoline the least and in what month was it the greatest? Estimate each price.

U.S. Average Retail Price Regular Unleaded Gasoline*

*Retail price includes state and federal taxes

Source: eia.gov

WRITING

111. Explain the difference between ten and one-tenth.

112. "The more digits a number contains, the larger it is." Is this statement true? Explain.

113. Explain why it is wrong to read 2.103 as *"two and one hundred and three thousandths."*

114. Signs.

  a. A sign in front of a fast-food restaurant had the cost of a hamburger listed as .99¢. Explain the error.

  b. The illustration below shows the unusual notation that some service stations use to express the price of a gallon of gasoline. Explain the error.

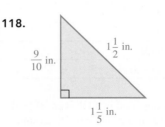

115. Write a definition for each of these words.

  decade   *decathlon*   *decimal*

116. Show that in the decimal numeration system, each place-value column for the fractional part of a decimal is $\frac{1}{10}$ of the value of the place directly to its left.

Find the perimeter and area of each figure below.

117.

3$\frac{1}{2}$ ft

2$\frac{3}{4}$ ft

118.

$\frac{9}{10}$ in.

1$\frac{1}{2}$ in.

1$\frac{1}{5}$ in.

OBJECTIVES

1 Add decimals.

2 Subtract decimals.

3 Add and subtract signed decimals.

4 Estimate sums and differences of decimals.

5 Solve application problems by adding and subtracting decimals.

SECTION 4.2 **Adding and Subtracting Decimals**

To add or subtract objects, they must be similar. The federal income tax form shown below has a vertical line to make sure that dollars are added to dollars and cents added to cents. In this section, we show how decimal numbers are added and subtracted using this type of vertical form.

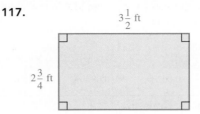

Form 1040EZ	Department of the Treasury—Internal Revenue Service **Income Tax Return for Single and Joint Filers With No Dependents** **2016**			
Income **Attach Form(s) W-2 here.** Enclose, but do not attach, any payment.	1	Wages, salaries, and tips. This should be shown in box 1 of your Form(s) W-2. Attach your Form(s) W-2.	1	21,056	89
	2	Taxable interest. If the total is over $1,500, you cannot use Form 1040EZ.	2	42	06
	3	Unemployment compensation and Alaska Permanent Fund dividends (see page 11).	3	200	00
	4	Add lines 1, 2, and 3. This is your **adjusted gross income.**	4	21,298	95

OBJECTIVE 1 **Add decimals.**

Adding decimals is similar to adding whole numbers. We use **vertical form** and stack the decimals with their corresponding place values and decimal points lined up. Then we add the digits in each column, working from right to left, making sure that hundredths are added to hundredths, tenths are added to tenths, ones are added to ones, and so on. We write the decimal point in the **sum** so that it lines up with the decimal points in the **addends**. For example, to find 4.21 + 1.23 + 2.45, we proceed as follows:

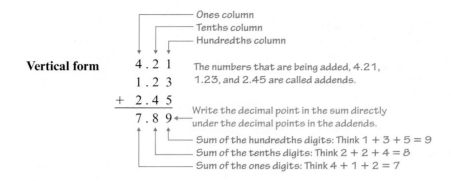

Vertical form

The numbers that are being added, 4.21, 1.23, and 2.45 are called addends.

Write the decimal point in the sum directly under the decimal points in the addends.

Sum of the hundredths digits: Think $1 + 3 + 5 = 9$
Sum of the tenths digits: Think $2 + 2 + 4 = 8$
Sum of the ones digits: Think $4 + 1 + 2 = 7$

The sum is 7.89.

In this example, each addend had two decimal places, tenths and hundredths. If the number of decimal places in the addends are different, we can insert additional zeros so that the number of decimal places match.

Adding Decimals

To add decimal numbers:

1. Write the numbers in vertical form with the decimal points lined up.
2. Add the numbers as you would add whole numbers, from right to left.
3. Write the decimal point in the result from Step 2 directly below the decimal points in the addends.

As with whole number addition, if the sum of the digits in any place-value column is greater than 9, we must **carry**.

EXAMPLE 1　Add: $31.913 + 5.6 + 68 + 16.78$

Strategy We will write the addition in vertical form so that the corresponding place values and decimal points of the addends are lined up. Then we will add the digits, column by column, working from right to left.

WHY We can only add digits with the same place value.

Solution To make the column additions easier, we will write two zeros after the 6 in the addend 5.6 and one zero after the 8 in the addend 16.78. Since whole numbers have an "understood" decimal point immediately to the right of their ones digit, we can write the addend 68 as 68.000 to help line up the columns.

$$
\begin{array}{r}
31.913 \\
5.600 \\
68.000 \\
+\ 16.780 \\
\end{array}
$$

Insert two zeros after the 6.
Insert a decimal point and three zeros: $68 = 68.000$.
Insert a zero after the 8.

Line up the decimal points.

Now we add, right to left, as we would whole numbers, writing the sum from each column below the horizontal bar.

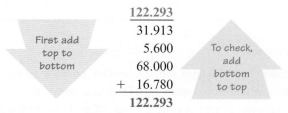

Carry a 2 (shown in blue) to the ones column.
Carry a 2 (shown in red) to the tens column.

Write the decimal point in the result directly below the decimal points in the addends.

Self Check 1

Add:
41.07 + 35 + 67.888 + 4.1

Now Try ➲ **Problem 19**

The sum is 122.293.

Success Tip In Example 1, the digits in each place-value column were added from *top to bottom*. To check the answer, we can instead add from *bottom to top*. Adding down or adding up should give the same result. If it does not, an error has been made and you should re-add.

First add
top to
bottom

$$
\begin{array}{r}
\underline{122.293} \\
31.913 \\
5.600 \\
68.000 \\
+\ 16.780 \\
\hline
122.293
\end{array}
$$

To check,
add
bottom
to top

Using Your Calculator ▶ **Adding Decimals**

The bar graph on the right shows the number of grams of fiber in a standard serving of each of several foods. It is believed that men can significantly cut their risk of heart attack by eating at least 25 grams of fiber a day. Does this diet meet or exceed the 25-gram requirement?

To find the total fiber intake, we add the fiber content of each of the foods. We can use a calculator to add the decimals.

3.1 $+$ 12.75 $+$ .9 $+$ 3.5 $+$ 1.1 $+$ 7.3 $=$ 28.65

On some calculators, the ENTER key is pressed to find the sum.

Since 28.65 > 25, this diet exceeds the daily fiber requirement of 25 grams.

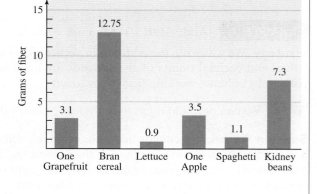

OBJECTIVE 2 Subtract decimals.

Subtracting decimals is similar to subtracting whole numbers. We use **vertical form** and stack the decimals with their corresponding place values and decimal points lined up so that we subtract similar objects—hundredths from hundredths, tenths from tenths, ones from ones, and so on. We write the decimal point in the **difference** so that it lines up with the decimal points in the **minuend** and **subtrahend**. For example, to find 8.59 − 1.27, we proceed as follows:

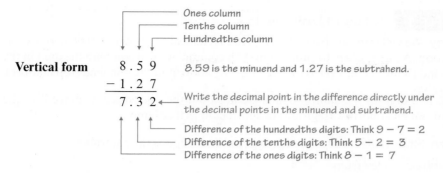

Vertical form

8 . 5 9   8.59 is the minuend and 1.27 is the subtrahend.
− 1 . 2 7
7 . 3 2 ◄—   Write the decimal point in the difference directly under
               the decimal points in the minuend and subtrahend.

Difference of the hundredths digits: Think $9 - 7 = 2$
Difference of the tenths digits: Think $5 - 2 = 3$
Difference of the ones digits: Think $8 - 1 = 7$

The difference is 7.32.

Subtracting Decimals

To subtract decimal numbers:

1. Write the numbers in vertical form with the decimal points lined up.
2. Subtract the numbers as you would subtract whole numbers from right to left.
3. Write the decimal point in the result from Step 2 directly below the decimal points in the minuend and the subtrahend.

As with whole numbers, if the subtraction of the digits in any place-value column requires that we subtract a larger digit from a smaller digit, we must **borrow** or **regroup**.

EXAMPLE 2   Subtract: $279.6 - 138.7$

Strategy As we prepare to subtract in each column, we will compare the digit in the subtrahend (bottom number) to the digit directly above it in the minuend (top number).

WHY If a digit in the subtrahend is greater than the digit directly above it in the minuend, we must borrow (regroup) to subtract in that column.

Solution Since 7 in the tenths column of 138.**7** is greater than 6 in the tenths column of 279.**6**, we cannot immediately subtract in that column because $6 - 7$ is *not* a whole number. To subtract in the tenths column, we must regroup by borrowing as shown below.

$$
\begin{array}{r}
\overset{8\ 16}{279.6} \\
-\ 138.7 \\
\hline
140.9
\end{array}
$$

To subtract in the tenths column, borrow 1 one in the form of 10 tenths from the ones column. Add 10 to the 6 in the tenths column to get 16 (shown in blue).

Recall from Section 1.3 that subtraction can be checked by addition. If a subtraction is done correctly, the sum of the difference and the subtrahend will equal the minuend: **Difference + subtrahend = minuend**.

Check:

$$
\begin{array}{r}
\overset{1}{140.9} \\
+\ 138.7 \\
\hline
279.6
\end{array}
$$

Difference
Subtrahend
Minuend

Since the sum of the difference and the subtrahend is the minuend, the subtraction is correct.

Some subtractions require borrowing from two (or more) place-value columns.

Self Check 2

Subtract: $382.5 - 227.1$

Now Try ➡ Problem 27

EXAMPLE 3　　Subtract 13.059 from 15.4.

Strategy We will translate the sentence to mathematical symbols and then perform the subtraction. As we prepare to subtract in each column, we will compare the digit in the subtrahend (bottom number) to the digit directly above it in the minuend (top number).

WHY If a digit in the subtrahend is greater than the digit directly above it in the minuend, we must borrow (regroup) to subtract in that column.

Solution Since 13.059 is the number to be subtracted, it is the subtrahend.

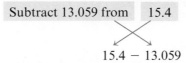

Subtract 13.059 from　　15.4

$$15.4 - 13.059$$

To find the difference, we write the subtraction in vertical form. To help with the column subtractions, we write two zeros to the right of 15.4 so that both numbers have three decimal places.

$$
\begin{array}{r}
15\,.\,400 \\
-\ 13\,.\,059 \\
\end{array}
$$
　　Insert two zeros after the 4 so that the decimal places match.

　　Line up the decimal points.

Since 9 in the thousandths column of 13.059 is greater than 0 in the thousandths column of 15.400, we cannot immediately subtract. It is not possible to borrow from the digit **0** in the hundredths column of 15.400. We can, however, borrow from the digit **4** in the tenths column of 15.400.

$$
\begin{array}{r}
^{3\ 10} \\
15\,.\,\cancel{4}\cancel{0}0 \\
-\ 13\,.\,059 \\
\end{array}
$$
　　Borrow 1 tenth in the form of 10 hundredths from 4 in the tenths column.
　　Add 10 to 0 in the hundredths column to get 10 (shown in blue).

Now we complete the two-column borrowing process by borrowing from the **10** in the hundredths column. Then we subtract, column-by-column, from the right to the left to find the difference.

$$
\begin{array}{r}
^{\quad 9} \\
^{3\ 10\ 10} \\
15\,.\,\cancel{4}\cancel{0}\cancel{0} \\
-\ 13\,.\,059 \\
\hline
2\,.\,341 \\
\end{array}
$$
　　Borrow 1 hundredth in the form of 10 thousandths from 10 in the hundredths column. Add 10 to 0 in the thousandths column to get 10 (shown in green).

When 13.059 is subtracted from 15.4, the difference is 2.341.

Check:

Self Check 3

Subtract 27.122 from 29.7.

Now Try ⟳ Problem 31

$$
\begin{array}{r}
^{1\ 1} \\
2.341 \\
+\ 13.059 \\
\hline
15.400 \\
\end{array}
$$
　　Since the sum of the difference and the subtrahend is the minuend, the subtraction is correct.

Using Your Calculator ▶　　Subtracting Decimals

A giant weather balloon is made of a flexible rubberized material that has an uninflated thickness of 0.011 inch. When the balloon is inflated with helium, the thickness becomes 0.0018 inch. To find the change in thickness, we need to subtract. We can use a calculator to subtract the decimals.

.011 [−] .0018 [=]

| 0.0092 |

On some calculators, the [**ENTER**] key is pressed to find the difference.

After the balloon is inflated, the rubberized material loses 0.0092 inch in thickness.

OBJECTIVE 3 **Add and subtract signed decimals.**

To add signed decimals, we use the same rules that we used for adding integers.

Adding Two Decimals That Have the Same (Like) Signs

1. To add two positive decimals, add them as usual. The final answer is positive.

2. To add two negative decimals, add their absolute values and make the final answer negative.

Adding Two Decimals That Have Different (Unlike) Signs

To add a positive decimal and a negative decimal, subtract the smaller absolute value from the larger.

1. If the positive decimal has the larger absolute value, the final answer is positive.

2. If the negative decimal has the larger absolute value, make the final answer negative.

EXAMPLE 4   Add: $-6.1 + (-4.7)$

Strategy We will use the rule for adding two decimals that have the same sign.

WHY Both addends, -6.1 and -4.7, are negative.

Solution Find the absolute values: $|-6.1| = 6.1$ and $|-4.7| = 4.7$.

$$-6.1 + (-4.7) = -10.8$$

Add the absolute values, 6.1 and 4.7, to get 10.8. Then make the final answer negative.

$$\begin{array}{r} 6.1 \\ +\ 4.7 \\ \hline 10.8 \end{array}$$

Self Check 4

Add: $-5.04 + (-2.32)$

Now Try ➡ **Problem 35**

EXAMPLE 5   Add: $5.35 + (-12.9)$

Strategy We will use the rule for adding two decimals that have different signs.

WHY One addend is positive and the other is negative.

Solution Find the absolute values: $|5.35| = 5.35$ and $|-12.9| = 12.9$.

$$5.35 + (-12.9) = -7.55$$

Subtract the smaller absolute value from the larger: $12.9 - 5.35 = 7.55$. Since the negative number, -12.9, has the larger absolute value, make the final answer negative.

$$\begin{array}{r} \overset{8\,10}{12.9\cancel{0}} \\ -\ 5.3\,5 \\ \hline 7.5\,5 \end{array}$$

Self Check 5

Add: $-21.4 + 16.75$

Now Try ➡ **Problem 39**

The rule for subtraction that was introduced in Section 2.3 can be used with signed decimals: *To subtract two decimals, add the first decimal to the opposite of the decimal to be subtracted.*

EXAMPLE 6 Subtract: $-35.6 - 5.9$

Strategy We will apply the rule for subtraction: Add the first decimal to the opposite of the decimal to be subtracted.

WHY It is easy to make an error when subtracting signed decimals. We will probably be more accurate if we write the subtraction as addition of the opposite.

Solution The number to be subtracted is 5.9. Subtracting 5.9 is the same as adding its opposite, -5.9.

Change the subtraction to addition.

$$-35.6 - 5.9 = -35.6 + (-5.9) = -41.5$$

Change the number being subtracted to its opposite.

Use the rule for adding two decimals with the same sign. Make the final answer negative.

$$\begin{array}{r} \overset{1\,1}{35.6} \\ +\ \ 5.9 \\ \hline 41.5 \end{array}$$

Self Check 6
Subtract: $-1.18 - 2.88$

Now Try Problem 43

EXAMPLE 7 Subtract: $-8.37 - (-16.2)$

Strategy We will apply the rule for subtraction: Add the first decimal to the opposite of the decimal to be subtracted.

WHY It is easy to make an error when subtracting signed decimals. We will probably be more accurate if we write the subtraction as addition of the opposite.

Solution The number to be subtracted is -16.2. Subtracting -16.2 is the same as adding its opposite, 16.2.

Add . . .

$$-8.37 - (-16.2) = -8.37 + 16.2 = 7.83$$

. . . the opposite

Use the rule for adding two decimals with different signs. Since 16.2 has the larger absolute value, the final answer is positive.

$$\begin{array}{r} \overset{5\ \overset{11}{\cancel{1}}10}{16.2\cancel{0}} \\ -\ 8.37 \\ \hline 7.83 \end{array}$$

Self Check 7
Subtract: $-2.56 - (-4.4)$

Now Try ➡ Problem 47

EXAMPLE 8 Evaluate: $-12.2 - (-14.5 + 3.8)$

Strategy We will perform the operation within the parentheses first.

WHY This is the first step of the order of operations rule.

Solution We perform the addition within the grouping symbols first.

$$-12.2 - (-14.5 + 3.8) = -12.2 - (-10.7) \qquad \text{Perform the addition.}$$
$$= -12.2 + 10.7 \qquad \text{Add the opposite of } -10.7.$$
$$= -1.5 \qquad \text{Perform the addition.}$$

$$\begin{array}{r} \overset{3\ 15}{14.\cancel{5}} \\ -\ 3.8 \\ \hline 10.7 \end{array}$$

$$\begin{array}{r} \overset{1\ 12}{12.\cancel{2}} \\ -10.7 \\ \hline 1.5 \end{array}$$

Self Check 8
Evaluate: $-4.9 - (-1.2 + 5.6)$

Now Try ➡ Problem 51

OBJECTIVE 4 Estimate sums and differences of decimals.

Estimation can be used to check the reasonableness of an answer to a decimal addition or subtraction. There are several ways to estimate, but the objective is the same: Simplify the numbers in the problem so that the calculations can be made easily and quickly.

EXAMPLE 9

a. Estimate by rounding the addends to the nearest ten: 261.76 + 432.94

b. Estimate using front-end rounding: 381.77 − 57.01

Strategy We will use rounding to approximate each addend, minuend, and subtrahend. Then we will find the sum or difference of the approximations.

WHY Rounding produces numbers that contain many 0's. Such numbers are easier to add or subtract.

Solution

a. $261.76 \rightarrow 260$ *Round to the nearest ten.*
      $+ 432.94 \rightarrow + 430$ *Round to the nearest ten.*
      ——————
                    690

The estimate is 690. If we compute 261.76 + 432.94, the sum is 694.7. We can see that the estimate is close; it's just 4.7 less than 694.7.

b. We use front-end rounding. Each number is rounded to its largest place value.

      $381.77 \rightarrow 400$ *Round to the nearest hundred.*
      $- 57.01 \rightarrow - 60$ *Round to the nearest ten.*
      ——————
                    340

The estimate is 340. If we compute 381.77 − 57.01, the difference is 324.76. We can see that the estimate is close; it's 15.24 more than 324.76.

Self Check 9

a. Estimate by rounding the addends to the nearest ten: 526.93 + 284.03

b. Estimate using front-end rounding: 512.33 − 36.47

Now Try ➲ Problems 55 and 57

OBJECTIVE 5 Solve application problems by adding and subtracting decimals.

To make a profit, a merchant must sell an item for more than she paid for it. The price at which the merchant sells the product, called the **retail price**, is the *sum* of the item's **cost** to the merchant plus the **markup**.

> **Retail price = cost + markup**

EXAMPLE 10   Pricing. Find the retail price of a Rubik's Cube if a game store owner buys them for $5.95 each and then marks them up $4.25 to sell in her store.

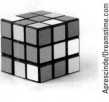
Aprescinde/Dreamstime.com

Analyze
- Rubik's Cubes cost the store owner $5.95 each.   *Given*
- She marks up the price $4.25.   *Given*
- What is the retail price of a Rubik's Cube?   *Find*

Form We translate the words of the problem to numbers and symbols.

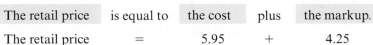

The retail price	is equal to	the cost	plus	the markup.
The retail price	=	5.95	+	4.25

Calculate Use vertical form to perform decimal addition:

$$\begin{array}{r} \overset{1\ 1}{5.95} \\ +\ \ 4.25 \\ \hline 10.20 \end{array}$$

State The retail price of a Rubik's Cube is $10.20.

Check We can estimate to check the result. If we use $6 to approximate the cost of a Rubik's Cube to the store owner and $4 to be the approximate markup, then the retail price is about $6 + $4 = $10. The result, $10.20, seems reasonable.

Self Check 10

Pricing. Find the retail price of a wool coat if a clothing outlet buys them for $109.95 each and then marks them up $99.95 to sell in its stores.

Now Try ➲ Problem 95

EXAMPLE 11   **Kitchen sinks.**   One model of kitchen sink is made of 18-gauge stainless steel that is 0.0500 inch thick. Another, less expensive, model is made from 20-gauge stainless steel that is 0.0375 inch thick. How much thicker is the 18-gauge?

Analyze

- The 18-gauge stainless steel is 0.0500 inch thick.   *Given*
- The 20-gauge stainless steel is 0.0375 inch thick.   *Given*
- How much thicker is the 18-gauge stainless steel?   *Find*

Form   Phrases such as *how much older, how much longer*, and, in this case, *how much thicker*, indicate subtraction. We translate the words of the problem to numbers and symbols.

How much thicker	is equal to	the thickness of the 18-gauge stainless steel	minus	the thickness of the 20-gauge stainless steel.
How much thicker	=	0.0500	−	0.0375

Calculate   Use vertical form to perform subtraction:

$$
\begin{array}{r}
\overset{\overset{9}{4\,10\,10}}{0.0\cancel{5}\,\cancel{0}\,\cancel{0}} \\
-\ 0.03\,7\,5 \\
\hline
0.01\,2\,5
\end{array}
$$

State   The 18-gauge stainless steel is 0.0125 inch thicker than the 20-gauge.

Check   We can add to check the subtraction:

$$
\begin{array}{r}
\overset{1\ 1}{0.0125} \\
+\ 0.0375 \\
\hline
0.0500
\end{array}
\quad
\begin{array}{l}
\text{Difference} \\
\text{Subtrahend} \\
\text{Minuend}
\end{array}
$$

The result checks.

Sometimes more than one operation is needed to solve a problem involving decimals.

Self Check 11

Aluminum.   How much thicker is 16-gauge aluminum that is 0.0508 inch thick than 22-gauge aluminum that is 0.0253 inch thick?

Now Try ➡ **Problem 101**

EXAMPLE 12   **Conditioning programs.**   A 350-pound football player lost 15.7 pounds during the first week of practice. During the second week, he gained 4.9 pounds. Find his weight after the first two weeks of practice.

Analyze

- The football player's beginning weight was 350 pounds.   *Given*
- The first week he lost 15.7 pounds.   *Given*
- The second week he gained 4.9 pounds.   *Given*
- What was his weight after two weeks of practice?   *Find*

Form   The word *lost* indicates subtraction. The word *gained* indicates addition. We translate the words of the problem to numbers and symbols.

The player's weight after two weeks of practice	is equal to	his beginning weight	minus	the first-week weight loss	plus	the second-week weight gain.
The player's weight after two weeks of practice	=	350	−	15.7	+	4.9

Calculate To evaluate $350 - 15.7 + 4.9$, we work from left to right and perform the subtraction first, then the addition.

$$
\begin{array}{r}
{\scriptstyle 9} \\
{\scriptstyle 4\ \cancel{10}\ 10} \\
3\cancel{5}\cancel{0}.\cancel{0} \\
-\ \ 1\,5.7 \\
\hline
3\,3\,4.3
\end{array}
$$

Write the whole number 350 as 350.0 and use a two-column borrowing process to subtract in the tenths column.

This is the player's weight after one week of practice.

Next, we add the 4.9-pound gain to the previous result to find the player's weight after two weeks of practice.

$$
\begin{array}{r}
{\scriptstyle 1} \\
334.3 \\
+\ \ \ 4.9 \\
\hline
339.2
\end{array}
$$

State The player's weight was 339.2 pounds after two weeks of practice.

Check We can estimate to check the result. The player lost about 16 pounds the first week and then gained back about 5 pounds the second week, for a net loss of 11 pounds. If we subtract the approximate 11-pound loss from his beginning weight, we get $350 - 11 = 339$ pounds. The result, 339.2 pounds, seems reasonable.

Self Check 12

Wrestling. A 195.5-pound wrestler had to lose 6.5 pounds to make his weight class. After the weigh-in, he gained back 3.7 pounds. What did he weigh then?

Now Try ⟳ Problem 107

Answers to Self Checks

1. 148.058 **2.** 155.4 **3.** 2.578 **4.** −7.36 **5.** −4.65 **6.** −4.06 **7.** 1.84 **8.** −9.3 **9. a.** 810 **b.** 460 **10.** $209.90 **11.** 0.0255 in. **12.** 192.7 lb

---

SECTION 4.2   STUDY SET

VOCABULARY

Fill in the blanks.

1. In the addition problem shown below, label each *addend* and the *sum*.

$$
\begin{array}{r}
1.72 \leftarrow \rule{2cm}{0.4pt} \\
4.68 \leftarrow \rule{2cm}{0.4pt} \\
+\ 2.02 \leftarrow \rule{2cm}{0.4pt} \\
\hline
8.42 \leftarrow \rule{2cm}{0.4pt}
\end{array}
$$

2. When using the vertical form to add decimals, if the addition of the digits in any one column produces a sum greater than 9, we must _____.

3. In the subtraction problem shown below, label the *minuend*, *subtrahend*, and the *difference*.

$$
\begin{array}{r}
12.9 \leftarrow \rule{2cm}{0.4pt} \\
-\ 4.3 \leftarrow \rule{2cm}{0.4pt} \\
\hline
8.6 \leftarrow \rule{2cm}{0.4pt}
\end{array}
$$

4. If the subtraction of the digits in any place-value column requires that we subtract a larger digit from a smaller digit, we must _____ or *regroup*.

5. To see whether the result of an addition is reasonable, we can round the addends and _____ the sum.

6. In application problems, phrases such as *how much older*, *how much longer*, and *how much thicker* indicate the operation of _____.

CONCEPTS

7. Check the following result. Use addition to determine if 15.2 is the correct difference.

$$
\begin{array}{r}
28.7 \\
-\ 12.5 \\
\hline
15.2
\end{array}
$$

8. Determine whether the *sign* of each result is positive or negative. *You do not have to find the sum.*

   a. $-7.6 + (-1.8)$

   b. $-24.99 + 29.08$

   c. $133.2 + (-400.43)$

9. Fill in the blank: To subtract signed decimals, add the _____ of the decimal that is being subtracted.

10. Apply the rule for subtraction and fill in the three blanks.

    $$3.6 - (-2.1) = 3.6 \;\boxed{}\,\boxed{}\,\boxed{} = \boxed{}$$

11. Fill in the blanks to rewrite each subtraction as addition of the opposite of the number being subtracted.

    a. $6.8 - 1.2 = 6.8 + (\;\boxed{}\;)$

    b. $29.03 - (-13.55) = 29.03 + \boxed{}$

    c. $-5.1 - 7.4 = -5.1 + (\;\boxed{}\;)$

12. Fill in the blanks to complete the estimation.

567.7	$\rightarrow$ $\boxed{}$	Round to the nearest ten.
$+\;214.3$	$\rightarrow$ $+\,\boxed{}$	Round to the nearest ten.
782.0	$\boxed{}$	

NOTATION

13. Copy the following addition problem. Insert a decimal point and additional zeros so that the number of decimal places in the addends match.

    $$\begin{array}{r} 46.6 \\ 11 \\ +\;15.702 \\ \hline \end{array}$$

14. Refer to the subtraction problem below. Fill in the blanks: To subtract in the _____ column, we borrow 1 tenth in the form of 10 hundredths from the 3 in the _____ column.

    $$\begin{array}{r} {}^{2\,1\,1}\!\!\!\!\!\! \\ 29.3\!\!\not{1} \\ -\;25.16 \\ \hline \end{array}$$

GUIDED PRACTICE

Add. **See Objective 1.**

15. $\begin{array}{r} 32.5 \\ +\;\;7.4 \\ \hline \end{array}$ 16. $\begin{array}{r} 16.3 \\ +\;\;3.5 \\ \hline \end{array}$

17. $\begin{array}{r} 3.04 \\ 4.12 \\ +\;1.43 \\ \hline \end{array}$ 18. $\begin{array}{r} 2.11 \\ 5.04 \\ +\;2.72 \\ \hline \end{array}$

Add. **See Example 1.**

19. $36.821 + 7.3 + 42 + 15.44$

20. $46.228 + 5.6 + 39 + 19.37$

21. $27.471 + 6.4 + 157 + 12.12$

22. $52.763 + 9.1 + 128 + 11.84$

Subtract. **See Objective 2.**

23. $\begin{array}{r} 6.83 \\ -\;3.52 \\ \hline \end{array}$ 24. $\begin{array}{r} 9.47 \\ -\;5.06 \\ \hline \end{array}$

25. $\begin{array}{r} 8.97 \\ -\;6.22 \\ \hline \end{array}$ 26. $\begin{array}{r} 7.56 \\ -\;2.33 \\ \hline \end{array}$

Subtract. **See Example 2.**

27. $\begin{array}{r} 495.4 \\ -\;153.7 \\ \hline \end{array}$ 28. $\begin{array}{r} 977.6 \\ -\;345.8 \\ \hline \end{array}$

29. $\begin{array}{r} 878.1 \\ -\;174.6 \\ \hline \end{array}$ 30. $\begin{array}{r} 767.2 \\ -\;614.7 \\ \hline \end{array}$

Perform the indicated operation. **See Example 3.**

31. Subtract 11.065 from 18.3.

32. Subtract 15.041 from 17.8.

33. Subtract 23.037 from 66.9.

34. Subtract 31.089 from 75.6.

Add. **See Example 4.**

35. $-6.3 + (-8.4)$ 36. $-9.2 + (-6.7)$

37. $-5.6 + (-5.6)$ 38. $-7.3 + (-5.4)$

Add. **See Example 5.**

39. $4.12 + (-18.8)$ 40. $9.95 + (-9.95)$

41. $6.45 + (-12.6)$ 42. $8.81 + (-14.9)$

Subtract. **See Example 6.**

43. $-1.9 - 0.2$ 44. $-56.1 - 8.6$

45. $-42.5 - 2.8$ 46. $-93.2 - 3.9$

Subtract. **See Example 7.**

47. $-4.49 - (-11.3)$ 48. $-1.35 - (-0.5)$

49. $-6.78 - (-24.6)$ 50. $-8.51 - (-27.4)$

Evaluate each expression. **See Example 8.**

51. $-11.1 - (-14.4 + 7.8)$

52. $-12.3 - (-13.6 + 7.9)$

53. $-16.4 - (-18.9 + 5.9)$

54. $-15.5 - (-19.8 + 5.7)$

Estimate each sum by rounding the addends to the nearest ten. **See Example 9.**

55. $510.65 + 279.19$ 56. $424.08 + 169.04$

Estimate each difference by using front-end rounding. **See Example 9.**

57. $671.01 - 88.35$ 58. $447.23 - 36.16$

TRY IT YOURSELF

Perform the indicated operations.

59. $-45.6 + 34.7$

60. $-6.95 + 8.2$

61. $-9.5 - 7.1$

62. $-1.5 - 0.35$

63. $15.95 + (-15.95)$

64. $34.7 + (-30.1)$

65.
$$\begin{array}{r} 21.88 \\ + 33.12 \\ \hline \end{array}$$

66.
$$\begin{array}{r} 19.05 \\ + 31.95 \\ \hline \end{array}$$

67. $2.55 - (-1.1)$

68. $143.3 - (-64.01)$

69. $645 + 9.90005 + 0.12 + 3.02002$

70. $505.0103 + 23 + 0.989 + 12.0704$

71. Subtract 47.5 from 0.

72. Subtract 5.9 from 7.001.

73. $(3.4 - 6.6) + 7.3$

74. $3.4 - (6.6 + 7.3)$

75. $247.9 + 40 + 0.56$

76. $0.0053 + 1.78 + 6$

77.
$$\begin{array}{r} 78.1 \\ - \;\; 7.81 \\ \hline \end{array}$$

78.
$$\begin{array}{r} 202.234 \\ - \;\; 19.34 \\ \hline \end{array}$$

79. $-7.8 + (-6.5)$

80. $-5.78 + (-33.1)$

81. $19.93 + (-20.21) + 1.37$

82. $-43 - (0.032 - 0.045)$

83. Find the sum of *two and forty-three hundredths* and *five and six tenths.*

84. Find the difference of *nineteen hundredths* and *six thousandths.*

85. $|-14.1 + 6.9| + 8$

86. $15 - |-2.3 + (-2.4)|$

87. $5 - 0.023$

88. $30 - 11.98$

89. $-2.002 - (-4.6)$

90. $-0.005 - (-8)$

LOOK ALIKES

Perform the indicated operations.

91. a. Subtract 5.65 from 11.37

 b. Find the difference of 5.65 and 11.37

92. a.
$$\begin{array}{r} 102.014 \\ + 99.927 \\ \hline \end{array}$$

 b.
$$\begin{array}{r} 102.014 \\ - 99.927 \\ \hline \end{array}$$

93. a. $3.86 - 2.13$

 b. $-3.86 - 2.13$

 c. $3.86 - (-2.13)$

 d. $-3.86 - (-2.13)$

94. a. $216.7 - (-50.5 - 32.8)$

 b. $216.7 - |-50.3 - 32.8|$

APPLICATIONS

95. Retailing. Find the retail price of each appliance listed in the table in the next column if a department store purchases them for the given costs and then marks them up as shown.

Appliance	Cost	Markup	Retail price
Refrigerator	$610.80	$205.00	
Washing machine	$389.50	$155.50	
Dryer	$363.99	$167.50	

96. Pricing. Find the retail price of a Kenneth Cole two-button suit if a men's clothing outlet buys them for $210.95 each and then marks them up $144.95 to sell in its stores.

97. Offshore drilling. A company needs to construct a pipeline from an offshore oil well to a refinery located on the coast. Company engineers have come up with two plans for consideration, as shown. Use the information in the illustration to complete the table that is shown below.

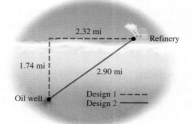

	Pipe underwater (mi)	Pipe underground (mi)	Total pipe (mi)
Design 1			
Design 2			

98. Driving directions. Find the total distance of the trip using the information in the printout shown below.

START	**1:** Start out going EAST on SUNKIST AVE.	0.0 mi
	2: Turn LEFT onto MERCED AVE.	0.4 mi
	3: Turn Right onto PUENTE AVE.	0.3 mi
WEST 10	**4:** Merge onto I-10 W toward LOS ANGELES.	2.2 mi
SOUTH 605	**5:** Merge onto I-605 S.	10.6 mi
SOUTH 5	**6:** Merge onto I-5 S toward SANTA ANA.	14.9 mi
110 A EXIT	**7:** Take the HARBOR BLVD exit, EXIT 110A.	0.3 mi
	8: Turn RIGHT onto S HARBOR BLVD.	0.1 mi
END	**9:** End at 1313 S Harbor Blvd Anaheim, CA.	

99. Pipe (PVC). Find the *outside* diameter of the plastic sprinkler pipe shown below if the thickness of the pipe wall is 0.218 inch and the inside diameter is 1.939 inches.

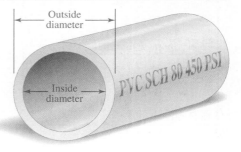

100. pH scale. The pH scale shown below is used to measure the strength of acids and bases in chemistry. Find the difference in pH readings between

   a. bleach and stomach acid.

   b. ammonia and coffee.

   c. blood and coffee.

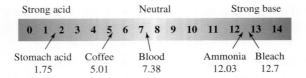

| Strong acid | | Neutral | | Strong base |

Stomach acid 1.75   Coffee 5.01   Blood 7.38   Ammonia 12.03   Bleach 12.7

101. Record holders. The late Florence Griffith-Joyner of the United States holds the women's world record in the 100-meter sprint: 10.49 seconds. Cate Campbell of Australia holds the women's world record in the 100-meter freestyle swim: 52.06 seconds. How much faster did Griffith-Joyner run the 100 meters than Campbell swam it? (Source: *The World Almanac and Book of Facts,* 2017)

102. Weather reports. Barometric pressure readings are recorded on the weather map below. In a low-pressure area (L on the map), the weather is often stormy. The weather is usually fair in a high-pressure area (H). What is the difference in readings between the areas of highest and lowest pressure?

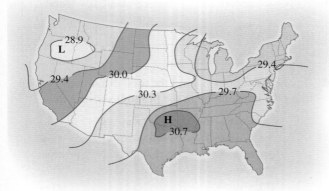

103. Banking. A businesswoman deposited several checks in her company's bank account, as shown on the deposit slip below. Find the *Subtotal* line on the slip by adding the amounts of the checks and total from the reverse side. If the woman wanted to get $25 in cash back from the teller, what should she write as the *Total deposit* on the slip?

Deposit slip

Cash		
Checks (properly endorsed)	116	10
	47	93
Total from reverse side	359	16
Subtotal		
Less cash	25	00
Total deposit		

104. Sports pages. Decimals are often used in the sports pages of newspapers. Two examples are given below.

   a. "German bobsledders set a world record today with a final run of 53.03 seconds, finishing ahead of the Italian team by only fourteen thousandths of a second." What was the time for the Italian bobsled team?

   b. "The women's figure skating title was decided by only thirty-three hundredths of a point." If the winner's point total was 102.71, what was the second-place finisher's total? (*Hint:* The highest score wins in a figure skating contest.)

Phovoir/Shutterstock.com

from Campus to Careers

Home Health Aide

105. Suppose certain portions of a patient's morning (A.M.) temperature chart were not filled in. Use the given information to complete the chart below. (*Hint:* 98.6°F is considered normal.)

Day of week	Patient's A.M. temperature	Amount above normal
Monday	99.7°	
Tuesday		2.5°
Wednesday	98.6°	
Thursday	100.0°	
Friday		0.9°

106. Quality control. An electronics company has strict specifications for the silicon chips it uses in its computers. The company only installs chips that are within 0.05 centimeter of the indicated thickness. The table below gives the specifications for two types of chips. Fill in the blanks to complete the chart.

Chip type	Thickness specification	Acceptable range	
		Low	High
A	0.78 cm		
B	0.643 cm		

107. Flight paths. Find the added distance a plane must travel to avoid flying through the storm.

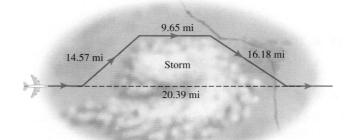

108. **Television.** The following illustration shows the six most-watched television series finales of all time.

   a. What was the combined total audience of all six shows?

   b. How many more people watched the last episode of *M*A*S*H* than watched the last episode of *Friends*?

   c. How many more people would have had to watch the last *Magnum, P.I.* to move it into a tie for fifth place?

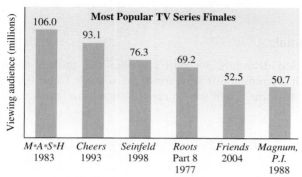

Most Popular TV Series Finales

Viewing audience (millions)

106.0	93.1	76.3	69.2	52.5	50.7
*M*A*S*H* 1983	*Cheers* 1993	*Seinfeld* 1998	*Roots* Part 8 1977	*Friends* 2004	*Magnum, P.I.* 1988

Source: businessinsider.com

109. **The Home Shopping Network.** The illustration shows a description of a cookware set that was sold on television.

   a. Find the difference between the manufacturer's suggested retail price (MSRP) and the sale price.

   b. Including shipping and handling (S & H), how much will the cookware set cost?

Item 229-442	
Continental 9-piece Cookware Set	
Stainless steel	
MSRP	$149.79
HSN Price	$59.85
On Sale	**$47.85**
S & H	$7.95

110. **Vehicle specifications.** Certain dimensions of a compact car are shown. Find the wheelbase of the car.

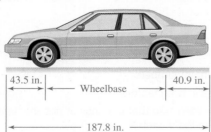

43.5 in.   Wheelbase   40.9 in.

187.8 in.

WRITING

111. Explain why we line up the decimal points and corresponding place-value columns when adding decimals.

112. Explain why we can write additional zeros to the right of a decimal such as 7.89 without affecting its value.

113. Explain what is wrong with the work shown below.

$$
\begin{array}{r}
203.56 \\
37 \\
+0.43 \\
\hline
204.36
\end{array}
$$

114. Consider the following addition:

$$
\begin{array}{r}
\overset{2}{23}.7 \\
41.9 \\
+12.8 \\
\hline
78.4
\end{array}
$$

Explain the meaning of the small red 2 written above the ones column.

115. Write a set of instructions that explains the two-column borrowing process shown below.

$$
\begin{array}{r}
\overset{9}{\overset{4\,10\,10}{2.6\,\cancel{5}\cancel{0}\cancel{0}}} \\
-1.3246 \\
\hline
1.3254
\end{array}
$$

116. Explain why it is easier to add the decimals 0.3 and 0.17 than the fractions $\frac{3}{10}$ and $\frac{17}{100}$.

REVIEW LOOK ALIKES

Perform the indicated operations.

117. a. $\dfrac{4}{5} + \dfrac{5}{12}$ b. $\dfrac{4}{5} - \dfrac{5}{12}$

   c. $\dfrac{4}{5} \cdot \dfrac{5}{12}$ d. $\dfrac{4}{5} \div \dfrac{5}{12}$

118. a. $\dfrac{3}{8} + \dfrac{1}{6}$ b. $\dfrac{3}{8} - \dfrac{1}{6}$

   c. $\dfrac{3}{8} \cdot \dfrac{1}{6}$ d. $\dfrac{3}{8} \div \dfrac{1}{6}$

OBJECTIVES

1 Multiply decimals.

2 Multiply decimals by powers of 10.

3 Multiply signed decimals.

4 Evaluate exponential expressions that have decimal bases.

5 Use the order of operations rule.

6 Evaluate formulas.

7 Estimate products of decimals.

8 Solve application problems by multiplying decimals.

SECTION 4.3 Multiplying Decimals

Since decimal numbers are *base-ten* numbers, multiplication of decimals is similar to multiplication of whole numbers. However, when multiplying decimals, there is one additional step—we must determine where to write the decimal point in the product.

OBJECTIVE 1 Multiply decimals.

To develop a rule for multiplying decimals, we will consider the multiplication $0.3 \cdot 0.17$ and find the product in a roundabout way. First, we write 0.3 and 0.17 as fractions and multiply them in that form. Then we express the resulting fraction as a decimal.

$$0.3 \cdot 0.17 = \frac{3}{10} \cdot \frac{17}{100} \quad \text{Express the decimals 0.3 and 0.17 as fractions.}$$

$$= \frac{3 \cdot 17}{10 \cdot 100} \quad \begin{array}{l} \text{Multiply the numerators.} \\ \text{Multiply the denominators.} \end{array}$$

$$= \frac{51}{1,000}$$

$$= 0.051 \quad \text{Write the resulting fraction } \tfrac{51}{1,000} \text{ as a decimal.}$$

From this example, we can make observations about multiplying decimals.

■ The digits in the answer are found by multiplying 3 and 17.

$$0.3 \quad \cdot \quad 0.17 \quad = \quad 0.051$$
$$3 \cdot 17 = 51$$

■ The answer has 3 decimal places. The *sum* of the number of decimal places in the factors 0.3 and 0.17 is also 3.

$$0.3 \quad \cdot \quad 0.17 \quad = \quad 0.051$$

1 decimal place   2 decimal places   3 decimal places

These observations illustrate the following rule for multiplying decimals.

Multiplying Decimals

To multiply two decimals:

1. Multiply the decimals as if they were whole numbers.

2. Find the total number of decimal places in both factors.

3. Insert a decimal point in the result from step 1 so that the answer has the same number of decimal places as the total found in step 2.

EXAMPLE 1   Multiply: $5.9 \cdot 3.4$

Strategy We will ignore the decimal points and multiply 5.9 and 3.4 as if they were whole numbers. Then we will write a decimal point in that result so that the final answer has two decimal places.

WHY Since the factor 5.9 has 1 decimal place, and the factor 3.4 has 1 decimal place, the product should have $1 + 1 = 2$ decimal places.

Solution We write the multiplication in vertical form and proceed as follows:

Vertical form

$$
\begin{array}{r}
5.9 \leftarrow \text{1 decimal place} \\
\times \quad 3.4 \leftarrow \text{1 decimal place} \\
\hline
236 \\
1770 \\
\hline
20.06
\end{array}
$$

The answer will have $1 + 1 = 2$ decimal places.

Move 2 places from the right to the left and insert a decimal point in the answer.

Thus, $5.9 \cdot 3.4 = 20.06$.

LANGUAGE OF MATHEMATICS

Recall the vocabulary of *multiplication*.

$$
\begin{array}{r}
5.9 \leftarrow \text{Factor} \\
\times \quad 3.4 \leftarrow \text{Factor} \\
\hline
236 \\
1770 \\
\hline
20.06 \leftarrow \text{Product}
\end{array}
$$

Partial products

Self Check 1

Multiply: $2.7 \cdot 4.3$

Now Try ➡ Problem 9

EXAMPLE 2   Multiply: $1.3(0.005)$

Strategy We will ignore the decimal points and multiply 1.3 and 0.005 as if they were whole numbers. Then we will write a decimal point in that result so that the final answer has four decimal places.

WHY Since the factor 1.3 has 1 decimal place, and the factor 0.005 has 3 decimal places, the product should have $1 + 3 = 4$ decimal places.

Solution Since many students find vertical form multiplication of decimals easier if the decimal with the smaller number of nonzero digits is written on the bottom, we will write 0.005 under 1.3.

$$
\begin{array}{r}
1.3 \leftarrow \text{1 decimal place} \\
\times \quad 0.005 \leftarrow \text{3 decimal places} \\
\hline
0.0065
\end{array}
$$

The answer will have $1 + 3 = 4$ decimal places.

Write 2 placeholder zeros in front of 6. Then move 4 places from the right to the left and insert a decimal point in the answer.

Thus, $1.3(0.005) = 0.0065$.

Self Check 2

Multiply: $(0.0002)7.2$

Now Try ➡ Problem 13

EXAMPLE 3   Multiply: $234(5.1)$

Strategy We will ignore the decimal point and multiply 234 and 5.1 as if they were whole numbers. Then we will write a decimal point in that result so that the final answer has one decimal place.

WHY Since the factor 234 has 0 decimal places, and the factor 5.1 has 1 decimal place, the product should have $0 + 1 = 1$ decimal place.

Solution We write the multiplication in vertical form, with 5.1 under 234.

$$
\begin{array}{r}
234 \leftarrow \text{No decimal places} \\
\times \quad 5.1 \leftarrow \text{1 decimal place} \\
\hline
23\,4 \\
1170\,0 \\
\hline
1193.4
\end{array}
$$

The answer will have $0 + 1 = 1$ decimal place.

Move 1 place from the right to the left and insert a decimal point in the answer.

Thus, $234(5.1) = 1{,}193.4$.

Success Tip When multiplying decimals, we do not need to line up the decimal points, as Example 2 and 3 illustrate.

Self Check 3

Multiply: $178(4.7)$

Now Try ➡ Problem 17

Using Your Calculator ▶ Multiplying Decimals

When billing a household, a gas company converts the amount of natural gas used to units of heat energy called therms. The number of therms used by a household in one month and the cost per therm are shown below.

Customer charge . . . . 39 therms @ $0.72264

To find the total charges for the month, we multiply the number of therms by the cost per therm: $39 \cdot 0.72264$.

39 ⊠ .72264 ▣ 28.18296 | 28.18296 |

On some calculator models, the [ENTER] key is pressed to display the product.
Rounding to the nearest cent, we see that the total charge is $28.18.

Think it Through ● OVERTIME

"Employees covered by the Fair Labor Standards Act must receive overtime pay for hours worked in excess of 40 in a workweek of at least 1.5 times their regular rates of pay."

—*United States Department of Labor*

The map of the United States shown below is divided into nine regions. The average hourly wage for private industry workers in each region is also listed in the legend below the map. Find the average hourly wage for the region where you live. Then calculate the corresponding average hourly overtime wage for that region.

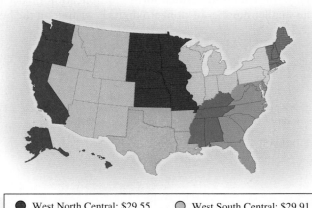

- ● West North Central: $29.55
- ○ Mountain: $29.15
- ● Pacific: $35.79
- ● East South Central: $25.28
- ○ East North Central: $30.06
- ○ West South Central: $29.91
- ● New England: $38.92
- ○ Middle Atlantic: $39.12
- ○ South Atlantic: $30.43

Source: Bureau of Labor Statistics

OBJECTIVE 2 **Multiply decimals by powers of 10.**

The numbers 10, 100, and 1,000 are called **powers of 10** because they are the results when we evaluate 10^1, 10^2, and 10^3. To develop a rule to find the product when multiplying a decimal by a power of 10, we multiply 8.675 by three different powers of 10.

Multiply: 8.675 · **10**

$$
\begin{array}{r}
8.675 \\
\times \quad 10 \\
\hline
0000 \\
86750 \\
\hline
86.750
\end{array}
$$

Multiply: 8.675 · **100**

$$
\begin{array}{r}
8.675 \\
\times \quad 100 \\
\hline
0000 \\
00000 \\
867500 \\
\hline
867.500
\end{array}
$$

Multiply: 8.675 · **1,000**

$$
\begin{array}{r}
8.675 \\
\times \quad 1000 \\
\hline
0000 \\
00000 \\
000000 \\
8675000 \\
\hline
8675.000
\end{array}
$$

When we inspect the answers, the decimal point in the first factor 8.675 appears to be moved to the right by the multiplication process. The number of decimal places it moves depends on the power of 10 by which 8.675 is multiplied.

One zero in 10

$8.675 \cdot 10 = 86{.}75$

It moves 1 place
to the right.

Two zeros in 100

$8.675 \cdot 100 = 867{.}5$

It moves 2 places
to the right.

Three zeros in 1,000

$8.675 \cdot 1{,}000 = 8675$

It moves 3 places
to the right.

These observations illustrate the following rule.

Multiplying a Decimal by 10, 100, 1,000, and So On

To find the product of a decimal and 10, 100, 1,000, and so on, move the decimal point to the right the same number of places as there are zeros in the power of 10.

EXAMPLE 4   Multiply: **a.** 2.81 · 10   **b.** 0.076(10,000)

Strategy For each multiplication, we will identify the factor that is a power of 10 and count the number of zeros that it has.

WHY To find the product of a decimal and a power of 10 that is greater than 1, we move the decimal point to the right the same number of places as there are zeros in the power of 10.

Solution

a. $2.81 \cdot 10 = 28.1$

Since 10 has one zero, move the decimal point in 2.81 one place to the right.

b. $0.076(10,000) = 00760.$

$= 760$

Since 10,000 has four zeros, move the decimal point in 0.076 four places to the right. Write a placeholder zero (shown in blue).

Self Check 4

Multiply:

a. 0.721 · 100

b. 6.08(1,000)

Now Try ➭ Problems 21 and 23

Numbers such as 10, 100, and 1,000 are powers of 10 that are *greater than 1*. There are also powers of 10 that are *less than 1*, such as 0.1, 0.01, and 0.001. To develop a rule to find the product when multiplying a decimal by one tenth, one hundredth, one thousandth, and so on, we will consider three examples:

Multiply: 5.19 · **0.1**

$$
\begin{array}{r}
5.19 \\
\times \quad 0.1 \\
\hline
0.519
\end{array}
$$

Multiply: 5.19 · **0.01**

$$
\begin{array}{r}
5.19 \\
\times \quad 0.01 \\
\hline
0.0519
\end{array}
$$

Multiply: 5.19 · **0.001**

$$
\begin{array}{r}
5.19 \\
\times \quad 0.001 \\
\hline
0.00519
\end{array}
$$

When we inspect the answers, the decimal point in the first factor 5.19 appears to be moved to the left by the multiplication process. The number of places that it moves depends on the power of 10 by which it is multiplied.

These observations illustrate the following rule.

Multiplying a Decimal by 0.1, 0.01, 0.001, and So On

To find the product of a decimal and 0.1, 0.01, 0.001, and so on, move the decimal point to the left the same number of decimal places as there are in the power of 10.

Self Check 5

Multiply:

a. 0.1(129.9)

b. 0.002 · 0.00001

Now Try ➲ Problems 25 and 27

EXAMPLE 5   Multiply: **a.** 145.8 · 0.01   **b.** 9.76(0.0001)

Strategy For each multiplication, we will identify the factor of the form 0.1, 0.01, and 0.001, and count the number of decimal places that it has.

WHY To find the product of a decimal and a power of 10 that is less than 1, we move the decimal point to the left the same number of decimal places as there are in the power of 10.

Solution

a. $145.8 \cdot 0.01 = 1.458$ Since 0.01 has *two decimal places*, move the decimal point in 145.8 *two* places to the left.

b. $9.76(0.0001) = 0.000976$ Since 0.0001 has *four decimal places*, move the decimal point in 9.76 *four* places to the left. This requires that three placeholder zeros (shown in blue) be inserted in front of the 9.

Quite often, newspapers, websites, and television programs present large numbers in a shorthand notation that involves a decimal in combination with a place-value column name. For example,

- In 2016, The Pokémon Company sold *11.4 million* Pokémon Sun/Moon units worldwide. (Source: statista.com)
- Boston's Big Dig was the most expensive single highway project in U.S. history. It cost about *$14.63 billion*. (Source: Roadtraffic-technology.com)
- The distance that light travels in one year is about *5.878 trillion* miles. (Source: *Encyclopedia Britannica*)

We can use the rule for multiplying a decimal by a power of 10 to write these large numbers in standard form.

EXAMPLE 6   Write each number in standard notation:

a. 14.6 million   **b.** 14.63 billion   **c.** 5.878 trillion

Strategy We will express each of the large numbers as the product of a decimal and a power of 10.

WHY Then we can use the rule for multiplying a decimal by a power of 10 to find their product. The result will be in the required standard form.

Solution

a. 14.6 million = **14.6 · 1 million**

= **14.6 · 1,000,000**   Write 1 million in standard form.

= **14,600,000**   Since 1,000,000 has *six zeros*, move the decimal point in 14.6 *six* places to the right.

b. 14.63 billion = 14.63 · **1 billion**

= 14.63 · 1,000,000,000   *Write 1 billion in standard form.*

= 14,630,000,000   *Since 1,000,000,000 has nine zeros, move the decimal point in 14.63 nine places to the right.*

c. 5.878 trillion = 5.878 · **1 trillion**

= 5.878 · 1,000,000,000,000   *Write 1 trillion in standard form.*

= 5,878,000,000,000   *Since 1,000,000,000,000 has twelve zeros, move the decimal point in 5.878 twelve places to the right.*

Self Check 6

Write each number in standard notation:

a. In 2016, the U.S. had 2.07 million farms. (Source: *The World Almanac and Book of Facts*, 2016)

b. Americans took 10.6 billion trips on public transportation in 2015. (Source: American Public Transit Association)

c. It would take about 1.818 trillion pennies to fill the Empire State Building. (Source: The Mega Penny Project)

Now Try ➭ Problem 113

OBJECTIVE 3 Multiply signed decimals.

The rules for multiplying integers also hold for multiplying signed decimals. The product of two decimals with like signs is positive, and the product of two decimals with unlike signs is negative.

EXAMPLE 7   Multiply: **a.** $-1.8(4.5)$ **b.** $(-1,000)(-59.08)$

Strategy In part a, we will use the rule for multiplying signed decimals that have different (unlike) signs. In part b, we will use the rule for multiplying signed decimals that have the same (like) signs.

WHY In part a, one factor is negative and one is positive. In part b, both factors are negative.

Solution

a. Find the absolute values: $|-1.8| = 1.8$ and $|4.5| = 4.5$. Since the decimals have unlike signs, their product is negative.

$-1.8(4.5) = -8.1$ *Multiply the absolute values, 1.8 and 4.5, to get 8.1. Then make the final answer negative.*

$$\begin{array}{r} 1.8 \\ \times\ 4.5 \\ \hline 90 \\ 720 \\ \hline 8.10 \end{array}$$

b. Find the absolute values: $|-1,000| = 1,000$ and $|-59.08| = 59.08$. Since the decimals have like signs, their product is positive.

$(-1,000)(-59.08) = 1,000(59.08)$

= 59080.   *Multiply the absolute values, 1,000 and 59.08. Since 1,000 has three zeros, move the decimal point in 59.08 three places to the right. Write a placeholder zero. The answer is positive.*

Self Check 7

Multiply:

a. $6.6(-5.5)$

b. $-44.968(-100)$

Now Try ➭ Problems 37 and 41

OBJECTIVE 4 Evaluate exponential expressions that have decimal bases.

We have evaluated exponential expressions that have whole number bases, integer bases, and fractional bases. The base of an exponential expression can also be a positive or a negative decimal.

EXAMPLE 8 Evaluate: **a.** $(2.4)^2$ **b.** $(-0.05)^2$

Strategy We will write each exponential expression as a product of repeated factors and then perform the multiplication. This requires that we identify the base and the exponent.

WHY The exponent tells the number of times the base is to be written as a factor.

Solution

a. $(2.4)^2 = 2.4 \cdot 2.4$ The base is 2.4 and the exponent is 2.
Write the base as a factor 2 times.

$\qquad\quad = 5.76$ Multiply the decimals.

$$\begin{array}{r} 2.4 \\ \times\ 2.4 \\ \hline 96 \\ 480 \\ \hline 5.76 \end{array}$$

Self Check 8

Evaluate:

a. $(-1.3)^2$

b. $(0.09)^2$

Now Try ➲ Problems 45 and 47

b. $(-0.05)^2 = (-0.05)(-0.05)$ The base is −0.05 and the exponent is 2. Write the base as a factor 2 times.

$\qquad\qquad\quad = 0.0025$ Multiply the decimals. The product of two decimals with like signs is positive.

$$\begin{array}{r} 0.05 \\ \times\ 0.05 \\ \hline 0.0025 \end{array}$$

OBJECTIVE 5 Use the order of operations rule.

Recall that the order of operations rule is used to evaluate expressions that involve more than one operation.

EXAMPLE 9 Evaluate: $-(0.6)^2 + 5|-3.6 + 1.9|$

Strategy The absolute value bars are grouping symbols. We will perform the addition within them first.

WHY By the order of operations rule, we must perform all calculations within parentheses and other grouping symbols (such as absolute value bars) first.

Solution

$-(0.6)^2 + 5|-3.6 + 1.9|$

$= -(0.6)^2 + 5|-1.7|$ Do the addition within the absolute value symbols. Use the rule for adding two decimals with different signs.

$= -(0.6)^2 + 5(1.7)$ Simplify: $|-1.7| = 1.7$.

$= -0.36 + 5(1.7)$ Evaluate: $(0.6)^2 = 0.36$.

$= -0.36 + 8.5$ Do the multiplication: $5(1.7) = 8.5$.

$= 8.14$ Use the rule for adding two decimals with different signs.

$$\begin{array}{r} \overset{2\ \ 16}{\cancel{3}.\cancel{6}} \\ -\ 1.9 \\ \hline 1.7 \end{array}$$

$$\begin{array}{r} \overset{3}{1.7} \\ \times\ 5 \\ \hline 8.5 \end{array}$$

$$\begin{array}{r} \overset{4\,10}{8.\cancel{5}\cancel{0}} \\ -0.36 \\ \hline 8.14 \end{array}$$

Self Check 9

Evaluate:
$-2|-4.4 + 5.6| + (-0.8)^2$

Now Try ➲ Problem 49

OBJECTIVE 6 Evaluate formulas.

Recall that to evaluate a formula, we replace the letters (called **variables**) with specific numbers and then use the order of operations rule.

EXAMPLE 10 Evaluate the formula $S = 6.28r(h + r)$ for $h = 3.1$ and $r = 6$.

Strategy In the given formula, we will replace the letter r with 6 and h with 3.1.

WHY Then we can use the order of operations rule to find the value of the expression on the right side of the = symbol.

Solution

$$S = 6.28r(h + r) \qquad \text{6.28}r(h + r) \text{ means } 6.28 \cdot r \cdot (h + r).$$

$$S = 6.28(6)(3.1 + 6) \qquad \text{Replace } r \text{ with 6 and } h \text{ with 3.1.}$$

$$S = 6.28(6)(9.1) \qquad \text{Do the addition within the parentheses.}$$

$$S = 37.68(9.1) \qquad \text{Do the multiplication: } 6.28(6) = 37.68.$$

$$S = 342.888 \qquad \text{Do the multiplication.}$$

$$\begin{array}{r} 37.68 \\ \times\ 9.1 \\ \hline 3768 \\ 339120 \\ \hline 342.888 \end{array}$$

Self Check 10

Evaluate $V = 1.3\pi r^3$ for $\pi \approx 3.14$ and $r = 3$.

Now Try ➋ Problem 53

OBJECTIVE 7 **Estimate products of decimals.**

Estimation can be used to check the reasonableness of an answer to a decimal multiplication. There are several ways to estimate, but the objective is the same: Simplify the numbers in the problem so that the calculations can be made easily and quickly.

EXAMPLE 11

a. Estimate using front-end rounding: $27 \cdot 6.41$

b. Estimate by rounding each factor to the nearest tenth: $13.91 \cdot 5.27$

c. Estimate by rounding: $0.1245(101.4)$

Strategy We will use rounding to approximate the factors. Then we will find the product of the approximations.

WHY Rounding produces factors that contain fewer digits. Such numbers are easier to multiply.

Solution

a. To estimate $27 \cdot 6.41$ by front-end rounding, we begin by rounding both factors to their *largest* place value.

$$\begin{array}{rcr} 27 & \longrightarrow & 30 \qquad \text{Round to the nearest ten.} \\ \times\ 6.41 & \longrightarrow & \times\ \ 6 \qquad \text{Round to the nearest one.} \\ \hline & & 180 \end{array}$$

The estimate is 180. If we calculate $27 \cdot 6.41$, the product is exactly 173.07. The estimate is close: It's about 7 more than 173.07.

b. To estimate $13.91 \cdot 5.27$, we will round both decimals to the nearest tenth.

$$\begin{array}{rcr} 13.91 & \longrightarrow & 13.9 \qquad \text{Round to the nearest tenth.} \\ \times\ 5.27 & \longrightarrow & \times\ 5.3 \qquad \text{Round to the nearest tenth.} \\ \hline & & 417 \\ & & 6950 \\ \hline & & 73.67 \end{array}$$

The estimate is 73.67. If we calculate $13.91 \cdot 5.27$, the product is exactly 73.3057. The estimate is close: It's just slightly more than 73.3057.

c. Since 101.4 is approximately 100, we can estimate $0.1245(101.4)$ using $0.1245(100)$.

$$0.1245(100) = 12.45 \qquad \text{Since 100 has two zeros, move the decimal point in}$$
$$\text{0.1245 two places to the right.}$$

The estimate is 12.45. If we calculate $0.1245(101.4)$, the product is exactly 12.6243. Note that the estimate is close: It's slightly less than 12.6243.

Self Check 11

a. Estimate using front-end rounding: $4.337 \cdot 65$

b. Estimate by rounding the factors to the nearest tenth: $3.092 \cdot 11.642$

c. Estimate by rounding: $0.7899(985.34)$

Now Try ➋ Problems 61 and 63

OBJECTIVE 8 Solve application problems by multiplying decimals.

Application problems that involve repeated addition are often more easily solved using multiplication.

EXAMPLE 12 **Coins.** Banks wrap pennies in rolls of 50 coins. If a penny is 1.55 millimeters thick, how tall is a stack of 50 pennies?

Analyze
- There are 50 pennies in a stack.   *Given*
- A penny is 1.55 millimeters thick.   *Given*
- How tall is a stack of 50 pennies?   *Find*

Form The height (in millimeters) of a stack of 50 pennies, each of which is 1.55 thick, is the sum of fifty 1.55's. This repeated addition can be calculated more simply by multiplication.

The height of a stack of pennies	is equal to	the thickness of one penny	times	the number of pennies in the stack.
The height of a stack of pennies	=	1.55	·	50

Calculate Use vertical form to perform the multiplication:

$$
\begin{array}{r}
1.55 \\
\times \quad 50 \\
\hline
000 \\
7750 \\
\hline
77.50
\end{array}
$$

State A stack of 50 pennies is 77.5 millimeters tall.

Check We can estimate to check the result. If we use 2 millimeters to approximate the thickness of one penny, then the height of a stack of 50 pennies is about 2 · 50 millimeters = 100 millimeters. The result, 77.5 mm, seems reasonable.

Self Check 12

Coins. Banks wrap nickels in rolls of 40 coins. If a nickel is 1.95 millimeters thick, how tall is a stack of 40 nickels?

Now Try ➡ **Problem 101**

Sometimes more than one operation is needed to solve a problem involving decimals.

EXAMPLE 13 **Weekly earnings.** A cashier's basic workweek is 40 hours. After his daily shift is over, he can work overtime at a rate 1.5 times his regular rate of $13.10 per hour. How much money will he earn in a week if he works 6 hours of overtime?

Analyze
- A cashier's basic workweek is 40 hours.   *Given*
- His overtime pay rate is 1.5 times his regular rate of $13.10 per hour. *Given*
- How much money will he earn in a week if he works his regular shift and 6 hours overtime?   *Find*

Form To find the cashier's overtime pay rate, we multiply 1.5 times his regular pay rate, $13.10.

$$
\begin{array}{r}
13.10 \\
\times \quad 1.5 \\
\hline
6550 \\
13100 \\
\hline
19.650
\end{array}
$$

The cashier's overtime pay rate is $19.65 per hour.

We now translate the words of the problem to numbers and symbols.

The total amount the cashier earns in a week	is equal to	40 hours	times	his regular pay rate	plus	the number of overtime hours	times	his overtime rate.
The total amount the cashier earns in a week	=	40	·	$13.10	+	6	·	$19.65

Calculate We will use the rule for the order of operations to evaluate the expression:

$40 \cdot 13.10 + 6 \cdot 19.65 = 524.00 + 117.90$ Do the multiplication first.

$= 641.90$ Do the addition.

```
        53 3
13.10   19.65
×  40  ×   6
0000   117.90
52400
524.00
```

```
  1
524.00
+117.90
641.90
```

State The cashier will earn a total of $641.90 for the week.

Check We can use estimation to check. The cashier works 40 hours per week for approximately $13 per hour to earn about $40 \cdot \$13 = \$520$. His 6 hours of overtime at approximately $20 per hour earns him about $6 \cdot \$20 = \$120$. His total earnings that week are about $\$520 + \$120 = \$640$. The result, $641.90, seems reasonable.

Self Check 13

Weekly earnings. A pharmacy assistant's basic workweek is 40 hours. After her daily shift is over, she can work overtime at a rate of 1.5 times her regular rate of $15.90 per hour. How much money will she earn in a week if she works 4 hours of overtime?

Now Try ➡ **Problem 117**

Answers to Self Checks

1. 11.61   **2.** 0.00144   **3.** 836.6   **4. a.** 72.1   **b.** 6,080   **5. a.** 12.99   **b.** 0.00000002   **6. a.** 2,070,000
b. 10,600,000,000   **c.** 1,818,000,000,000   **7. a.** −36.3   **b.** 4,496.8   **8. a.** 1.69   **b.** 0.0081   **9.** −1.76
10. 110.214   **11. a.** 280   **b.** 35.96   **c.** 789.9   **12.** 78 mm   **13.** $731.40

SECTION 4.3 STUDY SET

VOCABULARY

Fill in the blanks.

1. In the multiplication problem shown below, label each *factor*, the *partial products*, and the *product*.

```
  3.4 ←
× 2.6 ←
  204 ←
  680 ←
 8.84 ←
```

2. Numbers such as 10, 100, and 1,000 are called _____ of 10.

CONCEPTS

Fill in the blanks.

3. Insert a decimal point in the correct place for each product shown below. Write placeholder zeros, if necessary.

a.
```
  3.8
× 0.6
  228
```

b.
```
   1.79
×   8.1
   179
 14320
 14499
```

c.
```
  2.0
×   7
  140
```

d.
```
  0.013
× 0.02
 0026
```

4. Fill in the blanks.

a. To find the product of a decimal and 10, 100, 1,000, and so on, move the decimal point to the _____ the same number of places as there are zeros in the power of 10.

b. To find the product of a decimal and 0.1, 0.01, 0.001, and so on, move the decimal point to the _____ the same number of places as there are in the power of 10.

5. Determine whether the *sign* of each result is positive or negative. *You do not have to find the product.*

a. $-7.6(-1.8)$

b. $-4.09 \cdot 2.274$

6. a. When we move its decimal point to the right, does a decimal number get larger or smaller?

b. When we move its decimal point to the left, does a decimal number get larger or smaller?

NOTATION

7. a. List the first five powers of 10 that are greater than 1.

   b. List the first five powers of 10 that are less than 1.

8. Write each number in standard notation.

   a. one million

   b. one billion

   c. one trillion

GUIDED PRACTICE

Multiply. See Example 1.

9. $4.8 \cdot 6.2$

10. $3.5 \cdot 9.3$

11. $5.6(8.9)$

12. $7.2(8.4)$

Multiply. See Example 2.

13. $0.003(2.7)$

14. $0.002(2.6)$

15. $\begin{array}{r} 5.8 \\ \times\, 0.009 \end{array}$

16. $\begin{array}{r} 8.7 \\ \times\, 0.004 \end{array}$

Multiply. See Example 3.

17. $179(6.3)$

18. $225(4.9)$

19. $\begin{array}{r} 316 \\ \times\, 7.4 \end{array}$

20. $\begin{array}{r} 527 \\ \times\, 3.7 \end{array}$

Multiply. See Example 4.

21. $6.84 \cdot 100$

22. $2.09 \cdot 100$

23. $0.041(10,000)$

24. $0.034(10,000)$

Multiply. See Example 5.

25. $647.59 \cdot 0.01$

26. $317.09 \cdot 0.01$

27. $1.15(0.001)$

28. $2.83(0.001)$

Write each number in standard notation. See Example 6.

29. 14.2 million

30. 33.9 million

31. 98.2 billion

32. 80.4 billion

33. 1.421 trillion

34. 3.056 trillion

35. 657.1 billion

36. 422.7 billion

Multiply. See Example 7.

37. $-1.9(7.2)$

38. $-5.8(3.9)$

39. $-3.3(-1.6)$

40. $-4.7(-2.2)$

41. $(-10,000)(-44.83)$

42. $(-10,000)(-13.19)$

43. $678.231(-1,000)$

44. $491.565(-1,000)$

Evaluate each expression. See Example 8.

45. $(3.4)^2$

46. $(5.1)^2$

47. $(-0.03)^2$

48. $(-0.06)^2$

Evaluate each expression. See Example 9.

49. $-(-0.2)^2 + 4|-2.3 + 1.5|$

50. $-(-0.3)^2 + 6|-6.4 + 1.7|$

51. $-(-0.8)^2 + 7|-5.1 - 4.8|$

52. $-(-0.4)^2 + 6|-6.2 - 3.5|$

Evaluate each formula. See Example 10.

53. $A = P + Prt$ for $P = 85.50$, $r = 0.08$, and $t = 5$

54. $A = P + Prt$ for $P = 99.95$, $r = 0.05$, and $t = 10$

55. $A = lw$ for $l = 5.3$ and $w = 7.2$

56. $A = 0.5bh$ for $b = 7.5$ and $h = 6.8$

57. $P = 2l + 2w$ for $l = 3.7$ and $w = 3.6$

58. $P = a + b + c$ for $a = 12.91$, $b = 19$, and $c = 23.6$

59. $C = 2\pi r$ for $\pi \approx 3.14$ and $r = 2.5$

60. $A = \pi r^2$ for $\pi \approx 3.14$ and $r = 4.2$

Estimate each product using front-end rounding. See Example 11.

61. $46 \cdot 5.3$

62. $37 \cdot 4.29$

Estimate each product by rounding the factors to the nearest tenth. See Example 11.

63. $17.11 \cdot 3.85$

64. $18.33 \cdot 6.46$

TRY IT YOURSELF

Perform the indicated operations.

65. $-0.56 \cdot 0.33$

66. $(-3.7)(0.4)$

67. $(-0.1)^2$

68. $(-0.6)^2$

69. $(-0.7 - 0.5)(2.4 - 3.1)$

70. $(-8.1 - 7.8)(0.3 + 0.7)$

71. $\begin{array}{r} 0.008 \\ \times\, 0.09 \end{array}$

72. $\begin{array}{r} 0.003 \\ \times\, 0.09 \end{array}$

73. $-0.2 \cdot 1,000,000$

74. $-1,000,000 \cdot 1.9$

75. $(-5.6)(-2.2)$

76. $(-7.1)(-4.1)$

77. $-4.6(23.4 - 19.6)$

78. $6.9(9.8 - 8.9)$

79. $(-4.9)(-0.001)$

80. $(-0.001)(-7.09)$

81. $(-0.2)^2 + 2(7.1)$

82. $(-6.3)(3) - (1.2)^2$

83. $\begin{array}{r} 2.13 \\ \times\,4.05 \\ \hline \end{array}$

84. $\begin{array}{r} 3.06 \\ \times\,1.82 \\ \hline \end{array}$

85. $-7(8.1781)$

86. $-5(4.7199)$

87. $-1,000(0.02239)$

88. $-100(0.0897)$

89. $(0.5 + 0.6)^2(-3.2)$

90. $(-5.1)(4.9 - 3.4)^2$

91. $-0.2(306)(-0.4)$

92. $(-2.5)(-4)(-9)$

93. $-0.01(|-2.6 - 6.7|)^2$

94. $-0.01(|-8.16 + 9.9|)^2$

Complete each table.

95.

Decimal	Its square
0.1	
0.2	
0.3	
0.4	
0.5	
0.6	
0.7	
0.8	
0.9	

96.

Decimal	Its cube
0.1	
0.2	
0.3	
0.4	
0.5	
0.6	
0.7	
0.8	
0.9	

LOOK ALIKES

Write each number in standard notation.

97. a. 72.31 million **b.** 72.31 billion **c.** 72.31 trillion

Evaluate each expression.

98. a. $(0.4)^2$ **b.** $(0.4)^3$ **c.** $(0.4)^4$

Perform each operation.

99. a. $5.7 + 3.8$ **b.** $5.7 - 3.8$ **c.** $5.7 \cdot 3.8$

100. a. $68.2(1,000)$ **b.** $68.2(10)$ **c.** $68.2(0.01)$
   d. $68.2(100)$ **e.** $68.2(0.000001)$ **f.** $68.2(-10,000)$

APPLICATIONS

101. Reams of paper. Find the thickness of a 500-sheet ream of copier paper if each sheet is 0.0038 inch thick.

102. Mileage claims. Each month, a salesman is reimbursed by his company for any work-related travel that he does in his own car at the rate of $0.565 per mile. How much will the salesman receive if he traveled a total of 120 miles in his car on business in the month of June?

103. Salaries. Use the following formula to determine the annual salary of a recording engineer who works 38 hours per week at a rate of $37.35 per hour. Round the result to the nearest hundred dollars.

Annual salary	=	hourly rate	·	hours per week	·	52.2 weeks

104. Paychecks. If you are paid every other week, your monthly gross income is your gross income from one paycheck times 2.17. Find the monthly gross income of a supermarket clerk who earns $1,095.70 every two weeks. Round the result to the nearest cent.

105. Bakery supplies. A bakery buys various types of nuts as ingredients for cookies. Complete the table by filling in the cost of each purchase.

Type of nut	Price per pound	Pounds	Cost
Almonds	$9.99	16	
Walnuts	$4.95	25	

106. New homes. Find the cost to build the home shown below if construction costs are $92.55 per square foot.

House Plan #DP-2203
Square Feet: **2,291 Sq Ft.**
Stories: **Single Story**
Bedrooms: **3**
Bathrooms: **3**
Garage Bays: **2**

107. Biology. Cells contain DNA. In humans, it determines such traits as eye color, hair color, and height. A model of DNA appears below. If 1 Å (angstrom) = 0.000000004 inch, find the dimensions of 34 Å, 3.4 Å, and 10 Å, shown in the illustration.

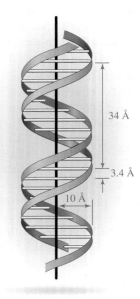

108. Tachometers.

a. Estimate the decimal number to which the tachometer needle points in the illustration below.

b. What engine speed (in rpm) does the tachometer indicate?

109. City planning. The streets shown in blue on the city map below are 0.35 mile apart. Find the distance of each trip between the two given locations.

a. The airport to the Convention Center

b. City Hall to the Convention Center

c. The airport to City Hall

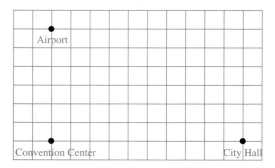

110. Retrofits. The illustration below shows the current widths of the three columns of a freeway overpass. A computer analysis indicated that the width of each column should actually be 1.4 times what it currently is to withstand the stresses of an earthquake. According to the analysis, how wide should each of the columns be?

111. Electric bills. When billing a household, a utility company charges for the number of kilowatt-hours used. A kilowatt-hour (kwh) is a standard measure of electricity. If the cost of 1 kwh is $0.14277, what is the electric bill for a household that uses 719 kwh in a month? Round the answer to the nearest cent.

112. Utility taxes. Some gas companies are required to tax the number of therms used each month by the customer. What are the taxes collected on a monthly usage of 31 therms if the tax rate is $0.00566 per therm? Round the answer to the nearest cent.

113. Write each highlighted number in standard form.

a. **Conservation.** The *19.6-million acre* Arctic National Wildlife Refuge is located in the northeast corner of Alaska. (Source: National Wildlife Federation)

b. **Population.** According to the website census.gov, at 03:07 P.M. eastern time on Saturday, March 4, 2017, the population of the Earth was *7.376 billion* people.

c. **Driving.** The U.S. Federal Highway Administration estimated that Americans drove a total of *3.148 trillion miles* in 2015.

114. Write each highlighted number in standard form.

a. **Mileage.** As of February 28, 2016, Irv Gordon of Long Island, New York, had driven a record *3.174 million miles* in his 1966 Volvo P-1800. (Source: newyork.cbslocal)

b. **E-Commerce.** Online spending during the 2016 holiday season (November 1 through December 23) was about *$63.1 billion*. (Source: comscore.com)

c. **Federal debt.** On March 4, 2017, at 3:16 P.M. eastern time, the U.S. national debt limit was *$19.981 trillion.*

115. Soccer. A soccer goal is rectangular and measures 24 feet wide by 8 feet high. Major league soccer officials are proposing to increase its width by 1.5 feet and increase its height by 0.75 foot.

a. What is the area of the goal opening now?

b. What would the area be if the proposal were adopted?

c. How much area would be added?

116. Salt intake. Studies done by the Centers for Disease Control and Prevention found that the average American eats 3.436 grams of salt each day. The recommended amount is 1.5 grams per day. How many more grams of salt does the average American eat in one week compared with what the Center recommends?

117. Concert seating. Two types of tickets were sold for a concert. Floor seating costs $12.50 a ticket, and balcony seats cost $15.75.

   a. Complete the following table and find the receipts from each type of ticket.

   b. Find the total receipts from the sale of both types of tickets.

Ticket type	Price	Number sold	Receipts
Floor		1,000	
Balcony		100	

118. Plumbing bills. A corner of the invoice for plumbing work is torn. What is the labor charge for the 4 hours of work? What is the total charge (standard service charge, parts, labor)?

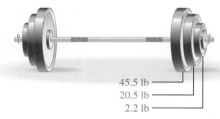

Carter Plumbing 100 W. Dalton Ave.		Invoice #210
Standard service charge	$	25.75
Parts	$	38.75
Labor: 4 hr @ $40.55/hr	$	
Total charges	$	

119. Weightlifting. The barbell is evenly loaded with iron plates. How much plate weight is loaded on the barbell?

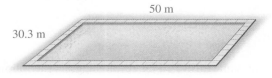

45.5 lb
20.5 lb
2.2 lb

120. Swimming pools. Long bricks, called *coping*, can be used to outline the edge of a swimming pool. How many meters of coping will be needed in the construction of the swimming pool shown?

50 m

30.3 m

121. Storm damage. After a rainstorm, the saturated ground under a hilltop house began to give way. A survey team noted that the house dropped 0.57 inch initially. In the next three weeks, the house fell 0.09 inch per week. How far did the house fall during this three-week period?

122. Water usage. In May, the water level of a reservoir reached its high mark for the year. During the summer months, as water usage increased, the level dropped. In the months of May and June, it fell 4.3 feet each month. In August and September, because of high temperatures, it fell another 8.7 feet each month. By the beginning of October, how far below the year's high mark had the water level fallen?

123. Donations. The graph below shows the daily donations the ALS Association received from the fundraising campaign called the *Ice Bucket Challenge*. This event raised awareness and funds to find treatments and a cure for amyotrophic lateral sclerosis (also know as Lou Gehrig's disease). Find the total amount of contributions from August 12 through August 27, 2014. Write the answer in standard notation.

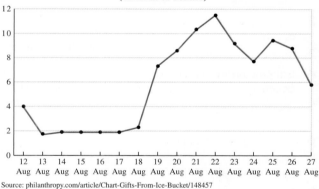

The Ice Bucket Challenge Donation to the ALS Association (millions of dollars)

Source: philanthropy.com/article/Chart-Gifts-From-Ice-Bucket/148457

124. Saving water. Several areas of the country are offering homeowners money to remove grass lawns and replace them with artificial turf or drought-resistant plants. One such program pays the homeowner $4.10 per square foot for the first 1,000 square feet of lawn removed and $2.25 per square foot for any more removed after that.

   a. Estimate the number of square feet of lawn in the landscape drawing shown below.

   b. Use your answer to part a to determine the amount of money the homeowner will receive from the program if all of the grass in the diagram is removed.

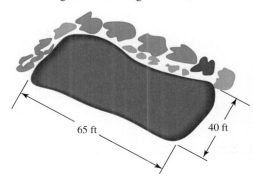

65 ft   40 ft

WRITING

125. Explain how to determine where to place the decimal point in the answer when multiplying two decimals.

126. List the similarities and differences between whole-number multiplication and decimal multiplication.

127. Explain how to multiply a decimal by a power of 10 that is greater than 1, and by a power of 10 that is less than 1.

128. Is it easier to multiply the decimals 0.4 and 0.16 or the fractions $\frac{4}{10}$ and $\frac{16}{100}$? Explain why.

129. Why do we have to line up the decimal points when adding but not when multiplying?

130. Which vertical form for the following multiplication do you like better? Explain why.

$$\begin{array}{r} 0.000003 \\ \times \quad 2.7 \\ \hline \end{array} \qquad \begin{array}{r} 2.7 \\ \times \ 0.000003 \\ \hline \end{array}$$

REVIEW

Find the prime factorization of each number. Use exponents in your answer, when helpful.

131. 220   **132.** 400

133. 162   **134.** 735

SECTION 4.4   Dividing Decimals

OBJECTIVES

1 Divide a decimal by a whole number.

2 Divide a decimal by a decimal.

3 Round a decimal quotient.

4 Estimate quotients of decimals.

5 Divide decimals by powers of 10.

6 Divide signed decimals.

7 Use the order of operations rule.

8 Evaluate formulas.

9 Solve application problems by dividing decimals.

In Chapter 1, we used a process called long division to divide whole numbers.

Long division form

$$\begin{array}{r} 2 \leftarrow \text{Quotient} \\ \text{Divisor} \rightarrow 5\overline{)10} \leftarrow \text{Dividend} \\ \underline{10} \\ 0 \leftarrow \text{Remainder} \end{array}$$

In this section, we consider division problems in which the divisor, the dividend, or both are decimals.

OBJECTIVE 1   Divide a decimal by a whole number.

To develop a rule for decimal division, let's consider the problem $47 \div 10$. If we rewrite the division as $\frac{47}{10}$, we can use the long division method from Chapter 3 for changing an improper fraction to a mixed number to find the answer:

$$\begin{array}{r} 4\frac{7}{10} \\ 10\overline{)47} \\ \underline{-40} \\ 7 \end{array}$$

Here the result is written in quotient $+ \dfrac{\text{remainder}}{\text{divisor}}$ form.

To perform this same division using decimals, we write 47 as 47.0 and divide as we would divide whole numbers.

Note that the decimal point in the quotient (answer) is placed directly above the decimal point in the dividend.

$$\begin{array}{r} 4.7 \\ 10\overline{)47.0} \\ \underline{-40} \downarrow \\ 7\,0 \\ \underline{-7\,0} \\ 0 \end{array}$$

After subtracting 40 from 47, bring down the 0 and continue to divide.

The remainder is 0.

Since $4\frac{7}{10} = 4.7$, either method gives the same answer. This result suggests the following method for dividing a decimal by a whole number.

Dividing a Decimal by a Whole Number

To divide a decimal by a whole number:

1. Write the problem in long division form and place a decimal point in the quotient (answer) directly above the decimal point in the dividend.

2. Divide as if working with whole numbers.

3. If necessary, additional zeros can be written to the right of the last digit of the dividend to continue the division.

EXAMPLE 1   Divide: $42.6 \div 6$. Check the result.

Strategy Since the divisor, 6, is a whole number, we will write the problem in long division form and place a decimal point directly above the decimal point in 42.6. Then we will divide as if the problem were $426 \div 6$.

WHY To divide a decimal by a whole number, we divide as if working with whole numbers.

Solution

Step 1

Place a decimal point in the quotient that lines up with the decimal point in the dividend.

$$6\overline{)42\,.\,6}$$

Step 2 Now divide using the four-step division process: **estimate**, **multiply**, **subtract**, and **bring down**.

$$\begin{array}{r} 7.1 \\ 6\overline{)42.6} \\ -42 \\ \hline 0\,6 \\ -6 \\ \hline 0 \end{array}$$

Ignore the decimal points and divide as if working with whole numbers.

After subtracting 42 from 42, bring down the 6 and continue to divide.

The remainder is 0.

In Section 1.5, we checked whole-number division using multiplication. Decimal division is checked in the same way: *The product of the quotient and the divisor should be the dividend.*

$$\begin{array}{r} 7.1 \leftarrow \text{Quotient} \\ \times6 \leftarrow \text{Divisor} \\ \hline 42.6 \; -\text{Dividend} \end{array} \qquad 6\overline{)42.6}$$

The check confirms that $42.6 \div 6 = 7.1$.

Self Check 1

Divide: $20.8 \div 4$. Check the result.

Now Try Problem 15

EXAMPLE 2   Divide: $71.68 \div 28$

Strategy Since the divisor is a whole number, 28, we will write the problem in long division form and place a decimal point directly above the decimal point in 71.68. Then we will divide as if the problem were $7,168 \div 28$.

WHY To divide a decimal by a whole number, we divide as if working with whole numbers.

Solution

Write the decimal point in the quotient (answer) directly
above the decimal point in the dividend.

$$
\begin{array}{r}
2.56 \\
28\overline{)71.68} \\
-56 \\
\hline
15\,6 \\
-14\,0 \\
\hline
1\,68 \\
-1\,68 \\
\hline
0
\end{array}
$$

Ignore the decimal points and divide as if working
with whole numbers.

After subtracting 56 from 71, bring down the 6
and continue to divide.

After subtracting 140 from 156, bring down the 8
and continue to divide.

The remainder is 0.

We can use multiplication to check this result.

$$
\begin{array}{r}
2.56 \\
\times\quad 28 \\
\hline
2048 \\
5120 \\
\hline
\mathbf{71.68}
\end{array}
\qquad
\begin{array}{r}
2.56 \\
28\overline{)71.68}
\end{array}
$$

Self Check 2

Divide: $101.44 \div 32$

Now Try ⮑ **Problem 19**

The check confirms that $71.68 \div 28 = 2.56$.

EXAMPLE 3   Divide: $19.2 \div 5$

Strategy We will write the problem in long division form, place a decimal point directly above the decimal point in 19.2, and divide. If necessary, we will write additional zeros to the right of the 2 in 19.2.

WHY Writing additional zeros to the right of the 2 allows us to continue the division process until we obtain a remainder of 0 or the digits in the quotient repeat in a pattern.

Solution

$$
\begin{array}{r}
3.8 \\
5\overline{)19.2} \\
-15 \\
\hline
4\,2 \\
-4\,0 \\
\hline
2
\end{array}
$$

After subtracting 15 from 19, bring down the 2 and continue to divide.

All the digits in the dividend have been used, but the remainder is not 0.

We can write a zero to the right of 2 in the dividend and continue the division process. Recall that writing additional zeros to the right of the decimal point does not change the value of the decimal. That is, $19.2 = 19.20$.

$$
\begin{array}{r}
3.84 \\
5\overline{)19.20} \\
-15 \\
\hline
4\,2 \\
-4\,0 \\
\hline
20 \\
-20 \\
\hline
0
\end{array}
$$

Write a zero to the right of the 2 and bring it down.

Continue to divide.

The remainder is 0.

Check:

$$
\begin{array}{r}
3.84 \\
\times \quad 5 \\
\hline
19.20 \\
\end{array}
$$ ← *Since this is the dividend, the result checks.*

Self Check 3

Divide: $42.8 \div 8$

Now Try ➜ **Problem 23**

OBJECTIVE 2 **Divide a decimal by a decimal.**

To develop a rule for division involving a decimal divisor, let's consider the problem $0.36\overline{)0.2592}$, where the divisor is the decimal 0.36. First, we express the division in fraction form.

$$0.36\overline{)0.2592} \qquad \text{can be represented by} \qquad \frac{0.2592}{0.36}$$

↑————————— Divisor —————————↑

To be able to use the rule for dividing decimals by a *whole number* discussed earlier, we need to move the decimal point in the divisor 0.36 two places to the right. This can be accomplished by multiplying it by 100. However, if the denominator of the fraction is multiplied by 100, the numerator must also be multiplied by 100 so that the fraction maintains the same value. It follows that $\frac{100}{100}$ is the form of 1 that we should use to build $\frac{0.2592}{0.36}$.

$$
\begin{aligned}
\frac{0.2592}{0.36} &= \frac{0.2592}{0.36} \cdot \frac{100}{100} && \text{Multiply by a form of 1.} \\[2mm]
&= \frac{0.2592 \cdot 100}{0.36 \cdot 100} && \text{Multiply the numerators.} \\
& && \text{Multiply the denominators.} \\[2mm]
&= \frac{25.92}{36} && \text{Multiplying both decimals by 100 moves} \\
& && \text{their decimal points two places to the right.}
\end{aligned}
$$

This fraction represents the division problem $36\overline{)25.92}$. From this result, we have the following observations.

- The division problem $0.36\overline{)0.2592}$ is equivalent to $36\overline{)25.92}$; that is, they have the same answer.

- The decimal points in *both* the divisor and the dividend of the first division problem have been moved two decimal places to the right to create the second division problem.

$$0.36\overline{)0.2592} \qquad \text{becomes} \qquad 36\overline{)25.92}$$

These observations illustrate the following rule for division with a decimal divisor.

Division with a Decimal Divisor

To divide with a decimal divisor:

1. Write the problem in long division form.

2. Move the decimal point of the divisor so that it becomes a whole number.

3. Move the decimal point of the dividend the same number of places to the right.

4. Write the decimal point in the quotient (answer) directly above the decimal point in the dividend. Divide as if working with whole numbers.

5. If necessary, additional zeros can be written to the right of the last digit of the dividend to continue the division.

EXAMPLE 4   Divide: $\dfrac{0.2592}{0.36}$

Strategy We will move the decimal point of the divisor, 0.36, two places to the right, and we will move the decimal point of the dividend, 0.2592, the same number of places to the right.

WHY We can then use the rule for dividing a decimal by a *whole number*.

Solution We begin by writing the problem in long division form.

$$0\,36\overline{)0\,25\,.\,92}$$

Move the decimal point two places to the right in the divisor and the dividend. Write the decimal point in the quotient (answer) directly above the decimal point in the dividend.

Since the divisor is now a whole number, we can use the rule for dividing a decimal by a whole number to find the quotient.

$$
\begin{array}{r}
0.72 \\
36\overline{)25.92} \\
-25\,2\downarrow \\
\hline
72 \\
-72 \\
\hline
0
\end{array}
$$

Now divide as with whole numbers.

Check:

$$
\begin{array}{r}
0.72 \\
\times\quad 36 \\
\hline
432 \\
2160 \\
\hline
25.92
\end{array}
$$

Since this is the dividend, the result checks.

OBJECTIVE 3   Round a decimal quotient.

In Example 4, the division process stopped after we obtained a 0 from the second subtraction. Sometimes when we divide, the subtractions never give a zero remainder, and the division process continues forever. In such cases, we can round the result.

EXAMPLE 5   Divide: $\dfrac{9.35}{0.7}$. Round the quotient to the nearest hundredth.

Strategy We will use the methods of this section to divide to the thousandths column.

WHY To round to the hundredths column, we need to continue the division process for one more decimal place, which is the thousandths column.

Solution We begin by writing the problem in long division form.

$$0\,7\overline{)93\,.\,5}$$

To write the divisor as a whole number, move the decimal point one place to the right. Do the same for the dividend. Place the decimal point in the quotient (answer) directly above the decimal point in the dividend.

We need to write two zeros to the right of the last digit of the dividend so that we can divide to the thousandths column.

$$7\overline{)93.500}$$

Success Tip When dividing decimals, moving the decimal points the same number of places to the right in *both* the divisor and the dividend does not change the answer.

After dividing to the thousandths column, we round to the hundredths column.

The rounding digit in the hundredths column is 5.
The test digit in the thousandths column is 7.

$$
\begin{array}{r}
13.357 \\
7{\overline{\smash{\big)}\,93.500}} \\
\underline{-\ 7} \\
23 \\
\underline{-\ 21} \\
2\ 5 \\
\underline{-\ 2\ 1} \\
40 \\
\underline{-\ 35} \\
50 \\
\underline{-\ 49} \\
1
\end{array}
$$

The division process can stop. We have divided to the thousandths column.

Since the test digit 7 is 5 or greater, we will round 13.357 up to approximate the quotient to the nearest hundredth.

$$\frac{9.35}{0.7} \approx 13.36 \qquad \text{Read} \approx \text{as "is approximately equal to."}$$

Check:

$$
\begin{array}{r}
13.36 \quad \leftarrow \text{The approximation of the quotient}\\
\times\ \ 0.7 \quad \leftarrow \text{The original divisor}\\
\hline
9.352 \quad \leftarrow \text{Since this is close to the original dividend, 9.35, the result seems}\\
\text{reasonable.}
\end{array}
$$

Success Tip To round a quotient to a certain decimal place value, continue the division process one more column to its right to find the *test digit*.

Self Check 5

Divide: $12.82 \div 0.9$. Round the quotient to the nearest hundredth.

Now Try ➲ **Problem 33**

Using Your Calculator ▶ Dividing Decimals

The nucleus of a cell contains vital information about the cell in the form of DNA. The nucleus is very small: A typical animal cell has a nucleus that is only 0.00023622 inch across. How many nuclei (plural of *nucleus*) would have to be laid end to end to extend to a length of 1 inch?

To find how many 0.00023622-inch lengths there are in 1 inch, we must use division: $1 \div 0.00023622$.

$$1\ \boxed{\div}\ .00023622\ \boxed{=} \qquad\qquad \boxed{4233.3418}$$

On some calculators, we press the $\boxed{\text{ENTER}}$ key to display the quotient.

It would take approximately 4,233 nuclei laid end to end to extend to a length of 1 inch.

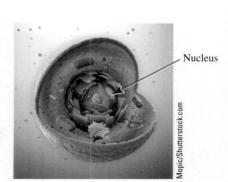

Nucleus

Mopic/Shutterstock.com

OBJECTIVE 4 **Estimate quotients of decimals.**

There are many ways to make an error when dividing decimals. Estimation is a helpful tool that can be used to determine whether or not an answer seems reasonable.

To estimate quotients, we use a method that approximates both the dividend and the divisor so that they divide easily. There is one rule of thumb for this method: If possible, round both numbers up or both numbers down.

EXAMPLE 6   Estimate the quotient: $248.687 \div 43.1$

Strategy We will round the dividend and the divisor down and find $240 \div 40$.

WHY The division can be made easier if the dividend and the divisor end with zeros. Also, 40 divides 240 exactly.

Solution

Self Check 6

Estimate the quotient:
$6,229.249 \div 68.9$

Now Try ➡ Problems 35 and 39

The dividend is approximately

$$248.687 \div 43.1 \qquad 240 \div 40 = 6$$

The divisor is approximately

To divide, drop one zero from 240 and from 40, and find $24 \div 4$.

The estimate is 6.

If we calculate $248.687 \div 43.1$, the quotient is exactly 5.77. Note that the estimate is close: It's just 0.23 more than 5.77.

OBJECTIVE 5   **Divide decimals by powers of 10.**

To develop a set of rules for division of decimals by a power of 10, we consider the problems $8.13 \div 10$ and $8.13 \div 0.1$.

$$
\begin{array}{r}
0.813 \\
10\overline{)8.130} \\
-\,8\,0\downarrow \\
\hline
13 \\
-\,10\downarrow \\
\hline
30 \\
-\,30 \\
\hline
0
\end{array}
$$
Write a zero to the right of the 3.

$$
\begin{array}{r}
81.3 \\
0\,1\overline{)81.3} \\
-\,8\downarrow \\
\hline
1 \\
-\,1\downarrow \\
\hline
3 \\
-\,3 \\
\hline
0
\end{array}
$$
Move the decimal points in the divisor and dividend one place to the right.

Note that the quotients, 0.813 and 81.3, and the dividend, 8.13, are the same except for the location of the decimal points. The first quotient, 0.813, can be easily obtained by moving the decimal point of the dividend one place to the left. The second quotient, 81.3, is easily obtained by moving the decimal point of the dividend one place to the right. These observations illustrate the following rules for dividing a decimal by a power of 10.

Dividing a Decimal by 10, 100, 1,000, and So On

To find the quotient of a decimal and 10, 100, 1,000, and so on, move the decimal point to the left the same number of places as there are zeros in the power of 10.

Dividing a Decimal by 0.1, 0.01, 0.001, and So On

To find the quotient of a decimal and 0.1, 0.01, 0.001, and so on, move the decimal point to the right the same number of decimal places as there are in the power of 10.

EXAMPLE 7   Find each quotient:

a. $16.74 \div 10$ **b.** $8.6 \div 10,000$ **c.** $\dfrac{290.623}{0.01}$

Strategy We will identify the divisor in each division. If it is a power of 10 greater than 1, we will count the number of zeros that it has. If it is a power of 10 less than 1, we will count the number of decimal places that it has.

WHY Then we will know how many places to the right or left to move the decimal point in the dividend to find the quotient.

Solution

a. $16.74 \div 10 = 1.674$ Since the divisor 10 has one zero, move the decimal point one place to the left.

b. $8.6 \div 10,000 = .00086$ Since the divisor 10,000 has four zeros, move the decimal point four places to the left. Write three placeholder zeros (shown in blue).

$= 0.00086$

c. $\dfrac{290.623}{0.01} = 29062.3$ Since the divisor 0.01 has *two* decimal places, move the decimal point in 290.623 *two* places to the right.

Self Check 7

Find each quotient:
a. $721.3 \div 100$

b. $\dfrac{1.07}{1,000}$

c. $19.4407 \div 0.0001$

Now Try ➲ Problems 43 and 49

OBJECTIVE 6 Divide signed decimals.

The rules for dividing integers also hold for dividing signed decimals. The quotient of two decimals with *like signs* is positive, and the quotient of two decimals with *unlike signs* is negative.

EXAMPLE 8   Divide: **a.** $-104.483 \div 16.3$ **b.** $\dfrac{-38.677}{-0.1}$

Strategy In part a, we will use the rule for dividing signed decimals that have different (unlike) signs. In part b, we will use the rule for dividing signed decimals that have the same (like) signs.

WHY In part a, the divisor is positive and the dividend is negative. In part b, both the dividend and divisor are negative.

Solution

a. First, we find the absolute values: $|-104.483| = 104.483$ and $|16.3| = 16.3$. Then we divide the absolute values, 104.483 by 16.3, using the methods of this section.

$$
\begin{array}{r}
6.41 \\
163\overline{)1044.83} \\
-978 \\
\hline
66\,8 \\
-65\,2 \\
\hline
1\,63 \\
-1\,63 \\
\hline
0
\end{array}
$$

Move the decimal point in the divisor and the dividend one place to the right.

Write the decimal point in the quotient (answer) directly above the decimal point in the dividend.

Divide as if working with whole numbers.

Since the signs of the original dividend and divisor are unlike, we make the final answer negative. Thus,

$$-104.483 \div 16.3 = -6.41$$

Check the result using multiplication.

Self Check 8

Divide:
a. $-100.624 \div 15.2$

b. $\dfrac{-23.9}{-0.1}$

Now Try ➡ Problems 51 and 55

b. We can use the rule for dividing a decimal by a power of 10 to find the quotient.

$$\frac{-38.677}{-0.1} = 386.\underset{\curvearrowright}{77}$$

Since the divisor 0.1 has one decimal place, move the decimal point in 38.677 one place to the right. Since the dividend and divisor have like signs, the quotient is positive.

OBJECTIVE 7 Use the order of operations rule.

Recall that the order of operations rule is used to evaluate expressions that involve more than one operation.

EXAMPLE 9 Evaluate: $\dfrac{2(0.351) + 0.5592}{0.2 - 0.6}$

Strategy We will evaluate the expression above and the expression below the fraction bar separately. Then we will do the indicated division, if possible.

WHY Fraction bars are grouping symbols. They group the numerator and denominator.

Solution

$$\frac{2(0.351) + 0.5592}{0.2 - 0.6}$$

$$= \frac{0.702 + 0.5592}{-0.4}$$ In the numerator, do the multiplication. In the denominator, do the subtraction.

$$= \frac{1.2612}{-0.4}$$ In the numerator, do the addition.

$$= -3.153$$ Do the division indicated by the fraction bar. The quotient of two numbers with unlike signs is negative.

$$\begin{array}{r} \overset{1}{0.351} \\ \times \quad 2 \\ \hline 0.702 \end{array} \qquad \begin{array}{r} \overset{1}{0.70}\overset{1}{2}0 \\ + 0.5592 \\ \hline 1.2612 \end{array}$$

$$\begin{array}{r} 3.153 \\ 4\overline{)12.612} \\ -\ 12 \\ \hline 6 \\ -4 \\ \hline 21 \\ -20 \\ \hline 12 \\ -12 \\ \hline 0 \end{array}$$

Self Check 9

Evaluate:
$$\frac{2.7756 + 3(-0.63)}{0.4 - 1.2}$$

Now Try ➡ Problem 59

OBJECTIVE 8 Evaluate formulas.

EXAMPLE 10 Evaluate the formula $b = \dfrac{2A}{h}$ for $A = 15.36$ and $h = 6.4$.

Strategy In the given formula, we will replace the letter A with 15.36 and h with 6.4.

WHY Then we can use the order of operations rule to find the value of the expression on the right side of the $=$ symbol.

Solution

$$b = \frac{2A}{h}$$ This is the given formula.

$$b = \frac{2(15.36)}{6.4}$$ Replace A with 15.36 and h with 6.4.

$$\begin{array}{r} \overset{1}{1}\overset{1}{5.36} \\ \times \quad 2 \\ \hline 30.72 \end{array}$$

$$b = \frac{30.72}{6.4}$$ *In the numerator, do the multiplication.*

$$b = 4.8$$ *Do the division indicated by the fraction bar.*

$$\begin{array}{r} 4.8 \\ 64\overline{)307.2} \\ -256 \\ \hline 51\,2 \\ -51\,2 \\ \hline 0 \end{array}$$

Self Check 10

Evaluate the formula $l = \frac{A}{w}$ for $A = 5.511$ and $w = 1.002$.

Now Try ➡ **Problem 63**

OBJECTIVE 9   Solve application problems by dividing decimals.

Recall that application problems that involve forming equal-sized groups can be solved by division.

EXAMPLE 11   **French bread.** A bread slicing machine cuts 25-inch-long loaves of French bread into 0.625-inch-thick slices. How many slices are there in one loaf?

Analyze

- 25-inch-long loaves of French bread are cut into slices.   *Given*
- Each slice is 0.625-inch thick.   *Given*
- How many slices are there in one loaf?   *Find*

Form   Cutting a loaf of French bread into equally thick slices indicates division. We translate the words of the problem to numbers and symbols.

The number of slices in a loaf of French bread	is equal to	the length of the loaf of French bread	divided by	the thickness of one slice.
The number of slices in a loaf of French bread	=	25	÷	0.625

Calculate   When we write 25 ÷ 0.625 in long division form, we see that the divisor is a decimal.

$$0.625\overline{)25.000}$$ *To write the divisor as a whole number, move the decimal point three places to the right. To move the decimal point three places to the right in the dividend, three placeholder zeros must be inserted (shown in blue).*

Now that the divisor is a whole number, we can perform the division.

$$\begin{array}{r} 40 \\ 625\overline{)25000} \\ -2500 \\ \hline 00 \\ -0 \\ \hline 0 \end{array}$$

State   There are 40 slices in one loaf of French bread.

Check   The multiplication below verifies that 40 slices, each 0.625-inch thick, makes a 25-inch-long loaf. The result checks.

$$\begin{array}{r} 0.625 \\ \times \quad 40 \\ \hline 0000 \\ 25000 \\ \hline 25.000 \end{array}$$

← *The thickness of one slice of bread (in inches)*
← *The number of slices in one loaf*

← *The length of one loaf of bread (in inches)*

Self Check 11

Fruitcakes. A 9-inch-long fruitcake loaf is cut into 0.25-inch-thick slices. How many slices are there in one fruitcake?

Now Try ➡ **Problem 99**

Recall that the **arithmetic mean**, or **average**, of several numbers is a value around which the numbers are grouped. We use addition and division to find the mean (average).

EXAMPLE 12   **Comparison shopping.**   An online travel website listed the following prices for a round trip airline ticket from Los Angeles International Airport to Honolulu Airport. What is the mean (average) price of a ticket?

| 7:45 a – 11:50 a   LAX–HNL $566.50 |
| 6 h 5 m   Nonstop |

| 8:35 a – 12:35 p   LAX–HNL $608.50 |
| 6 h 0 m   Nonstop |

| 8:25 a – 12:37 p   LAX–HNL $620.70 |
| 6 h 12 m   Nonstop |

| 8:22 a – 12:51 p   LAX–HNL $704.90 |
| 6 h 29 m   Nonstop |

Strategy   We will add 566.50, 608.50, 620.70, and 704.90 and divide the sum by 4.

WHY   To find the mean (average) of a set of values, we divide the sum of the values by the number of values.

Self Check 12

U.S. national parks. Use the following data to determine the average number of visitors per year to the national parks for the years 2012 through 2016. (Source: National Park Service)

Year	Visitors (millions)
2016	331.0
2015	307.2
2014	292.8
2013	273.6
2012	282.8

Now Try ➲ **Problem 107**

Solution

$$\text{Mean} = \frac{566.50 + 608.50 + 620.70 + 704.90}{4} \qquad \text{Since there are 4 prices, divide the sum by 4.}$$

$$\text{Mean} = \frac{2500.6}{4} \qquad \text{In the numerator, do the addition.}$$

$$\text{Mean} = 625.15 \qquad \text{Do the indicated division.}$$

$$\begin{array}{r} 1\,2\,2 \\ 566.50 \\ 608.50 \\ 620.70 \\ +704.90 \\ \hline 2500.60 \end{array}$$

$$\begin{array}{r} 625.15 \\ 4\overline{)2500.60} \\ -24 \\ \hline 10 \\ -8 \\ \hline 20 \\ -20 \\ \hline 6 \\ -4 \\ \hline 20 \\ -20 \\ \hline 0 \end{array}$$

The mean (average) price of a round trip airline ticket from Los Angeles to Honolulu is $625.15.

Think it Through ● GPA

"In considering all of the factors that are important to employers as they recruit students in colleges and universities nationwide, college major, grade point average, and work-related experience usually rise to the top of the list."
—Mary D. Feduccia, Ph.D., Career Services Director, Louisiana State University

A grade point average (GPA) is a weighted average based on the grades received and the number of units (credit hours) taken. A GPA for one semester (or term) is defined as

the quotient of the sum of the grade points earned for each class and the sum of the number of units taken. The number of grade points earned for a class is the product of the number of units assigned to the class and the value of the grade received in the class.

1. Use the table of grade values below to compute the GPA for the student whose semester grade report is shown. Round to the nearest hundredth.

Grade	Value
A	4
B	3
C	2
D	1
F	0

Class	Units	Grade
Geology	4	C
Algebra	5	A
Psychology	3	C
Spanish	2	B

2. If you were enrolled in school last semester (or term), list the classes taken, units assigned, and grades received like those shown in the grade report above. Then calculate your GPA.

Answers to Self Checks

1. 5.2 **2.** 3.17 **3.** 5.35 **4.** 0.93 **5.** 14.24 **6.** $6,300 \div 70 = 630 \div 7 = 90$ **7. a.** 7.213 **b.** 0.00107
c. 194,407 **8. a.** -6.62 **b.** 239 **9.** -1.107 **10.** 5.5 **11.** 36 slices **12.** 371.85 million visitors

---

SECTION 4.4   STUDY SET

VOCABULARY

Fill in the blanks.

1. In the division problem shown below, label the *dividend*, the *divisor*, and the *quotient*.

$$3.17 \leftarrow \underline{\qquad}$$
$$\underline{\qquad} \rightarrow 5)\overline{15.85} \;\nwarrow \underline{\qquad}$$

2. To perform the division $2.7)\overline{9.45}$, we move the decimal point of the divisor so that it becomes the _____ number 27. Then we move the decimal point of the _____ the same number of places to the right.

CONCEPTS

3. A decimal point is missing in each of the following quotients. Write a decimal point in the proper position.

   a. $\dfrac{526}{4)\overline{21.04}}$ **b.** $\dfrac{0008}{3)\overline{0.024}}$

4. **a.** How many places to the right must we move the decimal point in 6.14 so that it becomes a whole number?

   b. When the decimal point in 49.8 is moved three places to the right, what is the resulting number?

5. Move the decimal point in the divisor and the dividend the same number of places so that the divisor becomes a whole number. ***You do not have to find the quotient.***

   a. $1.3)\overline{10.66}$ **b.** $3.71)\overline{16.695}$

6. Fill in the blanks: To divide with a decimal divisor, write the problem in _____ division form. Move the decimal point of the divisor so that it becomes a _____ number. Then move the decimal point of the dividend the same number of places to the _____. Write the decimal point in the quotient directly _____ the decimal point in the dividend and divide as working with whole _____.

7. To perform the division $7.8\overline{)14.562}$, the decimal points in the divisor and dividend are moved 1 place to the right. This is equivalent to multiplying $\dfrac{14.562}{7.8}$ by what form of 1?

8. Use multiplication to check the following division. Is the result correct?

$$\frac{1.917}{0.9} = 2.13$$

9. When rounding a decimal to the hundredths column, to what other column must we look at first?

10. a. When 9.545 is divided by 10, is the answer smaller or larger than 9.545?

   b. When 9.545 is divided by 0.1, is the answer smaller or larger than 9.545?

11. Fill in the blanks.

   a. To find the quotient of a decimal and 10, 100, 1,000, and so on, move the decimal point to the _____ the same number of places as there are zeros in the power of 10.

   b. To find the quotient of a decimal and 0.1, 0.01, 0.001, and so on, move the decimal point to the _____ the same number of decimal places as there are in the power of 10.

12. Determine whether the *sign* of each result is positive or negative. *You do not have to find the quotient.*

   a. $-15.25 \div (-0.5)$

   b. $\dfrac{-25.92}{3.2}$

NOTATION

13. Explain what the red arrows are illustrating in the division problem below.

$467\overline{)3208.7}$

14. The division shown below is not finished. Why was the red 0 written after the 7 in the dividend?

$$\begin{array}{r} 2.3 \\ 2\overline{)4.70} \\ -4 \\ \hline 0\,7 \\ -6 \\ \hline 1 \end{array}$$

GUIDED PRACTICE

Divide. Check the result. **See Example 1.**

15. $12.6 \div 6$

16. $40.8 \div 8$

17. $3\overline{)27.6}$

18. $4\overline{)28.8}$

Divide. Check the result. **See Example 2.**

19. $98.21 \div 23$

20. $190.96 \div 28$

21. $37\overline{)320.05}$

22. $32\overline{)125.12}$

Divide. Check the result. **See Example 3.**

23. $13.4 \div 4$

24. $38.3 \div 5$

25. $5\overline{)22.8}$

26. $6\overline{)28.5}$

Divide. Check the result. **See Example 4.**

27. $\dfrac{0.1932}{0.42}$

28. $\dfrac{0.2436}{0.29}$

29. $0.29\overline{)0.1131}$

30. $0.58\overline{)0.1566}$

Divide. Round the quotient to the nearest hundredth. Check the result. **See Example 5.**

31. $\dfrac{11.83}{0.6}$

32. $\dfrac{16.43}{0.9}$

33. $\dfrac{17.09}{0.7}$

34. $\dfrac{13.07}{0.6}$

Estimate each quotient. **See Example 6.**

35. $289.842 \div 72.1$

36. $284.254 \div 91.4$

37. $383.76 \div 7.8$

38. $348.84 \div 5.7$

39. $3,883.284 \div 48.12$

40. $5,556.521 \div 67.89$

41. $6.1\overline{)15,819.74}$

42. $9.2\overline{)19,460.76}$

Find each quotient. **See Example 7.**

43. $451.78 \div 100$

44. $991.02 \div 100$

45. $\dfrac{30.09}{10,000}$

46. $\dfrac{27.07}{10,000}$

47. $1.25 \div 0.1$

48. $8.62 \div 0.01$

49. $\dfrac{545.2}{0.001}$

50. $\dfrac{67.4}{0.001}$

Divide. **See Example 8.**

51. $-110.336 \div 12.8$

52. $-121.584 \div 14.9$

53. $-91.304 \div (-22.6)$

54. $-66.126 \div (-32.1)$

55. $\dfrac{-20.3257}{-0.001}$

56. $\dfrac{-48.8933}{-0.001}$

57. $0.003 \div (-100)$

58. $0.008 \div (-100)$

Evaluate each expression. See Example 9.

59. $\dfrac{2(0.614) + 2.3854}{0.2 - 0.9}$ **60.** $\dfrac{2(1.242) + 0.8932}{0.4 - 0.8}$

61. $\dfrac{5.409 - 3(1.8)}{(0.3)^2}$ **62.** $\dfrac{1.674 - 5(0.222)}{(0.1)^2}$

Evaluate each formula. See Example 10.

63. $t = \dfrac{d}{r}$ for $d = 211.75$ and $r = 60.5$

64. $h = \dfrac{2A}{b}$ for $A = 9.62$ and $b = 3.7$

65. $r = \dfrac{d}{t}$ for $d = 219.375$ and $t = 3.75$

66. $\pi = \dfrac{C}{d}$ for $C = 14.4513$ and $d = 4.6$ (Round to the nearest hundredth.)

TRY IT YOURSELF

Perform the indicated operations. Round the result to the specified decimal place, when indicated.

67. $4.5\overline{)11.97}$ **68.** $4.1\overline{)14.637}$

69. $\dfrac{75.04}{10}$ **70.** $\dfrac{-1.3}{10}$

71. $8\overline{)0.036}$ **72.** $4\overline{)0.073}$

73. $9\overline{)2.889}$ **74.** $6\overline{)3.378}$

75. $\dfrac{-3(0.2) - 2(3.3)}{30(0.4)^2}$ **76.** $\dfrac{(-1.3)^2 + 9.2}{-2(0.2) - 0.5}$

77. Divide 1.2202 by -0.01.

78. Divide -0.4531 by -0.001.

79. $-5.714 \div 2.4$ (nearest tenth)

80. $-21.21 \div 3.8$ (nearest tenth)

81. $-39 \div (-4)$ **82.** $-26 \div (-8)$

83. $7.8915 \div 0.00001$ **84.** $23.025 \div 0.0001$

85. $\dfrac{-30.6}{5.1}$ **86.** $\dfrac{0.0092}{0.023}$

87. $12.243 \div 0.9$ (nearest hundredth)

88. $13.441 \div 0.6$ (nearest hundredth)

89. $1,000\overline{)34.8}$ **90.** $10,000\overline{)678.9}$

91. $\dfrac{40.7(3 - 8.3)}{0.4 - 0.61}$ (nearest hundredth)

92. $\dfrac{(0.5)^2 - (0.3)^2}{0.005 + 0.1}$ (nearest hundredth)

93. Divide 0.25 by 1.6   **94.** Divide 1.2 by 0.64

LOOK ALIKES

Perform each operation.

95. a. $4.08 - 1.7$ **b.** $4.08 \cdot 1.7$
   c. $4.08 + 1.7$ **d.** $4.08 \div 1.7$

96. a. $\dfrac{0.05}{0.008}$ **b.** $\dfrac{0.008}{0.05}$

97. a. $11.5\overline{)2.3}$ **b.** $2.3\overline{)11.5}$

98. a. $\dfrac{7.29}{1,000}$ **b.** $\dfrac{7.29}{0.1}$ **c.** $\dfrac{7.29}{1,000,000}$

   d. $\dfrac{7.29}{0.001}$ **e.** $\dfrac{7.29}{1,000}$ **f.** $\dfrac{7.29}{-10}$

APPLICATIONS

99. Butcher shops. A meat slicer trims 0.05-inch-thick pieces from a sausage. If the sausage is 14 inches long, how many slices are there in one sausage?

100. Electronics. The volume control on a computer is shown to the right. If the distance between the Low and High settings is 21 cm, how far apart are the equally spaced volume settings?

101. Computers. A computer can do an arithmetic calculation in 0.00003 second. How many of these calculations could it do in 60 seconds?

102. Powerball. In January of 2016, there was a three-way split of the world's largest lottery jackpot of $1.5864 billion. Winning tickets—with the numbers 4-8-19-27-34 and the Powerball number, 10—were sold in California, Florida, and Tennessee. Find each winner's share of the jackpot.

   a. Express your answer in billions of dollars.

   b. Express your answer in standard form.

103. Spray bottles. Each squeeze of the trigger of a spray bottle emits 0.017 ounce of liquid. How many squeezes are there in an 8.5-ounce bottle?

104. Car loans. See the loan statement below. How many more monthly payments must be made to pay off the loan?

American Finance Company	June
Monthly payment:	Paid to date: $547.30
$42.10	Loan balance: $631.50

105. Hiking. Refer to the illustration below to determine how long it will take the person shown to complete the hike. Then determine at what time of the day she will complete the hike.

106. Hourly pay. The graph below shows the average hours worked and the average weekly earnings of U.S. production workers for the years 2000 and 2015. What did the average production worker earn per hour

a. in 2000?   **b.** in 2015?

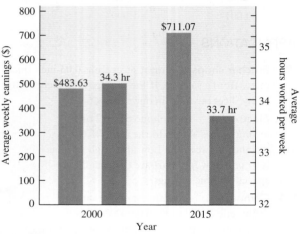

U.S. Production Workers in Manufacturing

Source: U. S. Department of Labor Statistic

107. Travel. The illustration shows the number of visitors to the U.S. for the years 2009–2015, as estimated by the National Travel and Tourism Office, International Trade Administration, U.S. Department of Commerce; World Tourism Organization. Find the average number of visitors per year for this period of time.

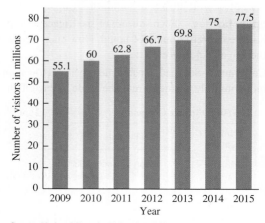

International Travel to the U.S.

Source: National Travel and Tourism Office, Intl. Trade Admin., U.S. Dept. of Commerce, World Tourism Organization

108. Oil wells. Geologists have mapped out the types of soil through which engineers must drill to reach an oil deposit. See the illustration below.

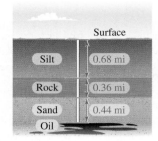

a. How far below the surface is the oil deposit?

b. What is the average depth that must be drilled each week if the drilling is to be a four-week project?

109. Reflexes. An online reaction time test is shown below. When the stoplight changes from red to green, the participant is to immediately click on the large green button. The program then displays the participant's reaction time in the table. After the participant takes the test five times, the *average* reaction time is found. Determine the average reaction time for the results shown below.

Test Number	Reaction Time (in seconds)	The stoplight to watch.	The button to click.
1	0.219		
2	0.233		Click here on green light
3	0.204		
4	0.297		
5	0.202		
AVG.	?		

110. Indy 500. Driver James Hincheliffe, of Canada, had the fastest average qualifying speed for the 2015 Indianapolis 500-mile race. This earned him the *pole position* to begin the race. The speeds for each of his four qualifying laps are shown below. What was his average qualifying speed?

Lap 1: 230.885 mph
Lap 2: 230.940 mph
Lap 3: 230.765 mph
Lap 4: 230.450 mph

Source: indianapolismotorspeedway.com

111. Customer satisfaction. An online store gives its customers an opportunity to rate an item that they purchased using a scale from 1 star (the lowest rating) to 5 stars (the highest rating). This type of feedback helps to inform other potential customers about the specific product. Twelve customer ratings for a steam iron are shown below. Find the average rating for the iron and then shade the correct number of stars on the right below to represent the average rating.

Customer Ratings

Average Rating

☆☆☆☆☆

112. Rating apps. The graph below shows the user ratings given by customers who downloaded a recently released iPhone app. The app is rated on a scale of 5 stars (highest) to 1 star (lowest). We see that 42 users rated it 5 stars, 8 users rated it 4 stars, 30 users rated it 3 stars, and so on.

User Ratings

a. How many customers rated the app?

b. Find the average rating.

113. Explain the process used to divide two numbers when both the divisor and the dividend are decimals. Give an example.

114. Explain why we must sometimes use rounding when we write the answer to a division problem.

115. The division $0.5\overline{)2.005}$ is equivalent to $5\overline{)20.05}$. Explain what equivalent means in this case.

116. In $3\overline{)0.7}$, why can additional zeros be placed to the right of 0.7 without affecting the result?

117. Explain how to estimate the following quotient: $0.75\overline{)2.415}$

118. Explain why multiplying $\dfrac{4.86}{0.2}$ by the form of 1 shown below moves the decimal points in the dividend, 4.86, and the divisor, 0.2, one place to the right.

$$\frac{4.86}{0.2} = \frac{4.86}{0.2} \cdot \frac{10}{10}$$

REVIEW

119. a. Find the GCF of 10 and 25.
b. Find the LCM of 10 and 25.

120. a. Find the GCF of 8, 12, and 16.
b. Find the LCM of 8, 12, and 16.

SECTION 4.5 Fractions and Decimals

In this section, we continue to explore the relationship between fractions and decimals.

OBJECTIVES

1 Write fractions as equivalent terminating decimals.

2 Write fractions as equivalent repeating decimals.

3 Round repeating decimals.

4 Graph fractions and decimals on a number line.

5 Compare fractions and decimals.

6 Evaluate expressions containing fractions and decimals.

7 Solve application problems involving fractions and decimals.

OBJECTIVE 1 **Write fractions as equivalent terminating decimals.**

A fraction and a decimal are said to be **equivalent** if they name the same number. Every fraction can be written in an equivalent decimal form by dividing the numerator by the denominator, as indicated by the fraction bar.

Writing a Fraction as a Decimal

To write a fraction as a decimal, divide the numerator of the fraction by its denominator.

EXAMPLE 1 Write each fraction as a decimal.

a. $\dfrac{3}{4}$ **b.** $\dfrac{5}{8}$ **c.** $\dfrac{7}{2}$

Strategy We will divide the numerator of each fraction by its denominator. We will continue the division process until we obtain a zero remainder.

WHY   We divide the numerator by the denominator because a fraction bar indicates division.

Solution

a. $\frac{3}{4}$ means $3 \div 4$. To find $3 \div 4$, we begin by writing it in long division form as $4\overline{)3}$. To proceed with the division, we must write the dividend 3 with a decimal point and some additional zeros. Then we use the procedure from Section 4.4 for dividing a decimal by a whole number.

$$
\begin{array}{r}
0.75 \\
4\overline{)3.00} \\
-2\,8\downarrow \\
\hline
20 \\
-20 \\
\hline
0
\end{array}
$$

Write a decimal point and two additional zeros to the right of 3.

$\leftarrow$ The remainder is 0.

Thus, $\frac{3}{4} = 0.75$. We say that the **decimal equivalent** of $\frac{3}{4}$ is 0.75.

We can check the result by writing 0.75 as a fraction in simplest form:

$$0.75 = \frac{75}{100}$$ 0.75 is seventy-five hundredths.

$$= \frac{3 \cdot \overset{1}{\cancel{25}}}{4 \cdot \underset{1}{\cancel{25}}}$$ To simplify the fraction, factor 75 as $3 \cdot 25$ and 100 as $4 \cdot 25$ and remove the common factor of 25.

$$= \frac{3}{4}$$ This is the original fraction.

b. $\frac{5}{8}$ means $5 \div 8$.

$$
\begin{array}{r}
0.625 \\
8\overline{)5.000} \\
-4\,8\downarrow \\
\hline
20 \\
-16 \\
\hline
40 \\
-40 \\
\hline
0
\end{array}
$$

Write a decimal point and three additional zeros to the right of 5.

$\leftarrow$ The remainder is 0.

Thus, $\frac{5}{8} = 0.625$.

c. $\frac{7}{2}$ means $7 \div 2$.

$$
\begin{array}{r}
3.5 \\
2\overline{)7.0} \\
-6\downarrow \\
\hline
1\,0 \\
-1\,0 \\
\hline
0
\end{array}
$$

Write a decimal point and one additional zero to the right of 7.

$\leftarrow$ The remainder is 0.

Thus, $\frac{7}{2} = 3.5$.

Self Check 1

Write each fraction as a decimal.

a. $\frac{1}{5}$

b. $\frac{3}{16}$

c. $\frac{9}{2}$

Now Try ➡ Problems 15, 17, and 21

Caution! A common error when finding a decimal equivalent for a fraction is to *incorrectly divide the denominator by the numerator.* An example of this is shown on the right, where the decimal equivalent of $\frac{5}{8}$ (a number less than 1) is incorrectly found to be 1.6 (a number greater than 1).

$$\begin{array}{r} 1.6 \\ 5\overline{)8.0} \\ -\,5 \\ \hline 3\,0 \\ -\,3\,0 \\ \hline 0 \end{array}$$

In parts a, b, and c of Example 1, the division process ended because a remainder of 0 was obtained. When such a division *terminates* with a remainder of 0, we call the resulting decimal a **terminating decimal**. Thus, 0.75, 0.625, and 3.5 are three examples of terminating decimals.

LANGUAGE OF MATHEMATICS

To ***terminate*** means to bring to an end. In the movie *The Terminator*, actor Arnold Schwarzenegger plays a heartless machine sent to Earth to bring an end to his enemies.

OBJECTIVE 2 Write fractions as equivalent repeating decimals.

Sometimes, when we are finding a decimal equivalent of a fraction, the division process never gives a remainder of 0. In this case, the result is a **repeating decimal**. Examples of repeating decimals are 0.4444 . . . and 1.373737 . . . . The three dots tell us that a block of digits repeats in the pattern shown. Repeating decimals can also be written using a bar over the repeating block of digits. For example, 0.4444 . . . can be written as $0.\overline{4}$, and 1.373737 . . . can be written as $1.\overline{37}$.

Caution! When using an **overbar** to write a repeating decimal, use the least number of digits necessary to show the repeating block of digits.

$0.333\ldots = 0.\overline{333}$ $\qquad$ $6.7454545\ldots = 6.7\overline{454}$

$0.333\ldots = 0.\overline{3}$ $\qquad$ $6.7454545\ldots = 6.7\overline{45}$

Some fractions can be written as decimals using an alternate approach. If the denominator of a fraction in simplified form has factors of only 2's or 5's, or a combination of both, it can be written as a decimal by multiplying it by a form of 1. The objective is to write the fraction in an equivalent form with a denominator that is a power of 10, such as 10, 100, 1,000, and so on.

EXAMPLE 2 Write each fraction as a decimal using multiplication by a form of 1: **a.** $\frac{4}{5}$ **b.** $\frac{11}{40}$

Strategy We will multiply $\frac{4}{5}$ by $\frac{2}{2}$ and we will multiply $\frac{11}{40}$ by $\frac{25}{25}$.

WHY The result of each multiplication will be an equivalent fraction with a denominator that is a power of 10. Such fractions are then easy to write in decimal form.

Solution

a. Since we need to multiply the denominator of $\frac{4}{5}$ by 2 to obtain a denominator of 10, it follows that $\frac{2}{2}$ should be the form of 1 that is used to build $\frac{4}{5}$.

$\frac{4}{5} = \frac{4}{5} \cdot \frac{2}{2}$ Multiply $\frac{4}{5}$ by 1 in the form of $\frac{2}{2}$.

$= \frac{8}{10}$ Multiply the numerators. Multiply the denominators.

$= 0.8$ Write the fraction as a decimal.

b. Since we need to multiply the denominator of $\frac{11}{40}$ by 25 to obtain a denominator of 1,000, it follows that $\frac{25}{25}$ should be the form of 1 that is used to build $\frac{11}{40}$.

Self Check 2

Write each fraction as a decimal using multiplication by a form of 1:

a. $\dfrac{2}{5}$

b. $\dfrac{8}{25}$

Now Try ➲ **Problems 27 and 29**

$$\frac{11}{40} = \frac{11}{40} \cdot \frac{25}{25} \qquad \text{Multiply } \frac{11}{40} \text{ by 1 in the form of } \frac{25}{25}.$$

$$= \frac{275}{1,000} \qquad \begin{array}{l}\text{Multiply the numerators.}\\ \text{Multiply the denominators.}\end{array}$$

$$= 0.275 \qquad \text{Write the fraction as a decimal.}$$

Mixed numbers can also be written in decimal form.

EXAMPLE 3 Write the mixed number $5\frac{7}{16}$ in decimal form.

Strategy We need only find the decimal equivalent for the fractional part of the mixed number.

WHY The whole-number part in the decimal form is the same as the whole-number part in the mixed number form.

Solution To write $\frac{7}{16}$ as a fraction, we find $7 \div 16$.

$$\begin{array}{r} 0.4375 \\ 16\overline{)7.0000} \\ -6\,4 \\ \hline 60 \\ -48 \\ \hline 120 \\ -112 \\ \hline 80 \\ -80 \\ \hline 0 \end{array}$$

Write a decimal point and four additional zeros to the right of 7.

←The remainder is 0.

Since the whole-number part of the decimal must be the same as the whole-number part of the mixed number, we have:

$$5\frac{7}{16} = 5.4375$$

We would have obtained the same result if we changed $5\frac{7}{16}$ to the improper fraction $\frac{87}{16}$ and divided 87 by 16.

Self Check 3

Write the mixed number $3\frac{17}{20}$ in decimal form.

Now Try ➲ **Problem 37**

EXAMPLE 4 Write $\dfrac{5}{12}$ as a decimal.

Strategy We will divide the numerator of the fraction by its denominator and watch for a repeating pattern of nonzero remainders.

WHY Once we detect a repeating pattern of remainders, the division process can stop.

Solution $\dfrac{5}{12}$ means $5 \div 12$.

```
        0.4166
    12)5.0000        Write a decimal point and four additional zeros to the right of 5.
      −4 8
         20
        −12
          80
         −72
           80
          −72       It is apparent that 8 will continue to reappear as the remainder. Therefore,
            8        6 will continue to reappear in the quotient. Since the repeating pattern is
                     now clear, we can stop the division.
```

We can use three dots to show that a repeating pattern of 6's appears in the quotient:

$$\frac{5}{12} = 0.416666\ldots$$

Or, we can use an overbar to indicate the repeating part (in this case, only the 6), and write the decimal equivalent in more compact form:

$$\frac{5}{12} = 0.41\overline{6}$$

Self Check 4

Write $\frac{1}{12}$ as a decimal.

Now Try ➲ Problem 41

EXAMPLE 5 Write $-\frac{6}{11}$ as a decimal.

Strategy To find the decimal equivalent for $-\frac{6}{11}$, we will first find the decimal equivalent for $\frac{6}{11}$. To do this, we will divide the numerator of $\frac{6}{11}$ by its denominator and watch for a repeating pattern of nonzero remainders.

WHY Once we detect a repeating pattern of remainders, the division process can stop.

Solution $\frac{6}{11}$ means $6 \div 11$.

```
        0.54545
    11)6.00000        Write a decimal point and five additional zeros to the right of 6.
      − 5 5
          50
        − 44
           60
         − 55
            50
          − 44
             60
           − 55       It is apparent that 6 and 5 will continue to reappear as remainders.
              5        Therefore, 5 and 4 will continue to reappear in the quotient. Since the
                       repeating pattern is now clear, we can stop the division process.
```

We can use three dots to show that a repeating pattern of 5 and 4 appears in the quotient:

$$\frac{6}{11} = 0.545454\ldots \text{ and therefore, } -\frac{6}{11} = -0.545454\ldots$$

Or, we can use an overbar to indicate the repeating part (in this case, 54), and write the decimal equivalent in more compact form:

$$\frac{6}{11} = 0.\overline{54} \text{ and therefore, } -\frac{6}{11} = -0.\overline{54}$$

Self Check 5

Write $-\frac{13}{33}$ as a decimal.

Now Try ➲ Problem 47

The repeating part of the decimal equivalent of some fractions is quite long. Here are some examples:

$$\frac{9}{37} = 0.\overline{243}$$ A block of three digits repeats.

$$\frac{13}{101} = 0.\overline{1287}$$ A block of four digits repeats.

$$\frac{6}{7} = 0.\overline{857142}$$ A block of six digits repeats.

Every fraction can be written as either a terminating decimal or a repeating decimal. For this reason, the set of fractions (**rational numbers**) form a subset of the set of decimals called the set of **real numbers**. The set of real numbers corresponds to all points on a number line.

Not all decimals are terminating or repeating decimals. For example,

$$0.2020020002\ldots$$

does not terminate, and it has no repeating block of digits. This decimal cannot be written as a fraction with an integer numerator and a nonzero integer denominator. Thus, it is not a rational number. It is an example from the set of **irrational numbers**.

OBJECTIVE 3 Round repeating decimals.

When a fraction is written in decimal form, the result is either a terminating or a repeating decimal. Repeating decimals are often rounded to a specified place value.

EXAMPLE 6   Write $\frac{1}{3}$ as a decimal and round to the nearest hundredth.

Strategy We will use the methods of this section to divide to the thousandths column.

WHY To round to the hundredths column, we need to continue the division process for one more decimal place, which is the thousandths column.

Solution $\frac{1}{3}$ means $1 \div 3$.

```
    0.333
 3)1.000      Write a decimal point and three additional zeros to the right of 1.
  − 9↓
    10
   − 9↓
     10
    − 9
      1        The division process can stop. We have divided to the thousandths column.
```

After dividing to the thousandths column, we round to the hundredths column.

┌─── The rounding digit in the hundredths column is 3.
│ ┌─ The test digit in the thousandths column is 3.
↓ ↓
0.333 . . .

Since 3 is less than 5, we round down, and we have

$$\frac{1}{3} \approx 0.33$$ Read ≈ as "is approximately equal to."

Self Check 6

Write $\frac{4}{9}$ as a decimal and round to the nearest hundredth.

Now Try Problem 51

EXAMPLE 7　Write $\frac{2}{7}$ as a decimal and round to the nearest thousandth.

Strategy　We will use the methods of this section to divide to the ten-thousandths column.

WHY　To round to the thousandths column, we need to continue the division process for one more decimal place, which is the ten-thousandths column.

Solution　$\frac{2}{7}$ means $2 \div 7$.

$$
\begin{array}{r}
0.2857 \\
7\overline{)2.0000} \\
-\ 14 \\
\hline
60 \\
-\ 56 \\
\hline
40 \\
-\ 35 \\
\hline
50 \\
-\ 49 \\
\hline
1
\end{array}
$$

Write a decimal point and four additional zeros to the right of 2.

The division process can stop.
We have divided to the ten-thousandths column.

After dividing to the ten-thousandths column, we round to the thousandths column.

The rounding digit in the thousandths column is 5.
The test digit in the ten-thousandths column is 7.

0.2857

Since 7 is greater than 5, we round up, and $\frac{2}{7} \approx 0.286$.

Self Check 7

Write $\frac{7}{24}$ as a decimal and round to the nearest thousandth.

Now Try ➡ Problem 61

Using Your Calculator ▶　The Fixed-Point Key

After performing a calculation, a scientific calculator can round the result to a given decimal place. This is done using the *fixed-point key*. As we did in Example 7, let's find the decimal equivalent of $\frac{2}{7}$ and round to the nearest thousandth. This time, we will use a calculator.

First, we set the calculator to round to the third decimal place (thousandths) by pressing $\boxed{\text{2nd}}$ $\boxed{\text{FIX}}$ 3. Then we press 2 $\boxed{\div}$ 7 $\boxed{=}$

$$\boxed{0.286}$$

Thus, $\frac{2}{7} \approx 0.286$. To round to the nearest tenth, we would fix 1; to round to the nearest hundredth, we would fix 2; and so on. After using the FIX feature, don't forget to remove it and return the calculator to the normal mode.

Graphing calculators can also round to a given decimal place. See the owner's manual for the required keystrokes.

OBJECTIVE 4　Graph fractions and decimals on a number line.

A number line can be used to show the relationship between fractions and their decimal equivalents. On the number line on the next page, sixteen equally spaced marks are used to scale from 0 to 1. Some commonly used fractions that have terminating decimal equivalents are shown. For example, we see that $\frac{1}{8} = 0.125$ and $\frac{13}{16} = 0.8125$.

On the next number line, six equally spaced marks are used to scale from 0 to 1. Some commonly used fractions and their repeating decimal equivalents are shown.

$$0.1\overline{6} \qquad 0.\overline{3} \qquad\qquad 0.\overline{6} \qquad 0.8\overline{3}$$

$$0 \qquad \frac{1}{6} \qquad \frac{1}{3} \qquad \frac{1}{2} \qquad \frac{2}{3} \qquad \frac{5}{6} \qquad 1$$

OBJECTIVE 5   Compare fractions and decimals.

To compare the size of a fraction and a decimal, it is helpful to write the fraction in its equivalent decimal form.

EXAMPLE 8   Place an $<$, $>$, or $=$ symbol in the box to make a true statement:

a. $\frac{4}{5}$ ☐ 0.91   **b.** 0.3$\overline{5}$ ☐ $\frac{1}{3}$ **c.** $\frac{9}{4}$ ☐ 22.25

Strategy In each case, we will write the given fraction as a decimal.

WHY Then we can use the procedure for comparing two decimals to determine which number is the larger and which is the smaller.

Solution

a. To write $\frac{4}{5}$ as a decimal, we divide 4 by 5.

$$\begin{array}{r} 0.8 \\ 5\overline{)4.0} \\ -40 \\ \hline 0 \end{array}$$ *Write a decimal point and one additional zero to the right of 4.*

Thus, $\frac{4}{5} = 0.8$.

To make the comparison of the decimals easier, we can write one zero after 8 so that they have the same number of digits to the right of the decimal point.

0.**8** 0   *This is the decimal equivalent for $\frac{4}{5}$.*

0.**9** 1

As we work from left to right, this is the first column in which the digits differ. Since $8 < 9$, it follows that $0.80 = \frac{4}{5}$ is less than 0.91, and we can write $\frac{4}{5} < 0.91$.

b. In Example 6, we saw that $\frac{1}{3} = 0.3333\ldots$. To make the comparison of these repeating decimals easier, we write them so that they have the same number of digits to the right of the decimal point.

0.3**5**55...   *This is $0.3\overline{5}$.*

0.3**3**33...   *This is the decimal equivalent of $\frac{1}{3}$.*

As we work from left to right, this is the first column in which the digits differ. Since $5 > 3$, it follows that $0.3555\ldots = 0.3\overline{5}$ is greater than $0.3333\ldots = \frac{1}{3}$, and we can write $0.3\overline{5} > \frac{1}{3}$.

c. To write $\frac{9}{4}$ as a decimal, we divide 9 by 4.

$$\begin{array}{r} 2.25 \\ 4\overline{)9.00} \\ -8 \\ \hline 1\,0 \\ -8 \\ \hline 20 \\ -20 \\ \hline 0 \end{array}$$

Write a decimal point and two additional zeros to the right of 9.

From the division, we see that $\frac{9}{4} = 2.25$.

Self Check 8

Place an $<$, $>$, or $=$ symbol in the box to make a true statement:

a. $\frac{3}{8}$ ☐ 0.305

b. $0.7\overline{6}$ ☐ $\frac{7}{9}$

c. $\frac{11}{4}$ ☐ 2.75

Now Try Problems 67, 69, and 71

EXAMPLE 9 Write the numbers in order from least to greatest:

$2.168,\ 2\frac{1}{6},\ \frac{20}{9}$

Strategy We will write $2\frac{1}{6}$ and $\frac{20}{9}$ in decimal form.

WHY Then we can do a column-by-column comparison of the numbers to determine the largest and smallest.

Solution From the number line on page 374, we see that $\frac{1}{6} = 0.1\overline{6}$. Thus, $2\frac{1}{6} = 2.1\overline{6}$. To write $\frac{20}{9}$ as a decimal, we divide 20 by 9.

$$\begin{array}{r} 2.222 \\ 9\overline{)20.000} \\ -18 \\ \hline 2\,0 \\ -1\,8 \\ \hline 20 \\ -18 \\ \hline 20 \\ -18 \\ \hline 2 \end{array}$$

Write a decimal point and two additional zeros to the right of 20.

Thus, $\frac{20}{9} = 2.222\ldots$.

To make the comparison of the three decimals easier, we stack them as shown below.

2.1680	This is 2.168 with an additional 0.
2.1666 . . .	This is $2\frac{1}{6} = 2.1\overline{6}$.
2.2222 . . .	This is $\frac{20}{9}$.

Working from left to right, this is the first column in which the digits differ. Since $2 > 1$, it follows that $2.222\ldots = \frac{20}{9}$ is the largest of the three numbers.

Working from left to right, this is the first column in which the top two numbers differ. Since $8 > 6$, it follows that 2.168 is the next largest number and that $2.1\overline{6} = 2\frac{1}{6}$ is the smallest.

Written in order from least to greatest, we have:

$2\frac{1}{6},\ 2.168,\ \frac{20}{9}$

Self Check 9

Write the numbers in order from least to greatest: 1.832, $\frac{9}{5}$, $1\frac{5}{6}$

Now Try Problem 75

OBJECTIVE 6 Evaluate expressions containing fractions and decimals.

Expressions can contain both fractions and decimals. In the following examples, we show two methods that can be used to evaluate expressions of this type. With the first method we find the answer by working in terms of fractions.

EXAMPLE 10 Evaluate $\frac{1}{3} + 0.27$ by working in terms of fractions.

Strategy We will begin by writing 0.27 as a fraction.

WHY Then we can use the methods of Chapter 3 for adding fractions with unlike denominators to find the sum.

Solution To write 0.27 as a fraction, it is helpful to read it aloud as "twenty-seven hundredths."

$$\frac{1}{3} + 0.27 = \frac{1}{3} + \frac{27}{100} \qquad \text{Replace 0.27 with } \frac{27}{100}.$$

$$= \frac{1}{3} \cdot \frac{100}{100} + \frac{27}{100} \cdot \frac{3}{3} \qquad \text{The LCD for } \frac{1}{3} \text{ and } \frac{27}{100} \text{ is 300. To build each fraction so that its denominator is 300, multiply by a form of 1.}$$

$$= \frac{100}{300} + \frac{81}{300} \qquad \text{Multiply the numerators.} \\ \text{Multiply the denominators.}$$

$$= \frac{181}{300} \qquad \text{Add the numerators and write the sum over the common denominator 300.}$$

Self Check 10

Evaluate by working in terms of fractions: $0.53 + \frac{1}{6}$

Now Try Problem 79

Now we will evaluate the expression from Example 10 by working in terms of decimals.

EXAMPLE 11 Estimate $\frac{1}{3} + 0.27$ by working in terms of decimals.

Strategy Since 0.27 has two decimal places, we will begin by finding a decimal approximation for $\frac{1}{3}$ to two decimal places.

WHY Then we can use the methods of this chapter for adding decimals to find the sum.

Solution We have seen that the decimal equivalent of $\frac{1}{3}$ is the repeating decimal 0.333 . . . . Rounded to the nearest hundredth: $\frac{1}{3} \approx 0.33$.

$$\frac{1}{3} + 0.27 \approx 0.33 + 0.27 \qquad \text{Approximate } \frac{1}{3} \text{ with the decimal 0.33.}$$

$$\approx 0.60 \qquad \text{Do the addition.}$$

$$\begin{array}{r} \overset{1}{0.33} \\ +0.27 \\ \hline 0.60 \end{array}$$

Self Check 11

Estimate the result by working in terms of decimals: $0.53 - \frac{1}{6}$

Now Try Problem 87

In Examples 10 and 11, we evaluated $\frac{1}{3} + 0.27$ in different ways. In Example 10, we obtained the exact answer, $\frac{181}{300}$. In Example 11, we obtained an approximation, 0.6. The results seem reasonable when we write $\frac{181}{300}$ in decimal form: $\frac{181}{300} = 0.60333$ . . . .

EXAMPLE 12   Evaluate: $\left(\dfrac{4}{5}\right)(1.35) + (0.5)^2$

Strategy We will find the decimal equivalent of $\frac{4}{5}$ and then evaluate the expression in terms of decimals.

WHY It is easier to perform multiplication and addition with the given decimals than it would be converting them to fractions.

Solution We use division to find the decimal equivalent of $\frac{4}{5}$.

$$
\begin{array}{r}
0.8 \\
5\overline{)4.0} \\
-4\,0 \\
\hline
0
\end{array}
$$
Write a decimal point and one additional zero to the right of the 4.

Now we use the order of operation rule to evaluate the expression.

$\left(\dfrac{4}{5}\right)(1.35) + (0.5)^2$

$= (\mathbf{0.8})(1.35) + (0.5)^2$ *Replace $\frac{4}{5}$ with its decimal equivalent, 0.8.*

$= (0.8)(1.35) + 0.25$ *Evaluate: $(0.5)^2 = 0.25$.*

$= 1.08 + 0.25$ *Do the multiplication: $(0.8)(1.35) = 1.08$.*

$= 1.33$ *Do the addition.*

$$
\begin{array}{r}
\overset{2}{0.5} \\
\times\ 0.5 \\
\hline
0.25
\end{array}
$$

$$
\begin{array}{r}
\overset{2\ 4}{1.35} \\
\times\ \ 0.8 \\
\hline
1.080
\end{array}
$$

$$
\begin{array}{r}
\overset{1}{1.08} \\
+0.25 \\
\hline
1.33
\end{array}
$$

Self Check 12

Evaluate:

$(-0.6)^2 + (2.3)\left(\dfrac{1}{8}\right)$

Now Try ➲ **Problem 99**

OBJECTIVE 7 Solve application problems involving fractions and decimals.

EXAMPLE 13   **Shopping.** A shopper purchased $\dfrac{3}{4}$ pound of fruit, priced at $0.88 a pound, and $\dfrac{1}{3}$ pound of fresh-ground coffee, selling for $6.60 a pound. Find the total cost of these items.

Analyze

- $\dfrac{3}{4}$ pound of fruit was purchased at $0.88 per pound.   *Given*

- $\dfrac{1}{3}$ pound of coffee was purchased at $6.60 per pound.   *Given*

- What was the total cost of the items?   *Find*

Form To find the total cost of each item, multiply the number of pounds purchased by the price per pound.

The total cost of the items	is equal to	the number of pounds of fruit	times	the price per pound	plus	the number of pounds of coffee	times	the price per pound.
The total cost of the items	=	$\dfrac{3}{4}$	·	$0.88	+	$\dfrac{1}{3}$	·	$6.60

Calculate   Because 0.88 is divisible by 4 and 6.60 is divisible by 3, we can work with the decimals and fractions in this form; no conversion is necessary.

$$\frac{3}{4} \cdot 0.88 + \frac{1}{3} \cdot 6.60$$

$$= \frac{3}{4} \cdot \frac{0.88}{1} + \frac{1}{3} \cdot \frac{6.60}{1} \qquad \text{Express 0.88 as } \tfrac{0.88}{1} \text{ and 6.60 as } \tfrac{6.60}{1}.$$

$$= \frac{2.64}{4} + \frac{6.60}{3} \qquad \text{Multiply the numerators.}$$
$$\qquad\qquad\qquad\qquad \text{Multiply the denominators.}$$

$$= 0.66 + 2.20 \qquad \text{Do each division.}$$

$$= 2.86 \qquad\qquad \text{Do the addition.}$$

$$\overset{2}{0.88}$$
$$\underline{\times \quad 3}$$
$$2.64$$

$$\begin{array}{r} 0.66 \\ 4\overline{)2.64} \\ -2\,4 \\ \hline 24 \\ -24 \\ \hline 0 \end{array} \qquad \begin{array}{r} 2.20 \\ 3\overline{)6.60} \\ -6 \\ \hline 06 \\ -6 \\ \hline 00 \\ -0 \\ \hline 0 \end{array}$$

$$\begin{array}{r} 0.66 \\ +2.20 \\ \hline 2.86 \end{array}$$

Self Check 13

Delicatessens.   A shopper purchased $\frac{2}{3}$ pound of Swiss cheese, priced at $2.19 per pound, and $\frac{3}{4}$ pound of sliced turkey, selling for $6.40 per pound. Find the total cost of these items.

Now Try ➲ **Problem 115**

State   The total cost of the items is $2.86.

Check   If approximately 1 pound of fruit, priced at approximately $1 per pound, was purchased, then about $1 was spent on fruit. If exactly $\frac{1}{3}$ of a pound of coffee, priced at approximately $6 per pound, was purchased, then about $\frac{1}{3} \cdot$ $6, or $2, was spent on coffee. Since the approximate cost of the items $1 + $2 = $3, is close to the result, $2.86, the result seems reasonable.

Answers to Self Checks

1. a. 0.2   **b.** 0.1875   **c.** 4.5   **2. a.** 0.4   **b.** 0.32   **3.** 3.85   **4.** $0.08\overline{3}$ **5.** $-0.\overline{39}$ **6.** 0.44   **7.** 0.292
8. a. >   **b.** <   **c.** =   **9.** $\frac{9}{5}$, 1.832, $1\frac{5}{6}$ **10.** $\frac{209}{300}$ **11.** approximately 0.36   **12.** 0.6475   **13.** $6.26

SECTION 4.5   STUDY SET

VOCABULARY

Fill in the blanks.

1. A fraction and a decimal are said to be _____ if they name the same number.

2. The _____ equivalent of $\frac{3}{4}$ is 0.75.

3. 0.75, 0.625, and 3.5 are examples of _____ decimals.

4. 0.3333 . . . and 1.666 . . . are examples of _____ decimals.

CONCEPTS

Fill in the blanks.

5. $\frac{7}{8}$ means 7 ▢ 8.

6. To write a fraction as a decimal, divide the _____ of the fraction by its denominator.

7. To perform the division shown below, a decimal point and two additional _____ were written to the right of 3.

$$4\overline{)3.00}$$

8. Sometimes, when finding the decimal equivalent of a fraction, the division process ends because a remainder of 0 is obtained. We call the resulting decimal a _____ decimal.

9. Sometimes, when we are finding the decimal equivalent of a fraction, the division process never gives a remainder of 0. We call the resulting decimal a _____ decimal.

10. If the denominator of a fraction in simplified form has factors of only 2's or 5's, or a combination of both, it can be written as a decimal by multiplying it by a form of ▢.

11. **a.** Round 0.3777 . . . to the nearest hundredth.

    b. Round 0.212121 . . . to the nearest thousandth.

12. **a.** When evaluating the expression $0.25 + \left(2.3 + \frac{2}{5}\right)^2$, would it be easier to work in terms of fractions or decimals?

    b. What is the first step that should be performed to evaluate the expression?

NOTATION

13. Write each decimal in fraction form.

   a. 0.7   **b.** 0.77

14. Write each repeating decimal in simplest form using an overbar.

   a. 0.888 . . .   **b.** 0.323232 . . .

   c. 0.56333 . . .   **d.** 0.8898989 . . .

GUIDED PRACTICE

Write each fraction as a decimal. See Example 1.

15. $\dfrac{1}{2}$ **16.** $\dfrac{1}{4}$

17. $\dfrac{7}{8}$ **18.** $\dfrac{3}{8}$

19. $\dfrac{11}{20}$ **20.** $\dfrac{17}{20}$

21. $\dfrac{13}{5}$ **22.** $\dfrac{15}{2}$

23. $\dfrac{9}{16}$ **24.** $\dfrac{3}{32}$

25. $-\dfrac{17}{32}$ **26.** $-\dfrac{15}{16}$

Write each fraction as a decimal using multiplication by a form of 1. See Example 2.

27. $\dfrac{3}{5}$ **28.** $\dfrac{13}{25}$

29. $\dfrac{9}{40}$ **30.** $\dfrac{7}{40}$

31. $\dfrac{19}{25}$ **32.** $\dfrac{21}{50}$

33. $\dfrac{1}{500}$ **34.** $\dfrac{1}{250}$

Write each mixed number in decimal form. See Example 3.

35. $3\dfrac{3}{4}$ **36.** $5\dfrac{4}{5}$

37. $12\dfrac{11}{16}$ **38.** $32\dfrac{9}{16}$

Write each fraction as a decimal. Use an overbar in your answer. See Example 4.

39. $\dfrac{1}{9}$ **40.** $\dfrac{8}{9}$

41. $\dfrac{7}{12}$ **42.** $\dfrac{11}{12}$

43. $\dfrac{7}{90}$ **44.** $\dfrac{1}{99}$

45. $\dfrac{1}{60}$ **46.** $\dfrac{1}{66}$

Write each fraction as a decimal. Use an overbar in your answer. See Example 5.

47. $-\dfrac{5}{11}$ **48.** $-\dfrac{7}{11}$

49. $-\dfrac{20}{33}$ **50.** $-\dfrac{16}{33}$

Write each fraction in decimal form. Round to the nearest hundredth. See Example 6.

51. $\dfrac{7}{30}$ **52.** $\dfrac{8}{9}$

53. $\dfrac{22}{45}$ **54.** $\dfrac{17}{45}$

55. $\dfrac{24}{13}$ **56.** $\dfrac{34}{11}$

57. $-\dfrac{13}{12}$ **58.** $-\dfrac{25}{12}$

Write each fraction in decimal form. Round to the nearest thousandth. See Example 7.

59. $\dfrac{5}{33}$ **60.** $\dfrac{5}{24}$

61. $\dfrac{10}{27}$ **62.** $\dfrac{17}{21}$

Graph the given numbers on a number line. See Objective 4.

63. $1\dfrac{3}{4}$, -0.75, $0.\overline{6}$, $-3.8\overline{3}$

64. $1\dfrac{2}{8}$, -2.375, $0.\overline{3}$, $4.1\overline{6}$

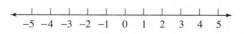

65. 3.875, $-3.\overline{5}$, $0.\overline{2}$, $-1\dfrac{4}{5}$

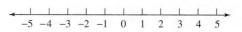

66. $1.375,\ -4\frac{1}{7},\ 0.\overline{1},\ -2.\overline{7}$

$$\xleftarrow{\quad\mid\ \mid\ \mid\ \mid\ \mid\ \mid\ \mid\ \mid\ \mid\ \mid\ \mid\quad}$$
$$-5\ -4\ -3\ -2\ -1\ \ 0\ \ 1\ \ 2\ \ 3\ \ 4\ \ 5$$

Place an <, >, *or* = *symbol in the box to make a true statement. See Example 8.*

67. $\frac{7}{8}$ ☐ 0.895

68. $\frac{3}{8}$ ☐ 0.381

69. $0.\overline{7}$ ☐ $\frac{17}{22}$

70. $0.\overline{45}$ ☐ $\frac{7}{16}$

71. $\frac{52}{25}$ ☐ 2.08

72. 4.4 ☐ $\frac{22}{5}$

73. $-\frac{11}{20}$ ☐ $-0.\overline{48}$

74. $-0.0\overline{9}$ ☐ $-\frac{1}{11}$

Write the numbers in order from least to greatest. See Example 9.

75. $6\frac{1}{2},\ 6.25,\ \frac{19}{3}$

76. $7\frac{3}{8},\ 7.08,\ \frac{43}{6}$

77. $-0.\overline{81},\ -\frac{8}{9},\ -\frac{6}{7}$

78. $-0.\overline{19},\ -\frac{1}{11},\ -0.1$

Evaluate each expression. Work in terms of fractions. See Example 10.

79. $\frac{1}{9} + 0.3$

80. $\frac{2}{3} + 0.1$

81. $0.9 - \frac{7}{12}$

82. $0.99 - \frac{5}{6}$

83. $\frac{5}{11}(0.3)$

84. $(0.9)\left(\frac{1}{27}\right)$

85. $\frac{1}{4}(0.25) + \frac{15}{16}$

86. $\frac{2}{5}(0.02) - (0.04)$

Estimate the value of each expression. Work in terms of decimals. See Example 11.

87. $0.24 + \frac{1}{3}$

88. $0.02 + \frac{5}{6}$

89. $5.69 - \frac{5}{12}$

90. $3.19 - \frac{2}{3}$

91. $0.43 - \frac{1}{12}$

92. $0.27 + \frac{5}{12}$

93. $\frac{1}{15} - 0.55$

94. $\frac{7}{30} - 0.84$

Evaluate each expression. Work in terms of decimals. See Example 12.

95. $(3.5 + 6.7)\left(-\frac{1}{4}\right)$

96. $\left(-\frac{5}{8}\right)\left(5.3 - 3\frac{9}{10}\right)$

97. $\left(\frac{1}{5}\right)^2 (1.7)$

98. $(2.35)\left(\frac{2}{5}\right)^2$

99. $7.5 - (0.78)\left(\frac{1}{2}\right)^2$

100. $8.1 - \left(\frac{3}{4}\right)^2 (0.12)$

101. $\frac{3}{8}(3.2) + \left(4\frac{1}{2}\right)\left(-\frac{1}{4}\right)$

102. $(-0.8)\left(\frac{1}{4}\right) + \left(\frac{1}{5}\right)(0.39)$

LOOK ALIKES

Write each fraction as a decimal.

103. a. $\frac{5}{8}$ b. $\frac{8}{5}$

104. a. $\frac{5}{16}$ b. $\frac{16}{5}$

Write each repeating decimal in simplest form using an overbar.

105. a. $0.44444...$ b. $0.4141414...$
  c. $0.414444...$ d. $0.14444...$

106. a. $-78.8888...$ b. $-78.878787...$
  c. $-78.87777...$ d. $-78.7778778778...$

APPLICATIONS

107. **Drafting.** The architect's scale shown below has several measuring edges. The edge marked 16 divides each inch into 16 equal parts. Find the decimal form for each fractional part of 1 inch that is highlighted with a red arrow.

108. Mileage signs. The freeway sign shown below gives the number of miles to the next three exits. Convert the mileages to decimal notation.

Barranca Ave.	$\frac{3}{4}$ mi
210 Freeway	$2\frac{1}{4}$ mi
Ada St.	$3\frac{1}{2}$ mi

109. Gardening. Two brands of replacement line for a lawn trimmer shown below are labeled in different ways. On one package, the line's thickness is expressed as a decimal; on the other, as a fraction. Which line is thicker?

NYLON LINE
Thickness: 0.065 in.

TRIMMER LINE
$\frac{3}{40}$ in. thick

110. Auto mechanics. While doing a tune-up, a mechanic checks the gap on one of the spark plugs of a car to be sure it is firing correctly. The owner's manual states that the gap should be $\frac{2}{125}$ inch. The gauge the mechanic uses to check the gap is in decimal notation; it registers 0.025 inch. Is the spark plug gap too large or too small?

111. Horse racing. In thoroughbred racing, the time a horse takes to run a given distance is measured using fifths of a second. For example, 23^2 (read "twenty-three and two") means $23\frac{2}{5}$ seconds. The illustration below lists four split times for a horse named *Speedy Flight* in a $1\frac{1}{16}$-mile race. Express each split time in decimal form.

Speedy Flight	Turfway Park, Ky	3-year–old
	17 May 2014	$1\frac{1}{16}$ mile
Splits	:23² :23⁴	:24¹ :32³

112. Geology. A geologist weighed a rock sample at the site where it was discovered and found it to weigh $17\frac{7}{8}$ lb. Later, a more accurate digital scale in the laboratory gave the weight as 17.671 lb. What is the difference in the two measurements?

113. Window replacements. The amount of sunlight that comes into a room depends on the area of the windows in the room. What is the area of the window shown below? (*Hint:* Use the formula $A = \frac{1}{2}bh$.)

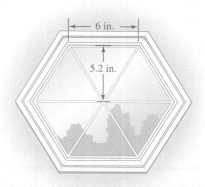

← 6 in. →
5.2 in.

114. Forestry. A command post asked each of three fire crews to estimate the length of the fire line they were fighting. Their reports came back in different forms, as shown. Find the perimeter of the fire. Round to the nearest tenth.

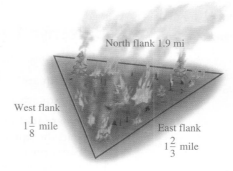

North flank 1.9 mi
West flank $1\frac{1}{8}$ mile
East flank $1\frac{2}{3}$ mile

115. Delicatessens. A shopper purchased $\frac{2}{3}$ pound of green olives, priced at $4.14 per pound, and $\frac{3}{4}$ pound of smoked ham, selling for $5.68 per pound. Find the total cost of these items.

116. Chocolate. A shopper purchased $\frac{3}{4}$ pound of dark chocolate, priced at $8.60 per pound, and $\frac{1}{3}$ pound of milk chocolate, selling for $5.25 per pound. Find the total cost of these items.

WRITING

117. Explain the procedure used to write a fraction in decimal form.

118. How does the terminating decimal 0.5 differ from the repeating decimal $0.\overline{5}$?

119. A student represented the repeating decimal 0.1333 . . . as $0.1\overline{333}$. Is this the best form? Explain why or why not.

120. Is 0.10100100010000 . . . a repeating decimal? Explain why or why not.

121. A student divided 19 by 25 to find the decimal equivalent of $\frac{19}{25}$ to be 0.76. Explain how she can check this result.

122. Explain the error in the following work to find the decimal equivalent for $\frac{5}{6}$.

$$
\begin{array}{r}
1.2 \\
5\overline{)6.0} \\
-5 \\
\hline
1\,0 \\
-1\,0 \\
\hline
0
\end{array}
\quad \text{Thus, } \frac{5}{6} = 1.2.
$$

REVIEW

123. Write each set of numbers.
   a. the first ten whole numbers
   b. the first ten prime numbers
   c. the integers

124. Evaluate: $(0.9)^2$

125. Evaluate: $(-0.1)^2$

126. Evaluate: $(0.8)^2$

OBJECTIVES

1 Find the square root of a perfect square.

2 Find the square root of fractions and decimals.

3 Evaluate expressions that contain square roots.

4 Evaluate formulas involving square roots.

5 Approximate square roots.

SECTION 4.6 Square Roots

We have discussed the relationships between addition and subtraction and between multiplication and division. In this section, we explore the relationship between raising a number to a power and finding a root. Decimals play an important role in this discussion.

OBJECTIVE 1 Find the square root of a perfect square.

When we raise a number to the second power, we are squaring it, or finding its **square**.

The square of 6 is 36 because $6^2 = 36$.

The square of -6 is 36 because $(-6)^2 = 36$.

The **square root** of a given number is a number whose square is the given number. For example, the square roots of 36 are 6 and -6 because either number, when squared, is 36.

Every positive number has two square roots. The number 0 has only one square root. In fact, it is its own square root because $0^2 = 0$.

Square Root

A number is a **square root** of a second number if the square of the first number equals the second number.

EXAMPLE 1 Find the two square roots of 49.

Strategy We will ask "What positive number and what negative number, when squared, is 49?"

WHY The square root of 49 is a number whose square is 49.

Solution

7 is a square root of 49 because $7^2 = 49$

and

-7 is a square root of 49 because $(-7)^2 = 49$.

Self Check 1

Find the two square roots of 64.

Now Try Problem 21

In Example 1, we saw that 49 has two square roots—one positive and one negative. The symbol $\sqrt{}$ is called a **radical symbol** and is used to indicate a positive square root of a nonnegative number. When reading this symbol, we usually drop the word *positive* and simply say *square root*. Since 7 is the positive square root of 49, we can write

$\sqrt{49} = 7$ $\sqrt{49}$ represents the positive number whose square is 49.
　　　　　　Read as "the square root of 49 is 7."

When a number, called the **radicand**, is written under a radical symbol, we have a **radical expression**.

Radical symbol $\rightarrow \sqrt{49} \leftarrow$ Radicand
Radical expression

Some other examples of radical expressions are:

$$\sqrt{36} \qquad \sqrt{100} \qquad \sqrt{144} \qquad \sqrt{81}$$

To evaluate (or simplify) a radical expression like those shown above, we need to find the positive square root of the radicand. For example, if we evaluate $\sqrt{36}$ (read as "the square root of 36"), the result is

$$\sqrt{36} = 6$$

because $6^2 = 36$.

The symbol $-\sqrt{}$ is used to indicate the **negative square root** of a positive number. It is the opposite of the positive square root. Since -6 is the negative square root of 36, we can write

$-\sqrt{36} = -6$ Read as "the negative square root of 36 is -6" or "the opposite of the square
　　　　　　　root of 36 is -6." $-\sqrt{36}$ represents the negative number whose square is 36.

If the number under the radical symbol is 0, we have $\sqrt{0} = 0$.

Numbers, such as 36 and 49, that are squares of whole numbers, are called **perfect squares**. To evaluate square root radical expressions, it is helpful to be able to identify perfect square radicands. You need to memorize the following list of perfect squares, shown in red.

> **Caution!** Remember that the radical symbol asks you to find only the *positive* square root of the radicand. It is incorrect, for example, to say that
>
> $\sqrt{36}$ is 6 and -6

Perfect Squares

$0 = 0^2$	$16 = 4^2$	$64 = 8^2$	$144 = 12^2$
$1 = 1^2$	$25 = 5^2$	$81 = 9^2$	$169 = 13^2$
$4 = 2^2$	$36 = 6^2$	$100 = 10^2$	$196 = 14^2$
$9 = 3^2$	$49 = 7^2$	$121 = 11^2$	$225 = 15^2$

A calculator is helpful in finding the square root of a perfect square that is larger than 225.

EXAMPLE 2   Evaluate each square root: **a.** $\sqrt{81}$ **b.** $-\sqrt{100}$

Strategy In each case, we will determine what positive number, when squared, produces the radicand.

WHY The radical symbol $\sqrt{}$ indicates that the positive square root of the number written under it should be found.

Solution
a. $\sqrt{81} = 9$ Ask: What positive number, when squared, is 81?
　　　　　　　The answer is 9 because $9^2 = 81$.
b. $-\sqrt{100}$ is the opposite (or negative) of the square root of 100. Since $\sqrt{100} = 10$, we have
$$-\sqrt{100} = -10$$

> **Self Check 2**
> Evaluate each square root:
> **a.** $\sqrt{144}$
> **b.** $-\sqrt{81}$
>
> **Now Try** Problems 25 and 29

Caution! Radical expressions such as

$$\sqrt{-36} \qquad \sqrt{-100} \qquad \sqrt{-144} \qquad \sqrt{-81}$$

do not represent real numbers because there are no real numbers that when squared give a negative number.

Be careful to note the difference between expressions such as $-\sqrt{36}$ and $\sqrt{-36}$. We have seen that $-\sqrt{36}$ is a real number: $-\sqrt{36} = -6$. In contrast, $\sqrt{-36}$ is not a real number.

Using Your Calculator ▶ Finding a Square Root

We use the square root key $\boxed{\sqrt{}}$ on a scientific calculator to find square roots. For example, to find $\sqrt{729}$, we enter the number and press the square root key.

729 $\boxed{\sqrt{}}$ $\boxed{27}$

We have found that $\sqrt{729} = 27$. To check this result, we need to square 27. This can be done by entering 27 and pressing the $\boxed{x^2}$ key. We obtain 729. Thus, 27 is the square root of 729.

Some calculator models require keystrokes of $\boxed{\text{2nd}}$ and then $\boxed{\sqrt{}}$ followed by the radicand to find a square root.

OBJECTIVE 2 **Find the square root of fractions and decimals.**

So far, we have found square roots of whole numbers. We can also find square roots of fractions and decimals.

EXAMPLE 3   Evaluate each square root: **a.** $\sqrt{\dfrac{25}{64}}$ **b.** $\sqrt{0.81}$

Strategy In each case, we will determine what positive number, when squared, produces the radicand.

WHY The radical symbol $\sqrt{}$ indicates that the positive square root of the number written under it should be found.

Solution

a. $\sqrt{\dfrac{25}{64}} = \dfrac{5}{8}$ Ask: What positive fraction, when squared, is $\frac{25}{64}$? The answer is $\frac{5}{8}$ because $\left(\frac{5}{8}\right)^2 = \frac{25}{64}$.

b. $\sqrt{0.81} = 0.9$ Ask: What positive decimal, when squared, is 0.81? The answer is 0.9 because $(0.9)^2 = 0.81$.

Self Check 3
Evaluate:

a. $\sqrt{\dfrac{16}{49}}$

b. $\sqrt{0.04}$

Now Try ⟹ Problems 37 and 43

OBJECTIVE 3 **Evaluate expressions that contain square roots.**

In Chapters 1, 2, and 3, we used the order of operations rule to evaluate expressions that involve more than one operation. If an expression contains any square roots, they are to be evaluated at the same stage in your solution as exponential expressions. (See Step 2 in the familiar order of operations rule on the next page.)

Order of Operations

1. Perform all calculations within parentheses and other grouping symbols following the order listed in Steps 2–4 below, working from the innermost pair of grouping symbols to the outermost pair.

2. Evaluate all exponential expressions and **square roots**.

3. Perform all multiplications and divisions as they occur from left to right.

4. Perform all additions and subtractions as they occur from left to right.

EXAMPLE 4   Evaluate: **a.** $\sqrt{64} + \sqrt{9}$ **b.** $-\sqrt{25} - \sqrt{225}$

Strategy We will scan the expression to determine what operations need to be performed. Then we will perform those operations, one at a time, following the order of operations rule.

WHY Every numerical expression has only one correct value. If we don't follow the correct order of operations, the expression can have more than one value.

Solution Since the expression does not contain any parentheses, we begin with Step 2 of the rules for the order of operations: Evaluate all exponential expressions and any square roots.

a. $\sqrt{64} + \sqrt{9} = 8 + 3$ Evaluate each square root first.
$\phantom{\sqrt{64} + \sqrt{9}} = 11$ Do the addition.

b. $-\sqrt{25} - \sqrt{225} = -5 - 15$ Evaluate each square root first.
$\phantom{-\sqrt{25} - \sqrt{225}} = -20$ Do the subtraction.

Self Check 4

Evaluate:
a. $\sqrt{121} + \sqrt{1}$
b. $-\sqrt{9} - \sqrt{196}$

Now Try ➲ Problems 49 and 53

EXAMPLE 5   Evaluate: **a.** $6\sqrt{100}$ **b.** $-5\sqrt{16} + 3\sqrt{9}$

Strategy We will scan the expression to determine what operations need to be performed. Then we will perform those operations, one at a time, following the order of operations rule.

WHY Every numerical expression has only one correct value. If we don't follow the correct order of operations, the expression can have more than one value.

Solution Since the expression does not contain any parentheses, we begin with Step 2 of the rules for the order of operations: Evaluate all exponential expressions and any square roots.

a. We note that $6\sqrt{100}$ means $6 \cdot \sqrt{100}$.

$\qquad 6\sqrt{100} = 6(10)$ Evaluate the square root first.
$\qquad \phantom{6\sqrt{100}} = 60$ Do the multiplication.

b. $-5\sqrt{16} + 3\sqrt{9} = -5(4) + 3(3)$ Evaluate each square root first.
$\phantom{-5\sqrt{16} + 3\sqrt{9}} = -20 + 9$ Do the multiplication.
$\phantom{-5\sqrt{16} + 3\sqrt{9}} = -11$ Do the addition.

Self Check 5

Evaluate:
a. $8\sqrt{121}$
b. $-6\sqrt{25} + 2\sqrt{36}$

Now Try ➲ Problems 57 and 61

EXAMPLE 6   Evaluate: $12 + 3[3^2 - (4 - 1)\sqrt{36}]$

Strategy We will work within the parentheses first and then within the brackets. Within each set of grouping symbols, we will follow the order of operations rule.

WHY By the order of operations rule, we must work from the *innermost* pair of grouping symbols to the *outermost*.

Solution

$$12 + 3\left[3^2 - (4 - 1)\sqrt{36}\right] = 12 + 3\left[3^2 - 3\sqrt{36}\right] \quad \text{Do the subtraction within the parentheses.}$$

$$= 12 + 3[9 - 3(6)] \quad \text{Within the brackets, evaluate the exponential expression and the square root.}$$

$$= 12 + 3[9 - 18] \quad \text{Do the multiplication within the brackets.}$$

$$= 12 + 3[-9] \quad \text{Do the subtraction within the brackets.}$$

$$= 12 + (-27) \quad \text{Do the multiplication.}$$

$$= -15 \quad \text{Do the addition.}$$

Self Check 6

Evaluate:
$10 - 4[2^2 - (3 + 2)\sqrt{4}]$

Now Try ➭ **Problems 65 and 69**

OBJECTIVE 4 Evaluate formulas involving square roots.

To evaluate formulas that involve square roots, we replace the letters with specific numbers and then use the order of operations rule.

EXAMPLE 7   Evaluate $c = \sqrt{a^2 + b^2}$ for $a = 3$ and $b = 4$.

Strategy In the given formula, we will replace the letter a with 3 and b with 4. Then we will use the order of operations rule to find the value of the radicand.

WHY We need to know the value of the radicand before we can find its square root.

Self Check 7

Evaluate $a = \sqrt{c^2 - b^2}$ for $c = 17$ and $b = 15$.

Now Try ➭ **Problem 81**

Solution

$$c = \sqrt{a^2 + b^2} \quad \text{This is the formula to evaluate.}$$

$$c = \sqrt{3^2 + 4^2} \quad \text{Replace } a \text{ with 3 and } b \text{ with 4.}$$

$$c = \sqrt{9 + 16} \quad \text{Evaluate the exponential expressions.}$$

$$c = \sqrt{25} \quad \text{Do the addition.}$$

$$c = 5 \quad \text{Evaluate the square root.}$$

OBJECTIVE 5 Approximate square roots.

In Examples 2–7, we have found square roots of perfect squares. If a number is not a perfect square, we can use the $\boxed{\sqrt{}}$ key on a calculator or a table of square roots to find its *approximate* square root. For example, to find $\sqrt{17}$ using a scientific calculator, we enter 17 and press the square root key:

$$17 \ \boxed{\sqrt{}}$$

The display reads

4.123105626

This result is an approximation because the exact value of $\sqrt{17}$ is a **nonterminating decimal** that never repeats. If we round to the nearest thousandth, we have

$$\sqrt{17} \approx 4.123 \quad \text{Read} \approx \text{ as "is approximately equal to."}$$

To check this approximation, we square 4.123.

$$(4.123)^2 = 16.999129$$

Since the result is close to 17, we know that $\sqrt{17} \approx 4.123$.

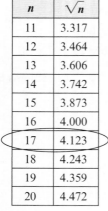

n	$\sqrt{n}$
11	3.317
12	3.464
13	3.606
14	3.742
15	3.873
16	4.000
17	4.123
18	4.243
19	4.359
20	4.472

A portion of the table of square roots from Appendix 3 on page A-23 is shown in the margin on the previous page. The table gives decimal approximations of square roots of whole numbers that are not perfect squares. To find an approximation of $\sqrt{17}$ to the nearest thousandth, we locate 17 in the n column of the table and scan directly right, to the $\sqrt{n}$ column, to find that $\sqrt{17} \approx 4.123$.

EXAMPLE 8   Use a calculator to approximate each square root. Round to the nearest hundredth.   **a.** $\sqrt{373}$ **b.** $\sqrt{56.2}$ **c.** $\sqrt{0.0045}$

Strategy We will identify the radicand and find the square root using the $\boxed{\sqrt{}}$ key. Then we will identify the digit in the thousandths column of the display.

WHY To round to the hundredths column, we must determine whether the digit in the thousandths column is less than 5, or greater than or equal to 5.

Solution
a. From the calculator, we get $\sqrt{373} \approx 19.31320792$. Rounded to the nearest hundredth, $\sqrt{373} \approx 19.31$.
b. From the calculator, we get $\sqrt{56.2} \approx 7.496665926$. Rounded to the nearest hundredth, $\sqrt{56.2} \approx 7.50$.
c. From the calculator, we get $\sqrt{0.0045} \approx 0.067082039$. Rounded to the nearest hundredth, $\sqrt{0.0045} \approx 0.07$.

Self Check 8

Use a scientific calculator to approximate each square root. Round to the nearest hundredth.

a. $\sqrt{153}$

b. $\sqrt{607.8}$

c. $\sqrt{0.076}$

Now Try ➲ Problems 87 and 91

Answers to Self Checks

1. 8 and -8 **2. a.** 12   **b.** -9 **3. a.** $\frac{4}{7}$ **b.** 0.2   **4. a.** 12   **b.** -17 **5. a.** 88   **b.** -18 **6.** 34
7. 8   **8. a.** 12.37   **b.** 24.65   **c.** 0.28

SECTION 4.6   STUDY SET

VOCABULARY

Fill in the blanks.

1. When we raise a number to the second power, we are squaring it, or finding its _____.

2. The square _____ of a given number is a number whose square is the given number.

3. The symbol $\sqrt{}$ is called a _____ symbol.

4. Label the *radicand,* the *radical expression,* and the *radical symbol* in the illustration below.

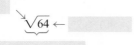

5. Whole numbers, such as 36 and 49, that are squares of whole numbers are called _____ squares.

6. The exact value of $\sqrt{17}$ is a _____ decimal that never repeats.

CONCEPTS

Fill in the blanks.

7. **a.** The square of 5 is ▢ because $5^2 = $ ▢ .

   b. The square of $\frac{1}{4}$ is ▢ because $\left(\frac{1}{4}\right)^2 = $ ▢ .

8. Complete the list of perfect squares: 1, 4, ▢ , 16, ▢ , 36, 49, 64, ▢ , 100, ▢ , 144, ▢ , 196, ▢ .

9. **a.** $\sqrt{49} = 7$ because ▢ $^2 = 49$.

   b. $\sqrt{4} = 2$ because ▢ $^2 = 4$.

10. **a.** $\sqrt{\dfrac{9}{16}} = $ ▢ because $\left(\dfrac{3}{4}\right)^2 = \dfrac{9}{16}$.

    b. $\sqrt{0.16} = $ ▢ because $(0.4)^2 = 0.16$.

11. Evaluate each square root.

    a. $\sqrt{1}$ **b.** $\sqrt{0}$

12. Evaluate each square root.

a. $\sqrt{121}$ **b.** $\sqrt{144}$ **c.** $\sqrt{169}$

d. $\sqrt{196}$ **e.** $\sqrt{225}$

13. In what step of the order of operations rule are square roots to be evaluated?

14. Graph $\sqrt{9}$ and $-\sqrt{4}$ on a number line.

15. Graph $-\sqrt{3}$ and $\sqrt{7}$ on a number line. (*Hint:* Use a calculator or square root table to approximate each square root first.)

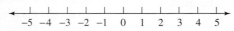

16. a. Between what two whole numbers would $\sqrt{19}$ be located when graphed on a number line? ____

     b. Between what two whole numbers would $\sqrt{50}$ be located when graphed on a number line?

NOTATION

Fill in the blanks.

17. a. The symbol $\sqrt{}$ is used to indicate a positive

     _____ _____.

     b. The symbol $-\sqrt{}$ is used to indicate the _____ square root of a positive number.

18. $4\sqrt{9}$ means $4 \boxed{} \sqrt{9}$.

Complete each step to evaluate the expression.

19. $-\sqrt{49} + \sqrt{64} = \boxed{} + \boxed{}$

               $= 1$

20. $2\sqrt{100} - 5\sqrt{25} = 2(\boxed{}) - 5(\boxed{})$

                    $= \boxed{} - 25$

                    $= -5$

GUIDED PRACTICE

Find the two square roots of each number. See Example 1.

21. 25   **22.** 1

23. 16   **24.** 144

Evaluate each square root without using a calculator. See Example 2.

25. $\sqrt{16}$ **26.** $\sqrt{64}$

27. $\sqrt{9}$ **28.** $\sqrt{25}$

29. $-\sqrt{144}$ **30.** $-\sqrt{121}$

31. $-\sqrt{49}$ **32.** $-\sqrt{81}$

Use a calculator to evaluate each square root. See Objective 1, Using Your Calculator.

33. $\sqrt{961}$ **34.** $\sqrt{841}$

35. $\sqrt{3,969}$ **36.** $\sqrt{5,625}$

Evaluate each square root without using a calculator. See Example 3.

37. $\sqrt{\dfrac{4}{25}}$ **38.** $\sqrt{\dfrac{36}{121}}$

39. $-\sqrt{\dfrac{16}{9}}$ **40.** $-\sqrt{\dfrac{64}{25}}$

41. $-\sqrt{\dfrac{1}{81}}$ **42.** $-\sqrt{\dfrac{1}{4}}$

43. $\sqrt{0.64}$ **44.** $\sqrt{0.36}$

45. $-\sqrt{0.81}$ **46.** $-\sqrt{0.49}$

47. $\sqrt{0.09}$ **48.** $\sqrt{0.01}$

Evaluate each expression without using a calculator. See Example 4.

49. $\sqrt{36} + \sqrt{1}$ **50.** $\sqrt{100} + \sqrt{16}$

51. $\sqrt{81} + \sqrt{49}$ **52.** $\sqrt{4} + \sqrt{36}$

53. $-\sqrt{144} - \sqrt{16}$ **54.** $-\sqrt{1} - \sqrt{196}$

55. $-\sqrt{225} + \sqrt{144}$ **56.** $-\sqrt{169} + \sqrt{16}$

Evaluate each expression without using a calculator. See Example 5.

57. $4\sqrt{25}$ **58.** $2\sqrt{81}$

59. $-10\sqrt{196}$ **60.** $-40\sqrt{4}$

61. $-4\sqrt{169} + 2\sqrt{4}$ **62.** $-6\sqrt{81} + 5\sqrt{1}$

63. $-8\sqrt{16} + 5\sqrt{225}$ **64.** $-3\sqrt{169} + 2\sqrt{225}$

Evaluate each expression without using a calculator. See Example 6.

65. $15 + 4\left[5^2 - (6 - 1)\sqrt{4}\right]$

66. $18 + 2\left[4^2 - (7 - 3)\sqrt{9}\right]$

67. $50 - \left[(6^2 - 24) + 9\sqrt{25}\right]$

68. $40 - \left[(7^2 - 40) + 7\sqrt{64}\right]$

69. $\sqrt{196} + 3\left(5^2 - 2\sqrt{225}\right)$

70. $\sqrt{169} + 2\left(7^2 - 3\sqrt{144}\right)$

71. $\dfrac{\sqrt{16} - 6(2^2)}{\sqrt{4}}$

72. $\dfrac{\sqrt{49} - 3(1^6)}{\sqrt{16} - \sqrt{64}}$

73. $\sqrt{\dfrac{1}{16}} - \sqrt{\dfrac{9}{25}}$

74. $\sqrt{\dfrac{25}{9}} - \sqrt{\dfrac{64}{81}}$

75. $5\left(-\sqrt{49}\right)(-2)^2$

76. $\left(-\sqrt{64}\right)(-2)(3)^3$

77. $(6^2)\sqrt{0.04} + 2.36$

78. $(5^2)\sqrt{0.25} + 4.7$

79. $-\left(-3\sqrt{1.44} + 5\right)$

80. $-\left(-2\sqrt{1.21} - 6\right)$

Evaluate each formula without using a calculator. See Example 7.

81. Evaluate $c = \sqrt{a^2 + b^2}$ for $a = 9$ and $b = 12$.

82. Evaluate $c = \sqrt{a^2 + b^2}$ for $a = 6$ and $b = 8$.

83. Evaluate $a = \sqrt{c^2 - b^2}$ for $c = 25$ and $b = 24$.

84. Evaluate $b = \sqrt{c^2 - a^2}$ for $c = 17$ and $a = 8$.

Use a calculator (or the square root table in Appendix 1) to complete each square root table. Round to the nearest thousandth when an answer is not exact. See Example 8.

85.

Number	Square Root
1	
2	
3	
4	
5	
6	
7	
8	
9	
10	

86.

Number	Square Root
10	
20	
30	
40	
50	
60	
70	
80	
90	
100	

Use a calculator (or a square root table) to approximate each of the following to the nearest hundredth. See Example 8.

87. $\sqrt{5}$

88. $\sqrt{51}$

89. $\sqrt{66}$

90. $\sqrt{204}$

Use a calculator to approximate each of the following to the nearest thousandth. See Example 8.

91. $\sqrt{24.05}$

92. $\sqrt{70.69}$

93. $-\sqrt{11.1}$

94. $\sqrt{0.145}$

LOOK ALIKES

Evaluate each expression.

95. a. $4\sqrt{9}$ **b.** $9\sqrt{4}$ **c.** $9\sqrt{49}$

   d. $49\sqrt{4}$ **e.** $49\sqrt{9}$ **f.** $49\sqrt{49}$

96. a. $16\sqrt{25}$ **b.** $25\sqrt{16}$ **c.** $16\sqrt{16}$

   d. $25\sqrt{25}$ **e.** $\sqrt{16} \cdot \sqrt{25}$ **f.** $\sqrt{25} \cdot \sqrt{25}$

97. a. $\sqrt{64} + 36$ **b.** $\sqrt{64 + 36}$ **c.** $\sqrt{64} + \sqrt{36}$

   d. $\sqrt{64} \cdot \sqrt{36}$ **e.** $\sqrt{64} - \sqrt{36}$ **f.** $\sqrt{36} - \sqrt{64}$

98. a. $\sqrt{4 \cdot 25}$ **b.** $\sqrt{4 \cdot 4}$ **c.** $\sqrt{25 \cdot 25}$

   d. $\sqrt{4 \cdot 4 \cdot 4}$ **e.** $4 \cdot \sqrt{25}$ **f.** $25 \cdot \sqrt{4}$

APPLICATIONS

In the following problems, some lengths are expressed as square roots. Solve each problem by evaluating any square roots. You may need to use a calculator. If so, round to the nearest tenth when an answer is not exact.

99. Carpentry.   Find the length of the slanted side of each roof truss shown below.

a.

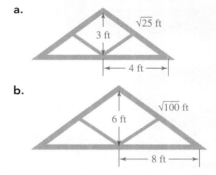

b.

100. Radio antennas.   Refer to the illustration below. How far from the base of the antenna is each guy wire anchored to the ground? (The measurements are in feet.)

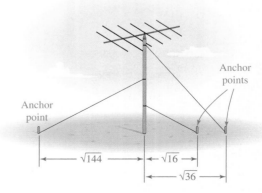

101. Baseball. The illustration below shows some dimensions of a major league baseball field. How far is it from home plate to second base?

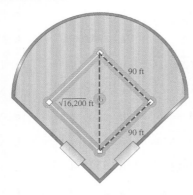

102. Surveying. Refer to the illustration below. Use the imaginary triangles set up by a surveyor to find the length of each lake. (The measurements are in meters.)

a.

Length: $\sqrt{318,096}$

b.

Length: $\sqrt{93,025}$

103. Flatscreen televisions. The picture screen on a television set is measured diagonally. What size screen is shown below?

$\sqrt{1,764}$ in.

104. Ladders. A painter's ladder is shown below. How long are the legs of the ladder?

$\sqrt{225}$ ft $\sqrt{169}$ ft

WRITING

105. When asked to find $\sqrt{16}$, a student answered 8. Explain his misunderstanding of the concept of square root.

106. Explain the difference between the *square* and the *square root* of a number. Give an example.

107. What is a *nonterminating* decimal? Use an example in your explanation.

108. **a.** How would you check whether $\sqrt{389} = 17$?
  b. How would you check whether $\sqrt{7} \approx 2.65$?

109. Explain why $\sqrt{-4}$ does not represent a real number.

110. Is there a difference between $-\sqrt{25}$ and $\sqrt{-25}$? Explain.

111. $\sqrt{6} \approx 2.449$. Explain why an $\approx$ symbol is used and not an $=$ symbol.

112. Without evaluating the following square roots, determine which is the largest and which is the smallest. Explain how you decided.

$$\sqrt{23}, \ \sqrt{27}, \ \sqrt{11}, \ \sqrt{6}, \ \sqrt{20}$$

REVIEW

113. Multiply: $6.75 \cdot 12.2$

114. Divide: $5.7\overline{)18.525}$

115. Evaluate: $(3.4)^3$

116. Add: $23.45 + 76 + 0.009 + 3.8$

4 Summary and Review

DEFINITIONS AND CONCEPTS	EXAMPLES
The place-value system for whole numbers can be extended to create the **decimal numeration system**. The place-value columns to the left of the decimal point form the **whole-number part** of the decimal number. The value of each of those columns is 10 times greater than the column directly to its right. The columns to the right of the decimal point form the **fractional part**. Each of those columns has a value that is $\frac{1}{10}$ of the value of the place directly to its left.	 The place value of the digit 3 is *3 hundredths*. The digit that tells the number of *ten-thousandths* is 1.
To write a decimal number in **expanded form** (**expanded notation**) means to write it as an addition of the place values of each of its digits.	Write 28.9341 in expanded notation: $$28.9341 = 20 + 8 + \frac{9}{10} + \frac{3}{100} + \frac{4}{1,000} + \frac{1}{10,000}$$
To read a decimal: 1. Look to the left of the decimal point and say the name of the whole number. 2. The decimal point is read as "and." 3. Say the fractional part of the decimal as a whole number followed by the name of the last place-value column of the digit that is the farthest to the right. We can use the steps for reading a decimal to **write it in words**.	Write the decimal in words and then as a fraction or mixed number: 28 .9341 *The whole-number part is 28. The fractional part is 9341. The digit the farthest to the right, 1, is in the ten-thousandths place.* *Twenty-eight and nine thousand three hundred forty-one ten-thousandths* Written as a mixed number, 28.9341 is $28\frac{9,341}{10,000}$. Write the decimal in words and then as a fraction or mixed number: 0 .079 *The whole-number part is 0. The fractional part is 79. The digit the farthest to the right, 9, is in the thousandths place.* *Seventy-nine thousandths* Written as a fraction, 0.079 is $\frac{79}{1,000}$.
The procedure for **reading a decimal** can be applied in reverse to convert from written-word form to standard form.	Write the decimal number in standard form: ***Negative twelve and sixty-five ten-thousandths*** -12.0065 *This is the ten-thousandths place-value column.* *Two placeholder 0's must be inserted here so that the last digit in 65 is in the ten-thousandths column.*
To compare two decimals: 1. Make sure both numbers have the same number of decimal places to the right of the decimal point. Write any additional zeros necessary to achieve this.	Compare 47.31572 and 47.31569. 47.31572 47.31569 *As we work from left to right, this is the first column in which the digits differ. Since 7 > 6, it follows that 47.31572 is greater than 47.31569.* Thus, 47.31572 > 47.31569.

2. Compare the digits of each decimal, column by column, working from left to right.

3. If the decimals are *positive*: When two digits differ, the decimal with the greater digit is the greater number.

If the decimals are *negative*: When two digits differ, the decimal with the smaller digit is the greater number.

Compare −6.418 and −6.41.

−6.41 8   These decimals are negative.

−6.41 0   Write a zero after 1 to help in the comparison.

⌐ As we work from left to right, this is the first column in which the digits differ. Since 0 < 8, it follows that −6.410 is greater than −6.418.

Thus, −6.41 > −6.418.

To **graph a decimal number** means to make a drawing that represents the number.

Graph −2.17, 0.6, −2.89, 3.99, and −0.5 on a number line.

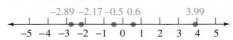

---

1. To **round a decimal** to a certain decimal place value, locate the **rounding digit** in that place.

2. Look at the **test digit** directly to the right of the rounding digit.

3. If the test digit is 5 or greater, round up by adding 1 to the rounding digit and dropping all the digits to its right. If the test digit is less than 5, round down by keeping the rounding digit and dropping all the digits to its right.

Round 33.41632 to the nearest thousandth.

Rounding digit: Keep the rounding digit:
thousandths column   Do not add 1.
      ↓   ↓

33.41632   33.41632
    ↑   ↑

Test digit: 3 is less than 5.   Drop the test digit and all digits to its right.

Thus, 33.41632 rounded to the nearest thousandth is 33.416.

Round 2.798 to the nearest hundredth.

Rounding digit: Add 1. Since 9 + 1 = 10,
hundredths column   write 0 in this column and
                                 carry 1 to the tenths column.
     ↓   ¹↓

2.798   2.798
    ↑   ↑

Test digit: 8 is 5 or greater.   Drop the test digit.

Thus, 2.798 rounded to the nearest hundredth is 2.80.

There are many situations in our daily lives that call for **rounding amounts of money**.

Rounded to the *nearest cent*, $0.14672 is $0.15.

Rounded to the *nearest dollar*, $142.39 is $142.

REVIEW EXERCISES

1. a. Represent the amount of the square region that is shaded, using a decimal and a fraction.

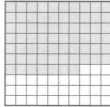

   b. Shade 0.8 of the region shown below.

2. Consider the decimal number 2,809.6735.

   a. What is the place value of the digit 7?

   b. Which digit tells the number of thousandths?

   c. Which digit tells the number of hundreds?

   d. What is the place value of the digit 5?

3. Write 16.4523 in expanded notation.

Write each decimal in words and then as a fraction or mixed number.

   4. 2.3   **5.** −615.59

   6. 0.0601   **7.** 0.00001

Write each number in standard form.

   8. One hundred and sixty-one hundredths

   9. Eleven and nine hundred ninety-seven thousandths

   10. Three hundred one and sixteen millionths

Place an < or an > symbol in the box to make a true statement.

   11. 5.68 ▢ 5.75   **12.** 106.8199 ▢ 106.82

   13. −78.23 ▢ −78.303   **14.** −555.098 ▢ −555.0991

   15. Graph: 1.55, −0.8, −2.1, and −2.7.

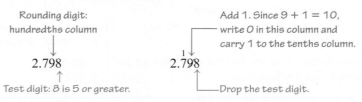

16. Determine whether each statement is true or false.

    a. $78 = 78.0$ **b.** $6.910 = 6.901$

    c. $-3.4700 = -3.470$ **d.** $0.008 = .00800$

Round each decimal to the indicated place value.

17. 4.578 nearest hundredth

18. 3,706.0815 nearest thousandth

19. −0.0614 nearest tenth

20. −88.12 nearest tenth

21. 6.702983 nearest ten-thousandth

22. 11.314964 nearest ten-thousandth

23. 0.2222282 nearest millionth

24. 0.635265 nearest hundred-thousandth

Round each given dollar amount.

25. $0.671456 to the nearest cent

26. $12.82 to the nearest dollar

27. Valedictorians. At the end of the school year, the five students listed were in the running to be class valedictorian (the student with the highest grade point average). Rank the students in order by GPA, beginning with the valedictorian. See the table in the next column.

Name	GPA
Diaz, Cielo	3.9809
Chou, Wendy	3.9808
Washington, Shelly	3.9865
Gerbac, Lance	3.899
Singh, Amani	3.9713

28. Allergy forecast. The graph below shows a four-day forecast of pollen levels for Las Vegas, Nevada. Determine the decimal-number forecast for each day.

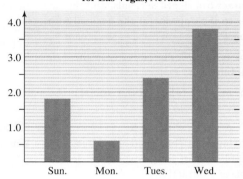

Allergy Alert 4-Day Forecast for Las Vegas, Nevada

SECTION 4.2 ▶ **Adding and Subtracting Decimals**

DEFINITIONS AND CONCEPTS	EXAMPLES
To add or subtract decimals: **1.** Write the numbers in **vertical form** with the decimal points lined up. **2.** Add (or subtract) as you would whole numbers. **3.** Write the decimal point in the result from Step 2 below the decimal points in the problem. If the number of decimal places in the problem are different, insert additional zeros so that the number of decimal places match. If the sum of the digits in any place-value column is greater than 9, we must **carry**. If the subtraction of the digits in any place-value column requires that we subtract a larger digit from a smaller digit, we must **borrow** or **regroup**.	Add: $15.82 + 19 + 32.995$ Write the problem in vertical form and add, column by column, working right to left. $\begin{array}{r} \overset{1\,1\;1}{15.820} \\ 19.000 \\ +\,32.995 \\ \hline 67.815 \end{array}$ Insert an extra zero. Insert a decimal point and extra zeros. Line up the decimal points. To **check** the result, add *bottom to top*. Subtract: $8.4 - 3.029$ Write the problem in **vertical form** and subtract, column by column, working right to left. $\begin{array}{r} \overset{9}{} \\ 8.\cancel{4}\cancel{0}\cancel{0} \\ -\,3.029 \\ \hline 5.371 \end{array}$ Insert extra zeros. First, borrow from the tenths column, then borrow from the hundredths column. To **check**: The sum of the difference and the subtrahend should equal the minuend. $\begin{array}{r} \overset{1\,1}{5.371} \leftarrow \text{Difference} \\ +\,3.029 \leftarrow \text{Subtrahend} \\ \hline 8.400 \leftarrow \text{Minuend} \end{array}$

To **add signed decimals**, we use the same rules that are used for adding integers.	Add: $-21.35 + (-64.52)$
	Find the absolute values: $\|-21.35\| = 21.35$ and $\|-64.52\| = 64.52$
With like signs: Add their absolute values and attach their common sign to the sum.	$-21.35 + (-64.52) = -85.87$ *Add the absolute values, 21.35 and 64.52, to get 85.87. Since both decimals are negative, make the final result negative.*
With unlike signs: Subtract their absolute values (the smaller from the larger). If the positive decimal has the larger absolute value, the final answer is positive. If the negative decimal has the larger absolute value, make the final answer negative.	Add: $-7.4 + 9.8$
	Find the absolute values: $\|-7.4\| = 7.4$ and $\|9.8\| = 9.8$
	$-7.4 + 9.8 = 2.4$ *Subtract the smaller absolute value from the larger: $9.8 - 7.4 = 2.4$. Since the positive number, 9.8, has the larger absolute value, the final answer is positive.*

To **subtract two signed decimals**, add the first decimal to the opposite of the decimal to be subtracted.	Subtract: $-8.62 - (-1.4)$
	The number to be subtracted is -1.4. Subtracting -1.4 is the same as adding its opposite, 1.4.
	Add . . . $-8.62 - (-1.4) = -8.62 + 1.4 = -7.22$ *Use the rule for adding two decimals with different signs.* *. . . the opposite*

Estimation can be used to check the reasonableness of an answer to a decimal addition or subtraction.	Estimate the sum by rounding the addends to the nearest ten: $328.99 + 459.02$
	$328.99 \longrightarrow 330$ *Round to the nearest ten.* $\underline{+459.02} \longrightarrow \underline{+460}$ *Round to the nearest ten.* $788.01 \qquad\quad 790$ *This is the estimate.*
	Estimate the difference by using **front-end rounding**: $302.47 - 36.9$
	Each number is rounded to its largest place value.
	$302.47 \longrightarrow 300$ *Round to the nearest hundred.* $\underline{-\ 36.9} \longrightarrow \underline{-\ 40}$ *Round to the nearest ten.* $265.57 \qquad\quad 260$ *This is the estimate.*

We can use addition and subtraction to solve many application problems that involve decimals.	See Examples 10–12 that begin on page 331 to review how to solve application problems by adding and subtracting decimals.

REVIEW EXERCISES

Perform each indicated operation.

29. $19.5 + 34.4 + 12.8$

30. $3.4 + 6.78 + 35 + 0.008$

31. $68.47 - 53.3$ **32.** $45.8 - 17.372$

33. $345.214 - 27.39$

34. $\begin{array}{r} 8.61 \\ 5.97 \\ +\,9.72 \\ \hline \end{array}$

35. $-16.1 + 8.4$ **36.** $-4.8 - (-7.9)$

37. $-3.55 + (-1.25)$ **38.** $-15.1 - 13.99$

Evaluate each expression.

39. $-8.8 + (-7.3 - 9.5)$

40. $(5 - 0.096) - (-0.035)$

41. a. Estimate the sum by rounding the addends to the nearest ten: $612.05 + 145.006$

    b. Estimate the difference by using front-end rounding: $289.43 - 21.86$

42. Coins.   The thicknesses of a penny, nickel, dime, quarter, half-dollar, and presidential $1 coin are 1.55 millimeters, 1.95 millimeters, 1.35 millimeters, 1.75 millimeters, 2.15 millimeters, and 2.00 millimeters, in that order. Find the height of a stack made from one of each type of coin.

43. Sale prices.   A calculator normally sells for $52.20. If it is being discounted $3.99, what is the sale price?

44. Microwave ovens.   A microwave oven is shown below. How tall is the window?

SECTION 4.3 ▶ Multiplying Decimals

DEFINITIONS AND CONCEPTS	EXAMPLES
To multiply two decimals: 1. Multiply the decimals as if they were whole numbers. 2. Find the total number of decimal places in both factors. 3. Insert a decimal point in the result from Step 1 so that the answer has the same number of decimal places as the total found in Step 2. When multiplying decimals, we *do not* need to line up the decimal points.	Multiply: $2.76 \cdot 4.3$ Write the problem in vertical form and multiply 2.76 and 4.3 as if they were whole numbers. $\begin{array}{r} 2.76 \\ \times\ 4.3 \\ \hline 828 \\ 11040 \\ \hline 11.868 \end{array}$ 2 decimal places · 1 decimal place — The answer will have $2 + 1 = 3$ decimal places. 11.868 *Move 3 places from right to left and insert a decimal point in the answer.* Thus, $2.76 \cdot 4.3 = 11.868$.
Multiplying a decimal by 10, 100, 1,000, and so on To find the product of a decimal and 10, 100, 1,000, and so on, move the decimal point to the right the same number of places as there are zeros in the power of 10. **Multiplying a decimal by 0.1, 0.01, 0.001, and so on** To find the product of a decimal and 0.1, 0.01, 0.001, and so on, move the decimal point to the left the same number of places as there are in the power of 10.	Multiply: $84.561 \cdot 10,000 = 845,610$ *Since 10,000 has four zeros, move the decimal point in 84.561 four places to the right. Write a placeholder zero (shown in blue).* Multiply: $32.67 \cdot 0.01 = 0.3267$ *Since 0.01 has two decimal places, move the decimal point in 32.67 two places to the left.*
The rules for multiplying integers also hold for **multiplying signed decimals**: The product of two decimals with **like signs** is positive, and the product of two decimals with **unlike signs** is negative.	Multiply: $(-0.03)(-4.1)$ Find the absolute values: $|-0.03| = 0.03$ and $|4.1| = 4.1$ Since the decimals have like signs, the product is positive. $(-0.03)(-4.1) = 0.123$ *Multiply the absolute values, 0.03 and 4.1, to get 0.123.* Multiply: $-5.7(0.4)$ Find the absolute values: $|-5.7| = 5.7$ and $|0.4| = 0.4$ Since the decimals have unlike signs, the product is negative. $-5.7(0.4) = -2.28$ *Multiply the absolute values, 5.7 and 0.4, to get 2.28. Make the final answer negative.*
We can use the rule for multiplying a decimal by a power of 10 to **write large numbers in standard form**.	Write *4.16 billion* in standard notation: $4.16 \text{ billion} = 4.16 \cdot \textbf{1 billion}$ $= 4.16 \cdot \textbf{1,000,000,000}$ *Write 1 billion in standard form.* $= 4,160,000,000$ *Since 1,000,000,000 has nine zeros, move the decimal point in 4.16 nine places to the right.*
The base of an **exponential expression** can be a positive or a negative decimal.	Evaluate: $(1.5)^2$ $(1.5)^2 = 1.5 \cdot 1.5$ *The base is 1.5 and the exponent is 2. Write the base as a factor 2 times.* $= 2.25$ *Multiply the decimals.* Evaluate: $(-0.02)^2$ $(-0.02)^2 = (-0.02)(-0.02)$ *The base is −0.02 and the exponent is 2. Write the base as a factor 2 times.* $= 0.0004$ *Multiply the decimals. The product of two decimals with like signs is positive.*

To **evaluate a formula**, we replace the letters with specific numbers and then use the order of operations rule.	Evaluate $P = 2l + 2w$ for $l = 4.9$ and $w = 3.4$. $P = 2l + 2w$ $\quad = 2(\mathbf{4.9}) + 2(\mathbf{3.4})$ *Replace l with 4.9 and w with 3.4.* $\quad = 9.8 + 6.8$ *Do the multiplication.* $\quad = 16.6$ *Do the addition.*

Estimation can be used to check the reasonableness of an answer to a decimal multiplication.	Estimate $37 \cdot 8.49$ by **front-end rounding**.
	
	$\begin{array}{ccc} 37 & \longrightarrow & 40 \\ \times 8.49 & \longrightarrow & \times\ 8 \\ \hline & & 320 \end{array}$ *Round to the nearest ten.*
	                           Round to the nearest one.
	                           This is the estimate.
	
	The estimate is 320. If we calculate $37 \cdot 8.49$, the product is exactly 314.13.

We can use multiplication to solve many application problems that involve decimals.	See Examples 12 and 13 that begin on page 346 to review how to solve application problems by multiplying decimals.

REVIEW EXERCISES

Multiply.

45. $2.3 \cdot 6.9$

46. $32.45(6.1)$

47. $\begin{array}{r} 1.7 \\ \times\,0.004 \\ \hline \end{array}$

48. $\begin{array}{r} 275 \\ \times\,8.4 \\ \hline \end{array}$

49. $1.5(-0.9)$

50. $(-0.003)(-0.02)$

51. $1{,}000(90.1452)$

52. $0.001(2.897)$

Evaluate each expression.

53. $(0.2)^2$

54. $(-0.15)^2$

55. $(0.6 + 0.7)^2 - (-3)(-4.1)$

56. $3(7.8) + 2(1.1)^2$

57. $(-3.3)^2(0.00001)$

58. $(0.1)^3 + 2|-45.63 - 12.24|$

Write each number in standard notation.

59. a. Geography. China is the third largest country in land area with territory that extends over *9.6 million* square kilometers. (Source: china.org)

    b. Motor travel. The number of miles driven by motorists on U.S. roads last year was a record *3.22 trillion* miles. (Source: U.S. DOT)

60. a. Estimate the product using front-end rounding: $193.28 \cdot 7.63$

    b. Estimate the product by rounding the factors to the nearest tenth: $12.42 \cdot 7.38$

61. Evaluate the formula $A = P + Prt$ for $P = 70.05$, $r = 0.08$, and $t = 5$.

62. Shopping. If crab meat sells for \$12.95 per pound, what would 1.5 pounds of crab meat cost? Round to the nearest cent.

63. Auto painting. A manufacturer uses a three-part process to finish the exterior of the cars it produces.

    Step 1: A 0.03-inch-thick rust-prevention undercoat is applied.

    Step 2: Three layers of color coat, each 0.015 inch thick, are sprayed on.

    Step 3: The finish is then buffed down, losing 0.005 inch of its thickness.

What is the resulting thickness of the automobile's finish?

64. Computers. The Page Setup screen for a document is shown. Find the area that can be filled with text on an 8.5 in. $\times$ 11 in. piece of paper if the margins are set as shown.

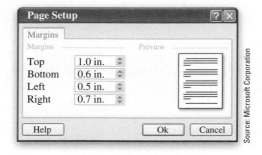

SECTION 4.4 ▶ Dividing Decimals

DEFINITIONS AND CONCEPTS	EXAMPLES
To divide a decimal by a whole number: 1. Write the problem in long division form and place a decimal point in the quotient (answer) directly above the decimal point in the dividend. 2. Divide as if working with whole numbers. 3. If necessary, additional zeros can be written to the right of the last digit of the dividend to continue the division.	Divide: $6.2 \div 4$ Place a decimal point in the quotient that lines up with the decimal point in the dividend. $$\begin{array}{r} 1.55 \\ 4)\overline{6.20} \\ -4\downarrow \\ \overline{2\,2} \\ -2\,0 \\ \overline{20} \\ -20 \\ \overline{0} \end{array}$$ Ignore the decimal points and divide as if working with whole numbers. Write a zero to the right of the 2 and bring it down. Continue to divide. The remainder is 0.
To **check** the result, we multiply the divisor by the quotient. The result should be the dividend.	***Check:*** $$\begin{array}{r} 1.55 \leftarrow \text{Quotient} \\ \times4 \leftarrow \text{Divisor} \\ \hline 6.20 \leftarrow \text{Dividend} \end{array}$$ The check confirms that $6.2 \div 4 = 1.55$.
To divide with a decimal divisor: 1. Write the problem in long division form. 2. Move the decimal point of the divisor so that it becomes a whole number. 3. Move the decimal point of the dividend the same number of places to the right. 4. Write a decimal point in the quotient (answer) directly above the decimal point in the dividend. Divide as if working with whole numbers. 5. If necessary, additional zeros can be written to the right of the last digit of the dividend to continue the division.	Divide: $\dfrac{1.462}{3.4}$ $$3.4)\overline{1.462}$$ Write the problem in long division form. Move the decimal point of the divisor, 3.4, one place to the right to make it a whole number. Move the decimal point of the dividend, 1.462, the same number of places to the right. Now use the rule for dividing a decimal by a *whole number*. $$\begin{array}{r} 0.4\,3 \\ 34)\overline{14.6\,2} \\ -13\,6\downarrow \\ \overline{1\,0\,2} \\ -1\,0\,2 \\ \overline{0} \end{array}$$ Write a decimal point in the quotient (answer) directly above the decimal point in the dividend. Divide as with whole numbers.
Sometimes when we divide decimals, the subtractions never give a zero remainder, and the division process continues forever. In such cases, we can **round the result**.	Divide: $0.77 \div 6$. Round the quotient to the nearest hundredth. To round to the hundredths column, we need to continue the division process for one more decimal place, which is the thousandths column. Rounding digit: hundredths column Test digit: Since 8 is 5 or greater, add 1 to the rounding digit and drop the test digit. $$\begin{array}{r} 0.128 \\ 6)\overline{0.770} \\ -6 \\ \overline{17} \\ -12 \\ \overline{50} \\ -48 \\ \overline{2} \end{array}$$ The remainder is still not 0. Thus, $0.77 \div 6 \approx 0.13$.

To **estimate quotients**, we use a method that approximates both the dividend and the divisor so that they divide easily. There is one rule of thumb for this method: If possible, round both numbers up or both numbers down.	Estimate the quotient: $337.96 \div 23.8$ The dividend is approximately $337.96 \div 23.8 \qquad 320 \div 20 = 16$ To divide, drop one zero from 320 and one zero from 20, and then find $32 \div 2$. The divisor is approximately The estimate is 16. (The exact answer is 14.2.)
Dividing a decimal by 10, 100, 1,000, and so on To find the quotient of a decimal and 10, 100, 1,000, and so on, move the decimal point to the left the same number of places as there are zeros in the power of 10.	Divide: $79.36 \div 10,000$ $79.36 \div 10,000 = 0.007936$ Since the divisor 10,000 has four zeros, move the decimal point four places to the left. Insert two placeholder zeros (shown in blue).
Dividing a decimal by 0.1, 0.01, 0.001, and so on To find the quotient of a decimal and 0.1, 0.01, 0.001, and so on, move the decimal point to the right the same number of decimal places as there are in the power of 10.	Divide: $\dfrac{1.6402}{0.001}$ $\dfrac{1.6402}{0.001} = 1,640.2$ Since the divisor 0.001 has *three* decimal places, move the decimal point in 1.6402 *three* places to the right.
The rules for dividing integers also hold for **dividing signed decimals**. The quotient of two decimals with *like signs* is positive, and the quotient of two decimals with *unlike signs* is negative.	Divide: $-1.53 \div 0.3 = -5.1$ Since the signs of the dividend and divisor are unlike, the final answer is negative. Divide: $\dfrac{-0.84}{-4.2} = 0.2$ Since the dividend and divisor have like signs, the quotient is positive.
We use the order of operations rule to **evaluate expressions** and **formulas**.	Evaluate: $\dfrac{37.8 - (1.2)^2}{0.1 + 0.3}$ $\dfrac{37.8 - (1.2)^2}{0.1 + 0.3} = \dfrac{37.8 - 1.44}{0.4}$ In the numerator, evaluate $(1.2)^2$. In the denominator, do the addition. $= \dfrac{36.36}{0.4}$ In the numerator, do the subtraction. $= 90.9$ Do the division indicated by the fraction bar.
We can use division to solve many application problems that involve decimals.	See Examples 11 and 12 that begin on page 361 to review how to solve application problems by dividing decimals.

REVIEW EXERCISES

Divide. Check the result.

65. $3\overline{)27.9}$ **66.** $41.8 \div 4$

67. $\dfrac{-29.67}{-23}$ **68.** $24.618 \div 0.6$

69. $-80.625 \div 12.9$ **70.** $\dfrac{0.0742}{1.4}$

71. $\dfrac{15.75}{0.25}$ **72.** $\dfrac{-0.003726}{-0.0046}$

73. $89.76 \div 1,000$ **74.** $\dfrac{0.0112}{-10}$

75. Divide -0.8765 by -0.001.

76. $77.021 \div 0.0001$

Estimate each quotient.

77. $4,983.01 \div 41.33$

78. $8.8\overline{)25,904.39}$

Divide and round each result to the specified decimal place.

79. $78.98 \div 6.1$ (nearest tenth)

80. $\dfrac{-5.438}{0.007}$ (nearest hundredth)

81. Evaluate: $\dfrac{(1.4)^2 - 2(-4.6)}{0.5 + 0.3}$

82. Evaluate the formula $C = \dfrac{5}{9}(F - 32)$ for $F = 68.9$.

83. Thanksgiving dinner. The cost of purchasing the food for a Thanksgiving dinner for a family of five was $24.95. What was the cost of the dinner per person?

84. Drinking water. Water samples from five wells were taken and tested for PCBs (polychlorinated biphenyls). The number of parts per billion (ppb) found in each sample is given below. Find the average number of parts per billion for these samples.

Sample #1:	0.44 ppb
Sample #2:	0.50 ppb
Sample #3:	0.46 ppb
Sample #4:	0.52 ppb
Sample #5:	0.63 ppb

85. Serving size. The illustration below shows the package labeling on a box of children's cereal. Use the information given to find the number of servings.

Nutrition Facts	
Serving size	1.1 ounce
Servings per container	?
Package weight	15.4 ounces

86. Telescopes. To change the position of a focusing mirror on a telescope, an adjustment knob is used. The mirror moves 0.025 inch with each revolution of the knob. The mirror needs to be moved 0.2375 inch to improve the sharpness of the image. How many revolutions of the adjustment knob does this require?

SECTION 4.5 ▶ Fractions and Decimals

DEFINITIONS AND CONCEPTS	EXAMPLES
A fraction and a decimal are said to be **equivalent** if they name the same number. To **write a fraction as a decimal**, divide the numerator of the fraction by its denominator. Sometimes, when finding the decimal equivalent of a fraction, the division process ends because a remainder of 0 is obtained. We call the resulting decimal a **terminating decimal**.	Write $\frac{3}{5}$ as a decimal. We divide the numerator by the denominator because a fraction bar indicates division: $\frac{3}{5}$ means $3 \div 5$. $$\begin{array}{r} 0.6 \\ 5\overline{)3.0} \\ -30 \\ \hline 0 \end{array}$$ *Write a decimal point and one additional zero to the right of 3.* ← *Since a zero remainder is obtained, the result is a terminating decimal.* Thus, $\frac{3}{5} = 0.6$. We say that 0.6 is the **decimal equivalent** of $\frac{3}{5}$.
If the denominator of a fraction in simplified form has factors of only 2's or 5's, or a combination of both, it can be written as a decimal by **multiplying it by a form of 1**. The objective is to write the fraction in an equivalent form with a denominator that is a power of 10, such as 10, 100, 1,000, and so on.	Write $\frac{3}{25}$ as a decimal. Since we need to multiply the denominator of $\frac{3}{25}$ by 4 to obtain a denominator of 100, it follows that $\frac{4}{4}$ should be the form of 1 that is used to build $\frac{3}{25}$. $\frac{3}{25} = \frac{3}{25} \cdot \frac{4}{4}$ *Multiply $\frac{3}{25}$ by 1 in the form of $\frac{4}{4}$.* $= \frac{12}{100}$ *Multiply the numerators.* *Multiply the denominators.* $= 0.12$ *Write the fraction as a decimal.*
Sometimes, when we are finding the decimal equivalent of a fraction, the division process never gives a remainder of 0. We call the resulting decimal a **repeating decimal**. An **overbar** can be used instead of the three dots . . . to represent the repeating pattern in a **repeating decimal**.	Write $\frac{5}{6}$ as a decimal. $$\begin{array}{r} 0.833 \\ 6\overline{)5.000} \\ -48 \\ \hline 20 \\ -18 \\ \hline 20 \\ -18 \\ \hline 2 \end{array}$$ *Write a decimal point and three additional zeros to the right of 5.* *It is apparent that 2 will continue to reappear as the remainder. Therefore, 3 will continue to reappear in the quotient. Since the repeating pattern is now clear, we can stop the division.* Thus, $\frac{5}{6} = 0.8333 \ldots$, or, using an overbar, we have $\frac{5}{6} = 0.8\overline{3}$.

When a fraction is written in decimal form, the result is either a terminating or repeating decimal. Repeating decimals are often **rounded** to a specified place value.	The decimal equivalent for $\dfrac{5}{11}$ is 0.454545 . . . . Round it to the nearest hundredth.

Rounding digit: hundredths column.
Test digit: Since 4 is less than 5, round down.

$$\frac{5}{11} = 0.454545\ldots$$

Thus, $\dfrac{5}{11} \approx 0.45$.

To write a mixed number in decimal form, we need only find the decimal equivalent for the fractional part of the mixed number. The whole-number part in the decimal form is the same as the whole-number part in the mixed-number form.	Whole-number part $4\dfrac{7}{8} = 4.875$ Write the fraction as a decimal.
A number line can be used to show the relationship between fractions and decimals.	Graph 3.125, $-4\frac{5}{7}$, $0.\overline{6}$, and $-1.\overline{09}$ on a number line. $-3.\overline{3}$ $-\dfrac{9}{10}$ 1.125 $2\dfrac{3}{4}$ $\begin{array}{ccccccccccc} & & & & & & & & & & \\ \hline -5 & -4 & -3 & -2 & -1 & 0 & 1 & 2 & 3 & 4 & 5 \end{array}$
To **compare the size of a fraction and a decimal**, it is helpful to write the fraction in its equivalent decimal form.	Place an $<$, $>$, or an $=$ symbol in the box to make a true statement: $\dfrac{3}{50}$ ☐ 0.07 To write $\dfrac{3}{50}$ as a decimal, divide 50 by 3: $\dfrac{3}{50} = 0.06$. Since 0.06 is less than 0.07, we have: $\dfrac{3}{50} < 0.07$.
To evaluate **expressions that can contain both fractions and decimals**, we can work in terms of decimals or in terms of fractions.	Evaluate: $\dfrac{1}{6} + 0.31$ If we work in terms of fractions, we have:

$$\frac{1}{6} + 0.31 = \frac{1}{6} + \frac{31}{100} \qquad \text{Write 0.31 in fraction form.}$$

$$= \frac{1}{6} \cdot \frac{\mathbf{50}}{\mathbf{50}} + \frac{31}{100} \cdot \frac{3}{3} \qquad \begin{array}{l}\text{The LCD is 300. Build each fraction by}\\ \text{multiplying by a form of 1.}\end{array}$$

$$= \frac{50}{300} + \frac{93}{300} \qquad \begin{array}{l}\text{Multiply the numerators.}\\ \text{Multiply the denominators.}\end{array}$$

$$= \frac{143}{300} \qquad \begin{array}{l}\text{Add the numerators and write the sum over}\\ \text{the common denominator 300.}\end{array}$$

If we work in terms of decimals, we have:

$$\frac{1}{6} + 0.31 \approx \mathbf{0.17} + 0.31 \qquad \text{Approximate } \tfrac{1}{6} \text{ with the decimal 0.17.}$$

$$\approx 0.48 \qquad \text{Do the addition.}$$

We can use the five-step **problem-solving strategy** to solve application problems that involve fractions and decimals.	See Example 13 on page 377 to review how to solve application problems involving fractions and decimals.

REVIEW EXERCISES

Write each fraction or mixed number as a decimal. Use an overbar when necessary.

87. $\dfrac{7}{8}$ **88.** $-\dfrac{2}{5}$ **89.** $\dfrac{9}{16}$ **90.** $\dfrac{3}{50}$

91. $\dfrac{6}{11}$ **92.** $-\dfrac{4}{3}$ **93.** $3\dfrac{7}{125}$ **94.** $\dfrac{26}{45}$

Write each fraction as a decimal. Round to the nearest hundredth.

95. $\dfrac{19}{33}$

96. $\dfrac{31}{30}$

Place an <, >, or an = symbol in the box to make a true statement.

97. $\dfrac{13}{25}$ ☐ 0.499

98. $-\dfrac{4}{15}$ ☐ $-0.2\overline{6}$

99. Write the numbers in order from least to greatest:

$\dfrac{10}{33}$, $0.\overline{3}$, 0.3

100. Graph 1.125, $-3.\overline{3}$, $2\frac{3}{4}$, and $-\frac{9}{10}$ on a number line.

$$-5 \quad -4 \quad -3 \quad -2 \quad -1 \quad 0 \quad 1 \quad 2 \quad 3 \quad 4 \quad 5$$

Evaluate each expression. Work in terms of fractions.

101. $\dfrac{1}{3} + 0.4$

102. $\dfrac{5}{6} + 0.19$

Evaluate each expression. Work in terms of decimals.

103. $\dfrac{4}{5}(-7.8)$

104. $\dfrac{1}{8}(7.3 - 5\frac{9}{10})$

105. $\dfrac{1}{2}(9.7 + 8.9)(10)$

106. $7.5 - (0.78)\left(\dfrac{1}{2}\right)^2$

107. Roadside emergency. What is the area of the reflector shown below?

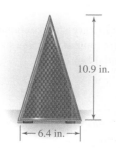

10.9 in.

6.4 in.

108. Seafood. A shopper purchased $\frac{3}{4}$ pound of crab meat, priced at \$13.80 per pound, and $\frac{1}{3}$ pound of lobster meat, selling for \$35.70 per pound. Find the total cost of these items.

SECTION 4.6 ▶ **Square Roots**

DEFINITIONS AND CONCEPTS	EXAMPLES

The **square root** of a given number is a number whose square is the given number.
 	
Every positive number has two square roots. The number 0 has only one square root.
	Find the two square roots of 81.
	   9 is a square root of 81 because $9^2 = 81$
	                    and
	   -9 is a square root of 81 because $(-9)^2 = 81$.

A **radical symbol** $\sqrt{}$ is used to indicate a positive square root. To **evaluate a radical expression** such as $\sqrt{4}$, find the positive square root of the radicand. Radical symbol $\sqrt{4}$ ← Radicand Read as "the square root Radical expression of 4." Numbers such as 4, 64, and 225, that are squares of whole numbers, are called **perfect squares**. To evaluate square root radical expressions, it is helpful to be able to identify **perfect square radicands**. Review the list of perfect squares on page 383.	Evaluate each square root: $\sqrt{4} = 2$ Ask: What positive number, when squared, is 4? The answer is 2 because $2^2 = 4$. $\sqrt{64} = 8$ Ask: What positive number, when squared, is 64? The answer is 8 because $8^2 = 64$. $\sqrt{225} = 15$ Ask: What positive number, when squared, is 225? The answer is 15 because $15^2 = 225$.
The symbol $-\sqrt{}$ is used to indicate the **negative square root** of a positive number. It is the opposite of the positive square root.	Evaluate: $-\sqrt{36}$ $-\sqrt{36}$ is the opposite (or negative) of the positive square root of 36. Since $\sqrt{36} = 6$, we have: $-\sqrt{36} = -6$

We can find the square root of fractions and decimals.

Evaluate each square root:

$$\sqrt{\frac{49}{100}}$$

Ask: What positive fraction, when squared, is $\frac{49}{100}$?

The answer is $\frac{7}{10}$ because $\left(\frac{7}{10}\right)^2 = \frac{49}{100}$.

$$\sqrt{0.25}$$

Ask: What positive decimal, when squared, is 0.25?
The answer is 0.5 because $(0.5)^2 = 0.25$.

When **evaluating an expression containing square roots**, evaluate square roots at the same stage in your solution as exponential expressions.

Evaluate: $20 + 6\left(2^3 - 4\sqrt{9}\right)$

Perform the operations within the parentheses first.

$20 + 6\left(2^3 - 4\sqrt{9}\right) = 20 + 6(8 - 4 \cdot 3)$ — Within the parentheses, evaluate the exponential expression and the square root.

$= 20 + 6(8 - 12)$ — Within the parentheses, do the multiplication.

$= 20 + 6(-4)$ — Within the parentheses, do the subtraction.

$= 20 + (-24)$ — Do the multiplication.

$= -4$ — Do the addition.

To **evaluate formulas that involve square roots**, we replace the letters with specific numbers and then use the order of operations rule.

Evaluate $a = \sqrt{c^2 - b^2}$ for $c = 25$ and $b = 20$.

$a = \sqrt{c^2 - b^2}$ — This is the formula to evaluate.

$a = \sqrt{25^2 - 20^2}$ — Replace c with 25 and b with 20.

$a = \sqrt{625 - 400}$ — Evaluate the exponential expressions.

$a = \sqrt{225}$ — Do the subtraction.

$a = 15$ — Evaluate the square root.

If a number is not a perfect square, we can use the square root key $\sqrt{}$ on a calculator (or a table of square roots) to find its **approximate square root**.

Approximate $\sqrt{149}$. Round to the nearest hundredth.

From a scientific calculator we get $\sqrt{149} \approx 12.20655562$. Rounded to the nearest hundredth,

$$\sqrt{149} \approx 12.21$$

REVIEW EXERCISES

109. Find the two square roots of 25.

110. Fill in the blanks: $\sqrt{49} = \boxed{}$ because $\boxed{}^2 = 49$

Evaluate each square root without using a calculator.

111. $\sqrt{49}$

112. $-\sqrt{16}$

113. $\sqrt{100}$

114. $\sqrt{0.09}$

115. $\sqrt{\dfrac{64}{169}}$

116. $\sqrt{0.81}$

117. $-\sqrt{\dfrac{1}{36}}$

118. $\sqrt{0}$

119. Graph each square root: $\sqrt{9}$, $-\sqrt{2}$, $\sqrt{3}$, $-\sqrt{16}$ (*Hint*: Use a calculator or square root table to approximate any square roots, when necessary.)

-5 -4 -3 -2 -1 0 1 2 3 4 5

120. Use a calculator to approximate each square root to the nearest hundredth.

a. $\sqrt{19}$
b. $\sqrt{598}$
c. $\sqrt{12.75}$

Evaluate each expression without using a calculator.

121. $-3\sqrt{100}$

122. $5\sqrt{196}$

123. $-3\sqrt{49} - \sqrt{36}$

124. $\sqrt{\dfrac{100}{9}} + \sqrt{225}$

125. $40 + 6[5^2 - (7 - 2)\sqrt{16}]$

126. $1 - 7[6^2 + (1 + 2)\sqrt{81}]$

127. Evaluate $b = \sqrt{c^2 - a^2}$ for $c = 17$ and $a = 15$.

128. **Sheet metal.** Find the length of the side of the range hood shown in the illustration below.

$\sqrt{1,089}$ in.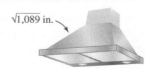

129. Between what two whole numbers would $\sqrt{83}$ be located when graphed on a number line?

130. $\sqrt{7} \approx 2.646$. Explain why an $\approx$ symbol is used and not an $=$ symbol.

1. Fill in the blanks.

   a. Copy the following addition. Label each *addend* and the *sum*.

   $$\begin{array}{r} 2.67 \leftarrow \rule{2cm}{0pt} \\ + 6.01 \leftarrow \rule{2cm}{0pt} \\ \hline 8.68 \leftarrow \rule{2cm}{0pt} \end{array}$$

   b. Copy the following subtraction. Label the *minuend*, the *subtrahend*, and the *difference*.

   $$\begin{array}{r} 9.6 \leftarrow \rule{2cm}{0pt} \\ - 6.2 \leftarrow \rule{2cm}{0pt} \\ \hline 3.4 \leftarrow \rule{2cm}{0pt} \end{array}$$

   c. Copy the following multiplication. Label the *factors* and the *product*.

   $$\begin{array}{r} 1.3 \leftarrow \rule{2cm}{0pt} \\ \times\ 7 \leftarrow \rule{2cm}{0pt} \\ \hline 9.1 \leftarrow \rule{2cm}{0pt} \end{array}$$

   d. Copy the following division. Label the *dividend*, the *divisor*, and the *quotient*.

   $$\begin{array}{r} 3.4 \leftarrow \rule{2cm}{0pt} \\ \rule{1.5cm}{0pt} \rightarrow 2\overline{)6.8} \leftarrow \rule{2cm}{0pt} \end{array}$$

   e. 0.6666 . . . and 0.8333 . . . are examples of _____ decimals.

   f. The $\sqrt{}$ symbol is called a _____ symbol.

2. Express the amount of the square region that is shaded using a fraction and a decimal.

   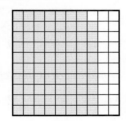

3. Consider the decimal number 629.471.

   a. What is the place value of the digit 1?
   b. Which digit tells the number of tenths?
   c. Which digit tells the number of hundreds?
   d. What is the place value of the digit 2?

4. **Water purity.**
   A county health department sampled the pollution content of tap water in five cities, with the results shown. Rank the cities in order, from dirtiest tap water to cleanest.

City	Pollution, parts per million
Monroe	0.0909
Covington	0.0899
Paston	0.0901
Cadia	0.0890
Selway	0.1001

5. Write *four thousand five hundred nineteen and twenty-seven ten-thousandths* in standard form.

6. Write each decimal in:
   ■ expanded form
   ■ words
   ■ as a fraction or mixed number. (You do not have to simplify the fraction.)

   a. **Skateboarding.** Kyle Wester of Denver, Colorado, set the skateboard speed record of 89.41 mph in 2016. (Source: skateboarding.transworld.net)

   b. **Money.** A dime weighs 0.08013 ounce.

7. Round each decimal number to the indicated place value.

   a. 461.728, nearest tenth
   b. 2,733.0495, nearest thousandth
   c. −1.9833732, nearest millionth

8. Round $0.648209 to the nearest cent.

Perform each operation.

9. $4.56 + 2 + 0.896 + 3.3$

10. Subtract 39.079 from 45.2

11. $(0.32)^2$

12. $\dfrac{0.1368}{0.24}$

13. $(-3.2)(0.6)$

14. $\begin{array}{r} 8.7 \\ \times\ 0.004 \\ \hline \end{array}$

15. $11\overline{)13}$

16. $-2.4 - (-1.6)$

17. Divide. Round the quotient to the nearest hundredth:

    $$\dfrac{12.146}{-5.3}$$

18. a. Estimate the product using front-end rounding: $34 \cdot 6.83$

    b. Estimate the quotient: $3,907.2 \div 19.3$

19. Perform each operation in your head.

    a. $567.909 \div 1,000$

    b. $0.00458 \cdot 100$

20. Write 61.4 billion in standard notation.

21. **Earthquake damage.** After an earthquake, geologists found that the ground on the west side of the fault line had dropped 0.834 inch. The next week, a strong aftershock caused the same area to sink 0.192 inch deeper. How far did the ground on the west side of the fault drop because of the earthquake and the aftershock?

22. New York City. Refer to the illustration on the right. Central Park, which lies in the middle of Manhattan, is the city's best-known park. If it is 2.5 miles long and 0.5 mile wide, what is its area?

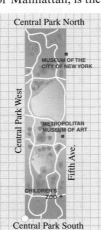

23. Telephone books. To print a telephone book, 565 sheets of paper were used. If the book is 2.26 inches thick, what is the thickness of each sheet of paper?

24. Accounting. At an ice-skating complex, receipts on Friday were $130.25 for indoor skating and $162.25 for outdoor skating. On Saturday, the corresponding amounts were $140.50 and $175.75. On which day, Friday or Saturday, were the receipts higher? How much higher?

25. Chemistry. In a lab experiment, a chemist mixed three compounds together to form a mixture weighing 4.37 grams. Later, she discovered that she had forgotten to record the weight of compound C in her notes. Find the weight of compound C used in the experiment.

	Weight
Compound A	1.86 grams
Compound B	2.09 grams
Compound C	?
Mixture total	4.37 grams

26. Weight of water. One gallon of water weighs 8.33 pounds. How much does the water in a $2\frac{1}{2}$-gallon jug weigh?

27. Evaluate the formula $C = 2\pi r$ for $\pi = 3.14$ and $r = 1.7$.

28. Write each fraction as a decimal.

a. $\dfrac{17}{50}$ b. $\dfrac{5}{12}$

Evaluate each expression.

29. $4.1 - (3.2)(0.4)^2$

30. $\left(\dfrac{2}{5}\right)^2 + 6\left|-6.2 - 3\dfrac{1}{4}\right|$

31. $8 - 2\left(2^4 - 60 + 6\sqrt{81}\right)$

32. $\dfrac{2}{3} + 0.7$ (Work in terms of fractions.)

33. a. Graph $\dfrac{3}{8}$, $\dfrac{2}{3}$, and $-\dfrac{4}{5}$ on the number line. Label each point using the decimal equivalent of the fraction.

b. Graph $\sqrt{16}$, $\sqrt{2}$, $-\sqrt{9}$, and $-\sqrt{5}$ on the number line below. (*Hint:* When necessary, use a calculator or square root table to approximate a square root.)

34. Salads. A shopper purchased $\frac{3}{4}$ pound of potato salad, priced at $5.60 per pound, and $\frac{1}{3}$ pound of coleslaw, selling for $4.35 per pound. Find the total cost of these items.

35. Use a calculator to evaluate $c = \sqrt{a^2 + b^2}$ for $a = 12$ and $b = 35$.

36. Write each number as a decimal.

a. $-\dfrac{27}{25}$ b. $2\dfrac{9}{16}$

37. Fill in the blank: $\sqrt{144} = \boxed{}$ because $\boxed{}^2 = 144$.

38. Place an $<$, $>$, or an $=$ symbol in the box to make a true statement.

a. $-6.78 \boxed{} -6.79$ b. $0.3 \boxed{} \dfrac{3}{8}$

c. $\sqrt{\dfrac{16}{81}} \boxed{} 0.\overline{4}$ d. $0.45 \boxed{} 0.\overline{45}$

Evaluate each expression without using a calculator.

39. $-2\sqrt{25} + 3\sqrt{49}$

40. $\sqrt{\dfrac{1}{36}} - \sqrt{\dfrac{1}{25}}$

41. Evaluate each square root without using a calculator.

a. $-\sqrt{0.04}$ b. $\sqrt{1.69}$
c. $\sqrt{225}$ d. $-\sqrt{121}$

42. Although the decimal 3.2999 contains more digits than 3.3, it is smaller than 3.3. Explain why this is so.

43. Wedding costs. A printer charges a setup fee of $24 and then 95 cents for each wedding announcement printed (tax included). If a couple has budgeted $100 for printing costs, how many announcements can they have printed?

Ola-la/Shutterstock.com

44. Recycling. At a recycling center, customers first drive their loaded vehicle onto a scale where it is weighed. (That is called the vehicle's *gross weight*.) Then they unload the materials and the empty vehicle is weighed again. (That is called the vehicle's *tare weight*.) Use the facts in the illustration to determine how much the recycler will earn for the cardboard if he is paid $0.025 per pound.

1. Write 154,302
   a. in words [Section 1.1]
   b. in expanded form [Section 1.1]

2. Use 3, 4, and 5 to express the associative property of addition. [Section 1.2]

3. Add: 9,339 + 471 + 6,883 [Section 1.2]

4. Subtract 199 from 301. [Section 1.3]

5. **Sudoku.** The world's largest Sudoku puzzle was carved into a hillside near Bristol, England. It measured 275 ft by 275 ft. Find the area covered by the square-shaped puzzle. (Source: joe-ks.com) [Section 1.4]

6. Divide: $43\overline{)1,203}$ [Section 1.5]

7. **The executive branch.** The annual salary of the President of the United States is $400,000, and the Vice President is paid $230,700 a year. How much more money does the President make than the Vice President during a four-year term? [Section 1.6]

8. List the factors of 20, from smallest to largest. [Section 1.7]

9. Find the prime factorization of 220. [Section 1.7]

10. Find the LCM and the GCF of 100 and 120. [Section 1.8]

11. Find the mean (average) of 7, 1, 8, 2, and 2. [Section 1.9]

12. Place an < or an > symbol in the box to make a true statement: $|-50|$ ☐ $-(-40)$ [Section 2.1]

13. Add: $-8 + (-5)$ [Section 2.2]

14. Fill in the blank: Subtraction is the same as _____ the opposite. [Section 2.3]

15. **Weather.** Marsha flew from her Minneapolis home to Hawaii for a week of vacation. She left blizzard conditions and a temperature of $-11°F$, and stepped off the airplane into $72°F$ weather. What temperature change did she experience? [Section 2.3]

16. Multiply: $-3(-5)(2)(-9)$ [Section 2.4]

17. Evaluate: $(-1)^5$ [Section 2.4]

18. **Submarines.** As part of a training exercise, the captain of a submarine ordered it to descend 350 feet, level off for 10 minutes, and then repeat the process several times. If the sub was on the ocean's surface at the beginning of the exercise, find its depth after the 6th dive. [Section 2.4]

19. Consider the division statement $\frac{-15}{-5} = 3$. What is its related multiplication statement? [Section 2.5]

20. Divide: $-420,000 \div (-7,000)$ [Section 2.5]

21. Complete the steps to evaluate the expression. [Section 2.6]

$$(-6)^2 - 2(5 - 4 \cdot 2) = (-6)^2 - 2(5 - \boxed{})$$
$$= (-6)^2 - 2(\boxed{})$$
$$= \boxed{} - 2(-3)$$
$$= 36 - (\boxed{})$$
$$= 36 + \boxed{}$$
$$= 42$$

22. Evaluate: $|-7(5)|$ [Section 2.6]

23. Estimate the value of $-3,887 + (-5,806) + 4,701$ by rounding each number to the nearest hundred. [Section 2.6]

24. **Flags.** What fraction of the stripes on a U.S. flag are white? [Section 3.1]

25. Although the fractions listed below look different, they all represent the same value. What concept does this illustrate? [Section 3.1]

$$\frac{1}{2} = \frac{2}{4} = \frac{3}{6} = \frac{4}{8} = \frac{5}{10} = \frac{6}{12}$$

26. Simplify: $\frac{90}{126}$ [Section 3.1]

Perform the operations. Simplify the result.

27. $\frac{3}{8} \cdot \frac{7}{16}$ [Section 3.2]

28. $-\frac{15}{8} \div 10$ [Section 3.3]

29. $\frac{1}{9} + \frac{5}{6}$ [Section 3.4]

30. $-4\frac{1}{4}\left(-4\frac{1}{2}\right)$ [Section 3.5]

31. $76\frac{1}{6} - 49\frac{7}{8}$ [Section 3.6]

32. $\dfrac{\frac{5}{27}}{-\frac{5}{9}}$ [Section 3.6]

33. What is $\frac{1}{4}$ of $\frac{7}{16}$? [Section 3.2]

34. Tape measures. Use the information shown in the illustration below to determine the inside length of the drawer. [Section 3.6]

35. Evaluate: $\left(\dfrac{9}{20} \div 2\frac{2}{5}\right) + \left(\dfrac{3}{4}\right)^2$ [Section 3.7]

36. Glass. Some electronic and medical equipment uses glass that is only 0.00098 inch thick. Round this number to the nearest thousandth. [Section 4.1]

37. Place an < or > symbol in the box to make a true statement. [Section 4.1]

356.1978 ⬚ 356.22

38. Graph $-3\frac{1}{4}$, 0.75, -1.5, $-\frac{9}{8}$, 3.8, and $\sqrt{4}$ on a number line. [Section 4.1]

-5 -4 -3 -2 -1 0 1 2 3 4 5

Perform the operations.

39. $56.228 + 5.6 + 39 + 29.37$ [Section 4.2]

40. $7.001 - 5.9$ [Section 4.2]

41. $-1.8(4.52)$ [Section 4.3]

42. $56.012(0.001)$ [Section 4.3]

43. $\dfrac{-21.28}{-3.8}$ [Section 4.4]

44. $\dfrac{0.897}{10,000}$ [Section 4.4]

45. Weekly schedules. Use the information in the illustration below to determine the number of hours during a week that the typical adult spends watching television. [Section 4.2]

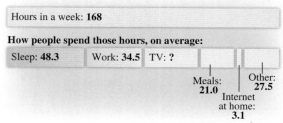

Source: National Sleep Foundation and the U.S. Bureau of Statistics

46. Kites. Find the area of the front of the kite shown below. [Section 4.3]

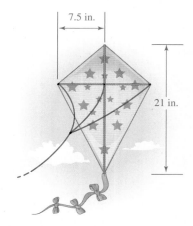

47. Evaluate the formula $C = \frac{5}{9}(F - 32)$ for $F = 451$. Round to the nearest tenth. [Section 4.4]

48. Write the fraction $\dfrac{5}{12}$ as a decimal. [Section 4.5]

49. Evaluate: $\dfrac{3}{8}(-3.2) + \left(4\frac{1}{2}\right)\left(-\frac{1}{4}\right)$ [Section 4.5]

50. Fill in the blanks: $\sqrt{64} = $ ⬚ because ⬚$^2 = 64$. [Section 4.6]

Evaluate each expression. [Section 4.6]

51. $\sqrt{49}$

52. $\sqrt{\dfrac{225}{16}}$

53. $-4\sqrt{36} + 2\sqrt{81}$

54. $\sqrt{169} + 2\left(7^2 - 3\sqrt{144}\right)$

5 Ratio, Proportion, and Measurement

from Campus to Careers

Chef

Chefs prepare and cook a wide range of foods—from soups, snacks, and salads to main dishes, side dishes, and desserts. They work in a variety of restaurants and food service kitchens. They measure, mix, and cook ingredients according to recipes, using a variety of equipment and tools. They are also responsible for directing the tasks of other kitchen workers, estimating food requirements, and ordering food supplies.

In **Problem 94** of **Study Set 5.2**, you will see how a chef can use proportions to determine the correct amounts of each ingredient needed to make a large batch of brownies.

JOB TITLE:
Chef

EDUCATION:
Training programs are available through culinary schools, two- or four-year college degree programs, and the armed forces.

JOB OUTLOOK:
Job opportunities are expected to be good; however, competition is going to be high.

ANNUAL EARNINGS:
The average (median) salary in 2015 was $41,500.

FOR MORE INFORMATION:
www.bls.gov/oes/current/oes351011.htm

OBJECTIVES

1 Write ratios as fractions.

2 Simplify ratios involving decimals and mixed numbers.

3 Convert units to write ratios.

4 Write rates as fractions.

5 Find unit rates.

6 Find the best buy based on unit price.

SECTION **5.1** Ratios and Rates

Ratios are often used to describe important relationships between two quantities. Here are three examples:

To prepare fuel for an outboard marine engine, gasoline must be mixed with oil in the ratio of 50 to 1.

To make 14-karat jewelry, gold is combined with other metals in the ratio of 14 to 10.

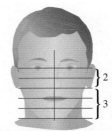

In this drawing, the eyes-to-nose distance and the nose-to-chin distance are drawn using a ratio of 2 to 3.

OBJECTIVE 1 **Write ratios as fractions.**

Ratios give us a way to compare two numbers or two quantities measured in the same units.

Ratios

A **ratio** is the quotient of two numbers or the quotient of two quantities that have the same units.

There are three ways to write a ratio. The most common way is as a fraction. Ratios can also be written as two numbers separated by the word *to*, or as two numbers separated by a colon. For example, the ratios described in the illustrations above can be expressed as:

$$\frac{50}{1}, \qquad 14 \text{ to } 10, \qquad \text{and} \qquad 2\!:\!3$$

- The fraction $\frac{50}{1}$ is read as "the ratio of 50 to 1."   *A fraction bar separates the numbers being compared.*

- 14 **to** 10 is read as "the ratio of 14 to 10."   *The word "to" separates the numbers being compared.*

- 2:3 is read as "the ratio of 2 to 3."   *A colon separates the numbers being compared.*

Writing a Ratio as a Fraction

To **write a ratio as a fraction**, write the first number (or quantity) mentioned as the numerator and the second number (or quantity) mentioned as the denominator. Then simplify the fraction, if possible.

EXAMPLE 1   Write each ratio as a fraction: **a.** 3 to 7   **b.** 10:11

Strategy We will identify the numbers before and after the word *to* and the numbers before and after the colon.

WHY The word *to* and the colon separate the numbers to be compared in a ratio.

Solution To write the ratio as a fraction, the first number mentioned is the numerator and the second number mentioned is the denominator.

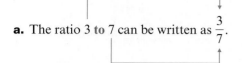

a. The ratio 3 **to** 7 can be written as $\dfrac{3}{7}$.   The fraction $\frac{3}{7}$ is in simplest form.

b. The ratio 10 : 11 can be written as $\dfrac{10}{11}$.   The fraction $\frac{10}{11}$ is in simplest form.

Self Check 1

Write each ratio as a fraction:
a. 4 to 9   **b.** 8:15

Now Try ⟳ Problem 13

When a ratio is written as a fraction, the fraction should be in simplest form. (Recall from Chapter 3 that a fraction is in simplest form, or lowest terms, when the numerator and denominator have no common factors other than 1.)

EXAMPLE 2   Write the ratio 35 to 10 as a fraction in simplest form.

Strategy We will translate the ratio from its given form in words to fractional form. Then we will look for any factors common to the numerator and denominator and remove them.

WHY We need to make sure that the numerator and denominator have no common factors other than 1. If that is the case, the ratio will be in *simplest form*.

Solution

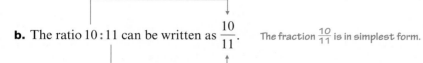

The ratio 35 **to** 10 can be written as $\dfrac{35}{10}$.   The fraction $\frac{35}{10}$ is not in simplest form.

Now we simplify the fraction, using the method discussed in Section 3.1.

$$\frac{35}{10} = \frac{\overset{1}{\cancel{5}} \cdot 7}{2 \cdot \underset{1}{\cancel{5}}} \qquad \text{Factor 35 as } 5 \cdot 7 \text{ and 10 as } 2 \cdot 5. \text{ Then remove the}$$
$$\text{common factor of 5 in the numerator and denominator.}$$
$$= \frac{7}{2}$$

The ratio 35 to 10 can be written as the fraction $\frac{35}{10}$, which simplifies to $\frac{7}{2}$ (read as "7 to 2"). Because the fractions $\frac{35}{10}$ and $\frac{7}{2}$ represent equal numbers, they are called **equal ratios**.

Caution! Since ratios are comparisons of two numbers, it would be *incorrect* in Example 2 to write the ratio $\frac{7}{2}$ as the mixed number $3\frac{1}{2}$. Ratios written as improper fractions are perfectly acceptable—just make sure the numerator and denominator have no common factors other than 1.

Self Check 2

Write the ratio 12 to 9 as a fraction in simplest form.

Now Try ⟳ Problems 17 and 23

To write a ratio in simplest form, we remove any common factors of the numerator and denominator as well as any common units.

EXAMPLE 3   **Carry-on luggage.**   An airline allows its passengers to carry a piece of luggage onto an airplane only if it will fit in the space shown below.

a. Write the ratio of the width of the space to its length as a fraction in simplest form.

b. Write the ratio of the length of the space to its width as a fraction in simplest form.

Strategy To write each ratio as a fraction, we will identify the quantity before the word *to* and the quantity after it.

WHY The first quantity mentioned is the numerator of the fraction and the second quantity mentioned is the denominator.

Solution

a. The ratio of the width of the space to its length is $\dfrac{10 \text{ inches}}{24 \text{ inches}}$.

To write a ratio in simplest form, we remove the common factors *and* the common units of the numerator and denominator.

$$\frac{10 \text{ inches}}{24 \text{ inches}} = \frac{\overset{1}{2} \cdot 5 \text{ inches}}{\underset{1}{2} \cdot 12 \text{ inches}}$$

Factor 10 as 2 · 5 and 24 as 2 · 12. Then remove the common factor of 2 and the common units of inches from the numerator and denominator.

$$= \frac{5}{12}$$

> **Caution!** Example 3 shows that order is important when writing a ratio. The width-to-length ratio is $\frac{5}{12}$ while the length-to-width ratio is $\frac{12}{5}$.

The width-to-length ratio of the carry-on space is $\dfrac{5}{12}$ (read as "5 to 12").

b. The ratio of the length of the space to its width is $\dfrac{24 \text{ inches}}{10 \text{ inches}}$.

$$\frac{24 \text{ inches}}{10 \text{ inches}} = \frac{\overset{1}{2} \cdot 12 \text{ inches}}{\underset{1}{2} \cdot 5 \text{ inches}}$$

Factor 24 and 10. Then remove the common factor of 2 and the common units of inches from the numerator and denominator.

$$= \frac{12}{5}$$

The length-to-width ratio of the carry-on space is $\dfrac{12}{5}$ (read as "12 to 5").

> **Self Check 3**
>
> **Carry-on luggage.**
> **a.** Write the ratio of the height to the length of the carry-on space shown in the illustration in Example 3 as a fraction in simplest form.
> **b.** Write the ratio of the length of the carry-on space to its height in simplest form.
>
> **Now Try** ➡ Problem 27

OBJECTIVE 2 Simplify ratios involving decimals and mixed numbers.

EXAMPLE 4   Write the ratio 0.3 to 1.8 as a fraction in simplest form.

Strategy After writing the ratio as a fraction, we will multiply it by a form of 1 to obtain an equivalent ratio of whole numbers.

WHY A ratio of whole numbers is easier to understand than a ratio of decimals.

Solution

The ratio 0.3 **to** 1.8 can be written as $\dfrac{0.3}{1.8}$.

To write this as a ratio of *whole numbers*, we need to move the decimal points in the numerator and denominator one place to the right. Recall that to find the product of a decimal and 10, we simply move the decimal point one place to the right. Therefore, it follows that $\frac{10}{10}$ is the form of 1 that we should use to build $\frac{0.3}{1.8}$ into an equivalent ratio.

$$\frac{0.3}{1.8} = \frac{0.3}{1.8} \cdot \frac{\mathbf{10}}{\mathbf{10}}$$ Multiply the ratio by a form of 1.

$$= \frac{0.3 \cdot 10}{1.8 \cdot 10}$$ Multiply the numerators.
Multiply the denominators.

$$= \frac{3}{18}$$ Do the multiplications by moving each decimal point one place to the right. $0.3 \cdot 10 = 3$ and $1.8 \cdot 10 = 18$.

$$= \frac{1}{6}$$ Simplify the fraction: $\frac{3}{18} = \frac{\overset{1}{\cancel{3}}}{\underset{1}{\cancel{3} \cdot 6}} = \frac{1}{6}$.

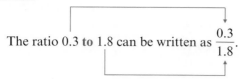

Self Check 4

Write the ratio 0.8 to 2.4 as a fraction in simplest form.

Now Try ➲ **Problems 29 and 33**

EXAMPLE 5   Write the ratio $4\frac{2}{3}$ to $1\frac{1}{6}$ as a fraction in simplest form.

Strategy After writing the ratio as a fraction, we will use the method for simplifying a complex fraction from Section 3.7 to obtain an equivalent ratio of whole numbers.

WHY A ratio of whole numbers is easier to understand than a ratio of mixed numbers.

Solution

The ratio of $4\frac{2}{3}$ **to** $1\frac{1}{6}$ can be written as $\dfrac{4\frac{2}{3}}{1\frac{1}{6}}$.

The resulting ratio is a complex fraction. To write the ratio in simplest form, we perform the division indicated by the main fraction bar (shown in red).

$$\frac{4\frac{2}{3}}{1\frac{1}{6}} = \frac{\frac{14}{3}}{\frac{7}{6}}$$ Write $4\frac{2}{3}$ and $1\frac{1}{6}$ as improper fractions.

$$= \frac{14}{3} \div \frac{7}{6}$$ Write the division indicated by the main fraction bar using a ÷ symbol.

$$= \frac{14}{3} \cdot \frac{6}{7}$$ Use the rule for dividing fractions: Multiply the first fraction by the reciprocal of $\frac{7}{6}$, which is $\frac{6}{7}$.

$$= \frac{14 \cdot 6}{3 \cdot 7}$$ Multiply the numerators.
Multiply the denominators.

$$= \frac{2 \cdot \overset{1}{\cancel{7}} \cdot 2 \cdot \overset{1}{\cancel{3}}}{\underset{1}{\cancel{3}} \cdot \underset{1}{\cancel{7}}}$$ To simplify the fraction, factor 14 as 2 · 7 and 6 as 2 · 3.
Then remove the common factors 3 and 7.

$$= \frac{4}{1}$$ Multiply the remaining factors in the numerator.
Multiply the remaining factors in the denominator.

Self Check 5

Write the ratio $3\frac{1}{3}$ to $1\frac{1}{9}$ as a fraction in simplest form.

Now Try ➡ **Problem 37**

We would normally simplify the result $\frac{4}{1}$ and write it as 4. But since a ratio compares two numbers, we leave the result in fractional form.

OBJECTIVE 3 Convert units to write ratios.

When a ratio compares two quantities, both quantities must be measured in the same units. For example, inches must be compared to inches, pounds to pounds, and seconds to seconds.

EXAMPLE 6 Write the ratio *12 ounces to 2 pounds* as a fraction in simplest form.

Strategy We will convert 2 pounds to ounces and write a ratio that compares ounces to ounces. Then we will simplify the ratio.

WHY A ratio compares two quantities that have the *same* units. When the units are different, it's usually easier to write the ratio using the smaller unit of measurement. Since ounces are smaller than pounds, we will compare in ounces.

Solution To express 2 pounds in ounces, we use the fact that there are 16 ounces in one pound.

$$2 \cdot 16 \text{ ounces} = 32 \text{ ounces}$$

We can now express the ratio *12 ounces to 2 pounds* using the same units:

12 **ounces** to 32 **ounces**

Next, we write the ratio in fraction form and simplify.

Self Check 6

Write the ratio *6 feet to 3 yards* as a fraction in simplest form. (*Hint:* 3 feet = 1 yard.)

Now Try ➡ **Problem 41**

$$\frac{12 \text{ ounces}}{32 \text{ ounces}} = \frac{3 \cdot \overset{1}{\cancel{4}} \cancel{\text{ounces}}}{\underset{1}{\cancel{4}} \cdot 8 \cancel{\text{ounces}}}$$ To simplify, factor 12 as 3 · 4 and 32 as 4 · 8. Then remove the common factor of 4 and the common units of ounces from the numerator and denominator.

$$= \frac{3}{8}$$

The ratio in simplest form is $\frac{3}{8}$.

OBJECTIVE 4 Write rates as fractions.

When we compare two quantities that have different units (and neither unit can be converted to the other), we call the comparison a **rate**, and we can write it as a fraction. For example, on the label of the can of paint shown on the right, we see that 1 quart of paint is needed for every 200 square feet to be painted. Writing this as a rate in fractional form, we have

$$\frac{1 \text{ quart}}{200 \text{ square feet}}$$ Read as "1 quart per 200 square feet."

The word **per** is associated with the operation of division, and it means "for each" or "for every." For example, when we say 1 quart of paint *per* 200 square feet, we mean 1 quart of paint *for every* 200 square feet.

Rates

A **rate** is a quotient of two quantities that have different units.

When writing a rate, always include the units. Some other examples of rates are:

- 16 computers **for** 75 students
- 1,550 feet **in** 4.5 seconds
- 88 tomatoes **from** 3 plants
- 250 miles **on** 2 gallons of gasoline

LANGUAGE OF MATHEMATICS

As seen on the left, words such as **per, for, in, from,** and **on** are used to separate the two quantities that are compared in a rate.

Writing a Rate as a Fraction

To **write a rate as a fraction**, write the first quantity mentioned as the numerator and the second quantity mentioned as the denominator, and then simplify, if possible. Write the units as part of the fraction.

EXAMPLE 7   **Snowfall.**   According to the *Guinness Book of World Records*, a total of 78 inches of snow fell at Mile 47 Camp, Cooper River Division, Arkansas, in a 24-hour period in 1963. Write the rate of snowfall as a fraction in simplest form.

Strategy We will use a fraction to compare the amount of snow that fell (in inches) to the amount of time in which it fell (in hours). Then we will simplify it.

WHY A rate is a quotient of two quantities with different units.

Solution

78 inches **in** 24 hours can be written as $\dfrac{78 \text{ inches}}{24 \text{ hours}}$.

Now, we simplify the fraction.

$$\frac{78 \text{ inches}}{24 \text{ hours}} = \frac{\overset{1}{\cancel{6}} \cdot 13 \text{ inches}}{4 \cdot \underset{1}{\cancel{6}} \text{ hours}}$$

To simplify, factor 78 as 6 · 13 and 24 as 4 · 6. Then remove the common factor of 6 from the numerator and denominator.

$$= \frac{13 \text{ inches}}{4 \text{ hours}}$$

Since the units are different, they cannot be removed.

The snow fell at a rate of 13 inches per 4 hours.

Self Check 7

Growth rates. The fastest-growing flowering plant on record grew 12 feet in 14 days. Write the rate of growth as a fraction in simplest form.

Now Try ➲ Problems 49 and 53

OBJECTIVE 5 Find unit rates.

Unit Rate

A **unit rate** is a rate in which the denominator is 1.

To illustrate the concept of a unit rate, suppose a driver makes the 354-mile trip from Pittsburgh to Indianapolis in 6 hours. Then the motorist's rate (or more specifically, rate of speed) is given by

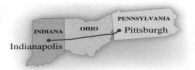

$$\frac{354 \text{ miles}}{6 \text{ hours}} = \frac{\overset{1}{\cancel{6}} \cdot 59 \text{ miles}}{\cancel{6} \cdot \text{hours}}$$ Factor 354 as 6 · 59 and remove the common factor of 6 from the numerator and denominator.

$$= \frac{59 \text{ miles}}{1 \text{ hour}}$$ Since the units are different, they cannot be removed. Note that the denominator is 1.

We can also find the unit rate by dividing 354 by 6.

Rate:

$$\frac{354 \text{ miles}}{6 \text{ hours}}$$

$$\begin{array}{r} 59 \\ 6\overline{)354} \\ -30 \\ \hline 54 \\ -54 \\ \hline 0 \end{array}$$ ← This quotient is the numerical part of the unit rate, written as a fraction. →

The numerical part of the denominator is always 1. →

Unit rate:

$$\frac{59 \text{ miles}}{1 \text{ hour}}$$

LANGUAGE OF MATHEMATICS

A *slash mark* **/** is often used to write a unit rate. In such cases, we read the slash mark as "per." For example, 33 pounds/gallon is read as 33 pounds *per* gallon.

The unit rate $\frac{59 \text{ miles}}{1 \text{ hour}}$ can be expressed in any of the following forms:

$$59 \frac{\text{miles}}{\text{hour}}, \quad 59 \text{ miles per hour}, \quad 59 \text{ miles/hour}, \quad \text{or} \quad 59 \text{ mph}$$

Writing a Rate as a Unit Rate

To **write a rate as a unit rate**, divide the numerator of the rate by the denominator.

EXAMPLE 8 **Coffee.** There are 384 calories in a 16-ounce cup of caramel Frappuccino blended coffee with whip cream. Write this rate as a unit rate. (*Hint:* Find the number of calories in 1 ounce.)

Strategy We will translate the rate from its given form in words to fractional form. Then we will perform the indicated division.

WHY To write a rate as a unit rate, we divide the numerator of the rate by the denominator.

Solution

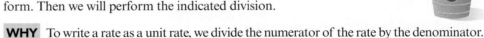

384 calories **in** 16 ounces can be written as $\dfrac{384 \text{ calories}}{16 \text{ ounces}}$.

To find the number of calories in 1 ounce of the coffee (the unit rate), we perform the division as indicated by the fraction bar:

Self Check 8

Nutrition. There are 204 calories in a 12-ounce can of cranberry juice. Write this rate as a unit rate. (*Hint:* Find the number of calories in 1 ounce.)

Now Try ➡ Problem 57

$$\begin{array}{r} 24 \\ 16\overline{)384} \\ -32 \\ \hline 64 \\ -64 \\ \hline 0 \end{array}$$ Divide the numerator of the rate by the denominator.

For the caramel Frappuccino blended coffee with whip cream, the unit rate is $\dfrac{24 \text{ calories}}{1 \text{ ounce}}$, which can be written as 24 calories per ounce or 24 calories/ounce.

EXAMPLE 9   **Part-time jobs.**   A student earns $74 for working 8 hours in a bookstore. Write this rate as a unit rate. (*Hint*: Find his hourly rate of pay.)

Strategy We will translate the rate from its given form in words to fractional form. Then we will perform the indicated division.

WHY To write a rate as a unit rate, we divide the numerator of the rate by the denominator.

Solution

$74 **for** working 8 hours can be written as $\dfrac{\$74}{8 \text{ hours}}$.

To find the rate of pay for 1 hour of work (the unit rate), we divide 74 by 8.

$$
\begin{array}{r}
9.25 \\
8)\overline{74.00} \\
-72 \\
\hline
2\,0 \\
-1\,6 \\
\hline
40 \\
-40 \\
\hline
0
\end{array}
$$
Write a decimal point and two additional zeros to the right of 4.

The unit rate of pay is $\dfrac{\$9.25}{1 \text{ hour}}$, which can be written as $9.25 per hour or $9.25/hr.

Self Check 9

Full-time jobs. Joan earns $436 per 40-hour week managing a dress shop. Write this rate as a unit rate. (*Hint*: Find her hourly rate of pay.)

Now Try ➔ **Problem 61**

OBJECTIVE 6 **Find the best buy based on unit price.**

If a grocery store sells a 5-pound package of hamburger for $18.75, a consumer might want to know what the hamburger costs per pound. When we find the cost of 1 pound of the hamburger, we are finding a **unit price**. To find the unit price of an item, we begin by comparing its price to the number of units.

$\dfrac{\$18.75}{5 \text{ pounds}}$ ⟵ Price
⟵ Number of units

Then we divide the price by the number of units.

$$
\begin{array}{r}
3.75 \\
5)\overline{18.75}
\end{array}
$$

The unit price of the hamburger is $3.75 per pound.
Other examples of unit prices are:

- $8.15 per ounce
- $200 per day
- $0.75 per foot

Unit Price

A **unit price** is a rate that tells how much is paid for *one* unit (or *one* item). It is the quotient of price to the number of units.

$$\text{Unit price} = \dfrac{\text{price}}{\text{number of units}}$$

When shopping, it is often difficult to determine the best buys because the items that we purchase come in so many different sizes and brands. Comparison shopping can be made easier by finding unit prices. *The best buy is the item that has the lowest unit price.*

EXAMPLE 10   **Comparison shopping.**   Olives come packaged in a 10-ounce jar, which sells for $2.49, or in a 6-ounce jar, which sells for $1.53. Which is the better buy?

Strategy   We will find the unit price for each jar of olives. Then we will identify which jar has the lower unit price.

WHY   The better buy is the jar of olives that has the lower unit price.

Solution   To find the unit price of each jar of olives, we write the quotient of its price and its weight and then perform the indicated division. Before dividing, we convert each price from dollars to cents so that the unit price can be expressed in cents per ounce.

The 10-ounce jar:

$$\frac{\$2.49}{10\ oz} = \frac{249¢}{10\ oz}$$ Write the rate: $\frac{price}{number\ of\ units}$.
Then change $2.49 to 249 cents.

$$= 24.9¢\ per\ oz$$ Divide 249 by 10 by moving the decimal point 1 place to the left.

The 6-ounce jar:

$$\frac{\$1.53}{6\ oz} = \frac{153¢}{6\ oz}$$ Write the rate: $\frac{price}{number\ of\ units}$.
Then change $1.53 to 153 cents.

$$= 25.5¢\ per\ oz$$ Do the division.

$$\begin{array}{r} 25.5 \\ 6\overline{)153.0} \\ -12 \\ \hline 33 \\ -30 \\ \hline 3\ 0 \\ -3\ 0 \\ \hline 0 \end{array}$$

One ounce for 24.9¢ is a better buy than one ounce for 25.5¢. The unit price is less when olives are packaged in 10-ounce jars, so that is the better buy.

Self Check 10

Comparison shopping.
A fast-food restaurant sells a 12-ounce soda for 72¢ and a 16-ounce soda for 99¢. Which is the better buy?

Now Try ⮕ Problems 65 and 109

Think it Through • UNIT COSTS

"You might not know it, but you probably have anywhere between 1.3 to 1.5 gallons of one of the most expensive liquids in the world on you right now. So what is the expensive liquid that you currently have in big supply? It just so happens to be the blood that runs inside your veins."
—**From providr.com "Top 10 Most Expensive Liquids in the World" by Brenden Mernagh**

Eight of the ten most expensive liquids in the world are listed below. Match each liquid in the left column with its approximate *cost per ounce* in the right column.

1. Black printer ink

2. Chanel No. 5 perfume

3. Human blood

4. Insulin
(It is very expensive to produce in its biosynthetic form)

5. King Cobra Venom
(It contains a unique protein that is used in a high-strength painkiller.)

6. Mercury
(It is used in vapor form in street lighting and fluorescent bulbs.)

7. Nail polish

8. Scorpion venom
(It is used to treat pain in those who suffer from multiple sclerosis.)

a. $55   **e.** $590

b. $95   **f.** $1,625

c. $170   **g.** $9,600

d. $215   **h.** $2,500,000

Charly Morlock/Shutterstock.com

Skynavin/Shutterstock.com

Answers to Self Checks

1. a. $\frac{4}{9}$ **b.** $\frac{8}{15}$ **2.** $\frac{4}{3}$ **3. a.** $\frac{2}{3}$ **b.** $\frac{3}{2}$ **4.** $\frac{1}{3}$ **5.** $\frac{3}{1}$ **6.** $\frac{2}{3}$ **7.** $\frac{6\ feet}{7\ days}$ **8.** 17 calories/oz
9. $10.90 per hour   **10.** the 12-oz soda

SECTION **5.1** STUDY SET

VOCABULARY

Fill in the blanks.

1. A _____ is the quotient of two numbers or the quotient of two quantities that have the same units.
2. A _____ is the quotient of two quantities that have different units.
3. A _____ rate is a rate in which the denominator is 1.
4. A unit _____ is a rate that tells how much is paid for one unit or one item.

CONCEPTS

5. To write the ratio $\frac{15}{24}$ in simplest form, we remove any common factors of the numerator and denominator. What common factor do they have?
6. Complete the steps used to write the ratio $\frac{14}{21}$ in simplest form.

$$\frac{14}{21} = \frac{2 \cdot 7}{\blacksquare \cdot \blacksquare} = \frac{2 \cdot \overset{1}{\cancel{7}}}{\blacksquare \cdot \underset{1}{\cancel{7}}} = \frac{\blacksquare}{\blacksquare}$$

7. Consider the ratio $\frac{0.5}{0.6}$. By what number should we multiply the numerator and denominator to make this a ratio of whole numbers?
8. What should be done to write the ratio $\frac{15 \text{ inches}}{22 \text{ inches}}$ in simplest form?
9. Write $\frac{11 \text{ minutes}}{1 \text{ hour}}$ so that it compares the same units and then simplify, if possible.
10. **a.** Consider the rate $\frac{\$248}{16 \text{ hours}}$. What division should be performed to find the unit rate in dollars per hour?
    b. Suppose 3 pairs of socks sell for $7.95: $\frac{\$7.95}{3 \text{ pairs}}$. What division should be performed to find the unit price of one pair of socks?

NOTATION

11. Write the ratio of the flag's length to its width using a fraction, using the word *to*, and using a colon.

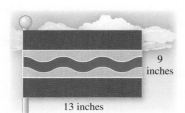

9 inches

13 inches

12. The rate $\frac{55 \text{ miles}}{1 \text{ hour}}$ can be expressed as

    ■ 55 _____ _____ _____ (in three words)

    ■ 55 _____ / _____ (in two words with a slash)

    ■ 55 ___ ___ ___ (in three letters)

GUIDED PRACTICE

Write each ratio as a fraction. See Example 1.

13. 5 to 8
14. 3 to 23
15. 11:16
16. 9:25

Write each ratio as a fraction in simplest form. See Example 2.

17. 25 to 15
18. 45 to 35
19. 63:36
20. 54:24
21. 22:33
22. 14:21
23. 17 to 34
24. 19 to 38

Write each ratio as a fraction in simplest form. See Example 3.

25. 4 ounces to 12 ounces
26. 3 inches to 15 inches
27. 24 miles to 32 miles
28. 56 yards to 64 yards

Write each ratio as a fraction in simplest form. See Example 4.

29. 0.3 to 0.9
30. 0.2 to 0.6
31. 0.65 to 0.15
32. 2.4 to 1.5
33. 3.8:7.8
34. 4.2:8.2
35. 7:24.5
36. 5:22.5

Write each ratio as a fraction in simplest form. See Example 5.

37. $2\frac{1}{3}$ to $4\frac{2}{3}$
38. $1\frac{1}{4}$ to $1\frac{1}{2}$
39. $10\frac{1}{2}$ to $1\frac{3}{4}$
40. $12\frac{3}{4}$ to $2\frac{1}{8}$

Write each ratio as a fraction in simplest form. See Example 6.

41. 12 minutes to 1 hour
42. 8 ounces to 1 pound
43. 3 days to 1 week
44. 4 inches to 1 yard
45. 18 months to 2 years
46. 8 feet to 4 yards
47. 21 inches to 3 feet
48. 32 seconds to 2 minutes

Write each rate as a fraction in simplest form. See Example 7.

49. 64 feet in 6 seconds
50. 45 applications for 18 openings
51. 75 days on 20 gallons of water
52. 3,000 students over a 16-year career
53. 84 made out of 100 attempts
54. 16 right compared to 34 wrong
55. 18 beats every 12 measures
56. 10 inches as a result of 30 turns

Write each rate as a unit rate. See Example 8.

57. 60 revolutions in 5 minutes
58. 14 trips every 2 months
59. $50,000 paid over 10 years
60. 245 presents for 35 children

Write each rate as a unit rate. See Example 9.

61. 12 errors in 8 hours
62. 114 times in a 12-month period
63. 4,007,500 people living in 12,500 square miles
64. 117.6 pounds of pressure on 8 square inches

Find the unit price of each item. See Example 10.

65. They charged $48 for 12 minutes.

66. 150 barrels cost $4,950.

67. Four sold for $272.

68. 7,020 pesos will buy six tickets.

69. 65 ounces sell for 78 cents.

70. For 7 dozen, you will pay $10.15.

71. $3.50 for 50 feet

72. $4 billion over a 5-month span

LOOK ALIKES

Simplify the ratio in part a. Then use that result to quickly write the ratio in part b in simplest form. No work is necessary.

73. a. 26 to 65   **b.** 65:26

74. a. 34 to 51   **b.** 51:34

75. a. 15 minutes to 2 hours   **b.** 2 hours to 15 minutes

76. a. 4 weeks to 12 days   **b.** 12 days to 4 weeks

APPLICATIONS

77. Gear ratios. Refer to the illustration below.
  a. Write the ratio of the number of teeth of the smaller gear to the number of teeth of the larger gear in simplest form.
  b. Write the ratio of the number of teeth of the larger gear to the number of teeth of the smaller gear in simplest form.

78. Cards. The suit of hearts from a deck of playing cards is shown below. What is the ratio of the number of face cards to the total number of cards in the suit? (*Hint:* A face card is a Jack, Queen, or King.)

79. Baking. A recipe for sourdough bread calls for $5\frac{1}{4}$ cups of all-purpose flour and $1\frac{3}{4}$ cups of water. Write the ratio of flour to water in simplest form.

80. Desserts. Refer to the recipe card shown below. Write the ratio of milk to sugar in simplest form.

> **Frozen Chocolate Slush**
> (Serves 8)
> Once frozen, this chocolate can be cut into cubes and stored in sealed plastic bags for a spur-of-the-moment dessert.
>
> $\frac{1}{2}$ cup Dutch cocoa powder, sifted
>
> $1\frac{1}{2}$ cups sugar
>
> $3\frac{1}{2}$ cups skim milk

81. Skin. Refer to the cross-section of human skin shown below. Write the ratio of the thickness of the stratum corneum to the thickness of the dermis in simplest form. (Source: Philips Research Laboratories)

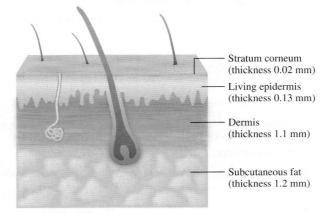

Stratum corneum (thickness 0.02 mm)

Living epidermis (thickness 0.13 mm)

Dermis (thickness 1.1 mm)

Subcutaneous fat (thickness 1.2 mm)

82. Painting. A 9.5-mil thick coat of fireproof paint is applied with a roller to a wall. (A *mil* is a unit of measure equal to 1/1,000 of an inch.) The coating dries to a thickness of 5.7 mils. Write the ratio of the thickness of the coating when wet to the thickness when dry in simplest form.

83. Bankruptcy. After declaring bankruptcy, a company could pay its creditors only 5¢ on the dollar. Write this as a ratio in simplest form.

84. Eggs. An average-sized ostrich egg weighs 3 pounds and an average-sized chicken egg weighs 2 ounces. Write the ratio of the weight of an ostrich egg to the weight of a chicken egg in simplest form.

85. CPR. A paramedic performed 165 compressions to 10 breaths on an adult with no pulse. What compressions-to-breaths rate did the paramedic use?

86. Faculty–Student ratios. At a college, there are 125 faculty members and 2,000 students. Find the rate of faculty to students. (This is often referred to as the faculty–student *ratio*, even though the units are different.)

87. Budgets. Refer to the circle graph below that shows a monthly budget for a family. Write each ratio in simplest form.

  a. Find the total amount for the monthly budget.

  b. Write the ratio of the amount budgeted for rent to the total budget.

  c. Write the ratio of the amount budgeted for food to the total budget.

  d. Write the ratio of the amount budgeted for the phone to the total budget.

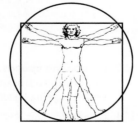

Rent $800
Food $600
Entertainment $80
Utilities $120
Phone $100
Transportation $100

88. Taxes. Refer to the list of tax deductions shown below. Write each ratio in simplest form.

  a. Write the ratio of the real estate tax deduction to the total deductions.

  b. Write the ratio of the charitable contributions to the total deductions.

  c. Write the ratio of the mortgage interest deduction to the union dues deduction.

Item	Amount
Medical expenses	$875
Real estate taxes	$1,250
Charitable contributions	$1,750
Mortgage interest	$4,375
Union dues	$500
Total deductions	$8,750

89. Art history. Leonardo da Vinci drew the human figure shown within a square. Write the ratio of the length of the man's horizontally outstretched arms to his height. (*Hint:* All four sides of a square are the same length.)

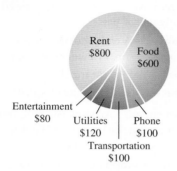

90. Flags. The checkered flag is composed of 24 equal-sized squares. What is the ratio of the width of the flag to its length? (*Hint:* All four sides of a square are the same length.)

91. Finger ratio. A number of studies have examined the ratio of the length of a person's second finger (index finger) to the length of the fourth finger (ring finger). The ratio is known as the *second to fourth digit ratio* and is written 2D:4D. Research has found a correlation between this ratio and various health and behavior traits. Match each ratio description below to the correct illustration.

  a. Equal 2D:4D ratio   **b.** Low 2D:4D ratio

  c. High 2D:4D ratio

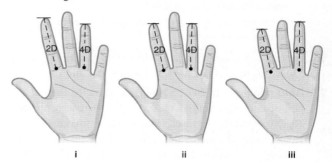

i   ii   iii

92. Screens. The *aspect ratio* is the ratio of the width W to the height H of a screen and it is often written in the form $W{:}H$. Label each type of screen shown below with the correct aspect ratio.

  a. iPad screen Aspect ratio 1.33:1

  b. High-definition TV Aspect ratio 1.78:1

  c. Most U.S. theaters screens Aspect ratio 1.85:1

  d. Wide-screen theaters Aspect ratio 2.39:1

i.
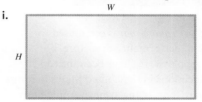

ii.
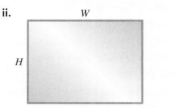

iii.
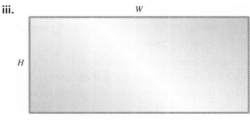

iv.
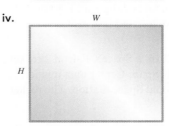

93. Airline complaints. An airline had 3.29 complaints for every 1,000 passengers. Write this as a rate of whole numbers.

94. Fingernails. On average, fingernails grow 0.02 inch per week. Write this rate using whole numbers.

95. Internet sales. A website determined that it had 112,500 hits in one month. Of those visiting the site, 4,500 made purchases.

   a. Those that visited the site, but did not make a purchase, are called *browsers*. How many browsers visited the website that month?

   b. What was the browsers-to-buyers unit rate for the website that month?

96. Typing. A secretary typed a document containing 330 words in 5 minutes. Write this rate as a unit rate.

97. Gold mining. In 2009, a geologist estimated that on a plot of land near Smith Creek, Alaska, there were about 35,500 cubic feet of gravel, from which approximately 1,065 ounces of gold could be mined. Find the predicted number of ounces of gold per cubic foot of gravel on the land. (Source: mrdata.usgs.gov)

George Allen Penton/Shutterstock.com

98. Apple Inc. *Sales per square foot* is often used to measure the financial performance of a store. For the last few years, Apple Stores have been the top performing retail locations in the United States in terms of revenue per square foot. If the average Apple Store has about 9,000 square feet of floor space and has average sales of about $49,950,000, what are the sales per square foot?

99. Unit prices. A 12-ounce can of soda sells for 84¢. Find the unit price in cents per ounce.

100. Day care. A day-care center charges $32 for 8 hours of supervised care. Find the unit price in dollars per hour for the day care.

101. Parking. A parking meter requires 25¢ for 20 minutes of parking. Find the unit price to park.

102. Lava. During a recent active lava flow on Hawaii's Big Island, civil defense officials warned residents of a residential area to prepare to evacuate. Authorities said the lava was 600 yards from the main road in town and that it could reach the road in anywhere from 30 to 40 hours. Determine the slower and faster rates of speed of the lava flow.

103. Rockets. On July 18, 2016, a SpaceX Falcon 9 rocket blasted off from Kennedy Space Center on a mission to resupply the International Space Station. The rocket carried a payload of about 5,500 pounds of experiments and supplies. If the launch cost about $133,000,000, what was the price per pound to put the payload into space?

104. Unit costs. A 24-ounce package of green beans sells for $1.29. Find the unit price in cents per ounce.

105. Draining tanks. An 11,880-gallon tank of water can be emptied in 27 minutes. Find the unit rate of flow of water out of the tank.

106. Pay rate. Ricardo worked for 27 hours to help insulate a hockey arena. For his work, he received $337.50. Find his hourly rate of pay.

107. Auto travel. A car's odometer reads 34,746 at the beginning of a trip. Five hours later, it reads 35,071.

   a. How far did the car travel?

   b. What was its average rate of speed?

108. Rates of speed. An airplane travels from Chicago to San Francisco, a distance of 1,883 miles, in 3.5 hours. Find the rate of speed of the plane.

109. Comparison shopping. A 6-ounce can of orange juice sells for 89¢, and an 8-ounce can sells for $1.19. Which is the better buy?

110. Comparison shopping. A 30-pound bag of planting mix costs $12.25, and an 80-pound bag costs $30.25. Which is the better buy?

111. Comparison shopping. A certain brand of cold and sinus medication is sold in 20-tablet boxes for $4.29 and in 50-tablet boxes for $9.59. Which is the better buy?

112. Comparison shopping. Which tire shown is the better buy?

ECONOMY $50.99 35,000-mile warranty   PREMIUM $59.50 40,000-mile warranty

113. Comparing speeds. A car travels 345 miles in 6 hours, and a truck travels 376 miles in 6.2 hours. Which vehicle is going faster?

114. Reading. One seventh-grader read a 54-page book in 40 minutes. Another read an 80-page book in 62 minutes. If the books were equally difficult, which student read faster?

115. Gas mileage. One car went 1,235 miles on 51.3 gallons of gasoline, and another went 1,456 miles on 55.78 gallons. Which car got the better gas mileage?

116. Electricity rates. In one community, a bill for 575 kilowatt-hours of electricity is $38.81. In a second community, a bill for 831 kWh is $58.10. In which community is electricity cheaper?

117. 2015 Crime rates. The U.S. crime statistics released each year by the FBI are often expressed as a *rate per 100,000 inhabitants*. Complete the second sentence to express the given crime rate in an equivalent form.

  a. The rate of violent crime was 372 offenses per 100,000 inhabitants.

            Or

    The rate of violent crime was 93 offenses per   inhabitants.

  b. The property crime rate was 2,487 offenses per 100,000 inhabitants.

            Or

    The property crime rate was   per 1,000,000 inhabitants.

118. Expensive art. In 2015, a colorful painting of two Tahitian women by French artist Paul Gauguin reportedly sold for $300 million to a museum in the Middle East. The painting, entitled *When Will You Marry?,* is 30 inches wide and 40 inches tall.

rook76/Shutterstock.com

  a. What is the area of the painting?

  b. What is the cost per square inch of the painting?

119. Are the ratios 3 to 1 and 1 to 3 the same? Explain why or why not.

120. Give three examples of ratios (or rates) that you have encountered in the past week.

121. How will the topics studied in this section make you a better shopper?

122. What is a unit rate? Give some examples.

123. Simplify: $\dfrac{48}{54}$

124. Simplify: $\dfrac{34}{153}$

125. Multiply: $\dfrac{3}{4} \cdot 16$

126. Multiply: $\dfrac{2}{3} \cdot 21$

SECTION 5.2   Proportions

OBJECTIVES

1 Write proportions.

2 Determine whether proportions are true or false.

3 Solve a proportion to find an unknown term.

4 Write proportions to solve application problems.

One of the most useful concepts in mathematics is the *equation*. Recall that an **equation** is a statement indicating that two expressions are equal. All equations contain an = symbol. Some examples of equations are:

$$4 + 4 = 8, \qquad 15.6 - 4.3 = 11.3, \qquad \frac{1}{2} \cdot 10 = 5, \qquad \text{and} \qquad -16 \div 8 = -2$$

Each of the equations shown above is true. Equations can also be false. For example,

$$3 + 2 = 6 \quad \text{and} \quad -40 \div (-5) = -8$$

are false equations.

    In this section, we will work with equations that state that two ratios (or rates) are equal.

OBJECTIVE 1 **Write proportions.**

Like any tool, a ladder can be dangerous if used improperly. When setting up an extension ladder, users should follow the *4-to-1 rule:* For every 4 feet of ladder height, position the legs of the ladder 1 foot away from the base of the wall. The 4-to-1 rule for ladders can be expressed using a ratio.

$$\frac{4 \text{ feet}}{1 \text{ foot}} = \frac{4 \text{ feet}}{1 \text{ foot}} = \frac{4}{1} \qquad \textit{Remove the common units of feet.}$$

The figure on the right shows how the 4-to-1 rule was used to properly position the legs of a ladder 3 feet from the base of a 12-foot-high wall. We can write a ratio comparing the ladder's height to its distance from the wall.

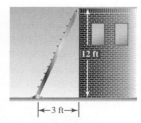

$$\frac{12 \text{ feet}}{3 \text{ feet}} = \frac{12 \,\cancel{\text{feet}}}{3 \,\cancel{\text{feet}}} = \frac{12}{3}$$ *Remove the common units of feet.*

Since this ratio satisfies the 4-to-1 rule, the two ratios $\frac{4}{1}$ and $\frac{12}{3}$ must be equal. Therefore, we have

$$\frac{4}{1} = \frac{12}{3}$$

Equations like this, which show that two ratios are equal, are called *proportions*.

Proportion

A **proportion** is a statement that two ratios (or rates) are equal.

Some examples of proportions are

- $\dfrac{1}{2} = \dfrac{3}{6}$ *Read as "1 is to 2 as 3 is to 6."*

- $\dfrac{3 \text{ waiters}}{7 \text{ tables}} = \dfrac{9 \text{ waiters}}{21 \text{ tables}}$ *Read as "3 waiters are to 7 tables as 9 waiters are to 21 tables."*

EXAMPLE 1   Write each statement as a proportion.

a. 22 is to 6 as 11 is to 3.

b. 1,000 administrators is to 8,000 teachers as 1 administrator is to 8 teachers.

Strategy We will locate the word *as* in each statement and identify the ratios (or rates) before and after it.

WHY The word *as* translates to the = symbol that is needed to write the statement as a proportion (equation).

Solution

a. This proportion states that two ratios are equal.

$$\underbrace{22 \text{ is \textbf{to} } 6}_{\dfrac{22}{6}} \quad \text{as} \quad \underbrace{11 \text{ is \textbf{to} } 3}_{\dfrac{11}{3}}$$

Recall that the word "to" is used to separate the numbers being compared.

b. This proportion states that two rates are equal.

$$\underbrace{1{,}000 \text{ administrators is \textbf{to} } 8{,}000 \text{ teachers}}_{\dfrac{1{,}000 \text{ administrators}}{8{,}000 \text{ teachers}}} \quad \text{as} \quad \underbrace{1 \text{ administrator is \textbf{to} } 8 \text{ teachers}}_{\dfrac{1 \text{ administrator}}{8 \text{ teachers}}}$$

When proportions involve rates, the units are often written outside of the proportion, as shown below:

$$\text{Administrators} \longrightarrow \frac{1{,}000}{8{,}000} = \frac{1}{8} \longleftarrow \text{Administrators}$$
$$\text{Teachers} \longrightarrow \phantom{\frac{1{,}000}{8{,}000}} \phantom{\frac{1}{8}} \longleftarrow \text{Teachers}$$

Self Check 1

Write each statement as a proportion.
a. 16 is to 28 as 4 is to 7.
b. 300 children is to 500 adults as 3 children is to 5 adults.

Now Try ➡ **Problems 17 and 19**

OBJECTIVE 2 **Determine whether proportions are true or false.**

Since a proportion is an equation, a proportion can be true or false. A proportion is true if its ratios (or rates) are equal and false if its ratios (or rates) are not equal. One way to determine whether a proportion is true is to use the fraction simplifying skills of Chapter 3.

EXAMPLE 2　　Determine whether each proportion is true or false by simplifying.

a. $\dfrac{3}{8} = \dfrac{21}{56}$　　**b.** $\dfrac{30}{4} = \dfrac{45}{12}$

Strategy We will simplify any ratios in the proportion that are not in simplest form. Then we will compare them to determine whether they are equal.

WHY If the ratios are equal, the proportion is true. If they are not equal, the proportion is false.

Solution

a. On the left side of the proportion $\frac{3}{8} = \frac{21}{56}$, the ratio $\frac{3}{8}$ is in simplest form. On the right side, the ratio $\frac{21}{56}$ can be simplified.

$$\frac{21}{56} = \frac{3 \cdot \overset{1}{\cancel{7}}}{\cancel{7} \cdot 8} = \frac{3}{8}$$　*Factor 21 and 56 and then remove the common factor of 7 in the numerator and denominator.*

Since the ratios on the left and right sides of the proportion are equal, the proportion is true.

b. Neither ratio in the proportion $\frac{30}{4} = \frac{45}{12}$ is in simplest form. To simplify each ratio, we proceed as follows:

$$\frac{30}{4} = \frac{\overset{1}{\cancel{2}} \cdot 15}{\underset{1}{\cancel{2}} \cdot 2} = \frac{15}{2} \qquad \frac{45}{12} = \frac{\overset{1}{\cancel{3}} \cdot 15}{\underset{1}{\cancel{3}} \cdot 4} = \frac{15}{4}$$

Since the ratios on the left and right sides of the proportion are not equal $\left(\frac{15}{2} \neq \frac{15}{4}\right)$, the proportion is false.

Self Check 2

Determine whether each proportion is true or false by simplifying.
a. $\dfrac{4}{5} = \dfrac{16}{20}$
b. $\dfrac{30}{24} = \dfrac{28}{16}$

Now Try ➲ **Problem 23**

There is another way to determine whether a proportion is true or false. Before we can discuss it, we need to introduce some more vocabulary of proportions.

Each of the four numbers in a proportion is called a **term**. The first and fourth terms are called the **extremes**, and the second and third terms are called the **means**.

First term (extreme) ⟶ $\dfrac{1}{2} = \dfrac{3}{6}$ ⟵ Third term (mean)
Second term (mean) ⟶ 　　　　⟵ Fourth term (extreme)

In the proportion shown above, the *product of the extremes is equal to the product of the means*.

$$1 \cdot 6 = 6 \qquad \text{and} \qquad 2 \cdot 3 = 6$$

These products can be found by multiplying diagonally in the proportion in such a way that one numerator is multiplied by the other denominator. We call $1 \cdot 6$ and $2 \cdot 3$ **cross products**.

Multiplication along one diagonal is shown in red.
Multiplication along the other diagonal is shown in blue.

Note that the cross products are equal. This example illustrates the following property of proportions.

> ### Cross-Products Property (Means-Extremes Property)
> To determine whether a proportion is true or false, first multiply along one diagonal, and then multiply along the other diagonal.
>
> ■ If the cross products are *equal*, the proportion is true.
> ■ If the cross products are *not equal,* the proportion is false.
>
> (If the product of the extremes is *equal* to the product of the means, the proportion is true. If the product of the extremes is *not equal* to the product of the means, the proportion is false.)

Caution! We cannot remove common factors "across" an = symbol. When this is done, the true proportion from Example 3 part a, $\frac{3}{7} = \frac{9}{21}$, is changed into the false proportion $\frac{1}{7} = \frac{9}{7}$.

$$\frac{\overset{1}{\cancel{3}}}{7} \diagup\!\!\!\!\diagdown \frac{9}{\underset{7}{\cancel{21}}}$$

Self Check 3

Determine whether the proportion

$$\frac{6}{13} = \frac{18}{39}$$

is true or false.

Now Try ➡ Problem 25

EXAMPLE 3   Determine whether each proportion is true or false.

a. $\dfrac{3}{7} = \dfrac{9}{21}$ **b.** $\dfrac{8}{3} = \dfrac{13}{5}$

Strategy We will check to see whether the cross products are equal (the product of the extremes is equal to the product of the means).

WHY If the cross products are equal, the proportion is true. If the cross products are not equal, the proportion is false.

Solution

a. $3 \cdot 21 = 63$ $7 \cdot 9 = 63$

$$\frac{3}{7} \diagup\!\!\!\!\diagdown \frac{9}{21} \qquad \textit{Each cross product is 63.}$$

Since the cross products are equal, the proportion is true.

b. $8 \cdot 5 = 40$ $3 \cdot 13 = 39$

$$\frac{8}{3} \diagup\!\!\!\!\diagdown \frac{13}{5} \qquad \textit{One cross product is 40 and the other is 39.}$$

Since the cross products are not equal, the proportion is false.

EXAMPLE 4   Determine whether each proportion is true or false.

a. $\dfrac{0.9}{0.6} = \dfrac{2.4}{1.5}$ **b.** $\dfrac{2\frac{1}{3}}{3\frac{1}{2}} = \dfrac{4\frac{2}{3}}{7}$

Strategy We will check to see whether the cross products are equal (the product of the extremes is equal to the product of the means).

WHY If the cross products are equal, the proportion is true. If the cross products are not equal, the proportion is false.

Solution

a.
$$\begin{array}{r} 1.5 \\ \times\ 0.9 \\ \hline 1.35 \end{array} \qquad \begin{array}{r} 2.4 \\ \times\ 0.6 \\ \hline 1.44 \end{array}$$

$$\frac{0.9}{0.6} \diagup\!\!\!\!\diagdown \frac{2.4}{1.5} \qquad \textit{One cross product is 1.35 and the other is 1.44.}$$

Since the cross products are not equal, the proportion is not true.

b.
$$3\frac{1}{2} \cdot 4\frac{2}{3} = \frac{7}{2} \cdot \frac{14}{3}$$

$$2\frac{1}{3} \cdot 7 = \frac{7}{3} \cdot \frac{7}{1} \qquad\qquad = \frac{7 \cdot \overset{1}{2} \cdot 7}{\underset{1}{2} \cdot 3}$$

$$= \frac{49}{3} \qquad\qquad\qquad = \frac{49}{3}$$

$$\frac{2\frac{1}{3}}{3\frac{1}{2}} \overset{\times}{=} \frac{4\frac{2}{3}}{7}$$ Each cross product is $\frac{49}{3}$.

Since the cross products are equal, the proportion is true.

When two pairs of numbers such as 2, 3 and 8, 12 form a true proportion, we say that they are **proportional**. To show that 2, 3 and 8, 12 are proportional, we check to see whether the equation

$$\frac{2}{3} = \frac{8}{12}$$

is a true proportion. To do so, we find the cross products.

$$2 \cdot 12 = 24 \qquad\qquad 3 \cdot 8 = 24$$

Since the cross products are equal, the proportion is true, and the numbers are proportional.

EXAMPLE 5   Determine whether 3, 7 and 36, 91 are proportional.

Strategy We will use the given pairs of numbers to write two ratios and form a proportion. Then we will find the cross products.

WHY If the cross products are equal, the proportion is true, and the numbers are proportional. If the cross products are not equal, the proportion is false, and the numbers are not proportional.

Solution The pair of numbers 3 and 7 form one ratio and the pair of numbers 36 and 91 form a second ratio. To write a proportion, we set the ratios equal. Then we find the cross products.

$$3 \cdot 91 = 273 \qquad\qquad 7 \cdot 36 = 252$$

$$\frac{3}{7} \overset{\times}{=} \frac{36}{91} \qquad \text{One cross product is 273 and the other is 252.}$$

Since the cross products are not equal, the numbers are not proportional.

OBJECTIVE 3 **Solve a proportion to find an unknown term.**

Suppose that we know three of the four terms in the following proportion.

$$\frac{?}{5} = \frac{24}{20}$$

In mathematics, we often let a letter represent an unknown number. We call such a letter a **variable**. To find the unknown term, we let the variable x represent it in the proportion and we can write:

$$\frac{x}{5} = \frac{24}{20}$$

If the proportion is to be true, the cross products must be equal.

$x \cdot 20 = 5 \cdot 24$ *Find the cross products for $\frac{x}{5} = \frac{24}{20}$ and set them equal.*

$x \cdot 20 = 120$ *To simplify the right side of the equation, do the multiplication: 5 · 24 = 120.*

$$\begin{array}{r} \overset{2}{2}4 \\ \times\ 5 \\ \hline 120 \end{array}$$

On the left side of the equation, the unknown number x is multiplied by 20. To undo the multiplication by 20 and isolate x, we divide both sides of the equation by 20.

$$\frac{x \cdot 20}{20} = \frac{120}{20}$$

$$\begin{array}{r} 6 \\ 20\overline{)120} \\ -\ 120 \\ \hline 0 \end{array}$$

We can simplify the fraction on the left side of the equation by removing the common factor of 20 from the numerator and denominator. On the right side, we perform the division indicated by the fraction bar.

$$\frac{x \cdot \overset{1}{\cancel{20}}}{\underset{1}{\cancel{20}}} = 6$$ *To simplify the left side of the equation, remove the common factor of 20 in the numerator and denominator.*
To simplify the right side of the equation, do the division: 120 ÷ 20 = 6.

Since the product of any number and 1 is that number, it follows that the numerator $x \cdot 1$ on the left side can be replaced by x.

$$\frac{x}{1} = 6$$

Since the quotient of any number and 1 is that number, it follows that $\frac{x}{1}$ on the left side of the equation can be replaced with x. Therefore,

$$x = 6$$

We have found that the unknown term in the proportion is 6 and we can write:

$$\frac{6}{5} = \frac{24}{20}$$

To check this result, we find the cross products.

Check:

$$\frac{6}{5} \overset{?}{=} \frac{24}{20} \qquad 6 \cdot 20 = 120$$
$$\qquad\qquad\qquad 5 \cdot 24 = 120$$

Since the cross products are equal, the result, 6, checks.

In the previous example, when we find the value of the variable x that makes the given proportion true, we say that we have *solved the proportion* to find the unknown term.

Solving a Proportion to Find an Unknown Term

1. Set the cross products equal to each other to form an equation.

2. Isolate the variable on one side of the equation by dividing both sides by the number that is multiplied by that variable.

3. Check by substituting the result into the original proportion and finding the cross products.

EXAMPLE 6   Solve the proportion: $\dfrac{12}{20} = \dfrac{3}{x}$

Strategy We will set the cross products equal to each other to form an equation.

WHY Then we can isolate the variable x on one side of the equation to find the unknown term in the proportion that it represents.

Solution

$$\dfrac{12}{20} = \dfrac{3}{x}$$ This is the proportion to solve.

$12 \cdot x = 20 \cdot 3$ Set the cross products equal to each other to form an equation.

$12 \cdot x = 60$ To simplify the right side of the equation, multiply: $20 \cdot 3 = 60$.

$$\dfrac{12 \cdot x}{12} = \dfrac{60}{12}$$ To undo the multiplication by 12 and isolate x, divide both sides by 12.

$$\begin{array}{r} 5 \\ 12\overline{)60} \\ -60 \\ \hline 0 \end{array}$$

$x = 5$ To simplify the left side, remove the common factor of 12. To simplify the right side of the equation, do the division: $60 \div 12 = 5$.

Thus, x is 5. To check this result, we substitute 5 for x in the original proportion.

Check:

$\dfrac{12}{20} \overset{?}{=} \dfrac{3}{5}$ $12 \cdot 5 = 60$
$20 \cdot 3 = 60$

Since the cross products are equal, the result, 5, checks.

Self Check 6

Solve the proportion:

$$\dfrac{15}{x} = \dfrac{20}{32}$$

Now Try ➲ **Problem 41**

EXAMPLE 7   Solve the proportion: $\dfrac{3.5}{7.2} = \dfrac{x}{15.84}$

Strategy We will set the cross products equal to each other to form an equation.

WHY Then we can isolate the variable x on one side of the equation to find the unknown term in the proportion that it represents.

Solution

$$\dfrac{3.5}{7.2} = \dfrac{x}{15.84}$$ This is the proportion to solve.

$3.5 \cdot 15.84 = 7.2 \cdot x$ Set the cross products equal to each other to form an equation.

$55.44 = 7.2 \cdot x$ To simplify the left side of the equation, multiply: $3.5 \cdot 15.84 = 55.44$.

$$\dfrac{55.44}{7.2} = \dfrac{7.2 \cdot x}{7.2}$$ To undo the multiplication by 7.2 and isolate x, divide both sides by 7.2.

$7.7 = x$ To simplify the left side of the equation, do the division: $55.44 \div 7.2 = 7.7$. To simplify the right side, remove the common factor of 7.2.

$$\begin{array}{r} 15.84 \\ \times 3.5 \\ \hline 7920 \\ 47520 \\ \hline 55.440 \end{array}$$

$$\begin{array}{r} 7.7 \\ 7.2\overline{)55.44} \\ -50\ 4 \\ \hline 5\ 04 \\ -5\ 04 \\ \hline 0 \end{array}$$

Thus, x is 7.7. Check the result in the original proportion.

Self Check 7

Solve the proportion:

$$\dfrac{6.7}{x} = \dfrac{33.5}{38}$$

Now Try ➲ **Problem 45**

EXAMPLE 8

Solve the proportion $\dfrac{x}{4\frac{1}{5}} = \dfrac{5\frac{1}{2}}{16\frac{1}{2}}$. Write the result as a mixed number.

Strategy We will set the cross products equal to each other to form an equation.

WHY Then we can isolate the variable x on one side of the equation to find the unknown term in the proportion that it represents.

Solution

$$\frac{x}{4\frac{1}{5}} = \frac{5\frac{1}{2}}{16\frac{1}{2}}$$ This is the proportion to solve.

$$x \cdot 16\frac{1}{2} = 4\frac{1}{5} \cdot 5\frac{1}{2}$$ Set the cross products equal to each other to form an equation.

$$x \cdot \frac{33}{2} = \frac{21}{5} \cdot \frac{11}{2}$$ Write each mixed number as an improper fraction.

$$\frac{x \cdot \frac{33}{2}}{\frac{33}{2}} = \frac{\frac{21}{5} \cdot \frac{11}{2}}{\frac{33}{2}}$$ To undo the multiplication by $\frac{33}{2}$ and isolate x, divide both sides by $\frac{33}{2}$.

$$x = \frac{21}{5} \cdot \frac{11}{2} \cdot \frac{2}{33}$$ To simplify the left side, remove the common factor of $\frac{33}{2}$ in the numerator and denominator. Perform the division on the right side indicated by the complex fraction bar. Multiply the numerator of the complex fraction by the reciprocal of $\frac{33}{2}$, which is $\frac{2}{33}$.

$$x = \frac{21}{5} \cdot \frac{11}{2} \cdot \frac{2}{33}$$ Multiply the numerators.
Multiply the denominators.

$$x = \frac{\overset{1}{2} \cdot \overset{1}{3} \cdot 7 \cdot \overset{1}{11}}{\underset{1}{3} \cdot \underset{1}{11} \cdot 5 \cdot \underset{1}{2}}$$ To simplify the fraction, factor 21 and 33, and then remove the common factors 2, 3, and 11 in the numerator and denominator.

$$x = \frac{7}{5}$$ Multiply the remaining factors in the numerator.
Multiply the remaining factors in the denominator.

$$x = 1\frac{2}{5}$$ Write the improper fraction as a mixed number.

Thus, x is $1\frac{2}{5}$. Check this result in the original proportion.

$$\begin{array}{r} 1 \\ 5{\overline{)7}} \\ -5 \\ \hline 2 \end{array}$$

Self Check 8

Solve the proportion:

$$\frac{x}{2\frac{1}{3}} = \frac{2\frac{1}{4}}{1\frac{1}{2}}$$

Write the result as a mixed number.

Now Try ➲ **Problem 49**

Using Your Calculator ▶   Solving Proportions with a Calculator

To solve the proportion in Example 7, we set the cross products equal and divided both sides by 7.2 to isolate the variable x.

$$\frac{3.5 \cdot 15.84}{7.2} = x$$

We can find x by entering these numbers and pressing these keys on a calculator.

3.5 $\boxed{\times}$ 15.84 $\boxed{\div}$ 7.2 $\boxed{=}$ $\boxed{7.7}$

Thus, x is 7.7.

OBJECTIVE 4 Write proportions to solve application problems.

Proportions can be used to solve application problems from a wide variety of fields such as medicine, accounting, construction, and business. It is easy to spot problems that can be solved using a proportion. You will be given a ratio (or rate) and asked to find the missing part of another ratio (or rate). It is helpful to revise the five-step problem-solving strategy seen earlier in the text and use the following **six-step approach**.

1. **Analyze the problem.**
2. **Assign a variable** to represent an unknown value in the problem.
   This means, in most cases, to let x equal what you are asked to find.
3. **Form an equation.**
4. **Solve the equation** formed in Step 3.
5. **State the conclusion clearly.**
6. **Check the result.**

EXAMPLE 9 **Shopping.** If 5 apples cost \$1.15, find the cost of 16 apples.

Analyze

■ We can express the fact that 5 apples cost \$1.15 using the rate: $\dfrac{5 \text{ apples}}{\$1.15}$.

■ What is the cost of 16 apples?

Assign We will let the variable c represent the unknown cost of 16 apples.

Form If we compare the number of apples to their cost, we know that the two rates must be equal and we can write a proportion.

5 apples is **to** \$1.15 **as** 16 apples is **to** \$c.

5 apples ⟶ $\dfrac{5}{1.15} = \dfrac{16}{c}$ ⟵ 16 apples
Cost of 5 apples ⟶ ⟵ Cost of 16 apples

The units can be written outside of the proportion.

Solve To find the cost of 16 apples, we solve the proportion for c.

$5 \cdot c = 1.15 \cdot 16$ Set the cross products equal to each other to form an equation.

$5 \cdot c = 18.4$ To simplify the right side of the equation, multiply: 1.15(16) = 18.4.

$\dfrac{5 \cdot c}{5} = \dfrac{18.4}{5}$ To undo the multiplication by 5 and isolate c, divide both sides by 5.

$c = 3.68$ To simplify the left side, remove the common factor of 5. On the right side, do the division: 18.4 ÷ 5 = 3.68.

$$\begin{array}{r} 3.68 \\ 5\overline{)18.40} \\ -15 \\ \hline 3\,4 \\ -3\,0 \\ \hline 40 \\ -40 \\ \hline 0 \end{array}$$

State Sixteen apples will cost \$3.68.

Check If 5 apples cost \$1.15, then 15 apples would cost 3 times as much: $3 \cdot \$1.15 = \$3.45$. It seems reasonable that 16 apples would cost \$3.68.

We could have compared the cost of the apples to the number of apples. If we solve this proportion for c, we obtain the same result: 3.68.

Cost of 5 apples ⟶ $\dfrac{1.15}{5} = \dfrac{c}{16}$ ⟵ Cost of 16 apples
5 apples ⟶ ⟵ 16 apples

Self Check 9

Concert tickets. If 9 tickets to a concert cost \$112.50, find the cost of 15 tickets.

Now Try Problem 77

Caution! When solving problems using proportions, make sure that the units of the numerators are the same and the units of the denominators are the same. For Example 9, it would be incorrect to write

Cost of 5 apples ⟶ $\cancel{\dfrac{1.15}{5}} = \cancel{\dfrac{16}{c}}$ ⟵ 16 apples
5 apples ⟶   ⟵ Cost of 16 apples

EXAMPLE 10   **Scale drawings.**   A **scale** is a ratio (or rate) that compares the size of a model, drawing, or map to the size of an actual object. The airplane shown below is drawn using a scale of 1 inch: 6 feet. This means that 1 inch on the drawing is actually 6 feet on the plane. The distance from wing tip to wing tip (the wingspan) on the drawing is 4.5 inches. What is the actual wingspan of the plane?

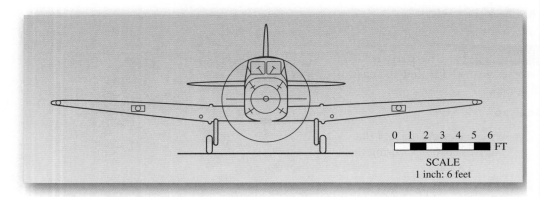

0 1 2 3 4 5 6
█░█░█░█ FT
SCALE
1 inch: 6 feet

Analyze
- The airplane is drawn using a scale of 1 inch: 6 feet, which can be written as a ratio in fraction form as: $\dfrac{1 \text{ inch}}{6 \text{ feet}}$.
- The wingspan of the airplane on the drawing is 4.5 inches.
- What is the actual wingspan of the plane?

Assign We will let *w* represent the unknown actual wingspan of the plane.

Form If we compare the measurements on the drawing to their actual measurement of the plane, we know that those two rates must be equal and we can write a proportion.

1 inch corresponds **to** 6 feet **as** 4.5 inches corresponds **to** *w* feet.

Measure on the drawing ⟶ $\dfrac{1}{6} = \dfrac{4.5}{w}$ ⟵ Measure on the drawing
Measure on the plane ⟶   ⟵ Measure on the plane

Self Check 10

Scale models. In a scale model of a city, a 300-foot-tall building is 4 inches high. An observation tower in the model is 9 inches high. How tall is the actual tower?

Now Try ⟹ Problem 87

Solve To find the actual wingspan of the airplane, we solve the proportion for *w*.

$1 \cdot w = 6 \cdot 4.5$ *Set the cross products equal to form an equation.*

$w = 27$ *To simplify each side of the equation, do the multiplication.*

$$\begin{array}{r} \overset{3}{4.5} \\ \times\ 6 \\ \hline 27.0 \end{array}$$

State The actual wingspan of the plane is 27 feet.

Check Every 1 inch on the scale drawing corresponds to an actual length of 6 feet on the plane. Therefore, a 5-inch measurement corresponds to an actual wingspan of 5 · 6 feet, or 30 feet. It seems reasonable that a 4.5-inch measurement corresponds to an actual wingspan of 27 feet.

EXAMPLE 11 **Baking.** A recipe for chocolate cake calls for $1\frac{1}{2}$ cups of sugar for every $2\frac{1}{4}$ cups of flour. If a baker has only $\frac{1}{2}$ cup of sugar on hand, how much flour should he add to it to make chocolate cake batter?

Analyze

- The sugar-to-flour rate can be expressed as: $\dfrac{1\frac{1}{2} \text{ cups sugar}}{2\frac{1}{4} \text{ cups flour}}$

- How much flour should be added to $\frac{1}{2}$ cup of sugar?

Assign We will let the variable f represent the unknown cups of flour.

Form If we compare the cups of sugar to the cups of flour, we know that the two ratios must be equal and we can write a proportion.

$1\frac{1}{2}$ cups of sugar is **to** $2\frac{1}{4}$ cups of flour **as** $\frac{1}{2}$ cup of sugar is **to** f cups of flour

Cups of sugar $\longrightarrow$ $\dfrac{1\frac{1}{2}}{2\frac{1}{4}} = \dfrac{\frac{1}{2}}{f}$ $\longleftarrow$ Cup of sugar
Cups of flour $\longrightarrow$ $\qquad\qquad\qquad\longleftarrow$ Cups of flour

Solve To find the amount of flour that is needed, we solve the proportion for f.

$$\dfrac{1\frac{1}{2}}{2\frac{1}{4}} = \dfrac{\frac{1}{2}}{f}$$
This is the proportion to solve.

$$1\frac{1}{2} \cdot f = 2\frac{1}{4} \cdot \frac{1}{2}$$
Set the cross products equal to each other to form an equation.

$$\frac{3}{2} \cdot f = \frac{9}{4} \cdot \frac{1}{2}$$
Write each mixed number as an improper fraction.

$$\dfrac{\frac{3}{2} \cdot f}{\frac{3}{2}} = \dfrac{\frac{9}{4} \cdot \frac{1}{2}}{\frac{3}{2}}$$
To undo the multiplication by $\frac{3}{2}$ and isolate f, divide both sides by $\frac{3}{2}$.

$$f = \frac{9}{4} \cdot \frac{1}{2} \cdot \frac{2}{3}$$
To simplify the left side, remove the common factor of $\frac{3}{2}$ in the numerator and denominator. Perform the division on the right side indicated by the complex fraction bar. Multiply the numerator of the complex fraction by the reciprocal of $\frac{3}{2}$, which is $\frac{2}{3}$.

$$f = \frac{9 \cdot 1 \cdot 2}{4 \cdot 2 \cdot 3}$$
Multiply the numerators.
Multiply the denominators.

$$f = \frac{\overset{1}{\cancel{3}} \cdot 3 \cdot 1 \cdot \overset{1}{\cancel{2}}}{4 \cdot \underset{1}{\cancel{2}} \cdot \underset{1}{\cancel{3}}}$$
To simplify the fraction, factor 9 and then remove the common factors 2 and 3 in the numerator and denominator.

$$f = \frac{3}{4}$$
Multiply the remaining factors in the numerator.
Multiply the remaining factors in the denominator.

State The baker should use $\frac{3}{4}$ cups of flour.

Check The rate of $1\frac{1}{2}$ cups of sugar for every $2\frac{1}{4}$ cups of flour is about 1 to 2. The rate of $\frac{1}{2}$ cup of sugar to $\frac{3}{4}$ cup flour is also about 1 to 2. The result, $\frac{3}{4}$, seems reasonable.

Self Check 11

Baking. See Example 11. How many cups of flour will be needed to make several chocolate cakes that will require a total of $12\frac{1}{2}$ cups of sugar?

Now Try ⟳ Problem 93

Success Tip In Example 11, an alternate approach would be to write each term of the proportion in its equivalent decimal form and then solve for f.

Fractions and mixed numbers   **Decimals**

$$\frac{1\frac{1}{2}}{2\frac{1}{4}} = \frac{\frac{1}{2}}{f} \longrightarrow \frac{1.5}{2.25} = \frac{0.5}{f}$$

Answers to Self Checks

1. a. $\frac{16}{28} = \frac{4}{7}$ **b.** $\frac{300 \text{ children}}{500 \text{ adults}} = \frac{3 \text{ children}}{5 \text{ adults}}$ **2. a.** true **b.** false **3.** true **4. a.** true **b.** true **5.** yes

6. 24 **7.** 7.6 **8.** $3\frac{1}{2}$ **9.** \$187.50 **10.** 675 ft **11.** $18\frac{3}{4}$ cups

---

SECTION 5.2 STUDY SET

VOCABULARY

Fill in the blanks.

1. A _____ is a statement that two ratios (or rates) are equal.

2. In $\frac{1}{2} = \frac{5}{10}$, the terms 1 and 10 are called the _____ of the proportion and the terms 2 and 5 are called the _____ of the proportion.

3. The _____ products for the proportion $\frac{4}{7} = \frac{36}{x}$ are $4 \cdot x$ and $7 \cdot 36$.

4. When two pairs of numbers form a proportion, we say that the numbers are _____ .

5. A letter that is used to represent an unknown number is called a _____ .

6. When we find the value of x that makes the proportion $\frac{3}{8} = \frac{x}{16}$ true, we say that we have _____ the proportion.

7. We solve proportions by writing a series of steps that result in an equation of the form x = a number or a number = x. We say that the variable x is _____ on one side of the equation.

8. A _____ is a ratio (or rate) that compares the size of a model, drawing, or map to the size of an actual object.

CONCEPTS

Fill in the blanks.

9. If the cross products of a proportion are equal, the proportion is _____. If the cross products are *not equal*, the proportion is _____.

10. The proportion $\frac{2}{5} = \frac{4}{10}$ will be true if the product ▨ · 10 is equal to the product ▨ · 4.

11. Complete the cross products.

▨ · 10 = ▨   2 · ▨ = ▨

$$\frac{9}{2} = \frac{45}{10}$$

12. In the equation $6 \cdot x = 2 \cdot 12$, to undo the multiplication by 6 and isolate x, _____ both sides of the equation by 6.

13. Label the missing units in the proportion.

Teacher's aides ⟶ $\dfrac{12}{100} = \dfrac{3}{25}$ ⟵
⟶ ⟵ Children

14. Consider the following problem: *For every 15 feet of chain link fencing, 4 support posts are used. How many support posts will be needed for 300 feet of chain link fencing?* Which of the proportions below could be used to solve this problem?

i. $\dfrac{15}{4} = \dfrac{300}{x}$ **ii.** $\dfrac{15}{4} = \dfrac{x}{300}$

iii. $\dfrac{4}{15} = \dfrac{300}{x}$ **iv.** $\dfrac{4}{15} = \dfrac{x}{300}$

NOTATION

Complete each step.

15. Solve the proportion: $\dfrac{2}{3} = \dfrac{x}{9}$

$2 \cdot 9 = $ ▨

▨ $= 3 \cdot x$

$\dfrac{18}{▨} = \dfrac{3 \cdot x}{▨}$

▨ $= x$

The solution is ▨.

16. Solve the proportion: $\dfrac{14}{x} = \dfrac{49}{17.5}$

$14 \cdot \rule{1cm}{0.4pt} = x \cdot 49$

$\rule{1cm}{0.4pt} = x \cdot 49$

$\dfrac{245}{\rule{0.8cm}{0.4pt}} = \dfrac{x \cdot 49}{\rule{0.8cm}{0.4pt}}$

$\rule{1cm}{0.4pt} = x$

The solution is $\rule{0.5cm}{0.4pt}$.

GUIDED PRACTICE

Write each statement as a proportion. **See Example 1.**

17. 20 is to 30 as 2 is to 3.

18. 9 is to 36 as 1 is to 4.

19. 400 sheets is to 100 beds as 4 sheets is to 1 bed.

20. 50 shovels is to 125 laborers as 2 shovels is to 5 laborers.

Determine whether each proportion is true or false by simplifying. **See Example 2.**

21. $\dfrac{7}{9} = \dfrac{70}{81}$ **22.** $\dfrac{2}{5} = \dfrac{8}{20}$

23. $\dfrac{21}{14} = \dfrac{18}{12}$ **24.** $\dfrac{42}{38} = \dfrac{95}{60}$

Determine whether each proportion is true or false by finding cross products. **See Example 3.**

25. $\dfrac{4}{32} = \dfrac{2}{16}$ **26.** $\dfrac{6}{27} = \dfrac{4}{18}$

27. $\dfrac{9}{19} = \dfrac{38}{80}$ **28.** $\dfrac{40}{29} = \dfrac{29}{22}$

Determine whether each proportion is true or false by finding cross products. **See Example 4.**

29. $\dfrac{0.5}{0.8} = \dfrac{1.1}{1.3}$ **30.** $\dfrac{0.6}{1.4} = \dfrac{0.9}{2.1}$

31. $\dfrac{1.2}{3.6} = \dfrac{1.8}{5.4}$ **32.** $\dfrac{3.2}{4.5} = \dfrac{1.6}{2.7}$

33. $\dfrac{1\frac{4}{5}}{3\frac{3}{7}} = \dfrac{2\frac{3}{16}}{4\frac{1}{6}}$ **34.** $\dfrac{2\frac{1}{2}}{1\frac{1}{5}} = \dfrac{3\frac{3}{4}}{2\frac{9}{10}}$

35. $\dfrac{\frac{1}{5}}{1\frac{1}{6}} = \dfrac{1\frac{1}{7}}{11\frac{2}{3}}$ **36.** $\dfrac{11\frac{1}{4}}{2\frac{1}{2}} = \dfrac{\frac{3}{4}}{\frac{1}{6}}$

Determine whether the numbers are proportional. **See Example 5.**

37. 18, 54 and 3, 9   **38.** 4, 3 and 12, 9

39. 8, 6 and 21, 16   **40.** 15, 7 and 13, 6

Solve each proportion. Check each result. **See Example 6.**

41. $\dfrac{5}{10} = \dfrac{3}{x}$ **42.** $\dfrac{7}{14} = \dfrac{2}{x}$

43. $\dfrac{2}{3} = \dfrac{x}{6}$ **44.** $\dfrac{3}{6} = \dfrac{x}{8}$

Solve each proportion. Check each result. **See Example 7.**

45. $\dfrac{0.6}{9.6} = \dfrac{x}{4.8}$ **46.** $\dfrac{0.4}{3.4} = \dfrac{x}{13.6}$

47. $\dfrac{2.75}{x} = \dfrac{1.5}{1.2}$ **48.** $\dfrac{9.8}{x} = \dfrac{2.8}{5.4}$

Solve each proportion. Check each result. Write each result as a fraction or mixed number. **See Example 8.**

49. $\dfrac{x}{1\frac{1}{2}} = \dfrac{10\frac{1}{2}}{4\frac{1}{2}}$ **50.** $\dfrac{x}{3\frac{1}{3}} = \dfrac{1\frac{1}{2}}{1\frac{9}{11}}$

51. $\dfrac{x}{1\frac{1}{6}} = \dfrac{2\frac{5}{8}}{3\frac{1}{2}}$ **52.** $\dfrac{x}{2\frac{2}{3}} = \dfrac{1\frac{1}{20}}{3\frac{1}{2}}$

TRY IT YOURSELF

Solve each proportion.

53. $\dfrac{4{,}000}{x} = \dfrac{3.2}{2.8}$ **54.** $\dfrac{0.4}{1.6} = \dfrac{96.7}{x}$

55. $\dfrac{12}{6} = \dfrac{x}{1\frac{1}{4}}$ **56.** $\dfrac{15}{10} = \dfrac{x}{1\frac{1}{3}}$

57. $\dfrac{x}{8} = \dfrac{900}{200}$ **58.** $\dfrac{x}{200} = \dfrac{1{,}800}{600}$

59. $\dfrac{x}{2.5} = \dfrac{3.7}{9.25}$

60. $\dfrac{8.5}{x} = \dfrac{4.25}{1.7}$

61. $\dfrac{0.8}{2} = \dfrac{x}{5}$

62. $\dfrac{0.9}{0.3} = \dfrac{6}{x}$

63. $\dfrac{x}{4\frac{1}{10}} = \dfrac{3\frac{3}{4}}{1\frac{7}{8}}$

64. $\dfrac{x}{2\frac{1}{4}} = \dfrac{\frac{1}{2}}{\frac{1}{5}}$

65. $\dfrac{340}{51} = \dfrac{x}{27}$

66. $\dfrac{480}{36} = \dfrac{x}{15}$

67. $\dfrac{0.4}{1.2} = \dfrac{6}{x}$

68. $\dfrac{5}{x} = \dfrac{2}{4.4}$

69. $\dfrac{4.65}{7.8} = \dfrac{x}{5.2}$

70. $\dfrac{8.6}{2.4} = \dfrac{x}{6}$

71. $\dfrac{\frac{3}{4}}{\frac{1}{2}} = \dfrac{0.25}{x}$

72. $\dfrac{\frac{7}{8}}{\frac{1}{2}} = \dfrac{0.25}{x}$

LOOK ALIKES

Solve each proportion in part a. Use your result to determine the solutions to the proportions in parts b and c quickly.

73. a. $\dfrac{y}{0.4} = \dfrac{2.7}{6}$ **b.** $\dfrac{0.4}{y} = \dfrac{6}{2.7}$ **c.** $\dfrac{6}{0.4} = \dfrac{2.7}{y}$

74. a. $\dfrac{x}{1\frac{1}{4}} = \dfrac{\frac{1}{2}}{3\frac{1}{8}}$ **b.** $\dfrac{3\frac{1}{8}}{1\frac{1}{4}} = \dfrac{\frac{1}{2}}{x}$ **c.** $\dfrac{x}{1.25} = \dfrac{0.5}{3.125}$

Solve each proportion and simplify each expression.

75. a. $\dfrac{2}{3} = \dfrac{x}{9}$ **b.** $\dfrac{2}{3} + \dfrac{1}{9}$

76. a. $\dfrac{10}{n} = \dfrac{2}{5}$ **b.** $\dfrac{10}{3} - \dfrac{2}{5}$

APPLICATIONS

To solve each problem, write and then solve a proportion.

77. School lunches. A manager of a school cafeteria orders 750 pudding cups. What will the order cost if she purchases them wholesale, 6 cups for $1.75?

78. Clothes shopping. As part of a spring clearance, a men's store put dress shirts on sale, 2 for $25.98. How much will a businessman pay if he buys five shirts?

79. Anniversary gifts. A florist sells a dozen long-stemmed red roses for $57.99. In honor of their 16th wedding anniversary, a man wants to buy 16 roses for his wife. What will the roses cost? (*Hint:* How many roses are in one dozen?)

80. Cooking. A recipe for spaghetti sauce requires four 16-ounce bottles of ketchup to make 2 gallons of sauce. How many bottles of ketchup are needed to make 10 gallons of sauce? (*Hint:* Read the problem very carefully.)

81. Business performance. The following bar graph shows the yearly costs and the revenue received for a business. Are the ratios of costs to revenue for 2016 and 2017 equal?

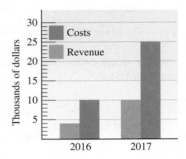

82. Ramps. Write a ratio of the rise to the run for each ramp shown. Set the ratios equal.

a. Is the resulting proportion true?

b. Is one ramp steeper than the other?

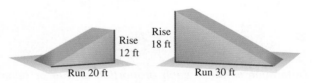

83. Mixing perfumes. A perfume is to be mixed in the ratio of 3 drops of pure essence to 7 drops of alcohol. How many drops of pure essence should be mixed with 56 drops of alcohol?

84. Making cologne. A cologne can be made by mixing 2 drops of pure essence with 5 drops of distilled water. How many drops of water should be used with 15 drops of pure essence?

85. Lab work. In a red blood cell count, a drop of the patient's diluted blood is placed on a grid like that shown on the next page. Instead of counting each and every red blood cell in the 25-square grid, a technician counts only the number of cells in the five highlighted squares. Then he or she uses a proportion to estimate the total red blood cell count. If there are 195 red blood cells in the blue squares, about how many red blood cells are in the entire grid?

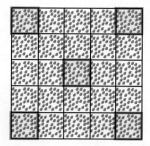

86. Dosages. The proper dosage of a certain medication for a 30-pound child is shown. At this rate, what would be the dosage for a 45-pound child?

87. Drafting. In a scale drawing, a 280-foot antenna tower is drawn 7 inches high. The building next to it is drawn 2 inches high. How tall is the actual building?

88. Blueprints. The scale for the drawing in the blueprint tells the reader that a $\frac{1}{4}$-inch length $\left(\frac{1}{4}''\right)$ on the drawing corresponds to an actual size of 1 foot (1′0″). Suppose the length of the kitchen is $2\frac{1}{2}$ inches on the blueprint. How long is the actual kitchen?

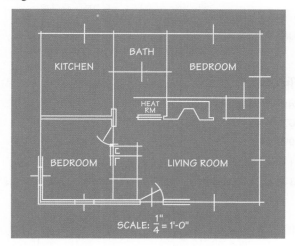

89. Model railroads. An HO-scale model railroad engine is 9 inches long. If HO scale is 87 feet to 1 foot, how long is a real engine? (*Hint:* Compare feet to inches. How many inches are in one foot?)

90. Model railroads. An N-scale model railroad caboose is 4 inches long. If N scale is 169 feet to 1 foot, how long is a real caboose? (*Hint:* Compare feet to inches. How many inches are in one foot?)

91. Carousels. The ratio in the illustration below indicates that 1 inch on the model carousel is equivalent to 160 inches on the actual carousel. How wide should the model be if the actual carousel is 35 feet wide? (*Hint:* Convert 35 feet to inches.)

Carousel ratio
1:160

92. Mixing fuels. The instructions on a can of oil intended to be added to lawn mower gasoline read as shown. Are these instructions correct? (*Hint:* There are 128 ounces in 1 gallon.)

Recommended	Gasoline	Oil
50 to 1	6 gal	16 oz

93. Making cookies. A recipe for chocolate chip cookies calls for $1\frac{1}{4}$ cups of flour and 1 cup of sugar. The recipe will make $3\frac{1}{2}$ dozen cookies. How many cups of flour will be needed to make 12 dozen cookies?

from **Campus to Careers**

Chef
94. Making brownies.
A recipe for brownies calls for 4 eggs and $1\frac{1}{2}$ cups of flour. If the recipe makes 15 brownies, how many cups of flour will be needed to make 130 brownies?

wong yu liang/shutterstock.com

95. Computer speed. Using an advanced mathematics program, a high-speed computer can perform a set of 15 complex calculations in 2.85 seconds. How long will it take the computer to perform 100 such calculations?

96. Quality control. Out of a sample of 500 men's shirts, 17 were rejected because of crooked collars. How many crooked collars would you expect to find in a run of 15,000 shirts?

97. Dogs. Refer to the illustration below. A Saint Bernard website lists the "ideal proportions for the *height at the withers* to *body length* as 5 : 6." What is the ideal height at the withers for a Saint Bernard whose body length is $37\frac{1}{2}$ inches?

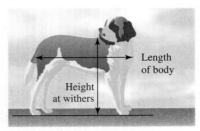

Length of body

Height at withers

98. Mileage. Under normal conditions, a Hummer can travel 325 miles on a full tank (25 gallons) of diesel. How far can it travel on its auxiliary tank, which holds 17 gallons of diesel?

99. Paychecks. Billie earns $412 for a 40-hour week. If she missed 10 hours of work last week, how much did she get paid?

100. Staffing. A school board has determined that there should be 3 teachers for every 50 students. Complete the table by filling in the number of teachers needed at each school.

	Glenwood High	Goddard Junior High	Sellers Elementary
Enrollment	2,700	1,900	850
Teachers			

101. Springs. A 5-pound weight stretches a spring 8 inches. How far will a 12-pound weight stretch the spring?

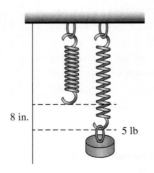

8 in.

5 lb

102. Recycling shingles. The Owens Corning Company has developed a program that reuses discarded asphalt shingles removed from homes that are being reroofed. The old shingles are taken to a recycling center where they are ground up and repurposed into asphalt paving mix. Shingles from 13 average-sized homes can help pave one-half mile of a two-lane highway. Old shingles from how many such homes would be needed to pave a 10-mile long stretch of two-lane highway?

jocic/Shutterstock.com

WRITING

103. Explain the difference between a ratio and a proportion.

104. The following paragraph is from a book about dollhouses. What concept from this section is mentioned?

Today, the internationally recognized scale for dollhouses and miniatures is 1 in. = 1 ft. This is small enough to be defined as a miniature, yet not too small for all details of decoration and furniture to be seen clearly.

105. Write a problem that could be solved using the following proportion.

Ounces of cashews $\longrightarrow \dfrac{4}{639} = \dfrac{10}{x} \longleftarrow$ Ounces of cashews

Calories $\longrightarrow$ $\hspace{1.5cm}$ $\longleftarrow$ Calories

106. Write a problem about a situation you encounter in your daily life that could be solved by using a proportion.

REVIEW

Perform each operation.

107. $7.4 + 6.78 + 35 + 0.008$

108. $29.5 + 34.4 + 12.8$

109. $48.8 - 17.372$

110. $78.47 - 53.3$

111. $-3.8 - (-7.9)$

112. $-17.1 + 8.4$

113. $-35.1 - 13.99$

114. $-5.55 + (-1.25)$

SECTION 5.3 American Units of Measurement

Two common systems of measurement are the **American** (or **English**) **system** and the **metric system.** We will discuss American units of measurement in this section and metric units in the next. Some common American units are *inches, feet, miles, ounces, pounds, tons, cups, pints, quarts,* and *gallons.* These units are used when measuring length, weight, and capacity.

A newborn baby is 20 inches long.

First-class postage for a letter that weighs less than 1 ounce is 46¢.

Milk is sold in gallon containers.

OBJECTIVE 1 Use a ruler to measure lengths in inches.

A ruler is one of the most common tools used for measuring distances or lengths. The figure below shows part of a ruler. Most rulers are 12 inches (1 foot) long. Since 12 inches = 1 foot, a ruler is divided into 12 equal lengths of 1 inch. Each inch is divided into halves of an inch, quarters of an inch, eighths of an inch, and sixteenths of an inch.

The left end of a ruler can be (but sometimes isn't) labeled with a 0. Each point on a ruler, like each point on a number line, has a number associated with it. That number is the distance between the point and 0. Several lengths on the ruler are shown below.

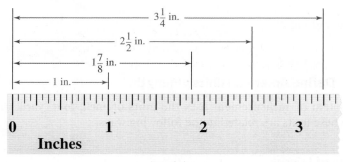

Actual size

EXAMPLE 1   Find the length of the paper clip shown here.

Strategy We will place a ruler below the paper clip, with the left end of the ruler (which could be thought of as 0) directly underneath one end of the paper clip.

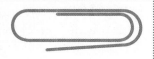

WHY Then we can find the length of the paper clip by identifying where its other end lines up on the tick marks printed in black on the ruler.

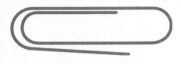

Self Check 1

Find the length of the jumbo paper clip.

Now Try ⟹ Problem 28

Solution Since the tick marks between 0 and 1 on the ruler create eight equal spaces, the ruler is scaled in eighths of an inch. The paper clip is $1\frac{3}{8}$ inches long.

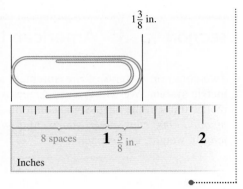

EXAMPLE 2 Find the length of the nail shown below.

Strategy We will place a ruler below the nail, with the left end of the ruler (which could be thought of as 0) directly underneath the head of the nail.

WHY Then we can find the length of the nail by identifying where its pointed end lines up on the tick marks printed in black on the ruler.

Solution Since the tick marks between 0 and 1 on the ruler create sixteen equal spaces, the ruler is scaled in sixteenths of an inch.

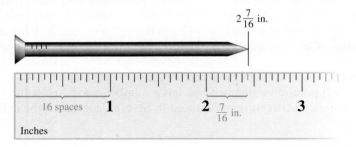

Self Check 2

Find the width of the circle.

Now Try ⟹ Problem 31

The nail is $2\frac{7}{16}$ inches long.

OBJECTIVE 2 Define American units of length.

The American system of measurement uses the units of **inch**, **foot**, **yard**, and **mile** to measure length. These units are related in the following ways.

American Units of Length

1 foot (ft) = 12 inches (in.)	1 yard (yd) = 36 inches
1 yard = 3 feet	1 mile (mi) = 5,280 feet

The abbreviation for each unit is written within parentheses.

OBJECTIVE 3 Convert from one American unit of length to another.

To convert from one unit of length to another, we use *unit conversion factors*. To find the unit conversion factor between yards and feet, we begin with this fact:

3 ft = 1 yd

If we divide both sides of this equation by 1 yard, we get

$$\frac{3 \text{ ft}}{1 \text{ yd}} = \frac{1 \text{ yd}}{1 \text{ yd}}$$

$$\frac{3 \text{ ft}}{1 \text{ yd}} = 1 \qquad \text{\small Simplify the right side. A nonzero number divided by itself is 1: } \frac{1 \text{ yd}}{1 \text{ yd}} = 1.$$

The fraction $\frac{3 \text{ ft}}{1 \text{ yd}}$ is called a **unit conversion factor**, because its value is 1. It can be read as "3 feet per yard." Since this fraction is equal to 1, multiplying a length by this fraction does not change its measure; it changes only the *units* of measure.

To convert units of length in the American system of measurement, we use the following unit conversion factors. Each conversion factor shown below is a form of 1.

To convert from	Use the unit conversion factor	To convert from	Use the unit conversion factor
feet to inches	$\frac{12 \text{ in.}}{1 \text{ ft}}$	inches to feet	$\frac{1 \text{ ft}}{12 \text{ in.}}$
yards to feet	$\frac{3 \text{ ft}}{1 \text{ yd}}$	feet to yards	$\frac{1 \text{ yd}}{3 \text{ ft}}$
yards to inches	$\frac{36 \text{ in.}}{1 \text{ yd}}$	inches to yards	$\frac{1 \text{ yd}}{36 \text{ in.}}$
miles to feet	$\frac{5,280 \text{ ft}}{1 \text{ mi}}$	feet to miles	$\frac{1 \text{ mi}}{5,280 \text{ ft}}$

EXAMPLE 3　　Convert 8 yards to feet.

Strategy We will multiply 8 yards by a carefully chosen unit conversion factor.

WHY If we multiply by the proper unit conversion factor, we can eliminate the unwanted units of yards and convert to feet.

Solution To convert from yards to feet, we must use a unit conversion factor that relates feet to yards. Since there are 3 feet per yard, we multiply 8 yards by the unit conversion factor $\frac{3 \text{ ft}}{1 \text{ yd}}$.

$$8 \text{ yd} = \frac{8 \text{ yd}}{1} \cdot \frac{3 \text{ ft}}{1 \text{ yd}} \qquad \text{\small Write 8 yd as a fraction: } 8 \text{ yd} = \frac{8 \text{ yd}}{1}$$
$$\text{\small Then multiply by a form of 1: } \frac{3 \text{ ft}}{1 \text{ yd}}$$

$$= \frac{8 \; \cancel{\text{yd}}}{1} \cdot \frac{3 \text{ ft}}{1 \; \cancel{\text{yd}}} \qquad \text{\small Remove the common units of yards from the numerator and}$$
$$\text{\small denominator. Notice that the units of feet remain.}$$

$$= 8 \cdot 3 \text{ ft} \qquad \text{\small Simplify.}$$

$$= 24 \text{ ft} \qquad \text{\small Multiply: } 8 \cdot 3 = 24.$$

8 yards is equal to 24 feet.

Success Tip Notice that in Example 3, we eliminated the units of yards and introduced the units of feet by multiplying by the appropriate unit conversion factor. In general, a unit conversion factor is a fraction with the following form:

$$\frac{\text{Unit we want to introduce}}{\text{Unit we want to eliminate}} \quad \begin{array}{l} \leftarrow \text{ Numerator} \\ \leftarrow \text{ Denominator} \end{array}$$

LANGUAGE OF MATHEMATICS

According to some sources, the *inch* was originally defined as the length from the tip of the thumb to the first knuckle. In some languages the word for *inch* is similar to or the same as *thumb*. For example, in Spanish, *pulgada* is inch and *pulgar* is thumb. In Swedish, *tum* is inch and *tumme* is thumb. In Italian, *pollice* is both inch and thumb.

Self Check 3

Convert 9 yards to feet.

Now Try ➨ Problem 35

EXAMPLE 4   Convert $1\frac{3}{4}$ feet to inches.

Strategy We will multiply $1\frac{3}{4}$ feet by a carefully chosen unit conversion factor.

WHY If we multiply by the proper unit conversion factor, we can eliminate the unwanted units of feet and convert to inches.

Solution To convert from feet to inches, we must choose a unit conversion factor whose numerator contains the units we want to introduce (inches), and whose denominator contains the units we want to eliminate (feet). Since there are 12 inches per foot, we will use

$\dfrac{12 \text{ in.}}{1 \text{ ft}}$ ← This is the unit we want to introduce.
     ← This is the unit we want to eliminate (the original unit).

To perform the conversion, we multiply.

$$1\frac{3}{4} \text{ ft} = \frac{7}{4} \text{ ft} \cdot \frac{12 \text{ in.}}{1 \text{ ft}}$$ Write $1\frac{3}{4}$ as an improper fraction: $1\frac{3}{4} = \frac{7}{4}$
     Then multiply by a form of 1: $\frac{12 \text{ in.}}{1 \text{ ft}}$

$$= \frac{7}{4} \cancel{\text{ft}} \cdot \frac{12 \text{ in.}}{1 \cancel{\text{ft}}}$$ Remove the common units of feet from the numerator and denominator. Notice that the units of inches remain.

$$= \frac{7 \cdot 12}{4 \cdot 1} \text{ in.}$$ Multiply the fractions.

$$= \frac{7 \cdot 3 \cdot \overset{1}{\cancel{4}}}{\underset{1}{\cancel{4}} \cdot 1} \text{ in.}$$ To simplify the fraction, factor 12. Then remove the common factor of 4 from the numerator and denominator.

$$= 21 \text{ in.}$$ Simplify.

$1\frac{3}{4}$ feet is equal to 21 inches.

Self Check 4

Convert $1\frac{1}{2}$ feet to inches.

Now Try **Problem 39**

Sometimes we must use two (or more) unit conversion factors to eliminate the given units while introducing the desired units. The following example illustrates this concept.

Caution! When converting lengths, if no common units appear in the numerator and denominator to remove, you have chosen the wrong conversion factor.

EXAMPLE 5   **Football.** A football field (including both end zones) is 120 yards long. Convert this length to miles. Give the exact answer and a decimal approximation, rounded to the nearest hundredth of a mile.

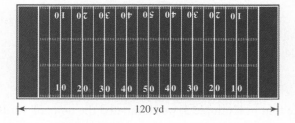

Strategy We will use a two-part multiplication process that converts 120 yards to feet and then converts that result to miles.

WHY We must use a two-part process because the table on page 441 does not contain a single unit conversion factor that converts from yards to miles.

Solution Since there are 3 feet per yard, we can convert 120 yards to feet by multiplying by the unit conversion factor $\frac{3 \text{ ft}}{1 \text{ yd}}$. Since there is 1 mile for every 5,280 feet, we can convert that result to miles by multiplying by the unit conversion factor $\frac{1 \text{ mi}}{5,280 \text{ ft}}$.

$$120 \text{ yd} = \frac{120 \text{ yd}}{1} \cdot \frac{3 \text{ ft}}{1 \text{ yd}} \cdot \frac{1 \text{ mi}}{5,280 \text{ ft}}$$

Write 120 yd as a fraction: $120 \text{ yd} = \frac{120 \text{ yd}}{1}$. Then multiply by two unit conversion factors: $\frac{3 \text{ ft}}{1 \text{ yd}} = 1$ and $\frac{1 \text{ mi}}{5,280 \text{ ft}} = 1$.

$$= \frac{120 \text{ y\!\!\!/d}}{1} \cdot \frac{3 \text{ f\!\!\!/t}}{1 \text{ y\!\!\!/d}} \cdot \frac{1 \text{ mi}}{5,280 \text{ f\!\!\!/t}}$$

Remove the common units of yards and feet in the numerator and denominator. Notice that all the units are removed except for miles.

$$= \frac{120 \cdot 3}{5,280} \text{ mi}$$

Multiply the fractions.

$$= \frac{\overset{1}{\cancel{2}} \cdot \overset{1}{\cancel{2}} \cdot \overset{1}{\cancel{2}} \cdot \overset{1}{\cancel{3}} \cdot \overset{1}{\cancel{5}} \cdot 3}{\underset{1}{\cancel{2}} \cdot \underset{1}{\cancel{2}} \cdot \underset{1}{\cancel{2}} \cdot 2 \cdot 2 \cdot \underset{1}{\cancel{3}} \cdot \underset{1}{\cancel{5}} \cdot 11} \text{ mi}$$

To simplify the fraction, prime factor 120 and 5,280, and remove the common factors 2, 3, and 5.

$$= \frac{3}{44} \text{ mi}$$

Multiply the remaining factors in the numerator. Multiply the remaining factors in the denominator.

A football field (including the end zones) is *exactly* $\frac{3}{44}$ miles long.

We can also present this conversion as a decimal. If we divide 3 by 44 (as shown on the right), and round the result to the nearest hundredth, we see that a football field (including the end zones) is *approximately* 0.07 mile long.

```
       0.068
44)3.000
     − 0
     3 00
   − 2 64
      360
    − 352
        8
```

Self Check 5

Marathons. The *marathon* is a long-distance race with an official distance of 26 miles 385 yards. Convert 385 yards to miles. Give the exact answer and a decimal approximation, rounded to the nearest hundredth of a mile.

Now Try ➲ **Problem 43**

OBJECTIVE 4 **Define American units of weight.**

The American system of measurement uses the units of **ounce**, **pound**, and **ton** to measure weight. These units are related in the following ways.

American Units of Weight

1 pound (lb) = 16 ounces (oz)　　　　1 ton (T) = 2,000 pounds

The abbreviation for each unit is written within parentheses.

OBJECTIVE 5 **Convert from one American unit of weight to another.**

To convert units of weight in the American system of measurement, we use the following unit conversion factors. Each conversion factor shown below is a form of 1.

To convert from	Use the unit conversion factor	To convert from	Use the unit conversion factor
pounds to ounces	$\frac{16 \text{ oz}}{1 \text{ lb}}$	ounces to pounds	$\frac{1 \text{ lb}}{16 \text{ oz}}$
tons to pounds	$\frac{2,000 \text{ lb}}{1 \text{ ton}}$	pounds to tons	$\frac{1 \text{ ton}}{2,000 \text{ lb}}$

EXAMPLE 6   Convert 40 ounces to pounds.

Strategy We will multiply 40 ounces by a carefully chosen unit conversion factor.

WHY If we multiply by the proper unit conversion factor, we can eliminate the unwanted units of ounces and convert to pounds.

Solution To convert from ounces to pounds, we must choose a unit conversion factor whose numerator contains the units we want to introduce (pounds), and whose denominator contains the units we want to eliminate (ounces). Since there is 1 pound for every 16 ounces, we will use

$\dfrac{1 \text{ lb}}{16 \text{ oz}}$ ← This is the unit we want to introduce.
      ← This is the unit we want to eliminate (the original unit).

To perform the conversion, we multiply.

$$40 \text{ oz} = \frac{40 \text{ oz}}{1} \cdot \boxed{\frac{1 \text{ lb}}{16 \text{ oz}}}$$

Write 40 oz as a fraction: $40 \text{ oz} = \frac{40 \text{ oz}}{1}$
Then multiply by a form of 1: $\frac{1 \text{ lb}}{16 \text{ oz}}$

$$= \frac{40 \cancel{\text{ oz}}}{1} \cdot \frac{1 \text{ lb}}{16 \cancel{\text{ oz}}}$$

Remove the common units of ounces from the numerator and denominator. Notice that the units of pounds remain.

$$= \frac{40}{16} \text{ lb}$$

Multiply the fractions.

There are two ways to complete the solution. First, we can remove any common factors of the numerator and denominator to simplify the fraction. Then we can write the result as a mixed number.

$$\frac{40}{16} \text{ lb} = \frac{5 \cdot \overset{1}{\cancel{8}}}{2 \cdot \underset{1}{\cancel{8}}} \text{ lb} = \frac{5}{2} \text{ lb} = 2\frac{1}{2} \text{ lb}$$

A second approach is to divide the numerator by the denominator and express the result as a decimal.

Self Check 6

Convert 60 ounces to pounds.

$$\frac{40}{16} \text{ lb} = 2.5 \text{ lb}$$ *Perform the division: 40 ÷ 16.*

Now Try Problem 47

40 ounces is equal to $2\frac{1}{2}$ lb (or 2.5 lb).

$$\begin{array}{r} 2.5 \\ 16\overline{)40.0} \\ -32 \\ \hline 8\,0 \\ -8\,0 \\ \hline 0 \end{array}$$

EXAMPLE 7   Convert 25 pounds to ounces.

Strategy We will multiply 25 pounds by a carefully chosen unit conversion factor.

WHY If we multiply by the proper unit conversion factor, we can eliminate the unwanted units of pounds and convert to ounces.

Solution To convert from pounds to ounces, we must choose a unit conversion factor whose numerator contains the units we want to introduce (ounces), and whose denominator contains the units we want to eliminate (pounds). Since there are 16 ounces per pound, we will use

$\dfrac{16 \text{ oz}}{1 \text{ lb}}$ ← This is the unit we want to introduce.
      ← This is the unit we want to eliminate (the original unit).

To perform the conversion, we multiply.

$$25 \text{ lb} = \frac{25 \text{ lb}}{1} \cdot \frac{\textbf{16 oz}}{\textbf{1 lb}}$$ Write 25 lb as a fraction: $25 \text{ lb} = \frac{25 \text{ lb}}{1}$
Then multiply by a form of 1: $\frac{16 \text{ oz}}{1 \text{ lb}}$

$$= \frac{25 \text{ lb}}{1} \cdot \frac{16 \text{ oz}}{1 \text{ lb}}$$ Remove the common units of pounds from the numerator and denominator. Notice that the units of ounces remain.

$$= 25 \cdot 16 \text{ oz}$$ Simplify.

$$= 400 \text{ oz}$$ Multiply: $25 \cdot 16 = 400$.

25 pounds is equal to 400 ounces.

$$\begin{array}{r} 25 \\ \times\, 16 \\ \hline 150 \\ 250 \\ \hline 400 \end{array}$$

Self Check 7
Convert 60 pounds to ounces.

Now Try ➲ **Problem 51**

OBJECTIVE 6 Define American units of capacity.

The American system of measurement uses the units of **ounce**, **cup**, **pint**, **quart**, and **gallon** to measure capacity. These units are related as follows.

American Units of Capacity	
1 cup (c) = 8 fluid ounces (fl oz)	1 pint (pt) = 2 cups
1 quart (qt) = 2 pints	1 gallon (gal) = 4 quarts

The abbreviation for each unit is written within parentheses.

OBJECTIVE 7 Convert from one American unit of capacity to another.

To convert units of capacity in the American system of measurement, we use the following unit conversion factors. Each conversion factor shown below is a form of 1.

To convert from	Use the unit conversion factor
cups to ounces	$\frac{8 \text{ fl oz}}{1 \text{ c}}$
pints to cups	$\frac{2 \text{ c}}{1 \text{ pt}}$
quarts to pints	$\frac{2 \text{ pt}}{1 \text{ qt}}$
gallons to quarts	$\frac{4 \text{ qt}}{1 \text{ gal}}$

To convert from	Use the unit conversion factor
ounces to cups	$\frac{1 \text{ c}}{8 \text{ fl oz}}$
cups to pints	$\frac{1 \text{ pt}}{2 \text{ c}}$
pints to quarts	$\frac{1 \text{ qt}}{2 \text{ pt}}$
quarts to gallons	$\frac{1 \text{ gal}}{4 \text{ qt}}$

LANGUAGE OF MATHEMATICS

The word *capacity* means the amount that can be contained. For example, a gas tank might have a *capacity* of 12 gallons.

EXAMPLE 8   **Cooking.**   If a recipe calls for 3 pints of milk, how many fluid ounces of milk should be used?

Strategy We will use a two-part multiplication process that converts 3 pints to cups and then converts that result to fluid ounces.

WHY We must use a two-part process because the table above does not contain a single unit conversion factor that converts from pints to fluid ounces.

Self Check 8

Convert 2.5 pints to fluid ounces.

Now Try ➡ **Problem 55**

Solution Since there are 2 cups per pint, we can convert 3 pints to cups by multiplying by the unit conversion factor $\frac{2 c}{1 pt}$. Since there are 8 fluid ounces per cup, we can convert that result to fluid ounces by multiplying by the unit conversion factor $\frac{8\,fl\,oz}{1 c}$.

$$3\ pt = \frac{3\ pt}{1} \cdot \frac{2\ c}{1\ pt} \cdot \frac{8\ fl\ oz}{1\ c}$$

Write 3 pt as a fraction: $3\ pt = \frac{3\ pt}{1}$
Multiply by two unit conversion factors:
$\frac{2 c}{1 pt} = 1$ and $\frac{8\,fl\,oz}{1 c} = 1$.

$$= \frac{3\ p\!\!\!/t}{1} \cdot \frac{2\ \cancel{c}}{1\ p\!\!\!/t} \cdot \frac{8\ fl\ oz}{1\ \cancel{c}}$$

Remove the common units of pints and cups in the numerator and denominator. Notice that all the units are removed except for fluid ounces.

$$= 3 \cdot 2 \cdot 8\ fl\ oz$$ Simplify.

$$= 48\ fl\ oz$$ Multiply.

Since 3 pints is equal to 48 fluid ounces, 48 fluid ounces of milk should be used.

OBJECTIVE 8 Define units of time.

The American system of measurement and the metric system both use the units of **second**, **minute**, **hour**, and **day** to measure time. These units are related as follows.

Units of Time

1 minute (min) = 60 seconds (sec)   1 hour (hr) = 60 minutes

1 day = 24 hours

The abbreviation for each unit is written within parentheses.

To convert units of time, we use the following unit conversion factors. Each conversion factor shown below is a form of 1.

To convert from	Use the unit conversion factor
minutes to seconds	$\frac{60\ sec}{1\ min}$
hours to minutes	$\frac{60\ min}{1\ hr}$
days to hours	$\frac{24\ hr}{1\ day}$

To convert from	Use the unit conversion factor
seconds to minutes	$\frac{1\ min}{60\ sec}$
minutes to hours	$\frac{1\ hr}{60\ min}$
hours to days	$\frac{1\ day}{24\ hr}$

OBJECTIVE 9 Convert from one unit of time to another.

EXAMPLE 9   **Astronomy.**   A lunar eclipse occurs when the Earth is between the sun and the moon in such a way that Earth's shadow darkens the moon. (See the figure below, which is not to scale.) A total lunar eclipse can last as long as 105 minutes. Express this time in hours.

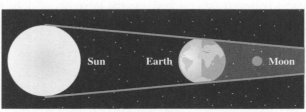

Strategy We will multiply 105 minutes by a carefully chosen unit conversion factor.

WHY If we multiply by the proper unit conversion factor, we can eliminate the unwanted units of minutes and convert to hours.

Solution To convert from minutes to hours, we must choose a unit conversion factor whose numerator contains the units we want to introduce (hours), and whose denominator contains the units we want to eliminate (minutes). Since there is 1 hour for every 60 minutes, we will use

$$\frac{1 \text{ hr}}{60 \text{ min}}$$ ← This is the unit we want to introduce.
← This is the unit we want to eliminate (the original unit).

To perform the conversion, we multiply.

$$105 \text{ min} = \frac{105 \text{ min}}{1} \cdot \frac{1 \text{ hr}}{60 \text{ min}}$$

Write 105 min as a fraction: $105 = \frac{105 \text{ min}}{1}$
Then multiply by a form of 1: $\frac{1 \text{ hr}}{60 \text{ min}}$

$$= \frac{105 \text{ min}}{1} \cdot \frac{1 \text{ hr}}{60 \text{ min}}$$

Remove the common units of minutes in the numerator and denominator. Notice that the units of hours remain.

$$= \frac{105}{60} \text{ hr}$$

Multiply the fractions.

$$= \frac{\overset{1}{3} \cdot \overset{1}{5} \cdot 7}{2 \cdot 2 \cdot \underset{1}{3} \cdot \underset{1}{5}} \text{ hr}$$

To simplify the fraction, prime factor 105 and 60. Then remove the common factors 3 and 5 in the numerator and denominator.

$$= \frac{7}{4} \text{ hr}$$

Multiply the remaining factors in the numerator.
Multiply the remaining factors in the denominator.

$$= 1\frac{3}{4} \text{ hr}$$

Write $\frac{7}{4}$ as a mixed number.

A total lunar eclipse can last as long as $1\frac{3}{4}$ hours.

Self Check 9

The sun. A solar eclipse (eclipse of the sun) can last as long as 450 seconds. Express this time in minutes.

Now Try ➡ Problem 59

Answers to Self Checks

1. $1\frac{7}{8}$ in. **2.** $1\frac{1}{4}$ in. **3.** 27 ft **4.** 18 in. **5.** $\frac{7}{32}$ mi ≈ 0.22 mi **6.** $3\frac{3}{4}$ lb = 3.75 lb **7.** 960 oz
8. 40 fl oz **9.** $7\frac{1}{2}$ min

SECTION 5.3 STUDY SET

VOCABULARY

Fill in the blanks.

1. A ruler is used for measuring _____.

2. Inches, feet, and miles are examples of American units of
_____.

3. $\frac{3 \text{ ft}}{1 \text{ yd}}$, $\frac{1 \text{ ton}}{2,000 \text{ lb}}$, and $\frac{4 \text{ qt}}{1 \text{ gal}}$ are examples of _____ conversion factors.

4. Ounces, pounds, and tons are examples of American units of _____.

5. Some examples of American units of _____ are cups, pints, quarts, and gallons.

6. Some units of _____ are seconds, minutes, hours, and days.

CONCEPTS

Fill in the blanks.

7. **a.** 12 inches = ▢ foot
   b. ▢ feet = 1 yard
   c. 1 yard = ▢ inches
   d. 1 mile = ▢ feet

8. **a.** ▢ ounces = 1 pound
   b. ▢ pounds = 1 ton

9. **a.** 1 cup = ▢ fluid ounces
   b. 1 pint = ▢ cups
   c. ▢ pints = 1 quart
   d. ▢ quarts = 1 gallon

10. **a.** 1 day = ▢ hours
    b. 2 hours = ▢ minutes

11. The value of any unit conversion factor is ▢.

12. In general, a unit conversion factor is a fraction with the following form:

$$\frac{\text{Unit that we want to } \rule{1cm}{0.4pt}}{\text{Unit that we want to } \rule{1cm}{0.4pt}} \begin{array}{l} \leftarrow \text{ Numerator} \\ \leftarrow \text{ Denominator} \end{array}$$

13. Consider the work shown below.

$$\frac{48 \text{ oz}}{1} \cdot \frac{1 \text{ lb}}{16 \text{ oz}}$$

   a. What units can be removed?
   b. What units remain?

14. Consider the work shown below.

$$\frac{600 \text{ yd}}{1} \cdot \frac{3 \text{ ft}}{1 \text{ yd}} \cdot \frac{1 \text{ mi}}{5,280 \text{ ft}}$$

   a. What units can be removed?
   b. What units remain?

15. Write a unit conversion factor to convert
   a. pounds to tons
   b. quarts to pints

16. Write the two unit conversion factors used to convert
   a. inches to yards
   b. days to minutes

17. Match each item with its proper measurement.

   a. Length of the U.S. coastline
   b. Height of a Barbie doll
   c. Span of the Golden Gate Bridge
   d. Width of a football field

   i. $11\frac{1}{2}$ in.
   ii. 4,200 ft
   iii. 53.5 yd
   iv. 12,383 mi

18. Match each item with its proper measurement.

   a. Weight of the men's shot put used in track and field
   b. Weight of an African elephant
   c. Amount of gold that is worth $2,000

   i. $1\frac{1}{2}$ oz
   ii. 16 lb
   iii. 7.2 tons

19. Match each item with its proper measurement.

   a. Amount of blood in an adult
   b. Size of the Exxon Valdez oil spill in 1989
   c. Amount of nail polish in a bottle
   d. Amount of milk to make 2 dozen pancakes

   i. $\frac{1}{2}$ fluid oz
   ii. 2 cups
   iii. 6 qt
   iv. 10,080,000 gal

20. Match each item with its proper measurement.

   a. Length of first U.S. manned space flight
   b. A leap year
   c. Time difference between New York and Fairbanks, Alaska
   d. Length of Wright Brothers' first flight

   i. 12 sec
   ii. 15 min
   iii. 4 hr
   iv. 366 days

NOTATION

21. What unit does each abbreviation represent?
   a. lb   **b.** oz
   c. fl oz

22. What unit does each abbreviation represent?
   a. qt   **b.** c
   c. pt

Complete each step.

23. Convert 2 yards to inches.

$$2 \text{ yd} = \frac{2 \text{ yd}}{1} \cdot \frac{\rule{0.6cm}{0.4pt} \text{ in.}}{1 \text{ yd}}$$

$$= 2 \cdot 36 \;\rule{0.6cm}{0.4pt}$$

$$= \rule{0.6cm}{0.4pt} \text{ in.}$$

24. Convert 24 pints to quarts.

$$24 \text{ pt} = \frac{24 \text{ pt}}{1} \cdot \frac{1 \text{ qt}}{\rule{0.6cm}{0.4pt} \text{ pt}}$$

$$= \frac{24}{1} \cdot \frac{1}{2} \;\rule{0.6cm}{0.4pt}$$

$$= \rule{0.6cm}{0.4pt} \text{ qt}$$

25. Convert 1 ton to ounces.

$$1 \text{ ton} = \frac{1 \text{ ton}}{1} \cdot \frac{\rule{0.8cm}{0.4pt} \text{ lb}}{1 \text{ ton}} \cdot \frac{\rule{0.6cm}{0.4pt} \text{ oz}}{1 \text{ lb}}$$

$$= 1 \cdot 2,000 \cdot 16 \;\rule{0.6cm}{0.4pt}$$

$$= \rule{0.8cm}{0.4pt} \text{ oz}$$

26. Convert 37,440 minutes to days.

$$37,440 \text{ min} = 37,440 \text{ min} \cdot \frac{1 \text{ hr}}{\rule{0.6cm}{0.4pt} \text{ min}} \cdot \frac{1 \text{ day}}{\rule{0.6cm}{0.4pt} \text{ hr}}$$

$$= \frac{37,440}{60 \cdot 24} \;\rule{0.6cm}{0.4pt}$$

$$= \rule{0.6cm}{0.4pt} \text{ days}$$

GUIDED PRACTICE

Refer to the given ruler to answer each question. **See Example 1.**

27. a. Each inch is divided into how many equal parts?
   b. Determine which measurements the arrows point to on the ruler.

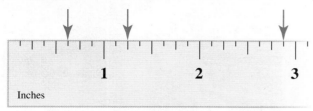

28. Find the length of the needle.

Refer to the given ruler to answer each question. **See Example 2.**

29. a. Each inch is divided into how many equal parts?
   b. Determine which measurements the arrows point to on the ruler.

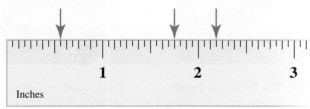

30. Find the length of the bolt.

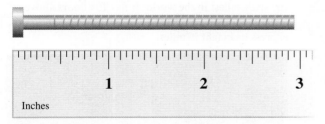

Use a ruler scaled in sixteenths of an inch to measure each object. **See Example 2.**

31. The width of a dollar bill
32. The length of a dollar bill
33. The length (top to bottom) of this page
34. The length of the word as printed here:
   supercalifragilisticexpialidocious

Perform each conversion. **See Example 3.**

35. 4 yards to feet
36. 6 yards to feet
37. 35 yards to feet
38. 33 yards to feet

Perform each conversion. **See Example 4.**

39. $3\frac{1}{2}$ feet to inches
40. $2\frac{2}{3}$ feet to inches
41. $5\frac{1}{4}$ feet to inches
42. $6\frac{1}{2}$ feet to inches

Use two unit conversion factors to perform each conversion. Give the exact answer and a decimal approximation, rounded to the nearest hundredth when necessary. **See Example 5.**

43. 105 yards to miles
44. 198 yards to miles
45. 1,540 yards to miles
46. 1,512 yards to miles

Perform each conversion. **See Example 6.**

47. 44 ounces to pounds
48. 24 ounces to pounds
49. 72 ounces to pounds
50. 76 ounces to pounds

Perform each conversion. **See Example 7.**

51. 50 pounds to ounces
52. 30 pounds to ounces
53. 87 pounds to ounces
54. 79 pounds to ounces

Perform each conversion. **See Example 8.**

55. 8 pints to fluid ounces
56. 5 pints to fluid ounces
57. 21 pints to fluid ounces
58. 30 pints to fluid ounces

Perform each conversion. **See Example 9.**

59. 165 minutes to hours
60. 195 minutes to hours
61. 330 minutes to hours
62. 80 minutes to hours

TRY IT YOURSELF

Perform each conversion.

63. 3 quarts to pints
64. 20 quarts to gallons
65. 7,200 minutes to days
66. 691,200 seconds to days
67. 56 inches to feet
68. 44 in. to ft
69. 4 ft to in.
70. 7 feet to inches
71. 16 pints to gallons
72. 3 gallons to fluid ounces
73. 80 ounces to pounds
74. 8 lbs to oz
75. 240 min to hr
76. 2,400 seconds to hours

77. 8 yards to inches

78. 324 inches to yards

79. 90 inches to yards

80. 12 yd to in.

81. 5 yards to feet

82. 21 feet to yards

83. 12.4 tons to pounds

84. 48,000 ounces to tons

85. 7 ft to yd

86. $4\frac{2}{3}$ yards to feet

87. 15,840 feet to miles

88. 2 mi to ft

89. $\frac{1}{2}$ mi to ft

90. 1,320 feet to miles

91. 7,000 pounds to tons

92. 2.5 tons to ounces

93. 32 fluid ounces to pints

94. 2 quarts to fluid ounces

LOOK ALIKES

Perform each conversion.

95. **a.** 24 yards to inches **b.** 24 inches to yards

96. **a.** 24 pounds to ounces **b.** 24 ounces to pounds

97. **a.** 24 quarts to fluid ounces **b.** 24 fluid ounces to quarts

98. **a.** 24 minutes to hours **b.** 24 hours to minutes

APPLICATIONS

99. **The Great Pyramid.** The Great Pyramid in Egypt is about 450 feet high. Express this distance in yards.

bumihills/Shutterstock.com

100. **The Wright Brothers.** In 1903, Orville Wright made the world's first sustained flight. It lasted 12 seconds, and the plane traveled 120 feet. Express the length of the flight in yards.

Everett Historical/Shutterstock.com

101. **The Great Sphinx.** The Great Sphinx of Egypt is 240 feet long. Express this in inches.

Plus Lee/Shutterstock.com

102. **Hoover Dam.** The Hoover Dam in Nevada is 726 feet high. Express this distance in inches.

Andrew Zarivny/Shutterstock.com

103. **The Freedom Tower.** One World Trade Center (also known as the Freedom Tower) is the main building of the rebuilt World Trade Center complex in New York City. It is the tallest building in the Western Hemisphere, and the sixth-tallest in the world. It has 104 floors above ground and is 1,776 feet tall. To the nearest hundredth, express its height in miles.

meunierd/Shutterstock.com

104. **NFL records.**
   a. Emmit Smith, the former Dallas Cowboys and Arizona Cardinals running back, holds the National Football League record for yards rushing in a career: 18,355. How many miles is this? Round to the nearest tenth of a mile.

b. Peyton Manning, the former Indianapolis Colt and Denver Bronco, holds the National Football League record for yards passing in a career. When he retired, his passing total was exactly 40.875 miles. How many yards is this?

105. Limos.   The world's longest limousine is the American Dream designed by Jay Ohrberg. It runs on 26 wheels and features a king-sized waterbed, swimming pool with a diving board, Jacuzzi tub, sun deck, and a helipad. It can actually bend in the middle when going around corners. It is $33\frac{1}{3}$ yards long. How many feet is this?

106. Lewis and Clark.   The trail traveled by the Lewis and Clark expedition is shown below. When the expedition reached the Pacific Ocean, Clark estimated that they had traveled 4,162 miles. (It was later determined that his guess was within 40 miles of the actual distance.) Express Clark's estimate of the distance in feet.

107. Weight of water.   One gallon of water weighs about 8 pounds. Express this weight in ounces.

108. Weight of a baby.   A newborn baby boy weighed 136 ounces. Express this weight in pounds.

109. Hippos.   An adult hippopotamus can weigh as much as 9,900 pounds. Express this weight in tons.

110. Elephants.   An adult elephant can consume as much as 495 pounds of grass and leaves in one day. How many ounces is this?

111. Buying paint.   A painter estimates that he will need 17 gallons of paint for a job. To take advantage of a closeout sale on quart cans, he decides to buy the paint in quarts. How many cans will he need to buy?

112. Catering.   How many cups of apple cider are there in a 10-gallon container of cider?

113. School lunches.   Each student attending Eagle River Elementary School receives 1 pint of milk for lunch each day. If 575 students attend the school, how many gallons of milk are used each day?

114. Radiators.   The radiator capacity of a piece of earth-moving equipment is 39 quarts. If the radiator is drained and new coolant put in, how many gallons of new coolant will be used?

115. Camping.   How many ounces of camping stove fuel will fit in the container shown?

116. Hiking.   A college student walks 11 miles in 155 minutes. To the nearest tenth, how many hours does he walk?

117. Space travel.   The astronauts of the *Apollo 8* mission, which was launched on December 21, 1968, were in space for 147 hours. How many days did the mission take?

118. Amelia Earhart.   In 1935, Amelia Earhart became the first woman to fly across the Atlantic Ocean alone, establishing a new record for the crossing: 13 hours and 30 minutes. How many minutes is this?

WRITING

119. a. Explain how to find the unit conversion factor that will convert feet to inches.

   b. Explain how to find the unit conversion factor that will convert pints to gallons.

120. Explain why the unit conversion factor $\frac{1\,lb}{16\,oz}$ is a form of 1.

REVIEW

121. Multiply: $2.09 \cdot 100$ **122.** Multiply: $0.0546 \cdot 1,000$

123. Multiply: $2.83(0.001)$ **124.** Multiply: $72.361(0.01)$

SECTION **5.4** Metric Units of Measurement; Medical Applications

The metric system is the system of measurement used by most countries in the world. All countries, including the United States, use it for scientific purposes. The metric system, like our decimal numeration system, is based on the number 10. For this reason, converting from one metric unit to another is easier than with the American system.

OBJECTIVE 1 Define metric units of length.

The basic metric unit of length is the **meter** (m). One meter is approximately 39 inches, which is slightly more than 1 yard. The figure below compares the length of a yardstick to a meterstick.

1 yard: 36 inches

1 meter: about 39 inches

Longer and shorter metric units of length are created by adding **prefixes** to the front of the basic unit, *meter*.

kilo means thousands	*deci* means tenths
hecto means hundreds	*centi* means hundredths
deka means tens	*milli* means thousandths

Metric Units of Length

Prefix	**kilo**meter	**hecto**meter	**deka**meter	**meter**	**deci**meter	**centi**meter	**milli**meter
Meaning	1,000 meters	100 meters	10 meters	1 meter	$\frac{1}{10}$ or 0.1 of a meter	$\frac{1}{100}$ or 0.01 of a meter	$\frac{1}{1,000}$ or 0.001 of a meter
Abbreviation	km	hm	dam	m	dm	cm	mm

LANGUAGE OF MATHEMATICS

It is helpful to memorize the **prefixes** listed above because they are also used with metric units of weight and capacity.

The most often used metric units of length are kilometers, meters, centimeters, and millimeters. It is important that you gain a practical understanding of metric lengths just as you have for the length of an inch, a foot, and a mile. Some examples of metric lengths are shown below.

1 kilometer is about the length of 60 train cars.

1 m

1 meter is about the distance from a doorknob to the floor.

1 cm

1 centimeter is about as wide as the nail on your little finger.

1 mm

1 millimeter is about the thickness of a dime.

OBJECTIVE 2 **Use a metric ruler to measure lengths.**

Parts of a metric ruler, scaled in centimeters, and a ruler scaled in inches are shown below. Several lengths on the metric ruler are highlighted.

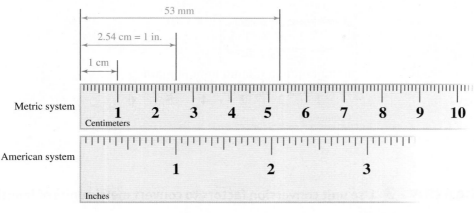

(Actual size)

EXAMPLE 1   Find the length of the nail shown below.

Strategy We will place a metric ruler below the nail, with the left end of the ruler (which could be thought of as 0) directly underneath the head of the nail.

WHY Then we can find the length of the nail by identifying where its pointed end lines up on the tick marks printed in black on the ruler.

Solution The longest tick marks on the ruler (those labeled with numbers) mark lengths in centimeters. Since the pointed end of the nail lines up on 6, the nail is 6 centimeters long.

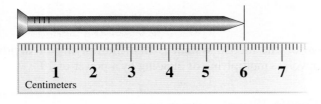

Self Check 1

To the nearest centimeter, find the width of the circle.

Now Try ⟳ Problem 24

EXAMPLE 2   Find the length of the paper clip shown below.

Strategy We will place a metric ruler below the paper clip, with the left end of the ruler (which could be thought of as 0) directly underneath one end of the paper clip.

WHY Then we can find the length of the paper clip by identifying where its other end lines up on the tick marks printed in black on the ruler.

Solution On the ruler, the shorter tick marks divide each centimeter into 10 millimeters, as shown. If we begin at the left end of the ruler and count by tens as we move right to 3, and then add an additional 6 millimeters to that result, we find that the length of the paper clip is 30 + 6 = 36 millimeters.

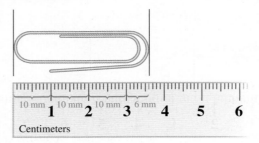

Self Check 2

Find the length of the jumbo paper clip.

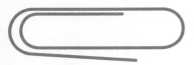

Now Try ➲ **Problem 26**

OBJECTIVE 3 Use unit conversion factors to convert metric units of length.

Metric units of length are related as shown in the following table.

Metric Units of Length	
1 kilometer (km) = 1,000 meters	1 meter = 10 decimeters (dm)
1 hectometer (hm) = 100 meters	1 meter = 100 centimeters (cm)
1 dekameter (dam) = 10 meters	1 meter = 1,000 millimeters (mm)
The abbreviation for each unit is written within parentheses.	

We can use the information in the table to write unit conversion factors that can be used to convert metric units of length. For example, in the table we see that

$$1 \text{ meter} = 100 \text{ centimeters}$$

From this fact, we can write two unit conversion factors.

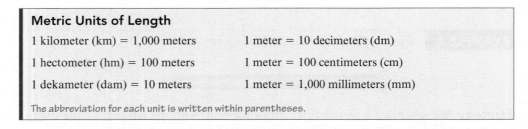

$$\frac{1 \text{ m}}{100 \text{ cm}} = 1 \qquad \text{and} \qquad \frac{100 \text{ cm}}{1 \text{ m}} = 1$$

To obtain the first unit conversion factor, divide both sides of the equation 1 m = 100 cm by 100 cm. To obtain the second unit conversion factor, divide both sides by 1 m.

One advantage of the metric system is that multiplying or dividing by a unit conversion factor involves multiplying or dividing by a power of 10.

EXAMPLE 3 Convert 350 centimeters to meters.

Strategy We will multiply 350 centimeters by a carefully chosen unit conversion factor.

WHY If we multiply by the proper unit conversion factor, we can eliminate the unwanted units of centimeters and convert to meters.

Solution To convert from centimeters to meters, we must choose a unit conversion factor whose numerator contains the units we want to introduce (meters), and whose denominator contains the units we want to eliminate (centimeters). Since there is 1 meter for every 100 centimeters, we will use

$$\frac{1 \text{ m}}{100 \text{ cm}}$$ ← This is the unit we want to introduce.
← This is the unit we want to eliminate (the original unit).

To perform the conversion, we multiply 350 centimeters by the unit conversion factor $\frac{1 \text{ m}}{100 \text{ cm}}$.

$$350 \text{ cm} = \frac{350 \text{ cm}}{1} \cdot \boxed{\frac{1 \text{ m}}{100 \text{ cm}}}$$
Write 350 cm as a fraction: $350 \text{ cm} = \frac{350 \text{ cm}}{1}$
Multiply by a form of 1: $\frac{1 \text{ m}}{100 \text{ cm}}$

$$= \frac{350 \text{ c\!\!\!\!/m}}{1} \cdot \frac{1 \text{ m}}{100 \text{ c\!\!\!\!/m}}$$
Remove the common units of centimeters from the numerator and denominator. Notice that the units of meter remain.

$$= \frac{350}{100} \text{ m}$$
Multiply the fractions.

$$= \frac{350.0}{100} \text{ m}$$
Write the whole number 350 as a decimal by placing a decimal point immediately to its right and entering a zero: $350 = 350.0$

$$= 3.5 \text{ m}$$
Divide 350.0 by 100 by moving the decimal point 2 places to the left: 3.500.

Thus, 350 centimeters = 3.5 meters.

Self Check 3

Convert 860 centimeters to meters.

Now Try ➲ **Problem 31**

OBJECTIVE 4 **Use a conversion chart to convert metric units of length.**

In Example 3, we converted 350 centimeters to meters using a unit conversion factor. We can also make this conversion by recognizing that all units of length in the metric system are powers of 10 of a meter.

To see this, review the table of metric units of length on page 452. Note that each unit has a value that is $\frac{1}{10}$ of the value of the unit immediately to its left and 10 times the value of the unit immediately to its right. Converting from one unit to another is as easy as multiplying (or dividing) by the correct power of 10 or simply moving a decimal point the correct number of places to the right (or left). For example, in the **conversion chart** below, we see that to convert from centimeters to meters, we move two places to the left.

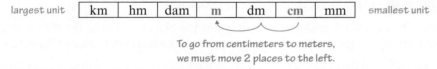

largest unit | km | hm | dam | **m** | dm | cm | mm | smallest unit

To go from centimeters to meters, we must move 2 places to the left.

If we write 350 centimeters as 350.0 centimeters, we can convert to meters by moving the decimal point two places to the left.

350.0 centimeters = 3.500 meters = 3.5 meters

Move 2 places to the left.

With the unit conversion factor method or the conversion chart method, we get 350 cm = 3.5 m.

Caution! When using a chart to help make a metric conversion, be sure to list the units from *largest to smallest* when reading from left to right.

EXAMPLE 4   Convert 2.4 meters to millimeters.

Strategy On a conversion chart, we will count the places and note the direction as we move from the original units of meters to the conversion units of millimeters.

WHY The decimal point in 2.4 must be moved the same number of places and in that same direction to find the conversion to millimeters.

Solution To construct a conversion chart, we list the metric units of length from largest (kilometers) to smallest (millimeters), working from left to right. Then we locate the original units of meters and move to the conversion units of millimeters, as shown below.

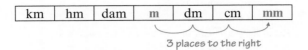

3 places to the right

We see that the decimal point in 2.4 should be moved three places to the right to convert from meters to millimeters.

2.4 meters = 2400. millimeters = 2,400 millimeters

Move 3 places to the right.

We can use the unit conversion factor method to confirm this result. Since there are 1,000 millimeters per meter, we multiply 2.4 meters by the unit conversion factor $\frac{1,000 \text{ mm}}{1 \text{ m}}$.

$$2.4 \text{ m} = \frac{2.4 \text{ m}}{1} \cdot \boxed{\frac{1,000 \text{ mm}}{1 \text{ m}}}$$

Write 2.4 m as a fraction: $2.4 \text{ m} = \frac{2.4 \text{ m}}{1}$
Multiply by a form 1: $\frac{1,000 \text{ mm}}{1 \text{ m}}$

$$= \frac{2.4 \text{ m\!\!\!/}}{1} \cdot \frac{1,000 \text{ mm}}{1 \text{ m\!\!\!/}}$$

Remove the common units of meters from the numerator and denominator. Notice that the units of millimeters remain.

$$= 2.4 \cdot 1,000 \text{ mm}$$

Multiply the fractions and simplify.

$$= 2,400 \text{ mm}$$

Multiply 2.4 by 1,000 by moving the decimal point 3 places to the right: 2 400.

Self Check 4

Convert 5.3 meters to millimeters.

Now Try Problem 35

EXAMPLE 5   Convert 3.2 centimeters to kilometers.

Strategy On a conversion chart, we will count the places and note the direction as we move from the original units of centimeters to the conversion units of kilometers.

WHY The decimal point in 3.2 must be moved the same number of places and in that same direction to find the conversion to kilometers.

Solution We locate the original units of centimeters on a conversion chart and then move to the conversion units of kilometers, as shown below.

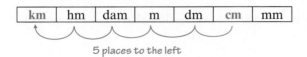

5 places to the left

We see that the decimal point in 3.2 should be moved five places to the left to convert centimeters to kilometers.

3.2 centimeters = 0.000032 kilometers = 0.000032 kilometers

Move 5 places to the left.

We can use the unit conversion factor method to confirm this result. To convert to kilometers, we must use two unit conversion factors so that the units of centimeters drop out and the units of kilometers remain. Since there is 1 meter for every 100 centimeters and 1 kilometer for every 1,000 meters, we multiply by $\frac{1 \text{ m}}{100 \text{ cm}}$ and $\frac{1 \text{ km}}{1,000 \text{ m}}$.

$$3.2 \text{ cm} = \frac{3.2 \text{ cm}}{1} \cdot \frac{1 \text{ m}}{100 \text{ cm}} \cdot \frac{1 \text{ km}}{1,000 \text{ m}}$$

Remove the common units of centimeters and meters. The units of km remain.

$$= \frac{3.2}{100 \cdot 1,000} \text{ km}$$

Multiply the fractions.

$$= 0.000032 \text{ km}$$

Divide 3.2 by 1,000 and 100 by moving the decimal point 5 places to the left.

Self Check 5

Convert 5.15 centimeters to kilometers.

Now Try ➡ **Problem 39**

OBJECTIVE 5 **Define metric units of mass.**

The **mass** of an object is a measure of the amount of material in the object. When an object is moved about in space, its mass does not change. The basic unit of mass in the metric system is the **gram** (g). A gram is defined to be the mass of water contained in a cube having sides 1 centimeter long. (See the figure below.)

1 cubic centimeter of water

1 g

Other units of mass are created by adding prefixes to the front of the basic unit, *gram*.

Metric Units of Mass

Prefix	kilo**gram**	hecto**gram**	deka**gram**	**gram**	deci**gram**	centi**gram**	milli**gram**
Meaning	1,000 grams	100 grams	10 grams	1 gram	$\frac{1}{10}$ or 0.1 of a gram	$\frac{1}{100}$ or 0.01 of a gram	$\frac{1}{1,000}$ or 0.001 of a gram
Abbreviation	kg	hg	dag	g	dg	cg	mg

The most often used metric units of mass are kilograms, grams, and milligrams. Some examples are shown below.

An average bowling ball weighs about 6 kilograms.

A raisin weighs about 1 gram.

A certain vitamin tablet contains 450 milligrams of calcium.

The **weight** of an object is determined by the Earth's gravitational pull on the object. Since gravitational pull on an object decreases as the object gets farther from Earth, the object weighs less as it gets farther from Earth's surface. This is why astronauts experience weightlessness in space. However, since most of us remain near Earth's surface, we will use the words *mass* and *weight* interchangeably. Thus, a mass of 30 grams is said to weigh 30 grams.

Metric units of mass are related as shown in the following table.

Metric Units of Mass	
1 kilogram (kg) = 1,000 grams	1 gram = 10 decigrams (dg)
1 hectogram (hg) = 100 grams	1 gram = 100 centigrams (cg)
1 dekagram (dag) = 10 grams	1 gram = 1,000 milligrams (mg)

The abbreviation for each unit is written within parentheses.

We can use the information in the table to write unit conversion factors that can be used to convert metric units of mass. For example, in the table we see that

1 kilogram = 1,000 grams

From this fact, we can write two unit conversion factors.

$$\frac{1 \text{ kg}}{1,000 \text{ g}} = 1 \quad \text{and} \quad \frac{1,000 \text{ g}}{1 \text{ kg}} = 1$$

To obtain the first unit conversion factor, divide both sides of the equation 1 kg = 1,000 g by 1,000 g. To obtain the second unit conversion factor, divide both sides by 1 kg.

Another metric unit of mass is the **microgram**, which is represented by the notation mcg. One microgram is related to the gram by

1 gram = 1,000,000 micrograms

This fact gives us the following conversion factors

$$\frac{1 \text{ g}}{1,000,000 \text{ mcg}} = 1 \quad \text{and} \quad \frac{1,000,000 \text{ mcg}}{1 \text{ g}} = 1$$

OBJECTIVE 6 Convert from one metric unit of mass to another.

EXAMPLE 6 Convert 7.86 kilograms to grams.

Strategy On a conversion chart, we will count the places and note the direction as we move from the original units of kilograms to the conversion units of grams.

WHY The decimal point in 7.86 must be moved the same number of places and in that same direction to find the conversion to grams.

Solution To construct a conversion chart, we list the metric units of mass from largest (kilograms) to smallest (milligrams), from left to right. Then we locate the original units of kilograms and move to the conversion units of grams, as shown below.

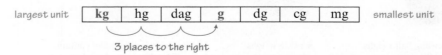

largest unit | kg | hg | dag | g | dg | cg | mg | smallest unit

3 places to the right

We see that the decimal point in 7.86 should be moved three places to the right to change kilograms to grams.

7.86 kilograms = 7860. grams = 7,860 grams

Move 3 places to the right.

We can use the unit conversion factor method to confirm this result. To convert to grams, we must choose a unit conversion factor such that the units of kilograms drop out and the units of grams remain. Since there are 1,000 grams per 1 kilogram, we multiply 7.86 kilograms by $\frac{1,000 \text{ g}}{1 \text{ kg}}$.

$$7.86 \text{ kg} = \frac{7.86 \text{ kg}}{1} \cdot \frac{1,000 \text{ g}}{1 \text{ kg}}$$

Remove the common units of kilograms in the numerator and denominator. The units of g remain.

$$= 7.86 \cdot 1,000 \text{ g}$$

Simplify.

$$= 7,860 \text{ g}$$

Multiply 7.86 by 1,000 by moving the decimal point 3 places to the right.

Self Check 6

Convert 5.83 kilograms to grams.

Now Try ➲ **Problem 43**

EXAMPLE 7   **Medications.** A bottle of Verapamil, a drug taken for high blood pressure, contains 30 tablets. If each tablet has 180 mg of active ingredient, how many grams of active ingredient are in the bottle?

sirtravelalot/Shutterstock.com

Strategy We will multiply the number of tablets in one bottle by the number of milligrams of active ingredient in each tablet.

WHY We need to know the total number of milligrams of active ingredient in one bottle before we can convert that number to grams.

```
   180
 × 30
   000
 5400
 5,400
```

Solution Since there are 30 tablets, and each one contains 180 mg of active ingredient, there are

$$30 \cdot 180 \text{ mg} = 5,400 \text{ mg}$$

of active ingredient in the bottle. To use a conversion chart to solve this problem, we locate the original units of milligrams and then move to the conversion units of grams, as shown below.

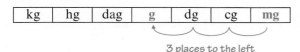

kg	hg	dag	g	dg	cg	mg

3 places to the left

We see that the understood decimal point in 5,400 should be moved three places to the left to convert from milligrams to grams.

$$5,400 \text{ milligrams} = 5.400 \text{ grams}$$ Think: 5,400 = 5,400.0.

Move 3 places to the left.

There are 5.4 grams of active ingredient in the bottle.

We can use the unit conversion factor method to confirm this result. To convert milligrams to grams, we multiply 5,400 milligrams by $\frac{1 \text{ g}}{1,000 \text{ mg}}$.

$$5,400 \text{ mg} = \frac{5,400 \text{ mg}}{1} \cdot \frac{1 \text{ g}}{1,000 \text{ mg}}$$

Remove the common units of milligrams from the numerator and denominator. The units of g remain.

$$= \frac{5,400}{1,000} \text{ g}$$

Multiply the fractions.

$$= 5.4 \text{ g}$$

Divide 5,400 by 1,000 by moving the understood decimal point in 5,400 three places to the left.

Self Check 7

Medications. A bottle of Isoptin (a drug taken for high blood pressure) contains 90 tablets, and each has 200 mg of active ingredient. How many grams of active ingredient are in the bottle?

Now Try ➲ **Problems 47 and 99**

OBJECTIVE 7 Define metric units of capacity.

In the metric system, the basic unit of capacity is the **liter** (L), which is defined to be the capacity of a cube with sides 10 centimeters long. Other units of capacity are created by adding prefixes to the front of the basic unit, liter.

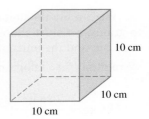

Metric Units of Capacity

Prefix	**kilo**liter	**hecto**liter	**deka**liter	**liter**	**deci**liter	**centi**liter	**milli**liter
Meaning	1,000 liters	100 liters	10 liters	1 liter	$\frac{1}{100}$ or 0.1 of a liter	$\frac{1}{100}$ or 0.01 of a liter	$\frac{1}{1,000}$ or 0.001 of a liter
Abbreviation	kL	hL	daL	L	dL	cL	mL

The most often used metric units of capacity are liters and milliliters. Here are some examples.

Soft drinks are sold in 2-liter plastic bottles.

The fuel tank of a minivan can hold about 75 liters of gasoline.

A teaspoon holds about 5 milliliters.

Metric units of capacity are related as shown in the following table.

Metric Units of Capacity

1 kiloliter (kL) = 1,000 liters   1 liter = 10 deciliters (dL)

1 hectoliter (hL) = 100 liters   1 liter = 100 centiliters (cL)

1 dekaliter (daL) = 10 liters   1 liter = 1,000 milliliters (mL)

The abbreviation for each unit is written within parentheses.

We can use the information in the table to write unit conversion factors that can be used to convert metric units of capacity. For example, in the table we see that

1 liter = 1,000 milliliters

From this fact, we can write two unit conversion factors.

$$\frac{1 \text{ L}}{1,000 \text{ mL}} = 1 \qquad \text{and} \qquad \frac{1,000 \text{ mL}}{1 \text{ L}} = 1$$

OBJECTIVE 8 **Convert from one metric unit of capacity to another.**

EXAMPLE 8 **Soft drinks.** How many milliliters are in three 2-liter bottles of soda?

Strategy We will multiply the number of bottles of soda by the number of liters of soda in each bottle.

WHY We need to know the total number of liters of soda before we can convert that number to milliliters.

Solution Since there are three bottles, and each contains 2 liters of soda, there are

$$3 \cdot 2 \text{ L} = 6 \text{ L}$$

of soda in the bottles. To construct a conversion chart, we list the metric units of capacity from largest (kiloliters) to smallest (milliliters), working from left to right. Then we locate the original units of liters and move to the conversion units of milliliters, as shown below.

largest unit | kL | hL | daL | L | dL | cL | mL | smallest unit

3 places to the right

We see that the understood decimal point in 6 should be moved three places to the right to convert from liters to milliliters.

$$6 \text{ liters} = 6000. \text{ milliliters} = 6{,}000 \text{ milliliters}$$

Move 3 places to the right.

Thus, there are 6,000 milliliters in three 2-liter bottles of soda.

We can use the unit conversion factor method to confirm this result. To convert to milliliters, we must choose a unit conversion factor such that liters drop out and the units of milliliters remain. Since there are 1,000 milliliters per 1 liter, we multiply 6 liters by the unit conversion factor $\frac{1{,}000 \text{ mL}}{1 \text{ L}}$.

$$6 \text{ L} = \frac{6 \cancel{\text{ L}}}{1} \cdot \frac{1{,}000 \text{ mL}}{1 \cancel{\text{ L}}}$$ Remove the common units of liters in the numerator and denominator. The units of mL remain.

$$= 6 \cdot 1{,}000 \text{ mL}$$ Simplify.

$$= 6{,}000 \text{ mL}$$ Multiply 6 by 1,000 by moving the understood decimal point in 6 three places to the right.

Self Check 8

Soft drinks. How many milliliters are in a case of twelve 2-liter bottles of soda?

Now Try ➲ Problems 51 and 101

OBJECTIVE 9 **Use conversions in medical applications.**

Another metric unit of capacity is the **cubic centimeter**, which is represented by the notation cm^3 or, more simply, cc. One milliliter and one cubic centimeter represent the same capacity.

$$1 \text{ mL} = 1 \text{ cm}^3 = 1 \text{ cc}$$

The units of cubic centimeters are used frequently in medicine. For example, when a nurse administers an injection containing 5 cc of medication, the dosage can also be expressed using milliliters.

$$5 \text{ cc} = 5 \text{ mL}$$

EXAMPLE 9   **Medical dosage.**   A doctor orders that a patient be given 1,000 cc of dextrose solution. How many liters were ordered?

Strategy On a conversion chart, we will count the places and note the direction as we move from the original units of cubic centimeters, which is the same as milliliters, to the conversion unit of liters.

WHY The decimal point in 1,000 must be moved the same number of places and in the same direction to find the conversion to liters.

Solution   The conversion chart for capacity is

kL	hL	daL	L	dL	cL	mL	cc

3 places to the right

We see the decimal point in 1,000 should be moved three places to the left to convert from milliliters to liters.

$$1,000 \text{ cc} = 1,000 \text{ mL} = 1,000 \text{ L} = 1 \text{ L}$$

Move 3 places to the left.

Self Check 9

Medical dosage. A doctor orders that a patient be given 2.5 liters of a 5% dextrose solution. How many cubic centimeters were ordered?

Now Try ➡ **Problem 106**

Answers to Self Checks

1. 3 cm   **2.** 47 mm   **3.** 8.6 m   **4.** 5,300 mm   **5.** 0.0000515 km   **6.** 5,830 g   **7.** 18 g
8. 24,000 mL   **9.** 2,500 cc

SECTION 5.4   **STUDY SET**

VOCABULARY

Fill in the blanks.

1. The meter, the gram, and the liter are basic units of measurement in the _____ system.

2. **a.** The basic unit of length in the metric system is the _____ .

   b. The basic unit of mass in the metric system is the _____ .

   c. The basic unit of capacity in the metric system is the _____ .

3. **a.** *Deka* means _____ .

   b. *Hecto* means _____ .

   c. *Kilo* means _____ .

4. **a.** *Deci* means _____ .

   b. *Centi* means _____ .

   c. *Milli* means _____ .

5. We can convert from one unit to another in the metric system using _____ conversion factors or a conversion _____ like that shown below.

km	hm	dam	m	dm	cm	mm

6. The _____ of an object is a measure of the amount of material in the object.

7. The _____ of an object is determined by the Earth's gravitational pull on the object.

8. Another metric unit of capacity is the cubic _____ , which is represented by the notation cm^3, or, more simply, cc.

CONCEPTS

Fill in the blanks.

9. **a.** 1 kilometer = _____ meters

   b. _____ centimeters = 1 meter

   c. _____ millimeters = 1 meter

10. **a.** 1 gram = _____ milligrams

    b. 1 kilogram = _____ grams

11. **a.** _____ milliliters = 1 liter

    b. 1 dekaliter = _____ liters

12. **a.** 1 milliliter = _____ cubic centimeter

    b. 1 liter = _____ cubic centimeters

13. Write a unit conversion factor to convert

    a. meters to kilometers

    b. grams to centigrams

    c. liters to milliliters

14. Use the chart to determine how many decimal places and in which direction to move the decimal point when converting the following.

a. Kilometers to centimeters

km	hm	dam	m	dm	cm	mm

b. Milligrams to grams

kg	hg	dag	g	dg	cg	mg

c. Hectoliters to centiliters

kL	hL	daL	L	dL	cL	mL

15. Match each item with its proper measurement.

a. Thickness of a phone book

b. Length of the Amazon River

c. Height of a soccer goal

i. 6,275 km

ii. 2 m

iii. 6 cm

16. Match each item with its proper measurement.

a. Weight of a giraffe

b. Weight of a paper clip

c. Active ingredient in an aspirin tablet

i. 800 kg

ii. 1 g

iii. 325 mg

17. Match each item with its proper measurement.

a. Amount of blood in an adult

b. Soda in an aluminum can

c. Kuwait's daily production of crude oil

i. 290,000 kL

ii. 5 L

iii. 355 mL

18. Of the objects shown below, which can be used to measure the following?

a. Millimeters

b. Milligrams

c. Milliliters

Balance

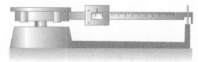

Beaker

Micrometer

Complete each step.

19. Convert 20 centimeters to meters.

$$20 \text{ cm} = \frac{20 \text{ cm}}{1} \cdot \frac{\boxed{} \text{ m}}{100 \text{ cm}}$$

$$= \frac{20}{\boxed{}} \text{ m}$$

$$= \boxed{} \text{ m}$$

20. Convert 3,000 milligrams to grams.

$$3{,}000 \text{ mg} = \frac{3{,}000 \text{ mg}}{1} \cdot \frac{1 \text{ g}}{1{,}000 \boxed{}}$$

$$= \frac{3{,}000}{1{,}000} \boxed{}$$

$$= \boxed{} \text{ g}$$

21. Convert 0.2 kilograms to milligrams.

$$0.2 \text{ kg} = \frac{0.2 \text{ kg}}{1} \cdot \frac{\boxed{} \text{ g}}{1 \text{ kg}} \cdot \frac{1{,}000 \text{ mg}}{\boxed{} \text{ g}}$$

$$= 0.2 \cdot 1{,}000 \cdot 1{,}000 \boxed{}$$

$$= \boxed{} \text{ mg}$$

22. Convert 400 milliliters to kiloliters.

$$400 \text{ mL} = \frac{400 \text{ mL}}{1} \cdot \frac{1 \text{ L}}{\boxed{} \text{ mL}} \cdot \frac{1}{1{,}000 \text{ L}}$$

$$= \frac{\boxed{}}{1{,}000 \cdot 1{,}000} \text{ kL}$$

$$= 0.0004 \text{ kL}$$

Refer to the given ruler to answer each question. See Example 1.

23. Determine which measurements the arrows point to on the metric ruler.

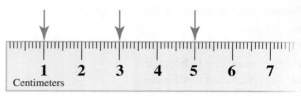

24. Find the length of the birthday candle (including the wick).

Refer to the given ruler to answer each question.
See Example 2.

25. a. Each centimeter is divided into how many equal parts? What is the length of one of those parts?

   b. Determine which measurements the arrows point to on the ruler.

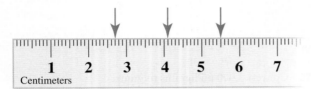

26. Find the length of the stick of gum.

Use a metric ruler scaled in millimeters to measure each object. See Example 2.

27. The length of a dollar bill

28. The width of a dollar bill

29. The length (top to bottom) of this page

30. The length of the word antidisestablishmentarianism as printed here.

Perform each conversion. See Example 3.

31. 380 centimeters to meters

32. 590 centimeters to meters

33. 120 centimeters to meters

34. 640 centimeters to meters

Perform each conversion. See Example 4.

35. 8.7 meters to millimeters

36. 1.3 meters to millimeters

37. 2.89 meters to millimeters

38. 4.06 meters to millimeters

Perform each conversion. See Example 5.

39. 4.5 centimeters to kilometers

40. 6.2 centimeters to kilometers

41. 0.3 centimeters to kilometers

42. 0.4 centimeters to kilometers

Perform each conversion. See Example 6.

43. 1.93 kilograms to grams

44. 8.99 kilograms to grams

45. 4.531 kilograms to grams

46. 6.077 kilograms to grams

Perform each conversion. See Example 7.

47. 6,000 milligrams to grams

48. 9,000 milligrams to grams

49. 3,500 milligrams to grams

50. 7,500 milligrams to grams

Perform each conversion. See Example 8.

51. 3 liters to milliliters

52. 4 liters to milliliters

53. 26.3 liters to milliliters

54. 35.2 liters to milliliters

TRY IT YOURSELF

Perform each conversion.

55. 0.31 decimeters to centimeters

56. 73.2 meters to decimeters

57. 500 milliliters to liters

58. 500 centiliters to milliliters

59. 2 kilograms to grams

60. 4,000 grams to kilograms

61. 0.074 centimeters to millimeters

62. 0.125 meters to millimeters

63. 1,000 kilograms to grams

64. 2 kilograms to centigrams

65. 658.23 liters to kiloliters

66. 0.0068 hectoliters to kiloliters

67. 4.72 cm to dm

68. 0.593 cm to dam

69. 10 mL = ▨ cc

70. 2,000 cc = ▨ L

71. 1,250 mcg to mg

72. 500 mcg to mg

73. 5,689 g to kg

74. 0.0579 km to mm

75. 453.2 cm to m

76. 675.3 cm to m

77. 0.325 dL to L

78. 0.0034 mL to L

79. 675 dam = ▨ cm

80. 76.8 hm = ▨ mm

81. 0.00777 cm = ▨ dam

82. 400 L = ▨ hL

83. 134 m to hm

84. 6.77 mm to cm

85. 65.78 km to dam

86. 5 g to cg

LOOK ALIKES

Perform each conversion.

87. a. 4.5 cm to m   **b.** 4.5 m to cm

88. a. 4.5 g to cg   **b.** 4.5 cg to g

89. a. 4.5 mL to L   **b.** 4.5 L to mL

90. a. 4.5 dm to mm   **b.** 4.5 mm to dm

APPLICATIONS

91. Speed skating. American Eric Heiden won an unprecedented five gold medals by capturing the men's 500 m, 1,000 m, 1,500 m, 5,000 m, and 10,000 m races at the 1980 Winter Olympic Games in Lake Placid, New York. Convert each race length to kilometers.

92. The Suez Canal. The 163-km-long Suez Canal connects the Mediterranean Sea with the Red Sea. It provides a shortcut for ships operating between European and American ports. Convert the length of the Suez Canal to meters.

93. Hair extensions. Custom clip-in women's hair extensions are available in lengths from 31 cm to 91 cm on a popular website. Convert these lengths to meters.

94. Weight of a baby. A baby weighs 4 kilograms. Give this weight in centigrams.

95. Health care. Blood pressure is measured by a sphygmomanometer *(as shown below)*. The measurement is read at two points and is expressed, for example, as 120/80. This indicates a *systolic* pressure of 120 millimeters of mercury and a *diastolic* pressure of 80 millimeters of mercury. Convert each measurement to centimeters of mercury.

96. Jewelry. A gold chain weighs 1,500 milligrams. Give this weight in grams.

97. Eye droppers. One drop from an eye dropper is 0.05 mL. Convert the capacity of one drop to liters.

98. Bottling. How many liters of wine are in a 750-mL bottle?

99. Medicine. A bottle of hydrochlorothiazide contains 60 tablets. If each tablet contains 50 milligrams of active ingredient, how many grams of active ingredient are in the bottle?

100. Ibuprofen. What is the total number of grams of ibuprofen contained in all the tablets in the box shown at the right?

101. Six packs. Some stores sell Fanta orange soda in 0.5-liter bottles. How many milliliters are there in a six pack of this size bottle?

102. Containers. How many deciliters of root beer are in two 2-liter bottles?

103. Olives. The net weight of a bottle of olives is 284 grams. Find the smallest number of bottles that must be purchased to have at least 1 kilogram of olives.

104. Injections. The illustration below shows a 3 cc syringe. Express its capacity using units of milliliters.

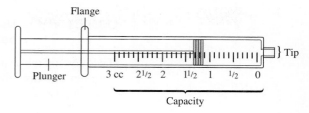

105. Medical dosage. A doctor ordered 1,500 cc of a saline solution from a pharmacy. How many liters of saline solution is this?

106. Medical dosage. A doctor ordered 2,000 cc of a saline (salt) solution from a pharmacy. How many liters of saline solution is this?

WRITING

107. To change 3.452 kilometers to meters, we can move the decimal point in 3.452 three places to the right to get 3,452 meters. Explain why.

108. To change 7,532 grams to kilograms, we can move the decimal point in 7,532 three places to the left to get 7.532 kilograms. Explain why.

109. A *centimeter* is one hundredth of a meter. Make a list of five other words that begin with the prefix *centi* or *cent* and write a definition for each.

110. List the advantages of the metric system of measurement as compared to the American system. There have been several attempts to bring the metric system into general use in the United States. Why do you think these efforts have been unsuccessful?

REVIEW

Write each fraction as a decimal. Use an overbar in your answer.

111. $\frac{8}{9}$

112. $\frac{11}{12}$

113. $\frac{7}{90}$

114. $\frac{1}{66}$

SECTION 5.5 Converting between American and Metric Units

It is often necessary to convert between American units and metric units. For example, we must convert units to answer the following questions.

- Which is higher: Pikes Peak (elevation 14,110 feet) or the Matterhorn (elevation 4,478 meters)?
- Does a 2-pound tub of butter weigh more than a 1-kilogram tub?
- Is a quart of soda pop more or less than a liter of soda pop?

In this section, we discuss how to answer such questions.

OBJECTIVE 1 **Use unit conversion factors to convert between American and metric units.**

The following table shows some conversions between American and metric units of length. In all but one case, the conversions are rounded approximations. An ≈ symbol is used to show this. The one exact conversion in the table is 1 inch = 2.54 centimeters.

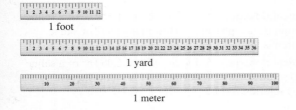

1 foot

1 yard

1 meter

Equivalent Lengths	
American to metric	**Metric to American**
1 in. = 2.54 cm	1 cm ≈ 0.39 in.
1 ft ≈ 0.30 m	1 m ≈ 3.28 ft
1 yd ≈ 0.91 m	1 m ≈ 1.09 yd
1 mi ≈ 1.61 km	1 km ≈ 0.62 mi

Unit conversion factors can be formed from the facts in the table to make specific conversions between American and metric units of length.

EXAMPLE 1 **Clothing labels.** The figure shows a label sewn into some pants made in Mexico that are for sale in the United States. Express the waist size to the nearest inch.

Strategy We will multiply 82 centimeters by a carefully chosen unit conversion factor.

WAIST: 82 cm
INSEAM: 76 cm
RN-80811
SEE REVERSE FOR CARE

MADE IN MEXICO

WHY If we multiply by the proper unit conversion factor, we can eliminate the unwanted units of centimeters and convert to inches.

Solution To convert from centimeters to inches, we must choose a unit conversion factor whose numerator contains the units we want to introduce (inches), and whose denominator contains the units we want to eliminate (centimeters). From the first row of the *Metric to American* column of the table, we see that there is approximately 0.39 inch per centimeter. Thus, we will use the unit conversion factor:

$$\frac{0.39 \text{ in.}}{1 \text{ cm}}$$

← This is the unit we want to introduce.

← This is the unit we want to eliminate (the original unit).

To perform the conversion, we multiply.

$$82 \text{ cm} \approx \frac{82 \text{ cm}}{1} \cdot \boxed{\frac{0.39 \text{ in.}}{1 \text{ cm}}}$$

Write 82 cm as a fraction: $82 \text{ cm} = \frac{82 \text{ cm}}{1}$

Multiply by a form of 1: $\frac{0.39 \text{ in.}}{1 \text{ cm}}$

$$\approx \frac{82 \cancel{\text{ cm}}}{1} \cdot \frac{0.39 \text{ in.}}{1 \cancel{\text{ cm}}}$$

Remove the common units of centimeters from the numerator and denominator. The units of inches remain.

$$\approx 82 \cdot 0.39 \text{ in.}$$ Simplify.

$$\approx 31.98 \text{ in.}$$ Do the multiplication.

$$\approx 32 \text{ in.}$$ Round to the nearest inch (ones column).

To the nearest inch, the waist size is 32 inches.

```
   0.39
 × 82
   78
 3120
 31.98
```

Self Check 1

Clothing labels. Refer to the figure in Example 1. What is the inseam length, to the nearest inch?

Now Try Problem 13

EXAMPLE 2   **Mountain elevations.**   Pikes Peak, one of the most famous peaks in the Rocky Mountains, has an elevation of 14,110 feet. The Matterhorn, in the Swiss Alps, rises to an elevation of 4,478 meters. Which mountain is higher?

Strategy We will convert the elevation of Pikes Peak, which is given in feet, to meters.

WHY Then we can compare the mountain's elevations in the same units, meters.

Solution To convert Pikes Peak elevation from feet to meters, we must choose a unit conversion factor whose numerator contains the units we want to introduce (meters) and whose denominator contains the units we want to eliminate (feet). From the second row of the *American to metric* column of the table, we see that there is approximately 0.30 meter per foot. Thus, we will use the unit conversion factor:

$$\frac{0.30 \text{ m}}{1 \text{ ft}} \quad \begin{array}{l} \leftarrow \text{ This is the unit we want to introduce.} \\ \leftarrow \text{ This is the unit we want to eliminate (the original unit).} \end{array}$$

To perform the conversion, we multiply.

$$14,110 \text{ ft} \approx \frac{14,110 \text{ ft}}{1} \cdot \boxed{\frac{0.30 \text{ m}}{1 \text{ ft}}}$$

Write 14,110 ft as a fraction: $14,110 \text{ ft} = \frac{14,110 \text{ ft}}{1}$

Multiply by a form of 1: $\frac{0.30 \text{ m}}{1 \text{ ft}}$

$$\approx \frac{14,110 \cancel{\text{ ft}}}{1} \cdot \frac{0.30 \text{ m}}{1 \cancel{\text{ ft}}}$$

Remove the common units of feet from the numerator and denominator. The units of meters remain.

$$\approx 14,110 \cdot 0.30 \text{ m}$$ Simplify.

$$\approx 4,233 \text{ m}$$ Do the multiplication.

```
   1
   14,110
 × 0.30
   000 00
 4233 00
 4233.00
```

Since the elevation of Pikes Peak is about 4,233 meters, we can conclude that the Matterhorn, with an elevation of 4,478 meters, is higher.

Self Check 2

Track and field. Which is longer: a 500-meter race or a 550-yard race?

Now Try Problem 17

We can convert between American units of weight and metric units of mass using the rounded approximations in the following table.

Equivalent Weights and Masses	
American to metric	**Metric to American**
1 oz ≈ 28.35 g	1 g ≈ 0.035 oz
1 lb ≈ 0.45 kg	1 kg ≈ 2.20 lb

1 pound

1 kilogram

EXAMPLE 3   Convert 50 pounds to grams.

Strategy We will use a two-part multiplication process that converts 50 pounds to ounces and then converts that result to grams.

WHY We must use a two-part process because the conversion table on page 467 does not contain a single unit conversion factor that converts from pounds to grams.

Solution Since there are 16 ounces per pound, we can convert 50 pounds to ounces by multiplying by the unit conversion factor $\frac{16\text{ oz}}{1\text{ lb}}$. Since there are approximately 28.35 g per ounce, we can convert that result to grams by multiplying by the unit conversion factor $\frac{28.35\text{ g}}{1\text{ oz}}$.

$$50\text{ lb} \approx \frac{50\text{ lb}}{1} \cdot \frac{16\text{ oz}}{1\text{ lb}} \cdot \frac{28.35\text{ g}}{1\text{ oz}}$$

Write 50 lb as a fraction: $50\text{ lb} = \frac{50\text{ lb}}{1}$
Multiply by two forms of 1: $\frac{16\text{ oz}}{1\text{ lb}}$ and $\frac{28.35\text{ g}}{1\text{ oz}}$

$$\approx \frac{50\text{ lb}}{1} \cdot \frac{16\text{ oz}}{1\text{ lb}} \cdot \frac{28.35\text{ g}}{1\text{ oz}}$$

Remove the common units of pounds and ounces from the numerator and denominator. The units of grams remain.

$$\approx 50 \cdot 16 \cdot 28.35\text{ g}$$ Simplify.

$$\approx 800 \cdot 28.35\text{ g}$$ Multiply: $50 \cdot 16 = 800$.

$$\approx 22{,}680\text{ g}$$ Do the multiplication.

$$\begin{array}{r} \overset{3}{16} \\ \times\ 50 \\ \hline 800 \end{array} \qquad \begin{array}{r} \overset{6\ 2\ 4}{28.35} \\ \times\ 800 \\ \hline 22680.00 \end{array}$$

Thus, 50 pounds ≈ 22,680 grams.

Self Check 3

Convert 68 pounds to grams. Round to the nearest gram.

Now Try ➡ **Problem 21**

EXAMPLE 4   **Packaging.** Does a 2.5-pound tub of butter weigh more than a 1.5-kilogram tub?

Strategy We will convert the weight of the 1.5-kilogram tub of butter to pounds.

WHY Then we can compare the weights of the tubs of butter in the same units, pounds.

Solution To convert 1.5 kilograms to pounds we must choose a unit conversion factor whose numerator contains the units we want to introduce (pounds), and whose denominator contains the units we want to eliminate (kilograms). From the second row of the *Metric to American* column of the table, we see that there are approximately 2.20 pounds per kilogram. Thus, we will use the unit conversion factor:

$$\frac{2.20\text{ lb}}{1\text{ kg}}$$ ← This is the unit we want to introduce.
   ← This is the unit we want to eliminate (the original unit).

To perform the conversion, we multiply.

$$1.5\text{ kg} \approx \frac{1.5\text{ kg}}{1} \cdot \frac{2.20\text{ lb}}{1\text{ kg}}$$

Write 1.5 kg as a fraction: $1.5\text{ kg} = \frac{1.5\text{ kg}}{1}$
Multiply by a form of 1: $\frac{2.20\text{ lb}}{1\text{ kg}}$

$$\approx \frac{1.5\text{ kg}}{1} \cdot \frac{2.20\text{ lb}}{1\text{ kg}}$$

Remove the common units of kilograms from the numerator and denominator. The units of pounds remain.

$$\approx 1.5 \cdot 2.20\text{ lb}$$ Simplify.

$$\approx 3.3\text{ lb}$$ Do the multiplication.

$$\begin{array}{r} 2.20 \\ \times\ 1.5 \\ \hline 1100 \\ 2200 \\ \hline 3.300 \end{array}$$

Self Check 4

Body weight. Who weighs more, a person who weighs 165 pounds or one who weighs 76 kilograms?

Now Try ➡ **Problem 25**

Since a 1.5-kilogram tub of butter weighs about 3.3 pounds, the 1.5-kilogram tub weighs more.

We can convert between American and metric units of capacity using the rounded approximations in the following table.

Equivalent Capacities	
American to metric	**Metric to American**
1 fl oz ≈ 29.57 mL	1 L ≈ 33.81 fl oz
1 pt ≈ 0.47 L	1 L ≈ 2.11 pt
1 qt ≈ 0.95 L	1 L ≈ 1.06 qt
1 gal ≈ 3.79 L	1 L ≈ 0.264 gal

1 liter   1 quart

Think it Through • **STUDYING IN OTHER COUNTRIES**

"The number of American college students studying abroad is now at an all-time high."
   —From the Institute of International Education

In 2015, a record number of 313,415 college students received credit for study abroad. Since students traveling to other countries are almost certain to come into contact with the metric system of measurement, they need to have a basic understanding of metric units.

   Suppose a student studying overseas needs to purchase the following school supplies. For each item in red, choose the appropriate metric units.

1. $8\frac{1}{2}$ in. × 11 in. notebook paper:

      216 meters × 279 meters   216 centimeters × 279 centimeters

               216 millimeters × 279 millimeters

2. A backpack that can hold **20 pounds** of books:

      9 kilograms   9 grams   9 milligrams

3. $\frac{3}{4}$ **fluid ounce** bottle of Liquid Paper correction fluid:

      22.5 hectoliters   2.5 liters   22.2 milliliters

EXAMPLE 5   **Cleaning supplies.**   A bottle of window cleaner contains 750 milliliters of solution. Convert this measure to quarts. Round to the nearest tenth.

Strategy We will use a two-part multiplication process that converts 750 milliliters to liters and then converts that result to quarts.

WHY We must use a two-part process because the conversion table at the top of this page does not contain a single unit conversion factor that converts from milliliters to quarts.

Solution Since there is 1 liter for every 1,000 milliliters, we can convert 750 milliliters to liters by multiplying by the unit conversion factor $\dfrac{1\text{ L}}{1,000\text{ mL}}$. Since there are approximately 1.06 quarts per liter, we can convert that result to quarts by multiplying by the unit conversion factor $\frac{1.06 \text{ qt}}{1 \text{ L}}$.

$$750 \text{ mL} \approx \frac{750 \text{ mL}}{1} \cdot \frac{1 \text{ L}}{1,000 \text{ mL}} \cdot \frac{1.06 \text{ qt}}{1 \text{ L}}$$

Write 750 mL as a fraction: $750 \text{ mL} = \frac{750 \text{ mL}}{1}$. Multiply by two forms of 1: $\frac{1 \text{ L}}{1,000 \text{ mL}}$ and $\frac{1.06 \text{ qt}}{1 \text{ L}}$

$$\approx \frac{750 \text{ mL}}{1} \cdot \frac{1 \text{ L}}{1,000 \text{ mL}} \cdot \frac{1.06 \text{ qt}}{1 \text{ L}}$$

Remove the common units of milliliters and liters from the numerator and denominator. The units of quarts remain.

$$\approx \frac{750 \cdot 1.06}{1,000} \text{ qt}$$

Multiply the fractions.

$$\approx \frac{795}{1,000} \text{ qt}$$

Multiply: $750 \cdot 1.06 = 795$.

$$\approx 0.795 \text{ qt}$$

Divide 795 by 1,000 by moving the decimal point 3 places to the left.

$$\approx 0.8 \text{ qt}$$

Round to the nearest tenth.

$$
\begin{array}{r}
750 \\
\times\, 1.06 \\
\hline
4500 \\
0000 \\
75000 \\
\hline
795.00
\end{array}
$$

The bottle contains approximately 0.8 qt of cleaning solution.

Self Check 5

Drinking water. A student bought a 360 mL bottle of water. Convert this measure to quarts. Round to the nearest tenth.

Now Try ➡ Problem 29

OBJECTIVE 2 Convert between Fahrenheit and Celsius temperatures.

In the American system, we measure temperature using **degrees Fahrenheit** (°F). In the metric system, we measure temperature using **degrees Celsius** (°C). These two scales are shown on the thermometers on the right. From the figures, we can see that

- $212°F = 100°C$ Water boils
- $32°F = 0°C$ Water freezes
- $5°F = -15°C$ A cold winter day
- $95°F = 35°C$ A hot summer day

There are formulas that enable us to convert from degrees Fahrenheit to degrees Celsius and from degrees Celsius to degrees Fahrenheit.

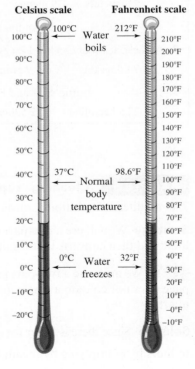

Conversion Formulas for Temperature

If F is the temperature in degrees Fahrenheit and C is the corresponding temperature in degrees Celsius, then

$$C = \frac{5}{9}(F - 32) \quad \text{and} \quad F = \frac{9}{5}C + 32$$

EXAMPLE 6   **Bathing.**   Warm bath water is 90°F. Express this temperature in degrees Celsius. Round to the nearest tenth of a degree.

Strategy   We will substitute 90 for F in the formula $C = \frac{5}{9}(F - 32)$.

WHY   Then we can use the rule for the order of operations to evaluate the right side of the equation and find the value of C, the temperature in degrees Celsius of the bath water.

Solution

$C = \dfrac{5}{9}(F - 32)$ This is the formula to find degrees Celsius.

$= \dfrac{5}{9}(90 - 32)$ Substitute 90 for F.

$= \dfrac{5}{9}(58)$ Do the subtraction within the parentheses first: 90 − 32 = 58.

$= \dfrac{5}{9}\left(\dfrac{58}{1}\right)$ Write 58 as a fraction: $58 = \frac{58}{1}$.

$= \dfrac{290}{9}$ Multiply the numerators. Multiply the denominators.

$= 32.222\ldots$ Do the division.

≈ 32.2 Round to the nearest tenth.

To the nearest tenth of a degree, the temperature of the bath water is 32.2°C.

$$\begin{array}{r} \overset{4}{58} \\ \times\ 5 \\ \hline 290 \end{array}$$

$$\begin{array}{r} 32.22 \\ 9\overline{)290.00} \\ -27 \\ \hline 20 \\ -18 \\ \hline 20 \\ -18 \\ \hline 20 \\ -18 \\ \hline 20 \\ -18 \\ \hline 2 \end{array}$$

Self Check 6

Coffee.   Hot coffee is 110°F. Express this temperature in degrees Celsius. Round to the nearest tenth of a degree.

Now Try ⮕ **Problem 33**

EXAMPLE 7   **Dishwashers.**   A dishwasher manufacturer recommends that dishes be rinsed in hot water with a temperature of 60°C. Express this temperature in degrees Fahrenheit.

Strategy   We will substitute 60 for C in the formula $F = \frac{9}{5}C + 32$.

WHY   Then we can use the rule for the order of operations to evaluate the right side of the equation and find the value of F, the temperature in degrees Fahrenheit of the water.

Solution

$F = \dfrac{9}{5}C + 32$ This is the formula to find degrees Fahrenheit.

$= \dfrac{9}{5}(60) + 32$ Substitute 60 for C.

$= \dfrac{540}{5} + 32$ Multiply: $\frac{9}{5}(60) = \frac{9}{5}\left(\frac{60}{1}\right) = \frac{540}{5}$.

$= 108 + 32$ Do the division.

$= 140$ Do the addition.

The manufacturer recommends that dishes be rinsed in 140°F water.

$$\begin{array}{r} 60 \\ \times\ 9 \\ \hline 540 \end{array}$$

$$\begin{array}{r} 108 \\ 5\overline{)540} \\ -5 \\ \hline 4 \\ -0 \\ \hline 40 \\ -40 \\ \hline 0 \end{array}$$

Self Check 7

Fevers.   To determine whether a baby has a fever, her mother takes her temperature with a Celsius thermometer. If the reading is 38.8°C, does the baby have a fever? (*Hint:* Normal body temperature is 98.6°F.)

Now Try ⮕ **Problem 37**

Answers to Self Checks

1. 30 in.   **2.** the 550-yard race   **3.** 30,845 g   **4.** the person who weighs 76 kg   **5.** 0.4 qt   **6.** 43.3°C   **7.** yes

SECTION 5.5   STUDY SET

VOCABULARY

Fill in the blanks.

1. In the American system, temperatures are measured in degrees _____. In the metric system, temperatures are measured in degrees _____.

2. **a.** Inches and centimeters are units used to measure _____.

   b. Pounds and grams are units used to measure _____ (weight).

   c. Gallons and liters are units used to measure _____.

CONCEPTS

3. Which is longer:
   a. A yard or a meter?
   b. A foot or a meter?
   c. An inch or a centimeter?
   d. A mile or a kilometer?

4. Which is heavier:
   a. An ounce or a gram?
   b. A pound or a kilogram?

5. Which is the greater unit of capacity:
   a. A pint or a liter?
   b. A quart or a liter?
   c. A gallon or a liter?

6. **a.** What formula is used for changing degrees Celsius to degrees Fahrenheit?
   b. What formula is used for changing degrees Fahrenheit to degrees Celsius?

7. Write a unit conversion factor to convert
   a. feet to meters
   b. pounds to kilograms
   c. gallons to liters

8. Write a unit conversion factor to convert
   a. centimeters to inches
   b. grams to ounces
   c. liters to fluid ounces

NOTATION

Complete each step.

9. Convert 4,500 feet to meters.

   $$4{,}500 \text{ ft} \approx \frac{4{,}500 \text{ ft}}{1} \cdot \frac{\boxed{}}{1 \text{ ft}}$$

   $$\approx 1{,}350 \;\boxed{}$$

10. Convert 8 liters to gallons.

    $$8 \text{ L} \approx \frac{8 \text{ L}}{1} \cdot \frac{\boxed{} \text{ gal}}{1 \text{ L}}$$

    $$\approx 2.112 \;\boxed{}$$

11. Convert 3 kilograms to ounces.

    $$3 \text{ kg} \approx \frac{3 \text{ kg}}{1} \cdot \frac{1{,}000 \text{ g}}{1 \text{ kg}} \cdot \frac{\boxed{} \text{ oz}}{1 \text{ g}}$$

    $$\approx 3 \cdot \boxed{} \cdot 0.035 \text{ oz}$$

    $$\approx 105 \;\boxed{}$$

12. Convert 70°C to degrees Fahrenheit.

    $$F = \frac{9}{5}C + 32$$

    $$= \frac{9}{5}(\boxed{}) + 32$$

    $$= \boxed{} + 32$$

    $$= 158$$

    Thus, $70°C = 158 \;\boxed{}$

GUIDED PRACTICE

Perform each conversion. Round to the nearest inch. See Example 1. Since most conversions are approximate, answers may vary slightly, depending on the method used.

13. 25 centimeters to inches
14. 35 centimeters to inches
15. 88 centimeters to inches
16. 91 centimeters to inches

Perform each conversion. See Example 2.

17. 8,400 feet to meters
18. 7,300 feet to meters
19. 25,115 feet to meters
20. 36,242 feet to meters

Perform each conversion. See Example 3.

21. 20 pounds to grams
22. 30 pounds to grams
23. 75 pounds to grams
24. 95 pounds to grams

Perform each conversion. See Example 4.

25. 6.5 kilograms to pounds
26. 7.5 kilograms to pounds

27. 300 kilograms to pounds

28. 800 kilograms to pounds

Perform each conversion. Round to the nearest tenth.
See Example 5.

29. 650 milliliters to quarts

30. 450 milliliters to quarts

31. 1,200 milliliters to quarts

32. 1,500 milliliters to quarts

Express each temperature in degrees Celsius. Round to
the nearest tenth of a degree. See Example 6.

33. 120°F   34. 110°F

35. 35°F   36. 45°F

Express each temperature in degrees Fahrenheit. See Example 7.

37. 75°C   38. 85°C

39. 10°C   40. 20°C

TRY IT YOURSELF

Perform each conversion. If necessary, round answers
to the nearest tenth. Since most conversions are
approximate, answers may vary slightly depending on the
method used.

41. 25 pounds to grams

42. 7.5 ounces to grams

43. 50°C to degrees Fahrenheit

44. 36.2°C to degrees Fahrenheit

45. 0.75 quart to milliliters

46. 3 pints to milliliters

47. 0.5 kilogram to ounces

48. 35 grams to pounds

49. 3.75 meters to inches

50. 2.4 kilometers to miles

51. 3 fluid ounces to liters

52. 2.5 pints to liters

53. 12 kilometers to feet

54. 3,212 centimeters to feet

55. 37 ounces to kilograms

56. 10 pounds to kilograms

57. −10°C to degrees Fahrenheit

58. −22.5°C to degrees Fahrenheit

59. 17 grams to ounces

60. 100 kilograms to pounds

61. 7.2 liters to fluid ounces

62. 5 liters to quarts

63. 3 feet to centimeters

64. 7.5 yards to meters

65. 500 milliliters to quarts

66. 2,000 milliliters to gallons

67. 50°F to degrees Celsius

68. 67.7°F to degrees Celsius

69. 5,000 inches to meters

70. 25 miles to kilometers

71. −5°F to degrees Celsius

72. −10°F to degrees Celsius

LOOK ALIKES

Perform each conversion.

73. **a.** 62 g to lb   **b.** 62 lb to g

74. **a.** 62 km to mi   **b.** 62 mi to km

75. **a.** 62 L to gal   **b.** 62 gal to L

76. **a.** 62°F to °C   **b.** 62°C to °F

APPLICATIONS

Since most conversions are approximate, answers may
vary slightly depending on the method used.

77. **The Middle East.** The distance between Jerusalem and Bethlehem is 8 kilometers. To the nearest mile, give this distance in miles.

78. **The Dead Sea.** The Dead Sea is 80 kilometers long. To the nearest mile, give this distance in miles.

79. **Cheetahs.** A cheetah can run 112 kilometers per hour. Express this speed in mph.

80. **Lions.** A lion can run 50 mph. Express this speed in kilometers per hour.

81. **Zip codes.** The website *Free Map Tools* has a feature where the user can enter two U.S. zip codes and it will find the distance between them when traveled by road. It also gives the distance between them "as the crow flies." This phrase means the distance measured in a *straight line* between the two zip codes. In the example below, the distance between 90210 (Beverly Hills) and 10013 (New York City) is given in miles. Determine the two distances that will be displayed if the user clicks on km. Round to the nearest km.

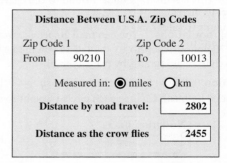

Distance Between U.S.A. Zip Codes	
Zip Code 1	Zip Code 2
From 90210	To 10013
Measured in: ◉ miles ○ km	
Distance by road travel:	2802
Distance as the crow flies	2455

82. **Track and field.** Track meets are held on an oval track. One lap around the track is usually 400 meters. However, some older tracks in the United States are 440-yard ovals. Are these two types of tracks the same length? If not, which is longer?

83. Hair growth. When hair is short, its rate of growth averages about $\frac{3}{4}$ inch per month. How many centimeters is this a month? Round to the nearest tenth of a centimeter.

84. Whales. An adult male killer whale can weigh as much as 12,000 pounds and be as long as 25 feet. Change these measurements to kilograms and meters.

85. Weightlifting. The table lists the personal best squat records for two of the world's best powerlifters. Change each metric weight to pounds. Round to the nearest pound.

Name	Hometown	Squat
Lee Ann Hewitt	Palm Beach, FL	262.5 kg
Ray Williams	Demopolis, AL	456 kg

Source: goheavy.net, t-nation.com

86. Words of wisdom. Refer to the wall hanging. Convert the first metric weight to ounces and the second to pounds. What famous saying results?

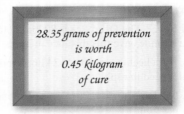

28.35 grams of prevention is worth 0.45 kilogram of cure

87. Ounces and fluid ounces.

a. There are 310 calories in 8 ounces of broiled chicken. Convert 8 ounces to grams.

b. There are 112 calories in a glass of fresh Valencia orange juice that holds 8 fluid ounces. Convert 8 fluid ounces to liters. Round to the nearest hundredth.

88. Track and field. A shot-put weighs 7.264 kilograms. Convert this weight to pounds. Round to the nearest pound.

89. Postal regulations. You can mail a package weighing up to 70 pounds via priority mail. Can you mail a package that weighs 32 kilograms by priority mail?

90. Nutrition. Refer to the nutrition label shown below for a packet of oatmeal. Change each circled weight to ounces.

Nutrition Facts
Serving Size: 1 Packet (46g)
Servings Per Container: 10

Amount Per Serving

Calories 170	Calories from Fat 20
	% Daily Value
Total fat 2g	**3%**
Saturated fat (0.5g)	**2%**
Polyunsaturated Fat 0.5g	
Monounsaturated Fat 1g	
Cholesterol 0mg	**0%**
Sodium (250mg)	**10%**
Total carbohydrate 35g	**12%**
Dietary fiber 3g	**12%**
Soluble Fiber 1g	
Sugars 16g	
Protein (4g)	

91. Hot springs. The thermal springs in Hot Springs National Park in central Arkansas emit water as warm as 143°F. Change this temperature to degrees Celsius.

92. Cooking meat. Meats must be cooked at temperatures high enough to kill harmful bacteria. According to the USDA and the FDA, the internal temperature for cooked roasts and steaks should be at least 145°F, and whole poultry should be 180°F. Convert these temperatures to degrees Celsius. Round up to the next degree.

93. Taking a shower. When you take a shower, which water temperature would you choose: 15°C, 38°C, or 50°C?

94. Drinking water. To get a cold drink of water, which temperature would you choose: −2°C, 10°C, or 25°C?

95. Snowy weather. At which temperatures might it snow: −5°C, 0°C, or 10°C?

96. Air conditioning. At which outside temperature would you be likely to run the air conditioner: 15°, 20°C, or 30°C?

97. Comparison shopping. Which is the better buy: 3 quarts of root beer for $4.50 or 2 liters of root beer for $3.60?

98. Comparison shopping. Which is the better buy: 3 gallons of antifreeze for $10.35 or 12 liters of antifreeze for $10.50?

99. Fire extinguishers. Refer to the instructions for using a fire extinguisher that are shown below. Convert the safe distance to remain away from the fire to feet.

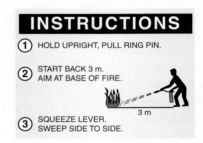

INSTRUCTIONS
① HOLD UPRIGHT, PULL RING PIN.
② START BACK 3 m. AIM AT BASE OF FIRE.
③ SQUEEZE LEVER. SWEEP SIDE TO SIDE.
3 m

100. Celsius poems. Over the years, many people have written rhymes and poems to help students have a better understanding of the Celsius temperature scale. One such "poem" is shown below. Convert each temperature to degrees Fahrenheit.

The Celsius Poem
30 is hot,
20 is nice,
10 put a coat on,
0 is ice.

101. Eye glasses. *Pupillary distance*, or PD as it is called, is the distance between the centers of the pupils in each eye. This measurement is made by an optometrist to make sure the patient's eyes match up with the optical center of the lenses when being fitted for glasses. The PD measurement shown below is in millimeters. Convert it to inches. Round to the nearest hundredth.

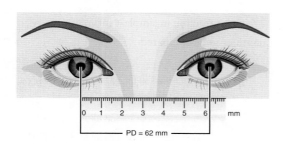

102. Oscars. Probably the most recognized trophy in the world is the Oscar Statuette. As of January 2017 a total of 3,048 of them have been awarded by the Academy of Motion Picture Arts and Sciences to recognize excellence in the film industry. Convert the height of the statue ($3\frac{1}{2}$ inches) to centimeters and its weight ($8\frac{1}{2}$ pounds) to kilograms. Round to the nearest tenth.

Featureflash Photo Agency/ Shutterstock.com

WRITING

103. Explain how to change kilometers to miles.

104. Explain how to change 50°C to degrees Fahrenheit.

105. The United States is the only industrialized country in the world that does not officially use the metric system. Some people claim this is costing American businesses money. Do you think so? Why?

106. What is meant by the phrase *a table of equivalent measures*?

REVIEW LOOK ALIKES

Perform each operation.

107. $\dfrac{3}{5} + \dfrac{4}{3}$

108. $\dfrac{3}{5} - \dfrac{4}{3}$

109. $\dfrac{3}{5} \cdot \dfrac{4}{3}$

110. $\dfrac{3}{5} \div \dfrac{4}{3}$

Perform each operation. Round to the nearest tenth when necessary.

111. $3.5 + 4.3$

112. $3.5 - 4.3$

113. $3.5 \cdot 4.3$

114. $4.3\overline{)3.5}$

⑤ Summary and Review

| SECTION 5.1 | ▶ Ratios and Rates |

DEFINITIONS AND CONCEPTS	EXAMPLES
Ratios are often used to describe important relationships between two quantities. A **ratio** is the quotient of two numbers or the quotient of two quantities that have the same units. Ratios are written in three ways: as fractions, in words separated by the word *to*, and using a colon.	The ratio 4 **to** 5 can be written as $\dfrac{4}{5}$. The ratio 5 : 12 can be written as $\dfrac{5}{12}$.

To **write a ratio as a fraction,** write the first number (or quantity) mentioned as the numerator and the second number (or quantity) mentioned as the denominator. Then simplify the fraction, if possible.	Write the ratio 30 to 36 as a fraction in simplest form. The word *to* separates the numbers to be compared. $\dfrac{30}{36} = \dfrac{5 \cdot \overset{1}{\cancel{6}}}{\underset{1}{\cancel{6}} \cdot 6}$ To simplify, factor 30 and 36. Then remove the common factor of 6 from the numerator and denominator. $\phantom{\dfrac{30}{36}} = \dfrac{5}{6}$
To write a **ratio in simplest form,** remove any common factors of the numerator and denominator as well as any common units.	Write the ratio *14 feet: 2 feet* as a fraction in simplest form. A colon separates the quantities to be compared. $\dfrac{14 \text{ feet}}{2 \text{ feet}} = \dfrac{\overset{1}{\cancel{2}} \cdot 7 \cancel{\text{ feet}}}{\underset{1}{\cancel{2}} \cancel{\text{ feet}}}$ To simplify, factor 14. Then remove the common factor of 2 and the common units of feet from the numerator and denominator. $\phantom{\dfrac{14 \text{ feet}}{2 \text{ feet}}} = \dfrac{7}{1}$ Since a ratio compares two numbers, we leave the result in fractional form. Do not simplify further.
To **simplify ratios involving decimals,** multiply the ratio by a form of 1 so that the numerator and denominator become whole numbers. Then simplify, if possible.	Write the ratio 0.23 to 0.71 as a fraction in simplest form. To write this as a ratio of *whole numbers*, we need to move the decimal points in the numerator and denominator two places to the right. This will occur if they are both multiplied by 100. $\dfrac{0.23}{0.71} = \dfrac{0.23}{0.71} \cdot \dfrac{100}{100}$ Multiply the ratio by a form of 1. $\phantom{\dfrac{0.23}{0.71}} = \dfrac{0.23 \cdot 100}{0.71 \cdot 100}$ Multiply the numerators. Multiply the denominators. $\phantom{\dfrac{0.23}{0.71}} = \dfrac{23}{71}$ To find the product of each decimal and 100, simply move the decimal point two places to the right. The resulting fraction is in simplest form.
To **simplify ratios involving mixed numbers,** use the method for simplifying complex fractions from Section 3.7. Perform the division indicated by the main fraction bar.	Write the ratio $3\dfrac{1}{3}$ to $4\dfrac{1}{6}$ as a fraction in simplest form. $\dfrac{3\frac{1}{3}}{4\frac{1}{6}} = \dfrac{\frac{10}{3}}{\frac{25}{6}}$ Write $3\frac{1}{3}$ and $4\frac{1}{6}$ as improper fractions. $\phantom{\dfrac{3\frac{1}{3}}{4\frac{1}{6}}} = \dfrac{10}{3} \div \dfrac{25}{6}$ Write the division indicated by the main fraction bar using a $\div$ symbol. $\phantom{\dfrac{3\frac{1}{3}}{4\frac{1}{6}}} = \dfrac{10}{3} \cdot \dfrac{6}{25}$ Use the rule for dividing fractions: Multiply the first fraction by the reciprocal of $\frac{25}{6}$, which is $\frac{6}{25}$. $\phantom{\dfrac{3\frac{1}{3}}{4\frac{1}{6}}} = \dfrac{10 \cdot 6}{3 \cdot 25}$ Multiply the numerators. Multiply the denominators. $\phantom{\dfrac{3\frac{1}{3}}{4\frac{1}{6}}} = \dfrac{2 \cdot \overset{1}{\cancel{5}} \cdot 2 \cdot \overset{1}{\cancel{3}}}{\underset{1}{\cancel{3}} \cdot \underset{1}{\cancel{5}} \cdot 5}$ To simplify the fraction, factor 10, 6, and 25. Then remove the common factors 3 and 5. $\phantom{\dfrac{3\frac{1}{3}}{4\frac{1}{6}}} = \dfrac{4}{5}$ Multiply the remaining factors in the numerator. Multiply the remaining factors in the denominator.

When a ratio compares two quantities, both quantities must be measured in the **same units.** When the units are different, it's usually easier to write the ratio using the smaller unit of measurement.	Write the ratio *5 inches to 2 feet* as a fraction in simplest form. Since inches are smaller than feet, compare in inches: 5 **inches** to 24 **inches** *Because 2 feet = 24 inches.* Next, write the ratio in fraction form and simplify. $$\frac{5 \text{ inches}}{24 \text{ inches}} = \frac{5}{24} \quad \text{\textit{Remove the common units of inches.}}$$
When we compare two quantities that have different units (and neither unit can be converted to the other), we call the comparison a **rate.** To **write a rate as a fraction,** write the first quantity mentioned as the numerator and the second quantity mentioned as the denominator, and then simplify, if possible. Write the units as part of the fraction. Words such as *per, for, in, from,* and *on* are used to separate the two quantities that are compared in a rate.	Write the rate *33 miles in 6 hours* as a fraction in simplest form. 33 miles **in** 6 hours can be written as $\dfrac{33 \text{ miles}}{6 \text{ hours}}$ $$\frac{33 \text{ miles}}{6 \text{ hours}} = \frac{\overset{1}{\cancel{3}} \cdot 11 \text{ miles}}{2 \cdot \underset{1}{\cancel{3}} \text{ hours}} \quad \begin{array}{l}\textit{To simplify, factor 33 and 6. Then}\\ \textit{remove the common factor of 3 from}\\ \textit{the numerator and denominator.}\end{array}$$ $$= \frac{11 \text{ miles}}{2 \text{ hours}} \quad \textit{Write the units as part of the rate.}$$ The rate can be written as 11 miles per 2 hours.
A **unit rate** is a rate in which the denominator is 1. To **write a rate as a unit rate,** divide the numerator of the rate by the denominator. A **slash mark /** is often used to write a unit rate.	Write as a unit rate: 2,490 apples from 6 trees. To find the unit rate, divide 2,490 by 6. $$\begin{array}{r} 415 \\ 6\overline{)2{,}490} \end{array}$$ The unit rate is $\frac{415 \text{ apples}}{1 \text{ tree}}$. This rate can also be expressed as: $415\frac{\text{apples}}{\text{tree}}$, 415 apples per tree, or 415 apples/tree.
A **unit price** is a rate that tells how much is paid for *one* unit (or *one* item). It is the quotient of price to the number of units. $$\text{Unit price} = \frac{\text{price}}{\text{number of units}}$$ Comparison shopping can be made easier by finding **unit prices.** The best buy is the item that has the lowest unit price.	Which is the better buy for shampoo? 12 ounces for \$3.84 or 16 ounces for \$4.64 To find the unit price of a bottle of shampoo, write the quotient of its price and its capacity, and then perform the indicated division. Before dividing, convert each price from dollars to cents so that the unit price can be expressed in cents per ounce. $$\frac{\$3.84}{12 \text{ oz}} = \frac{384¢}{12 \text{ oz}} \qquad \qquad \frac{\$4.64}{16 \text{ oz}} = \frac{464¢}{16 \text{ oz}}$$ $$= 32¢ \text{ per oz} \qquad \qquad = 29¢ \text{ per oz}$$ One ounce of shampoo for 29¢ is better than one ounce for 32¢. Thus, the 16-ounce bottle is the better buy.

REVIEW EXERCISES

Write each ratio as a fraction in simplest form.

1. 7 to 25

2. 15 : 16

3. 24 to 36

4. 21 : 14

5. 4 inches to 12 inches

6. 63 meters to 72 meters

7. 0.28 to 0.35

8. 5.1 : 1.7

9. $2\frac{1}{3}$ to $2\frac{2}{3}$

10. $4\frac{1}{6} : 3\frac{1}{3}$

11. 15 minutes : 3 hours

12. 8 ounces to 2 pounds

Write each rate as a fraction in simplest form.

13. 64 centimeters in 12 years

14. $15 for 25 minutes

Write each rate as a unit rate.

15. 600 tickets in 20 minutes

16. 45 inches every 3 turns

17. 195 feet in 6 rolls

18. 48 calories in 15 pieces

Find the unit price of each item.

19. 5 pairs cost $11.45.

20. $3 billion in a 12-month span

21. Aircraft. Specifications for a Boeing B-52 Stratofortress are shown below. What is the ratio of the airplane's wingspan to its length?

Crew: 6

Length: 160 ft
Wingspan: 185 ft
Maximum takeoff
weight: 488,000 lb
Maximum speed:
  595 mph
Maximum altitude: more than
50,000 ft
Range: 7,500 mi

22. Pay rates. Find the hourly rate of pay for a student who earned $333.25 for working 43 hours.

23. Crowd control. After a concert is over, it takes 48 minutes for a crowd of 54,000 people to exit a stadium. Find the unit rate of people exiting the stadium.

24. Comparison shopping. Mixed nuts come packaged in a 12-ounce can, which sells for $4.95, or an 8-ounce can, which sells for $3.25. Which is the better buy?

SECTION 5.2 ▶ Proportions

DEFINITIONS AND CONCEPTS	EXAMPLES
A **proportion** is a statement that two ratios or two rates are equal.	Write each statement as a proportion. $$\underset{\dfrac{6}{10}}{\underline{6 \text{ is to } 10}} \text{ as } \underset{\dfrac{3}{5}}{\underline{3 \text{ is to } 5}} = \quad \text{The word "to" is used to separate the numbers to be compared in a ratio (or rate).}$$ $$\dfrac{6}{10} = \dfrac{3}{5}$$ $$\underset{\dfrac{\$300}{500 \text{ minutes}}}{\underline{\$300 \text{ is to } 500 \text{ minutes}}} \text{ as } \underset{\dfrac{\$3}{5 \text{ minutes}}}{\underline{\$3 \text{ is to } 5 \text{ minutes}}}$$ $$\dfrac{\$300}{500 \text{ minutes}} = \dfrac{\$3}{5 \text{ minutes}}$$
Each of the four numbers in a proportion is called a **term.** The first and fourth terms are called the **extremes,** and the second and third terms are called the **means.**	First term *(extreme)* Third term *(mean)* $$\dfrac{1}{2} = \dfrac{3}{6}$$ Second term *(mean)* Fourth term *(extreme)*
Since a proportion is an equation, **a proportion can be true or false.** A proportion is true if its ratios (or rates) are equivalent and false if its ratios (or rates) are not equivalent.	Determine whether the proportion $\dfrac{3}{5} = \dfrac{15}{27}$ is true or false. *Method 1* Simplify any ratios in the proportion that are not in simplest form. Then compare them to determine whether they are equal.

One way to determine whether a proportion is true or false is to use the fraction simplifying skills of Chapter 3.

The two products found by multiplying diagonally in a proportion are called **cross products**.

Another way to determine whether a proportion is true or false involves the cross products. If the **cross products are equal**, the proportion is true. If the cross products are *not equal*, the proportion is false.

$$\frac{15}{27} = \frac{\overset{1}{\cancel{3}} \cdot 5}{\underset{1}{\cancel{3}} \cdot 9} = \frac{5}{9}$$ *Simplify the ratio on the right side.*

Since the ratios on the left and right sides of the proportion are not equal, the proportion is false.

Method 2 Check to see whether the cross products are equal.

Cross products

$$3 \cdot 27 = 81 \qquad 5 \cdot 15 = 75$$

$$\frac{3}{5} \bowtie \frac{15}{27}$$

Since the cross products are not equal, the proportion is not true.

When two pairs of numbers form a proportion, we say that they are **proportional.**

Determine whether 0.7, 0.3 and 2.1, 0.9 are proportional.

Write two ratios and form a proportion. Then find the cross products.

$$\frac{0.7}{0.3} = \frac{2.1}{0.9} \qquad 0.7 \cdot 0.9 = \mathbf{0.63} \qquad 0.3 \cdot 2.1 = \mathbf{0.63}$$

Since the cross products are equal, the numbers are proportional.

Solving a proportion to find an unknown term:

1. Set the cross products equal to each other to form an equation.

2. Isolate the variable on one side of the equation by dividing both sides by the number that is multiplied by that variable.

3. Check by substituting the result into the original proportion and finding the cross products.

Solve the proportion: $\dfrac{5}{37.5} = \dfrac{2}{x}$

$$\frac{5}{37.5} = \frac{2}{x}$$ *This is the proportion to solve.*

$$5 \cdot x = 37.5 \cdot 2$$ *Set the cross products equal to each other to form an equation.*

$$5 \cdot x = 75$$ *To simplify the right side of the equation, multiply: 37.5 · 2 = 75.*

$$\frac{5 \cdot x}{5} = \frac{75}{5}$$ *To undo the multiplication by 5 and isolate x, divide both sides by 5.*

$$x = 15$$ *To simplify the left side, remove the common factor of 5. To simplify the right side, do the division: 75 ÷ 5 = 15.*

Thus, x is 15. Check this result in the original proportion by finding the cross products.

Proportions can be used to **solve application problems.** It is easy to spot problems that can be solved using a proportion. You will be given a ratio (or rate) and asked to find the missing part of another ratio (or rate).

To solve such problems, it's helpful to revise the five-step problem-solving strategy seen earlier in the text and use the following **six-step approach.**

1. **Analyze the problem.**

2. **Assign a variable** to represent an unknown value in the problem. This means, in most cases, to let x equal what you are asked to find.

3. **Form an equation.**

4. **Solve the equation** formed in Step 3.

5. **State the conclusion clearly.**

6. **Check the result.**

Peanut butter. It takes 360 peanuts to make 8 ounces of peanut butter. How many peanuts does it take to make 12 ounces? (Source: National Peanut Board)

Analyze
- We can express the fact that it takes 360 peanuts to make 8 ounces of peanut butter as a rate: $\frac{360 \text{ peanuts}}{8 \text{ ounces}}$.
- How many peanuts does it take to make 12 ounces?

Assign We will let the variable p represent the unknown number of peanuts.

Form

360 peanuts is **to** 8 ounces **as** p peanuts is **to** 12 ounces.

$$\text{Number of peanuts} \longrightarrow \frac{360}{8} = \frac{p}{12} \longleftarrow \text{Number of peanuts}$$
$$\text{Ounces of peanut butter} \longrightarrow \qquad\qquad \longleftarrow \text{Ounces of peanut butter}$$

Solve To find the number of peanuts needed, solve the proportion for p.

$$360 \cdot 12 = 8 \cdot p \qquad \text{Set the cross products equal to each other to form an equation.}$$

$$4{,}320 = 8 \cdot p \qquad \text{To simplify the left side of the equation, multiply: } 360 \cdot 12 = 4{,}320.$$

$$\frac{4{,}320}{8} = \frac{8 \cdot p}{8} \qquad \text{To undo the multiplication by 8 and isolate } p, \text{ divide both sides by 8.}$$

$$540 = p \qquad \text{To simplify the left side, do the division: } 4{,}320 \div 8 = 540. \text{ To simplify the right side, remove the common factor of 8.}$$

State It takes 540 peanuts to make 12 ounces of peanut butter.

Check 16 ounces of peanut butter would require twice as many peanuts as 8 ounces: $2 \cdot 360$ peanuts $= 720$ peanuts. It seems reasonable that 12 ounces would require 540 peanuts.

REVIEW EXERCISES

25. Write each statement as a proportion.

   a. 20 is to 30 as 2 is to 3.

   b. 6 buses replace 100 cars as 36 buses replace 600 cars.

26. Complete the cross products.

$$\blacksquare \cdot 27 = \blacksquare \qquad 9 \cdot \blacksquare = \blacksquare$$

$$\frac{2}{9} \diagup\!\!\!\!\diagdown \frac{6}{27}$$

Determine whether each proportion is true or false by simplifying.

27. $\dfrac{8}{12} = \dfrac{3}{7}$

28. $\dfrac{4}{18} = \dfrac{10}{45}$

Determine whether each proportion is true or false by finding cross products.

29. $\dfrac{9}{27} = \dfrac{2}{6}$

30. $\dfrac{17}{7} = \dfrac{51}{21}$

31. $\dfrac{3.5}{9.3} = \dfrac{1.2}{3}$

32. $\dfrac{1\frac{1}{2}}{3\frac{1}{3}} = \dfrac{\frac{1}{4}}{1\frac{1}{7}}$

Determine whether the numbers are proportional.

33. 5, 9 and 20, 36

34. 7, 13 and 29, 54

Solve each proportion.

35. $\dfrac{12}{18} = \dfrac{3}{x}$

36. $\dfrac{4}{x} = \dfrac{2}{8}$

37. $\dfrac{4.8}{6.6} = \dfrac{x}{9.9}$

38. $\dfrac{0.08}{x} = \dfrac{0.04}{0.06}$

39. $\dfrac{1\frac{9}{11}}{x} = \dfrac{3\frac{1}{3}}{2\frac{3}{4}}$

40. $\dfrac{\frac{4}{5}}{1\frac{1}{20}} = \dfrac{2\frac{2}{3}}{x}$

41. $\dfrac{\frac{2}{3}}{\frac{1}{2}} = \dfrac{x}{0.25}$

42. $\dfrac{x}{300} = \dfrac{5{,}000}{1{,}500}$

43. Trucks. A Dodge Ram pickup truck can go 35 miles on 2 gallons of gas. How far can it go on 11 gallons?

44. Quality control. In a manufacturing process, 12 parts out of 66 were found to be defective. How many defective parts will be expected in a run of 1,650 parts?

45. Scale drawings. The illustration below shows an architect's drawing of a kitchen using a scale of $\frac{1}{8}$ inch to 1 foot $\left(\frac{1}{8}'' : 1'0''\right)$. On the drawing, the length of the kitchen is $1\frac{1}{2}$ inches. How long is the actual kitchen? (The symbol $''$ means inch and $'$ means foot.)

ELEVATION B-B
SCALE: $\frac{1}{8}''$ to $1'0''$

46. Dogs. The American Kennel Club website gives the ideal length-to-height proportions for a German Shepherd as $10 : 8\frac{1}{2}$. What is the ideal length of a German Shepherd that is $25\frac{1}{2}$ inches high at the shoulder?

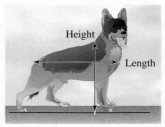

SECTION 5.3 ▶ American Units of Measurement

DEFINITIONS AND CONCEPTS	EXAMPLES
The **American system of measurement** uses the units of **inch, foot, yard,** and **mile** to measure **length.** A **ruler** is one of the most common tools for measuring lengths. Most rulers are 12 inches long. Each inch is divided into halves of an inch, quarters of an inch, eighths of an inch, and sixteenths of an inch.	1 ft = 12 in.　　　　　　1 yd = 3 ft 1 yd = 36 in.　　　　　　1 mi = 5,280 ft Since the black tick marks between 0 and 1 on the ruler create sixteen equal spaces, the ruler is scaled in sixteenths. $\frac{3}{16}$ in.　$\frac{1}{2}$ in.　　$1\frac{1}{4}$ in.　$1\frac{3}{4}$ in.　$2\frac{3}{8}$ in. 16 spaces　**1**　　　**2**　　　**3** Inches
To convert from one unit of length to another, we use **unit conversion factors.** They are called unit conversion factors because their value is 1. Multiplying a measurement by a unit conversion factor does not change the measure; it only changes the units of the measure. A list of unit conversion factors for American units of length is given on page 441.	Convert 4 yards to inches. To convert from yards to inches, we select a unit conversion factor that introduces the units of inches and eliminates the units of yards. Since there are 36 inches per yard, we will use: $\dfrac{36 \text{ in.}}{1 \text{ yd}}$ ← This is the unit we want to introduce. ← This is the unit we want to eliminate (the original unit). To perform the conversion, we multiply. $4 \text{ yd} = \dfrac{4 \text{ yd}}{1} \cdot \dfrac{36 \text{ in.}}{1 \text{ yd}}$　　Write 4 yd as a fraction. Then multiply by a form of 1: $\frac{36 \text{ in.}}{1 \text{ yd}}$ $= \dfrac{4 \cancel{\text{yd}}}{1} \cdot \dfrac{36 \text{ in.}}{1 \cancel{\text{yd}}}$　　Remove the common units of yards from the numerator and denominator. The units of inches remain. $= 4 \cdot 36 \text{ in.}$　　Simplify. $= 144 \text{ in.}$　　Do the multiplication. Thus, 4 yards = 144 inches.
The American system of measurement uses the units of **ounce, pound,** and **ton** to measure **weight.**	1 lb = 16 oz　　　　　　　1 ton = 2,000 lb
A list of unit conversion factors for American units of weight is given on page 443.	Convert 9,000 pounds to tons. To convert from pounds to tons, we select a unit conversion factor that introduces the units of tons and eliminates the units of pounds. Since there is 1 ton for every 2,000 pounds, we will use: $\dfrac{1 \text{ ton}}{2,000 \text{ lb}}$ ← This is the unit we want to introduce. ← This is the unit we want to eliminate (the original unit). To perform the conversion, we multiply. $9,000 \text{ lb} = \dfrac{9,000 \text{ lb}}{1} \cdot \dfrac{1 \text{ ton}}{2,000 \text{ lb}}$　　Write 9,000 lb as a fraction. Then multiply by a form of 1: $\frac{1 \text{ ton}}{2,000 \text{ lb}}$ $= \dfrac{9,000 \overset{1}{\cancel{\text{lb}}}}{1} \cdot \dfrac{1 \text{ ton}}{2,000 \underset{1}{\cancel{\text{lb}}}}$　　Remove the common units of pounds from the numerator and denominator. The units of tons remains. $= \dfrac{9,000}{2,000} \text{ ton}$　　Multiply the fractions.

	There are two ways to complete the solution. First, we can remove any common factors of the numerator and denominator to simplify the fraction. Then we can write the result as a mixed number.
	$$\frac{9{,}000}{2{,}000} \text{ tons} = \frac{9 \cdot \overset{1}{\cancel{1{,}000}}}{2 \cdot \underset{1}{\cancel{1{,}000}}} \text{ tons} = \frac{9}{2} \text{ tons} = 4\frac{1}{2} \text{ tons}$$
	A second approach is to divide the numerator by the denominator and express the result as a decimal.
	$$\frac{9{,}000}{2{,}000} \text{ tons} = 4.5 \text{ tons}$$
	Thus, 9,000 pounds is equal to $4\frac{1}{2}$ tons (or 4.5 tons).
The American system of measurement uses the units of **ounce, cup, pint, quart,** and **gallon** to measure **capacity.**	1 c = 8 fl oz 1 pt = 2 c 1 qt = 2 pt 1 gal = 4 qt
A list of unit conversion factors for American units of capacity is given on page 445. Some conversions require the use of **two** (or more) **unit conversion factors.**	Convert 5 gallons to pints. There is not a single unit conversion factor that converts from gallons to pints. We must use two unit conversion factors. Since there are 4 quarts per gallon, we can convert 5 gallons to quarts by multiplying by the unit conversion factor $\frac{4 \text{ qt}}{1 \text{ gal}}$. Since there are 2 pints per quart, we can convert that result to pints by multiplying by the unit conversion factor $\frac{2 \text{ pt}}{1 \text{ qt}}$.

$$5 \text{ gal} = \frac{5 \text{ gal}}{1} \cdot \boxed{\frac{4 \text{ qt}}{1 \text{ gal}}} \cdot \boxed{\frac{2 \text{ pt}}{1 \text{ qt}}}$$

$$= \frac{5 \text{ }\cancel{\text{gal}}}{1} \cdot \frac{4 \text{ }\cancel{\text{qt}}}{1 \text{ }\cancel{\text{gal}}} \cdot \frac{2 \text{ pt}}{1 \text{ }\cancel{\text{qt}}}$$

Remove the common units of gallons and quarts in the numerator and denominator. The units of pints remain.

$$= 40 \text{ pt}$$

Do the multiplication: $5 \cdot 4 \cdot 2 = 40$.

Thus, 5 gallons = 40 pints.

The American and metric systems of measurement both use the units of **seconds, minutes, hours,** and **days** to measure time.	1 min = 60 sec 1 hr = 60 min 1 day = 24 hr
A list of unit conversion factors for units of time is given on page 446.	Convert 240 minutes to hours. To convert from minutes to hours, we select a unit conversion factor that introduces the units of hours and eliminates the units of minutes. Since there is 1 hour for every 60 minutes, we will use:

$$\frac{1 \text{ hr}}{60 \text{ min}}$$

← *This is the unit we want to introduce.*

← *This is the unit we want to eliminate (the original unit).*

To perform the conversion, we multiply.

$$240 \text{ min} = \frac{240 \text{ min}}{1} \cdot \boxed{\frac{1 \text{ hr}}{60 \text{ min}}}$$

Write 240 min as a fraction. Then multiply by a form of 1: $\frac{1 \text{ hr}}{60 \text{ min}}$

$$= \frac{240 \text{ }\cancel{\text{min}}}{1} \cdot \frac{1 \text{ hr}}{60 \text{ }\cancel{\text{min}}}$$

Remove the common units of minutes from the numerator and denominator. The units of hours remain.

$$= \frac{240}{60} \text{ hr}$$

Multiply the fractions.

$$= 4 \text{ hr}$$

Do the division.

Thus, 240 minutes = 4 hours.

REVIEW EXERCISES

47. a. Refer to the ruler below. Each inch is divided into how many equal parts?

   b. Determine which measurements the arrows point to on the ruler.

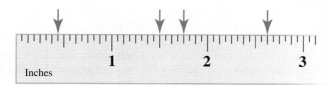

48. Use a ruler to measure the length of the computer mouse.

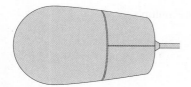

49. Write two unit conversion factors using the fact that 1 mile = 5,280 ft.

50. Consider the work shown below.

$$\frac{100 \text{ min}}{1} \cdot \frac{60 \text{ sec}}{1 \text{ min}}$$

   a. What units can be removed?

   b. What units remain?

Perform each conversion.

51. 5 yards to feet

52. 6 yards to inches

53. 66 in. to ft

54. 9,240 feet to miles

55. $4\frac{1}{2}$ feet to inches

56. 1 mi to yd

57. 32 ounces to pounds

58. 17.2 pounds to ounces

59. 3 tons to ounces

60. 4,500 pounds to tons

61. 5 pints to fluid ounces

62. 8 c to gal

63. 17 quarts to cups

64. 176 fluid ounces to quarts

65. 5 gallons to pints

66. 3.5 gal to c

67. 20 min to sec

68. 900 seconds to minutes

69. 200 hours to days

70. 6 hr to min

71. 4.5 days to hours

72. 1 day to seconds

73. Convert 210 yards to miles. Give the exact answer and a decimal approximation, rounded to the nearest hundredth.

74. Trucking. Large concrete trucks can carry roughly 40,500 pounds of concrete. Express this weight in tons.

Elemental Imaging/Shutterstock.com

75. Skyscrapers. The Willis Tower in Chicago is 1,454 feet high. Express this height in yards.

76. Bottling. A farmer sells apple cider in 2-quart bottles. How many bottles will he need to hold 50 gallons of cider?

SECTION 5.4 ▶ Metric Units of Measurement

DEFINITIONS AND CONCEPTS	EXAMPLES	
The basic metric unit of measurement is the **meter,** which is abbreviated **m.** Longer and shorter metric units are created by adding **prefixes** to the front of the basic unit, meter.	*kilo* means thousands *hecto* means hundreds *deka* means tens	*deci* means tenths *centi* means hundredths *milli* means thousandths *micro* means millionths
Common metric units of length are the **kilometer, hectometer, dekameter, decimeter, centimeter,** and **millimeter.** Abbreviations are often used when writing these units. See the table on page 452.	1 km = 1,000 m 1 hm = 100 m 1 dam = 10 m	1 m = 10 dm 1 m = 100 cm 1 m = 1,000 mm

A **metric ruler** can be used for measuring lengths. On most metric rulers, each centimeter is divided into 10 millimeters.	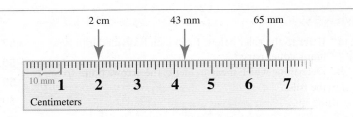

To convert from one metric unit of length to another, we use **unit conversion factors.**

Convert 4 meters to centimeters.

To convert from meters to centimeters, we select a unit conversion factor that introduces the units of centimeters and eliminates the units of meters. Since there are 100 centimeters per meter, we will use:

$$\frac{100 \text{ cm}}{1 \text{ m}} \begin{array}{l} \longleftarrow \text{ This is the unit we want to introduce.} \\ \longleftarrow \text{ This is the unit we want to eliminate (the original unit).} \end{array}$$

To perform the conversion, we multiply.

$$4 \text{ m} = \frac{4 \text{ m}}{1} \cdot \frac{100 \text{ cm}}{1 \text{ m}} \qquad \begin{array}{l} \text{Write 4 m as a fraction.} \\ \text{Then multiply by a form of 1: } \frac{100 \, cm}{1 \, m} \end{array}$$

$$= \frac{4 \text{ m}}{1} \cdot \frac{100 \text{ cm}}{1 \text{ m}} \qquad \begin{array}{l} \text{Remove the common units of meters from} \\ \text{the numerator and denominator. The units} \\ \text{of cm remain.} \end{array}$$

$$= 400 \text{ cm} \qquad \text{Multiply the fractions and simplify.}$$

Thus, 4 meters = 400 centimeters.

The **mass** of an object is a measure of the amount of material in the object.

Common metric units of mass are the **kilogram, hectogram, dekagram, decigram, centigram,** and **milligram.** Abbreviations are often used when writing these units. See the table on page 457.

1 kg = 1,000 g	1 g = 10 dg
1 hg = 100 g	1 g = 100 cg
1 dag = 10 g	1 g = 1,000 mg
	1 g = 1,000,000 mcg

Converting from one metric unit to another can be done using **unit conversion factors** or a **conversion chart.**

In a conversion chart, the units are listed from largest to smallest, reading left to right. We **count the places** and note the **direction** as we move from the original units to the conversion units.

Convert 820 grams to kilograms.

To use a conversion chart, locate the original units of grams and move to the conversion units of kilograms.

To go from grams to kilograms, we must move 3 places to the left.

If we write 820 grams as 820.0 grams, we can convert to kilograms by moving the decimal point three places to the left.

820.0 grams = 0.8200 kilograms = 0.82 kilograms

The unit conversion factor method gives the same result:

$$820 \text{ g} = \frac{820 \text{ g}}{1} \cdot \frac{1 \text{ kg}}{1,000 \text{ g}}$$

$$= \frac{820}{1,000} \text{ kg}$$

$$= 0.82 \text{ kg}$$

Thus, 820 grams = 0.82 kilograms.

Common metric units of capacity are the **kiloliter, hectoliter, dekaliter, deciliter, centiliter,** and **milliliter.** Abbreviations are often used when writing these units. See the table on page 460.	1 kL = 1,000 L 1 L = 10 dL 1 hL = 100 L 1 L = 100 cL 1 daL = 10 L 1 L = 1,000 mL
Converting from one metric unit to another can be done using **unit conversion factors** or a **conversion chart.**	Convert 0.7 kiloliters to milliliters. To use a conversion chart, locate the original units of kiloliters and move to the conversion units of milliliters.  *To go from kiloliters to milliliters, we must move 6 places to the right.* We can convert to milliliters by moving the decimal point six places to the right. 0.7 kiloliters = 0700000. milliliters = 700,000 milliliters The unit conversion factor method gives the same result: $0.7 \text{ kL} = \dfrac{0.7 \text{ kL}}{1} \cdot \dfrac{1{,}000 \text{ L}}{1 \text{ kL}} \cdot \dfrac{1{,}000 \text{ mL}}{1 \text{ L}}$ $= 0.7 \cdot 1{,}000 \cdot 1{,}000 \text{ mL}$ $= 700{,}000 \text{ mL}$ Thus, 0.7 kiloliters = 700,000 milliliters.
Another metric unit of capacity is the **cubic centimeter,** written cm^3, or, more simply, cc. One cc represents the same capacity as one milliliter. The units of cubic centimeters are used frequently in medicine.	1 milliliter = 1 cm^3 = 1 cc 5 milliliters = 5 cm^3 = 5 cc 0.6 milliliters = 0.6 cm^3 = 0.6 cc

REVIEW EXERCISES

77. a. Refer to the metric ruler below. Each centimeter is divided into how many equal parts? What is the length of one of those parts?

    b. Determine which measurements the arrows point to on the ruler.

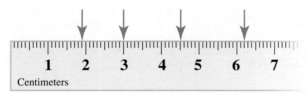

78. Use a metric ruler to measure the length of the computer mouse to the nearest centimeter.

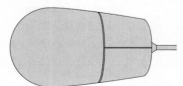

79. Write two unit conversion factors using the given fact.

    a. 1 km = 1,000 m

    b. 1 g = 100 cg

80. Use the chart to determine how many decimal places and in which direction to move the decimal point when converting from centimeters to kilometers.

km	hm	dam	m	dm	cm	mm

Perform each conversion.

81. 475 cm to m

82. 8 meters to millimeters

83. 165.7 kilometers to meters

84. 6,789 centimeters to decimeters

85. 25 micrograms to milligrams

86. 800 centigrams to grams

87. 5,425 grams to kilograms

88. 5,425 grams to milligrams

89. 150 cL to L

90. 3,250 liters to kiloliters

91. 400 mL to cL

92. 1 hectoliter to deciliters

93. The brain. The adult human brain weighs about 1,350 g. Convert the weight to kilograms.

94. Test tubes. A rack holds one dozen 20 mL test tubes. Find the total capacity of the test tubes in the rack in liters.

95. Tylenol. A bottle of Extra-Strength Tylenol contains 100 caplets of 500 milligrams each. How many grams of Tylenol are in the bottle?

96. Surgery. A dextrose solution is being administered to a patient intravenously as shown to the right. How many milliliters of solution does the IV bag hold?

SECTION 5.5 ▶ Converting between American and Metric Units

DEFINITIONS AND CONCEPTS	EXAMPLES
We **convert between American and metric units** of length using the facts on the right. In all but one case, the conversions are rounded approximations.	**American to metric** **Metric to American** 1 in. = 2.54 cm 1 cm ≈ 0.39 in. 1 ft ≈ 0.30 m 1 m ≈ 3.28 ft 1 yd ≈ 0.91 m 1 m ≈ 1.09 yd 1 mi ≈ 1.61 km 1 km ≈ 0.62 mi
Unit conversion factors can be formed from the facts in the table above right to make specific conversions between American and metric units of length.	Convert 15 inches to centimeters. To convert from inches to centimeters, we select a unit conversion factor that introduces the units of centimeters and eliminates the units of inches. Since there are 2.54 centimeters for every inch, we will use: $\dfrac{2.54 \text{ cm}}{1 \text{ in.}}$ ⟵ This is the unit we want to introduce. ⟵ This is the unit we want to eliminate (the original unit). To perform the conversion, we multiply. $15 \text{ in.} = \dfrac{15 \text{ in.}}{1} \cdot \dfrac{2.54 \text{ cm}}{1 \text{ in.}}$ Write 15 in. as a fraction. Then multiply by a form of 1: $\frac{2.54 \text{ cm}}{1 \text{ in.}}$ $= \dfrac{15 \cancel{\text{ in.}}}{1} \cdot \dfrac{2.54 \text{ cm}}{1 \cancel{\text{ in.}}}$ Remove the common units of inches from the numerator and denominator. The units of cm remain. $= 15 \cdot 2.54 \text{ cm}$ Simplify. $= 38.1 \text{ cm}$ Do the multiplication. Thus, 15 inches = 38.1 centimeters.
We **convert between American and metric units** of mass (weight) using the facts on the right. The conversions are rounded approximations.	**American to metric** **Metric to American** 1 oz ≈ 28.35 g 1 g ≈ 0.035 oz 1 lb ≈ 0.45 kg 1 kg ≈ 2.20 lb

Unit conversion factors can be formed from the facts in the table on the previous page, bottom right to make specific conversions between American and metric units of mass (weight).	Convert 6 kilograms to ounces. There is not a single unit conversion factor that converts from kilograms to ounces. We must use **two unit conversion factors**—one to convert kilograms to grams, and another to convert that result to ounces. $$6 \text{ kg} \approx \frac{6\text{kg}}{1} \cdot \frac{\mathbf{1{,}000\,g}}{\mathbf{1\,kg}} \cdot \frac{\mathbf{0.035\,oz}}{\mathbf{1\,g}}$$ $\approx \dfrac{6\text{ k\!\!\!\!/g}}{1} \cdot \dfrac{1{,}000\text{ \!\!\!\!/g}}{1\text{ k\!\!\!\!/g}} \cdot \dfrac{0.035\text{ oz}}{1\text{ \!\!\!\!/g}}$ *Remove the common units of kilograms and grams in the numerator and denominator. The units of oz remain.* $\approx 6 \cdot 1{,}000 \cdot 0.035 \text{ oz}$ *Simplify.* $\approx 6 \cdot 35 \text{ oz}$ *Multiply the last two factors: $1{,}000 \cdot 0.035 = 35$.* $\approx 210 \text{ oz}$ *Do the multiplication.* Thus, 6 kilograms $\approx$ 210 ounces.
We **convert between American and metric units** of capacity using the facts on the right. The conversions are rounded approximations.	**American to metric** $\qquad$ **Metric to American** 1 fl oz $\approx$ 29.57 mL $\qquad$ 1 L $\approx$ 33.81 fl oz 1 pt $\approx$ 0.47 L $\qquad\quad$ 1 L $\approx$ 2.11 pt 1 qt $\approx$ 0.95 L $\qquad\quad$ 1 L $\approx$ 1.06 qt 1 gal $\approx$ 3.79 L $\qquad\quad$ 1 L $\approx$ 0.264 gal
Unit conversion factors can be formed from the facts in the table above right to make specific conversions between American and metric units of capacity.	Convert 5 fluid ounces to milliliters. Round to the nearest tenth. To convert from fluid ounces to milliliters, we select a unit conversion factor that introduces the units of milliliters and eliminates the units of fluid ounces. Since there are 29.57 milliliters for every fluid ounce, we will use: $\dfrac{29.57 \text{ mL}}{1 \text{ fl oz}}$ $\leftarrow$ *This is the unit we want to introduce.* $\leftarrow$ *This is the unit we want to eliminate (the original unit).* To perform the conversion, we multiply. $$5 \text{ fl oz} \approx \frac{5 \text{ fl oz}}{1} \cdot \frac{\mathbf{29.57\,mL}}{\mathbf{1\,fl\,oz}}$$ *Write 5 fl oz as a fraction. Then multiply by a form of 1: $\frac{29.57 \text{ mL}}{1 \text{ fl oz}}$* $\approx \dfrac{5 \text{ f\!\!\!/l\,o\!\!\!/z}}{1} \cdot \dfrac{29.57 \text{ mL}}{1 \text{ f\!\!\!/l\,o\!\!\!/z}}$ *Remove the common units of fluid ounces from the numerator and denominator. The units of mL remain.* $\approx 5 \cdot 29.57 \text{ mL}$ *Simplify.* $\approx 147.85 \text{ mL}$ *Do the multiplication.* $\approx 147.9 \text{ mL}$ *Round to the nearest tenth.* Thus, 5 fluid ounces $\approx$ 147.9 milliliters.
In the American system, we measure temperature using **degrees Fahrenheit (°F)**. In the metric system, we measure temperature using **degrees Celsius (°C)**. If F is the temperature in degrees Fahrenheit and C is the corresponding temperature in degrees Celsius, then $$C = \frac{5}{9}(F - 32) \quad \text{and} \quad F = \frac{9}{5}C + 32$$	Convert 92°F to degrees Celsius. Round to the nearest tenth of a degree. $C = \dfrac{5}{9}(F - 32)$ *This is the formula to find degrees Celsius.* $= \dfrac{5}{9}(92 - 32)$ *Substitute 92 for F.* $= \dfrac{5}{9}(60)$ *Do the subtraction within the parentheses first.*

$$= \frac{5}{9}\left(\frac{60}{1}\right)$$ *Write 60 as a fraction.*

$$= \frac{300}{9}$$ *Multiply the numerators: 5 · 60 = 300.*
Multiply the denominators.

$$= 33.333\ldots$$ *Do the division.*

$$\approx 33.3$$ *Round to the nearest tenth.*

Thus, 92°F ≈ 33.3°C.

REVIEW EXERCISES

97. Swimming. Olympic-size swimming pools are 50 meters long. Express this distance in feet.

98. High-rise buildings. The Willis Tower in Chicago is 443 meters high, and the Empire State Building in New York is 1,250 feet high. Which building is taller?

99. Western settlers. The Oregon Trail was an overland route pioneers used from the 1840s through the 1870s to reach the Oregon Territory. It stretched 1,930 miles from Independence, Missouri, to Oregon City, Oregon. Find this distance to the nearest kilometer.

100. U.S. Presidents. Abraham Lincoln was the tallest of all presidents: 6 feet, 4 inches. Express his height in centimeters. Round to the nearest centimeter.

Perform each conversion. Since most conversions are approximate, answers may vary slightly depending on the method used.

101. 30 ounces to grams

102. 15 kilograms to pounds

103. 50 pounds to grams

104. 2,000 pounds to kilograms

105. Polar bears. At birth, polar bear cubs weigh less than human babies—about 910 grams. Convert this to pounds.

andamanec/Shutterstock.com

106. Bottled water. LaCroix bottled water can be purchased in bottles containing 17 fluid ounces. Mountain Valley water can be purchased in half-liter bottles. Which bottle contains more water?

107. Crude oil. There are 42 gallons in a barrel of crude oil. How many liters of crude oil is that?

108. Convert 105°C to degrees Fahrenheit.

109. Convert 77°F to degrees Celsius.

110. Recreation. Which water temperature is appropriate for swimming: 10°C, 30°C, 50°C, or 70°C?

1. Fill in the blanks.

   a. A _____ is the quotient of two numbers or the quotient of two quantities that have the same units.

   b. A _____ is the quotient of two quantities that have different units.

   c. A _____ is a statement that two ratios (or rates) are equal.

   d. The _____ products for the proportion $\frac{3}{8} = \frac{6}{16}$ are $3 \cdot 16$ and $8 \cdot 6$.

   e. *Deci* means _____, *centi* means _____, and *milli* means _____.

   f. The meter, the gram, and the liter are basic units of measurement in the _____ system.

   g. In the American system, temperatures are measured in degrees _____. In the metric system, temperatures are measured in degrees _____.

2. **Pianos.** A piano keyboard is made up of a total of eighty-eight keys, as shown below. What is the ratio of the number of black keys to white keys?

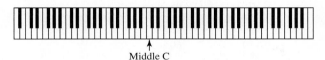

Middle C

Write each ratio as a fraction in simplest form.

3. 6 feet to 8 feet

4. 8 ounces to 3 pounds

5. 0.26 : 0.65

6. $3\frac{1}{3}$ to $3\frac{8}{9}$

7. Write the rate 54 feet in 36 seconds as a fraction in simplest form.

8. **Comparison shopping.** A 2-pound can of coffee sells for $3.38, and a 5-pound can of the same brand of coffee sells for $8.50. Which is the better buy?

9. **Utility costs.** A household used 675 kilowatt-hours of electricity during a 30-day month. Find the rate of electric usage in kilowatt-hours per day.

10. Write the following statement as a proportion: 15 billboards to 50 miles as 3 billboards to 10 miles.

11. Determine whether each proportion is true.

    a. $\frac{25}{33} = \frac{2}{3}$

    b. $\frac{2.2}{3.5} = \frac{1.76}{2.8}$

12. Are the numbers 7, 15 and 35, 75 proportional?

Solve each proportion.

13. $\frac{x}{3} = \frac{35}{7}$

14. $\frac{15.3}{x} = \frac{3}{12.4}$

15. $\dfrac{2\frac{2}{9}}{\frac{4}{3}} = \dfrac{x}{1\frac{1}{2}}$

16. $\dfrac{25}{\frac{1}{10}} = \dfrac{50}{x}$

17. **Shopping.** If 13 ounces of tea costs $2.79, how much would you expect to pay for 16 ounces of tea?

18. **Baking.** A recipe calls for $1\frac{2}{3}$ cup of sugar and 5 cups of flour. How much sugar should be used with 6 cups of flour?

19. a. Refer to the ruler below. Each inch is divided into how many equal parts?

    b. Determine which measurements the arrows point to on the ruler.

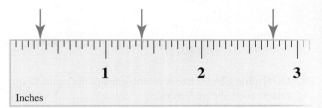

Inches

20. Fill in the blanks. In general, a unit conversion factor is a fraction with the following form:

Unit that we want to ⬚	← Numerator
Unit that we want to ⬚	← Denominator

21. Convert 180 inches to feet.

22. **Tools.** If a 25-foot tape measure is completely extended, how many yards does it stretch? Write your answer as a mixed number.

23. Convert $10\frac{3}{4}$ pounds to ounces.

24. **Automobiles.** A car weighs 1.6 tons. Find its weight in pounds.

25. **Containers.** How many fluid ounces are in a 1-gallon carton of milk?

26. **Literature.** An excellent work of early science fiction is the book *Around the World in 80 Days* by Jules Verne (1828–1905). Convert 80 days to minutes.

27. a. A quart and a liter of fruit punch are shown below. Which is the 1-liter carton: the larger one or the smaller one?

b. The figures below show the relative lengths of a yardstick and a meterstick. Which one represents the meterstick: the longer one or the shorter one?

c. One ounce and one gram weights are placed on a balance, as shown below. On which side is the gram: the left side or the right side?

28. Determine which measurements the arrows point to on the metric ruler shown below.

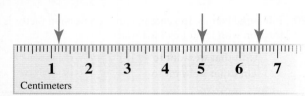

29. Speed skating. American Bonnie Blair won gold medals in the women's 500-meter speed skating competitions at the 1988, 1992, and 1994 Winter Olympic Games. Convert the race length to kilometers.

30. How many centimeters are in 5 meters?

31. Convert 5,000 micrograms to milligrams.

32. Convert 70 liters to milliliters.

33. Prescriptions. A bottle contains 50 tablets, each containing 150 mg of medicine. How many grams of medicine does the bottle contain?

34. Track. Which is the longer distance: a 100-yard race or an 80-meter race?

35. Body weight. Which person is heavier: Jim, who weighs 160 pounds, or Ricardo, who weighs 71 kilograms?

36. Convert 810 milliliters to quarts. Round to the nearest tenth.

37. Shoelaces. The proper shoelace length depends on how many *pairs* of eyelets the shoes have. The chart below shows the recommended shoelace length for athletic shoes. Convert the recommended length for the shoe shown below to the nearest centimeter.

Pairs of eyelets on a shoe	Recommended shoelace length
4	27 in.
5	36 in.
6–7	45 in.
8	54 in.
9–10	60 in.

38. Cooking meat. The USDA recommends that turkey be cooked to a temperature of 83°C. Change this to degrees Fahrenheit. To be safe, *round up* to the next degree. (*Hint:* $F = \frac{9}{5}C + 32$.)

39. What is a scale drawing? Give an example.

40. Explain the benefits of the metric system of measurement as compared to the American system.

1. Write 5,764,502:

   a. in words

   b. in expanded notation [Section 1.1]

2. **Basketball records.** On December 13, 1983, the Detroit Pistons and the Denver Nuggets played in the highest-scoring game in NBA history. See the game summary below. [Section 1.2]

   a. What was the final score?

   b. Which team won?

   c. What was the total number of points scored in the game?

	Quarter				Overtime			
	1	**2**	**3**	**4**	**1**	**2**	**3**	**Total**
Detroit	38	36	34	37	14	12	15	?
Denver	34	40	39	32	14	12	13	?

Source: ESPN.com

3. Subtract: $70,006 - 348$ [Section 1.3]

4. Multiply: $504 \cdot 729$ [Section 1.4]

5. Divide: $37\overline{)743}$ [Section 1.5]

6. **Discount lodging.** A hotel is offering rooms that normally go for $189 per night for only $109 a night. How many dollars would a traveler save if he stays in such a room for one week? [Section 1.6]

7. List the factors of 30, from smallest to largest. [Section 1.7]

8. Find the prime factorization of 360. [Section 1.7]

9. Find the LCM and the GCF of 20 and 28. [Section 1.8]

10. Evaluate: $81 + 9[7^2 - 7(11 - 4)]$ [Section 1.9]

11. Place an $<$ or an $>$ symbol in the box to make a true statement: $-(-10)$ ▨ $|-11|$ [Section 2.1]

12. Evaluate: $(-12 + 6) + (-6 + 8)$ [Section 2.2]

13. **Golf.** Tiger Woods won the 100th U.S. Open in June of 2000 by the largest margin in the history of that tournament. If he shot 12 under par (-12) and the second-place finisher, Miguel Ángel Jiménez, shot 3 over par ($+3$), what was Tiger's margin of victory? [Section 2.3]

14. Evaluate: -3^2 and $(-3)^2$ [Section 2.4]

15. Evaluate each expression, if possible. [Sections 2.5]

   a. $0 + (-8)$

   b. $\dfrac{-8}{0}$

   c. $0 - |-8|$

   d. $\dfrac{0}{-8}$

   e. $0 - (-8)$

   f. $0(-8)$

16. Evaluate: $\dfrac{3 + 3[5(-6) - (1 - 10)]}{-1 + (-1)}$ [Section 2.6]

17. Estimate the value of the following expression by rounding each number to the nearest hundred. [Section 2.6]

   $$-3,887 + (-5,806) + 4,701$$

18. Simplify: $-\dfrac{16}{20}$ [Section 3.1]

19. Express $\dfrac{9}{10}$ as an equivalent fraction with a denominator of 60. [Section 3.1]

20. **Geography.** Earth has a surface area of about 197,000,000 square miles. Use the information in the circle graph below to determine the number of square miles of Earth's surface covered by land. (Source: scienceclarified.com) [Section 3.2]

Land covers about $\dfrac{3}{10}$ of the Earth's surface

Water covers about $\dfrac{7}{10}$ of the Earth's surface

21. What is the formula for the area of a triangle? [Section 3.2]

22. Divide: $-\dfrac{7}{8} \div \dfrac{7}{8}$ [Section 3.3]

23. Subtract: $\dfrac{11}{12} - \dfrac{7}{15}$ [Section 3.4]

24. Determine which fraction is larger: $\dfrac{19}{15}$ or $\dfrac{5}{4}$ [Section 3.4]

25. **Hardware.** Find the length of the wood screw shown below. [Section 3.4]

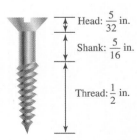

Head: $\dfrac{5}{32}$ in.

Shank: $\dfrac{5}{16}$ in.

Thread: $\dfrac{1}{2}$ in.

26. Multiply: $-15\dfrac{1}{3}\left(-\dfrac{9}{20}\right)$ [Section 3.5]

27. Motors. What is the difference in horsepower (hp) between the two motors shown? [Section 3.6]

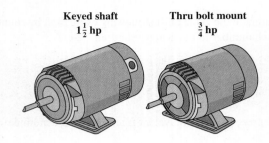

Keyed shaft
$1\frac{1}{2}$ hp

Thru bolt mount
$\frac{3}{4}$ hp

28. Simplify: $\dfrac{4 - \dfrac{3}{4}}{-1\dfrac{7}{8}}$ [Section 3.7]

29. Place an < or > symbol in the box to make a true statement. [Section 4.1]

-64.22 ▒ -64.238

30. Graph $-1\frac{3}{4}$, 2.25, -0.5, $\frac{11}{8}$, -3.2, and $\sqrt{9}$ on a number line. [Section 4.1]

$$\overset{\begin{array}{ccccccccccc} & & & & & & & & & & \\ -5 & -4 & -3 & -2 & -1 & 0 & 1 & 2 & 3 & 4 & 5 \end{array}}{\longleftrightarrow}$$

31. Add: $-20.04 + 2.4$ [Section 4.2]

32. Subtract: $-8.08 - 15.3$ [Section 4.2]

33. Multiply: $2.5 \cdot 100$ [Section 4.3]

34. Aquariums. One gallon of water weighs 8.33 pounds. What is the weight of the water in an aquarium that holds 55 gallons of water? [Section 4.3]

35. Divide: $2.5 \div 100$ [Section 4.4]

36. Evaluate the formula $t = \dfrac{d}{r}$ for $d = 107.95$ and $r = 8.5$. [Section 4.4]

37. Write $\dfrac{1}{12}$ as a decimal. [Section 4.5]

38. Lunch meats. A shopper purchased $\frac{3}{4}$ pound of barbequed beef, priced at $8.60 per pound, and $\frac{2}{3}$ pound of ham, selling for $5.25 per pound. Find the total cost of these items. [Section 4.5]

39. Evaluate: $3\sqrt{25} + 4\sqrt{4}$ [Section 4.6]

40. Express the phrase "3 inches to 15 inches" as a ratio in simplest form. [Section 5.1]

41. Building materials. Which is the better buy: a 94-pound bag of cement for $4.48 or a 100-pound bag of cement for $4.80? [Section 5.1]

42. Determine whether the proportion $\frac{25}{33} = \frac{12}{17}$ is true or false. [Section 5.2]

43. Caffeine. There are 55 milligrams of caffeine in 12 ounces of Mountain Dew. How many milligrams of caffeine are there in a super-size 44-ounce cup of Mountain Dew? Round to the nearest milligram. [Section 5.2]

44. Solve the proportion: $\dfrac{x}{3} = \dfrac{35}{7}$ [Section 5.2]

45. Survival guide [Section 5.3]

   a. A person can go without food for about 40 days. How many hours is this?

   b. A person can go without water for about 3 days. How many minutes is that?

   c. A person can go without breathing oxygen for about 8 minutes. How many seconds is that?

46. Convert 40 ounces to pounds. [Section 5.3]

47. Convert 2.4 meters to millimeters. [Section 5.4]

48. Convert 320 grams to kilograms. [Section 5.4]

49. a. Which holds more: a 2-liter bottle or a 1-gallon bottle? [Section 5.5]

   b. Which is longer: a meterstick or a yardstick?

50. Belts. A leather belt made in Mexico is 92 centimeters long. Express the length of the belt to the nearest inch. [Section 5.5]

6 Percent

Image Source/Getty Images

from **Campus to Careers**

Loan Officer

Loan officers help people apply for loans. *Commercial loan officers* work with businesses, *mortgage loan officers* work with people who want to buy a house or other real estate, and *consumer loan officers* work with people who want to buy a boat, a car, or need a loan for college. Loan officers analyze the applicant's financial history and often use banking formulas to determine the possibility of granting a loan.

In **Problem 43** of **Study Set 6.5**, you will see how a credit union loan officer calculates the interest to be charged on a loan.

JOB TITLE:
Loan Officer

EDUCATION:
Most have a degree in finance, economics, or a similar field. Mathematics and computer classes are good preparation for this job.

JOB OUTLOOK:
Employment of loan officers is expected to grow about as fast as the average (14%) for all jobs through 2020.

ANNUAL EARNINGS:
In 2015, the median salary for a loan officer was $63,430.

FOR MORE INFORMATION:
www.bls.gov/oes/current/oes132072.htm

CHAPTER OUTLINE

6.1 Percents, Decimals, and Fractions

6.2 Solving Percent Problems Using Percent Equations and Proportions

6.3 Applications of Percent

6.4 Estimation with Percent

6.5 Interest

CHAPTER SUMMARY AND REVIEW

CHAPTER TEST

CUMULATIVE REVIEW

OBJECTIVES

1 Explain the meaning of percent.

2 Write percents as fractions.

3 Write percents as decimals.

4 Write decimals as percents.

5 Write fractions as percents.

SECTION **6.1** Percents, Decimals, and Fractions

We see percents everywhere, every day. Stores use them to advertise discounts, manufacturers use them to describe the contents of their products, and banks use them to list interest rates for loans and savings accounts. Newspapers are full of information presented in percent form. In this section, we introduce percents and show how fractions, decimals, and percents are related.

OBJECTIVE 1 **Explain the meaning of percent.**

A percent tells us the number of parts per one hundred. You can think of a percent as the *numerator* of a fraction (or ratio) that has a denominator of 100.

Percent

Percent means parts per one hundred.

In the figure below, there are 100 equal-sized square regions, and 93 of them are shaded. Thus, $\frac{93}{100}$ or 93 percent of the figure is shaded. The word *percent* can be written using the symbol %, so we say that 93% of the figure is shaded.

Numerator

$$\frac{93}{100} = 93\%$$

Per 100

If the entire figure had been shaded, we would say that 100 out of the 100 square regions, or 100%, was shaded. Using this fact, we can determine what percent of the figure is *not* shaded by subtracting the percent of the figure that is shaded from 100%.

$$100\% - 93\% = 7\%$$

So 7% of the figure is *not* shaded.

To illustrate a percent greater than 100%, say 121%, we would shade one entire figure and 21 of the 100 square regions in a second, equal-sized grid.

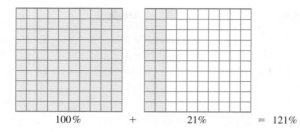

100%   +   21%   =   121%

EXAMPLE 1   **Tossing a coin.**   A coin was tossed 100 times and it landed heads up 51 times.

a. What percent of the time did the coin land heads up?

b. What percent of the time did it land tails up?

Strategy We will write a fraction that compares the number of times that the coin landed heads up (or tails up) to the total number of tosses.

WHY Since the denominator in each case will be 100, the numerator of the fraction will give the percent.

Solution

a. If a coin landed heads up 51 times after being tossed 100 times, then

$$\frac{51}{100} = 51\%$$

of the time it landed heads up.

b. The number of times the coin landed tails up is $100 - 51 = 49$ times. If a coin landed tails up 49 times after being tossed 100 times, then

$$\frac{49}{100} = 49\%$$

of the time it landed tails up.

Self Check 1

Board games.   A standard Scrabble game contains 100 tiles. There are 42 vowel tiles, 2 blank tiles, and the rest are consonant tiles.

a. What percent of the tiles are vowels?

b. What percent of the tiles are consonants?

Now Try   Problem 13

OBJECTIVE 2 Write percents as fractions.

We can use the definition of percent to write any percent in an equivalent fraction form.

Writing Percents as Fractions

To write a percent as a fraction, drop the % symbol and write the given number over 100. Then simplify the fraction, if possible.

EXAMPLE 2   **Earth.**   The chemical makeup of Earth's atmosphere is 78% nitrogen, 21% oxygen, and 1% other gases. Write each percent as a fraction in simplest form.

Strategy We will drop the % symbol and write the given number over 100. Then we will simplify the resulting fraction, if possible.

WHY *Percent* means parts per one hundred, and the word *per* indicates a ratio (fraction).

Solution   We begin with nitrogen.

$$78\% = \frac{78}{100}$$ *Drop the % symbol and write 78 over 100.*

$$= \frac{\overset{1}{\cancel{2}} \cdot 39}{\underset{1}{\cancel{2}} \cdot 50}$$ *To simplify the fraction, factor 78 as 2 · 39 and 100 as 2 · 50. Then remove the common factor of 2 from the numerator and denominator.*

$$= \frac{39}{50}$$

Nitrogen makes up $\frac{78}{100}$, or $\frac{39}{50}$, of Earth's atmosphere.

Oxygen makes up 21%, or $\frac{21}{100}$, of Earth's atmosphere. Other gases make up 1%, or $\frac{1}{100}$, of the atmosphere.

Self Check 2

Watermelons.   An average watermelon is 92% water. Write this percent as a fraction in simplest form.

Now Try ➨ **Problems 17 and 23**

EXAMPLE 3   **Unions.**   In 2016, 10.7% of the U.S. labor force belonged to a union. Write this percent as a fraction in simplest form. (Source: Bureau of Labor Statistics)

Strategy   We will drop the % symbol and write the given number over 100. Then we will multiply the resulting fraction by a form of 1 and simplify, if possible.

WHY   When writing a percent as a fraction, the numerator and denominator of the fraction should be whole numbers that have no common factors (other than 1).

Solution

$$10.7\% = \frac{10.7}{100}$$ *Drop the % symbol and write 10.7 over 100.*

To write this as an equivalent fraction of *whole numbers*, we need to move the decimal point in the numerator one place to the right. (Recall that to find the product of a decimal and 10, we simply move the decimal point one place to the right.) Therefore, it follows that $\frac{10}{10}$ is the form of 1 that we should use to build $\frac{10.7}{100}$.

$$\frac{10.7}{100} = \frac{10.7}{100} \cdot \mathbf{\frac{10}{10}}$$ *Multiply the fraction by a form of 1.*

$$= \frac{10.7 \cdot \mathbf{10}}{100 \cdot \mathbf{10}}$$ *Multiply the numerators. Multiply the denominators.*

$$= \frac{107}{1,000}$$ *Since 107 and 1,000 do not have any common factors (other than 1), the fraction is in simplest form.*

Self Check 3

Unions.   In 2012, 11.3% of the U.S. labor force belonged to a union. Write this percent as a fraction in simplest form. (Source: Bureau of Labor Statistics)

Now Try ➨ **Problems 27 and 31**

Thus, $10.7\% = \frac{107}{1,000}$. This means that 107 out of every 1,000 workers in the U.S. labor force belonged to a union in 2016.

EXAMPLE 4   Write $66\frac{2}{3}\%$ as a fraction in simplest form.

Strategy   We will drop the % symbol and write the given number over 100. Then we will perform the division indicated by the fraction bar and simplify, if possible.

WHY   When writing a percent as a fraction, the numerator and denominator of the fraction should be whole numbers that have no common factors (other than 1).

Solution

$$66\frac{2}{3}\% = \frac{66\frac{2}{3}}{100}$$ *Drop the % symbol and write $66\frac{2}{3}$ over 100.*

To write this as a fraction of whole numbers, we will perform the division indicated by the fraction bar.

$$\frac{66\frac{2}{3}}{100} = 66\frac{2}{3} \div 100 \qquad \text{\textit{The fraction bar indicates division.}}$$

$$= \frac{200}{3} \cdot \frac{1}{100} \qquad \begin{array}{l}\text{\textit{Write }} 66\frac{2}{3} \text{ \textit{as an improper fraction and then multiply}}\\ \text{\textit{by the reciprocal of 100.}}\end{array}$$

$$= \frac{200 \cdot 1}{3 \cdot 100} \qquad \begin{array}{l}\text{\textit{Multiply the numerators.}}\\ \text{\textit{Multiply the denominators.}}\end{array}$$

$$= \frac{2 \cdot \overset{1}{\cancel{100}} \cdot 1}{3 \cdot \underset{1}{\cancel{100}}} \qquad \begin{array}{l}\text{\textit{To simplify the fraction, factor 200 as }} 2 \cdot 100. \text{ \textit{Then remove the}}\\ \text{\textit{common factor of 100 from the numerator and denominator.}}\end{array}$$

$$= \frac{2}{3}$$

Self Check 4

Write $83\frac{1}{3}\%$ as a fraction in simplest form.

Now Try ➲ Problem 35

EXAMPLE 5   **a.** Write 175% as a fraction in simplest form.

b. Write 0.22% as a fraction in simplest form.

Strategy We will drop the % symbol and write each given number over 100. Then we will simplify the resulting fraction, if possible.

WHY *Percent* means parts per one hundred and the word *per* indicates a ratio (fraction).

Solution

a. $175\% = \dfrac{175}{100}$ *Drop the % symbol and write 175 over 100.*

$$= \frac{\overset{1}{\cancel{5}} \cdot \overset{1}{\cancel{5}} \cdot 7}{2 \cdot 2 \cdot \underset{1}{\cancel{5}} \cdot \underset{1}{\cancel{5}}} \qquad \begin{array}{l}\text{\textit{To simplify the fraction, prime factor 175}}\\ \text{\textit{and 100. Remove the common factors of}}\\ \text{\textit{5 from the numerator and denominator.}}\end{array} \qquad \begin{array}{ll}5\,\lfloor\underline{175} & 2\,\lfloor\underline{100}\\ 5\,\lfloor\underline{35} & 2\,\lfloor\underline{50}\\ \quad 7 & 5\,\lfloor\underline{25}\\ & \quad 5\end{array}$$

$$= \frac{7}{4}$$

Thus, $175\% = \dfrac{7}{4}$.

b. $0.22\% = \dfrac{0.22}{100}$ *Drop the % symbol and write 0.22 over 100.*

To write this as an equivalent fraction of *whole numbers*, we need to move the decimal point in the numerator two places to the right. (Recall that to find the product of a decimal and 100, we simply move the decimal point two places to the right.) Therefore, it follows that $\frac{100}{100}$ is the form of 1 that we should use to build $\frac{0.22}{100}$.

$$\frac{0.22}{100} = \frac{0.22}{100} \cdot \boxed{\frac{100}{100}} \qquad \text{\textit{Multiply the fraction by a form of 1.}}$$

$$= \frac{0.22 \cdot 100}{100 \cdot 100} \qquad \begin{array}{l}\text{\textit{Multiply the numerators.}}\\ \text{\textit{Multiply the denominators.}}\end{array}$$

$$= \frac{22}{10,000}$$

$$= \frac{\overset{1}{\cancel{2}} \cdot 11}{\underset{1}{\cancel{2}} \cdot 5,000} \qquad \begin{array}{l}\text{\textit{To simplify the fraction, factor 22 and 10,000.}}\\ \text{\textit{Remove the common factor of 2 from the}}\\ \text{\textit{numerator and denominator.}}\end{array}$$

$$= \frac{11}{5,000}$$

Thus, $0.22\% = \dfrac{11}{5,000}$.

Success Tip When percents that are greater than 100% are written as fractions, the fractions are greater than 1. When percents that are less than 1% are written as fractions, the fractions are less than $\frac{1}{100}$.

Self Check 5

a. Write 210% as a fraction in simplest form.

b. Write 0.54% as a fraction in simplest form.

Now Try ➲ Problems 39 and 43

OBJECTIVE 3 Write percents as decimals.

To write a percent as a decimal, recall that a percent can be written as a fraction with denominator 100 and that a denominator of 100 indicates division by 100.

For example, consider 14%, which means 14 parts per 100.

$$14\% = \frac{14}{100}$$ *Use the definition of percent: write 14 over 100.*

$$= 14 \div 100$$ *The fraction bar indicates division.*

$$= 14.0 \div 100$$ *Write the whole number 14 in decimal notation by placing a decimal point immediately to its right and entering a zero to the right of the decimal point.*

$$= .140$$ *Since the divisor 100 has two zeros, move the decimal point two places to the left.*

$$= 0.14$$ *Write a zero to the left of the decimal point.*

We have found that 14% = 0.14. This example suggests the following procedure.

Writing Percents as Decimals

To write a percent as a decimal, drop the % symbol and divide the given number by 100 by moving the decimal point two places to the left.

EXAMPLE 6   **Social networks.** The graph below shows the percent of market share of visits for the top five social networking sites.

a. Write the percent of market share for the Facebook site as a decimal.

b. Write the percent of market share for the Twitter site as a decimal.

Strategy We will drop the % symbol and divide each given number by 100 by moving the decimal point two places to the left.

WHY Recall from Section 4.4 that to find the quotient of a decimal and 10, 100, 1,000, and so on, move the decimal point to the left the same number of places as there are zeros in the power of 10.

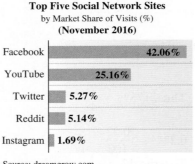

Top Five Social Network Sites
by Market Share of Visits (%)
(November 2016)

Facebook	**42.06%**
YouTube	**25.16%**
Twitter	**5.27%**
Reddit	**5.14%**
Instagram	**1.69%**

Source: dreamgrow.com

Solution

a. From the graph, we see that the percent market share of visits for Facebook is 42.06%. To write this percent as a decimal, we proceed as follows.

$$42.06\% = .4206$$ *Drop the % symbol and divide 42.06 by 100 by moving the decimal point two places to the left.*

$$= 0.4206$$ *Write a zero to the left of the decimal point.*

42.06%, written as a decimal, is 0.4206.

Self Check 6

Social networks. Refer to the graph on the right.

a. Write the percent of market share of visits for YouTube as a decimal.

b. Write the percent of market share of visits for Instagram as a decimal.

Now Try Problems 51 and 55

b. From the graph, we see that the percent market share of visits for Twitter is 5.27%. To write this percent as a decimal, we proceed as follows.

$5.27\% = .0527$ Drop the % symbol and divide 5.27 by 100 by moving the decimal point two places to the left. This requires that a placeholder zero (shown in blue) be inserted in front of the 5.

$\qquad = 0.0527$ Write a zero to the left of the decimal point.

5.27%, written as a decimal, is 0.0527.

EXAMPLE 7   **Population.**   The population of the state of Oregon is approximately $1\frac{1}{4}\%$ of the population of the United States. Write this percent as a decimal. (Source: U.S. Census Bureau)

Strategy We will write the mixed number $1\frac{1}{4}$ in decimal notation.

WHY With $1\frac{1}{4}$ in mixed-number form, we cannot apply the rule for writing a percent as a decimal; there is no decimal point to move two places to the left.

Solution To change a percent to a decimal, we drop the percent symbol and divide by 100 by moving the decimal point two places to the left. In this case, however, there is no decimal point to move in $1\frac{1}{4}\%$. Since $1\frac{1}{4} = 1 + \frac{1}{4}$, and since the decimal equivalent of $\frac{1}{4}$ is 0.25, we can write $1\frac{1}{4}\%$ in an equivalent form as 1.25%.

$1\dfrac{1}{4}\% = 1.25\%$ Write $1\frac{1}{4}$ as 1.25.

$\qquad = .0125$ Drop the % symbol and divide 1.25 by 100 by moving the decimal point two places to the left. This requires that a placeholder zero (shown in blue) be inserted in front of the 1.

$\qquad = 0.0125$ Write a zero to the left of the decimal point.

$1\frac{1}{4}\%$, written as a decimal, is 0.0125.

Self Check 7

Population. The population of the state of Ohio is approximately $3\frac{3}{4}\%$ of the population of the United States. Write this percent as a decimal. (Source: U.S. Census Bureau)

Now Try ➡ **Problem 57**

EXAMPLE 8   **a.** Write 310% as a decimal.   **b.** Write 0.9% as a decimal.

Strategy We will drop the % symbol and divide each given number by 100 by moving the decimal point two places to the left.

WHY Recall that to find the quotient of a decimal and 100, we move the decimal point to the left the same number of places as there are zeros in 100.

Solution

a. $310\% = 310.0\%$ Write the whole number 310 in decimal notation: $310 = 310.0$.

$\qquad = 3.100$ Drop the % symbol and divide 310 by 100 by moving the decimal point two places to the left.

$\qquad = 3.1$ Drop the unnecessary zeros to the right of the 1.

310%, written as a decimal, is 3.1.

b. $0.9\% = .009$ Drop the % symbol and divide 0.9 by 100 by moving the decimal point two places to the left. This requires that a placeholder zero (shown in blue) be inserted in front of the 0.

$\qquad = 0.009$ Write a zero to the left of the decimal point.

0.9%, written as a decimal, is 0.009.

Success Tip When percents that are greater than 100% are written as decimals, the decimals are greater than 1.0. When percents that are less than 1% are written as decimals, the decimals are less than 0.01.

Self Check 8

a. Write 600% as a decimal.

b. Write 0.8% as a decimal.

Now Try ➡ **Problems 63 and 67**

OBJECTIVE 4 Write decimals as percents.

To write a percent as a decimal, we drop the % symbol and move the decimal point two places to the left. To write a decimal as a percent, we do the opposite: we move the decimal point two places to the right and insert a % symbol.

Writing Decimals as Percents

To write a decimal as a percent, multiply the decimal by 100 by moving the decimal point two places to the right, and then insert a % symbol.

EXAMPLE 9 **Geography.** Land areas make up 0.291 of Earth's surface. Write this decimal as a percent.

Strategy We will multiply the decimal by 100 by moving the decimal point two places to the right, and then insert a % symbol.

WHY To write a *decimal as a percent*, we reverse the steps used to write a *percent as a decimal.*

Solution

$$0.291 = 029.1\%$$ Multiply 0.291 by 100 by moving the decimal point two places to the right, and then insert a % symbol.

$$= 29.1\%$$

0.291, written as a percent, is 29.1%

Self Check 9

Write 0.5343 as a percent.

Now Try ⟳ **Problems 71 and 75**

OBJECTIVE 5 Write fractions as percents.

We use a two-step process to write a fraction as a percent. First, we write the fraction as a decimal. Then we write that decimal as a percent.

$$\boxed{\text{Fraction}} \longrightarrow \boxed{\text{decimal}} \longrightarrow \boxed{\text{percent}}$$

Writing Fractions as Percents

To write a fraction as a percent:

1. Write the fraction as a decimal by dividing its numerator by its denominator.

2. Multiply the decimal by 100 by moving the decimal point two places to the right, and then insert a % symbol.

EXAMPLE 10 **Television.** The highest-rated television show of all time was a special episode of *M*A*S*H* that aired February 28, 1983. Surveys found that three out of every five American households watched this show. Express the rating as a percent. (Source: *The World Almanac and Book of Facts*, 2013)

Strategy First, we will translate the phrase *three out of every five* to fraction form and write that fraction as a decimal. Then we will write that decimal as a percent.

WHY A fraction-to-decimal-to-percent approach must be used to write a fraction as a percent.

Solution

Step 1 The phrase *three out of every five* can be expressed as $\frac{3}{5}$. To write this fraction as a decimal, we divide the numerator, 3, by the denominator, 5.

$$
\begin{array}{r}
0.6 \\
5\overline{)3.0} \\
-3\,0 \\
\hline
0
\end{array}
$$
Write a decimal point and one additional zero to the right of 3.

← The remainder is 0.

The result is a terminating decimal.

Step 2 To write 0.6 as a percent, we proceed as follows.

$$\frac{3}{5} = 0.6$$

$$0.6 = 0\underset{\curvearrowright}{60}.\%$$ Write a placeholder 0 to the right of the 6 (shown in blue). Multiply 0.60 by 100 by moving the decimal point two places to the right, and then insert a % symbol.

$$= 60\%$$

60% of American households watched the special episode of *M*A*S*H*.

EXAMPLE 11   Write $\frac{13}{4}$ as a percent.

Strategy We will write the fraction $\frac{13}{4}$ as a decimal. Then we will write that decimal as a percent.

WHY A fraction-to-decimal-to-percent approach must be used to write a fraction as a percent.

Solution

Step 1 To write $\frac{13}{4}$ as a decimal, we divide the numerator, 13, by the denominator, 4.

$$
\begin{array}{r}
3.25 \\
4\overline{)13.00} \\
-12\downarrow \\
\hline
1\,0 \\
-8\downarrow \\
\hline
20 \\
-20 \\
\hline
0
\end{array}
$$
Write a decimal point and two additional zeros to the right of 3.

← The remainder is 0.

The result is a terminating decimal.

Step 2 To write 3.25 as a percent, we proceed as follows.

$$3.25 = 3\underset{\curvearrowright}{25}.\%$$ Multiply 3.25 by 100 by moving the decimal point two places to the right, and then insert a % symbol.

$$= 325\%$$

The fraction $\frac{13}{4}$, written as a percent, is 325%.

In Examples 10 and 11, the result of the division was a terminating decimal. Sometimes when we write a fraction as a decimal, the result of the division is a repeating decimal.

*M*A*S*H* was a highly popular sitcom that aired on CBS from 1972 through 1983. It was set during the Korean War and focused on the relationships between members of a Mobile Army Surgical Hospital unit.

Self Check 10

Advertising. There is a commercial that says, "Four out of five dentists recommend sugarless gum for their patients who chew gum." Express this recommendation as a percent.

Now Try Problem 79

Success Tip When fractions that are greater than 1 are written as percents, the percents are greater than 100%.

Self Check 11

Write $\frac{5}{2}$ as a percent.

Now Try Problem 85

EXAMPLE 12 Write $\frac{5}{6}$ as a percent. Give the exact answer and an approximation to the nearest tenth of one percent.

Strategy We will write the fraction $\frac{5}{6}$ as a decimal. Then we will write that decimal as a percent.

WHY A fraction-to-decimal-to-percent approach must be used to write a fraction as a percent.

Solution

Step 1 To write $\frac{5}{6}$ as a decimal, we divide the numerator, 5, by the denominator, 6.

$$
\begin{array}{r}
0.8333 \\
6\overline{)5.0000} \\
-4\,8 \\
\hline
20 \\
-18 \\
\hline
20 \\
-18 \\
\hline
20 \\
-18 \\
\hline
2
\end{array}
$$

Write a decimal point and several zeros to the right of 5.

← The repeating pattern is now clear. We can stop the division.

The result is a repeating decimal.

Step 2 To write the decimal as a percent, we proceed as follows.

$$\frac{5}{6} = 0.8333\ldots$$

$$0.833\ldots = 0\,83.33\ldots\%$$

Multiply 0.8333... by 100 by moving the decimal point two places to the right, and then insert a % symbol.

$$= 83.33\ldots\%$$

We must now decide whether we want an exact answer or an approximation. For an exact answer, we can represent *the repeating part of the decimal using an equivalent fraction.* For an approximation, we can round 83.333 . . . % to a specific place value.

Exact answer:

$$\frac{5}{6} = 83.\underbrace{3333\ldots}\%$$

$$= 83\frac{1}{3}\%$$

Use the fraction $\frac{1}{3}$ to represent .3333....

Thus,

$$\frac{5}{6} = 83\frac{1}{3}\%$$

Approximation:

$$\frac{5}{6} = 83.33\ldots\%$$

$$\approx 83.3\%$$

Round to the nearest tenth.

Thus,

$$\frac{5}{6} \approx 83.3\%$$

Self Check 12

Write $\frac{2}{3}$ as a percent. Give the exact answer and an approximation to the nearest tenth of one percent.

Now Try Problem 91

Some percents occur so frequently that it is useful to memorize their fractional and decimal equivalents.

Percent	Decimal	Fraction	Percent	Decimal	Fraction
1%	0.01	$\frac{1}{100}$	$33\frac{1}{3}\%$	0.3333 . . .	$\frac{1}{3}$
10%	0.1	$\frac{1}{10}$	50%	0.5	$\frac{1}{2}$
$16\frac{2}{3}\%$	0.1666 . . .	$\frac{1}{6}$	$66\frac{2}{3}\%$	0.6666 . . .	$\frac{2}{3}$
20%	0.2	$\frac{1}{5}$	75%	0.75	$\frac{3}{4}$
25%	0.25	$\frac{1}{4}$	$83\frac{1}{3}\%$	0.8333 . . .	$\frac{5}{6}$

Answers to Self Checks

1. a. 42% **b.** 56%   **2.** $\frac{23}{25}$ **3.** $\frac{113}{1,000}$ **4.** $\frac{5}{6}$ **5. a.** $\frac{21}{10}$ **b.** $\frac{27}{5,000}$ **6. a.** 0.2516 **b.** 0.0169   **7.** 0.0375
8. a. 6 **b.** 0.008   **9.** 53.43%   **10.** 80%   **11.** 250%   **12.** $66\frac{2}{3}\% \approx 66.7\%$

SECTION **6.1** **STUDY SET**

VOCABULARY

Fill in the blanks.

1. _____ means parts per one hundred.
2. The word *percent* is formed from the prefix *per*, which means _____, and the suffix *cent*, which comes from the Latin word *centum*, meaning ____.

CONCEPTS

Fill in the blanks.

3. To write a percent as a fraction, drop the % symbol and write the given number over ____. Then _____ the fraction, if possible.
4. To write a percent as a decimal, drop the % symbol and divide the given number by 100 by moving the decimal point two places to the ____.
5. To write a decimal as a percent, multiply the decimal by 100 by moving the decimal point two places to the _____, and then insert a % symbol.
6. To write a fraction as a percent, first write the fraction as a _____. Then multiply the decimal by 100 by moving the decimal point two places to the right, and then insert a ___ symbol.

NOTATION

7. What does the symbol % mean?
8. Write the whole number 45 as a decimal.

GUIDED PRACTICE

What percent of the figure is shaded? What percent of the figure is not shaded? See Objective 1.

9.    **10.**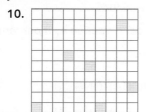

In the following illustrations, each set of 100 square regions represents 100%. What percent is shaded?

11.

12.

For Problems 13–16, see Example 1.

13. **The Internet.** The following sentence appeared on a technology blog: "Ask Internet users what they want from their service and 99 times out of 100 the answer will be the same: more speed." According to the blog, what percent of the time do Internet users give that answer?
14. **Basketball records.** In 1962, Wilt Chamberlain of the Philadelphia Warriors scored a total of 100 points in an NBA game. If 28 of his points came from made free throws, what percent of his point total came from free throws?
15. **Quilts.** A quilt is made from 100 squares of colored cloth.
    a. If 15 of the squares are blue, what percent of the squares in the quilt are blue?
    b. What percent of the squares are not blue?
16. **Divisibility.** Of the natural numbers from 1 through 100, only 14 of them are divisible by 7.
    a. What percent of the numbers are divisible by 7?
    b. What percent of the numbers are not divisible by 7?

Write each percent as a fraction. Simplify, if possible. See Example 2.

17. 17%   **18.** 31%

19. 91%   **20.** 89%

21. 4%   **22.** 5%

23. 60%   **24.** 40%

Write each percent as a fraction. Simplify, if possible. See Example 3.

25. 1.9%   **26.** 2.3%

27. 54.7%   **28.** 97.1%

29. 12.5%   **30.** 62.5%

31. 6.8%   **32.** 4.2%

Write each percent as a fraction. Simplify, if possible. See Example 4.

33. $1\frac{1}{3}\%$ **34.** $3\frac{1}{3}\%$

35. $14\frac{1}{6}\%$ **36.** $10\frac{5}{6}\%$

Write each percent as a fraction. Simplify, if possible.
See Example 5.

37. 130%

38. 160%

39. 220%

40. 240%

41. 0.35%

42. 0.45%

43. 0.25%

44. 0.75%

Write each percent as a decimal. See Objective 3.

45. 16%

46. 11%

47. 81%

48. 93%

Write each percent as a decimal. See Example 6.

49. 34.12%

50. 27.21%

51. 50.033%

52. 40.083%

53. 6.99%

54. 4.77%

55. 1.3%

56. 8.6%

Write each percent as a decimal. See Example 7.

57. $7\frac{1}{4}$%

58. $9\frac{3}{4}$%

59. $18\frac{1}{2}$%

60. $25\frac{1}{2}$%

Write each percent as a decimal. See Example 8.

61. 460%

62. 230%

63. 316%

64. 178%

65. 0.5%

66. 0.9%

67. 0.03%

68. 0.06%

Write each decimal or whole number as a percent.
See Example 9.

69. 0.362

70. 0.245

71. 0.98

72. 0.57

73. 1.71

74. 4.33

75. 4

76. 9

Write each fraction as a percent. See Example 10.

77. $\frac{2}{5}$

78. $\frac{1}{5}$

79. $\frac{4}{25}$

80. $\frac{9}{25}$

81. $\frac{5}{8}$

82. $\frac{3}{8}$

83. $\frac{7}{16}$

84. $\frac{9}{16}$

Write each fraction as a percent. See Example 11.

85. $\frac{9}{4}$

86. $\frac{11}{4}$

87. $\frac{21}{20}$

88. $\frac{33}{20}$

Write each fraction as a percent. Give the exact answer and an approximation to the nearest tenth of one percent. See Example 12.

89. $\frac{1}{6}$

90. $\frac{2}{9}$

91. $\frac{5}{3}$

92. $\frac{4}{3}$

TRY IT YOURSELF

Complete the table. Give an exact answer and an approximation to the nearest tenth of one percent when necessary. Round decimals to the nearest hundredth when necessary.

	Fraction	Decimal	Percent
93.		0.0314	
94.		0.0021	
95.			40.8%
96.			34.2%
97.			$5\frac{1}{4}$%
98.			$6\frac{3}{4}$%
99.	$\frac{7}{3}$		
100.	$\frac{7}{9}$		

LOOK ALIKES

Write each percent as a decimal.

101. a. 3.07%　　　　　　**b.** 30.7%

Write each decimal as a percent.

102. a. 0.845　　　　　　**b.** 8.45

Write each percent as a fraction.

103. a. 6.91%　　　　　　**b.** 69.1%

Write each fraction as a percent.

104. a. $\frac{5}{4}$　　　　　　**b.** $\frac{4}{5}$

APPLICATIONS

105. U.S. charities. For every dollar donated to The Salvation Army, 82 cents goes toward program services such as food for the hungry, relief for disaster victims, and assistance for the disabled and elderly. The remaining money goes to cover overhead costs like fundraising and management.

a. What percent of the donations goes toward program services?

b. What percent goes to cover overhead?

106. Biology. Blood is 92% water. Write the percent as a fraction in simplest form. (Source: waterinfo.org)

107. Regions of the country. The continental United States is divided into seven regions as shown below. What percent of the 50 states are

a. in the Rocky Mountain region?

b. in the Southwestern region?

c. not located in any of the seven regions shown here?

108. Road signs. Sometimes, signs like that shown below are posted to warn truckers when they are approaching a steep grade on the highway.

a. Write the grade shown on the sign as a fraction.

b. Write the grade shown on the sign as a decimal.

109. Interest rates. Write each interest rate for the following accounts as a decimal.

a. Home loan: 4.75%

b. Savings account: 1%

c. Credit card: 14.25%

110. Drunk driving. In most states, it is illegal to drive with a blood alcohol concentration of 0.08% or higher.

a. Write this percent as a fraction. Do not simplify.

b. Use your answer to part a to fill in the blanks: A blood alcohol concentration of 0.08% means ___ parts alcohol to ___ parts blood.

111. Human skin. The illustration below shows what percent of the total skin area that each section of the body covers. Find the missing percent for the torso, and then complete the bar graph.

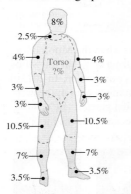

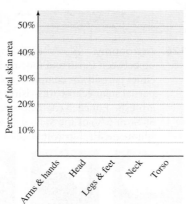

112. Recycling. The U.S. aluminum beverage can recycling rates for the years 2009 through 2015 are given in the table below as percents. For example, in 2009, 57.4% of the aluminum beverage cans that were used were recycled. Construct a line graph for the data.

2009	2010	2011	2012	2013	2014	2015
57.4%	58.1%	65.1%	67.0%	66.7%	66.5%	64.3%

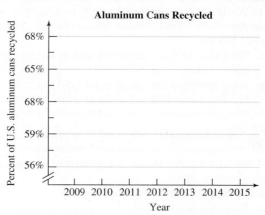

Source: cancentral.com

113. The UN Security Council. The United Nations has 193 members. The United States, Russia, the United Kingdom, France, and China, along with ten other nations, make up the Security Council.

a. What fraction of the members of the United Nations belong to the Security Council?

b. Write your answer to part a as a decimal. Round to the nearest ten-thousandth.

c. Write your answer to part b as a percent.

114. Soap. Ivory soap claims to be $99\frac{44}{100}\%$ pure. Write this percent as a decimal.

115. Logos. In the illustration, what part of the company's logo is shaded red? Express your answer as a percent (exact), a fraction, and a decimal (using an overbar).

Recycling Industries Inc.

116. The human spine. The human spine consists of a group of bones (vertebrae) as shown.

a. What fraction of the vertebrae are lumbar?

b. What percent of the vertebrae are lumbar? (Round to the nearest one percent.)

c. What percent of the vertebrae are cervical? (Round to the nearest one percent.)

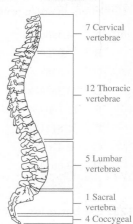

7 Cervical vertebrae

12 Thoracic vertebrae

5 Lumbar vertebrae

1 Sacral vertebra

4 Coccygeal vertebrae

117. Boxing. Oscar De La Hoya, also known as "the Golden Boy," generated over $600 million in pay-per-view revenue for his fights. He won 39 out of 45 of his professional fights.

a. What fraction of his fights did he win?

b. What percent of his fights did he win? Give the exact answer and an approximation to the nearest tenth of one percent.

118. Baseball history. In 1997, the Florida Marlins were the World Series champions. In 1998, they had the worst regular season record in major league baseball: 54 wins and 108 losses.

a. What was the total number of regular season games that the Marlins played in 1998?

b. What percent of the games played did the Marlins win in 1998? Give the exact answer and an approximation to the nearest tenth of one percent.

119. Economic forecasts. One economic indicator of the national economy is the number of orders placed by manufacturers. One month, the number of orders rose *one-fourth of 1 percent.*

a. Write this using a % symbol.

b. Express it as a fraction.

c. Express it as a decimal.

120. Taxes. Several years ago, Springfield, Missouri, voters approved a *one-eighth of one percent* sales tax to fund transportation projects in the city.

a. Write the percent as a decimal.

b. Write the percent as a fraction.

121. a. Birthdays. If the day of your birthday represents $\frac{1}{365}$ of a year, what percent of the year is it? Round to the nearest hundredth of a percent.

b. **Population.** As a fraction, each resident of the United States represents approximately $\frac{1}{325,000,000}$ of the U.S. population. Express this as a percent. Round to one nonzero digit.

122. The alphabet. To determine which letters of the alphabet are used the most often in forming the words of the English language, an analysis of all of the root words in the Oxford Dictionary was done. A chart showing the percent usage of each letter is shown below. List the 26 letters, in order, from greatest usage to least usage.

A	8.4966%	J	0.1965%	S	5.7351%
B	2.0720%	K	1.1016%	T	6.9509%
C	4.5388%	L	5.4893%	U	3.6308%
D	3.3844%	M	3.0129%	V	1.0074%
E	11.1607%	N	6.6544%	W	1.2899%
F	1.8121%	O	7.1635%	X	0.2902%
G	2.4705%	P	3.1671%	Y	1.7779%
H	3.0034%	Q	0.1962%	Z	0.2722%
I	7.5448%	R	7.5809%		

123. Car batteries. A voltmeter can be attached to a car's battery to measure its charge in volts. The range of typical voltmeter readings are shown in the next column. Mechanics recommend batteries that have less than a 75% charge should be recharged. Suppose the reading in the illustration was made at 50°F, does the car's battery need to be recharged?

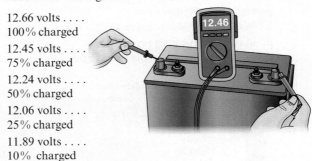

12.66 volts . . . .
100% charged

12.45 volts . . . .
75% charged

12.24 volts . . . .
50% charged

12.06 volts . . . .
25% charged

11.89 volts . . . .
10% charged

*These readings are at 80°F. Battery voltage readings drop with the temperature roughly 0.01 volt for every 10°F.

124. Copy paper. Hammermill® Copy Plus $8\frac{1}{2}$ in. by 11 in. paper comes with following Jam-Free™ guarantee:

We guarantee you will not experience more than one jam in 10,000 sheets on your high-speed digital equipment or we will replace your paper or refund your purchase price.

a. Express one in 10,000 as a fraction, decimal, and percent.

b. What percent of 10,000 sheets of paper will not jam?

125. Firefighting. A forest fire *containment* number is the percent of the fire's perimeter that is surrounded by a fire control line. For example, a fire is 100% contained when it's completely encircled by a dirt track. Refer to the illustration below. What percent of the fire is contained? Choose one: 10%, 25%, 50%, or 75%.

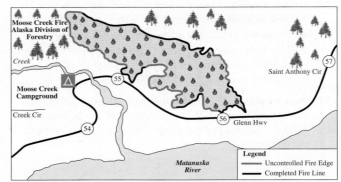

126. Astronomy. A *solar eclipse* occurs when the Moon lines up between Earth and the Sun in such a way that the Moon either partially or completely blocks out the Sun. In 2013, a partial solar eclipse was visible from many cites in the eastern U.S. For each city below, tell what percent (54%, 44%, 36%, or 6%) of the Sun's disk was covered by the moon's silhouette. Each percent can only be used once.

Akron, OH	Boston, MA	Miami, FL	Philadelphia, PA
(a)	(b)	(c)	(d)

WRITING

127. If you were writing advertising, which form do you think would attract more customers: "25% off" or "$\frac{1}{4}$ off"? Explain your reasoning.

128. Many coaches ask their players to give a 110% effort during practices and games. What do you think this means? Is it possible?

129. Explain how an amusement park could have an attendance that is 103% of capacity.

130. Won–lost records. In sports, when a team wins as many games as it loses, it is said to be playing "500 ball." Suppose in its first 40 games, a team wins 20 games and loses 20 games. Use the concepts in this section to explain why such a record could be called "500 ball."

REVIEW

131. The width of a rectangle is 6.5 centimeters and its length is 10.5 centimeters.
 a. Find its perimeter.
 b. Find its area.

132. The length of a side of a square is 9.8 meters.
 a. Find its perimeter.
 b. Find its area.

SECTION 6.2 Solving Percent Problems Using Percent Equations and Proportions

The articles on the front page of the newspaper below illustrate three types of percent problems.

OBJECTIVES

PERCENT EQUATIONS

1 Translate percent sentences to percent equations.

2 Solve percent equations to find the amount.

3 Solve percent equations to find the percent.

4 Solve percent equations to find the base.

PERCENT PROPORTIONS

1 Write percent proportions.

2 Solve percent proportions to find the amount.

3 Solve percent proportions to find the percent.

4 Solve percent proportions to find the base.

5 Read circle graphs.

Type 1 In the labor article, if we want to know how many union members voted to accept the new offer, we would ask:

What number is 84% of 500?

Type 2 In the article on drinking water, if we want to know what percent of the wells are safe, we would ask:

38 is what percent of 40?

Type 3 In the article on new appointees, if we want to know how many members are on the State Board of Examiners, we would ask:

6 is 75% of what number?

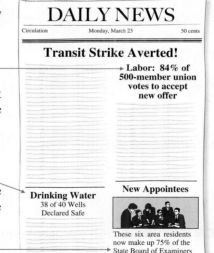

DAILY NEWS

Circulation   Monday, March 23   50 cents

Transit Strike Averted!

Labor: 84% of 500-member union votes to accept new offer

Drinking Water
38 of 40 Wells
Declared Safe

New Appointees

These six area residents now make up 75% of the State Board of Examiners

This section introduces two methods that can be used to solve the percent problems shown above. The first method involves writing and solving *percent equations*. The second method involves writing and solving *percent proportions*. If your instructor only requires you to learn the proportion method, then turn to page 514 and begin reading Objective 1.

METHOD 1: PERCENT EQUATIONS

OBJECTIVE 1 Translate percent sentences to percent equations.

The **percent sentences** highlighted in blue in the introduction above have three things in common.

■ Each contains the word *is*. Here, *is* can be translated as an = symbol.

■ Each contains the word *of*. In this case, *of* means multiply.

■ Each contains a phrase such as *what number* or *what percent*. In other words, there is an unknown number that can be represented by a variable.

These observations suggest that each percent sentence contains key words that can be translated to form an equation. The equation, called a **percent equation**, will contain three numbers (two known and one unknown represented by a variable), the operation of multiplication, and, of course, an = symbol.

The key words in a percent sentence translate as follows:

- *is* translates to an equal symbol =
- *of* translates to multiplication that is shown with a raised dot ·
- *what number* or *what percent* translates to an unknown number that is represented by a variable

---

EXAMPLE 1 Translate each percent sentence to a percent equation.

a. What number is 12% of 64?

b. What percent of 88 is 11?

c. 165% of what number is 366?

Strategy We will look for the key words *is*, *of*, and *what number* (or *what percent*) in each percent sentence.

WHY These key words translate to mathematical symbols that form the percent equation.

Solution In each case, we will let the variable x represent the unknown number. However, any letter can be used.

a. What number | is | 12% | of | 64?   *This is the given percent sentence.*

$$x = 12\% \cdot 64$$ *This is the percent equation.*

b. What percent | of | 88 | is | 11?   *This is the given percent sentence.*

$$x \cdot 88 = 11$$ *This is the percent equation.*

c. 165% | of | what number | is | 366?   *This is the given percent sentence.*

$$165\% \cdot x = 366$$ *This is the percent equation.*

Self Check 1

Translate each percent sentence to a percent equation.

a. What number is 33% of 80?

b. What percent of 55 is 6?

c. 172% of what number is 4?

Now Try Problem 17

OBJECTIVE 2 Solve percent equations to find the amount.

To solve the labor union percent problem (Type 1 from the newspaper), we translate the percent sentence into a percent equation and then find the unknown number.

EXAMPLE 2 What number is 84% of 500?

Strategy We will look for the key words *is*, *of*, and *what number* in the percent sentence and translate them to mathematical symbols to form a percent equation.

WHY Then it will be clear what operation should be performed to find the unknown number.

Solution First, we translate.

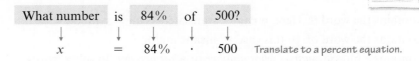

What number | is | 84% | of | 500?

$$x = 84\% \cdot 500$$ *Translate to a percent equation.*

Caution! When solving percent equations, always write the percent as a decimal (or a fraction) before performing any calculations. In Example 2, we wrote 84% as 0.84 before multiplying by 500.

Now we perform the multiplication on the right side of the equation.

$x = 0.84 \cdot 500$ Write 84% as a decimal: 84% = 0.84.

$x = 420$ Do the multiplication.

We have found that 420 is 84% of 500. That is, 420 union members mentioned in the newspaper article voted to accept the new offer.

Self Check 2
What number is 36% of 400?

Now Try ➲ **Problems 19 and 71**

Percent sentences involve a comparison of numbers. In the statement "420 is 84% of 500," the number 420 is called the **amount**, 84% is the **percent**, and 500 is called the **base**. Think of the base as the standard of comparison—it represents the **whole** of some quantity. The amount is a **part** of the base, but it can exceed the base when the percent is more than 100%. The percent, of course, has the % symbol.

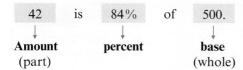

42	is	84%	of	500.
↓		↓		↓
Amount		**percent**		**base**
(part)				(whole)

In any percent problem, the relationship between the amount, the percent, and the base is as follows: *Amount is percent of base.* This relationship is shown below as the **percent equation** (also called the **percent formula**).

Percent Equation (Formula)

Any percent sentence can be translated to a percent equation that has the form:

Amount = percent · base   or   Part = percent · whole

EXAMPLE 3   What number is 160% of 15.8?

Strategy We will look for the key words *is*, *of*, and *what number* in the percent sentence and translate them to mathematical symbols to form a percent equation.

WHY Then it will be clear what operation needs to be performed to find the unknown number.

Solution First, we translate.

What number	is	160%	of	15.8?
↓	↓	↓	↓	↓
x	=	160%	·	15.8

x is the amount, 160% is the percent, and 15.8 is the base.

LANGUAGE OF MATHEMATICS

When we find the value of the variable that makes a percent equation true, we say that we have **solved the equation**. In Example 2, we solved $x = 84\% \cdot 500$ to find that x is 420.

Now we solve the equation by performing the multiplication on the right side.

$x = 1.6 \cdot 15.8$ Write 160% as a decimal: 160% = 1.6.

$x = 25.28$ Do the multiplication.

$$\begin{array}{r} 15.8 \\ \times\, 1.6 \\ \hline 948 \\ 1580 \\ \hline 25.28 \end{array}$$

Thus, 25.28 is 160% of 15.8. In this case, the amount exceeds the base because the percent is more than 100%.

Self Check 3
What number is 240% of 80.3?

Now Try ➲ **Problem 23**

OBJECTIVE 3 Solve percent equations to find the percent.

In the drinking water problem (Type 2 from the newspaper), we must find the percent. Once again, we translate the words of the problem into a percent equation and solve it.

EXAMPLE 4   38 is what percent of 40?

Strategy We will look for the key words *is*, *of*, and *what percent* in the percent sentence and translate them to mathematical symbols to form a percent equation.

WHY Then we can solve the equation to find the unknown percent.

Solution First, we translate.

38	is	what percent	of	40?
↓	↓	↓	↓	↓
38	=	x	·	40

38 is the amount, x is the percent, and 40 is the base.

$38 = x \cdot 40$ *This is the equation to solve.*

On the right side of the equation, the unknown number x is multiplied by 40. To undo the multiplication by 40 and isolate x, we divide both sides by 40.

$$\frac{38}{40} = \frac{x \cdot 40}{40}$$

We can simplify the fraction on the right side of the equation by removing the common factor of 40 from the numerator and denominator. On the left side, we perform the division indicated by the fraction bar.

$$0.95 = \frac{x \cdot \overset{1}{\cancel{40}}}{\underset{1}{\cancel{40}}}$$ *To simplify the left side, divide 38 by 40.*

$$0.95 = x$$

$$\begin{array}{r} 0.95 \\ 40\overline{)38.00} \\ -36\ 06 \\ \hline 2\ 00 \\ -2\ 00 \\ \hline 0 \end{array}$$

Since we want to find the percent, we need to write the decimal 0.95 as a percent.

$0\underset{\curvearrowright}{95.}\% = x$ *To write 0.95 as a percent, multiply it by 100 by moving the decimal point two places to the right, and then insert a % symbol.*

$95\% = x$

We have found that 38 is 95% of 40. That is, 95% of the wells mentioned in the newspaper article were declared safe.

EXAMPLE 5   14 is what percent of 32?

Strategy We will look for the key words *is*, *of*, and *what percent* in the percent sentence and translate them to mathematical symbols to form a percent equation.

WHY Then we can solve the equation to find the unknown percent.

Solution First, we translate.

14	is	what percent	of	32?
↓	↓	↓	↓	↓
14	=	x	·	32

14 is the amount, x is the percent, and 32 is the base.

LANGUAGE OF MATHEMATICS

We solve percent equations by writing a series of steps that result in an equation of the form $x =$ a number or a number $= x$. We say that the variable x is **isolated** on one side of the equation. Isolated means alone or by itself.

Self Check 4

4 is what percent of 80?

Now Try ➡ Problems 27 and 79

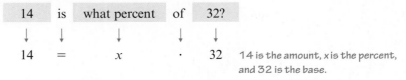

DAILY NEWS

Circulation   Monday, March 23   50 cents

Transit Strike Averted!

Labor: 84% of 500-member union votes to accept new offer

Drinking Water
38 of 40 Wells Declared Safe

New Appointees

These six area residents now make up 75% of the State Board of Examiners

$14 = x \cdot 32$ This is the equation to solve.

$\dfrac{14}{32} = \dfrac{x \cdot 32}{32}$ To undo the multiplication by 32 and isolate x on the right side of the equation, divide both sides by 32.

$0.4375 = x$ Do the division: 14 ÷ 32 = 0.4375.

$043.75\% = x$ To write the decimal 0.4375 as a percent, multiply it by 100 by moving the decimal point two places to the right, and then insert a % symbol.

$43.75\% = x$

Thus, 14 is 43.75% of 32.

$$\begin{array}{r} 0.4375 \\ 32\overline{)14.0000} \\ -12\,8 \\ \hline 1\,20 \\ -\,96 \\ \hline 240 \\ -224 \\ \hline 160 \\ -160 \\ \hline 0 \end{array}$$

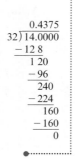

Self Check 5

9 is what percent of 16?

Now Try ➡ **Problem 31**

Using Your Calculator ▶   Cost of an Air Bag

An air bag is estimated to add $500 to the cost of a car. What percent of the $16,295 sticker price is the cost of the air bag?

First, we translate the words of the problem into a percent equation.

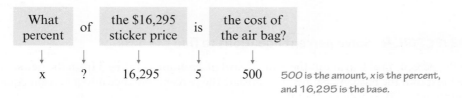

What percent	of	the $16,295 sticker price	is	the cost of the air bag?
↓	↓	↓	↓	↓
x	?	16,295	5	500

500 is the amount, x is the percent, and 16,295 is the base.

Then we solve the equation.

$16{,}295x = 500$ Write x · 16,295 as 16,295x.

$\dfrac{16{,}295x}{16{,}295} = \dfrac{500}{16{,}295}$ To undo the multiplication by 16,295 and isolate x on the left side, divide both sides of the equation by 16,295.

$x = \dfrac{500}{16{,}295}$

To perform the division on the right side using a scientific calculator, enter the following:

$500 \boxed{\div} 16295 \boxed{=}$ $\boxed{0.03068\mathsf{4}259}$

This display gives the answer in decimal form. To change it to a percent, we multiply the result by 100. This moves the decimal point two places to the right. (See the display.) Then we insert a % symbol. If we round to the nearest tenth of a percent, the cost of the air bag is about 3.1% of the sticker price.

$\boxed{3.068\mathsf{4}25898}$

EXAMPLE 6   What percent of 6 is 7.5?

Strategy We will look for the key words *is*, *of*, and *what percent* in the percent sentence and translate them to mathematical symbols to form a percent equation.

WHY Then we can solve the equation to find the unknown percent.

Solution First, we translate.

What percent   of   6   is   7.5?

↓   ↓   ↓   ↓

x ·   6   =   7.5

$x \cdot 6 = 7.5$ *This is the equation to solve.*

$\dfrac{x \cdot 6}{6} = \dfrac{7.5}{6}$ *To undo the multiplication by 6 and isolate x on the left side of the equation, divide both sides by 6.*

$\dfrac{x \cdot \overset{1}{\cancel{6}}}{\underset{1}{\cancel{6}}} = 1.25$ *To simplify the fraction on the left side of the equation, remove the common factor of 6 from the numerator and denominator. On the right side, divide 7.5 by 6.*

$x = 1.25$

$x = 125.\%$ *To write the decimal 1.25 as a percent, multiply it by 100 by moving the decimal point two places to the right, and then insert a % symbol.*

$x = 125\%$

$$\begin{array}{r} 1.25 \\ 6\overline{)7.50} \\ \underline{-6} \\ 1\,5 \\ \underline{-1\,2} \\ 30 \\ \underline{-30} \\ 0 \end{array}$$

Self Check 6

What percent of 5 is 8.5?

Now Try ➡ **Problem 35**

Thus, 7.5 is 125% of 6.

OBJECTIVE 4 Solve percent equations to find the base.

In the percent problem about the State Board of Examiners (Type 3 from the newspaper), we must find the base. As before, we translate the percent sentence into a percent equation and then find the unknown number.

EXAMPLE 7   6 is 75% of what number?

Strategy We will look for the key words *is*, *of*, and *what number* in the percent sentence and translate them to mathematical symbols to form a percent equation.

WHY Then we can solve the equation to find the unknown number.

Solution First, we translate.

6   is   75%   of   what number?

↓   ↓   ↓   ↓   ↓

6   =   75%   · x *6 is the amount, 75% is the percent, and x is the base.*

Now we solve the equation.

$6 = 0.75 \cdot x$ *Write 75% as a decimal: 75% = 0.75.*

$\dfrac{6}{0.75} = \dfrac{0.75 \cdot x}{0.75}$ *To undo the multiplication by 0.75 and isolate x on the right side, divide both sides of the equation by 0.75.*

$8 = \dfrac{\overset{1}{\cancel{0.75}} \cdot x}{\underset{1}{\cancel{0.75}}}$ *To simplify the fraction on the right side of the equation, remove the common factor of 0.75. On the left side, divide 6 by 0.75.*

$$\begin{array}{r} 8 \\ 75\overline{)600} \\ \underline{-600} \\ 0 \end{array}$$

$8 = x$

Self Check 7

3 is 5% of what number?

Now Try ➡ **Problem 39**

Thus, 6 is 75% of 8. That is, there are 8 members on the State Board of Examiners mentioned in the newspaper article.

DAILY NEWS

Circulation   Monday, March 23   50 cents

Transit Strike Averted!

Labor: 84% of 500-member union votes to accept new offer

Drinking Water
38 of 40 Wells Declared Safe

New Appointees

These six area residents now make up 75% of the State Board of Examiners

EXAMPLE 8   31.5 is $33\frac{1}{3}\%$ of what number?

Strategy We will look for the key words *is*, *of*, and *what number* in the percent sentence and translate them to mathematical symbols to form a percent equation.

WHY Then we can solve the equation to find the unknown number.

Solution First, we translate.

31.5	is	$33\frac{1}{3}\%$	of	what number?
↓	↓	↓	↓	↓

$$31.5 = 33\frac{1}{3}\% \cdot \qquad x$$

31.5 is the amount, $33\frac{1}{3}\%$ is the percent, and x is the base.

In this case, the calculations can be made easier by writing $33\frac{1}{3}\%$ as a fraction instead of as a repeating decimal.

$$31.5 = \frac{1}{3} \cdot x \qquad \text{Recall from Section 6.1 that } 33\% = \frac{1}{3}.$$

$$\frac{31.5}{\frac{1}{3}} = \frac{\frac{1}{3} \cdot x}{\frac{1}{3}}$$

To undo the multiplication by $\frac{1}{3}$ and isolate x on the right side of the equation, divide both sides by $\frac{1}{3}$.

$$\frac{31.5}{\frac{1}{3}} = \frac{\overset{1}{\cancel{\frac{1}{3}}} \cdot x}{\cancel{\frac{1}{3}}}$$

To simplify the fraction on the right side of the equation, remove the common factor of $\frac{1}{3}$ from the numerator and denominator.

$$31.5 \div \frac{1}{3} = x \qquad \text{On the left side, the fraction bar indicates division.}$$

$$\frac{31.5}{1} \cdot \frac{3}{1} = x$$

On the left side, write 31.5 as a fraction: $\frac{31.5}{1}$.
Then use the rule for dividing fractions:
Multiply by the reciprocal of $\frac{1}{3}$, which is $\frac{3}{1}$.

$$\begin{array}{r} \overset{1}{31.5} \\ \times\ 3 \\ \hline 94.5 \end{array}$$

$$94.5 = x \qquad \text{Do the multiplication.}$$

Thus, 31.5 is $33\frac{1}{3}\%$ of 94.5.

> **Success Tip** Sometimes the calculations to solve a percent problem are made easier if we write the percent as a fraction instead of a decimal. This is the case with percents that have *repeating* decimal equivalents such as $33\frac{1}{3}\%$, $66\frac{2}{3}\%$, and $16\frac{2}{3}\%$. You may want to review the table of percents and their fractional equivalents on page 502.

> **Self Check 8**
> 150 is $66\frac{2}{3}\%$ of what number?
>
> **Now Try** ⟳ Problems 43 and 83

To solve percent application problems, we often have to rewrite the facts of the problem in percent sentence form before we can translate to an equation.

EXAMPLE 9   **Rentals.** In an apartment complex, 198 of the units are currently occupied. If this represents an 88% occupancy rate, how many units are in the complex?

Strategy We will carefully read the problem and use the given facts to write them in the form of a percent sentence.

WHY Then we can translate the sentence into a percent equation and solve it to find the unknown number of units in the complex.

Solution An occupancy rate of 88% means that 88% of the units are occupied. Thus, the 198 units that are currently occupied are 88% of some unknown number of units in the complex, and we can write:

198	is	88%	of	what number?
↓	↓	↓	↓	↓

$$198 = 88\% \cdot \qquad x$$

198 is the amount, 88% is the percent, and x is the base.

Now we solve the equation.

$$198 = 88\% \cdot x$$

$$198 = 0.88 \cdot x \qquad \text{Write 88\% as a decimal: 88\% = 0.88.}$$

$$\frac{198}{\mathbf{0.88}} = \frac{0.88 \cdot x}{\mathbf{0.88}} \qquad \begin{array}{l}\text{To undo the multiplication by 0.88 and isolate x on the}\\ \text{right side, divide both sides of the equation by 0.88.}\end{array}$$

$$\frac{198}{0.88} = \frac{\overset{1}{\cancel{0.88}} \cdot x}{\underset{1}{\cancel{0.88}}} \qquad \begin{array}{l}\text{To simplify the fraction on the right side of the}\\ \text{equation, remove the common factor of 0.88}\\ \text{from the numerator and denominator. On the left}\\ \text{side, divide 198 by 0.88.}\end{array}$$

$$225 = x$$

$$\begin{array}{r} 225 \\ 88\overline{)19800} \\ -176 \\ \hline 220 \\ -176 \\ \hline 440 \\ -440 \\ \hline 0 \end{array}$$

The apartment complex has 225 units, of which 198, or 88%, are occupied.

If you are only learning the percent equation method for solving percent problems, turn to page 521 and pick up your reading at Objective 5.

METHOD 2: PERCENT PROPORTIONS

OBJECTIVE 1 Write percent proportions.

Another method to solve percent problems involves writing and then solving a proportion. To introduce this method, consider the figure on the right. The vertical line down its middle divides the figure into two equal-sized parts. Since 1 of the 2 parts is shaded red, the shaded portion of the figure can be described by the ratio $\frac{1}{2}$. We call this an **amount-to-base** (or **part-to-whole**) **ratio**.

Now consider the 100 equal-sized square regions within the figure. Since 50 of them are shaded red, we say that $\frac{50}{100}$, or 50% of the figure is shaded. The ratio $\frac{50}{100}$ is called a **percent ratio**.

Since the amount-to-base ratio, $\frac{1}{2}$, and the percent ratio, $\frac{50}{100}$, represent the same shaded portion of the figure, they must be equal, and we can write

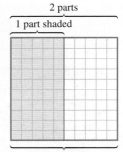

2 parts

1 part shaded

50 of the 100 parts shaded: 50% shaded

$$\underset{\text{The amount-to-base ratio}}{\Longrightarrow} \quad \frac{1}{2} = \frac{50}{100} \quad \underset{\text{The percent ratio}}{\Longleftarrow}$$

Recall from Section 5.2 that statements of this type stating that two ratios are equal are called *proportions*. We call $\frac{1}{2} = \frac{50}{100}$ a **percent proportion**. The four terms of a percent proportion are shown here.

Percent Proportion

To translate a percent sentence to a **percent proportion**, use the following form:

Amount is to base as percent is to 100.   *Part is to whole as percent is to 100.*

$$\frac{\text{amount}}{\text{base}} = \frac{\text{percent}}{100} \qquad\qquad \text{or} \qquad\qquad \frac{\text{part}}{\text{whole}} = \frac{\text{percent}}{100}$$

↑
This is always 100 because percent means parts per one hundred.

To write a percent proportion, you must identify three of the terms as you read the problem. (Remember, the fourth term of the proportion is always 100.) Here are some ways to identify those terms.

- The **percent** is easy to find. Look for the % symbol or the words *what percent*.
- The **base** (or **whole**) usually follows the word *of*.
- The **amount** (or **part**) is compared to the base (or whole).

Self Check 9

Capacity of a gym. A total of 784 people attended a graduation in a high school gymnasium. If this was 98% of capacity, what is the total capacity of the gym?

Now Try � **Problem 81**

EXAMPLE 1 Translate each percent sentence to a percent proportion.

a. What number is 12% of 64?
b. What percent of 88 is 11?
c. 165% of what number is 366?

Strategy A percent proportion has the form $\frac{\text{amount}}{\text{base}} = \frac{\text{percent}}{100}$. Since one of the terms of the percent proportion is always 100, we only need to identify three terms to write the proportion. We will begin by identifying the percent and the base in the given sentence.

WHY The remaining number (or unknown) must be the amount.

Solution

a. We will identify the terms in this order:

- *First:* the percent (next to the % symbol)
- *Second:* the base (usually after the word *of*)
- *Last:* the amount (the number that remains)

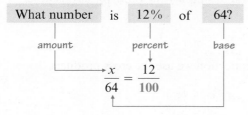

b.

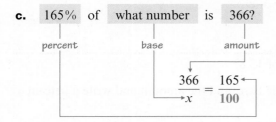

c.

Self Check 1

Translate each percent sentence to a percent proportion.

a. What number is 33% of 80?
b. What percent of 55 is 6?
c. 172% of what number is 4?

Now Try ⟶ Problem 17

OBJECTIVE 2 **Solve percent proportions to find the amount.**

Recall the labor union problem from the newspaper example in the introduction to this section. We can write and solve a percent proportion to find the unknown amount.

EXAMPLE 2   What number is 84% of 500?

Strategy We will identify the percent, the base, and the amount and write a percent proportion of the form $\frac{\text{amount}}{\text{base}} = \frac{\text{percent}}{100}$.

WHY Then we can solve the proportion to find the unknown number.

Solution First, we write the percent proportion.

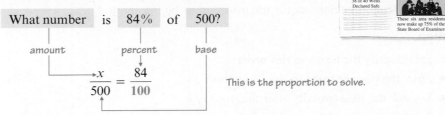

$$\frac{x}{500} = \frac{84}{100}$$ This is the proportion to solve.

To make the calculations easier, it is helpful to simplify the ratio $\frac{84}{100}$ at this time.

$$\frac{x}{500} = \frac{21}{25}$$ On the right side, simplify: $\frac{84}{100} = \frac{4 \cdot 21}{4 \cdot 25} = \frac{21}{25}$.

> **LANGUAGE OF MATHEMATICS**
>
> When we find the value of the variable that makes a percent proportion true, we say that we have **solved the proportion**. In Example 2, we solved $\frac{x}{500} = \frac{84}{100}$ to find that x is 420.

Recall from Section 5.2 that to solve a proportion we use the cross products.

$x \cdot 25 = 500 \cdot 21$ Find the cross products: $\frac{x}{500} = \frac{21}{25}$. Then set them equal.

$x \cdot 25 = 10{,}500$ To simplify the right side of the equation, do the multiplication: $500 \cdot 21 = 10{,}500$.

$\dfrac{x \cdot \overset{1}{\cancel{25}}}{\underset{1}{\cancel{25}}} = \dfrac{10{,}500}{25}$ To undo the multiplication by 25 and isolate x on the left side, divide both sides of the equation by 25. Then remove the common factor of 25 from the numerator and denominator.

$x = 420$ On the right side, divide 10,500 by 25.

$$\begin{array}{r} 500 \\ \times\ 21 \\ \hline 500 \\ 10\ 000 \\ \hline 10{,}500 \end{array}$$

$$\begin{array}{r} 420 \\ 25\overline{)10{,}500} \\ -10\ 0 \\ \hline 50 \\ -50 \\ \hline 00 \\ -0 \\ \hline 0 \end{array}$$

Self Check 2

What number is 36% of 400?

Now Try ➡ **Problems 19 and 75**

We have found that 420 is 84% of 500. That is, 420 union members mentioned in the newspaper article voted to accept the new offer.

EXAMPLE 3   What number is 160% of 15.8?

Strategy We will identify the percent, the base, and the amount and write a percent proportion of the form $\frac{\text{amount}}{\text{base}} = \frac{\text{percent}}{100}$.

WHY Then we can solve the proportion to find the unknown number.

Solution First, we write the percent proportion.

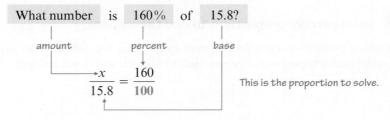

$$\frac{x}{15.8} = \frac{160}{100}$$ This is the proportion to solve.

To make the calculations easier, it is helpful to simplify the ratio $\frac{160}{100}$ at this time.

$$\frac{x}{15.8} = \frac{8}{5}$$ *On the right side, simplify:* $\frac{160}{100} = \frac{8 \cdot \overset{1}{\cancel{20}}}{5 \cdot \cancel{20}} = \frac{8}{5}$.

$$x \cdot 5 = 15.8 \cdot 8$$ *Find the cross products:* $\frac{x}{15.8} \diagup\!\!\!\!\diagup \frac{8}{5}$.
Then set them equal.

$$x \cdot 5 = 126.4$$ *To simplify the right side of the equation, do the*
multiplication: $15.8 \cdot 8 = 126.4$.

$$\frac{x \cdot \overset{1}{\cancel{5}}}{\cancel{5} \atop 1} = \frac{126.4}{5}$$ *To undo the multiplication by 5 and isolate x*
on the left side, divide both sides of the equation
by 5. Then remove the common factor of 5 from
the numerator and denominator.

$$x = 25.28$$ *On the right side, divide 126.4 by 5.*

Thus, 25.28 is 160% of 15.8.

$$\begin{array}{r} \overset{4\ 6}{15.8} \\ \times\ \ 8 \\ \hline 126.4 \end{array}$$

$$\begin{array}{r} 25.28 \\ 5)\overline{126.40} \\ -10 \\ \hline 26 \\ -25 \\ \hline 1\ 4 \\ -1\ 0 \\ \hline 40 \\ -40 \\ \hline 0 \end{array}$$

Self Check 3
What number is 240% of
80.3?

Now Try ➲ Problem 23

OBJECTIVE 3 Solve percent proportions to find the percent.

Recall the drinking water problem from the newspaper example in the introduction to this section. We can write and solve a percent proportion to find the unknown percent.

EXAMPLE 4   38 is what percent of 40?

Strategy We will identify the percent, the base, and the amount and write a percent proportion of the form
$\frac{\text{amount}}{\text{base}} = \frac{\text{percent}}{100}$.

WHY Then we can solve the proportion to find the unknown percent.

Solution First, we write the percent proportion.

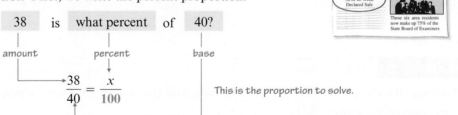

38	is	what percent	of	40?
amount		percent		base

$$\frac{38}{40} = \frac{x}{100}$$ *This is the proportion to solve.*

To make the calculations easier, it is helpful to simplify the ratio $\frac{38}{40}$ at this time.

$$\frac{19}{20} = \frac{x}{100}$$ *On the left side, simplify:* $\frac{38}{40} = \frac{\overset{1}{\cancel{2}} \cdot 19}{\cancel{2} \cdot 20} = \frac{19}{20}$.

$$19 \cdot 100 = 20 \cdot x$$ *To solve the proportion, find the cross products:* $\frac{19}{20} \diagup\!\!\!\!\diagup \frac{x}{100}$.
Then set them equal.

$$1{,}900 = 20 \cdot x$$ *To simplify the left side of the equation, do the*
multiplication: $19 \cdot 100 = 1{,}900$.

$$\frac{1{,}900}{20} = \frac{\overset{1}{\cancel{20}} \cdot x}{\cancel{20} \atop 1}$$ *To undo the multiplication by 20 and isolate x*
on the right side, divide both sides of the equation
by 20. Then remove the common factor of 20 from
the numerator and denominator.

$$95 = x$$ *On the left side, divide 1,900 by 20.*

$$\begin{array}{r} 95 \\ 20)\overline{1{,}900} \\ -1\ 80 \\ \hline 100 \\ -100 \\ \hline 0 \end{array}$$

Self Check 4
4 is what percent of 80?

We have found that 38 is 95% of 40. That is, 95% of the wells mentioned in the newspaper article were declared safe.

Now Try ➲ Problems 27
and 83

EXAMPLE 5   14 is what percent of 32?

Strategy We will identify the percent, the base, and the amount and write a percent proportion of the form $\frac{\text{amount}}{\text{base}} = \frac{\text{percent}}{100}$.

WHY Then we can solve the proportion to find the unknown percent.

Solution First, we write the percent proportion.

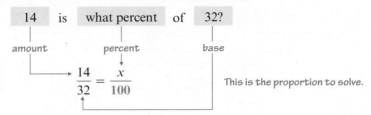

To make the calculations easier, it is helpful to simplify the ratio $\frac{14}{32}$ at this time.

$$\frac{7}{16} = \frac{x}{100} \qquad \text{On the left side, simplify: } \frac{14}{32} = \frac{\overset{1}{\cancel{2}} \cdot 7}{\underset{1}{\cancel{2}} \cdot 16} = \frac{7}{16}.$$

$$7 \cdot 100 = 16 \cdot x \qquad \text{To solve the proportion, find the cross products: } \frac{7}{16} = \frac{x}{100}.$$
$$\text{Then set them equal.}$$

$$700 = 16 \cdot x \qquad \text{To simplify the left side of the equation, do the}$$
$$\text{multiplication: } 7 \cdot 100 = 700.$$

$$\frac{700}{16} = \frac{\overset{1}{\cancel{16}} \cdot x}{\underset{1}{\cancel{16}}} \qquad \text{To undo the multiplication by 16 and isolate } x$$
$$\text{on the right side, divide both sides of the equation}$$
$$\text{by 16. Then remove the common factor of 16 from}$$
$$\text{the numerator and denominator.}$$

$$43.75 = x \qquad \text{On the left side, divide 700 by 16.}$$

$$
\begin{array}{r}
43.75 \\
16\overline{)700.00} \\
-64 \\
\hline
60 \\
-48 \\
\hline
12\,0 \\
-11\,2 \\
\hline
80 \\
-80 \\
\hline
0
\end{array}
$$

Self Check 5

9 is what percent of 16?

Now Try ➲ **Problem 31**

Thus, 14 is 43.75% of 32.

EXAMPLE 6   What percent of 6 is 7.5?

Strategy We will identify the percent, the base, and the amount and write a percent proportion of the form $\frac{\text{amount}}{\text{base}} = \frac{\text{percent}}{100}$.

WHY Then we can solve the proportion to find the unknown percent.

Solution First, we write the percent proportion.

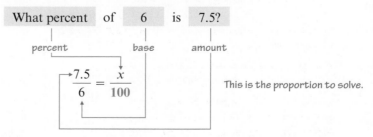

$7.5 \cdot 100 = 6 \cdot x$ To solve the proportion, find the cross products: $\frac{7.5}{6} = \frac{x}{100}$. Then set them equal.

$750 = 6 \cdot x$ To simplify the left side of the equation, do the multiplication: $7.5 \cdot 100 = 750$.

$\dfrac{750}{6} = \dfrac{\overset{1}{\cancel{6}} \cdot x}{\cancel{6}}$ To undo the multiplication by 6 and isolate x on the right side, divide both sides of the equation by 6. Then remove the common factor of 6 from the numerator and denominator.

$125 = x$ On the left side, divide 750 by 6.

$$
\begin{array}{r}
125 \\
6\overline{)750} \\
-6 \\
\hline
15 \\
-12 \\
\hline
30 \\
-30 \\
\hline
0
\end{array}
$$

Thus, 7.5 is 125% of 6.

Self Check 6

What percent of 5 is 8.5?

Now Try Problem 35

OBJECTIVE 4   **Solve percent proportions to find the base.**

Recall the State Board of Examiners problem from the newspaper example in the introduction to this section. We can write and solve a percent proportion to find the unknown base.

EXAMPLE 7   6 is 75% of what number?

Strategy We will identify the percent, the base, and the amount and write a percent proportion of the form $\frac{\text{amount}}{\text{base}} = \frac{\text{percent}}{100}$.

WHY Then we can solve the proportion to find the unknown number.

Solution First, we write the percent proportion.

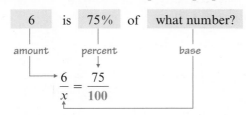

To make the calculations easier, it is helpful to simplify the ratio $\frac{75}{100}$ at this time.

$\dfrac{6}{x} = \dfrac{3}{4}$ Simplify: $\frac{75}{100} = \frac{3 \cdot \cancel{25}^{1}}{4 \cdot \cancel{25}_{1}} = \frac{3}{4}$.

$6 \cdot 4 = x \cdot 3$ To solve the proportion, find the cross products: $\frac{6}{x} = \frac{3}{4}$. Then set them equal.

$24 = x \cdot 3$ To simplify the left side of the equation, do the multiplication: $6 \cdot 4 = 24$.

$\dfrac{24}{3} = \dfrac{x \cdot \overset{1}{\cancel{3}}}{\cancel{3}}$ To undo the multiplication by 3 and isolate x on the right side, divide both sides of the equation by 3. Then remove the common factor of 3 from the numerator and denominator.

$8 = x$ On the left side, divide 24 by 3.

Thus, 6 is 75% of 8. That is, there are 8 members on the State Board of Examiners mentioned in the newspaper article.

Self Check 7

3 is 5% of what number?

Now Try Problem 39

EXAMPLE 8   31.5 is $33\frac{1}{3}\%$ of what number?

Strategy We will identify the percent, the base, and the amount and write a percent proportion of the form $\frac{\text{amount}}{\text{base}} = \frac{\text{percent}}{100}$.

WHY Then we can solve the proportion to find the unknown number.

Solution First, we write the percent proportion.

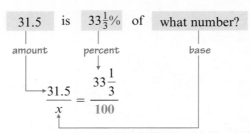

$$\frac{31.5}{x} = \frac{33\frac{1}{3}}{100}$$

To make the calculations easier, it is helpful to write the mixed number $33\frac{1}{3}$ as the improper fraction $\frac{100}{3}$.

$$\frac{31.5}{x} = \frac{\frac{100}{3}}{100}$$ Write $33\frac{1}{3}$ as $\frac{100}{3}$.

$$31.5 \cdot 100 = x \cdot \frac{100}{3}$$ To solve the proportion, find the cross products: $\frac{31.5}{x} \diagup \frac{\frac{100}{3}}{100}$. Then set them equal.

$$3,150 = \frac{100}{3}x$$ On the left side, do the multiplication: $31.5 \cdot 100 = 3,150$. On the right side, use the commutative property of multiplication to write $x \cdot \frac{100}{3}$ as $\frac{100}{3}x$.

$$3,150 = x \cdot \frac{100}{3}$$ To simplify the left side of the equation, do the multiplication: $31.5 \cdot 100 = 3,150$.

$$\frac{3,150}{\frac{100}{3}} = \frac{x \cdot \overset{1}{\cancel{\frac{100}{3}}}}{\cancel{\frac{100}{3}}_{1}}$$ To undo the multiplication by $\frac{100}{3}$ and isolate x on the right side, divide both sides of the equation by $\frac{100}{3}$. Then remove the common factor of $\frac{100}{3}$ from the numerator and denominator.

$$3,150 \div \frac{100}{3} = x$$ On the left side, the fraction bar indicates division.

$$\frac{3,150}{1} \cdot \frac{3}{100} = x$$ On the left side, write 3,150 as a fraction: $\frac{3,150}{1}$. Then use the rule for dividing fractions: Multiply by the reciprocal of $\frac{100}{3}$, which is $\frac{3}{100}$.

$$\frac{9,450}{100} = x$$ Multiply the numerators.
                                Multiply the denominators.

$$94.50 = x$$ Divide 9,450 by 100 by moving the understood decimal point in 9,450 two places to the left.

$$\begin{array}{r} \overset{1}{3,150} \\ \times\ 3 \\ \hline 9,450 \end{array}$$

Self Check 8

150 is $66\frac{2}{3}\%$ of what number?

Now Try ➡ Problems 43 and 87

Thus, 31.5 is $33\frac{1}{3}\%$ of 94.5.

To solve percent application problems, we often have to rewrite the facts of the problem in percent sentence form before we can translate to an equation.

EXAMPLE 9 **Rentals.** In an apartment complex, 198 of the units are currently occupied. If this represents an 88% occupancy rate, how many units are in the complex?

Strategy We will carefully read the problem and use the given facts to write them in the form of a percent sentence.

WHY Then we can write and solve a percent proportion to find the unknown number of units in the complex.

Solution An occupancy rate of 88% means that 88% of the units are occupied. Thus, the 198 units that are currently occupied are 88% of some unknown number of units in the complex, and we can write:

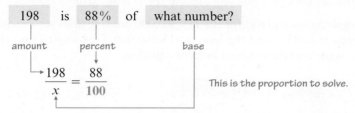

$$\frac{198}{x} = \frac{88}{100}$$ *This is the proportion to solve.*

To make the calculations easier, it is helpful to simplify the ratio $\frac{88}{100}$ at this time.

$$\frac{198}{x} = \frac{22}{25}$$ *On the right side, simplify:* $\frac{88}{100} = \frac{\overset{1}{\cancel{4} \cdot 22}}{\underset{1}{\cancel{4} \cdot 25}} = \frac{22}{25}.$

$$198 \cdot 25 = x \cdot 22$$ *Find the cross products. Then set them equal.*

$$4{,}950 = x \cdot 22$$ *To simplify the left side, do the multiplication:*
$$198 \cdot 25 = 4{,}950.$$

$$\frac{4{,}950}{22} = \frac{x \cdot \overset{1}{\cancel{22}}}{\underset{1}{\cancel{22}}}$$ *To undo the multiplication by 22 and isolate x on the right side, divide both sides of the equation by 22. Then remove the common factor of 22 from the numerator and denominator.*

$$225 = x$$ *On the left side, divide 4,950 by 22.*

$$\begin{array}{r} 198 \\ \times 25 \\ \hline 990 \\ 3960 \\ \hline 4{,}950 \end{array}$$

$$\begin{array}{r} 225 \\ 22\overline{)4{,}950} \\ -44 \\ \hline 55 \\ -44 \\ \hline 110 \\ -110 \\ \hline 0 \end{array}$$

The apartment complex has 225 units, of which 198, or 88%, are occupied.

Self Check 9

Capacity of a gym. A total of 784 people attended a graduation in a high school gymnasium. If this was 98% of capacity, what is the total capacity of the gym?

Now Try Problem 85

OBJECTIVE 5 **Read circle graphs.**

Percents are used with **circle graphs**, or **pie charts**, as a way of presenting data for comparison. In the figure below, the entire circle represents the total amount of electricity generated in the United States in 2015. The pie-shaped pieces of the graph show the relative sizes of the energy sources used to generate the electricity. For example, we see that the greatest amount of electricity (33%) was generated from each, coal and natural gas. Note that if we add the percents from all categories (33% + 1% + 6% + 33% + 20% + 7%), the sum is 100%.

The 100 tick marks equally spaced around the circle serve as a visual aid when constructing a circle graph. For example, to represent hydropower as 6%, a line was drawn from the center of the circle to a tick mark. Then we counted off 6 ticks and drew a second line from the center to that tick to complete the pie-shaped wedge.

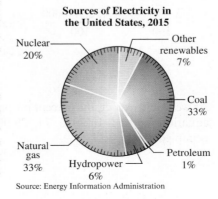

Sources of Electricity in the United States, 2015

Nuclear 20%
Other renewables 7%
Coal 33%
Natural gas 33%
Hydropower 6%
Petroleum 1%

Source: Energy Information Administration

EXAMPLE 10 **Housing.** In 2016, the U.S. Census Bureau estimates that there were 118,300,000 housing units in the United States. Use the information in the circle graph to find the number of housing units that were owner occupied at that time.

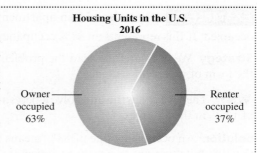

Housing Units in the U.S. 2016

Owner occupied 63%

Renter occupied 37%

Strategy We will rewrite the facts of the problem in percent sentence form.

WHY Then we can translate the sentence to a percent equation (or percent proportion) to find the number of owner occupied housing units in the U.S. in 2016.

Solution The circle graph shows that 63% of the 118,300,000 housing units were owner occupied. Thus, the percent is 63% and the base is 118,300,000. One way to find the unknown amount is to write and then solve a percent equation.

What number	is	63%	of	118,300,000?
↓	↓	↓	↓	↓
x	$=$	63%	·	118,300,000

Translate to a percent equation.

Now we perform the multiplication on the right side of the equation.

$x = 0.63 \cdot 118{,}300{,}000$ Write 63% as a decimal: 63% = 0.63.

$x = 74{,}529{,}000$ Do the multiplication.

$$\begin{array}{r} 118{,}300{,}000 \\ \times 0.63 \\ \hline 354900000 \\ 7098000000 \\ \hline 74{,}529{,}000.00 \end{array}$$

Thus, 74,529,000 of the 118,300,000 housing units were owner occupied in 2016.

Another way to find the unknown amount is to write and then solve a percent proportion.

What number	is	63%	of	118,300,000?
amount		percent		base

$$\frac{x}{118{,}300{,}000} = \frac{63}{100}$$ This is the proportion to solve.

$x \cdot 100 = 118{,}300{,}000 \cdot 63$ Find the cross products: $\frac{x}{118{,}300{,}000} = \frac{63}{100}$. Then set them equal.

$$\begin{array}{r} 118{,}300{,}000 \\ \times 63 \\ \hline 354900000 \\ 7098000000 \\ \hline 7{,}452{,}900{,}000 \end{array}$$

$x \cdot 100 = 7{,}452{,}900{,}000$ To simplify the right side, do the multiplication: 118,300,000 · 63

$$\frac{x \cdot \overset{1}{\cancel{100}}}{\underset{1}{\cancel{100}}} = \frac{7{,}452{,}900{,}000}{100}$$

To undo the multiplication by 100 and isolate x on the left side, divide both sides of the equation by 100. Then remove the common factor of 100 from the numerator and denominator.

$x = 74{,}529{,}000$ On the right side, divide 7,452,900,00 by 100.

As we would expect, the percent proportion method gives the same answer as the percent equation method. In 2016, 74,529,000 out of 118,300,000 housing units were owner occupied.

Now Try ➡ Problem 89

Self Check 10

Facebook. In 2017, there were a total of 214 million Facebook users in the United States. Use the information in the circle graph to find the number of female users.

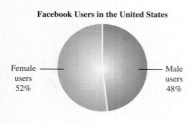

Facebook Users in the United States

Female users 52%

Male users 48%

Think it Through ● Community College Students

"When the history of American higher education is updated years from now, the story of our current times will highlight the pivotal role community colleges played in developing human capital and bolstering the nation's educational system."
—*Community College Survey of Student Engagement, 2007*

More than 453,000 students responded to the most recent Community College Survey of Student Engagement. Some results are shown below. Study each circle graph and then complete its legend.

Enrollment in Community Colleges	**Community College Students Who Work More Than 20 Hours per Week**	**Community College Students Who Discussed Their Grades or Assignments with an Instructor**

■ 55% are enrolled in college part time.
■ ?

■ 47% of the students work more than 20 hours per week.
■ ?

■ 49% often or very often
■ 42% sometimes
■ ?

Answers to Self Checks

1. a. $x = 33\% \cdot 80$ or $\frac{x}{80} = \frac{33}{100}$ **b.** $x \cdot 55 = 6$ or $\frac{6}{55} = \frac{x}{100}$ **c.** $172\% \cdot x = 4$ or $\frac{4}{x} = \frac{172}{100}$ **2.** 144   **3.** 192.72
4. 5%   **5.** 56.25%   **6.** 170%   **7.** 60   **8.** 225   **9.** 800 people   **10.** 111,280,000 female users of Facebook

---

SECTION 6.2 STUDY SET

VOCABULARY

Fill in the blanks.

1. We call "What number is 15% of 25?" a percent _____. It translates to the percent _____ $x = 15\% \cdot 25$.

2. The key words in a percent sentence translate as follows:
- ___ translates to an equal symbol =
- ___ translates to multiplication that is shown with a raised dot ·
- _____ *number* or _____ *percent* translates to an unknown number that is represented by a variable

3. When we find the value of the variable that makes a percent equation true, we say that we have _____ the equation.

4. In the percent sentence "45 is 90% of 50," 45 is the _____, 90% is the percent, and 50 is the _____.

5. The amount is _____ of the base. The base is the standard of comparison—it represents the _____ of some quantity.

6. a. *Amount is to base as percent is to 100:*

$$\frac{}{\text{base}} = \frac{\text{percent}}{}$$

b. *Part is to whole as percent is to 100:*

$$\frac{\text{part}}{} = \frac{}{100}$$

7. The _____ products for the proportion $\frac{24}{x} = \frac{36}{100}$ are $24 \cdot 100$ and $x \cdot 36$.

8. In a _____ graph, pie-shaped wedges are used to show the division of a whole quantity into its component parts.

CONCEPTS

9. Fill in the blanks to complete the percent equation (formula):

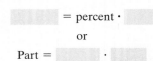

$$\boxed{} = \text{percent} \cdot \boxed{}$$

or

$$\text{Part} = \boxed{} \cdot \boxed{}$$

10. **a.** Without doing the calculation, tell whether 12% of 55 is more than 55 or less than 55.

   b. Without doing the calculation, tell whether 120% of 55 is more than 55 or less than 55.

11. **Pizza sales.** The circle graph shows what percent of total U.S. pizza sales the leading companies had in 2015. What percent of the sales did Pizza Hut have?

Percent of Total U.S. Pizza Sales, 2015

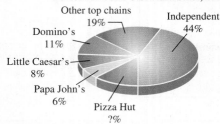

Other top chains 19%
Independent 44%
Domino's 11%
Little Caesar's 8%
Papa John's 6%
Pizza Hut ?%

Source: franchisedirect.com

12. **Smartphones.** The circle graph shows the percent U.S. market share for the leading smartphone subscribers in 2016. What was the market share for Samsung?

U.S. Smartphone Subscribers, 2016

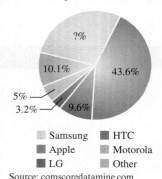

?%
10.1%
43.6%
5%
3.2%
9.6%

☐ Samsung   ■ HTC
■ Apple   ■ Motorola
■ LG   ☐ Other

Source: comscoredatamine.com

NOTATION

13. When computing with percents, we must change the percent to a decimal or a fraction. Change each percent to a decimal.

   a. 12%   **b.** 5.6%

   c. 125%   **d.** $\frac{1}{4}$%

14. When computing with percents, we must change the percent to a decimal or a fraction. Change each percent to a fraction.

   a. $33\frac{1}{3}$%   **b.** $66\frac{2}{3}$%

   c. $16\frac{2}{3}$%   **d.** $83\frac{1}{3}$%

GUIDED PRACTICE

Translate each percent sentence to a percent equation or percent proportion. Do not solve. **See Example 1.**

15. **a.** What number is 7% of 16?

   b. 125 is what percent of 800?

   c. 1 is 94% of what number?

16. **a.** What number is 28% of 372?

   b. 9 is what percent of 21?

   c. 4 is 17% of what number?

17. **a.** 5.4% of 99 is what number?

   b. 75.1% of what number is 15?

   c. What percent of 33.8 is 3.8?

18. **a.** 1.5% of 3 is what number?

   b. 49.2% of what number is 100?

   c. What percent of 100.4 is 50.2?

Translate to a percent equation or percent proportion and then solve to find the unknown number. **See Example 2.**

19. What is 34% of 200?

20. What is 48% of 600?

21. What is 88% of 150?

22. What number is 52% of 350?

Translate to a percent equation or percent proportion and then solve to find the unknown number. **See Example 3.**

23. What number is 224% of 7.9?

24. What number is 197% of 6.3?

25. What number is 105% of 23.2?

26. What number is 228% of 34.5?

Translate to a percent equation or percent proportion and then solve to find the unknown number. **See Example 4.**

27. 8 is what percent of 32?

28. 9 is what percent of 18?

29. 51 is what percent of 60?

30. 52 is what percent of 80?

Translate to a percent equation or percent proportion and then solve to find the unknown number. **See Example 5.**

31. 5 is what percent of 8?

32. 7 is what percent of 8?

33. 7 is what percent of 16?

34. 11 is what percent of 16?

Translate to a percent equation or percent proportion and then solve to find the unknown number. **See Example 6.**

35. What percent of 60 is 66?

36. What percent of 50 is 56?

37. What percent of 24 is 84?

38. What percent of 14 is 63?

Translate to a percent equation or percent proportion and then solve to find the unknown number. See Example 7.

39. 9 is 30% of what number?

40. 8 is 40% of what number?

41. 36 is 24% of what number?

42. 24 is 16% of what number?

Translate to a percent equation or percent proportion and then solve to find the unknown number. See Example 8.

43. 19.2 is $33\frac{1}{3}$% of what number?

44. 32.8 is $33\frac{1}{3}$% of what number?

45. 48.4 is $66\frac{2}{3}$% of what number?

46. 56.2 is $16\frac{2}{3}$% of what number?

TRY IT YOURSELF

Translate to a percent equation or percent proportion and then solve to find the unknown number.

47. What percent of 40 is 0.5?

48. What percent of 15 is 0.3?

49. 7.8 is 12% of what number?

50. 39.6 is 44% of what number?

51. $33\frac{1}{3}$% of what number is 33?

52. $66\frac{2}{3}$% of what number is 28?

53. What number is 36% of 250?

54. What number is 82% of 300?

55. 16 is what percent of 20?

56. 13 is what percent of 25?

57. What number is 0.8% of 12?

58. What number is 5.6% of 40?

59. 3.3 is 7.5% of what number?

60. 8.4 is 20% of what number?

61. What percent of 0.05 is 1.25?

62. What percent of 0.06 is 2.46?

63. 102% of 105 is what number?

64. 210% of 66 is what number?

65. $9\frac{1}{2}$% of what number is 5.7?

66. $\frac{1}{2}$% of what number is 5,000?

67. What percent of 8,000 is 2,500?

68. What percent of 3,200 is 1,400?

69. Find $7\frac{1}{4}$% of 600.

70. Find $1\frac{3}{4}$% of 800.

LOOK ALIKES

71. a. What number is 16% of 80?

 b. What percent of 80 is 16?

 c. 16% of what number is 80?

72. a. What number is 48% of 120?

 b. What percent of 120 is 48?

 c. 48% of what number is 120?

73. a. 1.5 is 75% of what number?

 b. 1.5 is what percent of 75?

 c. What is 75% of 1.5?

74. a. 2,000 is 2.5% of what number?

 b. 2,000 is what percent of 2.5?

 c. What is 2.5% of 2,000?

APPLICATIONS

75. Downloading. The message on the computer monitor screen shown below indicates that 24% of the 50MB of information that the user has decided to view have been downloaded to her computer at that time. Find the number of bytes of information that have been downloaded. (50MB stands for 50,000,000.)

76. Lumber. The rate of tree growth for walnut trees is about 3% per year. If a walnut tree has 400 board feet of lumber that can be cut from it, how many more board feet will it produce in a year? (Source: Iowa Department of Natural Resources)

77. Cash back. A credit card company offers a *cashback reward program* where 1.5% of the amount spent on the card is paid back to the card holder at the end of each month. A credit card statement for the month of March, 2017, is shown below. How large of a cash back reward can the card holder expect to receive for the month?

Date	Description	Amount
Mar 1, 2017	CVS Pharmacy	$47.23
Mar 8, 2017	Home Depot	$979.55
Mar 15, 2017	Toyota Financial Services	$550.10
Mar 21, 2017	Staples	$123.09
Mar 22, 2017	City Wide Florist	$89.15
Mar 31, 2017	Macy's	$1,020.88

78. Price guarantees. To assure its customers of low prices, the Home Club offers a "10% Plus" guarantee. If the customer finds the same item selling for less somewhere else, he or she receives the difference in price, plus 10% of the difference. A woman bought miniblinds at the Home Club for $120 but later saw the same blinds on sale for $98 at another store.

 a. What is the difference in the prices of the miniblinds?

 b. What is 10% of the difference in price?

 c. How much money can the woman expect to receive if she takes advantage of the "10% Plus" guarantee from the Home Club?

79. Carnival games. The result of a 2015 survey of game-booth managers at some of the country's busiest state fairs, boardwalks, and carnivals found that there is a 57% chance of winning the popular balloon dart game. Suppose a game operator charges $5 to play and the stuffed animal prizes cost him $3 each. If 200 people play the game in one day, how much profit will the game owner make? (Source: *Men's Health* magazine July/August 2015 edited by Ben Paynter)

Kent Weakley/Shutterstock.com

80. Copy Machines.

 a. The reduce feature on a copier is set at 98%, and a 2-inch wide picture is to be copied. What will be the width of the reduced picture?

 b. The enlarge feature on a copier is set at 180%, and a 1.5-inch wide picture is to be copied. What will be the width of the enlarged picture?

81. Driver's license. On the written part of his driving test, a man answered 28 out of 40 questions correctly. If 70% correct is passing, did he pass the test?

82. Housing. A general budget rule of thumb is that your rent or mortgage payment should be less than 30% of your income. Together, a couple earns $4,500 per month and they pay $1,260 in rent. Are they following the budget rule of thumb for housing?

83. Insurance. The cost to repair a car after a collision was $4,000. The automobile insurance policy paid the entire bill except for a $200 deductible, which the driver paid. What percent of the cost did he pay?

84. Floor space. A house has 1,200 square feet on the first floor and 800 square feet on the second floor.

 a. What is the total square footage of the house?

 b. What percent of the square footage of the house is on the first floor?

85. Child care. After the first day of registration, 84 children had been enrolled in a new day care center. That represented 70% of the available slots. What was the maximum number of children the center could enroll?

86. Racing programs. One month before a stock car race, the sale of ads for the official race program was slow. Only 12 pages, or 60% of the available pages, had been sold. What was the total number of pages devoted to advertising in the program?

87. Antibiotics. A report published in the *Journal of the American Medical Association* estimated that $33\frac{1}{3}\%$ of the prescriptions written each year in doctor's offices and emergency rooms for antibiotics are unnecessary. That is because they are prescribed for such illnesses as common colds, viral sore throats, bronchitis, and ear infections that do not respond to antibiotics. If an estimated 153 million prescriptions for antibiotics are written annually, how many of them are likely not appropriate?

88. Remodeling. A remodeling website claims that, on average, $66\frac{2}{3}\%$ of the cost of a major kitchen remodel will be recouped when the home is sold. The website lists the average amount spent nationally on such a remodel is $63,000. How much of that can a homeowner expect to recover upon the sale of the home?

89. Government spending. The circle graph below shows the breakdown of federal spending for fiscal year 2015. If the total spending was approximately $3.603 trillion, how many dollars were spent on social programs, Social Security, Medicare, and other retirement programs?

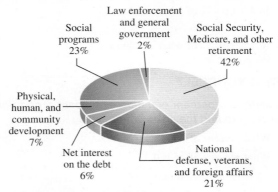

Note: Numbers may not total to 100% due to rounding.
Source: 2016 Federal Income Tax Form 1040 Instructions

90. Waste. The circle graph below shows the types of trash U.S. residents, businesses, and institutions generated in 2014. If the total amount of trash produced that year was about 258 million tons, how many million tons of yard trimmings were there?

U.S. Trash Generation by Material
Before Recycling, 2014
(258 Million Tons)

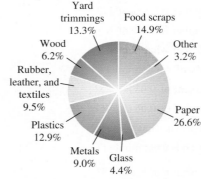

Source: Environmental Protection Agency

91. Product promotion. To promote sales, a free 6-ounce bottle of shampoo is packaged with every large bottle. Use the information on the package to find how many ounces of shampoo the large bottle contains.

92. Nutrition facts. The nutrition label on a package of corn chips is shown.

a. How many milligrams of sodium are in one serving of chips?

b. According to the label, what percent of the daily value of sodium is this?

c. What daily value of sodium intake is considered healthy?

Nutrition Facts
Serving Size: 1 oz. (28g/About 29 chips)
Servings Per Container: About 11

Amount Per Serving
Calories 160   Calories from Fat 90

% Daily Value

Total fat 10g	**15%**
Saturated fat 1.5 g	**7%**
Cholesterol 0mg	**0%**
Sodium 240mg	**12%**
Total carbohydrate 15g	**5%**
Dietary fiber 1g	**4%**
Sugars less than 1g	
Protein 2g	

93. Mixtures. Complete the table to find the number of gallons of sulfuric acid in each of two storage tanks.

	Gallons of solution in tank	% Sulfuric acid	Gallons of sulfuric acid in tank
Tank 1	60	50%	
Tank 2	40	30%	

94. The alphabet. What percent of the English alphabet do the vowels a, e, i, o, and u make up? (Round to the nearest one percent.)

95. Tips. In August of 2006, a customer left Applebee's employee Cindy Kienow of Hutchinson, Kansas, a $10,000 tip for a bill that was approximately $25. What percent tip is this? (Source: cbsnews.com)

96. Elections. In Los Angeles City Council races, if no candidate receives more than 50% of the vote, a runoff election is held between the first- and second-place finishers.

a. How many total votes were cast?

b. Determine whether there must be a runoff election for District 10.

City council	District 10
Nate Holden	8,501
Madison T. Shockley	3,614
Scott Suh	2,630
Marsha Brown	2,432

97. Renewable energy. On Sunday, August 7, 2016, the winds were strong enough in Scotland to supply all of its electricity needs for that day and more. That day, wind turbines provided 39,432 megawatt-hours of electricity, which is 106% of Scotland's total electrical needs on a Sunday. How many megawatt-hours of electricity does Scotland require on a Sunday?

zhu difeng/Shutterstock.com

98. Home loans. *Origination points* are a fee borrowers pay to lenders in order to compensate them for the role they play in evaluating, processing, and approving home mortgage loans. Each origination point represents 1% of the mortgage loan. What would a couple have to pay if they are borrowing $446,000 to purchase a home and the lender is charging them 2 loan orientation points?

99. Online dating. The *eHarmony* website claims an average of 219 of their members marry every day in the U.S. as a result of being matched by the online dating service.

a. If this claim is true, how many eHarmony marriages would there be in one year? Round the result to the nearest thousand. (There are 365 days in a year.)

b. There are about 2,000,000 marriages in the United States each year. What percent of those marriages does eHarmony claim is a result of their service?

100. Optical scanners. The U.S. Postal Service uses optical character recognition (OCR) technology to automate the mail handling process. Scanners have the ability to read many styles of machine print and handwriting on envelopes. The Postal Service has specified the exact size of the areas on an envelope that are scanned by the OCR machines. Find the dimensions (length and height) of the return address area shown below. Round to the nearest hundredth of an inch.

Number 10 Envelope
Length 9.5 in. × Height 4.125 in.

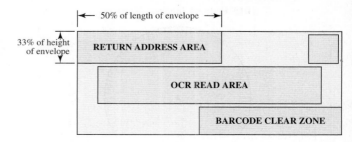

Use a circle graph to illustrate the given data. A circle divided into 100 sections is provided to help in the graphing process.

101. Paying the college bill. Draw a circle graph to show what percent of the payment for college was provided by each source.

Paying the College Bill

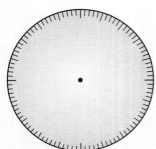

Student Borrowing	23%
Parent Borrowing	16%
Parent Income/Savings	32%
Grants/Scholarships	15%
Student Income/ Savings	10%
Friends/Relatives support	4%

Source: www.tieronetutors.com

102. Greenhouse gases. Draw a circle graph to show what percent of the total U.S. greenhouse gas emissions in 2014 came from each economic sector.

Electric power	30%
Transportation	27%
Industry	20%
Agriculture	10%
Commercial	7%
Residential	6%

Source: Environmental Protection Agency

103. Government income. Complete the following table by finding what percent of total federal government income in 2015 each source provided. Then draw a circle graph for the data.

Total Income, Fiscal Year 2015: $3,250 Billion

Source of income	Amount	Percent of total
Social Security, Medicare, unemployment taxes	$942.5 billion	
Personal income taxes	$1,365 billion	
Corporate income taxes	$292.5 billion	
Excise, estate, customs taxes	$260 billion	
Borrowing to cover deficit	$390 billion	

Source: 2016 Federal Income Tax Form 1040 Instructions

2015 Federal Income, Sources

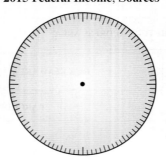

104. Water usage. The per-person indoor water use in the typical single family home is about 70 gallons per day. Complete the following table. Then draw a circle graph for the data.

Use	Gallons per person per day	Percent of total daily use
Showers	11.9	
Clothes washer	15.4	
Dishwasher	0.7	
Toilets	18.9	
Baths	1.4	
Leaks	9.8	
Faucets	10.5	
Other	1.4	

Source: American Water Works Association

Daily Water Use per Person

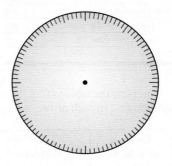

WRITING

105. Write a real-life situation that can be described by "9 is what percent of 20?"

106. Write a real-life situation that can be translated to $15 = 25\% \cdot x$.

107. Explain why 150% of a number is more than the number.

108. Explain why each of the following problems is easy to solve.

a. What is 9% of 100?

b. 16 is 100% of what number?

c. 27 is what percent of 27?

109. When solving percent problems, when is it best to write a given percent as a fraction instead of as a decimal?

110. Explain how to identify the amount, the percent, and the base in a percent problem.

REVIEW

111. Add: $2.78 + 6 + 9.09 + 0.3$

112. Evaluate: $\sqrt{64} + 3\sqrt{9}$

113. On the number line, which is closer to 5: the number 4.9 or the number 5.001?

114. Multiply: $34.5464 \cdot 1{,}000$

115. Evaluate: $(0.2)^3$

116. Evaluate: $15 - 2(3) + 7$

SECTION 6.3 Applications of Percent

In this section, we discuss applications of percent. Three of them (taxes, commissions, and discounts) are directly related to purchasing. A solid understanding of these concepts will make you a better shopper and consumer. The fourth uses percent to describe increases or decreases of such things as population and unemployment.

OBJECTIVE 1 Calculate sales taxes, total cost, and tax rates.

The department store sales receipt shown below gives a detailed account of what items were purchased, how many of each were purchased, and the price of each item.

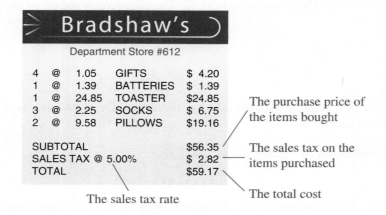

The receipt shows that the $56.35 purchase price (labeled *subtotal*) was taxed at a rate of 5%. Sales tax of $2.82 was charged.

This example illustrates the following sales tax formula. Notice that the formula is based on the percent equation discussed in Section 6.2.

Finding the Sales Tax

The sales tax on an item is a percent of the purchase price of the item.

$$\text{Sales tax} = \text{sales tax rate} \cdot \text{purchase price}$$
$$\text{amount} = \text{percent} \cdot \text{base}$$

Sales tax rates are usually expressed as a percent. When necessary, sales tax dollar amounts are rounded to the nearest cent.

EXAMPLE 1   **Sales tax.**   Find the sales tax on a purchase of $56.35 if the sales tax rate is 5%. (This is the purchase on the sales receipt shown above.)

Strategy We will identify the sales tax rate and the purchase price.

WHY Then we can use the sales tax formula to find the unknown sales tax.

Solution The sales tax rate is 5% and the purchase price is $56.35.

$$\text{Sales tax} = \textbf{sales tax rate} \cdot \textbf{purchase price} \quad \text{This is the sales tax formula.}$$

$$\text{Sales tax} = \quad 5\% \quad \cdot \quad \$56.35 \qquad \text{Substitute 5\% for the sales tax rate and } \$56.35 \text{ for the purchase price.}$$

$$\text{Sales tax} = 0.05 \cdot \$56.35 \qquad \text{Write 5\% as a decimal: 5\% = 0.05.}$$

$$\text{Sales tax} = \$2.8175 \qquad \text{Do the multiplication.}$$

$$\begin{array}{r} \overset{3\,1}{5}\overset{2}{6}.35 \\ \times\,0.05 \\ \hline 2.8175 \end{array}$$

The rounding digit in the hundredths column is 1.

↓

$$\text{Sales tax} = \$2.8\overset{}{1}75 \qquad \begin{array}{l}\text{Prepare to round the sales tax to the nearest}\\\text{cent (hundredth) by identifying the rounding}\\\text{digit and test digit.}\end{array}$$

↑

The test digit is 7.

$$\text{Sales tax} \approx \$2.82 \qquad \text{Since the test digit is 5 or greater, round up.}$$

The sales tax on the $56.35 purchase is $2.82. The sales receipt shown on the previous page is correct.

Self Check 1

Sales tax. What would the sales tax be if the $56.35 purchase were made in a state that has a 6.25% state sales tax?

Now Try ➡ **Problem 13**

Success Tip It is helpful to see the sales tax problem in Example 1 as a type of percent problem from Section 6.2.

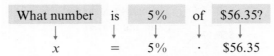

What number	is	5%	of	$56.35?
↓	↓	↓	↓	↓
x	=	5%	·	$56.35

Look at the department store sales receipt once again. Note that the sales tax was added to the purchase price to get the total cost. This example illustrates the following formula for total cost.

Finding the Total Cost

The total cost of an item is the sum of its purchase price and the sales tax on the item.

Total cost = purchase price + sales tax

EXAMPLE 2 **Total cost.** Find the total cost of the child's car seat shown on the right if the sales tax rate is 7.2%.

Strategy First, we will find the sales tax on the child's car seat.

WHY Then we can add the purchase price and the sales tax to find the total of the car seat.

Safety-T First

Child's
Car
Seat

$249.50
Buy today!

Ships next business day

Solution The sales tax rate is 7.2% and the purchase price is $249.50.

Sales tax = **sales tax rate · purchase price**   This is the sales tax formula.

Sales tax =   **7.2%**   ·   **$249.50**   Substitute 7.2% for the sales tax rate and $249.50 for the purchase price.

Sales tax = 0.072 · $249.50   Write 7.2% as a decimal: 7.2% = 0.072.

Sales tax = $17.964   Do the multiplication.

The rounding digit in the hundredths column is 6.

Sales tax = $17.96̂4   Prepare to round the sales tax to the nearest cent (hundredth) by identifying the rounding digit and test digit.

The test digit is 4.

Sales tax ≈ $17.96   Since the test digit is less than 5, round down.

$$\begin{array}{r} 249.50 \\ \times\ 0.072 \\ \hline 49900 \\ 1746500 \\ \hline 17.96400 \end{array}$$

Thus, the sales tax on the $249.50 purchase is $17.96. The total cost of the car seat is the sum of its purchase price and the sales tax.

Total cost = **purchase price + sales tax**   This is the formula for the total cost.

Total cost = $249.50 + $17.96   Substitute $249.50 for the purchase price and $17.96 for the sales tax.

Total cost = $267.46   Do the addition.

$$\begin{array}{r} \overset{1\ 1}{249.50} \\ +\ 17.96 \\ \hline 267.46 \end{array}$$

In addition to sales tax, we pay many other taxes in our daily lives. Income tax, gasoline tax, and Social Security tax are just a few. To find such tax rates, we can use an approach like that discussed in Section 6.2.

Self Check 2

Total cost. Find the total cost of a $179.95 baby stroller if the sales tax rate on the purchase is 3.2%.

Now Try ➲ **Problem 17**

EXAMPLE 3   **Withholding tax.**   A waitress found that $11.04 was deducted from her weekly gross earnings of $240 for federal income tax. What withholding tax rate was used?

Strategy We will carefully read the problem and use the given facts to write them in the form of a percent sentence.

WHY   Then we can translate the sentence into a percent equation (or percent proportion) and solve it to find the unknown withholding tax rate.

Solution There are two methods that can be used to solve this problem.

The percent equation method: Since the withholding tax of $11.04 is some unknown percent of her weekly gross earnings of $240, the percent sentence is:

$11.04   is   what percent   of $240?

11.04   = x ·   240   This is the percent equation to solve.

$$\frac{11.04}{240} = \frac{x \cdot 240}{240}$$ To isolate x on the right side of the equation, divide both sides by 240.

$$0.046 = \frac{x \cdot \overset{1}{\cancel{240}}}{\underset{1}{\cancel{240}}}$$ To simplify the fraction on the right side of the equation, remove the common factor of 240 from the numerator and denominator. On the left side, divide 11.04 by 240.

$$0.046 = x$$

$$0\,04.6\% = x$$ To write the decimal 0.046 as a percent, multiply it by 100 by moving the decimal point two places to the right, and then insert a % symbol.

$$4.6\% = x$$

$$\begin{array}{r} 0.046 \\ 240\overline{)11.0400} \\ -0 \\ \hline 11\;04 \\ -\;9\;60 \\ \hline 1\;440 \\ -\;1\;440 \\ \hline 0 \end{array}$$

The withholding tax rate was 4.6%.

The percent proportion method: Since the withholding tax of $11.04 is some unknown percent of her weekly gross earnings of $240, the percent sentence is:

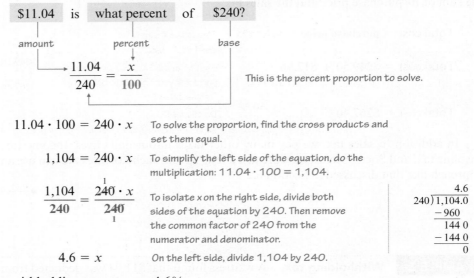

$11.04   is   what percent   of $240?

amount   percent   base

$$\frac{11.04}{240} = \frac{x}{100}$$ This is the percent proportion to solve.

$$11.04 \cdot 100 = 240 \cdot x$$ To solve the proportion, find the cross products and set them equal.

$$1{,}104 = 240 \cdot x$$ To simplify the left side of the equation, do the multiplication: 11.04 · 100 = 1,104.

$$\frac{1{,}104}{240} = \frac{\overset{1}{\cancel{240}} \cdot x}{\underset{1}{\cancel{240}}}$$ To isolate x on the right side, divide both sides of the equation by 240. Then remove the common factor of 240 from the numerator and denominator.

$$4.6 = x$$ On the left side, divide 1,104 by 240.

$$\begin{array}{r} 4.6 \\ 240\overline{)1{,}104.0} \\ -\;960 \\ \hline 144\;0 \\ -\;144\;0 \\ \hline 0 \end{array}$$

The withholding tax rate was 4.6%.

Self Check 3

Inheritance tax. A tax of $5,250 was paid on an inheritance of $15,000. What was the inheritance tax rate?

Now Try ⮕ **Problem 21**

OBJECTIVE 2   Calculate commissions and commission rates.

Instead of working for a salary or getting paid at an hourly rate, many salespeople are paid on **commission**. They earn a certain percent of the total dollar amount of the goods or services that they sell. The following formula to calculate a commission is based on the percent equation discussed in Section 6.2.

Finding the Commission

The amount of commission paid is a percent of the total dollar sales of goods or services.

Commission   =   commission rate   ·   sales

amount   =   percent   ·   base

EXAMPLE 4   **Appliance sales.**   The commission rate for a salesperson at an appliance store is 16.5%. Find his commission from the sale of a refrigerator that costs $500.

Strategy We will identify the commission rate and the dollar amount of the sale.

WHY Then we can use the commission formula to find the unknown amount of the commission.

Solution The commission rate is 16.5% and the dollar amount of the sale is $500.

Commission = **commission rate** · **sales**   *This is the commission formula.*

Commission =   **16.5%**   · **$500**   *Substitute 16.5% for the commission rate and $500 for the sales.*

Commission = 0.165 · $500   *Write 16.5% as a decimal: 16.5% = 0.165.*

Commission = $82.50   *Do the multiplication.*

$$\begin{array}{r} \overset{3\,2}{0.165} \\ \times\ 500 \\ \hline 82.500 \end{array}$$

The commission earned on the sale of the $500 refrigerator is $82.50.

Self Check 4

Selling insurance. An insurance salesperson receives a 4.1% commission on each $120 premium paid by a client. What is the amount of the commission on this premium?

Now Try ➲ Problem 25

EXAMPLE 5   **Jewelry sales.**   A jewelry salesperson earned a commission of $448 for selling a diamond ring priced at $5,600. Find the commission rate.

Strategy We will identify the commission and the dollar amount of the sale.

WHY Then we can use the commission formula to find the unknown commission rate.

Solution The commission is $448 and the dollar amount of the sale is $5,600.

Commission = commission rate · sales   *This is the commission formula.*

$448   = x · $5,600   *Substitute $448 for the commission and $5,600 for the sales. Let x represent the unknown commission rate.*

$$\frac{448}{5,600} = \frac{x \cdot 5,600}{5,600}$$ *We can drop the dollar signs. To undo the multiplication by 5,600 and isolate x on the right side of the equation, divide both sides by 5,600.*

$$0.08 = \frac{x \cdot 5,\overset{1}{\cancel{600}}}{\underset{1}{\cancel{5,600}}}$$ *On the right side, remove the common factor of 5,600 from the numerator and denominator. On the left side, divide 448 by 5,600.*

$$\begin{array}{r} 0.08 \\ 5,600\overline{)448.00} \\ -448\ 00 \\ \hline 0 \end{array}$$

$0\underset{\curvearrowright}{08}\% = x$ *To write the decimal 0.08 as a percent, multiply it by 100 by moving the decimal point two places to the right, and then insert a % symbol.*

$8\% = x$

The commission rate paid to the salesperson on the sale of the diamond ring was 8%.

Self Check 5

Selling electronics. If the commission on a $430 digital video camera is $21.50, what is the commission rate?

Now Try ➲ Problem 29

OBJECTIVE 3 **Find the percent of increase or decrease.**

Percents can be used to describe how a quantity has changed. For example, consider the table on the right, which shows the number of television channels that the average U.S. home received in 2010 and 2016.

Year	Number of television channels that the average U.S. home received
2010	135
2016	206

Source: videomind.ooyala.com

From the table, we see that the number of television channels received increased considerably from 2010 to 2016. To describe this increase using a percent, we first subtract to find the **amount of increase**.

$$206 - 135 = 71$$ *Subtract the number of TV channels received in 2010 from the number received in 2016.*

Thus, the number of channels received increased by 71 from 2010 to 2016.

Next, we find what percent of the *original* 135 channels received in 2010 that the 71 channel increase represents. To do this, we translate the problem into a percent equation (or percent proportion) and solve it.

The percent equation method:

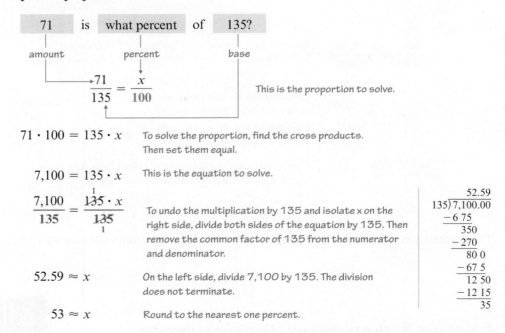

71	is	what percent	of	135?

$$71 = x \cdot 135$$ *This is the percent equation to solve.*

$$7{,}100 = 135 \cdot x$$ *To simplify the left side, do the multiplication: 71·100 = 7,100.*

$$\frac{7{,}100}{135} = \frac{\overset{1}{\cancel{135}} \cdot x}{\underset{1}{\cancel{135}}}$$ *To undo the multiplication by 135 and isolate x on the right side, divide both sides of the equation by 135. Then remove the common factor of 135 from the numerator and denominator.*

$$52.59\% \approx x$$ *To write the decimal 0.5259 as a percent, multiply it by 100 by moving the decimal point two places to the right, and then insert a % symbol.*

$$53\% \approx x$$ *Round to the nearest one percent.*

```
        0.5259
135)71.0000
   - 67 5
     3 50
   - 2 70
       800
     - 675
      1250
    - 1215
        35
```

The percent proportion method:

71	is	what percent	of	135?
amount		percent		base

$$\frac{71}{135} = \frac{x}{100}$$ *This is the proportion to solve.*

$$71 \cdot 100 = 135 \cdot x$$ *To solve the proportion, find the cross products. Then set them equal.*

$$7{,}100 = 135 \cdot x$$ *This is the equation to solve.*

$$\frac{7{,}100}{135} = \frac{\overset{1}{\cancel{135}} \cdot x}{\underset{1}{\cancel{135}}}$$ *To undo the multiplication by 135 and isolate x on the right side, divide both sides of the equation by 135. Then remove the common factor of 135 from the numerator and denominator.*

$$52.59 \approx x$$ *On the left side, divide 7,100 by 135. The division does not terminate.*

$$53 \approx x$$ *Round to the nearest one percent.*

```
        52.59
135)7,100.00
   - 6 75
      350
    - 270
      80 0
    - 67 5
      12 50
    - 12 15
        35
```

With either method, we see that there was a 53% increase in the number of television channels received by the average American home from 2010 to 2016.

EXAMPLE 6   **JFK.**   A 1996 auction included an oak rocking chair used by President John F. Kennedy in the Oval Office. The chair, originally valued at $5,000, sold for $453,500. Find the percent of increase in the value of the rocking chair.

Strategy We will begin by finding the amount of increase in the value of the rocking chair.

WHY Then we can calculate what percent of the original $5,000 value of the chair that the increase represents.

Solution First, we find the amount of increase in the value of the rocking chair.

Paul Schutzer/The LIFE Picture Collection/Getty Images

$$453{,}500 - 5{,}000 = 448{,}500$$ Subtract the original value from the price paid at auction.

The rocking chair increased in value by $448,500. Next, we find what percent of the original $5,000 value of the rocking chair the $448,500 increase represents by translating the problem into a percent equation (or percent proportion) and solving it.

The percent equation method:

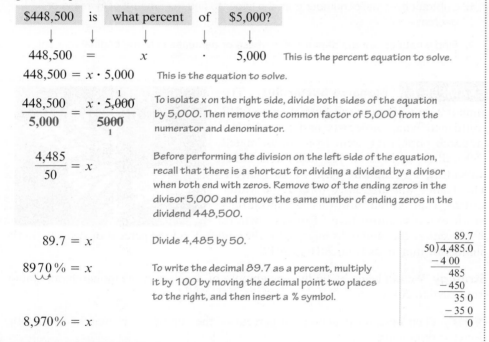

$448{,}500$ is   what percent   of $5{,}000?

$$448{,}500 \quad = \quad x \quad \cdot \quad 5{,}000$$ This is the percent equation to solve.

$$448{,}500 = x \cdot 5{,}000$$ This is the equation to solve.

$$\frac{448{,}500}{5{,}000} = \frac{x \cdot \overset{1}{\cancel{5{,}000}}}{\underset{1}{\cancel{5000}}}$$ To isolate x on the right side, divide both sides of the equation by 5,000. Then remove the common factor of 5,000 from the numerator and denominator.

$$\frac{4{,}485}{50} = x$$ Before performing the division on the left side of the equation, recall that there is a shortcut for dividing a dividend by a divisor when both end with zeros. Remove two of the ending zeros in the divisor 5,000 and remove the same number of ending zeros in the dividend 448,500.

$$89.7 = x$$ Divide 4,485 by 50.

$$89\underset{\curvearrowright}{7\,0}\% = x$$ To write the decimal 89.7 as a percent, multiply it by 100 by moving the decimal point two places to the right, and then insert a % symbol.

$$8{,}970\% = x$$

$$\begin{array}{r} 89.7 \\ 50)\overline{4{,}485.0} \\ \underline{-4\ 00} \\ 485 \\ \underline{-450} \\ 35\ 0 \\ \underline{-35\ 0} \\ 0 \end{array}$$

The percent proportion method:

$448{,}500$ is   what percent   of $5{,}000?$

amount   percent   base

$$\frac{448{,}500}{5{,}000} = \frac{x}{100}$$ This is the proportion to solve.

$$448{,}500 \cdot 100 = 5{,}000 \cdot x$$ To solve the proportion, find the cross products. Then set them equal.

$$44{,}850{,}000 = 5{,}000 \cdot x$$ To simplify the left side of the equation, do the multiplication: $448{,}500 \cdot 100 = 44{,}850{,}000$.

Caution! The percent of increase (or decrease) is a percent of the *original number*—that is, the number before the change occurred. Thus, in Example 6, it would be incorrect to write a percent sentence that compares the increase to the *new value* of the Kennedy rocking chair.

$448,500 is what percent of ~~$453,500?~~

$$\frac{44,850,000}{5,000} = \frac{\overset{1}{\cancel{5,000}} \cdot x}{\underset{1}{\cancel{5,000}}}$$

To isolate *x* on the right side, divide both sides of the equation by 5,000. Then remove the common factor of 5,000 from the numerator and denominator.

$$\frac{44,850,000}{5,000} = x$$

Before performing the division on the left side of the equation, recall that there is a shortcut for dividing a dividend by a divisor when both end with zeros.

$$\frac{44,850}{5} = x$$

Remove the three ending zeros in the divisor 5,000 and remove the same number of ending zeros in the dividend 44,850,000.

$$8,970 = x$$

Divide 44,850 by 5.

```
        8 970
    5)44,850
     -40
       4 8
      -4 5
        35
       -35
         0
        -0
         0
```

Self Check 6

News/talk radio stations.
From 2006 to 2016 the number of news/talk radio stations increased from 1,335 to 1,353. Find the percent of increase. Round to the nearest tenth percent.

Now Try ➡ **Problem 33**

With either method, we see that there was an amazing 8,970% increase in the value of the Kennedy rocking chair.

Finding the Percent of Increase or Decrease

To find the percent of increase or decrease:

1. Subtract the smaller number from the larger to find the amount of increase or decrease.

2. Find what percent the amount of increase or decrease is of the original amount.

EXAMPLE 7   **Monarch butterflies.**   The annual count of monarch butterflies in 2017 confirmed what biologists had feared. The monarch population went from an estimated 150 million in 2016 to 109 million in 2017, indicating an ongoing risk of extinction. The decline was attributed to extreme winter storms that killed millions of monarchs in Mexico's mountain forests, where 99 percent of the monarchs migrate for the winter. Find the percent of decrease in the number of monarchs from 2016 to 2017.

James Laurie/Shutterstock.com

Strategy We will begin by finding the amount of decrease in the monarch population from 2016 to 2017.

WHY Then we can calculate what percent of the number of monarchs in 2016 the decrease represents.

Solution Subtraction is used to find the amount of decrease in the monarch population. To keep from having to write a lot of zeros, we will work in millions.

$$150 - 109 = 41$$ Subtract the 2017 monarch population (in millions) from the 2016 population (in millions).

The monarch population decreased by 41 million.

Next, we find what percent of the 2016 monarch population the 41 million decrease represents by translating the problem into a percent equation (or a percent proportion) and solving it.

The percent equation method:

41	is	what percent	of	150?
↓	↓	↓	↓	↓
41	=	x	·	150

This is the percent equation to solve.

$$41 = 150 \cdot x \qquad \text{This is the equation to solve.}$$

$$\frac{41}{150} = \frac{\overset{1}{\cancel{150}} \cdot x}{\underset{1}{\cancel{150}}} \qquad$$ To undo the multiplication by 150 and isolate x on the right side, divide both sides of the equation by 150. Then remove the common factor of 150 from the numerator and denominator.

$$0.2733 = x \qquad$$ On the left side, divide 41 by 150. The division does not terminate.

$$27.33\% = x \qquad$$ To write the decimal 0.2733 as a percent, multiply it by 100 by moving the decimal point two places to the right, and then insert a % symbol.

$$27\% = x \qquad$$ Round to the nearest one percent.

```
        0.2733
150) 41.0000
    − 30 0
      11 00
    − 10 50
        500
      − 450
        500
      − 450
         50
```

The percent proportion method:

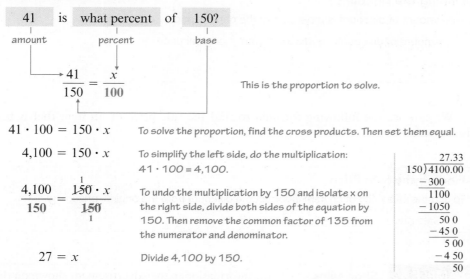

$$41 \cdot 100 = 150 \cdot x \qquad$$ To solve the proportion, find the cross products. Then set them equal.

$$4,100 = 150 \cdot x \qquad$$ To simplify the left side, do the multiplication: $41 \cdot 100 = 4,100$.

$$\frac{4,100}{150} = \frac{\overset{1}{\cancel{150}} \cdot x}{\underset{1}{\cancel{150}}} \qquad$$ To undo the multiplication by 150 and isolate x on the right side, divide both sides of the equation by 150. Then remove the common factor of 135 from the numerator and denominator.

$$27 = x \qquad$$ Divide 4,100 by 150.

```
        27.33
150) 4100.00
    − 300
     1100
    − 1050
       50 0
     − 45 0
        5 00
      − 4 50
         50
```

With either method, we see that there was a 27% decrease in the number of monarchs from 2016 to 2017.

Self Check 7

Reducing fat intake. One serving of the original Jif peanut butter has 16 grams of fat per serving. The new Jif Reduced Fat product contains 12 grams of fat per serving. What is the percent decrease in the number of grams of fat per serving?

Now Try ➲ Problem 37

OBJECTIVE 4 Calculate the amount of discount, the sale price, and the discount rate.

While shopping, you have probably noticed that many stores display signs advertising sales. Store managers have found that offering discounts attracts more customers. To be a smart shopper, it is important to know the vocabulary of discount sales.

The difference between the **original price** and the **sale price** of an item is called the **amount of discount**, or simply the **discount**. If the discount is expressed as a percent of the selling price, it is called the **discount rate**.

If we know the original price and the sale price of an item, we can use the following formula to find the amount of discount.

Finding the Discount

The amount of discount is the difference between the original price and the sale price.

Amount of discount = original price − sale price

If we know the original price of an item and the discount rate, we can use the following formula to find the amount of discount. Like several other formulas in this section, it is based on the percent equation discussed in Section 6.2.

Finding the Discount

The amount of discount is a percent of the original price.

Amount of discount = discount rate · original price

amount = percent · base

We can use the following formula to find the sale price of an item that is being discounted.

Finding the Sale Price

To find the sale price of an item, subtract the discount from the original price.

Sale price = original price − discount

EXAMPLE 8   **Shoe sales.**   Use the information in the advertisement shown on the previous page to find the amount of the discount on the pair of men's basketball shoes. Then find the sale price.

Strategy We will identify the discount rate and the original price of the shoes and use a formula to find the amount of the discount.

WHY Then we can subtract the discount from the original price to find the sale price of the shoes.

Solution From the advertisement, we see that the discount rate on the men's shoes is 25% and the original price is $89.80.

Amount of discount = **discount rate · original price**   This is the amount of discount formula.

Amount of discount =   **25%**   ·   **$89.80**   Substitute 25% for the discount rate and $89.80 for the original price.

$$\begin{array}{r} 89.80 \\ \times\,0.25 \\ \hline 44900 \\ 179600 \\ \hline 22.4500 \end{array}$$

Amount of discount = 0.25 · $89.80   Write 25% as a decimal: 25% = 0.25.

Amount of discount = $22.45   Do the multiplication.

The discount on the men's shoes is $22.45. To find the sale price, we use subtraction.

Sale price = **original price** − **discount**   *This is the sale price formula.*

Sale price =   **$89.8**   −   **$22.45**   *Substitute $89.80 for the original price and $22.45 for the discount.*

$$\begin{array}{r} 710 \\ 89.8\cancel{0} \\ -22.45 \\ \hline 67.35 \end{array}$$

Sale price = $67.35   *Do the subtraction.*

The sale price of the men's basketball shoes is $67.35.

Self Check 8

Sunglasses sales. Sunglasses, regularly selling for $30.80, are discounted 15%. Find the amount of the discount. Then find the sale price.

Now Try ➭ Problem 41

EXAMPLE 9   **Discounts.**   Find the discount rate on the ladies' cross trainer shoes shown in the advertisement on page 537. Round to the nearest one percent.

Strategy We will think of this as a percent-of-decrease problem.

WHY We want to find what percent of the $59.99 original price the amount of discount represents.

Solution From the advertisement, we see that the original price of the women's shoes is $59.99 and the sale price is $33.99. The discount (decrease in price) is found using subtraction.

$59.99 − $33.99 = $26   *Use the formula:*
Amount of discount = original price − sale price.

The shoes are discounted $26. Now we find what percent of the original price the $26 discount represents.

Amount of discount = discount rate · original price   *This is the amount of discount formula.*

   26   = x ·   **$59.99**   *Substitute 26 for the amount of discount and $59.99 for the original price. Let x represent the unknown discount rate.*

$$\frac{26}{59.99} = \frac{x \cdot 59.99}{59.99}$$ *We can drop the dollar signs. To undo the multiplication by 59.99 and isolate x on the right side of the equation, divide both sides by 59.99.*

$$0.433 \approx \frac{x \cdot \overset{1}{\cancel{59.99}}}{\underset{1}{\cancel{59.99}}}$$ *To simplify the fraction on the right side of the equation, remove the common factor of 59.99 from the numerator and denominator. On the left side, divide 26 by 59.99.*

043.3% ≈ x *To write the decimal 0.433 as a percent, multiply it by 100 by moving the decimal point two places to the right, and then insert a % symbol.*

43% ≈ x *Round to the nearest one percent.*

$$\begin{array}{r} 0.433 \\ 59.99{\overline{\smash{\big)}\,260\,0.000}} \\ -23\,99\,6 \\ \hline 200\,40 \\ -179\,97 \\ \hline 20\,430 \\ -17\,997 \\ \hline 2\,433 \end{array}$$

To the nearest one percent, the discount rate on the women's shoes is 43%.

Self Check 9

Dining out. An early-bird special at a restaurant offers a $20.99 prime rib dinner for only $18.79 if it is ordered before 6 P.M. Find the rate of discount. Round to the nearest one percent.

Now Try ➭ Problem 45

Answers to Self Checks

   1. $3.52   **2.** $185.71   **3.** 35%   **4.** $4.92   **5.** 5%   **6.** 1.3%   **7.** 25%   **8.** $4.62, $26.18   **9.** 10%

VOCABULARY

Fill in the blanks.

1. Instead of working for a salary or getting paid at an hourly rate, some salespeople are paid on _____. They earn a certain percent of the total dollar amount of the goods or services they sell.

2. Sales tax _____ are usually expressed as a percent.

3. **a.** When we use percent to describe how a quantity has increased compared to its original value, we are finding the percent of _____.

   b. When we use percent to describe how a quantity has decreased compared to its _____ value, we are finding the percent of decrease.

4. Refer to the advertisement below for a ceiling fan on sale.

   a. The _____ price of the ceiling fan was $199.99.

   b. The amount of the _____ is $40.00.

   c. The discount _____ is 20%.

   d. The _____ price of the ceiling fan is $159.99.

Ceiling Fan

Hampton Bay
52 in.
Quick install
Antique Brass

20% OFF

Was: $199.99
−40.00
Now: **$159.99**

CONCEPTS

Fill in the blanks in each of the following formulas.

5. Sales tax = sales tax rate · ▢▢▢▢

6. Total cost = ▢▢▢▢ + sales tax

7. Commission = commission rate · ▢▢

8. **a.** Amount of discount = original price − ▢▢▢

   b. Amount of discount = ▢▢▢ · original price

   c. Sale price = ▢▢▢ − discount

9. **a.** The sales tax on an item priced at $59.32 is $4.75. What is the total cost of the item?

   b. The original price of an item is $150.99. The amount of discount is $15.99. What is the sale price of the item?

10. Round each dollar amount to the nearest cent.

    a. $168.257

    b. $57.234

    c. $3.396

11. Fill in the blanks: To find the percent decrease, _____ the smaller number from the larger number to find the amount of decrease. Then find what percent that difference is of the _____ amount.

12. **U.S. armed services.** The table below shows the number of personnel on active duty in the United States Navy and Marine Corps in 2015 and 2016.

Active Duty Personnel

	Navy	Marine Corps
2015	326,504	184,587
2016	330,556	183,370

Source: *The World Almanac and Book of Facts*, 2017

a. What was the *amount of increase* in the number of active duty personnel in the Navy?

b. What was the *amount of decrease* in the number of active duty personnel in the Marine Corps?

GUIDED PRACTICE

Solve each problem to find the sales tax. **See Example 1.**

13. Find the sales tax on a purchase of $92.70 if the sales tax rate is 4%.

14. Find the sales tax on a purchase of $33.60 if the sales tax rate is 8%.

15. Find the sales tax on a purchase of $83.90 if the sales tax rate is 5%.

16. Find the sales tax on a purchase of $234.80 if the sales tax rate is 2%.

Solve each problem to find the total cost. **See Example 2.**

17. Find the total cost of a $68.24 purchase if the sales tax rate is 3.8%.

18. Find the total cost of an $86.56 purchase if the sales tax rate is 4.3%.

19. Find the total cost of a $60.18 purchase if the sales tax rate is 6.4%.

20. Find the total cost of a $70.73 purchase if the sales tax rate is 5.9%.

Solve each problem to find the tax rate. **See Example 3.**

21. **Sales tax.** The purchase price for a blender is $140. If the sales tax is $7.28, what is the sales tax rate?

22. **Sales tax.** The purchase price for a camping tent is $180. If the sales tax is $8.64, what is the sales tax rate?

23. **Self-employment tax.** A business owner paid self-employment tax of $4,590 on a taxable income of $30,000. What is the self-employment tax rate?

24. **Capital gains tax.** A couple paid $3,000 in capital gains tax on a profit of $20,000 made from the sale of some shares of stock. What is the capital gains tax rate?

Solve each problem to find the commission. See Example 4.

25. Selling shoes.
A shoe salesperson earns a 12% commission on all sales. Find her commission if she sells a pair of casual shoes for $95.

26. Selling cars. A used car salesperson earns an 11% commission on all sales. Find his commission if he sells a 2007 Chevy Malibu for $4,800.

27. Employment agencies. An employment counselor receives a 35% commission on the first week's salary of anyone that she places in a new job. Find her commission if one of her clients is hired as a secretary at $480 per week.

28. Pharmaceutical sales. A medical sales representative is paid an 18% commission on all sales. Find her commission if she sells $75,000 of Coumadin, a blood-thinning drug, to a pharmacy chain.

Solve each problem to find the commission rate. See Example 5.

29. Auctions. An auctioneer earned a $15 commission on the sale of an antique chair for $750. What is the commission rate?

30. Selling tires. A tire salesman was paid a $28 commission after one of his customers purchased a set of new tires for $560. What is the commission rate?

31. Selling electronics. If the commission on a $500 laptop computer is $20, what is the commission rate?

32. Selling clocks. If the commission on a $600 grandfather clock is $54, what is the commission rate?

Solve each problem to find the percent of increase. See Example 6.

33. Clubs. The number of members of a service club increased from 80 to 88. What was the percent of increase in club membership?

34. Savings accounts. The amount of money in a savings account increased from $2,500 to $3,000. What was the percent of increase in the amount of money saved?

35. Raises. After receiving a raise, the salary of a secretary increased from $300 to $345 dollars per week. What was the percent of increase in her salary?

36. Tuition. The tuition at a community college increased from $2,500 to $2,650 per semester. What was the percent of increase in the tuition?

Solve each problem to find the percent of decrease. See Example 7.

37. Travel time. After a new freeway was completed, a commuter's travel time to work decreased from 30 minutes to 24 minutes. What was the percent of decrease in travel time?

38. Layoffs. A printing company reduced the number of employees from 300 to 246. What was the percent of decrease in the number of employees?

39. Enrollment. Thirty-six of the 40 students originally enrolled in an algebra class completed the course. What was the percent of decrease in the number of students in the class?

40. Declining sales. One year, a pumpkin patch sold 1,200 pumpkins. The next year, they only sold 900 pumpkins. What was the percent of decrease in the number of pumpkins sold?

Solve each problem to find the amount of the discount and the sale price. See Example 8.

41. Dinnerware sales. Find the amount of the discount on a six-place dinnerware set if it regularly sells for $90, but is on sale for 33% off. Then find the sale price of the dinnerware set.

42. Bedding sales. Find the amount of the discount on a $130 bedspread that is now selling for 20% off. Then find the sale price of the bedspread.

43. Men's clothing sales. 501 Levi jeans that regularly sell for $58 are now discounted 15%. Find the amount of the discount. Then find the sale price of the jeans.

44. Book sales. At a bookstore, the list price of $23.50 for the *Merriam-Webster's Collegiate Dictionary* is crossed out, and a 30% discount sticker is pasted on the cover. Find the amount of the discount. Then find the sale price of the dictionary.

Solve each problem to find the discount rate. See Example 9.

45. Ladder sales. Find the discount rate on an aluminum ladder regularly priced at $79.95 that is on sale for $64.95. Round to the nearest one percent.

46. Office supplies sales. Find the discount rate on an electric pencil sharpener regularly priced at $49.99 that is on sale for $45.99. Round to the nearest one percent.

47. Discount tickets. The price of a one-way airline ticket from Atlanta to New York City was reduced from $209 to $179. Find the discount rate. Round to the nearest one percent.

48. Discount hotels. The cost of a one-night stay at a hotel was reduced from $245 to $200. Find the discount rate. Round to the nearest one percent.

APPLICATIONS

49. Sales tax. The Massachusetts state sales tax rate is 6.25%. Find the sales tax on a dining room set that sells for $900.

50. Sales tax. Find the sales tax on a pair of jeans costing $40 if they are purchased in Missouri, which has a state sales tax rate of 4.225%.

51. Sales receipts. Complete the sales receipt below by finding the subtotal, the sales tax, and the total cost of the purchase.

NURSERY CENTER

Your one-stop garden supply

3 @	2.99	PLANTING MIX	$ 8.97
1 @	9.87	GROUND COVER	$ 9.87
2 @	14.25	SHRUBS	$28.50

SUBTOTAL	$
SALES TAX @ 6.00%	$
TOTAL	$

52. Sales receipts. Complete the sales receipt below by finding all three prices, the subtotal, the sales tax, and the total cost of the purchase.

McCOY'S FURNITURE

1 @	450.00	SOFA	$
2 @	90.00	END TABLES	$
1 @	350.00	LOVE SEAT	$

SUBTOTAL	$
SALES TAX @ 4.20%	$
TOTAL	$

53. Room tax. After checking out of a hotel, a man noticed that the hotel bill included an additional charge labeled *room tax*. If the price of the room was $129 plus a room tax of $10.32, find the room tax rate.

54. Tanning. As part of the Affordable Care Act of 2010, indoor tanning services are subject to a federal excise tax. If a customer pays $25 for a 10-minute tanning session, and the tax is $2.50, what is the excise tax rate?

55. Gambling. For state authorized wagers (bets) placed with legal bookmakers and lottery operators, there is a federal excise tax on the wager. What is the excise tax rate if there is an excise tax of $5 on a $2,000 bet?

56. Buying fishing equipment. There are federal excise taxes on the retail price when purchasing fishing equipment. The taxes are intended to help pay for parks and conservation. What is the federal excise tax rate if there is an excise tax of $17.50 on a fishing rod and reel that has a retail price of $175?

57. Tax hikes. In order to raise more revenue, some states raise the sales tax rate. How much additional money will be collected on the sale of a $24,950 car if the sales tax rate is raised 1%?

58. Foreign travel. Value-added tax (VAT) is a consumer tax on goods and services. Currently, VAT systems are in place all around the world. (The United States is one of the few nations not using a value-added tax system.) Complete the table in the next column by determining the VAT a traveler would pay in each country on a dinner that cost $25. Round to the nearest cent.

Country	VAT rate	Tax on a $25 dinner
Mexico	16%	
Germany	19%	
Bahamas	7.5%	
Sweden	25%	

Source: www.worldwide-tax.com

59. Paychecks. Use the information on the paycheck stub to find:

a. the tax rate for the federal withholding, worker's compensation, Medicare, and Social Security taxes that were deducted from the gross pay.

b. Many paychecks have a tax column labeled FICA. FICA stands for the Federal Insurance Contribution Act tax. It can get complicated, but for most employees the FICA tax is simply the combination of the Social Security tax and the Medicare tax. What is the amount of FICA tax paid by this employee?

c. What is the FICA tax *rate* paid by this employee?

6286244

Issue date: 03-27-17

GROSS PAY	$360.00
TAXES	
FED. TAX	$ 28.80
WORK. COMP.	$ 13.50
MEDICARE	$ 5.22
SOCIAL SECURITY	$ 22.32
NET PAY	$290.16

60. Gasoline tax. In one state, a gallon of unleaded gasoline sells for $2.52. This price includes federal and state taxes that total approximately $0.764. Therefore, the price of a gallon of gasoline, before taxes, is $1.756. What is the tax rate on gasoline? Round to the nearest one percent.

61. Police force. A police department plans to increase its 80-person force to 84 persons. Find the percent increase in the size of the police force.

62. Cost-of-living increases. A woman making $32,000 a year receives a cost-of-living increase that raises her salary to $32,768 per year. Find the percent of increase in her yearly salary.

63. Lake shorelines.
Because of a heavy spring runoff, the shoreline of a lake increased from 5.8 miles to 7.6 miles. What was the percent of increase in the length of the shoreline? Round to the nearest one percent.

iStock.com/TT

64. Crop damage. After flooding damaged much of the crop, the cost of a head of lettuce jumped from $0.99 to $2.20. What percent of increase is this? Round to the nearest one percent.

65. Overtime. From May to June, the number of overtime hours for employees at a printing company increased from 42 to 106. What is the percent of increase in the number of overtime hours? Round to the nearest percent.

66. Tourism. The graph below shows the number of international visitors (travelers) to the United States each year from 2008 to 2015.

   a. The greatest percent of increase in the number of travelers was between 2013 and 2014. Find the percent increase. Round to the nearest one percent.

   b. The only decrease in the number of travelers was between 2008 and 2009. Find the percent decrease. Round to the nearest one percent.

International Travel to the U.S.

Source: U.S. Department of Commerce

67. Reduced calories. A company advertised its new, improved chips as having 96 calories per serving. The original style contained 150 calories. What percent of decrease in the number of calories per serving is this?

68. Car insurance. A student paid a car insurance premium of $420 every three months. Then the premium dropped to $370 because she qualified for a good-student discount. What was the percent of decrease in the premium? Round to the nearest percent.

69. Bus passes. To increase the number of riders, a bus company reduced the price of a monthly pass from $112 to $98. What was the percent of decrease in the cost of a bus pass?

70. Baseball. The illustration below shows the path of a baseball hit 110 mph, with a launch angle of 35 degrees, at sea level and at Coors Field, home of the Colorado Rockies. What is the percent of increase in the distance the ball travels at Coors Field?

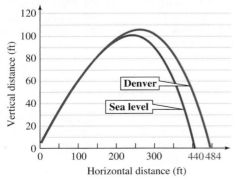

Source: *Los Angeles Times*, September 16, 1996

71. Earth moving. The illustration below shows the typical soil volume change during earth moving. (One cubic yard of soil fits in a cube that is 1 yard long, 1 yard wide, and 1 yard high.)

   a. Find the percent of increase in the soil volume as it goes through step 1 of the process.

   b. Find the percent of decrease in the soil volume as it goes through step 2 of the process.

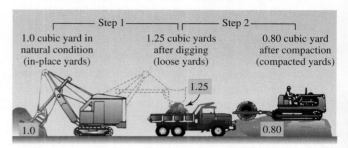

Source: U.S. Department of the Army

72. Parking. The management of a mall has decided to increase the parking area. The plans are shown below. What will be the percent of increase in the parking area when the project is completed?

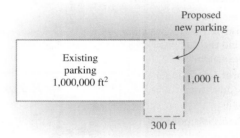

73. Real estate. After selling a house for $198,500, a real estate agent split the 6% commission with another agent. How much did each person receive?

74. Commissions. A salesperson for a medical supplies company is paid a regular commission of 9%. For orders exceeding $8,000, she receives an additional 2% in commission on the total amount. What is her commission on a sale of $14,600?

75. Sports agents. A sports agent charges her clients a fee to represent them during contract negotiations. The fee is based on a percent of the contract amount. If the agent earned $37,500 when her client signed a $2,500,000 professional football contract, what rate did she charge for her services?

76. Art galleries. An art gallery displays paintings for artists and receives a commission from the artist when a painting is sold. What is the commission rate if a gallery received $135.30 when a painting was sold for $820?

77. Whole life insurance. For the first 12 months, insurance agents earn a very large commission on the monthly premium of any whole life policy that they sell. After that, the commission rate is lowered significantly. Suppose on a new policy with monthly premiums of $160, an agent is paid monthly commissions of $144. Find the commission rate.

78. Term insurance. For the first 12 months, insurance agents earn a large commission on the monthly premium of any term life policy that they sell. After that, the commission rate is lowered significantly. Suppose on a new policy with monthly premiums of $180, an agent is paid monthly commissions of $81. Find the commission rate.

79. Concert parking. A concert promoter gets a commission of $33\frac{1}{3}\%$ of the revenue an arena receives from parking the night of the performance. How much can the promoter make if 6,000 cars are expected and parking costs $6 a car?

80. Parties. A homemaker invited her neighbors to a kitchenware party to show off cookware and utensils. As party hostess, she received 12% of the total sales. How much was purchased if she received $41.76 for hosting the party?

81. Watch sale. Refer to the advertisement to the right.
  a. Find the amount of the discount on the watch.
  b. Find the sale price of the watch.

82. Scooter sale. Refer to the advertisement to the right.
  a. Find the amount of the discount on the scooter.
  b. Find the sale price of the scooter.

83. Hover boards. Find the discount rate on a Halo Rover hover board shown in the advertisement. Round to the nearest one percent.

84. Printers. A printer regularly priced at $160, is on sale for $116. What is the discount rate?

85. Disc players. What are the sale price and the discount rate for a Blu-ray disc player that regularly sells for $399.97 and is being discounted $50? Round to the nearest one percent.

86. Camcorder sale. What are the sale price and the discount rate for a camcorder that regularly sells for $559.97 and is being discounted $80? Round to the nearest one percent.

87. Rebates. Find the discount rate and the new price for a case of motor oil if a shopper receives the manufacturer's rebate mentioned in the advertisement. Round to the nearest one percent.

Regular price $15.48/case
Mfr's rebate: $3.60

88. Double coupons. Find the discount, the discount rate, and the reduced price for a box of cereal that normally sells for $3.29 if a shopper presents the coupon at a store that doubles the value of the coupon.

89. TV shopping. Determine the Home Shopping Network (HSN) price of the ring described in the illustration if it sells it for 55% off of the retail price. Ignore shipping and handling costs.

90. Infomercials. The host of a TV infomercial says that the suggested retail price of a rotisserie grill is $249.95 and that it is now offered "for just 4 easy payments of only $39.95." What is the discount, and what is the discount rate?

Item 169-117
2.75 lb ctw
10K
Blue Topaz
Ring
6, 7, 8, 9, 10
Retail value $170
HSN Price
$??.??
S&H $5.95

91. Ring sale. What does a ring regularly sell for if it has been discounted 20% and is on sale for $149.99? (*Hint*: The ring is selling for 80% of its regular price.)

92. Blinds sale. What do vinyl blinds regularly sell for if they have been discounted 55% and are on sale for $49.50? (*Hint:* The blinds are selling for 45% of their regular price.)

93. Pickers. In the History channel TV show *American Pickers*, Mike Wolfe and Frank Fritz travel the country looking to buy antiques and collectibles that they can resell for a profit. In one episode, Mike bought (or "picked," as they say) a 92-year-old Martin guitar for $1,050 and he planned to resell it for $1,500.

   a. By how much did Mike markup the price of the guitar?

   b. What is the percent of markup on the price of the guitar? Round to the nearest percent.

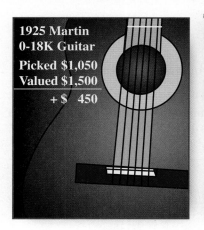

**1925 Martin
0-18K Guitar**

Picked $1,050
Valued $1,500
+ $ 450

94. Pawn shops. In an episode of the History channel TV show *Pawn Stars*, pawn shop owner Rick Harrison bought an 1800's dueling pistol for $850 from a man who was originally asking $1,100.

   a. By how much did the value of the pistol decrease in the eyes of its owner?

   b. What was the percent decrease in value of the pistol that the owner experienced? Round to the nearest percent.

WRITING

95. Explain the difference between a sales tax and a sales tax rate.

96. List the pros and cons of working on commission.

97. Suppose the price of an item increases $25 from $75 to $100. Explain why the following percent sentence *cannot* be used to find the percent of increase in the price of the item.

     25   is   what percent   of   100?

98. Explain how to find the sale price of an item if you know the regular price and the discount rate.

REVIEW

99. Multiply: $-5(-5)(-2)$

100. Divide: $\dfrac{-320}{40}$

101. Subtract: $-4 - (-7)$

102. Add: $-17 + 6 + (-12)$

103. Evaluate: $|-5 - 8|$

104. Evaluate: $\sqrt{25} - \sqrt{16}$

SECTION 6.4 Estimation with Percent

Estimation can be used to find approximations when exact answers aren't necessary. For example, when dining at a restaurant, it's helpful to be able to estimate the amount of the tip. When shopping, the ability to estimate a discount or the sale price of an item also comes in handy. In this section, we will discuss some estimation methods that can be used to make quick calculations involving percents.

OBJECTIVE 1 Estimate answers to percent problems involving 1% and 10%.

There is an easy way to find 1% of a number that does not require any calculations. First, recall that $1\% = \frac{1}{100} = 0.01$. Thus, to find 1% of a number, we multiply it by 0.01, and a quick way to multiply the number by 0.01 is to move its decimal point *two places to the left*.

Finding 1% of a Number

To find 1% of a number, move the decimal point in the number two places to the left.

EXAMPLE 1   What is 1% of 423.1? Find the exact answer and an estimate using front-end rounding.

Strategy To find the exact answer, we will move the decimal point in 423.1 two places to the left. To find an estimate, we will move the decimal point in an approximation of 423.1 two places to the left.

WHY We move the decimal point *two* places to the left because 1% of a number means 0.01 of (times) the number.

Solution
Exact answer:

$$1\% \text{ of } 423.1 = 4.231 \qquad \textit{Move the decimal point in 423.1 two places to the left.}$$

Estimate: Recall from Chapter 1 that with **front-end rounding**, a number is rounded to its largest place value so that all but its first digit is zero. To estimate 1% of 423.1, we can front-end round 423.1 to 400 and find 1% of 400. If we move the understood decimal point in 400 two places to the left, we get 4. Thus,

$$1\% \text{ of } 423.1 \approx 4 \qquad \textit{Because 1% of 400 = 4.}$$

Self Check 1

What is 1% of 519.3? Find the exact answer and an estimate using front-end rounding.

Now Try ⟹ Problem 11

Success Tip To quickly find 2% of a number, find 1% of the number by moving the decimal point two places to the left, and then double (multiply by 2) the result. In Example 1, we found that 1% of 423.1 is **4.231**. Thus, 2% of 423.1 is 2 · **4.231** = 8.462. A similar approach can be used to find 3% of a number, 4% of a number, and so on.

There is also an easy way to find 10% of a number that doesn't require any calculations. First, recall that $10\% = \frac{10}{100} = \frac{1}{10}$. Thus, to find 10% of a number, we multiply the number by 0.1, and a quick way to multiply the number by 0.1 is to move its decimal point *one place to the left*.

Finding 10% of a Number

To find 10% of a number, move the decimal point in the number one place to the left.

EXAMPLE 2   What is 10% of 6,872 feet? Find the exact answer and an estimate using front-end rounding.

Strategy To find the exact answer, we will move the decimal point in 6,872 one place to the left. To find an estimate, we will move the decimal point in an approximation of 6,872 one place to the left.

WHY We move the decimal point *one* place to the left because 10% of a number means 0.10 of (times) the number.

Solution
Exact answer:

10% of 6,872 feet = 687.2 feet   *Move the understood decimal point in 6,872 one place to the left.*

Estimate: To estimate 10% of 6,872 feet, we can front-end round 6,872 to 7,000 and find 10% of 7,000 feet. If we move the understood decimal point in 7,000 one place to the left, we get 700. Thus,

10% of 6,872 feet ≈ 700 feet   *Because 10% of 7,000 = 700.*

The rule for finding 10% of a number can be extended to help us quickly find multiples of 10% of a number.

Finding 20%, 30%, 40%, . . . of a Number

To find 20% of a number, find 10% of the number by moving the decimal point one place to the left, and then double (multiply by 2) the result. A similar approach can be used to find 30% of a number, 40% of a number, and so on.

EXAMPLE 3   Estimate the answer: What is 20% of 416?

Strategy We will estimate 10% of 416 and double (multiply by 2) the result.

WHY 20% of a number is twice as much as 10% of a number.

Solution Since 10% of 416 is 41.6 (or about **42**), it follows that 20% of 416 is about $2 \cdot \mathbf{42}$, which is 84. Thus,

20% of 416 ≈ 84   *Because 10% of 416 = 41.6 ≈ 42 and 2 · 42 = 84.*

OBJECTIVE 2   Estimate answers to percent problems involving 50%, 25%, 5%, and 15%.

There is an easy way to find 50% of a number. First, recall that $50\% = \frac{50}{100} = \frac{1}{2}$. Thus, to find 50% of a number means to find $\frac{1}{2}$ of that number, and to find $\frac{1}{2}$ of a number we simply divide it by 2.

Self Check 2
What is 10% of 3,536 pounds? Find the exact answer and an estimate using front-end rounding.

Now Try ⬤ Problem 15

Caution! In Examples 1 and 2, *front-end rounding* was used to find estimates of answers to percent problems. Since there are other ways to approximate (round) the numbers involved in a percent problem, the answers to estimation problems may vary.

Self Check 3
Estimate the answer: What is 20% of 129?

Now Try ⬤ Problem 19

> **Finding 50% of a Number**
> To find 50% of a number, divide the number by 2.

EXAMPLE 4   Estimate the answer: What is 50% of 2,595,603?

Strategy We will divide an approximation of 2,595,603 by 2.

WHY To find 50% of a number, we divide the number by 2.

Solution To estimate 50% of 2,595,603, we will find 50% of 2,600,000. We use 2,600,000 as an approximation because it is close to 2,595,603, because it is even, and, therefore, divisible by 2, and because it ends with many zeros.

$$50\% \text{ of } 2{,}595{,}603 \approx 1{,}300{,}000 \quad \textit{Because 50\% of 2,600,000} = \frac{2{,}600{,}000}{2} = 1{,}300{,}000.$$

Self Check 4

Estimate the answer: What is 50% of 14,272,549?

Now Try ➥ **Problem 23**

There is also an easy way to find 25% of a number. First, find 50% of the number by dividing the number by 2. Then, since 25% is one-half of 50%, divide that result by 2. Or, to save time, simply divide the original number by 4.

> **Finding 25% of a Number**
> To find 25% of a number, divide the number by 4.

EXAMPLE 5   Estimate the answer: What is 25% of 43.02?

Strategy We will divide an approximation of 43.02 by 4.

WHY To find 25% of a number, divide the number by 4.

Solution To estimate 25% of 43.02, we will find 25% of 44. We use 44 as an approximation because it is close to 43.02 and because it is divisible by 4.

$$25\% \text{ of } 43.02 \approx 11 \quad \textit{Because 25\% of 44} = \frac{44}{4} = 11.$$

Self Check 5

Estimate the answer: What is 25% of 27.16?

Now Try ➥ **Problem 27**

There is a quick way to find 5% of a number. First, find 10% of the number by moving the decimal point in the number one place to the left. Then, since 5% is one-half of 10%, divide that result by 2.

> **Finding 5% of a Number**
> To find 5% of a number, find 10% of the number by moving the decimal point in the number one place to the left. Then, divide that result by 2.

EXAMPLE 6   **Electricity usage.** The average U.S. household uses 10,812 kilowatt-hours of electricity each year. Several energy conservation groups would like each household to take steps to reduce its electricity usage by 5%. Estimate 5% of 10,812 kilowatt-hours. (Source: Energy Information Administration)

Strategy We will find 10% of 10,812. Then, we will divide an approximation of that result by 2.

WHY 5% of a number is one-half of 10% of a number.

Solution First, we find 10% of 10,812.

10% of 10,812 = 1,081.2   *Move the understood decimal point in 10,812 one place to the left.*

Gravity Images/Taxi/Getty Images

We will use 1,080 as an approximation of this result because it is close to 1,081.2 and because it is even, and, therefore, divisible by 2. Next, we divide the approximation by 2 to estimate 5% of 10,812.

$$\frac{1,080}{2} = 540$$ *Divide the approximation of 10% of 10,812 by 2.*

Thus, 5% of 10,812 ≈ 540. A 5% reduction in electricity usage by the average U.S. household is about 540 kilowatt-hours.

Self Check 6

Estimate the answer: What is 5% of 24,198?

Now Try Problem 31

We can use the shortcuts for finding 10% and 5% of a number to find 15% of a number.

Finding 15% of a Number

To find 15% of a number, find the sum of 10% of the number and 5% of the number.

EXAMPLE 7 **Tipping.** As a general rule, if the service in a restaurant is acceptable, a tip of 15% of the total bill should be left for the server. Estimate the 15% tip on a $77.55 dinner bill.

tetra images/First Light

Strategy We will find 10% and 5% of an approximation of $77.55. Then we will add those results.

WHY To find 15% of a number, find the sum of 10% of the number and 5% of the number.

Solution To simplify the calculations, we will estimate the cost of the $77.55 dinner to be $80. Then, to estimate the tip, we find 10% of $80 and 5% of $80, and add.

10% of $80 is $8 ⟶ $8
5% of $80 (half as much as 10% of $80) ⟶ + $4
$12   *Add to get the estimated tip.*

The tip should be $12.

Self Check 7

Tipping. Estimate the 15% tip on a $29.55 breakfast bill.

Now Try Problems 35 and 79

OBJECTIVE 3 **Estimate answers to percent problems involving 200%.**

Since 100% of a number is the number itself, it follows that 200% of a number would be twice the number. We can extend this rule to quickly find multiples of 100% of a number.

Finding 200%, 300%, 400%, . . . of a Number

To find 200% of a number, multiply the number by 2. A similar approach can be used to find 300% of a number, 400% of a number, and so on.

EXAMPLE 8   Estimate the answer: What is 200% of 5.673?

Strategy We will multiply an approximation of 5.673 by 2.

WHY To find 200% of a number, multiply the number by 2.

Solution To estimate 200% of 5.673, we will find 200% of 6. We use 6 as an approximation because it is close to 5.673 and it makes the multiplication by 2 easy.

200% of 5.673 ≈ 12   *Because 200% of 6 = 2 · 6 = 12.*

Self Check 8

Estimate the answer: What is 200% of 12.437?

Now Try **Problem 43**

OBJECTIVE 4 **Use estimation to solve percent application problems.**

In the previous examples of this section, we were given the percent (1%, 10%, 50%, 25%, 5%, 15%, or 200%), we approximated the base, and then we estimated the amount. Sometimes we must approximate the percent, as well, to estimate an answer.

EXAMPLE 9   **Music education.**   Of the 350 children attending an elementary school, 24% of them are enrolled in the instrumental music program. Estimate the number of children taking instrumental music.

Strategy We will use the rule from this section for finding 25% of a number.

WHY 24% is approximately 25%, and there is a quick way to find 25% of a number.

Solution 24% of the 350 children in the school are taking instrumental music. To estimate 24% of 350, we will find 25% of 360. We use 360 as an approximation because it is close to 350 and it is divisible by 4.

24% of 350 ≈ 90   *Because 25% of 360 = $\frac{360}{4}$ = 90.*

There are approximately 90 children in the school taking instrumental music.

Self Check 9

Student drivers. Of the 1,550 students attending a high school, 26% of them drive to school. Estimate the number of students that drive to school.

Now Try **Problem 89**

Answers to Self Checks

1. 5.193, 5   **2.** 353.6 lb, 400 lb   **3.** 26   **4.** 7,000,000   **5.** 7   **6.** 1,210   **7.** $4.50   **8.** 24
9. 400 students

SECTION 6.4 **STUDY SET**

VOCABULARY

Fill in the blanks.

1. _____ can be used to find approximations when exact answers aren't necessary.

2. With _____-end rounding, a number is rounded to its largest place value so that all but its first digit is zero.

CONCEPTS

Fill in the blanks.

3. To find 1% of a number, move the decimal point in the number _____ places to the left.

4. To find 10% of a number, move the decimal point in the number _____ place to the left.

5. To find 20% of a number, find 10% of the number by moving the decimal point one place to the left, and then double (multiply by ___) the result.

6. To find 50% of a number, divide the number by ___.

7. To find 25% of a number, divide the number by ___.

8. To find 5% of a number, find 10% of the number by moving the decimal point one place to the left. Then, divide that result by ___.

9. To find 15% of a number, find the sum of ___% of the number and ___% of the number.

10. To find 200% of a number, multiply the number by ___.

GUIDED PRACTICE

What is 1% of the given number? Find the exact answer and an estimate using front-end rounding. See Example 1.

11. 275.1   **12.** 460.9

13. 12.67   **14.** 92.11

What is 10% of the given number? Find the exact answer and an estimate using front-end rounding. See Example 2.

15. 4,059 pounds

16. 7,435 hours

17. 691.4 minutes

18. 881.2 kilometers

Estimate each answer. (Answers may vary.) See Example 3.

19. What is 20% of 346?

20. What is 20% of 409?

21. What is 20% of 67?

22. What is 20% of 32?

Estimate each answer. (Answers may vary.) See Example 4.

23. What is 50% of 4,195,898?

24. What is 50% of 6,802,117?

25. What is 50% of 397,020?

26. What is 50% of 793,288?

Estimate each answer. (Answers may vary.) See Example 5.

27. What is 25% of 15.49?

28. What is 25% of 7.02?

29. What is 25% of 49.33?

30. What is 25% of 39.74?

Estimate each answer. (Answers may vary because of the approximation used.) See Example 6.

31. What is 5% of 16,359?

32. What is 5% of 44,191?

33. What is 5% of 394.182?

34. What is 5% of 176.001?

Estimate a 15% tip on each dollar amount. (Answers may vary.) See Example 7.

35. $58.99   **36.** $38.60

37. $27.16   **38.** $49.05

39. $115.75   **40.** $135.88

41. $9.74   **42.** $11.75

Estimate each answer. (Answers may vary.) See Example 8.

43. What is 200% of 4.212?

44. What is 200% of 5.189?

45. What is 200% of 35.77?

46. What is 200% of 80.32?

TRY IT YOURSELF

Find the exact answer using methods from this section.

47. What is 2% of 600?

48. What is 3% of 700?

49. What is 30% of 18?

50. What is 40% of 45?

Estimate each answer. (Answers may vary.)

51. What is 300% of 59.2?

52. What is 400% of 203.77?

53. What is 5% of 4,605?

54. What is 5% of 8,401?

55. What is 1% of 628.21?

56. What is 1% of 12,847.9?

57. What is 15% of 119?

58. What is 15% of 237?

59. What is 10% of 67.0056?

60. What is 10% of 94.2424?

61. What is 25% of 275?

62. What is 25% of 313?

63. What is 50% of 23,898?

64. What is 25% of 56,716?

65. What is 200% of 0.9123?

66. What is 200% of 0.4189?

Find the exact answer.

67. What is 1% of 50% of 98?

68. What is 10% of 25% of 20?

69. What is 15% of 20% of 400?

70. What is 5% of 10% of 30?

LOOK ALIKES

71. What is…
  a. 1% of 1?   **b.** 10% of 1?   **c.** 100% of 1?

72. What is…
  a. 2% of 2?   **b.** 20% of 2?   **c.** 200% of 2?

73. What is…
  a. 10% of 10?   **b.** 20% of 10?   **c.** 30% of 10?

74. What is…
  a. 30% of 30?   **b.** 15% of 30?   **c.** 60% of 30?

APPLICATIONS

Estimate each answer unless stated otherwise. (Answers may vary.)

75. College courses. 20% of the 815 students attending a small college were enrolled in a science course. How many students is this?

76. Special offers. In the grocery store, a 65-ounce bottle of window cleaner was marked "25% free." How many ounces are free?

77. Discounts. By how much is the price of a coat discounted if the regular price of $196.88 is reduced by 30%?

78. Signs. The nation's largest electronic billboard is at the south intersection of Times Square in New York City. It has 12,000,000 LED lights. If just 1% of these lights burnt out, how many lights would have to be replaced? Give the exact answer.

79. Tipping. A restaurant tip is normally 15% of the cost of the meal. Find the tip on a dinner costing $38.64.

80. Visa receipts. Refer to the receipt below. Estimate the 15% gratuity (tip) and then find the total.

CLARK'S SEAFOOD
OKLAHOMA CITY, OK

Date:
Card Type: VISA
Acct Num: ★★★★★★★★★★★★0241
Exp Date: ★★/★★
Customer: WONG/TOM
Server: 209 Colleen

Amount: $58.47
Gratuity: ?
Total: ?

81. Dining out. A couple went out to eat at a restaurant. The food they ordered cost $28.55 and the drinks they ordered cost $19.75. Estimate a 15% tip on the total bill.

82. Splitting the tip. The total bill for three businessmen who went out to eat at a Chinese restaurant was $121.10. If they split the tip equally, estimate each person's share.

83. Fire damage. An insurance company paid 25% of the $118,000 it cost to rebuild a home that was destroyed by fire. How much did the insurance company pay?

84. Safety inspections. Of the 2,513 vehicles inspected at a safety checkpoint, 10% had code violations. How many cars had code violations?

85. Weightlifting. A 158-pound weightlifter can bench press 200% of his body weight. How many pounds can he bench press?

86. Testing. On a 60-question true/false test, 5% of a student's answers were wrong. How many questions did she miss?

87. Traffic studies. According to an electronic traffic monitor, 30% of the 690 motorists who passed it were speeding. How many of these motorists were speeding?

88. Selling a home. A homeowner has been told she will get back 50% of her $6,125 investment if she paints her home before selling it. How much will she get back if she paints her home?

Approximate the percent and then estimate each answer. (Answers may vary.)

89. No-shows. The attendance at a seminar was only 24% of what the organizers had anticipated. If 875 people were expected, how many actually attended the seminar?

90. Honor roll. Of the 900 students in a school, 16% were on the principal's honor roll. How many students were on the honor roll?

91. Internet surveys. The illustration shows an online survey question. How many people voted yes?

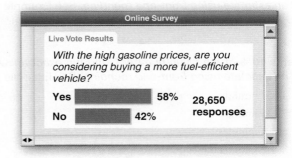

92. Sales tax. The state sales tax rate in Nebraska is 5.5%. Estimate the sales tax on a purchase of $596.

93. Voting. On election day, 48% of the 6,200 workers at the polls were volunteers. How many volunteers helped with the election?

94. Budgets. Each department at a college was asked to cut its budget by 21%. By how much money should the mathematics department budget be reduced if it is currently $4,715?

WRITING

95. Explain why 200% of a number is twice the number.

96. If you know 10% of a number, explain how you can find 30% of the same number.

97. If you know 10% of a number, explain how you can find 5% of the same number.

98. Explain why 25% of a number is the same as $\frac{1}{4}$ of the number.

REVIEW LOOK ALIKES

Perform each operation and simplify, if possible.

99. a. $\frac{5}{6} + \frac{1}{2}$ **b.** $\frac{5}{6} - \frac{1}{2}$

　　 c. $\frac{5}{6} \cdot \frac{1}{2}$ **d.** $\frac{5}{6} \div \frac{1}{2}$

100. a. $\frac{7}{15} + \frac{7}{18}$ **b.** $\frac{7}{15} - \frac{7}{18}$

　　 c. $\frac{7}{15} \cdot \frac{7}{18}$ **d.** $\frac{7}{15} \div \frac{7}{18}$

SECTION 6.5   Interest

When money is borrowed, the lender expects to be paid back the amount of the loan plus an additional charge for the use of the money. The additional charge is called **interest**. When money is deposited in a bank, the depositor is paid for the use of the money. The money the deposit earns is also called interest. In general, *interest is money that is paid for the use of money.*

OBJECTIVE 1   Calculate simple interest.

Interest is calculated in one of two ways: either as **simple interest** or as **compound interest**. We begin by discussing simple interest. First, we need to introduce some key terms associated with borrowing or lending money.

- **Principal:** the amount of money that is invested, deposited, loaned, or borrowed.
- **Interest rate:** a percent that is used to calculate the amount of interest to be paid. The interest rate is assumed to be per year (annual interest) unless otherwise stated.
- **Time:** the length of time that the money is invested, deposited, or borrowed.

The amount of interest to be paid depends on the principal, the rate, and the time. That is why all three are usually mentioned in advertisements for bank accounts, investments, and loans, as shown below.

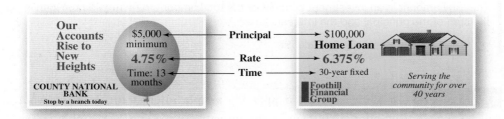

Simple interest is interest earned only on the original principal. It is found using the following formula.

Simple Interest Formula

     Interest = principal · rate · time   or $I = P \cdot r \cdot t$

where the rate r is expressed as an annual (yearly) rate and the time t is expressed in years. This formula can be written more simply without the multiplication raised dots as

     $I = Prt$

EXAMPLE 1   If \$3,000 is invested for 1 year at a rate of 5%, how much simple interest is earned?

Strategy We will identify the principal, rate, and time for the investment.

WHY Then we can use the formula $I = Prt$ to find the unknown amount of simple interest earned.

Solution   The principal is $3,000, the interest rate is 5%, and the time is 1 year.

$$P = \$3,000 \qquad r = 5\% = 0.05 \qquad t = 1$$

$I = \mathbf{Prt}$	This is the simple interest formula.
$I = \mathbf{\$3,000 \cdot 0.05 \cdot 1}$	Substitute the values for P, r, and t. Remember to write the rate r as a decimal.
$I = \$3,000 \cdot 0.05$	Multiply: $0.05 \cdot 1 = 0.05$.
$I = \$150$	Complete the multiplication.

$$\begin{array}{r} 3,000 \\ \times\, 0.05 \\ \hline 150.00 \end{array}$$

Self Check 1

If $4,200 is invested for 2 years at a rate of 4%, how much simple interest is earned?

Now Try   Problem 17

The simple interest earned in 1 year is $150.

The information given in this problem and the result can be presented in a table.

Principal	Rate	Time	Interest earned
$3,000	5%	1 year	$150

If no money is withdrawn from an investment, the investor receives the principal *and* the interest at the end of the time period. Similarly, a borrower must repay the principal *and* the interest when taking out a loan. In each case, the **total amount** of money involved is given by the following formula.

Finding the Total Amount

The total amount in an investment account or the total amount to be repaid on a loan is the sum of the principal and the interest.

Total amount = principal + interest

Caution!   When we use the formula $I = Prt$, the time must be expressed in years. If the time is given in days or months, we rewrite it as a fractional part of a year. For example, a 30-day investment lasts $\frac{30}{365}$ of a year, since there are 365 days in a year. For a 6-month loan, we express the time as $\frac{6}{12}$ or $\frac{1}{2}$ of a year, since there are 12 months in a year.

EXAMPLE 2   If $800 is invested at 4.5% simple interest for 3 years, what will be the total amount of money in the investment account at the end of the 3 years?

Strategy   We will find the simple interest earned on the investment and add it to the principal.

WHY   At the end of 3 years, the total amount of money in the account is the sum of the principal and the interest earned.

Solution   The principal is $800, the interest rate is 4.5%, and the time is 3 years. To find the interest the investment earns, we use multiplication.

$$P = \$800 \qquad r = 4.5\% = 0.045 \qquad t = 3$$

$I = \mathbf{Prt}$	This is the simple interest formula.
$I = \mathbf{\$800 \cdot 0.045 \cdot 3}$	Substitute the values for P, r, and t. Remember to write the rate r as a decimal.
$I = \$36 \cdot 3$	Multiply: $\$800 \cdot 0.045 = \$36$.
$I = \$108$	Complete the multiplication.

$$\begin{array}{r} \overset{4}{0.045} \\ \times\ 800 \\ \hline 36.000 \end{array} \qquad \begin{array}{r} \overset{1}{36} \\ \times 3 \\ \hline 108 \end{array}$$

Self Check 2

If $600 is invested at 2.5% simple interest for 4 years, what will be the total amount of money in the investment account at the end of the 4 years?

Now Try   Problem 21

The simple interest earned in 3 years is $108. To find the total amount of money in the account, we add.

Total amount = **principal** + **interest**	This is the total amount formula.
Total amount = **$800** + **$108**	Substitute $800 for the principal and $108 for the interest.
Total amount = $908	Do the addition.

At the end of 3 years, the total amount of money in the account will be $908.

EXAMPLE 3   **Education costs.** A student borrowed $920 at 3% for 9 months to pay some college tuition expenses. Find the simple interest that must be paid on the loan.

Strategy We will rewrite 9 months as a fractional part of a year, and then we will use the formula $I = Prt$ to find the unknown amount of simple interest to be paid on the loan.

WHY To use the formula $I = Prt$, the time must be expressed in years, or as a fractional part of a year.

Solution Since there are 12 months in a year, we have

$$9 \text{ months} = \frac{9}{12} \text{ year} = \frac{\overset{1}{\cancel{3}} \cdot 3}{\cancel{3} \cdot 4} \text{ year} = \frac{3}{4} \text{ year}$$

Simplify the fraction $\frac{9}{12}$ by removing a common factor of 3 from the numerator and denominator.

The time of the loan is $\frac{3}{4}$ year. To find the amount of interest, we multiply.

$$P = \$920 \qquad r = 3\% = 0.03 \qquad t = \frac{3}{4}$$

$I = Prt$ This is the simple interest formula.

$I = \$920 \cdot 0.03 \cdot \dfrac{3}{4}$ Substitute the values for P, r, and t. Remember to write the rate r as a decimal.

$I = \dfrac{\$920}{1} \cdot \dfrac{0.03}{1} \cdot \dfrac{3}{4}$ Write $920 and 0.03 as fractions.

$I = \dfrac{\$82.80}{4}$ Multiply the numerators. Multiply the denominators.

$I = \$20.70$ Do the division: 82.80 ÷ 4 = 20.70.

The simple interest to be paid on the loan is $20.70.

$$\begin{array}{r} 920 \\ \times 0.03 \\ \hline 27.60 \end{array} \qquad \begin{array}{r} \overset{2\ 1}{27.60} \\ \times\quad 3 \\ \hline 82.80 \end{array}$$

$$\begin{array}{r} 20.70 \\ 4\overline{)82.80} \\ -8 \\ \hline 02 \\ -0 \\ \hline 2\ 8 \\ -2\ 8 \\ \hline 00 \\ -0 \\ \hline 0 \end{array}$$

Self Check 3

Short-term loans. Find the simple interest on a loan of $810 at 9% for 8 months.

Now Try ➜ Problem 25

EXAMPLE 4   **Short-term business loans.** To start a business, a couple borrowed $5,500 for 90 days to purchase equipment and supplies. If the loan has a 14% simple interest rate, find the total amount they must repay at the end of the 90-day period.

Strategy We will rewrite 90 days as a fractional part of a year, and then we will use the formula $I = Prt$ to find the unknown amount of simple interest to be paid on the loan.

WHY To use the formula $I = Prt$, the time must be expressed in years, or as a fractional part of a year.

Solution Since there are 365 days in a year, we have

$$90 \text{ days} = \frac{90}{365} \text{ year} = \frac{\overset{1}{\cancel{5}} \cdot 18}{\cancel{5} \cdot 73} \text{ year} = \frac{18}{73} \text{ year}$$

Simplify the fraction $\frac{90}{365}$ by removing a common factor of 5 from the numerator and denominator.

The time of the loan is $\frac{18}{73}$ year. To find the amount of interest, we multiply.

$$P = \$5{,}500 \qquad r = 14\% = 0.14 \qquad t = \frac{90}{365} = \frac{18}{73}$$

$I = \mathbf{Prt}$ This is the simple interest formula.

$I = \mathbf{\$5{,}500} \cdot 0.14 \cdot \dfrac{18}{73}$ Substitute the values for P, r, and t.

$I = \dfrac{\$5{,}500}{1} \cdot \dfrac{0.14}{1} \cdot \dfrac{18}{73}$ Write $5,500 and 0.14 as fractions.

$I = \dfrac{\$13{,}860}{73}$ Multiply the numerators.
Multiply the denominators.

$I \approx \$189.86$ Divide 13,860 by 73. The division does not terminate. Round to the nearest cent.

5,500	770
×0.14	×18
22000	6160
55000	7700
770.00	13,860

Self Check 4

Accounting. To cover payroll expenses, a small business owner borrowed $3,200 at a simple interest rate of 15%. Find the total amount he must repay at the end of 120 days.

Now Try ➲ **Problem 29**

The interest on the loan is $189.86. To find how much they must pay back, we add.

Total amount = **principal** + interest   This is the total amount formula.

       = **$5,500** + $189.86   Substitute $5,500 for the principal and $189.86 for the interest.

       = $5,689.86   Do the addition.

The couple must pay back $5,689.86 at the end of 90 days.

OBJECTIVE 2   **Calculate compound interest.**

Most savings accounts and investments pay *compound interest* rather than simple interest. We have seen that simple interest is paid only on the original principal. **Compound interest** is paid on the principal and *previously earned interest.* To illustrate this concept, suppose that $2,000 is deposited in a savings account at a rate of 5% for 1 year. We can use the formula $I = Prt$ to calculate the interest earned at the end of 1 year.

$I = \mathbf{Prt}$ This is the simple interest formula.

$I = \$2{,}000 \cdot 0.05 \cdot 1$ Substitute for P, r, and t.

$I = \$100$ Do the multiplication.

Interest of $100 was earned. At the end of the first year, the account contains the interest ($100) plus the original principal ($2,000), for a balance of $2,100.

    Suppose that the money remains in the savings account for another year at the same interest rate. For the second year, interest will be paid on a principal of $2,100. That is, during the second year, we earn *interest on the interest* as well as on the original $2,000 principal. Using $I = Prt$, we can find the interest earned in the second year.

$I = \mathbf{Prt}$ This is the simple interest formula.

$I = \$2{,}100 \cdot 0.05 \cdot 1$ Substitute for P, r, and t.

$I = \$105$ Do the multiplication.

In the second year, $105 of interest is earned. The account now contains that interest plus the $2,100 principal, for a total of $2,205.

    As the figure on the next page shows, we calculated the simple interest two times to find the compound interest.

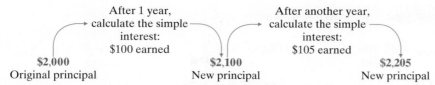

After 1 year, calculate the simple interest: $100 earned

After another year, calculate the simple interest: $105 earned

$2,000
Original principal

$2,100
New principal

$2,205
New principal

If we compute only the *simple interest* on $2,000, at 5% for 2 years, the interest earned is $I = \$2,000 \cdot 0.05 \cdot 2 = \$200$. Thus, the account balance would be $2,200. Comparing the balances, we find that the account earning compound interest will contain $5 more than the account earning simple interest.

In the previous example, the interest was calculated at the end of each year, or **annually**. When compounding, we can compute the interest in other time spans, such as **semiannually** (twice a year), **quarterly** (four times a year), or even **daily**.

EXAMPLE 5   **Compound interest.** As a special gift for her newborn granddaughter, a grandmother opens a $1,000 savings account in the baby's name. The interest rate is 4.2%, compounded quarterly. Find the amount of money the child will have in the bank on her first birthday.

Strategy We will use the simple interest formula $I = Prt$ four times in a series of steps to find the amount of money in the account after 1 year. Each time, the time t is $\frac{1}{4}$.

WHY The interest is compounded *quarterly*.

Solution If the interest is compounded quarterly, the interest will be computed four times in one year. To find the amount of interest $1,000 will earn in the first quarter of the year, we use the simple interest formula, where t is $\frac{1}{4}$ of a year.

Interest earned in the first quarter:

$$P_{\text{1st Qtr}} = \$1,000 \qquad r = 4.2\% = 0.042 \qquad t = \frac{1}{4}$$

$I = Prt$ This is the simple interest formula.

$I = \$1,000 \cdot 0.042 \cdot \dfrac{1}{4}$ Substitute for P, r, and t.

$I = \$42 \cdot \dfrac{1}{4}$ Multiply: $\$1,000 \cdot 0.042 = \$42$.

$I = \dfrac{\$42}{4}$ Do the multiplication. Think of 42 as $\frac{42}{1}$.

$I = \$10.50$ Do the division: $42 \div 4 = 10.5$.

$$
\begin{array}{r}
10.5 \\
4\overline{)42.0} \\
-4 \\
\hline
02 \\
-0 \\
\hline
2\,0 \\
-2\,0 \\
\hline
0
\end{array}
$$

The interest earned in the first quarter is $10.50. This now becomes part of the principal for the second quarter.

$P_{\text{2nd Qtr}} = \$1,000 + \$10.50 = \$1,010.50$ Add the original principal and the interest that it earned to find the second-quarter principal.

To find the amount of interest $1,010.50 will earn in the second quarter of the year, we use the simple interest formula, where t is again $\frac{1}{4}$ of a year.

Interest earned in the second quarter:

$$P_{\text{2nd Qtr}} = \$1{,}010.50 \qquad r = 0.042 \qquad t = \frac{1}{4}$$

$I = Prt$ *This is the simple interest formula.*

$$I = \$1{,}010.50 \cdot 0.042 \cdot \frac{1}{4}$$ *Substitute for P, r, and t.*

$$I = \frac{\$1{,}010.50 \cdot 0.042 \cdot 1}{4}$$ *Multiply.*

$I \approx \$10.61$ *Use a calculator. Round to the nearest cent (hundredth).*

The interest earned in the second quarter is \$10.61. This becomes part of the principal for the third quarter.

$$P_{\text{3rd Qtr}} = \$1{,}010.50 + \$10.61 = \$1{,}021.11$$ *Add the second-quarter principal and the interest that it earned to find the third-quarter principal.*

To find the interest \$1,021.11 will earn in the third quarter of the year, we proceed as follows.

Interest earned in the third quarter:

$$P_{\text{3rd Qtr}} = \$1{,}021.11 \qquad r = 0.042 \qquad t = \frac{1}{4}$$

$I = Prt$ *This is the simple interest formula.*

$$I = \$1{,}021.11 \cdot 0.042 \cdot \frac{1}{4}$$ *Substitute for P, r, and t.*

$$I = \frac{\$1{,}021.11 \cdot 0.042 \cdot 1}{4}$$ *Multiply.*

$I \approx \$10.72$ *Use a calculator. Round to the nearest cent (hundredth).*

The interest earned in the third quarter is \$10.72. This now becomes part of the principal for the fourth quarter.

$$P_{\text{4th Qtr}} = \$1{,}021.11 + \$10.72 = \$1{,}031.83$$ *Add the third-quarter principal and the interest that it earned to find the fourth-quarter principal.*

To find the interest \$1,031.83 will earn in the fourth quarter, we again use the simple interest formula.

Interest earned in the fourth quarter:

$$P_{\text{4th Qtr}} = \$1{,}031.83 \qquad r = 0.042 \qquad t = \frac{1}{4}$$

$I = Prt$ *This is the simple interest formula.*

$$I = \$1{,}031.83 \cdot 0.042 \cdot \frac{1}{4}$$ *Substitute for P, r, and t.*

$$I = \frac{\$1{,}031.83 \cdot 0.042 \cdot 1}{4}$$ *Multiply.*

$I \approx \$10.83$ *Use a calculator. Round to the nearest cent (hundredth).*

The interest earned in the fourth quarter is $10.83. Adding this to the existing principal, we get

Total amount = $1,031.83 + $10.83 = $1,042.66   *Add the fourth-quarter principal and the interest that it earned.*

The total amount in the account after four quarters, or 1 year, is $1,042.66.

Calculating compound interest by hand can take a long time. The **compound interest formula** can be used to find the total amount of money that an account will contain at the end of the term quickly.

Self Check 5

Compound interest.
Suppose $8,000 is deposited in an account that earns 2.3% compounded quarterly. Find the amount of money in an account at the end of the first year.

Now Try ⮕ **Problem 33**

Compound Interest Formula

The total amount A in an account can be found using the formula

$$A = P\left(1 + \frac{r}{n}\right)^{nt}$$

where P is the principal, r is the annual interest rate expressed as a decimal, t is the length of time in years, and n is the number of compoundings in one year.

A calculator is very helpful in performing the operations on the right side of the compound interest formula.

Using Your Calculator ▶ **Compound Interest**

A businessperson invests $9,250 at 7.6% interest, to be compounded monthly. To find what the investment will be worth in 3 years, we use the compound interest formula with the following values.

$P = \$9,250$ $r = 7.6\% = 0.076$ $t = 3$ years $n = 12$ times a year (monthly)

$$A = P\left(1 + \frac{r}{n}\right)^{nt}$$ *This is the compound interest formula.*

$$A = 9,250\left(1 + \frac{0.076}{12}\right)^{12(3)}$$ *Substitute the values of P, r, t, and n. In the exponent, nt means n·t.*

$$A = 9,250\left(1 + \frac{0.076}{12}\right)^{36}$$ *Evaluate the exponent: 12(3) = 36.*

To evaluate the expression on the right-hand side of the equation using a calculator, we enter these numbers and press these keys.

9250 × (1 + .076 ÷ 12) y^x 36 = | 11610.43875 |

On some calculator models, the ^ key is used in place of the y^x key. Also, the | ENTER | key is pressed instead of the = key for the result to be displayed.

Rounded to the nearest cent, the amount in the account after 3 years will be $11,610.44.

If your calculator does not have parenthesis keys, calculate the sum within the parentheses first. Then find the power. Finally, multiply by 9,250.

EXAMPLE 6　**Compounding daily.**　An investor deposited $50,000 in a long-term account at 6.8% interest, compounded daily. How much money will he be able to withdraw in 7 years if the principal is to remain in the bank?

Strategy　We will use the compound interest formula to find the *total amount* in the account after 7 years. Then we will subtract the original principal from that result.

WHY　When the investor withdraws money, he does not want to touch the original $50,000 principal in the account.

Solution　"Compounded daily" means that compounding will be done 365 times in a year for 7 years.

$$P = \$50,000 \qquad r = 6.8\% = 0.068 \qquad t = 7 \qquad n = 365$$

$$A = P\left(1 + \frac{r}{n}\right)^{nt} \qquad \text{This is the compound interest formula.}$$

$$A = 50,000\left(1 + \frac{0.068}{365}\right)^{365(7)} \qquad \begin{array}{l}\text{Substitute the values of } P, r, t, \text{ and } n. \\ \text{In the exponent, } nt \text{ means } n \cdot t.\end{array} \qquad \begin{array}{r}{\scriptstyle 4\,3} \\ 365 \\ \times\ \ 7 \\ \hline 2,555\end{array}$$

$$A = 50,000\left(1 + \frac{0.068}{365}\right)^{2,555} \qquad \text{Evaluate the exponent: } 365 \cdot 7 = 2,555.$$

$$A \approx 80,477.58 \qquad \text{Use a calculator. Round to the nearest cent.}$$

The account will contain $80,477.58 at the end of 7 years. To find how much money the man can withdraw, we must subtract the original principal of $50,000 from the total amount in the account.

$$80,477.58 - 50,000 = 30,477.58$$

The man can withdraw $30,477.58 without having to touch the $50,000 principal.

Self Check 6

Compounding daily.
Find the amount of interest $25,000 will earn in 10 years if it is deposited in an account at 5.99% interest, compounded daily.

Now Try　Problem 37

Answers to Self Checks

1. $336　**2.** $660　**3.** $48.60　**4.** $3,357.81　**5.** $8,185.59　**6.** $20,505.20

SECTION 6.5　STUDY SET

VOCABULARY

Fill in the blanks.

1. In general, _____ is money that is paid for the use of money.
2. In banking, the original amount of money invested, deposited, loaned, or borrowed is known as the _____.
3. The percent that is used to calculate the amount of interest to be paid is called the interest ____.
4. _____ interest is interest earned only on the original principal.
5. The _____ amount in an investment account is the sum of the principal and the interest.
6. _____ interest is interest paid on the principal and previously earned interest.

CONCEPTS

7. Refer to the home loan advertisement below.

Loans.com
Great mortgage rates

Home Loan　**5%**　30-year fixed

$125,000 available online

 a. What is the principal?
 b. What is the interest rate?
 c. What is the time?

8. Refer to the investment advertisement below.

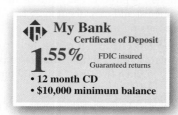

My Bank
Certificate of Deposit

1.55% FDIC insured
Guaranteed returns

• **12 month CD**
• **$10,000 minimum balance**

    a. What is the minimum principal?

    b. What is the interest rate?

    c. What is the time?

9. When making calculations involving percents, they must be written as decimals or fractions. Change each percent to a decimal.

    a. 7% **b.** 9.8% **c.** $6\frac{1}{4}\%$

10. Express each of the following as a fraction of a year. Simplify the fraction.

    a. 6 months   **b.** 90 days

    c. 120 days   **d.** 1 month

11. Complete the table by finding the simple interest earned.

Principal	Rate	Time	Interest earned
$10,000	6%	3 years	

12. Determine how many times a year the interest on a savings account is calculated if the interest is compounded

    a. annually   **b.** semiannually

    c. quarterly   **d.** daily

    e. monthly

13. **a.** What concept studied in this section is illustrated by the diagram below?

    b. What was the original principal?

    c. How many times was the interest found?

    d. How much interest was earned on the first compounding?

    e. For how long was the money invested?

    ⌒ **1st qtr** ⌄⌒ **2nd qtr** ⌄⌒ **3rd qtr** ⌄⌒ **4th qtr** ⌒

    $1,000 $1,050 $1,102.50 $1,157.63 $1,215.51

14. $3,000 is deposited in a savings account that earns 10% interest compounded annually. Complete the series of calculations in the illustration below to find how much money will be in the account at the end of 2 years.

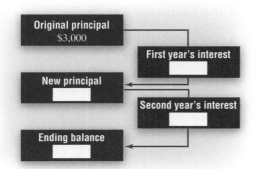

Original principal
$3,000

First year's interest

New principal

Second year's interest

Ending balance

15. Write the simple interest formula $I = P \cdot r \cdot t$ without the multiplication raised dots.

16. In the formula $A = P\left(1 + \dfrac{r}{n}\right)^{nt}$, how many operations must be performed to find A?

GUIDED PRACTICE

Calculate the simple interest earned. **See Example 1.**

17. If $2,000 is invested for 1 year at a rate of 5%, how much simple interest is earned?

18. If $6,000 is invested for 1 year at a rate of 7%, how much simple interest is earned?

19. If $700 is invested for 4 years at a rate of 9%, how much simple interest is earned?

20. If $800 is invested for 5 years at a rate of 8%, how much simple interest is earned?

Calculate the total amount in each account. **See Example 2.**

21. If $500 is invested at 2.5% simple interest for 2 years, what will be the total amount of money in the investment account at the end of the 2 years?

22. If $400 is invested at 6.5% simple interest for 6 years, what will be the total amount of money in the investment account at the end of the 6 years?

23. If $1,500 is invested at 1.2% simple interest for 5 years, what will be the total amount of money in the investment account at the end of the 5 years?

24. If $2,500 is invested at 4.5% simple interest for 8 years, what will be the total amount of money in the investment account at the end of the 8 years?

Calculate the simple interest. **See Example 3.**

25. Find the simple interest on a loan of $550 borrowed at 4% for 9 months.

26. Find the simple interest on a loan of $460 borrowed at 9% for 9 months.

27. Find the simple interest on a loan of $1,320 borrowed at 7% for 4 months.

28. Find the simple interest on a loan of $1,250 borrowed at 10% for 3 months.

Calculate the total amount that must be repaid at the end of each short-term loan. **See Example 4.**

29. $12,600 is loaned at a simple interest rate of 18% for 90 days. Find the total amount that must be repaid at the end of the 90-day period.

30. $45,000 is loaned at a simple interest rate of 12% for 90 days. Find the total amount that must be repaid at the end of the 90-day period.

31. $40,000 is loaned at 10% simple interest for 45 days. Find the total amount that must be repaid at the end of the 45-day period.

32. $30,000 is loaned at 20% simple interest for 60 days. Find the total amount that must be repaid at the end of the 60-day period.

Calculate the total amount in each account. **See Example 5.**

33. Suppose $2,000 is deposited in a savings account that pays 3% interest, compounded quarterly. How much money will be in the account in one year?

34. Suppose $3,000 is deposited in a savings account that pays 2% interest, compounded quarterly. How much money will be in the account in one year?

35. If $5,400 earns 4% interest, compounded quarterly, how much money will be in the account at the end of one year?

36. If $10,500 earns 8% interest, compounded quarterly, how much money will be in the account at the end of one year?

Use a calculator to solve the following problems. **See Example 6.**

37. A deposit of $30,000 is placed in a savings account that pays 4.8% interest, compounded daily. How much money can be withdrawn at the end of 6 years if the principal is to remain in the bank?

38. A deposit of $12,000 is placed in a savings account that pays 5.6% interest, compounded daily. How much money can be withdrawn at the end of 8 years if the principal is to remain in the bank?

39. If 8.55% interest, compounded daily, is paid on a deposit of $55,250, how much money will be in the account at the end of 4 years?

40. If 4.09% interest, compounded daily, is paid on a deposit of $39,500, how much money will be in the account at the end of 9 years?

APPLICATIONS

41. **Retirement income.**   A retiree invests $5,000 in a savings plan that pays a simple interest rate of 6%. What will the account balance be at the end of the first year?

42. **Investments.**   A developer promised a return of 8% simple interest on an investment of $15,000 in her company. How much could an investor expect to make in the first year?

from **Campus to Careers**

Loan Officer

43. A member of a credit union was loaned $1,200 to pay for car repairs. The loan was made for 3 years at a simple interest rate of 5.5%. Find the interest due on the loan.

44. **Remodeling.**   A homeowner borrows $8,000 to pay for a kitchen remodeling project. The terms of the loan are 9.2% simple interest and repayment in 2 years. How much interest will be paid on the loan?

45. **Smoke damage.**   The owner of a café borrowed $4,500 for 2 years at 12% simple interest to pay for the cleanup after a kitchen fire. Find the total amount due on the loan.

46. **Alternative fuels.**   To finance the purchase of a fleet of natural-gas-powered vehicles, a city borrowed $200,000 for 4 years at a simple interest rate of 3.5%. Find the total amount due on the loan.

47. **Short-term loans.**   A loan of $1,500 at 12.5% simple interest is paid off in 3 months. What is the interest charged?

48. **Farm loans.**   An apple orchard owner borrowed $7,000 from a farmers' co-op bank. The money was loaned at 8.8% simple interest for 18 months. How much money did the co-op charge him for the use of the money?

49. **Meeting payrolls.**   In order to meet end-of-the-month payroll obligations, a small business had to borrow $4,200 for 30 days. How much did the business have to repay if the simple interest rate was 18%?

50. **Car loans.**   To purchase a car, a man takes out a loan for $2,000 for 120 days. If the simple interest rate is 9% per year, how much interest will he have to pay at the end of the 120-day loan period?

51. **Savings accounts.**   Find the interest earned on $10,000 at $7\frac{1}{4}$% for 2 years. Use the table to organize your work.

P	r	t	I

52. **Tuition.**   A student borrows $300 from an educational fund to pay for books for spring semester. If the loan is for 45 days at $3\frac{1}{2}$% annual interest, what will the student owe at the end of the loan period?

53. **Loan applications.**   Complete the following loan application.

Loan Application Worksheet

1. Amount of loan (principal) ___$1,200.00___

2. Length of loan (time) ___2 YEARS___

3. Annual percentage rate ___8%___
   (simple interest)
4. Interest charged _____

5. Total amount to be repaid _____

6. Check method of repayment:
   ☐ 1 lump sum   ☑ monthly payments

   Borrower agrees to pay ___24___ equal
   payments of _____ to repay loan.

54. Loan applications. Complete the following loan application.

Loan Application Worksheet

1. Amount of loan (principal) ___$810.00___

2. Length of loan (time) ___9 mos.___

3. Annual percentage rate ___12%___
   (simple interest)
4. Interest charged _____

5. Total amount to be repaid _____

6. Check method of repayment:
   ☐ 1 lump sum   ☑ monthly payments

   Borrower agrees to pay ___9___ equal
   payments of _____ to repay loan.
---

55. Low-interest loans. An underdeveloped country receives a low-interest loan from a bank to finance the construction of a water treatment plant. What must the country pay back at the end of $3\frac{1}{2}$ years if the loan is for $18 million at 2.3% simple interest?

56. Redevelopment. A city is awarded a low-interest loan to help renovate the downtown business district. The $40-million loan, at 1.75% simple interest, must be repaid in $2\frac{1}{2}$ years. How much interest will the city have to pay?

A calculator will be helpful in solving the following problems.

57. Compounding annually. If $600 is invested in an account that earns 8%, compounded annually, what will the account balance be after 3 years?

58. Compounding semiannually. If $600 is invested in an account that earns annual interest of 8%, compounded semiannually, what will the account balance be at the end of 3 years?

59. College funds. A ninth-grade student opens a savings account that locks her money in for 4 years at an annual rate of 6%, compounded daily. If the initial deposit is $1,000, how much money will be in the account when she begins college in 4 years?

60. Certificate of deposits. A 3-year certificate of deposit pays an annual rate of 5%, compounded daily. The maximum allowable deposit is $90,000. What is the most interest a depositor can earn from the CD?

61. Tax refunds. A couple deposits an income tax refund check of $545 in an account paying an annual rate of 4.6%, compounded daily. What will the size of the account be at the end of 1 year?

62. Inheritance. After receiving an inheritance of $11,000, a man deposits the money in an account paying an annual rate of 7.2%, compounded daily. How much money will be in the account at the end of 1 year?

63. Lottery. Suppose you won $500,000 in the lottery and deposited the money in a savings account that paid an annual rate of 6% interest, compounded daily. How much interest would you earn in a year?

64. Cash gifts. After receiving a $250,000 cash gift, a university decides to deposit the money in an account paying an annual rate of 5.88%, compounded quarterly. How much money will the account contain in 5 years?

65. Withdrawing only interest. A financial adviser invested $90,000 in a long-term account at 5.1% interest, compounded daily. How much money will she be able to withdraw in 20 years if the principal is to remain in the account?

66. Living on the interest. A couple sold their home and invested the profit of $490,000 in an account at 6.3% interest, compounded daily. How much money will they be able to withdraw in 2 years if they don't want to touch the principal?

67. Paycheck advance. A *paycheck advance* is a small, short-term loan that is to be paid back at the borrower's next payday. These loans (usually for about $1,000 or less) can be obtained in a matter of minutes, but there is a catch. The interest rates charged are extremely high. A typical two-week payday advance costs $15 per $100. To determine the annual simple interest rate, solve the following equation for r.

$$15 = 100 \cdot r \cdot \frac{2}{52}$$

68. Car title loans. A *car title loan* is a small, short-term, high-rate loan that uses the title on a vehicle as collateral. These loans are usually for 15 or 30 days and have a *monthly percentage rate* of 25%. Find the interest charged on a one-month-long $5,000 car title loan at that rate.

WRITING

69. What is the difference between simple and compound interest?

70. Explain this statement: *Interest is the amount of money paid for the use of money.*

71. On some accounts, banks charge a penalty if the depositor withdraws the money before the end of the term. Why would a bank do this?

72. Explain why it is better for a depositor to open a savings account that pays 5% interest compounded daily than one that pays 5% interest compounded monthly.

REVIEW

73. Evaluate: $\sqrt{\dfrac{1}{4}}$

74. Evaluate: $\left(\dfrac{1}{4}\right)^2$

75. Add: $\dfrac{3}{7} + \dfrac{2}{5}$

76. Subtract: $\dfrac{3}{7} - \dfrac{2}{5}$

77. Multiply: $2\dfrac{1}{2} \cdot 3\dfrac{1}{3}$

78. Divide: $-12\dfrac{1}{2} \div 5$

79. Evaluate: -6^2

80. Evaluate: $(0.2)^2 - (0.3)^2$

---

6 Summary and Review

SECTION 6.1 ▸ Percents, Decimals, and Fractions

DEFINITIONS AND CONCEPTS	EXAMPLES
Percent means parts per one hundred. The word *percent* can be written using the symbol %.	In the figure below, there are 100 equal-sized square regions, and 37 of them are shaded. We say that $\frac{37}{100}$, or 37%, of the figure is shaded. Numerator $\dfrac{37}{100} \quad = \quad 37\%$ Per 100
To **write a percent as a fraction**, drop the % symbol and write the given number over 100. Then simplify the fraction, if possible.	Write 22% as a fraction. $22\% = \dfrac{22}{100}$ Drop the % symbol and write 22 over 100. $= \dfrac{\overset{1}{2} \cdot 11}{\underset{1}{2} \cdot 50}$ To simplify the fraction, factor 22 and 100. Then remove the common factor of 2 from the numerator and denominator. Thus, $22\% = \frac{11}{50}$.
Percents such as 9.1% and 36.23% can be written as fractions of whole numbers by multiplying the numerator and denominator by a power of 10.	Write 9.1% as a fraction. $9.1\% = \dfrac{9.1}{100}$ Drop the % symbol and write 9.1 over 100. $= \dfrac{9.1}{100} \cdot \dfrac{10}{10}$ To obtain an equivalent fraction of *whole numbers*, we need to move the decimal point in the numerator one place to the right. Choose $\frac{10}{10}$ as the form of 1 to build the fraction. $= \dfrac{91}{1,000}$ Multiply the numerators. Multiply the denominators. Thus, $9.1\% = \frac{91}{1,000}$.

Mixed number percents, such as $2\frac{1}{3}\%$ and $23\frac{5}{6}\%$, can be written as fractions of whole numbers by performing the indicated division.	Write $2\frac{1}{3}\%$ as a fraction. $2\frac{1}{3}\% = \dfrac{2\frac{1}{3}}{100}$ *Drop the % symbol and write $2\frac{1}{3}$ over 100.* $= 2\frac{1}{3} \div 100$ *The fraction bar indicates division.* $= \dfrac{7}{3} \cdot \dfrac{1}{100}$ *Write $2\frac{1}{3}$ as an improper fraction and then multiply by the reciprocal of 100.* $= \dfrac{7}{300}$ *Multiply the numerators. Multiply the denominators.* Thus, $2\frac{1}{3}\% = \frac{7}{300}$.
When **percents that are greater than 100%** are written as fractions, the fractions are greater than 1.	Write 170% as a fraction. $170\% = \dfrac{170}{100}$ *Drop the % symbol and write 170 over 100.* $= \dfrac{\overset{1}{\cancel{10}} \cdot 17}{\underset{1}{\cancel{10}} \cdot 10}$ *To simplify the fraction, factor 170 and 100. Then remove the common factor of 10 from the numerator and denominator.* Thus, $170\% = \frac{17}{10}$.
When **percents that are less than 1%** are written as fractions, the fractions are less than $\frac{1}{100}$.	Write 0.03% as a fraction. $0.03\% = \dfrac{0.03}{100}$ *Drop the % symbol and write 0.03 over 100.* $= \dfrac{0.03}{100} \cdot \dfrac{\mathbf{100}}{\mathbf{100}}$ *To obtain an equivalent fraction of whole numbers, we need to move the decimal point in the numerator two places to the right. Choose $\frac{100}{100}$ as the form of 1 to build the fraction.* $= \dfrac{3}{10,000}$ *Multiply the numerators and multiply the denominators. Since the numerator and denominator do not have any common factors (other than 1), the fraction is in simplified form.* Thus, $0.03\% = \frac{3}{10,000}$.
To **write a percent as a decimal**, drop the % symbol and divide the given number by 100 by moving the decimal point two places to the left.	Write each percent as a decimal. $14\% = 14.0\% = 0.14$ *Write a decimal point and 0 to the right of the 4 in 14%.* $9.35\% = 0.0935$ *Write a placeholder 0 (shown in blue) to the left of the 9.* $198\% = 198.0\% = 1.98$ *Write a decimal point and 0 to the right of the 8 in 198%.* $0.75\% = 0.0075$ *Write two placeholder zeros.*
Mixed number percents, such as $1\frac{3}{4}\%$ and $10\frac{1}{2}\%$, can be written as decimals by writing the fractional part of the mixed number in its equivalent decimal form.	Write $1\frac{3}{4}\%$ as a decimal. There is no decimal point to move in $1\frac{3}{4}\%$. Since $1\frac{3}{4} = 1 + \frac{3}{4}$ and since the decimal equivalent of $\frac{3}{4}$ is 0.75, we can write $1\frac{3}{4}\%$ as 1.75% $1\frac{3}{4}\% = 1.75\% = 0.0175$ *Write a placeholder 0 (shown in blue) to the left of the 1.*
To **write a decimal as a percent**, multiply the decimal by 100 by moving the decimal point two places to the right, and then insert a % symbol.	Write each decimal as a percent. $0.501 = 50.1\%$ $3.66 = 366\%$ $0.002 = 000.2\% = 0.2\%$

To **write a fraction as a percent,**
1. Write the fraction as a decimal by dividing its numerator by its denominator.
2. Multiply the decimal by 100 by moving the decimal point two places to the right, and then insert a % symbol.

Fraction ⟶ decimal ⟶ percent

Write $\dfrac{3}{4}$ as a percent.

Step 1 Divide the numerator by the denominator.

$$
\begin{array}{r}
0.75 \\
4\overline{)3.00} \\
-2\,8 \\
\hline
20 \\
-20 \\
\hline
0
\end{array}
$$

Write a decimal point and some additional zeros to the right of 3.

The remainder is 0.

Step 2 Write the decimal 0.75 as a percent.

$$\frac{3}{4} = 0.75 = 75\%$$

Sometimes, when we want to write a fraction as a percent, the result of the division is a **repeating decimal**. In such cases, we can give an **exact answer** or an **approximate answer**.

Write $\dfrac{2}{3}$ as a percent.

Step 1 Divide the numerator by the denominator.

$$
\begin{array}{r}
0.666 \\
3\overline{)2.000} \\
-1\,8 \\
\hline
20 \\
-18 \\
\hline
20 \\
-18 \\
\hline
2
\end{array}
$$

Write a decimal point and some additional zeros to the right of 2.

The repeating pattern is now clear. We can stop the division.

Step 2 Write the decimal 0.6666 . . . as a percent.

$$0.6666 = 66.66\ldots\%$$

Exact Answer:

Use $\frac{2}{3}$ to represent 0.666. . . .

$$\frac{2}{3} = 66.66\ldots\%$$

$$= 66\frac{2}{3}\%$$

Approximation:

Round to the nearest tenth.

$$\frac{2}{3} = 66.66\ldots\%$$

$$\approx 66.7\%$$

REVIEW EXERCISES

Express the amount of each figure that is shaded as a percent, as a decimal, and as a fraction. Each set of squares represents 100 %.

1.

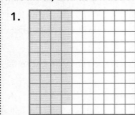

2.

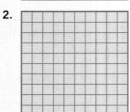

3. In Problem 1, what percent of the figure is not shaded?

4. The Internet. The following sentence once appeared on a technology blog: "54 out of the top 100 websites failed Yahoo's performance test."

   a. What percent of the websites failed the test?

   b. What percent of the websites passed the test?

Write each percent as a fraction.

   5. 15%   **6.** 120%   **7.** $9\frac{1}{4}\%$ **8.** 0.2%

Write each percent as a decimal.

   9. 27%   **10.** 8%   **11.** 655%   **12.** $1\frac{4}{5}\%$

   13. 0.75%   **14.** 0.23%

Write each decimal or whole number as a percent.

15. 0.83   **16.** 1.625   **17.** 0.051   **18.** 6

Write each fraction as a percent.

19. $\dfrac{1}{2}$ **20.** $\dfrac{4}{5}$ **21.** $\dfrac{7}{8}$ **22.** $\dfrac{1}{16}$

Write each fraction as a percent. Give the exact answer and an approximation to the nearest tenth of a percent.

23. $\dfrac{1}{3}$ **24.** $\dfrac{5}{6}$ **25.** $\dfrac{11}{12}$ **26.** $\dfrac{15}{9}$

27. Water distribution. The oceans contain 97.2% of all the water on Earth. (Source: National Ground Water Association)

   a. Write this percent as a decimal.

   b. Write this percent as a fraction in simplest form.

28. Bill of Rights. There are 27 amendments to the Constitution of the United States. The first 10 are known as the Bill of Rights. What percent of the amendments were adopted after the Bill of Rights? (Round to the nearest one percent.)

29. Taxes. The city of Grand Prairie, Texas, had a special election in 2013 to continue a *one-fourth of one percent* sales tax to help fund park improvements.

   a. Write this percent as a decimal.

   b. Write this percent as a fraction.

30. Social security. If your full retirement age is 66, your Social Security benefits are reduced by $\frac{1}{15}$ if you retire at age 65. Write this fraction as a percent. Give the exact answer and an approximation to the nearest tenth of a percent. (Source: Social Security Administration)

SECTION 6.2 ▶ Solving Percent Problems Using Percent Equations and Proportions

DEFINITIONS AND CONCEPTS	EXAMPLES
The key words in a **percent sentence** can be translated to a percent equation. ■ Each *is* translates to an equal symbol = ■ *of* translates to multiplication that is shown with a raised dot · ■ *what* number or *what* percent translates to an unknown number that is represented by a variable.	Translate the percent sentence to a percent equation. What number is 26% of 180? ↓ ↓ ↓ ↓ ↓ x = 26% · 180 This is the percent equation.

Percent sentences involve a comparison of numbers. The relationship between the **base** (the standard of comparison, the whole), the **amount** (a part of the base), and the **percent** is:
 	
    Amount = percent · base
 	
 or
 	
    Part = percent · whole
	8 is 12.5% of 64.
	   ↑ ↑ ↑
	 Amount **percent** **base**
	 (part) (whole)

The percent equation method: We can translate percent sentences to percent equations and solve to **find the amount**. **Caution!** When solving percent equations, always write the percent as a decimal (or fraction) before performing any calculations.	What number is 45% of 120? ↓ ↓ ↓ ↓ ↓ x = 45% · 120 Translate. Now, solve the percent equation. $x = 0.45 \cdot 120$ Write 45% as a decimal. $x = 54$ Do the multiplication. Thus, 54 is 45% of 120.

We can translate percent sentences to percent equations and solve to **find the percent**.

12	is	what percent	of	192?
↓	↓	↓	↓	↓
12	=	x	·	192

Translate.

Now, solve the percent equation.

$$12 = x \cdot 192$$

$$\frac{12}{192} = \frac{x \cdot \overset{1}{\cancel{192}}}{\underset{1}{\cancel{192}}}$$

To isolate x on the right side of the equation, divide both sides by 192. Then remove the common factor of 192 in the numerator and denominator.

$$0.0625 = x$$

On the left side, divide 12 by 192.

$$0\underset{\smile}{6.25}\% = x$$

To write 0.0625 as a percent, multiply it by 100 by moving the decimal point two places to the right, and then insert a % symbol.

Thus, 12 is 6.25% of 192.

We can translate percent sentences to percent equations and solve to **find the base**.

Caution! Sometimes the calculations to solve a percent problem are made easier if we **write the percent as a fraction instead of a decimal.** This is the case with percents that have *repeating* decimal equivalents such as $33\frac{1}{3}\%$, $66\frac{2}{3}\%$, and $16\frac{2}{3}\%$.

8.2	is	$33\frac{1}{3}\%$	of	what number?
↓	↓	↓		↓
8.2	=	$33\frac{1}{3}\%$	·	x

Translate.

Now, solve the percent equation.

$$8.2 = \frac{1}{3} \cdot x$$

Write the percent as a fraction: $33\frac{1}{3}\% = \frac{1}{3}$.

$$\frac{8.2}{\frac{1}{3}} = \frac{\overset{1}{\cancel{\frac{1}{3}}} \cdot x}{\underset{1}{\cancel{\frac{1}{3}}}}$$

To isolate x on the right side of the equation, divide both sides by $\frac{1}{3}$. Then remove the common factor of $\frac{1}{3}$ in the numerator and denominator.

$$8.2 \div \frac{1}{3} = x$$

On the left side, the fraction bar indicates division.

$$\frac{8.2}{1} \cdot \frac{3}{1} = x$$

On the left side, write 8.2 as a fraction. Then use the rule for dividing fractions: Multiply by the reciprocal of $\frac{1}{3}$, which is $\frac{3}{1}$.

$$24.6 = x$$

Do the multiplication.

Thus, 8.2 is $33\frac{1}{3}\%$ of 24.6.

The percent proportion method:

We can translate percent sentences to percent proportions and solve to **find the amount**.

To translate a percent sentence to a **percent proportion**, use the following form:

Amount is to base as percent is to 100:

$$\frac{\text{amount}}{\text{base}} = \frac{\text{percent}}{100}$$

or

Part is to whole as percent is to 100:

$$\frac{\text{part}}{\text{whole}} = \frac{\text{percent}}{100}$$

What number	is	45%	of	120?
amount		percent		base

$$\frac{x}{120} = \frac{45}{100}$$

This is the proportion to solve.

To make the calculations easier, simplify the ratio $\frac{45}{100}$.

$$\frac{x}{120} = \frac{9}{20}$$

Simplify: $\frac{45}{100} = \frac{\overset{1}{\cancel{5}} \cdot 9}{\underset{1}{\cancel{5}} \cdot 20} = \frac{9}{20}$.

To solve the proportion we use the cross products.

$$x \cdot 20 = 120 \cdot 9$$

Find the cross products and set them equal.

$$x \cdot 20 = 1{,}080$$

To simplify the right side, do the multiplication: $120 \cdot 9 = 1{,}080$.

$$\frac{x \cdot \overset{1}{\cancel{20}}}{\underset{1}{\cancel{20}}} = \frac{1{,}080}{20}$$

To isolate x on the left side, divide both sides of the equation by 20. Then remove the common factor of 20 from the numerator and denominator.

$$x = 54$$

On the right side, divide 1,080 by 20.

Thus, 54 is 45% of 120.

We can translate percent sentences to percent proportions and solve to **find the percent**.

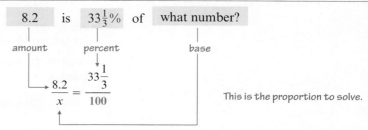

12 is what percent of 192?

amount   percent   base

$$\frac{12}{192} = \frac{x}{100}$$

This is the proportion to solve.

To make the calculations easier, simplify the ratio $\frac{12}{192}$ first.

$$\frac{1}{16} = \frac{x}{100}$$

Simplify: $\frac{12}{192} = \frac{\overset{1}{\cancel{2}} \cdot \overset{1}{\cancel{2}} \cdot \overset{1}{\cancel{3}}}{\underset{1}{\cancel{2}} \cdot \underset{1}{\cancel{2}} \cdot 2 \cdot 2 \cdot 2 \cdot \underset{1}{\cancel{3}}} = \frac{1}{16}$.

$1 \cdot 100 = 16 \cdot x$ Find the cross products and set them equal.

$100 = 16 \cdot x$ On the left side, do the multiplication: $1 \cdot 100 = 100$.

$\dfrac{100}{16} = \dfrac{\overset{1}{\cancel{16}} \cdot x}{\underset{1}{\cancel{16}}}$ To isolate x on the right side, divide both sides of the equation by 16. Then remove the common factor of 16 from the numerator and denominator.

$6.25 = x$ On the left side, divide 100 by 16.

Thus, 12 is 6.25% of 192.

We can translate percent sentences to percent proportions and solve to **find the base**.

8.2 is $33\frac{1}{3}$% of what number?

amount   percent   base

$$\frac{8.2}{x} = \frac{33\frac{1}{3}}{100}$$

This is the proportion to solve.

To make the calculations easier, write the mixed number $33\frac{1}{3}$ as the improper fraction $\frac{100}{3}$.

$\dfrac{8.2}{x} = \dfrac{\frac{100}{3}}{100}$ Write $33\frac{1}{3}$ as $\frac{100}{3}$.

$8.2 \cdot 100 = x \cdot \dfrac{100}{3}$ To solve the proportion, find the cross products and set them equal.

$820 = x \cdot \dfrac{100}{3}$ To simplify the left side, do the multiplication: $8.2 \cdot 100 = 820$.

$\dfrac{820}{\frac{100}{3}} = \dfrac{x \cdot \frac{\overset{1}{\cancel{100}}}{\cancel{3}}}{\underset{1}{\frac{\cancel{100}}{\cancel{3}}}}$ To isolate x on the right side, divide both sides of the equation by $\frac{100}{3}$. Then remove the common factor of $\frac{100}{3}$ from the numerator and denominator.

$820 \div \dfrac{100}{3} = x$ On the left side, the fraction bar indicates division.

$\dfrac{820}{1} \cdot \dfrac{3}{100} = x$ On the left side, write 820 as a fraction. Use the rule for dividing fractions: Multiply by the reciprocal of $\frac{100}{3}$.

$\dfrac{2,460}{100} = x$ Multiply the numerators. Multiply the denominators.

$24.6 = x$ Divide 2,460 by 100 by moving the understood decimal point in 2,460 two places to the left.

Thus, 8.2 is $33\frac{1}{3}$% of 24.6.

A **circle graph** is a way of presenting data for comparison. The pie-shaped pieces of the graph show the relative sizes of each category.

The 100 tick marks equally spaced around the circle serve as a visual aid when constructing a circle graph.

To solve percent application problems, we often have to **rewrite the facts** of the problem in percent sentence form before we can translate to an equation.

Facebook. As of the spring of 2017, Facebook had approximately 1,860,000,000 users worldwide. Use the information in the circle graph on the right to find how many of them were male.

The circle graph shows that 47% of the 1,860,000,000 users of Facebook were male.

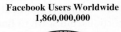
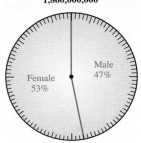

Facebook Users Worldwide
1,860,000,000

Male 47%

Female 53%

Source: omnicoreagency.com

Method 1: To find the unknown amount write and then solve a **percent equation**.

What number	is	47%	of	1,860,000,000?
↓	↓	↓	↓	↓
x	$=$	47%	$\cdot$	1,860,000,000 Translate.

Now, solve the percent equation.

$x = 0.47 \cdot 1,860,000,000$ Write 47% as a decimal: 47% = 0.47.

$x = 874,200,000$ Do the multiplication.

In the spring of 2017, there were approximately 874,200,000 male users of Facebook worldwide.

Method 2: To find the unknown amount write and then solve a **percent proportion**.

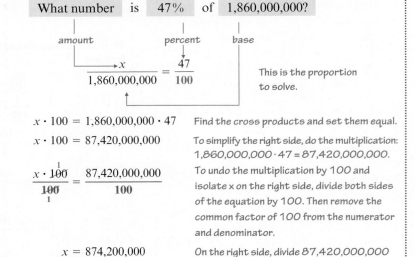

What number   is   47%   of   1,860,000,000?

amount   percent   base

$$\frac{x}{1,860,000,000} = \frac{47}{100}$$ This is the proportion to solve.

$x \cdot 100 = 1,860,000,000 \cdot 47$ Find the cross products and set them equal.

$x \cdot 100 = 87,420,000,000$ To simplify the right side, do the multiplication: 1,860,000,000 · 47 = 87,420,000,000.

$$\frac{x \cdot \overset{1}{\cancel{100}}}{\underset{1}{\cancel{100}}} = \frac{87,420,000,000}{100}$$ To undo the multiplication by 100 and isolate x on the right side, divide both sides of the equation by 100. Then remove the common factor of 100 from the numerator and denominator.

$x = 874,200,000$ On the right side, divide 87,420,000,000 by 100.

In the spring of 2017, there were approximately 874,200,000 million male users of Facebook worldwide.

REVIEW EXERCISES

31. a. Identify the amount, the base, and the percent in the statement "15 is $33\frac{1}{3}$% of 45."

b. Fill in the blanks to complete the percent equation (formula):

_____ = percent · _____

or

Part = _____ · whole

32. When computing with percents, we must change the percent to a decimal or a fraction. Change each percent to a decimal.

a. 13%   **b.** 7.1%

c. 195%   **d.** $\frac{1}{4}$%

Change each percent to a fraction.

e. $33\frac{1}{3}$%   **f.** $66\frac{2}{3}$%

g. $16\frac{2}{3}$%

33. Translate each percent sentence into a *percent equation*. **Do not solve.**

    a. What number is 32% of 96?

    b. 64 is what percent of 135?

    c. 9 is 47.2% of what number?

34. Translate each percent sentence into a *percent proportion*. **Do not solve.**

    a. What number is 32% of 96?

    b. 64 is what percent of 135?

    c. 9 is 47.2% of what number?

Translate to a percent equation or percent proportion and then solve to find the unknown number.

35. What number is 40% of 500?

36. 16% of what number is 20?

37. 1.4 is what percent of 80?

38. $66\frac{2}{3}$% of 3,150 is what number?

39. Find 220% of 55.

40. What is 0.05% of 60,000?

41. 3.88 is 2.5% of what number?

42. What percent of 0.08 is 4.24?

43. Racing. The nitro–methane fuel mixture used to power some experimental cars is 96% nitro and 4% methane. How many gallons of methane are needed when filling a 15-gallon fuel tank?

44. Home sales. After the first day on the market, 51 homes in a new subdivision had already sold. This was 75% of the total number of homes available. How many homes were originally for sale?

45. Hurricane damage. In a mobile home park, 96 of the 110 trailers were either damaged or destroyed by hurricane winds. What percent is this? (Round to the nearest 1 percent.)

46. Tipping. The cost of dinner for a family of five at a restaurant was $54.30. Find the amount of the tip if it should be 20% of the cost of dinner.

47. Alternative fuels. *Biodiesel* is a clean burning renewable fuel made from vegetable oils and fats. It is used as a replacement for petroleum diesel and can be used with no modifications to the diesel engine or the vehicle. More than 1.5 billion gallons of biodiesel were produced in the U.S. in 2016. The table shows the percent of each type of oil and fat used in the production of biodiesel. Draw a circle graph for the data.

Oils and Fats Used in the Production of Biodiesel

Soybean oil	52%	Used cooking oil	13%
Canola oil	8%	Animal fats	14%
Corn oil	11%	Other	2%

Oils and Fats Used in the Production of Biodiesel

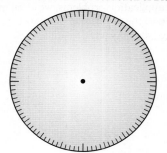

48. Earth's surface. The surface of Earth is approximately 196,800,000 square miles. Use the information in the circle graph to determine the number of square miles of Earth's surface that are covered with water.

Water 70.9%
Land 29.1%

SECTION 6.3 ▶ **Applications of Percent**

DEFINITIONS AND CONCEPTS	EXAMPLES
The **sales tax** on an item is a percent of the purchase price of the item. Sales tax = sales tax rate · purchase price ↓ ↓ ↓ ↓ ↓ Amount = percent · base Notice that the formula is based on the percent equation discussed in Section 6.2. **Sales tax dollar amounts** are rounded to the nearest cent (hundredth). The **total cost** of an item is the sum of its purchase price and the sales tax on the item. Total cost = purchase price + sales tax	**Shopping.** Find the sales tax and total cost of a $50.95 purchase if the sales tax rate is 8%. Sales tax = **sales tax rate · purchase price** Sales tax = 8% · $50.95 Sales tax = 0.08 · $50.95 *Write 8% as a decimal: 8% = 0.08.* Sales tax = $4.076 *Do the multiplication.* Sales tax ≈ $4.08 *Round the sales tax to the nearest cent (hundredth).* Thus, the sales tax is $4.08. The total cost is the sum of its purchase price and the sales tax. Total cost = **purchase price + sales tax rate** Total cost = $50.95 + **$4.08** Total cost = $55.03 *Do the addition.* The total cost of the purchase is $55.03.

Sales tax rates are usually expressed as a percent.

Two methods can be used to find the unknown sales tax rate:

- The percent equation method
- The percent proportion method

Appliances. The purchase price of a toaster is $82. If the sales tax is $5.33, what is the sales tax rate?

The sales tax of $5.33 is some unknown percent of the purchase price of $82. Two methods can be used to solve this problem.

The percent equation method:

5.33 is what percent of $82?$

$$5.33 \quad = \quad x \quad \cdot \quad 82 \qquad \text{Translate.}$$

Now, solve the percent equation.

$$\frac{5.33}{82} = \frac{x \cdot 82}{82} \qquad \text{To isolate } x \text{ on the right side of the equation, divide both sides by 82.}$$

$$0.065 = \frac{x \cdot \overset{1}{\cancel{82}}}{\underset{1}{\cancel{82}}} \qquad \text{On the right side of the equation, remove the common factor of 82 from the numerator and denominator. On the left side, divide 5.33 by 82.}$$

$$0.065 = x$$

$$006.5\% = x \qquad \text{Write the decimal 0.065 as a percent.}$$

$$6.5\% = x$$

The sales tax rate is 6.5%.

The percent proportion method:

5.33 is what percent of $82?$

amount   percent   base

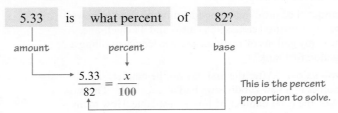

$$\frac{5.33}{82} = \frac{x}{100} \qquad \text{This is the percent proportion to solve.}$$

$$5.33 \cdot 100 = 82 \cdot x \qquad \text{To solve the proportion, find the cross products and set them equal.}$$

$$533 = 82 \cdot x \qquad \text{Do the multiplication on the left side of the equation.}$$

$$\frac{533}{82} = \frac{\overset{1}{\cancel{82}} \cdot x}{\underset{1}{\cancel{82}}} \qquad \text{To isolate } x \text{ on the right side, divide both sides of the equation by 82. Then remove the common factor of 82 from the numerator and denominator.}$$

$$6.5 = x \qquad \text{On the left side, divide 533 by 82.}$$

The sales tax rate is 6.5%.

Instead of working for a salary or getting paid at an hourly rate, many salespeople are paid on **commission**.

The **amount of commission** paid is a percent of the total dollar sales of goods or services.

$$\text{Commission} = \text{commission rate} \cdot \text{sales}$$

Commissions. A salesperson earns an 11% commission on all appliances that she sells. If she sells a $450 dishwasher, what is her commission?

$$\text{Commission} = \textbf{commission rate} \quad \cdot \quad \textbf{sales}$$

$$\text{Commission} = \quad\quad 11\% \quad\quad \cdot \quad \$450$$

$$\text{Commission} = 0.11 \cdot \$450 \qquad \text{Write 11\% as a decimal.}$$

$$\text{Commission} = \$49.50 \qquad \text{Do the multiplication.}$$

The commission earned on the sale of the $450 dishwasher is $49.50.

The **commission rate** is usually expressed as a percent.

Telemarketing. A telemarketer made a commission of $600 in one week on sales of $4,000. What is his commission rate?

$$\text{Commission} = \text{commission rate} \quad \cdot \quad \text{sales}$$

$$\$600 \quad = \quad x \quad\quad \cdot \; \$4,000 \qquad \text{Let } x \text{ represent the unknown commission rate.}$$

$$\frac{600}{4{,}000} = \frac{x \cdot 4{,}000}{4{,}000}$$

We can drop the dollar signs. To isolate x on the right side of the equation, divide both sides by 4,000.

$$0.15 = \frac{x \cdot \overset{1}{\cancel{4{,}000}}}{\underset{1}{\cancel{4{,}000}}}$$

Remove the common factor of 4,000 from the numerator and denominator. On the left side, divide 600 by 4,000.

$$0\underset{\curvearrowright}{.}15\% = x$$

Write the decimal 0.15 as a percent.

The commission rate is 15%.

To find **percent of increase or decrease**:

1. Subtract the smaller number from the larger to find the amount of increase or decrease.

2. Find what percent the amount of increase or decrease is of the original amount.

Two methods can be used to find the unknown percent of increase (or decrease):

- The percent equation method
- The percent proportion method

Caution! The percent of increase (or decrease) is a percent of the *original number*—that is, the number before the change occurred.

Watching television. According to the Nielsen Company, the average American watched 154 minutes of TV a day in 2005. That increased to 166 minutes per day in 2015. Find the percent of increase. Round to the nearest one percent.

First, subtract to find the amount of increase.

$$166 - 154 = 12 \quad \text{\textit{Subtract the smaller number from the larger number.}}$$

The number of minutes watched per day increased by 12.

Next, find what percent of the *original* 154 minutes the 12 minute increase represents.

The percent equation method:

12	is	what percent	of	154?
↓	↓	↓	↓	↓
12	=	x	·	154

Translate.

Now, solve the percent equation.

$$12 = 154 \cdot x$$

$$\frac{12}{154} = \frac{\overset{1}{\cancel{154}} \cdot x}{\underset{1}{\cancel{154}}}$$

To undo the multiplication by 154 and isolate x on the right side, divide both sides of the equation by 154. Then remove the common factor of 100 from the numerator and denominator.

$$0.078 \approx x$$

On the left side, divide 12 by 154. The division does not terminate.

$$0\underset{\curvearrowright}{0}7.8\% \approx x$$

Write the decimal 0.078 as a percent.

$$8\% \approx x$$

Round to the nearest one percent.

Between 2005 and 2015, the number of minutes of television watched by the average American each day increased by 8%.

If the **percent proportion method** is used, solve the following proportion for x to find the percent of increase.

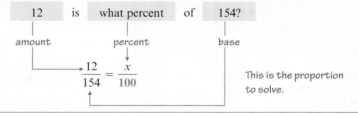

The **amount of discount** is a percent of the original price.

$$\underset{\substack{\uparrow \\ \text{amount}}}{\text{Amount of}\atop\text{discount}} = \underset{\substack{\uparrow \\ \text{percent}}}{\text{discount}\atop\text{rate}} \cdot \underset{\substack{\uparrow \\ \text{base}}}{\text{original}\atop\text{price}}$$

Notice that the formula is based on the percent equation discussed in Section 6.2.

Tool sales. Find the amount of the discount on a tool kit if it is normally priced at $89.95 but is currently on sale for 35% off. Then find the sale price.

Amount of discount = **discount rate** · **original price**

Amount of discount =   35%   · $89.95

Amount of discount = 0.35 · $89.95   *Write 35% as a decimal.*

Amount of discount = $31.4825   *Do the multiplication.*

Amount of discount ≈ $31.48   *Round to the nearest cent (hundredth).*

To find the **sale price** of an item, subtract the discount from the original price.

Sale price = original price − discount

The discount on the tool kit is \$31.48. To find the sale price, we use subtraction.

Sale price = **original price** − **discount**

Sale price = **\$89.95** − **\$31.48**

Sale price = \$58.47 *Do the subtraction.*

The sale price of the tool kit is \$58.47.

The difference between the original price and the sale price is the **amount of discount**.

$$\text{Amount of discount} = \text{original price} - \text{sale price}$$

Furniture sales. Find the discount rate on a living room set regularly priced at \$2,500 that is on sale for \$1,870. Round to the nearest one percent.

We will think of this as a *percent-of-decrease problem.* The discount (decrease in price) is found using subtraction.

\$2,500 − \$1,870 = \$630 *Discount = original price − sale price*

The living room set is discounted \$630. Now we find what percent of the original price the \$630 discount represents.

Amount of discount = discount rate · **original price**

\$630 = x · **\$2,500**

$$\frac{630}{2{,}500} = \frac{x \cdot 2{,}500}{2{,}500}$$ *Drop the dollar signs. To isolate x on the right side of the equation, divide both sides by 2,500.*

$$0.252 = \frac{x \cdot \overset{1}{\cancel{2{,}500}}}{\underset{1}{\cancel{2{,}500}}}$$ *On the right side of the equation, remove the common factor of 2,500 from the numerator and denominator. On the left side, divide 630 by 2,500.*

$025.2\% = x$ *Write the decimal 0.252 as a percent.*

$25\% \approx x$ *Round to the nearest one percent.*

To the nearest one percent, the discount rate on the living room set is 25%.

REVIEW EXERCISES

49. Sales receipts. Complete the sales receipt shown below by finding the sales tax and total cost of the camera.

CAMERA CENTER

35mm Canon Camera	\$59.99
SUBTOTAL	\$59.99
SALES TAX @ 5.5%	?
TOTAL	?

50. Sales tax rates. Find the sales tax rate if the sales tax is \$492 on the purchase of an automobile priced at \$12,300.

51. Commissions. If the commission rate is 6%, find the commission earned by an appliance salesperson who sells a washing machine for \$369.97 and a dryer for \$299.97.

52. Selling medical supplies. A salesperson made a commission of \$646 on a \$15,200 order of antibiotics. What is her commission rate?

53. T-shirt sales. A stadium owner earns a commission of $33\frac{1}{3}\%$ of the T-shirt sales from any concert or sporting event. How much can the owner make if 12,000 T-shirts are sold for \$25 each at a soccer match?

54. Fill in the blank: The percent of increase (or decrease) is a percent of the ________ number—that is, the number before the change occurred.

55. The United Nations. In 2016, the United Nations expanded its peacekeeping force in South Sudan from 13,000 to 17,000 troops. Find the percent of increase in the number of troops. Round to the nearest one percent. (Source: washingtonpost.com)

56. Gas mileage. A woman found that the gas mileage fell from 18.8 to 17.0 miles per gallon when she experimented with a new brand of gasoline in her truck. Find the percent of decrease in her mileage. Round to the nearest tenth of one percent.

57. Fill in the blanks.

a. Sales tax = sales tax rate · ________

b. Total cost = purchase price + ________

c. Commission = ________ · sales

58. Fill in the blanks.

a. Amount of discount = original price − ________

b. Amount of discount = discount rate · ________

c. Sale price = original price − ________

59. Tool chests. Use the information in the advertisement below to find the discount, the original price, and the discount rate on the tool chest.

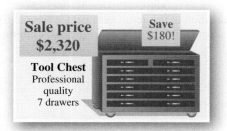

Sale price
$2,320

Save
$180!

Tool Chest
Professional
quality
7 drawers

60. Rents. Find the discount rate if the monthly rent for an apartment is reduced from $980 to $931 per month.

SECTION 6.4 ▶ **Estimation with Percent**

DEFINITIONS AND CONCEPTS	EXAMPLES
Estimation can be used to find approximations when exact answers aren't necessary. To find **1% of a number**, move the decimal point in the number two places to the left.	What is 1% of 291.4? Find the exact answer and an estimate using front-end rounding. **Exact answer:** 1% of 291.4 = 2.914 *Move the decimal point two places to the left.* **Estimate:** 291.4 front-end rounds to 300. If we move the understood decimal point in 300 two places to the left, we get 3. Thus 1% of 291.4 ≈ 3 *Because 1% of 300 = 3.*
To find **10% of a number**, move the decimal point in the number one place to the left.	What is 10% of 40,735 pounds? Find the exact answer and an estimate using front-end rounding. **Exact answer:** 10% of 40,735 = 4,073.5 *Move the decimal point one place to the left.* **Estimate:** 40,735 front-end rounds to 40,000. If we move the understood decimal point in 40,000 one place to the left, we get 4,000. Thus 10% of 40,735 ≈ 4,000 *Because 10% of 40,000 = 4,000.*
To find **20% of a number**, find 10% of the number by moving the decimal point one place to the left, and then double (multiply by 2) the result. A similar approach can be used to find 30% of a number, 40% of a number, and so on.	Estimate the answer: What is 20% of 809? Since 10% of 809 is 80.9 (or about **81**), it follows that 20% of 809 is about 2 · **81**, which is 162. Thus, 20% of 809 ≈ 162 *Because 10% of 809 ≈ 81.*
To find **50% of a number**, divide the number by 2.	Estimate the answer: What is 50% of 1,442,957? We use 1,400,000 as an approximation of 1,442,957 because it is even, divisible by 2, and ends with many zeros. 50% of 1,442,957 ≈ 700,000 *Because 50% of 1,400,000 = $\frac{1,400,000}{2}$ = 700,000.*
To find **25% of a number**, divide the number by 4.	Estimate the answer: What is 25% of 21.004? We use 20 as an approximation because it is close to 21.004 and because it is divisible by 4. 25% of 21.004 ≈ 5 *Because 25% of 20 = $\frac{20}{4}$ = 5.*

To find **5% of a number**, find 10% of the number by moving the decimal point in the number one place to the left. Then, divide that result by 2.	Estimate the answer: What is 5% of 36,150?
	First, we find 10% of 36,150:
	$$10\% \text{ of } 36,150 = 3,615$$
	We use 3,600 as an approximation of this result because it is close to 3,615 and because it is even, and therefore divisible by 2. Next, we divide the approximation by 2 to estimate 5% of 36,150.
	$$\frac{3,600}{2} = 1,800$$
	Thus, 5% of 36,150 ≈ 1,800.
To find **15% of a number**, find the sum of 10% of the number and 5% of the number.	**Tipping.** Estimate the 15% tip on a dinner costing $88.55.
	To simplify the calculations, we will estimate the cost of the $88.55 dinner to be $90. Then, to estimate the tip, we find 10% of $90 and 5% of $90, and add.
	10% of $90 is $9 ⟶ $9 5% of $90 (half as much as 10% of $90) ⟶ + $4.50 $13.50
	The tip should be $13.50.
To find **200% of a number**, multiply the number by 2. A similar approach can be used to find 300% of a number, 400% of a number, and so on.	Estimate the answer: What is 200% of 3.509?
	To estimate 200% of 3.509, we will find 200% of 4. We use 4 as an approximation because it is close to 3.509 and it makes the multiplication by 2 easy.
	200% of 3.509 ≈ 8 *Because 200% of 4 = 2 · 4 = 8.*
Sometimes we must **approximate the percent** to estimate an answer.	**Quality control.** In a production run of 145,350 ceramic tiles, 3% were found to be defective. Estimate the number of defective tiles.
	To estimate 3% of 145,350, we will find 1% of 150,000, and multiply the result by 3. We use 150,000 as the approximation because it is close to 145,350 and it ends with several zeros.
	3% of 145,350 ≈ 4,500 *Because 1% of 150,000 = 1,500 and* *3 · 1,500 = 4,500.*
	There were about 4,500 defective tiles in the production run.

REVIEW EXERCISES

What is 1% of the given number? Find the exact answer and an estimate using front-end rounding.

61. 342.03

62. 8,687

What is 10% of the given number? Find the exact answer and an estimate using front-end rounding.

63. 43.4 seconds

64. 10,900 liters

Estimate each answer. (Answers may vary.)

65. What is 20% of 63?

66. What is 20% of 612?

67. What is 50% of 279,985?

68. What is 50% of 327?

69. What is 25% of 13.02?

70. What is 25% of 39.9?

71. What is 5% of 7,150?

72. What is 5% of 19,359?

73. What is 200% of 29.78?

74. What is 200% of 1.125?

Estimate a 15% tip on each dollar amount. (Answers may vary.)

75. $243.55

76. $46.99

Estimate each answer. (Answers may vary.)

77. Special offers. A home improvement store sells a 50-fluid-ounce pail of asphalt driveway sealant that is labeled "25% free." How many ounces are free?

78. Job training. 15% of the 785 people attending a job training program had a college degree. How many people is this?

Approximate the percent and then estimate each answer. (Answers may vary.)

79. Seat belts. A state trooper survey on an interstate highway found that of the 3,850 cars that passed the inspection point, 6% of the drivers were not wearing a seat belt. Estimate the number not wearing a seat belt.

80. Down payments. Estimate the amount of an 11% down payment on a house that is selling for $279,950.

SECTION 6.5 ▶ **Interest**

DEFINITIONS AND CONCEPTS	EXAMPLES
Interest is money that is paid for the use of money. **Simple interest** is interest earned on the original principal and is found using the formula $\quad I = Prt$ where P is the principal, r is the annual (yearly) interest rate, and t is the length of time in years.	If \$4,000 is invested for 3 years at a rate of 7.2%, how much simple interest is earned? $\quad P = \$4,000 \qquad r = 7.2\% = 0.072 \qquad t = 3$ $\quad I = Prt$ — This is the simple interest formula. $\quad I = \$4,000 \cdot 0.072 \cdot 3$ — Substitute the values for P, r, and t. Remember to write the rate r as a decimal. $\quad I = \$288 \cdot 3$ — Multiply: $\$4,000 \cdot 0.072 = \$288$. $\quad I = \$864$ — Complete the multiplication. The simple interest earned in 3 years is \$864.
The **total amount** in an investment account or the total amount to be repaid on a loan is the sum of the principal and the interest. $\quad$ Total amount = principal + interest	**Home repairs.** A homeowner borrowed \$5,600 for 2 years at 10% simple interest to pay for a new concrete driveway. Find the total amount due on the loan. $\quad P = \$5,600 \qquad r = 10\% = 0.10 \qquad t = 2$ $\quad I = Prt$ — This is the simple interest formula. $\quad I = \$5,600 \cdot 0.10 \cdot 2$ — Write the rate r as a decimal. $\quad I = \$560 \cdot 2$ — Multiply: $\$5,600 \cdot 0.10 = \$560$. $\quad I = \$1,120$ — Complete the multiplication. The interest due in 2 years is \$1,120. To find the total amount of money due on the loan, we add. $\quad$ Total amount = **principal** + **interest** $\qquad\qquad\quad = \ \ \textbf{\$5,600} \ \ + \ \ \textbf{\$1,120}$ $\qquad\qquad\quad = \$6,720$ — Do the addition. At the end of 2 years, the total amount of money due on the loan is \$6,720.
When using the formula $I = Prt$, the **time must be expressed in years**. If the time is given in days or months, rewrite it as a fractional part of a year. Here are two examples: ▪ Since there are 365 days in a year, $\quad 60 \text{ days} = \dfrac{60}{365} \text{ year} = \dfrac{\overset{1}{\cancel{5}} \cdot 12}{\underset{1}{\cancel{5}} \cdot 73} \text{ year} = \dfrac{12}{73} \text{ year}$ ▪ Since there are 12 months in a year, $\quad 4 \text{ months} = \dfrac{4}{12} \text{ year} = \dfrac{\overset{1}{\cancel{4}}}{3 \cdot \underset{1}{\cancel{4}}} \text{ year} = \dfrac{1}{3} \text{ year}$	**Fines.** A man borrowed \$300 at 15% for 45 days to get his car out of an impound parking garage. Find the simple interest that must be paid on the loan. Since there are 365 days in a year, we have $\quad 45 \text{ days} = \dfrac{45}{365} \text{ year} = \dfrac{\overset{1}{\cancel{5}} \cdot 9}{\underset{1}{\cancel{5}} \cdot 73} \text{ year} = \dfrac{9}{73} \text{ year}$ — Simplify the fraction. The time of the loan is $\frac{9}{73}$ year. To find the amount of interest, we multiply. $\quad P = \$300 \qquad r = 15\% = 0.15 \qquad t = \dfrac{9}{73}$ $\quad I = Prt$ — This is the simple interest formula. $\quad I = \$300 \cdot 0.15 \cdot \dfrac{9}{73}$ — Write the rate r as a decimal. $\quad I = \dfrac{\$300}{1} \cdot \dfrac{0.15}{1} \cdot \dfrac{9}{73}$ — Write \$300 and 0.15 as fractions. $\quad I = \dfrac{\$405}{73}$ — Multiply the numerators. Multiply the denominators. $\quad I \approx \$5.55$ — Do the division. Round to the nearest cent. The simple interest that must be paid on the loan is \$5.55.

Compound interest is interest earned on the original principal and previously earned interest.

When compounding, we can calculate interest:

- **annually:** once a year
- **semiannually:** twice a year
- **quarterly:** four times a year
- **daily:** 365 times a year

Compound interest.   Suppose $10,000 is deposited in an account that earns 6.5% compounded semiannually. Find the amount of money in an account at the end of the first year.

The word *semiannually* means that the interest will be compounded two times in one year. To find the amount of interest $10,000 will earn in the first half of the year, use the simple interest formula, where t is $\frac{1}{2}$ of a year.

Interest earned in the first half of the year:

$$P = \$10,000 \qquad r = 6.5\% = 0.065 \qquad t = \frac{1}{2}$$

$I = Prt$	This is the simple interest formula.
$I = \$10,000 \cdot 0.065 \cdot \dfrac{1}{2}$	Write the rate r as a decimal.
$I = \dfrac{\$10,000}{1} \cdot \dfrac{0.065}{1} \cdot \dfrac{1}{2}$	Write $10,000 and 0.065 as fractions.
$I = \dfrac{\$650}{2}$	Multiply the numerators. Multiply the denominators.
$I = \$325$	Do the division.

The interest earned in the first half of the year is $325. The original principal and this interest now become the principal for the second half of the year.

$$\$10,000 + \$325 = \$10,325$$

To find the amount of interest $10,325 will earn in the second half of the year, use the simple interest formula, where t is again $\frac{1}{2}$ of a year.

Interest earned in the second half of the year:

$$P = \$10,325 \qquad r = 6.5\% = 0.065 \qquad t = \frac{1}{2}$$

$I = Prt$	This is the simple interest formula.
$I = \$10,325 \cdot 0.065 \cdot \dfrac{1}{2}$	Write the rate r as a decimal.
$I = \dfrac{\$10,325}{1} \cdot \dfrac{0.065}{1} \cdot \dfrac{1}{2}$	Write $10,325 and 0.065 as fractions.
$I = \dfrac{\$671.125}{2}$	Multiply the numerators. Multiply the denominators.
$I \approx \$335.56$	Do the division. Round to the nearest cent.

The interest earned in the second half of the year is $335.56. Adding this to the principal for the second half of the year, we get

$$\$10,325 + \$335.56 = \$10,660.56$$

The total amount in the account after one year is $10,660.56

Computing compound interest by hand can take a long time. The **compound interest formula** can be used to find the amount of money that an account will contain at the end of the term.

$$A = P\left(1 + \frac{r}{n}\right)^{nt}$$

where A is the amount in the account, P is the principal, r is the annual interest rate, n is the number of compoundings in one year, and t is the length of time in years.

A **calculator** is helpful in performing the operations on the right side of the compound interest formula.

Compounding daily. A mini-mall developer promises investors in his company $3\frac{1}{4}\%$ interest, compounded daily. If a businessman decides to invest $80,000 with the developer, how much money will be in his account in 8 years?

Compounding daily means the compounding will be done 365 times a year.

$$P = \$80,000 \qquad r = 3\frac{1}{4}\% = 0.0325 \qquad t = 8 \qquad n = 365$$

$$A = P\left(1 + \frac{r}{n}\right)^{nt} \qquad \text{This is the compound interest formula.}$$

$$A = 80,000\left(1 + \frac{0.0325}{365}\right)^{365(8)} \qquad \text{Substitute for } P, r, n, \text{ and } t.$$

$$A = 80,000\left(1 + \frac{0.0325}{365}\right)^{2,920} \qquad \text{Evaluate the exponent: } 365 \cdot 8 = 2,920.$$

$$A \approx 103,753.21 \qquad \text{Use a calculator. Round to the nearest cent.}$$

There will be $103,753.21 in the account in 8 years.

REVIEW EXERCISES

81. Investments. Find the simple interest earned on $6,000 invested at 8% for 2 years. Use the following table to organize your work.

P	*r*	*t*	*I*

82. Investment accounts. If $24,000 is invested at a simple interest rate of 4.5% for 3 years, what will be the total amount of money in the investment account at the end of the term?

83. Emergency loans. A teacher's credit union loaned a client $2,750 at a simple interest rate of 11% so that he could pay an overdue medical bill. How much interest does the client pay if the loan must be paid back in 3 months?

84. Code violations. A business was ordered to correct safety code violations in a production plant. To pay for the needed corrections, the company borrowed $10,000 at 12.5% simple interest for 90 days. Find the total amount that had to be paid after 90 days.

Duplass/Shutterstock.com

85. Monthly payments. A couple borrows $1,500 for 1 year at a simple interest rate of $7\frac{3}{4}\%$.

a. How much interest will they pay on the loan?

b. What is the total amount they must repay on the loan?

c. If the couple decides to repay the loan by making 12 equal monthly payments, how much will each monthly payment be?

86. Savings accounts. Find the amount of money that will be in a savings account at the end of 1 year if $2,000 is the initial deposit and the interest rate of 7% is compounded semi-annually.

87. Savings accounts. Find the amount that will be in a savings account at the end of 3 years if a deposit of $5,000 earns interest at a rate of $6\frac{1}{2}\%$, compounded daily.

88. Cash grants. Each year a cash grant is given to a deserving college student. The grant consists of the interest earned that year on a $500,000 savings account. What is the cash award for the year if the money is invested at a rate of 8.3%, compounded daily?

1. Fill in the blanks.

   a. _____ means parts per one hundred.

   b. The key words in a percent sentence translate as follows:
      - ___ translates to an equal symbol =
      - ___ translates to multiplication that is shown with a raised dot ·
      - _____ *number* or _____ *percent* translates to an unknown number that is represented by a variable

   c. In the percent sentence "5 is 25% of 20," 5 is the _____, 25% is the percent, and 20 is the _____.

   d. When we use percent to describe how a quantity has increased compared to its original value, we are finding the percent of _____.

   e. _____ interest is interest earned only on the original principal. _____ interest is interest paid on the principal and previously earned interest.

2. a. Express the amount of the figure that is shaded as a percent, as a fraction, and as a decimal.

   b. What percent of the figure is not shaded?

   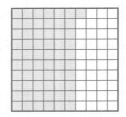

3. In the illustration below, each set of 100 square regions represents 100%. Express as a percent the amount of the figure that is shaded. Then express that percent as a fraction and as a decimal.

   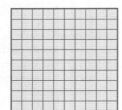 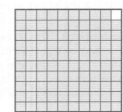

4. Write each percent as a decimal.

   a. 67%   b. 12.3%   c. $9\frac{3}{4}\%$

5. Write each percent as a decimal.

   a. 0.06%   b. 210%   c. 55.375%

6. Write each fraction as a percent.

   a. $\frac{1}{4}$ b. $\frac{5}{8}$ c. $\frac{28}{25}$

7. Write each decimal as a percent.

   a. 0.19   b. 3.47   c. 0.005

8. Write each decimal or whole number as a percent.

   a. 0.667   b. 2   c. 0.9

9. Write each percent as a fraction. Simplify, if possible.

   a. 55%   b. 0.01%   c. 125%

10. Write each percent as a fraction. Simplify, if possible.

    a. $6\frac{2}{3}\%$ b. 37.5%   c. 8%

11. Write each fraction as a percent. Give the exact answer and an approximation to the nearest tenth of a percent.

    a. $\frac{1}{30}$ b. $\frac{16}{9}$

12. 65 is what percent of 1,000?

13. What percent of 14 is 35?

14. **Fugitives.** As of March 2017, exactly 481 of the 512 fugitives who have appeared on the FBI's Ten Most Wanted list have been captured or located. What percent is this? Round to the nearest hundredth of one percent. (Source: www.fbi.gov/wanted)

    WANTED BY THE FBI

15. **Swimming workouts.** A swimmer was able to complete 18 laps before a shoulder injury forced him to stop. This was only 20% of a typical workout. How many laps does he normally complete during a workout?

16. **College employees.** The 700 employees at a community college fall into three major categories, as shown in the circle graph. How many employees are in administration?

    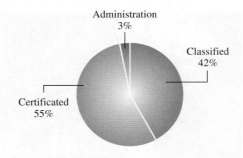

17. What number is 224% of 60?

18. 2.6 is $33\frac{1}{3}\%$ of what number?

19. Shrinkage. See the following label from a new pair of jeans. The measurements are in inches. (*Inseam* is a measure of the length of the jeans.)

WAIST	INSEAM
33	**34**

Expect shrinkage of approximately **3%** in length after the jeans are washed.

   a. How much length will be lost due to shrinkage?

   b. What will be the length of the jeans after being washed?

20. Total cost. Find the total cost of a $25.50 purchase if the sales tax rate is 2.9%.

21. Sales tax. The purchase price for a watch is $90. If the sales tax is $2.70, what is the sales tax rate?

22. Population increases. After a new freeway was completed, the population of a town it passed through increased from 2,800 to 3,444 in two years. Find the percent of increase.

23. Insurance. An automobile insurance salesperson receives a 4% commission on the annual premium of any policy she sells. Find her commission on a policy if the annual premium is $898.

24. Telemarketing. A telemarketer earned a commission of $528 on $4,800 worth of new business that she obtained over the telephone. Find her rate of commission.

Neustockimages/E+/Getty Images

25. Cost-of-living. A teacher earning $40,000 just received a cost-of-living increase of 3.6%. What is the teacher's new salary?

26. Auto care. Refer to the advertisement below. Find the discount, the sale price, and the discount rate on the car waxing kit.

SAVE! SAVE! SAVE! SAVE!
CAR WAX KIT
$9 OFF
CLEAN & SHINE
COMPLETE
Regularly $75.00

27. Towel sales. Find the amount of the discount on a beach towel if it regularly sells for $20 but is on sale for 33% off. Then find the sale price of the towel.

28. Fill in the blanks.

   a. To find 1% of a number, move the decimal point in the number _____ places to the _____.

   b. To find 10% of a number, move the decimal point in the number _____ place to the _____.

29. Estimate each answer. (Answers may vary slightly.)

   a. What is 20% of 396?

   b. What is 50% of 6,189,034?

   c. What is 200% of 21.2?

30. Credit ranges. A FICO credit score is a powerful measure of a person's creditworthiness as a lender might see it. The score can range from 300 to 850. The graph below shows the percent of consumers with scores in a particular range. What percent of consumers have a credit score from 750 to 799?

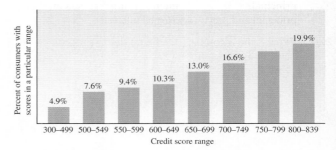

Source: Fico score & data as of April 2015

31. Shade Figure b so that Figure a and Figure b have the same percent shaded.

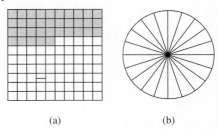

(a)   (b)

32. Brake inspections. Of the 1,920 trucks inspected at a safety checkpoint, 5% had problems with their brakes. Estimate the number of trucks that had brake problems.

Dmitry Kalinovsky/Shutterstock.com

33. Tipping.   Estimate the amount of a 15% tip on a lunch costing $28.40.

34. Car shows.   24% of 63,400 people that attended a five-day car show were female. Estimate the number of females that attended the car show.

35. Interest charges.   Find the simple interest on a loan of $3,000 at 5% per year for 1 year.

36. Investments.   If $23,000 is invested at $4\frac{1}{2}\%$ simple interest for 5 years, what will be the total amount of money in the investment account at the end of the 5 years?

37. Short-term loans.   Find the simple interest on a loan of $2,000 borrowed at 8% for 90 days.

38. Use the formula $A = P\left(1 + \dfrac{r}{n}\right)^{nt}$ to find the amount of interest earned on an investment of $24,000 paying an annual rate of 6.4% interest, compounded daily for 3 years.

1. Write 6,054,346 [Section 1.1]

   a. in words   **b.** in expanded notation

2. **Weather.** The cloudiest major city in the United States is Juneau, Alaska. The tables below show the average number of cloudy days in Juneau each month. Find the total number of cloudy days in a year. (Source: Western Regional Climate Center) [Section 1.2]

Jan	Feb	Mar	Apr	May	June
23	21	23	22	23	22

July	Aug	Sept	Oct	Nov	Dec
23	22	24	27	24	25

3. Subtract: $50,055 - 7,899$ [Section 1.3]

4. Multiply: $308 \cdot 75$ [Section 1.4]

5. Divide: $37\overline{)561}$ [Section 1.5]

6. **Bottled Water.** How many 8-ounce servings are there in a 5-gallon bottle of water? (*Hint*: There are 128 fluid ounces in one gallon.) [Section 1.6]

7. List the factors of 40, from smallest to largest. [Section 1.7]

8. Find the prime factorization of 294. [Section 1.7]

9. Find the LCM and the GCF of 24 and 30. [Section 1.8]

10. Evaluate: $\dfrac{39 + 3[4^3 - 2(2^2 - 3)]}{4 \cdot 2^2 - 1}$ [Section 1.9]

11. Place an $<$ or an $>$ symbol in the box to make a true statement: $|-8|$ ☐ $-(-5)$ [Section 2.1]

12. Evaluate: $(-20 + 9) + (-13 + 24)$ [Section 2.2]

13. **Overdraft protection.** A student forgot that she had only \$55 in her bank account and wrote a check for \$75, used an ATM to get \$60 cash, and used her debit card to buy \$25 worth of groceries. On each of the three transactions, the bank charged her a \$10 overdraft protection fee. Find the new account balance. [Section 2.3]

14. Evaluate: -6^2 and $(-6)^2$ [Section 2.4]

15. Evaluate each expression, if possible. [Sections 2.2–2.5]

   a. $\dfrac{-14}{0}$ **b.** $\dfrac{0}{-12}$

   c. $-3(-4)(-5)(0)$ **d.** $0 - (-14)$

16. Evaluate: $\dfrac{3 + 3[5(-6) - (1 - 10)]}{1 - (-1)}$ [Section 2.6]

17. Estimate the following sum by rounding each number to the nearest hundred. [Section 2.6]

   $-5,684 + (-2,270) + 3,404 + 2,689$

18. Simplify: $\dfrac{54}{60}$ [Section 3.1]

19. Express $\dfrac{4}{5}$ as an equivalent fraction with a denominator of 45. [Section 3.1]

20. What is $\dfrac{1}{4}$ of -240? [Section 3.2]

21. **Kites.** Find the number of square inches of nylon cloth used to make the kite shown below. (*Hint:* Find the area.) [Section 4.2]

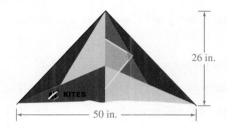

22. Divide: $\dfrac{4}{9} \div \left(-\dfrac{16}{27}\right)$ [Section 3.3]

23. Subtract: $\dfrac{9}{10} - \dfrac{3}{14}$ [Section 3.4]

24. Determine which fraction is larger: $\dfrac{23}{20}$ or $\dfrac{7}{6}$ [Section 3.4]

25. **Hamburgers.** What is the difference in weight between a $\frac{1}{4}$-pound and a $\frac{1}{3}$-pound hamburger? [Section 3.4]

26. Multiply: $\left(-3\dfrac{3}{4}\right)(8)$ [Section 3.5]

27. **Belts.** Refer to the belt shown below. What is the maximum waist size that the belt will fit if it is fastened using the last hole? [Section 3.6]

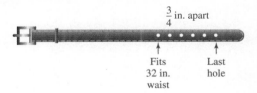

28. Simplify: $\dfrac{\frac{1}{3} - \frac{3}{4}}{\frac{1}{6} + \frac{1}{3}}$ [Section 3.7]

29. Round each decimal. [Section 4.1]

   a. Round 452.0298 to the nearest hundredth.

   b. Round 452.0298 to the nearest thousandth.

30. Evaluate: $3.4 - (6.6 + 7.3) + 5$ [Section 4.2]

31. Weekly earnings. A welder's basic work week is 40 hours. After his daily shift is over, he can work overtime at a rate of 1.5 times his regular rate of $15.90 per hour. How much money will he earn in a week if he works 4 hours of overtime? [Section 4.3]

32. Divide: $0.58\overline{)0.1566}$ [Section 4.4]

33. Write $\dfrac{11}{15}$ as a decimal. Use an overbar. [Section 4.5]

34. Evaluate: $3\sqrt{81} - 8\sqrt{49}$ [Section 4.6]

35. Write the ratio $1\frac{1}{4}$ to $1\frac{1}{2}$ as a fraction in simplest form. [Section 5.1]

36. Solve the proportion: $\dfrac{7}{14} = \dfrac{2}{x}$ [Section 5.2]

37. How many days are in 960 hours? [Section 5.3]

38. Convert 2,400 millimeters to meters. [Section 5.4]

39. Convert 6.5 kilograms to pounds. [Section 5.5]

40. Complete the table. [Section 6.1]

Percent	Decimal	Fraction
	0.29	
47.3%		
		$\frac{7}{8}$

41. 16% of what number is 20? [Section 6.2]

42. Genealogy. Through an extensive computer search, a genealogist determined that worldwide, 180 out of every 10 million people had his last name. What percent is this? [Section 6.2]

43. Health clubs. The number of members of a health club increased from 300 to 534. What was the percent of increase in club membership? [Section 6.3]

44. Guitar sale. What are the regular price, the sale price, the discount, and discount rate for the guitar shown in the advertisement below? [Section 6.3]

Save on the Standard Strat

Fender

Now Only
$321⁰⁰
Save $107

45. Tipping. Refer to the sales receipt below. [Section 6.4]

  a. Estimate the 15% tip.

  b. Find the total.

STEAK STAMPEDE	
Bloomington, MN	
Server #12\ AT	
VISA	67463777288
NAME	DALTON/ LIZ
AMOUNT	$78.18
GRATUITY $	_____
TOTAL $	_____

46. Investments. Find the simple interest earned on $10,000 invested for 2 years at 7.25%. [Section 6.5]

7 Graphs and Statistics

from **Campus to Careers**

Postal Service Mail Carrier

Mail carriers follow schedules as they collect and deliver mail to homes and businesses. They must have the ability to quickly and accurately compare similarities and differences among sets of letters, numbers, objects, pictures, and patterns. They also need to have strong problem-solving skills to redirect mislabeled letters and packages. Mail carriers weigh items on postal scales and make calculations with money as they read postage rate tables.

In **Problem 19** of **Study Set 7.1**, you will see how a mail carrier must be able to read a postal rate table and know American units of weight to determine the cost to send a package using priority mail.

JOB TITLE:
Postal Service Mail Carrier

EDUCATION:
A high school diploma (or equivalent) and a passing score on a written exam are required.

JOB OUTLOOK:
Competition for jobs is high since positions usually come open only upon retirement of current mail carriers.

ANNUAL EARNINGS:
Average (mean) salary in 2015 was $51,130

FOR MORE INFORMATION:
http://stats.bls.gov/oes/current/oes/435052.htm

SECTION 7.1   Reading Graphs and Tables

We live in an information age. Never before have so many facts and figures been right at our fingertips. Since information is often presented in the form of tables or graphs, we need to be able to read and make sense of data displayed in that way.

The following **table**, **bar graph**, and **circle graph** (or **pie chart**) show the results of a shopper survey. A large sample of adults were asked how far in advance they typically shop for a gift. In the bar graph, the length of a bar represents the percent of responses for a given shopping method. In the circle graph, the size of a colored region represents the percent of responses for a given shopping method.

Shopper Survey

How far in advance gift givers typically shop

Survey Responses: A Table

Time in advance	Percent
A month or longer	8%
Within a month	12%
Within 3 weeks	12%
Within 2 weeks	23%
Within a week	41%
The same day as giving it	4%

Survey Responses: A Bar Graph

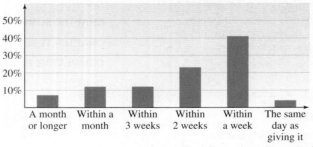

Survey Responses: A Circle Graph

Source: Harris interactive online study via QuickQuery for Gifts.com

It is often said that a picture is worth a thousand words. That is the case here, where the graphs display the results of the survey more clearly than the table. It's easy to see from the graphs that most people shop within a week of when they need to purchase a gift. It is also apparent that same-day shopping for a gift was the least popular response. That information also appears in the table, but it is just not as obvious.

OBJECTIVE 1 Read tables.

Data are often presented in tables, with information organized in **rows** and **columns.** To read a table, we must find the *intersection* of the row and column that contains the desired information.

EXAMPLE 1   **Postal rates.**   Refer to the table of priority mail postal rates from 2017 below. Find the cost of mailing an $8\frac{1}{2}$-pound package by priority mail to postal zone 4.

Priority Mail 2017	Zone 1 & 2	Zone 3	Zone 4	Zone 5	Zone 6	Zone 7	Zone 8	Zone 9
1 lb	$5.75	$6.26	$6.37	$6.53	$6.66	$6.71	$7.04	$9.59
2 lbs	$6.33	$6.40	$6.63	$7.91	$8.91	$9.69	$9.97	$14.69
3 lbs	$6.41	$7.16	$7.62	$9.22	$11.46	$12.39	$14.45	$19.92
4 lbs	$6.62	$7.39	$8.46	$10.19	$13.36	$15.13	$17.08	$23.99
5 lbs	$6.96	$7.77	$8.91	$10.62	$15.21	$17.39	$19.80	$27.86
6 lbs	$7.29	$8.18	$9.26	$13.23	$17.06	$19.83	$22.67	$31.92
7 lbs	$7.60	$8.59	$9.60	$14.68	$18.89	$22.36	$25.46	$35.84
8 lbs	$7.96	$9.00	$10.64	$15.87	$20.77	$24.61	$28.59	$40.24
9 lbs	$8.32	$9.57	$10.75	$16.96	$22.59	$26.66	$31.79	$44.75
10 lbs	$8.82	$9.97	$10.88	$18.24	$24.41	$29.30	$34.57	$48.66

Strategy We will read the number at the intersection of the 9th row and the column labeled Zone 4.

WHY Since $8\frac{1}{2}$ pounds is more than 8 pounds, we cannot use the 8th row. Since $8\frac{1}{2}$ pounds does not exceed 9 pounds, we use the 9th row of the table.

Solution

The number at the intersection of the 9th row (in red) and the column labeled Zone 4 (in blue) is 10.75 (in purple). This means it would cost $10.75 to mail the $8\frac{1}{2}$-pound package by priority mail.

Self Check 1

Postal rates. Refer to the table of priority mail postal rates. Find the cost of mailing a 3.75-pound package by priority mail to postal zone 8.

Now Try Problem 17

OBJECTIVE 2   **Read bar graphs.**

Another popular way to display data is to use a **bar graph** with bars drawn vertically or horizontally. The relative heights (or lengths) of the bars make for easy comparisons of values. A horizontal or vertical line used for reference in a bar graph is called an **axis**. The **horizontal axis** and the **vertical axis** of a bar graph serve to frame the graph, and they are scaled in units such as years, dollars, minutes, pounds, and percent.

EXAMPLE 2   **Speed of animals.**   The following bar graph shows the maximum speeds for several animals over a given distance.

a. What animal in the graph has the fastest maximum speed?

b. What animal in the graph has the slowest maximum speed?

c. How much greater is the maximum speed of a lion compared to that of a coyote?

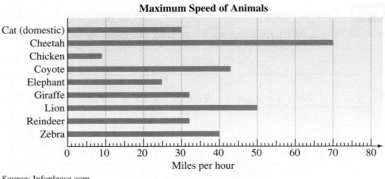

Source: Infoplease.com

Strategy We will locate the name of each desired animal on the vertical axis and move right to the end of its corresponding bar.

WHY Then we can extend downward and read the animal's maximum speed on the horizontal axis scale.

Solution

a. The longest bar in the graph has a length of 70 units and corresponds to a cheetah. Of all the animals listed in the graph, the cheetah has the fastest maximum speed at 70 mph.

b. The shortest bar in the graph has a length of approximately 9 units and corresponds to a chicken. Of all the animals listed in the graph, the chicken has the slowest maximum speed at 9 mph.

c. The length of the bar that represents a lion's maximum speed is 50 units long and the length of the bar that represents a coyote's maximum speed appears to be 43 units long. To find how much greater is the maximum speed of a lion compared to that of a coyote, we subtract

$$50 \text{ mph} - 43 \text{ mph} = 7 \text{ mph}$$ Subtract the coyote's maximum speed from the lion's maximum speed.

The maximum speed of a lion is about 7 mph faster than the maximum speed of a coyote.

Self Check 2

Speed of animals. Refer to the bar graph for Example 2.
a. What is the maximum speed of a giraffe?
b. How much greater is the maximum speed of a coyote compared to that of a reindeer?
c. Which animals listed in the graph have a maximum speed that is slower than that of a domestic cat?

Now Try Problem 21

To compare sets of related data, groups of two (or three) bars can be shown. For **double-bar** or **triple-bar graphs**, a **key** is used to explain the meaning of each type of bar in a group.

EXAMPLE 3 **The U.S. economy.** The following bar graph shows the total income generated by three sectors of the U.S. economy in each of three years.

a. What income was generated by retail sales in 2010?

b. Which sector of the economy consistently generated the most income?

c. By what amount did the income from the wholesale sector increase from 2010 to 2014?

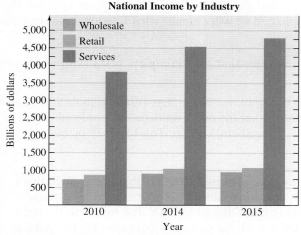

National Income by Industry

Source: *The World Almanac and Book of Facts*, 2016

Strategy To answer questions about years, we will locate the correct colored bar and look at the *horizontal axis* of the graph. To answer questions about the income, we will locate the correct colored bar and extend to the left to look at the *vertical axis* of the graph.

WHY The years appear on the horizontal axis. The height of each bar, representing income in billions of dollars, is measured on the scale on the vertical axis.

Solution

a. The first group of bars indicates income in the year 2010. According to the color key, the blue bar of that group shows the retail sales. Since the vertical axis is scaled in units of $\frac{\$500\text{ billion}}{2} = \$250$ billion, the height of that bar is approximately 750 plus one-half of 250, or 125. Thus, the height of the blue bar is approximately $750 + 125 = 875$, which represents $875 billion in retail sales in 2010.

b. In each group, the green bar is the tallest. That bar, according to the color key, represents the income from the services sector of the economy. Thus, services consistently generated the most income.

c. According to the color key, the orange bar in each group shows income from the wholesale sector. That sector generated about $725 billion of income in 2010 and $900 billion in income in 2014. The amount of increase is the difference of these two quantities.

$$\$900 \text{ billion} - \$725 \text{ billion} = \$175 \text{ billion}$$ *Subtract the 2010 wholesale income from the 2014 wholesale income.*

Wholesale income increased by about $175 billion between 2010 and 2014.

OBJECTIVE 3 Read pictographs.

A **pictograph** is like a bar graph, but the bars are made from pictures or symbols. A **key** tells the meaning (or value) of each symbol.

EXAMPLE 4 **Pizza deliveries.** The pictograph on the right shows the number of pizzas delivered to the three residence halls on a college campus during final exam week. In the graph, what information does the top row of pizzas give?

Pizzas ordered during final exam week

Men's residence hall

Women's residence hall

Co-ed residence hall

= 12 pizzas

Self Check 3

The U.S. economy. Refer to the bar graph for Example 3.

a. What income was generated by retail sales in 2015?

b. What income was generated by the wholesale sector in 2015?

c. In 2010, by what amount did the income from the services sector exceed the income from the retail sector?

Now Try ⭢ Problems 25 and 31

Strategy We will count the number of complete pizza symbols that appear in the top row of the graph, and we will estimate what fractional part of a pizza symbol also appears in that row.

WHY The key indicates that each complete pizza symbol represents one dozen (12) pizzas.

Solution The top row contains 3 complete pizza symbols and what appears to be $\frac{1}{4}$ of another. This means that the men's residence hall ordered 3 · 12, or 36 pizzas, plus approximately $\frac{1}{4}$ of 12, or about 3 pizzas. This totals 39 pizzas.

Caution! One drawback of a pictograph is that it can be difficult to determine what fractional amount is represented by a portion of a symbol. For example, if the BLU-RAY disc shown to the right represents 1,000 units sold, we can only estimate that the partial BLU-RAY disc symbol represents about 600 units sold.

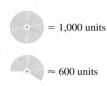

= 1,000 units

≈ 600 units

OBJECTIVE 4 **Read circle graphs.**

In a **circle graph**, regions called **sectors** are used to show what part of the whole each quantity represents.

EXAMPLE 5   **Silver production.** The circle graph to the right gives information about silver production. The entire circle represents the world's total production of 887 million ounces in 2015. Use the graph to answer the following questions.

a. What percent of the total was the combined production of the United States and Mexico?

b. What percent of the total production came from sources other than those listed?

c. To the nearest tenth of a million, how many ounces of silver did China produce in 2015?

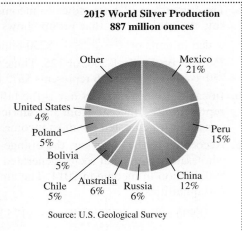

2015 World Silver Production
887 million ounces

Source: U.S. Geological Survey

Strategy We will look for the key words in each problem.

WHY Key words tell us what operation (addition, subtraction, multiplication, or division) must be performed to answer each question.

Solution

a. The key word *combined* indicates addition. According to the graph, the United States produced 4% and Mexico produced 21% of the total amount of silver in 2015. Together, they produced 4% + 21%, or 25% of the total.

b. The phrase *from sources other than those listed* indicates subtraction. To find the percent of silver produced by countries that are not listed, we add the contributions of all the listed sources and subtract that total from 100%.

$100\% - (21\% + 15\% + 12\% + 6\% + 6\% + 5\% + 5\% + 5\% + 4\%) = 100\% - 79\% = 21\%$

Countries that are not listed in the graph produced 21% of the world's total production of silver in 2015.

c. From the graph we see that China produced 12% of the world's silver in 2015. To find the number of ounces produced by China (the amount), we use the method for solving percent problems from Section 6.2.

What number	is	12%	of	887?

This is the percent sentence. The units are millions of troy ounces.

$$x \quad = \quad 12\% \quad \cdot \quad 887$$

Translate to a percent equation.

Now we perform the multiplication on the right side of the equation.

$x = 0.12 \cdot 887$ Write 12% as a decimal: 12% = 0.12.

$x = 106.44$ Do the multiplication.

$$\begin{array}{r} 887 \\ \times\ 0.12 \\ \hline 1774 \\ 8870 \\ \hline 106.44 \end{array}$$

Rounded to the nearest tenth of a million, China produced 106.4 million ounces of silver in 2015.

Self Check 5

Silver production.   Refer to the circle graph for Example 5. To the nearest tenth of a million, how many ounces of silver did Russia produce in 2015?

Now Try ➔ Problems 37, 41, and 43

OBJECTIVE 5   Read line graphs.

A **line graph** is used to show how quantities change with time. From such a graph, we can determine when a quantity is increasing and when it is decreasing.

EXAMPLE 6   **ATMs.**   The line graph below shows the number of automated teller machines (ATMs) in the United States for the years 2002 through 2016. Use the graph to answer the following questions.

a. How many ATMs were there in the United States in 2005?

b. Between which two years was there the greatest increase in the number of ATMs?

c. When did the number of ATMs decrease for the first time?

d. Between which two years did the number of ATMs remain about the same?

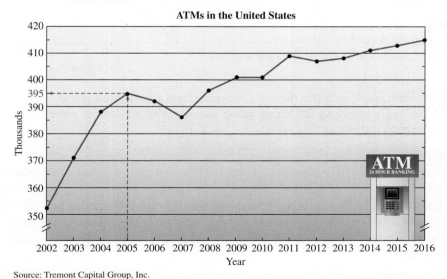

Source: Tremont Capital Group, Inc.

LANGUAGE OF MATHEMATICS

The symbol $\frac{\ell}{\ell}$ is used to show a break in the scale on an axis. It enables us to omit large portions of empty space on a graph.

Strategy We will determine whether the graph is rising, falling, or horizontal.

WHY When the graph rises as we read from left to right, the number of ATMs is increasing. When the graph falls as we read from left to right, the number of ATMs is decreasing. If the graph is horizontal, there is no change in the number of ATMs.

Solution

a. To find the number of ATMs in 2005, we follow the dashed blue line from the label 2005 on the horizontal axis straight up to the line graph. Then we extend directly over to the scale on the vertical axis, where the arrowhead points to 395. Since the vertical scale is in thousands of ATMs, there were about 395,000 ATMs in 2005 in the United States.

b. This line graph is composed of fourteen line segments that connect pairs of consecutive years. The steepest of those seven segments represents the greatest increase in the number of ATMs. Since that segment is between 2002 and 2003, the greatest increase in the number of ATMs occurred between 2002 and 2003.

c. The first line segment of the graph that falls as we read from left to right is the segment connecting the data points for the years 2005 and 2006. Thus, the number of ATMs decreased from 2005 to 2006.

d. The line segment connecting the data points for the years 2009 and 2010 appears to be horizontal. Since there is little or no change in the number of ATMs for those years, the number of ATMs remained about the same from 2009 to 2010.

Self Check 6

ATMs.   Refer to the line graph for Example 6.
a. Find the increase in the number of ATMs between 2007 and 2008.
b. How many more ATMs were there in the United States in 2016 as compared to 2002?

Now Try ➡ Problems 45, 47, and 51

Two quantities that are changing with time can be compared by drawing both lines on the same graph.

EXAMPLE 7   **Trains.**   The line graph below shows the movements of two trains. The horizontal axis represents time, and the vertical axis represents the distance that the trains have traveled.

a. How are the trains moving at time A?
b. At what time (A, B, C, D, or E) are both trains stopped?
c. At what times have both trains traveled the same distance?

Strategy We will determine whether the graphs are rising or are horizontal. We will also consider the relative positions of the graphs for a given time.

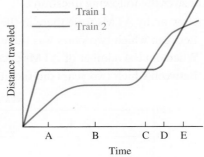

WHY A rising graph indicates the train is moving, and a horizontal graph means it is stopped. For any given time, the higher graph indicates that the train it represents has traveled the greater distance.

Solution

The movement of train 1 is represented by the red line, and that of train 2 is represented by the blue line.

a. At time A, the blue line is rising. This shows that the distance traveled by train 2 is increasing. Thus, at time A, train 2 is moving. At time A, the red line is horizontal. This indicates that the distance traveled by train 1 is not changing: At time A, train 1 is stopped.

b. To find the time at which both trains are stopped, we find the time at which both the red and the blue lines are horizontal. At time B, both trains are stopped.

c. At any time, the height of a line gives the distance a train has traveled. Both trains have traveled the same distance whenever the two lines are the same height—that is, at any time when the lines intersect. This occurs at times C and E.

Self Check 7

Trains.   In the graph for Example 7, what is train 1 doing at time D?

Now Try ➡ Problems 53, 55, and 59

OBJECTIVE 6 Read histograms and frequency polygons.

A company that makes vitamins is sponsoring a program on a cable TV channel. The marketing department must choose from three advertisements to show during the program.

1. Children talking about a chewable vitamin that the company makes.

2. A college student talking about an active-life vitamin that the company makes.

3. A grandmother talking about a multivitamin that the company makes.

A survey of the viewing audience records the age of each viewer, counting the number in the 6-to-15-year-old age group, the 16-to-25-year-old age group, and so on. The graph of the data is displayed in a special type of bar graph called a **histogram**, as shown on the right. The vertical axis, labeled **Frequency**, indicates the number of viewers in each age group. For example, the histogram shows that 105 viewers are in the 36-to-45-year-old age group.

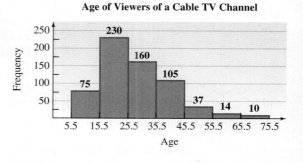

Age of Viewers of a Cable TV Channel

A histogram is a bar graph with three important features.

1. The bars of a histogram touch.

2. Data values never fall at the edge of a bar.

3. The widths of the bars are equal and represent a range of values.

The width of each bar of a histogram represents a range of numbers called a **class interval**. The histogram above has seven class intervals, each representing an age span of 10 years. Since most viewers are in the 16-to-25-year-old age group, the marketing department decides to advertise the active-life vitamins in commercials that appeal to young adults.

EXAMPLE 8 **Carry-on luggage.** An airline weighed the carry-on luggage of 2,260 passengers. The data are displayed in the histogram below.

a. How many passengers carried luggage in the 8-to-11-pound range?

b. How many carried luggage in the 12-to-19-pound range?

Strategy We will examine the scale on the horizontal axis of the histogram and identify the interval that contains the given range of weight for the carry-on luggage.

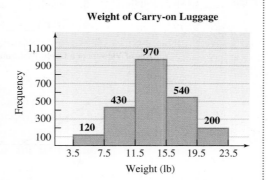

Weight of Carry-on Luggage

WHY Then we can read the height of the corresponding bar to answer the question.

Solution

a. The second bar, with edges at 7.5 and 11.5 pounds, corresponds to the 8-to-11-pound range. Use the height of the bar (or the number written there) to determine that 430 passengers carried such luggage.

b. The 12-to-19-pound range is covered by two bars. The total number of passengers with luggage in this range is 970 + 540, or 1,510.

Self Check 8

Carry-on luggage. Refer to the histogram for Example 8. How many passengers carried luggage in the 20-to-23-pound range?

Now Try ➥ Problem 61

A special line graph, called a **frequency polygon**, can be constructed from the carry-on luggage histogram by joining the center points at the top of each bar. (See the graphs below.) On the horizontal axis, we write the middle value of each bar. After erasing the bars, we get the frequency polygon shown on the right below.

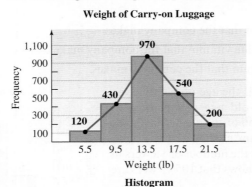

Histogram

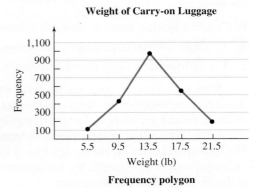

Frequency polygon

Answers to Self Checks

1. $17.08   **2. a.** 32 mph   **b.** 11 mph   **c.** a chicken and an elephant   **3. a.** about $1,050 billion
b. about $990 billion   **c.** about $3,000 billion   **4.** 33 pizzas were delivered to the co-ed residence hall.
5. 53.2 million ounces   **6. a.** about 10,000   **b.** about 63,000   **7.** Train 1, which had been stopped, is beginning to move.   **8.** 200

SECTION 7.1   STUDY SET

VOCABULARY

For problems 1–6, refer to graphs a through f below. Fill in the blanks with the correct letter. Each letter is used only once.

1. Graph _____ is a bar graph. **2.** Graph _____ is a circle graph.

3. Graph _____ is a pictograph.

4. Graph _____ is a line graph.

5. Graph _____ is a histogram.

6. Graph _____ is a frequency polygon.

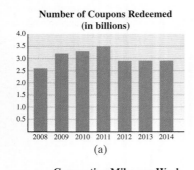

(a)

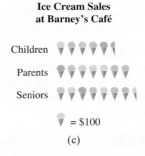

(b)

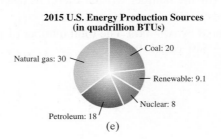

(c)

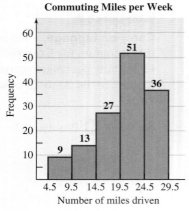

(d)

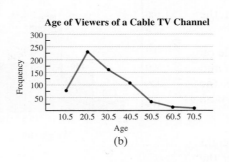

(e)

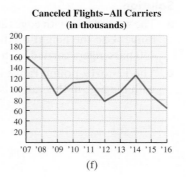

(f)

7. A horizontal or vertical line used for reference in a bar graph is called an _____.

8. A circle graph uses slice-of-pie-shaped regions called _____ to show what part of the whole each quantity represents.

CONCEPTS

Fill in the blanks.

9. To read a table, we must find the _____ of the row and column that contains the desired information.

10. The _____ axis and the vertical axis of a bar graph serve to frame the graph; they are scaled in units such as years, dollars, minutes, pounds, and percent.

11. A pictograph is like a bar graph, but the bars are made from _____ or symbols.

12. Line graphs are often used to show how a quantity changes with _____. On such graphs, we can easily see when a quantity is increasing and when it is _____.

13. A histogram is a bar graph with three important features.
   - The _____ of a histogram touch.
   - Data values never fall at the _____ of a bar.
   - The widths of the bars are _____ and represent a range of values.

14. A frequency polygon can be constructed from a histogram by joining the _____ points at the top of each bar.

NOTATION

15. If the symbol  = 1,000 buses, estimate what the symbol ![bus] represents.

16. Fill in the blank: The symbol ⌇ is used when graphing to show a _____ in the scale on an axis.

GUIDED PRACTICE

Refer to the postal rate table on page 587 to answer the following questions. **See Example 1.**

17. Find the cost of using priority mail to send a package weighing $7\frac{1}{4}$ pounds to zone 3.

18. Find the cost of sending a package weighing $2\frac{1}{4}$ pounds to zone 5 by priority mail.

---

from **Campus to Careers**

Postal Service Mail Carrier

19. A woman wants to send a birthday gift and an anniversary gift to her brother, who lives in zone 6, using priority mail. One package weighs 2 pounds 9 ounces, and the other weighs 3 pounds 8 ounces. Suppose you are the woman's mail carrier and she asks you how much money will be saved by sending both gifts as one package instead of two. Make the necessary calculations to answer her question. (*Hint*: 16 ounces = 1 pound.)

---

20. Juan wants to send a package weighing 11 pounds 15 ounces to a friend living in zone 2. Express (overnight) mail service would cost $39.50. How much more would it cost to send the package by express mail instead of priority mail?

Refer to the bar graph below to answer the following questions. **See Example 2.**

21. List the top three most commonly owned pets in the United States.

22. There are three types of pets that are owned in approximately equal numbers. What are they?

23. The number of fish is less than the combined total of which two types of pets?

24. How many more pet cats are there than pet dogs?

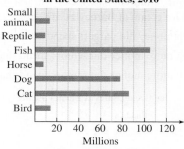

Total Number of Pets Owned in the United States, 2016

Source: National Pet Owners Survey, AAPA

Refer to the bar graph on the next page to answer the following questions. **See Example 3.**

Zinc and lead are two commonly used metals. A zinc coating stops steel and iron from rusting, and zinc is combined with copper to form brass. Lead is used in batteries and in aprons to shield patients during X-rays.

25. For the years shown in the graph, has the production of zinc always exceeded the production of lead?

26. Estimate how many times greater the amount of zinc produced in 2000 was compared to the amount of lead produced that year?

27. What is the sum of the amounts of lead produced in 1990, 2000, and 2015?

28. For which metal, lead or zinc, and for which two years, did the production remain about the same?

29. In what years was the amount of zinc produced at least twice that of lead?

30. Find the difference between the amount of zinc produced in 2015 and the amount produced in 2000.

31. By how many metric tons did the amount of zinc produced increase between 1990 and 2015?

32. Between which two years did the production of lead decrease?

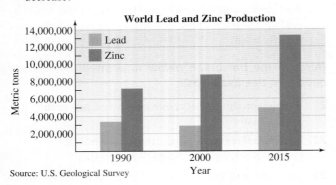

Source: U.S. Geological Survey

Refer to the pictograph below to answer the following questions. See Example 4.

33. Which group (children, parents, or seniors) spent the most money on ice cream at Barney's Café?

34. How much money did parents spend on ice cream?

35. How much more money did seniors spend than parents?

36. How much more money did seniors spend than children?

Ice Cream Sales at Barney's Café

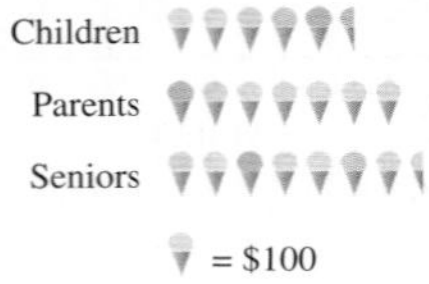

Children
Parents
Seniors

= $100

Refer to the circle graph in the next column to answer the following questions. See Example 5.

37. Of the languages in the graph, which is spoken by the greatest number of people?

38. Do more people speak Spanish or French?

39. Together, do more people speak English, French, Spanish, Russian, and German combined than Chinese?

40. One pair of languages shown in the graph are spoken by groups of the same size. Which pair of languages are they?

41. What percent of the world's population speak a language other than the eight shown in the graph?

42. What percent of the world's population speak Russian or English?

43. To the nearest one million, how many people in the world speak Chinese?

44. To the nearest one million, how many people in the world speak Arabic?

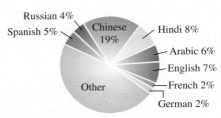

World Languages
Percents of the world population that speak them

Estimated world population 2015: 7,200,000,000

Source: ethnologue.com, U.S. Census Bureau

Refer to the line graph below to answer the following questions. See Example 6.

45. How many U.S. ski resorts were in operation in 2004?

46. How many U.S. ski resorts were in operation in 2008?

47. Between which two years was there a decrease in the number of ski resorts in operation? (*Hint:* there is more than one answer.)

48. Between which two years was there an increase in the number of ski resorts in operation? (*Hint:* there is more than one answer.)

49. For which two years was the number of ski resorts in operation exactly the same? (*Hint:* there is more than one answer.)

50. Find the difference in the number of ski resorts in operation in 2001 and 2008.

51. Between which two years was there the greatest decrease in the number of ski resorts in operation? What was the decrease?

52. Between which two years was there the greatest increase in the number of ski resorts in operation? What was the increase?

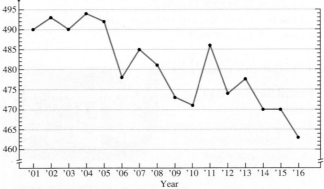

Number of U.S. Ski Resorts in Operation

Source: National Ski Area Assn.

Refer to the line graph below to answer the following questions. See Example 7.

53. Which runner ran faster at the start of the race?

54. At time A, which runner was ahead in the race?

55. At what time during the race were the runners tied for the lead?

56. Which runner stopped to rest first?

57. Which runner dropped his watch and had to go back to get it?

58. At which of these times (A, B, C, D, E) was runner 1 stopped and runner 2 running?

59. Describe what was happening at time E. Who was running? Who was stopped?

60. Which runner won the race?

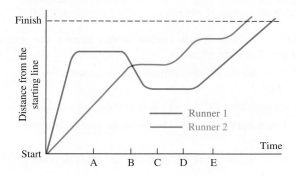

Refer to the histogram and frequency polygon below to answer the following questions. See Example 8.

61. Commuting miles. An insurance company collected data on the number of miles its employees drive to and from work. The data are presented in the histogram below.

   a. How many employees have a commute that is in the range of 15 to 19 miles per week?

   b. How many employees commute 14 miles or less per week?

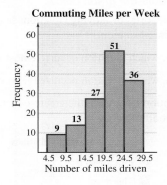

62. Night shift staffing. A hospital administrator surveyed the medical staff to determine the number of room calls during the night. She constructed the frequency polygon below.

   a. On how many nights were there about 30 room calls?

   b. On how many nights were there about 60 room calls?

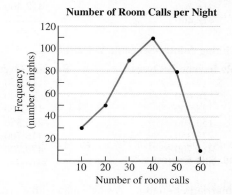

TRY IT YOURSELF

Refer to the 2017 federal income tax tables on the next page.

63. Filing a single return. Herb is single and has taxable income of $79,250. Compute his federal income tax.

64. Millionaires. Compute the federal income tax on a taxable income of $1,000,000 for:

   a. a single adult

   b. a married couple filing jointly

65. Lotto winnings. In 2016, a married couple from Tennessee won the Powerball jackpot and took it as a lump sum of $328,000,000. What is the federal income tax on their winnings?

66. Filing a joint return. Raul and his wife have a combined taxable income of $57,100. Compute their federal income tax if they file jointly.

67. Tax-saving strategy. Angelina is single and has a taxable income of $53,000. If she marries someone with no income, she will gain other deductions that will reduce her income by $3,900, and she can file a joint return.

   a. Compute her federal income tax if she remains single.

   b. Compute her federal income tax if she gets married.

   c. How much will she save in federal income tax by getting married?

Unmarried Individuals-Schedule X 2017

If Taxable Income Is Over	But Not Over	The Tax Is	Plus	Of the Amount Over
$ 0	$ 9,325	$ 0	10%	$ 0
9,325	37,950	932.50	15%	9,325
37,950	91,900	5,226.25	25%	37,950
91,900	191,650	18,713.75	28%	91,900
191,650	416,700	46,643.75	33%	191,650
416,700	418,400	120,910.25	35%	416,700
418,400	----------	121,500.25	39.6%	418,400

Married Filing Jointly-Schedule Y-1 2017

If Taxable Income Is Over	But Not Over	The Tax Is	Plus	Of the Amount Over
$ 0	$ 18,650	$ 0	10%	$ 0
18,650	75,900	1,865.00	15%	18,650
75,900	153,100	10,452.50	25%	75,900
153,100	233,350	29,752.50	28%	153,100
233,350	416,700	52,222.50	33%	233,350
416,700	470,700	112,728.00	35%	416,700
470,700	----------	131,628.00	39.6%	470,700

68. The marriage penalty. A single man with a taxable income of $80,000 is dating a single woman with a taxable income of $75,000.

  a. Find the amount of federal income tax each of them must pay.

  b. Add the results from part a.

  c. If they get married and file a joint return, how much federal income tax will they have to pay on their combined incomes?

  d. Will they owe more federal income taxes if they get married or if they stay single? Find the amount of the "marriage penalty."

Refer to the bar graph in the next column.

69. In which year was the largest percent of flights canceled? Estimate the percent.

70. In which year was the smallest percent of flights canceled? Estimate the percent.

71. Did the percent of canceled flights increase or decrease between 2010 and 2012? By how much?

72. Did the percent of canceled flights increase or decrease between 2008 and 2009? By how much?

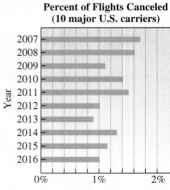

Percent of Flights Canceled
(10 major U.S. carriers)

Source: Bureau of Transportation Statistics

Refer to the line graph below, which shows the altitude of a small private airplane.

73. How did the plane's altitude change between times B and C?

74. At what time did the pilot first level off the airplane?

75. When did the pilot first begin his descent to land the airplane?

76. How did the plane's altitude change between times D and E?

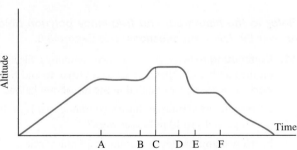

Refer to the double-bar graph below.

77. In which categories of moving violations have violations decreased since last month?

78. Last month, which violation occurred most often?

79. This month, which violation occurred least often?

80. Which violation has shown the greatest decrease in number since last month?

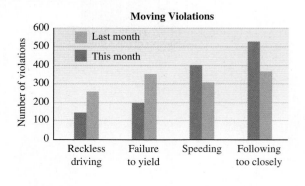

Moving Violations

Refer to the line graph below.

81. What were the average weekly earnings in mining for the year 1980?

82. What were the average weekly earnings in construction for the year 1980?

83. Were the average weekly earnings in mining and construction ever the same?

84. What was the difference in a miner's and a construction worker's weekly earnings in 2015?

85. In the period between 2005 and 2010, which occupation's weekly earnings were increasing more rapidly, the miner's or the construction worker's?

86. Did the weekly earnings of a miner or a construction worker ever decrease over a five-year span?

87. In the period from 1980 to 2015, which group of workers received the greater increase in weekly earnings?

88. In what five-year span was the miner's increase in weekly earnings the smallest?

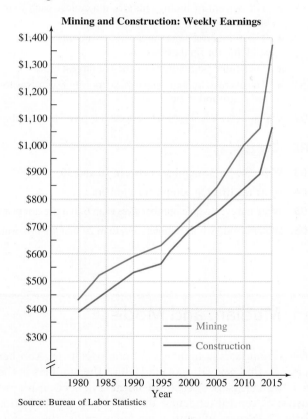

Mining and Construction: Weekly Earnings

Source: Bureau of Labor Statistics

Refer to the pictograph in the next column.

89. What is the daily parking rate for Honolulu?

90. What is the daily parking rate for Boston?

91. How much more would it cost to park a car for five days in Boston compared to five days in San Francisco?

92. How much more would it cost to park a car for five days in Honolulu compared to five days in Boston?

Daily Parking Rates

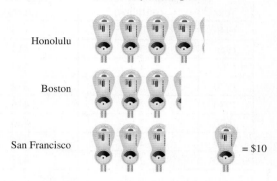

Source: Colliers International

Refer to the circle graph below.

93. What percent of the bank robberies were known to have occurred on a Tuesday? Round to the nearest percent.

94. What percent of the bank robberies were known to have occurred on Thursday? Round to the nearest percent.

95. What percent of the bank robberies were known to have occurred on a weekend? Round to the nearest percent.

96. By what percent does the number of bank robberies known to have occurred on a Friday exceed the number known to have occurred on a Monday?

Number of U.S. Bank Robberies by Day of the Week, 2015

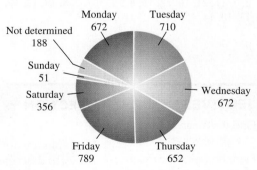

Total: 4,090

Source: U.S. Department of Justice

97. Number of U.S. farms. Use the data in the table on the next page to construct a bar graph showing the number of U.S. farms for selected years from 1940 through 2010.

98. Size of U.S. farms. Use the data in the table below to construct a line graph showing the average acreage of U.S. farms for selected years from 1940 through 2010.

Year	Number of farms (millions)	Average size farm (acres)
1940	6.3	174
1950	5.6	213
1960	4.0	297
1970	2.9	374
1980	2.4	426
1990	2.1	460
2000	2.2	436
2010	2.2	418

Source: U.S. Dept. of Agriculture

99. Coupons. Each coupon value shown in the table below provides savings for shoppers. Construct a line graph that relates the original price (in dollars, on the horizontal axis) to the sale price (in dollars, on the vertical axis).

Coupon value: amount saved	Original price of the item
$10	$100, but less than $250
$25	$250, but less than $500
$50	$500 or more

100. Dentistry. To study the effect of fluoride in preventing tooth decay, researchers counted the number of fillings in the teeth of 28 patients and recorded these results:

3, 7, 11, 21, 16, 22, 18, 8, 12, 3, 7, 2, 8, 19, 12, 19, 12, 10, 13, 10, 14, 15, 14, 14, 9, 10, 12, 13

Tally the results by completing the table. Then construct a histogram. The first bar extends from 0.5 to 5.5, the second bar from 5.5 to 10.5, and so on.

Number of fillings	Frequency
1–5	
6–10	
11–15	
16–20	
21–25	

WRITING

101. What kind of presentation (table, bar graph, line graph, circle graph, pictograph, or histogram) is most appropriate for displaying each type of information? Explain your choices.
- The percent of students at a college, classified by major
- The percent of biology majors at a college each year since 1970
- The number of hours a group of students spent studying for final exams
- The ethnic populations of the ten largest cities
- The average annual salary of corporate executives for ten major industries

102. Explain why a histogram is a special type of bar graph.

REVIEW

103. Write the prime numbers between 10 and 30.

104. Write the first ten composite numbers.

105. Write the even whole numbers less than 6 that are not prime.

106. Write the odd whole numbers less than 20 that are not prime.

OBJECTIVES

1 Find the mean (average) of a set of values.

2 Find the weighted mean of a set of values.

3 Find the median of a set of values.

4 Find the mode of a set of values.

5 Use the mean, median, mode, and range to describe a set of values.

SECTION 7.2   Mean, Median, and Mode

Graphs are not the only way of describing sets of numbers in a compact form. Another way to describe a set of numbers is to find *one* value around which the numbers in the set are grouped. We call such a value a **measure of central tendency**. In Section 1.9, we studied the most popular measure of central tendency, the *mean* or *average*. In this section we will examine two other measures of central tendency, called the *median* and the *mode*.

OBJECTIVE 1 Find the mean (average) of a set of values.

Recall that the *mean* or *average* of a set of values gives an indication of the "center" of the set of values. To review this concept, let's consider the case of a student who has taken five tests this semester in a history class scoring 87, 73, 89, 92, and 84. To find out how well she is doing, she calculates the mean, or the average, of these scores, by finding their sum and then dividing it by 5.

$$\text{Mean} = \frac{87 + 73 + 89 + 92 + 84}{5} \quad \longleftarrow \text{The sum of the test scores}$$
$$\longleftarrow \text{The number of test scores}$$

$$= \frac{425}{5} \quad \text{In the numerator, do the addition.}$$

$$= 85 \quad \text{Do the division.}$$

$$
\begin{array}{r}
\overset{2}{8}7 \\
73 \\
89 \\
92 \\
+\,84 \\
\hline
425
\end{array}
\qquad
\begin{array}{r}
85 \\
5\overline{)425} \\
-\,40 \\
\hline
25 \\
-\,25 \\
\hline
0
\end{array}
$$

The mean is 85. Some scores were better and some were worse, but 85 is a good indication of her performance in the class.

Finding the Mean (Arithmetic Average)

The **mean**, or the **average**, of a set of values is given by the formula:

$$\text{Mean (average)} = \frac{\text{the sum of the values}}{\text{the number of values}}$$

EXAMPLE 1   **Store sales.**   One week's sales in men's, women's, and children's departments of the Clothes Shoppe are given in the table below. Find the mean of the daily sales in the women's department for the week.

Strategy We will add $3,135, $2,310, $3,206, $2,115, $1,570, and $2,100 and divide the sum by 6.

Total Daily Sales by Department—Clothes Shoppe			
Day	**Men's department**	**Women's department**	**Children's department**
Monday	$2,315	$3,135	$1,110
Tuesday	2,020	2,310	890
Wednesday	1,100	3,206	1,020
Thursday	2,000	2,115	880
Friday	955	1,570	1,010
Saturday	850	2,100	1,000

WHY To find the mean (average) of a set of values, we divide the sum of the values by the number of values. In this case, there are 6 days of sales (Monday through Saturday).

Solution Since there are 6 days of sales, divide the sum by 6.

$$\text{Mean} = \frac{\$3,135 + \$2,310 + \$3,206 + \$2,115 + \$1,570 + \$2,100}{6}$$

$$= \frac{\$14,436}{6} \quad \text{In the numerator, do the addition.}$$

$$= \$2,406 \quad \text{Do the division.}$$

The mean of the week's daily sales in the women's department is $2,406.

$$
\begin{array}{r}
\overset{1\ \ 11}{3,135} \\
2,310 \\
3,206 \\
2,115 \\
1,570 \\
+\,2,100 \\
\hline
14,436
\end{array}
\qquad
\begin{array}{r}
2406 \\
6\overline{)14,436} \\
-\,12 \\
\hline
2\,4 \\
-\,2\,4 \\
\hline
03 \\
-\,0 \\
\hline
36 \\
-\,36 \\
\hline
0
\end{array}
$$

Self Check 1

Store sales. Find the mean of the daily sales in the men's department of the Clothes Shoppe for the week.

Now Try ⟹ Problems 9 and 45

Using Your Calculator ▶   Finding the Mean

Most scientific calculators do statistical calculations and can easily find the mean of a set of numbers. To use a scientific calculator in statistical mode to find the mean in Example 1, try these keystrokes:

- Set the calculator to statistical mode.
- Reset the calculator to clear the *statistical registers*.
- Enter each number, followed by the $\boxed{\Sigma +}$ key instead of the $\boxed{+}$ key. That is, enter 3135, press $\boxed{\Sigma +}$, enter 2310, press $\boxed{\Sigma +}$, and so on.
- When all data are entered, find the mean by pressing the $\boxed{\bar{x}}$ key. You may need to press $\boxed{\text{2nd}}$ first. The mean is 2,406.

Because keystrokes vary among calculator brands, you might have to check the owner's manual if these instructions don't work.

EXAMPLE 2   Driving.   In the month of January, a trucker drove a total of 4,805 miles. On the average, how many miles did he drive per day?

January						
S	M	T	W	T	F	S
				1	2	3	4
5	6	7	8	9	10	11
12	13	14	15	16	17	18
19	20	21	22	23	24	25
26	27	28	29	30	31	

Strategy We will divide 4,805 by 31 (the number of days in the month of January).

WHY We do not have to find the sum of the miles driven each day in January. That total is given in the problem as 4,805 miles.

Solution

$$\text{Average number of miles driven per day} = \frac{\text{the total miles driven}}{\text{the number of days}}$$

$$= \frac{4,805}{31} \quad \begin{array}{l}\text{← This is given.}\\ \text{← January has 31 days.}\end{array}$$

$$= 155 \qquad \text{Do the division.}$$

On average, the trucker drove 155 miles per day.

$$\begin{array}{r} 155 \\ 31\overline{)4,805} \\ -31 \\ \hline 1\,70 \\ -1\,55 \\ \hline 155 \\ -155 \\ \hline 0 \end{array}$$

Self Check 2

Trucking. If a trucker drove 3,360 miles in February, how many miles did he drive per day, on average? (Assume it is not a leap year.)

Now Try ⮕ **Problem 47**

OBJECTIVE 2 Find the weighted mean of a set of values.

When a value in a set appears more than once, that value has a greater "influence" on the mean than another value that only occurs a single time. To simplify the process of finding a mean, any value that appears more than once can be "weighted" by multiplying it by the number of times it occurs. A mean that is found in this way is called a **weighted mean**.

EXAMPLE 3   Hotel reservations.   A hotel electronically recorded the number of times the reservation desk telephone rang before it was answered by a receptionist. The results of the weeklong survey are shown in the table on the next page. Find the average number of times the phone rang before a receptionist answered.

Strategy First, we will determine the total number of times the reservation desk telephone rang during the week before it was answered. Then we will divide that result by the total number of calls received.

Number of rings	Number of calls
1	11
2	46
3	45
4	28
5	20

WHY To find the average of a set of values, we divide the sum of the values by the number of values.

Solution To find the total number of times the reservation desk telephone rang during the week before it was answered, we multiply each number of rings (1, 2, 3, 4, and 5) by the number of times it occurred and add those results to get 450. The calculations are highlighted in blue in the "Weighted number of rings" column.

Number of rings	Number of calls	Weighted number of rings	
1	11	$1 \cdot 11 \rightarrow$	11
2	46	$2 \cdot 46 \rightarrow$	92
3	45	$3 \cdot 45 \rightarrow$	135
4	28	$4 \cdot 28 \rightarrow$	112
5	+ 20	$5 \cdot 20 \rightarrow$	+ 100
Totals	**150**		**450**

To find the total number of calls received, we add the values in the "Number of calls" column of the table and get 150, as highlighted in red. To find the average, we divide.

$$\text{Average} = \frac{450}{150} \xleftarrow{\text{The total number of rings}}_{\text{The total number of calls}}$$

$$= 3 \qquad \text{Do the division.}$$

$$\begin{array}{r} 3 \\ 150\overline{)450} \\ -450 \\ \hline 0 \end{array}$$

The average number of times the phone rang before it was answered was 3.

Now Try ➡ Problem 49

Self Check 3

Quiz results. The class results on a five-question true-or-false Spanish quiz are shown in the table below. Find the average number of incorrect answers on the quiz.

Total number of incorrect answers on the quiz	Number of students
0	8
1	8
2	5
3	15
4	3
5	1

Finding the Weighted Mean

To find the weighted mean of a set of values:

1. Multiply each value by the number of times it occurs.

2. Find the sum of products from step 1.

3. Divide the sum from step 2 by the total number of individual values.

Another example of a weighted mean is a **grade point average (GPA)**. To find a GPA, we divide:

$$\text{GPA} = \frac{\text{total number of grade points}}{\text{total number of credit hours}}$$

EXAMPLE 4   **Finding GPAs.**   Find the semester grade point average for a student that received the following grades. Round to the nearest hundredth.

Course	Grade	Credits
Speech	C	2
Basic Mathematics	A	4
French	B	4
Business Law	D	3
Study Skills	A	1

Strategy   First, we will determine the total number of grade points earned by the student. Then we will divide that result by the total number of credits.

WHY   To find the mean of a set of values, we divide the sum of the values by the number of values.

Solution

The point values of grades that are used at most colleges and universities are:

A: 4 pts   **B**: 3 pts   **C**: 2 pts   **D**: 1 pt   **F**: 0 pt

To find the total number of grade points that the student earned, we multiply the number of credits for each course by the point value of the grade received. Then we add those results to find that the total number of grade points is 39. The calculations are highlighted in blue in the "Weighted grade points" column below.

To find the total number of credits, we add the values in that column (highlighted in red), to get 14.

Course	Grade	Credits	Weighted grade points	
Speech	C	2	$2 \cdot 2 \rightarrow$	4
Basic Mathematics	A	4	$4 \cdot 4 \rightarrow$	16
French	B	4	$3 \cdot 4 \rightarrow$	12
Business Law	D	3	$1 \cdot 3 \rightarrow$	3
Study Skills	A	+1	$4 \cdot 1 \rightarrow$	+ 4
Totals		**14**		**39**

LANGUAGE OF MATHEMATICS

Some schools assign a certain number of **credit hours (credits)** to a course while others assign a certain number of **units**. For example, at San Antonio College, the Basic Mathematics course is 3 credit hours while the same course at Los Angeles City College is 3 units.

To find the GPA, we divide.

$$\text{GPA} = \frac{39}{14} \quad \begin{array}{l} \leftarrow \text{ The total number of grade points} \\ \leftarrow \text{ The total number of credits} \end{array}$$

≈ 2.785 *Do the division.*

$= 2.79$ *Round 2.785 to the nearest hundredth.*

The student's semester GPA is 2.79.

$$\begin{array}{r} 2.785 \\ 14\overline{)39.000} \\ -28 \\ \hline 11\,0 \\ -9\,8 \\ \hline 1\,20 \\ -1\,12 \\ \hline 80 \\ -70 \\ \hline 10 \end{array}$$

Self Check 4

Finding GPAs.   Find the semester grade point average for a student that received the following grades.

Course	Grade	Credits
MATH 130	A	4
ENG 101	D	3
PHY 080	B	4
SWIM 100	C	1

Now Try Problem 55

OBJECTIVE 3 **Find the median of a set of values.**

The mean is not always the best measure of central tendency. It can be affected by very high or very low values. For example, suppose the weekly earnings of four workers in a small company are $280, $300, $380, and $240, and the owner pays himself $5,000 a week. At that company, the mean salary per week is

$$\text{Mean} = \frac{\$280 + \$300 + \$380 + \$240 + \$5,000}{5}$$ There are 4 employees plus the owner: 4 + 1 = 5.

$$= \frac{\$6,200}{5}$$ In the numerator, do the addition.

$$= \$1,240$$ Do the division.

The owner could say, "Our employees earn an average of $1,240 per week." Clearly, the mean does not fairly represent the typical worker's salary there.

A better measure of the company's typical salary is the *median*: the salary in the middle when all of them are arranged by size.

Smallest $240 $280   **$300** $380 $5,000   Largest

Two salaries   The middle salary   Two salaries

The typical worker earns $300 per week, far less than the mean salary.

The Median

The **median** of a set of values is the middle value. To find the median:

1. Arrange the values in increasing order.

2. If there is an odd number of values, the median is the middle value.

3. If there is an even number of values, the median is the mean (average) of the middle two values.

EXAMPLE 5   Find the median of the following set of values:

7.5   20.9   9.9   4.4   9.8   5.3   6.2   7.5   4.9

Strategy We will arrange the nine values in increasing order.

WHY It is easier to find the middle value when they are written in that way.

Solution
Since there is an odd number of values, the median is the middle value.

Smallest   4.4   4.9   5.3   6.2   **7.5**   7.5   9.8   9.9   20.9   Largest

Four values   The middle value   Four values

The median is 7.5.

If there is an even number of values in a set, there is no middle value. In that case, the median is the mean (average) of the two values closest to the middle.

Success Tip The median is a single value that is "typical" of a set of values. It can be, but is not necessarily, one of the values in the set. In Example 5, you will see the median, 7.5, was one of the given values. In Example 6, you will see the median exam score, 64, was not in the given set of exam scores.

Self Check 5

Find the median of the following set of values:

$1\frac{7}{8}$ $2\frac{1}{2}$ $3\frac{3}{5}$ $\frac{1}{2}$ $2\frac{3}{4}$

Now Try Problems 17 and 21

EXAMPLE 6   **Grade distributions.**   On an exam, there were three scores of 59, four scores of 77, and scores of 43, 47, 53, 60, 68, 82, and 97. Find the median score.

Strategy We will arrange the fourteen exam scores in increasing order.

WHY It is easier to find the two middle scores when they are written in that way.

Solution
Since there is an even number of exam scores, we need to identify the two middle scores.

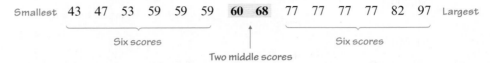

Since there is an even number of scores, the median is the average (mean) of the two scores closest to the middle: the 60 and the 68.

$$\text{Median} = \frac{60 + 68}{2} = \frac{128}{2} = 64$$

The median is 64.

Self Check 6

Grade distributions. On a mathematics exam, there were four scores of 68, five scores of 83, and scores of 72, 78, and 90. Find the median score.

Now Try Problems 25 and 29

OBJECTIVE 4 Find the mode of a set of values.

The mean and the median are not always the best measure of central tendency. For example, suppose a hardware store displays 20 outdoor thermometers. Ten of them read 80°, and the other ten all have different readings.

To choose an accurate thermometer, should we choose one with a reading that is closest to the *mean* of all 20, or to their *median*? Neither. Instead, we should choose one of the 10 that all read the same, figuring that any of those that agree will likely be correct.

By choosing that temperature that appears most often, we have chosen the *mode* of the 20 values.

> **The Mode**
>
> The **mode** of a set of values is the single value that occurs most often. The mode of several values is also called the **modal value**.

LANGUAGE OF MATHEMATICS

In Example 7, the set of values has one mode. If a set of values has two modes (exactly two values that occur an equal number of times and more often than any other value), it is said to be *bimodal*. If no value in a set occurs more often than another, then there is *no mode*.

EXAMPLE 7   Find the mode of these values:

3 6 5 7 3 7 2 4 3 5 3 7 8 7 3 7 6 3 4

Strategy We will determine how many times each of the values, 2, 3, 4, 5, 6, 7, and 8 occurs.

WHY We need to know which values occur most often.

Solution
It is not necessary to list the values in increasing order. Instead, we can make a chart and use **tally marks** to keep track of the number of times that the values 2, 3, 4, 5, 6, 7, and 8 occur.

2	3	4	5	6	7	8	← These values appear in the list.
/	///// /	//	//	//	/////	/	← Tally marks

Because 3 occurs more times than any other value, it is the mode.

Self Check 7

Find the mode of these values:
2 3 4 6 2 4
3 4 3 4 2 5

Now Try Problems 33 and 37

OBJECTIVE 5 **Use the mean, median, mode, and range to describe a set of values.**

Another measure that is used to describe a set of values is the **range**. It indicates the spread of the values.

The Range

The **range** of a set of values is the difference between the largest value and the smallest value.

EXAMPLE 8 **Machinist's tools.** The diameters (distances across) of eight stainless steel bearings were found using the calipers shown below. Find **a.** the mean, **b.** the median, **c.** the mode, and **d.** the range of the set of measurements listed below.

3.43 cm   3.25 cm   3.48 cm   3.39 cm   3.54 cm   3.48 cm   3.23 cm   3.24 cm

Calipers

Stainless
steel bearing

Strategy We will determine the sum of the measurements, the number of measurements, the middle measurement(s), the most often occurring measurement, and the difference between the largest and smallest measurement.

WHY We need to know that information to find the mean, median, mode, and range.

Solution

a. To find the mean, we add the measurements and divide by the number of values, which is 8.

$$\text{Mean} = \frac{3.43 + 3.25 + 3.48 + 3.39 + 3.54 + 3.48 + 3.23 + 3.24}{8}$$

$$= \frac{27.04}{8} \quad \text{In the numerator, do the addition.}$$

$$= 3.38 \quad \text{Do the division.}$$

$$
\begin{array}{r}
\overset{3\ 4}{3.43} \\
3.25 \\
3.48 \\
3.39 \\
3.54 \\
3.48 \\
3.23 \\
+\ 3.24 \\
\hline
27.04
\end{array}
$$

$$
\begin{array}{r}
3.38 \\
8\overline{)27.04} \\
-24 \\
\hline
3\,0 \\
-2\,4 \\
\hline
64 \\
-64 \\
\hline
0
\end{array}
$$

The mean is 3.38 cm.

b. To find the median, we first arrange the eight measurements in increasing order.

Smallest   3.23   3.24   3.25   **3.39   3.43**   3.48   3.48   3.54   *Largest*

↑
Two middle measurements

Because there is an even number of measurements, the median is the average of the two middle values.

$$\text{Median} = \frac{\mathbf{3.39 + 3.43}}{2} = \frac{6.82}{2} = 3.41 \text{ cm}$$

c. Since the measurement 3.48 cm occurs most often (twice), it is the mode.

d. In part b, we see that the smallest value is 3.23 and the largest value is 3.54. To find the range, we subtract the smallest value from the largest value.

$$\text{Range} = 3.54 - 3.23 = 0.31 \text{ cm}$$

$$
\begin{array}{r}
3.54 \\
-3.23 \\
\hline
0.31
\end{array}
$$

Self Check 8

Mobile phones. The weights of eight different makes of mobile phones are: 4.37 oz, 5.98 oz, 4.36 oz, 4.95 oz, 5.05 oz, 5.95 oz, 4.95 oz, and 5.27 oz. Find the mean, median, and mode weight. Then find the range of the weights.

Now Try Problem 51

Think it Through ● THE VALUE OF AN EDUCATION

"If we want America to lead in the 21st century, nothing is more important than giving everyone the best education possible."
—President Barack Obama, 2012

As college costs increase, some people wonder if it is worth it to spend years working toward a degree when that same time could be spent earning money. The following median income data makes it clear that, over time, additional education is well worth the investment. Use the given facts to complete the bar graph.

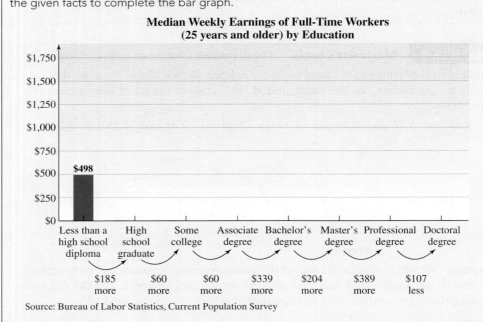

Median Weekly Earnings of Full-Time Workers
(25 years and older) by Education

Source: Bureau of Labor Statistics, Current Population Survey

Answers to Self Checks

1. $1,540   **2.** 120 miles per day   **3.** 2 incorrect answers   **4.** 2.75   **5.** $2\frac{1}{2}$ **6.** 80.5   **7.** 4
8. mean: 5.11 oz; median: 5.00 oz; mode: 4.95 oz; range: 1.62 oz

---

SECTION 7.2   STUDY SET

VOCABULARY

Fill in the blanks.

1. The _____ (average) of a set of values is the sum of the values divided by the number of values in the set.

2. The _____ of a set of values written in increasing order is the middle value.

3. The _____ of a set of values is the single value that occurs most often.

4. The _____ of a set of values is the difference between the largest value and the smallest value.

CONCEPTS

5. Fill in the blank. The mean of a set of values is given by the formula

$$\text{Mean} = \frac{\text{the sum of the values}}{\rule{2cm}{0.4pt}}$$

6. Consider the following set of values written in increasing order:

    3   6   8   10   11   15   16

a. Is there an even or an odd number of values?

b. What is the middle number of the list?

c. What is the median of the set of values?

d. What is the largest value? What is the smallest value? What is the range of the values?

7. Consider the following set of values written in increasing order:

    4   5   5   6   8   9   9   15

a. Is there an even or odd number of values?

b. What are the middle numbers of the set of values?

c. Fill in the blanks:

$$\text{Median} = \frac{\boxed{} + \boxed{}}{2} = \frac{\boxed{}}{2} = \boxed{}$$

d. What is the largest value? What is the smallest value? What is the range of the values?

8. Consider the following set of values:

> 1 6 8 6 10 9 10 2 6

a. What value occurs the most often? How many times does it occur?

b. What is the mode of the set of values?

c. What is the largest value? What is the smallest value? What is the range of the values?

GUIDED PRACTICE

Find the mean of each set of values. See Example 1.

9. 3 4 7 7 8 11 16

10. 13 15 17 17 15 13

11. 5 9 12 35 37 45 60 77

12. 0 0 3 4 7 9 12

13. 15 7 12 19 27 17 19 35 20

14. 45 67 42 35 86 52 91 102

15. 4.2 3.6 7.1 5.9 8.2

16. 19.1 12.8 16.5 20.0

Find the median of each set of values. See Example 5.

17. 29 5 1 9 11 17 2

18. 20 4 3 2 9 8 1

19. 7 5 4 7 3 6 7 4 1

20. 0 0 3 4 0 0 3 4 5

21. 15.1 44.9 19.7 13.6 17.2

22. 22.4 22.1 50.5 22.3 22.2

23. $\dfrac{1}{100}$ $\dfrac{999}{1,000}$ $\dfrac{16}{15}$ $\dfrac{1}{3}$ $\dfrac{5}{8}$

24. $\dfrac{1}{30}$ $\dfrac{17}{30}$ $\dfrac{7}{30}$ $\dfrac{29}{30}$ $\dfrac{11}{30}$

Find the median of each set of values. See Example 6.

25. 8 10 16 63 6 7

26. 7 2 11 5 4 17

27. 39 1 50 41 51 47

28. 47 18 35 29 27 16

29. 1.8 1.7 2.0 9.0 2.1 2.3 2.1 2.0

30. 5.0 1.3 5.0 2.3 4.3 5.6 3.2 4.5

31. $\dfrac{1}{5}$ $\dfrac{11}{5}$ $\dfrac{13}{5}$ $\dfrac{2}{5}$ $\dfrac{3}{5}$ $\dfrac{7}{5}$

32. $\dfrac{1}{9}$ $\dfrac{2}{9}$ $\dfrac{7}{9}$ $\dfrac{11}{9}$ $\dfrac{13}{9}$ $\dfrac{29}{9}$

Find the mode (if any) of each set of values. See Example 7.

33. 3 5 7 3 5 4 6 7 2 3 1 4

34. 12 12 17 17 12 13 17 12

35. −6 −7 −6 −4 −3 −6 −7

36. 0 3 0 2 7 0 6 0 3 4 2 0

37. 23.1 22.7 23.5 22.7 34.2 22.7

38. 21.6 19.3 1.3 19.3 1.6 9.3 2.6

39. $\dfrac{1}{2}$ $\dfrac{1}{3}$ $\dfrac{1}{3}$ 2 $\dfrac{1}{2}$ 2 $\dfrac{1}{5}$ $\dfrac{1}{2}$ 5 $\dfrac{1}{3}$

40. 5 9 12 35 37 45 60

LOOK ALIKES

41. Find the mean, median, mode, and range of the values:
3 4 5 5 8

42. Find the mean, median, mode, and range of the values:
2.5 5 5 5 7.5

43. Find the mean, median, and mode of each set of values:

a. 1 2 3 3 3 4 5

b. 1 2 3 3 3 3 4 5

44. Find the mean, median, and mode of each set of values:

a. −1 9 10 10 10 11 21

b. −1 9 10 10 10 10 11 21

APPLICATIONS

45. Semester grades.   Frank's algebra grade is based on the average of four exams, which count equally. His grades are 75, 80, 90, and 85.

a. Find his average exam score.

b. If Frank's professor decided to count the fourth exam double, what would Frank's average be?

46. Hurricanes.   The table lists the number of major hurricanes to strike the mainland of the United States by decade. Find the average (mean) number per decade. Round to the nearest one.

Decade	Number	Decade	Number
1891–1900	8	1951–1960	9
1901–1910	4	1961–1970	6
1911–1920	7	1971–1980	4
1921–1930	5	1981–1990	4
1931–1940	8	1991–2000	5
1941–1950	10	2001–2010	7

Source: National Hurricane Center NOAA Technical Memorandum, 2011

47. Fleet mileage. An insurance company's sales force uses 37 cars. Last June, those cars logged a total of 98,790 miles.

   a. On average, how many miles did each car travel that month?

   b. Find the average number of miles driven daily for each car.

48. Budgets. A family of five spent $519 on groceries last April.

   a. On average, how much was spent on groceries each day?

   b. What is the average spent for groceries for one family member for one day?

49. Cash awards. A contest is to be part of a promotional kickoff for a new children's cereal. The prizes to be awarded are shown.

   a. How much money will be awarded in the promotion?

   b. How many cash prizes will be awarded?

   c. What is the average cash prize?

Coloring Contest

Grand prize: Disney World vacation plus $2,500
Four 1st place prizes of $500
Thirty-five 2nd place prizes of $150
Eighty-five 3rd place prizes of $25

---

50. Surveys. Some students were asked to rate their college cafeteria food on a scale from 1 to 5. The responses are shown on the tally sheet. Find the average rating.

Poor		Fair		Excellent
1	2	3	4	5
	|||	|||	┼╫	┼╫ ||||

51. Candy bars. The prices (in cents) of the different types of candy bars sold in a drug store are: 50, 60, 50, 50, 70, 75, 50, 45, 50, 50, 60, 75, 60, 75, 100, 50, 80, 75, 100, 75.

   a. Find the mean price of a candy bar.

   b. Find the median price for a candy bar.

   c. Find the mode of the prices of the candy bars.

   d. Find the range of the candy bar prices.

52. Computer supplies. Several computer stores reported differing prices for toner cartridges for a laser printer (in dollars): 51, 55, 73, 75, 72, 70, 53, 59, 75.

   a. Find the mean price of a toner cartridge.

   b. Find the median price for a toner cartridge.

   c. Find the mode of the prices for a toner cartridge.

   d. Find the range of the toner cartridge prices.

53. Temperature changes. Temperatures were recorded at hourly intervals and listed in the table in the next column. Find the average temperature of the period from midnight to 11:00 A.M.

Time	Temperature	Time	Temperature
12:00 A.M.	53	12:00 noon	71
1:00	54	1:00 P.M.	73
2:00	57	2:00	76
3:00	58	3:00	77
4:00	59	4:00	78
5:00	59	5:00	71
6:00	61	6:00	70
7:00	62	7:00	64
8:00	64	8:00	61
9:00	66	9:00	59
10:00	68	10:00	53
11:00	71	11:00	51

54. Average temperatures. Find the average temperature for the 24-hour period shown in the table in Exercise 53.

For Exercises 55–58, find the semester grade point average for a student that received the following grades. Round to the nearest hundredth, when necessary.

55.

Course	Grade	Credits
MATH 210	C	5
ACCOUNTING 175	A	3
HEALTH 090	B	1
JAPANESE 010	D	4

56.

Course	Grade	Credits
NURSING 101	D	3
READING 150	B	4
PAINTING 175	A	2
LATINO STUDIES 090	C	3

57.

Course	Grade	Credits
PHOTOGRAPHY	D	3
MATH 020	B	4
CERAMICS 175	A	1
ELECTRONICS 090	C	3
SPANISH 130	B	5

58.

Course	Grade	Credits
ANTHROPOLOGY 050	D	3
STATISTICS 100	A	4
ASTRONOMY 100	C	1
FORESTRY 130	B	5
CHOIR 130	C	1

59. Exam averages. Roberto received the same score on each of five exams, and his mean score is 85. Find his median score and the mode of his scores.

60. Exam scores. The scores on the first exam of the students in a history class were 57, 59, 61, 63, 63, 63, 70, 87, 89, 95, 99, and 100. Kia, who scored the 70, claims that "70 is better than average." Which of the three measures of central tendency (mean, median, or mode) is Kia's score better than?

61. Comparing grades. A student received scores of 37, 53, and 78 on three quizzes. His sister received scores of 53, 57, and 58. Who had the better average? Whose grades were more consistent?

62. What is the average of all of the integers from −100 to 100, inclusive?

63. Octuplets. In December 1998, Nkem Chukwu gave birth to eight babies in Texas Children's Hospital. Find the mean, the median, and the range of their birth weights listed below.

Ebuka (girl)	24 oz	Odera (girl)	11.2 oz
Chidi (girl)	27 oz	Ikem (boy)	17.5 oz
Echerem (girl)	28 oz	Jioke (boy)	28.5 oz
Chima (girl)	26 oz	Gorom (girl)	18 oz

64. Comparison shopping. A survey of grocery stores found the price of a 14-ounce box of Cheerios cereal ranging from $3.89 to $4.39, as shown below. What are the mean, median, mode, and range of the prices listed?

$4.29 $3.89 $4.29 $4.09 $4.24 $3.99
$3.98 $4.19 $4.19 $4.39 $3.97 $4.29

65. Earthquakes. The table below lists the 17 most powerful earthquakes in 2016. Consider the depths of the earthquakes that are given in kilometers. Find the mean (round to the nearest tenth), the median, mode, and the range.

Magnitude	Location	Depth (km)
7.9	Papua New Guinea	103.2
7.8	Ecuador	20.6
7.8	Indonesia	24.0
7.8	New Zealand	22.0
7.8	Solomon Islands	41.0
7.7	North Mariana Islands	212.4
7.6	Chile	35.2
7.4	South Sandwich Islands	10.0
7.2	Russia	163.2
7.2	South Sandwich Islands	72.7
7.2	New Caledonia	9.9
7.1	Alaska	125.6
7.1	Ascension Island	10.0
7.0	Japan	10.0
7.0	Vanuatu	27.2
7.0	New Zealand	19.0
7.0	El Salvador	10.3

Source: Infoplease.com

66. Fuel efficiency. The ten most fuel-efficient cars in 2017, omitting electric cars, based on manufacturer's estimated city and highway average miles per gallon (mpg), are shown in the table below.

   a. Find the mean, median, and mode of the city mileage.

   b. Find the mean, median, and mode of the highway mileage.

Model	mpg city/hwy
Toyota Prius 3	43/59
Toyota Prius C	37/48
Chevy Malibu Hybrid	33/49
Toyota Prius V	33/47
Lexus CT 200h	31/47
Ford Fusion	35/41
Hyundai Sonata	31/46
Toyota Camry	32/43
Ford C-Max	35/38
Mitsubishi Mirage	28/47

Source: Consumer Reports.

67. Sport fishing. The report shown below lists the fishing conditions at Pyramid Lake for a Saturday in January. Find the median, the mode, and the range of the weights of the striped bass caught at the lake.

   Pyramid Lake—Some striped bass are biting but are on the small side. Striking jigs and plastic worms. Water is cold: 38°. Weights of fish caught (lb): 6, 9, 4, 7, 4, 3, 3, 5, 6, 9, 4, 5, 8, 13, 4, 5, 4, 6, 9

68. Nutrition. Refer to the table below.

   a. Find the mean number of calories in one serving of the meats shown.

   b. Find the median.

   c. Find the mode.

   d. Find the range of the number of calories.

NUTRITIONAL COMPARISONS
Per 3.5 oz. serving of cooked meat

Species	Calories
Bison	143
Beef (Choice)	283
Beef (Select)	201
Pork	212
Chicken (Skinless)	190
Sockeye Salmon	216

Source: The National Bison Association

69. Explain how to find the mean, the median, the mode, and the range of a set of values.

70. The mean, median, and mode are used to measure the central tendency of a set of values. What is meant by central tendency?

71. Which measure of central tendency—mean, median, or mode—do you think is the best for describing the salaries at a large company? Explain your reasoning.

72. When is the mode a better measure of central tendency than the mean or the median? Give an example and explain why.

Translate to a percent equation (or percent proportion) and then solve to find the unknown number.

73. 52 is what percent of 80?

74. What percent of 50 is 56?

75. $66\frac{2}{3}\%$ of what number is 28?

76. 56.2 is $16\frac{2}{3}\%$ of what number?

77. 5 is what percent of 8?

78. What number is 52% of 350?

79. Find $7\frac{1}{4}\%$ of 600.

80. $\frac{1}{2}\%$ of what number is 5,000?

OBJECTIVES

1 Identify the sample space for an experiment.

2 Identify an event of an experiment.

3 Find the probability of an event.

SECTION **7.3**   Probability

If we flip a coin, it can land in one of two equally likely ways—either heads or tails. Because one of these two results is heads, we say the *probability* of obtaining heads in a single toss is $\frac{1}{2}$. If records show that out of 100 days with weather conditions like today's, 70 have received rain, we say that there is a $\frac{70}{100}$, or more simply, a $\frac{7}{10}$ probability of rain today. If there are two chances in three that a basketball player will make a free throw, we say that the probability that the player will make a free throw is $\frac{2}{3}$.

In this section, we will discuss the concept of probability. But before we can do that, we need to discuss some important vocabulary.

OBJECTIVE 1 **Identify the sample space for an experiment.**

An **experiment** is any process for which the result is uncertain. Some examples of experiments are

- Flipping a coin
- Rolling dice
- Drawing a card from a standard deck of playing cards
- Selecting a marble from a bag of colored marbles

The results of an experiment are called **outcomes**.

In case you are not familiar with dice and a deck of cards, here are some descriptions of those items that will help you better understand some of the probability problems in this section.

A **die** (the singular form of the word *dice*) is a cube with six faces, each containing a number of dots from one to six.

A **standard deck** of 52 playing cards has two red **suits** (hearts and diamonds) and two black suits (clubs and spades). Each suit has 13 cards, including a king, queen, jack (called **face cards**), an ace, and cards numbered from 2 to 10. Assume that the card decks we work with in this section are well shuffled.

When performing an experiment, we often speak of *favorable outcomes*. These are the outcomes of an experiment that satisfy the requirements of a particular event. For the experiment of rolling a die once and the event "getting a prime number," the outcomes that are considered *favorable* to the event are 2, 3, and 5.

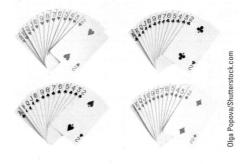

Olga Popova/Shutterstock.com

For any experiment, the set of all possible outcomes is called a **sample space**. When writing a sample space, we list all possible outcomes within **set braces** { }. For example, the sample space for flipping a single coin one time is

{H, T}   H represents the outcome "heads," and T represents the outcome "tails."

The sample space for flipping a coin twice is a set of four ordered pairs:

{(H, H), (H, T), (T, H), (T, T)}   The pair (H, T), for example, represents the outcome "heads on the first coin and tails on the second coin."

One way to determine all possible outcomes of a multistep experiment is to draw a **tree diagram**. For the experiment of flipping a coin twice, we have:

```
First toss    Second toss    Outcome
                   H ———————— (H, H)
        H <
                   T ———————— (H, T)
                   H ———————— (T, H)
        T <
                   T ———————— (T, T)
```

Alexey Stiop/Shutterstock.com

EXAMPLE 1   Write out the sample space of the experiment "rolling two dice one time."

Strategy We will write all the possible outcomes of this experiment as ordered pairs.

WHY The outcome from rolling the white die can be any number from 1 to 6, and the outcome from rolling the red die can be any number from 1 to 6.

Solution We will let the first number in each ordered pair be the result on the first die and the second number in each ordered pair be the result on the second die.

The sample space contains 36 ordered pairs:

		White Die					
		1	**2**	**3**	**4**	**5**	**6**
	1	(1, 1)	(2, 1)	(3, 1)	(4, 1)	(5, 1)	(6, 1)
	2	(1, 2)	(2, 2)	(3, 2)	(4, 2)	(5, 2)	(6, 2)
Red	**3**	(1, 3)	(2, 3)	(3, 3)	(4, 3)	(5, 3)	(6, 3)
Die	**4**	(1, 4)	(2, 4)	(3, 4)	(4, 4)	(5, 4)	(6, 4)
	5	(1, 5)	(2, 5)	(3, 5)	(4, 5)	(5, 5)	(6, 5)
	6	(1, 6)	(2, 6)	(3, 6)	(4, 6)	(5, 6)	(6, 6)

Self Check 1

Each of the letters in the word *truck* is written on a card and placed in a jar. Then one card is drawn from the jar. List the sample space for the experiment.

Now Try ➔ Problem 15

OBJECTIVE 2 Identify an event of an experiment.

An **event** is one or more outcomes of an experiment. Events are subsets of the sample space and are usually labeled with the capital letter *E*. For the experiment of rolling a die, some events are

- Getting an even number: $E = \{2, 4, 6\}$
- Getting a number greater than 2: $E = \{3, 4, 5, 6\}$

For the experiment of tossing a coin twice, some events are

- Getting at least one heads: $E = \{(H, H), (H, T), (T, H)\}$
- Getting no tails: $E = \{(H, H)\}$

EXAMPLE 2   What is the event "getting a prime number" for the experiment of rolling a die once?

Strategy We will consider which of the numbers 1, 2, 3, 4, 5, and 6 are prime numbers.

WHY The only possible outcomes when rolling a die once are the numbers 1 through 6.

Solution Recall that a prime number is a whole number greater than 1 that has only 1 and itself as factors. Thus:

$$E = \{2, 3, 5\} \quad \text{1, 4, and 6 are not prime numbers.}$$

Self Check 2

What is the event "getting at least one tails" for the experiment of tossing a coin twice?

Now Try ➲ Problem 19

OBJECTIVE 3 Find the probability of an event.

Let's again consider the experiment of tossing a coin twice. We have seen that the sample space contains four ordered pairs:

{(H, H), (H, T), (T, H), (T, T)}

We have also seen that the event E "getting at least one heads" contains three outcomes:

$E = \{(H, H), (H, T), (T, H)\}$

Because the outcome of getting at least one heads can occur in 3 ways out of a total of 4 possible outcomes, we say that the probability of event E is $\frac{3}{4}$. In symbols, we write

$$P(E) = P(\text{at least one heads}) = \frac{3}{4} \quad \text{Read P(E) as "the probability of event E."}$$

In general, the **probability** of an event is a measure of the likelihood (or chance) of it occurring. We define the probability of an event as follows.

> **Probability of an Event**
>
> If a sample space of an experiment has n distinct and equally likely outcomes, and E is an event that occurs in s favorable ways, the **probability of event E** is
>
> $$P(E) = \frac{\text{number of favorable outcomes}}{\text{number of possible outcomes}} = \frac{s}{n}$$

Events that have a less than even chance of occurring are classified as **unlikely**. Events that have a more than even chance of occurring are classified as **likely**. These facts are illustrated graphically on the probability graph below.

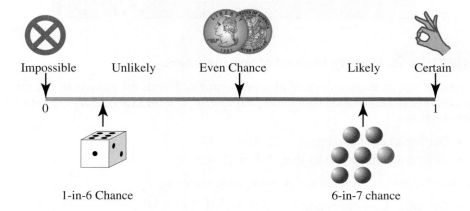

1-in-6 Chance   6-in-7 chance

Probability is always a number from 0 to 1

Because $0 \leq s \leq n$, it follows that $0 \leq \dfrac{s}{n} \leq 1$. This implies that all probabilities have values from 0 to 1, including 0 and 1. An event that cannot happen has a probability of 0. An event that is certain to happen has a probability of 1.

EXAMPLE 3 Find the probability of drawing an ace from a standard card deck.

Strategy First, we will determine how many ways the required event (drawing an ace) can occur. Then, we will determine how many equally likely outcomes there are when drawing one card from a standard deck.

WHY The probability of the event is the ratio (fraction) of those two numbers.

Solution
Since the sample space is 52 cards, $n = 52$. We let E be the set of outcomes that give an ace:

   $E = \{$ace of hearts, ace of diamonds, ace of spades, ace of clubs$\}$

Since there are 4 ways to draw an ace, $s = 4$. The probability of drawing an ace is the ratio of the number of favorable outcomes, s, to the number of possible outcomes, n.

$$P(\text{ace}) = \frac{s}{n} = \frac{4}{52} = \frac{1}{13} \quad \text{Factor 52 as } 4 \cdot 13 \text{ and simplify the fraction: } \dfrac{\overset{1}{\cancel{4}}}{\underset{1}{\cancel{4} \cdot 13}}.$$

The probability of drawing an ace from a standard card deck is $\frac{1}{13}$.

Self Check 3

Find the probability of drawing a diamond from a standard card deck.

Now Try ➲ Problem 23

EXAMPLE 4 Find the probability of the event "rolling a sum of 7 on one roll of two dice."

Strategy First, we will determine how many ways the required event (rolling a sum of 7) can occur. Then, we will determine how many equally likely outcomes there are when rolling two dice.

WHY The probability of the event is the ratio (fraction) of those two numbers.

Solution
The sample space for this experiment was listed in Example 1. Since there are 36 possible outcomes, $n = 36$.
   We let E be the set of outcomes that give a sum of 7:

   $E = \{(1, 6), (2, 5), (3, 4), (4, 3), (5, 2), (6, 1)\}$

Since there are 6 favorable outcomes, $s = 6$.
   The probability of rolling a sum of 7 is the ratio of the number of favorable outcomes, s, to the number of possible outcomes, n.

$$P(\text{rolling a 7}) = \frac{s}{n} = \frac{6}{36} = \frac{1}{6} \quad \text{Always simplify the fraction, if possible.}$$

The probability of rolling a sum of 7 on one roll of two dice is $\frac{1}{6}$.

Self Check 4

Find the probability of the event "rolling a sum of 4 on one roll of two dice."

Now Try ➲ Problem 27

EXAMPLE 5 **Balloons.** A bag contains 8 red, 4 blue, 2 white, 3 green, and 3 yellow balloons. If a child reaches in to the bag and randomly selects one balloon, what is the probability that it is

a. blue

b. red or yellow

c. neither white nor green

d. not yellow

Strategy First, we will determine how many ways the event can occur. Then we will determine how many equally likely outcomes there are when selecting one balloon from the bag.

WHY The probability of each event is the ratio (fraction) of those two numbers.

Solution

The bag contains a total of $8 + 4 + 2 + 3 + 3 = 20$ balloons. Therefore, there are 20 equally likely outcomes when selecting one balloon from the bag.

a. Since the bag contains 4 blue balloons, there are 4 favorable outcomes. To find the probability, we compare the favorable to possible outcomes.

$$P(\text{blue}) = \frac{\textbf{number of favorable outcomes}}{\textbf{total number of possible outcomes}} = \frac{4}{20} = \frac{\overset{1}{\cancel{4}}}{\cancel{4} \cdot 5} = \frac{1}{5} \quad \text{Simplify the fraction.}$$

The probability of selecting a blue balloon is $\frac{1}{5}$.

b. Since the bag contains 8 red and 3 yellow balloons, there are $8 + 3 = 11$ favorable outcomes. To find the probability, we compare the favorable to possible outcomes.

$$P(\text{red or yellow}) = \frac{\textbf{number of favorable outcomes}}{\textbf{total number of possible outcomes}} = \frac{11}{20} \quad \text{The fraction does not simplify.}$$

The probability of selecting a red or yellow balloon is $\frac{11}{20}$.

c. If the balloon selected is to be neither white nor green, it must be red, blue, or yellow. Since there is a total of $8 + 4 + 3 = 15$ balloons with those colors, there are 15 favorable outcomes. To find the probability, we compare the favorable to possible outcomes. Remember to simplify the resulting fraction.

$$P(\text{neither white nor green}) = \frac{\textbf{number of favorable outcomes}}{\textbf{total number of possible outcomes}} = \frac{15}{20} = \frac{3 \cdot \overset{1}{\cancel{5}}}{4 \cdot \cancel{5}} = \frac{3}{4}$$

The probability of selecting neither a white nor green balloon is $\frac{3}{4}$.

d. Since the bag contains 20 balloons, and 3 are yellow, there are $20 - 3 = 17$ balloons that are not yellow. To find the probability, we compare the favorable to possible outcomes.

$$P(\text{not yellow}) = \frac{\textbf{number of favorable outcomes}}{\textbf{total number of possible outcomes}} = \frac{17}{20} \quad \text{The fraction does not simplify.}$$

The probability of selecting a balloon that is not yellow is $\frac{17}{20}$.

Self Check 5

Refer to Example 5. If a child reaches into the bag and randomly selects one balloon, what is the probability that it is

a. white

b. not white

c. neither red nor yellow

d. blue or red

Now Try ➲ Problem 59

SECTION 7.3   STUDY SET

VOCABULARY

Fill in the blanks.

1. An _____ is any process for which the result is uncertain.
2. The results of an experiment are called _____.
3. For any experiment, the set of all possible outcomes is called a _____ _____.
4. An _____ is one or more outcomes of an experiment.
5. _____ outcomes of an experiment satisfy the requirements of that particular event.
6. The _____ of an event is a measure of the likelihood (or chance) of its occurring.

CONCEPTS

Fill in the blanks.

7. If a sample space of an experiment has n distinct and equally likely outcomes, and E is an event that occurs in s favorable ways, the probability of event E is $P(E) = $ _____ .
8. **a.** The probability of an event is always a number from _____ to _____ , including _____ and _____ .
   b. If an event cannot happen, its probability is _____ .
   c. If an event is certain to happen, its probability is _____ .
9. How many outcomes are there when rolling a die? List them.
10. The chart below organizes all of the possible sums from 2 through 12 when tossing two dice. Draw the correct number of dots on the dice of the missing outcomes.

11. Some of the cards from the suit of diamonds shown below are not shown. Draw the missing cards.

12. In a standard deck of playing cards,
   a. How many cards are there?
   b. How many red cards are there?
   c. How many clubs are there?
   d. How many face cards are there in each suit?

NOTATION

Fill in the blanks.

13. When writing a sample space, we list all possible outcomes within _____ { } .
14. We read $P(E)$ as "the _____ of _____ E."

GUIDED PRACTICE

List the sample space of each experiment. **See Example 1.**

15. Rolling a die and tossing a coin
16. Tossing three coins
17. Selecting a letter of the alphabet that is a vowel
18. Guessing a one-digit number

List each event. **See Example 2.**

19. What is the event "getting an even number" for the experiment of rolling a die once?
20. What is the event "getting a red face card" for the experiment of drawing a card from a standard deck?
21. What is the event "getting at least one heads" for the experiment of tossing a coin twice?
22. What is the event "getting a tails and an odd number" for the experiment of tossing a coin and then rolling a die?

Find the probability of each event. **See Example 3.**

23. Drawing a black card on one draw from a standard card deck

24. Drawing a diamond on one draw from a standard card deck

25. Drawing a face card on one draw from a standard card deck

26. Drawing a jack or a king on one draw from a standard card deck

Two dice are rolled. Find the probability of each event. **See Example 4.**

27. Getting a sum of 6

28. Getting a sum of 10

29. Getting a sum of 13

30. Getting a sum of less than 13

A basket contains 40 red poker chips, 20 blue poker chips, and 30 black poker chips. One chip is chosen at random. Find each probability.

31. The chip chosen is not red.

32. The chip chosen is neither red nor black.

33. The chip chosen is not blue.

34. The chip chosen is red or blue.

TRY IT YOURSELF

A die is rolled once. Find the probability of each event.

35. Rolling a 2

36. Rolling a number greater than 4

37. Rolling a number larger than 1 but less than 6

38. Rolling a number that is an odd number

Balls numbered from 1 to 42 are placed in a jar and stirred. If one is drawn at random, find the probability of each result.

39. The number is less than 20.

40. The number is less than 50.

41. The number is a prime number.

42. The number is less than 10 or greater than 40.

Find the probability of each event.

43. Drawing a red egg from a basket containing 5 red eggs and 7 blue eggs

44. Drawing an orange cube from a bowl containing 5 orange cubes and 1 beige cube

45. Obtaining a tails and an even number when a coin is flipped and then a die is rolled

46. Obtaining a heads and a number greater than 2 when a coin is flipped and then a die is rolled

If the spinner shown below is spun, find the probability of each event. Assume that the spinner never stops on a line.

47. The spinner stops on red.

48. The spinner stops on green.

49. The spinner stops on brown.

50. The spinner stops on yellow.

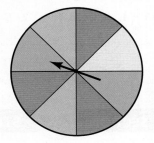

LOOK ALIKES

51. a. When a die is tossed once, what is the probability of rolling a 1? What is the probability of *not* rolling a 1?

   b. Add your results from part a.

52 a. When a single card is chosen from a standard deck of playing cards, what is the probability of drawing the ace of spades? What is the probability of *not* drawing the ace of spades?

   b. Add your results from part a.

53. a. When two dice are tossed, what is the probability of getting a sum of 2? What is the probability of getting a sum that is *not* 2?

   b. Add your results from part a.

54. a. When a die is tossed, what is the probability of rolling a number between 1 and 6, inclusive? What is the probability of rolling a number that is not between 1 and 6, inclusive?

   b. Add your results from part a.

APPLICATIONS

55. Aeronautics. An aircraft has 4 engines. Assume that an engine has an equal chance of surviving or failing during a test.

   a. Find the sample space. Let S represent an engine that survives and F represent an engine that fails. A tree diagram is helpful.

   b. Find the probability that all engines will survive the test.

   c. Find the probability that exactly 1 engine will survive.

   d. Find the probability that exactly 2 engines will survive.

   e. Find the probability that exactly 3 engines will survive.

   f. Find the probability that no engines will survive.

56. Find the sum of the probabilities in problem 55. What do you discover?

57. Opinion surveys. A survey of 282 people is taken to determine the opinions of doctors, teachers, and lawyers on a proposed piece of legislation, with the results shown in the table. A person is chosen at random from those surveyed. Refer to the table to find the probability of each event.

a. The person favors the legislation.

b. A doctor opposes the legislation.

c. A person who opposes the legislation is a lawyer.

d. The person has no opinion.

	Number that favor	Number that oppose	Number with no opinion	Total
Doctors	70	32	17	119
Teachers	83	24	10	117
Lawyers	23	15	8	46
Total	176	71	35	282

58. TV game shows. On *The Price is Right*, contestants can take up to two spins of a large wheel to get as close to $1.00 as possible, without going over. After the first spin, the player can choose to stay with what he/she landed on or spin again. The wheel has the following values (in cents) displayed on it: 5, 100, 15, 80, 35, 60, 20, 40, 75, 55, 95, 50, 85, 30, 65, 10, 45, 70, 25, 90. What is the probability that the wheel lands on 100 on a contestant's first spin?

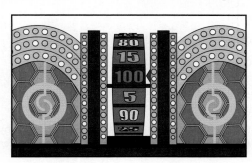

59. TV game shows. On the *Wheel of Fortune*, contestants spin a roulette-style wheel to begin each turn. See the illustration in the next column. What is the probability of landing on "Bankrupt"?

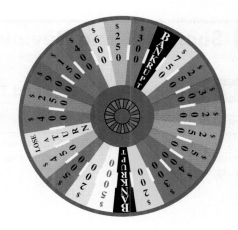

60. Testing.

a. Suppose you don't know the answer to a true false question on a test. If you guess, what is the probability that your answer is correct?

b. Suppose you do not know the answer to the first multiple-choice question on a test. If you randomly select choice D, as shown below, what is the probability that your answer is correct?

—	1	[A]	[B]	[C]	[≣]	[E]
—	2	[A]	[B]	[C]	[D]	[E]
—	3	[A]	[B]	[C]	[D]	[E]

WRITING

61. Explain why all probability values range from 0 to 1.

62. Explain the concept of probability.

63. a. Give an example of a likely event.

  b. Give an example of an unlikely event.

64. a. Give an example of an impossible event.

  b. Give an example of a certain event.

REVIEW LOOK ALIKES

65. $(-8 - 5) - 3$ **66.** $-8 - (5 - 3)$

67. $(-8 - 5)(-3)$ **68.** $-(8 - 5) - 3$

7 Summary and Review

SECTION 7.1 ▶ **Reading Graphs and Tables**

DEFINITIONS AND CONCEPTS	EXAMPLES
To read a **table** and locate a specific fact in it, we find the *intersection* of the correct row and column that contains the desired information.	**Salary schedules.** Find the annual salary for a teacher with a master's degree plus 15 additional units of study who is beginning her 4th year of teaching.

Teacher Salary Schedule

Step	BA	BA+15	BA+30	BA+45	MA	MA+15	MA+30
1	37,295	38,362	39,416	40,480	41,556	42,612	43,669
2	38,504	39,581	40,652	41,728	42,812	43,879	44,952
3	39,716	40,802	41,885	42,973	44,066	45,147	46,234
4	40,926	42,021	43,120	44,220	45,321	46,417	47,514
5	42,135	43,240	44,356	45,465	46,577	47,682	48,795
6	44,458	45,567	46,683	47,782	48,897	50,010	51,113
7	46,780	47,891	49,003	50,115	51,226	52,330	53,438

The annual salary is $46,417. It can be found by looking on the fourth row (labeled Step 4) in the 6th column (labeled MA + 15).

A **bar graph** presents data using vertical or horizontal bars. A **horizontal axis** and **vertical axis** serve to frame the graph, and they are scaled in units such as years, dollars, minutes, pounds, and percent.	**Cancer deaths.** Refer to the graph. Estimate how many more deaths were caused by lung cancer than by colon cancer in the United States in 2016.

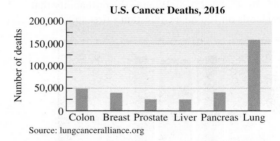

From the graph, we see that there were about 160,000 deaths caused by lung cancer and about 50,000 deaths from colon cancer. To find the difference, we subtract:

$$160,000 - 50,000 = 110,000$$

There were about 110,000 more deaths caused by lung cancer than deaths caused by colon cancer in the United States in 2016.

To compare sets of related data, groups of two (or three) bars can be shown. For **double-bar** or **triple-bar graphs**, a key is used to explain the meaning of each type of bar in a group.	**Seat belts.** Refer to the graph. How did the percent of male high school students that rarely or never wore seat belts change from 2001 to 2015?

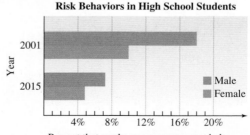

Source: *The World Almanac and Book of Facts*, 2003, 2016

From the graph, we see that in 2001 about 18% of male high school students rarely or never wore seat belts. By 2015, the percent was about 7%, a decrease of 18% − 7%, or 11%.

A **pictograph** is like a bar graph, but the bars are made from pictures or symbols. A **key** tells the meaning (or value) of each symbol.

Medical schools. Refer to the pictograph below. In 2017, how many students were enrolled in California medical schools?

Total Medical School Enrollment by State, 2017

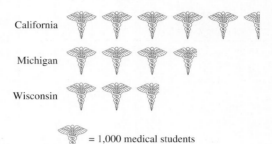

= 1,000 medical students

Source: Association of American Medical Colleges

The California row contains five complete symbols and a large part of another. This means that there were 5 · 1,000, or 5,000 medical students, plus approximately 500 more. In 2017, about 5,500 students were enrolled in California medical schools.

In a **circle graph**, regions called *sectors* (they look like slices of pizza) are used to show what part of the whole each quantity represents.

Checking email. The circle graph to the right shows the results of a survey of adults who were asked how many personal email addresses they regularly check. What percent of the adults surveyed check 4 or more email addresses regularly?

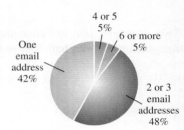

Source: Ipsos for Habeas

We add the percent of the responses for 4 or 5 email addresses and the percent of the responses for 6 or more email addresses:

$$5\% + 5\% = 10\%$$

Thus, 10% of the adults surveyed check 4 or more email addresses regularly.

Use the survey results to predict the number of adults in a group of 5,000 that would check only one email address regularly.

In the survey, 42% said they check only one email address. We need to find:

What number	is	42%	of	5,000?
↓	↓	↓	↓	↓
x	=	42%	·	5,000

$x = 0.42 \cdot 5,000$ Write 42% as a decimal.

$x = 2,100$ Do the multiplication.

According to the survey, about 2,100 of the 5,000 adults would check only one email address regularly.

A **line graph** is used to show how quantities change with time. From such a graph, we can determine when a quantity is increasing and when it is decreasing.

Snowboarding.   The line graph below shows the number of people who participated in snowboarding in the United States for the years 1997–2016.

Estimated Number of Active U.S. Snowboarders

Source: NSAA Demographic Study

When did the popularity of snowboarding seem to peak?

The year with the highest participation was 2010.

Between which two years was there the greatest decrease in the number of snowboarding participants?

The line segment with the greatest "fall" as we read left to right is the segment connecting the data points for the years 2011 and 2012. Thus, the greatest decrease in the number of snowboarding participants occurred between 2011 and 2012.

Two quantities that are changing with time can be compared by **drawing both lines on the same graph**.

Skateboarding.   Refer to the line graphs below that show the results of a skateboarding race.

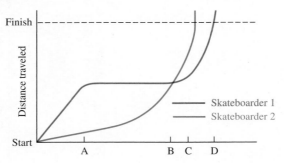

Observations:

- Since the red graph is well above the blue graph at time A, skateboarder 1 was well ahead of skateboarder 2 at that stage of the race.
- Since the red graph is horizontal from time A to time B, skateboarder 1 had stopped.
- Since the blue graph crosses the red graph at time B, at that instant, the skateboarders are tied for the lead.
- Since the blue graph crosses the dashed finish line at time C, which is sooner than time D, skateboarder 2 won the race.

A **histogram** is a bar graph with these features:

1. The bars of the histogram touch.
2. Data values never fall at the edge of a bar.
3. The widths of the bars are equal and represent a range of values.

Sleep.   A group of parents of junior high students were surveyed and asked to estimate the number of hours that their children slept each night. The results are displayed in the histogram on the next page. How many children sleep 6 to 9 hours a night?

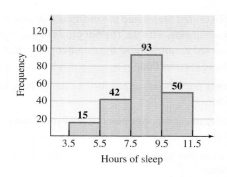

The bar with edges 5.5 and 7.5 corresponds to the 6- to 7-hour range. The height of that bar indicates that 42 children sleep 6 to 7 hours. The bar with edges 7.5 and 9.5 corresponds to the 8- to 9-hour range. The height of that bar indicates that 93 children sleep 8 to 9 hours. The total number of children sleeping 6 to 9 hours is found using addition:

$$42 + 93 = 135$$

135 of the junior high children sleep 6 to 9 hours a night.

A **frequency polygon** is a special line graph formed from a histogram by joining the center points at the top of each bar. On the horizontal axis, we write the coordinate of the middle value of each **class interval**. Then we erase the bars.

Frequency polygon

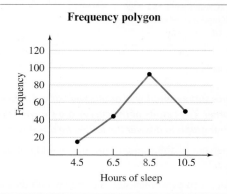

REVIEW EXERCISES

Refer to the table to the right to answer the following questions.

Determining the Windchill Temperature

1. **Windchill temperatures.**
   a. Find the windchill temperature on a 10°F day when a 15-mph wind is blowing.
   b. Find the windchill temperature on a −15°F day when a 30-mph wind is blowing.

2. **Wind speeds.**
   a. The windchill temperature is −25°F, and the actual outdoor temperature is 15°F. How fast is the wind blowing?
   b. The windchill temperature is −38°F, and the actual outdoor temperature is −5°F. How fast is the wind blowing?

Wind speed	Actual temperature							
	20°F	15°F	10°F	5°F	0°F	−5°F	−10°F	−15°F
5 mph	16°	12°	7°	0°	−5°	−10°	−15°	−21°
10 mph	3°	−3°	−9°	−15°	−22°	−27°	−34°	−40°
15 mph	−5°	−11°	−18°	−25°	−31°	−38°	−45°	−51°
20 mph	−10°	−17°	−24°	−31°	−39°	−46°	−53°	−60°
25 mph	−15°	−22°	−29°	−36°	−44°	−51°	−59°	−66°
30 mph	−18°	−25°	−33°	−41°	−49°	−56°	−64°	−71°
35 mph	−20°	−27°	−35°	−43°	−52°	−58°	−67°	−74°

As of 2016, the United States had the most nuclear power plants in operation worldwide, with 99. The bar graph below shows the rest of the top ten countries and the number of nuclear power plants they have in operation.

3. How many nuclear power plants does Korea have in operation?

4. How many nuclear power plants does France have in operation?

5. Which countries have fewer than 20 nuclear power plants in operation?

6. How many more nuclear power plants in operation does Japan have than Canada?

Number of Nuclear Power Plants in Operation, 2016

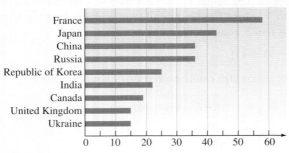

Source: euronuclear.org

In a workplace survey, employed adults were asked if they would date a coworker. The results of the survey are shown below. Use the double-bar graph to answer the following questions.

7. What percent of the women said they would not date a coworker?

8. Did more men or women say that they would date a coworker? What percent more?

9. When asked, were more men or more women unsure if they would date a coworker?

10. Which of the three responses to the survey was given by approximately the same percent of men and women?

Responses to the Survey: Would you date a coworker?

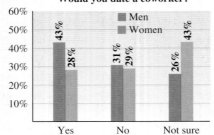

Source: Spherion Workplace Survey

Refer to the pictograph in the next column to answer the following questions.

11. How many animals are there at the San Diego Zoo?

12. Which of the zoos listed has the most animals? How many?

13. How many animals would have to be added to the Phoenix Zoo for it to have the same number as the San Diego Zoo?

14. Find the total number of animals in all three zoos.

America's Best Zoos Number of Animals

Source: zoo.findthebest.com

Refer to the circle graph below to answer the following questions.

15. What element makes up the largest percent of the body weight of a human?

16. Elements other than oxygen, carbon, hydrogen, and nitrogen account for what percent of the weight of a human body?

17. Hydrogen accounts for how much of the body weight of a 135-pound woman?

18. Oxygen and carbon account for how much of the body weight of a 200-pound man?

Elements in the Human Body (by weight)

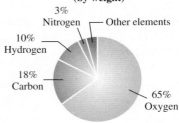

Source: General Chemistry Online

Refer to the line graph on the next page to answer the following questions.

19. How many eggs were produced in Nebraska in 2003?

20. How many eggs were produced in North Carolina in 2008?

21. In what year was the egg production of Nebraska equal to that of North Carolina? How many eggs?

22. What was the total egg production of Nebraska and North Carolina in 2005?

23. Between what two years did the egg production in North Carolina increase dramatically?

24. Between what two years did the egg production in Nebraska decrease dramatically?

25. How many more eggs did North Carolina produce in 2009 compared to Nebraska?

26. How many more eggs did North Carolina produce in 2015 compared to Nebraska?

Total Egg Production

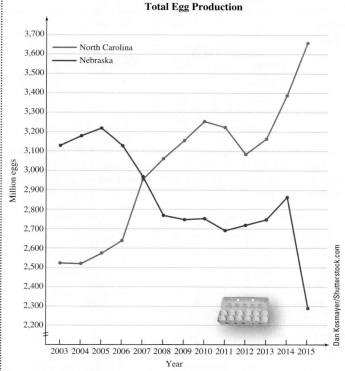

Source: U.S. Department of Agriculture and Agricultural Statistics Service

A survey of the weekly television viewing habits of 320 households produced the histogram in the next column. Use the graph to answer the following questions.

27. How many households watch between 1 and 5 hours of TV each week?

28. How many households watch between 6 and 15 hours of TV each week?

29. How many households watch 11 hours or more each week?

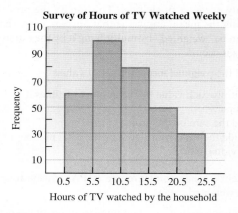

30. Create a frequency polygon from the histogram shown above.

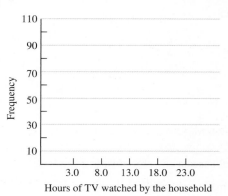

SECTION 7.2 ▶ **Mean, Median, and Mode**

DEFINITIONS AND CONCEPTS	EXAMPLES
It is often useful to find one number to represent the "center" of all the numbers in a set of data. There are three measures of **central tendency**: mean, median, mode. The **mean** of a set of values is given by the formula $$\text{Mean} = \frac{\text{sum of the values}}{\text{number of values}}$$	Find the mean of the following set of values: 6 8 3 5 9 8 10 7 8 5 To find the mean, we divide the sum of the values by the number of values, which is 10. $$\frac{6 + 8 + 3 + 5 + 9 + 8 + 10 + 7 + 8 + 5}{10} = \frac{69}{10}$$ $$= 6.9$$ Thus, 6.9 is the mean.

When a value in a set appears more than once, that value has a greater "influence" on the mean than another value that only occurs a single time. To simplify the process of finding the mean, any value that appears more than once can be "weighted" by multiplying it by the number of times it occurs.

To find the **weighted mean** of a set of values:

1. Multiply each value by the number of times it occurs.
2. Find the sum of the products from step 1.
3. Divide the sum from step 2 by the total number of individual values.

A student's **grade point average (GPA)** can be found using a weighted mean.

Some schools assign a certain number of **credit hours** to a course while others assign a certain number of **units**.

GPAs. Find the semester grade point average for a student that received the following grades. (The point values are A = 4, B = 3, C = 2, D = 1, and F = 0.)

Course	Grade	Credits
Algebra	A	5
History	C	3
Art	D	4

Multiply the number of credits for each course by the point value of the grade received. Add the results (as shown highlighted in blue) to get the total number of grade points. To find the total number of credits, add as shown highlighted in red.

Course	Grade	Credits	Weighted grade points	
Algebra	A	5	$4 \cdot 5 \rightarrow$	20
History	C	3	$2 \cdot 3 \rightarrow$	6
Art	D	+ 4	$1 \cdot 4 \rightarrow +$	4
Totals		12		30

To find the GPA, we divide.

$$\text{GPA} = \frac{30}{12} \begin{array}{l} \leftarrow \text{ The total number of grade points} \\ \leftarrow \text{ The total number of credits} \end{array}$$

$$= 2.5 \quad \text{Do the division.}$$

The student's semester GPA is 2.5.

To find the **median** of a set of values:

1. Arrange the values in increasing order.

2. If there is an odd number of values, the median is the middle value.

3. If there is an even number of values, the median is the mean (average) of the middle two values.

To find the median of

6 8 3 5 9 8 10 7 8 5

arrange them in increasing order:

Smallest   Largest

3 5 5 6 7 8 8 8 9 10   There are 10 values.

           ↑

     Middle two values

Since there is an even number of values, the median is the mean (average) of the two middle values:

$$\frac{7 + 8}{2} = \frac{15}{2} = 7.5$$

Thus, 7.5 is the median.

The **mode** of a set of values is the single value that occurs most often.

To find the mode of

6 8 3 5 9 8 10 7 8 5

we find the value that occurs most often.

6 **8** 3 5 9 **8** 10 7 **8** 5

3 times

Since 8 occurs the most times, it is the mode.

The **range** of a set of values is the difference between the largest value and the smallest value.

The range of the data listed above is:

range = 10 − 3 = 7   Subtract the smallest value, 3, from the largest value, 10.

When a collection of values has two modes, it is called **bimodal**.	The collection of values 1 2 **3 3** 4 5 **6 6** 7 8 has two modes: 3 and 6.

REVIEW EXERCISES

31. Grades. Jose worked hard this semester, earning grades of 87, 92, 97, 100, 100, 98, 90, and 98. If he needs a 95 average to earn an A in the class, did he make it?

32. Grade summaries. The students in a mathematics class had final averages of 43, 83, 40, 100, 40, 36, 75, 39, and 100. When asked how well her students did, their teacher answered, "43 was typical." What measure was the teacher using: mean, median, or mode?

33. Pretzel packaging. Samples of SnacPak pretzels were weighed to find out whether the package claim "Net weight 1.2 ounces" was accurate. The tally appears in the table. Find the mode of the weights.

Weights of SnacPak Pretzels

Ounces	Number
0.9	1
1.0	6
1.1	18
1.2	23
1.3	2

Africa Studio/Shutterstock.com

34. Find the mean weight and the range of the weights of the samples in Exercise 33.

35. Blood samples. A medical laboratory technician examined a blood sample under a microscope and measured the sizes (in microns) of the white blood cells. The data are listed below. Find the mean, median, mode, and range.

7.8   6.9   7.9   6.7   6.8   8.0   7.2   6.9   7.5

36. Summer reading. A paperback version of the classic *Gone with the Wind* is 960 pages long. If a student wants to read the entire book during the month of June, how many pages must she average per day?

37. Walk-a-thons. Use the data in the table to find the mean (average) donation to a charity walk-a-thon.

Donation amount	$5	$10	$20	$50	$100
Number received	20	65	25	5	10

38. GPAs. Find the semester grade point average for a student who received the grades shown below. Round to the nearest hundredth. (Assume the following standard point values for the letter grades: A = 4, B = 3, C = 2, D = 1, and F = 0.)

Course	Grade	Credits
Chemistry	A	5
Sociology	C	3
Economics	D	4
Archery	A	1

SECTION 7.3 ▶ **Probability**

DEFINITIONS AND CONCEPTS	EXAMPLES
An **experiment** is any process for which the result is uncertain. The results of an experiment are called **outcomes.**	Some examples of experiments are ■ Flipping a coin ■ Rolling dice ■ Drawing a card from a standard deck ■ Selecting a marble from a bag of colored marbles
For any experiment, the set of all possible outcomes is called a **sample space**.	The sample space for flipping a coin twice is a set of four ordered pairs: $\{(H,H), (H,T), (T,H), (T,T)\}$

An **event** is one or more outcomes of an experiment. Events are subsets of the sample space and are usually labeled with the capital letter E.	For the experiment of rolling a die, some events are ■ Getting an even number: $E = \{2, 4, 6\}$ ■ Getting a number greater than 2: $E = \{3, 4, 5, 6\}$
When performing an experiment, we often speak of **favorable outcomes**. These are the outcomes of an experiment that satisfy the requirements of a particular event.	For the experiment of rolling a die once and the event "getting a prime number," the outcomes that are considered *favorable* to the event are 2, 3, and 5.
The **probability** of an event is a measure of the likelihood (or chance) of its occurring. If a sample space of an experiment has n distinct and equally likely outcomes, and E is an event that occurs in s favorable ways, the **probability of event E** is $P(E) = \dfrac{number\ of\ favorable\ outcomes}{number\ of\ possible\ outcomes} = \dfrac{s}{n}$	To find the probability of drawing a queen from a standard card deck, we note that ■ Since the sample space is 52 cards, $n = 52$. ■ Since there are 4 ways to draw a queen, $s = 4$. $P(queen) = \dfrac{s}{n} = \dfrac{4}{52} = \dfrac{1}{13}$ Simplify the fraction.
An event that cannot happen has a probability of 0. An event that is certain to happen has a probability of 1. All other events have probabilities between 0 and 1.	The probability of the event "rolling a 13" on one roll of two dice is $P(E) = 0$. The probability of the event "rolling a number less than 7" on one roll of a die is $P(E) = 1$.

REVIEW EXERCISES

In Exercises 39–41, assume that a dart is randomly thrown at the colored chart.

39. What is the probability that the dart lands in a blue area?

40. What is the probability that the dart lands in an even-numbered area?

41. What is the probability that the dart lands in an area whose number is greater than 2?

1	2	3	4
5	6	7	8
9	10	11	12
13	14	15	16

42. Find the probability of rolling a sum of 11 on one roll of two dice?

43. Find the probability of getting a 7 on one roll of a die.

44. Find the probability of drawing a 10 from a standard deck of cards.

45. Find the probability of getting a number that is not prime for the experiment of rolling a die once.

46. Balls numbered 1 through 50 are placed in a jar and stirred. One ball is drawn at random. Find the probability that the number drawn is less than 55.

Fill in the blanks.

1. **a.** A horizontal or vertical line used for reference in a bar graph is called an _____.

   b. The _____ (average) of a set of values is the sum of the values divided by the number of values in the set.

   c. The _____ of a set of values written in increasing order is the middle value.

   d. The _____ of a set of values is the single value that occurs most often.

   e. The mean, median, and mode are three measures of _____ tendency.

2. **Workouts.** Refer to the table below to answer the following questions.

   a. How many calories will a 155-pound person burn if she runs for one hour at a rate of 5 mph?

   b. In one hour, how many more calories will a 190-pound person burn if he runs at a rate of 7 mph instead of 6 mph?

   c. At what rate does a 130-pound person have to run for one hour to burn approximately 800 calories?

Number of Calories Burned While Running for One Hour

Running Speed (mph)	Body Weight		
	130 lb	155 lb	190 lb
5	472	563	690
6	590	704	863
7	679	809	992
8	797	950	1,165
9	885	1,056	1,294

Source: nutristrategy.com

3. **Moving.** Refer to the bar graph below to answer the following questions.

   a. Which piece of furniture shown in the graph requires the greatest number of feet of bubble wrap? How much?

   b. How many more feet of bubble wrap is needed to wrap a desk than a coffee table?

   c. How many feet of bubble wrap is needed to cover a bedroom set that has a headboard, a dresser, and two end tables?

Amount of Bubble Wrap Needed to Wrap Pieces of Furniture When Moving

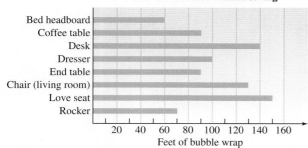

Source: transitsystems.com

4. **Cancer survival rates.** Refer to the graph below to answer the following questions.

   a. What was the survival rate (in percent) from breast cancer in 1976?

   b. By how many percent did the cancer survival rate for breast cancer increase by 2012?

   c. Which type of cancer shown in the graph has the lowest survival rate?

   d. Which type of cancer has had the greatest increase in survival rate from 1976 to 2012? How much of an increase?

Five-Year Survival Rates

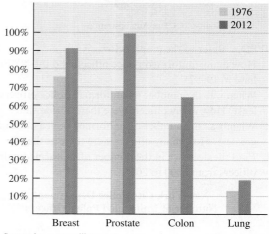

Source: lungcanceralliance.org

5. **Energy drinks.** Refer to the pictograph below to answer the following questions.

   a. How many grams of sugar are there in 12 ounces of Big Red?

   b. For a 12-ounce serving, how many more grams of sugar are there in Monster Energy Drink than in Starbucks Tall Caffè Mocha?

Sugar Content in Energy Drinks and Coffee (12-ounce serving)

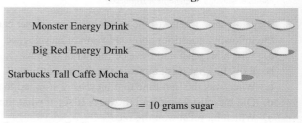

Source: energyfiend.com

6. **Fires.** Refer to the graph below to answer the following questions.
   a. In 2015, what percent of the fires in the United States were vehicle fires?
   b. In 2015, there were a total of 1,345,500 fires in the United States. How many were structure fires?

Where Fires Occurred, 2015

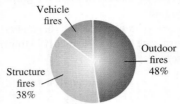

Source: U.S. Fire Administration

7. **NYPD.** Refer to the graph below.
   a. How many uniformed police officers did the NYPD have in 1999?
   b. When was the number of uniformed police officers the least? How many officers were there at that time?
   c. When was the number of uniformed police officers the greatest? How many officers were there at that time?
   d. Find the decrease in the number of uniformed police officers from 2000 to 2016.

New York City Police Department Number of Uniformed Police Officers

Source: Mayor's Management Reports

8. **Bicycle race.** Refer to the graph below to answer each of the following questions about a two-man bicycle race.
   a. Which bicyclist had traveled farther at time A?
   b. Explain what was happening in the race at time B.
   c. When was the first time that bicyclist 2 stopped to rest?
   d. Did bicyclist 2 ever lead the race? If so, at what time?
   e. Which bicyclist won the race?

Ten-Mile Bicycle Race

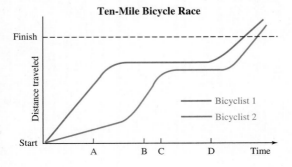

9. **Commuting time.** A school district collected data on the number of minutes it took its employees to drive to work in the morning. The results are presented in the histogram below.
   a. How many employees have a commute time that is in the 7-to-10-minute range?
   b. How many employees have a commute time that is less than 10 minutes?
   c. How many employees have a commute that takes 15 minutes or more each day?

School District Employees' Commute

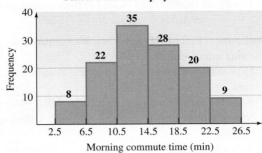

10. **Volunteer service.** The number of hours served last month by each of the volunteers at a homeless shelter are listed below:

    4 6 8 2 8 10 11 9 5 12 5 18 7 5 1 9

    a. Find the mean (average) of the hours of volunteer service.
    b. Find the median of the hours of volunteer service.
    c. Find the mode of the hours of volunteer service.
    d. Find the range of the number of hours of volunteer service.

11. **Rating movies.** An online DVD rental company allows members to rate movies using a 5-star system. The table below shows a tally of the ratings that a group of college students gave a movie. Find the mean (average) rating of the movie.

Number of Stars	Comments	Tally
★★★★★	Loved it	|||
★★★★	Really liked it	||||
★★★	Liked it	⊮⊩
★★	Didn't like it	⊮⊩	
★	Hated it	||

12. GPAs. Find the semester grade point average for a student who received the following grades. Round to the nearest hundredth.

Course	Grade	Credits
WEIGHT TRAINING	C	1
TRIGONOMETRY	A	3
GOVERNMENT	B	2
PHYSICS	A	4
PHYSICS LAB	D	1

13. Ratings. The eight top-rated cable television programs for Friday, March 4, 2017, are given below. What are the mean, median, mode, and range of the viewer data?

Show/day/time/network	Millions of viewers
Gold Rush, 9 P.M., DISC	4.37
Gold Rush: Dirt Aftershow, 10:30 P.M., DISC	3.29
Gold Rush: The Dirt, 8 P.M., DISC	2.23
Family Guy, 10:30 P.M., ADSM	1.85
Descendants, 8 P.M., DSNY	1.84
SpongeBob, 4:30 P.M., NICK	1.62
Alvinnn!!! and the Chipmunks, 5:30 P.M., NICK	1.54
Alvinnn!!! and the Chipmunks, 5:30 P.M, NICK	1.53

Source: The Nielsen Company

14. Real estate. In December 2016, the median sales price of an existing single-family home in the United States was $233,800. Explain what is meant by the median sales price. (Source: National Association of Realtors)

Find each probability.

15. Rolling a 5 on one roll of a die.

16. Drawing a heart from a standard card deck.

17. Getting a sum of 7 when rolling 2 dice.

18. Tossing 2 heads in 2 tosses of a fair coin.

19. Shade an appropriate number of pie-shaped sections of a circle so that the probability of a spinner stopping on an unshaded section is $\frac{9}{16}$.

20. Fill in the blanks:

  a. The probability of an event that cannot happen is ____.

  b. The probability of an event that is guaranteed to happen is ____.

1. **Automobiles.** In 2016, a total of 72,105,435 cars were produced in the world. Write this number in words and in expanded notation. (Source: worldometers.info) [Section 1.1]

2. Round 59,999 to the nearest hundred. [Section 1.1]

Perform each operation.

3. $\begin{array}{r} 48,908 \\ +\ 5,696 \end{array}$ [Section 1.2]

4. $\begin{array}{r} 8,700 \\ -\ 5,491 \end{array}$ [Section 1.3]

5. $\begin{array}{r} 408 \\ \times\ 67 \end{array}$ [Section 1.4]

6. $87)\overline{2,001}$ [Section 1.5]

7. Explain how to check the following result using addition. [Section 1.3]

$\begin{array}{r} 2,142 \\ -\ 459 \\ \hline 1,683 \end{array}$

8. **Geometry.** Find the perimeter and the area of the rectangle shown below. [Section 1.4]

6 in.

14 in.

9. **The Vietnamese calendar.** An animal represents each Vietnamese lunar year. Recent Years of the Cat are listed below. If the cycle continues, what year will be the next Year of the Cat? [Section 1.6]

1927 1939 1951 1963 1975 1987 1999 2011

10. **a.** Find the factors of 36. [Section 1.7]
    b. Find the prime factorization of 36. [Section 1.7]

11. Write the first ten prime numbers. [Section 1.7]

12. **a.** Find the LCM of 8 and 12. [Section 1.8]
    b. Find the GCF of 8 and 12. [Section 1.8]

Evaluate each expression. [Section 1.9]

13. $15 + 5\left[12 - (2^2 + 4)\right]$

14. $\dfrac{12 + 5 \cdot 3}{3^2 - 2 \cdot 3}$

15. Graph the integers greater than -3 but less than 4. [Section 2.1]

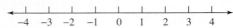

$\quad -4 \quad -3 \quad -2 \quad -1 \quad 0 \quad 1 \quad 2 \quad 3 \quad 4$

16. **a.** Simplify: $-(-6)$ [Section 2.1]
    b. Find the absolute value: $|-5|$
    c. Is the statement $-12 > -10$ true or false?

17. *Perform each operation.*
    a. $-35 + 5$ [Section 2.2]
    b. $-35 - (-5)$ [Section 2.3]

c. $-35(5)$ [Section 2.4]

d. $\dfrac{-35}{-5}$ [Section 2.5]

18. **Planets.** Mercury orbits closer to the sun than does any other planet. Temperatures on Mercury can get as high as 810°F and as low as -290°F. What is the temperature range? [Section 2.3]

Evaluate each expression. [Section 2.6]

19. $\dfrac{(-6)^2 - 1^5}{-4 - 3}$

20. $-3 + 3(-4 - 4 \cdot 2)^2$

21. $-\left|\dfrac{45}{-9} - (-9)\right|$

22. $-10^2 - (-10)^2$

23. Simplify each fraction. [Section 3.1]
    a. $\dfrac{60}{108}$
    b. $\dfrac{24}{16}$

24. Simplify, if possible. [Section 3.1]
    a. $\dfrac{0}{64}$
    b. $\dfrac{27}{0}$

Perform each operation. Simplify, if possible.

25. $\dfrac{4}{5} \cdot \dfrac{2}{7}$ [Section 3.2]

26. $\dfrac{8}{63} \div \dfrac{2}{7}$ [Section 3.3]

27. Subtract $\dfrac{2}{3}$ from $\dfrac{1}{2}$. [Section 3.4]

28. $\dfrac{11}{12} + \dfrac{1}{30}$ [Section 3.4]

29. **Class time.** In a chemistry course, students spend a total of 300 minutes in lab and lecture each week. If $\dfrac{7}{15}$ of the time is spent in lab each week, how many minutes are spent in lecture each week? [Section 3.3]

30. Divide: $2\dfrac{4}{5} \div \left(-2\dfrac{2}{3}\right)$ [Section 3.5]

31. **Tennis.** Find the length of the handle on the tennis racquet shown below. [Section 3.6]

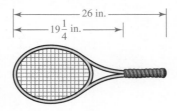

26 in.

$19\dfrac{1}{4}$ in.

32. Evaluate the formula $A = \frac{1}{2}h(a + b)$ for $a = 4\frac{1}{2}$, $b = 5\frac{1}{2}$, and $h = 2\frac{1}{8}$. [Section 3.7]

33. Simplify the complex fraction: $\dfrac{-\dfrac{1}{5}}{\dfrac{8}{15}}$ [Section 3.7]

34. Write $400 + 20 + 8 + \frac{9}{10} + \frac{1}{100}$ as a decimal. [Section 4.1]

35. Checkbooks. Find the total dollar amount of the checks written in the register shown below. [Section 4.2]

DATE		CHECK NUMBER	TRANSACTION DESCRIPTION		✓ T	AMOUNT OF PAYMENT OR DEBIT (•)	
3	17	703	TO:	Albertsons		$ 213	16
			FOR:	Groceries			
3	19	704	TO:	Brian Auto		$1,504	80
			FOR:	Car Repair			
3	19	705	TO:	Nordstrom		$ 89	73
			FOR:	Sweater			
3	21	706	TO:	Girl Scouts		$ 7	50
			FOR:	Cookies			

36. Perform each operation in your head. [Section 4.3]

   a. Multiply: $3.45 \cdot 100$ **b.** Divide: $3.45 \div 10{,}000$

Perform each operation.

37. Subtract: $\begin{array}{r} 760.2 \\ -\ 614.7 \end{array}$ [Section 4.2]

38. Multiply: $(-0.31)(2.4)$ [Section 4.3]

39. Divide: $0.72\overline{)536.4}$ [Section 4.4]

40. Divide: $4\overline{)0.073}$ [Section 4.4]

41. Write $\frac{8}{11}$ as a decimal. [Section 4.5]

42. Evaluate: $15 + \sqrt{16}\left[5^2 - \left(\sqrt{9} + 2\right)\sqrt{4}\right]$ [Section 4.6]

43. Express the phrase "8 feet to 4 yards" as a ratio in simplest form. [Section 5.1]

44. Clothes shopping. As part of a summer clearance, a women's store put turtleneck sweaters on sale, 3 for $35.97. How much will five turtleneck sweaters cost? [Section 5.2]

45. Solve the proportion: $\dfrac{\frac{7}{8}}{\frac{1}{2}} = \dfrac{\frac{1}{4}}{x}$ [Section 5.2]

46. Convert 8 pints to fluid ounces. [Section 5.3]

47. Convert 640 centimeters to meters. [Section 5.4]

48. Convert 67.7°F to degrees Celsius. Round to the nearest tenth. [Section 5.3]

49. Complete the table below. [Section 6.1]

Fraction	Decimal	Percent
		3%
$\frac{9}{4}$		
	0.041	

50. 90 is what percent of 525? Round to the nearest one percent. [Section 6.2]

51. What number is 105% of 23.2? [Section 6.2]

52. 19.2 is $33\frac{1}{3}\%$ of what number? [Section 6.2]

53. Shark Tank. On the TV show *Shark Tank*, entrepreneur-contestants make business presentations to a panel of "shark" investors who then decide whether or not to invest in the contestant's company. One contestant offered the "sharks" a 40% share of his young company for a $90,000 investment from one of them. What would this offer indicate that total value of the contestant's company is? [Section 6.2]

Kathy Hutchins/Shutterstock.com

54. Selling electronics. If the commission on a $1,500 laptop computer is $240, what is the commission rate? [Section 6.3]

55. Tipping. Estimate the 15% tip on a $77.55 dinner bill. [Section 6.4]

56. Remodeling. A homeowner borrows $18,000 to pay for a kitchen remodeling project. The terms of the loan are 9.2% annual simple interest and repayment in 2 years. How much interest will be paid on the loan? [Section 6.5]

57. Loans. $12,600 is loaned at an annual simple interest rate of 18%. Find the total amount that must be repaid at the end of a 90-day period. [Section 6.5]

58. Spinal cord injuries. Refer to the circle graph below. [Section 7.1]

   a. What percent of spinal cord injuries are caused by sports accidents?

   b. If there are approximately 17,000 new cases of spinal cord injury each year, according to the graph, how many of them were caused by vehicle crashes?

Causes of Spinal Cord Injury in the United States

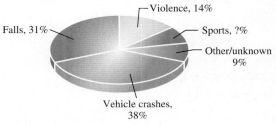

Violence, 14%
Falls, 31%
Sports, ?%
Other/unknown 9%
Vehicle crashes, 38%

Source: National Spinal Cord Injury Statistical Center

59. Avalanches. The bar graph below shows the number of deaths from avalanches in the United States for the winter seasons ending in the years 2003 to 2016. Use the graph to answer the following questions. [Section 7.1]

   a. In which year(s) were there the most deaths from avalanches? How many deaths were there?

   b. Between what two years was there the greatest increase in the number of deaths from avalanches? What was the increase?

   c. Between what two years was there the greatest decrease in the number of deaths from avalanches? What was the decrease?

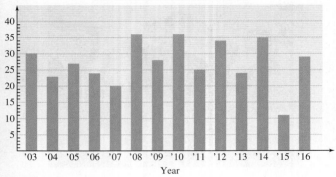

U.S. Annual Avalanche Deaths

Source: Northwest Weather and Avalanche Center

60. Team GPA. The grade point averages of the players on a badminton team are listed below. Find the mean, median, mode, and range of the team's GPAs. [Section 7.2]

   3.04   4.00   2.75   3.23   3.87   2.21

   3.02   2.25   2.98   2.56   3.58   2.75

61. What is the probability of drawing a face card from a standard card deck? [Section 7.3]

62. What is the probability of rolling a number less than 3 on one roll of a die? [Section 7.3]

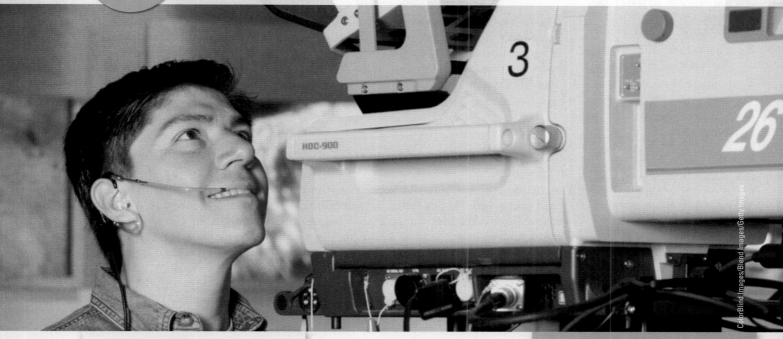

8 An Introduction to Algebra

from **Campus to Careers**

Broadcasting Technician

It takes many people behind the scenes at television stations to make what we see and hear over the airwaves possible. Broadcast technicians set up, operate, and maintain transmitter and sound equipment. They also use computers to edit audio and video recordings. This job requires skills in programming and operating electronic equipment, and the mathematical ability to analyze data.

In **Problem 41** of **Study Set 8.5**, you will see how a broadcast technician determines the amount of commercial time and program time he should schedule for a 30-minute time slot.

JOB TITLE:
Broadcasting Technician

EDUCATION:
Broadcasting jobs in large markets are usually offered to individuals who have an associate's degree or vocational certificate.

JOB OUTLOOK:
Employment in broadcasting is expected to increase about 7 percent over the 2014–2029 period.

ANNUAL EARNINGS:
The 2015 median annual salary was $41,780.

FOR MORE INFORMATION:
http://www.bls.gov/ooh/media-and-communication

SECTION 8.1 The Language of Algebra

The first seven chapters of this textbook have been an in-depth study of arithmetic. It's now time to begin the move toward *algebra*. **Algebra** is the language of mathematics. It can be used to solve many types of problems. In this chapter, you will learn more about thinking and writing in the language of algebra using its most important component—a variable.

THE LANGUAGE OF ALGEBRA

The word *algebra* comes from the title of the book *Ihm Al-jabr wa'l muqābalah*, written by an Arabian mathematician around A.D. 800.

THE LANGUAGE OF ALGEBRA

The word *variable* is based on the root word *vary*, which means change or changing. For example, the length and width of rectangles *vary*, and the unknown numbers in percent problems *vary*.

OBJECTIVE 1 Use variables to state properties of addition, multiplication, and division.

One of the major differences between arithmetic and algebra is the use of *variables*. Recall that a **variable** is a letter (or symbol) that stands for a number. In this course, we have used variables on several occasions. For example, in Chapter 1, we let l stand for the length and w stand for the width in the formula for the area of a rectangle: $A = lw$. In Chapter 6, we let x represent the unknown number in percent problems.

Many symbols used in arithmetic are also used in algebra. For example, a plus symbol $+$ is used to indicate addition, a minus symbol $-$ is used to indicate subtraction, and an $=$ symbol means *is equal to*.

Since the letter x is often used in algebra and could be confused with the multiplication symbol $\times$, we usually write multiplication using a **raised dot** or **parentheses**. When multiplying a variable by a number, or a variable by another variable, we can omit the symbol for multiplication. For example,

$2b$ means $2 \cdot b$ xy means $x \cdot y$ $8abc$ means $8 \cdot a \cdot b \cdot c$

In the notation $2b$, the number 2 is an example of a **constant** because it does not change value.

Many of the patterns that we have seen while working with whole numbers, integers, fractions, and decimals can be generalized and stated in symbols using variables. Here are some familiar properties of addition written in a very compact form, where the variables a and b represent any numbers.

■ **The Commutative Property of Addition**

$$a + b = b + a$$

Changing the order when adding does not affect the answer.

■ **The Associative Property of Addition**

$$(a + b) + c = a + (b + c)$$

Changing the grouping when adding does not affect the answer.

■ **Addition Property of 0 (Identity Property of Addition)**

$$a + 0 = a \quad \text{and} \quad 0 + a = a$$

When 0 is added to any number, the result is the same number.

Here are several familiar properties of multiplication stated using variables.

■ **The Commutative Property of Multiplication**

$$ab = ba$$

Changing the order when multiplying does not affect the answer.

■ **The Associative Property of Multiplication**

$$(ab)c = a(bc)$$

Changing the grouping when multiplying does not affect the answer.

■ **Multiplication Property of 0**

$$0 \cdot a = 0 \quad \text{and} \quad a \cdot 0 = 0 \quad \text{The product of 0 and any number is 0.}$$

■ **Multiplication Property of 1**

$$1 \cdot a = a \quad \text{and} \quad a \cdot 1 = a \quad \begin{array}{l}\text{The product of 1 and any number is that}\\ \text{number.}\end{array}$$

Here are two familiar properties of division stated using a variable.

■ **Division Properties**

$$\frac{a}{1} = a \quad \text{and} \quad \frac{a}{a} = 1 \quad \text{provided } a \neq 0 \quad \begin{array}{l}\text{Any number divided by 1}\\ \text{is the number itself.}\\ \text{Any number (except 0)}\\ \text{divided by itself is 1.}\end{array}$$

OBJECTIVE 2 **Identify terms and coefficients of terms.**

When we combine variables and numbers using arithmetic operations, the result is an *algebraic expression*.

Algebraic Expressions

Variables and/or numbers can be combined with the operations of addition, subtraction, multiplication, and division to create *algebraic expressions*.

Here are some examples of algebraic expressions.

THE LANGUAGE OF ALGEBRA

We often refer to *algebraic expressions* as simply *expressions*.

$4a + 7$ — This expression is a combination of the numbers 4 and 7, the variable a, and the operations of multiplication and addition.

$\dfrac{10 - y}{3}$ — This expression is a combination of the numbers 10 and 3, the variable y, and the operations of subtraction and division.

$15mn(2m)$ — This expression is a combination of the numbers 15 and 2, the variables m and n, and the operation of multiplication.

Addition symbols separate expressions into parts called *terms*. For example, the expression $x + 8$ has two terms.

$$\underset{\text{First term}}{x} \quad + \quad \underset{\text{Second term}}{8}$$

Since subtraction can be written as addition of the opposite, the expression $a^2 - 3a - 9$ has three terms.

$$a^2 - 3a - 9 = \quad \underset{\text{First term}}{a^2} \quad + \quad \underset{\text{Second term}}{(-3a)} \quad + \quad \underset{\text{Third term}}{(-9)}$$

In general, a **term** is a product or quotient of numbers and/or variables. A single number or variable is also a term. Examples of terms are

$$4, \quad y, \quad 6r, \quad -w^3, \quad 3.7x^5, \quad \frac{3}{n}, \quad -15ab^2$$

The numerical factor of a term is called the **coefficient** of the term. For instance, the term $6r$ has a coefficient of 6 because $6r = 6 \cdot r$. The coefficient of $-15ab^2$ is -15 because $-15ab^2 = -15 \cdot ab^2$. More examples are shown below.

Caution! By the commutative property of multiplication, $r6 = 6r$ and $-15b^2a = -15ab^2$. However, when writing terms, we usually write the numerical factor first and the variable factors in alphabetical order.

A term, such as 4, that consists of a single number, is called a **constant term**.

Term	Coefficient
$8y^2$	8
$-0.9pq$	-0.9
$\dfrac{3}{4}b$	$\dfrac{3}{4}$
$-\dfrac{x}{6}$	$-\dfrac{1}{6}$
x	1
$-t$	-1
27	27

This term could be written $\frac{3b}{4}$.

Because $-\frac{x}{6} = -\frac{1x}{6} = -\frac{1}{6} \cdot x$

Because $x = 1x$

Because $-t = -1t$

The coefficient of a constant term is that constant.

EXAMPLE 1 Identify the coefficient of each term in the expression: $7x^2 - x + 6$

Strategy We will begin by writing the subtraction as addition of the opposite. Then we will determine the numerical factor of each term.

WHY Addition symbols separate expressions into terms.

Solution If we write $7x^2 - x + 6$ as $7x^2 + (-x) + 6$, we see that it has three terms: $7x^2$, $-x$, and 6. The numerical factor of each term is its coefficient.

The coefficient of $7x^2$ is **7** because $7x^2$ means $7 \cdot x^2$.

The coefficient of $-x$ is **−1** because $-x$ means $-1 \cdot x$.

The coefficient of the constant 6 is 6.

Self Check 1

Identify the coefficient of each term in the expression: $p^3 - 12p^2 + 3p - 4$

Now Try Problem 23

It is important to be able to distinguish between the *terms* of an expression and the *factors* of a term.

EXAMPLE 2 Is m used as a *factor* or a *term* in each expression?
a. $m + 6$ **b.** $8m$

Strategy We will begin by determining whether m is involved in an addition or a multiplication.

WHY Addition symbols separate expressions into *terms*. A *factor* is a number being multiplied.

Solution
a. Since m is added to 6, m is a term of $m + 6$.
b. Since m is multiplied by 8, m is a factor of $8m$.

Self Check 2

Is b used as a *factor* or a *term* in each expression?
a. $-27b$
b. $5a + b$

Now Try Problems 27 and 29

OBJECTIVE 3 Translate word phrases to algebraic expressions.

The tables below show how key phrases can be translated into algebraic expressions.

Addition	
the sum of a and 8	$a + 8$
4 plus c	$4 + c$
16 added to m	$m + 16$
4 more than t	$t + 4$
20 greater than F	$F + 20$
T increased by r	$T + r$
exceeds y by 35	$y + 35$

Subtraction	
the difference of 23 and P	$23 - P$
550 minus h	$550 - h$
18 less than w	$w - 18$
7 decreased by j	$7 - j$
M reduced by x	$M - x$
12 subtracted from L	$L - 12$
5 less f	$5 - f$

Multiplication	
the product of 4 and x	$4x$
20 times B	$20B$
twice r	$2r$
double the amount a	$2a$
triple the profit P	$3P$
three-fourths of m	$\frac{3}{4}m$

Division	
the quotient of R and 19	$\frac{R}{19}$
s divided by d	$\frac{s}{d}$
k split into 4 equal parts	$\frac{k}{4}$
the ratio of c to d	$\frac{c}{d}$

Caution! Be careful when translating subtraction. Order is important. For example, when a translation involves the phrase *less than*, note how the terms are reversed.

18 less than w

$w - 18$

Caution! Be careful when translating division. As with subtraction, order is important. For example, s divided by d is *not* written $\frac{d}{s}$.

EXAMPLE 3   Write each phrase as an algebraic expression of:

a. one-half of the profit P

b. 5 less than the capacity c

c. the product of the weight w and 2,000, increased by 300

Strategy We will begin by identifying any key words or phrases.

WHY Key words or phrases can be translated to mathematical symbols.

Solution

a. Key phrase: *one-half of*   **Translation:** multiplication by $\frac{1}{2}$

The algebraic expression is: $\frac{1}{2}P$.

b. Key phrase: *less than*   **Translation:** subtraction

Sometimes thinking in terms of specific numbers makes translating easier. Suppose the capacity was 100. Then 5 *less than* 100 would be $100 - 5$. If the capacity is c, then we need to make c 5 less. The algebraic expression is: $c - 5$.

Caution! $5 < c$ is the translation of the statement 5 *is less than* the capacity c and not 5 *less than* the capacity c.

c. Key phrase: *product of*   **Translation:** multiplication
   Key phrase: *increased by*   **Translation:** addition

In the given wording, the comma after 2,000 means w is first multiplied by 2,000; then 300 is added to that product. The algebraic expression is: $2{,}000w + 300$.

Self Check 3

Write each phrase as an algebraic expression:

a. 80 less than the total, t

b. $\frac{2}{3}$ of the time T

c. the difference of twice a and 15, squared

Now Try Problems 15, 17, 23, 31, 33, and 39

To solve application problems, we let a variable stand for an unknown quantity.

Self Check 4

Lottos.   The payoff for a winning lottery ticket is to be split equally among fifteen friends. Write an algebraic expression that represents each person's share of the prize (in dollars).

Now Try Problem 67

EXAMPLE 4   **Swimming.**   A pool is to be sectioned into eight equally wide swimming lanes. Write an algebraic expression that represents the width of each lane.

Strategy There are two unknowns—the width of the pool and the width of each lane. We will begin by letting w = the width of the pool (in feet), as shown in the illustration.

WHY The width of each lane is related to (based on) the width of the pool.

Solution The width of the pool is sectioned into *eight equally wide lanes.*

   Key phrase: *eight equally wide lanes*   **Translation:** division by 8

Therefore, the width of each lane is $\dfrac{w}{8}$ feet.

Self Check 5

Scholarships.   Part of a $900 donation to a college went to the scholarship fund, the rest to the building fund. Choose a variable to represent the amount donated to one of the funds. Then write an expression that represents the amount donated to the other fund.

Now Try Problem 12

EXAMPLE 5   **Painting.**   A 10-inch-long paintbrush has two parts: a handle and bristles. Choose a variable to represent the length of one of the parts. Then write an algebraic expression to represent the length of the other part.

Strategy There are two approaches. We can let h = the length of the handle or we can let b = the length of the bristles.

WHY Both the length of the handle and the length of the bristles are unknown.

Solution Refer to the first drawing on the right. If we let h = the length of the handle (in inches), then the length of the bristles is $10 - h$.

   Now refer to the second drawing. If we let b = the length of the bristles (in inches), then the length of the handle is $10 - b$.

Self Check 6

Elections.   In an election, the incumbent received 55 fewer votes than three times the challenger's votes. Choose a variable to represent the number of votes received by one candidate. Then write an algebraic expression that represents the number of votes received by the other.

Now Try Problem 91

EXAMPLE 6   **Enrollments.**   Second-semester enrollment in a nursing program was 32 more than twice that of the first semester. Choose a variable to represent the enrollment for one of the semesters. Then write an algebraic expression that represents the enrollment for the other semester.

Strategy There are two unknowns—the enrollment for the first semester and the enrollment for the second semester. We will begin by letting x = the enrollment for the first semester.

WHY The second-semester enrollment is related to (based on) the first-semester enrollment.

Solution

   Key phrase: *more than*   **Translation:** addition
   Key phrase: *twice that*   **Translation:** multiplication by 2

The second-semester enrollment was $2x + 32$.

OBJECTIVE 4 Evaluate algebraic expressions.

To **evaluate an algebraic expression**, we substitute given numbers for each variable and perform the necessary calculations in the proper order.

EXAMPLE 7   Evaluate each expression for $x = 3$ and $y = -4$:

a. $y^3 + y^2$ **b.** $-y - x$ **c.** $|5xy - 7|$ **d.** $\dfrac{y - 0}{x - (-1)}$

Strategy We will replace each x and y in the expression with the given value of the variable, and evaluate the expression using the order of operation rules.

WHY To *evaluate an expression* means to find its numerical value, once we know the value of its variable(s).

Solution

a. $y^3 + y^2 = (-4)^3 + (-4)^2$ Substitute -4 for each y. We must write -4 within parentheses so that it is the base of each exponential expression.

$\quad\quad\quad = -64 + 16$ Evaluate each exponential expression.

$\quad\quad\quad = -48$ Do the addition.

$$\begin{array}{r} \overset{5\,1\,4}{64} \\ -\ 16 \\ \hline 48 \end{array}$$

b. $-y - x = -(-4) - 3$ Substitute -4 for y and 3 for x. Don't forget to write the $-$ sign in front of (-4).

$\quad\quad\quad = 4 - 3$ Simplify: $-(-4) = 4$.

$\quad\quad\quad = 1$

c. $|5xy - 7| = |5(3)(-4) - 7|$ Substitute 3 for x and -4 for y.

$\quad\quad\quad = |-60 - 7|$ Do the multiplication: $5(3)(-4) = -60$.

$\quad\quad\quad = |-67|$ Do the subtraction: $-60 - 7 = -60 + (-7) = -67$.

$\quad\quad\quad = 67$ Find the absolute value of -67.

d. $\dfrac{y - 0}{x - (-1)} = \dfrac{-4 - 0}{3 - (-1)}$ Substitute 3 for x and -4 for y.

$\quad\quad\quad = \dfrac{-4}{4}$ In the denominator, do the subtraction: $3 - (-1) = 3 + 1 = 4$.

$\quad\quad\quad = -1$ Do the division.

Caution! When replacing a variable with its numerical value, we must often write the replacement number within parentheses to convey the proper meaning.

Self Check 7

Evaluate each expression for $a = -2$ and $b = 5$:

a. $|a^3 + b^2|$

b. $-a + 2ab$

c. $\dfrac{a + 2}{b - 3}$

Now Try ➲ Problems 73 and 85

Answers to Self Checks

1. $1, -12, 3, -4$ **2. a.** factor **b.** term   **3. a.** $t - 80$ **b.** $\dfrac{2}{3}T$ **c.** $(2a - 15)^2$ **4.** $x =$ the lottery payoff in dollars; $\dfrac{x}{15} =$ each person's share in dollars   **5.** $s =$ the amount donated to the scholarship fund (in dollars); $900 - s =$ the amount donated to the building fund (in dollars)   **6.** $x =$ the number of votes received by the challenger; $3x - 55 =$ the number of votes received by the incumbent   **7. a.** 17 **b.** -18 **c.** 0

VOCABULARY

Fill in the blanks.

1. _____ are letters (or symbols) that stand for numbers.

2. The word _____ comes from the title of a book written by an Arabian mathematician around A.D. 800.

3. Variables and/or numbers can be combined with the operations of arithmetic to create algebraic _____ .

4. A _____ is a product or quotient of numbers and/or variables. Examples are: $8x$, $\frac{t}{2}$, and $-cd^3$.

5. Addition symbols separate algebraic expressions into parts called _____ .

6. A term, such as 27, that consists of a single number is called a _____ term.

7. The _____ of the term $10x$ is 10.

8. To _____ $4x - 3$ for $x = 5$, we substitute 5 for x and perform the necessary calculations in order.

9. **Cutlery.** The knife shown below is 12 inches long. Write an algebraic expression that represents the length of the blade (in inches).

10. **Savings accounts.** A student inherited $5,000 and deposits x dollars in American Savings. Write an expression that represents the amount of money left to deposit in a City Mutual account.

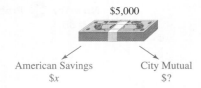

$5,000

American Savings   City Mutual
$x $?

11. **a. Mixing solutions.** Solution 1 is poured into solution 2. Write an expression that represents the number of ounces in the mixture.

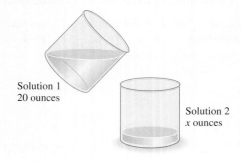

Solution 1
20 ounces

Solution 2
x ounces

b. Snacks. Cashews were mixed with p pounds of peanuts to make 100 pounds of a mixture. Write an expression that represents the number of pounds of cashews that were used.

PEANUTS   CASHEWS

p pounds   ? pounds

MIX

100 pounds

12. **Building materials.**

   a. Let b = the length of the beam shown below (in feet). Write an expression that represents the length of the pipe.

   b. Let p = the length of the pipe (in feet). Write an expression that represents the length of the beam.

15 ft

NOTATION

Complete each step. Evaluate each expression for $a = 5$, $x = -2$, and $y = 4$.

13. $9a - a^2 = 9(\ \ ) - (5)^2$

$= 9(5) - \boxed{}$

$= \boxed{} - 25$

$= 20$

14. $-x + 6y = -(\ \ ) + 6(\ \ )$

$= \boxed{} + 24$

$= \boxed{}$

Write each expression without using a multiplication symbol or parentheses.

15. $4 \cdot x$ 16. $P \cdot r \cdot t$

17. $2(w)$ 18. $(x)(y)$

GUIDED PRACTICE

Use the following variables to write each property of addition and multiplication. **See Objective 1.**

19. **a.** Write the commutative property of addition using the variables x and y.

   b. Write the associative property of addition using the variables r, s, and t.

20. **a.** Write the commutative property of multiplication using the variables *m* and *n*.

   b. Write the associative property of multiplication using the variables *x*, *y*, and *z*.

21. Write the multiplication property of zero using the variable *s*.

22. Write the multiplication property of 1 using the variable *b*.

Answer the following questions about terms and coefficients. **See Example 1.**

23. Consider the expression $3x^3 + 11x^2 - x + 9$.

   a. How many terms does the expression have?

   b. What is the coefficient of each term?

24. Consider the expression $4a^2 - 6a - 1$.

   a. How many terms does the expression have?

   b. What is the coefficient of each term?

25. Complete the following table.

Term	$6m$	$-75t$	w	$\frac{1}{2}bh$	$\frac{x}{5}$	t
Coefficient						

26. Complete the following table.

Term	$4a$	$-2r$	c	$\frac{3}{4}lw$	$\frac{d}{9}$	$-x$
Coefficient						

Determine whether the variable c is used as a factor or as a term. **See Example 2.**

27. $c + 32$ 28. $-24c + 6$

29. $5c$ 30. $a + b + c$

Translate each phrase to an algebraic expression. If no variable is given, use x as the variable. **See Example 3.**

31. The sum of the length *l* and 15

32. The difference of a number and 10

33. The product of a number and 50

34. Three-fourths of the population *p*

35. The ratio of the amount won *w* and lost *l*

36. The tax *t* added to *c*

37. *P* increased by two-thirds of *p*

38. 21 less than the total height *h*

39. The square of *k*, minus 2,005

40. *s* subtracted from *S*

41. 1 less than twice the attendance *a*

42. *J* reduced by 500

43. 1,000 split *n* equal ways

44. Exceeds the cost *c* by 25,000

45. 90 more than twice the current price *p*

46. 64 divided by the cube of *y*

47. Three times the total of 35, *h*, and 300

48. Decrease *x* by -17

49. 680 fewer than the entire population *p*

50. Triple the number of expected participants

51. The product of *d* and 4, decreased by 15

52. The quotient of *y* and 6, cubed

53. Twice the sum of 200 and *t*

54. The square of the quantity 14 less than *x*

55. The absolute value of the difference of *a* and 2

56. The absolute value of *a*, decreased by 2

57. One-tenth of the distance *d*

58. Double the difference of *x* and 18

Translate each algebraic expression into words. (Answers may vary.) **See Example 3.**

59. $\frac{3}{4}r$ 60. $\frac{2}{3}d$

61. $t - 50$

62. $c + 19$

63. xyz

64. $10ab$

65. $2m + 5$

66. $2s - 8$

Answer with an algebraic expression. **See Example 4.**

67. **Modeling.** A model's skirt is *x* inches long. The designer then lets the hem down 2 inches. What is the length of the altered skirt?

Reza Syathir/Shutterstock.com

68. **Production lines.** A soft drink manufacturer produced *c* cans of cola during the morning shift. Write an expression for how many six-packs of cola can be assembled from the morning shift's production.

69. Pants. The tag on a new pair of 36-inch-long jeans warns that after washing, they will shrink x inches in length. What is the length of the jeans after they are washed?

70. Road trips. A caravan of b cars, each carrying 5 people, traveled to the state capital for a political rally. How many people were in the caravan?

Evaluate each expression for $x = 3$, $y = -2$, and $z = -4$. See Example 7.

71. $-y$

72. $-z$

73. $-z + 3x$

74. $-y - 5x$

75. $3y^2 - 6y - 4$

76. $-z^2 - z - 12$

77. $(3 + x)y$

78. $(4 + z)y$

79. $(x + y)^2 - |z + y|$

80. $[(z - 1)(z + 1)]^2$

81. $-\dfrac{2x + y^3}{y + 2z}$

82. $-\dfrac{2z^2 - x}{2x - y^2}$

Evaluate each expression. See Example 7.

83. $b^2 - 4ac$ for $a = -1, b = 5$, and $c = -2$

84. $(x - a)^2 + (y - b)^2$ for $x = -2, y = 1, a = 5,$ and $b = -3$

85. $a^2 + 2ab + b^2$ for $a = -5$ and $b = -1$

86. $\dfrac{a - x}{y - b}$ for $x = -2, y = 1, a = 5,$ and $b = 2$

87. $\dfrac{n}{2}[2a + (n - 1)d]$ for $n = 10, a = -4.2,$ and $d = 6.6$

88. $\dfrac{a(1 - r^n)}{1 - r}$ for $a = -5, r = 2,$ and $n = 3$

89. $(27c^2 - 4d^2)^3$ for $c = \frac{1}{3}$ and $d = \frac{1}{2}$

90. $\dfrac{-b^2 + 16a^2 + 1}{2}$ for $a = \frac{1}{4}$ and $b = -10$

LOOK ALIKES

Translate each phrase to mathematical symbols. Let x represent the unknown number.

91. a. 23 less than a number
   b. 23 is less than a number

92. a. The sum of a number and 7 squared
   b. The sum of a number and 7, squared

93. a. 6 times a number increased by 8
   b. 6 times a number, increased by 8

94. a. Twice a number, decreased by 3
   b. Twice a number decreased by 3

APPLICATIONS

95. Vehicle weights. A Hummer H2 weighs 340 pounds less than twice a Honda Element.
   a. Let x represent the weight of one of the vehicles. Write an expression for the weight of the other vehicle.
   b. If the weight of the Element is 3,370 pounds, what is the weight of the Hummer?

96. Sod farms. The expression $20,000 - 3s$ gives the number of square feet of sod that are left in a field after s strips have been removed. Suppose a city orders 7,000 strips of sod. Evaluate the expression and explain the result.

Strips of sod, cut and ready to be loaded on a truck for delivery

97. Computer companies. IBM was founded 80 years before Apple Computer. Dell Computer Corporation was founded 9 years after Apple.
   a. Let x represent the age (in years) of one of the companies. Write expressions to represent the ages (in years) of the other two companies.
   b. On April 1, 2008, Apple Computer Company was 32 years old. How old were the other two computer companies then?

98. Thrill rides. The distance in feet that an object will fall in t seconds is given by the expression $16t^2$. Find the distance that riders on "Drop Zone" will fall during the times listed in the table.

Ellen McKnight/Alamy

Time (seconds)	Distance (feet)
1	
2	
3	
4	

WRITING

99. What is a variable? Give an example of how variables are used.

100. What is an algebraic expression? Give some examples.

101. Explain why *2 less than x* does not translate to $2 < x$.

102. In this section, we substituted a number for a variable. List some other uses of the word *substitute* that you encounter in everyday life.

REVIEW

103. Find the LCD for $\frac{5}{12}$ and $\frac{1}{15}$.

104. Simplify: $\frac{3 \cdot 3 \cdot 5}{3 \cdot 5 \cdot 5 \cdot 11}$

105. Evaluate: $\left(\frac{2}{3}\right)^3$

106. Find the result when $\frac{7}{8}$ is multiplied by its reciprocal.

SECTION 8.2 Simplifying Algebraic Expressions

In algebra, we frequently replace one algebraic expression with another that is equivalent and simpler in form. That process, called *simplifying an algebraic expression*, often involves the use of one or more properties of real numbers.

OBJECTIVES

1 Simplify products.

2 Use the distributive property.

3 Identify like terms.

4 Combine like terms.

OBJECTIVE 1 Simplify products.

The commutative and associative properties of multiplication can be used to simplify certain products. For example, let's simplify $8(4x)$.

$$8(4x) = 8 \cdot (4 \cdot x) \quad \text{Rewrite } 4x \text{ as } 4 \cdot x.$$

$$= (8 \cdot 4) \cdot x \quad \text{Use the associative property of multiplication to group 4 with 8.}$$

$$= 32x \quad \text{Do the multiplication within the parentheses.}$$

We have found that $8(4x) = 32x$. We say that $8(4x)$ and $32x$ are **equivalent expressions** because for each value of x, they represent the same number. For example, if $x = 10$, both expressions have a value of 320. If $x = -3$, both expressions have a value of -96.

If $x = 10$		**If $x = -3$**	
$8(4x) = 8[4(10)]$ $32x = 32(10)$		$8(4x) = 8[4(-3)]$ $32x = 32(-3)$	
$= 8(40)$ $= 320$		$= 8(-12)$ $= -96$	
$= 320$		$= -96$	

same result   same result

THE LANGUAGE OF ALGEBRA

Be careful when using the words *simplify* and *solve*. In mathematics, we **simplify** expressions and we **solve** equations.

Success Tip By the commutative property of multiplication, we can change the order of factors. By the associative property of multiplication, we can change the grouping of factors.

EXAMPLE 1   Simplify:

a. $9(3b)$ **b.** $15a(6)$ **c.** $3(7p)(-5)$ **d.** $\dfrac{8}{3} \cdot \dfrac{3}{8}r$ **e.** $35\left(\dfrac{4}{5}x\right)$

Strategy We will use the commutative and associative properties of multiplication to reorder and regroup the factors in each expression.

WHY We want to group all of the numerical factors of an expression together so that we can find their product.

Solution

a. $9(3b) = (9 \cdot 3)b$ Use the associative property of multiplication to regroup the factors.

      $= 27b$ Do the multiplication within the parentheses.

b. $15a(6) = 15(6)a$ Use the commutative property of multiplication to reorder the factors.

      $= 90a$ Do the multiplication: $15(6) = 90$.

$$\begin{array}{r} \overset{3}{15} \\ \times\ 6 \\ \hline 90 \end{array}$$

c. $3(7p)(-5) = [3(7)(-5)]p$ Use the commutative and associative properties of multiplication to reorder and regroup the factors.

      $= [21(-5)]p$ Multiply within the brackets.

      $= -105p$ Complete the multiplication within the brackets.

$$\begin{array}{r} 21 \\ \times\ 5 \\ \hline 105 \end{array}$$

Self Check 1

Simplify:

a. $9 \cdot 6s$

b. $-4(6u)(-2)$

c. $\dfrac{2}{3} \cdot \dfrac{3}{2}m$

d. $36\left(\dfrac{2}{9}y\right)$

Now Try ➡ Problems 15, 25, 29, and 31

d. $\dfrac{8}{3} \cdot \dfrac{3}{8}r = \left(\dfrac{8}{3} \cdot \dfrac{3}{8}\right)r$ Use the associative property of multiplication to group the factors.

      $= 1r$ Multiply within the parentheses.

              The product of a number and its reciprocal is 1: $\dfrac{\overset{1}{\cancel{8}}}{\underset{1}{\cancel{3}}} \cdot \dfrac{\overset{1}{\cancel{3}}}{\underset{1}{\cancel{8}}} = 1$.

      $= r$ The coefficient 1 need not be written.

e. $35\left(\dfrac{4}{5}x\right) = \left(\dfrac{35}{1} \cdot \dfrac{4}{5}\right)x$ Use the associative property of multiplication to regroup the factors.

      $= \left(\dfrac{\overset{1}{\cancel{5}} \cdot 7 \cdot 4}{1 \cdot \underset{1}{\cancel{5}}}\right)x$ Factor 35 as $5 \cdot 7$ and then remove the common factor 5.

      $= 28x$ Do the multiplication and then simplify: $\dfrac{28}{1} = 28$.

OBJECTIVE 2 Use the distributive property.

Another property that is often used to simplify algebraic expressions is the **distributive property**. To introduce it, we will evaluate $4(5 + 3)$ in two ways.

Method 1	*Method 2*
Use the order of operations:	***Distribute the multiplication:***
$4(5 + 3) = 4(8)$	$4(5 + 3) = 4(5) + 4(3)$
$= 32$	$= 20 + 12$
	$= 32$

THE LANGUAGE OF ALGEBRA

To **distribute** means to give from one to several. You have probably *distributed* candy to children coming to your door on Halloween.

Each method gives a result of 32. This observation suggests the following property.

The Distributive Property

For any numbers a, b, and c,

$$a(b + c) = ab + ac$$

To illustrate one use of the distributive property, let's consider the expression $5(x + 3)$. Since we are not given the value of x, we cannot add x and 3 within the parentheses. However, we can distribute the multiplication by the factor of 5 that is outside the parentheses to x and to 3 and add those products.

$$5(x + 3) = 5(x) + 5(3) \quad \text{Distribute the multiplication by 5.}$$
$$= 5x + 15 \quad \text{Do the multiplication.}$$

> **THE LANGUAGE OF ALGEBRA**
>
> Formally, it is called the *distributive property of multiplication over addition*. When we use it to write a product, such as $5(x + 2)$, as a sum, $5x + 10$, we say that we have *removed* or *cleared* the parentheses.

EXAMPLE 2

Multiply: **a.** $8(m + 9)$ **b.** $-12(4t + 1)$ **c.** $6\left(\dfrac{x}{3} + \dfrac{9}{2}\right)$

Strategy In each case, we will distribute the multiplication by the factor *outside* the parentheses over each term *within* the parentheses.

WHY In each case, we cannot simplify the expression within the parentheses. To multiply, we must use the distributive property.

Solution

a. We read $8(m + 9)$ as "eight times the *quantity* of m plus nine." The word *quantity* alerts us to the grouping symbols in the expression.

$$8(m + 9) = 8 \cdot m + 8 \cdot 9 \quad \text{Distribute the multiplication by 8.}$$
$$= 8m + 72 \quad \text{Do the multiplications. Try to go directly to this step.}$$

b. $-12(4t + 1) = -12(4t) + (-12)(1) \quad \text{Distribute the multiplication by } -12.$

$$= -48t + (-12) \quad \text{Do the multiplications.}$$

$$= -48t - 12 \quad \begin{array}{l}\text{Write the result in simpler form. Recall that}\\ \text{adding } -12 \text{ is the same as subtracting 12.}\end{array}$$

$$\begin{array}{r} 12 \\ \times\ 4 \\ \hline 48 \end{array}$$

c. $6\left(\dfrac{x}{3} + \dfrac{9}{2}\right) = 6 \cdot \dfrac{x}{3} + 6 \cdot \dfrac{9}{2} \quad \text{Distribute the multiplication by 6.}$

$$= \dfrac{2 \cdot \overset{1}{\cancel{3}} \cdot x}{\underset{1}{\cancel{3}}} + \dfrac{\overset{1}{\cancel{2}} \cdot 3 \cdot 9}{\underset{1}{\cancel{2}}} \quad \begin{array}{l}\text{Factor 6 as } 2 \cdot 3 \text{ and then remove}\\ \text{the common factors 3 and 2.}\end{array}$$

$$= 2x + 27$$

> **Self Check 2**
>
> Multiply:
> **a.** $7(m + 2)$
> **b.** $-80(8x + 3)$
> **c.** $24\left(\dfrac{y}{6} + \dfrac{3}{8}\right)$
>
> **Now Try** ➡ Problems 35, 37, and 39

Since subtraction is the same as adding the opposite, the distributive property also holds for subtraction.

$$a(b - c) = ab - ac$$

EXAMPLE 3

Multiply: **a.** $3(3b - 4)$ **b.** $-6(-3y - 8)$ **c.** $-1(t - 9)$

Strategy In each case, we will distribute the multiplication by the factor *outside* the parentheses over each term *within* the parentheses.

WHY In each case, we cannot simplify the expression within the parentheses. To multiply, we must use the distributive property.

Solution

a. $3(3b - 4) = 3(3b) - 3(4) \quad \text{Distribute the multiplication by 3.}$

$$= 9b - 12 \quad \text{Do the multiplications. Try to go directly to this step.}$$

Caution! A common mistake is to forget to distribute the multiplication over each of the terms within the parentheses.

$$3(3b - 4) = 9b - 4$$

b. $-6(-3y - 8) = -6(-3y) - (-6)(8)$ *Distribute the multiplication by −6.*

$$= 18y - (-48) \qquad \text{*Do the multiplications.*}$$

$$= 18y + 48 \qquad \text{*Write the result in simpler form.*}$$
  Add the opposite of −48.

Another approach is to write the subtraction within the parentheses as addition of the opposite. Then we distribute the multiplication by -6 over the addition.

$$-6(-3y - 8) = -6[-3y + (-8)] \qquad \text{*Add the opposite of 8.*}$$

$$= -6(-3y) + (-6)(-8) \qquad \text{*Distribute the multiplication by −6.*}$$

$$= 18y + 48 \qquad \text{*Do the multiplications.*}$$

c. $-1(t - 9) = -1(t) - (-1)(9)$ *Distribute the multiplication by −1.*

$$= -t - (-9) \qquad \text{*Do the multiplications.*}$$

$$= -t + 9 \qquad \text{*Write the result in simpler form. Add the opposite of −9.*}$$

Notice that distributing the multiplication by -1 *changes the sign* of each term within the parentheses.

Self Check 3

Multiply:
a. $5(2x - 1)$
b. $-9(-y - 4)$
c. $-1(c - 22)$

Now Try ➡ **Problems 43, 47, and 49**

Caution! The distributive property does not apply to every expression that contains parentheses—only those in which multiplication is distributed over addition (or subtraction). For example, to simplify $6(5x)$, we do not use the distributive property.

Correct	**Incorrect**
$6(5x) = (6 \cdot 5)x = 30x$	$6(5x) = 30 \cdot 6x = 180x$

The distributive property can be extended to several other useful forms. Since multiplication is commutative, we have:

$$(b + c)a = ba + ca \qquad\qquad (b - c)a = ba - ca$$

For situations in which there are more than two terms within parentheses, we have:

$$a(b + c + d) = ab + ac + ad \qquad a(b - c - d) = ab - ac - ad$$

EXAMPLE 4 Multiply:

a. $(6x + 4)\dfrac{1}{2}$ **b.** $2(a - 3b)8$ **c.** $-0.3(3a - 4b + 7)$

Strategy We will multiply each term within the parentheses by the factor (or factors) outside the parentheses.

WHY In each case, we cannot simplify the expression within the parentheses. To multiply, we use the distributive property.

Solution

a. $(6x + 4)\dfrac{1}{2} = (6x)\dfrac{1}{2} + (4)\dfrac{1}{2}$ Distribute the multiplication by $\frac{1}{2}$.

$\qquad\qquad\quad = 3x + 2$ Do the multiplications.

b. This expression contains three factors.   Use the commutative property of
$\quad 2(a - 3b)8 = 2 \cdot 8(a - 3b)$ multiplication to reorder the factors.

$\qquad\qquad = 16(a - 3b)$ Multiply 2 and 8 to get 16.

$\qquad\qquad = 16a - 48b$ Distribute the multiplication by 16.

c. $-0.3(3a - 4b + 7) = -0.3(3a) - (-0.3)(4b) + (-0.3)(7)$

$\qquad\qquad\qquad = -0.9a + 1.2b - 2.1$ Do each multiplication.

$$\begin{array}{cc} & \overset{1}{} \\ 0.3 & 0.3 \\ \times\,3 & \times\,4 \\ \hline 0.9 & 1.2 \end{array}$$

$$\begin{array}{c} \overset{2}{0.3} \\ \times\,7 \\ \hline 2.1 \end{array}$$

Self Check 4

Multiply:

a. $(-6x - 24)\dfrac{1}{3}$

b. $6(c - 2d)9$

c. $-0.7(2r + 5s - 8)$

Now Try ➭ Problems 53, 55, and 57

We can use the distributive property to find the opposite of a sum. For example, to find $-(x + 10)$, we interpret the $-$ symbol as a factor of -1, and proceed as follows:

$-(x + 10) = -1(x + 10)$ Replace the $-$ symbol with -1.

$\qquad\qquad = -1(x) + (-1)(10)$ Distribute the multiplication by -1.

$\qquad\qquad = -x - 10$

In general, we have the following property.

The Opposite of a Sum

The opposite of a sum is the sum of the opposites.

For any numbers a and b,

$$-(a + b) = -a + (-b)$$

EXAMPLE 5   Simplify: $-(-9s - 3)$

Strategy We will multiply each term within the parentheses by -1.

WHY The $-$ outside the parentheses represents a factor of -1 that is to be distributed.

Solution

$-(-9s - 3) = -1(-9s - 3)$ Replace the $-$ symbol in front
$\qquad\qquad\qquad$ of the parentheses with -1.

$\qquad\qquad = -1(-9s) - (-1)(3)$ Distribute the multiplication by -1.

$\qquad\qquad = 9s + 3$ Do the multiplications. Try to go
$\qquad\qquad\qquad$ directly to this step.

Self Check 5

Simplify: $-(-5x + 18)$

Now Try ➭ Problem 59

OBJECTIVE 3 Identify like terms.

Before we can discuss methods for simplifying algebraic expressions involving addition and subtraction, we need to introduce some new vocabulary.

> **Like Terms**
>
> **Like terms** are terms containing exactly the same variables raised to exactly the same powers. Any constant terms in an expression are considered to be like terms. Terms that are not like terms are called **unlike terms**.

Success Tip When looking for like terms, don't look at the coefficients of the terms. Consider only the variable factors of each term. If two terms are like terms, only their coefficients may differ.

Here are several examples.

Like terms	*Unlike terms*	
$4x$ and $7x$	$4x$ and $7y$	The variables are not the same.
$-10p^2$ and $25p^2$	$-10p$ and $25p^2$	Same variable, but different powers.
$\frac{1}{3}c^3d$ and c^3d	$\frac{1}{3}c^3d$ and c^3	The variables are not the same.

EXAMPLE 6   Identify the like terms in each expression:
a. $7r + 5 + 3r$ **b.** $6x^4 - 6x^2 - 6x$ **c.** $-17m^3 + 3 - 2 + m^3$

Strategy First, we will identify the terms of the expression. Then we will look for terms that contain the same variables raised to exactly the same powers.

WHY If two terms contain the same variables raised to the same powers, they are like terms.

Self Check 6

Identify the like terms:
a. $2x - 2y + 7y$
b. $5p^2 - 12 + 17p^2 + 2$

Now Try Problem 63

Solution

a. $7r + 5 + 3r$ contains the like terms $7r$ and $3r$.

b. Since the exponents on x are different, $6x^4 - 6x^2 - 6x$ contains no like terms.

c. $-17m^3 + 3 - 2 + m^3$ contains two pairs of like terms: $-17m^3$ and m^3 are like terms, and the constant terms, 3 and -2, are like terms.

OBJECTIVE 4 Combine like terms.

To add or subtract objects, they must be similar. For example, fractions that are to be added must have a common denominator. When adding decimals, we align columns to be sure to add tenths to tenths, hundredths to hundredths, and so on. The same is true when working with terms of an algebraic expression. They can be added or subtracted only if they are like terms.

This expression can be simplified because it contains like terms.	This expression *cannot* be simplified because its terms are not like terms.
$3x + 4x$	$3x + 4y$

Recall that the distributive property can be written in the following forms:

$$(b + c)a = ba + ca \qquad (b - c)a = ba - ca$$

Success Tip Just as 3 apples plus 4 apples is 7 apples, $3x + 4x = 7x$

We can use these forms of the distributive property in reverse to simplify a sum or difference of like terms. For example, we can simplify $3x + 4x$ as follows:

$$3x + 4x = (3 + 4)x \qquad \text{Use the form: } ba + ca = (b + c)a.$$
$$= 7x$$

We can simplify $15m^2 - 9m^2$ in a similar way:

$$15m^2 - 9m^2 = (15 - 9)m^2 \quad \text{Use the form: } ba - ca = (b - c)a.$$
$$= 6m^2$$

These examples suggest the following general rule.

THE LANGUAGE OF ALGEBRA

Simplifying a sum or difference of like terms is called *combining like terms*.

Combining Like Terms

Like terms can be combined by adding or subtracting the coefficients of the terms and keeping the same variables with the same exponents.

EXAMPLE 7   Simplify by combining like terms, if possible:

a. $2x + 9x$ **b.** $-8p + (-2p) + 4p$ **c.** $0.5s^3 - 0.3s^3$ **d.** $4w + 6$ **e.** $\dfrac{4}{9}b + \dfrac{7}{9}b$

Strategy We will use the distributive property in reverse to add (or subtract) the coefficients of the like terms. We will keep the same variables raised to the same powers.

WHY To *combine like terms* means to add or subtract the like terms in an expression.

Solution

a. Since $2x$ and $9x$ are like terms with the common variable x, we can combine them.

$\qquad 2x + 9x = 11x \qquad$ Think: $(2 + 9)x = 11x$.

b. $-8p + (-2p) + 4p = -6p \quad$ Think: $[-8 + (-2) + 4]p = -6p$.

c. $0.5s^3 - 0.3s^3 = 0.2s^3 \qquad$ Think: $(0.5 - 0.3)s^3 = 0.2s^3$.

d. Since $4w$ and 6 are not like terms, they cannot be combined. The expression $4w + 6$ doesn't simplify.

e. $\dfrac{4}{9}b + \dfrac{7}{9}b = \dfrac{11}{9}b \quad$ Think: $\left(\dfrac{4}{9} + \dfrac{7}{9}\right)b = \dfrac{11}{9}b$.

Self Check 7

Simplify, if possible:
a. $3x + 5x$
b. $-6y + (-6y) + 9y$
c. $4.4s^4 - 3.9s^4$
d. $4a - 2$
e. $\dfrac{10}{7}c - \dfrac{4}{7}c$

Now Try ➲ Problems 67, 71, 79, and 83

EXAMPLE 8   Simplify by combining like terms:

a. $16t - 15t$ **b.** $16t - t$ **c.** $15t - 16t$ **d.** $16t + t$

Strategy As we combine like terms, we must be careful when working with terms such as t and $-t$.

WHY Coefficients of 1 and -1 are usually not written.

Solution

a. $16t - 15t = t \qquad$ Think: $(16 - 15)t = 1t = t$.

b. $16t - t = 15t \qquad$ Think: $16t - 1t = (16 - 1)t = 15t$.

c. $15t - 16t = -t \qquad$ Think: $(15 - 16)t = -1t = -t$.

d. $16t + t = 17t \qquad$ Think: $16t + 1t = (16 + 1)t = 17t$.

Self Check 8

Simplify:
a. $9h - h$ **b.** $9h + h$
c. $9h - 8h$ **d.** $8h - 9h$

Now Try ➲ Problems 73 and 77

EXAMPLE 9 Simplify: $6a^2 + 54a - 4a - 36$

Strategy First we will identify any like terms in the expression. Then we will use the distributive property in reverse to combine them.

WHY To *simplify* an expression, we use properties of real numbers to write an equivalent expression in simpler form.

Solution
We can combine the like terms that involve the variable a.

$$6a^2 + 54a - 4a - 36 = 6a^2 + 50a - 36 \quad \text{Think: } (54 - 4)a = 50a.$$

Self Check 9

Simplify:
$7y^2 + 21y - 2y - 6$

Now Try Problem 93

EXAMPLE 10 Simplify: $4(x + 5) - 5 - (2x - 4)$

Strategy First, we will remove the parentheses. Then we will identify any like terms and combine them.

WHY To *simplify* an expression, we use properties of real numbers, such as the distributive property, to write an equivalent expression in simpler form.

Solution
Here, the distributive property is used both *forward* (to remove parentheses) and in *reverse* (to combine like terms).

$$4(x + 5) - 5 - (2x - 4) = 4(x + 5) - 5 - 1(2x - 4) \quad \begin{array}{l}\text{Replace the } - \text{ symbol}\\ \text{in front of } (2x - 4)\\ \text{with } -1.\end{array}$$

$$= 4x + 20 - 5 - 2x + 4 \quad \begin{array}{l}\text{Distribute the multiplication}\\ \text{by 4 and } -1.\end{array}$$

$$= 2x + 19 \quad \begin{array}{l}\text{Think: } (4 - 2)x = 2x.\\ \text{Think: } (20 - 5 + 4) = 19.\end{array}$$

Self Check 10

Simplify:
$6(3y - 1) + 2 - (-3y + 4)$

Now Try Problem 99

Answers to Self Checks

1. a. $54s$ **b.** $48u$ **c.** m **d.** $8y$ **2. a.** $7m + 14$ **b.** $-640x - 240$ **c.** $4y + 9$ **3. a.** $10x - 5$
b. $9y + 36$ **c.** $-c + 22$ **4. a.** $-2x - 8$ **b.** $54c - 108d$ **c.** $-1.4r - 3.5s + 5.6$ **5.** $5x - 18$
6. a. $-2y$ and $7y$ **b.** $5p^2$ and $17p^2$; -12 and 2 **7. a.** $8x$ **b.** $-3y$ **c.** $0.5s^4$ **d.** does not simplify
e. $\frac{6}{7}c$ **8. a.** $8h$ **b.** $10h$ **c.** h **d.** $-h$ **9.** $7y^2 + 19y - 6$ **10.** $21y - 8$

SECTION 8.2 STUDY SET

VOCABULARY

Fill in the blanks.

1. To _____ the expression $5(6x)$ means to write it in simpler form: $5(6x) = 30x$.

2. $5(6x)$ and $30x$ are _____ expressions because for each value of x, they represent the same number.

3. To perform the multiplication $2(x + 8)$, we use the _____ property.

4. We call $-(c + 9)$ the _____ of a sum.

5. Terms such as $7x^2$ and $5x^2$, which have the same variables raised to exactly the same power, are called _____ terms.

6. When we write $9x + x$ as $10x$, we say we have _____ like terms.

CONCEPTS

7. a. Fill in the blanks to simplify the expression.

$$4(9t) = (\quad \cdot 9)t = \quad t$$

b. What property did you use in part a?

8. a. Fill in the blanks to simplify the expression.

$$-6y \cdot 2 = \quad \cdot 2 \cdot y = \quad y$$

b. What property did you use in part a?

9. Fill in the blanks.

a. $2(x + 4) = 2x \quad 8$

b. $2(x - 4) = 2x \quad 8$

c. $-2(x + 4) = -2x \quad 8$

d. $-2(-x - 4) = 2x \quad 8$

10. Fill in the blanks to combine like terms.

a. $4m + 6m = (\quad + \quad)m = \quad m$

b. $30n^2 - 50n^2 = (\quad - \quad)n^2 = \quad n^2$

c. $12 + 32d + 15 = 32d + \quad$

d. Like terms can be combined by adding or subtracting the _____ of the terms and keeping the same _____ with the same exponents.

11. Simplify each expression, if possible.

a. $5(2x)$ **b.** $5 + 2x$

c. $6(-7x)$ **d.** $6 - 7x$

e. $2(3x)(3)$ **f.** $2 + 3x + 3$

12. Fill in the blanks: Distributing multiplication by -1 changes the _____ of each term within the parentheses.

$$-(x + 10) = \quad (x + 10) = -x \quad 10$$

NOTATION

13. Translate to symbols.

a. Six times the quantity of h minus four.

b. The opposite of the sum of z and sixteen.

14. Write an equivalent expression for the given expression using fewer symbols.

a. $1x$ **b.** $-1d$ **c.** $0m$

d. $5x - (-1)$ **e.** $16t + (-6)$

GUIDED PRACTICE

Simplify. **See Example 1.**

15. $3 \cdot 4t$ **16.** $9 \cdot 3s$

17. $9(7m)$ **18.** $12n(8)$

19. $5(-7q)$ **20.** $-7(5t)$

21. $5t \cdot 60$ **22.** $70a \cdot 10$

23. $(-5.6x)(-2)$ **24.** $(-4.4x)(-3)$

25. $5(4c)(3)$ **26.** $9(2h)(2)$

27. $-4(-6)(-4m)$ **28.** $-5(-9)(-4n)$

29. $\dfrac{5}{3} \cdot \dfrac{3}{5}g$ **30.** $\dfrac{9}{7} \cdot \dfrac{7}{9}k$

31. $12\left(\dfrac{5}{12}x\right)$ **32.** $15\left(\dfrac{4}{15}w\right)$

33. $8\left(\dfrac{3}{4}y\right)$ **34.** $27\left(\dfrac{2}{3}x\right)$

Multiply. **See Example 2.**

35. $5(x + 3)$ **36.** $4(x + 2)$

37. $-3(4x + 9)$ **38.** $-5(8x + 9)$

39. $45\left(\dfrac{x}{5} + \dfrac{2}{9}\right)$ **40.** $35\left(\dfrac{y}{5} + \dfrac{8}{7}\right)$

41. $0.4(x + 4)$ **42.** $2.2(2q + 1)$

Multiply. **See Example 3.**

43. $6(6c - 7)$ **44.** $9(9d - 3)$

45. $-6(13c - 3)$ **46.** $-2(10s - 11)$

47. $-15(-2t - 6)$ **48.** $-20(-4z - 5)$

49. $-1(-4a + 1)$ **50.** $-1(-2x + 3)$

Multiply. **See Example 4.**

51. $(3t + 2)8$ **52.** $(2q + 1)9$

53. $(3w - 6)\dfrac{2}{3}$ **54.** $(2y - 8)\dfrac{1}{2}$

55. $4(7y + 4)2$ **56.** $8(2a - 3)4$

57. $25(2a - 3b + 1)$ **58.** $5(9s - 12t - 3)$

Simplify. **See Example 5.**

59. $-(x - 7)$ **60.** $-(y + 1)$

61. $-(-5.6y + 7)$ **62.** $-(-4.8a - 3)$

Identify the like terms in each expression, if any.
See Example 6.

63. $3x + 2 - 2x$

64. $3y + 4 - 11y + 6$

65. $-12m^4 - 3m^3 + 2m^2 - m^3$

66. $6x^3 + 3x^2 + 6x$

Simplify by combining like terms. **See Examples 7 and 8.**

67. $3x + 7x$ **68.** $12y - 15y$

69. $-4x + 4x$ **70.** $-16y + 16y$

71. $-7b^2 + 27b^2$ **72.** $-2c^3 + 12c^3$

73. $13r - 12r$ **74.** $25s + s$

75. $36y + y - 9y$ **76.** $32a - a + 5a$

77. $43s^3 - 44s^3$ **78.** $8j^3 - 9j^3$

79. $-9.8c + 6.2c$ **80.** $-5.7m + 4.3m$

81. $-0.2r - (-0.6r)$

82. $-1.1m - (-2.4m)$

83. $\frac{3}{5}t + \frac{1}{5}t$

84. $\frac{3}{16}x - \frac{5}{16}x$

85. $-\frac{7}{16}x - \frac{3}{16}x$

86. $-\frac{5}{18}x - \frac{7}{18}x$

Simplify by combining like terms, if possible. See Example 9.

87. $15y - 10 - y - 20y$

88. $9z - 7 - z - 19z$

89. $3x + 4 - 5x + 1$

90. $4b + 9 - 9b + 9$

91. $6m^2 - 6m + 6$

92. $9a^2 + 9a - 9$

93. $4x^2 + 5x - 8x + 9$

94. $10y^2 - 8y + y - 7$

Simplify. See Example 10.

95. $2z + 5(z - 3)$

96. $12(m + 11) - 11$

97. $2(s^2 - 7) - (s^2 - 2)$

98. $4(d^2 - 3) - (d^2 - 1)$

99. $-9(3r - 9) - 7(2r - 7)$

100. $-6(3t - 6) - 3(11t - 3)$

101. $36\left(\frac{2}{9}x - \frac{3}{4}\right) + 36\left(\frac{1}{2}\right)$

102. $40\left(\frac{3}{8}y - \frac{1}{4}\right) + 40\left(\frac{4}{5}\right)$

TRY IT YOURSELF

Simplify each expression.

103. $6 - 4(-3c - 7)$

104. $10 - 5(-5g - 1)$

105. $-4r - 7r + 2r - r$

106. $-v - 3v + 6v + 2v$

107. $24\left(-\frac{5}{6}r\right)$

108. $\frac{3}{4} \cdot \frac{1}{2}g$

109. $a + a + a$

110. $t - t - t - t$

111. $60\left(\frac{3}{20}r - \frac{4}{15}\right)$

112. $72\left(\frac{7}{8}f - \frac{8}{9}\right)$

113. $5(-1.2x)$

114. $5(-6.4c)$

115. $-(c + 7) + 2(c - 3)$

116. $-(z + 2) + 5(3 - z)$

117. $a^3 + 2a^2 + 4a - 2a^2 - 4a - 8$

118. $c^3 - 3c^2 + 9c + 3c^2 - 9c + 27$

LOOK ALIKES

Simplify each expression.

119. a. $3(5x)4$ **b.** $3(5x + 4)$

120. a. $6(t - 11)$ **b.** $(t - 11)6$

121. a. $-(4s - 8)$ **b.** $-(-4s - 8)$

122. a. $7(-2x - 1)4$ **b.** $4(-2x - 1)7$

APPLICATIONS

In Exercises 123–126, recall that the perimeter of a figure is equal to the sum of the lengths of its sides.

123. The Red Cross.   In 1891, Clara Barton founded the Red Cross. Its symbol is a white flag bearing a red cross. If each side of the cross has length x, write an expression that represents the perimeter of the cross.

124. Billiards.   Billiard tables vary in size, but all tables are twice as long as they are wide.

a. If the billiard table is x feet wide, write an expression that represents its length.

b. Write an expression that represents the perimeter of the table.

125. Ping-Pong.   Write an expression that represents the perimeter of the Ping-Pong table.

126. Sewing.   Write an expression that represents the length of the purple trim needed to outline a pennant with the given side lengths.

WRITING

127. Explain why the distributive property applies to $2(3 + x)$ but not to $2(3x)$.

128. Explain how to combine like terms. Give an example.

REVIEW

Evaluate each expression for $x = -3$, $y = -5$, and $z = 0$.

129. $\dfrac{x - y^2}{2y - 1 + x}$

130. $\dfrac{2y + 1}{x} - x$

SECTION 8.3   Solving Equations Using Properties of Equality

In this section, we introduce four properties of equality that are used to solve equations.

OBJECTIVES

1 Determine whether a number is a solution.

2 Use the addition property of equality.

3 Use the subtraction property of equality.

4 Use the multiplication property of equality.

5 Use the division property of equality.

OBJECTIVE 1   Determine whether a number is a solution.

An **equation** is a statement indicating that two expressions are equal. All equations contain an equal symbol $=$. An example is $x + 5 = 15$. The equal symbol $=$ separates the equation into two parts: The expression $x + 5$ is the **left side**, and 15 is the **right side**. The letter x is the **variable** (or the **unknown**). The sides of an equation can be reversed, so we can write $x + 5 = 15$ or $15 = x + 5$.

- An equation can be true: $6 + 3 = 9$
- An equation can be false: $2 + 4 = 7$
- An equation can be neither true nor false. For example, $x + 5 = 15$ is neither true nor false because we don't know what number x represents.

An equation that contains a variable is made true or false by substituting a number for the variable. If we substitute 10 for x in $x + 5 = 15$, the resulting equation is true: $10 + 5 = 15$. If we substitute 1 for x, the resulting equation is false: $1 + 5 = 15$. A number that makes an equation true when substituted for the variable is called a **solution**; it is said to **satisfy** the equation. Therefore, 10 is a solution of $x + 5 = 15$, and 1 is not.

> **THE LANGUAGE OF ALGEBRA**
>
> It is important to know the difference between an **equation** and an **expression**. An equation contains an $=$ symbol; an expression does not.

EXAMPLE 1   Is 9 a solution of $3y - 1 = 2y + 7$?

Strategy We will substitute 9 for each y in the equation and evaluate the expression on the left side and the expression on the right side separately.

WHY If a true statement results, 9 is a solution of the equation. If we obtain a false statement, 9 is not a solution.

Solution

Evaluate the expression on the left side.

Evaluate the expression on the right side.

$$3y - 1 = 2y + 7$$
$$3(9) - 1 \stackrel{?}{=} 2(9) + 7 \qquad \text{Read } \stackrel{?}{=} \text{ as "is possibly equal to."}$$
$$27 - 1 \stackrel{?}{=} 18 + 7$$
$$26 = 25$$

Since $26 = 25$ is false, 9 is *not* a solution of $3y - 1 = 2y + 7$.

Self Check 1

Is 25 a solution of $10 - x = 35 - 2x$?

Now Try Problem 19

OBJECTIVE 2   Use the addition property of equality.

To **solve an equation** means to find all values of the variable that make the equation true. We can develop an understanding of how to solve equations by referring to the scales shown on the right.

The first scale represents the equation $x - 2 = 3$. The scale is in balance because the weights on the left side and right side are equal. To find x, we must add 2 to the left side. To keep the scale in balance, we must also add 2 to the right side. After doing this, we see that x is balanced by 5. Therefore, x must be 5. We say that we have solved the equation $x - 2 = 3$ and that the solution is 5.

In this example, we solved $x - 2 = 3$ by transforming it to a simpler *equivalent equation*, $x = 5$.

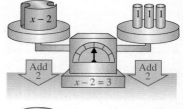

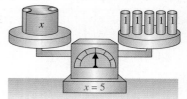

> ### Equivalent Equations
> Equations with the same solutions are called **equivalent equations.**

The procedure that we used to solve $x - 2 = 3$ illustrates the following property of equality.

> ### Addition Property of Equality
> Adding the same number to both sides of an equation does not change its solution.
>   For any numbers a, b, and c,
>   if $a = b$, then $a + c = b + c$

When we use this property of equality, the resulting equation is *equivalent to the original one.* We will now show how it is used to solve $x - 2 = 3$.

EXAMPLE 2   Solve: $x - 2 = 3$

Strategy We will use a property of equality to isolate the variable on one side of the equation.

WHY To solve the original equation, we want to find a simpler equivalent equation of the form $x = $ **a number**, whose solution is obvious.

Solution
We will use the addition property of equality to isolate x on the left side of the equation. We can undo the subtraction of 2 by adding 2 to both sides.

$$x - 2 = 3 \qquad \text{This is the equation to solve.}$$
$$x - 2 + 2 = 3 + 2 \qquad \text{Add 2 to both sides.}$$
$$x + 0 = 5 \qquad \begin{array}{l}\text{On the left side, the sum of a number and its opposite is zero:}\\ -2 + 2 = 0.\text{ On the right side, add: } 3 + 2 = 5.\end{array}$$
$$x = 5 \qquad \begin{array}{l}\text{On the left side, when 0 is added to a number, the result}\\ \text{is the same number.}\end{array}$$

Since 5 is obviously the solution of the equivalent equation $x = 5$, the solution of the original equation, $x - 2 = 3$, is also 5. To check this result, we substitute 5 for x in the original equation and simplify.

Self Check 2

Solve: $n - 16 = 33$

Now Try ⟳ Problem 37

$$x - 2 = 3 \qquad \text{This is the original equation.}$$
$$5 - 2 \stackrel{?}{=} 3 \qquad \text{Substitute 5 for } x.$$
$$3 = 3 \qquad \text{True}$$

Since the statement $3 = 3$ is true, 5 is the solution of $x - 2 = 3$.

THE LANGUAGE OF ALGEBRA

We solve equations by writing a series of steps that result in an equivalent equation of the form

$x = $ *a number* or *a number* $= x$

We say the variable is **isolated** on one side of the equation. Isolated means alone or by itself.

EXAMPLE 3   Solve: **a.** $-19 = y - 7$ **b.** $-27 + y = -3$

Strategy We will use a property of equality to isolate the variable on one side of the equation.

WHY To solve the original equation, we want to find a simpler equivalent equation of the form $y = $ **a number** or **a number** $= y$, whose solution is obvious.

Solution

a. To isolate y on the right side, we use the addition property of equality. We can undo the subtraction of 7 by adding 7 to both sides.

$$-19 = y - 7 \qquad \text{This is the equation to solve.}$$
$$-19 + 7 = y - 7 + 7 \qquad \text{Add 7 to both sides.}$$
$$-12 = y \qquad \text{On the left side, add. On the right side, the sum of a number and its opposite is zero: } -7 + 7 = 0.$$

Check: $-19 = y - 7 \qquad \text{This is the original equation.}$
$$-19 \stackrel{?}{=} -12 - 7 \qquad \text{Substitute } -12 \text{ for } y.$$
$$-19 = -19 \qquad \text{True}$$

Since the statement $-19 = -19$ is true, the solution is -12.

Caution! We may solve an equation so that the variable is isolated on either side of the equation. Note that $-12 = y$ is equivalent to $y = -12$.

b. To isolate y, we use the addition property of equality. We can eliminate -27 on the left side by adding its opposite to both sides.

$$-27 + y = -3 \qquad \text{The equation to solve.}$$
$$-27 + y + 27 = -3 + 27 \qquad \text{Add 27 to both sides.}$$
$$y = 24 \qquad \text{The sum of a number and its opposite is zero: } -27 + 27 = 0.$$

Check: $-27 + y = -3 \qquad \text{This is the original equation.}$
$$-27 + 24 \stackrel{?}{=} -3 \qquad \text{Substitute 24 for } y.$$
$$-3 = -3 \qquad \text{True}$$

The solution of $-27 + y = -3$ is 24.

Caution! After checking a result, be careful when stating your conclusion. Here, it would be incorrect to say:

The solution is -3.

The number we were checking was 24, not -3.

Self Check 3

Solve:
a. $-5 = b - 38$
b. $-20 + n = 29$

Now Try ⟶ Problems **39**
and **43**

OBJECTIVE 3 **Use the subtraction property of equality.**

To introduce another property of equality, consider the first scale shown on the right, which represents the equation $x + 3 = 5$. The scale is in balance because the weights on the left and right sides are equal. To find x, we need to remove 3 from the left side. To keep the scale in balance, we must also remove 3 from the right side. After doing this, we see that x is balanced by 2. Therefore, x must be 2. We say that we have solved the equation $x + 3 = 5$ and that the solution is 2. This example illustrates the following property of equality.

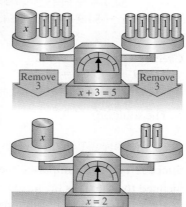

> **Subtraction Property of Equality**
>
> Subtracting the same number from both sides of an equation does not change its solution.
>
> For any numbers a, b, and c,
>
> if $a = b$, then $a - c = b - c$

When we use this property of equality, the resulting equation is equivalent to the original one.

EXAMPLE 4   Solve: **a.** $x + \dfrac{1}{8} = \dfrac{7}{4}$ **b.** $54.9 + x = 45.2$

Strategy We will use a property of equality to isolate the variable on one side of the equation.

WHY To solve the original equation, we want to find a simpler equivalent equation of the form $x = $ **a number**, whose solution is obvious.

Solution

a. To isolate x, we use the subtraction property of equality. We can undo the addition of $\frac{1}{8}$ by subtracting $\frac{1}{8}$ from both sides.

$$x + \frac{1}{8} = \frac{7}{4} \qquad \text{This is the equation to solve.}$$

$$x + \frac{1}{8} - \frac{1}{8} = \frac{7}{4} - \frac{1}{8} \qquad \text{Subtract } \tfrac{1}{8} \text{ from both sides.}$$

$$x = \frac{7}{4} - \frac{1}{8} \qquad \text{On the left side, } \tfrac{1}{8} - \tfrac{1}{8} = 0.$$

$$x = \frac{7}{4} \cdot \frac{2}{2} - \frac{1}{8} \qquad \text{On the right side, build } \tfrac{7}{4} \text{ so that it has a denominator of 8.}$$

$$x = \frac{14}{8} - \frac{1}{8} \qquad \text{Multiply the numerators and multiply the denominators.}$$

$$x = \frac{13}{8} \qquad \text{Subtract the numerators. Write the result over the common denominator 8.}$$

The solution is $\dfrac{13}{8}$. Check by substituting it for x in the original equation.

b. To isolate x, we use the subtraction property of equality. We can undo the addition of 54.9 by subtracting 54.9 from both sides.

$$54.9 + x = 45.2 \qquad \text{This is the equation to solve.}$$

$$54.9 + x - \mathbf{54.9} = 45.2 - \mathbf{54.9} \qquad \text{Subtract 54.9 from both sides.}$$

$$x = -9.7 \qquad \text{On the left side, } 54.9 - 54.9 = 0.$$

$$\begin{array}{r} ^{4\ 14} \\ 5\!\!\!/4.9 \\ -\ 45.2 \\ \hline 9.7 \end{array}$$

Check: $54.9 + x = 45.2 \qquad \text{This is the original equation.}$

$54.9 + (-9.7) \overset{?}{=} 45.2 \qquad \text{Substitute } -9.7 \text{ for } x.$

$45.2 = 45.2 \qquad \text{True}$

$$\begin{array}{r} ^{4\ 14} \\ 5\!\!\!/4.9 \\ -\ 9.7 \\ \hline 45.2 \end{array}$$

The solution is -9.7.

Self Check 4

Solve:

a. $x + \dfrac{4}{15} = \dfrac{11}{5}$

b. $0.7 + a = 0.2$

Now Try ➡ Problems 49 and 51

Success Tip In Example 4a, the solution, $\frac{13}{8}$, is an improper fraction. If you were inclined to write it as the mixed number $1\frac{5}{8}$, that is not necessary. It is common practice in algebra to leave such solutions in improper fraction form. Just make sure that they are simplified (the numerator and denominator have no common factors other than 1).

OBJECTIVE 4   Use the multiplication property of equality.

The first scale shown on the right represents the equation $\frac{x}{3} = 25$. The scale is in balance because the weights on the left side and right side are equal. To find x, we must triple (multiply by 3) the weight on the left side. To keep the scale in balance, we must also triple the weight on the right side. After doing this, we see that x is balanced by 75. Therefore, x must be 75.

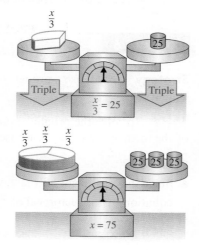

The procedure that we used to solve $\frac{x}{3} = 25$ illustrates the following property of equality.

Multiplication Property of Equality

Multiplying both sides of an equation by the same nonzero number does not change its solution.

For any numbers a, b, and c, where c is not 0,

if $a = b$, then $ca = cb$

When we use this property, the resulting equation is equivalent to the original one. We will now show how it is used to solve $\frac{x}{3} = 25$ algebraically.

EXAMPLE 5   Solve: $\dfrac{x}{3} = 25$

Strategy We will use a property of equality to isolate the variable on one side of the equation.

WHY To solve the original equation, we want to find a simpler equivalent equation of the form $x = $ **a number**, whose solution is obvious.

Solution
To isolate x, we use the multiplication property of equality. We can undo the division by 3 by multiplying both sides by 3.

$$\frac{x}{3} = 25 \qquad \text{This is the equation to solve.}$$

$$3 \cdot \frac{x}{3} = 3 \cdot 25 \qquad \text{Multiply both sides by 3.}$$

$$\frac{3}{1} \cdot \frac{x}{3} = 3 \cdot 25 \qquad \text{Write 3 as } \tfrac{3}{1}.$$

$$\frac{3x}{3} = 75 \qquad \text{Do the multiplications.}$$

$$1x = 75 \qquad \text{Simplify } \tfrac{3x}{3} \text{ by removing the common factor of 3 in the numerator and denominator: } \frac{\overset{1}{\cancel{3}}x}{\underset{1}{\cancel{3}}} = 1.$$

$$x = 75 \qquad \text{The coefficient 1 need not be written since } 1x = x.$$

$$\begin{array}{r} 25 \\ \times\ 3 \\ \hline 75 \end{array}$$

If we substitute 75 for x in $\frac{x}{3} = 25$, we obtain the true statement $25 = 25$. This verifies that 75 is the solution.

Self Check 5

Solve: $\dfrac{b}{24} = 3$

Now Try ⟿ **Problem 53**

Since the product of a number and its reciprocal is 1, we can solve equations such as $\frac{2}{3}x = 6$, where the coefficient of the variable term is a fraction, as follows.

EXAMPLE 6   Solve: **a.** $\frac{2}{3}x = 6$ **b.** $-\frac{5}{4}x = 3$

Strategy We will use a property of equality to isolate the variable on one side of the equation.

WHY To solve the original equation, we want to find a simpler equivalent equation of the form $x = \textbf{a number}$, whose solution is obvious.

Solution

a. Since the coefficient of x is $\frac{2}{3}$, we can isolate x by multiplying both sides of the equation by the reciprocal of $\frac{2}{3}$, which is $\frac{3}{2}$.

$$\frac{2}{3}x = 6 \qquad \text{This is the equation to solve.}$$

$$\frac{3}{2} \cdot \frac{2}{3}x = \frac{3}{2} \cdot 6 \qquad \text{To undo the multiplication by } \frac{2}{3}, \text{ multiply both sides}$$
$$\text{by the reciprocal of } \frac{2}{3}.$$

$$\left(\frac{3}{2} \cdot \frac{2}{3}\right)x = \frac{3}{2} \cdot 6 \qquad \text{Use the associative property of multiplication to group } \frac{3}{2} \text{ and } \frac{2}{3}.$$

$$1x = 9 \qquad \text{On the left side, the product of a number and its reciprocal is 1: } \frac{3}{2} \cdot \frac{2}{3} = 1.$$
$$\text{On the right side, } \frac{3}{2} \cdot 6 = \frac{18}{2} = 9.$$

$$x = 9 \qquad \text{The coefficient 1 need not be written since } 1x = x.$$

Check: $\frac{2}{3}x = 6 \qquad \text{This is the original equation.}$

$$\frac{2}{3}(9) \stackrel{?}{=} 6 \qquad \text{Substitute 9 for } x \text{ in the original equation.}$$

$$6 = 6 \qquad \text{On the left side, } \frac{2}{3}(9) = \frac{18}{3} = 6.$$

Since the statement $6 = 6$ is true, 9 is the solution of $\frac{2}{3}x = 6$.

b. To isolate x, we multiply both sides by the reciprocal of $-\frac{5}{4}$, which is $-\frac{4}{5}$.

$$-\frac{5}{4}x = 3 \qquad \text{This is the equation to solve.}$$

$$-\frac{4}{5}\left(-\frac{5}{4}x\right) = -\frac{4}{5}(3) \qquad \text{To undo the multiplication by } -\frac{5}{4}, \text{ multiply both sides}$$
$$\text{by the reciprocal of } -\frac{5}{4}.$$

$$1x = -\frac{12}{5} \qquad \text{On the left side, the product of a number and its reciprocal is 1:}$$
$$-\frac{4}{5}\left(-\frac{5}{4}\right) = 1.$$

$$x = -\frac{12}{5} \qquad \text{The coefficient 1 need not be written since } 1x = x.$$

The solution is $-\frac{12}{5}$. Verify that this is correct by checking.

Self Check 6

Solve:

a. $\frac{7}{2}x = 21$

b. $-\frac{3}{8}b = 2$

Now Try Problems 61 and 67

OBJECTIVE 5 Use the division property of equality.

To introduce a fourth property of equality, consider the first scale shown on the next page, which represents the equation $2x = 6$. The scale is in balance because the weights on the left and right sides are equal. To find x, we need to split the amount of weight on the left side in half. To keep the scale in balance, we must split the amount of weight in half on the right side. After doing this, we see that x is balanced by 3. Therefore, x must be 3. We say that we have solved the equation $2x = 6$ and that the solution is 3. This example illustrates the following property of equality.

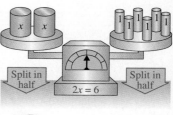

Division Property of Equality

Dividing both sides of an equation by the same nonzero number does not change its solution.

For any numbers a, b, and c, where c is not 0,

$$\text{if } a = b, \text{ then } \frac{a}{c} = \frac{b}{c}$$

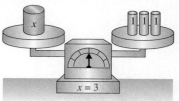

When we use this property of equality, the resulting equation is equivalent to the original one.

EXAMPLE 7   Solve: **a.** $2t = 80$ **b.** $-6.02 = -8.6t$

Strategy We will use a property of equality to isolate the variable on one side of the equation.

WHY To solve the original equation, we want to find a simpler equivalent equation of the form $t = \textbf{a number}$ or $\textbf{a number} = t$, whose solution is obvious.

Solution

a. To isolate t on the left side, we use the division property of equality. We can undo the multiplication by 2 by dividing both sides of the equation by 2.

$2t = 80$ This is the equation to solve.

$\dfrac{2t}{2} = \dfrac{80}{2}$ Divide both sides by 2.

$1t = 40$ On the left side, simplify $\frac{2t}{2}$ by removing the common factor of 2 in the numerator and denominator: $\frac{\overset{1}{\cancel{2}}t}{\underset{1}{\cancel{2}}} = 1t$. On the right side, do the division.

$t = 40$ The coefficient 1 need not be written since $1t = t$.

If we substitute 40 for t in $2t = 80$, we obtain the true statement $80 = 80$. This verifies that 40 is the solution.

Since division by 2 is the same as multiplication by $\frac{1}{2}$, we can also solve $2t = 80$ using the multiplication property of equality. We could also isolate t by multiplying both sides by the *reciprocal* of 2, which is $\frac{1}{2}$:

$$\frac{1}{2} \cdot 2t = \frac{1}{2} \cdot 80$$

b. To isolate t on the right side, we use the division property of equality. We can undo the multiplication by -8.6 by dividing both sides by -8.6.

$-6.02 = -8.6t$ This is the equation to solve.

$\dfrac{-6.02}{-8.6} = \dfrac{-8.6t}{-8.6}$ Use the division property of equality: Divide both sides by -8.6.

$0.7 = t$ On the left side, do the division. The quotient of two negative numbers is positive. On the right side, simplify by removing the common factor of -8.6 from the numerator and denominator: $\frac{\overset{1}{\cancel{-8.6}}t}{\underset{1}{\cancel{-8.6}}} = t$.

$$8\,6\,)\overline{60.2} \quad \begin{array}{r} 0.7 \\ \hline \end{array}$$
$$\underline{-60\ 2}$$
$$0$$

The solution is 0.7. Verify that this is correct by checking.

Self Check 7

Solve:
a. $16x = 176$
b. $10.04 = -0.4r$

Now Try ➡ Problems 69 and 79

Success Tip It is usually easier to multiply on each side if the coefficient of the variable term is a *fraction*, and divide on each side if the coefficient is an *integer* or *decimal*.

EXAMPLE 8 Solve: $-x = 3$

Strategy The variable x is not isolated because there is a $-$ sign in front of it. Since the term $-x$ has an understood coefficient of -1, the equation can be written as $-1x = 3$. We need to select a property of equality and use it to isolate the variable on one side of the equation.

WHY To find the solution of the original equation, we want to find a simpler equivalent equation of the form $x = $ **a number**, whose solution is obvious.

Solution To isolate x, we can either multiply or divide both sides by -1.

Multiply both sides by -1		*Divide both sides by -1*	
$-x = 3$	*The equation to solve*	$-x = 3$	*The equation to solve*
$-1x = 3$	*Write: $-x = -1x$*	$-1x = 3$	*Write: $-x = -1x$*
$(-1)(-1x) = (-1)3$		$\dfrac{-1x}{-1} = \dfrac{3}{-1}$	
$1x = -3$		$1x = -3$	*On the left side, $\frac{-1}{-1} = 1$.*
$x = -3$		$x = -3$	

Check: $\quad -x = 3 \quad$ *This is the original equation.*

$-(-3) \overset{?}{=} 3 \quad$ *Substitute -3 for x.*

$3 = 3 \quad$ *On the left side, the opposite of -3 is 3.*

Since the statement $3 = 3$ is true, -3 is the solution of $-x = 3$.

Self Check 8

Solve: $-h = -12$

Now Try ➡ Problem 81

Answers to Self Checks

1. yes **2.** 49 **3. a.** 33 **b.** 49 **4. a.** $\frac{29}{15}$ **b.** -0.5 **5.** 72 **6. a.** 6 **b.** $-\frac{16}{3}$ **7. a.** 11
b. -25.1 **8.** 12

SECTION 8.3 STUDY SET

VOCABULARY

Fill in the blanks.

1. An _____, such as $x + 1 = 7$, is a statement indicating that two expressions are equal.

2. Any number that makes an equation true when substituted for the variable is said to _____ the equation. Such numbers are called _____.

3. To _____ an equation means to find all values of the variable that make the equation true.

4. To solve an equation, we _____ the variable on one side of the equal symbol.

5. Equations with the same solutions are called _____ equations.

6. To _____ the solution of an equation, we substitute the value for the variable in the original equation and determine whether the result is a true statement.

CONCEPTS

7. Consider $x + 6 = 12$.
   a. What is the left side of the equation?
   b. Is this equation true or false?
   c. Is 5 a solution?
   d. Does 6 satisfy the equation?

8. For each equation, determine what operation is performed on the variable. Then explain how to undo that operation to isolate the variable.
   a. $x - 8 = 24$
   b. $x + 8 = 24$
   c. $\dfrac{x}{8} = 24$
   d. $8x = 24$

9. Complete the following properties of equality.

    a. If $a = b$, then

$$a + c = b + \boxed{} \quad \text{and} \quad a - c = b - \boxed{}$$

    b. If $a = b$, then

$$ca = \boxed{}\, b \quad \text{and} \quad \frac{a}{c} = \frac{b}{\boxed{}} \quad (c \neq 0)$$

10. **a.** To solve $\frac{h}{10} = 20$, do we multiply both sides of the equation by 10 or 20?

    b. To solve $4k = 16$, do we subtract 4 from both sides of the equation or divide both sides by 4?

11. Simplify each expression.

    a. $x + 7 - 7$ **b.** $y - 2 + 2$

    c. $\dfrac{5t}{5}$ **d.** $6 \cdot \dfrac{h}{6}$

12. **a.** To solve $-\frac{4}{5}x = 8$, we can multiply both sides by the reciprocal of $-\frac{4}{5}$. What is the reciprocal of $-\frac{4}{5}$?

    b. What is $-\frac{5}{4}\left(-\frac{4}{5}\right)$?

NOTATION

Complete each step to solve the equation.

13. $x - 5 = 45$ **Check:** $x - 5 = 45$

    $x - 5 + \boxed{} = 45 + \boxed{}$ $\boxed{} - 5 \boxed{} 45$

          $x = \boxed{}$ $\boxed{} = 45$ True

                                $\boxed{}$ is the solution.

14. $8x = 40$ **Check:** $8x = 40$

    $\dfrac{8x}{\boxed{}} = \dfrac{40}{\boxed{}}$ $8(\boxed{}) \overset{?}{=} 40$

      $x = \boxed{}$ $\boxed{} = 40$ True

                                  $\boxed{}$ is the solution.

15. **a.** What does the symbol $\overset{?}{=}$ mean?

    b. If you solve an equation and obtain $50 = x$, can you write $x = 50$?

16. Fill in the blank: $-x = \boxed{}\, x$

GUIDED PRACTICE

Check to determine whether the number in red is a solution of the equation. **See Example 1.**

17. 6, $x + 12 = 28$ 18. 110, $x - 50 = 60$

19. -8, $2b + 3 = -15$ 20. -2, $5t - 4 = -16$

21. 5, $0.5x = 2.9$ 22. 3.5, $1.2 + x = 4.7$

23. -6, $33 - \dfrac{x}{2} = 30$ 24. -8, $\dfrac{x}{4} + 98 = 100$

25. -2, $|c - 8| = 10$ 26. -45, $|30 - r| = 15$

27. 12, $3x - 2 = 4x - 5$ 28. 5, $5y + 8 = 3y - 2$

29. -3, $x^2 - x - 6 = 0$ 30. -2, $y^2 + 5y - 3 = 0$

31. 1, $\dfrac{2}{a+1} + 5 = \dfrac{12}{a+1}$ 32. 4, $\dfrac{2t}{t-2} - \dfrac{4}{t-2} = 1$

33. $\dfrac{3}{4}$, $x - \dfrac{1}{8} = \dfrac{5}{8}$ 34. $\dfrac{7}{3}$, $-4 = a + \dfrac{5}{3}$

35. -3, $(x - 4)(x + 3) = 0$ 36. 5, $(2x + 1)(x - 5) = 0$

Use a property of equality to solve each equation. Then check the result. **See Examples 2–4.**

37. $a - 5 = 66$ 38. $x - 34 = 19$

39. $9 = p - 9$ 40. $3 = j - 88$

41. $x - 1.6 = -2.5$ 42. $y - 1.2 = -1.3$

43. $-3 + a = 0$ 44. $-1 + m = 0$

45. $d - \dfrac{1}{9} = \dfrac{7}{9}$ 46. $\dfrac{7}{15} = b - \dfrac{1}{15}$

47. $x + 7 = 10$ 48. $y + 15 = 24$

49. $s + \dfrac{1}{5} = \dfrac{4}{25}$ 50. $\dfrac{1}{6} = h + \dfrac{4}{3}$

51. $3.5 + f = 1.2$ 52. $9.4 + h = 8.1$

Use a property of equality to solve each equation. Then check the result. **See Example 5.**

53. $\dfrac{x}{15} = 3$ 54. $\dfrac{y}{7} = 12$

55. $0 = \dfrac{v}{11}$ 56. $\dfrac{d}{49} = 0$

57. $\dfrac{d}{-7} = -3$ 58. $\dfrac{c}{-2} = -11$

59. $\dfrac{y}{0.6} = -4.4$ 60. $\dfrac{y}{0.8} = -2.9$

Use a property of equality to solve each equation. **See Example 6.**

61. $\dfrac{4}{5}t = 16$ 62. $\dfrac{11}{15}y = 22$

63. $\dfrac{2}{3}c = 10$ 64. $\dfrac{9}{7}d = 81$

65. $-\dfrac{7}{2}r = 21$ 66. $-\dfrac{4}{5}s = 36$

67. $-\dfrac{5}{4}h = -5$ 68. $-\dfrac{3}{8}t = -3$

Use a property of equality to solve each equation. See Example 7.

69. $4x = 16$

70. $5y = 45$

71. $63 = 9c$

72. $40 = 5t$

73. $23b = 23$

74. $16 = 16h$

75. $-8h = 48$

76. $-9a = 72$

77. $-100 = -5g$

78. $-80 = -5w$

79. $-3.4y = -1.7$

80. $-2.1x = -1.26$

Use a property of equality to solve each equation. See Example 8.

81. $-x = 18$

82. $-y = 50$

83. $-n = \dfrac{4}{21}$

84. $-w = \dfrac{11}{16}$

TRY IT YOURSELF

Solve each equation. Then check the result.

85. $8.9 = -4.1 + t$

86. $7.7 = -3.2 + s$

87. $-2.5 = -m$

88. $-1.8 = -b$

89. $-\dfrac{9}{8}x = 3$

90. $-\dfrac{14}{3}c = 7$

91. $r - \dfrac{1}{7} = \dfrac{5}{7}$

92. $\dfrac{4}{13} = a - \dfrac{1}{13}$

93. $-100 = -5g$

94. $-80 = -5w$

95. $-10 = n - 5$

96. $-8 = t - 2$

97. $\dfrac{h}{-40} = 5$

98. $\dfrac{x}{-7} = 12$

99. $x - 9 = 12$

100. $18 = x - 3$

LOOK ALIKES

Solve each equation. Then check the result.

101. a. $4 + x = 20$ **b.** $4x = 20$

102. a. $-7 + x = 14$ **b.** $-7x = 14$

103. a. $\dfrac{1}{3}x = 12$ **b.** $3x = 12$

104. a. $-5x = 30$ **b.** $-\dfrac{1}{5}x = 30$

APPLICATIONS

105. Synthesizers.
To find the unknown angle measure, which is represented by x, solve the equation $x + 115 = 180$.

106. Stop signs.   To find the measure of one angle of the stop sign, which is represented by x, solve the equation $8x = 1{,}080$.

107. Lotteries.   The largest lottery payout in the United States was a Mega Millions jackpot won on March 30, 2012. Three tickets had the winning numbers, and each received a $158 million cash payout. To find the total amount of the cash payout (in millions of dollars), which is represented by x, solve the equation $\dfrac{x}{3} = 158$. (Source: http://articles.latimes.com/2012/apr/10/nation/la-na-nn-maryland-mega-millions-school-employees-20120410)

108. Tennis.   Chris Evert won 154 Women's Tennis Association tournament titles in her career. This is 13 less than the all-time leader, Martina Navratilova. To find the number of titles won by Navratilova, which is represented by x, solve the equation $154 = x - 13$. (Source: tennis-x.com)

WRITING

109. What does it mean to solve an equation?

110. When solving an equation, we *isolate* the variable on one side of the equation. Write a sentence in which the word *isolate* is used in a different context.

111. Explain the error in the following work.

Solve: $x + 2 = 40$

$x + 2 - 2 = 40$

$x = 40$

112. After solving an equation, how do we check the result?

REVIEW

113. Evaluate $-9 - 3x$ for $x = -3$.

114. Evaluate: $-5^2 + (-5)^2$

115. Translate to symbols: Subtract x from 45

116. Evaluate: $\dfrac{2^3 + 3(5 - 3)}{15 - 4 \cdot 2}$

SECTION 8.4   More About Solving Equations

We have solved simple equations by using properties of equality. We will now expand our equation-solving skills by considering more complicated equations.

OBJECTIVES

1 Use more than one property of equality to solve equations.

2 Simplify expressions to solve equations.

OBJECTIVE 1   Use more than one property of equality to solve equations.

Sometimes we must use several properties of equality to solve an equation. For example, on the left side of $2x + 6 = 10$, the variable x is multiplied by 2, and then 6 is added to that product. To isolate x, we use the order of operations rules in reverse. First, we undo the addition of 6, and then we undo the multiplication by 2.

$2x + 6 = 10$ *This is the equation to solve.*

$2x + 6 - 6 = 10 - 6$ *To undo the addition of 6, subtract 6 from both sides.*

$2x = 4$ *Do the subtractions.*

$\dfrac{2x}{2} = \dfrac{4}{2}$ *To undo the multiplication by 2, divide both sides by 2.*

$x = 2$ *On the left side, simplify by removing the common factor of 2 from the numerator and denominator:*
                          $\dfrac{\overset{1}{\cancel{2}}x}{\underset{1}{\cancel{2}}} = 1x = x.$ *On the right side, do the division.*

The solution is 2.

EXAMPLE 1   Solve: $12x + 5 = 17$

Strategy First we will use a property of equality to isolate the *variable term* on one side of the equation. Then we will use a second property of equality to isolate the *variable* itself.

WHY To solve the original equation, we want to find a simpler equivalent equation of the form $x = $ **a number**, whose solution is obvious.

Solution

On the left side of the equation, x is multiplied by 12, and then 5 is added to that product. To isolate x, we undo the operations in the opposite order.

- To isolate the variable term, $12x$, we subtract 5 from both sides to undo the addition of 5.
- To isolate the variable, x, we divide both sides by 12 to undo the multiplication by 12.

$12x + 5 = 17$ *This is the equation to solve. First, we want to isolate the variable term, 12x.*

$12x + 5 - 5 = 17 - 5$ *Use the subtraction property of equality: Subtract 5 from both sides to isolate 12x.*

$12\,x = 12$ *Do the subtractions: 5 − 5 = 0 and 17 − 5 = 12. Now we want to isolate the variable, x.*

$\dfrac{12x}{12} = \dfrac{12}{12}$ *Use the division property of equality: Divide both sides by 12 to isolate x.*

$x = 1$ *On the left side simplify: $\dfrac{\overset{1}{\cancel{12}}x}{\underset{1}{\cancel{12}}} = x.$*
                          On the right side, do the division.

THE LANGUAGE OF ALGEBRA

In Example 1, we subtract 5 from both sides to isolate the **variable term**, *12x*. Then we divide both sides by 12 to isolate the variable, *x*.

Check: $12x + 5 = 17$ This is the original equation.

$12(\mathbf{1}) + 5 \stackrel{2}{=} 17$ Substitute 1 for x.

$12 + 5 \stackrel{2}{=} 17$ Do the multiplication on the left side.

$17 = 17$ True

The solution is 1.

Self Check 1

Solve: $8x - 13 = 43$

Now Try ➡ **Problem 15**

Caution!   When checking solutions, always use the original equation.

EXAMPLE 2   Solve: $10 = -5s - 60$

Strategy First we will use a property of equality to isolate the *variable term* on one side of the equation. Then we will use a second property of equality to isolate the *variable* itself.

WHY   To solve the original equation, we want to find a simpler equivalent equation of the form **a number** $= s$, whose solution is obvious.

Solution
On the right side of the equation, 5 is multiplied by -5, and then 60 is subtracted from that product. To isolate s, we undo the operations in the opposite order.

- To isolate the variable term, $-5s$, we add 60 to both sides to undo the subtraction of 60.
- To isolate the variable, s, we divide both sides by -5 to undo the multiplication by -5.

$10 = \boxed{-5s} - 60$ This is the equation to solve. First, we want to isolate the variable term, $-5s$.

$10 + \mathbf{60} = -5s - 60 + \mathbf{60}$ Use the addition property of equality: Add 60 to both sides to isolate $-5s$.

$70 = -5\boxed{s}$ Do the additions: $10 + 60 = 70$ and $-60 + 60 = 0$. Now we want to isolate the variable, s.

$\dfrac{70}{\mathbf{-5}} = \dfrac{-5s}{\mathbf{-5}}$ Use the division property of equality: Divide both sides by -5 to isolate s.

On the left side, do the division. The quotient of a positive and a negative number is negative.

$-14 = s$ On the right side, simplify: $\dfrac{\overset{1}{\cancel{-5}}s}{\underset{1}{\cancel{-5}}} = s$.

$$\begin{array}{r} 14 \\ 5\overline{)70} \\ -5 \\ \hline 20 \\ -20 \\ \hline 0 \end{array}$$

Check: $10 = -5s - 60$ This is the original equation.

$10 \stackrel{2}{=} -5(\mathbf{-14}) - 60$ Substitute -14 for s.

$10 \stackrel{2}{=} 70 - 60$ Do the multiplication on the right side.

$10 = 10$ True

$$\begin{array}{r} \overset{2}{1}4 \\ \times\ 5 \\ \hline 70 \end{array}$$

Self Check 2

Solve: $40 = -4d - 8$

Now Try ➡ **Problem 25**

The solution is -14.

EXAMPLE 3   Solve: $\dfrac{5}{8}m - 2 = -12$

Strategy We will use properties of equality to isolate the variable on one side of the equation.

WHY To solve the original equation, we want to find a simpler equivalent equation of the form **m = a number**, whose solution is obvious.

Solution

We note that the coefficient of m is $\frac{5}{8}$ and proceed as follows.

- To isolate the variable term $\frac{5}{8}m$, we add 2 to both sides to undo the subtraction of 2.
- To isolate the variable, m, we multiply both sides by $\frac{8}{5}$ to undo the multiplication by $\frac{5}{8}$.

$$\dfrac{5}{8}m - 2 = -12$$

This is the equation to solve. First, we want to isolate the variable term, $\frac{5}{8}m$.

$$\dfrac{5}{8}m - 2 + 2 = -12 + 2$$

Use the addition property of equality: Add 2 to both sides to isolate $\frac{5}{8}m$.

$$\dfrac{5}{8}m = -10$$

Do the additions: $-2 + 2 = 0$ and $-12 + 2 = -10$. Now we want to isolate the variable, m.

$$\dfrac{8}{5}\left(\dfrac{5}{8}m\right) = \dfrac{8}{5}(-10)$$

Use the multiplication property of equality: Multiply both sides by $\frac{8}{5}$ (which is the reciprocal of $\frac{5}{8}$) to isolate m.

$$m = -16$$

On the left side: $\frac{8}{5}\left(\frac{5}{8}\right) = 1$ and $1m = m$.

On the right side: $\frac{8}{5}(-10) = -\frac{8 \cdot 2 \cdot \overset{1}{\cancel{5}}}{\cancel{5}} = -16$.

The solution is -16. Check by substituting it into the original equation.

Self Check 3

Solve: $\dfrac{7}{12}a - 6 = -27$

Now Try ➡ **Problem 29**

EXAMPLE 4   Solve: $-0.2 = -0.8 - y$

Strategy First, we will use a property of equality to isolate the variable term on one side of the equation. Then we will use a second property of equality to isolate the variable itself.

WHY To solve the original equation, we want to find a simpler equivalent equation of the form **a number = y**, whose solution is obvious.

Solution

To isolate the variable term $-y$ on the right side, we eliminate -0.8 by adding 0.8 to both sides.

$$-0.2 = -0.8 - y$$

This is the equation to solve. First, we want to isolate the variable term, $-y$.

$$-0.2 + 0.8 = -0.8 - y + 0.8$$

Add 0.8 to both sides to isolate $-y$.

$$0.6 = -y$$

Do the additions.

$$\begin{array}{r} 0.8 \\ -0.2 \\ \hline 0.6 \end{array}$$

Since the term $-y$ has an understood coefficient of -1, the equation can be written as $0.6 = -1y$. To isolate y, we can either multiply both sides or divide both sides by -1. If we choose to divide both sides by -1, we proceed as follows.

$$0.6 = -1y$$

Now we want to isolate the variable y.

$$\dfrac{0.6}{-1} = \dfrac{-1y}{-1}$$

On the left side, do the division. The quotient of a positive and a negative number is negative. On the right side, simplify: $\frac{\overset{1}{\cancel{-1}}y}{\cancel{-1}} = y$.

$$-0.6 = y$$

The solution is -0.6. Check this by substituting it into the original equation.

Self Check 4

Solve: $-6.6 - m = -2.7$

Now Try ➡ **Problem 35**

OBJECTIVE 2 Simplify expressions to solve equations.

When solving equations, we should simplify the expressions that make up the left and right sides before applying any properties of equality. Often, that involves removing parentheses and/or combining like terms.

EXAMPLE 5 Solve: **a.** $3(k + 1) - 5k = 0$ **b.** $8a - 2(a - 7) = 68$

Strategy We will use the distributive property along with the process of combining like terms to simplify the left side of each equation.

WHY It's best to simplify each side of an equation before using a property of equality.

Solution

a.

$3(k + 1) - 5k = 0$	This is the equation to solve.
$3k + 3 - 5k = 0$	Distribute the multiplication by 3.
$-2k + 3 = 0$	Combine like terms: $3k - 5k = -2k$. First, we want to isolate the variable term, $-2k$.
$-2k + 3 - 3 = 0 - 3$	To undo the addition of 3, subtract 3 from both sides. This isolates $-2k$.
$-2k = -3$	Do the subtractions: $3 - 3 = 0$ and $0 - 3 = -3$. Now we want to isolate the variable, k.
$\dfrac{-2k}{-2} = \dfrac{-3}{-2}$	To undo the multiplication by -2, divide both sides by -2. This isolates k.
$k = \dfrac{3}{2}$	On the right side, simplify: $\dfrac{-3}{-2} = \dfrac{3}{2}$.

Check: $3(k + 1) - 5k = 0$	This is the original equation.
$3\left(\dfrac{3}{2} + 1\right) - 5\left(\dfrac{3}{2}\right) \stackrel{?}{=} 0$	Substitute $\dfrac{3}{2}$ for k.
$3\left(\dfrac{5}{2}\right) - 5\left(\dfrac{3}{2}\right) \stackrel{?}{=} 0$	Do the addition within the parentheses. Think of 1 as $\dfrac{2}{2}$ and then add: $\dfrac{3}{2} + \dfrac{2}{2} = \dfrac{5}{2}$.
$\dfrac{15}{2} - \dfrac{15}{2} \stackrel{?}{=} 0$	Do the multiplications.
$0 = 0$	True

The solution is $\dfrac{3}{2}$.

Caution! To check a result, we evaluate each side of the equation following the order of operations rule. For the check shown above, perform the addition within parentheses first. *Don't distribute the multiplication by 3.*

$$\underbrace{3\left(\dfrac{3}{2} + 1\right)}_{\text{Add first}}$$

b.

$8a - 2(a - 7) = 68$	This is the equation to solve.
$8a - 2a + 14 = 68$	Distribute the multiplication by -2.
$6a + 14 = 68$	Combine like terms: $8a - 2a = 6a$. First, we want to isolate the variable term, $6a$.
$6a + 14 - 14 = 68 - 14$	To undo the addition of 14, subtract 14 from both sides. This isolates $6a$.

$$\begin{array}{r} 68 \\ -\ 14 \\ \hline 54 \end{array}$$

$6a = 54$ Do the subtractions. Now we want to isolate the variable, a.

$\dfrac{6a}{6} = \dfrac{54}{6}$ To undo the multiplication by 6, divide both sides by 6. This isolates a.

$a = 9$ On the left side, simplify: $\dfrac{\overset{1}{\cancel{6}}a}{\underset{1}{\cancel{6}}} = a$.

On the right side, do the division.

The solution is 9. Use a check to verify this.

Self Check 5

Solve:
a. $4(a + 2) - a = 11$
b. $9x - 5(x - 9) = 1$

Now Try ➲ Problems 39 and 45

When solving an equation, if variables appear on both sides, we can use the addition (or subtraction) property of equality to get all variable terms on one side and all constant terms on the other.

EXAMPLE 6 Solve: $3x - 15 = 4x + 36$

Strategy There are variable terms ($3x$ and $4x$) on both sides of the equation. We will eliminate $3x$ from the left side of the equation by subtracting $3x$ from both sides.

WHY To solve for x, all the terms containing x must be on the same side of the equation.

Solution

$3x - 15 = 4x + 36$ This is the equation to solve. There are variable terms on both sides of the equation.

$3x - 15 - 3x = 4x + 36 - 3x$ Subtract 3x from both sides to isolate the variable term on the right side.

$-15 = x + 36$ Combine like terms: $3x - 3x = 0$ and $4x - 3x = x$. Now we want to isolate the variable, x.

$-15 - 36 = x + 36 - 36$ To undo the addition of 36, subtract 36 from both sides. This isolates x.

$-51 = x$ Do the subtractions.

$\begin{array}{r} \overset{1}{1}5 \\ + 36 \\ \hline 51 \end{array}$

Check: $3x - 15 = 4x + 36$ This is the original equation.

$3(-51) - 15 \overset{?}{=} 4(-51) + 36$ Substitute −51 for x.

$-153 - 15 \overset{?}{=} -204 + 36$ Do the multiplications.

$-168 = -168$ True

$\begin{array}{cc} 51 & 51 \\ \times\,3 & \times\,4 \\ \hline 153 & 204 \end{array}$

$\begin{array}{cc} & \overset{9}{\overset{1\cancel{1}0}{2\cancel{0}\cancel{4}}} \\ 153 & \\ + 15 & -36 \\ \hline 168 & 168 \end{array}$

The solution is -51.

Self Check 6

Solve: $30 + 6n = 4n - 2$

Now Try ➲ Problem 57

Success Tip In Example 6, we could have eliminated $4x$ from the right side by subtracting $4x$ from both sides:

$3x - 15 - 4x = 4x + 36 - 4x$

$-x - 15 = 36$ Note that the coefficient of x is negative.

However, it is usually easier to isolate the variable term on the side that will result in a *positive* coefficient.

EXAMPLE 7   Solve: $3(4x - 80) + 6x = 2(x + 40)$

Strategy We will use the distributive property on each side of the equation to remove the parentheses. Then we will combine any like terms.

WHY It is easiest to simplify the expressions that make up the left and right sides of the equation before using the properties of equality to isolate the variable.

Solution

$3(4x - 80) + 6x = 2(x + 40)$	This is the equation to solve.
$12x - 240 + 6x = 2x + 80$	Distribute the multiplication by 3 and by 2.
$18x - 240 = 2x + 80$	On the left side, combine like terms: $12x + 6x = 18x$. There are variable terms on both sides.
$18x - 240 - 2x = 2x + 80 - 2x$	To eliminate the term $2x$ on the right side, subtract $2x$ from both sides.
$16x - 240 = 80$	Combine like terms on each side: $18x - 2x = 16x$ and $2x - 2x = 0$.
$16x - 240 + 240 = 80 + 240$	To isolate the variable term, $16x$, on the left side, add 240 to both sides to undo the subtraction of 240.
$16x = 320$	Do the addition on each side: $-240 + 240 = 0$ and $80 + 240 = 320$. Now we want to isolate the variable, x.
$\dfrac{16x}{16} = \dfrac{320}{16}$	To isolate x on the left side, divide both sides by 16 to undo the multiplication by 16.
$x = 20$	On the left side, simplify, $\dfrac{16x}{16} = x$. On the right side, do the division.

$$\begin{array}{r} \overset{1}{}240 \\ +\ 80 \\ \hline 320 \end{array}$$

$$\begin{array}{r} 20 \\ 16\overline{)320} \\ -\ 32 \\ \hline 00 \\ -\ 0 \\ \hline 0 \end{array}$$

The solution is 20. Check by substituting it in the original equation.

Self Check 7

Solve: $6(5x - 30) - 2x = 8(x + 50)$

Now Try ➡ **Problem 57**

The previous examples suggest the following strategy for solving equations. You won't always have to use all four steps to solve a given equation. If a step doesn't apply, skip it and go to the next step.

Strategy for Solving Equations

1. **Simplify each side of the equation:** Use the distributive property to remove parentheses, and then combine like terms on each side.
2. **Isolate the variable term on one side:** Add (or subtract) to get the variable term on one side of the equation and a number on the other using the addition (or subtraction) property of equality.
3. **Isolate the variable:** Multiply (or divide) to isolate the variable using the multiplication (or division) property of equality.
4. **Check the result:** Substitute the possible solution for the variable in the *original* equation to see if a true statement results.

Answers to Self Checks

   1. 7   **2.** -12 **3.** -36 **4.** -3.9 **5. a.** 1 **b.** -11 **6.** -16 **7.** 29

SECTION **8.4**   **STUDY SET**

VOCABULARY

Fill in the blanks.

1. To _____ an equation means to find all values of the variable that make the equation true.

2. The equation $6x + 3 = 4x + 1$ has variable terms on _____ sides.

3. When solving equations, _____ the expressions that make up the left and right sides of the equation before using the properties of equality to isolate the variable.

4. When we write the expression $9x + x$ as $10x$, we say we have _____ like terms.

CONCEPTS

5. On the left side of the equation $4x + 9 = 25$, the variable x is multiplied by ☐, and then ☐ is added to that product.

6. On the right side of the equation $16 = -5t - 1$, the variable t is multiplied by ☐, and then ☐ is subtracted from that product.

Fill in the blanks.

7. To solve $3x - 5 = 1$, we first undo the _____ of 5 by adding 5 to both sides. Then we undo the _____ by 3 by dividing both sides by 3.

8. To solve $\frac{x}{2} + 3 = 5$, we can undo the _____ of 3 by subtracting 3 from both sides. Then we can undo the _____ by 2 by multiplying both sides by 2.

9. a. Combine like terms on the left side of
$6x - 8 - 8x = -24$.

b. Distribute and then combine like terms on the right side of $-20 = 4(3x - 4) - 9x$.

10. Distribute on both sides of the equation shown below. **Do not solve.**
$$7(3x + 2) = 4(x - 3)$$

11. Use a check to determine whether -2 is a solution of the equation.
a. $6x + 5 = 7$ **b.** $8(x + 3) = 8$

12. a. Simplify: $3x + 5 - x$
b. Solve: $3x + 5 = 9$
c. Evaluate $3x + 5 - x$ for $x = 9$.
d. Check: Is -1 a solution of $3x + 5 - x = 9$?

NOTATION

Complete each step.

13. Solve:
$$2x - 7 = 21$$
$$2x - 7 + \boxed{} = 21 + \boxed{}$$
$$2x = 28$$
$$\frac{2x}{\boxed{}} = \frac{28}{\boxed{}}$$
$$x = 14$$

Check:
$$2x - 7 = 21$$
$$2(\boxed{}) - 7 \quad 21$$
$$\boxed{} - 7 \overset{?}{=} 21$$
$$\boxed{} = 21$$
$\boxed{}$ is the solution.

14. Fill in the blank: $-y = \boxed{} y$

GUIDED PRACTICE

Solve each equation and check the result. See Example 1.

15. $2x + 5 = 17$ **16.** $4p + 3 = 43$

17. $5q - 2 = 23$ **18.** $3x - 5 = 13$

19. $-33 = 5t + 2$ **20.** $-55 = 3w + 5$

21. $0.7 + 4y = 1.7$ **22.** $0.3 + 2x = 0.9$

23. $-5 - 2d = 0$ **24.** $-8 - 3c = 0$

Solve each equation and check the result. See Example 2.

25. $12 = -7a - 9$ **26.** $15 = -8b - 1$

27. $-3 = -3p + 7$ **28.** $-1 = -2r + 8$

Solve each equation and check the result. See Example 3.

29. $\frac{2}{3}t + 2 = 6$ **30.** $\frac{3}{5}x - 6 = -12$

31. $\frac{5}{6}k - 5 = 10$ **32.** $\frac{2}{5}c - 12 = 2$

33. $-\frac{7}{16}h + 28 = 21$ **34.** $-\frac{5}{8}h + 25 = 15$

Solve each equation and check the result. See Example 4.

35. $-1.7 = 1.2 - x$ **36.** $0.6 = 4.1 - x$

37. $-6 - y = -2$ **38.** $-1 - h = -9$

Solve each equation and check the result. See Example 5.

39. $3(2y - 2) - y = 5$

40. $2(-3a + 2) + a = 2$

41. $9(x + 11) + 5(13 - x) = 0$

42. $20b + 2(6b - 1) = -34$

43. $-(4 - m) = -10$

44. $-(6 - t) = -12$

45. $10.08 = 4(0.5x + 2.5)$

46. $-3.28 = 8(1.5y - 0.5)$

47. $6a - 3(3a - 4) = 30$

48. $16y - 8(3y - 2) = -24$

49. $-(19 - 3s) - (8s + 1) = 35$

50. $6x - 5(3x + 1) = 58$

Solve each equation and check the result. **See Example 6.**

51. $5x = 4x + 7$

52. $3x = 2x + 2$

53. $8y + 44 = 4y$

54. $9y + 36 = 6y$

55. $60r - 50 = 15r - 5$

56. $100f - 75 = 50f + 75$

57. $8y - 2 = 4y + 16$

58. $7 + 3w = 4 + 9w$

Solve each equation and check the result. **See Example 7.**

59. $3(A + 2) + 4A = 2(A - 7)$

60. $9(T - 1) + 18T = 6(T + 2)$

61. $2 - 3(x - 5) = 4(x - 1)$

62. $2 - (4x + 7) = 3 + 2(x + 2)$

TRY IT YOURSELF

Solve each equation. Check the result.

63. $3x - 8 - 4x - 7x = -2 - 8$

64. $-6t - 7t - 5t - 1 = 12 - 3$

65. $4(d - 5) + 20 = 5 - 2d$

66. $1 - t = 5(t - 2) + 10$

67. $30x - 12 = 1{,}338$

68. $40y - 19 = 1{,}381$

69. $-7 = \dfrac{3}{7}r + 14$

70. $21 = \dfrac{2}{5}f - 19$

71. $10 - 2y = 8$

72. $7 - 7x = -21$

73. $9 + 5(r + 3) = 6 + 3(r - 2)$

74. $2 + 3(n - 6) = 4(n + 2) - 21$

75. $-\dfrac{2}{3}z + 4 = 8$

76. $-\dfrac{7}{5}x + 9 = -5$

77. $-2(9 - 3s) - (5s + 2) = -25$

78. $4(x - 5) - 3(12 - x) = 7$

79. $9a - 2.4 = 7a + 4.6$

80. $4c - 1.6 = 7c + 3.2$

LOOK ALIKES

Solve each equation. Then check the result.

81. a. $5(4x) = 40$ **b.** $5(4 + x) = 40$

82. a. $2(-7 + x) = 28$ **b.** $2(-7x) = 28$

83. a. $\dfrac{1}{3}(6x) = 12$ **b.** $\dfrac{1}{3}(x + 6) = 12$

84. a. $-5(x - 4) = 30$ **b.** $-\dfrac{1}{5}(10x) = 30$

WRITING

85. To solve $3x - 4 = 5x + 1$, one student began by subtracting $3x$ from both sides. Another student solved the same equation by first subtracting $5x$ from both sides. Will the students get the same solution? Explain why or why not.

86. Explain the error in the following solution.

$$\text{Solve:} \qquad 2x + 4 = 30$$

$$\cancel{\dfrac{2x}{2} + 4 = \dfrac{30}{2}}$$

$$\cancel{x + 4 = 15}$$

$$\cancel{x + 4 - 4 = 15 - 4}$$

$$\cancel{x = 11}$$

REVIEW

Name the property that is used.

87. $x \cdot 9 = 9x$

88. $x + 99 = 99 + x$

89. $(x + 1) + 2 = x + (1 + 2)$

90. $2(30y) = (2 \cdot 30)y$

OBJECTIVES

1 Solve application problems to find one unknown.

2 Solve application problems to find two unknowns.

SECTION 8.5 Using Equations to Solve Application Problems

Throughout this course, we have used the steps *Analyze, Form, Calculate, State, and Check* as a strategy to solve application problems. Now that you have had an introduction to algebra, we can modify that strategy and make use of your newly learned skills.

OBJECTIVE 1 Solve application problems to find one unknown.

To become a good problem solver, you need a plan to follow, such as the following six-step strategy. You will notice that the steps are quite similar to the strategy first introduced in Chapter 1. However, this new approach uses the concept of variable, the translation skills from Section 8.1, and the equation-solving methods of Sections 8.3 and 8.4.

Strategy for Problem Solving

1. **Analyze the problem** by reading it carefully to understand the given facts. What information is given? What are you asked to find? What vocabulary is given? Often, a diagram will help you understand the facts of the problem.
2. **Assign a variable** to represent an unknown value in the problem. This means, in most cases, to let $x =$ what you are asked to find. If there are other unknown values, represent each of them using an algebraic expression that involves the variable.
3. **Form an equation** by translating the words of the problem into mathematical symbols.
4. **Solve the equation** formed in Step 3.
5. **State the conclusion** clearly. Be sure to include the units (such as feet, seconds, or pounds) in your answer.
6. **Check the result** using the original wording of the problem, not the equation that was formed in Step 3.

EXAMPLE 1   **Systems analysis.** A company's telephone use would have to increase by 350 calls per hour before the system would reach the maximum capacity of 1,500 calls per hour. Currently, how many calls are being made each hour on the system?

Analyze

Neustockimages/E+/Getty Images

- If the number of calls increases by 350, the system will reach capacity. *Given*
- The maximum capacity of the system is 1,500 calls per hour. *Given*
- How many calls are currently being made each hour? *Find*

Caution! Unlike an arithmetic approach, you *do not* have to determine whether to add, subtract, multiply, or divide at this stage. Simply translate the words of the problem to mathematical symbols to form an equation that describes the situation. Then solve the equation.

Assign Let $n =$ the number of calls currently being made each hour.

Form To form an equation involving n, we look for a key word or phrase in the problem.

   Key phrase: *increase by 350*   **Translation:** addition

The key phrase tells us to add 350 to the current number of calls to obtain an expression for the maximum capacity of the system. Now we translate the words of the problem into a **verbal model** and then into an equation.

The current number of calls per hour	increased by	350	equals	the maximum capacity of the system.
n	$+$	350	$=$	$1{,}500$

Solve

$$n + 350 = 1{,}500$$ We need to isolate n on the left side.

$$n + 350 - 350 = 1{,}500 - 350$$ To isolate n, subtract 350 from both sides to undo the addition of 350.

$$n = 1{,}150$$ Do the subtraction.

$$\begin{array}{r} {\scriptstyle 4\,10} \\ 1{,}5\cancel{0}\cancel{0} \\ -\ 350 \\ \hline 1{,}150 \end{array}$$

State Currently, 1,150 calls per hour are being made.

Check If the number of calls currently being made each hour is 1,150, and we increase that number by 350, we should obtain the maximum capacity of the system.

$$\begin{array}{r} 1{,}150 \\ +\ 350 \\ \hline 1{,}500 \end{array}$$ ← *This is the maximum capacity.*

The result, 1,150, checks.

Caution! Always check the result in the original wording of the problem, not by substituting it into the equation. Why? The equation may have been solved correctly, but the danger is that you may have formed it incorrectly.

Self Check 1

Apartment buildings. Owners of a newly constructed apartment building would have to sell 34 more units before all of the 510 units were sold. How many of the apartment units have been sold to date?

Now Try ➲ **Problem 19**

EXAMPLE 2 **Small businesses.** Last year, a stylist lost 17 customers who moved away. If she now has 73 customers, how many did she have originally?

Analyze
- She lost 17 customers.   *Given*
- She now has 73 customers.   *Given*
- How many customers did she originally have?   *Find*

Assign We can let $c =$ the original number of customers.

Form To form an equation involving c, we look for a key word or phrase in the problem.

    Key phrase: *moved away*   **Translation:** *subtraction*

Now we translate the words of the problem into an equation.

The original number of customers	minus	17	is	the number of customers she has now.
c	$-$	17	$=$	73

This is called the verbal model.

Solve

$$c - 17 = 73 \qquad \text{We need to isolate } c \text{ on the left side.}$$
$$c - 17 + \mathbf{17} = 73 + \mathbf{17} \qquad \begin{array}{l}\text{To isolate } c,\text{ add 17 to both sides to undo the}\\ \text{subtraction of 17.}\end{array}$$
$$c = 90 \qquad \text{Do the addition.}$$

$$\begin{array}{r} \overset{1}{7}3 \\ +\ 17 \\ \hline 90 \end{array}$$

State She originally had 90 customers.

Check If the hair stylist originally had 90 customers, and we decrease that number by the 17 that moved away, we should obtain the number of customers she has now.

$$\begin{array}{r} \overset{8\ 10}{9\cancel{0}} \\ -\ 17 \\ \hline 73 \end{array}$$ *This is the number of customers the hair stylist now has.*

The result, 90, checks.

Self Check 2

Gasoline storage. A tank currently contains 1,325 gallons of gasoline. If 450 gallons have recently been pumped from the tank, how many gallons did it originally contain?

Now Try ➲ **Problem 20**

EXAMPLE 3   **Traffic fines.**   For speeding in a construction zone, a motorist had to pay a fine of $592. The violation occurred on a highway posted with signs like the one shown on the right. What would the fine have been if such signs were not posted?

> **TRAFFIC FINES
> DOUBLED IN
> CONSTRUCTION ZONE**

Analyze
- For speeding, the motorist was fined $592.   *Given*
- The fine was double what it would normally have been.   *Given*
- What would the fine have been, had the sign not been posted?   *Find*

Assign We can let f = the amount that the fine would normally have been.

Form To form an equation, we look for a key word or phrase in the problem or analysis.

Key word: *double*   **Translation:** *multiply by 2*

Now we translate the words of the problem into an equation.

Two	times	the normal speeding fine	is	the new fine.
2	·	f	=	592

Solve $2f = 592$ We need to isolate f on the left side.

$$\frac{2f}{2} = \frac{592}{2}$$ To isolate f, divide both sides by 2 to undo the multiplication by 2.

$$f = 296$$ Do the division.

$$\begin{array}{r} 296 \\ 2\overline{)592} \\ -4 \\ \hline 19 \\ -18 \\ \hline 12 \\ -12 \\ \hline 0 \end{array}$$

State The fine would normally have been $296.

Check If the normal fine was $296, and we double it, we should get the new fine.

$$\begin{array}{r} \overset{1\,1}{296} \\ \times\ 2 \\ \hline 592 \end{array}$$ ←—This is the new fine.

The result, $296, checks.

Self Check 3

Speed-reading. A Speed-reading course claims it can teach a person to read four times faster. After taking the course, a student can now read 700 words per minute. If the company's claims are true, what was the student's reading rate before taking the course?

Now Try ➲ Problem 21

EXAMPLE 4   **Entertainment costs.**   A five-piece band worked on New Year's Eve. If each player earned $120, what fee did the band charge?

Analyze
- There were 5 players in the band.   *Given*
- Each player made $120.   *Given*
- What fee did the band charge?   *Find*

Assign We can let f = the band's fee.

Form To form an equation, we look for a key word or phrase. In this case, we find it in the analysis of the problem. If each player earned the same amount ($120), the band's fee must have been divided into 5 equal parts.

Key phrase: *divided into 5 equal parts*   **Translation:** *division*

Now we translate the words of the problem into an equation.

The band's fee	divided by	the number of players in the band	is	each person's share.
f	$\div$	5	=	120

Solve

$$\frac{f}{5} = 120 \qquad \textit{We need to isolate } f \textit{ on the left side.}$$

$$5 \cdot \frac{f}{5} = 5 \cdot 120 \qquad \textit{To isolate f, multiply both sides by 5 to undo the division by 5.}$$

$$f = 600 \qquad \textit{Do the multiplication.}$$

$$\begin{array}{r} \overset{1}{120} \\ \times \quad 5 \\ \hline 600 \end{array}$$

State The band's fee was $600.

Check If the band's fee was $600, and we divide it into 5 equal parts, we should get the amount that each player earned.

$$\begin{array}{r} 120 \\ 5)\overline{600} \\ -5 \\ \hline 10 \\ -10 \\ \hline 00 \\ -0 \\ \hline 0 \end{array}$$ ←—This is the amount each band member earned.

The result, $600, checks.

Self Check 4

Classical music. A woodwind quartet was hired to play at an art exhibit. If each member made $85 for the performance, what fee did the quartet charge?

Now Try ➲ **Problem 22**

EXAMPLE 5 **Volunteer service hours.** To receive a degree in child development, students at one college must complete 135 hours of volunteer service by working 3-hour shifts at a local preschool. If a student has already volunteered 87 hours, how many more 3-hour shifts must she work to meet the service requirement for her degree?

Analyze
- Students must complete 135 hours of volunteer service.   *Given*
- Students work 3-hour shifts.   *Given*
- A student has already completed 87 hours of service.   *Given*
- How many more 3-hour shifts must she work?   *Find*

Assign Let x = the number of shifts needed to complete the service requirement.

Form Since each shift is 3 hours long, multiplying 3 by the number of shifts will give the number of additional hours the student needs to volunteer.

The number of hours she has already completed	plus	3 times	the number of shifts yet to be completed	is	the number of hours required.
87	+	3 ·	x	=	135

Solve

$$87 + 3x = 135$$

We need to isolate x on the left side.

$$87 + 3x - 87 = 135 - 87$$

To isolate the variable term 3x, subtract 87 from both sides to undo the addition of 87.

$$3x = 48$$

Do the subtraction.

$$\frac{3x}{3} = \frac{48}{3}$$

To isolate x, divide both sides by 3 to undo the multiplication by 3.

$$x = 16$$

Do the division.

$$\begin{array}{r} \overset{12}{\underset{}{\cancel{2}}}\,15 \\ \cancel{135} \\ -\ 87 \\ \hline 48 \end{array}$$

$$\begin{array}{r} 16 \\ 3\overline{)48} \\ -\ 3 \\ \hline 18 \\ -\ 18 \\ \hline 0 \end{array}$$

State The student needs to complete 16 more 3-hour shifts of volunteer service.

Check The student has already completed 87 hours. If she works 16 more shifts, each 3 hours long, she will have $16 \cdot 3 = 48$ more hours. Adding the two sets of hours, we get:

$$\begin{array}{r} 87 \\ +\ 48 \\ \hline 135 \end{array} \longleftarrow \text{This is the total number of hours needed.}$$

The result, 16, checks.

Self Check 5

Service clubs. To become a member of a service club, students at one college must complete 72 hours of volunteer service by working 4-hour shifts at the tutoring center. If a student has already volunteered 48 hours, how many more 4-hour shifts must she work to meet the service requirement for membership in the club?

Now Try ➭ Problem 23

EXAMPLE 6   **Attorney's fees.** In return for her services, an attorney and her client split the jury's cash award equally. After paying her assistant $1,000, the attorney ended up making $10,000 from the case. What was the amount of the award?

Analyze
- The attorney and client split the award equally.   *Given*
- The attorney's assistant was paid $1,000.   *Given*
- The attorney made $10,000.   *Given*
- What was the amount of the award?   *Find*

Assign Let x = the amount of the award.

Form Two key phrases in the problem help us form an equation.

    Key phrase: *split the award equally*   **Translation:** divide by 2

    Key phrase: *paying her assistant $1,000*   **Translation:** subtract $1,000

Now we translate the words of the problem into an equation.

The award split in half	minus	the amount paid to the assistant	is	the amount the attorney makes.
$\frac{x}{2}$	$-$	$1,000$	$=$	$10,000$

Solve

$$\frac{x}{2} - 1,000 = 10,000$$

We need to isolate x on the left side.

$$\frac{x}{2} - 1,000 + \mathbf{1,000} = 10,000 + \mathbf{1,000}$$

To isolate the variable term $\frac{x}{2}$, add 1,000 to both sides to undo the subtraction of 1,000.

$$\frac{x}{2} = 11,000$$

Do the addition.

$$\mathbf{2} \cdot \frac{x}{2} = \mathbf{2} \cdot 11,000$$

To isolate the variable x, multiply both sides by 2 to undo the division by 2.

$$\begin{array}{r} 11,000 \\ \times\ \ \ \ 2 \\ \hline 22,000 \end{array}$$

$$x = 22,000$$

Do the multiplication.

Yard sales. A husband and wife split equally the money that they made on a yard sale. The husband gave $75 of his share to charity, leaving him with $210. How much money did the couple make at their yard sale?

Now Try ➲ Problem 24

State The amount of the award was $22,000.

Check If the award of $22,000 is split in half, the attorney's share is $11,000. If $1,000 is paid to her assistant, we subtract to get:

$11,000
$- 1,000$
$\overline{\$10,000}$ ⟵ *This is what the attorney made.*

The result, $22,000, checks.

OBJECTIVE 2 Solve application problems to find two unknowns.

When solving application problems, we usually let the variable stand for the quantity we are asked to find. In the next two examples, each problem contains a second unknown quantity. We will look for a key word or phrase in the problem to help us describe it using an algebraic expression.

EXAMPLE 7 **Civil service.** A candidate for a position with the FBI scored 12 points higher on the written part of the civil service exam than she did on her interview. If her combined score was 92, what were her scores on the interview and on the written part of the exam?

Analyze
- She scored 12 points higher on the written part than on the interview. *Given*
- Her combined score was 92. *Given*
- What were her scores on the interview and on the written part? *Find*

Assign Since we are told that her score on the written part was related to her score on the interview, we let x = her score on the interview.

There is a second unknown quantity—her score on the written part of the exam. We look for a key phrase to help us decide how to represent that score using an algebraic expression.

Key phrase: 12 points *higher* on the written part than on the interview

Translation: add 12 points to the interview score

So $x + 12$ = her score on the written part of the test.

Form Now we translate the words of the problem into an equation.

The score on the interview	plus	the score on the written part	is	the overall score.
x	$+$	$x + 12$	$=$	92

Solve

$$x + x + 12 = 92$$ *We need to isolate x on the left side.*

$$2x + 12 = 92$$ *On the left side, combine like terms: x + x = 2x.*

$$2x + 12 - 12 = 92 - 12$$ *To isolate the variable term 2x, subtract 12 from both sides to undo the addition of 12.*

$$2x = 80$$ *Do the subtraction.*

$$\frac{2x}{2} = \frac{80}{2}$$ *To isolate the variable x, divide both sides by 2 to undo the multiplication by 2.*

$$x = 40$$ *Do the division. This is her score on the interview.*

To find the second unknown, we substitute 40 for x in the expression that represents her score on the written part.

$$x + 12 = \mathbf{40} + 12$$
$$= 52 \quad \text{\textit{This is her score on the written part.}}$$

State Her score on the interview was 40, and her score on the written part was 52.

Check Her score of 52 on the written exam was 12 points higher than her score of 40 on the interview. Also, if we add the two scores, we get:

$$\begin{array}{r} 40 \\ + \ 52 \\ \hline 92 \end{array}$$ ← *This is her combined score.*

The results, 40 and 52, check.

Self Check 7

Civil service. A candidate for a position with the IRS scored 15 points higher on the written part of the civil service exam than he did on his interview. If his combined score was 155, what were his scores on the interview and on the written part?

Now Try ➲ **Problem 25**

EXAMPLE 8   **Playgrounds.** After receiving a donation of 400 feet of chain link fencing, the staff of a preschool decided to use it to enclose a playground that is rectangular. Find the length and the width of the playground if the length is three times the width.

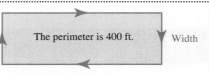

The perimeter is 400 ft.   Width

The length is three times the width.

Analyze

- The perimeter is 400 ft.   Given
- The length is three times the width.   Given
- What is the length and what is the width of the rectangle?   Find

Assign We will let w = the width of the playground. There is a second unknown quantity: the length of the playground. We look for a key phrase to help us decide how to represent it using an algebraic expression.

     Key phrase: length *three times* the width   **Translation:** multiply width by 3

So $3w$ = the length of the playground.

Form The formula for the perimeter of a rectangle is $P = 2l + 2w$. In words, we can write

$2 \cdot$	the length of the playground	plus	$2 \cdot$	the width of the playground	is	the perimeter.
$2 \cdot$	$3w$	$+$	$2 \cdot$	w	$=$	400

Solve

$$2 \cdot 3w + 2w = 400 \quad \text{\textit{We need to isolate w on the left side.}}$$
$$6w + 2w = 400 \quad \text{\textit{Do the multiplication: } } 2 \cdot 3w = 6w.$$
$$8w = 400 \quad \text{\textit{On the left side, combine like terms: } } 6w + 2w = 8w.$$
$$\frac{8w}{8} = \frac{400}{8} \quad \text{\textit{To isolate w, divide both sides by 8 to undo the multiplication by 8.}}$$
$$w = 50 \quad \text{\textit{Do the division.}}$$

$$\begin{array}{r} 50 \\ 8\overline{)400} \\ -\ 40 \\ \hline 00 \\ -\ 0 \\ \hline 0 \end{array}$$

Self Check 8

Crime scenes. Police used 800 feet of yellow tape to fence off a rectangular lot for an investigation. Each width used 50 feet less tape than each length. Find the length and the width of the lot.

Now Try ➡ **Problem 26**

To find the second unknown, we substitute 50 for w in the expression that represents the length of the playground.

$$3w = 3(50) \quad \text{Substitute 50 for } w.$$
$$ = 150 \quad \text{This is the length of the playground.}$$

State The width of the playground is 50 feet, and the length is 150 feet.

Check If we add two lengths and two widths, we get $2(150) + 2(50) = 300 + 100 = 400$. Also, the length (150 ft) is three times the width (50 ft). The results check.

Answers to Self Checks

1. 476 units have been sold.   **2.** The tank originally contained 1,775 gallons of gasoline.   **3.** The student used to read 175 words per minute.   **4.** The quartet charged $340 for the performance.   **5.** The student needs to complete 6 more 4-hour shifts of volunteer service.   **6.** The couple made $570 at the yard sale.   **7.** His score on the interview was 70, and his score on the written part was 85.   **8.** The length of the lot is 225 feet, and the width of the lot is 175 feet.

SECTION 8.5   STUDY SET

VOCABULARY

Fill in the blanks.

1. The six-step problem-solving strategy is:
   - _____ the problem.
   - _____ a variable.
   - Form an _____ .
   - _____ the equation.
   - State the _____ .
   - _____ the result.

2. Words such as *doubled* and *tripled* indicate the operation of _____ .

3. Phrases such as *distributed equally* and *sectioned off uniformly* indicate the operation of _____ .

4. Words such as *trimmed*, *removed*, and *melted* indicate the operation of _____ .

5. Words such as *extended* and *reclaimed* indicate the operation of _____ .

6. A letter (or symbol) that is used to represent a number is called a _____ .

CONCEPTS

In each of the following problems, find the key word or phrase and tell how it translates. You do not have to solve the problem.

7. **Fast food.** The franchise fee and startup costs for a Taco Bell restaurant total $1,324,300. If an entrepreneur has $550,000 to invest, how much money will she need to borrow to open her own Taco Bell restaurant?

   Key word: _____

   Translation: _____

8. **Graduation announcements.** Six of Tom's graduation announcements were returned by the post office stamped "no longer at this address," but 27 were delivered. How many announcements did he send?

   Key word: _____

   Translation: _____

9. **Working in groups.** When a history teacher had the students in her class form equal-size discussion groups, there were seven complete groups, with five students in each group. How many students were in the class?

   Key word: _____

   Translation: _____

10. **Self-help books.** An author claimed that the information in his book could double a salesperson's monthly income. If a medical supplies salesperson currently earns $5,000 a month, what monthly income can she expect to make after reading the book?

    Key word: _____

    Translation: _____

11. **Scholarships.** See the illustration. How many scholarships were awarded this year?

Last year, *s* scholarships were awarded.   Six more scholarships were awarded this year than last year.

12. Ocean travel. See the illustration. How many miles did the passenger ship travel?

The freighter traveled *m* miles.

The passenger ship traveled 3 times farther than the freighter.

13. Service stations. See the illustration. How many gallons does the smaller tank hold?

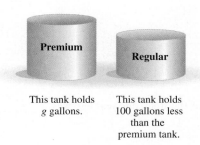

This tank holds *g* gallons.

This tank holds 100 gallons less than the premium tank.

14. Complete this statement about the perimeter of the rectangle shown.

$$2 \cdot \boxed{} + 2 \cdot \boxed{} = 240$$

The perimeter is 240 ft.   *w*

5*w*

In Exercises 15–18, complete the steps to answer the question.

15. History. A 1,700-year-old scroll is 425 years older than the clay jar in which it was found. How old is the jar?

Analyze

■ The scroll is _____ years old.

■ The scroll is _____ years older than the jar.

■ How old is the ____?

Assign

Let *x* = the ____ of the jar.

Form

Now we look for a key phrase in the problem.

  Key phrase: older than   **Translation:** _____

Now we translate the words of the problem into an equation.

The age of the scroll	is	425 years	plus	the age of the jar.
	=	425	+	

Solve

$$\boxed{} = 425 + x$$
$$1{,}700 - \boxed{} = 425 + x - \boxed{}$$
$$\boxed{} = x$$

State

The jar is _____ years old.

Check

  $\boxed{}$
  $+\ 425$
  ————
  $\boxed{}$ ⟵ This is the age of the scroll.

The result checks.

16. Banking. After a student wrote a $1,500 check to pay for a car, he had a balance of $750 in his account. How much did he have in the account before he wrote the check?

Analyze

■ A _____ check was written.

■ The new balance in the account was _____.

■ How much did he have in the account _____ he wrote the check?

Assign

Let *x* = the account balance _____ he wrote the check.

Form Now we look for a key phrase in the problem.

  Key phrase: wrote a check   **Translation:** _____

Now we translate the words of the problem into an equation.

The account balance before writing the check	minus	the amount of the check	is	the new balance.
$\boxed{}$	−	1,500	=	$\boxed{}$

Solve

$$\boxed{} - 1{,}500 = 750$$
$$x - 1{,}500 + \boxed{} = 750 + \boxed{}$$
$$x = \boxed{}$$

State

The account balance before writing the check was _____.

Check

  $\boxed{}$
  $-\ 1{,}500$
  ————
  $\boxed{}$ ⟵ This is the new balance.

The result checks.

17. Airline seating.   An 88-seat passenger plane has ten times as many economy seats as first-class seats. Find the number of first-class seats and the number of economy seats.

Analyze
- There are ▢ seats on the plane.
- There are ▢ times as many economy as first-class seats.
- Find the number of _____ seats and the number of _____ seats.

Assign

Since the number of economy seats is related to the number of first-class seats, we let x = the number of _____ seats.

To represent the number of economy seats, look for a key phrase in the problem.
Key phrase: ten times as many   **Translation:** multiply by ____

So ▢ = the number of economy seats.

Form

The number of first-class seats	plus	the number of economy seats	is	88.
x	$+$	▢	$=$	88

Solve

$$x + 10x = ▢$$
$$▢ = 88$$
$$\frac{11x}{▢} = \frac{88}{▢}$$
$$x = ▢$$

State

There are ▢ first-class seats and $10 \cdot 8 = ▢$ economy seats.

Check

The number of economy seats, 80, is ▢ times the number of first-class seats, 8. Also, if we add the numbers of seats, we get:

▢
$+ 8$

▢ ← This is the total number of seats.

The results check.

18. The stock market.   An investor has seen the value of his stock double in the last 12 months. If the current value of his stock is $274,552, what was its value one year ago?

Analyze
- The value of the stock _____ in 12 months.
- The current value of the stock is _____.
- What was the _____ of the stock one year ago?

Assign

We can let x = the _____ of the stock one year ago.

Form

We now look for a key word in the problem.

Key phrase: double   **Translation:** _____ by 2

Now we translate the words of the problem into an equation.

2	times	the value of the stock one year ago	is	the current value of the stock.
2	$\cdot$	▢	$=$	274,552

Solve

$$2x = ▢$$
$$\frac{2x}{▢} = \frac{274,552}{▢}$$
$$x = ▢$$

State

The value of the stock one year ago was _____.

Check

▢
$\times \quad 2$

▢ ← This is the current value of the stock.

The result checks.

GUIDED PRACTICE

Form an equation and solve it to answer each question. See Example 1.

19. Fast food.   The franchise fee and startup costs for a Pizza Hut restaurant are $316,500. If an entrepreneur has $68,500 to invest, how much money will she need to borrow to open her own Pizza Hut restaurant?

See Example 2.

20. Party invitations.   Three of Mia's party invitations were lost in the mail, but 59 were delivered. How many invitations did she send?

See Example 3.

21. Speed-reading.   An advertisement for a speed-reading program claimed that successful completion of the course could triple a person's reading rate. After taking the course, Alicia can now read 399 words per minute. If the company's claims are true, what was her reading rate before taking the course?

See Example 4.

22. **Physical education.** A high school PE teacher had the students in her class form three-person teams for a basketball tournament. Thirty-two teams participated in the tournament. How many students were in the PE class?

See Example 5.

23. **Business.** After beginning a new position with 15 established accounts, a salesman made it his objective to add 5 new accounts every month. His goal was to reach 100 accounts. At this rate, how many months would it take to reach his goal?

See Example 6.

24. **Tax refunds.** After receiving their tax refund, a husband and wife split the refunded money equally. The husband then gave $50 of his money to charity, leaving him with $70. What was the amount of the tax refund check?

See Example 7.

25. **Scholarships.** Because of increased giving, a college scholarship program awarded six more scholarships this year than last year. If a total of 20 scholarships were awarded over the last two years, how many were awarded last year and how many were awarded this year?

See Example 8.

26. **Geometry.** The perimeter of a rectangle is 150 inches. Find the length and the width if the length is four times the width.

APPLICATIONS

Form an equation and solve it to answer each question.

27. **Loans.** A student plans to pay back a $600 loan with monthly payments of $30. How many payments has she made if she now only owes $420?

28. **Antiques.** A woman purchases 8 antique spoons each year. She now owns 56 spoons. In how many years will she have 200 spoons in her collection?

29. **Hip Hop.** *Forbes* magazine estimates that in 2011, Shawn "Jay-Z" Carter earned $38 million. If this was $72 million less than Andre "Dr. Dre" Young's earnings, how much did Dr. Dre earn in 2011?

30. **Buying golf clubs.** A man needs $345 for a new set of golf clubs. How much more money does he need if he now has $317?

31. **Interior decorating.** As part of redecorating, crown molding was installed around the ceiling of a room. Sixty feet of molding was needed for the project. Find the length and the width of the room if its length is twice the width.

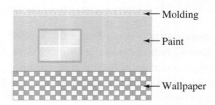

32. **Sprinkler systems.** A landscaper buried a water line around a rectangular lawn to serve as a supply line for a sprinkler system. The length of the lawn is 5 times its width. If 240 feet of pipe was used to do the job, what is the length and the width of the lawn?

33. **Gravity.** The weight of an object on Earth is 6 times greater than what it is on the moon. The situation shown below took place on Earth. If it took place on the moon, what weight would the scale register?

34. Infomercials. The number of orders received each week by a company selling skin care products increased fivefold after a Hollywood celebrity was added to the company's infomercial. After adding the celebrity, the company received about 175 orders each week. How many orders were received each week before the celebrity took part?

35. Theater. The play *Romeo and Juliet,* by William Shakespeare, has 5 acts and a total of 24 scenes. The second act has the most scenes, 6. The third and fourth acts both have 5 scenes. The last act has the least number of scenes, 3. How many scenes are in the first act?

36. U.S. presidents. As of December 31, 1999, there had been 42 presidents of the United States. George Washington and John Adams were the only presidents in the 18th century (1700–1799). During the 19th century (1800–1899), there were 23 presidents. How many presidents were there during the 20th century (1900–1999)?

37. Help wanted. From the following ad from the classified section of a newspaper, determine the value of the benefit package. ($45K means $45,000.)

> **★ACCOUNTS PAYABLE★**
> 2-3 yrs exp as supervisor. Degree a +. High vol company. Good pay, $45K & xlnt benefits; total compensation worth $52K. Fax resume.

38. Power outages. The electrical system in a building automatically shuts down when the meter shown reads 85. By how much must the current reading increase to cause the system to shut down?

39. Video games. After a week of playing *Angry Birds,* a boy scored 105,880 points in one game—an improvement of 77,420 points over the very first time he played. What was the score for his first game?

40. Auto repair. A woman paid $29 less to have her car fixed at a muffler shop than she would have paid at a gas station. At the gas station, she would have paid $219. How much did she pay to have her car fixed?

from Campus to Careers

41. For a half-hour time slot on television, a producer scheduled 18 minutes more time for the program than time for the commercials. How many minutes of commercials and how many minutes of the program were there in that time slot? (*Hint:* How many minutes are there in a half hour?)

42. Service stations. At a service station, the underground tank storing regular gas holds 100 gallons less than the tank storing premium gas. If the total storage capacity of the tanks is 700 gallons, how much does the premium gas tank and how much does the regular gas tank hold?

43. Class time. In a biology course, students spend a total of 250 minutes in lab and lecture each week. The lab time is 50 minutes shorter than the lecture time. How many minutes do the students spend in lecture and how many minutes do students spend in lab per week?

44. Ocean travel. At noon, a passenger ship and a freighter left a port traveling in opposite directions. By midnight, the passenger ship was 3 times farther from port than the freighter was. How far was the freighter and how far was the passenger ship from port if the distance between the ships was 84 miles?

45. Animal shelters. The number of phone calls to an animal shelter quadrupled after the evening news aired a segment explaining the services the shelter offered. Before the publicity, the shelter received 8 calls a day. How many calls did the shelter receive each day after being featured on the news?

46. Open houses. The attendance at an elementary school open house was only half of what the principal had expected. If 120 people visited the school that evening, how many had she expected to attend?

47. Bus riders. A man had to wait 20 minutes for a bus today. Three days ago, he had to wait 15 minutes longer than he did today, because four buses passed by without stopping. How long did he wait three days ago?

48. Hit records. The oldest artist to have a number one single was Louis Armstrong, with the song *Hello Dolly.* He was 55 years older than the youngest artist to have a number one single, 12-year-old Jimmy Boyd, with *I Saw Mommy Kissing Santa Claus.* How old was Louis Armstrong when he had the number one song? (Source: *The Top 10 of Everything,* 2000.)

Library of Congress, Prints & Photographs Division, Reproduction number LC-USZ62-127236 (b&w film copy neg.)

49. Cost overruns. Lengthy delays and skyrocketing costs caused a rapid-transit construction project to go over budget by a factor of 10. The final audit showed the project costing $540 million. What was the initial cost estimate?

50. Lotto winners. The grocery store employees listed below pooled their money to buy $120 worth of lottery tickets each week, with the understanding that they would split the prize equally if they happened to win. One week they did have the winning ticket and won $480,000. What was each employee's share of the winnings?

Sam M. Adler	Ronda Pellman	Manny Fernando
Lorrie Jenkins	Tom Sato	Sam Lin
Kiem Nguyen	H. R. Kinsella	Tejal Neeraj
Virginia Ortiz	Libby Sellez	Alicia Wen

51. Rentals. In renting an apartment with two other friends, Enrique agreed to pay the security deposit of $100 himself. The three of them agreed to contribute equally toward the monthly rent. Enrique's first check to the apartment owner was for $425. What was the monthly rent for the apartment?

52. Bottled water delivery. A truck driver left the plant carrying 300 bottles of drinking water. His delivery route consisted of office buildings, each of which was to receive 3 bottles of water. The driver returned to the plant at the end of the day with 117 bottles of water on the truck. To how many office buildings did he deliver water?

53. Construction. To get a heavy-equipment operator's certificate, 48 hours of on-the-job training are required. If a woman has completed 24 hours, and the training sessions last for 6 hours, how many more sessions must she take to get the certificate?

54. The Bermuda Triangle. The Bermuda Triangle is a triangular region in the Atlantic Ocean where many ships and airplanes have disappeared. The perimeter of the triangle is about 3,075 miles. It is formed by three imaginary lines. The first, 1,100 miles long, is from Melbourne, Florida, to Puerto Rico. The second, 1,000 miles long, stretches from Puerto Rico to Bermuda. The third extends from Bermuda back to Florida. Find its length.

WRITING

55. What is the most difficult step of the six-step problem-solving strategy for you? Explain why it is.

56. Give ten words or phrases that indicate subtraction.

57. What does the word *translate* mean?

58. Unlike an arithmetic approach, you *do not* have to determine whether to add, subtract, multiply, or divide to solve the application problems in this section. That decision is made for you when you solve the equation that mathematically describes the situation. Explain.

59. Write a problem that could be represented by the following equation.

Age of father	plus	age of son	is	50.
x	$+$	$x - 20$	$=$	50

60. Write a problem that could be represented by the following equation.

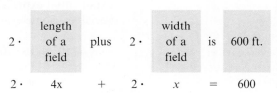

	length of a field	plus		width of a field	is	600 ft.
$2 \cdot$	$4x$	$+$	$2 \cdot$	x	$=$	600

REVIEW

Find the LCM and the GCF of the given numbers.

61. 100, 120

62. 120, 180

63. 14, 140

64. 15, 300

65. 8, 9, 49

66. 9, 16, 25

67. 66, 198, 242

68. 52, 78, 130

OBJECTIVES

1 Identify bases and exponents.

2 Multiply exponential expressions that have like bases.

3 Raise exponential expressions to a power.

4 Find powers of products.

SECTION 8.6 **Multiplication Rules for Exponents**

In this section, we will use the definition of exponent to develop some rules for simplifying expressions that contain exponents.

OBJECTIVE 1 **Identify bases and exponents.**

Recall that an **exponent** indicates repeated multiplication. It indicates how many times the base is used as a factor. For example, 3^5 represents the product of five 3's.

Exponent ⟶ 5 factors of 3

$$3^5 = 3 \cdot 3 \cdot 3 \cdot 3 \cdot 3$$

Base ⟶

In general, we have the following definition.

Natural-Number Exponents

A natural-number* exponent tells how many times its base is to be used as a factor. For any number x and any natural number n,

n factors of x

$$x^n = x \cdot x \cdot x \cdot \, \cdots \, \cdot x$$

*The set of natural numbers is {1, 2, 3, 4, 5, … }.

Expressions of the form x^n are called **exponential expressions**. The base of an exponential expression can be a number, a variable, or a combination of numbers and variables. Some examples are:

$$10^5 = 10 \cdot 10 \cdot 10 \cdot 10 \cdot 10$$
The base is 10. The exponent is 5. Read as "10 to the fifth power."

$$y^2 = y \cdot y$$
The base is y. The exponent is 2. Read as "y squared."

$$(-2s)^3 = (-2s)(-2s)(-2s)$$
The base is $-2s$. The exponent is 3. Read as "negative $2s$ raised to the third power" or "negative $2s$ cubed."

Caution! Bases that contain a − sign *must* be written within parentheses.

$$(-2s)^3 \leftarrow \text{Exponent}$$

Base

$$-8^4 = -(8 \cdot 8 \cdot 8 \cdot 8)$$
Since the − sign is not written within parentheses, the base is 8. The exponent is 4. Read as "the opposite (or the negative) of 8 to the fourth power."

When an exponent is 1, it is usually not written. For example, $4 = 4^1$ and $x = x^1$.

EXAMPLE 1 Identify the base and the exponent in each expression:

a. 8^5 **b.** $7a^3$ **c.** $(7a)^3$

Strategy To identify the base and exponent, we will look for the form ▪▪.

WHY The exponent is the small raised number to the right of the base.

Solution

a. In 8^5, the base is 8 and the exponent is 5.

b. $7a^3$ means $7 \cdot a^3$. Thus, the base is a, not $7a$. The exponent is 3.

c. Because of the parentheses in $(7a)^3$, the base is $7a$ and the exponent is 3.

Self Check 1

Identify the base and the exponent:

a. $3y^4$

b. $(3y)^4$

Now Try ⟹ Problems 13 and 17

EXAMPLE 2   Write each expression in an equivalent form using an exponent:
a. $b \cdot b \cdot b \cdot b$ **b.** $5 \cdot t \cdot t \cdot t$

Strategy We will look for repeated factors and count the number of times each appears.

WHY We can use an exponent to represent repeated multiplication.

Solution

a. Since there are four repeated factors of b in $b \cdot b \cdot b \cdot b$, the expression can be written as b^4.

b. Since there are three repeated factors of t in $5 \cdot t \cdot t \cdot t$, the expression can be written as $5t^3$.

Self Check 2

Write as an exponential expression:
$(x + y)(x + y)(x + y)(x + y)$
$(x + y)$

Now Try ➲ Problems 25 and 29

OBJECTIVE 2 Multiply exponential expressions that have like bases.

To develop a rule for multiplying exponential expressions that have the same base, we consider the product $6^2 \cdot 6^3$. Since 6^2 means that 6 is to be used as a factor two times, and 6^3 means that 6 is to be used as a factor three times, we have

$$6^2 \cdot 6^3 = \overbrace{6 \cdot 6}^{2 \text{ factors of } 6} \cdot \overbrace{6 \cdot 6 \cdot 6}^{3 \text{ factors of } 6}$$

$$= \overbrace{6 \cdot 6 \cdot 6 \cdot 6 \cdot 6}^{5 \text{ factors of } 6}$$

$$= 6^5$$

We can quickly find this result if we keep the common base 6 and add the exponents on 6^2 and 6^3.

$$6^2 \cdot 6^3 = 6^{2+3} = 6^5$$

This example illustrates the following rule for exponents.

Product Rule for Exponents

To multiply exponential expressions that have the same base, keep the common base and add the exponents.

    For any number x and any natural numbers m and n,

$$x^m \cdot x^n = x^{m+n}$$ Read as "x to the mth power times x to the nth power equals x to the m plus nth power."

EXAMPLE 3   Simplify:

a. $9^5(9^6)$ **b.** $x^3 \cdot x^4$ **c.** $y^2 y^4 y$ **d.** $(c^2 d^3)(c^4 d^5)$

Strategy In each case, we want to write an equivalent expression using one base and one exponent. We will use the product rule for exponents to do this.

WHY The product rule for exponents is used to multiply exponential expressions that have the same base.

Solution

a. $9^5(9^6) = 9^{5+6} = 9^{11}$ *Keep the common base, 9, and add the exponents. Since 9^{11} is a very large number, we will leave the answer in this form. We won't evaluate it.*

Caution! Don't make the mistake of multiplying the bases when using the product rule. Keep the *same* base.

$$9^5(9^6) \neq 81^{11}$$

b. $x^3 \cdot x^4 = x^{3+4} = x^7$ *Keep the common base, x, and add the exponents.*

c. $y^2 y^4 y = y^2 y^4 y^1$ *Write y as y^1.*

$= y^{2+4+1}$ *Keep the common base, y, and add the exponents.*

$= y^7$

d. $(c^2 d^3)(c^4 d^5) = (c^2 c^4)(d^3 d^5)$ *Use the commutative and associative properties of multiplication to group like bases together.*

$= (c^{2+4})(d^{3+5})$ *Keep the common base, c, and add the exponents. Keep the common base, d, and add the exponents.*

$= c^6 d^8$

Self Check 3

Simplify:
a. $7^8(7^7)$
b. $x^2 x^3 x$
c. $(y-1)^5 (y-1)^5$
d. $(s^4 t^3)(s^4 t^4)$

Now Try Problems 33, 35, and 37

Caution! We cannot use the product rule to simplify expressions like $3^2 \cdot 2^3$, where the bases are not the same. However, we can simplify this expression by doing the arithmetic:

$$3^2 \cdot 2^3 = 9 \cdot 8 = 72 \qquad 3^2 = 3 \cdot 3 = 9 \text{ and } 2^3 = 2 \cdot 2 \cdot 2 = 8.$$

Recall that *like terms* are terms with exactly the same variables raised to exactly the same powers. To add or subtract exponential expressions, they must be like terms. To multiply exponential expressions, only the bases need to be the same.

$x^5 + x^2$ *These are not like terms; the exponents are different. We cannot add.*

$x^2 + x^2 = 2x^2$ *These are like terms; we can add. Recall that $x^2 = 1x^2$.*

$x^5 \cdot x^2 = x^7$ *The bases are the same; we can multiply.*

OBJECTIVE 3 **Raise exponential expressions to a power.**

To develop another rule for exponents, we consider $(5^3)^4$. Here, an exponential expression, 5^3, is raised to a power. Since 5^3 is the base and 4 is the exponent, $(5^3)^4$ can be written as $5^3 \cdot 5^3 \cdot 5^3 \cdot 5^3$. Because each of the four factors of 5^3 contains three factors of 5, there are $4 \cdot 3$ or 12 factors of 5.

12 factors of 5

$$(5^3)^4 = 5^3 \cdot 5^3 \cdot 5^3 \cdot 5^3 = 5 \cdot 5 \cdot 5 \cdot 5 \cdot 5 \cdot 5 \cdot 5 \cdot 5 \cdot 5 \cdot 5 \cdot 5 \cdot 5 = 5^{12}$$

$5^3 \quad 5^3 \quad 5^3 \quad 5^3$

We can quickly find this result if we keep the common base of 5 and multiply the exponents.

$$(5^3)^4 = 5^{3 \cdot 4} = 5^{12}$$

This example illustrates the following rule for exponents.

Power Rule for Exponents

To raise an exponential expression to a power, keep the base and multiply the exponents.

For any number x and any natural numbers m and n,

$$(x^m)^n = x^{m \cdot n} = x^{mn}$$ Read as "the quantity of x to the mth power raised to the nth power equals x to the mnth power."

EXAMPLE 4   Simplify: **a.** $(2^3)^7$ **b.** $[(-6)^2]^5$ **c.** $(z^8)^8$

Strategy In each case, we want to write an equivalent expression using one base and one exponent. We will use the power rule for exponents to do this.

WHY Each expression is a power of a power.

Solution

a. $(2^3)^7 = 2^{3 \cdot 7} = 2^{21}$ Keep the base, 2, and multiply the exponents. Since 2^{21} is a very large number, we will leave the answer in this form.

b. $[(-6)^2]^5 = (-6)^{2 \cdot 5} = (-6)^{10}$ Keep the base, −6, and multiply the exponents. Since $(-6)^{10}$ is a very large number, we will leave the answer in this form.

c. $(z^8)^8 = z^{8 \cdot 8} = z^{64}$ Keep the base, z, and multiply the exponents.

THE LANGUAGE OF ALGEBRA

An exponential expression raised to a power, such as $(2^3)^7$, is also called a *power of a power*.

Self Check 4
Simplify:
a. $(4^6)^5$
b. $(y^5)^2$

Now Try Problems 49, 51, and 53

EXAMPLE 5   Simplify: **a.** $(x^2 x^5)^2$ **b.** $(z^2)^4(z^3)^3$

Strategy In each case, we want to write an equivalent expression using one base and one exponent. We will use the product and power rules for exponents to do this.

WHY The expressions involve multiplication of exponential expressions that have the same base and they involve powers of powers.

Solution

a. $(x^2 x^5)^2 = (x^7)^2$ Within the parentheses, keep the common base, x, and add the exponents: $2 + 5 = 7$.

$\quad = x^{14}$ Keep the base, x, and multiply the exponents: $7 \cdot 2 = 14$.

b. $(z^2)^4(z^3)^3 = z^8 z^9$ For each power of z raised to a power, keep the base and multiply the exponents: $2 \cdot 4 = 8$ and $3 \cdot 3 = 9$.

$\quad = z^{17}$ Keep the common base, z, and add the exponents: $8 + 9 = 17$.

Self Check 5
Simplify:
a. $(a^4 a^3)^3$
b. $(a^3)^3(a^4)^2$

Now Try Problems 57 and 61

OBJECTIVE 4 Find powers of products.

To develop another rule for exponents, we consider the expression $(2x)^3$, which is a *power of the product* of 2 and x.

$$(2x)^3 = 2x \cdot 2x \cdot 2x \qquad \text{Write the base } 2x \text{ as a factor 3 times.}$$
$$= (2 \cdot 2 \cdot 2)(x \cdot x \cdot x) \quad \text{Change the order of the factors and group like bases.}$$
$$= 2^3 x^3 \qquad \text{Write each product of repeated factors in exponential form.}$$
$$= 8x^3 \qquad \text{Evaluate: } 2^3 = 8.$$

This example illustrates the following rule for exponents.

Power of a Product

To raise a product to a power, raise each factor of the product to that power.
    For any numbers x and y, and any natural number n,

$$(xy)^n = x^n y^n$$

EXAMPLE 6   Simplify: **a.** $(3c)^4$ **b.** $(x^2 y^3)^5$

Strategy In each case, we want to write the expression in an equivalent form in which each base is raised to a single power. We will use the power of a product rule for exponents to do this.

WHY Within each set of parentheses is a product, and each of those products is raised to a power.

Solution

a. $(3c)^4 = 3^4 c^4$ Raise each factor of the product $3c$ to the 4th power.
$$= 81c^4 \quad \text{Evaluate: } 3^4 = 81.$$

b. $(x^2 y^3)^5 = (x^2)^5 (y^3)^5$ Raise each factor of the product $x^2 y^3$ to the 5th power.
$$= x^{10} y^{15} \qquad \text{For each power of a power, keep each base, } x \text{ and } y, \text{ and}$$
$$\text{multiply the exponents: } 2 \cdot 5 = 10 \text{ and } 3 \cdot 5 = 15.$$

Self Check 6

Simplify:

a. $(2t)^4$

b. $(c^3 d^4)^6$

Now Try ⮕ Problems 65 and 69

EXAMPLE 7   Simplify: $(2a^2)^2 (4a^3)^3$

Strategy We want to write an equivalent expression using one base and one exponent. We will begin the process by using the power of a product rule for exponents.

WHY Within each set of parentheses is a product, and each product is raised to a power.

Solution

$$(2a^2)^2 (4a^3)^3 = 2^2 (a^2)^2 \cdot 4^3 (a^3)^3 \qquad \text{Raise each factor of the product } 2a^2 \text{ to the 2nd power. Raise}$$
$$\text{each factor of the product } 4a^3 \text{ to the 3rd power.}$$

$$= 4a^4 \cdot 64a^9 \qquad \text{Evaluate: } 2^2 = 4 \text{ and } 4^3 = 64. \text{ For each power of a power,}$$
$$\text{keep each base and multiply the exponents: } 2 \cdot 2 = 4$$
$$\text{and } 3 \cdot 3 = 9.$$

$$= (4 \cdot 64)(a^4 \cdot a^9) \qquad \text{Group the numerical factors. Group}$$
$$\text{the factors that have the same base.}$$

$$= 256a^{13} \qquad \text{Do the multiplication: } 4 \cdot 64 = 256. \text{ Keep the}$$
$$\text{common base } a \text{ and add the exponents: } 4 + 9 = 13.$$

$$\begin{array}{r} \overset{1}{64} \\ \times\ 4 \\ \hline 256 \end{array}$$

Self Check 7

Simplify: $(4y^3)^2 (3y^4)^3$

Now Try ⮕ Problem 73

The rules for natural-number exponents are summarized as follows.

Rules for Exponents

If m and n represent natural numbers and there are no divisions by zero, then

Exponent of 1	Product rule	Power rule	Power of a product
$x^1 = x$	$x^m x^n = x^{m+n}$	$(x^m)^n = x^{mn}$	$(xy)^n = x^n y^n$

Answers to Self Checks

1. a. base: y, exponent: 4 **b.** base: $3y$, exponent: 4 **2.** $(x + y)^5$ **3. a.** 7^{15} **b.** x^6 **c.** $(y - 1)^{10}$ **d.** s^8t^7 **4. a.** 4^{30} **b.** y^{10} **5. a.** a^{21} **b.** a^{17} **6. a.** $16t^4$ **b.** $c^{18}d^{24}$ **7.** $432y^{18}$

SECTION 8.6 STUDY SET

VOCABULARY

Fill in the blank.

1. Expressions such as x^4, 10^3, and $(5t)^2$ are called ___________ expressions.

2. Match each expression with the proper description.
   $(a^4b^2)^5$ $\quad (a^8)^4 \quad a^5 \cdot a^3$
   a. Product of exponential expressions with the same base
   b. Power of an exponential expression
   c. Power of a product

CONCEPTS

Fill in the blanks.

3. a. $(3x)^4 = \square \cdot \square \cdot \square \cdot \square$
   b. $(-5y)(-5y)(-5y) = \square$

4. a. $x = x^{\square}$ b. $x^m x^n = \square$
   c. $(xy)^n = \square$ d. $(a^b)^c = \square$

5. To simplify each expression, determine whether you add, subtract, multiply, or divide the exponents.
   a. $b^6 \cdot b^9$
   b. $(n^8)^4$
   c. $(a^4b^2)^5$

6. To simplify $(2y^3z^2)^4$, what factors within the parentheses must be raised to the fourth power?

Simplify each expression, if possible.

7. a. $x^2 + x^2$ b. $x^2 \cdot x^2$
8. a. $x^2 + x$ b. $x^2 \cdot x$
9. a. $x^3 - x^2$ b. $x^3 \cdot x^2$
10. a. $4^2 \cdot 2^4$ b. $x^3 \cdot y^2$

NOTATION

Complete each step to simplify each expression.

11. $(x^4x^2)^3 = (\square)^3$
    $= x^{\square}$

12. $(x^4)^3(x^2)^3$
    $= x^{\square} \cdot x^6$
    $= x^{\square}$

GUIDED PRACTICE

Identify the base and the exponent in each expression. See Example 1.

13. 4^3
14. $(-8)^2$
15. x^5
16. $\left(\frac{5}{x}\right)^3$
17. $(-3x)^2$
18. $(2xy)^{10}$
19. $-\frac{1}{3}y^6$
20. $-x^4$
21. $9m^{12}$
22. $3.14r^4$
23. $(y + 9)^4$
24. $(z - 2)^3$

Write each expression in an equivalent form using an exponent. See Example 2.

25. $m \cdot m \cdot m \cdot m \cdot m$
26. $r \cdot r \cdot r \cdot r \cdot r \cdot r$
27. $4t \cdot 4t \cdot 4t \cdot 4t$
28. $-5u(-5u)(-5u)(-5u)(-5u)$
29. $4 \cdot t \cdot t \cdot t \cdot t \cdot t$
30. $5 \cdot u \cdot u \cdot u$
31. $a \cdot a \cdot b \cdot b \cdot b$
32. $m \cdot m \cdot m \cdot n \cdot n$

Use the product rule for exponents to simplify each expression. Write the results using exponents. See Example 3.

33. $5^3 \cdot 5^4$
34. $3^4 \cdot 3^6$
35. $a^3 \cdot a^3$
36. $m^7 \cdot m^7$
37. bb^2b^3
38. aa^3a^5
39. $(c^5)(c^8)$
40. $(d^4)(d^{20})$
41. $(a^2b^3)(a^3b^3)$
42. $(u^3v^5)(u^4v^5)$
43. $cd^4 \cdot cd$
44. $ab^3 \cdot ab^4$

45. $x^2 \cdot y \cdot x \cdot y^{10}$ **46.** $x^3 \cdot y \cdot x \cdot y^{12}$

47. $m^{100} \cdot m^{100}$ **48.** $n^{600} \cdot n^{600}$

Use the power rule for exponents to simplify each expression. Write the results using exponents. **See Example 4.**

49. $(3^2)^4$ **50.** $(4^3)^3$

51. $[(-4.3)^3]^8$ **52.** $[(-1.7)^9]^8$

53. $(m^{50})^{10}$ **54.** $(n^{25})^4$

55. $(y^5)^3$ **56.** $(b^3)^6$

Use the product and power rules for exponents to simplify each expression. **See Example 5.**

57. $(x^2x^3)^5$ **58.** $(y^3y^4)^4$

59. $(p^4p^5)^3$ **60.** $(r^3r^4)^2$

61. $(t^3)^4(t^2)^3$ **62.** $(b^2)^5(b^3)^2$

63. $(u^4)^2(u^3)^2$ **64.** $(v^5)^2(v^3)^4$

Use the power of a product rule for exponents to simplify each expression. **See Example 6.**

65. $(6a)^2$ **66.** $(3b)^3$

67. $(5y)^4$ **68.** $(4t)^4$

69. $(3a^4b^7)^3$ **70.** $(5m^9n^{10})^2$

71. $(-2r^2s^3)^3$ **72.** $(-2x^2y^4)^5$

Use the power of a product rule for exponents to simplify each expression. **See Example 7.**

73. $(2c^3)^3(3c^4)^2$ **74.** $(5b^4)^2(3b^8)^2$

75. $(10d^7)^2(4d^9)^3$ **76.** $(2x^7)^3(4x^8)^2$

TRY IT YOURSELF

Simplify each expression.

77. $(7a^9)^2$ **78.** $(12b^6)^2$

79. $t^4 \cdot t^5 \cdot t$ **80.** $n^4 \cdot n \cdot n^3$

81. $y^3y^2y^4$ **82.** y^4yy^6

83. $(-6a^3b^2)^3$ **84.** $(-10r^3s^2)^2$

85. $(n^4n)^3(n^3)^6$ **86.** $(y^3y)^2(y^2)^2$

87. $(b^2b^3)^{12}$ **88.** $(s^3s^3)^3$

89. $(2b^4b)^5(3b)^2$ **90.** $(2aa^7)^3(3a)^3$

91. $(c^2)^3(c^4)^2$ **92.** $(t^5)^2(t^3)^3$

93. $(3s^4t^3)^3(2st)^4$ **94.** $(2a^3b^5)^2(4ab)^3$

95. $x \cdot x^2 \cdot x^3 \cdot x^4 \cdot x^5$ **96.** $x^{10} \cdot x^9 \cdot x^8 \cdot x^7$

LOOK ALIKES

Simplify each expression, if possible.

97. a. $x^3 \cdot x^3$ **b.** $(x^3)^3$ **c.** $x^3 + x^3$

98. a. $(x^5)^7$ **b.** $x^5 \cdot x^7$ **c.** $x^5 + x^7$

99. a. $(n^2n^4)^6$ **b.** $(n^2)^6(n^4)^6$ **c.** $n^2n^4n^6$

100. a. $ac^8 \cdot ac$ **b.** $ac^8 - ac$ **c.** $(ac^8) \cdot (ac)$

APPLICATIONS

101. Art history.   Leonardo da Vinci's drawing relating a human figure to a square and a circle is shown below. Find an expression for the area of the square if the man's height is $5x$ feet.

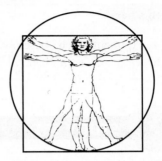

102. Packaging.   Find an expression for the volume of the box shown below.

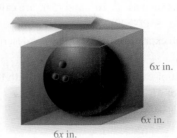

$6x$ in.

$6x$ in.

$6x$ in.

WRITING

103. Explain the mistake in the following work.

$$2^3 \cdot 2^2 = 4^5 = 1{,}024$$

104. Explain why we can simplify $x^4 \cdot x^5$, but cannot simplify $x^4 + x^5$.

REVIEW

105. Jewelry.   A lot of what we refer to as gold jewelry is actually made of a combination of gold and another metal. For example, 18-karat gold is $\frac{18}{24}$ gold by weight. Simplify this ratio.

106. After evaluation, what is the sign of $(-13)^5$?

107. Divide: $\dfrac{-25}{-5}$

108. How much did the temperature change if it went from $-4°F$ to $-17°F$?

109. Evaluate: $2\left(\dfrac{12}{-3}\right) + 3(5)$

110. Solve: $-10 = x + 1$

111. Solve: $-x = -12$

112. Divide: $\dfrac{0}{10}$

8   Summary and Review

DEFINITIONS AND CONCEPTS	EXAMPLES
A **variable** is a letter (or symbol) that stands for a number. Since numbers do not change value, they are called **constants**.	Variables: x, a, and y Constants: 8, -10, $2\frac{3}{5}$, and 3.14
When multiplying a variable by a number, or a variable by another variable, we can omit the symbol for multiplication.	$3x$ means $3 \cdot x$ ab means $a \cdot b$ $4rst$ means $4 \cdot r \cdot s \cdot t$
Many of the properties that we have seen while working with whole numbers, integers, fractions, and decimals can be generalized and stated in symbols using variables.	**The Commutative Property of Addition** $$a + b = b + a$$ **The Associative Property of Multiplication** $$(ab)c = a(bc)$$
Variables and/or numbers can be combined with the operations of addition, subtraction, multiplication, and division to create **algebraic expressions**. We often refer to *algebraic expressions* as simply **expressions**.	Expressions: $5y + 7$ $\dfrac{12 - x}{5}$ $8a(b - 3)$
A **term** is a product or quotient of numbers and/or variables. A single number or variable is also a term. A term, such as 4, that consists of a single number is called a **constant term**.	Terms: 4, y, $6r$, $-w^3$, $3.7x^5$, $\dfrac{3}{n}$, $-15ab^2$
Addition symbols separate expressions into parts called **terms**. The numerical factor of a term is called the **coefficient** of the term.	Since $6a^2 + a - 5$ can be written as $6a^2 + a + (-5)$, it has three terms. <table><tr><th>Term</th><th>Coefficient</th></tr><tr><td>$6a^2$</td><td>6</td></tr><tr><td>a</td><td>1</td></tr><tr><td>-5</td><td>-5</td></tr></table>
It is important to be able to distinguish between the **terms of an expression** and the **factors of a term**.	$x + 6$ $6x$ ↑ ↑ *x is a term.* *x is a factor.*
Key words and **key phrases** can be translated into algebraic expressions.	*5 more than x* can be expressed as $x + 5$. *25 less than twice y* can be expressed as $2y - 25$. One-half of the cost c can be expressed as $\dfrac{1}{2}c$.

To **evaluate algebraic expressions,** we substitute the values of its variables and apply the rules for the order of operations.

Evaluate $\dfrac{x^2 - y^2}{x + y}$ for $x = 2$ and $y = -3$.

$$\dfrac{x^2 - y^2}{x + y} = \dfrac{2^2 - (-3)^2}{2 + (-3)}$$ Substitute 2 for x and -3 for y.

$$= \dfrac{4 - 9}{-1}$$ In the numerator, evaluate the exponential expressions. In the denominator, add.

$$= \dfrac{-5}{-1}$$ In the numerator, subtract.

$$= 5$$ Do the division.

REVIEW EXERCISES

1. Write each expression without using a multiplication symbol or parentheses.

  a. $6 \cdot b$

  b. $x \cdot y \cdot z$

  c. $2(t)$

2. a. Write the commutative property of addition using the variables c and d.

  b. Write the associative property of multiplication using the variables r, s, and t.

3. Determine whether the variable h is used as a *term* or as a *factor*.

  a. $5h + 9$ **b.** $h + 16$

4. How many terms does each expression have?

  a. $3x^2 + 2x - 5$ **b.** $-12xyz$

5. Identify the coefficient of each term of the given expression.

  a. $16x^2 - x + 25$ **b.** $\dfrac{x}{2} + y$

6. Translate the expression $m - 500$ into words.

7. Translate each phrase to an algebraic expression.

  a. 25 more than the height h

  b. 100 reduced by twice the cutoff score s

  c. 6 less than one-half of the time t

  d. The absolute value of the difference of 2 and the square of a.

8. Hardware.   Refer to the illustration in the next column.

  a. Let n represent the length of the nail (in inches). Write an algebraic expression that represents the length of the bolt (in inches).

  b. Let b represent the length of the bolt (in inches). Write an algebraic expression that represents the length of the nail (in inches).

4 in.

9. a. Clothes designers.   The legs on a pair of pants are x inches long. The designer then lets the hem down 1 inch. Write an algebraic expression that represents the length of the altered pants legs.

  b. Butchers.   A roast weighs p pounds. A butcher trimmed the roast into 8 equal-sized servings. Write an algebraic expression that represents the weight of one serving.

10. Sports equipment.   An NBA basketball weighs 2 ounces more than twice the weight of a volleyball.

  a. Let x represent the weight of one of the balls. Write an expression for the weight of the other ball.

  b. If the weight of the volleyball is 10 ounces, what is the weight of the NBA basketball?

Evaluate each algebraic expression for the given values of the variables.

11. $2x^2 + 3x + 7$ for $x = 5$

12. $(x - 7)^2$ for $x = -1$

13. $b^2 - 4ac$ for $b = -10$, $a = 3$, and $c = 5$

14. $\dfrac{x + y}{-x - z}$ for $x = 19$, $y = 17$, and $z = -18$

SECTION 8.2 ▶ Simplifying Algebraic Expressions

DEFINITIONS AND CONCEPTS	EXAMPLES
We often use the *commutative property of multiplication* to reorder factors and the *associative property of multiplication* to regroup factors when **simplifying expressions**.	Simplify: $-5(3y) = (-5 \cdot 3)y = -15y$ Simplify: $-45b\left(\dfrac{5}{9}\right) = \left(-45 \cdot \dfrac{5}{9}\right)b = -\dfrac{5 \cdot \overset{1}{\cancel{9}} \cdot 5}{\underset{1}{\cancel{9}}}b = -25b$
The **distributive property** can be used to remove parentheses: $a(b + c) = ab + ac \qquad a(b - c) = ab - ac$ $a(b + c + d) = ab + ac + ad$	Multiply: $7(x + 3) = 7 \cdot x + 7 \cdot 3$ $\qquad\qquad\qquad = 7x + 21$ Multiply: $-0.2(4m - 5n - 7) = -0.2(4m) - (-0.2)(5n) - (-0.2)(7)$ $\qquad\qquad\qquad\qquad = -0.8m + n + 1.4$
Like terms are terms with exactly the same variables raised to exactly the same powers.	$3x$ and $-5x$ are like terms. $-4t^3$ and $3t^2$ are unlike terms because the variable t has different exponents. $0.5xyz$ and $3.7xy$ are unlike terms because they have different variables.
Simplifying the sum or difference of like terms is called **combining like terms**. Like terms can be combined by adding or subtracting the coefficients of the terms and keeping the same variables with the same exponents.	Simplify: $4a + 2a = 6a$ Think: $(4 + 2)a = 6a$. Simplify: $5p^2 + p - p^2 - 9p = 4p^2 - 8p$ Think: $(5 - 1)p^2 = 4p^2$ and $\qquad\qquad\qquad\qquad\qquad\qquad\qquad\qquad\qquad (1 - 9)p = -8p.$ Simplify: $2(k - 1) - 3(k + 2) = 2k - 2 - 3k - 6$ $\qquad\qquad\qquad\qquad\qquad = -k - 8$

REVIEW EXERCISES

Simplify each expression.

15. $4(7w)$

16. $3(-2x)(-4)$

17. $0.4(5.2f)$

18. $\dfrac{7}{2} \cdot \dfrac{2}{7}r$

Use the distributive property to remove parentheses.

19. $5(x + 3)$

20. $-(2x + 3 - y)$

21. $\dfrac{3}{4}(4c - 8)$

22. $2(3c + 7)(2.1)$

List the like terms in each expression.

23. $7a + 3 + 9a$

24. $2x^2 + 2x + 3x^2 - x$

Simplify each expression by combining like terms, if possible.

25. $8p + 5p - 4p$

26. $-5m + 2 - 2m - 2$

27. $n + n + n + n$

28. $5(p - 2) - 2(3p + 4)$

29. $55.7k^2 - 55.6k^2$

30. $8a^3 + 4a^3 + 2a - 4a^3 - 2a - 1$

31. $10x + 10y$

32. $4x^3 - 4x^2 - 4x - 4$

33. $\dfrac{3}{5}w - \left(-\dfrac{2}{5}w\right)$

34. $36\left(\dfrac{1}{9}h - \dfrac{3}{4}\right) + 36\left(\dfrac{1}{3}\right)$

35. Write an equivalent expression for the given expression using fewer symbols.

    a. $1x$ **b.** $-1x$

    c. $4x - (-1)$ **d.** $4x + (-1)$

36. **Geometry.** Write an algebraic expression in simplified form that represents the perimeter of the triangle.

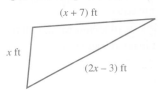

SECTION 8.3 ▶ Solving Equations Using Properties of Equality

DEFINITIONS AND CONCEPTS	EXAMPLES
An **equation** is a statement indicating that two expressions are equal. All equations contain an equal symbol. The equal symbol $=$ separates an equation into two parts: the left side and the right side.	Equations: $$2x + 4 = 10 \qquad -5(a + 4) = -11a \qquad \frac{3}{2}t + 6 = t - \frac{1}{3}$$
A number that makes an equation a true statement when substituted for the variable is called a **solution** of the equation.	Determine whether 2 is a solution of $x + 4 = 3x$. **Check:** $\quad x + 4 = 3x$ $\qquad 2 + 4 \stackrel{?}{=} 3(2) \qquad$ Substitute 2 for each x. $\qquad\quad 6 = 6 \qquad$ True Since the resulting statement, $6 = 6$, is true, 2 is a solution of $x + 4 = 3x$.
Equivalent equations have the same solutions.	$x - 2 = 6$ and $x = 8$ are equivalent equations because they have the same solution, 8.
To **solve an equation**, isolate the variable on one side of the equation by undoing the operations performed on it using properties of equality. **Addition (subtraction) property of equality:** If the same number is added to (or subtracted from) both sides of an equation, the result is an equivalent equation.	Solve: $\quad x - 5 = 7 \qquad\qquad$ Solve: $\quad c + 9 = 16$ $\quad x - 5 + 5 = 7 + 5 \qquad\qquad c + 9 - 9 = 16 - 9$ $\qquad\qquad x = 12 \qquad\qquad\qquad\qquad c = 7$
Multiplication (division) property of equality: If both sides of an equation are multiplied (or divided) by the same nonzero number, the result is an equivalent equation.	Solve: $\quad \dfrac{m}{3} = 2 \qquad\qquad\qquad$ Solve: $\quad 10y = 50$ $\quad 3\left(\dfrac{m}{3}\right) = 3(2) \qquad\qquad\qquad \dfrac{10y}{10} = \dfrac{50}{10}$ $\qquad\quad m = 6 \qquad\qquad\qquad\qquad\qquad y = 5$

REVIEW EXERCISES

Use a check to determine whether the number in red is a solution of the equation.

37. 84, $x - 34 = 50$

38. 3, $5y + 2 = 12$

39. -30, $\dfrac{x}{5} = 6$

40. 2, $a^2 - a - 1 = 0$

41. -3, $5b - 2 = 3b - 8$

42. 1, $\dfrac{2}{y + 1} = \dfrac{12}{y + 1} - 5$

Fill in the blanks.

43. An _____ is a statement indicating that two expressions are equal.

44. To solve $x - 8 = 10$ means to find all the values of the variable that make the equation a _____ statement.

Solve each equation. Check the result.

45. $x - 9 = 12$

46. $-y = -32$

47. $a + 3.7 = -16.9$

48. $100 = -7 + r$

49. $120 = 5c$

50. $t - \dfrac{1}{2} = \dfrac{3}{2}$

51. $\dfrac{4}{3}t = -12$

52. $3 = \dfrac{q}{-2.6}$

53. $6b = 0$

54. $\dfrac{15}{16}s = -3$

SECTION 8.4 ▶ More About Solving Equations

DEFINITIONS AND CONCEPTS	EXAMPLES
A strategy for solving equations:	Solve: $6x + 2 = 14$
1. *Simplify* each side. Use the distributive property and combine like terms when necessary.	To isolate the variable, we use the order of operations rule in reverse. ■ To isolate the variable term, $6x$, we subtract 2 from both sides to undo the addition of 2.
2. *Isolate the variable term.* Use the addition and subtraction properties of equality.	■ To isolate the variable, x, we divide both sides by 6 to undo the multiplication by 6.
3. *Isolate the variable.* Use the multiplication and division properties of equality.	$6x + 2 - 2 = 14 - 2$ Subtract 2 from both sides to isolate 6x.
4. *Check* the result in the original equation.	$6x = 12$ Do the subtractions.
	$\dfrac{6x}{6} = \dfrac{12}{6}$ Divide both sides by 6 to isolate x.
	$x = 2$
	The solution is 2. Check by substituting it into the original equation.
When solving equations, we should simplify the expressions that make up the left and right sides before applying any properties of equality.	Solve: $2(y + 2) + 4y = 11 - y$
	$2y + 4 + 4y = 11 - y$ Distribute the multiplication by 2.
	$6y + 4 = 11 - y$ Combine like terms: 2y + 4y = 6y.
	$6y + 4 + y = 11 - y + y$ To eliminate −y on the right, add y to both sides.
	$7y + 4 = 11$ Combine like terms.
	$7y + 4 - 4 = 11 - 4$ To isolate the variable term 7y, subtract 4 from both sides.
	$7y = 7$ Simplify each side of the equation.
	$\dfrac{7y}{7} = \dfrac{7}{7}$ To isolate y, divide both sides by 7.
	$y = 1$
	The solution is 1. Check by substituting it into the original equation.

REVIEW EXERCISES

Solve each equation. Check the result.

55. $5x + 4 = 14$

56. $98.6 - t = 129.2$

57. $\dfrac{n}{5} - 2 = 4$

58. $\dfrac{3}{4}c + 10 = -11$

59. $12a - 9 = 4a + 15$

60. $8t + 3.2 = 4t - 1.6$

61. $5(2x - 4) - 5x = 0$

62. $-2(x - 5) = 5(-3x + 4) + 3$

63. $2(m + 40) - 6m = 3(4m - 80)$

64. $-8(1.5r - 0.5) = 3.28$

DEFINITIONS AND CONCEPTS	EXAMPLES
To solve application problems, use the six-step problem-solving strategy.	**Nobel Prize.** In 1998, three Americans, Louis Ignarro, Robert Furchgott, and Fred Murad, were awarded the Nobel Prize for medicine. They shared the prize money equally. If each person received $318,500, what was the amount of the cash award for the Nobel Prize for medicine? (Source: nobelprize.org)

To solve application problems, use the six-step problem-solving strategy.

1. **Analyze the problem:** What information is given? What are you asked to find?

2. **Assign a variable:** Let a variable represent an unknown value in the problem.

3. **Form an equation:** Translate the words of the problem into an equation.

4. **Solve the equation.**

5. **State the conclusion clearly:** Be sure to include the units (such as feet, seconds, or pounds) in your answer.

6. **Check the result:** Use the original wording of the problem, not the equation that was formed in Step 3 from the words.

Nobel Prize. In 1998, three Americans, Louis Ignarro, Robert Furchgott, and Fred Murad, were awarded the Nobel Prize for medicine. They shared the prize money equally. If each person received $318,500, what was the amount of the cash award for the Nobel Prize for medicine? (Source: nobelprize.org)

Analyze

- 3 people shared the cash award equally.   *Given*

- Each person received $318,500.   *Given*

- What was the amount of the cash award?   *Find*

Assign

Let a = the amount of the cash award for the Nobel Prize.

Form

Look for a key word or phrase in the problem.

   Key phrase: shared the prize money equally

   Translation: division

Translate the words of the problem into an equation.

The amount of the cash award	divided by	the number of people that shared it equally	was	$318,500.
a	$\div$	3	$=$	318,500

Solve

$\dfrac{a}{3} = 318,500$ *We need to isolate a on the left side.*

$3 \cdot \dfrac{a}{3} = 3 \cdot 318,500$ *To isolate a, undo the division by 3 by multiplying both sides by 3.*

$a = 955,500$ *Do the multiplication.*

$$\begin{array}{r} \overset{2\,1}{318,500} \\ \times\quad 3 \\ \hline 955,500 \end{array}$$

State

The amount of the cash award for the Nobel Prize in medicine was $955,500.

Check

If the cash prize was $955,500, then the amount that each winner received can be found using division:

$$3\overline{)955,500}^{\,318,500} \quad \longleftarrow \text{ This is the amount each prize winner received.}$$

The result, $955,500 checks.

The six-step problem-solving strategy can be used to solve application problems to find **two unknowns**.

Sound systems.   A 45-foot-long speaker wire is cut into two pieces. One piece is 9 feet longer than the other. Find the length of each piece of wire.

Analyze

- A 45-foot-long wire is cut into two pieces.   *Given*
- One piece is 9 feet longer than the other.   *Given*
- What is the length of the shorter piece and the length of the longer piece of wire?   *Find*

Assign

Since we are told that the length of the longer piece of wire is related to the length of the shorter piece, let $x =$ the length of the shorter piece of wire.

There is a second unknown quantity. Look for a key phrase to help represent the length of the longer piece of wire using an algebraic expression.

Key phrase: 9 feet longer   **Translation:** addition

So $x + 9 =$ the length of the longer piece of wire.

Form

Now translate the words of the problem into an equation.

The length of the shorter piece	plus	the length of the longer piece	is	45 feet.
x	$+$	$x + 9$	$=$	45

Solve

$$x + x + 9 = 45 \qquad \text{We need to isolate } x \text{ on the left side.}$$
$$2x + 9 = 45 \qquad \text{Combine like terms: } x + x = 2x.$$
$$2x + 9 - 9 = 45 - 9 \qquad \text{To isolate } 2x, \text{ subtract 9 from both sides.}$$
$$2x = 36 \qquad \text{Do the subtraction.}$$
$$\frac{2x}{2} = \frac{36}{2} \qquad \text{To isolate } x, \text{ undo the multiplication by 2 by dividing both sides by 2.}$$
$$x = 18 \qquad \text{Do the division.}$$

To find the second unknown, we substitute 18 for x in the expression that represents the length of the longer piece of wire.

$$x + 9 = \mathbf{18} + 9 = 27$$

State

The length of the shorter piece of wire is 18 feet, and the length of the longer piece is 27 feet.

Check

The length of the longer piece of wire, 27 feet, is 9 feet more than the length of the shorter piece, 18 feet. Adding the two lengths, we get

$$
\begin{array}{r}
18 \\
+\ 27 \\
\hline
45
\end{array}
$$

45 ◄—— This is the original length of the wire, before It was cut into two pieces.

The results, 18 ft and 27 ft, check.

REVIEW EXERCISES

Form an equation and solve it to answer each question.

65. Financing.   A newly married couple made a $25,000 down payment on a house priced at $122,750. How much did they need to borrow?

66. Patient lists.   After moving his office, a doctor lost 53 patients. If he had 672 patients left, how many did he have originally?

67. Construction delays.   Because of a shortage of materials, the final cost of a construction project was three times greater than the original estimate. Upon completion, the project cost $81 million. What was the original cost estimate?

68. Social work.   A human services program assigns each of its social workers a caseload of 80 clients. How many clients are served by 45 social workers?

69. Cold storage.   A meat locker lowers the temperature of a product 7° Fahrenheit every hour. If freshly ground hamburger is placed in the locker, how long will it take to go from room temperature of 71°F to 29°F?

Jorge Royan/Alamy Stock Photo

70. Moving expenses.   Tom and his friend split the cost of renting a U-Haul trailer equally. Tom also agreed to pay the $4 to rent a refrigerator dolly. In all, Tom paid $20. What did it cost to rent the trailer?

71. Fitness.   The midweek workout for a fitness instructor consists of walking and running. She walks 3 fewer miles than she runs. If her workout covers a total of 15 miles, how many miles does she run and how many miles does she walk?

72. Rodeos.   Attendance during the first day of a two-day rodeo was low. On the second day, attendance doubled. If a total of 6,600 people attended the show, what was the attendance on the first day and what was the attendance on the second day?

73. Parking lots.   A rectangular parking lot is 4 times as long as it is wide. If the perimeter of the parking lot is 250 feet, what is its length and width?

74. Space travel.   The 364-foot-tall *Saturn V* rocket carried the first astronauts to the moon. Its first, second, and third stages were 138, 98, and 46 feet tall (in that order). Atop the third stage was a lunar module, and from it extended a 28-foot escape tower. How tall was the lunar module? (Source: NASA)

SECTION 8.6 ▶ Multiplication Rules for Exponents

DEFINITIONS AND CONCEPTS	EXAMPLES

An **exponent** indicates repeated multiplication. It tells how many times the base is to be used as a factor.
 	
Exponent ⌐
                   n factors of x
       $x^n = x \cdot x \cdot x \cdot \cdots \cdot x$
Base ⌐
	Identify the base and the exponent in each expression.
	
	$2^6 = 2 \cdot 2 \cdot 2 \cdot 2 \cdot 2 \cdot 2$ *2 is the base and 6 is the exponent.*
	
	$(-xy)^3 = (-xy)(-xy)(-xy)$ *Because of the parentheses, −xy is the base and 3 is the exponent.*
	
	$5t^4 = 5 \cdot t \cdot t \cdot t \cdot t$ *The base is t and 4 is the exponent.*
	
	$8^1 = 8$ *The base is 8 and 1 is the exponent.*

Rules for Exponents If *m* and *n* represent integers, **Product rule:** $x^m x^n = x^{m+n}$ **Power rule:** $(x^m)^n = x^{m \cdot n} = x^{mn}$ **Power of a product rule:** $(xy)^m = x^m y^m$	Simplify each expression: $5^2 5^7 = 5^{2+7} = 5^9$ *Keep the common base, 5, and add the exponents.* $(6^3)^7 = 6^{3 \cdot 7} = 6^{21}$ *Keep the base, 6, and multiply the exponents.* $(2p)^5 = 2^5 p^5 = 32p^5$ *Raise each factor of the product 2p to the 5th power.*

To simplify some expressions, we must apply two (or more) rules for exponents.	Simplify: $(c^2 c^5)^4 = (c^7)^4$ *Within the parentheses, keep the common base, c, and add the exponents: 2 + 5 = 7.*
	
	          $= c^{28}$ *Keep the base, c, and multiply the exponents: 7 · 4 = 28.*
	
	Simplify: $(t^2)^4(t^3)^3 = t^8 t^9$ *For each power of t raised to a power, keep the base and multiply the exponents: 2 · 4 = 8 and 3 · 3 = 9.*
	
	          $= t^{17}$ *Keep the common base, t, and add the exponents: 8 + 9 = 17.*

REVIEW EXERCISES

75. Identify the base and the exponent in each expression.

     a. n^{12} **b.** $(2x)^6$

     c. $3r^4$ **d.** $(y - 7)^3$

76. Write each expression in an equivalent form using an exponent.

     a. $m \cdot m \cdot m \cdot m \cdot m$ **b.** $-3 \cdot x \cdot x \cdot x \cdot x$

     c. $a \cdot a \cdot b \cdot b \cdot b \cdot b$ **d.** $(pq)(pq)(pq)$

77. Simplify, if possible.

     a. $x^2 \cdot x^2$ **b.** $x^2 + x^2$

     c. $x \cdot x^2$ **d.** $x + x^2$

78. Explain each error.

     a. $3^2 \cdot 3^4 = 9^6$ **b.** $(3^2)^4 = 3^6$

Simplify each expression.

79. $7^4 \cdot 7^8$ **80.** $mmnn^2$

81. $(y^7)^3$ **82.** $(3x)^4$

83. $(6^3)^{12}$ **84.** $-b^3 b^4 b^5$

85. $(-16s^3)^2 s^4$ **86.** $(2.1x^2 y)^2$

87. $[(-9)^3]^5$ **88.** $(a^5)^3 (a^2)^4$

89. $(2x^2 x^3)^3$ **90.** $(m^2 m^3)^2 (n^2 n^4)^3$

91. $(3a^4)^2 (2a^3)^3$ **92.** $x^{100} \cdot x^{100}$

93. $(4m^3)^3 (2m^2)^2$ **94.** $(3t^4)^3 (2t^5)^2$

Fill in the blanks.

1. a. _____ are letters (or symbols) that stand for numbers.

  b. To perform the multiplication $3(x + 4)$, we use the _____ property.

  c. Terms such as $7x^2$ and $5x^2$, which have the same variables raised to exactly the same power, are called _____ terms.

  d. When we write $4x + x$ as $5x$, we say we have _____ like terms.

  e. The _____ of the term $9y$ is 9.

  f. To evaluate $y^2 + 9y - 3$ for $y = -5$, we _____ -5 for y and apply the order of operations rule.

  g. Variables and/or numbers can be combined with the operations of arithmetic to create algebraic _____ .

  h. An _____ is a statement indicating that two expressions are equal.

  i. To _____ an equation means to find all values of the variable that make the equation true.

  j. To _____ the solution of an equation, we substitute the value for the variable in the original equation and determine whether the result is a true statement.

2. Use the following variables to state each property in symbols.

  a. Write the associative property of addition using the variables b, c, and d.

  b. Write the multiplication property of 1 using the variable t.

3. Fish. Refer to the illustration below. Let the variable s represent the length of the salmon (in inches). Write an algebraic expression that represents the length of the trout (in inches).

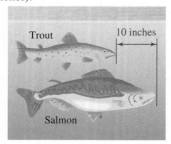

Trout   10 inches

Salmon

4. Translate to symbols

  a. 2 less than r

  b. The product of 3, x, and y

  c. The cost c split three equal ways

  d. 7 more than twice the width w

5. Translate the algebraic expression $\frac{3}{4}t$ into words.

6. Retaining walls. Refer to the illustration below. Let $h =$ the height of the retaining wall (in feet).

  a. Write an algebraic expression to represent the length of the upper base of the brick retaining wall.

  b. Write an algebraic expression to represent the length of the lower base of the brick retaining wall.

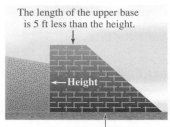

The length of the upper base is 5 ft less than the height.

←Height

The length of the lower base is 3 ft less than twice the height.

7. Determine whether a is used as a factor or as a term.

  a. $5ab$ **b.** $8b + a + 6$

8. Consider the expression $x^3 + 8x^2 - x - 6$.

  a. How many terms does the expression have?

  b. What is the coefficient of each term?

9. Evaluate $\dfrac{x - 16}{x}$ for $x = 4$.

10. Evaluate $a^2 + 2ab + b^2$ for $a = -5$ and $b = -1$.

11. Simplify each expression.

  a. $9 \cdot 4s$ **b.** $-10(12t)$

  c. $18\left(\dfrac{2}{3}x\right)$ **d.** $-4(-6)(-3m)$

12. Multiply.

  a. $5(5x + 1)$ **b.** $-6(7 - x)$

  c. $-(6y + 4)$ **d.** $0.3(2a + 3b - 7)$

  e. $\dfrac{1}{2}(2m - 8)$ **f.** $(2r + 1)9$

13. Identify the like terms in the following expression:
  $$12m^2 - 3m + 2m^2 + 3$$

14. Simplify by combining like terms, if possible.

  a. $20y - 8y$

  b. $34a - a + 7a$

  c. $-8b^2 + 29b^2$

  d. $9z - 6 + 2z + 19$

15. Simplify: $4(2y + 3) - 5(y + 3)$

16. Use a check to determine whether 7 is a solution of $2y - 1 = y + 8$.

Solve each equation and check the result.

17. $x + 6 = 10$

18. $1.8 = y - 1.3$

19. $5t = 55$

20. $\dfrac{q}{3} = -27$

21. $d - \dfrac{1}{3} = \dfrac{1}{6}$

22. $\dfrac{7}{8}n = 21$

23. $15a - 10 = 20$

24. $8x + 6 = 3x + 7$

25. $3.6 - r = 9.8$

26. $2(4x - 1) = 3(4 - 3x) + 3x$

27. $-\dfrac{15}{16}x + 15 = 0$

28. $-b = 15$

Form an equation and solve it to answer each question.

29. Hearing protection. When an airplane mechanic wears earmuffs, the sound intensity that he experiences from a jet engine is only 81 decibels. If the earmuffs reduce sound intensity by 29 decibels, what is the actual sound intensity of a jet engine?

30. Parking. After many student complaints, a college has decided to triple the number of parking spaces on campus by constructing a parking structure. That increase will bring the total number of spaces up to 6,240. How many parking spaces does the college have at this time?

31. Orchestras. A 98-member orchestra is made up of a woodwind section with 19 musicians, a brass section with 23 players, a 2-person percussion section, and a large string section. How many musicians make up the string section of the orchestra?

32. Recreation. A developer donated a large plot of land to a city for a park. Half of the acres will be used for sports fields. From the other half, 4 acres will be used for parking. This will leave 18 acres for a nature habitat. How many acres of land did the developer donate to the city?

33. Number problem. The sum of two numbers is 63. One number is 17 more than the other. What are the numbers?

34. Picture framing. A rectangular picture frame is twice as long as it is wide. If 144 inches of framing material were used to make it, what is the width and what is the length of the frame?

35. Identify the base and the exponent of each expression.

   a. 6^5 **b.** $7b^4$

36. Simplify each expression, if possible.

   a. $x^2 + x^2$ **b.** $x^2 \cdot x^2$

   c. $x^2 + x$ **d.** $x^2 \cdot x$

37. Simplify each expression.

   a. $h^2 h^4$ **b.** $(m^{10})^2$

   c. $b^2 \cdot b \cdot b^5$ **d.** $(x^3)^4(x^2)^3$

   e. $(a^2 b^3)(a^4 b^7)$ **f.** $(12a^9 b)^2$

   g. $(2x^2)^3(3x^3)^3$ **h.** $(t^2 t^3)^3$

38. Explain what is wrong with the following work:

$$5^4 \cdot 5^3 = 25^7$$

1. Round 7,535,670 [Section 1.1]

  a. to the nearest hundred.

  b. to the nearest ten thousand.

2. Gasoline. In 2011, the United States produced three billion, one hundred ninety-four million, seven hundred fifty-four thousand barrels of finished motor gasoline. Write this number in standard notation. (Source: U.S. Energy Information Administration). [Section 1.1]

Perform each operation.

3. $5,679 + 68 + 109 + 3,458$ [Section 1.2]

4. Subtract 4,375 from 7,697. [Section 1.3]

5. $5,345 \cdot 46$ [Section 1.4]

6. $35\overline{)30,625}$ [Section 1.5]

7. Refer to the illustration of the rectangular swimming pool below.

  a. Find the perimeter of the pool. [Section 1.2]

  b. Find the area of the surface of the pool. [Section 1.5]

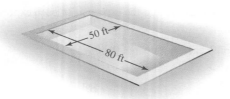

8. Discount lodging. A hotel is offering rooms that normally go for $99 per night for only $65 a night. How many dollars would a traveler save if he stays in such a room for 5 nights? [Section 1.6]

9. a. Find the factors of 20. [Section 1.7]

  b. Find the prime factorization of 20.

10. a. Find the LCM of 14 and 21. [Section 1.8]

  b. Find the GCF of 14 and 21.

Evaluate each expression. [Section 1.9]

11. $6 + 5[20 - (3^2 + 1)]$ **12.** $\dfrac{25 - (2 \cdot 3 - 1)}{2 \cdot 9 - 8}$

13. Graph the integers greater than -3 but less than 6. [Section 2.1]

        $-6\ -5\ -4\ -3\ -2\ -1\ \ 0\ \ 1\ \ 2\ \ 3\ \ 4\ \ 5\ \ 6$

14. a. Simplify: $-(-11)$ [Section 2.1]

  b. Find the absolute value: $|-11|$

  c. Is the statement $-11 > -10$ true or false?

15. Perform each operation.

  a. $-16 + 11$ [Section 2.2]

  b. $21 - (-17)$ [Section 2.3]

  c. $-6(40)$ [Section 2.4]

  d. $\dfrac{-80}{-10}$ [Section 2.5]

16. The Gateway City. The record high temperature for St. Louis, Missouri, is 107°F. The record low temperature is -18°F. Find the temperature range for these extremes. (Source: *The World Almanac and Book of Facts*, 2013) [Section 2.3]

Evaluate each expression. [Section 2.6]

17. $\dfrac{(-6)^2 - 1^5}{-4 - 3}$

18. $-10^2 - (-10)^2$

19. Simplify: $\dfrac{36}{96}$ [Section 3.1]

20. Write $\dfrac{5}{6}$ as an equivalent fraction with denominator 54. [Section 3.1]

Perform the operations.

21. $\dfrac{10}{21} \cdot \dfrac{3}{10}$ [Section 3.2]   **22.** $\dfrac{22}{25} \div \dfrac{11}{5}$ [Section 3.3]

23. $\dfrac{1}{9} + \dfrac{5}{6}$ [Section 3.4]

24. $-20\dfrac{1}{4} \div \left(-1\dfrac{11}{16}\right)$ [Section 3.5]

25. $58\dfrac{4}{11} - 15\dfrac{1}{2}$ [Section 3.6]

26. $\dfrac{\frac{2}{5} + \frac{1}{4}}{\frac{2}{5} - \frac{1}{4}}$ [Section 3.7]

27. Reading. A student has read $\frac{2}{3}$ of a novel. He plans to read one-half of the remaining pages by this evening. [Section 3.3]

  a. What fraction of the book will he have read by this evening?

  b. What fraction of the book will he have left to read?

28. Consider the decimal number: 304.817 [Section 4.1]

  a. What is the place value of the digit 1?

  b. Which digit tells the number of thousandths?

  c. Which digit tells the number of hundreds?

  d. What is the place value of the digit 7?

  e. Round 304.817 to the nearest hundredth.

Perform the operations.

29. $645 + 9.90005 + 0.12 + 3.02002$ [Section 4.2]

30. $202.234 - 19.34$ [Section 4.2]

31. $-5.8(3.9)(100)$ [Section 4.3]

32. $-(-0.2)^2 + 4|-2.3 + 1.5|$ [Section 4.3]

33. Divide -0.4531 by -0.001. [Section 4.4]

34. $12.243 \div 0.9$ (nearest hundredth) [Section 4.4]

35. Estimate the quotient: $284.254 \div 91.4$ [Section 4.4]

36. Coins. Banks wrap dimes in rolls of 50 coins. If a dime is 1.35 millimeters thick, how tall is a stack of 50 dimes? [Section 4.3]

37. Write each fraction as a decimal. [Section 4.5]

   a. $\dfrac{19}{25}$
   b. $\dfrac{1}{66}$ (use an overbar)

38. Evaluate: $50 - [(6^2 - 24) + 9\sqrt{25}]$ [Section 4.6]

39. Write the ratio $45:35$ as a fraction in simplest form. [Section 5.1]

40. Anniversary gifts. A florist sells a dozen long-stemmed red roses for $45. In honor of their 25th wedding anniversary, a man wants to buy 25 roses for his wife. What will the roses cost? (*Hint:* How many roses are in one dozen?) [Section 5.2]

41. Solve the proportion: $\dfrac{9.8}{x} = \dfrac{2.8}{5.4}$ [Section 5.2]

42. Convert 80 minutes to hours. [Section 5.3]

43. Convert 7,500 milligrams to grams [Section 5.4]

44. Track and field. A shot-put weighs 7.264 kilograms. Give this weight in pounds. [Section 5.5]

45. Complete the table below. [Section 6.1]

Fraction	Decimal	Percent
	0.25	
$\dfrac{1}{3}$		$33\dfrac{1}{3}\%$
		4.2%

46. 13 is what percent of 25? [Section 6.2]

47. 7.8 is 12% of what number? [Section 6.2]

48. Instructional equipment. Find the amount of the discount and the sale price of the overhead projector shown below. [Section 6.3]

OVERHEAD PROJECTORS
SALE! 15% OFF
Reg Price: $248⁰⁰
Hi/low switch
15 ft cord
360 W Halogen lamp

49. Estimate: What is 5% of 16,359? [Section 6.4]

50. Loans. If $400 is invested at 6.5% simple interest for 6 years, what will be the total amount of money in the investment account at the end of the 6 years? [Section 6.5]

51. Space travel. A Gallup Poll conducted July 10–12, 2009, asked a group of adults whether the U.S. space program had brought enough benefits to the country to justify the costs. The results are shown in the bar graph below. [Section 7.1]

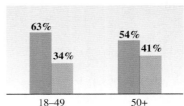

It's now 40 years since the United States first landed men on the moon. Do you think the space program has brought enough benefits to this country to justify its costs, or don't you think so?

By age
■ Yes ■ No

63% 34% 54% 41%
18–49   50+

Source: gallup.com

   a. Which age group felt more positive about the benefits of the space program?

   b. If 800 people in the survey were in the 50+ age group, how many of them responded that the benefits of the space program did not justify the costs?

52. Find the mean, median, and mode of the following set of values. [Section 7.2]

   10   4   5   7   10   3   2   3   10

53. Find the probability of drawing a black card on one draw from a standard card deck. [Section 7.3]

54. Two dice are rolled. Find the probability of getting a sum of 10. [Section 7.3]

55. Evaluate $3x - x^3$ for $x = 4$. [Section 8.1]

56. Translate each phrase to an algebraic expression. [Section 8.1]

   a. 4 less than x

   b. Twice the weight w increased by 50

57. Simplify each expression. [Section 8.2]

   a. $-3(5x)$ **b.** $-4x(-7x)$

58. Multiply. [Section 8.2]

   a. $-2(3x - 4)$ **b.** $5(3x - 2y + 4)$

59. Combine like terms. [Section 8.2]

   a. $8x - 3x$ **b.** $4a^2 + 6a^2 + 3a^2 - a^2$

   c. $4x - 3y - 5x + 2y$ **d.** $9(3x - 4) + 2x$

60. Use a check to determine whether 4 is a solution of $3x - 1 = x + 8$. [Section 8.3]

Solve each equation and check the result. [Section 8.4]

61. $3x + 2 = -13$

62. $\dfrac{y}{4} - 1 = -5$

63. $3(3y - 8) = -2(y - 4) + 3y$

64. $8 - y = -10$

Form an equation and solve it to answer each question.

65. Observation hours. To get a master's degree in elementary education, a graduate student must have 100 hours of observation time. If a student has already observed for 37 hours, how many more 3-hour shifts must she observe? [Section 8.5]

Michael Newman/PhotoEdit

66. Geometry. The perimeter of a rectangle is 210 feet. If the length is four times the width, what is the length and width of the rectangle? [Section 8.5]

67. Identify the base and the exponent of each expression. [Section 8.6]

   a. 8^9 **b.** $2a^3$

68. Simplify each expression. [Section 8.6]

   a. p^3pp^5 **b.** $(t^5)^3$

   c. $(x^2y^3)(x^3y^4)$ **d.** $(3a^2)^4$

   e. $(2p^3)^2(3p^2)^3$ **f.** $[(-2.6)^2]^8$

9 An Introduction to Geometry

from **Campus to Careers**

Surveyor

Surveyors measure distances, directions, elevations (heights), contours (curves), and angles between lines on Earth's surface. Surveys are also done in the air and underground. Surveyors often work in teams. They use a variety of instruments and electronics, including the Global Positioning System (GPS). In general, people who like surveying also like math—primarily geometry and trigonometry. The field attracts people with geology, forestry, history, engineering, computer science, and astronomy backgrounds, too.

In **Problem 83** of **Study Set 9.5,** you will see how a surveyor, using geometry, can stay on dry land and yet measure the width of a river.

JOB TITLE:
Surveyor

EDUCATION:
Courses in algebra, geometry, trigonometry, and computer science are required.

JOB OUTLOOK:
Job growth is expected to be 25% through 2020—much faster than the average for all occupations.

ANNUAL EARNINGS:
In 2015, the median annual income was $58,020.

FOR MORE INFORMATION:
http://www.bls.gov/oes/current/oes171022.htm

OBJECTIVES

1 Identify and name points, lines, and planes.

2 Identify and name line segments and rays.

3 Identify and name angles.

4 Use a protractor to measure angles.

5 Solve problems involving adjacent angles.

6 Use the property of vertical angles to solve problems.

7 Solve problems involving complementary and supplementary angles.

Euclid

Science Source/Science Source

SECTION **9.1**   Basic Geometric Figures; Angles

Geometry is a branch of mathematics that studies the properties of two- and three-dimensional figures such as triangles, circles, cylinders, and spheres. The word *geometry* comes from the Greek words *geo* (meaning earth) and *metron* (meaning measure). More than 5,000 years ago, Egyptian surveyors used geometry to measure areas of land in the flooded plains of the Nile River after heavy spring rains. Even today, engineers marvel at the Egyptians' use of geometry in the design and construction of the pyramids. History records many other practical applications of geometry made by Babylonians, Chinese, Indians, and Romans.

Many scholars consider **Euclid** (330?–275? BCE) to be the greatest of the Greek mathematicians. His book *The Elements* is an impressive study of geometry and number theory. It presents geometry in a highly structured form that begins with several simple assumptions and then expands on them using logical reasoning. For more than 2,000 years, *The Elements* was the textbook that students all over the world used to learn geometry.

OBJECTIVE 1   Identify and name points, lines, and planes.

Geometry is based on three undefined words: *point*, *line*, and *plane*. Although we will make no attempt to define these words formally, we can think of a **point** as a geometric figure that has position but no length, width, or depth. Points can be represented on paper by drawing small dots, and they are labeled with capital letters. For example, point *A* is shown in figure (a) below.

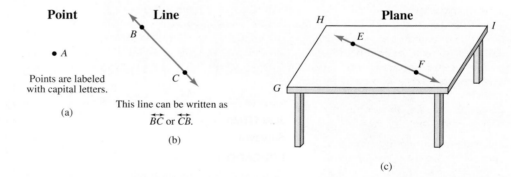

Lines are made up of points. A line extends infinitely far in both directions, but has no width or depth. A line can be represented on paper by drawing a straight line with arrowheads at either end. We can name a line using any two points on the line. In figure (b) above, the line that passes through points *B* and *C* is written as $\overleftrightarrow{BC}$.

Planes are also made up of points. A plane is a flat surface, extending infinitely far in every direction, that has length and width but no depth. The top of a table, a floor, or a wall is part of a plane. We can name a plane using any three points that lie in the plane. In figure (c) above, $\overleftrightarrow{EF}$ lies in plane *GHI*.

As figure (b) illustrates, points *B* and *C* determine exactly one line, the line $\overleftrightarrow{BC}$. In figure (c), the points *E* and *F* determine exactly one line, the line $\overleftrightarrow{EF}$. In general, any two different points determine exactly one line.

As figure (c) illustrates, points *G*, *H*, and *I* determine exactly one plane. In general, any three different points that do not lie on a line determine exactly one plane.

Other geometric figures can be created by using parts or combinations of points, lines, and planes.

OBJECTIVE 2 Identify and name line segments and rays.

> **Line Segment**
>
> The **line segment** *AB*, written as $\overline{AB}$, is the part of a line that consists of points *A* and *B* and all points in between (see the figure below). Points *A* and *B* are the **endpoints** of the segment.

Line segment

This line segment can be written as $\overline{AB}$ or $\overline{BA}$.

Every line segment has a **midpoint**, which divides the segment into two parts of equal length. In the figure below, *M* is the midpoint of segment $\overline{AB}$, because the measure of $\overline{AM}$, which is written as m($\overline{AM}$), is equal to the measure of $\overline{MB}$ which is written as m($\overline{MB}$).

$$m(\overline{AM}) = 4 - 1$$
$$m(\overline{AM}) = 3$$

and

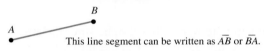

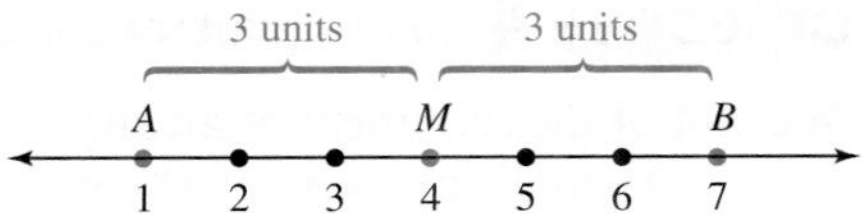

$$m(\overline{MB}) = 7 - 4$$
$$m(\overline{MB}) = 3$$

Since the measure of both segments is 3 units, we can write m($\overline{AM}$) = m($\overline{MB}$).

When two line segments have the same measure, we say that they are **congruent**. Since m($\overline{AM}$) = m($\overline{MB}$), we can write

$$\overline{AM} \cong \overline{MB}$$ Read the symbol $\cong$ as "is congruent to."

Another geometric figure is the ray, as shown below.

> **Ray**
>
> A **ray** is the part of a line that begins at some point (say, *A*) and continues forever in one direction. Point *A* is the **endpoint** of the ray.

Ray

Ray *AB* is written as $\overrightarrow{AB}$. The endpoint of the ray is always listed first.

To name a ray, we list its endpoint and then one other point on the ray. Sometimes it is possible to name a ray in more than one way. For example, in the figure on the right, $\overrightarrow{DE}$ and $\overrightarrow{DF}$ name the same ray. This is because both have point *D* as their endpoint and extend forever in the same direction. In contrast, $\overrightarrow{DE}$ and $\overrightarrow{ED}$ are not the same ray. They have different endpoints and point in opposite directions.

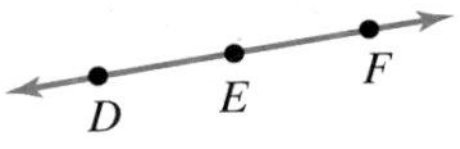

OBJECTIVE 3 Identify and name angles.

> **Angle**
>
> An **angle** is a figure formed by two rays with a common endpoint. The common endpoint is called the **vertex**, and the rays are called **sides**.

The angle shown below can be written as $\angle BAC$, $\angle CAB$, $\angle A$, or $\angle 1$. The symbol $\angle$ means angle.

Angle

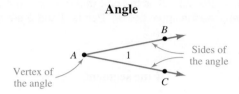

When using three letters to name an angle, be sure the letter name of the vertex is the middle letter. Furthermore, we can only name an angle using a single vertex letter when there is no possibility of confusion. For example, in the figure on the right, we cannot refer to any of the angles as simply $\angle X$, because we would not know if that meant $\angle WXY$, $\angle WXZ$, or $\angle YXZ$.

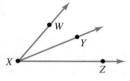

OBJECTIVE 4 Use a protractor to measure angles.

One unit of measurement of an angle is the **degree**. The symbol for degree is a small raised circle, °. An angle measure of 1° (read as "one degree") means that one side of an angle is rotated $\frac{1}{360}$ of a complete revolution about the vertex from the other side of the angle. The measure of $\angle ABC$, shown below, is 1°. We can write this in symbols as m($\angle ABC$) = 1°.

This side of the angle is rotated $\frac{1}{360}$ of a complete
revolution from the other side of the angle.

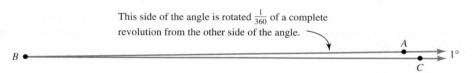

The following figures show the measures of several other angles. An angle measure of 90° is equivalent to $\frac{90}{360} = \frac{1}{4}$ of a complete revolution. An angle measure of 180° is equivalent to $\frac{180}{360} = \frac{1}{2}$ of a complete revolution, and an angle measure of 60° is equivalent to $\frac{60}{360} = \frac{1}{6}$ of a complete revolution.

$$\text{m}(\angle FED) = 90° \qquad \text{m}(\angle IHG) = 180° \qquad \text{m}(\angle JKL) = 60°$$

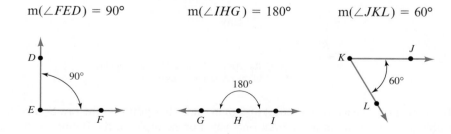

We can use a **protractor** to measure angles. To begin, we place the center of the protractor at the vertex of the angle, with the edge of the protractor aligned with one side of the angle, as shown below. The angle measure is found by determining where the other side of the angle crosses the scale. Be careful to use the appropriate scale, inner or outer, when reading an angle measure.

If we read the protractor from right to left, using the outer scale, we see that m($\angle ABC$) = 30°. If we read the protractor from left to right, using the inner scale, we can see that m($\angle GBF$) = 30°.

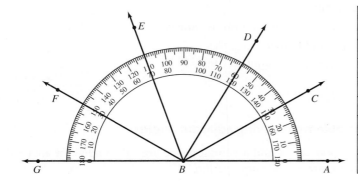

Angle	Measure in degrees
$\angle ABC$	30°
$\angle ABD$	60°
$\angle ABE$	110°
$\angle ABF$	150°
$\angle ABG$	180°
$\angle GBF$	30°
$\angle GBC$	150°

When two angles have the same measure, we say that they are **congruent**. Since m($\angle ABC$) = 30° and m($\angle GBF$) = 30°, we can write

$$\angle ABC \cong \angle GBF$$ Read the symbol $\cong$ as "is congruent to."

We classify angles according to their measure.

Classifying Angles

Acute angles: Angles whose measures are greater than 0° but less than 90°.
Right angles: Angles whose measures are 90°.
Obtuse angles: Angles whose measures are greater than 90° but less than 180°.
Straight angles: Angles whose measures are 180°.

40°	90°	130°	180°
Acute angle	**Right angle**	**Obtuse angle**	**Straight angle**

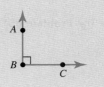

Self Check 1

Classify ∠*EFG*, ∠*DEF*, ∠1, and ∠*GED* in the figure as an acute angle, a right angle, an obtuse angle, or a straight angle.

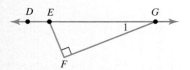

Now Try ➡ Problems 57, 59, and 61

EXAMPLE 1   Classify each angle in the figure as an acute angle, a right angle, an obtuse angle, or a straight angle.

Strategy We will determine how each angle's measure compares to 90° or to 180°.

WHY Acute, right, obtuse, and straight angles are defined with respect to 90° and 180° angle measures.

Solution

Since m(∠1) < 90°, it is an acute angle.

Since m(∠2) > 90° but less than 180°, it is an obtuse angle.

Since m(∠*BDE*) = 90°, it is a right angle.

Since m(∠*ABC*) = 180°, it is a straight angle.

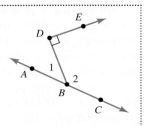

OBJECTIVE 5 Solve problems involving adjacent angles.

Two angles that have a common vertex and a common side are called **adjacent angles** if they are side by side and their interiors do not overlap.

Success Tip We can use the algebra concepts of variable and equation that were introduced in Chapter 8 to solve many types of geometry problems.

EXAMPLE 2   Two angles with degree measures of *x* and 35° are adjacent angles, as shown. Use the information in the figure to find *x*.

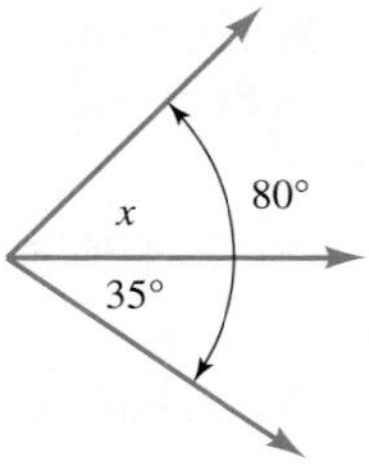

Strategy We will write an equation involving *x* that mathematically models the situation.

WHY We can then solve the equation to find the unknown angle measure.

Adjacent angles

Self Check 2

Use the information in the figure to find *x*.

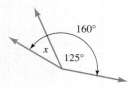

Now Try Problem 65

Solution

Since the sum of the measures of the two adjacent angles is 80°, we have

$$x + 35° = 80°$$ *The word sum indicates addition.*

$$x + 35° - 35° = 80° - 35°$$ *To isolate x, undo the addition of 35° by subtracting 35° from both sides.*

$$x = 45°$$ *Do the subtractions: 35° − 35° = 0° and 80° − 35° = 45°.*

$$\begin{array}{r} {\scriptstyle 7\,10} \\ 8\cancel{0} \\ -35 \\ \hline 45 \end{array}$$

Thus, *x* is 45°. As a check, we see that 45° + 35° = 80°.

In the figure for Example 2, we used the variable *x* to represent an unknown angle measure. In such cases, we will assume that the variable "carries" with it the associated units of degrees. That means we do not have to write a ° symbol next to the variable. Furthermore, if *x* represents an unknown number of degrees, then expressions such as 3*x*, *x* + 15°, and 4*x* − 20° also have units of degrees.

OBJECTIVE 6 Use the property of vertical angles to solve problems.

When two lines intersect, pairs of nonadjacent angles are called **vertical angles**. In the following figure, ∠1 and ∠3 are vertical angles and ∠2 and ∠4 are vertical angles.

Vertical angles

- ∠1 and ∠3
- ∠2 and ∠4

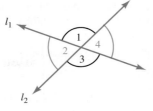

LANGUAGE OF MATHEMATICS

When we work with two (or more) lines at one time, we can use **subscripts** to name the lines. The prefix *sub* means below or beneath, as in *sub*marine or *sub*way. To name the first line in the figure to the left, we use l_1, which is read as "*l* sub one." To name the second line, we use l_2, which is read as "*l* sub two."

To illustrate that vertical angles always have the same measure, refer to the figure below, with angles having measures of x, y, and 30°. Since the measure of any straight angle is 180°, we have

$$30° + x = 180° \qquad \text{and} \qquad 30° + y = 180°$$
$$x = 150° \qquad\qquad\qquad y = 150°$$

To undo the addition of 30°, subtract 30° from both sides.

Since x and y are both 150°, we conclude that $x = y$.

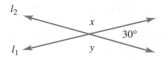

Note that the angles having measures x and y are vertical angles.

The previous example illustrates that vertical angles have the same measure. Recall that when two angles have the same measure, we say that they are *congruent*. Therefore, we have the following important fact.

Property of Vertical Angles

Vertical angles are congruent (have the same measure).

EXAMPLE 3   Refer to the figure. Find:
a. m(∠1)   **b.** m(∠ABF)

Strategy To answer part a, we will use the property of vertical angles. To answer part b, we will write an equation involving m(∠ABF) that mathematically models the situation.

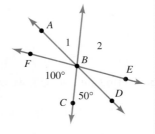

WHY For part a, we note that $\overleftrightarrow{AD}$ and $\overleftrightarrow{BC}$ intersect to form vertical angles. For part b, we can solve the equation to find the unknown, m(∠ABF).

Solution
a. If we ignore $\overleftrightarrow{FE}$ for the moment, we see that $\overleftrightarrow{AD}$ and $\overleftrightarrow{BC}$ intersect to form the pair of vertical angles ∠CBD and ∠1. By the property of vertical angles,

$$\angle CBD \cong \angle 1 \quad \text{Read as "angle } CBD \text{ is congruent to angle one."}$$

Since congruent angles have the same measure,

$$\text{m}(\angle CBD) = \text{m}(\angle 1)$$

In the figure, we are given m(∠CBD) = 50°. Thus, m(∠1) is also 50°, and we can write m(∠1) = 50°.

Self Check 3

Refer to the figure for
Example 3. Find:

a. m($\angle 2$)

b. m($\angle DBE$)

Now Try 🠖 **Problems 69
and 71**

b. Since $\angle ABD$ is a straight angle, the sum of the measures of $\angle ABF$, the 100° angle, and the 50° angle is 180°. If we let $x = \text{m}(\angle ABF)$, we have

$$x + 100° + 50° = 180° \qquad \text{The word } \textit{sum} \text{ indicates addition.}$$
$$x + 150° = 180° \qquad \text{On the left side, combine like terms: } 100° + 50° = 150°.$$
$$x = 30° \qquad \text{To isolate x, undo the addition of 150° by subtracting 150°}$$
$$\text{from both sides: } 180° - 150° = 30°.$$

Thus, m($\angle ABF$) = 30°

EXAMPLE 4   In the figure on the right, find:

a. x **b.** m($\angle ABC$)   **c.** m($\angle CBE$)

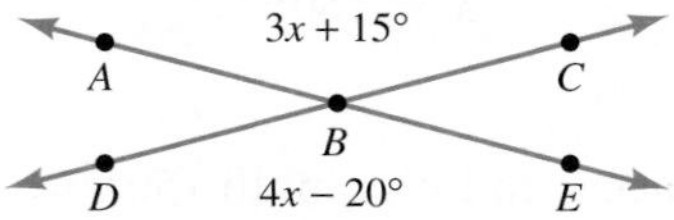

Strategy We will use the property of vertical angles to write an equation that mathematically models the situation.

WHY $\overleftrightarrow{AE}$ and $\overleftrightarrow{DC}$ intersect to form two pairs of vertical angles.

Solution

a. In the figure, two vertical angles have degree measures that are represented by the algebraic expressions $4x - 20°$ and $3x + 15°$. Since the angles are vertical angles, they have equal measures.

$$4x - 20° = 3x + 15° \qquad \text{Set the algebraic expressions equal.}$$
$$4x - 20° - 3x = 3x + 15° - 3x \qquad \text{To eliminate 3x from the right side, subtract 3x}$$
$$\text{from both sides.}$$
$$x - 20° = 15° \qquad \text{Combine like terms: } 4x - 3x = x$$
$$\text{and } 3x - 3x = 0.$$
$$x = 35° \qquad \text{To isolate x, undo the subtraction of}$$
$$20° \text{ by adding 20° to both sides.}$$

Thus, x is 35°.

Self Check 4

In the figure below, find:

a. y

b. m($\angle XYZ$)

c. m($\angle MYX$)

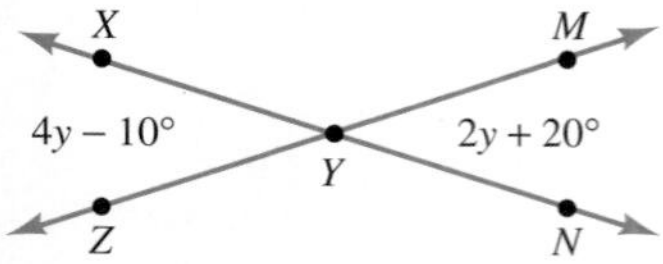

Now Try 🠖 **Problem 75**

b. To find m($\angle ABC$), we evaluate the expression $3x + 15°$ for $x = 35°$.

$$3x + 15° = 3(35°) + 15° \qquad \text{Substitute 35° for x.}$$
$$= 105° + 15° \qquad \text{Do the multiplication.}$$
$$= 120° \qquad \text{Do the addition.}$$

$$\begin{array}{r} 1 \\ 35 \\ \times\, 3 \\ \hline 105 \end{array}$$

Thus, m($\angle ABC$) = 120°.

c. $\angle ABE$ is a straight angle. Since the measure of a straight angle is 180° and m($\angle ABC$) = 120°, m($\angle CBE$) must be 180° − 120°, or 60°.

OBJECTIVE 7 **Solve problems involving complementary
and supplementary angles.**

Complementary and Supplementary Angles

Two angles are **complementary angles** when the sum of their measures is 90°.

Two angles are **supplementary angles** when the sum of their measures is 180°.

In figure (a) below, $\angle ABC$ and $\angle CBD$ are complementary angles because the sum of their measures is 90°. Each angle is said to be the **complement** of the other. In figure (b) below, $\angle X$ and $\angle Y$ are also complementary angles, because $m(\angle X) + m(\angle Y) = 90°$. Figure (b) illustrates an important fact: Complementary angles need not be adjacent angles.

Complementary angles

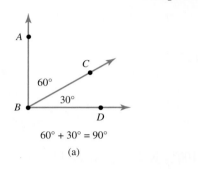

$60° + 30° = 90°$

(a)

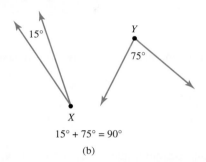

$15° + 75° = 90°$

(b)

In figure (a) below, $\angle MNO$ and $\angle ONP$ are supplementary angles because the sum of their measures is 180°. Each angle is said to be the **supplement** of the other. Supplementary angles need not be adjacent angles. For example, in figure (b) below, $\angle G$ and $\angle H$ are supplementary angles, because $m(\angle G) + m(\angle H) = 180°$.

Supplementary angles

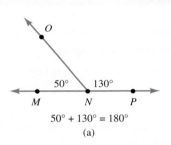

$50° + 130° = 180°$

(a)

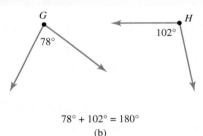

$78° + 102° = 180°$

(b)

EXAMPLE 5

a. Find the complement of a 35° angle.

b. Find the supplement of a 105° angle.

Strategy We will use the definitions of complementary and supplementary angles to write equations that mathematically model each situation.

WHY We can then solve each equation to find the unknown angle measure.

Solution

a. It is helpful to draw a figure, as shown to the right. Let x represent the measure of the complement of the 35° angle. Since the angles are complementary, we have

$x + 35° = 90°$ *The sum of the angles' measures must be 90°.*

$x = 55°$ *To isolate x, undo the addition of 35° by subtracting 35° from both sides: 90° − 35° = 55°.*

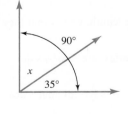

The complement of a 35° angle has measure 55°.

Caution! The definition of supplementary angles requires that the sum of *two* angles be 180°. Three angles of 40°, 60°, and 80° are not supplementary even though their sum is 180°.

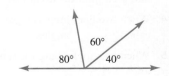

b. It is helpful to draw a figure, as shown on the right. Let y represent the measure of the supplement of the 105° angle. Since the angles are supplementary, we have

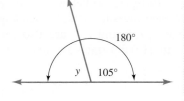

$$y + 105° = 180°$$ The sum of the angles' measures must be 180°.

$$y = 75°$$ To isolate y, undo the addition of 105° by subtracting 105° from both sides: $180° − 105° = 75°$.

The supplement of a 105° angle has measure 75°.

Answers to Self Checks

   1. right angle, obtuse angle, acute angle, straight angle   **2.** 35°   **3. a.** 100° **b.** 30°   **4. a.** 15° **b.** 50°
   c. 130°   **5. a.** 40° **b.** 130°

SECTION **9.1**   **STUDY SET**

VOCABULARY

Fill in the blanks.

1. Three undefined words in geometry are _____, _____, and _____.

2. A line _____ has two endpoints.

3. A _____ divides a line segment into two parts of equal length.

4. A _____ is the part of a line that begins at some point and continues forever in one direction.

5. An _____ is formed by two rays with a common endpoint.

6. An angle is measured in _____.

7. A _____ is used to measure angles.

8. The measure of an _____ angle is less than 90°.

9. The measure of a _____ angle is 90°.

10. The measure of an _____ angle is greater than 90° but less than 180°.

11. The measure of a straight angle is _____.

12. When two segments have the same length, we say that they are _____.

13. _____ angles have the same vertex, are side by side, and their interiors do not overlap.

14. When two lines intersect, pairs of nonadjacent angles are called _____ angles.

15. When two angles have the same measure, we say that they are _____.

16. The word *sum* indicates the operation of _____.

17. The sum of two complementary angles is _____.

18. The sum of two _____ angles is 180°.

CONCEPTS

19. a. Given two points (say, M and N), how many different lines pass through these two points?

   b. Fill in the blank: In general, two different points determine exactly one _____.

20. Refer to the figure.

   a. Name $\overrightarrow{NM}$ in another way.

   b. Do $\overrightarrow{MN}$ and $\overrightarrow{NM}$ name the same ray?

21. Consider the acute angle shown below.

   a. What two rays are the sides of the angle?

   b. What point is the vertex of the angle?

   c. Name the angle in four ways.

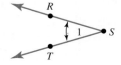

22. Estimate the measure of each angle. Do not use a protractor.

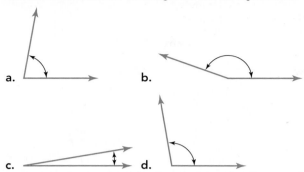

a.   b.

c.   d.

23. Draw an example of each type of angle.

    a. an acute angle   **b.** an obtuse angle

    c. a right angle   **d.** a straight angle

24. Fill in the blanks with the correct symbol.

    a. If $m(\overline{AB}) = m(\overline{CD})$, then $\overline{AB}$ ___ $\overline{CD}$.

    b. If $\angle ABC \cong \angle DEF$, then $m(\angle ABC)$ ___ $m(\angle DEF)$.

25. **a.** Draw a pair of adjacent angles. Label them $\angle ABC$ and $\angle CBD$.

    b. Draw two intersecting lines. Label them lines l_1 and l_2. Label one pair of vertical angles that are formed as $\angle 1$ and $\angle 2$.

    c. Draw two adjacent complementary angles.

    d. Draw two adjacent supplementary angles.

26. Fill in the blank:

    If $\angle MNO \cong \angle BFG$, then $m(\angle MNO)$ ___ $m(\angle BFG)$.

27. Fill in the blank:

    The vertical angle property: Vertical angles are _____.

28. Refer to the figure below. Fill in the blanks.

    a. $\angle XYZ$ and $\angle$ _____ are vertical angles.

    b. $\angle XYZ$ and $\angle ZYW$ are _____ angles.

    c. $\angle ZYW$ and $\angle XYV$ are _____ angles.

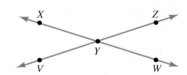

29. Refer to the figure below and tell whether each statement is true.

    a. $\angle AGF$ and $\angle BGC$ are vertical angles.

    b. $\angle FGE$ and $\angle BGA$ are adjacent angles.

    c. $m(\angle AGB) = m(\angle BGC)$.

    d. $\angle AGC \cong \angle DGF$.

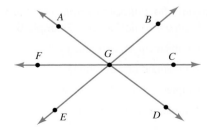

30. Refer to the figure below and tell whether the angles are congruent.

    a. $\angle 1$ and $\angle 2$ **b.** $\angle FGB$ and $\angle CGE$

    c. $\angle AGF$ and $\angle FGE$ **d.** $\angle CGD$ and $\angle CGB$

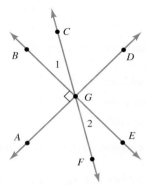

Refer to the figure above and tell whether each statement is true.

31. $\angle 1$ and $\angle CGD$ are adjacent angles.

32. $\angle FGA$ and $\angle AGC$ are supplementary.

33. $\angle AGB$ and $\angle BGC$ are complementary.

34. $\angle AGF$ and $\angle 2$ are complementary.

NOTATION

Fill in the blanks.

35. The symbol $\overleftrightarrow{AB}$ is read as "_____ AB."

36. The symbol $\overline{AB}$ is read as "_____ AB."

37. The symbol $\overrightarrow{AB}$ is read as "_____ AB."

38. We read $m(\overline{AB})$ as "the _____ of segment AB."

39. We read $\angle ABC$ as "_____ ABC."

40. We read $m(\angle ABC)$ as "the _____ of angle ABC."

41. The symbol for _____ is a small raised circle, °.

42. The symbol ⌐ indicates a _____ angle.

43. The symbol $\cong$ is read as "is _____ to."

44. The symbol l_1 can be used to name a line. It is read as "line l _____ one."

GUIDED PRACTICE

45. Draw each geometric figure and label it completely. **See Objective 1.**

    a. Point T **b.** $\overleftrightarrow{JK}$

    c. Plane ABC

46. Draw each geometric figure and label it completely. **See Objectives 2 and 3.**

    a. $\overline{RS}$ **b.** $\overrightarrow{PQ}$

    c. $\angle XYZ$ **d.** $\angle L$

47. Refer to the figure and find the length of each segment.
See Objective 2.

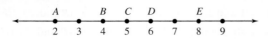

a. $\overline{AB}$ **b.** $\overline{CE}$

c. $\overline{DC}$ **d.** $\overline{EA}$

48. Refer to the figure above and find each midpoint. **See Objective 2.**

a. Find the midpoint of $\overline{AD}$.

b. Find the midpoint of $\overline{BE}$.

c. Find the midpoint of $\overline{EA}$.

Use the protractor to find each angle measure listed below. See Objective 4.

49. m($\angle GDE$)   **50.** m($\angle ADE$)

51. m($\angle EDS$)   **52.** m($\angle EDR$)

53. m($\angle CDR$)   **54.** m($\angle CDA$)

55. m($\angle CDG$)   **56.** m($\angle CDS$)

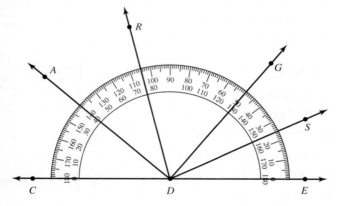

Classify the following angles in the figure as an acute angle, a right angle, an obtuse angle, or a straight angle. See Example 1.

57. $\angle MNO$ **58.** $\angle OPN$

59. $\angle NOP$ **60.** $\angle POS$

61. $\angle MPQ$ **62.** $\angle PNO$

63. $\angle QPO$ **64.** $\angle MNQ$

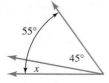

Find x. See Example 2.

65.

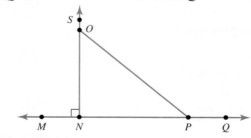

66.

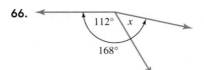

67.   **68.**

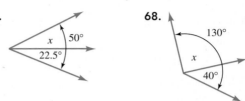

Refer to the figure below. Find the measure of each angle. See Example 3.

69. $\angle 1$ **70.** $\angle MYX$

71. $\angle NYZ$ **72.** $\angle 2$

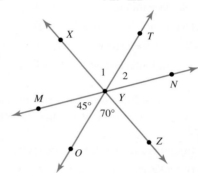

First find x. Then find m($\angle ABD$) and m($\angle DBE$). See Example 4.

73.   **74.**

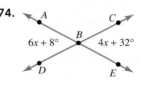

First find x. Then find m($\angle ZYQ$) and m($\angle PYQ$). See Example 4.

75.   **76.**

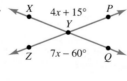

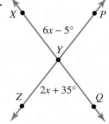

Let x represent the unknown angle measure. Write an equation and solve it to find x. See Example 5.

77. Find the complement of a 30° angle.

78. Find the supplement of a 30° angle.

79. Find the supplement of a 105° angle.

80. Find the complement of a 75° angle.

TRY IT YOURSELF

81. Refer to the figure below and tell whether each statement is true. If a statement is false, explain why.

   a. $\overrightarrow{GF}$ has point G as its endpoint.

   b. $\overline{AG}$ has no endpoints.

   c. $\overleftrightarrow{CD}$ has three endpoints.

   d. Point D is the vertex of $\angle DGB$.

   e. $m(\angle AGC) = m(\angle BGD)$

   f. $\angle AGF \cong \angle BGE$

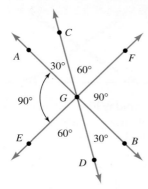

82. Refer to the figure for Problem 81 and tell whether each angle is an acute angle, a right angle, an obtuse angle, or a straight angle.

   a. $\angle AGC$ **b.** $\angle EGA$

   c. $\angle FGD$ **d.** $\angle BGA$

Use a protractor to measure each angle.

83.   **84.**

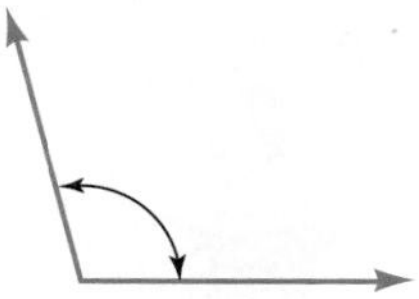

85.   **86.**

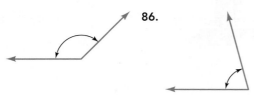

87. Refer to the figure below, in which $m(\angle 1) = 50°$. Find the measure of each angle or sum of angles.

   a. $\angle 3$

   b. $\angle 4$

   c. $m(\angle 1) + m(\angle 2) + m(\angle 3)$

   d. $m(\angle 2) + m(\angle 4)$

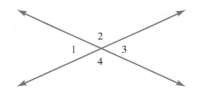

88. Refer to the figure below, in which $m(\angle 1) + m(\angle 3) + m(\angle 4) = 180°$, $\angle 3 \cong \angle 4$, and $\angle 4 \cong \angle 5$. Find the measure of each angle.

   a. $\angle 1$ **b.** $\angle 2$

   c. $\angle 3$ **d.** $\angle 6$

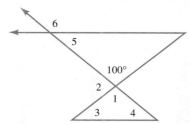

89. Refer to the figure below where $\angle 1 \cong \angle ACD$, $\angle 1 \cong \angle 2$, and $\angle BAC \cong \angle 2$.

   a. What is the complement of $\angle BAC$?

   b. What is the supplement of $\angle BAC$?

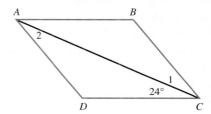

90. Refer to the figure below where $\angle EBS \cong \angle BES$.

   a. What is the measure of $\angle AEF$?

   b. What is the supplement of $\angle AET$?

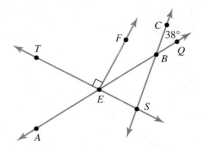

91. Find the supplement of the complement of a 51° angle.

92. Find the complement of the supplement of a 173° angle.

93. Find the complement of the complement of a 1° angle.

94. Find the supplement of the supplement of a 6° angle.

LOOK ALIKES

95. If we are working with an angle named $\angle ABC$, can we also call it $\angle CBA$?

96. If we are working with an angle named $\angle ABC$, can we also call it $\angle BCA$?

97. If we are working with ray named $\overrightarrow{AB}$, can we also call it $\overrightarrow{BA}$?

98. If M is the midpoint of $\overline{AB}$, can we also say it is the midpoint of $\overline{BA}$?

APPLICATIONS

99. Musical instruments. Suppose that you are a beginning band teacher describing the correct posture needed to play various instruments. Using the diagrams shown below, approximate the angle measure (in degrees) at which each instrument should be held in relation to the student's body.

a. flute   **b.** clarinet   **c.** trumpet

  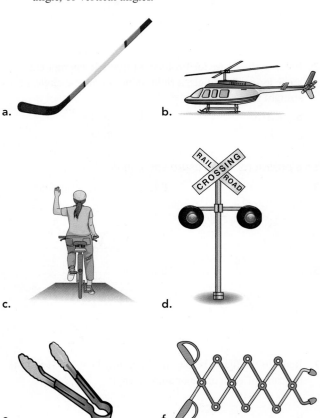

100. Planets. The figures below show the direction of rotation of several planets in our solar system. They also show the angle of tilt of each planet.

    a. Which planets have an angle of tilt that is an acute angle?

    b. Which planets have an angle of tilt that is an obtuse angle?

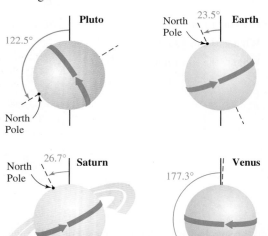

101. a. Aviation. How many degrees from the horizontal position are the wings of the airplane?

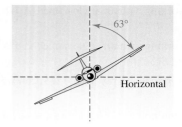

b. Gardening. What angle does the handle of the lawn mower make with the ground?

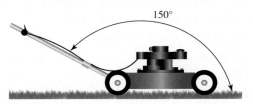

102. Synthesizer. Find x and y.

103. Angles in everyday life. Classify each illustration as an example of an acute angle, an obtuse angle, a right angle, or vertical angles.

a.   b.

c.   d.

e.   f.

104. Playground equipment. What geometric concept studied in this section is illustrated by the perfectly balanced playground seesaw shown in the diagram?

105. Phrases. Explain what you think each of these phrases means. How is geometry involved?

   a. The president did a complete 180-degree flip on the subject of a tax cut.

   b. The rollerblader did a "360" as she jumped off the ramp.

106. In the statements below, the ° symbol is used in two different ways. Explain the difference.

$$m(\angle A) = 85° \quad \text{and} \quad 85°F$$

107. Can two angles that are complementary be equal? Explain.

108. Explain why the angles highlighted below are not vertical angles.

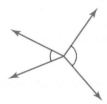

109. Solve: $x + 3x + 20 = 180$

110. Solve: $3x + 4 = 5x - 40$

111. Solve: $5x = 6x - 15$

112. Solve: $50 + x + 45 = 180$

SECTION 9.2   Parallel and Perpendicular Lines

In this section, we will consider *parallel* and *perpendicular* lines. Since parallel lines are always the same distance apart, the railroad tracks shown in figure (a) illustrate one application of parallel lines. Figure (b) shows one of the events of men's gymnastics, the parallel bars. Since perpendicular lines meet and form right angles, the monument and the ground shown in figure (c) illustrate one application of perpendicular lines.

OBJECTIVES

1. Identify and define parallel and perpendicular lines.

2. Identify corresponding angles, interior angles, and alternate interior angles.

3. Use properties of parallel lines cut by a transversal to find unknown angle measures.

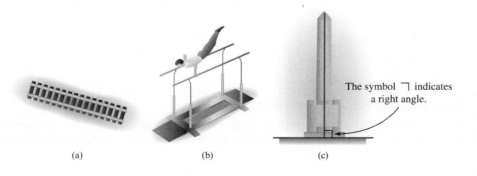

The symbol ⌐ indicates a right angle.

   (a)   (b)   (c)

OBJECTIVE 1   **Identify and define parallel and perpendicular lines.**

If two lines lie in the same plane, they are called **coplanar**. Two coplanar lines that do not intersect are called **parallel lines**. See figure (a) on the next page. If two lines do not lie in the same plane, they are called noncoplanar. Two noncoplanar lines that do not intersect are called **skew lines**.

Parallel lines   **Perpendicular lines**

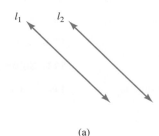

(a)   (b)

Parallel Lines

Parallel lines are coplanar lines that do not intersect.

Some lines that intersect are perpendicular. See figure (b) above.

Perpendicular Lines

Perpendicular lines are lines that intersect and form right angles.

OBJECTIVE 2 **Identify corresponding angles, interior angles, and alternate interior angles.**

A line that intersects two coplanar lines in two distinct (different) points is called a **transversal**. For example, line l_1 in the figure to the right is a transversal intersecting lines l_2 and l_3.

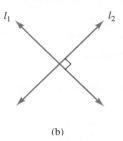

When two lines are cut by a transversal, all eight angles that are formed are important in the study of parallel lines. Descriptive names are given to several pairs of these angles.

In the figure below, four pairs of **corresponding angles** are formed.

Corresponding angles

- $\angle 1$ and $\angle 5$
- $\angle 3$ and $\angle 7$
- $\angle 2$ and $\angle 6$
- $\angle 4$ and $\angle 8$

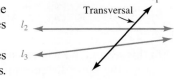

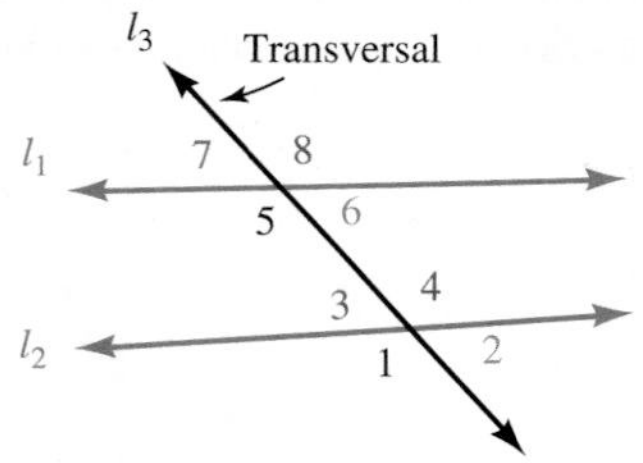

Corresponding Angles

If two lines are cut by a transversal, then the angles on the same side of the transversal and in corresponding positions with respect to the lines are called corresponding angles.

In the figure below, four **interior angles** are formed.

Interior angles

- $\angle 3$, $\angle 4$, $\angle 5$, and $\angle 6$

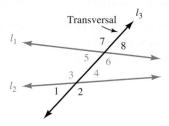

In the figure below, two pairs of **alternate interior angles** are formed.

Alternate interior angles

- ∠4 and ∠5
- ∠3 and ∠6

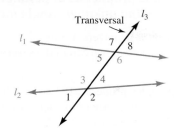

Alternate Interior Angles

If two lines are cut by a transversal, then the nonadjacent angles on opposite sides of the transversal and on the interior of the two lines are called alternate interior angles.

Success Tip Alternate interior angles are easily spotted because they form a Z-shape or a backward Z-shape, as shown below.

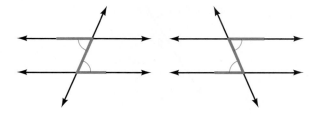

EXAMPLE 1   Refer to the figure. Identify:

a. all pairs of corresponding angles

b. all interior angles

c. all pairs of alternate interior angles

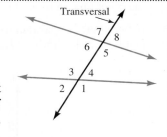

Strategy When two lines are cut by a transversal, eight angles are formed. We will consider the relative position of the angles with respect to the two lines and the transversal.

WHY There are four pairs of corresponding angles, four interior angles, and two pairs of alternate interior angles.

Solution

a. To identify corresponding angles, we examine the angles to the right of the transversal and the angles to the left of the transversal. The pairs of corresponding angles in the figure are

- ∠1 and ∠5   - ∠4 and ∠8
- ∠2 and ∠6   - ∠3 and ∠7

b. To identify the interior angles, we determine the angles inside the two lines cut by the transversal. The interior angles in the figure are

∠3, ∠4, ∠5, and ∠6

c. Alternate interior angles are nonadjacent angles on opposite sides of the transversal inside the two lines. Thus, the pairs of alternate interior angles in the figure are

- ∠3 and ∠5   - ∠4 and ∠6

Self Check 1

Refer to the figure below. Identify:

a. all pairs of corresponding angles

b. all interior angles

c. all pairs of alternate interior angles

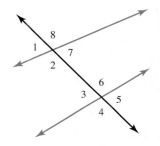

Now Try ➲ **Problem 21**

OBJECTIVE 3 **Use properties of parallel lines cut by a transversal to find unknown angle measures.**

Lines that are cut by a transversal may or may not be parallel. When a pair of parallel lines are cut by a transversal, we can make several important observations about the angles that are formed.

1. **Corresponding angles property:** If two parallel lines are cut by a transversal, each pair of corresponding angles are congruent. In the figure below, if $l_1 \parallel l_2$, then $\angle 1 \cong \angle 5$, $\angle 3 \cong \angle 7$, $\angle 2 \cong \angle 6$, and $\angle 4 \cong \angle 8$.

2. **Alternate interior angles property:** If two parallel lines are cut by a transversal, alternate interior angles are congruent. In the figure below, if $l_1 \parallel l_2$, then $\angle 3 \cong \angle 6$ and $\angle 4 \cong \angle 5$.

3. **Interior angles property:** If two parallel lines are cut by a transversal, interior angles on the same side of the transversal are supplementary. In the figure below, if $l_1 \parallel l_2$, then $\angle 3$ is supplementary to $\angle 5$ and $\angle 4$ is supplementary to $\angle 6$.

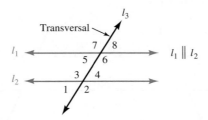

4. If a transversal is perpendicular to one of two parallel lines, it is also perpendicular to the other line. In figure (a) below, if $l_1 \parallel l_2$ and $l_3 \perp l_1$, then $l_3 \perp l_2$.

5. If two lines are parallel to a third line, they are parallel to each other. In figure (b) below, if $l_1 \parallel l_2$ and $l_1 \parallel l_3$, then $l_2 \parallel l_3$.

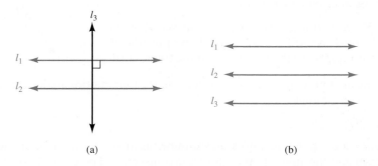

(a)   (b)

EXAMPLE 2 Refer to the figure. If $l_1 \parallel l_2$ and $m(\angle 3) = 120°$, find the measures of the other seven angles that are labeled.

Strategy We will look for vertical angles, supplementary angles, and alternate interior angles in the figure.

WHY The facts that we have studied about vertical angles, supplementary angles, and alternate interior angles enable us to use known angle measures to find unknown angle measures.

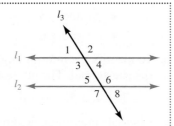

Solution

$m(\angle 1) = 60°$ *∠3 and ∠1 are supplementary: m(∠3) + m(∠1) = 180°.*

$m(\angle 2) = 120°$ *Vertical angles are congruent: m(∠2) = m(∠3).*

$m(\angle 4) = 60°$ *Vertical angles are congruent: m(∠4) = m(∠1).*

$m(\angle 5) = 60°$ *If two parallel lines are cut by a transversal, alternate interior angles are congruent: m(∠5) = m(∠4).*

$m(\angle 6) = 120°$ *If two parallel lines are cut by a transversal, alternate interior angles are congruent: m(∠6) = m(∠3).*

$m(\angle 7) = 120°$ *Vertical angles are congruent: m(∠7) = m(∠6).*

$m(\angle 8) = 60°$ *Vertical angles are congruent: m(∠8) = m(∠5).*

Self Check 2

Refer to the figure for Example 2. If $l_1 \parallel l_2$ and $m(\angle 8) = 50°$, find the measures of the other seven angles that are labeled.

Now Try ➲ Problem 23

Some geometric figures contain two transversals.

EXAMPLE 3 Refer to the figure. If $\overline{AB} \parallel \overline{DE}$, which pairs of angles are congruent?

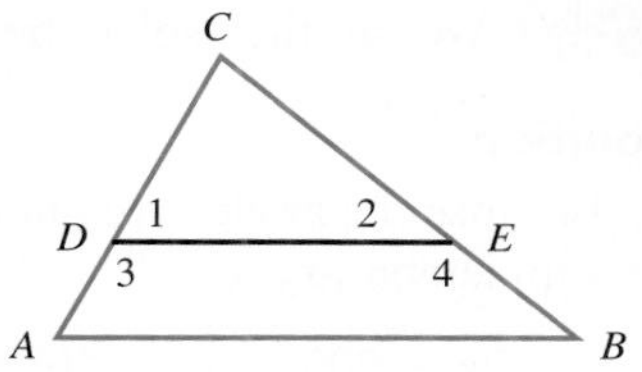

Strategy We will use the corresponding angles property twice to find two pairs of congruent angles.

WHY Both $\overleftrightarrow{AC}$ and $\overleftrightarrow{BC}$ are transversals cutting the parallel line segments $\overline{AB}$ and $\overline{DE}$.

Solution Since $\overline{AB} \parallel \overline{DE}$, and $\overleftrightarrow{AC}$ is a transversal cutting them, corresponding angles are congruent. So we have

$$\angle A \cong \angle 1$$

Since $\overline{AB} \parallel \overline{DE}$ and $\overleftrightarrow{BC}$ is a transversal cutting them, corresponding angles must be congruent. So we have

$$\angle B \cong \angle 2$$

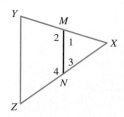

Self Check 3

See the figure below. If $\overline{YZ} \parallel \overline{MN}$, which pairs of angles are congruent?

Now Try ➲ Problem 25

EXAMPLE 4 In the figure, $l_1 \parallel l_2$. Find x.

Strategy We will use the corresponding angles property to write an equation that mathematically models the situation.

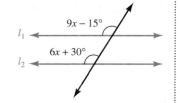

WHY We can then solve the equation to find x.

Solution In the figure, two corresponding angles have degree measures that are represented by the algebraic expressions $9x - 15°$ and $6x + 30°$. Since $l_1 \parallel l_2$, the two corresponding angles are congruent.

In the figure below, $l_1 \parallel l_2$. Find y.

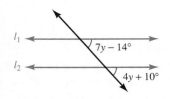

Now Try ➡ **Problem 27**

$9x - 15° = 6x + 30°$ *Since the angles are congruent, their measures are equal.*

$3x - 15° = 30°$ *To eliminate 6x from the right side, subtract 6x from both sides.*

$3x = 45°$ *To isolate the variable term 3x, undo the subtraction of 15° by adding 15° to both sides: 30° + 15° = 45°.*

$x = 15°$ *To isolate x, undo the multiplication by 3 by dividing both sides by 3.*

Thus, x is 15°.

EXAMPLE 5   In the figure, $l_1 \parallel l_2$.

a. Find x.

b. Find the measures of both angles labeled in the figure.

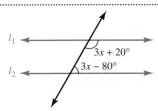

Strategy We will use the interior angles property to write an equation that mathematically models the situation.

WHY We can then solve the equation to find x.

Solution

a. Because the angles are interior angles on the same side of the transversal, they are supplementary.

$3x - 80° + 3x + 20° = 180°$ *The sum of the measures of two supplementary angles is 180°.*

$6x - 60° = 180°$ *Combine like terms: 3x + 3x = 6x.*

$6x = 240°$ *To undo the subtraction of 60°, add 60° to both sides: 180° + 60° = 240°.*

$x = 40°$ *To isolate x, undo the multiplication by 6 by dividing both sides by 6.*

Thus, x is 40°.

This problem may be solved using a different approach. In the figure below, we see that $\angle 1$ and the angle with measure $3x - 80°$ are corresponding angles.

Because l_1 and l_2 are parallel, all pairs of corresponding angles are congruent. Therefore,

$$m(\angle 1) = 3x - 80°$$

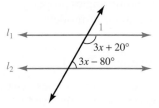

In the figure below, $l_1 \parallel l_2$.

a. Find x.

b. Find the measures of both angles labeled in the figure.

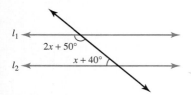

Now Try ➡ **Problem 29**

In the figure, we also see that $\angle 1$ and the angle with measure $3x + 20°$ are supplementary. That means that the sum of their measures must be 180°. We have

$$m(\angle 1) + 3x + 20° = 180°$$
$$3x - 80° + 3x + 20° = 180°$$ *Replace m(∠1) with 3x − 80°.*

This is the same equation that we obtained in the previous solution. When it is solved, we find that x is 40°.

b. To find the measures of the angles in the figure, we evaluate the expressions $3x + 20°$ and $3x - 80°$ for $x = 40°$.

$$3x + 20° = 3(\mathbf{40°}) + 20° \qquad 3x - 80° = 3(\mathbf{40°}) - 80°$$
$$= 120° + 20° \qquad\qquad\quad = 120° - 80°$$
$$= 140° \qquad\qquad\qquad\quad = 40°$$

The measures of the angles labeled in the figure are 140° and 40°.

Answers to Self Checks

1. a. $\angle 1$ and $\angle 3$, $\angle 2$ and $\angle 4$, $\angle 8$ and $\angle 6$, $\angle 7$ and $\angle 5$ **b.** $\angle 2$, $\angle 7$, $\angle 3$, and $\angle 6$ **c.** $\angle 2$ and $\angle 6$, $\angle 7$ and $\angle 3$
2. $m(\angle 5) = 50°$, $m(\angle 7) = 130°$, $m(\angle 6) = 130°$, $m(\angle 3) = 130°$, $m(\angle 4) = 50°$, $m(\angle 1) = 50°$, and $m(\angle 2) = 130°$
3. $\angle 1 \cong \angle Y$, $\angle 3 \cong \angle Z$ **4.** $8°$ **5. a.** $30°$ **b.** $110°$, $70°$

SECTION 9.2 STUDY SET

VOCABULARY

Fill in the blanks.

1. Two lines that lie in the same plane are called _____.
Two lines that lie in different planes are called _____.

2. Two coplanar lines that do not intersect are called
_____ lines. Two noncoplanar lines that do not intersect
are called _____ lines.

3. _____ lines are lines that intersect and form right
angles.

4. A line that intersects two coplanar lines in two distinct
(different) points is called a _____.

5. In the figure below, $\angle 4$ and $\angle 6$ are _____ interior angles.

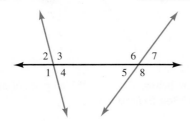

6. In the figure above, $\angle 2$ and $\angle 6$ are _____ angles.

CONCEPTS

7. a. Draw two parallel lines. Label them l_1 and l_2.
b. Draw two lines that are not parallel. Label them l_1 and l_2.

8. a. Draw two perpendicular lines. Label them l_1 and l_2.
b. Draw two lines that are not perpendicular. Label them
l_1 and l_2.

9. a. Draw two parallel lines cut by a transversal. Label the
lines l_1 and l_2 and label the transversal l_3.

b. Draw two lines that are not parallel and cut by a
transversal. Label the lines l_1 and l_2 and label the
transversal l_3.

10. Draw three parallel lines. Label them l_1, l_2, and l_3.

In Problems 11–14, two parallel lines are cut by a transversal.
Fill in the blanks.

11. In the figure below, on the left, $\angle ABC \cong \angle BEF$.
When two parallel lines are cut by a transversal,
_____ angles are congruent.

12. In the figure below, on the right, $\angle 1 \cong \angle 2$. When two
parallel lines are cut by a transversal, _____ _____
angles are congruent.

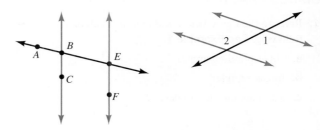

13. In the figure on the next page, on the left, $m(\angle ABC) +$
$m(\angle BCD) = 180°$. When two parallel lines are cut by
a transversal, _____ angles on the same side of the
transversal are supplementary.

14. In the figure below, on the right, $\angle 8 \cong \angle 6$. When two parallel lines are cut by a transversal, _____ angles are congruent.

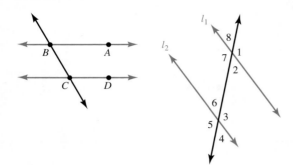

15. In the figure below, on the left, $l_1 \parallel l_2$. What can you conclude about l_1 and l_3?

16. In the figure below, on the right, $l_1 \parallel l_2$ and $l_2 \parallel l_3$. What can you conclude about l_1 and l_3?

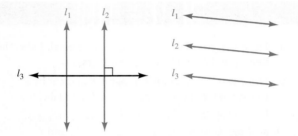

NOTATION

Fill in the blanks.

17. The symbol ⌐ indicates a _____ angle.

18. The symbol $\parallel$ is read as "is _____ to."

19. The symbol $\perp$ is read as "is _____ to."

20. The symbol l_1 is read as "line l _____ one."

GUIDED PRACTICE

21. Refer to the figure below and identify each of the following. **See Example 1.**

a. corresponding angles

b. interior angles

c. alternate interior angles

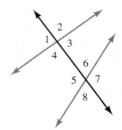

22. Refer to the figure below and identify each of the following. **See Example 1.**

a. corresponding angles

b. interior angles

c. alternate interior angles

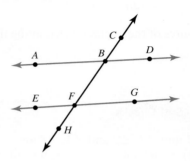

23. In the figure below, $l_1 \parallel l_2$ and $m(\angle 4) = 130°$. Find the measures of the other seven angles that are labeled. **See Example 2.**

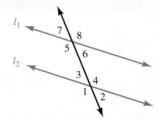

24. In the figure below, $l_1 \parallel l_2$ and $m(\angle 2) = 40°$. Find the measures of the other angles. **See Example 2.**

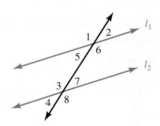

25. In the figure below, $\overline{YM} \parallel \overline{XN}$. Which pairs of angles are congruent? **See Example 3.**

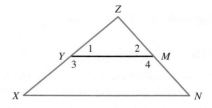

26. In the figure below, $\overline{AE} \parallel \overline{BD}$. Which pairs of angles are congruent? **See Example 3.**

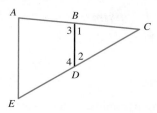

In Problems 27 and 28, $l_1 \parallel l_2$. First find x. Then determine the measure of each angle that is labeled in the figure. See Example 4.

27.

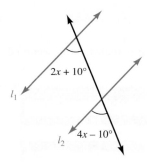

28.

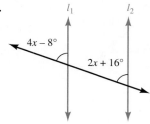

In Problems 29 and 30, $l_1 \parallel l_2$. First find x. Then determine the measure of each angle that is labeled in the figure. See Example 5.

29.

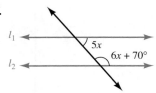

30.

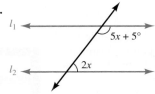

31. In the figure below, $l_1 \parallel \overrightarrow{AB}$. Find:

a. m($\angle 1$), m($\angle 2$), m($\angle 3$), and m($\angle 4$)

b. m($\angle 3$) + m($\angle 4$) + m($\angle ACD$)

c. m($\angle 1$) + m($\angle ABC$) + m($\angle 4$)

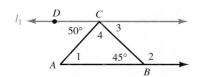

32. In the figure below, $\overline{AB} \parallel \overline{DE}$. Find m($\angle B$), m($\angle E$), and m($\angle 1$).

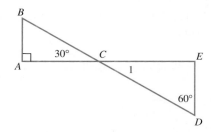

33. In the figure below, $\overline{AB} \parallel \overline{DE}$. What pairs of angles are congruent? Explain your reasoning.

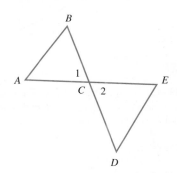

34. In the figure below, $l_1 \parallel l_2$. Find x.

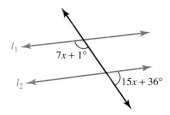

In Problems 35–38, first find x. Then determine the measure of each angle that is labeled in the figure.

35. $l_1 \parallel \overline{CA}$

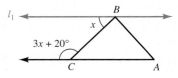

36. $\overline{AB} \parallel \overline{DE}$

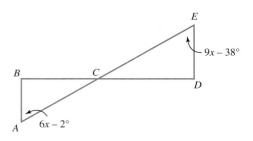

37. $\overline{AB} \parallel \overline{DE}$

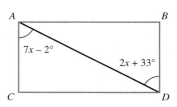

38. $\overline{AC} \parallel \overline{BD}$

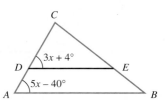

LOOK ALIKES

In Problems 39–42, $l_1 \parallel l_2$. First find x. Then determine the measure of each angle that is labeled in the figure.

39. a.

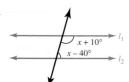

b.

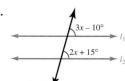

40. a.

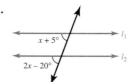

b.

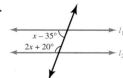

41. a.

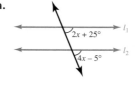

b.

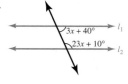

42. a.

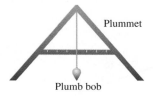

b.

APPLICATIONS

43. Constructing pyramids. The Egyptians used a device called a **plummet** to tell whether stones were properly leveled. A plummet (shown below) is made up of an A-frame and a plumb bob suspended from the peak of the frame. How could a builder use a plummet to tell that the two stones on the left are not level and that the three stones on the right are level?

44. Greece. What two concepts studied in this section does the flag of Greece illustrate?

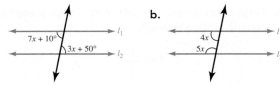

45. Beauty tips. The figure to the right shows how one website illustrated the "geometry" of the ideal eyebrow. If $l_1 \parallel l_2$ and m($\angle DCF$) = 130°, find m($\angle ABE$).

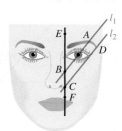

46. Painting signs. For many sign painters, the most difficult letter to paint is a capital E because of all the right angles involved. How many right angles are there?

47. Hanging wallpaper. Explain why the concepts of *perpendicular* and *parallel* are both important when hanging wallpaper.

48. Tools. What geometric concepts are seen in the design of the rake shown here?

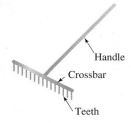

Handle
Crossbar
Teeth

49. Seismology. The figure shows how an earthquake fault occurs when two blocks of earth move apart and one part drops down. Determine the measures of ∠1, ∠2, and ∠3.

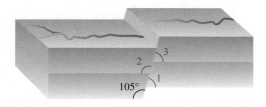

2   3
1
105°

50. Carpentry. A carpenter cross braced three 2 × 4's as shown below and then used a tool to measure the three highlighted angles in red. Are the 2 × 4's parallel? Explain your answer.

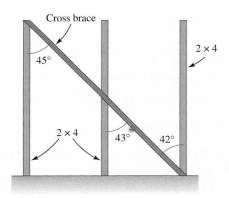

Cross brace
2 × 4
45°
2 × 4
43°   42°

51. Skiing. What concept studied in this section describes the position of the skis shown in the illustration below?

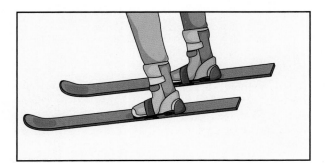

52. Solar energy. A solar panel collects the most solar radiation when the sun's rays are perpendicular to the panel's surface. Since the angle of the sun varies throughout the year, the ideal panel tilt angle for the spring differs from the ideal tilt angle for the summer. The first illustration below shows the ideal angle tilt for a roof-mounted solar panel for March. Draw the ideal tilt angle for a roof-mounted panel for the month of June in the second illustration.

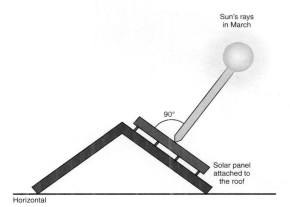

Sun's rays in March
90°
Solar panel attached to the roof
Horizontal

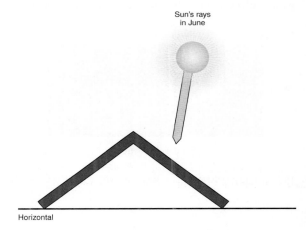

Sun's rays in June
Horizontal

53. Line of sight. Refer to the illustration below. When the observer in the lighthouse looks down at the fisherman, her line of sight from the horizontal creates an *angle of depression*. When the fisherman in the boat looks upward at the observer in the lighthouse, his line of sight from the horizontal creates an *angle of elevation*. Explain why those two angles are congruent.

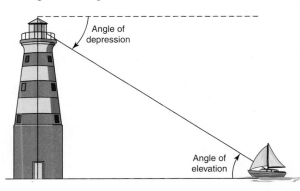

Angle of depression
Angle of elevation

54. Rowing. The four-person rowing crew shown below moves their oars in unison to propel the boat in a straight line. Why are the measures of the angles labeled a, b, c, and d congruent?

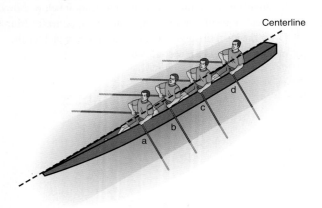

Centerline

WRITING

55. Parking design. Using terms from this section, write a paragraph describing the parking layout shown below.

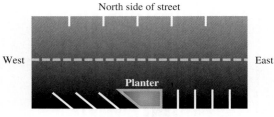

North side of street

West — — — East

Planter

South side of street

56. In the figure below, $l_1 \parallel l_2$. Explain why m($\angle BDE$) = 91°.

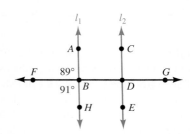

57. In the figure below, $l_1 \parallel l_2$. Explain why m($\angle FEH$) = 100°.

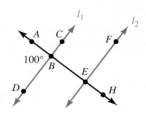

58. In the figure below, $l_1 \parallel l_2$. Explain why the figure must be mislabeled.

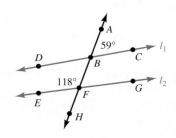

59. Are pairs of alternate interior angles always congruent? Explain.

60. Are pairs of interior angles on the same side of a transversal always supplementary? Explain.

REVIEW

61. Find 60% of 120.

62. 80% of what number is 400?

63. What percent of 500 is 225?

64. Simplify: $3.45 + 7.37 \cdot 2.98$

65. Is every whole number an integer?

66. Multiply: $2\dfrac{1}{5} \cdot 4\dfrac{3}{7}$

67. Express the phrase as a ratio in lowest terms: 4 ounces to 12 ounces

68. Convert 5,400 milligrams to kilograms.

SECTION 9.3 Polygons and Triangles

OBJECTIVES

1 Classify polygons.
2 Classify triangles.
3 Identify isosceles triangles.
4 Find unknown angle measures of triangles.

We will now discuss geometric figures called *polygons*. We see these shapes every day. For example, the walls of most buildings are rectangular in shape. Some tile and vinyl floor patterns use the shape of a pentagon or a hexagon. Stop signs are in the shape of an octagon.

In this section, we will focus on one specific type of polygon called a *triangle.* Triangular shapes are especially important because triangles contribute strength and stability to walls and towers. The gable roofs of houses are triangular, as are the sides of many ramps.

OBJECTIVE 1 Classify polygons.

> **Polygon**
>
> A **polygon** is a closed geometric figure with at least three line segments for its sides.

The House of the Seven Gables, Salem, Massachusetts

Polygons are formed by fitting together line segments in such a way that

- no two of the segments intersect, except at their endpoints, and
- no two line segments with a common endpoint lie on the same line.

The line segments that form a polygon are called its **sides**. The point where two sides of a polygon intersect is called a **vertex** of the polygon (plural **vertices**). The polygon shown to the right has five sides and five vertices.

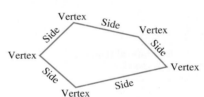

Polygons are classified according to the number of sides that they have. For example, in the figure below, we see that a polygon with four sides is called a *quadrilateral*, and a polygon with eight sides is called an *octagon*. If a polygon has sides that are all the same length and angles that are the same measure, we call it a **regular polygon**.

	Triangle 3 sides	Quadrilateral 4 sides	Pentagon 5 sides	Hexagon 6 sides	Heptagon 7 sides	Octagon 8 sides	Nonagon 9 sides	Decagon 10 sides	Dodecagon 12 sides
Polygons									
Regular polygons									

EXAMPLE 1 Give the number of vertices of:

a. a triangle   **b.** a hexagon

Strategy We will determine the number of angles that each polygon has.

WHY The number of its vertices is equal to the number of its angles.

Success Tip From the results of Example 1, we see that *the number of vertices of a polygon is equal to the number of its sides.*

Self Check 1

Give the number of vertices of:

a. a quadrilateral

b. a pentagon

Now Try Problems 25 and 27

Solution

a. From the figure on the previous page, we see that a triangle has three angles and therefore three vertices.

b. From the figure on the previous page, we see that a hexagon has six angles and therefore six vertices.

OBJECTIVE 2 Classify triangles.

A **triangle** is a polygon with three sides (and three vertices). Recall that in geometry points are labeled with capital letters. We can use the capital letters that denote the vertices of a triangle to name the triangle. For example, when referring to the triangle below, with vertices A, B, and C, we can use the notation $\triangle ABC$ (read as "triangle ABC").

When naming a triangle, we may begin with any vertex. Then we move around the figure in a clockwise (or counterclockwise) direction as we list the remaining vertices. Other ways of naming the triangle shown here are $\triangle ACB$, $\triangle BCA$, $\triangle BAC$, $\triangle CAB$, and $\triangle CBA$.

The figures below show how triangles can be classified according to the lengths of their sides. The single **tick marks** drawn on each side of the equilateral triangle indicate that the sides are of equal length. The double tick marks drawn on two of the sides of the isosceles triangle indicate that they have the same length. Each side of the scalene triangle has a different number of tick marks to indicate that the sides have different lengths.

LANGUAGE OF MATHEMATICS

Since every angle of an equilateral triangle has the same measure, an equilateral triangle is also *equiangular*. Since equilateral triangles have at least two sides of equal length, they are also isosceles. However, isosceles triangles are not necessarily equilateral.

Equilateral triangle
(all sides equal length)

Isosceles triangle
(at least two sides of equal length)

Scalene triangle
(no sides of equal length)

Triangles may also be classified by their angles, as shown below.

Acute triangle
(has three acute angles)

Obtuse triangle
(has an obtuse angle)

Right triangle
(has one right angle)

Right triangles have many real-life applications. For example, in figure (a) below, we see that a right triangle is formed when a ladder leans against the wall of a building.

The longest side of a right triangle is called the **hypotenuse**, and the other two sides are called **legs**. The hypotenuse of a right triangle is always opposite the 90° (right) angle. The legs of a right triangle are adjacent to (next to) the right angle, as shown in figure (b).

Right triangles

(a)

(b)

OBJECTIVE 3 Identify isosceles triangles.

In an isosceles triangle, the angles opposite the sides of equal length are called **base angles**, the sides of equal length form the **vertex angle**, and the third side is called the **base**. Two examples of isosceles triangles are shown below.

Isosceles triangles

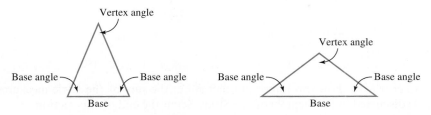

We have seen that isosceles triangles have two sides of equal length. The **isosceles triangle theorem** states that such triangles have one other important characteristic: Their base angles are congruent.

> ### Isosceles Triangle Theorem
> If two sides of a triangle are congruent, then the angles opposite those sides are congruent.

Tick marks can be used to denote the sides of a triangle that have the same length. They can also be used to indicate the angles of a triangle with the same measure. For example, we can show that the base angles of the isosceles triangle below are congruent by using single tick marks.

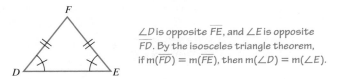

$\angle D$ is opposite $\overline{FE}$, and $\angle E$ is opposite $\overline{FD}$. By the isosceles triangle theorem, if $m(\overline{FD}) = m(\overline{FE})$, then $m(\angle D) = m(\angle E)$.

If a mathematical statement is written in the form *if p* . . . , *then q* . . . , we call the statement *if q* . . . , *then p* . . . its **converse**. The converses of some statements are true, while the converses of other statements are false. It is interesting to note that the converse of the isosceles triangle theorem is true.

> ### Converse of the Isosceles Triangle Theorem
> If two angles of a triangle are congruent, then the sides opposite the angles have the same length, and the triangle is isosceles.

EXAMPLE 2 Is the triangle shown here an isosceles triangle?

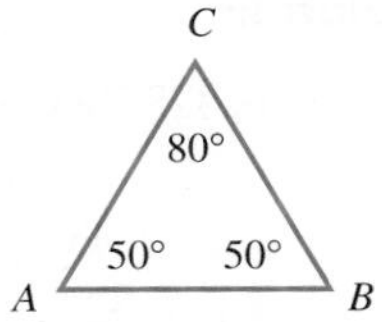

Strategy We will consider the measures of the angles of the triangle.

WHY If two angles of a triangle are congruent, then the sides opposite the angles have the same length, and the triangle is isosceles.

Solution $\angle A$ and $\angle B$ have the same measure, 50°. By the converse of the isosceles triangle theorem, if $m(\angle A) = m(\angle B)$, we know that $m(\overline{BC}) = m(\overline{AC})$ and that $\triangle ABC$ is isosceles.

Self Check 2

Is the triangle shown below an isosceles triangle?

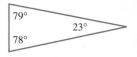

Now Try Problems 33 and 35

OBJECTIVE 4 **Find unknown angle measures of triangles.**

If you draw several triangles and carefully measure each angle with a protractor, you will find that the sum of the angle measures of each triangle is 180°. Two examples are shown below.

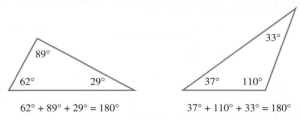

$$62° + 89° + 29° = 180°$$ $$37° + 110° + 33° = 180°$$

Another way to show this important fact about the sum of the angle measures of a triangle is discussed in Problem 90 of the Study Set at the end of this section.

> **Angles of a Triangle**
> The sum of the angle measures of any triangle is 180°.

EXAMPLE 3 In the figure, find x.

Strategy We will use the fact that the sum of the angle measures of any triangle is 180° to write an equation that models the situation.

WHY We can then solve the equation to find the unknown angle measure, x.

Self Check 3
In the figure, find y.

Now Try Problem 37

Solution Since the sum of the angle measures of any triangle is 180°, we have

$$x + 40° + 90° = 180°$$ The ⌐ symbol indicates that the measure of the angle is 90°.

$$x + 130° = 180°$$ Do the addition.

$$x = 50°$$ To isolate x, undo the addition of 130° by subtracting 130° from both sides.

$$\begin{array}{r} 90 \\ +40 \\ \hline 130 \end{array}$$

Thus, x is 50°.

EXAMPLE 4 In the figure, find the measure of each angle of $\triangle ABC$.

Strategy We will use the fact that the sum of the angle measures of any triangle is 180° to write an equation that models the situation.

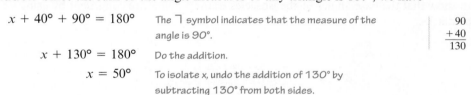

WHY We can then solve the equation to find the unknown angle measure x, and use it to evaluate the expressions $2x$ and $x + 32°$.

Solution

$$x + 32° + x + 2x = 180°$$ The sum of the angle measures of any triangle is 180°.

$$4x + 32° = 180°$$ Combine like terms: $x + x + 2x = 4x$.

$$4x + 32° - 32° = 180° - 32°$$ To isolate the variable term, 4x, subtract 32° from both sides.

$$4x = 148°$$ Do the subtractions.

$$\frac{4x}{4} = \frac{148°}{4}$$ To isolate x, divide both sides by 4.

$$x = 37°$$ Do the divisions. This is the measure of $\angle B$.

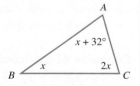

To find the measures of $\angle A$ and $\angle C$, we evaluate the expressions $x + 32°$ and $2x$ for $x = 37°$.

$$x + 32° = 37° + 32°$$ Substitute 37 for x. $$\quad\quad 2x = 2(37°)$$ Substitute 37 for x.

$$= 69°$$ $$= 74°$$

The measure of $\angle B$ is 37°, the measure of $\angle A$ is 69°, and the measure of $\angle C$ is 74°.

EXAMPLE 5 If one base angle of an isosceles triangle measures 70°, what is the measure of the vertex angle?

Strategy We will use the isosceles triangle theorem and the fact that the sum of the angle measures of any triangle is 180° to write an equation that models the situation.

WHY We can then solve the equation to find the unknown angle measure.

Solution By the isosceles triangle theorem, if one of the base angles measures 70°, so does the other. (See the figure on the right.) If we let x represent the measure of the vertex angle, we have

$$x + 70° + 70° = 180°$$ The sum of the measures of the angles of a triangle is 180°.

$$x + 140° = 180°$$ Combine like terms: 70° + 70° = 140°.

$$x = 40°$$ To isolate x, undo the addition of 140° by subtracting 140° from both sides.

The vertex angle measures 40°.

EXAMPLE 6 If the vertex angle of an isosceles triangle measures 99°, what are the measures of the base angles?

Strategy We will use the fact that the base angles of an isosceles triangle have the same measure and the sum of the angle measures of any triangle is 180° to write an equation that mathematically models the situation.

WHY We can then solve the equation to find the unknown angle measures.

Solution The base angles of an isosceles triangle have the same measure. If we let x represent the measure of one base angle, the measure of the other base angle is also x. (See the figure to the right.) Since the sum of the measures of the angles of any triangle is 180°, the sum of the measures of the base angles and of the vertex angle is 180°. We can use this fact to form an equation.

$$x + x + 99° = 180°$$

$$2x + 99° = 180°$$ Combine like terms: x + x = 2x.

$$2x = 81°$$ To isolate the variable term, 2x, undo the addition of 99° by subtracting 99° from both sides.

$$\frac{2x}{2} = \frac{81°}{2}$$ To isolate x, undo the multiplication by 2 by dividing both sides by 2.

$$x = 40.5°$$

The measure of each base angle is 40.5°.

$$\begin{array}{r} 40.5 \\ 2\overline{)81.0} \\ -8 \\ \hline 01 \\ -0 \\ \hline 1\,0 \\ -1\,0 \\ \hline 0 \end{array}$$

SECTION **9.3** **STUDY SET**

VOCABULARY

Fill in the blanks.

1. A _____ is a closed geometric figure with at least three line segments for its sides.

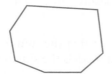

2. The polygon shown to the right has seven _____ and seven vertices.

3. A point where two sides of a polygon intersect is called a _____ of the polygon.

4. A _____ polygon has sides that are all the same length and angles that all have the same measure.

5. A triangle with three sides of equal length is called an _____ triangle. An _____ triangle has at least two sides of equal length. A _____ triangle has no sides of equal length.

6. An _____ triangle has three acute angles. An _____ triangle has one obtuse angle. A _____ triangle has one right angle.

7. The longest side of a right triangle is called the _____. The other two sides of a right triangle are called _____.

8. The _____ angles of an isosceles triangle have the same measure. The sides of equal length of an isosceles triangle form the _____ angle.

9. In this section, we discussed the sum of the measures of the angles of a triangle. The word *sum* indicates the operation of _____.

10. Complete the table.

Number of Sides	Name of Polygon
3	
4	
5	
6	
7	
8	
9	
10	
12	

CONCEPTS

11. Draw an example of each type of regular polygon.

    a. hexagon   **b.** octagon

    c. quadrilateral   **d.** triangle

    e. pentagon   **f.** decagon

12. Refer to the triangle below.

    a. What are the names of the vertices of the triangle?

    b. How many sides does the triangle have? Name them.

    c. Use the vertices to name this triangle in three ways.

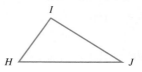

13. Draw an example of each type of triangle.

    a. isosceles   **b.** equilateral

    c. scalene   **d.** obtuse

    e. right   **f.** acute

14. Classify each triangle as an acute, an obtuse, or a right triangle.

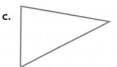

    a.   **b.** 90°

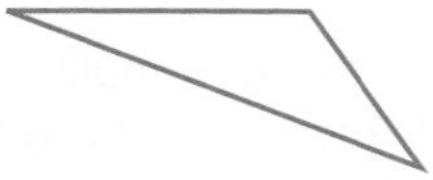

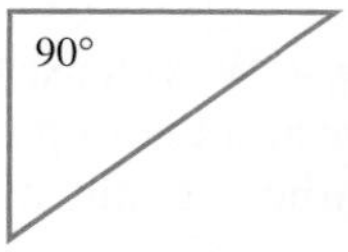

    c.   **d.** 91°

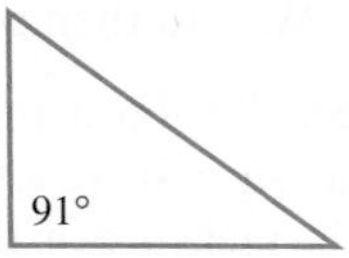

15. Refer to the triangle shown to the right.

    a. What is the measure of ∠B?

    b. What type of triangle is it?

    c. What two line segments form the legs?

    d. What line segment is the hypotenuse?

    e. Which side of the triangle is the longest?

    f. Which side is opposite ∠B?

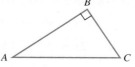

16. Fill in the blanks.

    a. The sides of a right triangle that are adjacent to the right angle are called the _____.

    b. The hypotenuse of a right triangle is the side _____ the right angle.

17. Fill in the blanks.

    a. The _____ triangle theorem states that if two sides of a triangle are congruent, then the angles opposite those sides are congruent.

    b. The _____ of the isosceles triangle theorem is true and states that if two angles of a triangle are congruent, then the sides opposite the angles have the same length, and the triangle is isosceles.

18. Refer to the given triangle.

    a. What two sides are of equal length?

    b. What type of triangle is $\triangle XYZ$?

    c. Name the base angles.

    d. Which side is opposite $\angle X$?

    e. What is the vertex angle?

    f. Which angle is opposite side $\overline{XY}$?

    g. Which two angles are congruent?

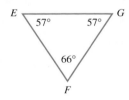

19. Refer to the triangle below.

    a. What do we know about $\overline{EF}$ and $\overline{GF}$?

    b. What type of triangle is $\triangle EFG$?

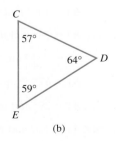

20. a. Find the sum of the measures of the angles of $\triangle JKL$, shown in figure (a).

    b. Find the sum of the measures of the angles of $\triangle CDE$, shown in figure (b).

    c. What is the sum of the measures of the angles of *any* triangle?

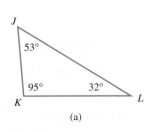

        (a)   (b)

NOTATION

Fill in the blanks.

21. The symbol $\triangle$ means _____.

22. The symbol $m(\angle A)$ means the _____ of angle A.

Refer to the triangle below.

23. What fact about the sides of $\triangle ABC$ do the tick marks indicate?

24. What fact about the angles of $\triangle ABC$ do the tick marks indicate?

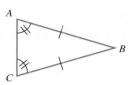

GUIDED PRACTICE

For each polygon, give the number of sides it has, give its name, and then give the number of vertices that it has. See Example 1.

25. a.   **b.**

26. a.   **b.**

27. a.   **b.**

28. a.   **b.**

Classify each triangle as an equilateral triangle, an isosceles triangle, or a scalene triangle. See Objective 2.

29. a.    **b.**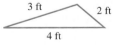

30. a.   **b.**

31. a.   **b.**

32. a.

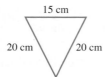

b.

State whether each of the triangles is an isosceles triangle. See Example 2.

33.

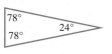

34.

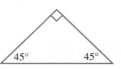

35.

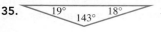

36.

Find y. See Example 3.

37.

38.

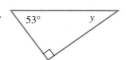

39.

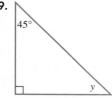

40.

The degree measures of the angles of a triangle are represented by algebraic expressions. First find x. Then determine the measure of each angle of the triangle. See Example 4.

41.

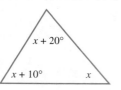

42.

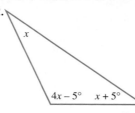

43.

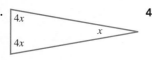

44.

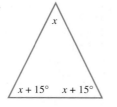

Find the measure of the vertex angle of each isosceles triangle given the following information. **See Example 5.**

45. The measure of one base angle is 56°.

46. The measure of one base angle is 68°.

47. The measure of one base angle is 85.5°.

48. The measure of one base angle is 4.75°.

Find the measure of one base angle of each isosceles triangle given the following information. **See Example 6.**

49. The measure of the vertex angle is 102°.

50. The measure of the vertex angle is 164°.

51. The measure of the vertex angle is 90.5°.

52. The measure of the vertex angle is 2.5°.

TRY IT YOURSELF

Find the measure of each vertex angle.

53.

54.

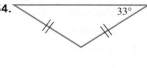

55.

56.

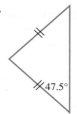

The measures of two angles of $\triangle ABC$ are given. Find the measure of the third angle.

57. $m(\angle A) = 30°$ and $m(\angle B) = 60°$; find $m(\angle C)$.

58. $m(\angle A) = 45°$ and $m(\angle C) = 105°$; find $m(\angle B)$.

59. $m(\angle B) = 100°$ and $m(\angle A) = 35°$; find $m(\angle C)$.

60. $m(\angle B) = 33°$ and $m(\angle C) = 77°$; find $m(\angle A)$.

61. $m(\angle A) = 25.5°$ and $m(\angle B) = 63.8°$; find $m(\angle C)$.

62. $m(\angle B) = 67.25°$ and $m(\angle C) = 72.5°$; find $m(\angle A)$.

63. $m(\angle A) = 29°$ and $m(\angle C) = 89.5°$; find $m(\angle B)$.

64. $m(\angle A) = 4.5°$ and $m(\angle B) = 128°$; find $m(\angle C)$.

In Problems 65–68, find x.

65.

66.  **67.**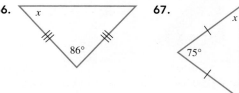

68.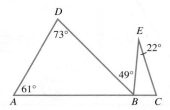

69. One angle of an isosceles triangle has a measure of 39°. What are the possible measures of the other angles?

70. One angle of an isosceles triangle has a measure of 2°. What are the possible measures of the other angles?

71. Find m(∠C).

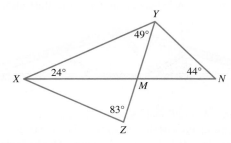

72. Find:

   a. m(∠MXZ)

   b. m(∠MYN)

73. Find m(∠NOQ).

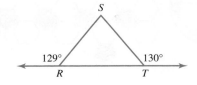

74. Find m(∠S).

In Problems 75–78, find x.

75. a. 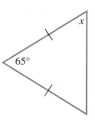 **b.**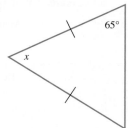

76. a. 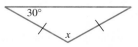 **b.**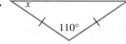

77. a.  **b.**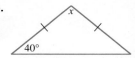

78. a. 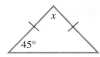 **b.**

79. Polygons in nature. As seen below, a starfish fits the shape of a pentagon. What polygon shape do you see in each of the other objects?

Starfish   a. Lemon   b. Apple

c. Honey comb   d. Chili pepper   e. Flower

80. Chemistry. Polygons are used to represent the chemical structure of compounds. In the figure below, what types of polygons are used to represent methylprednisolone, the active ingredient in an anti-inflammatory medication?

Methylprednisolone

81. Automobile jack. Refer to the figure below. No matter how high the jack is raised, it always forms two isosceles triangles. Explain why.

82. Easels. Refer to the figure below. What type of triangle studied in this section is used in the design of the legs of the easel?

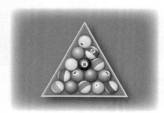

83. Pool. The rack shown below is used to set up the billiard balls when beginning a game of pool. Which type of triangle is it?

84. Drafting. Among the tools used in drafting are the two clear plastic triangles shown in the next column. Classify each according to the lengths of its sides and then according to its angle measures.

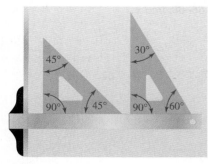

85. Shipping. A *boom* is a strong pole attached to a ship's mast that is used in combination with pulleys, wires, and a hoist motor to load and unload cargo. Find the measures of the angles labeled *a*, *b*, *c*, and *d* in the illustration below.

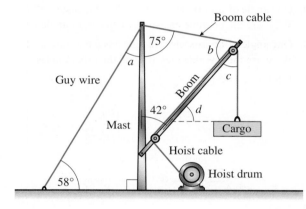

86. Trusses. A truss is a supporting structure that is often used in the design of roofs and bridges. Find the measures of the angles labeled *a*, *b*, *c*, and *d* in the illustration below.

87. Soccer. A traditional soccer ball is made up of two types of polygon panels as shown below. When the panels are sewn together and inflated, they make a nearly perfect sphere.

a. What specific types of polygons are used to construct a soccer ball?

b. How many of each type are used?

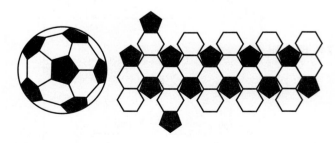

88. Geodesic domes. A geodesic (pronounced gee•o •dess•ick) dome, as shown below, is a structure made from a clever arrangement of polygons. These domes are quite stable and can support very heavy loads for their size. What type of polygon is the basic element of a geodesic dome?

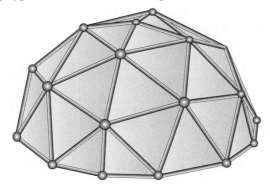

WRITING

89. In this section, we discussed the definition of a pentagon. What is *the* Pentagon? Why is it named that?

90. A student cut a triangular shape out of blue construction paper and labeled the angles $\angle 1$, $\angle 2$, and $\angle 3$, as shown in figure (a) in the next column. Then she tore off each of the three corners and arranged them as shown in figure (b).

Explain what important geometric concept this model illustrates.

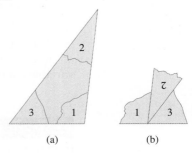

(a)   (b)

91. Explain why a triangle cannot have two right angles.

92. Explain why a triangle cannot have two obtuse angles.

REVIEW

93. Find 20% of 110.

94. Find 15% of 50.

95. What percent of 200 is 80?

96. 20% of what number is 500?

97. Evaluate: $0.85 \div 2(0.25)$

98. First aid. When checking an accident victim's pulse, a paramedic counted 13 beats during a 15-second span. How many beats would be expected in 60 seconds?

SECTION 9.4 The Pythagorean Theorem

OBJECTIVES

1 Use the Pythagorean theorem to find the exact length of a side of a right triangle.

2 Use the Pythagorean theorem to approximate the length of a side of a right triangle.

3 Use the converse of the Pythagorean theorem.

A **theorem** is a mathematical statement that can be proven. In this section, we will discuss one of the most widely used theorems of geometry—the Pythagorean theorem. It is named after Pythagoras, a Greek mathematician who lived about 2,500 years ago. He is thought to have been the first to develop a proof of it. The Pythagorean theorem expresses the relationship between the lengths of the sides of any right triangle.

OBJECTIVE 1 **Use the Pythagorean theorem to find the exact length of a side of a right triangle.**

Recall that a right triangle is a triangle that has a right angle (an angle with measure 90°). In a right triangle, the longest side is called the **hypotenuse**. It is the side opposite the right angle. The other two sides are called **legs**. It is common practice to let the variable c represent the length of the hypotenuse and the variables a and b represent the lengths of the legs, as shown on the right.

If we know the lengths of any two sides of a right triangle, we can find the length of the third side using the **Pythagorean theorem**.

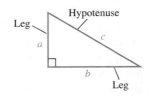

Pythagorean Theorem

If a and b are the lengths of two legs of a right triangle and c is the length of the hypotenuse, then

$$a^2 + b^2 = c^2$$

Pythagoras

In words, the Pythagorean theorem is expressed as follows:

In a right triangle, the sum of the squares of the lengths of the two legs is equal to the square of the length of the hypotenuse.

When using the **Pythagorean equation** $a^2 + b^2 = c^2$, we can let a represent the length of either leg of the right triangle. We then let b represent the length of the other leg. The variable c must always represent the length of the hypotenuse.

EXAMPLE 1   Find the length of the hypotenuse of the right triangle shown here.

Strategy We will use the Pythagorean theorem to find the length of the hypotenuse.

WHY If we know the lengths of any two sides of a right triangle, we can find the length of the third side using the Pythagorean theorem.

Solution We will let $a = 3$ and $b = 4$, and substitute into the Pythagorean equation to find c.

$a^2 + b^2 = c^2$ This is the Pythagorean equation.

$3^2 + 4^2 = c^2$ Substitute 3 for a and 4 for b.

$9 + 16 = c^2$ Evaluate each exponential expression.

$25 = c^2$ Do the addition.

$c^2 = 25$ Reverse the sides of the equation so that c^2 is on the left.

To find c, we must find a number that, when squared, is 25. There are two such numbers, one positive and one negative; they are the square roots of 25. Since c represents the length of a side of a triangle, c cannot be negative. For this reason, we need only find the positive square root of 25 to get c.

$c = \sqrt{25}$ The symbol $\sqrt{}$ is used to indicate the positive square root of a number.

$c = 5$ $\sqrt{25} = 5$ because $5^2 = 25$.

The length of the hypotenuse is 5 in.

Success Tip The Pythagorean theorem is used to find the lengths of sides of right triangles. A calculator with a square root key $\boxed{\sqrt{}}$ is often helpful in the final step of the solution process when we must find the positive square root of a number.

Self Check 1

Find the length of the hypotenuse of the right triangle shown below.

Now Try Problem 13

EXAMPLE 2   **Firefighting.** To fight a forest fire, the forestry department plans to clear a rectangular fire break around the fire, as shown in the figure to the right. Crews are equipped with mobile communications that have a 3,000-yard range. Can crews at points A and B remain in radio contact?

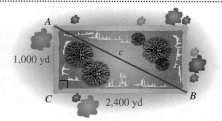

Strategy We will use the Pythagorean theorem to find the distance between points A and B.

WHY If the distance is less than 3,000 yards, the crews can communicate by radio. If it is greater than 3,000 yards, they cannot.

Solution The line segments connecting points A, B, and C form a right triangle. To find the distance c from point A to point B, we can use the Pythagorean equation, substituting 2,400 for a and 1,000 for b and solving for c.

$$a^2 + b^2 = c^2$$ This is the Pythagorean equation.

$$2,400^2 + 1,000^2 = c^2$$ Substitute for *a* and *b*.

$$5,760,000 + 1,000,000 = c^2$$ Evaluate each exponential expression.

$$6,760,000 = c^2$$ Do the addition.

$$c^2 = 6,760,000$$ Reverse the sides of the equation so that c^2 is on the left.

$$c = \sqrt{6,760,000}$$ If $c^2 = 6,760,000$, then *c* must be a square root of 6,760,000. Because *c* represents a length, it must be the positive square root of 6,760,000.

$$c = 2,600$$ Use a calculator to find the square root.

The two crews are 2,600 yards apart. Because this distance is less than the 3,000-yard range of the radios, they can communicate by radio.

Self Check 2

In Example 2, can the crews communicate by radio if the distance from point *B* to point *C* remains the same but the distance from point *A* to point *C* increases to 2,520 yards?

Now Try **Problems 19 and 47**

EXAMPLE 3   The lengths of two sides of a right triangle are given in the figure. Find the missing side length.

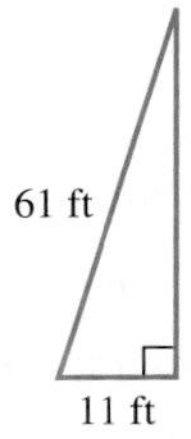

Strategy We will use the Pythagorean theorem to find the missing side length.

WHY If we know the lengths of any two sides of a right triangle, we can find the length of the third side using the Pythagorean theorem.

Solution We may substitute 11 for either *a* or *b*, but 61 must be substituted for the length *c* of the hypotenuse. If we substitute 11 for *b*, we can find the unknown side length *a* as follows.

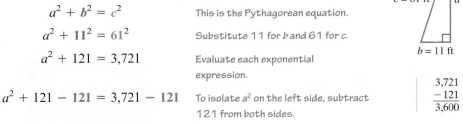

$$a^2 + b^2 = c^2$$ This is the Pythagorean equation.

$$a^2 + 11^2 = 61^2$$ Substitute 11 for *b* and 61 for *c*.

$$a^2 + 121 = 3,721$$ Evaluate each exponential expression.

$$a^2 + 121 - 121 = 3,721 - 121$$ To isolate a^2 on the left side, subtract 121 from both sides.

$$a^2 = 3,600$$ Do the subtraction.

$$a = \sqrt{3,600}$$ If $a^2 = 3,600$, then *a* must be a square root of 3,600. Because *a* represents a length, it must be the positive square root of 3,600.

$$a = 60$$ Use a calculator, if necessary, to find the square root.

The missing side length is 60 ft.

```
  3,721
-   121
  3,600
```

Self Check 3

The lengths of two sides of a right triangle are given. Find the missing side length.

65 in.

33 in.

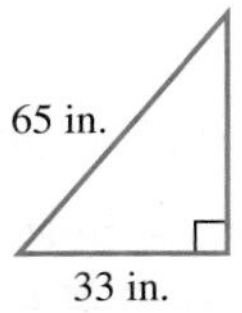

Now Try **Problem 23**

OBJECTIVE 2   **Use the Pythagorean theorem to approximate the length of a side of a right triangle.**

When we use the Pythagorean theorem to find the length of a side of a right triangle, the solution is sometimes the square root of a number that is not a perfect square. In that case, we can use a calculator to *approximate* the square root.

EXAMPLE 4   Refer to the right triangle shown here. Find the missing side length. Give the exact answer and an approximation to the nearest hundredth.

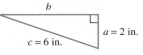

Strategy We will use the Pythagorean theorem to find the missing side length.

WHY If we know the lengths of any two sides of a right triangle, we can find the length of the third side using the Pythagorean theorem.

Solution We may substitute 2 for either a or b, but 6 must be substituted for the length c of the hypotenuse. If we choose to substitute 2 for a, we can find the unknown side length b as follows.

$$a^2 + b^2 = c^2 \quad \text{This is the Pythagorean equation.}$$
$$2^2 + b^2 = 6^2 \quad \text{Substitute 2 for } a \text{ and 6 for } c.$$
$$4 + b^2 = 36 \quad \text{Evaluate each exponential expression.}$$
$$4 + b^2 - 4 = 36 - 4 \quad \text{To isolate } b^2 \text{ on the left side, undo the addition of 4 by subtracting 4 from both sides.}$$
$$b^2 = 32 \quad \text{Do the subtraction.}$$

We must find a number that, when squared, is 32. Since b represents the length of a side of a triangle, we consider only the positive square root.

$$b = \sqrt{32} \quad \text{This is the exact length.}$$

The missing side length is exactly $\sqrt{32}$ inches long. Since 32 is not a perfect square, its square root is not a whole number. We can use a calculator to *approximate* $\sqrt{32}$. To the nearest hundredth, the missing side length is 5.66 inches.

$$\sqrt{32} \text{ in.} \approx 5.66 \text{ in.}$$

Self Check 4

Refer to the triangle below. Find the missing side length. Give the exact answer and an approximation to the nearest hundredth.

Now Try ⮕ **Problem 35**

Using Your Calculator ▶   **Finding the Width of a TV Screen**

The size of a television screen is the diagonal measure of its rectangular screen. To find the width of a 27-inch screen that is 17 inches high, we use the Pythagorean theorem with $c = 27$ and $b = 17$.

$$c^2 = a^2 + b^2$$
$$27^2 = a^2 + 17^2$$
$$27^2 - 17^2 = a^2$$

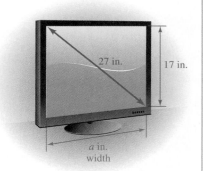

Since the variable a represents the width of the television screen, it must be positive. To find a, we find the positive square root of the result when 17^2 is subtracted from 27^2.

Using a radical symbol to indicate this, we have

$$\sqrt{27^2 - 17^2} = a$$

We can evaluate the expression on the left side by entering:

(27 x^2 − 17 x^2) √ | 20.97617696 |

To the nearest inch, the width of the television screen is 21 inches.

OBJECTIVE 3 Use the converse of the Pythagorean theorem.

If a mathematical statement is written in the form *if p . . . , then q . . .* , we call the statement *if q . . . , then p . . .* its **converse**. The converses of some statements are true, while the converses of other statements are false. It is interesting to note that the converse of the Pythagorean theorem is true.

Converse of the Pythagorean Theorem

If a triangle has three sides of lengths a, b, and c, such that $a^2 + b^2 = c^2$, then the triangle is a right triangle.

EXAMPLE 5   Is the triangle shown here a right triangle?

Strategy We will substitute the side lengths, 6, 8, and 11, into the Pythagorean equation $a^2 + b^2 = c^2$.

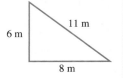

WHY By the converse of the Pythagorean theorem, the triangle is a right triangle if a true statement results. The triangle is not a right triangle if a false statement results.

Solution We must substitute the longest side length, 11, for c, because it is the possible hypotenuse. The lengths of 6 and 8 may be substituted for either a or b.

$$a^2 + b^2 = c^2 \qquad \text{This is the Pythagorean equation.}$$
$$6^2 + 8^2 \stackrel{?}{=} 11^2 \qquad \text{Substitute 6 for a, 8 for b, and 11 for c.}$$
$$36 + 64 \stackrel{?}{=} 121 \qquad \text{Evaluate each exponential expression.}$$
$$100 = 121 \qquad \text{This is a false statement.}$$

Since $100 \neq 121$, the triangle is not a right triangle.

$$\begin{array}{r} \frac{1}{36} \\ +64 \\ \hline 100 \end{array}$$

Self Check 5

Is the triangle below a right triangle?

Now Try Problem 39

Answers to Self Checks

1. 13 ft   **2.** no   **3.** 56 in.   **4.** $\sqrt{24}$ m ≈ 4.90 m   **5.** yes

---

SECTION 9.4 STUDY SET

VOCABULARY

1. In a right triangle, the side opposite the 90° angle is called the _____ . The other two sides are called _____ .

2. The Pythagorean theorem is named after the Greek mathematician _____ , who is thought to have been the first to prove it.

3. The _____ theorem states that in any right triangle, the square of the length of the hypotenuse is equal to the sum of the squares of the lengths of the two legs.

4. $a^2 + b^2 = c^2$ is called the Pythagorean _____ .

CONCEPTS

Fill in the blanks.

5. If a and b are the lengths of two legs of a right triangle and c is the length of the hypotenuse, then ▢ + ▢ = ▢ .

6. The two solutions of $c^2 = 36$ are $c =$ ▢ or $c =$ ▢. If c represents the length of the hypotenuse of a right triangle, then we can discard the solution ▢ .

7. The converse of the Pythagorean theorem: If a triangle has three sides of lengths a, b, and c, such that $a^2 + b^2 = c^2$, then the triangle is a _____ triangle.

8. Use a protractor to draw an example of a right triangle.

9. Refer to the triangle on the right.

   a. What side is the hypotenuse?

   b. What side is the longer leg?

   c. What side is the shorter leg?

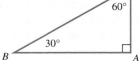

10. What is the first step when solving the equation $25 + b^2 = 81$ for b?

NOTATION

Complete the steps to solve the equation, where $a > 0$ and $c > 0$.

11. $8^2 + 6^2 = c^2$

$\boxed{} + 36 = c^2$

$\boxed{} = c^2$

$\sqrt{\boxed{}} = c$

$10 = c$

12. $a^2 + 15^2 = 17^2$

$a^2 + \boxed{} = \boxed{}$

$a^2 + 225 - \boxed{} = 289 - \boxed{}$

$a^2 = \boxed{}$

$a = \sqrt{\boxed{}}$

$a = \boxed{}$

GUIDED PRACTICE

Find the length of the hypotenuse of the right triangle shown below if it has the given side lengths. See Examples 1 and 2.

13. $a = 6$ ft and $b = 8$ ft

14. $a = 12$ mm and $b = 9$ mm

15. $a = 5$ m and $b = 12$ m

16. $a = 16$ in. and $b = 12$ in.

17. $a = 48$ mi and $b = 55$ mi

18. $a = 80$ ft and $b = 39$ ft

19. $a = 88$ cm and $b = 105$ cm

20. $a = 132$ mm and $b = 85$ mm

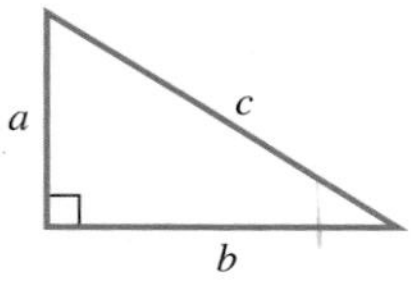

Refer to the right triangle below. See Example 3.

21. Find b if $a = 10$ cm and $c = 26$ cm.

22. Find b if $a = 14$ in. and $c = 50$ in.

23. Find a if $b = 18$ m and $c = 82$ m.

24. Find a if $b = 9$ yd and $c = 41$ yd.

25. Find a if $b = 21$ m and $c = 29$ m.

26. Find a if $b = 16$ yd and $c = 34$ yd.

27. Find b if $a = 180$ m and $c = 181$ m.

28. Find b if $a = 630$ ft and $c = 650$ ft.

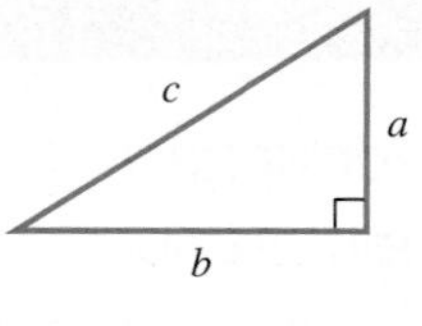

The lengths of two sides of a right triangle are given. Find the missing side length. Give the exact answer and an approximation to the nearest hundredth. See Example 4.

29. $a = 5$ cm and $c = 6$ cm

30. $a = 4$ in. and $c = 8$ in.

31. $a = 12$ m and $b = 8$ m

32. $a = 10$ ft and $b = 4$ ft

33. $a = 9$ in. and $b = 3$ in.

34. $a = 5$ mi and $b = 7$ mi

35. $b = 4$ in. and $c = 6$ in.

36. $b = 9$ mm and $c = 12$ mm

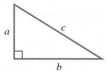

Is a triangle with the following side lengths a right triangle? See Example 5.

37. 12, 14, 15

38. 15, 16, 22

39. 33, 56, 65

40. 20, 21, 29

LOOK ALIKES

Find the length of the hypotenuse of each right triangle. Leave your answer as a square root and you need not include any units.

41.

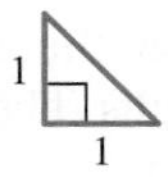

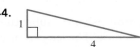

42.

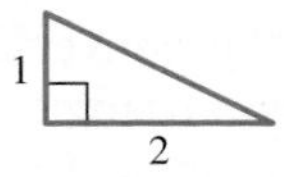

43.

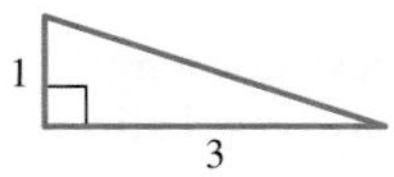

44.

APPLICATIONS

45. Adjusting ladders. A 20-foot ladder reaches a window 16 feet above the ground. How far from the wall is the base of the ladder?

46. Length of guy wires. A 30-foot tower is to be fastened by three guy wires attached to the top of the tower and to the ground at positions 20 feet from its base. How much wire is needed? Round to the nearest tenth.

47. Picture frames. After gluing and nailing two pieces of picture frame molding together, a frame maker checks her work by making a diagonal measurement. If the sides of the frame form a right angle, what measurement should the frame maker read on the yardstick?

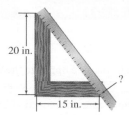

48. Carpentry. The gable end of the roof shown is divided in half by a vertical brace, 8 feet in height. Find the length of the roof line.

49. Baseball. A baseball diamond is a square with each side 90 feet long. How far is it from home plate to second base? Round to the nearest hundredth.

50. Sandwiches. A sandwich is made from 4-inch square slices of bread. It is diagonally cut into two pieces as shown below. How long is the diagonal side of the sandwich? Round to the nearest tenth.

gowithstock/Shutterstock.com

51. Firefighting. The base of the 37-foot ladder shown in the figure on the right is 9 feet from the wall. Will the top reach a window ledge that is 35 feet above the ground? Explain how you arrived at your answer.

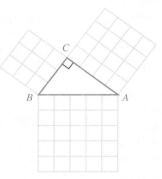

52. Wind damage. A tree was blown over in a wind storm. Find the height of the tree when it was standing vertically upright.

WRITING

53. State the Pythagorean theorem in your own words.

54. When the lengths of the sides of the triangle shown below are substituted into the equation $a^2 + b^2 = c^2$, the result is a false statement. Explain why.

$$a^2 + b^2 = c^2$$
$$2^2 + 4^2 = 5^2$$
$$4 + 16 = 25$$
$$20 = 25$$

55. In the figure below, equal-sized squares have been drawn on the sides of right triangle $\triangle ABC$. Explain how this figure demonstrates that $3^2 + 4^2 = 5^2$.

56. In the movie *The Wizard of Oz*, the scarecrow was in search of a brain. To prove that he had found one, he recited the following:

"The sum of the square roots of any two sides of an isosceles triangle is equal to the square root of the remaining side."

Unfortunately, this statement is not true. Correct it so that it states the Pythagorean theorem.

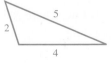

Pictorial Press Ltd/Alamy Stock Photo

REVIEW

Use a check to determine whether the given number in red is a solution of the equation.

57. $2b + 3 = -15, \mathbf{-8}$

58. $5t - 4 = -16, \mathbf{-2}$

59. $0.5x = 2.9, \mathbf{5}$

60. $1.2 + x = 4.7, \mathbf{3.5}$

61. $33 - \dfrac{x}{2} = 30, \mathbf{-6}$

62. $\dfrac{x}{4} + 98 = 100, \mathbf{-8}$

63. $3x - 2 = 4x - 5, \mathbf{12}$

64. $5y + 8 = 3y - 2, \mathbf{5}$

OBJECTIVES

1 Identify corresponding parts of congruent triangles.

2 Use congruence properties to prove that two triangles are congruent.

3 Determine whether two triangles are similar.

4 Use similar triangles to find unknown lengths in application problems.

SECTION **9.5** ## Congruent Triangles and Similar Triangles

In our everyday lives, we see many types of triangles. Triangular-shaped kites, sails, roofs, tortilla chips, and ramps are just a few examples. In this section, we will discuss how to compare the size and shape of two given triangles. From this comparison, we can make observations about their side lengths and angle measures.

OBJECTIVE 1 Identify corresponding parts of congruent triangles.

Simply put, two geometric figures are congruent if they have the same shape and size. For example, if $\triangle ABC$ and $\triangle DEF$ shown below are congruent, we can write

$$\triangle ABC \cong \triangle DEF \qquad \text{Read as "Triangle ABC is congruent to triangle DEF."}$$

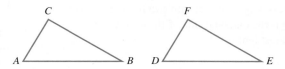

One way to determine whether two triangles are congruent is to see if one triangle can be moved onto the other triangle in such a way that it fits exactly. When we write $\triangle ABC \cong \triangle DEF$, we are showing how the vertices of one triangle are matched to the vertices of the other triangle to obtain a "perfect fit." We call this matching of points a **correspondence**.

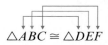

$$\triangle ABC \cong \triangle DEF$$

$A \leftrightarrow D$ Read as "Point A corresponds to point D."

$B \leftrightarrow E$ Read as "Point B corresponds to point E."

$C \leftrightarrow F$ Read as "Point C corresponds to point F."

When we establish a correspondence between the vertices of two congruent triangles, we also establish a correspondence between the angles and the sides of the triangles. Corresponding angles and corresponding sides of congruent triangles are called **corresponding parts**. *Corresponding parts of congruent triangles are always congruent.* That is, corresponding parts of congruent triangles always have the same measure. For the congruent triangles shown above, we have

$$\text{m}(\angle A) = \text{m}(\angle D) \qquad \text{m}(\angle B) = \text{m}(\angle E) \qquad \text{m}(\angle C) = \text{m}(\angle F)$$
$$\text{m}(\overline{BC}) = \text{m}(\overline{EF}) \qquad \text{m}(\overline{AC}) = \text{m}(\overline{DF}) \qquad \text{m}(\overline{AB}) = \text{m}(\overline{DE})$$

Congruent Triangles

Two triangles are congruent if and only if their vertices can be matched so that the corresponding sides and the corresponding angles are congruent.

EXAMPLE 1   Refer to the figure below, where $\triangle XYZ \cong \triangle PQR$.

a. Name the six congruent corresponding parts of the triangles.

b. Find $\text{m}(\angle P)$.

c. Find $\text{m}(\overline{XZ})$.

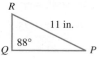

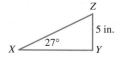

Strategy We will establish the correspondence between the vertices of △*XYZ* and the vertices of △*PQR*.

WHY This will, in turn, establish a correspondence between the congruent corresponding angles and sides of the triangles.

Solution

a. The correspondence between the vertices is

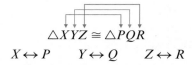

$$\triangle XYZ \cong \triangle PQR$$

$$X \leftrightarrow P \qquad Y \leftrightarrow Q \qquad Z \leftrightarrow R$$

Corresponding parts of congruent triangles are congruent. Therefore, the congruent corresponding angles are

$$\angle X \cong \angle P \qquad \angle Y \cong \angle Q \qquad \angle Z \cong \angle R$$

The congruent corresponding sides are

$$\overline{YZ} \cong \overline{QR} \qquad \overline{XZ} \cong \overline{PR} \qquad \overline{XY} \cong \overline{PQ}$$

b. From the figure, we see that m($\angle X$) = 27°. Since $\angle X \cong \angle P$, it follows that m($\angle P$) = 27°.

c. From the figure, we see that m($\overline{PR}$) = 11 inches. Since $\overline{XZ} \cong \overline{PR}$, it follows that m($\overline{XZ}$) = 11 inches.

Self Check 1

Refer to the figure below, where △*ABC* ≅ △*EDF*.

a. Name the six congruent corresponding parts of the triangles.

b. Find m($\angle C$).

c. Find m($\overline{FE}$).

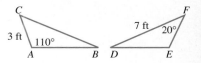

Now Try ➲ Problem 33

OBJECTIVE 2 **Use congruence properties to prove that two triangles are congruent.**

Sometimes it is possible to conclude that two triangles are congruent without having to show that three pairs of corresponding angles are congruent and three pairs of corresponding sides are congruent. To do so, we apply one of the following properties.

SSS Property

If three sides of one triangle are congruent to three sides of a second triangle, the triangles are congruent.

We can show that the triangles shown below are congruent by the SSS property:

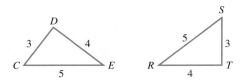

$\overline{CD} \cong \overline{ST}$ Since m($\overline{CD}$) = 3 and m($\overline{ST}$) = 3, the segments are congruent.

$\overline{DE} \cong \overline{TR}$ Since m($\overline{DE}$) = 4 and m($\overline{TR}$) = 4, the segments are congruent.

$\overline{EC} \cong \overline{RS}$ Since m($\overline{EC}$) = 3 and m($\overline{RS}$) = 5, the segments are congruent.

Therefore, △*CDE* ≅ △*STR*.

SAS Property

If two sides and the angle between them in one triangle are congruent, respectively, to two sides and the angle between them in a second triangle, the triangles are congruent.

We can show that the triangles shown below are congruent by the SAS property:

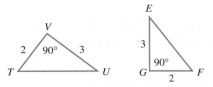

$\overline{TV} \cong \overline{FG}$ Since m($\overline{TV}$) = 2 and m($\overline{FG}$) = 2, the segments are congruent.

$\angle V \cong \angle G$ Since m($\angle V$) = 90° and m($\angle G$) = 90°, the angles are congruent.

$\overline{UV} \cong \overline{EG}$ Since m($\overline{UV}$) = 3 and m($\overline{EG}$) = 3, the segments are congruent.

Therefore, $\triangle TVU \cong \triangle FGE$.

ASA Property

If two angles and the side between them in one triangle are congruent, respectively, to two angles and the side between them in a second triangle, the triangles are congruent.

We can show that the triangles shown below are congruent by the ASA property:

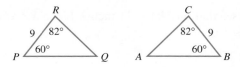

$\angle P \cong \angle B$ Since m($\angle P$) = 60° and m($\angle B$) = 60°, the angles are congruent.

$\overline{PR} \cong \overline{BC}$ Since m($\overline{PR}$) = 9 and m($\overline{BC}$) = 9, the segments are congruent.

$\angle R \cong \angle C$ Since m($\angle R$) = 82° and m($\angle C$) = 82°, the angles are congruent.

Therefore, $\triangle PQR \cong \triangle BAC$.

There is no SSA property. To illustrate this, consider the triangles shown below. Two sides and an angle of $\triangle ABC$ are congruent to two sides and an angle of $\triangle DEF$. But the congruent angle is not between the congruent sides.

We refer to this situation as SSA. Obviously, the triangles are not congruent because they are not the same shape and size.

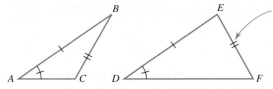

The tick marks indicate congruent parts. That is, the sides with one tick mark are the same length, the sides with two tick marks are the same length, and the angles with one tick mark have the same measure.

EXAMPLE 2   Explain why the triangles in the figure on the following page are congruent.

Strategy We will show that two sides and the angle between them in one triangle are congruent, respectively, to two sides and the angle between them in a second triangle.

WHY Then we know that the two triangles are congruent by the SAS property.

Solution Since vertical angles are congruent,

$$\angle 1 \cong \angle 2$$

From the figure, we see that

$$\overline{AC} \cong \overline{EC} \quad \text{and} \quad \overline{BC} \cong \overline{DC}$$

Since two sides and the angle between them in one triangle are congruent, respectively, to two sides and the angle between them in a second triangle, $\triangle ABC \cong \triangle EDC$ by the SAS property.

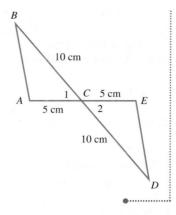

Self Check 2

Are the triangles in the figure below congruent? Explain why or why not.

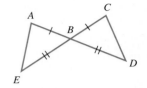

Now Try ➥ **Problem 35**

EXAMPLE 3 Are $\triangle RST$ and $\triangle RUT$ in the figure on the right congruent?

Strategy We will show that two angles and the side between them in one triangle are congruent, respectively, to two angles and the side between them in a second triangle.

WHY Then we know that the two triangles are congruent by the ASA property.

Solution From the markings on the figure, we know that two pairs of angles are congruent.

$$\angle SRT \cong \angle URT \quad$$ These angles are marked with 1 tick mark, which indicates that they have the same measure.

$$\angle STR \cong \angle UTR \quad$$ These angles are marked with 2 tick marks, which indicates that they have the same measure.

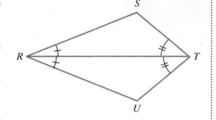

From the figure, we see that the triangles have side $\overline{RT}$ in common. Furthermore, $\overline{RT}$ is between each pair of congruent angles listed above. Since every segment is congruent to itself, we also have

$$\overline{RT} \cong \overline{RT}$$

Knowing that two angles and the side between them in $\triangle RST$ are congruent, respectively, to two angles and the side between them in $\triangle RUT$, we can conclude that $\triangle RST \cong \triangle RUT$ by the ASA property.

Self Check 3

Are the triangles in the following figure congruent? Explain why or why not.

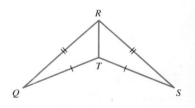

Now Try ➥ **Problem 37**

OBJECTIVE 3 **Determine whether two triangles are similar.**

We have seen that congruent triangles have the same shape and size. **Similar triangles** have the same shape, but not necessarily the same size. That is, one triangle is an exact scale model of the other triangle. If the triangles in the figure below are similar, we can write $\triangle ABC \sim \triangle DEF$ (read the symbol $\sim$ as "is similar to").

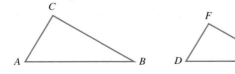

Success Tip Note that congruent triangles are always similar, but similar triangles are not always congruent.

The formal definition of similar triangles requires that we establish a correspondence between the vertices of the triangles. The definition also involves the word *proportional*.

Recall that a **proportion** is a mathematical statement that two ratios (fractions) are equal. An example of a proportion is

$$\frac{1}{2} = \frac{4}{8}$$

In this case, we say that $\frac{1}{2}$ and $\frac{4}{8}$ are *proportional*.

Similar Triangles

Two triangles are similar if and only if their vertices can be matched so that corresponding angles are congruent and the lengths of corresponding sides are proportional.

EXAMPLE 4   Refer to the figure below. If $\triangle PQR \sim \triangle CDE$, name the congruent angles and the sides that are proportional.

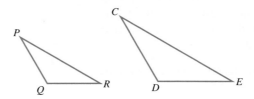

Strategy We will establish the correspondence between the vertices of $\triangle PQR$ and the vertices of $\triangle CDE$.

WHY This will, in turn, establish a correspondence between the congruent corresponding angles and proportional sides of the triangles.

Solution When we write $\triangle PQR \sim \triangle CDE$, a correspondence between the vertices of the triangles is established.

$$\triangle PQR \sim \triangle CDE$$

Since the triangles are similar, corresponding angles are congruent:

$$\angle P \cong \angle C \qquad \angle Q \cong \angle D \qquad \angle R \cong \angle E$$

The lengths of the corresponding sides are proportional. To simplify the notation, we will now let $PQ = \text{m}(\overline{PQ})$, $CD = \text{m}(\overline{CD})$, $QR = \text{m}(\overline{QR})$, and so on.

$$\frac{PQ}{CD} = \frac{QR}{DE} \qquad \frac{QR}{DE} = \frac{PR}{CE} \qquad \frac{PQ}{CD} = \frac{PR}{CE}$$

Written in a more compact way, we have

$$\frac{PQ}{CD} = \frac{QR}{DE} = \frac{PR}{CE}$$

Self Check 4

If $\triangle GEF \sim \triangle IJH$, name the congruent angles and the sides that are proportional.

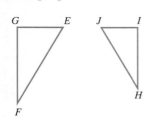

Now Try Problem 39

Property of Similar Triangles

If two triangles are similar, all pairs of corresponding sides are in proportion.

It is possible to conclude that two triangles are similar without having to show that all three pairs of corresponding angles are congruent and that the lengths of all three pairs of corresponding sides are proportional.

AAA Similarity Theorem

If the angles of one triangle are congruent to corresponding angles of another triangle, the triangles are similar.

EXAMPLE 5   In the figure on the right, $\overline{PR} \parallel \overline{MN}$. Are $\triangle PQR$ and $\triangle NQM$ similar triangles?

Strategy We will show that the angles of one triangle are congruent to corresponding angles of another triangle.

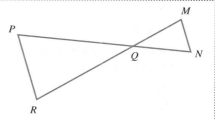

WHY Then we know that the two triangles are similar by the AAA property.

Solution Since vertical angles are congruent,

$$\angle PQR \cong \angle NQM \qquad \text{This is one pair of congruent corresponding angles.}$$

In the figure, we can view $\overleftrightarrow{PN}$ as a transversal cutting parallel line segments $\overline{PR}$ and $\overline{MN}$. Since alternate interior angles are then congruent, we have:

$$\angle RPQ \cong \angle MNQ \qquad \text{This is a second pair of congruent corresponding angles.}$$

Furthermore, we can view $\overleftrightarrow{RM}$ as a transversal cutting parallel line segments $\overline{PR}$ and $\overline{MN}$. Since alternate interior angles are then congruent, we have:

$$\angle QRP \cong \angle QMN \qquad \text{This is a third pair of congruent corresponding angles.}$$

These observations are summarized in the figure on the right. We see that corresponding angles of $\triangle PQR$ are congruent to corresponding angles of $\triangle NQM$. By the AAA similarity theorem, we can conclude that

$$\triangle PQR \sim \triangle NQM$$

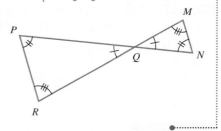

Self Check 5

In the figure below, $\overline{YA} \parallel \overline{ZB}$. Are $\triangle XYA$ and $\triangle XZB$ similar triangles?

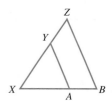

Now Try ➥ **Problems 41 and 43**

EXAMPLE 6   In the figure below, $\triangle RST \sim \triangle JKL$. Find: **a.** x **b.** y

Strategy To find x, we will write a proportion of corresponding sides so that x is the only unknown. Then we will solve the proportion for x. We will use a similar method to find y.

WHY Since $\triangle RST \sim \triangle JKL$, we know that the lengths of corresponding sides of $\triangle RST$ and $\triangle JKL$ are proportional.

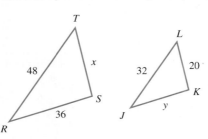

Solution

a. When we write $\triangle RST \sim \triangle JKL$, a correspondence between the vertices of the two triangles is established.

The lengths of corresponding sides of these similar triangles are proportional.

$$\frac{RT}{JL} = \frac{ST}{KL}$$ Each fraction is a ratio of a side length of $\triangle RST$ to its corresponding side length of $\triangle JKL$.

$$\frac{48}{32} = \frac{x}{20}$$ Substitute: $RT = 48$, $JL = 32$, $ST = x$, and $KL = 20$.

$$48(20) = 32x$$ Find each cross product and set them equal.

$$960 = 32x$$ Do the multiplication.

$$30 = x$$ To isolate x, undo the multiplication by 32 by dividing both sides by 32.

$$
\begin{array}{r}
48 \\
\times\,20 \\
\hline
960
\end{array}
$$

$$
\begin{array}{r}
30 \\
32\overline{)960} \\
-96 \\
\hline
00 \\
-00 \\
\hline
0
\end{array}
$$

Thus, x is 30.

b. To find y, we write a proportion of corresponding side lengths in such a way that y is the only unknown.

$$\frac{RT}{JL} = \frac{RS}{JK}$$

$$\frac{48}{32} = \frac{36}{y}$$ Substitute: $RT = 48$, $JL = 32$, $RS = 36$, and $JK = y$.

$$48y = 32(36)$$ Find each cross product and set them equal.

$$48y = 1{,}152$$ Do the multiplication.

$$y = 24$$ To isolate y, undo the multiplication by 48 by dividing both sides by 48.

$$
\begin{array}{r}
36 \\
\times\,32 \\
\hline
72 \\
1080 \\
\hline
1152
\end{array}
$$

$$
\begin{array}{r}
24 \\
48\overline{)1{,}152} \\
-96 \\
\hline
192 \\
-192 \\
\hline
0
\end{array}
$$

Thus, y is 24.

Self Check 6

In the figure below, $\triangle DEF \sim \triangle GHI$. Find:

a. x **b.** y

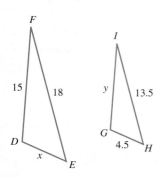

Now Try ➡ **Problem 53**

OBJECTIVE 4   **Use similar triangles to find unknown lengths in application problems.**

Similar triangles and proportions can be used to find lengths that would normally be difficult to measure. For example, we can use the reflective properties of a mirror to calculate the height of a flagpole while standing safely on the ground.

EXAMPLE 7   To determine the height of a flagpole, a woman walks to a point 20 feet from its base, as shown below. Then she takes a mirror from her purse, places it on the ground, and walks 2 feet farther away, where she can see the top of the pole reflected in the mirror. Find the height of the pole.

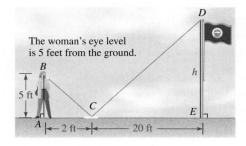

Strategy We will show that $\triangle ABC \sim \triangle EDC$.

WHY Then we can write a proportion of corresponding sides so that h is the only unknown and we can solve the proportion for h.

Solution To show that $\triangle ABC \sim \triangle EDC$, we begin by applying an important fact about mirrors. When a beam of light strikes a mirror, it is reflected at the same angle as it hits the mirror. Therefore, $\angle BCA \cong \angle DCE$. Furthermore, $\angle A \cong \angle E$ because the woman and the flagpole are perpendicular to the ground. Finally, if two pairs of corresponding angles are congruent, it follows that the third pair of corresponding angles are also congruent: $\angle B \cong \angle D$. By the AAA similarity theorem, we conclude that $\triangle ABC \sim \triangle EDC$.

Since the triangles are similar, the lengths of their corresponding sides are in proportion. If we let h represent the height of the flagpole, we can find h by solving the following proportion.

Height of the flagpole → $\dfrac{h}{5} = \dfrac{20}{2}$ ← Distance from flagpole to mirror
Height of the woman → ← Distance from woman to mirror

$$2h = 5(20) \quad \text{Find each cross product and set them equal.}$$
$$2h = 100 \quad \text{Do the multiplication.}$$
$$h = 50 \quad \text{To isolate } h, \text{ divide both sides by 2.}$$

The flagpole is 50 feet tall.

Self Check 7

In the figure below, $\triangle ABC \sim \triangle EDC$. Find h.

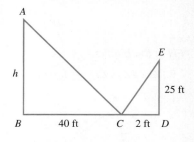

Now Try ➡ **Problem 85**

Answers to Self Checks

1. a. $\angle A \cong \angle E$, $\angle B \cong \angle D$, $\angle C \cong \angle F$, $\overline{AB} \cong \overline{ED}$, $\overline{BC} \cong \overline{DF}$, $\overline{CA} \cong \overline{FE}$　**b.** $20°$　**c.** 3 ft　**2.** yes, by the SAS property　**3.** yes, by the SSS property　**4.** $\angle G \cong \angle I$, $\angle E \cong \angle J$, $\angle F \cong \angle H$; $\dfrac{EG}{JI} = \dfrac{GF}{IH} = \dfrac{FE}{HJ}$　**5.** yes, by the AAA similarity theorem: $\angle X \cong \angle X$, $\angle XYA \cong \angle XZB$, $\angle XAY \cong \angle XBZ$　**6. a.** 6　**b.** 11.25　**7.** 500 ft

---

SECTION 9.5　STUDY SET

VOCABULARY

Fill in the blanks.

1. _____ triangles are the same size and the same shape.

2. When we match the vertices of $\triangle ABC$ with the vertices of $\triangle DEF$, as shown below, we call this matching of points a _____.

$$A \leftrightarrow D \qquad B \leftrightarrow E \qquad C \leftrightarrow F$$

3. Two angles or two line segments with the same measure are said to be _____.

4. Corresponding _____ of congruent triangles are congruent.

5. If two triangles are _____, they have the same shape but not necessarily the same size.

6. A mathematical statement that two ratios (fractions) are equal, such as $\dfrac{x}{18} = \dfrac{4}{9}$, is called a _____.

CONCEPTS

7. Refer to the triangles below.

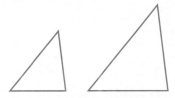

a. Do these triangles appear to be congruent? Explain why or why not.

b. Do these triangles appear to be similar? Explain why or why not.

8. a. Draw a triangle that is congruent to △*CDE* shown below. Label it △*ABC*.

   b. Draw a triangle that is similar to, but not congruent to, △*CDE*. Label it △*MNO*.

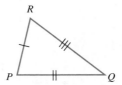

Fill in the blanks.

9. △*XYZ* ≅ △ _____

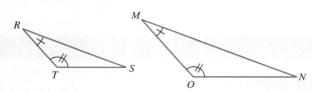

10. △ _____ ≅ △*DEF*

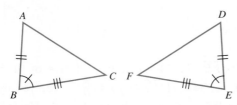

11. △*RST* ~ △ _____

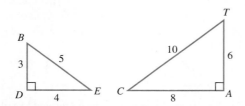

12. △ _____ ~ △*TAC*

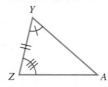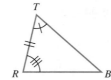

13. Name the six corresponding parts of the congruent triangles shown below.

14. Name the six corresponding parts of the congruent triangles shown below.

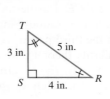

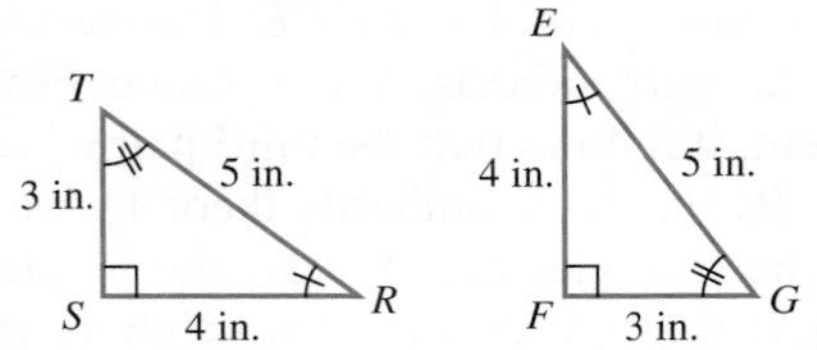

Fill in the blanks.

15. Two triangles are _____ if and only if their vertices can be matched so that the corresponding sides and the corresponding angles are congruent.

16. SSS property: If three _____ of one triangle are congruent to three _____ of a second triangle, the triangles are congruent.

17. SAS property: If two sides and the _____ between them in one triangle are congruent, respectively, to two sides and the _____ between them in a second triangle, the triangles are congruent.

18. ASA property: If two angles and the _____ between them in one triangle are congruent, respectively, to two angles and the _____ between them in a second triangle, the triangles are congruent.

Solve each proportion.

19. $\dfrac{x}{15} = \dfrac{20}{3}$

20. $\dfrac{5}{8} = \dfrac{35}{x}$

21. $\dfrac{h}{2.6} = \dfrac{27}{13}$

22. $\dfrac{11.2}{4} = \dfrac{h}{6}$

Fill in the blanks.

23. Two triangles are similar if and only if their vertices can be matched so that corresponding angles are congruent and the lengths of corresponding sides are _____.

24. If the angles of one triangle are congruent to corresponding angles of another triangle, the triangles are _____.

25. Congruent triangles are always similar, but similar triangles are not always _____.

26. For certain application problems, similar triangles and _____ can be used to find lengths that would normally be difficult to measure.

NOTATION

Fill in the blanks.

27. The symbol ≅ is read as "___ _____ ___."

28. The symbol ~ is read as "___ _____ ___."

29. Use tick marks to show the congruent parts of the triangles shown below.

$$\angle K \cong \angle H \qquad \overline{KR} \cong \overline{HJ} \qquad \angle M \cong \angle E$$

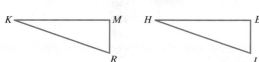

30. Use tick marks to show the congruent parts of the triangles shown below.

$$\angle P \cong \angle T \qquad \overline{LP} \cong \overline{RT} \qquad \overline{FP} \cong \overline{ST}$$

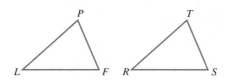

GUIDED PRACTICE

Name the six corresponding parts of the congruent triangles. **See Objective 1.**

31. $\overline{AC} \cong$ _____
$\overline{DE} \cong$ _____
$\overline{BC} \cong$ _____
$\angle A \cong$ _____
$\angle E \cong$ _____
$\angle F \cong$ _____

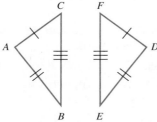

32. $\overline{AB} \cong$ _____
$\overline{EC} \cong$ _____
$\overline{AC} \cong$ _____
$\angle D \cong$ _____
$\angle B \cong$ _____
$\angle 1 \cong$ _____

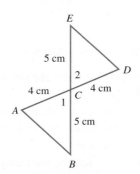

33. Refer to the figure below, where $\triangle BCD \cong \triangle MNO$. **See Example 1.**

  a. Name the six congruent corresponding parts of the triangles.

  b. Find m($\angle N$).

  c. Find m($\overline{MO}$).

  d. Find m($\overline{CD}$).

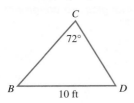

34. Refer to the figure below, where $\triangle DCG \cong \triangle RST$. **See Example 1.**

  a. Name the six congruent corresponding parts of the triangles.

  b. Find m($\angle R$).

  c. Find m($\overline{DG}$).

  d. Find m($\overline{ST}$).

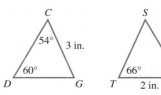

Determine whether each pair of triangles is congruent. If they are, tell why. **See Examples 2 and 3.**

35.

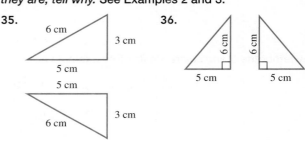

36.

37.

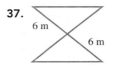

38.

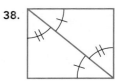

39. Refer to the similar triangles shown below. **See Example 4.**

  a. Name three pairs of congruent angles.

  b. Complete each proportion.

$$\frac{LM}{HJ} = \frac{}{JE} \qquad \frac{MR}{JE} = \frac{}{HE} \qquad \frac{}{HJ} = \frac{LR}{HE}$$

  c. We can write the answer to part b in a more compact form:

$$\frac{LM}{} = \frac{MR}{} = \frac{}{HE}$$

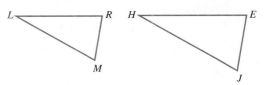

40. Refer to the similar triangles shown below. **See Example 4.**

a. Name three pairs of congruent angles.

b. Complete each proportion.

$$\frac{WY}{DF} = \frac{\boxed{}}{FE} \qquad \frac{WX}{\boxed{}} = \frac{YX}{FE} \qquad \frac{\boxed{}}{EF} = \frac{WY}{DF}$$

c. We can write the answer to part b in a more compact form:

$$\frac{\boxed{}}{DF} = \frac{YX}{\boxed{}} = \frac{WX}{\boxed{}}$$

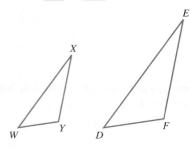

Tell whether the triangles are similar. **See Example 5.**

41.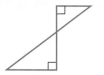

42.

43.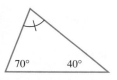

44.

45.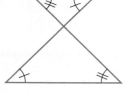

70° 40°   40° 70°

46.

47.

48.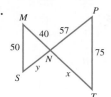

49. $\overline{XY} \parallel \overline{ZD}$

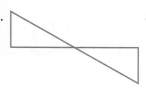

50. $\overline{QR} \parallel \overline{TU}$

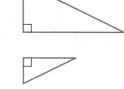

51.

52.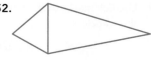

In Problems 53 and 54, $\triangle MSN \sim \triangle TPR$. *Find x and y.* **See Example 6.**

53.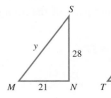

54.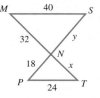

In Problems 55 and 56, $\triangle MSN \sim \triangle TPN$. *Find x and y.* **See Example 6.**

55.

56.

M   40   S

32   N   y

18   x

P   24   T

TRY IT YOURSELF

Tell whether each statement is true. If a statement is false, tell why.

57. If three sides of one triangle are the same length as the corresponding three sides of a second triangle, the triangles are congruent.

58. If two sides of one triangle are the same length as two sides of a second triangle, the triangles are congruent.

59. If two sides and an angle of one triangle are congruent, respectively, to two sides and an angle of a second triangle, the triangles are congruent.

60. If two angles and the side between them in one triangle are congruent, respectively, to two angles and the side between them in a second triangle, the triangles are congruent.

Determine whether each pair of triangles are congruent. If they are, tell why.

61.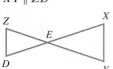

62.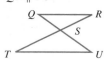
40°

40°

63.

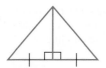

64.

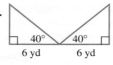

65. $\overline{AB} \parallel \overline{DE}$

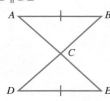

66. $\overline{XY} \parallel \overline{ZQ}$

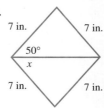

67.

68.

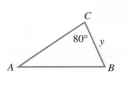

In Problems 69 and 70, △ ABC ≅ △ DEF. *Find x and y.*

69.

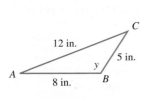

70.

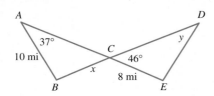

In Problems 71 and 72, find x and y.

71. △ABC ≅ △ABD

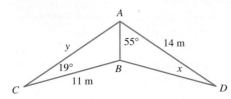

72. △ABC ≅ △DEC

In Problems 73–76, find x.

73.

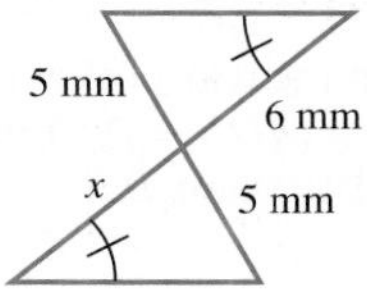

74.

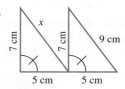

75.

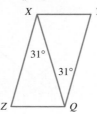

76.

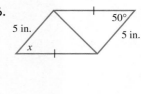

77. If $\overline{DE}$ in the figure below is parallel to $\overline{AB}$, △ABC will be similar to △DEC. Find x.

78. If $\overline{SU}$ in the figure below is parallel to $\overline{TV}$, △SRU will be similar to △TRV. Find x.

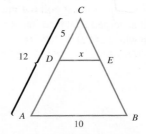

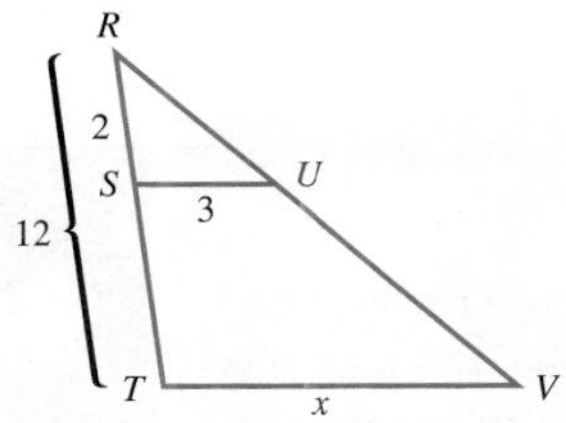

79. If $\overline{DE}$ in the figure below is parallel to $\overline{CB}$, △EAD will be similar to △BAC. Find x.

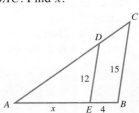

80. If $\overline{HK}$ in the figure below is parallel to $\overline{AB}$, △HCK will be similar to △ACB. Find x.

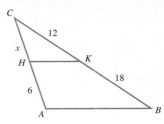

APPLICATIONS

81. Sewing. The pattern that is sewn on the rear pocket of a pair of blue jeans is shown below. If $\triangle AOB \cong \triangle COD$, how long is the stitching from point A to point D?

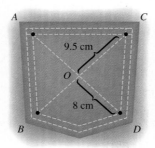

82. Camping. The base of the tent pole is placed at the midpoint between the stake at point A and the stake at point B, and it is perpendicular to the ground, as shown below. Explain why $\triangle ACD \cong \triangle BCD$.

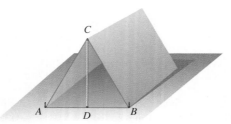

Surveyor

83. A surveying crew needs to find the width of the river shown in the illustration below. Because of a dangerous current, they decide to stay on the west side of the river and use geometry to find its width. Their approach is to create two similar right triangles on dry land. Then they write and solve a proportion to find w. What is the width of the river?

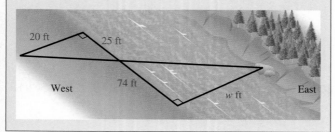

84. Height of a building. A man places a mirror on the ground and sees the reflection of the top of a building, as shown below. Find the height of the building.

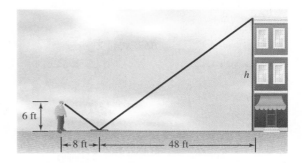

85. Height of a tree. The tree shown below casts a shadow 24 feet long when a man 6 feet tall casts a shadow 4 feet long. Find the height of the tree.

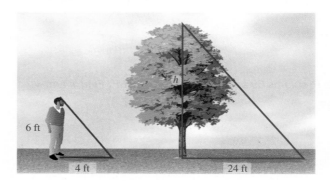

86. Washington, D.C. The Washington Monument casts a shadow of $166\frac{1}{2}$ feet at the same time as a 5-foot-tall tourist casts a shadow of $1\frac{1}{2}$ feet. Find the height of the monument.

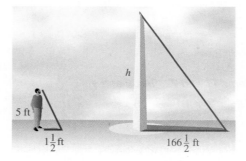

87. Height of a tree. A tree casts a shadow of 29 feet at the same time as a vertical yardstick casts a shadow of 2.5 feet. Find the height of the tree.

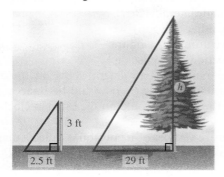

88. Geography. The diagram below shows how a laser beam was pointed over the top of a pole to the top of a mountain to determine the elevation of the mountain. Find h.

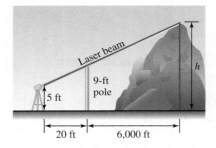

89. Flight path. An airplane ascends 200 feet as it flies a horizontal distance of 1,000 feet, as shown in the following figure. How much altitude is gained as it flies a horizontal distance of 1 mile? (*Hint:* 1 mile = 5,280 feet.)

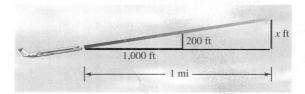

WRITING

90. Tell whether the statement is true or false. Explain your answer.

   a. Congruent triangles are always similar.

   b. Similar triangles are always congruent.

91. Explain why there is no SSA property for congruent triangles.

REVIEW

Find the LCM of the given numbers.

92. 16, 20   **93.** 21, 27

Find the GCF of the given numbers.

94. 18, 96   **95.** 63, 84

SECTION 9.6 Quadrilaterals and Other Polygons

Recall from Section 9.3 that a polygon is a closed geometric figure with at least three line segments for its sides. In this section, we will focus on polygons with four sides, called *quadrilaterals*. One type of quadrilateral is the *square*. The game boards for Monopoly and Scrabble have a square shape. Another type of quadrilateral is the *rectangle*. Most picture frames and many mirrors are rectangular. Utility knife blades and swimming fins have shapes that are examples of a third type of quadrilateral called a *trapezoid*.

macroworld/iStock/Getty Images

OBJECTIVES

1 Classify quadrilaterals.

2 Use properties of rectangles to find unknown angle measures and side lengths.

3 Find unknown angle measures of trapezoids.

4 Use the formula for the sum of the angle measures of a polygon.

OBJECTIVE 1 Classify quadrilaterals.

A **quadrilateral** is a polygon with four sides. Some common quadrilaterals are shown below.

   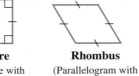

Parallelogram	**Rectangle**	**Square**	**Rhombus**	**Trapezoid**
(Opposite sides parallel)	(Parallelogram with four right angles)	(Rectangle with sides of equal length)	(Parallelogram with sides of equal length)	(Exactly two sides parallel)

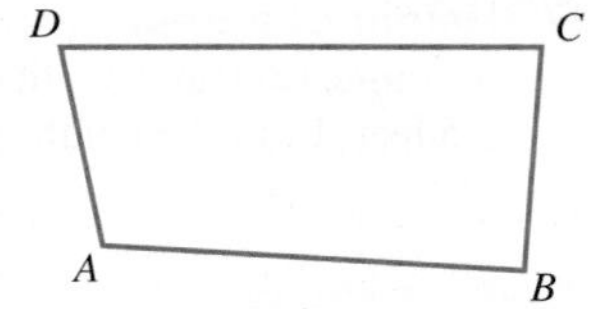

We can use the capital letters that label the vertices of a quadrilateral to name it. For example, when referring to the quadrilateral shown on the right, with vertices A, B, C, and D, we can use the notation quadrilateral $ABCD$.

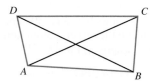

Quadrilateral $ABCD$

When naming a quadrilateral (or any other polygon), we may begin with any vertex. Then we move around the figure in a clockwise (or counterclockwise) direction as we list the remaining vertices. Some other ways of naming the quadrilateral above are quadrilateral $ADCB$, quadrilateral $CDAB$, and quadrilateral $DABC$. It would be unacceptable to name it as quadrilateral $ACDB$, because the vertices would not be listed in clockwise (or counterclockwise) order.

A segment that joins two nonconsecutive vertices of a polygon is called a **diagonal** of the polygon. Quadrilateral $ABCD$ shown below has two diagonals, $\overline{AC}$ and $\overline{BD}$.

OBJECTIVE 2 Use properties of rectangles to find unknown angle measures and side lengths.

Recall that a **rectangle** is a quadrilateral with four right angles. The rectangle is probably the most common and recognizable of all geometric figures. For example, most doors and windows are rectangular in shape. The boundaries of soccer fields and basketball courts are rectangles. Even our paper currency, such as the $1, $5, and $20 bills, is in the shape of a rectangle. Rectangles have several important characteristics.

Properties of Rectangles

In any rectangle:

1. All four angles are right angles.

2. Opposite sides are parallel.

3. Opposite sides have equal length.

4. The diagonals have equal length.

5. The diagonals intersect at their midpoints.

EXAMPLE 1 In the figure, quadrilateral $WXYZ$ is a rectangle. Find each measure:

a. m($\angle YXW$) **b.** m($\overline{XY}$) **c.** m($\overline{WY}$) **d.** m($\overline{XZ}$)

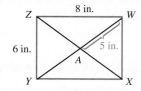

Strategy We will use properties of rectangles to find the unknown angle measure and the unknown measures of the line segments.

WHY Quadrilateral $WXYZ$ is a rectangle.

Solution

a. In any rectangle, all four angles are right angles. Therefore, $\angle YXW$ is a right angle, and m($\angle YXW$) = 90°.

b. $\overline{XY}$ and $\overline{WZ}$ are opposite sides of the rectangle, so they have equal length. Since the length of $\overline{WZ}$ is 8 inches, m($\overline{XY}$) is also 8 inches.

c. $\overline{WY}$ and $\overline{ZX}$ are diagonals of the rectangle, and they intersect at their midpoints. That means that point A is the midpoint of $\overline{WY}$. Since the length of $\overline{WA}$ is 5 inches, m($\overline{WY}$) is 2 · 5 inches, or 10 inches.

d. The diagonals of a rectangle are of equal length. In part c, we found that the length of $\overline{WY}$ is 10 inches. Therefore, m($\overline{XZ}$) is also 10 inches.

Self Check 1

In rectangle *RSTU* shown below, the length of $\overline{RT}$ is 13 ft. Find each measure:

a. m($\angle SRU$)

b. m($\overline{ST}$)

c. m($\overline{TG}$)

d. m($\overline{SG}$)

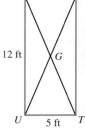

Now Try Problem 27

We have seen that if a quadrilateral has four right angles, it is a rectangle. The following statements establish some conditions that a parallelogram must meet to ensure that it is a rectangle.

Parallelograms That Are Rectangles

1. If a parallelogram has one right angle, then the parallelogram is a rectangle.

2. If the diagonals of a parallelogram are congruent, then the parallelogram is a rectangle.

EXAMPLE 2   **Construction.**   A carpenter wants to build a shed with a 9-foot-by-12-foot base. How can he make sure that the foundation has four right-angle corners?

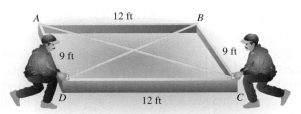

Strategy The carpenter should find the lengths of the diagonals of the foundation.

WHY If the diagonals are congruent, then the foundation is rectangular in shape and the corners are right angles.

Solution The four-sided foundation, which we will label as parallelogram *ABCD*, has opposite sides of equal length. The carpenter can use a tape measure to find the lengths of the diagonals $\overline{AC}$ and $\overline{BD}$. If these diagonals are of equal length, the foundation will be a rectangle and have right angles at its four corners. This process is commonly referred to as "squaring a foundation." Picture framers use a similar process to make sure their frames have four 90° corners.

Now Try Problem 59

OBJECTIVE 3 **Find unknown angle measures of trapezoids.**

A **trapezoid** is a quadrilateral with exactly two sides parallel. For the trapezoid shown on the next page, the parallel sides $\overline{AB}$ and $\overline{DC}$ are called **bases**. To distinguish between the two bases, we will refer to $\overline{AB}$ as the **upper base** and $\overline{DC}$ as the **lower base**. The angles on either side of the upper base are called **upper base angles**, and the angles on either side of the lower base are called **lower base angles**. The nonparallel sides are called **legs**.

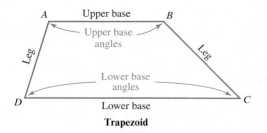

Trapezoid

   In the figure above, we can view $\overleftrightarrow{AD}$ as a transversal cutting the parallel lines $\overleftrightarrow{AB}$ and $\overleftrightarrow{DC}$. Since $\angle A$ and $\angle D$ are interior angles on the same side of a transversal, they are supplementary. Similarly, $\overleftrightarrow{BC}$ is a transversal cutting the parallel lines $\overleftrightarrow{AB}$ and $\overleftrightarrow{DC}$. Since $\angle B$ and $\angle C$ are interior angles on the same side of a transversal, they are also supplementary. These observations lead us to the conclusion that *there are always two pairs of supplementary angles in any trapezoid.*

EXAMPLE 3   Refer to trapezoid *KLMN* below, with $\overline{KL} \parallel \overline{NM}$. Find x and y.

Strategy We will use the interior angles property twice to write two equations that mathematically model the situation.

WHY We can then solve the equations to find x and y.

Solution $\angle K$ and $\angle N$ are interior angles on the same side of transversal $\overleftrightarrow{KN}$ that cuts the parallel lines segments $\overline{KL}$ and $\overline{NM}$. Similarly, $\angle L$ and $\angle M$ are interior angles on the same side of transversal $\overleftrightarrow{LM}$ that cuts the parallel lines segments $\overline{KL}$ and $\overline{NM}$. Recall that if two parallel lines are cut by a transversal, interior angles on the same side of the transversal are supplementary. We can use this fact twice—once to find x and a second time to find y.

Self Check 3

Refer to trapezoid *HIJK* below, with $\overline{HI} \parallel \overline{KJ}$. Find x and y.

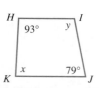

Now Try Problem 29

$m(\angle K) + m(\angle N) = 180°$	The sum of the measures of supplementary angles is 180°.
$x + 82° = 180°$	Substitute x for $m(\angle K)$ and 82° for $m(\angle N)$.
$x = 98°$	To isolate x, subtract 82° from both sides.

$$
\begin{array}{r}
\overset{17}{\cancel{\overset{7\,10}{180}}} \\
-\;82 \\
\hline
98
\end{array}
$$

Thus, x is 98°.

$m(\angle L) + m(\angle M) = 180°$	The sum of the measures of supplementary angles is 180°.
$121° + y = 180°$	Substitute 121° for $m(\angle L)$ and y for $m(\angle M)$.
$y = 59°$	To isolate y, subtract 121° from both sides.

$$
\begin{array}{r}
\overset{7\,10}{180} \\
-\;121 \\
\hline
59
\end{array}
$$

Thus, y is 59°.

   If the nonparallel sides of a trapezoid are the same length, it is called an **isosceles trapezoid**. The figure on the right shows isosceles trapezoid *DEFG* with $\overline{DG} \cong \overline{EF}$. In an isosceles trapezoid, *both pairs of base angles are congruent.* In the figure, $\angle D \cong \angle E$ and $\angle G \cong \angle F$.

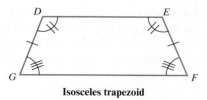

Isosceles trapezoid

EXAMPLE 4   **Landscaping.**   A cross section of a drainage ditch shown below is an isosceles trapezoid with $\overline{AB} \parallel \overline{DC}$. Find x and y.

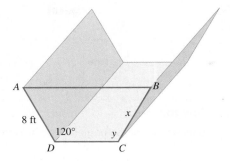

Strategy We will compare the nonparallel sides and compare a pair of base angles of the trapezoid to find each unknown.

 WHY   The nonparallel sides of an isosceles trapezoid have the same length and both pairs of base angles are congruent.

Solution Since $\overline{AD}$ and $\overline{BC}$ are the nonparallel sides of an isosceles trapezoid, m($\overline{AD}$) and m($\overline{BC}$) are equal, and x is 8 ft.
   Since $\angle D$ and $\angle C$ are a pair of base angles of an isosceles trapezoid, they are congruent and m($\angle D$) = m($\angle C$). Thus, y is 120°.

Self Check 4

Refer to the isosceles trapezoid shown below with $\overline{RS} \parallel \overline{UT}$. Find x and y.

Now Try ➡ **Problem 31**

OBJECTIVE 4 **Use the formula for the sum of the angle measures of a polygon.**

In the figure shown below, a protractor was used to find the measure of each angle of the quadrilateral. When we add the four angle measures, the result is 360°.

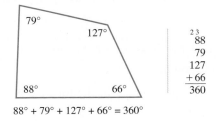

$$88° + 79° + 127° + 66° = 360°$$

   This illustrates an important fact about quadrilaterals: The sum of the measures of the angles of *any* quadrilateral is 360°. This can be shown using the diagram in figure (a) on the following page. In the figure, the quadrilateral is divided into two triangles. Since the sum of the angle measures of any triangle is 180°, the sum of the measures of the angles of the quadrilateral is 2 · 180°, or 360°.
   A similar approach can be used to find the sum of the measures of the angles of any pentagon or any hexagon. The pentagon in figure (b) is divided into three triangles. The sum of the measures of the angles of the pentagon is 3 · 180°, or 540°. The hexagon in figure (c) is divided into four triangles. The sum of the measures of the angles of the hexagon is 4 · 180°, or 720°. In general, a polygon with n sides can be divided into $n - 2$ triangles. Therefore, the sum of the angle measures of a polygon can be found by multiplying 180° by $n - 2$.

Quadrilateral	Pentagon	Hexagon

  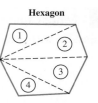

$2 \cdot 180° = 360°$	$3 \cdot 180° = 540°$	$4 \cdot 180° = 720°$
(a)	(b)	(c)

Sum of the Angles of a Polygon

The sum S, in degrees, of the measures of the angles of a polygon with n sides is given by the formula

$$S = (n - 2)180°$$

EXAMPLE 5   Find the sum of the angle measures of a 13-sided polygon.

Strategy We will substitute 13 for n in the formula $S = (n - 2)180°$ and evaluate the right side.

WHY The variable S represents the unknown sum of the measures of the angles of the polygon.

Self Check 5

Find the sum of the angle measures of the polygon shown below.

Solution

$S = (n - 2)180°$	This is the formula for the sum of the measures of the angles of a polygon.	$\begin{array}{r} 180 \\ \times 11 \\ \hline 180 \\ 1800 \\ \hline 1,980 \end{array}$
$S = (\mathbf{13} - 2)180°$	Substitute 13 for n, the number of sides.	
$S = (11)180°$	Do the subtraction within the parentheses.	
$S = 1{,}980°$	Do the multiplication.	

Now Try ➡ **Problem 33**

The sum of the measures of the angles of a 13-sided polygon is 1,980°.

EXAMPLE 6   The sum of the measures of the angles of a polygon is 1,080°. Find the number of sides the polygon has.

Strategy We will substitute 1,080° for S in the formula $S = (n - 2)180°$ and solve for n.

WHY The variable n represents the unknown number of sides of the polygon.

Solution

$S = (n - 2)180°$	This is the formula for the sum of the measures of the angles of a polygon.	
$\mathbf{1{,}080°} = (n - 2)180°$	Substitute 1,080° for S, the sum of the measures of the angles.	
$1{,}080° = 180°n - 360°$	Distribute the multiplication by 180°.	
$1{,}080° + 360° = 180°n - 360° + 360°$	To isolate 180°n, add 360° to both sides.	
$1{,}440° = 180°n$	Do the additions.	$\begin{array}{r} \overset{1}{1{,}080} \\ + \ 360 \\ \hline 1{,}440 \end{array}$ $\quad$ $\begin{array}{r} 8 \\ 180\overline{)1{,}440} \\ -1\ 440 \\ \hline 0 \end{array}$
$\dfrac{1{,}440°}{180°} = \dfrac{180°n}{180°}$	To isolate n, divide both sides by 180°.	
$8 = n$	Do the division.	

Self Check 6

The sum of the measures of the angles of a polygon is 1,620°. Find the number of sides the polygon has.

Now Try ➡ **Problem 43**

The polygon has 8 sides. It is an octagon.

SECTION 9.6 STUDY SET

VOCABULARY

Fill in the blanks.

1. A _____ is a polygon with four sides.

2. A _____ is a quadrilateral with opposite sides parallel.

3. A _____ is a quadrilateral with four right angles.

4. A rectangle with all sides of equal length is a _____.

5. A _____ is a parallelogram with four sides of equal length.

6. A segment that joins two nonconsecutive vertices of a polygon is called a _____ of the polygon.

7. A _____ has two sides that are parallel and two sides that are not parallel. The parallel sides are called _____. The legs of an _____ trapezoid have the same length.

8. A _____ polygon has sides that are all the same length and angles that are all the same measure.

CONCEPTS

9. Refer to the polygon below.

    a. How many vertices does it have? List them.
    b. How many sides does it have? List them.
    c. How many diagonals does it have? List them.
    d. Tell which of the following are acceptable ways of naming the polygon.

    quadrilateral *ABCD*
    quadrilateral *CDBA*
    quadrilateral *ACBD*
    quadrilateral *BADC*

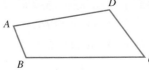

10. Draw an example of each type of quadrilateral.

    a. rhombus   b. parallelogram
    c. trapezoid   d. square
    e. rectangle   f. isosceles trapezoid

11. A parallelogram is shown below. Fill in the blanks.

    a. $\overline{ST} \parallel$ ____   b. $\overline{SV}$ ____ $\overline{TU}$

12. Refer to the rectangle in the next column.

    a. How many right angles does the rectangle have? List them.
    b. Which sides are parallel?

c. Which sides are of equal length?

d. Copy the figure and draw the diagonals. Call the point where the diagonals intersect point *X*. How many diagonals does the figure have? List them.

13. Fill in the blanks. In any rectangle:

    a. All four angles are _____ angles.
    b. Opposite sides are _____.
    c. Opposite sides have equal _____.
    d. The diagonals have equal _____.
    e. The diagonals intersect at their _____.

14. Refer to the figure below.

    a. What is m($\overline{CD}$)?   b. What is m($\overline{AD}$)?

15. In the figure below, $\overline{TR} \parallel \overline{DF}$, $\overline{DT} \parallel \overline{FR}$, and m($\angle D$) = 90°. What type of quadrilateral is *DTRF*?

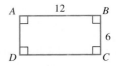

16. Refer to the parallelogram shown below. If m($\overline{GI}$) = 4 and m($\overline{HJ}$) = 4, what type of figure is quadrilateral *GHIJ*?

17. a. Is every rectangle a square?

    b. Is every square a rectangle?
    c. Is every parallelogram a rectangle?
    d. Is every rectangle a parallelogram?
    e. Is every rhombus a square?
    f. Is every square a rhombus?

18. Trapezoid *WXYZ* is shown below. Which sides are parallel?

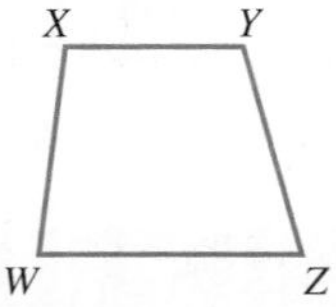

19. Trapezoid *JKLM* is shown below.

   a. What type of trapezoid is this?

   b. Which angles are the lower base angles?

   c. Which angles are the upper base angles?

   d. Fill in the blanks:

     m(∠*J*) = m(   )

     m(∠*K*) = m(   )

     m($\overline{JK}$) = m(   )

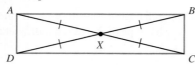

20. Find the sum of the measures of the angles of the hexagon below.

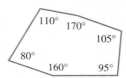

NOTATION

21. What do the tick marks in the figure indicate?

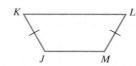

22. Rectangle *ABCD* is shown below. What do the tick marks indicate about point *X*?

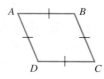

23. In the formula $S = (n - 2)180°$, what does S represent? What does n represent?

24. Suppose $n = 12$. What is $(n - 2)180°$?

In Problems 25 and 26, classify each quadrilateral as a rectangle, a square, a rhombus, or a trapezoid. Some figures may be correctly classified in more than one way. See Objective 1.

25. a.
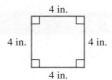

   b.

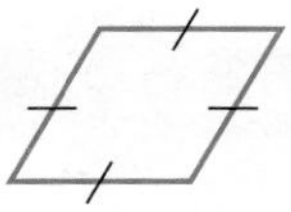

   c.
      d.

26. a. 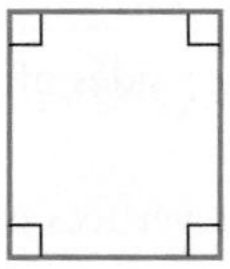
   b. 8 cm

   c. 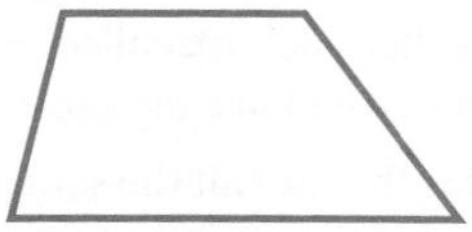
      d.

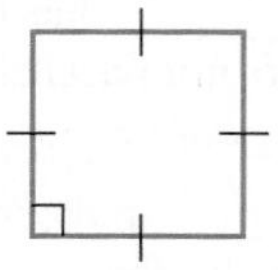

27. Rectangle *ABCD* is shown below. **See Example 1.**

   a. What is m(∠*DCB*)?

   b. What is m($\overline{AX}$)?

   c. What is m($\overline{AC}$)?

   d. What is m($\overline{BD}$)?

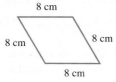

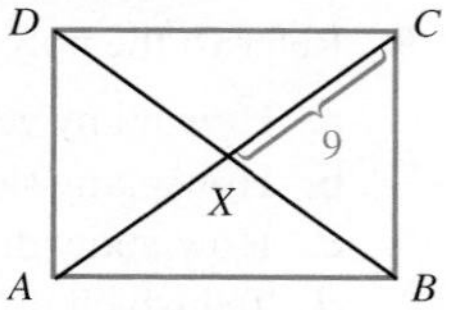

28. Refer to rectangle *EFGH* shown below. **See Example 1.**

   a. Find m(∠*EHG*).   **b.** Find m($\overline{FH}$).

   c. Find m($\overline{EI}$).   **d.** Find m($\overline{EG}$).

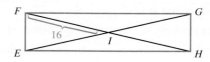

29. Refer to the trapezoid shown below. **See Example 3.**

   a. Find *x*.   **b.** Find *y*.

30. Refer to trapezoid *MNOP* shown below. **See Example 3.**

   a. Find m(∠*O*).   **b.** Find m(∠*M*).

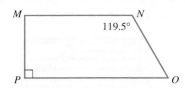

31. Refer to the isosceles trapezoid shown below. **See Example 4.**

   a. Find m($\overline{BC}$).   **b.** Find *x*.
   c. Find *y*.   **d.** Find *z*.

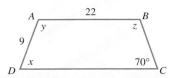

32. Refer to the trapezoid shown below. **See Example 4.**

   a. Find m(∠*T*).
   b. Find m(∠*R*).
   c. Find m(∠*S*).

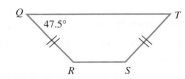

Find the sum of the angle measures of the polygon.
See Example 5.

33. a 14-sided polygon

34. a 15-sided polygon

35. a 20-sided polygon

36. a 22-sided polygon

37. an octagon   **38.** a decagon

39. a dodecagon   **40.** a nonagon

Find the number of sides a polygon has if the sum of its angle measures is the given number. **See Example 6.**

41. 540°   **42.** 720°

43. 900°   **44.** 1,620°

45. 1,980°   **46.** 1,800°

47. 2,160°   **48.** 3,600°

49. Refer to rectangle *ABCD* shown here.

   a. Find m(∠1).
   b. Find m(∠3).
   c. Find m(∠2).
   d. If m($\overline{AC}$) is 8 cm, find m($\overline{BD}$).
   e. Find m($\overline{PD}$).

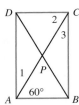

50. The following problem appeared on a quiz. Explain why the instructor must have made an error when typing the problem.

   The sum of the measures of the angles of a polygon is 1,000°. How many sides does the polygon have?

*For Problems 51 and 52, find *x*. Then find the measure of each angle of the polygon.*

51.

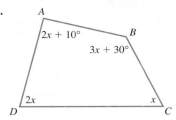

52.

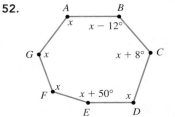

53. Quadrilaterals in everyday life. What quadrilateral shape do you see in each of the following objects?

   a. podium (upper portion)   **b.** checkerboard

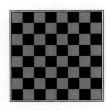

   c. dollar bill   **d.** swimming fin

   e. camper shell window

54. Flowchart. A flowchart shows a sequence of steps to be performed by a computer to solve a given problem. When designing a flowchart, the programmer uses a set of standardized symbols to represent various operations to be performed by the computer. Locate a rectangle, a rhombus, and a parallelogram in the flowchart shown to the right.

55. Baseball. Refer to the figure to the right. Find the sum of the measures of the angles of home plate.

56. Tools. The utility knife blade shown below has the shape of an isosceles trapezoid. Find x, y, and z.

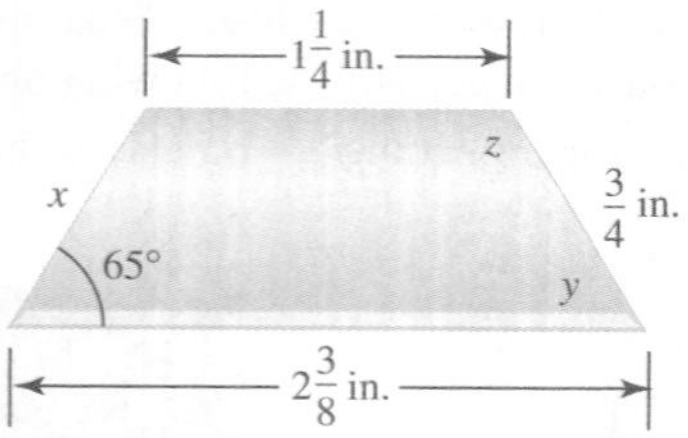

57. Basketball courts. The *free throw lane* for an NBA court and one for an international basketball court are shown below in yellow. Sometimes they are referred to as the "key" or "the paint." What type of quadrilateral is each?

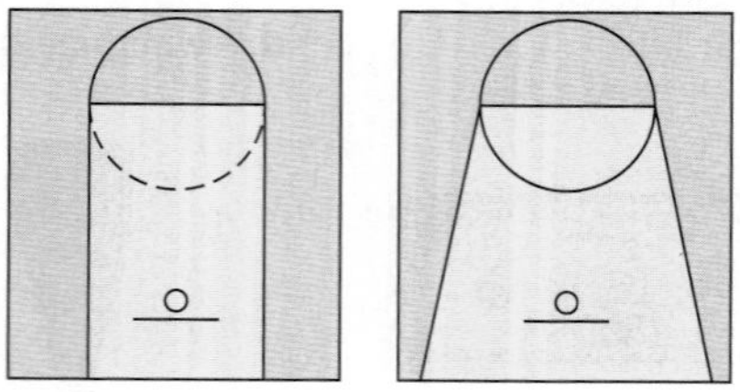

NBA key   **International key**

58. Working overhead. A *scissor lift* is a portable, hydraulic-powered piece of equipment with a platform that is raised upward above its base. What type of quadrilateral do the criss-crossing steel arms form?

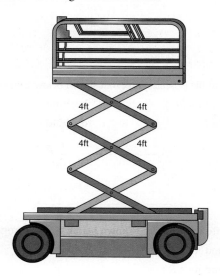

WRITING

59. Explain why a square is a rectangle.

60. Explain why a trapezoid is not a parallelogram.

61. Making a frame.
After gluing and nailing the pieces of a picture frame together, it didn't look right to a frame maker. (See the figure to the right.) How can she use a tape measure to make sure the corners are 90° (right) angles?

62. A decagon is a polygon with ten sides. What could you call a polygon with one hundred sides? With one thousand sides? With one million sides?

REVIEW

Write each number in words.

63. 254,309

64. 504,052,040

65. 82,000,415

66. 51,000,201,078

SECTION 9.7   Perimeters and Areas of Polygons

In this section, we will discuss how to find perimeters and areas of polygons. Finding perimeters is important when estimating the cost of fencing a yard or installing crown molding in a room. Finding area is important when calculating the cost of carpeting, painting a room, or fertilizing a lawn.

OBJECTIVES

1 Find the perimeter of a polygon.

2 Find the area of a polygon.

3 Find the area of figures that are combinations of polygons.

OBJECTIVE 1   **Find the perimeter of a polygon.**

The **perimeter** of a polygon is the distance around it. To find the perimeter P of a polygon, we simply add the lengths of its sides.

Triangle	Quadrilateral	Pentagon

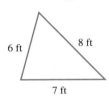

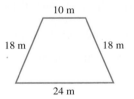

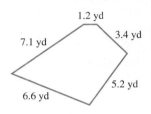

$P = 6 + 7 + 8$
$\quad = 21$

$P = 10 + 18 + 24 + 18$
$\quad = 70$

$P = 1.2 + 7.1 + 6.6 + 5.2 + 3.4$
$\quad = 23.5$

The perimeter is 21 ft.   The perimeter is 70 m.   The perimeter is 23.5 yd.

For some polygons, such as a square and a rectangle, we can simplify the computations by using a perimeter formula. Since a square has four sides of equal length s, its perimeter P is $s + s + s + s$, or $4s$.

Perimeter of a Square

If a square has a side of length s, its perimeter P is given by the formula

$$P = 4s$$

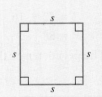

EXAMPLE 1   Find the perimeter of a square whose sides are 7.5 meters long.

Strategy We will substitute 7.5 for s in the formula $P = 4s$ and evaluate the right side.

WHY The variable P represents the unknown perimeter of the square.

Solution

$P = 4s$	This is the formula for the perimeter of a square.
$P = 4(7.5)$	Substitute 7.5 for s, the length of one side of the square.
$P = 30$	Do the multiplication.

The perimeter of the square is 30 meters.

$$\begin{array}{r} \overset{2}{7.5} \\ \times\ 4 \\ \hline 30.0 \end{array}$$

Self Check 1

A Scrabble game board has a square shape with sides of length 38.5 cm. Find the perimeter of the game board.

Now Try ➲ Problems 17 and 19

Since a rectangle has two lengths *l* and two widths *w*, its perimeter *P* is given by *l* + *w* + *l* + *w*, or 2*l* + 2*w*.

Caution! When finding the perimeter of a polygon, the lengths of the sides must be expressed in the same units.

Perimeter of a Rectangle

If a rectangle has length *l* and width *w*, its perimeter *P* is given by the formula

$$P = 2l + 2w$$

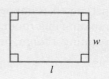

EXAMPLE 2   Find the perimeter of the rectangle shown on the right, in inches.

Strategy We will express the length of the rectangle in inches and then use the formula $P = 2l + 2w$ to find the perimeter of the figure.

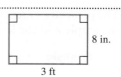

WHY We can only add quantities that are measured in the same units.

Solution Since 1 foot = 12 inches, we can convert 3 feet to inches by multiplying 3 feet by the unit conversion factor $\frac{12 \text{ in.}}{1 \text{ foot}}$.

$$3 \text{ ft} = 3 \text{ ft} \cdot \frac{12 \text{ in.}}{1 \text{ ft}} \quad \text{Multiply by 1: } \frac{12 \text{ in.}}{1 \text{ ft}} = 1.$$

$$= \frac{3 \cancel{\text{ft}}}{1} \cdot \frac{12 \text{ in.}}{1 \cancel{\text{ft}}} \quad \text{Write 3 ft as a fraction. Remove the common units of feet from the numerator and denominator. The units of inches remain.}$$

$$= 36 \text{ in.} \quad \text{Do the multiplication.}$$

The length of the rectangle is 36 inches. We can now substitute 36 for *l*, the length, and 8 for *w*, the width, in the formula for the perimeter of a rectangle.

Self Check 2

Find the perimeter of the triangle shown below, in inches.

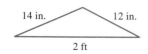

Now Try Problem 21

$$P = 2l + 2w \quad \text{This is the formula for the perimeter of a rectangle.}$$
$$P = 2(36) + 2(8) \quad \text{Substitute 36 for the length, and 8 for the width.}$$
$$P = 72 + 16 \quad \text{Do the multiplication.}$$
$$P = 88 \quad \text{Do the addition.}$$

The perimeter of the rectangle is 88 inches.

$$\begin{array}{r} \overset{1}{36} \\ \times 2 \\ \hline 72 \end{array} \qquad \begin{array}{r} 72 \\ + 16 \\ \hline 88 \end{array}$$

EXAMPLE 3   **Structural engineering.** The truss shown below is made up of three parts that form an isosceles triangle. If 76 linear feet of lumber were used to make the truss, how long is the base of the truss?

Analyze

- The truss is in the shape of an isosceles triangle.   Given
- One of the sides of equal length is 20 feet long.   Given
- The perimeter of the truss is 76 feet.   Given
- What is the length of the base of the truss?   Find

Assign We can let b equal the length of the base of the truss (in feet). At this stage, it is helpful to draw a sketch. (See the figure on the right.) If one of the sides of equal length is 20 feet long, so is the other.

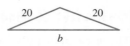

Form Because 76 linear feet of lumber were used to make the triangular-shaped truss,

The length of the base of the truss	plus	the length of one side	plus	the length of the other side	equals	the perimeter of the truss.
b	+	20	+	20	=	76

Solve

$$b + 20 + 20 = 76$$
$$b + 40 = 76 \quad \text{Combine like terms.}$$
$$b = 36 \quad \text{To isolate } b, \text{ subtract 40 from both sides.}$$

$$\begin{array}{r} 76 \\ -40 \\ \hline 36 \end{array}$$

State The length of the base of the truss is 36 ft.

Check If we add the lengths of the parts of the truss, we get 36 ft + 20 ft + 20 ft = 76 ft. The result checks.

Self Check 3

The perimeter of an isosceles triangle is 58 meters. If one of its sides of equal length is 15 meters long, how long is its base?

Now Try ➥ **Problem 25**

Using Your Calculator ▶ **Perimeters of Figures That Are Combinations of Polygons**

To find the perimeter of the figure shown below, we need to know the values of x and y. Since the figure is a combination of two rectangles, we can use a calculator to see that

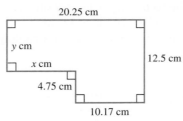

$$x = 20.25 - 10.17 \quad \text{and} \quad y = 12.5 - 4.75$$
$$x = 10.08 \text{ cm} \qquad\qquad y = 7.75 \text{ cm}$$

The perimeter P of the figure is

$$P = 20.25 + 12.5 + 10.17 + 4.75 + x + y$$
$$P = 20.25 + 12.5 + 10.17 + 4.75 + \mathbf{10.08} + \mathbf{7.75}$$

We can use a scientific calculator to make this calculation.

$$20.25 \boxed{+} 12.5 \boxed{+} 10.17 \boxed{+} 4.75 \boxed{+} 10.08 \boxed{+} 7.75 \boxed{=} \qquad \boxed{65.5}$$

The perimeter is 65.5 centimeters.

OBJECTIVE 2 Find the area of a polygon.

Caution! Do not confuse the concepts of perimeter and area. Perimeter is the distance around a polygon. It is measured in linear units, such as centimeters, feet, or miles. Area is a measure of the surface enclosed within a polygon. It is measured in square units, such as square centimeters, square feet, or square miles.

The **area** of a polygon is the measure of the amount of surface it encloses. Area is measured in square units, such as square inches or square centimeters, as shown below.

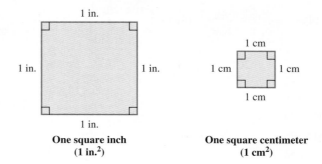

One square inch
(1 in.²)

One square centimeter
(1 cm²)

In everyday life, we often use areas. For example,

- To carpet a room, we buy square yards.
- A can of paint will cover a certain number of square feet.
- To measure vast amounts of land, we often use square miles.
- We buy house roofing by the "square." One square is 100 square feet.

The rectangle shown below has a length of 10 centimeters and a width of 3 centimeters. If we divide the rectangular region into square regions as shown in the figure, each square has an area of 1 square centimeter—a surface enclosed by a square measuring 1 centimeter on each side. Because there are 3 rows with 10 squares in each row, there are 30 squares. Since the rectangle encloses a surface area of 30 squares, its area is 30 square centimeters, which can be written as 30 cm².

This example illustrates that to find the area of a rectangle, we multiply its length by its width.

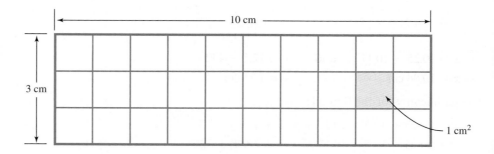

In practice, we do not find areas of polygons by counting squares. Instead, we use formulas to find areas of geometric figures.

Figure	Name	Formula for Area
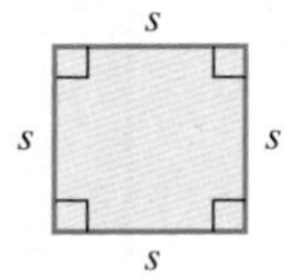	Square	$A = s^2$, where s is the length of one side.
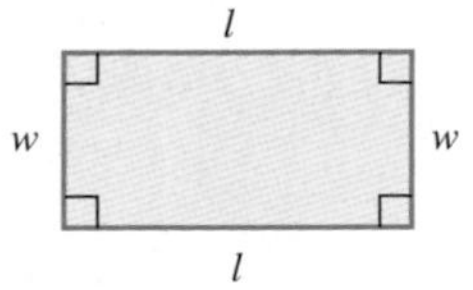	Rectangle	$A = lw$, where l is the length and w is the width.
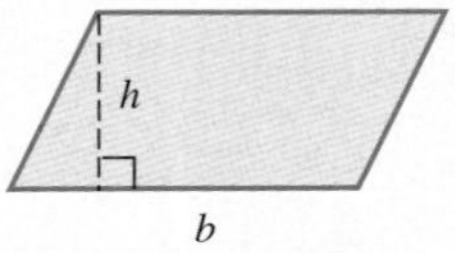	Parallelogram	$A = bh$, where b is the length of the base and h is the height. (A height is always perpendicular to the base.)
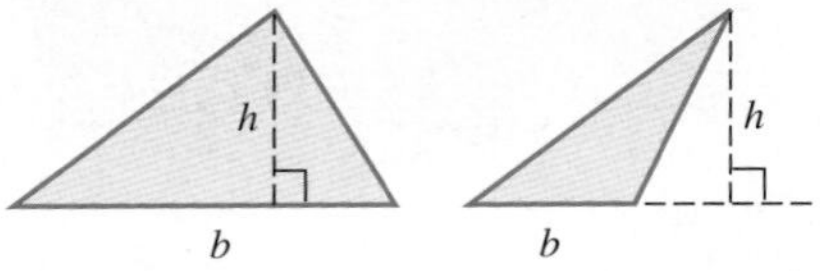	Triangle	$A = \frac{1}{2}bh$, where b is the length of the base and h is the height. The segment perpendicular to the base and representing the height (shown here using a dashed line) is called an **altitude**.
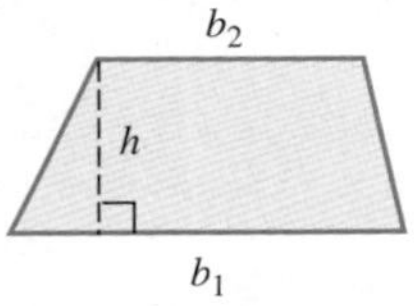	Trapezoid	$A = \frac{1}{2}h(b_1 + b_2)$, where h is the height of the trapezoid and b_1 and b_2 represent the lengths of the bases.

EXAMPLE 4   Find the area of the square shown on the right.

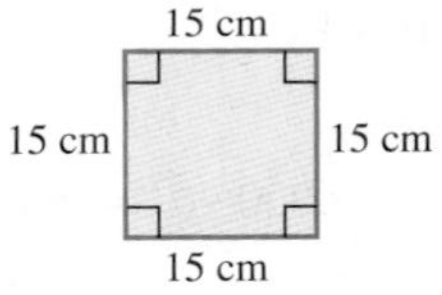

Strategy We will substitute 15 for s in the formula $A = s^2$ and evaluate the right side.

WHY The variable A represents the unknown area of the square.

Solution

$A = s^2$ This is the formula for the area of a square.

$A = 15^2$ Substitute 15 for s, the length of one side of the square.

$A = 225$ Evaluate the exponential expression.

$$\begin{array}{r} 15 \\ \times\, 15 \\ \hline 75 \\ 150 \\ \hline 225 \end{array}$$

Recall that area is measured in square units. Thus, the area of the square is 225 square centimeters, which can be written as 225 cm².

Self Check 4

Find the area of the square shown below.

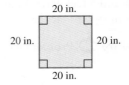

Now Try Problems 29 and 31

EXAMPLE 5   Find the number of square feet in 1 square yard.

Strategy A figure is helpful to solve this problem. We will draw a square yard and divide each of its sides into 3 equally long parts.

WHY Since a square yard is a square with each side measuring 1 yard, each side also measures 3 feet.

Self Check 5
Find the number of square centimeters in 1 square meter.

Now Try ⟶ **Problems 33 and 39**

Solution

$$1 \text{ yd}^2 = (1 \text{ yd})^2$$
$$= (3 \text{ ft})^2 \qquad \text{Substitute 3 feet for 1 yard.}$$
$$= (3 \text{ ft})(3 \text{ ft})$$
$$= 9 \text{ ft}^2$$

There are 9 square feet in 1 square yard.

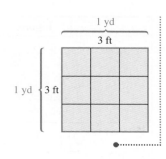

EXAMPLE 6   Women's sports. Field hockey is a team sport in which players use sticks to try to hit a ball into their opponents' goal. Find the area of the rectangular field shown on the right. Give the answer in square feet.

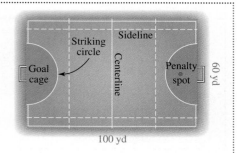

Strategy We will substitute 100 for l and 60 for w in the formula $A = lw$ and evaluate the right side.

WHY The variable A represents the unknown area of the rectangle.

Solution

$$A = lw \qquad \text{This is the formula for the area of a rectangle.}$$
$$A = 100(60) \qquad \text{Substitute 100 for } l, \text{ the length, and 60 for } w, \text{ the width.}$$
$$A = 6{,}000 \qquad \text{Do the multiplication.}$$

The area of the rectangle is 6,000 square yards. Since there are 9 square feet per square yard, we can convert this number to square feet by multiplying 6,000 square yards by $\frac{9 \text{ ft}^2}{1 \text{ yd}^2}$.

Self Check 6

Ping-Pong. A regulation-size Ping-Pong table is 9 feet long and 5 feet wide. Find its area in square inches.

Now Try ⟶ **Problem 41**

$$6{,}000 \text{ yd}^2 = 6{,}000 \text{ yd}^2 \cdot \frac{9 \text{ ft}^2}{1 \text{ yd}^2} \qquad \text{Multiply by the unit conversion factor: } \frac{9 \text{ ft}^2}{1 \text{ yd}^2} = 1.$$
$$= 6{,}000 \cdot 9 \text{ ft}^2 \qquad \text{Remove the common units of square yards in the numerator and denominator. The units of ft}^2 \text{ remain.}$$
$$= 54{,}000 \text{ ft}^2 \qquad \text{Multiply: } 6{,}000 \cdot 9 = 54{,}000.$$

The area of the field is 54,000 ft².

Think it Through • DORM ROOMS

"The United States has more than 4,000 colleges and universities, with 2.3 million students living in college dorms."
—*The New York Times, 2007*

The average dormitory room in a residence hall has about 180 square feet of floor space. The rooms are usually furnished with the following items having the given dimensions:

- 2 extra-long twin beds (each is 39 in. wide × 80 in. long × 24 in. high)
- 2 dressers (each is 18 in. wide × 36 in. long × 48 in. high)
- 2 bookcases (each is 12 in. wide × 24 in. long × 40 in. high)
- 2 desks (each is 24 in. wide × 48 in. long × 28 in. high)

How many square feet of floor space are left?

EXAMPLE 7   Find the area of the triangle shown on the right.

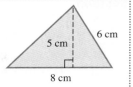

Strategy We will substitute 8 for b and 5 for h in the formula $A = \frac{1}{2}bh$ and evaluate the right side. (The side having length 6 cm is additional information that is not used to find the area.)

WHY The variable A represents the unknown area of the triangle.

Solution

$A = \dfrac{1}{2}bh$ This is the formula for the area of a triangle.

$A = \dfrac{1}{2}(8)(5)$ Substitute 8 for b, the length of the base, and 5 for h, the height.

$A = 4(5)$ Do the first multiplication: $\frac{1}{2}(8) = 4$.

$A = 20$ Complete the multiplication.

The area of the triangle is 20 cm².

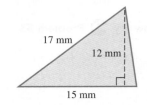

Self Check 7

Find the area of the triangle shown below.

Now Try ➡ Problem 45

EXAMPLE 8   Find the area of the triangle shown on the right.

Strategy We will substitute 9 for b and 13 for h in the formula $A = \frac{1}{2}bh$ and evaluate the right side. (The side having length 15 cm is additional information that is not used to find the area.)

WHY The variable A represents the unknown area of the triangle.

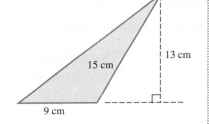

Solution In this case, the altitude falls outside the triangle.

$A = \dfrac{1}{2}bh$ This is the formula for the area of a triangle.

$A = \dfrac{1}{2}(9)(13)$ Substitute 9 for b, the length of the base, and 13 for h, the height.

$A = \dfrac{1}{2}\left(\dfrac{9}{1}\right)\left(\dfrac{13}{1}\right)$ Write 9 as $\frac{9}{1}$ and 13 as $\frac{13}{1}$.

$A = \dfrac{117}{2}$ Multiply the fractions.

$A = 58.5$ Do the division.

The area of the triangle is 58.5 cm².

$$\begin{array}{r} \overset{2}{13} \\ \times 9 \\ \hline 117 \end{array} \qquad \begin{array}{r} 58.5 \\ 2\overline{)117.0} \\ -10 \\ \hline 17 \\ -16 \\ \hline 1\,0 \\ -1\,0 \\ \hline 0 \end{array}$$

Self Check 8

Find the area of the triangle shown below.

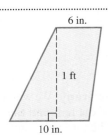

Now Try ➡ Problem 49

EXAMPLE 9   Find the area of the trapezoid shown on the right.

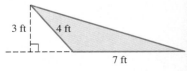

Strategy We will express the height of the trapezoid in inches and then use the formula $A = \frac{1}{2}h(b_1 + b_2)$ to find the area of the figure.

WHY The height of 1 foot must be expressed as 12 inches to be consistent with the units of the bases.

Find the area of the trapezoid shown below.

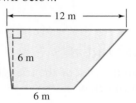

Now Try → **Problem 53**

Solution

$$A = \frac{1}{2}h(b_1 + b_2)$$ This is the formula for the area of a trapezoid.

$$A = \frac{1}{2}(12)(10 + 6)$$ Substitute 12 for h, the height; 10 for b_1, the length of the lower base; and 6 for b_2, the length of the upper base.

$$A = \frac{1}{2}(12)(16)$$ Do the addition within the parentheses.

$$A = 6(16)$$ Do the first multiplication: $\frac{1}{2}(12) = 6$.

$$A = 96$$ Complete the multiplication.

The area of the trapezoid is 96 in².

$$\begin{array}{r} \overset{3}{16} \\ \times 6 \\ \hline 96 \end{array}$$

EXAMPLE 10   The area of the parallelogram shown on the right is 360 ft². Find the height.

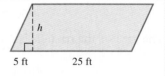

Strategy To find the height of the parallelogram, we will substitute the given values in the formula $A = bh$ and solve for h.

WHY The variable h represents the unknown height.

The area of the parallelogram below is 96 cm². Find its height.

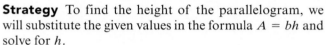

Now Try → **Problem 57**

Solution From the figure, we see that the length of the base of the parallelogram is

$$5 \text{ feet} + 25 \text{ feet} = 30 \text{ feet}$$

$$A = bh$$ This is the formula for the area of a parallelogram.

$$360 = 30h$$ Substitute 360 for A, the area, and 30 for b, the length of the base.

$$\frac{360}{30} = \frac{30h}{30}$$ To isolate h, undo the multiplication by 30 by dividing both sides by 30.

$$12 = h$$ Do the division.

The height of the parallelogram is 12 feet.

$$\begin{array}{r} 12 \\ 30\overline{)360} \\ -30 \\ \hline 60 \\ 60 \\ \hline 0 \end{array}$$

OBJECTIVE 3 **Find the area of figures that are combinations of polygons.**

To find the area of an irregular shape, break up the shape into familiar polygons. Find the area of each polygon and then add the results.

EXAMPLE 11   Find the area of one side of the tent shown below.

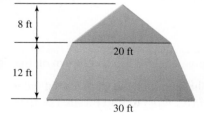

Strategy We will use the formula $A = \frac{1}{2}h(b_1 + b_2)$ to find the area of the lower portion of the tent and the formula $A = \frac{1}{2}bh$ to find the area of the upper portion of the tent. Then we will combine the results.

WHY The side of the tent is a combination of a trapezoid and a triangle.

Solution To find the area of the lower portion of the tent, we proceed as follows.

$A_{\text{trap.}} = \dfrac{1}{2}h(b_1 + b_2)$ This is the formula for the area of a trapezoid.

$A_{\text{trap.}} = \dfrac{1}{2}(12)(30 + 20)$ Substitute 30 for b_1, 20 for b_2, and 12 for h.

$A_{\text{trap.}} = \dfrac{1}{2}(12)(50)$ Do the addition within the parentheses.

$A_{\text{trap.}} = 6(50)$ Do the first multiplication: $\frac{1}{2}(12) = 6$.

$A_{\text{trap.}} = 300$ Complete the multiplication.

The area of the trapezoid is 300 ft^2.
To find the area of the upper portion of the tent, we proceed as follows.

$A_{\text{triangle}} = \dfrac{1}{2}bh$ This is the formula for the area of a triangle.

$A_{\text{triangle}} = \dfrac{1}{2}(20)(8)$ Substitute 20 for b and 8 for h.

$A_{\text{triangle}} = 80$ Do the multiplications, working from left to right: $\frac{1}{2}(20) = 10$ and then $10(8) = 80$.

The area of the triangle is 80 ft^2.
To find the total area of one side of the tent, we add:

$A_{\text{total}} = A_{\text{trap.}} + A_{\text{triangle}}$

$A_{\text{total}} = 300 \text{ ft}^2 + 80 \text{ ft}^2$

$A_{\text{total}} = 380 \text{ ft}^2$

The total area of one side of the tent is 380 ft^2.

Self Check 11

Find the area of the shaded figure below.

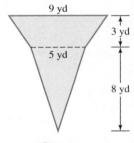

Now Try ➲ **Problem 65**

EXAMPLE 12 Find the area of the shaded region shown on the right.

Strategy We will subtract the unwanted area of the square from the area of the rectangle.

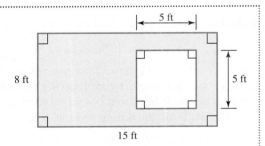

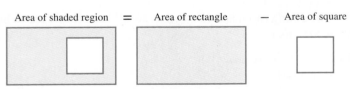

Area of shaded region = Area of rectangle — Area of square

WHY The area of the rectangular-shaped shaded figure does not include the square region inside of it.

Find the area of the shaded region shown below.

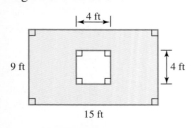

Solution

$$A_{\text{shaded}} = lw - s^2 \qquad \text{The formula for the area of a rectangle is } A = lw.$$
$$\text{The formula for the area of a square is } A = s^2.$$

$$A_{\text{shaded}} = 15(8) - 5^2 \qquad \text{Substitute 15 for the length } l \text{ and 8 for the}$$
$$\text{width } w \text{ of the rectangle. Substitute 5 for}$$
$$\text{the length } s \text{ of a side of the square.}$$

$$A_{\text{shaded}} = 120 - 25$$
$$A_{\text{shaded}} = 95$$

The area of the shaded region is 95 ft².

$$\begin{array}{r} \overset{4}{15} \\ \times 8 \\ \hline 120 \end{array}$$

$$\begin{array}{r} \overset{11}{\cancel{1}\cancel{2}\cancel{0}} \\ -25 \\ \hline 95 \end{array}$$

Now Try Problem 69

EXAMPLE 13 **Carpeting a room.** A living room/dining room has the floor plan shown in the figure. If carpet costs $29 per square yard, including pad and installation, how much will it cost to carpet both rooms? (Assume no waste.)

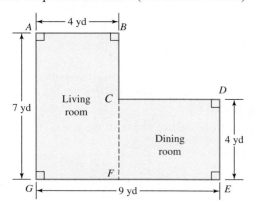

Strategy We will find the number of square yards of carpeting needed and multiply the result by $29.

WHY Each square yard costs $29.

Solution First, we must find the total area of the living room and the dining room:

$$A_{\text{total}} = A_{\text{living room}} + A_{\text{dining room}}$$

Since $\overline{CF}$ divides the space into two rectangles, the areas of the living room and the dining room are found by multiplying their respective lengths and widths. Therefore, the area of the living room is 4 yd · 7 yd = **28 yd²**.

The width of the dining room is given as 4 yd. To find its length, we subtract:

$$\text{m}(\overline{CD}) = \text{m}(\overline{GE}) - \text{m}(\overline{AB}) = 9 \text{ yd} - 4 \text{ yd} = 5 \text{ yd}$$

Thus, the area of the dining room is 5 yd · 4 yd = **20 yd²**. The total area to be carpeted is the sum of these two areas.

$$A_{\text{total}} = A_{\text{living room}} + A_{\text{dining room}}$$
$$A_{\text{total}} = 28 \text{ yd}^2 + 20 \text{ yd}^2$$
$$A_{\text{total}} = 48 \text{ yd}^2$$

$$\begin{array}{r} 48 \\ \times 29 \\ \hline 432 \\ 960 \\ \hline 1{,}392 \end{array}$$

Now Try Problem 73

At $29 per square yard, the cost to carpet both rooms will be 48 · $29, or $1,392.

1. 154 cm   **2.** 50 in.   **3.** 28 m   **4.** 400 in.²   **5.** 10,000 cm²   **6.** 6,480 in.²   **7.** 90 mm²
8. 10.5 ft²   **9.** 54 m²   **10.** 8 cm   **11.** 41 yd²   **12.** 119 ft²

SECTION 9.7   STUDY SET

VOCABULARY

Fill in the blanks.

1. The distance around a polygon is called the _____.

2. The _____ of a polygon is measured in linear units such as inches, feet, and miles.

3. The measure of the surface enclosed by a polygon is called its _____.

4. If each side of a square measures 1 foot, the area enclosed by the square is 1 _____ foot.

5. The _____ of a polygon is measured in square units.

6. The segment that represents the height of a triangle is called an _____.

CONCEPTS

7. The figure below shows a kitchen floor that is covered with 1-foot-square tiles. Without counting *all* of the squares, determine the area of the floor.

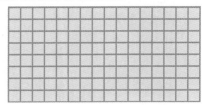

8. Tell which concept applies, perimeter or area.
   a. The length of a walk around New York's Central Park
   b. The amount of office floor space in the White House
   c. The amount of fence needed to enclose a playground
   d. The amount of land in Yellowstone National Park

9. Give the formula for the perimeter of a
   a. square   b. rectangle

10. Give the formula for the area of a
    a. square   b. rectangle
    c. triangle   d. trapezoid
    e. parallelogram

11. For each figure below, draw the altitude to the base *b*.
    a.   b.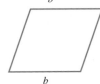

    c. 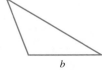   d.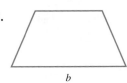

12. For each figure below, label the base *b* for the given altitude.
    a. 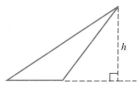   b.

    c.   d.

13. The shaded figure below is a combination of what two types of geometric figures?

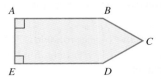

14. Explain how you would find the area of the following shaded figure.

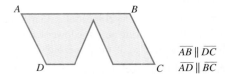

$\overline{AB} \parallel \overline{DC}$
$\overline{AD} \parallel \overline{BC}$

NOTATION

Fill in the blanks.

15. a. The symbol 1 in.² means one _____ _____.
    b. One square meter is expressed as _____.

16. In the figure below, the symbol ⌐ indicates that the dashed line segment, called an *altitude*, is _____ to the base.

GUIDED PRACTICE

Find the perimeter of each square. See Example 1.

17. 8 in.
    8 in.   8 in.
    8 in.

18. 93 in.
    93 in.   93 in.
    93 in.

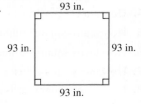

19. A square with sides 5.75 miles long

20. A square with sides 3.4 yards long

Find the perimeter of each rectangle, in inches. **See Example 2.**

21. 2 ft / 7 in.

22. 6 ft / 2 in.

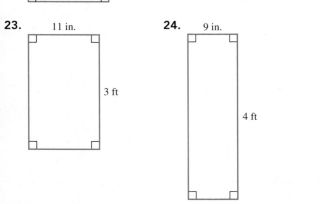

23. 11 in. / 3 ft

24. 9 in. / 4 ft

Write and then solve an equation to answer each problem. **See Example 3.**

25. The perimeter of an isosceles triangle is 35 feet. One of the sides of equal length is 10 feet long. Find the length of the base of the triangle.

26. The perimeter of an isosceles triangle is 94 feet. One of the sides of equal length is 42 feet long. Find the length of the base of the triangle.

27. The perimeter of an isosceles trapezoid is 35 meters. The upper base is 10 meters long, and the lower base is 15 meters long. How long is each leg of the trapezoid?

28. The perimeter of an isosceles trapezoid is 46 inches. The upper base is 12 inches long, and the lower base is 16 inches long. How long is each leg of the trapezoid?

Find the area of each square. **See Example 4.**

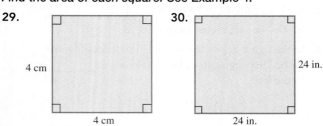

29. 4 cm / 4 cm

30. 24 in. / 24 in.

31. A square with sides 2.5 meters long

32. A square with sides 6.8 feet long

For Problems 33–40, **see Example 5.**

33. How many square inches are in 1 square foot?

34. How many square inches are in 1 square yard?

35. How many square millimeters are in 1 square meter?

36. How many square decimeters are in 1 square meter?

37. How many square feet are in 1 square mile?

38. How many square yards are in 1 square mile?

39. How many square meters are in 1 square kilometer?

40. How many square dekameters are in 1 square kilometer?

Find the area of each rectangle. Give the answer in square feet. **See Example 6.**

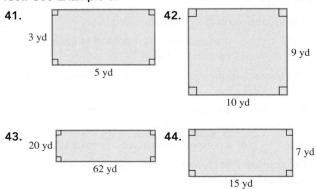

41. 3 yd / 5 yd

42. 9 yd / 10 yd

43. 20 yd / 62 yd

44. 7 yd / 15 yd

Find the area of each triangle. **See Example 7.**

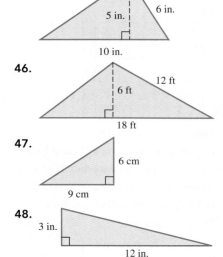

45. 5 in. / 6 in. / 10 in.

46. 12 ft / 6 ft / 18 ft

47. 6 cm / 9 cm

48. 3 in. / 12 in.

Find the area of each triangle. **See Example 8.**

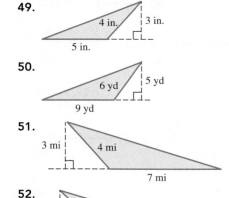

49. 4 in. / 3 in. / 5 in.

50. 6 yd / 5 yd / 9 yd

51. 3 mi / 4 mi / 7 mi

52. 5 ft / 7 ft / 11 ft

Find the area of each trapezoid. **See Example 9.**

53.

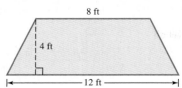

54.

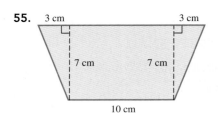

55.

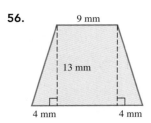

56.

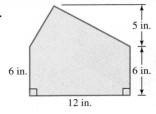

Solve each problem. **See Example 10.**

57. The area of a parallelogram is 60 m², and its height is 15 m. Find the length of its base.

58. The area of a parallelogram is 95 in.², and its height is 5 in. Find the length of its base.

59. The area of a rectangle is 36 cm², and its length is 3 cm. Find its width.

60. The area of a rectangle is 144 mi², and its length is 6 mi. Find its width.

61. The area of a triangle is 54 m², and the length of its base is 3 m. Find the height.

62. The area of a triangle is 270 ft², and the length of its base is 18 ft. Find the height.

63. The perimeter of a rectangle is 64 mi, and its length is 21 mi. Find its width.

64. The perimeter of a rectangle is 26 yd, and its length is 10.5 yd. Find its width.

Find the area of each shaded figure. **See Example 11.**

65.

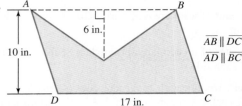

66.

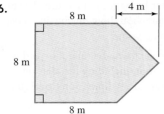

67.

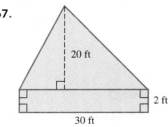

68.

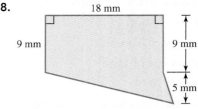

Find the area of each shaded figure. **See Example 12.**

69.

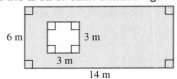

70.

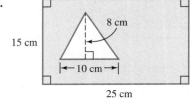

71.

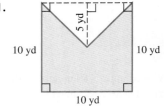

72.

$\overline{AB} \parallel \overline{DC}$
$\overline{AD} \parallel \overline{BC}$

Solve each problem. **See Example 13.**

73. Flooring. A rectangular family room is 8 yards long and 5 yards wide. At $30 per square yard, how much will it cost to put down vinyl sheet flooring in the room? (Assume no waste.)

74. Carpeting. A rectangular living room measures 10 yards by 6 yards. At $32 per square yard, how much will it cost to carpet the room? (Assume no waste.)

75. Fences. A man wants to enclose a rectangular yard with fencing that costs $12.50 a foot, including installation. Find the cost of enclosing the yard if its dimensions are 110 ft by 85 ft.

76. Frames. Find the cost of framing a rectangular picture with dimensions of 24 inches by 30 inches if framing material costs $0.75 per inch.

TRY IT YOURSELF

Sketch and label each of the figures.

77. Two different rectangles, each having a perimeter of 40 in.

78. Two different rectangles, each having an area of 40 in.²

79. A square with an area of 25 m²

80. A square with a perimeter of 20 m

81. A parallelogram with an area of 15 yd²

82. A triangle with an area of 20 ft²

83. A figure consisting of a combination of two rectangles, whose total area is 80 ft²

84. A figure consisting of a combination of a rectangle and a square, whose total area is 164 ft²

Find the area of each parallelogram.

85.

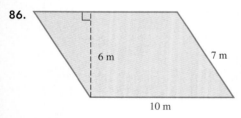

86.

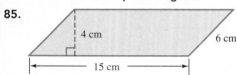

87. The perimeter of an isosceles triangle is 80 meters. If the length of one of the congruent sides is 22 meters, how long is the base?

88. The perimeter of a square is 35 yards. How long is a side of the square?

89. The perimeter of an equilateral triangle is 85 feet. Find the length of each side.

90. An isosceles triangle with congruent sides of length 49.3 inches has a perimeter of 121.7 inches. Find the length of the base.

Find the perimeter of the figure.

91.

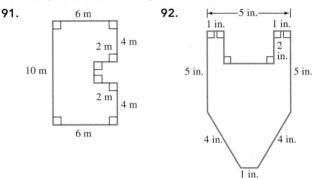

92.

93.

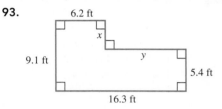

Find x and y. Then find the perimeter of the figure.

94.

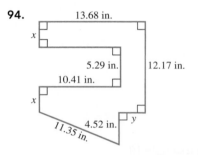

APPLICATIONS

95. Landscaping. A woman wants to plant a pine-tree screen around three sides of her rectangular-shaped backyard. (See the figure below.) If she plants the trees 3 feet apart, how many trees will she need?

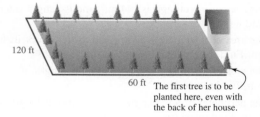

120 ft

60 ft   The first tree is to be planted here, even with the back of her house.

96. Gardening. A gardener wants to plant a border of marigolds around the garden shown below, to keep out rabbits. How many plants will she need if she allows 6 inches between plants?

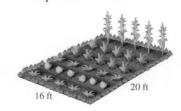

16 ft   20 ft

97. Comparison shopping. Which is more expensive: a ceramic-tile floor costing $3.75 per square foot or vinyl costing $34.95 per square yard?

98. Comparison shopping. Which is cheaper: a hardwood floor costing $6.95 per square foot or a carpeted floor costing $37.50 per square yard?

99. Tiles. A rectangular basement room measures 14 by 20 feet. Vinyl floor tiles that are 1 ft² cost $1.29 each. How much will the tile cost to cover the floor? (Assume no waste.)

100. Painting. The north wall of a barn is a rectangle 23 feet high and 72 feet long. There are five windows in the wall, each 4 by 6 feet. If a gallon of paint will cover 300 ft², how many gallons of paint must the painter buy to paint the wall?

101. Sails. If nylon is $12 per square yard, how much would the fabric cost to make a triangular sail with a base of 12 feet and a height of 24 feet?

102. Remodeling. The gable end of a house is an isosceles triangle with a height of 4 yards and a base of 23 yards. It will require one coat of primer and one coat of finish to paint the triangle. Primer costs $17 per gallon, and the finish paint costs $23 per gallon. If one gallon of each type of paint covers 300 square feet, how much will it cost to paint the gable, excluding labor?

103. Geography. Use the dimensions of the trapezoid that is superimposed over the state of Nevada to estimate the area of the "Silver State."

104. Solar covers. A swimming pool has the shape shown below. How many square feet of a solar blanket material will be needed to cover the pool? How much will the cover cost if it is $1.95 per square foot? (Assume no waste.)

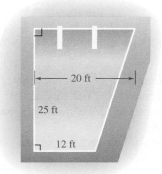

105. Carpentry. How many sheets of 4-foot-by-8-foot sheetrock are needed to drywall the inside walls on the first floor of the barn shown below? (Assume that the carpenters will cover each wall entirely before cutting out areas for the doors.)

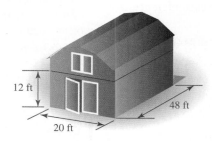

106. Carpentry. If it costs $90 per square foot to build a one-story home in northern Wisconsin, find the cost of building the house with the floor plan shown below.

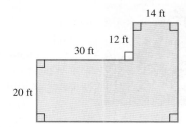

WRITING

107. Explain the difference between perimeter and area.

108. Why is it necessary that area be measured in square units?

109. A student expressed the area of the square in the figure below as 25^2 ft. Explain his error.

110. Refer to the figure below. What must be done before we can use the formula to find the area of this rectangle?

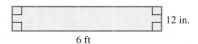

REVIEW

Simplify each expression.

111. $8\left(\dfrac{3}{4}t\right)$

112. $27\left(\dfrac{2}{3}m\right)$

113. $-\dfrac{2}{3}(3w - 6)$

114. $(2y - 8)\dfrac{1}{2}$

115. $-\dfrac{7}{16}x - \dfrac{3}{16}x$

116. $-\dfrac{5}{18}x - \dfrac{7}{18}x$

117. $60\left(\dfrac{3}{20}r - \dfrac{4}{15}\right)$

118. $72\left(\dfrac{7}{8}f - \dfrac{8}{9}\right)$

OBJECTIVES

1 Define circle, radius, chord, diameter, and arc.

2 Find the circumference of a circle.

3 Find the area of a circle.

In this section, we will discuss the circle, one of the most useful geometric figures of all. In fact, the discoveries of fire and the circular wheel are two of the most important events in the history of the human race. We will begin our study by introducing some basic vocabulary associated with circles.

OBJECTIVE 1 Define circle, radius, chord, diameter, and arc.

> **Circle**
>
> A **circle** is the set of all points in a plane that lie a fixed distance from a point called its **center**.

A segment drawn from the center of a circle to a point on the circle is called a **radius**. (The plural of *radius* is *radii*.) From the definition, it follows that all radii of the same circle are the same length.

A **chord** of a circle is a line segment that connects two points on the circle. A **diameter** is a chord that passes through the center of the circle. Since a diameter D of a circle is twice as long as a radius r, we have

$$D = 2r$$

Each of the previous definitions is illustrated in figure (a) below, in which O is the center of the circle.

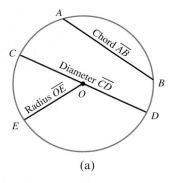   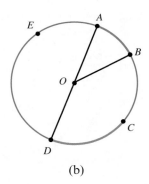

(a)   (b)

Any part of a circle is called an **arc**. In figure (b) above, the part of the circle from point A to point B that is highlighted in blue is $\overparen{AB}$, read as "arc AB." $\overparen{CD}$ is the part of the circle from point C to point D that is highlighted in green. An arc that is half of a circle is a **semicircle**.

> **Semicircle**
>
> A **semicircle** is an arc of a circle whose endpoints are the endpoints of a diameter.

Success Tip It is often possible to name a major arc in more than one way. For example, in figure (b), major arc $\overparen{ABE}$ is the part of the circle from point A to point E that includes point B. Two other names for the same major arc are $\overparen{ACE}$ and $\overparen{ADE}$.

If point O is the center of the circle in figure (b), $\overline{AD}$ is a diameter and $\overparen{AED}$ is a semicircle. The middle letter E distinguishes semicircle $\overparen{AED}$ (the part of the circle from point A to point D that includes point E) from semicircle $\overparen{ABD}$ (the part of the circle from point A to point D that includes point B).

An arc that is shorter than a semicircle is a **minor arc**. An arc that is longer than a semicircle is a **major arc**. In figure (b),

$\overparen{AE}$ is a minor arc   and $\overparen{ABE}$ is a major arc.

OBJECTIVE 2 Find the circumference of a circle.

Since early history, mathematicians have known that the ratio of the distance around a circle (the circumference) to the length of its diameter is approximately 3. First Kings, Chapter 7, of the Bible describes a round bronze tank that was 15 feet from brim to brim and 45 feet in circumference, and $\frac{45}{15} = 3$. Today, we use a more precise value for this ratio, known as π (pi). If C is the circumference of a circle and D is the length of its diameter, then

$$\pi = \frac{C}{D} \quad \text{where } \pi = 3.141592653589\ldots \quad \tfrac{22}{7} \text{ and } 3.14 \text{ are often used as estimates of } \pi.$$

If we multiply both sides of $\pi = \frac{C}{D}$ by D, we have the following formula.

Circumference of a Circle

The circumference of a circle is given by the formula

$C = \pi D$ where C is the circumference and D is the length of the diameter

---

Since a diameter of a circle is twice as long as a radius r, we can substitute $2r$ for D in the formula $C = \pi D$ to obtain another formula for the circumference C:

$C = 2\pi r$ The notation $2\pi r$ means $2 \cdot \pi \cdot r$.

EXAMPLE 1   Find the circumference of the circle shown on the right. Give the exact answer and an approximation.

Strategy We will substitute 5 for r in the formula $C = 2\pi r$ and evaluate the right side.

WHY The variable C represents the unknown circumference of the circle.

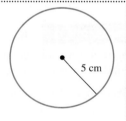
5 cm

Solution

$C = 2\pi r$ This is the formula for the circumference of a circle.

$C = 2\pi(\mathbf{5})$ Substitute 5 for r, the radius.

$C = 2(5)\pi$ When a product involves π, we usually rewrite it so that π is the last factor.

$C = 10\pi$ Do the first multiplication: $2(5) = 10$. This is the exact answer.

The circumference of the circle is exactly 10π cm. If we replace π with 3.14, we get an approximation of the circumference.

$C = 10\pi$

$C \approx 10(\mathbf{3.14})$ To multiply by 10, move the decimal point in 3.14 one place to the right.

$C \approx 31.4$

The circumference of the circle is approximately 31.4 cm.

Self Check 1

Find the circumference of the circle shown below. Give the exact answer and an approximation.

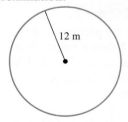
12 m

Now Try Problem 25

Using Your Calculator ▶   **Calculating Revolutions of a Tire**

When the $\boxed{\pi}$ key on a scientific calculator is pressed (on some models, the $\boxed{\text{2nd}}$ key must be pressed first), an approximation of π is displayed. To illustrate how to use this key, consider the following problem. How many times does the tire shown to the right revolve when a car makes a 25-mile trip?

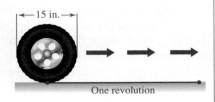
One revolution

We first find the circumference of the tire. From the figure, we see that the diameter of the tire is 15 inches. Since the circumference of a circle is the product of π and the length of its diameter, the tire's circumference is $\pi \cdot 15$ inches, or 15π inches. (Normally, we rewrite a product such as $\pi \cdot 15$ so that π is the second factor.)

We then change the 25 miles to inches using two unit conversion factors.

$$\frac{25 \text{ miles}}{1} \cdot \frac{5{,}280 \text{ feet}}{1 \text{ mile}} \cdot \frac{12 \text{ inches}}{1 \text{ foot}} = 25 \cdot 5{,}280 \cdot 12 \text{ inches}$$ The units of miles and feet can be removed.

The length of the trip is $25 \cdot 5{,}280 \cdot 12$ inches.

Finally, we divide the length of the trip by the circumference of the tire to get

$$\text{The number of revolutions of the tire} = \frac{25 \cdot 5{,}280 \cdot 12}{15\pi}$$

We can use a scientific calculator to make this calculation.

$\boxed{(}\ 25\ \boxed{\times}\ 5280\ \boxed{\times}\ 12\ \boxed{)}\ \boxed{\div}\ \boxed{(}\ 15\ \boxed{\times}\ \boxed{\pi}\ \boxed{)}\ \boxed{=}$ $\boxed{\text{33613.52398}}$

The tire makes about 33,614 revolutions.

EXAMPLE 2   **Architecture.**   A Norman window is constructed by adding a semicircular window to the top of a rectangular window. Find the perimeter of the Norman window shown here.

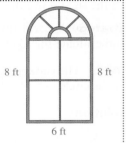
8 ft   8 ft

6 ft

Strategy We will find the perimeter of the rectangular part and the circumference of the circular part of the window and add the results.

 WHY The window is a combination of a rectangle and a semicircle.

Solution The perimeter of the rectangular part is

$$P_{\text{rectangular part}} = 8 + 6 + 8 = 22$$ Add only 3 sides of the rectangle.

The perimeter of the semicircle is one-half of the circumference of a circle that has a 6-foot diameter.

$$P_{\text{semicircle}} = \frac{1}{2}C$$ This is the formula for the circumference of a semicircle.

$$P_{\text{semicircle}} = \frac{1}{2}\pi D$$ Since we know the diameter, replace C with πD. We could also have replaced C with $2\pi r$.

$$P_{\text{semicircle}} = \frac{1}{2}\pi(6)$$ Substitute 6 for D, the diameter.

$$P_{\text{semicircle}} \approx 9.424777961$$ Use a calculator to do the multiplication.

The total perimeter is the sum of the two parts.

$$P_{\text{total}} = P_{\text{rectangular part}} + P_{\text{semicircle}}$$
$$P_{\text{total}} \approx 22 + 9.424777961$$
$$P_{\text{total}} \approx 31.424777961$$

To the nearest hundredth, the perimeter of the window is 31.42 feet.

Self Check 2

Find the perimeter of the figure shown below. Round to the nearest hundredth. (Assume the arc is a semicircle.)

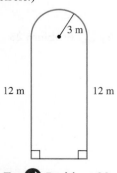
3 m

12 m   12 m

Now Try ⟳ **Problem 29**

OBJECTIVE 3 Find the area of a circle.

If we divide the circle shown in figure (a) below into an even number of pie-shaped pieces and then rearrange them as shown in figure (b), we have a figure that looks like a parallelogram. The figure has a base b that is one-half the circumference of the circle, and its height h is about the same length as a radius of the circle.

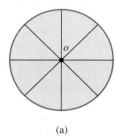
(a)

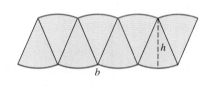
(b)

If we divide the circle into more and more pie-shaped pieces, the figure will look more and more like a parallelogram, and we can find its area by using the formula for the area of a parallelogram.

$A = bh$

$A = \dfrac{1}{2}Cr$ Substitute $\frac{1}{2}$ of the circumference for b, the length of the base of the "parallelogram." Substitute r for the height of the "parallelogram."

$A = \dfrac{1}{2}(2\pi r)r$ Substitute $2\pi r$ for C.

$A = \pi r^2$ Simplify: $\frac{1}{2} \cdot 2 = 1$ and $r \cdot r = r^2$.

This result gives the following formula.

Area of a Circle

The area A of a circle with radius r is given by the formula.

$A = \pi r^2$

EXAMPLE 3   Find the area of the circle shown on the right. Give the exact answer and an approximation to the nearest tenth.

Strategy We will find the radius of the circle, substitute that value for r in the formula $A = \pi r^2$, and evaluate the right side.

WHY The variable A represents the unknown area of the circle.

Solution Since the length of the diameter is 10 centimeters and the length of a diameter is twice the length of a radius, the length of the radius is 5 centimeters.

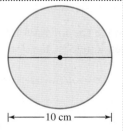
10 cm

$A = \pi r^2$ This is the formula for the area of a circle.
$A = \pi(5)^2$ Substitute 5 for r, the radius of the circle. The notation πr^2 means $\pi \cdot r^2$.
$A = \pi(25)$ Evaluate the exponential expression.
$A = 25\pi$ Write the product so that π is the last factor.

The exact area of the circle is 25π cm². We can use a calculator to approximate the area.

$A \approx 78.53981634$ Use a calculator to do the multiplication: $25 \cdot \pi$.

To the nearest tenth, the area is 78.5 cm².

Self Check 3

Find the area of a circle with a diameter of 12 feet. Give the exact answer and an approximation to the nearest tenth.

Now Try Problem 33

Using Your Calculator ▶   **Painting a Helicopter Landing Pad**

Orange paint is available in gallon containers at $39 each, and each gallon will cover 375 ft². To calculate how much the paint will cost to cover a circular helicopter landing pad 60 feet in diameter, we first calculate the area of the helicopter pad.

$A = \pi r^2$ *This is the formula for the area of a circle.*

$A = \pi(30)^2$ *Substitute one-half of 60 for r, the radius of the circular pad.*

$A = 30^2\pi$ *Write the product so that π is the last factor.*

The area of the pad is exactly $30^2\pi$ ft². Since each gallon of paint will cover 375 ft², we can find the number of gallons of paint needed by dividing $30^2\pi$ by 375.

$$\text{Number of gallons needed} = \frac{30^2\pi}{375}$$

We can use a scientific calculator to make this calculation.

30 $\boxed{x^2}$ $\boxed{\times}$ $\boxed{\pi}$ $\boxed{=}$ $\boxed{\div}$ 375 $\boxed{=}$ $\boxed{7.539822369}$

Because paint comes only in full gallons, the painter will need to purchase 8 gallons. The cost of the paint will be 8 · $39, or $312.

EXAMPLE 4   Find the area of the shaded figure on the right. Round to the nearest hundredth.

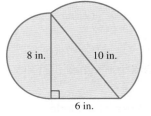

Strategy We will find the area of the entire shaded figure using the following approach:

$$A_{\text{total}} = A_{\text{triangle}} + A_{\text{smaller semicircle}} + A_{\text{larger semicircle}}$$

WHY The shaded figure is a combination of a triangular region and two semicircular regions.

Solution The area of the triangle is

$$A_{\text{triangle}} = \frac{1}{2}bh = \frac{1}{2}(6)(8) = \frac{1}{2}(48) = 24$$

Since the formula for the area of a circle is $A = \pi r^2$, the formula for the area of a semicircle is $A = \frac{1}{2}\pi r^2$. Thus, the area enclosed by the smaller semicircle is

$$A_{\text{smaller semicircle}} = \frac{1}{2}\pi r^2 = \frac{1}{2}\pi(4)^2 = \frac{1}{2}\pi(16) = 8\pi$$

The area enclosed by the larger semicircle is

$$A_{\text{larger semicircle}} = \frac{1}{2}\pi r^2 = \frac{1}{2}\pi(5)^2 = \frac{1}{2}\pi(25) = 12.5\pi$$

The total area is the sum of the three results:

$$A_{\text{total}} = 24 + 8\pi + 12.5\pi \approx 88.4026494 \quad \text{Use a calculator to perform the operations.}$$

$$\begin{array}{r} 12.5 \\ 2\overline{)25.0} \\ -2 \\ \hline 05 \\ -4 \\ \hline 1\,0 \\ -1\,0 \\ \hline 0 \end{array}$$

To the nearest hundredth, the area of the shaded figure is 88.40 in.².

Self Check 4

Find the area of the shaded figure below. Round to the nearest hundredth.

[figure: shaded figure with 10 yd, 26 yd, 24 yd labels]

10 yd

26 yd

24 yd

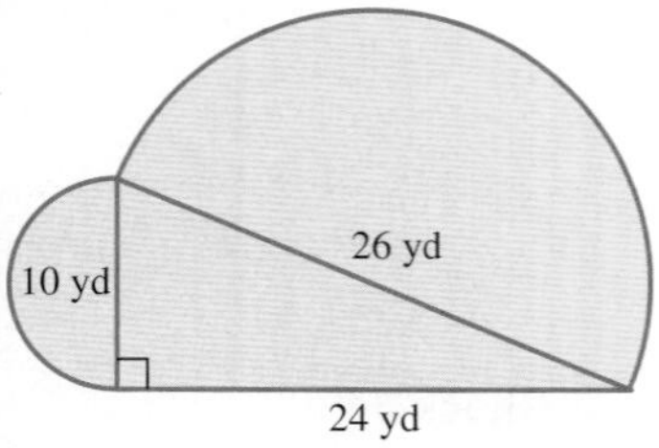

Now Try ⟳ **Problem 37**

Answers to Self Checks

   1. 24π m ≈ 75.4 m   **2.** 39.42 m   **3.** 36π ft² ≈ 113.1 ft²   **4.** 424.73 yd²

SECTION 9.8   STUDY SET

VOCABULARY

Fill in the blanks.

1. A segment drawn from the center of a circle to a point on the circle is called a _____ .

2. A segment joining two points on a circle is called a _____ .

3. A _____ is a chord that passes through the center of a circle.

4. An arc that is one-half of a complete circle is a _____ .

5. The distance around a circle is called its _____ .

6. The surface enclosed by a circle is called its _____ .

7. A diameter of a circle is _____ as long as a radius.

8. Suppose the exact circumference of a circle is 3π feet. When we write $C \approx 9.42$ feet, we are giving an _____ of the circumference.

CONCEPTS

Refer to the figure below, where point 0 is the center of the circle.

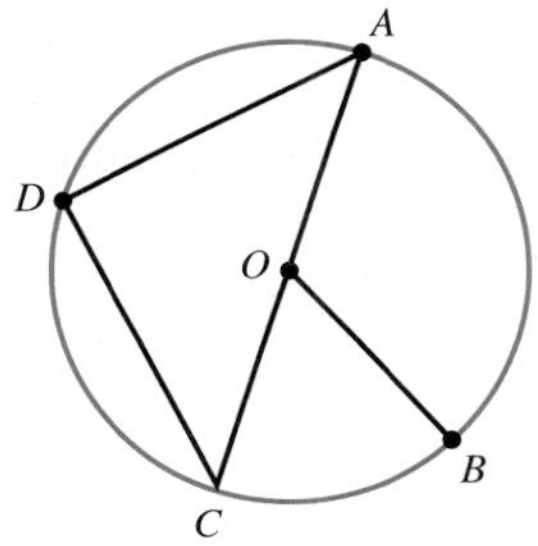

9. Name each radius.

10. Name a diameter.

11. Name each chord.

12. Name each minor arc.

13. Name each semicircle.

14. Name major arc $\overset{\frown}{ABD}$ in another way.

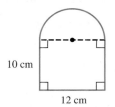

15. **a.** If you know the radius of a circle, how can you find its diameter?

    b. If you know the diameter of a circle, how can you find its radius?

16. **a.** What are the two formulas that can be used to find the circumference of a circle?

    b. What is the formula for the area of a circle?

17. If C is the circumference of a circle and D is its diameter, then $\dfrac{C}{D} =$ ▢ .

18. If D is the diameter of a circle and r is its radius, then $D =$ ▢ r.

19. When evaluating $\pi(6)^2$, what operation should be performed first?

20. Round $\pi = 3.141592653589 \ldots$ to the nearest hundredth.

NOTATION

Fill in the blanks.

21. The symbol $\overset{\frown}{AB}$ is read as "_____ _____."

22. To the nearest hundredth, the value of π is _____ .

23. **a.** In the expression $2\pi r$, what operations are indicated?

    b. In the expression πr^2, what operations are indicated?

24. Write each expression in better form. Leave π in your answer.

    a. $\pi(8)$ **b.** $2\pi(7)$ **c.** $\pi \cdot \dfrac{25}{3}$

GUIDED PRACTICE

The answers to the problems in this Study Set may vary slightly, depending on which approximation of π is used.

Find the circumference of the circle shown below. Give the exact answer and an approximation to the nearest tenth. See Example 1.

25.

4 ft

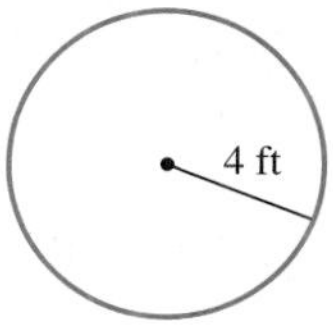

26.

8 in.

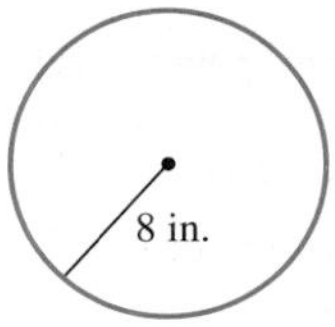

27.

6 m

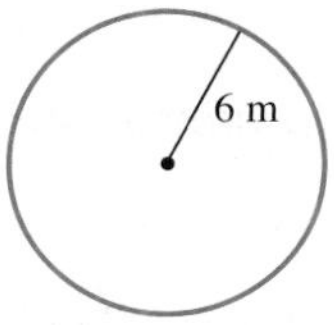

28.

10 mm

Find the perimeter of each figure. Assume each arc is a semicircle. Round to the nearest hundredth. See Example 2.

29.

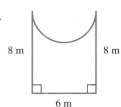
10 cm

12 cm

30.

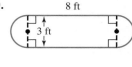
8 ft

3 ft

31.

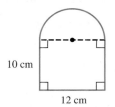
8 m   8 m

6 m

32.

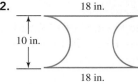
18 in.

10 in.

18 in.

Find the area of each circle given the following information. Give the exact answer and an approximation to the nearest tenth. See Example 3.

33.

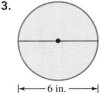
6 in.

34.

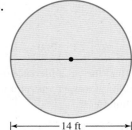
14 ft

35. Find the area of a circle with diameter 18 inches.

36. Find the area of a circle with diameter 20 meters.

Find the total area of each figure. Assume each arc is a semicircle. Round to the nearest tenth. **See Example 4.**

37.
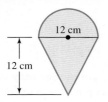
12 cm

12 cm

38.
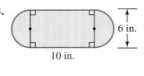
6 in.

10 in.

39.
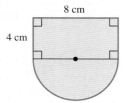
8 cm

4 cm

40.
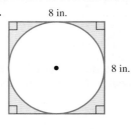
4 in.

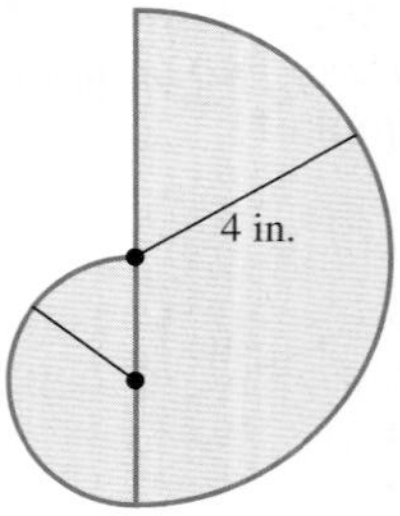

TRY IT YOURSELF

Find the area of each shaded region. Round to the nearest tenth.

41.
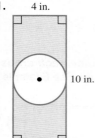
4 in.

10 in.

42.
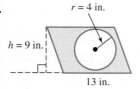
8 in.

8 in.

43.
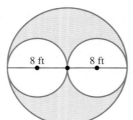
$r = 4$ in.

$h = 9$ in.

13 in.

44.
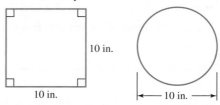
8 ft   8 ft

45. Find the circumference of the circle shown below. Give the exact answer and an approximation to the nearest hundredth.

50 yd

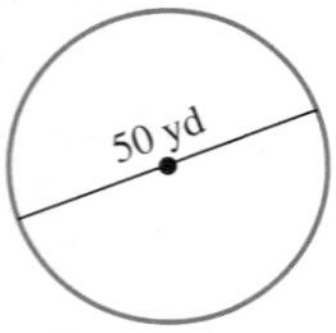

46. Find the circumference of the semicircle shown below. Give the exact answer and an approximation to the nearest hundredth.

25 cm

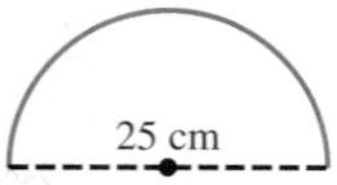

47. Find the circumference of the circle shown below if the square has sides of length 6 inches. Give the exact answer and an approximation to the nearest tenth.

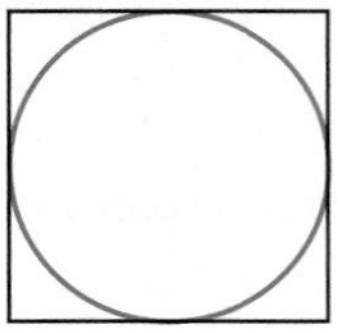

48. Find the circumference of the semicircle shown below if the length of the rectangle in which it is enclosed is 8 feet. Give the exact answer and an approximation to the nearest tenth.

8 ft

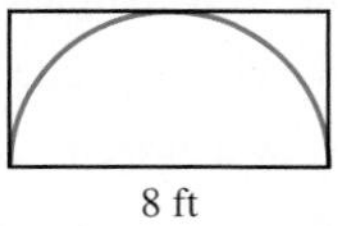

49. Find the area of the circle shown below if the square has sides of length 9 millimeters. Give the exact answer and an approximation to the nearest tenth.

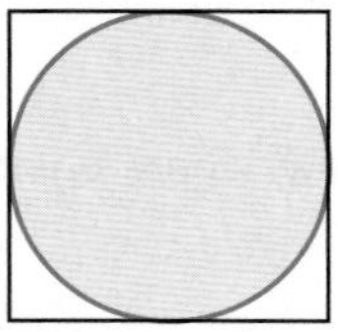

50. Find the area of the shaded semicircular region shown below. Give the exact answer and an approximation to the nearest tenth.

6.5 mi

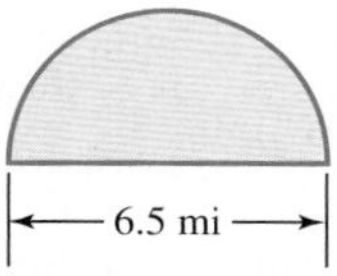

LOOK ALIKES

51. Find the perimeter or circumference of each figure below. Round to the nearest tenth, when necessary.

52. Find the area of each figure below. Round to the nearest tenth, when necessary.

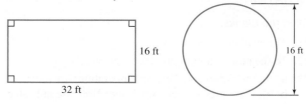
10 in.

10 in.

10 in.

53. Find the perimeter or circumference of each figure below. Round to the nearest tenth, when necessary.

54. Find the area of each figure below. Round to the nearest tenth, when necessary.

16 ft

32 ft

16 ft

APPLICATIONS

55. Suppose the two "legs" of the compass shown here are adjusted so that the distance between the pointed ends is 1 inch. Then a circle is drawn.

  a. What will the radius of the circle be?

  b. What will the diameter of the circle be?

  c. What will the circumference of the circle be? Give an exact answer and an approximation to the nearest hundredth.

  d. What will the area of the circle be? Give an exact answer and an approximation to the nearest hundredth.

56. Suppose we find the distance around a can and the distance across the can using a measuring tape, as shown to the right. Then we make a comparison, in the form of a ratio:

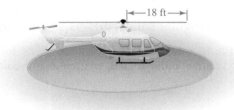

$$\frac{\text{The distance around the can}}{\text{The distance across the top of the can}}$$

After we do the indicated division, the result will be close to what number?

When appropriate, give the exact answer and an approximation to the nearest hundredth. Answers may vary slightly, depending on which approximation of π is used.

57. Lakes. Round Lake has a circular shoreline that is 2 miles in diameter. Find the area of the lake.

58. Helicopters. Refer to the figure below. How far does a point on the tip of a rotor blade travel when it makes one complete revolution?

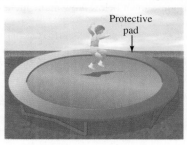

59. Giant sequoia. The largest sequoia tree is the General Sherman Tree in Sequoia National Park in California. In fact, it is considered to be the largest living thing in the world. According to the *Guinness Book of World Records*, it has a diameter of 32.66 feet, measured $4\frac{1}{2}$ feet above the ground. What is the circumference of the tree at that height?

60. Trampoline. See the figure in the next column. The distance from the center of the trampoline to the edge of its steel frame is 7 feet. The protective padding covering the springs is 18 inches wide. Find the area of the circular jumping surface of the trampoline, in square feet.

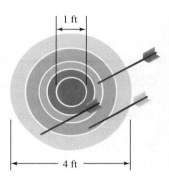

61. Jogging. Joan wants to jog 10 miles on a circular track $\frac{1}{4}$ mile in diameter. How many times must she circle the track? Round to the nearest lap.

62. Carpeting. A state capitol building has a circular floor 100 feet in diameter. The legislature wishes to have the floor carpeted. The lowest bid is $83 per square yard, including installation. How much must the legislature spend for the carpeting project? Round to the nearest dollar.

63. Archery. See the figure on the right. Find the area of the entire target and the area of the bull's eye. What percent of the area of the target is the bull's eye?

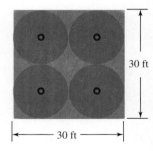

64. Landscape design. See the figure on the right. How many square feet of lawn does not get watered by the four sprinklers at the center of each circle?

65. Fire protection. When a fire sprinkler is activated by intense heat, it releases a water spray that covers a circular area on the floor directly beneath it. Determine the amount of floor space covered by the spray shown in the illustration below. Round to the nearest square foot.

66. Curling irons. A woman separated a section of her hair and wrapped it around a curling iron three times. If she used the model shown below, how many inches of her hair was curled?

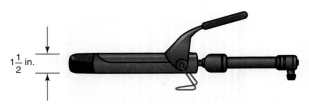

$1\frac{1}{2}$ in.

67. Explain what is meant by the circumference of a circle.

68. Explain what is meant by the area of a circle.

69. Explain the meaning of π.

70. Explain what it means for a car to have a small *turning radius*.

71. Write $\frac{9}{10}$ as a percent.

72. Write $\frac{7}{8}$ as a percent.

73. Write 0.827 as a percent.

74. Write 0.036 as a percent.

75. Unit costs. A 24-ounce package of green beans sells for $1.29. Give the unit cost in cents per ounce.

76. Mileage. One car went 1,235 miles on 51.3 gallons of gasoline, and another went 1,456 on 55.78 gallons. Which car got the better gas mileage?

77. How many sides does a pentagon have?

78. What is the sum of the measures of the angles of a triangle?

OBJECTIVES

1 Find the volume of rectangular solids, prisms, and pyramids.

2 Find the volume of cylinders, cones, and spheres.

SECTION **9.9** Volume

We have studied ways to calculate the perimeter and the area of two-dimensional figures that lie in a plane, such as rectangles, triangles, and circles. Now we will consider three-dimensional figures that occupy space, such as rectangular solids, cylinders, and spheres. In this section, we will introduce the vocabulary associated with these figures as well as the formulas that are used to find their volume. Volumes are measured in cubic units, such as cubic feet, cubic yards, or cubic centimeters. For example,

- We measure the capacity of a refrigerator in cubic feet.
- We buy gravel or topsoil by the cubic yard.
- We often measure amounts of medicine in cubic centimeters.

OBJECTIVE 1 **Find the volume of rectangular solids, prisms, and pyramids.**

The **volume** of a three-dimensional figure is a measure of its capacity. The following illustration shows two common units of volume: cubic inches, written as in.^{3}, and cubic centimeters, written as cm^3.

1 cubic inch: 1 in.^{3}

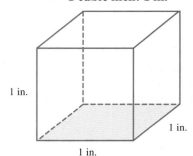

1 in.

1 in.

1 in.

1 cubic centimeter: 1 cm^3

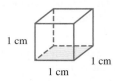

1 cm

1 cm

1 cm

The volume of a figure can be thought of as the number of cubic units that will fit within its boundaries. If we divide the figure shown in black below into cubes, each cube represents a volume of 1 cm³. Because there are 2 levels with 12 cubes on each level, the volume of the prism is 24 cm³.

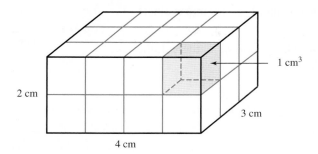

EXAMPLE 1   How many cubic inches are there in 1 cubic foot?

Strategy A figure is helpful to solve this problem. We will draw a cube and divide each of its sides into 12 equally long parts.

WHY Since a cubic foot is a cube with each side measuring 1 foot, each side also measures 12 inches.

Solution The figure on the right helps us understand the situation. Note that each level of the cubic foot contains $12 \cdot 12$ cubic inches and that the cubic foot has 12 levels. We can use multiplication to count the number of cubic inches contained in the figure. There are

$$12 \cdot 12 \cdot 12 = 1,728$$

cubic inches in 1 cubic foot. Thus, $1 \text{ ft}^3 = 1,728 \text{ in.}^3$.

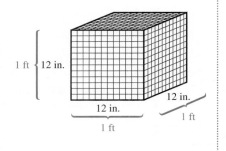

Self Check 1

How many cubic centimeters are in 1 cubic meter?

Now Try Problem 25

In practice, we do not find volumes of three-dimensional figures by counting cubes. Instead, we use the formulas shown in the table on the next page. Note that several of the volume formulas involve the variable B. It represents the area of the base of the figure.

EXAMPLE 2   **Storage tanks.** An oil storage tank is in the form of a rectangular solid. Find its volume.

Strategy We will substitute 17 for l, 10 for w, and 8 for h in the formula $V = lwh$ and evaluate the right side.

WHY The variable V represents the volume of a rectangular solid.

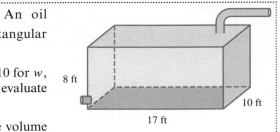

8 ft

10 ft

17 ft

Solution

Self Check 2

Find the volume of a rectangular solid with dimensions 8 meters by 12 meters by 20 meters.

Now Try ➡ **Problem 29**

$$V = lwh$$ This is the formula for the volume of a rectangular solid.

$$V = 17(10)(8)$$ Substitute 17 for l, the length, 10 for w, the width, and 8 for h, the height of the tank.

$$V = 1,360$$ Do the multiplication.

The volume of the tank is 1,360 ft³.

$$\overset{5}{170}$$
$$\underline{\times\ \ 8}$$
$$1,360$$

Cube	**Rectangular Solid**	**Sphere**
	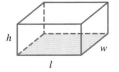	
$V = s^3$	$V = lwh$	$V = \dfrac{4}{3}\pi r^3$
where s is the length of a side	where l is the length, w is the width, and h is the height	where r is the radius

---

Prism	**Pyramid**
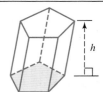 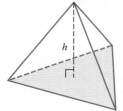	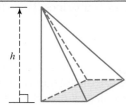
$V = Bh$	$V = \dfrac{1}{3}Bh$
where B is the area of the base and h is the height	where B is the area of the base and h is the height

---

Cylinder	**Cone**
 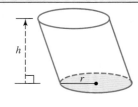	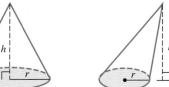
$V = Bh$ or $V = \pi r^2 h$ where B is the area of the base, h is the height, and r is the radius of the base	$V = \dfrac{1}{3}Bh$ or $V = \dfrac{1}{3}\pi r^2 h$ where B is the area of the base, h is the height, and r is the radius of the base

EXAMPLE 3   Find the volume of the prism shown on the right.

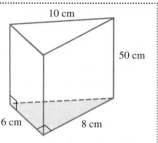

Strategy First, we will find the area of the base of the prism.

WHY To use the volume formula $V = Bh$, we need to know B, the area of the prism's base.

Solution The area of the triangular base of the prism is $\frac{1}{2}(6)(8) = 24$ square centimeters. To find its volume, we proceed as follows:

$V = Bh$ This is the formula for the volume of a triangular prism.

$V = 24(50)$ Substitute 24 for B, the area of the base, and 50 for h, the height.

$V = 1{,}200$ Do the multiplication.

$$\begin{array}{r} \overset{2}{24} \\ \times\ 50 \\ \hline 1{,}200 \end{array}$$

The volume of the triangular prism is 1,200 cm³.

Caution! Note that the 10 cm measurement was not used in the calculation of the volume.

Self Check 3

Find the volume of the prism shown below.

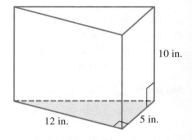

Now Try ➲ **Problem 33**

EXAMPLE 4   Find the volume of the pyramid shown on the right.

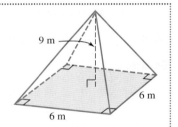

Strategy First, we will find the area of the square base of the pyramid.

WHY The volume of a pyramid is $\frac{1}{3}$ of the product of the area of its base and its height.

Solution Since the base is a square with each side 6 meters long, the area of the base is $(6\text{ m})^2$, or 36 m^2. To find the volume of the pyramid, we proceed as follows:

$V = \dfrac{1}{3}Bh$ This is the formula for the volume of a pyramid.

$V = \dfrac{1}{3}(36)(9)$ Substitute 36 for B, the area of the base, and 9 for h, the height.

$V = 12(9)$ Multiply: $\frac{1}{3}(36) = \frac{36}{3} = 12$.

$V = 108$ Complete the multiplication.

$$\begin{array}{r} \overset{1}{12} \\ \times\ 9 \\ \hline 108 \end{array}$$

The volume of the pyramid is 108 m³.

Self Check 4

Find the volume of the pyramid shown below.

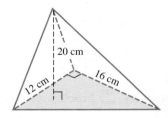

Now Try ➲ **Problem 37**

OBJECTIVE 2 Find the volume of cylinders, cones, and spheres.

EXAMPLE 5   Find the volume of the cylinder shown on the next page. Give the exact answer and an approximation to the nearest hundredth.

Strategy First, we will find the radius of the circular base of the cylinder.

WHY To use the formula for the volume of a cylinder, $V = \pi r^2 h$, we need to know r, the radius of the base.

Find the volume of the cylinder shown below. Give the exact answer and an approximation to the nearest hundredth.

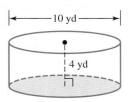

Now Try Problem 45

Solution Since a radius is one-half of the diameter of the circular base, $r = \frac{1}{2} \cdot 6$ cm $= 3$ cm. From the figure, we see that the height of the cylinder is 10 cm. To find the volume of the cylinder, we proceed as follows.

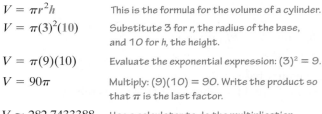

$V = \pi r^2 h$	This is the formula for the volume of a cylinder.
$V = \pi (3)^2 (10)$	Substitute 3 for r, the radius of the base, and 10 for h, the height.
$V = \pi (9)(10)$	Evaluate the exponential expression: $(3)^2 = 9$.
$V = 90\pi$	Multiply: $(9)(10) = 90$. Write the product so that π is the last factor.
$V \approx 282.7433388$	Use a calculator to do the multiplication.

The exact volume of the cylinder is 90π cm³. To the nearest hundredth, the volume is 282.74 cm³.

Caution! The height of a geometric solid is always measured along a line perpendicular to its base.

Find the volume of the cone shown below. Give the exact answer and an approximation to the nearest hundredth.

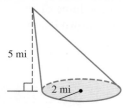

Now Try Problem 49

EXAMPLE 6 Find the volume of the cone shown on the right. Give the exact answer and an approximation to the nearest hundredth.

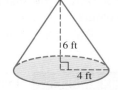

Strategy We will substitute 4 for r and 6 for h in the formula $V = \frac{1}{3}\pi r^2 h$ and evaluate the right side.

WHY The variable V represents the volume of a cone.

Solution

$V = \dfrac{1}{3}\pi r^2 h$	This is the formula for the volume of a cone.
$V = \dfrac{1}{3}\pi (4)^2 (6)$	Substitute 4 for r, the radius of the base, and 6 for h, the height.
$V = \dfrac{1}{3}\pi (16)(6)$	Evaluate the exponential expression: $(4)^2 = 16$.
$V = 2\pi (16)$	Multiply: $\frac{1}{3}(6) = 2$.
$V = 32\pi$	Multiply: $2(16) = 32$. Write the product so that π is the last factor.
$V \approx 100.5309649$	Use a calculator to do the multiplication.

The exact volume of the cone is 32π ft³. To the nearest hundredth, the volume is 100.53 ft³.

EXAMPLE 7 **Water towers.** How many cubic feet of water are needed to fill the spherical water tank shown on the right? Give the exact answer and an approximation to the nearest tenth.

Strategy We will substitute 15 for r in the formula $V = \frac{4}{3}\pi r^3$ and evaluate the right side.

WHY The variable V represents the volume of a sphere.

Solution

$$V = \frac{4}{3}\pi r^3 \qquad \text{This is the formula for the volume of a sphere.}$$

$$V = \frac{4}{3}\pi (15)^3 \qquad \text{Substitute 15 for } r, \text{ the radius of the sphere.}$$

$$V = \frac{4}{3}\pi (3{,}375) \qquad \text{Evaluate the exponential expression: } (15)^3 = 3{,}375.$$

$$V = \frac{13{,}500}{3}\pi \qquad \text{Multiply: } 4(3{,}375) = 13{,}500.$$

$$V = 4{,}500\pi \qquad \text{Divide: } \frac{13{,}500}{3} = 4{,}500. \text{ Write the product so that } \pi \text{ is the last factor.}$$

$$V \approx 14{,}137.16694 \qquad \text{Use a calculator to do the multiplication.}$$

$$\begin{array}{r} {\scriptstyle 1\,3\,2} \\ 3375 \\ \times\ \ 4 \\ \hline 13{,}500 \end{array}$$

The tank holds exactly $4{,}500\pi$ ft³ of water. To the nearest tenth, this is 14,137.2 ft³.

Self Check 7

Find the volume of a spherical water tank with a radius of 7 meters. Give the exact answer and an approximation to the nearest tenth.

Now Try Problem 53

Using Your Calculator ▶ Volume of a Silo

A silo is a structure used for storing grain. The silo shown on the right is a cylinder 50 feet tall topped with a dome in the shape of a hemisphere. To find the volume of the silo, we add the volume of the cylinder to the volume of the dome.

$$\text{Volume}_{\text{cylinder}} + \text{Volume}_{\text{dome}} = (\text{Area}_{\text{cylinder's base}})(\text{Height}_{\text{cylinder}}) + \frac{1}{2}(Volume_{sphere})$$

$$\text{Volume}_{\text{cylinder}} + \text{Volume}_{\text{dome}} = \pi r^2 h + \frac{1}{2}\left(\frac{4}{3}\pi r^3\right)$$

$$\text{Volume}_{\text{cylinder}} + \text{Volume}_{\text{dome}} = \pi r^2 h + \frac{2\pi r^3}{3} \qquad \text{Multiply and simplify: } \frac{1}{2}\left(\frac{4}{3}\pi r^3\right) = \frac{4}{6}\pi r^3 = \frac{2\pi r^3}{3}.$$

$$\text{Volume}_{\text{cylinder}} + \text{Volume}_{\text{dome}} = \pi (10)^2 (50) + \frac{2\pi (10)^3}{3} \qquad \text{Substitute 10 for } r \text{ and 50 for } h.$$

We can use a scientific calculator to make this calculation.

$$\boxed{\pi}\ \boxed{\times}\ 10\ \boxed{x^2}\ \boxed{\times}\ 50\ \boxed{+}\ \boxed{(}\ 2\ \boxed{\times}\ \boxed{\pi}\ \boxed{\times}\ 10\ \boxed{y^x}\ 3\ \boxed{)}\ \boxed{\div}\ 3\ \boxed{=}$$

$$\boxed{17802.35837}$$

The volume of the silo is approximately 17,802 ft³.

WoodyUpstate/E+/Getty Images

50 ft

10 ft

Answers to Self Checks

1. 1,000,000 cm³   **2.** 1,920 m³   **3.** 300 in.³   **4.** 640 cm³   **5.** 100π yd³ $\approx$ 314.16 yd³   **6.** $\frac{20}{3}\pi$ mi³ $\approx$ 20.94 mi³

7. $\frac{1{,}372}{3}\pi$ m³ $\approx$ 1,436.8 m³

SECTION 9.9 **STUDY SET**

VOCABULARY

Fill in the blanks.

1. The _____ of a three-dimensional figure is a measure of its capacity.

2. The volume of a figure can be thought of as the number of _____ units that will fit within its boundaries.

Give the name of each figure.

3.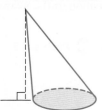

4.

5.

6.

7.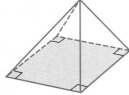

8.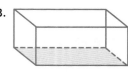

CONCEPTS

9. Draw a cube. Label a side s.

10. Draw a cylinder. Label the height h and radius r.

11. Draw a pyramid. Label the height h and the base b.

12. Draw a cone. Label the height h and radius r.

13. Draw a sphere. Label the radius r.

14. Draw a rectangular solid. Label the length l, the width w, and the height h.

15. Which of the following are acceptable units with which to measure volume?

ft²	mi³	seconds	days
cubic inches	mm	square yards	in.
pounds	cm²	meters	m³

16. In the figure on the right, the unit of measurement of length used to draw the figure is the inch.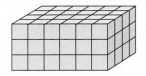

    a. What is the area of the base of the figure?

    b. What is the volume of the figure?

17. Which geometric concept (perimeter, circumference, area, or volume) should be applied when measuring each of the following?

    a. The distance around a checkerboard

    b. The size of a trunk of a car

    c. The amount of paper used for a postage stamp

    d. The amount of storage in a cedar chest

    e. The amount of beach available for sunbathing

    f. The distance the tip of a propeller travels

18. Complete the table.

Figure	Volume formula
Cube	
Rectangular solid	
Prism	
Cylinder	
Pyramid	
Cone	
Sphere	

19. Evaluate each expression. Leave π in the answer.

    a. $\dfrac{1}{3}\pi(25)6$

    b. $\dfrac{4}{3}\pi(125)$

20. a. Evaluate $\frac{1}{3}\pi r^2 h$ for $r = 2$ and $h = 27$. Leave π in the answer.

    b. Approximate your answer to part a to the nearest tenth.

NOTATION

21. a. What does "in.³" mean?

    b. Write "one cubic centimeter" using symbols.

22. In the formula $V = \frac{1}{3}Bh$, what does B represent?

23. In a drawing, what does the symbol ⌐ indicate?

24. Redraw the figure below using dashed lines to show the hidden edges.

GUIDED PRACTICE

Convert from one unit of measurement to another.
See Example 1.

25. How many cubic feet are in 1 cubic yard?

26. How many cubic decimeters are in 1 cubic meter?

27. How many cubic meters are in 1 cubic kilometer?

28. How many cubic inches are in 1 cubic yard?

Find the volume of each figure. **See Example 2.**

29.
7 ft, 2 ft, 4 ft

30.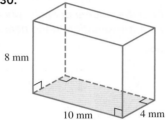
8 mm, 10 mm, 4 mm

31.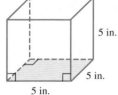
5 in., 5 in., 5 in.

32.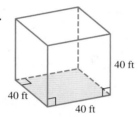
40 ft, 40 ft, 40 ft

Find the volume of each figure. **See Example 3.**

33.
5 cm, 0.2 m, 3 cm, 4 cm

34.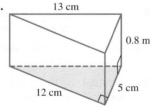
13 cm, 0.8 m, 12 cm, 5 cm

35.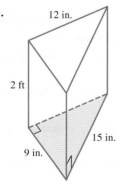
12 in., 2 ft, 9 in., 15 in.

36.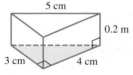
10 in., 24 in., 0.5 ft, 26 in.

Find the volume of each figure. **See Example 4.**

37.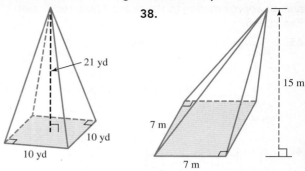
21 yd, 10 yd, 10 yd

38.
7 m, 15 m, 7 m

39.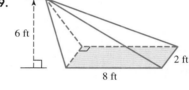
6 ft, 2 ft, 8 ft

40.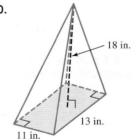
18 in., 13 in., 11 in.

41.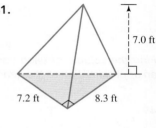
7.0 ft, 7.2 ft, 8.3 ft

42.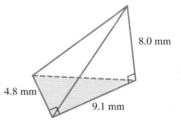
8.0 mm, 4.8 mm, 9.1 mm

43.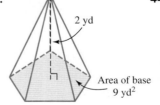
2 yd, Area of base 9 yd²

44.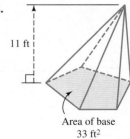
11 ft, Area of base 33 ft²

Find the volume of each cylinder. Give the exact answer and an approximation to the nearest hundredth. See Example 5.

45.

4 ft

12 ft

46.

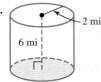

2 mi

6 mi

47.

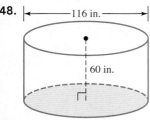

← 30 cm →

14 cm

48.

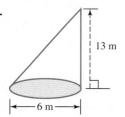

← 116 in. →

60 in.

Find the volume of each cone. Give the exact answer and an approximation to the nearest hundredth. See Example 6.

49.

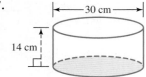

13 m

← 6 m →

50.

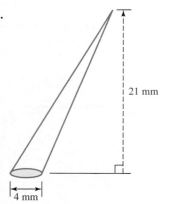

21 mm

4 mm

51.

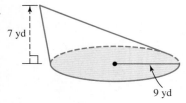

7 yd

9 yd

52.

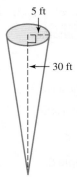

5 ft

30 ft

Find the volume of each sphere. Give the exact answer and an approximation to the nearest tenth. See Example 7.

53.

6 in.

54.

9 ft

55.

← 4 cm →

56.

← 10 in. →

TRY IT YOURSELF

Find the volume of each figure. If an exact answer contains π, approximate to the nearest hundredth.

57. A hemisphere with a radius of 9 inches
(*Hint:* a **hemisphere** is an exact half of a sphere.)

58. A hemisphere with a diameter of 22 feet
(*Hint:* a **hemisphere** is an exact half of a sphere.)

59. A cylinder with a height of 12 meters and a circular base with a radius of 6 meters

60. A cylinder with a height of 4 meters and a circular base with a diameter of 18 meters

61. A rectangular solid with dimensions of 3 cm by 4 cm by 5 cm

62. A rectangular solid with dimensions of 5 m by 8 m by 10 m

63. A cone with a height of 12 centimeters and a circular base with a diameter of 10 centimeters

64. A cone with a height of 3 inches and a circular base with a radius of 4 inches

65. A pyramid with a square base 10 meters on each side and a height of 12 meters

66. A pyramid with a square base 6 inches on each side and a height of 4 inches

67. A prism whose base is a right triangle with legs 3 meters and 4 meters long and whose height is 8 meters

68. A prism whose base is a right triangle with legs 5 feet and 12 feet long and whose height is 25 feet

Find the volume of each figure. Give the exact answer and, when needed, an approximation to the nearest hundredth.

69.

3 cm

8 cm

8 cm

8 cm

70.

10 in.

20 in.

8 in.

71.

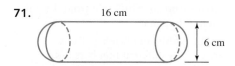

16 cm

6 cm

72.

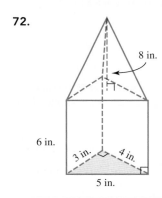

8 in.

6 in.

3 in.   4 in.

5 in.

APPLICATIONS

Solve each problem. If an exact answer contains π, approximate the answer to the nearest hundredth.

73. Sweeteners. A sugar cube is $\frac{1}{2}$ inch on each edge. How much volume does it occupy?

74. Ventilation. A classroom is 40 feet long, 30 feet wide, and 9 feet high. Find the number of cubic feet of air in the room.

75. Moving and storage. The largest container that Portable On Demand Storage (PODS) offers is shown below. In their description of its features, they state that it has approximately 857 cubic feet of interior packing space. What is the difference between the volume of an 8 ft by 8 ft by 16 ft rectangular solid and the interior packing space of a POD container.

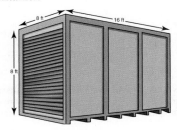

8 ft

16 ft

8 ft

76. Refrigerators. The largest refrigerator advertised in a JCPenney catalog has a capacity of 25.2 cubic feet. How many cubic inches is this?

77. Tanks. A cylindrical oil tank has a diameter of 6 feet and a length of 7 feet. Find the volume of the tank.

78. Desserts. A restaurant serves pudding in a conical dish that has a diameter of 3 inches. If the dish is 4 inches deep, how many cubic inches of pudding are in each dish?

79. Hot-air balloons. The lifting power of a spherical balloon depends on its volume. How many cubic feet of gas will a balloon hold if it is 40 feet in diameter?

80. Cereal boxes. A box of cereal measures 3 inches by 8 inches by 10 inches. The manufacturer plans to market a smaller box that measures $2\frac{1}{2}$ by 7 by 8 inches. By how much will the volume be reduced?

81. Engines. The *compression ratio* of an engine is the volume in one cylinder with the piston at bottom-dead-center (B.D.C.), divided by the volume with the piston at top-dead-center (T.D.C.). From the data given in the following figure, what is the compression ratio of the engine? Use a colon to express your answer.

Volume before compression: 30.4 in.³   Volume after compression: 3.8 in.³

T.D.C.

B.D.C.

82. Geography. Earth is not a perfect sphere but is slightly pear-shaped. To estimate its volume, we will assume that it is spherical, with a diameter of about 7,926 miles. What is its volume, to the nearest one billion cubic miles?

83. Birdbaths.
a. The bowl of the birdbath shown on the right is in the shape of a hemisphere (half of a sphere). Find its volume.
b. If 1 gallon of water occupies 231 cubic inches of space, how many gallons of water does the birdbath hold? Round to the nearest tenth.

30 in.

84. Concrete blocks. Find the number of cubic inches of concrete used to make the hollow, cube-shaped block shown below.

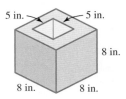

5 in.   5 in.

8 in.

8 in.   8 in.

85. Gourmet cooking. One type of knife cut that gourmet cooks often use when preparing celery, carrots, and potatoes is called *julienne*. The official measurements of a julienne cut are $\frac{1}{8}$ in. by $\frac{1}{8}$ in. by 3 in. What is the volume of a julienne carrot stick? Express the answer as a fraction and as a decimal.

Robyn Mackenzie/Shutterstock.com

86. AC ductwork. A 12-foot-long piece of air conditioning duct has an inside diameter of 8 inches. Find the number of cubic feet of air that the duct holds. Round to the nearest tenth. (*Hint:* Note that the units are different.)

WRITING

87. What is meant by the *volume* of a cube?

88. The stack of 3 × 5 index cards shown in figure (a) forms a right rectangular prism, with a certain volume. If the stack is pushed to lean to the right, as in figure (b), a new prism is formed. How will its volume compare to the volume of the right rectangular prism? Explain your answer.

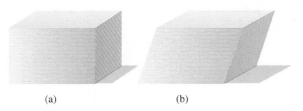

(a)   (b)

89. Are the units used to measure area different from the units used to measure volume? Explain.

90. The dimensions (length, width, and height) of one rectangular solid are entirely different numbers from the dimensions of another rectangular solid. Would it be possible for the rectangular solids to have the same volume? Explain.

REVIEW

91. Evaluate: $-5(5 - 2)^2 + 3$

92. Buying pencils. Carlos bought 6 pencils at $0.60 each and a notebook for $1.25. He gave the clerk a $5 bill. How much change did he receive?

93. Solve: $-x = 4$

94. 38 is what percent of 40?

95. Express the phrase "3 inches to 15 inches" as a ratio in simplest form.

96. Convert 40 ounces to pounds.

97. Convert 2.4 meters to millimeters.

98. State the Pythagorean equation.

---

⑨ Summary and Review

DEFINITIONS AND CONCEPTS	EXAMPLES
The word **geometry** comes from the Greek words *geo* (meaning Earth) and *metron* (meaning measure). Geometry is based on three undefined words: **point**, **line**, and **plane**.	**Point** **Line $\overleftrightarrow{BC}$** **Plane *EFG*** • A •B •E •G •C •F Points are labeled with capital letters. We can name a line using any two points on it. Floors, walls, and table tops are all parts of planes.

A **line segment** is a part of a line with two endpoints. Every line segment has a **midpoint**, which divides the segment into two parts of equal length. The notation $m(\overline{AM})$ is read as "the measure of line segment $\overline{AM}$." When two line segments have the same measure, we say that they are **congruent**. Read the symbol $\cong$ as "is congruent to." A **ray** is a part of a line with one **endpoint**.	**Line segment $\overline{AB}$** **Ray $\overrightarrow{CD}$** $m(\overline{AM}) = m(\overline{MB})$ $\overline{AM} \cong \overline{MB}$ 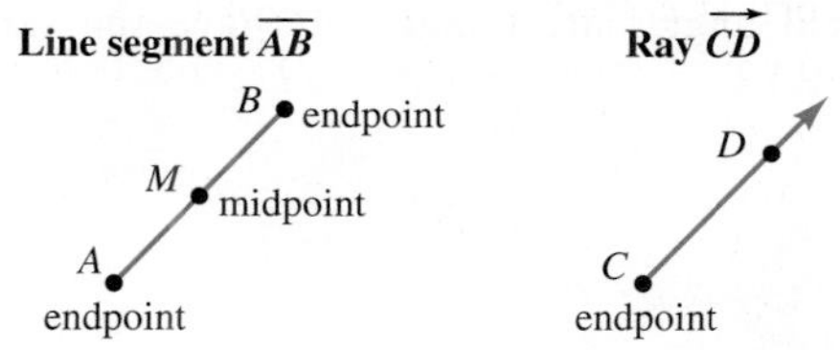 B endpoint M midpoint A endpoint D C endpoint
An **angle** is a figure formed by two rays (called **sides**) with a common endpoint. The common endpoint is called the **vertex** of the angle. We read the symbol $\angle$ as "angle."	The angle below can be written as $\angle BAC$, $\angle CAB$, $\angle A$, or $\angle 1$. 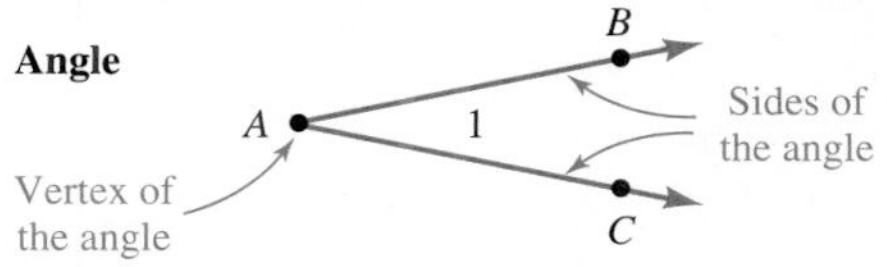 **Angle** B A 1 Sides of the angle Vertex of the angle C
When two angles have the same measure, we say that they are **congruent**. A **protractor** is used to find the measure of an angle. One unit of measurement of an angle is the **degree**. The notation $m(\angle DEF)$ is read as "the measure of $\angle DEF$."	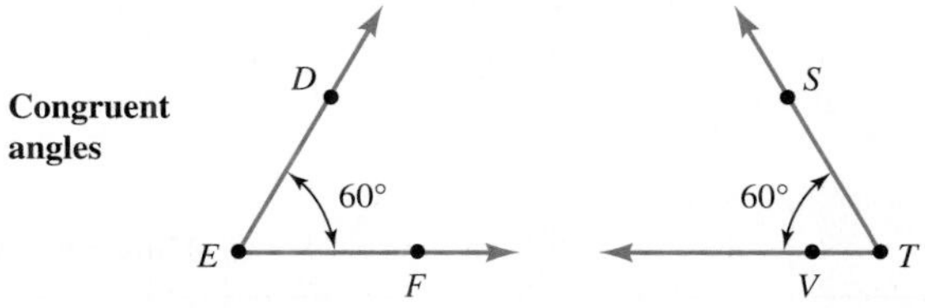 **Congruent angles** D S $60°$ $60°$ E F V T Since $m(\angle DEF) = m(\angle STV)$, we say that $\angle DEF \cong \angle STV$.
An **acute angle** has a measure that is greater than 0° but less than 90°. An **obtuse angle** has a measure that is greater than 90° but less than 180°. A **straight angle** measures 180°.	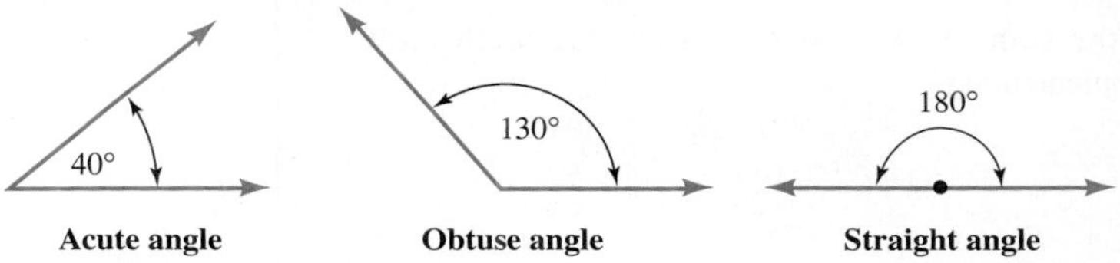 $40°$ $130°$ $180°$ **Acute angle** **Obtuse angle** **Straight angle**
A **right angle** measures 90°.	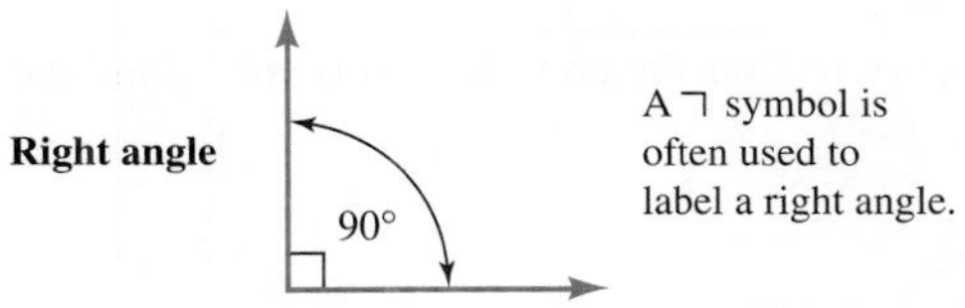 **Right angle** $90°$ A ⌐ symbol is often used to label a right angle.
Two angles that have the same vertex and are side by side are called **adjacent angles**.	Two angles with degree measures of x and 21° are adjacent angles, as shown here. Use the information in the figure to find x. **Adjacent angles** 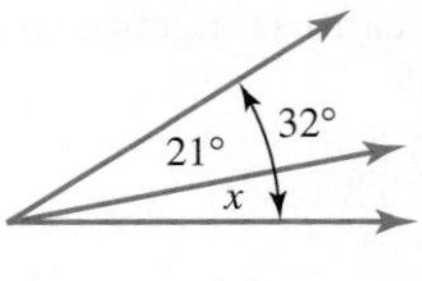 $21°$ $32°$ x
We can use the algebra concepts of variable and equation to solve many types of geometry problems.	The sum of the measures of the two adjacent angles is 32°: $x + 21° = 32°$ The word *sum* indicates addition. $x + 21° - 21° = 32° - 21°$ Subtract 21° from both sides. $x = 11°$ Do the subtraction. Thus, x is 11°.
When two lines intersect, pairs of nonadjacent angles are called **vertical angles**.	**Vertical angles** 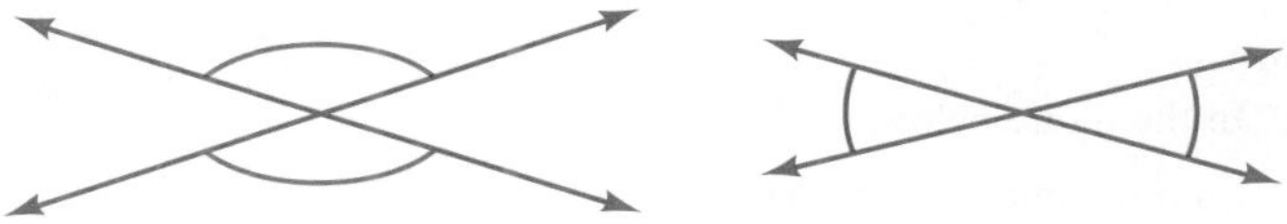

Vertical angles are congruent (have the same measure).	Refer to the figure below. Find x and m($\angle XYZ$).

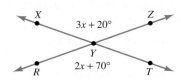

Since the angles are vertical angles, they have equal measures.

$3x + 20° = 2x + 70°$	Set the expressions equal.
$3x + 20° - 2x = 2x + 70° - 2x$	Eliminate 2x from the right side.
$x + 20° = 70°$	Combine like terms.
$x = 50°$	Subtract 20° from both sides.

Thus, x is 50°. To find m($\angle XYZ$), evaluate the expression $3x + 20°$ for $x = 50°$.

$3x + 20° = 3(50°) + 20°$	Substitute 50° for x.
$= 150° + 20°$	Do the multiplication.
$= 170°$	Do the addition.

Thus, m($\angle XYZ$) = 170°.

If the sum of two angles is 90°, the angles are **complementary**. If the sum of two angles is 180°, the angles are **supplementary**.	**Complementary angles** $63° + 27° = 90°$ 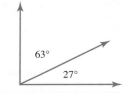 **Supplementary angles** $146° + 34° = 180°$ 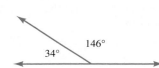

We can use algebra to find the complement of an angle.	Find the complement of an 11° angle.

Let x = the measure of the complement (in degrees).

$x + 11° = 90°$	The sum of the angles' measures must be 90°.
$x = 79°$	To isolate x, subtract 11° from both sides.

The complement of an 11° angle has measure 79°.

We can use algebra to find the supplement of an angle.	Find the supplement of a 68° angle.

Let x = the measure of the supplement (in degrees).

$x + 68° = 180°$	The sum of the angles' measures must be 180°.
$x = 112°$	To isolate x, subtract 68° from both sides.

The supplement of a 68° angle has measure 112°.

REVIEW EXERCISES

1. In the illustration, give the name of a point, a line, and a plane.

2. a. In the figure below, find m($\overline{AG}$).

   b. Find the midpoint of $\overline{BH}$.

   c. Is $\overline{AC} \cong \overline{GE}$?

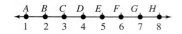

3. Give four ways to name the angle shown below.

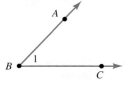

4. a. Is the angle shown above acute or obtuse?

   b. What is the vertex of the angle?

   c. What rays form the sides of the angle?

   d. Use a protractor to find the measure of the angle.

5. Identify each acute angle, right angle, obtuse angle, and straight angle in the figure below.

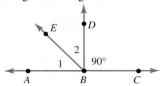

6. In the figure above, is $\angle ABD \cong \angle CBD$?

7. In the figure above, are $\overrightarrow{AC}$ and $\overrightarrow{AB}$ the same ray?

8. The measures of several angles are given below. Identify each angle as an acute angle, a right angle, an obtuse angle, or a straight angle.

    a. $m(\angle A) = 150°$

    b. $m(\angle B) = 90°$

    c. $m(\angle C) = 180°$

    d. $m(\angle D) = 25°$

9. The two angles shown here are adjacent angles. Find x.

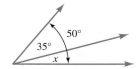

10. Line AB is shown in the figure below. Find y.

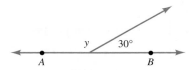

11. Refer to the figure on the right.

    a. Find $m(\angle 1)$.

    b. Find $m(\angle 2)$.

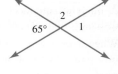

12. Refer to the figure below with lines $\overleftrightarrow{AE}$, $\overleftrightarrow{GD}$, and $\overleftrightarrow{CF}$.

    a. What is $m(\angle ABG)$?

    b. What is $m(\angle FBE)$?

    c. What is $m(\angle CBD)$?

    d. What is $m(\angle FBG)$?

    e. Are $\angle CBD$ and $\angle DBE$ complementary angles?

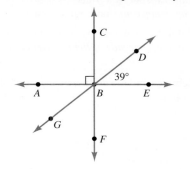

13. Refer to the figure with lines $\overleftrightarrow{EH}$ and $\overleftrightarrow{GI}$.

    a. Find x.

    b. What is $m(\angle HFI)$?

    c. What is $m(\angle GFH)$?

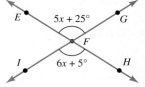

14. Find the complement of a 71° angle.

15. Find the supplement of a 143° angle.

16. Are angles measuring 30°, 60°, and 90° supplementary?

SECTION 9.2 ▶ **Parallel and Perpendicular Lines**

DEFINITIONS AND CONCEPTS	EXAMPLES
If two lines lie in the same plane, they are called **coplanar**. **Parallel lines** are coplanar lines that do not intersect. We read the symbol ‖ as "is parallel to." **Perpendicular lines** are lines that intersect and form right angles. We read the symbol ⊥ as "is perpendicular to."	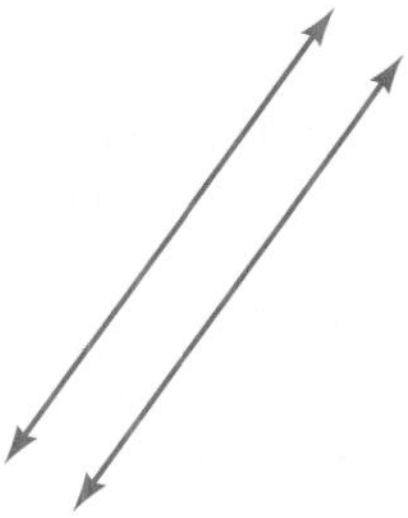 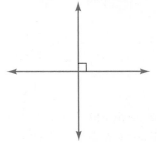

A line that intersects two coplanar lines in two distinct (different) points is called a **transversal**. When a transversal intersects two coplanar lines, four pairs of **corresponding angles** are formed. If two parallel lines are cut by a transversal, *corresponding angles are congruent* (have equal measures).	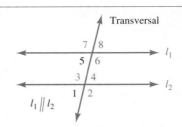 **Corresponding angles** • $\angle 1 \cong \angle 5$ • $\angle 2 \cong \angle 6$ • $\angle 3 \cong \angle 7$ • $\angle 4 \cong \angle 8$
When a transversal intersects two coplanar lines, two pairs of **interior angles** and two pairs of **alternate interior angles** are formed. If two parallel lines are cut by a transversal, *alternate interior angles are congruent* (have equal measures). If two parallel lines are cut by a transversal, *interior angles on the same side of the transversal are supplementary.*	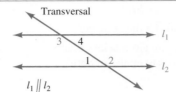 **Alternate interior angles** • $\angle 1 \cong \angle 4$ • $\angle 2 \cong \angle 3$ **Interior angles** $m(\angle 1) + m(\angle 3) = 180°$ $m(\angle 2) + m(\angle 4) = 180°$
We can use algebra to find the unknown measures of corresponding angles.	In the figure, $l_1 \parallel l_2$. Find x and the measure of each angle that is labeled. 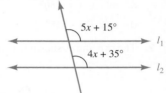 Since the lines are parallel, and the angles are corresponding angles, the angles are congruent. $\quad 5x + 15° = 4x + 35° \qquad$ The angle measures are equal. $\quad\quad x + 15° = 35° \qquad\quad$ Subtract $4x$ from both sides. $\quad\quad\quad\quad x = 20° \qquad\quad\quad$ To isolate x, subtract $15°$ from both sides. Thus, x is 20°. To find the measures of the angles labeled in the figure, we evaluate each expression for $x = 20°$. $\quad 5x + 15° = 5(\mathbf{20°}) + 15° \quad\mid\quad 4x + 35° = 4(\mathbf{20°}) + 35°$ $\quad\quad\quad\quad\quad = 100° + 15° \quad\mid\quad\quad\quad\quad\quad = 80° + 35°$ $\quad\quad\quad\quad\quad = 115° \quad\quad\quad\mid\quad\quad\quad\quad\quad = 115°$ The measure of each angle is 115°.
We can use algebra to find the unknown measures of interior angles.	In the figure, $l_1 \parallel l_2$. Find x and the measure of each angle that is labeled. 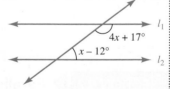 Since the angles are interior angles on the same side of the transversal, they are supplementary. $\quad 4x + 17° + x - 12° = 180° \qquad$ The sum of the measures of two supplementary angles is 180°. $\quad\quad\quad 5x + 5° = 180° \qquad\qquad$ Combine like terms. $\quad\quad\quad\quad\quad 5x = 175° \qquad\qquad$ Subtract 5° from both sides. $\quad\quad\quad\quad\quad\quad x = 35° \qquad\qquad$ Divide both sides by 5. Thus, x is 35°. To find the measures of the angles in the figure, we evaluate the expressions for $x = 35°$. $\quad 4x + 17° = 4(\mathbf{35°}) + 17° \quad\mid\quad x - 12° = 35° - 12°$ $\quad\quad\quad\quad\quad = 140° + 17° \quad\quad\mid\quad\quad\quad\quad = 23°$ $\quad\quad\quad\quad\quad = 157° \quad\quad\quad\quad\mid$ The measures of the angles labeled in the figure are 157° and 23°.

REVIEW EXERCISES

17. a. Lines l_1 and l_2 shown in figure (a) below do not intersect and are coplanar. What word describes the lines?

b. In figure (a), line l_3 intersects lines l_1 and l_2 in two distinct (different) points. What is the name given to line l_3?

c. What word describes the two lines shown in figure (b) below?

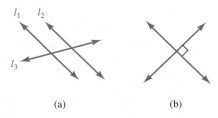

       (a)   (b)

18. Identify all pairs of alternate interior angles shown in the figure below.

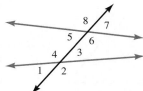

19. Refer to the figure in Problem 18. Identify all pairs of corresponding angles.

20. Refer to the figure in Problem 18. Identify all pairs of vertical angles.

21. In the figure below, $l_1 \parallel l_2$. Find the measure of each angle.

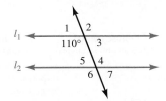

22. In the figure on the right, $\overline{DC} \parallel \overline{AB}$. Find the measure of each angle that is labeled.

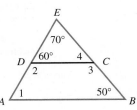

23. In the figure below, $l_1 \parallel l_2$.

a. Find x.

b. Find the measure of each angle that is labeled.

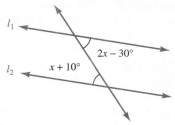

24. In the figure below, $l_1 \parallel l_2$.

a. Find x.

b. Find the measure of each angle that is labeled.

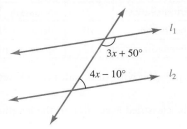

25. In the figure below, $\overline{AB} \parallel \overline{DC}$.

a. Find x.

b. Find the measure of each angle that is labeled.

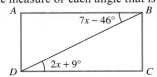

26. In the figure below, $\overline{EF} \parallel \overline{HI}$.

a. Find x.

b. Find the measure of each angle that is labeled.

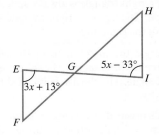

SECTION 9.3 ▶ Polygons and Triangles

DEFINITIONS AND CONCEPTS	EXAMPLES
A **polygon** is a closed geometric figure with at least three line segments for its sides. The points at which the sides intersect are called **vertices**. A **regular polygon** has sides that are all the same length and angles that are all the same measure. The number of vertices of a polygon is equal to the number of sides it has.	**Polygon** **Regular polygon**

Classifying Polygons	**Quadrilateral** **Hexagon** **Octagon** (4 sides) (6 sides) (8 sides)

Number of Sides	Name of Polygon	Number of Sides	Name of Polygon
3	triangle	8	octagon
4	quadrilateral	9	nonagon
5	pentagon	10	decagon
6	hexagon	12	dodecagon

A **triangle** is a polygon with three sides (and three vertices). Triangles can be classified according to the lengths of their sides. **Tick marks** indicate sides that are of equal length.	**Equilateral triangle** **Isosceles triangle** **Scalene triangle** (all sides of (at least two sides of (no sides of equal length) equal length) equal length)
Triangles can be classified by their angles.	**Acute triangle** **Obtuse triangle** **Right triangle** (has three acute angles) (has an obtuse angle) (has one right angle)
The longest side of a right triangle is called the **hypotenuse**, and the other two sides are called **legs**. The hypotenuse of a right triangle is always opposite the 90° (right) angle. The legs of a right triangle are adjacent to (next to) the right angle.	**Right triangle** Hypotenuse (longest side) Leg Leg
In an isosceles triangle, the angles opposite the sides of equal length are called **base angles**. The third angle is called the **vertex angle**. The third side is called the **base**. **Isosceles triangle theorem:** If two sides of a triangle are congruent, then the angles opposite those sides are congruent. **Converse of the isosceles triangle theorem:** If two angles of a triangle are congruent, then the sides opposite the angles have the same length, and the triangle is isosceles.	**Isosceles triangles** Vertex angle Base angle — — Base angle Base

The **sum of the measures of the angles** of any triangle is 180°.

We can use algebra to find unknown angle measures of a triangle.

Find the measure of each angle of △ABC.

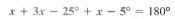

The sum of the angle measures of any triangle is 180°:

$$x + 3x - 25° + x - 5° = 180°$$
$$5x - 30° = 180° \quad \text{Combine like terms.}$$
$$5x = 210° \quad \text{Add 30° to both sides.}$$
$$x = 42° \quad \text{Divide both sides by 5.}$$

To find the measures of ∠B and ∠C, we evaluate the expressions $3x - 25°$ and $x - 5°$ for $x = 42°$.

$$3x - 25° = 3(42°) - 25° \qquad x - 5° = 42° - 5°$$
$$= 126° - 25° \qquad\qquad\quad = 37°$$
$$= 101°$$

Thus, m(∠A) = 42°, m(∠B) = 101°, and m(∠C) = 37°.

We can use algebra to find unknown angle measures of an isosceles triangle.

If the vertex angle of an isosceles triangle measures 26°, what is the measure of each base angle?

If we let x represent the measure of one base angle, the measure of the other base angle is also x. (See the figure.) Since the sum of the measures of the angles of any triangle is 180°, we have

$$x + x + 26° = 180°$$
$$2x + 26° = 180° \quad \text{On the left side, combine like terms.}$$
$$2x = 154° \quad \text{To isolate 2x, subtract 26° from both sides.}$$
$$x = 77° \quad \text{To isolate x, divide both sides by 2.}$$

The measure of each base angle is 77°.

REVIEW EXERCISES

27. For each of the following polygons, give the number of sides it has, tell its name, and then give the number of vertices that it has.

a.

b.

c.

d.

e.

f.

28. Classify each of the following triangles as an equilateral triangle, an isosceles triangle, or a scalene triangle. Some figures may be correctly classified in more than one way.

a.

8 in.   8 in.

b.
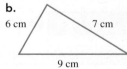
6 cm   7 cm
9 cm

c.

5 m   5 m
5 m

d.

44°
44°

29. Classify each of the following triangles as an acute, an obtuse, or a right triangle.

a.

b.
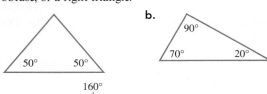

c.
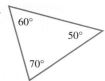

d.
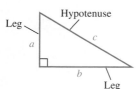

30. Refer to the triangle shown here.

   a. What is the measure of $\angle X$?

   b. What type of triangle is it?

   c. What two line segments are the legs?

   d. What line segment is the hypotenuse?

   e. Which side of the triangle is the longest?

   f. Which side is opposite $\angle X$?

In each triangle shown below, find x.

31.

32.
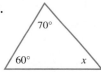

33. In $\triangle ABC$, $m(\angle B) = 32°$ and $m(\angle C) = 77°$. Find $m(\angle A)$.

34. For the triangle shown below, find x. Then determine the measure of each angle of the triangle.

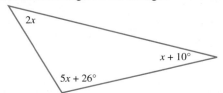

35. One base angle of an isosceles triangle measures 65°. Find the measure of the vertex angle.

36. The measure of the vertex angle of an isosceles triangle is 68°. Find the measure of each base angle.

37. Find the measure of $\angle C$ of the triangle shown here.

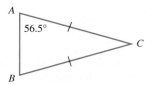

38. Refer to the figure shown here. Find $m(\angle C)$.

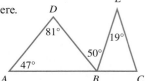

<table>
<tr><td colspan="2">**SECTION 9.4** ▶ **The Pythagorean Theorem**</td></tr>
<tr><td>DEFINITIONS AND CONCEPTS</td><td>EXAMPLES</td></tr>
</table>

Pythagorean theorem

If a and b are the lengths of the legs of a right triangle and c is the length of the hypotenuse, then

$$a^2 + b^2 = c^2$$

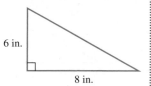

$a^2 + b^2 = c^2$ is called the **Pythagorean equation**.

Find the length of the hypotenuse of the right triangle shown here.

We will let $a = 6$ and $b = 8$, and substitute into the Pythagorean equation to find c.

$a^2 + b^2 = c^2$ This is the Pythagorean equation.

$6^2 + 8^2 = c^2$ Substitute 6 for a and 8 for b.

$36 + 64 = c^2$ Evaluate the exponential expressions.

$100 = c^2$ Do the addition.

$c^2 = 100$ Reverse the sides of the equation so that c^2 is on the left.

To find c, we must find a number that, when squared, is 100. There are two such numbers, one positive and one negative; they are the square roots of 100. Since c represents the length of a side of a triangle, c cannot be negative. For this reason, we need only find the positive square root of 100 to get c.

$$c = \sqrt{100}$$ The symbol $\sqrt{}$ is used to indicate the positive square root of a number.

$$c = 10$$ Because $10^2 = 100$.

The length of the hypotenuse of the triangle is 10 in.

When we use the Pythagorean theorem to find the length of a side of a right triangle, the solution is sometimes the square root of a number that is not a perfect square. In that case, we can use a calculator to *approximate* the square root.

The lengths of two sides of a right triangle are shown here. Find the missing side length.

We may substitute 9 for either a or b, but 11 must be substituted for the length c of the hypotenuse. If we substitute 9 for a, we can find the unknown side length b as follows.

$$a^2 + b^2 = c^2$$ This is the Pythagorean equation.

$$9^2 + b^2 = 11^2$$ Substitute 9 for a and 11 for c.

$$81 + b^2 = 121$$ Evaluate each exponential expression.

$$81 + b^2 - 81 = 121 - 81$$ To isolate b^2 on the left side, subtract 81 from both sides.

$$b^2 = 40$$

We must find a number that, when squared, is 40. Since b represents the length of a side of a triangle, we consider only the positive square root.

$$b = \sqrt{40}$$ This is the exact length.

The missing side length is exactly $\sqrt{40}$ feet long. Since 40 is not a perfect square, we use a calculator to approximate $\sqrt{40}$. To the nearest hundredth, the missing side length is 6.32 ft.

The converse of the Pythagorean theorem:
If a triangle has sides of lengths a, b, and c, such that $a^2 + b^2 = c^2$, then the triangle is a right triangle.

Is the triangle shown here a right triangle?

We must substitute the longest side length, 12, for c, because it is the possible hypotenuse. The lengths of 8 and 10 may be substituted for either a or b.

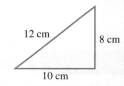

$$a^2 + b^2 = c^2$$ This is the Pythagorean equation.

$$8^2 + 10^2 \overset{?}{=} 12^2$$ Substitute 8 for a, 10 for b, and 12 for c.

$$64 + 100 \overset{?}{=} 144$$ Evaluate each exponential expression.

$$164 = 144$$ This is a false statement.

Since $164 \neq 144$, the triangle is not a right triangle.

REVIEW EXERCISES

Refer to the right triangle in the next column.

39. Find c, if $a = 5$ cm and $b = 12$ cm.

40. Find c, if $a = 8$ ft and $b = 15$ ft.

41. Find a, if $b = 77$ in. and $c = 85$ in.

42. Find b, if $a = 21$ ft and $c = 29$ ft.

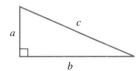

The lengths of two sides of a right triangle are given. Find the missing side length. Give the exact answer and an approximation to the nearest hundredth.

43.

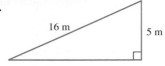

16 m   5 m

44.

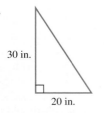

30 in.

20 in.

45. High-ropes adventure courses.   A builder of a high-ropes adventure course wants to secure a pole by attaching a support cable from the anchor stake 55 inches from the pole's base to a point 48 inches up the pole. See the illustration below. How long should the cable be?

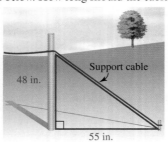

Support cable

48 in.

55 in.

46. TV screens.   Find the height of the television screen shown. Give the exact answer and an approximation to the nearest inch.

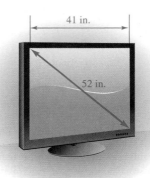

41 in.

52 in.

Determine whether each triangle shown here is a right triangle.

47.

8   11

15

48.

9   2

7

▶ **Congruent Triangles and Similar Triangles**

DEFINITIONS AND CONCEPTS	EXAMPLES
If two triangles have the same size and the same shape, they are **congruent triangles**.	C ... F $\triangle ABC \cong \triangle DEF$ A ... B D ... E
Corresponding parts of congruent triangles are congruent (have the same measure).	There are six pairs of congruent parts: three pairs of congruent angles and three pairs of congruent sides. ■ m($\angle A$) = m($\angle D$) ■ m($\overline{BC}$) = m($\overline{EF}$) ■ m($\angle B$) = m($\angle E$) ■ m($\overline{AC}$) = m($\overline{DF}$) ■ m($\angle C$) = m($\angle F$) ■ m($\overline{AB}$) = m($\overline{DE}$)

Three ways to show that two triangles are congruent are:

1. The SSS property: If three sides of one triangle are congruent to three sides of a second triangle, the triangles are congruent.

$\triangle MNO \cong \triangle RST$ by the SSS property.

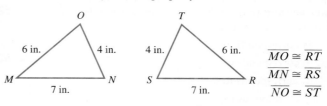

$\overline{MO} \cong \overline{RT}$
$\overline{MN} \cong \overline{RS}$
$\overline{NO} \cong \overline{ST}$

2. The SAS property: If two sides and the angle between them in one triangle are congruent, respectively, to two sides and the angle between them in a second triangle, the triangles are congruent.

$\triangle DEF \cong \triangle XYZ$ by the SAS property.

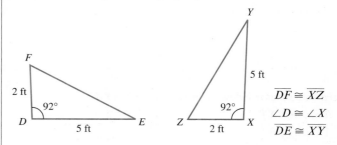

$\overline{DF} \cong \overline{XZ}$
$\angle D \cong \angle X$
$\overline{DE} \cong \overline{XY}$

3. The ASA property: If two angles and the side between them in one triangle are congruent, respectively, to two angles and the side between them in a second triangle, the triangles are congruent.

$\triangle ABC \cong \triangle TUV$ by the ASA property.

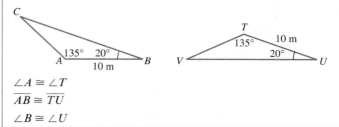

$\angle A \cong \angle T$
$\overline{AB} \cong \overline{TU}$
$\angle B \cong \angle U$

Similar triangles have the same shape, but not necessarily the same size.

We read the symbol ~ as "is similar to."

AAA similarity theorem
If the angles of one triangle are congruent to corresponding angles of another triangle, the triangles are similar.

$\triangle EFG \sim \triangle WXY$ by the AAA similarity theorem.

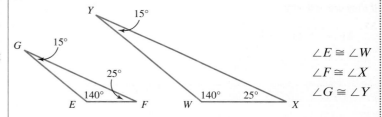

$\angle E \cong \angle W$
$\angle F \cong \angle X$
$\angle G \cong \angle Y$

Property of similar triangles
If two triangles are similar, all pairs of corresponding sides are in proportion.

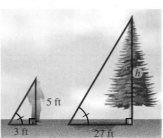

Similar triangles are determined by the tree and its shadow and the man and his shadow. Since the triangles are similar, the lengths of their corresponding sides are in proportion.

Landscaping. A tree casts a shadow 27 feet long at the same time as a man 5 feet tall casts a shadow 3 feet long. Find the height of the tree.

If we let $h =$ the height of the tree, we can find h by solving the following proportion.

The height of the tree → $\dfrac{h}{5} = \dfrac{27}{3}$ ← The length of the tree's shadow
The height of the man →   ← The length of the man's shadow

$3h = 5(27)$ Find each cross product and set them equal.

$3h = 135$ Do the multiplication.

$\dfrac{3h}{3} = \dfrac{135}{3}$ To isolate h, divide both sides by 3.

$h = 45$ Do the division.

The tree is 45 feet tall.

REVIEW EXERCISES

49. Two congruent triangles are shown below. Complete the list of corresponding parts.

a. ∠A corresponds to ____ .

b. ∠B corresponds to ____ .

c. ∠C corresponds to ____ .

d. $\overline{AC}$ corresponds to ____ .

e. $\overline{AB}$ corresponds to ____ .

f. $\overline{BC}$ corresponds to ____ .

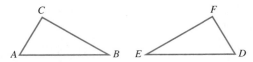

50. Refer to the figure below, where △ABC ≅ △XYZ.

a. Find m(∠X).

b. Find m(∠B).

c. Find m($\overline{YZ}$).

d. Find m($\overline{AB}$).

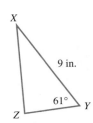

Determine whether the triangles in each pair are congruent. If they are, tell why.

51.

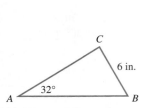

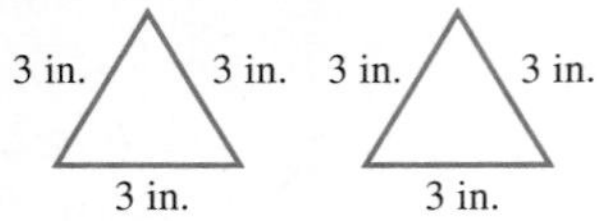

52.

53.

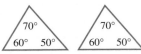

54.

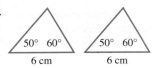

Determine whether the triangles are similar.

55.   **56.**

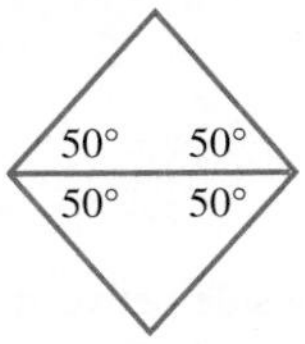

57. In the figure below, △RST ~ △MNO. Find x and y.

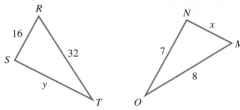

58. Height of a tree. A tree casts a 26-foot shadow at the same time a woman 5 feet tall casts a 2-foot shadow. What is the height of the tree? (*Hint:* Draw a diagram first and label the side lengths of the similar triangles.)

SECTION 9.6 ▶ **Quadrilaterals and Other Polygons**

DEFINITIONS AND CONCEPTS	EXAMPLES
A **quadrilateral** is a polygon with four sides. Use the capital letters that label the vertices of a quadrilateral to name it. A segment that joins two nonconsecutive vertices of a polygon is called a **diagonal** of the polygon.	**Quadrilateral WXYZ** 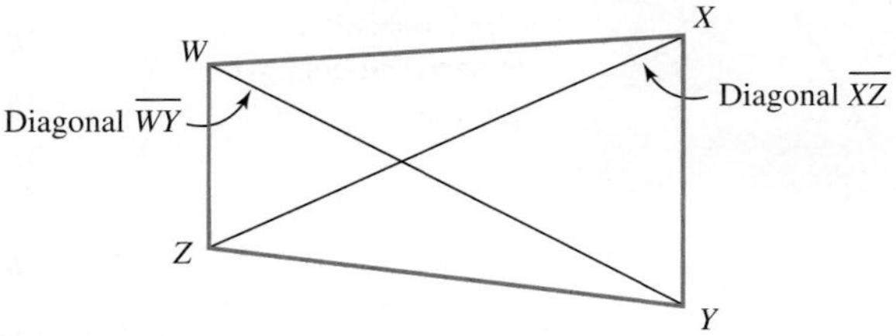 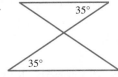 Diagonal $\overline{WY}$ Diagonal $\overline{XZ}$

Some special types of quadrilaterals are shown on the right.

Parallelogram
(Opposite sides parallel)

Rectangle
(Parallelogram with four right angles)

Square
(Rectangle with sides of equal length)

Rhombus
(Parallelogram with sides of equal length)

Trapezoid
(Exactly two sides parallel)

A **rectangle** is a quadrilateral with four right angles.

Rectangle *ABCD*

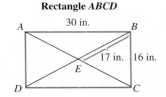

Properties of rectangles:

1. All four angles are right angles.

2. Opposite sides are parallel.

3. Opposite sides have equal length.

4. Diagonals have equal length.

5. The diagonals intersect at their midpoints.

1. $m(\angle DAB) = m(\angle ABC) = m(\angle BCD) = m(\angle CDA) = 90°$

2. $\overline{AD} \parallel \overline{BC}$ and $\overline{AB} \parallel \overline{DC}$

3. $m(\overline{AD}) = 16$ in. and $m(\overline{DC}) = 30$ in.

4. $m(\overline{DB}) = m(\overline{AC}) = 34$ in.

5. $m(\overline{DE}) = m(\overline{AE}) = m(\overline{EC}) = 17$ in.

Conditions that a parallelogram must meet to ensure that it is a rectangle:

1. If a parallelogram has one right angle, then the parallelogram is a rectangle.

2. If the diagonals of a parallelogram are congruent, then the parallelogram is a rectangle.

Read Example 2 on page 765 to see how these two conditions are used in construction to "square a foundation."

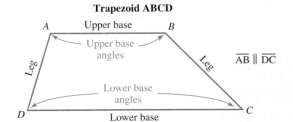

A **trapezoid** is a quadrilateral with exactly two sides parallel.

The parallel sides of a trapezoid are called **bases**. The nonparallel sides are called **legs**.

If the legs (the nonparallel sides) of a trapezoid are of equal length, it is called an **isosceles trapezoid**.

In an isosceles trapezoid, both pairs of **base angles** are congruent.

Trapezoid ABCD

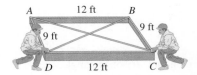

$\overline{AB} \parallel \overline{DC}$

The sum S, in degrees, of the measures of the angles of a polygon with n sides is given by the formula

$$S = (n - 2)180°$$

Find the sum of the angle measures of a hexagon.

Since a hexagon has six sides, we will substitute 6 for n in the formula.

$S = (n - 2)180°$

$S = (6 - 2)180°$ Substitute 6 for *n*, the number of sides.

$S = (4)180°$ Do the subtraction within the parentheses.

$S = 720°$ Do the multiplication.

The sum of the measures of the angles of a hexagon is 720°.

We can use the formula $S = (n - 2)180°$ to find the number of sides a polygon has.

The sum of the measures of the angles of a polygon is 2,340°. Find the number of sides the polygon has.

$$S = (n - 2)180°$$

$$2,340° = (n - 2)180°$$ — Substitute 2,340° for S. Now solve for n.

$$2,340° = 180°n - 360°$$ — Distribute the multiplication by 180°.

$$2,340° + 360° = 180°n - 360° + 360°$$ — Add 360° to both sides.

$$2,700° = 180°n$$ — Do the addition.

$$\frac{2,700°}{180°} = \frac{180°n}{180°}$$ — Divide both sides by 180°.

$$15 = n$$ — Do the division.

The polygon has 15 sides.

REVIEW EXERCISES

59. Classify each of the following quadrilaterals as a parallelogram, a rectangle, a square, a rhombus, or a trapezoid. Some figures may be correctly classified in more than one way.

a.

b.

2 cm
2 cm   2 cm
2 cm

c.
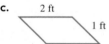
2 ft
1 ft

d.

e.

f.

60. The length of diagonal $\overline{AC}$ of rectangle $ABCD$ shown below is 15 centimeters. Find each measure.

a. m($\overline{BD}$)
b. m($\angle 1$)
c. m($\angle 2$)
d. m($\overline{EC}$)
e. m($\overline{AB}$)

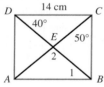
14 cm
D   C
40°
E   50°
2
1
A   B

61. Refer to rectangle $WXYZ$ below. Tell whether each statement is true or false.

a. m($\overline{WX}$) = m($\overline{ZY}$)   **b.** m($\overline{ZE}$) = m($\overline{EX}$)
c. Triangle WEX is isosceles.
d. m($\overline{WY}$) = m($\overline{WX}$)

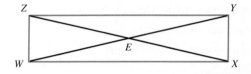
Z   Y
E
W   X

62. Refer to isosceles trapezoid $ABCD$ below. Find each measure.

a. m($\angle B$)
b. m($\angle C$)
c. m($\overline{CB}$)

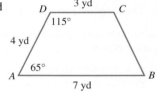
D   3 yd   C
115°
4 yd
65°
A   B
7 yd

63. Find the sum of the angle measures of an octagon.

64. The sum of the measures of the angles of a polygon is 3,240°. Find the number of sides the polygon has.

SECTION 9.7 ▶ **Perimeters and Areas of Polygons**

DEFINITIONS AND CONCEPTS	EXAMPLES
The **perimeter** of a polygon is the distance around it.	Find the perimeter of the triangle shown below.

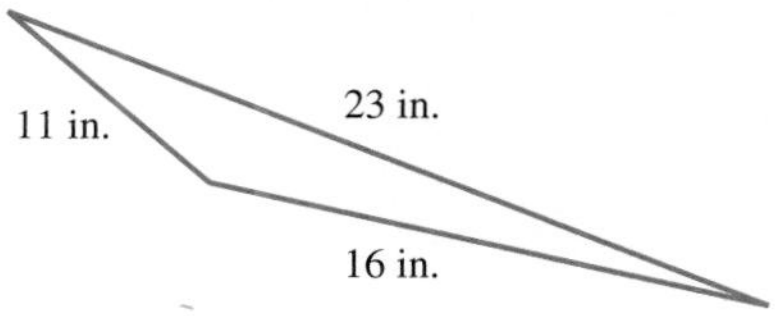

Figure	Perimeter Formula
Square	$P = 4s$
Rectangle	$P = 2l + 2w$
Triangle	$P = a + b + c$

$$P = a + b + c \qquad \text{This is the formula for the perimeter of a triangle.}$$

$$P = 11 + 16 + 23 \qquad \text{Substitute 11 for } a, 16 \text{ for } b, \text{ and } 23 \text{ for } c.$$

$$P = 50 \qquad \text{Do the addition.}$$

The perimeter of the triangle is 50 inches.

The **area** of a polygon is the measure of the amount of surface it encloses.

Figure	Area Formulas
Square	$A = s^2$
Rectangle	$A = lw$
Parallelogram	$A = bh$
Triangle	$A = \frac{1}{2}bh$
Trapezoid	$A = \frac{1}{2}h(b_1 + b_2)$

Find the area of the triangle shown here.

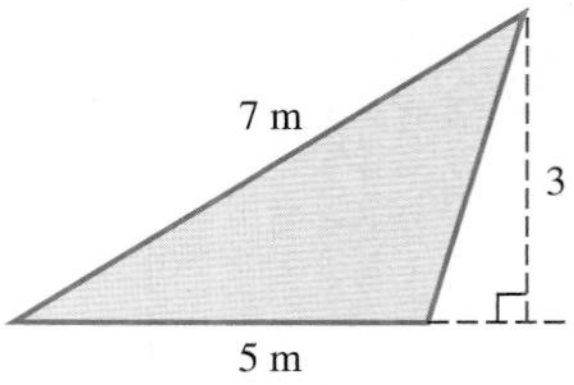

$$A = \frac{1}{2}bh \qquad \text{This is the formula for the area of a triangle.}$$

$$A = \frac{1}{2}(5)(3) \qquad \begin{array}{l}\text{Substitute 5 for } b, \text{ the length of the base,}\\ \text{and 3 for } h, \text{ the height. Note that the side length}\\ \text{7 m is not used in the calculation.}\end{array}$$

$$A = \frac{1}{2}\left(\frac{5}{1}\right)\left(\frac{3}{1}\right) \qquad \text{Write 5 as } \tfrac{5}{1} \text{ and 3 as } \tfrac{3}{1}.$$

$$A = \frac{15}{2} \qquad \begin{array}{l}\text{Multiply the numerators.}\\ \text{Multiply the denominators.}\end{array}$$

$$A = 7.5 \qquad \text{Do the division.}$$

The area of the triangle is 7.5 m².

To find the perimeter or area of a polygon, all the measurements must be in the **same units**. If they are not, use unit conversion factors to change them to the same unit.

To find the perimeter or area of the rectangle shown here, we need to express the length in inches.

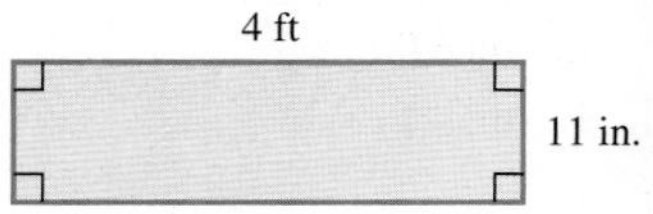

$$4 \text{ ft} = \frac{4 \text{ ft}}{1} \cdot \frac{12 \text{ in.}}{1 \text{ ft}} \qquad \begin{array}{l}\text{Convert 4 feet to inches using a unit}\\ \text{conversion factor.}\end{array}$$

$$= 4 \cdot 12 \text{ in.} \qquad \begin{array}{l}\text{Remove the common units of feet in the numerator}\\ \text{and denominator. The unit of inches remain.}\end{array}$$

$$= 48 \text{ in.} \qquad \text{Do the multiplication.}$$

The length of the rectangle is 48 inches. Now we can find the perimeter (in inches) or area (in in.²) of the rectangle.

If we know the area of a polygon, we can often use algebra to find an unknown measurement	The area of the parallelogram shown here is 208 ft². Find the height. 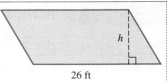

26 ft

$A = bh$ — This is the formula for the area of a parallelogram.

$208 = 26h$ — Substitute 208 for A, the area, and 26 for b, the length of the base.

$\dfrac{208}{26} = \dfrac{26h}{26}$ — To isolate h, undo the multiplication by 26 by dividing both sides by 26.

$8 = h$ — Do the division.

The height of the parallelogram is 8 feet.

To find the area of an irregular shape, break up the shape into familiar polygons. Find the area of each polygon, and then add the results.	Find the area of the shaded figure shown here. 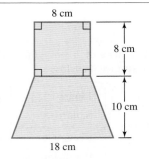

We will find the area of the lower portion of the figure (the trapezoid) and the area of the upper portion (the square) and then add the results.

$A_{trapezoid} = \dfrac{1}{2}h(b_1 + b_2)$ — This is the formula for the area of a trapezoid.

$A_{trapezoid} = \dfrac{1}{2}(10)(8 + 18)$ — Substitute 8 for b_1, 18 for b_2, and 10 for h.

$A_{trapezoid} = \dfrac{1}{2}(10)(26)$ — Do the addition within the parentheses.

$A_{trapezoid} = 130$ — Do the multiplication.

The area of the trapezoid is 130 cm².

$A_{square} = s^2$ — This is the formula for the area of a square.

$A_{square} = 8^2$ — Substitute 8 for s.

$A_{square} = 64$ — Evaluate the exponential expression.

The area of the square is 64 cm².

The total area of the shaded figure is

$A_{total} = A_{trapezoid} + A_{square}$

$A_{total} = 130 \text{ cm}^2 + 64 \text{ cm}^2$

$A_{total} = 194 \text{ cm}^2$

The area of the shaded figure is 194 cm².

To find the area of an irregular shape, we must sometimes use subtraction.	To find the area of the shaded figure below, we subtract the area of the triangle *from* the area of the rectangle. 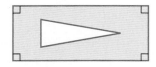 $A_{shaded} = A_{rectangle} - A_{triangle}$

REVIEW EXERCISES

65. Find the perimeter of a square with sides 18 inches long.

66. Find the perimeter (in inches) of a rectangle that is 7 inches long and 3 feet wide.

Find the perimeter of each polygon.

67.

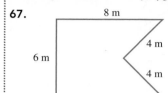

68.

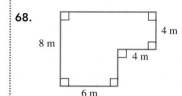

69. The perimeter of an isosceles triangle is 107 feet. If one of the congruent sides is 34 feet long, how long is the base?

70. a. How many square feet are there in 1 square yard?

    b. How many square inches are in 1 square foot?

Find the area of each polygon.

71.

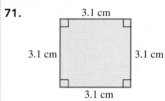

72.

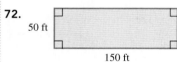

73.

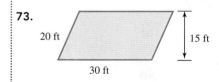

74.

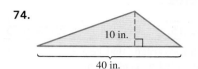

75.

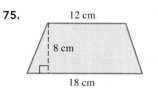

76.

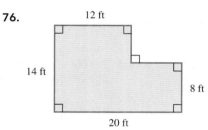

77.

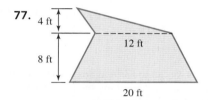

78.

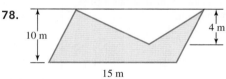

79. The area of a parallelogram is 240 ft². If the length of the base is 30 feet, what is its height?

80. The perimeter of a rectangle is 48 mm and its width is 6 mm. Find its length.

81. Fences. A man wants to enclose a rectangular front yard with chain link that costs $8.50 a foot (the price includes installation). Find the cost of enclosing the yard if its dimensions are 115 ft by 78 ft.

82. Lawns. A family is going to have artificial turf installed in their rectangular backyard that is 36 feet long and 24 feet wide. If the turf costs $48 per *square yard*, and the installation is free, what will this project cost? (Assume no waste.)

DEFINITIONS AND CONCEPTS	EXAMPLES
A **circle** is the set of all points in a plane that lie a fixed distance from a point called its **center**. The fixed distance is the circle's **radius**. A **chord** of a circle is a line segment connecting two points on the circle. A **diameter** is a chord that passes through the circle's center. Any part of a circle is called an **arc**. A **semicircle** is an arc of a circle whose endpoints are the endpoints of a diameter.	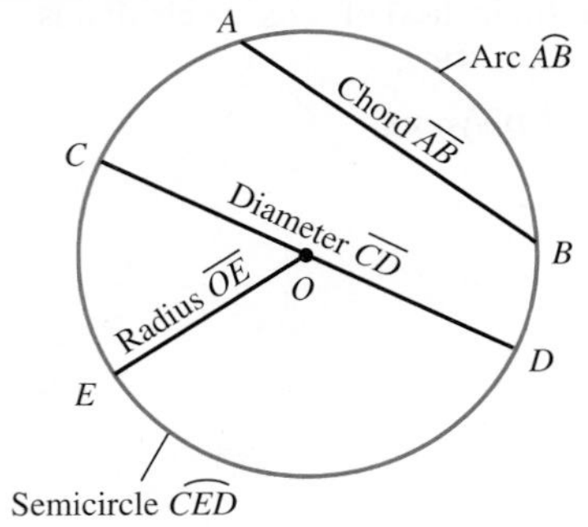 Arc $\widehat{AB}$ Chord $\overline{AB}$ Diameter $\overline{CD}$ Radius $\overline{OE}$ Semicircle $\widehat{CED}$
The **circumference** (perimeter) of a circle is given by the formulas $$C = \pi D \quad \text{or} \quad C = 2\pi r$$ where $\pi = 3.14159\ldots.$	Find the circumference of the circle shown here. Give the exact answer and an approximation. 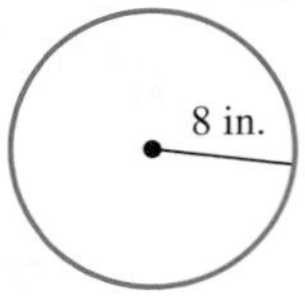 8 in. $C = 2\pi r$ This is the formula for the circumference of a circle. $C = 2\pi(8)$ Substitute 8 for r, the radius. $C = 2(8)\pi$ Rewrite the product so that π is the last factor. $C = 16\pi$ Do the first multiplication: 2(8) = 16. This is the exact answer.
If an exact answer contains π, we can use 3.14 as an approximation and complete the calculations by hand. Or, we can use a calculator that has a pi key $\boxed{\pi}$ to find an approximation.	The circumference of the circle is exactly 16π inches. If we replace π with 3.14, we get an approximation of the circumference. $C = 16\pi$ $C \approx 16(3.14)$ Substitute 3.14 for π. $C \approx 50.24$ Do the multiplication. The circumference of the circle is approximately 50.2 inches. We can also use a calculator to approximate 16π. $C \approx 50.26548246$
The **area** of a circle is given by the formula $$A = \pi r^2$$	Find the area of the circle shown here. Give the exact answer and an approximation to the nearest tenth. 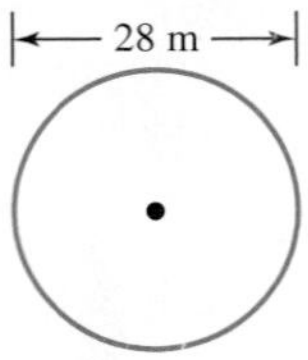 $\leftarrow$ 28 m $\rightarrow$ Since the diameter is 28 meters, the radius is half of that, or 14 meters. $A = \pi r^2$ This is the formula for the area of a circle. $A = \pi(14)^2$ Substitute 14 for r, the radius of the circle. $A = \pi(196)$ Evaluate the exponential expression. $A = 196\pi$ Write the product so that π is the last factor. The exact area of the circle is 196π m². We can use a calculator to approximate the area. $A \approx 615.7521601$ Use a calculator to do the multiplication. To the nearest tenth, the area is 615.8 m².
To find the area of an irregular shape, break it up into familiar figures.	To find the area of the shaded figure shown here, find the area of the triangle and the area of the semicircle, and then add the results. 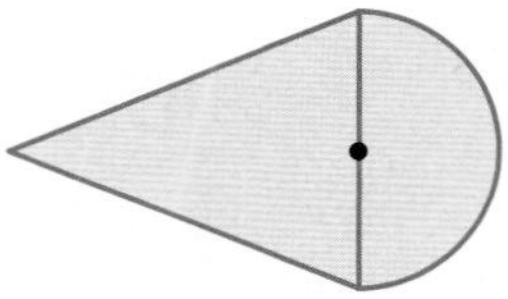 $$A_{\text{shaded figure}} = A_{\text{triangle}} + A_{\text{semicircle}}$$

REVIEW EXERCISES

83. Refer to the figure.

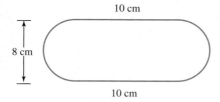

   a. Name each chord.

   b. Name each diameter.

   c. Name each radius.

   d. Name the center.

84. Find the circumference of a circle with a diameter of 21 feet. Give the exact answer and an approximation to the nearest hundredth.

85. Find the perimeter of the figure shown below. Round to the nearest tenth.

10 cm

8 cm

10 cm

86. Find the area of a circle with a diameter of 18 inches. Give the exact answer and an approximation to the nearest hundredth.

87. Find the area of the figure shown in Problem 85. Round to the nearest tenth.

88. Find the area of the shaded region shown on the right. Round to the nearest tenth.

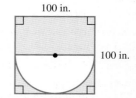

100 in.

100 in.

<table>
<tr><td>**SECTION 9.9**</td><td>▶ **Volume**</td></tr>
</table>

DEFINITIONS AND CONCEPTS	EXAMPLES
The volume of a figure can be thought of as the number of **cubic units** that will fit within its boundaries. Two common units of volume are cubic inches (in.³) and cubic centimeters (cm³).	**1 cubic inch: 1 in.³ 1 cubic centimeter: 1 cm³** 1 in. 1 cm 1 in. 1 cm 1 in. 1 cm

The **volume** of a solid is a measure of the space it occupies.

Figure	Volume Formula
Cube	$V = s^3$
Rectangular solid	$V = lwh$
Prism	$V = Bh$*
Pyramid	$V = \frac{1}{3}Bh$*
Cylinder	$V = \pi r^2 h$
Cone	$V = \frac{1}{3}\pi r^2 h$
Sphere	$V = \frac{4}{3}\pi r^3$

*B represents the area of the base.

Carry-on luggage. The largest carry-on bag that Alaska Airlines allows on board a flight is shown on the right. Find the volume of space that a bag that size occupies.

Height: 17 in.
Width: 10 in.
Length: 24 in.

$V = lwh$ This is the formula for the volume of a rectangular solid.

$V = 24(10)(17)$ Substitute 24 for l, the length, 10 for w, the width, and 17 for h, the height of the bag.

$V = 4,080$ Do the multiplication.

The volume of the space that the bag occupies is 4,080 in.³.

Caution! When finding the volume of a figure, only use the measurements that are called for in the formula. Sometimes a figure may be labeled with measurements that are not used.

Find the volume of the prism shown here.

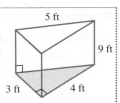

The area of the triangular base of the prism is $\frac{1}{2}(3)(4) = 6$ square feet. (The 5-ft measurement is not used.) To find the volume of the prism, proceed as follows:

$V = Bh$	This is the formula for the volume of a prism.
$V = 6(9)$	Substitute 6 for B, the area of the base, and 9 for h, the height.
$V = 54$	Do the multiplication.

The volume of the triangular prism is 54 ft³.

The letter B appears in two of the volume formulas. It represents the area of the base of the figure.

Note that the volume formulas for a pyramid and a cone contain a factor of $\frac{1}{3}$.

Cone: $V = \dfrac{1}{3}\pi r^2 h$

Pyramid: $V = \dfrac{1}{3}Bh$

Find the volume of the pyramid shown here.

Since the base is a square with each side 5 centimeters long, the area of the base is $5 \cdot 5 = 25$ cm².

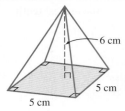

$V = \dfrac{1}{3}Bh$	This is the formula for the volume of a pyramid.
$V = \dfrac{1}{3}(25)(6)$	Substitute 25 for B, the area of the base, and 6 for h, the height.
$V = 25(2)$	Multiply the first and third factors: $\frac{1}{3}(6) = 2$.
$V = 50$	Complete the multiplication by 25.

The volume of the pyramid is 50 cm³.

Note that the volume formulas for a cone, a cylinder, and a sphere contain a factor of π.

Cone: $V = \dfrac{1}{3}\pi r^2 h$

Cylinder: $V = \pi r^2 h$

Sphere: $V = \dfrac{4}{3}\pi r^3$

Find the volume of the cylinder shown here. Give the exact answer and an approximation to the nearest hundredth.

Since a radius is one-half of the diameter of the circular base, $r = \frac{1}{2} \cdot 8$ yd = 4 yd. To find the volume of the cylinder, proceed as follows:

$V = \pi r^2 h$	This is the formula for the volume of a cylinder.
$V = \pi(4)^2(3)$	Substitute 4 for r, the radius of the base, and 3 for h, the height.
$V = \pi(16)(3)$	Evaluate the exponential expression.
$V = 48\pi$	Write the product so that π is the last factor.
$V \approx 150.7964474$	Use a calculator to do the multiplication.

The exact volume of the cylinder is 48π yd³. To the nearest hundredth, the volume is 150.80 yd³.

If an exact answer contains π, we can use 3.14 as an approximation and complete the calculations by hand. Or, we can use a calculator that has a pi key $\boxed{\pi}$ to find an approximation.

Find the volume of the sphere shown here. Give the exact answer and an approximation to the nearest tenth.

$$V = \frac{4}{3}\pi r^3$$ This is the formula for the volume of a sphere.

$$V = \frac{4}{3}\pi(6)^3$$ Substitute 6 for r, the radius of the sphere.

$$V = \frac{4}{3}\pi(216)$$ Evaluate the exponential expression.

$$V = \frac{864}{3}\pi$$ Multiply: 4(216) = 864.

$$V = 288\pi$$ Divide: $\frac{864}{3} = 288$.

$$V \approx 904.7786842$$ Use a calculator to do the multiplication.

The volume of the sphere is exactly 288π ft³. To the nearest tenth, this is 904.8 ft³.

REVIEW EXERCISES

Find the volume of each figure. If an exact answer contains π, approximate to the nearest hundredth.

89.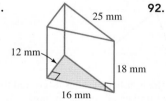
5 cm, 5 cm, 5 cm

90.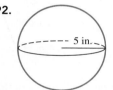
8 m, 10 m, 6 m

91.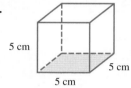
25 mm, 12 mm, 18 mm, 16 mm

92.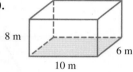
5 in.

93.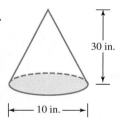
30 in., 10 in.

94.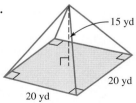
15 yd, 20 yd, 20 yd

95.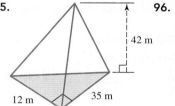
42 m, 12 m, 35 m

96.
16 in.

97. Farming. Find the volume of the corn silo shown below. Round to the nearest one cubic foot.

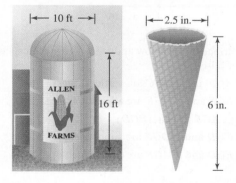
10 ft, 16 ft, 2.5 in., 6 in.

98. Waffle cones. Find the volume of the ice cream cone shown above. Give the exact answer and an approximation to the nearest tenth.

99. How many cubic inches are there in 1 cubic foot?

100. How many cubic feet are there in 2 cubic yards?

1. Estimate each angle measure. Then tell whether it is an acute, right, obtuse, or straight angle.

a.
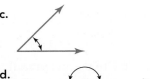

b.

c.
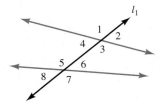

d.

2. Fill in the blanks.

a. If $\angle ABC \cong \angle DEF$, then the angles have the same _____.

b. Two congruent segments have the same _____.

c. Two different points determine one _____.

d. Two angles are called _____ if the sum of their measures is 90°.

3. Refer to the figure below. What is the midpoint of $\overline{BE}$?

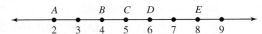

4. Refer to the figure below and tell whether each statement is true or false.

a. $\angle AGF$ and $\angle BGC$ are vertical angles.

b. $\angle EGF$ and $\angle DGE$ are adjacent angles.

c. $m(\angle AGB) = m(\angle EGD)$.

d. $\angle CGD$ and $\angle DGF$ are supplementary angles.

e. $\angle EGD$ and $\angle AGB$ are complementary angles.

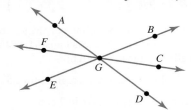

5. Find x. Then find $m(\angle ABD)$ and $m(\angle CBE)$.

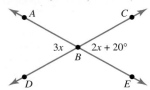

6. Find the supplement of a 47° angle.

7. Refer to the figure below. Fill in the blanks.

a. l_1 intersects two coplanar lines. It is called a _____.

b. $\angle 4$ and _____ are alternate interior angles.

c. $\angle 3$ and _____ are corresponding angles.

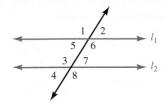

8. In the figure below, $l_1 \parallel l_2$ and $m(\angle 2) = 25°$. Find the measures of the other numbered angles.

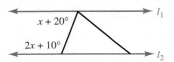

9. In the figure below, $l_1 \parallel l_2$. Find x. Then determine the measure of each angle that is labeled in the figure.

10. For each polygon, give the number of sides it has, tell its name, and then give the number of vertices it has.

a.

b.

c.

d.

11. Classify each triangle as an equilateral triangle, an isosceles triangle, or a scalene triangle.

a.

b. 4 in. 5 in. 6 in.

c.

d. 56° 56°

12. Find x.

13. The measure of the vertex angle of an isosceles triangle is 12°. Find the measure of each base angle.

14. Refer to rectangle *EFGH* shown below.

   a. Find m($\overline{HG}$).   **b.** Find m($\overline{FH}$).

   c. Find m($\angle FGH$).   **d.** Find m($\overline{EH}$).

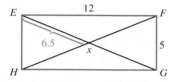

15. Refer to isosceles trapezoid *QRST* shown below.

   a. Find m($\overline{RS}$).   **b.** Find x.

   c. Find y.   **d.** Find z.

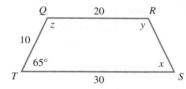

16. Find the sum of the measures of the angles of a decagon.

17. Find the perimeter of the figure shown below.

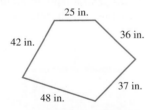

18. The perimeter of an equilateral triangle is 45.6 m. Find the length of each side.

19. Find the area of the shaded part of the figure shown below.

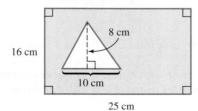

20. Decorating. A patio has the shape of a trapezoid, as shown on the right. If indoor/outdoor carpeting sells for $18 a *square yard* installed, how much will it cost to carpet the patio?

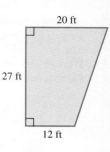

21. How many square inches are in one square foot?

22. Find the area of the rectangle shown below in square inches.

23. Refer to the figure below, where *O* is the center of the circle.

   a. Name each chord.

   b. Name a diameter.

   c. Name each radius.

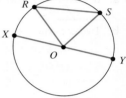

24. Fill in the blank: If C is the circumference of a circle and D is the length of its diameter, then $\frac{C}{D} = $ ___ .

In Problems 25–27, when appropriate, give the exact answer and an approximation to the nearest tenth.

25. Find the circumference of a circle with a diameter of 21 feet.

26. Find the perimeter of the figure shown below. Assume that the arcs are semicircles.

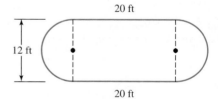

27. History. Stonehenge is a prehistoric monument in England, believed to have been built by the Druids. The site, 30 meters in diameter, consists of a circular arrangement of stones, as shown below. What area does the monument cover?

28. See the figure below, where $\triangle MNO \cong \triangle RST$. Name the six corresponding parts of the congruent triangles.

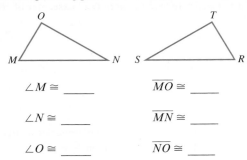

$\angle M \cong$ _____ $\overline{MO} \cong$ _____

$\angle N \cong$ _____ $\overline{MN} \cong$ _____

$\angle O \cong$ _____ $\overline{NO} \cong$ _____

29. Tell whether each pair of triangles are congruent. If they are, tell why.

a.
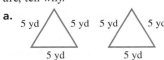
5 yd 5 yd 5 yd 5 yd
5 yd   5 yd

b.
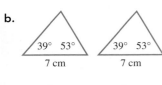
39° 53°   39° 53°
7 cm   7 cm

c.
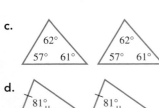
62°   62°
57° 61°   57° 61°

d.
81°   81°

30. Refer to the figure below, in which $\triangle ABC \cong \triangle DEF$.

a. Find m($\overline{DE}$).   **b.** Find m($\angle E$).

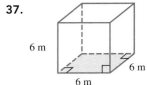

31. Tell whether the triangles in each pair are similar.

a.
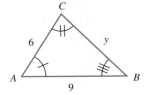
43° 43°
43° 43°

b.
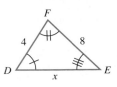
29°
29°

32. Refer to the triangles below. The units are meters.

a. Find x.   **b.** Find y.

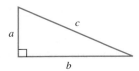

33. Shadows. If a tree casts a 7-foot shadow at the same time as a man 6 feet tall casts a 2-foot shadow, how tall is the tree?

34. Refer to the right triangle below. Find the missing side length. Approximate any exact answers that contain a square root to the nearest tenth.

a. Find c if $a = 10$ cm and $b = 24$ cm.
b. Find b if $a = 6$ in. and $c = 8$ in.

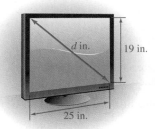

35. Televisions. To the nearest tenth of an inch, what is the diagonal measurement of the television screen shown below?

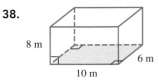

36. How many cubic inches are there in 1 cubic foot?

Find the volume of each figure. Give the exact answer and an approximation to the nearest hundredth if an answer contains π.

37.
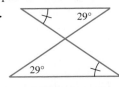
6 m
6 m
6 m

38.
8 m
6 m
10 m

39.

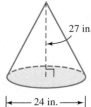

27 in.

|← 24 in. →|

40.

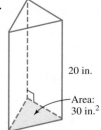

20 in.

Area: 30 in.²

41.

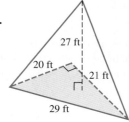

27 ft
20 ft
21 ft
29 ft

42.

3 yd

7 yd

43.

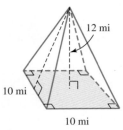

12 mi

10 mi

10 mi

44.

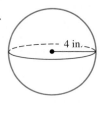

4 in.

45. Farming. A silo is used to store wheat and corn. Find the volume of the silo shown below. Give the exact answer and an approximation to the nearest cubic foot.

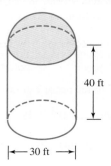

40 ft

|← 30 ft →|

46. Give a real-life example in which the concept of perimeter is used. Do the same for area and for volume. Be sure to discuss the type of units used in each case.

1. **Used cars.** The following ad appeared in *The Car Trader*. (O.B.O. means "or best offer.") If offers of $8,750, $8,875, $8,900, $8,850, $8,800, $7,995, $8,995, and $8,925 were received, what was the selling price of the car? [Section 1.1]

> 1969 Ford Mustang. New tires
> Must sell!!!! $10,500 O.B.O.

2. Round 2,109,567 to the nearest thousand. [Section 1.1]

3. Add: $458 + 8,099 + 23,419 + 58$ [Section 1.2]

4. Subtract: $35,021 - 23,999$ [Section 1.3]

5. **Parking.** The length of a rectangular parking lot is 204 feet, and its width is 97 feet. [Section 1.4]

   a. Find the perimeter of the lot.

   b. Find the area of the lot.

6. Divide: $1,363 \div 41$ [Section 1.5]

7. **Painting.** One gallon of paint covers 350 square feet. How many gallons are needed if the total area of walls and ceilings to be painted is 8,400 square feet, and if two coats must be applied? [Section 1.6]

8. a. Prime factor 220. [Section 1.7]

   b. Find all the factors of 12. [Section 1.7]

9. a. Find the LCM of 16 and 24. [Section 1.8]

   b. Find the GCF of 16 and 24.

10. Evaluate: $\dfrac{(3 + 5)^2 + 2}{2(8 - 5)}$ [Section 1.9]

11. a. Write the set of integers. [Section 2.1]

    b. Simplify: $-(-3)$ [Section 2.1]

12. Perform the operations.

    a. $-16 + 4$ [Section 2.2]

    b. $16 - (-4)$ [Section 2.3]

    c. $-16(4)$ [Section 2.4]

    d. $\dfrac{-16}{-4}$ [Section 2.5]

    e. -4^2 [Section 2.4]

    f. $(-4)^2$ [Section 2.4]

13. **Overdraft protection.** A student forgot that she had only $30 in her bank account and wrote a check for $55 and used her debit card to buy $75 worth of groceries. On each of the two transactions, the bank charged her a $20 overdraft protection fee. Find the new account balance. [Section 2.3]

14. Evaluate: $10 - 4|6 - (-3)^2|$ [Section 2.6]

15. a. Simplify: $\dfrac{35}{28}$ [Section 3.1]

    b. Write $\dfrac{3}{8}$ as an equivalent fraction with denominator 48. [Section 3.1]

    c. What is the reciprocal of $\dfrac{9}{8}$? [Section 3.3]

    d. Write $7\dfrac{1}{2}$ as an improper fraction. [Section 3.5]

16. **Gravity.** Objects on the moon weigh only one-sixth as much as on Earth. If a rock weighs 54 ounces on the Earth, how much does it weigh on the moon? [Section 3.2]

Perform the operations.

17. $-\dfrac{5}{77}\left(\dfrac{33}{50}\right)$ [Section 3.2]

18. $\dfrac{15}{16} \div \dfrac{45}{8}$ [Section 3.3]

19. $\dfrac{3}{4} - \dfrac{3}{5}$ [Section 3.4]

20. $-\dfrac{6}{25}\left(2\dfrac{7}{24}\right)$ [Section 3.5]

21. $45\dfrac{2}{3} + 96\dfrac{4}{5}$ [Section 3.6]

22. $\dfrac{7 - \dfrac{2}{3}}{4\dfrac{5}{6}}$ [Section 3.7]

23. **Pet medication.** A pet owner was told to use an eye dropper to administer medication to his sick kitten. The cup shown below contains 8 doses of the medication. Determine the size of a single dose. [Section 3.3]

24. **Baking.** A bag of all-purpose flour contains $17\dfrac{1}{2}$ cups. A baker uses $3\dfrac{3}{4}$ cups. How many cups of flour are left? [Section 3.6]

25. Evaluate: $\dfrac{3}{4} + \left(-\dfrac{1}{3}\right)^2\left(\dfrac{5}{4}\right)$ [Section 3.7]

26. a. Round the number pi to the nearest ten thousandth: $\pi = 3.141592654\ldots$ [Section 4.1]

   b. Place the proper symbol ($>$ or $<$) in the blank: 154.34 ▒ 154.33999.   [Section 4.1]

   c. Write 6,510,345.798 in words.   [Section 4.1]

   d. Write 7,498.6461 in expanded notation.   [Section 4.1]

Perform the operations.

27. $3.4 + 106.78 + 35 + 0.008$ [Section 4.2]

28. $5{,}091.5 - 1{,}287.89$ [Section 4.2]

29. $-8.8 + (-7.3 - 9.5)$ [Section 4.2]

30. $-5.5(-3.1)$ [Section 4.3]

31. $\dfrac{0.0742}{1.4}$ [Section 4.4]

32. $\dfrac{7}{8}(9.7 + 15.8)$ [Section 4.5]

33. Paychecks.   If you are paid every other week, your *monthly gross income* is your gross income from one paycheck times 2.17. Find the monthly gross income of a secretary who earns \$1,250 every two weeks.   [Section 4.3]

34. Perform each operation in your head.

   a. $(89.9708)(10{,}000)$ [Section 4.3]

   b. $\dfrac{89.9708}{100}$ [Section 4.4]

35. Estimate the quotient: $9.2\overline{)18{,}460.76}$ [Section 4.4]

36. Evaluate $\dfrac{(-1.3)^2 + 6.7}{-0.9}$ and round the result to the nearest hundredth.   [Section 4.4]

37. Write $\dfrac{2}{15}$ as a decimal. Use an overbar.   [Section 4.5]

38. Evaluate each expression.   [Section 4.6]

   a. $2\sqrt{121} - 3\sqrt{64}$

   b. $\sqrt{\dfrac{49}{81}}$

39. Graph each number on the number line.   [Section 4.6]

$$\left\{-4\dfrac{5}{8},\ \sqrt{17},\ 2.89,\ \dfrac{2}{3},\ -0.1,\ -\sqrt{9},\ \dfrac{3}{2}\right\}$$

$$\overset{}{\underset{-5\ \ -4\ \ -3\ \ -2\ \ -1\ \ \ 0\ \ \ 1\ \ \ 2\ \ \ 3\ \ \ 4\ \ \ 5}{\longleftarrow\!\mid\,\mid\,\mid\,\mid\,\mid\,\mid\,\mid\,\mid\,\mid\,\mid\,\mid\!\longrightarrow}}$$

40. Write each phrase as a ratio (fraction) in simplest form.   [Section 5.1]

   a. 3 centimeters to 7 centimeters

   b. 13 weeks to 1 year

41. Comparison shopping.   A dry-erase whiteboard with an area of 400 in.² sells for \$24. A larger board, with an area of 600 in.², sells for \$42. Which board is the better buy?   [Section 5.1]

42. Solve the proportion: $\dfrac{x}{14} = \dfrac{13}{28}$ [Section 5.2]

43. Insurance claims.   In one year, an auto insurance company had 3 complaints per 1,000 policies. If a total of 375 complaints were filed that year, how many policies did the company have?   [Section 5.2]

44. Scale drawings.   On the scale drawing below, $\frac{1}{4}$-inch represents an actual length of 3 feet. The length of the house on the drawing is $6\frac{1}{4}$ inches. What is the actual length of the house?   [Section 5.2]

Scale $\frac{1}{4}$ in. : 3 ft

45. Make each conversion.   [Section 5.3]

   a. Convert 168 inches to feet.

   b. Convert 212 ounces to pounds.

   c. Convert 30 gallons to quarts.

   d. Convert 12.5 hours to minutes.

46. Make each conversion.   [Section 5.4]

   a. Convert 1.538 kilograms to grams.

   b. Convert 500 milliliters to liters.

   c. Convert 0.3 centimeters to kilometers.

47. The Amazon. The Amazon River enters the Atlantic Ocean through a broad estuary, roughly estimated at 240,000 m in width. Convert the width to kilometers. [Section 5.4]

48. Ocean liners. In making ocean crossings from England to America, the *Queen Mary* consumed fuel at a rate of 13 feet to the gallon. [Section 5.5]

   a. How many meters per gallon is this?

   b. The fuel capacity of the ship was 3,000,000 gallons. How many liters is this?

49. Cooking. What is the weight of a 10-pound ham in kilograms? [Section 5.5]

50. Convert 75°C to degrees Fahrenheit. [Section 5.5]

51. Complete the table. [Section 6.1]

Percent	Decimal	Fraction
57%		
	0.001	
		$\frac{1}{3}$

52. Refer to the figure on the right. [Section 6.1]

   a. What percent of the figure is shaded?

   b. What percent is not shaded?

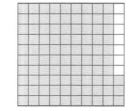

53. What number is 15% of 450? [Section 6.2]

54. 24.6 is 20.5% of what number? [Section 6.2]

55. 51 is what percent of 60? [Section 6.2]

56. Clothing sales. Find the amount of the discount and the sale price of the coat shown below. [Section 6.3]

Men's Open Range Coat
Save 25%
Regularly $820 ⁰⁰
Winter Coats on Sale!
Genuine leather

57. Sales tax. If the sales tax rate is $6\frac{1}{4}$%, how much sales tax will be added to the price of a new car selling for $18,550? [Section 6.3]

58. Collectibles. A porcelain figurine, which was originally purchased for $125, was sold by a collector ten years later for $750. What was the percent increase in the value of the figurine? [Section 6.3]

59. Tipping. Estimate a 15% tip on a dinner that cost $135.88. [Section 6.4]

60. Paying off loans. To pay for tuition, a college student borrows $1,500 for six months. If the annual interest rate is 9%, how much will the student have to repay when the loan comes due? [Section 6.5]

61. Freeways. Refer to the pictograph below to answer the following questions. [Section 7.1]

Freeway Traffic
Average number of vehicles daily

I-405 Los Angeles 🚗🚗🚗🚗🚗🚗🚗🚗🚗

I-95 Virginia 🚗🚗🚗🚗🚗🚗🚗

US-59 Houston 🚗🚗🚗🚗🚗🚗🚗🚗

I-270 Maryland 🚗🚗🚗🚗🚗🚗

🚗 = 50,000 vehicles

Source: Office of Highway Policy Information, Federal Highway Administration

   a. Estimate the number of vehicles that travel the I-405 Freeway in Los Angeles each day.

   b. Estimate the number of vehicles that travel the I-270 Freeway in Maryland each day.

   c. Estimate how many more vehicles travel the US-59 in Houston than the I-95 in Virginia each day.

62. Vegetarians. The graph below gives the results of a recent study by *Vegetarian Times*. [Section 7.1]

Survey Results: Ages of Adult Vegetarians in the United States

Over 55 yrs old
41% 35–54 yrs old
42% 18–34 yrs old

Source: *Vegetarian Times*

   a. According to the study, what percent of the adult vegetarians in the United States are over 55 years old?

   b. The study estimated that there were 7,300,000 adult vegetarians in the United States. How many of them are 35 to 54 years old?

63. Spending on pets. Refer to the bar graph below to answer the following questions. [Section 7.1]

a. In what category was the most money spent on pets? Estimate how much.

b. Estimate how much money was spent on purchasing pets.

c. Estimate how much more money was spent on vet care than on grooming and boarding.

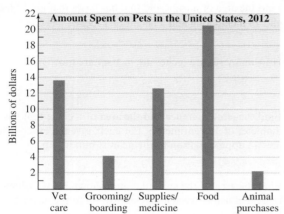

Source: americanpetproducts.org

64. Table tennis. The weights (in ounces) of 8 Ping-Pong balls that are to be used in a tournament are as follows: 0.85, 0.87, 0.88, 0.88, 0.85, 0.86, 0.84, and 0.85. Find the mean, median, and mode of the weights. [Section 7.2]

65. Find the probability of each event. [Section 7.3]

a. drawing a face card on one draw from a standard deck

b. rolling a number greater than 4 on one roll of a die

c. getting a sum of 6 when 2 dice are rolled

66. Evaluate the expression $\dfrac{2x + 3y}{z - y}$ for $x = 2$, $y = -3$, and $z = -4$. [Section 8.1]

67. Translate each phrase into an algebraic expression. [Section 8.1]

a. 16 less than twice x

b. the product of 75 and s, increased by 6

68. Simplify each expression. [Section 8.2]

a. $12(4a)$ **b.** $-2b(-7)(3)$

c. $9(3t - 10)$ **d.** $8(4x - 5y + 1)$

69. Combine like terms. [Section 8.2]

a. $10x - 7x$

b. $c^2 + 4c^2 + 2c^2 - c^2$

c. $4m - n - 12m + 7n$

d. $4x - 2(3x - 4) - 5(2x)$

70. Check to determine whether -6 is a solution of $5x + 9 = x + 16$. [Section 8.3]

Solve each equation and check the result. [Section 8.4]

71. $\dfrac{x}{8} - 2 = -5$ **72.** $4x - 40 = -20$

73. $3(2p + 15) = 3p - 4(11 - p)$

74. $-x + 2 = 13$

75. Observation hours. To pass a teacher education course, a student must have 90 hours of classroom observation time. If a student has already observed for 48 hours, how many 6-hour classroom visits must she make to meet the requirement? (*Hint:* Form an equation and solve it to answer the question.) [Section 8.5]

76. Identify the base and the exponent of each expression. [Section 8.6]

a. 4^8 **b.** $3s^4$

77. Simplify each expression. [Section 8.6]

a. $s^4 \cdot s^5 \cdot s$ **b.** $(a^5)^7$

c. $(r^2 t^4)(r^3 t^5)$ **d.** $(2b^3 c^6)^3$

e. $(y^5)^2(y^4)^3$ **f.** $\left[(-5.5)^3\right]^{12}$

78. Fill in the blanks. [Section 9.1]

a. The measure of an _____ angle is less than 90°.

b. The measure of a _____ angle is 90°.

c. The measure of an _____ angle is greater than 90° but less than 180°.

d. The measure of a straight angle is _____.

79. a. Find the supplement of an angle of 105°. [Section 9.1]

b. Find the complement of an angle of 75°. [Section 9.1]

80. Refer to the figure below, where $l_1 \parallel l_2$. Find the measure of each angle. [Section 9.2]

  a. m($\angle 1$)   **b.** m($\angle 3$)

  c. m($\angle 2$)   **d.** m($\angle 4$)

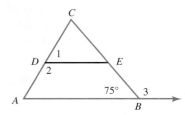

81. Refer to the figure below, where $AB \parallel DE$ and m($\overline{AC}$) = m($\overline{BC}$). Find the measure of each angle. [Section 9.3]

  a. m($\angle 1$)   **b.** m($\angle C$)

  c. m($\angle 2$)   **d.** m($\angle 3$)

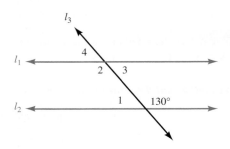

82. Javelin throw. Refer to the illustration below. Determine x and y. [Section 9.3]

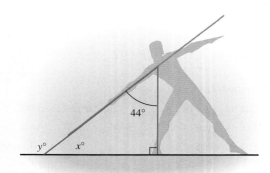

83. If the vertex angle of an isosceles triangle measures 34°, what is the measure of each base angle? [Section 9.3]

84. If the legs of a right triangle measure 10 meters and 24 meters, how long is the hypotenuse? [Section 9.4]

85. Determine whether a triangle with sides of length 16 feet, 63 feet, and 65 feet is a right triangle. [Section 9.4]

86. Shadows. If a tree casts a 35-foot shadow at the same time as a man 6 feet tall casts a 5-foot shadow, how tall is the tree? [Section 9.5]

87. Find the sum of the angles of a pentagon. [Section 9.6]

88. Find the perimeter and the area of a square that has sides each 12 meters long. [Section 9.7]

89. Find the area of a triangle with a base that is 14 feet long and a height of 18 feet. [Section 9.7]

90. Find the area of a trapezoid that has bases that are 12 inches and 14 inches long and a height of 7 inches. [Section 9.7]

91. How many square inches are in 1 square foot? [Section 9.7]

92. Find the circumference and the area of a circle that has a diameter of 14 centimeters. For each, give the exact answer and an approximation to the nearest hundredth. [Section 9.8]

93. Find the area of the shaded region shown below, which is created using two semicircles. Round to the nearest hundredth. [Section 9.8]

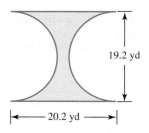

94. Ice. Find the volume of a block of ice that is in the shape of a rectangular solid with dimensions 15 in. × 24 in. × 18 in. [Section 9.9]

95. Find the volume of a sphere that has a diameter of 18 inches. Give the exact answer and an approximation to the nearest hundredth. [Section 9.9]

96. Find the volume of a cone that has a circular base with a radius of 4 meters and a height of 9 meters. Give the exact answer and an approximation to the nearest hundredth. [Section 9.9]

97. Find the volume of a cylindrical pipe that is 20 feet long and has a radius of 1 foot. Give the exact answer and an approximation to the nearest hundredth. [Section 9.9]

98. How many cubic inches are there in 1 cubic foot? [Section 9.9]

APPENDIX 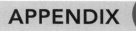 Addition and Multiplication Facts

Table of Basic Addition Facts

+	0	1	2	3	4	5	6	7	8	9
0	0	1	2	3	4	5	6	7	8	9
1	1	2	3	4	5	6	7	8	9	10
2	2	3	4	5	6	7	8	9	10	11
3	3	4	5	6	7	8	9	10	11	12
4	4	5	6	7	8	9	10	11	12	13
5	5	6	7	8	9	10	11	12	13	14
6	6	7	8	9	10	11	12	13	14	15
7	7	8	9	10	11	12	13	14	15	16
8	8	9	10	11	12	13	14	15	16	17
9	9	10	11	12	13	14	15	16	17	18

Fifty Addition Facts

1. $3 + 2$	**2.** $1 + 1$	**3.** $2 + 5$	**4.** $5 + 4$
5. $7 + 7$	**6.** $1 + 8$	**7.** $6 + 6$	**8.** $9 + 4$
9. $3 + 8$	**10.** $0 + 4$	**11.** $6 + 3$	**12.** $5 + 1$
13. $2 + 8$	**14.** $4 + 7$	**15.** $1 + 6$	**16.** $7 + 2$
17. $8 + 9$	**18.** $4 + 3$	**19.** $7 + 0$	**20.** $1 + 3$
21. $4 + 6$	**22.** $8 + 6$	**23.** $9 + 9$	**24.** $5 + 9$
25. $0 + 8$	**26.** $2 + 2$	**27.** $7 + 6$	**28.** $8 + 8$
29. $1 + 2$	**30.** $4 + 2$	**31.** $4 + 4$	**32.** $5 + 6$
33. $3 + 3$	**34.** $9 + 7$	**35.** $2 + 6$	**36.** $6 + 9$
37. $0 + 6$	**38.** $8 + 5$	**39.** $7 + 3$	**40.** $5 + 5$
41. $1 + 0$	**42.** $4 + 1$	**43.** $3 + 5$	**44.** $8 + 4$
45. $9 + 2$	**46.** $3 + 9$	**47.** $7 + 8$	**48.** $1 + 9$
49. $5 + 7$	**50.** $7 + 1$		

Fifty Subtraction Facts

1. $8 - 5$	**2.** $8 - 7$	**3.** $4 - 2$	**4.** $4 - 3$
5. $7 - 3$	**6.** $14 - 7$	**7.** $12 - 8$	**8.** $11 - 5$
9. $12 - 3$	**10.** $10 - 8$	**11.** $18 - 9$	**12.** $8 - 6$
13. $10 - 4$	**14.** $6 - 3$	**15.** $15 - 9$	**16.** $9 - 5$
17. $2 - 0$	**18.** $10 - 5$	**19.** $15 - 7$	**20.** $10 - 1$
21. $17 - 8$	**22.** $7 - 1$	**23.** $13 - 6$	**24.** $9 - 0$
25. $16 - 8$	**26.** $12 - 5$	**27.** $7 - 5$	**28.** $11 - 7$
29. $14 - 5$	**30.** $16 - 7$	**31.** $5 - 0$	**32.** $6 - 4$
33. $12 - 6$	**34.** $14 - 6$	**35.** $5 - 3$	**36.** $11 - 3$
37. $13 - 8$	**38.** $7 - 0$	**39.** $9 - 1$	**40.** $2 - 1$
41. $3 - 2$	**42.** $9 - 3$	**43.** $13 - 9$	**44.** $11 - 2$
45. $10 - 3$	**46.** $6 - 1$	**47.** $4 - 0$	**48.** $8 - 4$
49. $9 - 2$	**50.** $5 - 4$		

SECTION A1.2 Multiplication Table and One Hundred Multiplication and Division Facts

Table of Basic Multiplication Facts

×	0	1	2	3	4	5	6	7	8	9
0	**0**	0	0	0	0	0	0	0	0	0
1	0	**1**	2	3	4	5	6	7	8	9
2	0	2	**4**	6	8	10	12	14	16	18
3	0	3	6	**9**	12	15	18	21	24	27
4	0	4	8	12	**16**	20	24	28	32	36
5	0	5	10	15	20	**25**	30	35	40	45
6	0	6	12	18	24	30	**36**	42	48	54
7	0	7	14	21	28	35	42	**49**	56	63
8	0	8	16	24	32	40	48	56	**64**	72
9	0	9	18	27	36	45	54	63	72	**81**

Fifty Multiplication Facts

1. 4 ×4	**2.** 1 ×4	**3.** 6 ×3	**4.** 9 ×7
5. 5 ×7	**6.** 0 ×8	**7.** 5 ×2	**8.** 1 ×2
9. 7 ×8	**10.** 4 ×0	**11.** 3 ×3	**12.** 9 ×3
13. 5 ×6	**14.** 7 ×2	**15.** 3 ×5	**16.** 8 ×8
17. 1 ×8	**18.** 3 ×2	**19.** 0 ×7	**20.** 6 ×4
21. 8 ×6	**22.** 9 ×9	**23.** 6 ×0	**24.** 1 ×3
25. 4 ×8	**26.** 8 ×2	**27.** 9 ×1	**28.** 7 ×7
29. 9 ×6	**30.** 1 ×5	**31.** 9 ×0	**32.** 4 ×5
33. 8 ×3	**34.** 7 ×6	**35.** 6 ×2	**36.** 7 ×1
37. 5 ×8	**38.** 4 ×3	**39.** 7 ×4	**40.** 1 ×1
41. 9 ×5	**42.** 2 ×2	**43.** 7 ×3	**44.** 2 ×4
45. 6 ×6	**46.** 9 ×2	**47.** 5 ×5	**48.** 6 ×1
49. 8 ×9	**50.** 9 ×4		

Fifty Division Facts

1. $4\overline{)20}$	**2.** $8\overline{)56}$	**3.** $3\overline{)6}$	**4.** $1\overline{)8}$
5. $9\overline{)45}$	**6.** $7\overline{)42}$	**7.** $5\overline{)25}$	**8.** $3\overline{)24}$
9. $5\overline{)5}$	**10.** $7\overline{)21}$	**11.** $9\overline{)81}$	**12.** $3\overline{)0}$
13. $8\overline{)32}$	**14.** $6\overline{)18}$	**15.** $9\overline{)0}$	**16.** $2\overline{)10}$
17. $4\overline{)8}$	**18.** $3\overline{)27}$	**19.** $1\overline{)1}$	**20.** $6\overline{)30}$
21. $1\overline{)7}$	**22.** $4\overline{)16}$	**23.** $7\overline{)63}$	**24.** $5\overline{)0}$
25. $7\overline{)35}$	**26.** $3\overline{)3}$	**27.** $5\overline{)15}$	**28.** $8\overline{)48}$
29. $7\overline{)0}$	**30.** $2\overline{)16}$	**31.** $3\overline{)9}$	**32.** $6\overline{)12}$
33. $9\overline{)72}$	**34.** $8\overline{)0}$	**35.** $4\overline{)28}$	**36.** $8\overline{)64}$
37. $6\overline{)24}$	**38.** $6\overline{)54}$	**39.** $7\overline{)49}$	**40.** $7\overline{)14}$
41. $6\overline{)36}$	**42.** $1\overline{)9}$	**43.** $3\overline{)12}$	**44.** $4\overline{)36}$
45. $2\overline{)4}$	**46.** $8\overline{)40}$	**47.** $2\overline{)2}$	**48.** $1\overline{)4}$
49. $9\overline{)18}$	**50.** $6\overline{)6}$		

SECTION A2.1 Multiplication Rules for Exponents

OBJECTIVES

1 Identify bases and exponents.

2 Multiply exponential expressions that have like bases.

3 Raise exponential expressions to a power.

4 Find powers of products.

In this section, we will use the definition of exponent to develop some rules for simplifying expressions that contain exponents.

OBJECTIVE 1 Identify bases and exponents.

Recall that an **exponent** indicates repeated multiplication. It indicates how many times the base is used as a factor. For example, 3^5 represents the product of five 3's.

$$\text{Exponent} \longrightarrow \overbrace{3^5 = 3 \cdot 3 \cdot 3 \cdot 3 \cdot 3}^{\text{5 factors of 3}}$$
$$\text{Base} \longrightarrow$$

In general, we have the following definition.

Natural-Number Exponents

A natural-number* exponent tells how many times its base is to be used as a factor. For any number x and any natural number n,

$$x^n = \overbrace{x \cdot x \cdot x \cdot \cdots \cdot x}^{n \text{ factors of } x}$$

*The set of natural numbers is $\{1, 2, 3, 4, 5, \ldots\}$.

Expressions of the form x^n are called **exponential expressions**. The base of an exponential expression can be a number, a variable, or a combination of numbers and variables. Some examples are:

$10^5 = 10 \cdot 10 \cdot 10 \cdot 10 \cdot 10$ The base is 10. The exponent is 5. Read as "10 to the fifth power."

$y^2 = y \cdot y$ The base is y. The exponent is 2. Read as "y squared."

$(-2s)^3 = (-2s)(-2s)(-2s)$ The base is $-2s$. The exponent is 3. Read as "negative 2s raised to the third power" or "negative 2s cubed."

$-8^4 = -(8 \cdot 8 \cdot 8 \cdot 8)$ Since the $-$ sign is not written within parentheses, the base is 8. The exponent is 4. Read as "the opposite (or the negative) of 8 to the fourth power."

Caution! Bases that contain a $-$ sign *must* be written within parentheses.

$$(-2s)^3 \longleftarrow \text{Exponent}$$
$$\text{Base}$$

When an exponent is 1, it is usually not written. For example, $4 = 4^1$ and $x = x^1$.

EXAMPLE 1 Identify the base and the exponent in each expression:

a. 8^5 **b.** $7a^3$ **c.** $(7a)^3$

Strategy To identify the base and exponent, we will look for the form ▢▪.

WHY The exponent is the small raised number to the right of the base.

Self Check 1

Identify the base and the exponent:

a. $3y^4$

b. $(3y)^4$

Now Try ➲ **Problems 13 and 17**

Solution

a. In 8^5, the base is 8 and the exponent is 5.

b. $7a^3$ means $7 \cdot a^3$. Thus, the base is a, not $7a$. The exponent is 3.

c. Because of the parentheses in $(7a)^3$, the base is $7a$ and the exponent is 3.

EXAMPLE 2 Write each expression in an equivalent form using an exponent:

a. $b \cdot b \cdot b \cdot b$ **b.** $5 \cdot t \cdot t \cdot t$

Strategy We will look for repeated factors and count the number of times each appears.

WHY We can use an exponent to represent repeated multiplication.

Self Check 2

Write as an exponential expression:
$(x + y)(x + y)(x + y)(x + y)$
$(x + y)$

Now Try ➲ **Problems 25 and 29**

Solution

a. Since there are four repeated factors of b in $b \cdot b \cdot b \cdot b$, the expression can be written as b^4.

b. Since there are three repeated factors of t in $5 \cdot t \cdot t \cdot t$, the expression can be written as $5t^3$.

OBJECTIVE 2 **Multiply exponential expressions that have like bases.**

To develop a rule for multiplying exponential expressions that have the same base, we consider the product $6^2 \cdot 6^3$. Since 6^2 means that 6 is to be used as a factor two times, and 6^3 means that 6 is to be used as a factor three times, we have

$$\overset{\text{2 factors of 6}}{\overbrace{\qquad}} \quad \overset{\text{3 factors of 6}}{\overbrace{\qquad}}$$
$$6^2 \cdot 6^3 = \quad 6 \cdot 6 \quad \cdot \quad 6 \cdot 6 \cdot 6$$
$$\underset{\text{5 factors of 6}}{\underbrace{\qquad\qquad\qquad}}$$
$$= 6 \cdot 6 \cdot 6 \cdot 6 \cdot 6$$
$$= 6^5$$

We can quickly find this result if we keep the common base 6 and add the exponents on 6^2 and 6^3.

$$6^2 \cdot 6^3 = 6^{2+3} = 6^5$$

This example illustrates the following rule for exponents.

Product Rule for Exponents

To multiply exponential expressions that have the same base, keep the common base and add the exponents.

For any number x and any natural numbers m and n,

$$x^m \cdot x^n = x^{m+n}$$ Read as "x to the mth power times x to the nth power equals x to the m plus nth power."

EXAMPLE 3 Simplify:

a. $9^5(9^6)$ **b.** $x^3 \cdot x^4$ **c.** y^2y^4y **d.** $(c^2d^3)(c^4d^5)$

Strategy In each case, we want to write an equivalent expression using one base and one exponent. We will use the product rule for exponents to do this.

WHY The product rule for exponents is used to multiply exponential expressions that have the same base.

Solution

a. $9^5(9^6) = 9^{5+6} = 9^{11}$ *Keep the common base, 9, and add the exponents. Since 9^{11} is a very large number, we will leave the answer in this form. We won't evaluate it.*

Caution! Don't make the mistake of multiplying the bases when using the product rule. Keep the *same* base.

$$9^5(9^6) \neq 81^{11}$$

b. $x^3 \cdot x^4 = x^{3+4} = x^7$ *Keep the common base, x, and add the exponents.*

c. $y^2y^4y = y^2y^4y^1$ *Write y as y^1.*

$$= y^{2+4+1}$$ *Keep the common base, y, and add the exponents.*

$$= y^7$$

d. $(c^2d^3)(c^4d^5) = (c^2c^4)(d^3d^5)$ *Use the commutative and associative properties of multiplication to group like bases together.*

$$= (c^{2+4})(d^{3+5})$$ *Keep the common base, c, and add the exponents. Keep the common base, d, and add the exponents.*

$$= c^6d^8$$

Self Check 3

Simplify:

a. $7^8(7^7)$

b. x^2x^3x

c. $(y-1)^5(y-1)^5$

d. $(s^4t^3)(s^4t^4)$

Now Try ⟳ Problems 33, 35, and 37

Caution! We cannot use the product rule to simplify expressions like $3^2 \cdot 2^3$, where the bases are not the same. However, we can simplify this expression by doing the arithmetic:

$$3^2 \cdot 2^3 = 9 \cdot 8 = 72 \qquad 3^2 = 3 \cdot 3 = 9 \text{ and } 2^3 = 2 \cdot 2 \cdot 2 = 8.$$

Recall that *like terms* are terms with exactly the same variables raised to exactly the same powers. To add or subtract exponential expressions, they must be like terms. To multiply exponential expressions, only the bases need to be the same.

$$x^5 + x^2$$ *These are not like terms; the exponents are different. We cannot add.*

$$x^2 + x^2 = 2x^2$$ *These are like terms; we can add. Recall that $x^2 = 1x^2$.*

$$x^5 \cdot x^2 = x^7$$ *The bases are the same; we can multiply.*

OBJECTIVE 3 **Raise exponential expressions to a power.**

To develop another rule for exponents, we consider $(5^3)^4$. Here, an exponential expression, 5^3, is raised to a power. Since 5^3 is the base and 4 is the exponent, $(5^3)^4$ can be written as $5^3 \cdot 5^3 \cdot 5^3 \cdot 5^3$. Because each of the four factors of 5^3 contains three factors of 5, there are $4 \cdot 3$ or 12 factors of 5.

12 factors of 5

$$(5^3)^4 = 5^3 \cdot 5^3 \cdot 5^3 \cdot 5^3 = 5 \cdot 5 \cdot 5 \cdot 5 \cdot 5 \cdot 5 \cdot 5 \cdot 5 \cdot 5 \cdot 5 \cdot 5 \cdot 5 = 5^{12}$$

$$5^3 \qquad 5^3 \qquad 5^3 \qquad 5^3$$

We can quickly find this result if we keep the common base of 5 and multiply the exponents.

$$(5^3)^4 = 5^{3 \cdot 4} = 5^{12}$$

This example illustrates the following rule for exponents.

Power Rule for Exponents

To raise an exponential expression to a power, keep the base and multiply the exponents.

For any number x and any natural numbers m and n,

$$(x^m)^n = x^{m \cdot n} = x^{mn}$$ Read as "the quantity of x to the mth power raised to the nth power equals x to the mnth power."

THE LANGUAGE OF ALGEBRA

An exponential expression raised to a power, such as $(2^3)^7$, is also called a **power of a power**.

EXAMPLE 4   Simplify: **a.** $(2^3)^7$ **b.** $[(-6)^2]^5$ **c.** $(z^8)^8$

Strategy In each case, we want to write an equivalent expression using one base and one exponent. We will use the power rule for exponents to do this.

WHY Each expression is a power of a power.

Solution

a. $(2^3)^7 = 2^{3 \cdot 7} = 2^{21}$ Keep the base, 2, and multiply the exponents. Since 2^{21} is a very large number, we will leave the answer in this form.

b. $[(-6)^2]^5 = (-6)^{2 \cdot 5} = (-6)^{10}$ Keep the base, -6, and multiply the exponents. Since $(-6)^{10}$ is a very large number, we will leave the answer in this form.

c. $(z^8)^8 = z^{8 \cdot 8} = z^{64}$ Keep the base, z, and multiply the exponents.

Self Check 4

Simplify:

a. $(4^6)^5$

b. $(y^5)^2$

Now Try Problems 49, 51, and 53

EXAMPLE 5   Simplify: **a.** $(x^2x^5)^2$ **b.** $(z^2)^4(z^3)^3$

Strategy In each case, we want to write an equivalent expression using one base and one exponent. We will use the product and power rules for exponents to do this.

WHY The expressions involve multiplication of exponential expressions that have the same base and they involve powers of powers.

Solution

a. $(x^2x^5)^2 = (x^7)^2$ Within the parentheses, keep the common base, x, and add the exponents: $2 + 5 = 7$.

$= x^{14}$ Keep the base, x, and multiply the exponents: $7 \cdot 2 = 14$.

b. $(z^2)^4(z^3)^3 = z^8z^9$ For each power of z raised to a power, keep the base and multiply the exponents: $2 \cdot 4 = 8$ and $3 \cdot 3 = 9$.

$= z^{17}$ Keep the common base, z, and add the exponents: $8 + 9 = 17$.

Self Check 5

Simplify:

a. $(a^4a^3)^3$

b. $(a^3)^3(a^4)^2$

Now Try Problems 57 and 61

OBJECTIVE 4 **Find powers of products.**

To develop another rule for exponents, we consider the expression $(2x)^3$, which is a *power of the product* of 2 and x.

$$(2x)^3 = 2x \cdot 2x \cdot 2x \qquad \text{Write the base } 2x \text{ as a factor 3 times.}$$
$$= (2 \cdot 2 \cdot 2)(x \cdot x \cdot x) \qquad \text{Change the order of the factors and group like bases.}$$
$$= 2^3 x^3 \qquad \text{Write each product of repeated factors in exponential form.}$$
$$= 8x^3 \qquad \text{Evaluate: } 2^3 = 8.$$

This example illustrates the following rule for exponents.

Power of a Product

To raise a product to a power, raise each factor of the product to that power.
 For any numbers x and y, and any natural number n,

$$(xy)^n = x^n y^n$$

EXAMPLE 6 Simplify: **a.** $(3c)^4$ **b.** $(x^2 y^3)^5$

Strategy In each case, we want to write the expression in an equivalent form in which each base is raised to a single power. We will use the power of a product rule for exponents to do this.

 WHY Within each set of parentheses is a product, and each of those products is raised to a power.

Solution

a. $(3c)^4 = 3^4 c^4$ Raise each factor of the product 3c to the 4th power.
$$= 81c^4 \quad \text{Evaluate: } 3^4 = 81.$$

b. $(x^2 y^3)^5 = (x^2)^5 (y^3)^5$ Raise each factor of the product $x^2 y^3$ to the 5th power.
$$= x^{10} y^{15} \quad \text{For each power of a power, keep each base, } x \text{ and } y, \text{ and}$$
$$\text{multiply the exponents: } 2 \cdot 5 = 10 \text{ and } 3 \cdot 5 = 15.$$

Self Check 6

Simplify:

a. $(2t)^4$

b. $(c^3 d^4)^6$

Now Try ➡ **Problems 65 and 69**

EXAMPLE 7 Simplify: $(2a^2)^2 (4a^3)^3$

Strategy We want to write an equivalent expression using one base and one exponent. We will begin the process by using the power of a product rule for exponents.

 WHY Within each set of parentheses is a product, and each product is raised to a power.

Solution

$$(2a^2)^2 (4a^3)^3 = 2^2 (a^2)^2 \cdot 4^3 (a^3)^3 \quad \text{Raise each factor of the product } 2a^2 \text{ to the 2nd power. Raise}$$
$$\text{each factor of the product } 4a^3 \text{ to the 3rd power.}$$

$$= 4a^4 \cdot 64a^9 \quad \text{Evaluate: } 2^2 = 4 \text{ and } 4^3 = 64. \text{ For each power of a power,}$$
$$\text{keep each base and multiply the exponents: } 2 \cdot 2 = 4$$
$$\text{and } 3 \cdot 3 = 9.$$

$$= (4 \cdot 64)(a^4 \cdot a^9) \quad \text{Group the numerical factors. Group}$$
$$\text{the factors that have the same base.}$$

$$= 256a^{13} \quad \text{Do the multiplication: } 4 \cdot 64 = 256. \text{ Keep the}$$
$$\text{common base } a \text{ and add the exponents: } 4 + 9 = 13.$$

$$\begin{array}{r} \overset{1}{64} \\ \times\ 4 \\ \hline 256 \end{array}$$

Self Check 7

Simplify: $(4y^3)^2 (3y^4)^3$

Now Try ➡ **Problem 73**

The rules for natural-number exponents are summarized as follows.

Rules for Exponents

If m and n represent natural numbers and there are no divisions by zero, then

Exponent of 1	Product rule	Power rule	Power of a product
$x^1 = x$	$x^m x^n = x^{m+n}$	$(x^m)^n = x^{mn}$	$(xy)^n = x^n y^n$

Answers to Self Checks

1. a. base: y, exponent: 4 **b.** base: $3y$, exponent: 4 **2.** $(x + y)^5$ **3. a.** 7^{15} **b.** x^6 **c.** $(y - 1)^{10}$ **d.** $s^8 t^7$
4. a. 4^{30} **b.** y^{10} **5. a.** a^{21} **b.** a^{17} **6. a.** $16t^4$ **b.** $c^{18} d^{24}$ **7.** $432y^{18}$

SECTION 2.1 STUDY SET

VOCABULARY

Fill in the blank.

1. Expressions such as x^4, 10^3, and $(5t)^2$ are called _____ expressions.

2. Match each expression with the proper description.

   $(a^4 b^2)^5$ $(a^8)^4$ $a^5 \cdot a^3$

   a. Product of exponential expressions with the same base
   b. Power of an exponential expression
   c. Power of a product

CONCEPTS

Fill in the blanks.

3. **a.** $(3x)^4 = $ ▢ $\cdot$ ▢ $\cdot$ ▢ $\cdot$ ▢
   b. $(-5y)(-5y)(-5y) = $ ▢

4. **a.** $x = x^{▢}$ **b.** $x^m x^n = $ ▢
   c. $(xy)^n = $ ▢   **d.** $(a^b)^c = $ ▢

5. To simplify each expression, determine whether you add, subtract, multiply, or divide the exponents.
   a. $b^6 \cdot b^9$
   b. $(n^8)^4$
   c. $(a^4 b^2)^5$

6. To simplify $(2y^3 z^2)^4$, what factors within the parentheses must be raised to the fourth power?

Simplify each expression, if possible.

7. **a.** $x^2 + x^2$ **b.** $x^2 \cdot x^2$
8. **a.** $x^2 + x$ **b.** $x^2 \cdot x$
9. **a.** $x^3 - x^2$ **b.** $x^3 \cdot x^2$
10. **a.** $4^2 \cdot 2^4$ **b.** $x^3 \cdot y^2$

NOTATION

Complete each step to simplify each expression.

11. $(x^4 x^2)^3 = ($ ▢ $)^3$
    $= x^{▢}$

12. $(x^4)^3 (x^2)^3$
    $= x^{▢} \cdot x^6$
    $= x^{▢}$

GUIDED PRACTICE

Identify the base and the exponent in each expression. See Example 1.

13. 4^3 14. $(-8)^2$
15. x^5 16. $\left(\dfrac{5}{x}\right)^3$
17. $(-3x)^2$ 18. $(2xy)^{10}$
19. $-\dfrac{1}{3} y^6$ 20. $-x^4$
21. $9m^{12}$ 22. $3.14r^4$
23. $(y + 9)^4$ 24. $(z - 2)^3$

Write each expression in an equivalent form using an exponent. See Example 2.

25. $m \cdot m \cdot m \cdot m \cdot m$
26. $r \cdot r \cdot r \cdot r \cdot r \cdot r$
27. $4t \cdot 4t \cdot 4t \cdot 4t$
28. $-5u(-5u)(-5u)(-5u)(-5u)$
29. $4 \cdot t \cdot t \cdot t \cdot t \cdot t$
30. $5 \cdot u \cdot u \cdot u$
31. $a \cdot a \cdot b \cdot b \cdot b$
32. $m \cdot m \cdot m \cdot n \cdot n$

Use the product rule for exponents to simplify each expression. Write the results using exponents. See Example 3.

33. $5^3 \cdot 5^4$

34. $3^4 \cdot 3^6$

35. $a^3 \cdot a^3$

36. $m^7 \cdot m^7$

37. bb^2b^3

38. aa^3a^5

39. $(c^5)(c^8)$

40. $(d^4)(d^{20})$

41. $(a^2b^3)(a^3b^3)$

42. $(u^3v^5)(u^4v^5)$

43. $cd^4 \cdot cd$

44. $ab^3 \cdot ab^4$

45. $x^2 \cdot y \cdot x \cdot y^{10}$

46. $x^3 \cdot y \cdot x \cdot y^{12}$

47. $m^{100} \cdot m^{100}$

48. $n^{600} \cdot n^{600}$

Use the power rule for exponents to simplify each expression. Write the results using exponents. See Example 4.

49. $(3^2)^4$

50. $(4^3)^3$

51. $[(-4.3)^3]^8$

52. $[(-1.7)^9]^8$

53. $(m^{50})^{10}$

54. $(n^{25})^4$

55. $(y^5)^3$

56. $(b^3)^6$

Use the product and power rules for exponents to simplify each expression. See Example 5.

57. $(x^2x^3)^5$

58. $(y^3y^4)^4$

59. $(p^4p^5)^3$

60. $(r^3r^4)^2$

61. $(t^3)^4(t^2)^3$

62. $(b^2)^5(b^3)^2$

63. $(u^4)^2(u^3)^2$

64. $(v^5)^2(v^3)^4$

Use the power of a product rule for exponents to simplify each expression. See Example 6.

65. $(6a)^2$

66. $(3b)^3$

67. $(5y)^4$

68. $(4t)^4$

69. $(3a^4b^7)^3$

70. $(5m^9n^{10})^2$

71. $(-2r^2s^3)^3$

72. $(-2x^2y^4)^5$

Use the power of a product rule for exponents to simplify each expression. See Example 7.

73. $(2c^3)^3(3c^4)^2$

74. $(5b^4)^2(3b^8)^2$

75. $(10d^7)^2(4d^9)^3$

76. $(2x^7)^3(4x^8)^2$

TRY IT YOURSELF

Simplify each expression.

77. $(7a^9)^2$

78. $(12b^6)^2$

79. $t^4 \cdot t^5 \cdot t$

80. $n^4 \cdot n \cdot n^3$

81. $y^3y^2y^4$

82. y^4yy^6

83. $(-6a^3b^2)^3$

84. $(-10r^3s^2)^2$

85. $(n^4n)^3(n^3)^6$

86. $(y^3y)^2(y^2)^2$

87. $(b^2b^3)^{12}$

88. $(s^3s^3)^3$

89. $(2b^4b)^5(3b)^2$

90. $(2aa^7)^3(3a)^3$

91. $(c^2)^3(c^4)^2$

92. $(t^5)^2(t^3)^3$

93. $(3s^4t^3)^3(2st)^4$

94. $(2a^3b^5)^2(4ab)^3$

95. $x \cdot x^2 \cdot x^3 \cdot x^4 \cdot x^5$

96. $x^{10} \cdot x^9 \cdot x^8 \cdot x^7$

LOOK ALIKES

Simplify each expression, if possible.

97. a. $x^3 \cdot x^3$ **b.** $(x^3)^3$ **c.** $x^3 + x^3$

98. a. $(x^5)^7$ **b.** $x^5 \cdot x^7$ **c.** $x^5 + x^7$

99. a. $(n^2n^4)^6$ **b.** $(n^2)^6(n^4)^6$ **c.** $n^2n^4n^6$

100. a. $ac^8 \cdot ac$ **b.** $ac^8 - ac$ **c.** $(ac^8) \cdot (ac)$

APPLICATIONS

101. Art history.   Leonardo da Vinci's drawing relating a human figure to a square and a circle is shown below. Find an expression for the area of the square if the man's height is $5x$ feet.

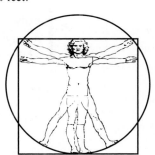

102. Packaging.   Find an expression for the volume of the box shown below.

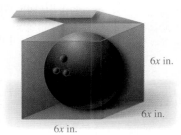

$6x$ in.

$6x$ in.

$6x$ in.

WRITING

103. Explain the mistake in the following work.

$$2^3 \cdot 2^2 = 4^5 = 1{,}024$$

104. Explain why we can simplify $x^4 \cdot x^5$, but cannot simplify $x^4 + x^5$.

REVIEW

105. Jewelry. A lot of what we refer to as gold jewelry is actually made of a combination of gold and another metal. For example, 18-karat gold is $\frac{18}{24}$ gold by weight. Simplify this ratio.

106. After evaluation, what is the sign of $(-13)^5$?

107. Divide: $\dfrac{-25}{-5}$

108. How much did the temperature change if it went from $-4°F$ to $-17°F$?

109. Evaluate: $2\left(\dfrac{12}{-3}\right) + 3(5)$

110. Solve: $-10 = x + 1$

111. Solve: $-x = -12$

112. Divide: $\dfrac{0}{10}$

OBJECTIVES

1 Know the vocabulary for polynomials.

2 Evaluate polynomials.

SECTION A2.2 Introduction to Polynomials

OBJECTIVE 1 Know the vocabulary for polynomials.

Recall that an **algebraic term**, or simply a **term**, is a number or a product of a number and one or more variables, which may be raised to powers. Some examples of terms are

$$17, \qquad 5x, \qquad 6t^2, \qquad \text{and} \qquad -8z^3$$

The *coefficients* of these terms are 17, 5, 6, and -8, in that order.

> **Polynomials**
>
> A **polynomial** is a single term or a sum of terms in which all variables have whole-number exponents and no variable appears in the denominator.

Some examples of polynomials are

$$141, \qquad 8y^2, \qquad 2x + 1, \qquad 4y^2 - 2y + 3, \qquad \text{and} \qquad 7a^3 + 2a^2 - a - 1$$

The polynomial $8y^2$ has one term. The polynomial $2x + 1$ has two terms, $2x$ and 1. Since $4y^2 - 2y + 3$ can be written as $4y^2 + (-2y) + 3$, it is the sum of three terms, $4y^2$, $-2y$, and 3.

We classify some polynomials by the number of terms they contain. A polynomial with one term is called a **monomial**. A polynomial with two terms is called a **binomial**. A polynomial with three terms is called a **trinomial**. Some examples of these polynomials are shown in the table below.

Monomials	Binomials	Trinomials
$5x^2$	$2x - 1$	$5t^2 + 4t + 3$
$-6x$	$18a^2 - 4a$	$27x^3 - 6x + 2$
29	$-27z^4 + 7z^2$	$32r^2 + 7r - 12$

EXAMPLE 1   Classify each polynomial as a monomial, a binomial, or a trinomial:
a. $3x + 4$ **b.** $3x^2 + 4x - 12$ **c.** $25x^3$

Strategy We will count the number of terms in the polynomial.

WHY The number of terms determines the type of polynomial.

Solution

a. Since $3x + 4$ has two terms, it is a binomial.

b. Since $3x^2 + 4x - 12$ has three terms, it is a trinomial.

c. Since $25x^3$ has one term, it is a monomial.

The monomial $7x^3$ is called a **monomial of third degree**, or a **monomial of degree 3**, because the variable occurs three times as a factor.

- $5x^2$ is a monomial of degree 2.   *Because the variable occurs two times as a factor: $x^2 = x \cdot x$.*

- $-8a^4$ is a monomial of degree 4.   *Because the variable occurs four times as a factor: $a^4 = a \cdot a \cdot a \cdot a$.*

- $\dfrac{1}{2}m^5$ is a monomial of degree 5.   *Because the variable occurs five times as a factor: $m^5 = m \cdot m \cdot m \cdot m \cdot m$.*

- 8 is a monomial of degree 0.   *The degree of a nonzero constant is 0.*

We define the degree of a polynomial by considering the degrees of each of its terms.

> ### Degree of a Polynomial
> The **degree of a polynomial** is equal to the highest degree of any term of the polynomial.

For example,

- $x^2 + 5x$ is a binomial of degree 2, because the degree of its term with largest degree (x^2) is 2.
- $4y^3 + 2y - 7$ is a trinomial of degree 3, because the degree of its term with largest degree ($4y^3$) is 3.
- $\dfrac{1}{2}z + 3z^4 - 2z^2$ is a trinomial of degree 4, because the degree of its term with largest degree ($3z^4$) is 4.

EXAMPLE 2   Find the degree of each polynomial:

a. $-2x + 4$ **b.** $5t^3 + t^4 - 7$ **c.** $3 - 9z + 6z^2 - z^3$

Strategy We will determine the degree of each term of the polynomial.

WHY The degree of the term with the highest degree gives the degree of the polynomial.

Solution

a. Since $-2x$ can be written as $-2x^1$, the degree of the term with largest degree is 1. Thus, the degree of the polynomial $-2x + 4$ is 1.

b. In $5t^3 + t^4 - 7$, the degree of the term with largest degree (t^4) is 4. Thus, the degree of the polynomial is 4.

c. In $3 - 9z + 6z^2 - z^3$, the degree of the term with largest degree ($-z^3$) is 3. Thus, the degree of the polynomial is 3.

Self Check 1

Classify each polynomial as a monomial, a binomial, or a trinomial:

a. $8x^2 + 7$

b. $5x$

c. $x^2 - 2x - 1$

Now Try ➜ Problems 11, 13, and 17

Self Check 2

Find the degree of each polynomial:

a. $7p^3$

b. $17r^4 + 2r^8 - r$

c. $-2g^5 - 7g^6 + 12g^7$

Now Try ➜ Problems 23, 25, and 29

OBJECTIVE 2 **Evaluate polynomials.**

When a number is substituted for the variable in a polynomial, the polynomial takes on a numerical value. Finding this value is called **evaluating the polynomial**.

EXAMPLE 3 Evaluate each polynomial for $x = 3$:

a. $3x - 2$ **b.** $-2x^2 + x - 3$

Strategy We will substitute the given value for each x in the polynomial and follow the order of operations rule.

WHY To *evaluate a polynomial* means to find its numerical value, once we know the value of its variable.

Solution

a. $3x - 2 = 3(3) - 2$ Substitute 3 for x.

$= 9 - 2$ Multiply: 3(3) = 9.

$= 7$ Subtract.

b. $-2x^2 + x - 3 = -2(3)^2 + 3 - 3$ Substitute 3 for x.

$= -2(9) + 3 - 3$ Evaluate the exponential expression.

$= -18 + 3 - 3$ Multiply: −2(9) = −18.

$= -15 - 3$ Add: −18 + 3 = −15.

$= -18$ Subtract: −15 − 3 = −15 + (−3) = −18.

Self Check 3

Evaluate each polynomial for $x = -1$:

a. $-2x^2 - 4$

b. $3x^2 - 4x + 1$

Now Try ➲ Problems 35 and 45

EXAMPLE 4   **Height of an object.**   The polynomial $-16t^2 + 28t + 8$ gives the height (in feet) of an object t seconds after it has been thrown into the air. Find the height of the object after 1 second.

Strategy We will substitute 1 for t and evaluate the polynomial.

WHY The variable t represents the time since the object was thrown into the air.

Solution
To find the height at 1 second, we evaluate the polynomial for $t = 1$.

$-16t^2 + 28t + 8 = -16(1)^2 + 28(1) + 8$ Substitute 1 for t.

$= -16(1) + 28(1) + 8$ Evaluate the exponential expression.

$= -16 + 28 + 8$ Multiply: −16(1) = −16 and 28(1) = 28.

$= 12 + 8$ Add: −16 + 28 = 12.

$= 20$ Add.

At 1 second, the height of the object is 20 feet.

Self Check 4

Refer to Example 4. Find the height of the object after 2 seconds.

Now Try ➲ Problems 51 and 53

Answers to Self Checks

1. a. binomial **b.** monomial **c.** trinomial **2. a.** 3 **b.** 8 **c.** 7 **3. a.** -6 **b.** 8 **4.** 0 ft

SECTION 2.2 STUDY SET

VOCABULARY

Fill in the blanks.

1. A _____ is a single term or a sum of terms in which all variables have whole-number exponents and no variable appears in the denominator.

2. A polynomial with one term is called a _____.

3. A polynomial with three terms is called a _____.

4. A polynomial with two terms is called a _____.

CONCEPTS

5. How many terms does each of the following polynomials have?
   a. $x^2 + 3x$
   b. $3a^4$
   c. $-4r^2 + 9r + 11$

6. Fill in the blank so that $10c^{}$ has degree 3.

7. Fill in the blank. The degree of a polynomial is equal to the _____ degree of any term of the polynomial.

8. What is the degree of each term of the polynomial $4x^3 + x^2 - 7x + 5$?

NOTATION

Complete each step.

9. Evaluate $3a^2 + 2a - 7$ for $a = 2$.
$$3a^2 + 2a - 7 = 3(\ \ )^2 + 2(\ \ ) - 7$$
$$= 3(\ \ ) + 2(2) - 7$$
$$= 12 + \ \ - 7$$
$$= \ \ - 7$$
$$= 9$$

10. Evaluate $-q^2 - 3q + 2$ for $q = -1$.
$$-q^2 - 3q + 2 = -(\ \ )^2 - 3(\ \ ) + 2$$
$$= -(\ \ ) - 3(-1) + 2$$
$$= -1 + \ \ + 2$$
$$= \ \ + 2$$
$$= 4$$

GUIDED PRACTICE

Classify each polynomial as a monomial, a binomial, or a trinomial. See Example 1.

11. $3x^2 - 4$
12. $5t^2 - t + 1$
13. $17e^4$
14. $x^2 + x + 7$
15. $25u^2$
16. $x^2 - 9$
17. $q^5 + q^2 + 1$
18. $4d^3 - 3d^2$
19. $81x^3 - 27$
20. $125m^3 - 8$
21. $4c^2 - 8c + 12$
22. $16n^4 - 8n^2 + n$

Find the degree of each polynomial. See Example 2.

23. $5x^3$
24. $3t^5 + 3t^2$
25. $2x^2 - 3x + 2$
26. $\frac{1}{2}p^4 - p^2$
27. $2m$
28. $7q - 5$
29. $25w^6 + 5w^7$
30. $p^6 - p^8$
31. $a^2 - 9$
32. $b^2 - 25$
33. $-m + 3m^4 + 5m^2$
34. $-r + 6r^8 - r^7$

Evaluate each polynomial for the given value. See Example 3.

35. $3x + 4$ for $x = 3$
36. $5n - 10$ for $n = 6$
37. $2x^2 + 4$ for $x = -1$
38. $9r^2 + 12$ for $r = -3$
39. $\frac{1}{2}x - 3$ for $x = -6$
40. $-\frac{1}{2}x^2 - 1$ for $x = 2$
41. $0.5t^3 - 1$ for $t = 4$
42. $0.75a^2 + 2.5a + 2$ for $a = 0$
43. $\frac{2}{3}b^2 - b + 1$ for $b = 3$
44. $\frac{3}{2}n^2 - n + 2$ for $n = 2$
45. $-2s^2 - 2s + 1$ for $s = -1$
46. $-4r^2 - 3r - 1$ for $r = -2$

LOOK ALIKES

47. a. Evaluate: $7m + 3$ for $m = -11$
    b. Solve: $7m + 3 = -11$

48. a. Evaluate: $-3r + 2$ for $r = -7$
    b. Solve: $-3r + 2 = -7$

49. a. Evaluate: $0.25x - 4.75$ for $x = 10$
    b. Solve: $0.25x - 4.75 = 10$

50. a. Evaluate: $0.4x + 1.4$ for $x = 100$
    b. Solve: $0.4x + 1.4 = 100$

APPLICATIONS

The height h (in feet) of a ball shot straight up with an initial velocity of 64 feet per second is given by the equation h = −16t² + 64t. Find the height of the ball after the given number of seconds.

51. 0 second
52. 1 second
53. 2 seconds
54. 4 seconds

from Campus to Careers

Police Officer

55. The number of feet that a car travels before stopping depends on the driver's reaction time and the braking distance, as shown below. For one driver, the stopping distance d is given by the equation

$$d = 0.04v^2 + 0.9v$$

where v is the velocity (speed) of the car. Find the stopping distance when the driver is traveling at 30 mph.

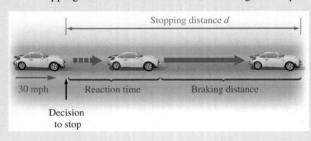

Stopping distance d

30 mph   Reaction time   Braking distance

Decision
to stop

In Problems 56–58, refer to Problem 55. Then find the stopping distance for each of the following speeds.

56. 50 mph
57. 60 mph
58. 70 mph

WRITING

59. Explain how to find the degree of the polynomial
    $2x^3 + 5x^5 - 7x$.
60. Explain how to evaluate the polynomial $-2x^2 - 3$ for $x = 5$.

REVIEW

Perform the operations.

61. $\dfrac{2}{3} + \dfrac{4}{3}$ 62. $\dfrac{36}{7} - \dfrac{23}{7}$

63. $\dfrac{5}{12} \cdot \dfrac{18}{5}$ 64. $\dfrac{23}{25} \div \dfrac{46}{5}$

Solve each equation.

65. $x - 4 = 12$ 66. $4z = 108$
67. $2(x - 3) = 6$ 68. $3(a - 5) = 4(a + 9)$

OBJECTIVES

1 Add polynomials.
2 Subtract polynomials.

SECTION A2.3 Adding and Subtracting Polynomials

Polynomials can be added, subtracted, and multiplied just like numbers in arithmetic. In this section, we show how to find sums and differences of polynomials.

OBJECTIVE 1 Add polynomials.

Recall that like terms have exactly the same variables and the same exponents. For example, the monomials

$3z^2$ and $-2z^2$ are like terms   Both have the same variable, z, with the same exponent, 2.

However, the monomials

$7b^2$ and $8a^2$ are not like terms   They have different variables.
$32p^2$ and $25p^3$ are not like terms   The exponents of p are different.

Also recall that we use the distributive property in reverse to simplify a sum or difference of like terms. We **combine like terms** by adding their coefficients and keeping the same variables and exponents. For example,

$$2y + 5y = (2 + 5)y \qquad \text{and} \qquad -3x^2 + 7x^2 = (-3 + 7)x^2$$
$$= 7y \qquad\qquad\qquad\qquad\qquad = 4x^2$$

These examples suggest the following rule.

Adding Polynomials

To add polynomials, combine their like terms.

EXAMPLE 1   Add: $5x^3 + 7x^3$

Strategy We will use the distributive property in reverse and add the coefficients of the terms.

WHY $5x^3$ and $7x^3$ are like terms and therefore can be added.

Solution

$$5x^3 + 7x^3 = 12x^3 \quad \text{Think: } (5 + 7)x^3 = 12x^3.$$

Self Check 1

Add: $7y^3 + 12y^3$

Now Try ➲ Problem 15

EXAMPLE 2   Add: $\dfrac{3}{2}t^2 + \dfrac{5}{2}t^2 + \dfrac{7}{2}t^2$

Strategy We will use the distributive property in reverse and add the coefficients of the terms.

WHY $\frac{3}{2}t^2$, $\frac{5}{2}t^2$, and $\frac{7}{2}t^2$ are like terms and therefore can be added.

Solution
Since the three monomials are like terms, we add the coefficients and keep the variables and exponents.

$$\frac{3}{2}t^2 + \frac{5}{2}t^2 + \frac{7}{2}t^2 = \left(\frac{3}{2} + \frac{5}{2} + \frac{7}{2}\right)t^2 \quad \begin{array}{l}\text{To add the fractions, add the numerators}\\\text{and keep the denominator: } 3 + 5 + 7 = 15.\end{array}$$

$$= \frac{15}{2}t^2$$

Self Check 2

Add:

$$\frac{1}{9}a^3 + \frac{2}{9}a^3 + \frac{5}{9}a^3$$

Now Try ➲ Problem 21

To add two polynomials, we write a + sign between them and combine like terms.

EXAMPLE 3   Add: $2x + 3$ and $7x - 1$

Strategy We will reorder and regroup to get the like terms together. Then we will combine like terms.

WHY To add polynomials means to combine their like terms.

Solution

$$(2x + 3) + (7x - 1) \qquad \text{Write a + sign between the binomials.}$$

$$= (2x + 7x) + (3 - 1) \qquad \begin{array}{l}\text{Use the associative and commutative properties to group}\\\text{like terms together.}\end{array}$$

$$= 9x + 2 \qquad \text{Combine like terms.}$$

Self Check 3

Add:
$5y - 2$ and $-3y + 7$

Now Try ➲ Problem 25

The binomials in Example 3 can be added by writing the polynomials so that like terms are in columns.

$$\begin{array}{r} 2x + 3 \\ + \underline{7x - 1} \\ 9x + 2 \end{array} \quad \text{Add the like terms, one column at a time.}$$

EXAMPLE 4   Add: $(5x^2 - 2x + 4) + (3x^2 - 5)$

Strategy We will combine the like terms of the trinomial and binomial.

WHY To add polynomials, we combine like terms.

Solution

$(5x^2 - 2x + 4) + (3x^2 - 5)$

$$= (5x^2 + 3x^2) + (-2x) + (4 - 5) \quad \text{Use the associative and commutative properties to group like terms together.}$$

$$= 8x^2 - 2x - 1 \quad \text{Combine like terms.}$$

Self Check 4

Add:
$(2b^2 - 4b) + (b^2 + 3b - 1)$

Now Try ➡ **Problem 29**

The polynomials in Example 4 can be added by writing the polynomials so that like terms are in columns.

$$\begin{array}{r} 5x^2 - 2x + 4 \\ + \underline{3x^2 \qquad - 5} \\ 8x^2 - 2x - 1 \end{array} \quad \text{Add the like terms, one column at a time.}$$

EXAMPLE 5   Add: $(3.7x^2 + 4x - 2) + (7.4x^2 - 5x + 3)$

Strategy We will combine the like terms of the two trinomials.

WHY To add polynomials, we combine like terms.

Solution

$(3.7x^2 + 4x - 2) + (7.4x^2 - 5x + 3)$

$$= (3.7x^2 + 7.4x^2) + (4x - 5x) + (-2 + 3) \quad \text{Use the associative and commutative properties to group like terms together.}$$

$$= 11.1x^2 - x + 1 \quad \text{Combine like terms.}$$

Self Check 5

Add:
$(s^2 + 1.2s - 5) + (3s^2 - 2.5s + 4)$

Now Try ➡ **Problem 31**

The trinomials in Example 5 can be added by writing them so that like terms are in columns.

$$\begin{array}{r} 3.7x^2 + 4x - 2 \\ + \underline{7.4x^2 - 5x + 3} \\ 11.1x^2 - \ x + 1 \end{array} \quad \text{Add the like terms, one column at a time.}$$

OBJECTIVE 2   **Subtract polynomials.**

To subtract one monomial from another, we add the opposite of the monomial that is to be subtracted. In symbols, $x - y = x + (-y)$.

EXAMPLE 6   Subtract: $8x^2 - 3x^2$

Strategy We will add the opposite of $3x^2$ to $8x^2$.

WHY To subtract monomials, we add the opposite of the monomial that is to be subtracted.

Solution

$$8x^2 - 3x^2 = 8x^2 + (-3x^2) \quad \text{Add the opposite of } 3x^2.$$

$$= 5x^2 \quad \text{Add the coefficients and keep the same variable and exponent.}$$
$$\text{Think: } [8 + (-3)]x^2 = 5x^2$$

Self Check 6

Subtract: $6y^3 - 9y^3$

Now Try ➡ **Problem 39**

Recall from Chapter 1 that we can use the distributive property to find the opposite of several terms enclosed within parentheses. For example, we consider $-(2a^2 - a + 9)$.

$$-(2a^2 - a + 9) = -1(2a^2 - a + 9) \qquad \text{Replace the } - \text{ symbol in front}$$
$$\text{of the parentheses with } -1.$$

$$= -2a^2 + a - 9 \qquad \text{Use the distributive property}$$
$$\text{to remove parentheses.}$$

This example illustrates the following method of subtracting polynomials.

Subtracting Polynomials

To subtract two polynomials, change the signs of the terms of the polynomial being subtracted, drop the parentheses, and combine like terms.

EXAMPLE 7   Subtract: $(3x - 4.2) - (5x + 7.2)$

Strategy We will change the signs of the terms of $5x + 7.2$, drop the parentheses, and combine like terms.

WHY This is the method for subtracting two polynomials.

Solution

$$(3x - 4.2) - (5x + 7.2)$$
$$= 3x - 4.2 - 5x - 7.2 \qquad \text{Change the signs of each term of } 5x + 7.2 \text{ and drop the}$$
$$\text{parentheses.}$$
$$= -2x - 11.4 \qquad \text{Combine like terms: Think: } (3 - 5)x = -2x$$
$$\text{and } (-4.2 - 7.2) = -11.4.$$

Self Check 7

Subtract:
$(3.3a - 5) - (7.8a + 2)$

Now Try ➥ Problem 43

The binomials in Example 7 can be subtracted by writing them so that like terms are in columns.

$$\begin{array}{r} 3x - 4.2 \\ -(5x + 7.2) \end{array} \longrightarrow \begin{array}{r} 3x - 4.2 \\ + \underline{-5x - 7.2} \\ -2x - 11.4 \end{array} \qquad \text{Change signs and add, column by column.}$$

EXAMPLE 8   Subtract: $(3x^2 - 4x - 6) - (2x^2 - 6x + 12)$

Strategy We will change the signs of the three terms of $2x^2 - 6x + 12$, drop the parentheses, and combine like terms.

WHY This is the method for subtracting two polynomials.

Solution

$$(3x^2 - 4x - 6) - (2x^2 - 6x + 12)$$
$$= 3x^2 - 4x - 6 - 2x^2 + 6x - 12 \qquad \text{Change the signs of each term of}$$
$$2x^2 - 6x + 12 \text{ and drop the parentheses.}$$
$$= x^2 + 2x - 18 \qquad \text{Combine like terms: Think: } (3 - 2)x^2 = x^2,$$
$$(-4 + 6)x = 2x, \text{ and } (-6 - 12) = -18.$$

Self Check 8

Subtract:
$(5y^2 - 4y + 2) - (3y^2 + 2y - 1)$

Now Try ➥ Problem 47

The trinomials in Example 8 can be subtracted by writing them so that like terms are in columns.

$$\begin{array}{r} 3x^2 - 4x - 6 \\ -(2x^2 - 6x + 12) \end{array} \longrightarrow \begin{array}{r} 3x^2 - 4x - 6 \\ + \underline{-2x^2 + 6x - 12} \\ x^2 + 2x - 18 \end{array} \qquad \begin{array}{l}\text{Change signs and add,} \\ \text{column by column.}\end{array}$$

Answers to Self Checks

1. $19y^3$ 2. $\dfrac{8}{9}a^3$ 3. $2y + 5$ 4. $3b^2 - b - 1$ 5. $4s^2 - 1.3s - 1$ 6. $-3y^3$ 7. $-4.5a - 7$ 8. $2y^2 - 6y + 3$

SECTION 2.3 STUDY SET

VOCABULARY

Fill in the blanks.

1. If two algebraic terms have exactly the same variables and exponents, they are called _____ terms.
2. Because the exponents on x are different, $3x^3$ and $3x^2$ are _____ terms.

CONCEPTS

Fill in the blanks.

3. To add two like term monomials, we add the _____ and keep the same _____ and exponents.
4. To subtract one monomial from another, we add the _____ of the monomial that is to be subtracted.

Determine whether the monomials are like terms. If they are, combine them.

5. $3y, 4y$ 6. $3x^2, 5x^2$
7. $3x, 3y$ 8. $3x^2, 6x$
9. $3x^3, 4x^3, 6x^3$ 10. $-2y^4, -6y^4, 10y^4$
11. $-5x^2, 13x^2, 7x^2$ 12. $23, 12x, 25x$

NOTATION

Complete each step.

13. $(3x^2 + 2x - 5) + (2x^2 - 7x)$
$= (3x^2 + \boxed{}) + (2x - \boxed{}) + (-5)$
$= \boxed{} + (-5x) - 5$
$= 5x^2 - 5x - 5$

14. $(3x^2 + 2x - 5) - (2x^2 - 7x)$
$= 3x^2 + 2x - 5 \boxed{} 2x^2 \boxed{} 7x$
$= (\boxed{} - \boxed{}) + (2x + 7x) - 5$
$= x^2 + 9x - 5$

GUIDED PRACTICE

Add. See Example 1.

15. $4y + 5y$ 16. $-2x + 3x$
17. $8t^2 + 4t^2$ 18. $15x^2 + 10x^2$

Add. See Example 2.

19. $\dfrac{1}{8}a + \dfrac{3}{8}a + \dfrac{5}{8}a$ 20. $\dfrac{1}{4}b + \dfrac{3}{4}b + \dfrac{1}{4}b$

21. $\dfrac{2}{3}c^2 + \dfrac{1}{3}c^2 + \dfrac{2}{3}c^2$ 22. $\dfrac{4}{9}d^3 + \dfrac{1}{9}d^3 + \dfrac{3}{9}d^3$

Add. See Example 3.

23. $3x + 7$ and $4x - 3$ 24. $2y - 3$ and $4y + 7$
25. $2x^2 + 3$ and $5x^2 - 10$ 26. $-4a^2 + 1$ and $5a^2 - 1$

Add. See Example 4.

27. $(5x^3 - 42x) + (7x^3 - 107x)$
28. $(-43a^3 + 25a) + (58a^3 - 10a)$
29. $(3x^2 + 2x - 4) + (5x^2 - 17)$
30. $(5a^2 - 2a) + (-2a^2 + 3a + 4)$

Add. See Example 5.

31. $(2.5a^2 + 3a - 9) + (3.6a^2 + 7a - 10)$
32. $(1.9b^2 - 4b + 10) + (3.7b^2 - 3b - 11)$
33. $(3n^2 - 5.8n + 7) + (-n^2 + 5.8n - 2)$
34. $(-3t^2 - t + 3.4) + (3t^2 + 2t - 1.8)$

35. $\begin{array}{r} 3x^2 + 4x + 5 \\ + 2x^2 - 3x + 6 \\ \hline \end{array}$ 36. $\begin{array}{r} 2x^2 - 3x + 5 \\ + -4x^2 - x - 7 \\ \hline \end{array}$

37. $\begin{array}{r} -3x^2 \quad\quad - 7 \\ + -4x^2 - 5x + 6 \\ \hline \end{array}$ 38. $\begin{array}{r} 4x^2 - 4x + 9 \\ + \quad\quad\quad 9x - 3 \\ \hline \end{array}$

Subtract. See Example 6.

39. $32u^3 - 16u^3$ 40. $25y^2 - 7y^2$
41. $18x^5 - 11x^5$ 42. $17x^6 - 22x^6$

Subtract. See Example 7.

43. $(4.5a + 3.7) - (2.9a - 4.3)$
44. $(5.1b - 7.6) - (3.3b + 5.9)$
45. $(7.2x^2 - 3.1x) - (9.4x^2 + 6.8x)$
46. $(3.7y^3 + 9.8y^2) - (2.4y^3 - 1.1y^2)$

Subtract. See Example 8.

47. $(2b^2 + 3b - 5) - (2b^2 - 4b - 9)$
48. $(3a^2 - 2a + 4) - (a^2 - 3a + 7)$
49. $(5p^2 - p + 71) - (4p^2 + p + 71)$
50. $(10m^2 - m - 19) - (6m^2 + m - 19)$

51. $\begin{array}{r} 3x^2 + 4x - 5 \\ - (-2x^2 - 2x + 3) \\ \hline \end{array}$ 52. $\begin{array}{r} 3y^2 - 4y + 7 \\ - (6y^2 - 6y - 13) \\ \hline \end{array}$

53. $\begin{array}{r} -2x^2 - 4x + 12 \\ - (10x^2 + 9x - 24) \\ \hline \end{array}$ 54. $\begin{array}{r} 25x^3 - 45x^2 + 31x \\ - (12x^3 + 27x^2 - 17x) \\ \hline \end{array}$

TRY IT YOURSELF

Perform the operations.

55. $(30x^2 - 4) - (11x^2 + 1)$

56. $(5x^3 - 8) - (2x^3 + 5)$

57. $(7y^2 + 5y) + (y^2 - y - 2)$

58. $(4p^2 - 4p + 5) + (6p - 2)$

59. $(3x^2 - 3x - 2) + (3x^2 + 4x - 3)$

60. $(4c^2 + 3c - 2) + (3c^2 + 4c + 2)$

61. $(m^2 - m - 5) - (m^2 + 5.5m - 75)$

62. $(3.7y^2 - 5) - (2y^2 - 3.1y + 4)$

63. $(t^2 - 4.5t + 5) - (2t^2 - 3.1t - 1)$

64. $(a^4 - 5.1a^3 + 1.1a) - (3a^4 - 6.7a^3 + 0.1a)$

65.
$$\begin{array}{r} -3x^2 + 4x + 25.4 \\ + \underline{5x^2 - 3x - 12.5} \end{array}$$

66.
$$\begin{array}{r} -6x^3 - 4.2x^2 + 7 \\ + \underline{-7x^3 + 9.7x^2 - 21} \end{array}$$

67. $3s^2 + 4s^2 + 7s^2$

68. $-2a^3 + 7a^3 - 3a^3$

69. $\dfrac{1}{3}b^4 + \dfrac{2}{3}b^4 - \dfrac{5}{3}b^4$

70. $\dfrac{4}{5}n^6 - \dfrac{1}{5}n^6 - \dfrac{2}{5}n^6$

71.
$$\begin{array}{r} z^3 + 6z^2 - 7z + 16 \\ + \underline{9z^3 - 6z^2 + 8z - 18} \end{array}$$

72.
$$\begin{array}{r} 3x^3 + 4x^2 - 3x + 5 \\ + \underline{3x^3 - 4x^2 - x - 7} \end{array}$$

73. $(-4h^3 + 5h^2 + 15) - (h^3 - 15)$

74. $(-c^5 + 5c^4 - 12) - (2c^5 - c^4)$

75. $0.6x^3 + 0.8x^4 + 0.7x^3 + (-0.8x^4)$

76. $1.9m^4 - 2.4m^6 - 3.7m^4 + 2.8m^6$

77. $(12.1h^3 - 9.9h^2 + 9.5) + (7.3h^3 - 1.2h^2 - 10.1)$

78. $(7.1a^2 + 2.2a - 5.8) - (3.4a^2 - 3.9a + 11.8)$

79.
$$\begin{array}{r} 4x^3 - 3x + 10 \\ - \underline{(5x^3 - 4x - 4)} \end{array}$$

80.
$$\begin{array}{r} 3x^3 + 4x^2 + 12 \\ - \underline{(-4x^3 + 6x^2 - 3)} \end{array}$$

LOOK ALIKES

Simplify each expression.

81. a. $(6m - 24) + (4m - 3)$
  b. $(6m - 24) - (4m - 3)$

82. a. $(1.2n^2 - 3.0n) + (4.7n^2 - 2.9n)$
  b. $(1.2n^2 - 3.0n) - (4.7n^2 - 2.9n)$

83. a. $(3x^2 + 2x - 6) + (x^2 - 4x + 1)$
  b. $(3x^2 + 2x - 6) - (x^2 - 4x + 1)$

84. a. $(8x^3 - 4x^2 + x) + (x^3 - 9x^2 + 11x)$
  b. $(8x^3 - 4x^2 + x) - (x^3 - 9x^2 + 11x)$

APPLICATIONS

85. Billiards. Billiard tables vary in size, but all tables are twice as long as they are wide.

  a. If the billiard table is x feet wide, write an expression that represents its length.

  b. Write an expression that represents the perimeter of the table.

x ft

86. Gardening. Find a polynomial that represents the length of the wooden handle of the shovel.

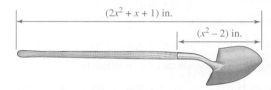

$(2x^2 + x + 1)$ in.

$(x^2 - 2)$ in.

87. Reading blueprints.

  a. What is the difference between the length and width of the one-bedroom apartment shown below?

  b. Find the perimeter of the apartment.

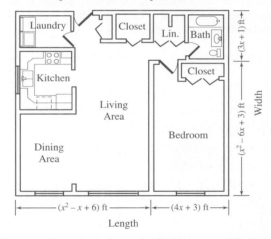

Laundry   Closet   Lin. Bath
Kitchen
Closet
Living Area
Bedroom
Dining Area

$(3x + 1)$ ft
$(x^2 - 6x + 3)$ ft   Width

$(x^2 - x + 6)$ ft $(4x + 3)$ ft

Length

88. Piñatas. Find the polynomial that represents the length of the rope used to hold up the piñata.

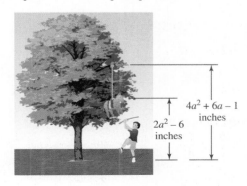

$4a^2 + 6a - 1$ inches

$2a^2 - 6$ inches

89. Greek architecture.   Find a polynomial that represents the difference in the heights of the columns shown here.

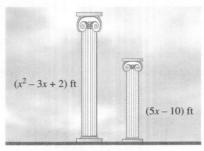

$(x^2 - 3x + 2)$ ft

$(5x - 10)$ ft

90. Greek architecture.   Find a polynomial that represents the height of the columns shown in problem 89 if they were stacked one atop the other.

WRITING

91. What are *like terms*?

92. Explain how to add two polynomials.

93. Explain how to subtract two polynomials.

94. When two binomials are added, is the result always a binomial? Explain.

REVIEW

95. Life insurance.   Refer to the chart in the next column. Determine the difference in the monthly premiums for a male and female (each 50 years old) to purchase a one-half-million dollar life insurance policy.

Monthly Premiums / Female & Male Nonsmoker						
Insurance	Age 40		Age 50		Age 60	
Amount	Female	Male	Female	Male	Female	Male
$100,000	$10.85	$11.64	$17.68	$19.43	$30.71	$41.91
$250,000	$15.09	$16.84	$29.97	$34.78	$57.09	$83.78
$500,000	$24.06	$27.56	$55.75	$63.44	$105.88	$154.88

96. Aerobics.   The number of calories burned when doing step aerobics depends on the step height. How many more calories are burned during a 10-minute workout using an 8-inch step instead of a 4-inch step?

Step height (in.)	Calories burned per minute
4	4.5
6	5.5
8	6.4
10	7.2

Source: *Reebok Instructor News* (Vol. 4, No. 3, 1991)

n	n^2	$\sqrt{n}$	n	n^2	$\sqrt{n}$
1	1	1.000	51	2,601	7.141
2	4	1.414	52	2,704	7.211
3	9	1.732	53	2,809	7.280
4	16	2.000	54	2,916	7.348
5	25	2.236	55	3,025	7.416
6	36	2.449	56	3,136	7.483
7	49	2.646	57	3,249	7.550
8	64	2.828	58	3,364	7.616
9	81	3.000	59	3,481	7.681
10	100	3.162	60	3,600	7.746
11	121	3.317	61	3,721	7.810
12	144	3.464	62	3,844	7.874
13	169	3.606	63	3,969	7.937
14	196	3.742	64	4,096	8.000
15	225	3.873	65	4,225	8.062
16	256	4.000	66	4,356	8.124
17	289	4.123	67	4,489	8.185
18	324	4.243	68	4,624	8.246
19	361	4.359	69	4,761	8.307
20	400	4.472	70	4,900	8.367
21	441	4.583	71	5,041	8.426
22	484	4.690	72	5,184	8.485
23	529	4.796	73	5,329	8.544
24	576	4.899	74	5,476	8.602
25	625	5.000	75	5,625	8.660
26	676	5.099	76	5,776	8.718
27	729	5.196	77	5,929	8.775
28	784	5.292	78	6,084	8.832
29	841	5.385	79	6,241	8.888
30	900	5.477	80	6,400	8.944
31	961	5.568	81	6,561	9.000
32	1,024	5.657	82	6,724	9.055
33	1,089	5.745	83	6,889	9.110
34	1,156	5.831	84	7,056	9.165
35	1,225	5.916	85	7,225	9.220
36	1,296	6.000	86	7,396	9.274
37	1,369	6.083	87	7,569	9.327
38	1,444	6.164	88	7,744	9.381
39	1,521	6.245	89	7,921	9.434
40	1,600	6.325	90	8,100	9.487
41	1,681	6.403	91	8,281	9.539
42	1,764	6.481	92	8,464	9.592
43	1,849	6.557	93	8,649	9.644
44	1,936	6.633	94	8,836	9.695
45	2,025	6.708	95	9,025	9.747
46	2,116	6.782	96	9,216	9.798
47	2,209	6.856	97	9,409	9.849
48	2,304	6.928	98	9,604	9.899
49	2,401	7.000	99	9,801	9.950
50	2,500	7.071	100	10,000	10.000

THINK IT THROUGH (page 9)

1. c **2.** b **3.** e **4.** d **5.** a

STUDY SET SECTION 1.1 (page 10)

1. digits **3.** standard **5.** expanded **7.** inequality

9.

11. a. forty **b.** ninety **c.** sixty-eight **d.** fifteen

13.

15.

17.

19.

21. braces **23. a.** 3 tens **b.** 7 **c.** 6 hundreds **d.** 5
25. a. 2 ten millions **b.** 6 **c.** 3 thousands **d.** 9
27. ninety-three **29.** seven hundred thirty-two
31. one hundred fifty-four thousand, three hundred two
33. fourteen million, four hundred thirty-two thousand, five
hundred **35.** nine hundred seventy billion, thirty-one million,
five hundred thousand, one hundred four
37. eighty-two million, four hundred fifteen **39.** 3,737
41. 930 **43.** 7,021 **45.** 26,000,432 **47.** 200 + 40 + 5
49. 3,000 + 600 + 9 **51.** 70,000 + 2,000 + 500 + 30 + 3
53. 100,000 + 4,000 + 400 + 1
55. 8,000,000 + 400,000 + 3,000 + 600 + 10 + 3
57. 20,000,000 + 6,000,000 + 100 + 50 + 6
59. a. > **b.** < **61. a.** > **b.** < **63.** 98,150
65. 512,970 **67.** 8,400 **69.** 32,400 **71.** 66,000
73. 2,581,000 **75.** 53,000; 50,000 **77.** 77,000; 80,000
79. 816,000; 820,000 **81.** 297,000; 300,000 **83. a.** 79,590
b. 79,600 **c.** 80,000 **d.** 80,000 **85. a.** $419,160
b. $419,200 **c.** $419,000 **d.** $420,000 **87.** 40,025
89. 202,036 **91.** 27,598 **93.** 10,700,506
95. a. 1,000,600,000,000 **b.** 1,000,600,000 **c.** 1,000,600
97. a. 9,000,000,000 **b.** 9,000,000,000 **99.** Aisha
101. a. under $25,000 **b.** $100,000 and over
c. 12 million **d.** 3 million

103.

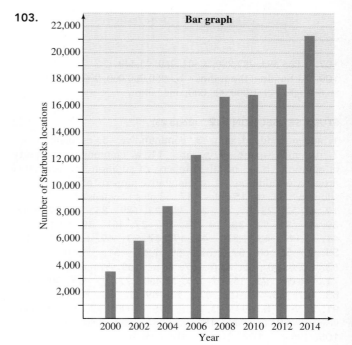

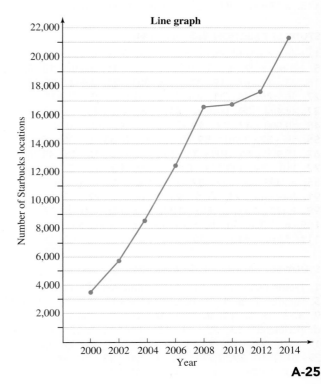

105. a.

b.

107. 1,865,593; 482,880; 1,503; 269; 43,449
109. a. hundred thousands   **b.** 980,000,000; 900,000,000 + 80,000,000   **c.** 1,000,000,000; one billion

STUDY SET SECTION **1.2**   (page 23)

1. addend, addend, sum   **3.** commutative   **5.** estimate
7. rectangle, square   **9.** square   **11. a.** commutative property of addition   **b.** associative property of addition   **c.** associative property of addition   **d.** commutative property of addition   **13.** 0   **15.** plus   **17.** 33 plus 12 equals 45
19. 47, 52   **23.** 17, 29   **21.** 38   **23.** 689   **25.** 76   **27.** 876
29. 35   **31.** 92   **33.** 70   **35.** 75   **37.** 461   **39.** 8,937
41. 18,143   **43.** 1,810   **45.** 19   **47.** 33   **49.** 137   **51.** 241
53. 30   **55.** 60   **57.** 1,615   **59.** 1,207   **61.** 37,500
63. 1,020,000   **65.** 88 ft   **67.** 68 in.   **69.** 376 mi
71. 186 cm   **73.** 15,907   **75.** 56,460   **77.** 65   **81.** 30,000
83. 121   **85.** 11,312   **87.** 50   **89. a.** 398   **b.** 398
91. a. 999   **b.** 999   **93.** 91 ft   **95.** 1,090   **97.** 243,547,000 visitors   **99.** 610,749   **101.** $34,752   **103.** 3,507,414,063
105. 196 in.   **107.** 216 in.   **109.** 250 ft
115. a. 3,000 + 100 + 20 + 5   **b.** 60,000 + 30 + 7

STUDY SET SECTION **1.3**   (page 34)

1. minuend, subtrahend, difference   **3.** subtraction
5. Estimate   **7.** 4, 3, 7   **9.** Left, right   **11.** minus   **13.** 83 − 30
15. 23   **17.** 61   **19.** 224   **21.** 303   **23.** 7,642   **25.** 2,562
27. 36   **29.** 48   **31.** 8,457   **33.** 6,483   **35.** 51,677
37. 44,444   **39.** correct   **41.** Incorrect   **43.** 66,000
45. 50,000   **47.** 29   **49.** 37   **51.** 608   **53.** 1,048
55. 59   **57.** 2,901   **59.** 102   **61.** 20   **63.** 65   **65.** 30
67. 19,929   **69.** 197   **71.** 10,457   **73.** 303   **75.** 48,760
77. 110   **79.** 143,559   **81.** 119,299   **83. a.** 527   **b.** 269
85. a. 495   **b.** 999   **87.** 2,272   **89.** 2,136 markets
91. 1,495 mi   **93.** $425   **95.** 28 points   **97.** 1,764
99. 1,472   **101.** $1,513   **103. a.** $49,565   **b.** $1,322
109. a. 5,370,650   **b.** 5,370,000   **c.** 5,400,000
111. 52 in.   **113.** 5,530

STUDY SET SECTION **1.4**   (page 48)

1. factor, factor, product   **3.** commutative, associative
5. square   **7. a.** 4 · 8   **b.** 15 + 15 + 15 + 15 + 15 + 15 + 15
9. a. 3   **b.** 5   **11. a.** area   **b.** perimeter   **c.** area
d. perimeter   **13.** ×, ·, ( )   **15.** $A = l · w$ or $A = lw$
17. 105   **19.** 272   **21.** 3,700   **23.** 750   **25.** 1,070,000
27. 512,000   **29.** 2,720   **31.** 11,200   **33.** 390,000
35. 108,000,000   **37.** 9,344   **39.** 18,368   **41.** 408,758

43. 16,868,238   **45.** 1,800   **47.** 135,000   **49.** 18,000
51. 400,000   **53.** 84 in.²   **55.** 144 in.²   **57.** 1,491
59. 68,948   **61.** 7,623   **63.** 0   **65.** 1,590   **67.** 44,486
69. 8,945,912   **71.** 374,644   **73.** 9,900   **75.** 2,400,000
77. 355,712   **79.** 166,500   **81. a.** 1,520   **b.** 15,200
c. 152,000   **83. a.** 3,528   **b.** 35,280   **c.** 352,800
85. 72 cups   **87.** 204 grams   **89.** 3,900 times   **91.** 63,360 in.
93. 77,000 words   **95.** $6,264,000   **97.** 72 entries
99. no   **101.** 18 hr   **103.** 162,000 attacks   **105.** 84 tablets
107. 54 ft²   **109.** 1,260 mi, 97,200 mi²   **113.** 20,642

STUDY SET SECTION **1.5**   (page 62)

1. 12: dividend, 4: divisor, 3: quotient; 4: divisor, 3: quotient, 12: dividend; 12: dividend, 4: divisor, 3: quotient   **3.** long
5. divisible   **7. a.** 7   **b.** 5, 2   **9. a.** 1   **b.** 6   **c.** undefined
d. 0   **11. a.** 2   **b.** 6   **c.** 3   **d.** 5   **13.** 37; 333   **15. a.** 0, 5
b. 2, 3   **c.** sum   **d.** 10   **17.** ÷,)‾   , −   **19.** 5, 9, 45
21. 4, 11, 44   **23.** 7 · 3 = 21   **25.** 6 · 12 = 72   **27.** 16
29. 29   **31.** 325   **33.** 218   **35.** 504   **37.** 602   **39.** 39 R 15
41. 21 R 33   **43.** 47 R 86   **45.** 19 R 132   **47.** 2, 3, 4, 5, 6, 10
49. 3, 5, 9   **51.** none   **53.** 2, 3, 4, 5, 6, 10   **55.** 70   **57.** 22
59. 9,000   **61.** 50   **63.** 4,325   **65.** 6   **67.** 8 R 25   **69.** 160
71. 106 R 3   **73.** 509   **75.** 3,080   **77.** 5   **79.** 23 R 211
81. 30 R 13   **83.** 89   **85.** 7 R 1   **87. a.** 36,800
b. 3,680   **c.** 368   **89. a.** 50 R 7   **b.** 50 R 8   **c.** 50 R 6
91. 625 tickets   **93.** 27 trips   **95.** 2 cartons, 4 cartons
97. 9 times, 28 ounces   **99.** 14,500 lb   **101.** $105
103. 5 mi   **105.** 13 dozen   **107.** 9 girls
109. $4,059, $4,353, $3,882   **115. a.** 276   **b.** 268   **c.** 1,088
d. 68   **117. a.** 1,150   **b.** 1,058   **c.** 50,784   **d.** 24

STUDY SET SECTION **1.6**   (page 72)

1. strategy   **3.** subtraction   **5.** multiplication
7. addition   **9.** multiplication   **11.** division
13. Analyze, Form, Calculate, State, Check   **15.** 40
17. $194,445   **19.** 179 episodes   **21.** 14 daily servings
23. 24 scenes   **25.** 26 full-size rolls   **27.** 68 files   **29.** 1,197,283 square miles   **31.** $4,376 million   **33.** 53,029 people applied
35. $462   **37.** 56 gal   **39.** Used:21 GB; available: 11 GB
41. 426 ft   **43.** 10,080 min   **45.** 14 fireplaces, 172 bricks left over
47. 179 squares   **49.** $730   **51.** approximately 71,334 mi
53. 7 $20 bills, $3 change   **55.** 113 points   **57.** 388 ft²
59. 4,900 cents = $49   **65.** Upward: 12,787. The sum is not correct.   **67.** Estimate: 4,200. The product does not seem reasonable.

STUDY SET SECTION **1.7**   (page 83)

1. factors   **3.** prime   **5.** prime   **7.** base, exponent
9. 45, 15, 9; 1, 3, 5, 9, 15, 45   **11.** yes   **13. a.** even, odd
b. 0, 2, 4, 6, 8, 10, 12, 14, 16, 18   **c.** 1, 3, 5, 7, 9, 11, 13, 15, 17, 19
15. 5, 6, 2; 2, 3, 5, 5   **17.** 2, 25, 2, 3, 5, 5   **19. a.** base: 7, exponent: 6   **b.** base: 15, exponent: 1   **21.** 1, 2, 5, 10
23. 1, 2, 4, 5, 8, 10, 20, 40   **25.** 1, 2, 3, 6, 9, 18   **27.** 1, 2, 4, 11, 22, 44   **29.** 1, 7, 11, 77   **31.** 1, 2, 4, 5, 10, 20, 25, 50, 100
33. 2 · 4   **35.** 3 · 9   **37.** 7 · 7   **39.** 2 · 10 or 4 · 5
41. 2 · 3 · 5   **43.** 3 · 3 · 7   **45.** 2 · 3 · 9 or 3 · 3 · 6
47. 2 · 3 · 10 or 2 · 2 · 15 or 2 · 5 · 6 or 3 · 4 · 5   **49.** 1 and 11
51. 1 and 37   **53.** yes   **55.** no, (9 · 11)   **57.** no, (3 · 17)
59. yes   **61.** 2 · 3 · 5   **63.** 3 · 13   **65.** 3 · 3 · 11 or 3^2 · 11
67. 2 · 3 · 3 · 3 · 3 or 2 · 3^4 **69.** 2 · 2 · 2 · 2 · 2 · 2 or 2^6
71. 3 · 7 · 7 or 3 · 7^2 **73.** 2 · 2 · 5 · 11 or 2^2 · 5 · 11
75. 2 · 3 · 17   **77.** 2^5 **79.** 5^4 **81.** $4^2(8^3)$ **83.** 7^7 · 9^2

85. a. 81 **b.** 64 **87. a.** 32 **b.** 25 **89. a.** 343 **b.** 2,187
91. a. 9 **b.** 1 **93.** 90 **95.** 847 **97.** 225 **99.** 2,808
101. 1, 2, 4, 7, 14, 28, 1 + 2 + 4 + 7 + 14 = 28
103. 2² square units, 3² square units, 4² square units
109. 125 band members

STUDY SET SECTION 1.8 (page 94)

1. multiples **3.** divisible **5. a.** 12 **b.** smallest **7. a.** 20
b. 20 **9. a.** two **b.** two **c.** one **d.** 2, 2, 3, 3, 5, 180
11. a. two **b.** three **c.** 2, 3, 108 **13. a.** 2, 3, 5 **b.** 30
15. a. GCF **b.** LCM **17.** 4, 8, 12, 16, 20, 24, 28, 32
19. 11, 22, 33, 44, 55, 66, 77, 88 **21.** 8, 16, 24, 32, 40, 48, 56, 64
23. 20, 40, 60, 80, 100, 120, 140, 160 **25.** 15 **27.** 24
29. 55 **31.** 28 **33.** 12 **35.** 30 **37.** 80 **39.** 150
41. 315 **43.** 600 **45.** 72 **47.** 60 **49.** 2 **51.** 3 **53.** 11
55. 15 **57.** 6 **59.** 14 **61.** 1 **63.** 1 **65.** 4 **67.** 36
69. 600; 20 **71.** 140; 14 **73.** 2,178; 22 **75.** 3,528; 1
77. 3,000; 5 **79.** 204; 34 **81.** 138; 23 **83.** 4,050; 1
85. a. 2 **b.** 20 **87. a.** 2 **b.** 2 **89.** 15,000 mi,
22,500 mi, 30,000 mi, 37,500 mi, 45,000 mi **91.** 180 min or 3 hr
93. 6 packages of hot dogs and 5 packages of buns
95. 12 pieces **97. a.** $7 **b.** 1st day: 4 students, 2nd day:
3 students, 3rd day: 9 students **103.** 11,110 **105.** 15,250

THINK IT THROUGH (page 103)

1. $678 **2.** $738 **3.** $798 **4.** $1,137 **5.** $1,341
6. $1,730 **7.** $1,623

STUDY SET SECTION 1.9 (page 103)

1. expressions **3.** parentheses, brackets **5.** inner, outer
7. a. square, multiply, subtract **b.** multiply, cube, add,
subtract **c.** square, multiply **d.** multiply, square
9. multiply, square **11.** the fraction bar, the numerator and
the denominator **13.** quantity **15.** 4, 20, 8 **17.** 9, 36, 16, 20
19. 47 **21.** 13 **23.** 38 **25.** 36 **27.** 24 **29.** 12
31. a. 33 **b.** 15 **33. a.** 43 **b.** 27 **35.** 100 **37.** 512
39. 64 **41.** 203 **43.** 73 **45.** 81 **47.** 3 **49.** 4 **51.** 6
53. 5 **55.** 16 **57.** 4 **59.** 5 **61.** 162 **63.** 27 **65.** 10
67. 3 **69.** 5,239 **71.** 15 **73.** 25 **75.** 22 **77.** 53 **79.** 2
81. 1 **83.** 25 **85.** 813 **87.** 49 **89.** 11 **91.** 191 **93.** 34
95. 323 **97.** undefined **99.** 14 **101.** 192 **103.** 74
105. a. 2 **b.** 50 **c.** 50 **d.** 2 **107. a.** 0 **b.** undefined
109. 3(7) + 4(4) + 2(3), $43 **111.** 3(8 + 7 + 8 + 8 + 7), 114
113. brick: 3(3) + 1 + 1 + 3 + 3(5), 29;
aphid: 3[1 + 2(3) + 4 + 1 + 2], 42
115. 2² + 3² + 5² + 7² = 4 + 9 + 25 + 49 = 87
117. 79° **119.** 31 therms **121.** 300 calories
123. a. 125 **b.** $11,875 **c.** $95
129. two hundred fifty-four thousand, three hundred nine

CHAPTER 1 REVIEW (page 107)

1. 6 **2.** 7 **3.** 1 billion **4.** 8 **5. a.** ninety-seven thousand,
two hundred eighty-three **b.** five billion, four hundred forty-
four million, sixty thousand, seventeen
6. a. 3,207 **b.** 23,253,412 **c.** 61,204
7. 500,000 + 70,000 + 300 + 2
8. 30,000,000 + 7,000,000 + 300,000 + 9,000 + 100 + 50 + 4

9.

10.

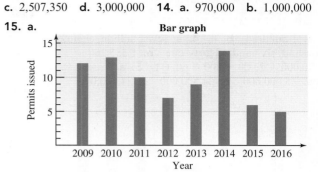

11. > **12.** < **13. a.** 2,507,300 **b.** 2,510,000
c. 2,507,350 **d.** 3,000,000 **14. a.** 970,000 **b.** 1,000,000

15. a.

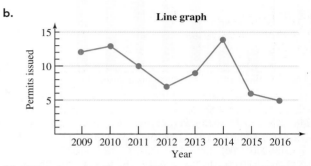

16. Nile, Amazon, Yangtze, Mississippi-Missouri, Ob-Irtysh
17. 463 **18.** 59 **19.** 6,000 **20.** 50 **21.** 12,601
22. 152,169 **23.** 59,400 **24. a.** 61 + 24 **b.** (9 + 91) + 29
25. 253,377,614 passengers **26.** no **27.** 14,661
28. 779,666 **29.** 2,467 square feet **30.** 2,746 ft **31.** 61
32. 217 **33.** 505 **34.** 2,075 **35.** incorrect
36. 12 + 8 = 20 **37.** 160,000 **38.** 2,772,459 square miles
39. $13,445 **40.** 54 days **41.** 423 **42.** 210 **43.** 720,000
44. 9,263 **45.** 1,580,344 **46.** 230,418 **47.** 2,800,000
48. a. 5 · 7 **b.** 2t **c.** mn **49. a.** 0 **b.** 7
50. a. associative property of multiplication
b. commutative property of multiplication **51.** 32 cm²
52. 6,084 in.² **53. a.** 2,555 hr **b.** 3,285 hr **54.** 330 members
55. Santiago **56.** 14,400 eggs **57.** 18 **58.** 37 **59.** 307
60. 19 R 6 **61.** 0 **62.** undefined **63.** 42 R 13
64. 380 **65.** 40 · 4 = 160 **66.** It is not correct.
67. It is divisible by 3, 5, and 9. **68.** 4,000 **69.** 16; 25
70. 34 cars **71.** 185° F **72.** 324 drive-in theaters **73.** 900 lbs
74. 1,200 cars **75.** 2,500 boxes **76.** 68 hats, 12 yards of thread
left over **77.** 147 cattle **78.** 96 children
79. 1, 2, 3, 6, 9, 18 **80.** 1, 3, 5, 15, 25, 75
81. 2 · 10 or 4 · 5 **82.** 2 · 3 · 9 or 3 · 3 · 6
83. a. prime **b.** composite **c.** neither **d.** neither
e. composite **f.** prime **84. a.** odd **b.** even **c.** even
d. odd **85.** 2 · 3 · 7 **86.** 3 · 5² **87.** 2² · 5 · 11
88. 2² · 5 · 7 **89.** 6⁴ **90.** 5³ · 13² **91.** 125 **92.** 121
93. 784 **94.** 2,700 **95.** 9, 18, 27, 36, 45, 54, 63, 72, 81, 90
96. a. 24, 48 **b.** 1, 2 **97.** 12 **98.** 12 **99.** 45 **100.** 36
101. 126 **102.** 360 **103.** 140 **104.** 84 **105.** 4 **106.** 3
107. 10 **108.** 15 **109.** 21 **110.** 28 **111.** 24 **112.** 44
113. 42 days **114. a.** 8 arrangements **b.** 4 red carnations,
3 white carnations, 2 blue carnations **115.** 45 **116.** 23
117. 243 **118.** 4 **119.** 32 **120.** 72 **121.** 8
122. 8 **123.** 1 **124.** 3 **125.** 28 **126.** 9
127. 77 **128.** 60.

CHAPTER 1 TEST (page 122)

1. a. whole **b.** inequality **c.** value **d.** area **e.** divisible
f. parentheses, brackets **g.** prime
2.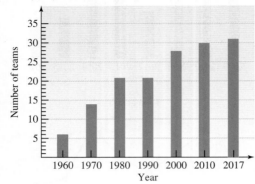
    0 1 2 3 4 5 6 7 8 9

3. a. 1 hundred **b.** 0 **4. a.** seven million, eighteen
thousand, six hundred forty-one **b.** 1,385,266
c. 90,000 + 2,000 + 500 + 60 + 1 **5. a.** > **b.** <
6. a. 35,000,000 **b.** 34,800,000 **c.** 34,760,000
7.

8. 248, 248 + 287 = 535 **9.** 225,164 **10.** 942 **11.** 424
12. 41,588 **13.** 72 **14.** 114 R 57, (73 • 114) + 57 = 8,379
15. 13,800,000 **16.** 250 **17.** 43,000 **18.** 2,168 in.
19. 529 cm² **20. a.** 1, 2, 3, 4, 6, 12 **b.** 4, 8, 12, 16, 20, 24
c. 8 • 5 **21.** 2² • 3² • 5 • 7 **22.** 32 teeth **23.** 4,933 tails
24. 96 students **25.** 4,085 ft² **26.** 414 mi **27.** $331,000
28. a. associative property of multiplication **b.** commutative
property of addition **29. a.** 0 **b.** 0 **c.** 1 **d.** undefined
30. 90 **31.** 72 **32.** 6 **33.** 4 **34. a.** 40 in. **b.** rice: 5
boxes, potatoes: 4 boxes **35.** It is divisible by 2, 3, 4, 5, 6, and 10.
36. 58 **37.** 29 **38.** 762 **39.** 44 **40.** 1

THINK IT THROUGH (page 129)

$4,621, $1,073, $3,325

STUDY SET SECTION 2.1 (page 132)

1. Positive, negative **3.** graph **5.** absolute value
7. a. −$225 **b.** −10 sec **c.** −3° **d.** −$12,000 **e.** −1 mi
9. a. The spacing is not uniform. **b.** The numbering is not
uniform. **c.** Zero is missing. **d.** The arrowheads are not
drawn. **11. a.** −4 **b.** −2 **13. a.** −7 **b.** 8
15. a. 15 > −12 **b.** −5 < −4

17.

Number	Opposite	Absolute value
−25	25	25
39	−39	39
0	0	0

19. a. −(−8) **b.** |−8| **c.** 8 − 8 **d.** −|−8|
21. a. greater, equal **b.** less, equal

23. ├────┼──┼──┼──┼──┼──┼──┼──┼──┼──┤
    −5 −4 −3 −2 −1 0 1 2 3 4 5

25. ├──●──┼──●──┼──●──┼──┼──●──┼──┤
    −5 −4 −3 −2 −1 0 1 2 3 4 5

27. ├──●──┼──┼──┼──┼──┼──●──┼──┤
    −5 −4 −3 −2 −1 0 1 2 3 4 5

29. ├──┼──●──●──●──┼──┼──●──┼──┼──┤
    −5 −4 −3 −2 −1 0 1 2 3 4 5

31. < **33.** < **35.** > **37.** > **39.** true **41.** true
43. false **45.** false **47.** 9 **49.** 8 **51.** 14 **53.** 180
55. 11 **57.** 4 **59.** 102 **61.** 561 **63.** −20 **65.** −6
67. −253 **69.** 0 **71.** > **73.** < **75.** > **77.** <
79. −52, −22, −12, 12, 52, 82 **81.** −3, −5, −7
83. a. 18 **b.** −18 **85. a.** 44 **b.** −44
87. a. True **b.** False **c.** False **d.** True
89. a. False **b.** True **c.** False **d.** True
91. −31 lengths **93.** 0, 20, 5, −40, −120 **95.** peaks:
2, 4, 0; valleys: −3, −5, −2 **97. a.** −1 (1 below par)
b. −3 (3 below par) **c.** Most of the scores are below
par. **99. a.** −20° to −10° **b.** 40° **c.** 10°
101. a. 200 yr **b.** A.D. **c.** B.C. **d.** the birth of Jesus

103.

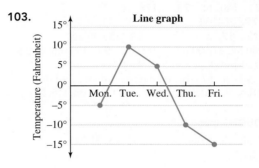

113. 23,500 **115.** 761
117. associative property of multiplication

THINK IT THROUGH (page 141)

decrease expenses, increase income, decrease expenses, increase
income, increase income, increase income, decrease expenses,
decrease expenses, increase income, decrease expenses

STUDY SET SECTION 2.2 (page 145)

1. like **3.** identity **5.** Commutative
7. a. |10| = 10, |−12| = 12 **b.** −12 **c.** 2 **d.** −2
9. subtract, larger **11. a.** yes **b.** yes **c.** no **d.** no **13. a.** 0
b. 0 **15.** −18, −19 **17.** 5, 2 **19.** −9 **21.** −10 **23.** −62
25. −96 **27.** −379 **29.** −874 **31.** −3 **33.** 1
35. −22 **37.** 48 **39.** 357 **41.** −60 **43.** 7 **45.** −4
47. −10 **49.** 41 **51.** 3 **53.** −6 **55.** 3 **57.** −7
59. 9 **61.** −562 **63.** 2 **65.** 0 **67.** −8 **69.** −2
71. −1 **73.** −3 **75.** −1,032 **77.** −21 **79.** −8,348
81. −20 **83. a.** 84 **b.** −98 **c.** −84 **d.** 98 **85. a.** −316
b. 148 **87.** 112°F, 115°F **89. a.** −15,720 ft **b.** −12,500 ft
91. a. −9 ft **b.** 2 ft above flood stage **93.** 195°
95. 5, 4% risk **97.** 3,250 m **99.** ($967) **101. a.** January,
April, June, September **b.** deficit, −$587,412,000000
c. negative five hundred eighty-seven billion, four hundred
twelve million dollars **111. a.** 16 ft **b.** 15 ft² **113.** 2 • 5³

STUDY SET SECTION 2.3 (page 155)

1. opposite **3.** value **5.** opposite **7.** −3
9. change **11. a.** 3 **b.** −12 **13.** +, 6, 9
15. a. −8 − (−4) **b.** −4 − (−8) **17.** −3, 2, 0

19. −2, −10, 6, −4 **21.** −7 **23.** −10 **25.** 9 **27.** 18
29. −18 **31.** −50 **33. a.** −10 **b.** 10
35. a. 25 **b.** −25 **37.** −15 **39.** 9 **41.** −2
43. −10 **45.** 9 **47.** −12 **49.** −8 **51.** 0 **53.** 32
55. −26 **57.** −2,447 **59.** 43,900 **61.** −5 **63.** 10
65. 8 **67.** 5 **69.** 3 **71.** −1 **73.** −9 **75.** −22
77. 9 **79.** −4 **81.** 0 **83.** −18 **85.** 8 **87.** −25
89. a. −46 **b.** −34 **91. a.** −85 **b.** −113
93. −2,200 ft **95.** 1,129 ft **97.** −8 **99.** −4 yd
101. −$140 **103.** Portland, Barrow, Kansas City, Atlantic
City, Norfolk **105.** 470°F **107.** 16-point increase
109. a. 2 minutes and 19 seconds of the song are left to play
b. 3 minutes and 46 seconds **115. a.** 24,090 **b.** 6,000
117. 156

STUDY SET SECTION 2.4 (page 165)

1. factor, factor, product **3.** unlike **5.** Associative
7. positive, negative **9.** negative **11.** unlike/different
13. 0 **15. a.** 3 **b.** 12 **17. a.** base: 8, exponent: 4
b. base: −7, exponent: 9 **19.** 6, −24 **21.** −15 **23.** −18
25. −72 **27.** −126 **29.** −1,665 **31.** −94,000 **33.** 56
35. 7 **37.** 156 **39.** 276 **41.** 1,947 **43.** 72,000,000
45. 90 **47.** 150 **49.** −384 **51.** −336 **53.** −48
55. −81 **57.** 36 **59.** 144 **61.** −27 **63.** −32
65. 625 **67.** 1 **69.** 49, −49 **71.** 144, −144 **73.** −60
75. 0 **77.** −64 **79.** −20 **81.** −18 **83.** 60 **85.** −48
87. −8,400,000 **89.** −625 **91.** 144 **93.** 1 **95.** −120
97. a. −2,407 **b.** 2,407 **c.** −2,407 **d.** 2,407 **99. a.** 4,352
b. −4,352 **c.** 4,352 **d.** −4,352 **101. a.** 8 **b.** 22 **c.** −105
103. a. −7 **b.** 1,394 **c.** −75 **105.** −2,000 ft **107. a.** high:
2, low: −3 **b.** high: 4, low: −6 **109. a.** −402,000 jobs
b. −423,000 jobs **c.** −581,000 jobs **d.** −528,000 jobs
111. −324°F **113.** −$1,200 **115.** −18 ft
117. −$226,524 **123.** 2, 3, 5, 7, 11, 13, 17, 19, 23, 29
125. 19 R 47

STUDY SET SECTION 2.5 (page 174)

1. 12: dividend, −4: divisor, −3: quotient; 12: dividend,
−4: divisor, −3: quotient **3.** by, of **5. a.** −5(5) = −25
b. 6(−6) = −36 **c.** 0(−15) = 0 **7. a.** positive
b. negative **9. a.** 0 **b.** undefined **11. a.** always true
b. sometimes true **c.** always true **13.** −7 **15.** −4
17. −6 **19.** −8 **21.** −22 **23.** −39 **25.** −30 **27.** −50
29. 2 **31.** 5 **33.** 9 **35.** 4 **37.** 16 **39.** 21 **41.** 40
43. 500 **45. a.** undefined **b.** 0 **47. a.** 0 **b.** undefined
49. 3 **51.** −17 **53.** 0 **55.** −5 **57.** −5 **59.** undefined
61. −19 **63.** 1 **65.** −20 **67.** −1 **69.** 10 **71.** −24
73. −30 **75.** −4 **77.** −542 **79.** −1,634 **81. a.** −16
b. −20 **c.** −36 **d.** −9 **83. a.** −64 **b.** 576 **c.** 9 **d.** −80
85. −$35 per week **87.** −1,010 ft **89.** −7° per min
91. −6 (6 games behind) **93.** −$15 **95.** −$17
103. 5 **105.** associative property of addition **107.** no

STUDY SET SECTION 2.6 (page 182)

1. order **3.** inner, outer **5. a.** square, multiplication,
subtraction **b.** multiplication, cube, subtraction, addition
c. subtraction, multiplication, addition **d.** square, multiplication

7. parentheses, brackets, absolute value symbols, fraction bar
9. 4, 20, −20, −28 **11.** −8, −1, −5, −14
13. −10 **15.** −62 **17.** 15 **19.** 12 **21.** −12
23. −80 **25.** −72 **27.** −200 **29.** 4 **31.** 28 **33.** 17
35. 71 **37.** 21 **39.** 50 **41.** −6 **43.** −12 **45. a.** 12
b. 5 **47. a.** 60 **b.** 14 **49.** −2 **51.** −3 **53.** −770
55. −5,000 **57.** −7 **59.** 1 **61.** 17 **63.** −21 **65.** 19
67. −7 **69.** 12 **71.** −14 **73.** −11 **75.** −9 **77.** −5
79. −3 **81.** −5 **83.** 166 **85.** 0 **87.** −14 **89.** 112
91. 22 **93.** 8 **95.** −3 **97. a.** 22 **b.** −13
99. a. −80 **b.** −20 **101.** −78,000 people **103.** 19
105. −$8 million **107.** It's better to refer to the last four
years, because there was an average budget surplus of
$16 billion. **109. a.** 90 ft below sea level (−90)
b. $600 lost (−600) **c.** −400 ft **115. a.** −3 **b.** −4 **117.** no

CHAPTER 2 REVIEW (page 186)

1. {. . . , −5, −4, −3, −2, −1, 0, 1, 2, 3, 4, 5, . . .}
2. a. −$1,200 **b.** −10 sec **3.** −33 ft
4. a.

b.

5. a. > **b.** < **6. a.** false **b.** true **7. a.** 5 **b.** 43 **c.** 0
8. a. −8 **b.** 8 **c.** 0 **9. a.** −12 **b.** 12 **c.** 0
10. a. negative **b.** the opposite of **c.** negative **d.** minus
11.

Position	Player	Score to par
1	Cristie Kerr	−20
2	Mirim Lee	−18
3	Lydia Ko	−17
4	Alison Lee	−16
5	Inbee Park	−15
6	Hyo Joo Kim	−14

12. a. 1998, $60 billion **b.** 2000, $230 billion **c.** 2009,
−$1,420 billion **13.** −10 **14.** −9 **15.** 32 **16.** 73 **17.** 0
18. 0 **19.** −8 **20.** −3 **21.** 10 **22.** 8 **23.** −4
24. −20 **25.** −76 **26.** −31 **27.** −374 **28.** 3,128
29. a. 11 **b.** −4 **c.** yes **d.** yes **e.** no **f.** no **30. a**
decrease of 2,202 people **31. a.** −100 ft **b.** −66 ft **32.** 134°F
33. opposite **34. a.** −9 − (−1) **b.** −6 − (−10)
35. −3 **36.** −21 **37.** 4 **38.** −6 **39.** −112 **40.** −8
41. −37 **42.** 30 **43.** 16 **44.** −24 **45.** −4 **46.** 22
47. 6 **48.** −8 **49.** −62 **50.** 103 **51.** 75
52. a. −77 **b.** 77 **53.** −225 ft **54.** 180°, 140°
55. 44 points **56.** −$80 **57.** −14 **58.** −376 **59.** 322
60. 25 **61.** −25 **62.** −204 **63.** −68,000,000
64. 30,000,000 **65.** −36 **66.** −36 **67.** 120 **68.** 100
69. 450 **70.** 48 **71.** −260, −390 **72.** −540 ft **73.** −125
74. −32 **75.** 4,096 **76.** 256 **77.** negative **78.** In the
first expression, the base is 9. In the second expression, the base
is −9, −81, 81 **79.** −3, 5, −15 **80.** The answer is
incorrect: 18(−8) ≠ −152 **81.** −5 **82.** −2 **83.** 8
84. −8 **85.** 10 **86.** 1 **87.** −50 **88.** 400
89. 23 **90.** −17 **91.** 0 **92.** undefined **93.** −32
94. 5 **95.** −2 min **96.** −4,729 ft **97.** −41

98. 4 **99.** 40 **100.** 8 **101.** 41 **102.** 0 **103.** −13
104. 32 **105.** 12 **106.** −16 **107.** −4
108. −34 **109.** −1 **110.** −4 **111.** 5 **112.** 55
113. 2,300 **114.** −2

CHAPTER 2 TEST (page 194)

1. a. integers **b.** inequality **c.** absolute value
d. opposites **e.** base, exponent **f.** solve **g.** check
2. a. > **b.** < **c.** < **3. a.** true **b.** true **c.** false
d. false **e.** true **4.** Poly
5.

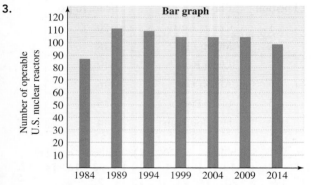

6. a. −2 **b.** −145 **c.** −1 **d.** −32 **e.** −3 **7. a.** −13
b. −1 **c.** 191 **d.** −15 **e.** −40 **8. a.** −70 **b.** 32
c. 48 **d.** 54 **e.** −26,000,000 **9.** 5(−4) = −20 **10. a.** −8
b. −8 **c.** 9 **d.** −34 **e.** −80 **11. a.** −12 **b.** 18 **c.** 4
d. −80 **12. a.** commutative property of addition
b. commutative property of multiplication **c.** adding
13. a. undefined **b.** −5 **c.** 0 **d.** 1 **14. a.** 16 **b.** −16
15. 1 **16.** −27 **17.** −34 **18.** 88 **19.** 6 **20.** 48
21. −24 **22.** 58 **23.** −72°F **24.** $203 lost (−203)
25. 154 ft **26.** −350 ft **27.** −15 **28.** −$60 million

CHAPTERS 1–2 CUMULATIVE REVIEW (page 196)

1. a. 7 millions **b.** 3 **c.** 7,326,500 **d.** 7,330,000
2. CRF Cable

3.

Bar graph

Source: allcountries.org and The World Almanac and Book of Facts, 2017

4. 360 **5.** 1,854 **6.** 24,388 **7.** 3,806 **8.** 4,684 **9.** 37,777
10. 1,432 **11.** no **12.** 65 wooden chairs **13.** 11,745
14. 5,528,166 **15.** 21,700,000 **16.** 864 tennis balls
17. 104 ft, 595 ft^2 **18.** 25; 144; 10,000 **19.** 87 R 5 **20.** 13
21. 467 **22.** 28 **23.** yes **24.** 10 times, 20 ounces
25. 60 rolls **26.** 1, 2, 3, 6, 9, 18 **27. a.** prime number,
odd number **b.** composite number, even number
c. neither, even number **d.** neither, odd number
28. $2^3 \cdot 3^2 \cdot 7$ **29.** 11^4 **30.** 175 **31.** 24 **32.** 30 **33.** 6
34. 27 **35.** −56 **36.** 10 **37.** 2 **38.** 41 mph
39. a.

b.

40. −3 **41.** 21 **42.** The new account balance will be −$79.
43. The melting point of helium is −273° Celsius.
44. Each investor was responsible for $55,000 of debt.
45. −37 **46.** 70 **47.** −3 **48.** 4 **49.** 129 **50.** 1
51. −23 **52.** 0 **53.** −4 **54.** −3 **55.** −100 ft
56. −$4,000,000

STUDY SET SECTION 3.1 (page 208)

1. fraction **3.** proper, improper **5.** equivalent
7. building **9.** equivalent fractions: $\frac{2}{6} = \frac{1}{3}$
11. a. improper fraction **b.** proper fraction **c.** proper
fraction **d.** improper fraction **13.** 5 **15.** numerators
17. $\frac{-7}{8}, -\frac{7}{8}$ **19.** 3, 1, 3, 18 **21.** numerator: 4; denominator: 5
23. numerator: 17; denominator: 10 **25.** $\frac{3}{4}, \frac{1}{4}$ **27.** $\frac{5}{8}, \frac{3}{8}$
29. $\frac{1}{4}, \frac{3}{4}$ **31.** $\frac{7}{12}, \frac{5}{12}$ **33. a.** 4 **b.** 1 **c.** 0 **d.** undefined
35. a. undefined **b.** 0 **c.** 1 **d.** 75 **37.** $\frac{35}{40}$ **39.** $\frac{12}{27}$
41. $\frac{45}{54}$ **43.** $\frac{4}{14}$ **45.** $\frac{15}{30}$ **47.** $\frac{22}{32}$ **49.** $\frac{35}{28}$
51. $\frac{48}{45}$ **53.** $\frac{36}{9}$ **55.** $\frac{48}{8}$ **57.** $\frac{15}{5}$ **59.** $\frac{28}{2}$
61. a. no **b.** yes **63. a.** yes **b.** no **65.** $\frac{2}{3}$ **67.** $\frac{4}{5}$
69. $\frac{1}{3}$ **71.** $\frac{1}{24}$ **73.** $\frac{3}{8}$ **75.** in simplest form
77. $\frac{5}{9}$ **79.** $\frac{10}{11}$ **81.** in simplest form **83.** $\frac{6}{7}$ **85.** $\frac{17}{13}$
87. $\frac{5}{2}$ **89.** $\frac{35}{12}$ **91.** $-\frac{1}{17}$ **93.** $-\frac{6}{7}$ **95.** $-\frac{8}{13}$
97. a. not equivalent **b.** equivalent **99.** $\frac{2}{12}$
101. a. $\frac{1}{4}$ **b.** $\frac{6}{24}$ **103. a.** $\frac{7}{10}$ **b.** $\frac{21}{30}$
105. a. 32 **b.** $\frac{5}{32}$ **107. a.** 16 **b.** $\frac{5}{8}$ **109. a.** 16, 33, 1
b. $\frac{16}{50} = \frac{8}{25}$ **c.** $\frac{33}{50}$ **d.** $\frac{1}{50}$ **111. a.** 20 **b.** $\frac{2}{5}, \frac{3}{5}$
113. Begin: April 1; end: June 30 **115.** $\frac{1}{4}$ **123.** $2,307

STUDY SET SECTION 3.2 (page 220)

1. multiplication **3.** simplify **5.** area **7.** numerators,
denominators, simplify **9. a.** negative **b.** positive
c. positive **d.** negative **11. a.** base, height, $\frac{1}{2}bh$
b. square **13. a.** $\frac{4}{1}$ **b.** $-\frac{3}{1}$ **c.** $\frac{1}{1}$
15. 7, 15, 2, 3, 5, 5, 24 **17.** $\frac{1}{8}$ **19.** $\frac{1}{45}$ **21.** $\frac{14}{27}$
23. $\frac{24}{77}$ **25.** $-\frac{4}{15}$ **27.** $-\frac{35}{72}$ **29.** $\frac{9}{8}$ **31.** $\frac{5}{2}$ **33.** $\frac{1}{2}$
35. $\frac{1}{7}$ **37.** $\frac{1}{10}$ **39.** $\frac{2}{15}$ **41. a.** $\frac{9}{25}$ **b.** $\frac{9}{25}$
43. a. $-\frac{1}{36}$ **b.** $-\frac{1}{216}$
45. $\frac{15}{32}$ **47.** 9 **49.** 15 ft^2 **51.** 63 in.^{2} **53.** 6 m^2 **55.** 60 ft^2

57.

·	$\frac{1}{2}$	$\frac{1}{3}$	$\frac{1}{4}$	$\frac{1}{5}$	$\frac{1}{6}$
$\frac{1}{2}$	$\frac{1}{4}$	$\frac{1}{6}$	$\frac{1}{8}$	$\frac{1}{10}$	$\frac{1}{12}$
$\frac{1}{3}$	$\frac{1}{6}$	$\frac{1}{9}$	$\frac{1}{12}$	$\frac{1}{15}$	$\frac{1}{18}$
$\frac{1}{4}$	$\frac{1}{8}$	$\frac{1}{12}$	$\frac{1}{16}$	$\frac{1}{20}$	$\frac{1}{24}$
$\frac{1}{5}$	$\frac{1}{10}$	$\frac{1}{15}$	$\frac{1}{20}$	$\frac{1}{25}$	$\frac{1}{30}$
$\frac{1}{6}$	$\frac{1}{12}$	$\frac{1}{18}$	$\frac{1}{24}$	$\frac{1}{30}$	$\frac{1}{36}$

59. $-\frac{1}{5}$ **61.** $\frac{21}{128}$ **63.** $\frac{1}{24}$ **65.** -15 **67.** $-\frac{27}{64}$

69. 1 **71.** $\frac{8}{3}$ **73.** $-\frac{3}{2}$ **75.** $\frac{2}{9}$ **77.** $-\frac{25}{81}$ **79.** $\frac{2}{3}$

81. $\frac{5}{6}$ **83.** $\frac{77}{60}$ **85.** $\frac{1}{2}$ **87. a.** $\frac{1}{4}$ **b.** $\frac{1}{8}$ **c.** $\frac{1}{16}$

d. $\frac{1}{32}$ **89. a.** $\frac{9}{50}$ **b.** $-\frac{9}{50}$ **91.** 60 votes

93. 18 in., 6 in., and 2 in. **95.** $\frac{3}{8}$ cup sugar, $\frac{1}{6}$ cup molasses

97.

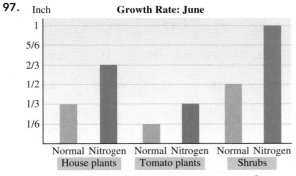

Growth Rate: June

99. 27 ft² **101.** 42 ft² **103.** 9,646 mi² **105.** $\frac{3}{4}$ in.

107. $\frac{1}{4}$ in. **115.** -2 **117.** 4

STUDY SET SECTION 3.3 (page 231)

1. reciprocal **3.** quotient **5. a.** multiply, reciprocal

b. $\cdot$, $\frac{3}{2}$ **7. a.** negative **b.** positive **9. a.** 1 **b.** 1

11. 27, 27, 8, 9, 2, 4, 4, 9, 3 **13. a.** $\frac{7}{6}$ **b.** $-\frac{8}{15}$ **c.** $\frac{1}{10}$

15. a. $\frac{8}{11}$ **b.** -14 **c.** $-\frac{1}{63}$ **17.** $\frac{3}{16}$ **19.** $\frac{14}{23}$ **21.** $\frac{35}{8}$

23. $\frac{3}{4}$ **25.** 45 **27.** 320 **29.** -4 **31.** $-\frac{7}{2}$ **33.** $\frac{4}{55}$

35. $\frac{3}{23}$ **37.** 50 **39.** $\frac{5}{6}$ **41.** $\frac{2}{3}$ **43.** 1 **45.** $-\frac{5}{8}$

47. 36 **49.** $\frac{2}{15}$ **51.** $\frac{1}{192}$ **53.** $-\frac{27}{8}$ **55.** $-\frac{15}{2}$

57. $\frac{27}{16}$ **59.** $-\frac{1}{64}$ **61.** $\frac{3}{14}$ **63.** $-\frac{5}{12}$ **65.** $\frac{13}{32}$ **67.** $\frac{2}{9}$

69. -6 **71.** $\frac{11}{6}$ **73.** $\frac{15}{28}$ **75.** $-\frac{5}{2}$ **77. a.** $\frac{15}{28}$

b. $\frac{21}{20}$ **79. a.** $\frac{1}{10}$ **b.** $-\frac{128}{45}$ **81.** 4 applications

83. 6 one-eighth cups **85. a.** 30 days **b.** 15 mi

c. 25 days **d.** route 2 **87. a.** 16 **b.** $\frac{3}{4}$ in.

c. $\frac{1}{120}$ in. **89.** 7,855 sections **97.** is less than

99. Zero **101.** 18

THINK IT THROUGH   (page 244)

$\frac{7}{20}$

STUDY SET SECTION 3.4 (page 244)

1. common **3.** build, $\frac{2}{2}$ **5.** numerators, common, Simplify

7. larger **9.** $\frac{9}{9}$ **11. a.** once **b.** twice **c.** three times

13. 2, 2, 3, 3, 5, 180

15. 7, 7, 14, 35, 14, 5, 19 **17.** $\frac{5}{9}$ **19.** $\frac{1}{2}$ **21.** $\frac{4}{15}$ **23.** $\frac{2}{5}$

25. $-\frac{3}{5}$ **27.** $-\frac{5}{21}$ **29.** $\frac{3}{8}$ **31.** $\frac{7}{11}$ **33.** $\frac{10}{21}$ **35.** $\frac{23}{45}$

37. $\frac{1}{20}$ **39.** $\frac{13}{28}$ **41.** $\frac{1}{4}$ **43.** $\frac{1}{2}$ **45.** $-\frac{13}{9}$ **47.** $-\frac{3}{4}$

49. $\frac{19}{24}$ **51.** $\frac{31}{36}$ **53.** $\frac{24}{35}$ **55.** $\frac{9}{20}$ **57.** $\frac{3}{8}$ **59.** $\frac{4}{5}$ **61.** $\frac{11}{12}$

63. $\frac{7}{6}$ **65.** $\frac{2}{3}$ **67.** $\frac{11}{10}$ **69.** $\frac{1}{3}$ **71.** $\frac{22}{15}$ **73.** $\frac{2}{5}$

75. $-\frac{11}{20}$ **77.** $-\frac{3}{16}$ **79.** $\frac{1}{4}$ **81.** $\frac{23}{10}$ **83.** $\frac{5}{12}$

85. $\frac{341}{400}$ **87.** $\frac{9}{20}$ **89.** $\frac{20}{103}$ **91.** $-\frac{23}{4}$ **93.** $\frac{17}{54}$ **95.** $-\frac{1}{50}$

97. $\frac{5}{36}$ **99.** $-\frac{17}{60}$ **101. a.** $\frac{3}{8}$ **b.** $\frac{1}{8}$ **c.** $\frac{1}{32}$ **d.** 2

103. a. $\frac{2}{5}$ **b.** $-\frac{1}{5}$ **c.** $\frac{3}{100}$ **d.** $\frac{1}{3}$ **105. a.** $\frac{31}{32}$ in.

b. $\frac{11}{32}$ in. **107.** $\frac{11}{16}$ in. **109. a.** $\frac{3}{8}$ **b.** $\frac{2}{6} = \frac{1}{3}$

c. $\frac{17}{24}$ of a pizza was left **111.** $\frac{1}{16}$ lb, undercharge

113. $\frac{13}{32}$ in. **115.** $\frac{7}{10}$ of the full-time students

study 2 or more hours a day. **117.** no **119. a.** RR: right rear

b. LR: left rear **123.** $\frac{3}{10}$ **125.** $-\frac{5}{3}$

STUDY SET SECTION 3.5 (page 258)

1. mixed **3.** improper **5. a.** $5\frac{1}{3}°$ **b.** $-6\frac{7}{8}$ in.

7. Multiply, Add, denominator **9.** $-\frac{4}{5}, -\frac{2}{5}, \frac{1}{5}$

11. improper   **13.** not reasonable: $4\frac{1}{5} \cdot 2\frac{5}{7} \approx 4 \cdot 3 = 12$

15. a. and, sixteenths   **b.** negative, two   **17.** 8, 4, 8, 4, 4, 4, 6, 6   **19.** $\frac{19}{8}, 2\frac{3}{8}$ **21.** $\frac{34}{25}, 1\frac{9}{25}$ **23.** $\frac{13}{2}$ **25.** $\frac{104}{5}$

27. $-\frac{68}{9}$ **29.** $-\frac{26}{3}$ **31.** $3\frac{1}{4}$ **33.** $5\frac{3}{5}$ **35.** $4\frac{2}{3}$ **37.** $10\frac{1}{2}$

39. 4   **41.** 2   **43.** $-8\frac{2}{7}$ **45.** $-3\frac{1}{3}$

47.

$-2\frac{8}{9}$ $-\frac{1}{2}$ $1\frac{2}{3}$ $\frac{16}{5}=3\frac{1}{5}$

$-5\ -4\ -3\ -2\ -1\ \ 0\ \ 1\ \ 2\ \ 3\ \ 4\ \ 5$

49.

$-\frac{10}{3}=-3\frac{1}{3}$ $-\frac{98}{99}$ $\frac{3}{2}=1\frac{1}{2}$ $3\frac{1}{7}$

$-5\ -4\ -3\ -2\ -1\ \ 0\ \ 1\ \ 2\ \ 3\ \ 4\ \ 5$

51. $8\frac{1}{6}$ **53.** $7\frac{2}{5}$ **55.** 8   **57.** -10 **59.** $\frac{4}{9}$ **61.** $6\frac{9}{10}$

63. $2\frac{1}{3}$ **65.** $1\frac{10}{21}$ **67.** $-13\frac{3}{4}$ **69.** $-\frac{9}{10}$ **71.** $\frac{25}{9}=2\frac{7}{9}$

73. $2\frac{1}{2}$ **75.** 12   **77.** 14   **79.** -2 **81.** $-8\frac{1}{3}$ **83.** $7\frac{3}{4}$

85. $2\frac{1}{2}$ **87.** $-1\frac{1}{4}$ **89.** $-\frac{64}{27} = -2\frac{10}{27}$ **91. a.** $\frac{15}{2}=7\frac{1}{2}$

b. $\frac{8}{15}$ **93. a.** $-\frac{11}{2}=-5\frac{1}{2}$ **b.** $-\frac{81}{22}=-3\frac{15}{22}$

95. a. $3\frac{2}{3}$ **b.** $\frac{11}{3}$ **97.** $2\frac{1}{2}$ **99. a.** $2\frac{2}{3}$ **b.** $-1\frac{1}{3}$

101. size 14, slim cut   **103.** $76\frac{9}{16}$ in.^{2} **105.** $42\frac{5}{8}$ in.^{2}

107. 64 calories   **109.** $357¢ = \$3.57$ **111.** $1\frac{1}{4}$ cups

113. 600 people   **115.** $8\frac{1}{2}$ furlongs   **119.** 60   **121.** 4

THINK IT THROUGH   (page 270)

workday: $6\frac{1}{2}$ hr; non-workday: $7\frac{1}{5}$ hr; $\frac{7}{10}$ hr

STUDY SET SECTION 3.6   (page 271)

1. mixed   **3.** fractions, whole   **5.** carry   **7. a.** 76, $\frac{3}{4}$

b. $76 + \frac{3}{4}$ **9. a.** 12   **b.** 30   **c.** 18   **d.** 24   **11.** 5, 5, 21, 35, 31, 35   **13.** $3\frac{7}{12}$ **15.** $6\frac{11}{15}$ **17.** $-2\frac{3}{8}$ **19.** $-3\frac{1}{6}$

21. $376\frac{17}{21}$ **23.** $714\frac{19}{20}$ **25.** $59\frac{28}{45}$ **27.** $132\frac{29}{33}$ **29.** $121\frac{9}{10}$

31. $147\frac{8}{9}$ **33.** $102\frac{13}{24}$ **35.** $129\frac{28}{45}$ **37.** $10\frac{1}{4}$ **39.** $13\frac{8}{15}$

41. $31\frac{14}{33}$ **43.** $71\frac{43}{56}$ **45.** $579\frac{4}{15}$ **47.** $62\frac{23}{32}$ **49.** $11\frac{1}{30}$

51. $5\frac{11}{30}$ **53.** $9\frac{3}{10}$ **55.** $3\frac{7}{8}$ **57.** $5\frac{2}{3}$ **59.** $10\frac{7}{16}$

61. $397\frac{5}{12}$ **63.** $-1\frac{11}{24}$ **65.** $7\frac{1}{2}$ **67.** $-5\frac{1}{4}$ **69.** $6\frac{1}{3}$

71. $53\frac{5}{12}$ **73.** $2\frac{1}{2}$ **75.** $-5\frac{7}{8}$ **77.** $3\frac{5}{8}$ **79.** $4\frac{1}{3}$

81. $461\frac{1}{8}$ **83.** $\frac{1}{4}$ **85.** 167, 45, 212; not reasonable

87. 616, 489, 127; reasonable   **89. a.** $\frac{32}{7}$ **b.** $\frac{38}{7}$ **c.** $-\frac{15}{7}$

d. $-\frac{35}{3}$ **91. a.** $5\frac{3}{7}$ **b.** $1\frac{1}{7}$ **c.** $-7\frac{2}{49}$ **d.** $-\frac{23}{15}$

93. $5\frac{1}{4}$ hr   **95.** $7\frac{1}{6}$ cups   **97.** $20\frac{1}{16}$ lb   **99.** $108\frac{1}{2}$ in.

101. $2\frac{3}{4}$ mi   **103.** $48\frac{1}{2}$ ft   **105. a.** 20¢ per gallon

b. 20¢ per gallon   **107.** $3\frac{1}{4}$ in.   **109.** about 60 in.

117. a. $\frac{3}{4}$ **b.** $\frac{1}{4}$ **c.** $\frac{1}{8}$ **d.** 2

STUDY SET SECTION 3.7   (page 283)

1. operations   **3.** complex   **5.** raising to a power (exponent), multiplication, and addition

7. $\left(\frac{2}{3} - \frac{1}{10}\right) + 1\frac{2}{15}$ **9.** $\frac{2}{3} \div \frac{1}{5}$ **11.** $\dfrac{\frac{1}{8} - \frac{3}{16}}{\frac{23}{4}}$

13. 3, 6, 2, 2, 2, 5   **15.** $\frac{17}{20}$ **17.** $-\frac{1}{6}$ **19.** $-\frac{7}{26}$ **21.** $-\frac{1}{12}$

23. $5\frac{13}{30}$ **25.** $2\frac{2}{3}$ **27.** $26\frac{1}{4}$ **29.** 18   **31.** $\frac{5}{32}$ **33.** $\frac{5}{6}$

35. $\frac{5}{18}$ **37.** $-\frac{1}{2}$ **39.** $\frac{50}{13}$ **41.** $\frac{25}{26}$ **43.** $-1\frac{27}{40}$ **45.** $-1\frac{1}{3}$

47. 36   **49.** $\frac{1}{3}$ **51.** $\frac{1}{8}$ **53.** 5   **55.** $14\frac{5}{24}$ **57.** 11

59. $-1\frac{1}{6}$ **61.** $\frac{3}{7}$ **63.** $\frac{3}{10}$ **65.** $44\frac{1}{3}$ **67.** $8\frac{1}{2}$ **69.** $\frac{4}{9}$

71. $1\frac{37}{70}$ **73.** $\frac{7}{8}$ **75.** $8\frac{4}{15}$ **77. a.** $\frac{22}{27}$ **b.** $\frac{8}{9}$ **79. a.** $\frac{15}{2}$

b. $\frac{55}{6}$ **81.** $91\frac{1}{4}$ in.   **83.** yes   **85.** $3\frac{1}{4}$ hr   **87.** 9 parts

89. 7 full tubes; $\frac{2}{3}$ of a tube is left over   **91.** 7 yd^2 **93.** 6 sec

99. 2,248   **101.** 20,217   **103.** 1, 2, 3, 4, 6, 8, 12, 24

CHAPTER 3 REVIEW   (page 288)

1. numerator: 11, denominator: 16; proper fraction

2. $\frac{4}{7}, \frac{3}{7}$ **3.** The figure is not divided into equal parts.

4. $-\frac{2}{3}, \frac{-2}{3}$ **5. a.** 1   **b.** 0   **c.** 18   **d.** undefined

6. equivalent fractions: $\frac{6}{8} = \frac{3}{4}$ **7.** $\frac{12}{18}$ **8.** $\frac{6}{16}$ **9.** $\frac{21}{45}$

10. $\frac{65}{60}$ **11.** $\frac{45}{9}$ **12. a.** no   **b.** yes   **13.** $\frac{1}{3}$ **14.** $\frac{5}{12}$

15. $\frac{11}{18}$ **16.** $\frac{9}{16}$ **17.** in simplest form   **18.** equivalent

19. $\frac{7}{24}, \frac{17}{24}$ **20. a.** The fraction $\frac{5}{8}$ is being expressed as an equivalent fraction with a denominator of 16. To build the fraction, multiply $\frac{5}{8}$ by 1 in the form of $\frac{2}{2}$. **b.** The fraction $\frac{4}{6}$ is being simplified. To simplify the fraction, remove the common factors of 2 from the numerator and denominator. This removes a factor equal to 1: $\frac{2}{2} = 1$. **21.** numerators, denominators, simplify **22.** $\frac{5}{6} \cdot \frac{2}{3}$ **23.** $\frac{1}{6}$ **24.** $-\frac{14}{45}$

25. $\frac{5}{12}$ **26.** $-\frac{1}{25}$ **27.** $\frac{21}{5}$ **28.** $\frac{9}{4}$ **29.** 1 **30.** 1

31. $-\frac{9}{16}$ **32.** $-\frac{125}{8}$ **33.** $-\frac{8}{125}$ **34.** $\frac{4}{9}$ **35.** 2 mi

36. 30 lb **37.** 60 in.² **38.** 165 ft² **39. a.** 8 **b.** $-\frac{12}{11}$

c. $\frac{1}{5}$ **d.** $\frac{7}{8}$ **40.** multiply, reciprocal **41.** $\frac{25}{66}$

42. $-\frac{7}{8}$ **43.** $\frac{6}{5}$ **44.** $\frac{30}{7}$ **45.** $-\frac{3}{2}$ **46.** $\frac{8}{5}$

47. $-\frac{1}{180}$ **48.** 1 **49.** 12 pins **50.** 30 pillow cases

51. $\frac{5}{7}$ **52.** $\frac{1}{2}$ **53.** $\frac{5}{4}$ **54.** $-\frac{6}{5}$ **55. a.** $\frac{5}{8}$ **b.** $\frac{1}{5}$

56. 2, 3, 3, 5, 90 **57.** $\frac{5}{6}$ **58.** $-\frac{31}{40}$ **59.** $\frac{19}{48}$ **60.** $\frac{20}{7}$

61. $-\frac{23}{36}$ **62.** $\frac{7}{12}$ **63.** $-\frac{23}{6}$ **64.** $\frac{47}{60}$ **65.** $\frac{7}{32}$ in.

66. $\frac{3}{4}$ **67.** the second hour: $\frac{3}{11} > \frac{2}{9}$ **68.** $\frac{1}{250}$

69. $4\frac{1}{4} = \frac{17}{4}$

70.

71. $3\frac{1}{5}$ **72.** $-3\frac{11}{12}$ **73.** 17 **74.** $2\frac{1}{3}$ **75.** $\frac{75}{8}$ **76.** $-\frac{11}{5}$

77. $\frac{53}{14}$ **78.** $\frac{199}{100}$ **79.** $2\frac{1}{10}$ **80.** $-\frac{21}{22}$ **81.** 40 **82.** $2\frac{1}{2}$

83. 16 **84.** $-40\frac{4}{5}$ **85.** $7\frac{9}{16}$ **86.** $6\frac{2}{9}$ **87.** $48\frac{1}{8}$ in.

88. 87 in.² **89.** 40 posters **90.** 9 loads **91.** $3\frac{23}{40}$

92. $6\frac{1}{6}$ **93.** $1\frac{1}{12}$ **94.** $1\frac{5}{16}$ **95.** $255\frac{19}{20}$ **96.** $23\frac{32}{35}$

97. $83\frac{1}{18}$ **98.** $113\frac{7}{20}$ **99.** $31\frac{11}{24}$ **100.** $316\frac{3}{4}$

101. $20\frac{1}{2}$ **102.** $34\frac{3}{8}$ **103.** $39\frac{11}{12}$ gal **104.** $\frac{5}{8}$ in.

105. $\frac{8}{9}$ **106.** $\frac{19}{72}$ **107.** $8\frac{8}{15}$ **108.** $-3\frac{5}{8}$ **109.** $-\frac{12}{17}$

110. $\frac{26}{29}$ **111.** $-\frac{2}{5}$ **112.** $\frac{63}{17}$ **113.** $2\frac{23}{40}$ **114.** $14\frac{1}{16}$

115. $8\frac{1}{3}$ **116.** $11\frac{1}{6}$ **117.** 5 full tubes, $\frac{9}{10}$ of a tube is left over

118. 8 in.

CHAPTER 3 TEST (page 304)

1. a. numerator, denominator **b.** equivalent **c.** simplest **d.** simplify **e.** reciprocal **f.** mixed **g.** complex

2. a. $\frac{4}{5}$ **b.** $\frac{1}{5}$ **3.** $\frac{13}{6} = 2\frac{1}{6}$

4.

5. yes **6. a.** $\frac{36}{45}$ **b.** $\frac{21}{24}$ **7. a.** 0 **b.** undefined

8. a. $\frac{3}{4}$ **b.** $\frac{2}{5}$ **9.** $\frac{5}{8}$ **10.** $-\frac{3}{20}$ **11.** 6 **12.** $\frac{11}{20}$

13. $\frac{11}{7}$ **14.** $\frac{1}{3}$ **15.** $\frac{9}{10}$ **16.** 40 **17.** $\frac{24}{25}$

18. a. $9\frac{1}{6}$ **b.** $\frac{39}{21}$ **19.** $261\frac{1}{6}$ **20.** $37\frac{5}{12}$

21. $1\frac{2}{3}$ **22. a.** Foreman, $39\frac{1}{2}$ lb **b.** Foreman, $5\frac{1}{2}$ in.

c. Ali, $\frac{1}{4}$ in. **23.** $\frac{8}{9}$ **24.** $\$1\frac{1}{2}$ million **25.** $11\frac{3}{4}$ in.

26. perimeter: $53\frac{1}{3}$ in., area: $106\frac{2}{3}$ in.² **27.** 60 calories

28. 12 servings **29.** $\frac{13}{24}$ **30.** $\frac{3}{10}$ **31.** $\frac{20}{21}$ **32.** $-\frac{5}{3}$

33. When we multiply a number, such as $\frac{3}{4}$, and its reciprocal, $\frac{4}{3}$, the result is 1: $\frac{3}{4} \cdot \frac{4}{3} = 1$

34. a. removing a common factor from the numerator and denominator (simplifying a fraction) **b.** equivalent fractions **c.** multiplying a fraction by a form of 1 (building an equivalent fraction)

CHAPTERS 1–3 CUMULATIVE REVIEW (page 306)

1. a. 5 **b.** 8 hundred thousands **c.** 5,896,600 **d.** 5,900,000 **2.** trillions **3.** Miami-Dade, Kings, Dallas, Riverside, Queens, San Bernadino **4. a.** 450 ft **b.** 11,250 ft²

5. 30,996 **6.** 16,544, 16,544 + 3,456 = 20,000

7. 2,400 stickers **8.** 299,320 **9.** 991, 991 · 35 = 34,685

10. $160 **11.** 1, 2, 3, 4, 6, 8, 12, 24 **12.** $2 \cdot 3^2 \cdot 5^2$ **13.** 80

14. 21 **15.** 35 **16.** $156,000 **17.** {..., −5, −4, −3, −2, −1, 0, 1, 2, 3, 4, 5, ...} **18.** true **19.** −15 **20.** 10 **21.** −200 ft

22. −11°F per hour **23.** 20 **24.** −35 **25.** 1 **26.** 2

27. $\frac{3}{4}$ **28.** $\frac{5}{2}$ **29.** $-\frac{4}{5}$ **30.** $\frac{1}{2}$ **31.** $1\frac{5}{12}$ **32.** $-\frac{1}{35}$

33. 30 sec **34.** $\frac{11}{16}$ in. **35.** $10\frac{5}{7}$ **36.** $-\frac{53}{8}$ **37.** $7\frac{2}{5}$

38. $6\frac{9}{10}$ **39.** $9\frac{11}{12}$ **40.** $5\frac{11}{15}$ **41.** width: 28 in., height: 6 in.

42. $274\frac{1}{4}$ gal **43.** $3\frac{5}{12}$ ft **44.** $-\frac{3}{64}$ **45.** $\frac{5}{6}$ **46.** $-\frac{2}{49}$

STUDY SET SECTION 4.1 (page 319)

1. point **3.** expanded **5.** Thousands, Hundreds, Tens, Ones, Tenths, Hundredths, Thousandths, Ten-thousandths

7. a. 10 **b.** $\frac{1}{10}$ **9. a.** $\frac{7}{10}, 0.7$ **b.** $\frac{47}{100}, 0.47$

11. Whole-number part, Fractional part **13.** ths

15. 79,816.0245 **17. a.** 9 tenths **b.** 6 **c.** 4 **d.** 5 ones

19. a. 8 millionths **b.** 0 **c.** 5 **d.** 6 ones

21. $30 + 7 + \frac{8}{10} + \frac{9}{100}$

23. $100 + 20 + 4 + \frac{5}{10} + \frac{7}{100} + \frac{5}{1,000}$

25. $7,000 + 400 + 90 + 8 + \frac{6}{10} + \frac{4}{100} + \frac{6}{1,000} + \frac{8}{10,000}$

27. $6 + \frac{4}{10} + \frac{9}{1,000} + \frac{4}{10,000} + \frac{1}{100,000}$

29. three tenths, $\frac{3}{10}$

31. fifty and forty-one hundredths, $50\frac{41}{100}$ **33.** nineteen and five hundred twenty-nine thousandths, $19\frac{529}{1,000}$

35. three hundred four and three ten-thousandths, $304\frac{3}{10,000}$

37. negative one hundred thirty-seven hundred-thousandths, $-\frac{137}{100,000}$ **39.** negative one thousand seventy-two and four hundred ninety-nine thousandths, $-1,072\frac{499}{1,000}$ **41.** 6.187

43. 10.0056 **45.** -16.39 **47.** 104.000004 **49.** > **51.** < **53.** > **55.** > **57.** < **59.** >

61.

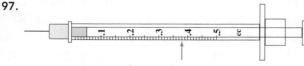

63.

$$\begin{array}{c} -4.25 \ -3.29 \ -1.84 \ -1.21 \qquad\quad 2.75 \\ \overleftrightarrow{ -5 \ -4 \ -3 \ -2 \ -1 \ 0 \ 1 \ 2 \ 3 \ 4 \ 5} \end{array}$$

65. 506.2 **67.** 33.08 **69.** 4.234 **71.** 0.3656 **73.** -0.14 **75.** -2.7 **77.** 3.150 **79.** 1.414213 **81.** 16.100 **83.** 290.30350 **85.** $0.28 **87.** $27,842 **89. a.** 8,500 **b.** 8,506.29 **91. a.** 100 **b.** 0.01 **c.** 1,000 **d.** 0.001 **e.** 1,000,000 **f.** 0.000001 **g.** 1,000,000,000 **h.** 0.000000001 **93.** -0.7 **95.** $1,025.78

97.

99. two-thousandths, $\frac{2}{1,000} = \frac{1}{500}$ **101.** $0.16, $1.02, $1.20, $0.00, $0.10 **103.** candlemaking, crafts, hobbies, folk dolls, modern art **105.** Cylinder 2, Cylinder 4 **107.** bacterium, plant cell, animal cell, asbestos fiber **109.** June 3, 4, 5, 6, 7

117. Perimeter: $12\frac{1}{2}$ ft; Area: $9\frac{5}{8}$ ft^2

STUDY SET SECTION 4.2 (page 333)

1. addend, addend, addend, sum **3.** minuend, subtrahend, difference **5.** estimate **7.** It is not correct:

$15.2 + 12.5 \ne 28.7$ **9.** opposite **11. a.** -1.2 **b.** 13.55 **c.** -7.4 **13.** 46.600, 11.000 **15.** 39.9 **17.** 8.59 **19.** 101.561 **21.** 202.991 **23.** 3.31 **25.** 2.75 **27.** 341.7 **29.** 703.5 **31.** 7.235 **33.** 43.863 **35.** -14.7 **37.** -11.2 **39.** -14.68 **41.** -6.15 **43.** -2.1 **45.** -45.3 **47.** 6.81 **49.** 17.82 **51.** -4.5 **53.** -3.4 **55.** 790 **57.** 610 **59.** -10.9 **61.** -16.6 **63.** 0 **65.** 55.00 **67.** 3.65 **69.** 658.04007 **71.** -47.5 **73.** 4.1 **75.** 288.46 **77.** 70.29 **79.** -14.3 **81.** 1.09 **83.** 8.03 **85.** 15.2 **87.** 4.977 **89.** 2.598 **91. a.** 5.72 **b.** -5.72 **93. a.** 1.73 **b.** -5.99 **c.** 5.99 **d.** -1.73 **95.** $815.80, $545.00, $531.49 **97.** 1.74, 2.32, 4.06; 2.90, 0, 2.90 **99.** 2.375 in. **101.** 41.57 sec **103.** $523.19, $498.19 **105.** 1.1°, 101.1°, 0°, 1.4°, 99.5° **107.** 20.01 mi **109. a.** $101.94 **b.** $55.80

117. a. $\frac{73}{60} = 1\frac{13}{60}$ **b.** $\frac{23}{60}$ **c.** $\frac{1}{3}$ **d.** $\frac{48}{25} = 1\frac{23}{25}$

STUDY SET SECTION 4.3 (page 347)

1. factor, factor, partial product, partial product, product **3. a.** 2.28 **b.** 14.499 **c.** 14.0 **d.** 0.00026 **5. a.** positive **b.** negative **7. a.** 10, 100, 1,000, 10,000, 100,000 **b.** 0.1, 0.01, 0.001, 0.0001, 0.00001 **9.** 29.76 **11.** 49.84 **13.** 0.0081 **15.** 0.0522 **17.** 1,127.7 **19.** 2,338.4 **21.** 684 **23.** 410 **25.** 6.4759 **27.** 0.00115 **29.** 14,200,000 **31.** 98,200,000,000 **33.** 1,421,000,000,000 **35.** 657,100,000,000 **37.** -13.68 **39.** 5.28 **41.** 448,300 **43.** $-678,231$ **45.** 11.56 **47.** 0.0009 **49.** 3.16 **51.** 68.66 **53.** 119.70 **55.** 38.16 **57.** 14.6 **59.** 15.7 **61.** 250 **63.** 66.69 **65.** -0.1848 **67.** 0.01 **69.** 0.84 **71.** 0.00072 **73.** $-200,000$ **75.** 12.32 **77.** -17.48 **79.** 0.0049 **81.** 14.24 **83.** 8.6265 **85.** -57.2467 **87.** -22.39 **89.** -3.872 **91.** 24.48 **93.** -0.8649 **95.** 0.01, 0.04, 0.09, 0.16, 0.25, 0.36, 0.49, 0.64, 0.81 **97. a.** 72,310,000 **b.** 72,310,000,000 **c.** 72,310,000,000,000 **99. a.** 9.5 **b.** 1.9 **c.** 21.66 **101.** 1.9 in. **103.** $74,100 **105.** $159.84, $123.75 **107.** 0.000000136 in., 0.0000000136 in., 0.00000004 in. **109. a.** 2.1 mi **b.** 3.5 mi **c.** 5.6 mi **111.** $102.65 **113. a.** 19,600,000 acres **b.** 7,376,000,000 **c.** 3,148,000,000,000 miles **115. a.** 192 ft^2 **b.** 223.125 ft^2 **c.** 31.125 ft^2 **117. a.** $12.50, $12,500, $15.75, $1,575 **b.** $14,075 **119.** 136.4 lb **121.** 0.84 in. **123.** $94.3 million = $94,300,000 **131.** $2^2 \cdot 5 \cdot 11$ **133.** $2 \cdot 3^4$

THINK IT THROUGH (page 363)

1. 2.86

STUDY SET SECTION 4.4 (page 363)

1. divisor, quotient, dividend **3. a.** 5.26 **b.** 0.008 **5. a.** $13\overline{)106.6}$ **b.** $371\overline{)1669.5}$ **7.** $\frac{10}{10}$ **9.** thousandths **11. a.** left **b.** right **13.** moving the decimal points in the divisor and dividend 2 places to the right **15.** 2.1 **17.** 9.2 **19.** 4.27 **21.** 8.65 **23.** 3.35 **25.** 4.56 **27.** 0.46 **29.** 0.39 **31.** 19.72 **33.** 24.41 **35.** $280 \div 70 = 28 \div 7 = 4$ **37.** $400 \div 8 = 50$ **39.** $4,000 \div 50 = 400 \div 5 = 80$ **41.** $15,000 \div 5 = 3,000$ **43.** 4.5178 **45.** 0.003009 **47.** 12.5 **49.** 545,200 **51.** -8.62 **53.** 4.04 **55.** 20,325.7 **57.** -0.00003 **59.** -5.162 **61.** 0.1 **63.** 3.5 **65.** 58.5 **67.** 2.66 **69.** 7.504 **71.** 0.0045 **73.** 0.321 **75.** -1.5

77. -122.02 **79.** -2.4 **81.** 9.75 **83.** 789,150 **85.** -6
87. 13.60 **89.** 0.0348 **91.** 1,027.19 **93.** 0.15625
95. a. 2.38 **b.** 6.936 **c.** 5.78 **d.** 2.4 **97. a.** 0.2 **b.** 5
99. 280 slices **101.** 2,000,000 calculations
103. 500 squeezes **105.** 11 hr, 6 P.M. **107.** 66.7 million
109. 0.231 sec **111.** 3.5 stars **119. a.** 5 **b.** 50

STUDY SET SECTION 4.5 (page 378)

1. equivalent **3.** terminating **5.** $\div$ **7.** zeros **9.** repeating
11. a. 0.38 **b.** 0.212 **13. a.** $\dfrac{7}{10}$ **b.** $\dfrac{77}{100}$ **15.** 0.5
17. 0.875 **19.** 0.55 **21.** 2.6 **23.** 0.5625 **25.** -0.53125
27. 0.6 **29.** 0.225 **31.** 0.76 **33.** 0.002 **35.** 3.75
37. 12.6875 **39.** $0.\overline{1}$ **41.** $0.58\overline{3}$ **43.** $0.0\overline{7}$ **45.** $0.01\overline{6}$
47. $-0.\overline{45}$ **49.** $-0.\overline{60}$ **51.** 0.23 **53.** 0.49 **55.** 1.85
57. -1.08 **59.** 0.152 **61.** 0.370
63.

65.

67. $<$ **69.** $>$ **71.** $=$ **73.** $<$ **75.** 6.25, $\dfrac{19}{3}$, $6\dfrac{1}{2}$
77. $-\dfrac{8}{9}, -\dfrac{6}{7}, -0.\overline{81}$ **79.** $\dfrac{37}{90}$ **81.** $\dfrac{19}{60}$ **83.** $\dfrac{3}{22}$ **85.** 1
87. 0.57 **89.** 5.27 **91.** 0.35 **93.** -0.48 **95.** -2.55
97. 0.068 **99.** 7.305 **101.** 0.075 **103. a.** 0.625 **b.** 1.6
105. a. $0.\overline{4}$ **b.** $0.\overline{41}$ **c.** $0.41\overline{4}$ **d.** $0.1\overline{4}$
107. 0.0625, 0.375, 0.5625, 0.9375
109. $\dfrac{3}{40}$ in. **111.** 23.4 sec, 23.8 sec, 24.2 sec, 32.6 sec
113. 93.6 in.² **115.** \$7.02 **123. a.** $\{0, 1, 2, 3, 4, 5, 6, 7, 8, 9\}$
b. $\{2, 3, 5, 7, 11, 13, 17, 19, 23, 29\}$ **c.** $\{\ldots, -3, -2, -1, 0, 1, 2, 3, \ldots\}$ **125.** 0.01

STUDY SET SECTION 4.6 (page 387)

1. square **3.** radical **5.** perfect **7. a.** 25, 25 **b.** $\dfrac{1}{16}, \dfrac{1}{16}$
9. a. 7 **b.** 2 **11. a.** 1 **b.** 0 **13.** Step 2: Evaluate all exponential expressions and any square roots.
15.

17. a. square root **b.** negative **19.** $-7, 8$ **21.** 5 and -5
23. 4 and -4 **25.** 4 **27.** 3 **29.** -12 **31.** -7 **33.** 31
35. 63 **37.** $\dfrac{2}{5}$ **39.** $-\dfrac{4}{3}$ **41.** $-\dfrac{1}{9}$ **43.** 0.8 **45.** -0.9
47. 0.3 **49.** 7 **51.** 16 **53.** -16 **55.** -3 **57.** 20
59. -140 **61.** -48 **63.** 43 **65.** 75 **67.** -7 **69.** -1
71. -10 **73.** $-\dfrac{7}{20}$ **75.** -140 **77.** 9.56 **79.** -1.4
81. 15 **83.** 7 **85.** 1, 1.414, 1.732, 2, 2.236, 2.449, 2.646, 2.828, 3, 3.162 **87.** 2.24 **89.** 8.12 **91.** 4.904 **93.** -3.332
95. a. 12 **b.** 18 **c.** 63 **d.** 98 **e.** 147 **f.** 343
97. a. 44 **b.** 10 **c.** 14 **d.** 48 **e.** 2 **f.** -2
99. a. 5 ft **b.** 10 ft **101.** 127.3 ft **103.** 42-inch screen
113. 82.35 **115.** 39.304

CHAPTER 4 REVIEW (page 391)

1. a. 0.67, $\dfrac{67}{100}$ **b.**

2. a. 7 hundredths **b.** 3 **c.** 8 **d.** 5 ten-thousandths
3. $10 + 6 + \dfrac{4}{10} + \dfrac{5}{100} + \dfrac{2}{1,000} + \dfrac{3}{10,000}$ **4.** two and three tenths, $2\dfrac{3}{10}$ **5.** negative six hundred fifteen and fifty-nine hundredths, $-615\dfrac{59}{100}$ **6.** six hundred one ten-thousandths, $\dfrac{601}{10,000}$ **7.** one hundred-thousandth, $\dfrac{1}{100,000}$ **8.** 100.61
9. 11.997 **10.** 301.000016 **11.** $<$ **12.** $<$ **13.** $>$ **14.** $>$
15.

16. a. true **b.** false **c.** true **d.** true **17.** 4.58
18. 3,706.082 **19.** -0.1 **20.** -88.1 **21.** 6.7030
22. 11.3150 **23.** 0.222228 **24.** 0.63527 **25.** \$0.67 **26.** \$13
27. Washington, Diaz, Chou, Singh, Gerbac **28.** Sun: 1.8, Mon: 0.6, Tues: 2.4, Wed: 3.8 **29.** 66.7 **30.** 45.188
31. 15.17 **32.** 28.428 **33.** 317.824 **34.** 24.30 **35.** -7.7
36. 3.1 **37.** -4.8 **38.** -29.09 **39.** -25.6 **40.** 4.939
41. a. 760 **b.** 280 **42.** 10.75 mm **43.** \$48.21 **44.** 8.15 in.
45. 15.87 **46.** 197.945 **47.** 0.0068 **48.** 2,310 **49.** -1.35
50. 0.00006 **51.** 90,145.2 **52.** 0.002897 **53.** 0.04
54. 0.0225 **55.** -10.61 **56.** 25.58 **57.** 0.0001089
58. 115.741 **59. a.** 9,600,000 km² **b.** 3,220,000,000,000 m
60. a. 1,600 **b.** 91.76 **61.** 98.07 **62.** \$19.43 **63.** 0.07 in.
64. 68.62 in.² **65.** 9.3 **66.** 10.45 **67.** 1.29 **68.** 41.03
69. -6.25 **70.** 0.053 **71.** 63 **72.** 0.81 **73.** 0.08976
74. -0.00112 **75.** 876.5 **76.** 770,210
77. $4,800 \div 40 = 480 \div 4 = 120$ **78.** $27,000 \div 9 = 3,000$
79. 12.9 **80.** -776.86 **81.** 13.95 **82.** 20.5 **83.** \$4.99
84. 0.51 ppb **85.** 14 servings **86.** 9.5 revolutions
87. 0.875 **88.** -0.4 **89.** 0.5625 **90.** 0.06 **91.** $0.\overline{54}$
92. $-1.\overline{3}$ **93.** 3.056 **94.** $0.5\overline{7}$ **95.** 0.58 **96.** 1.03
97. $>$ **98.** $=$ **99.** 0.3, $\dfrac{10}{33}$, $0.\overline{3}$
100.

101. $\dfrac{11}{15}$ **102.** $\dfrac{307}{300} = 1\dfrac{7}{300}$ **103.** -6.24 **104.** 0.175
105. 93 **106.** 7.305 **107.** 34.88 in.² **108.** \$22.25
109. 5 and -5 **110.** 7, 7 **111.** 7 **112.** -4
113. 10 **114.** 0.3 **115.** $\dfrac{8}{13}$ **116.** 0.9 **117.** $-\dfrac{1}{6}$ **118.** 0
119.

120. a. 4.36 **b.** 24.45 **c.** 3.57 **121.** -30 **122.** 70
123. -27 **124.** $18\dfrac{1}{3}$ **125.** 70 **126.** -440 **127.** 8
128. 33 in. **129.** 9 and 10 **130.** Since $(2.646)^2 = 7.001316$, we cannot use an $=$ symbol.

CHAPTER 4 TEST (page 403)

1. **a.** addend, addend, sum **b.** minuend, subtrahend, difference **c.** factor, factor, product **d.** divisor, quotient, dividend **e.** repeating **f.** radical **2.** $\dfrac{79}{100}$, 0.79

3. **a.** 1 thousandth **b.** 4 **c.** 6 **d.** 2 tens
4. Selway, Monroe, Paston, Covington, Cadia **5.** 4,519.0027
6. **a.** $80 + 9 + \dfrac{4}{10} + \dfrac{1}{100}$, eighty-nine and forty-one hundredths,

$89\dfrac{41}{100}$ **b.** $\dfrac{8}{100} + \dfrac{1}{10,000} + \dfrac{3}{100,000}$, eight thousand

thirteen one hundred-thousandths, $\dfrac{8,013}{100,000}$ **7. a.** 461.7

b. 2,733.050 **c.** −1.983373 **8.** $0.65 **9.** 10.756
10. 6.121 **11.** 0.1024 **12.** 0.57 **13.** −1.92 **14.** 0.0348
15. $1.1\overline{8}$ **16.** −0.8 **17.** −2.29 **18. a.** 210
b. $4,000 \div 20 = 400 \div 2 = 200$ **19. a.** 0.567909 **b.** 0.458
20. 61,400,000,000 **21.** 1.026 in. **22.** 1.25 mi²
23. 0.004 in. **24.** Saturday, $23.75 **25.** 0.42 gram
26. 20.825 lb **27.** 10.676 **28. a.** 0.34 **b.** $0.41\overline{6}$ **29.** 3.588
30. 56.86 **31.** −12 **32.** $\dfrac{41}{30}$

33. **a.**

b.

34. $5.65 **35.** 37 **36. a.** −1.08 **b.** 2.5625 **37.** 12, 12
38. a. > **b.** < **c.** = **d.** < **39.** 11 **40.** $-\dfrac{1}{30}$
41. a. −0.2 **b.** 1.3 **c.** 15 **d.** −11

CHAPTERS 1–4 CUMULATIVE REVIEW (page 406)

1. **a.** one hundred fifty-four thousand, three hundred two
b. $100,000 + 50,000 + 4,000 + 300 + 2$
2. $(3 + 4) + 5 = 3 + (4 + 5)$ **3.** 16,693 **4.** 102
5. 75,625 ft² **6.** 27 R 42 **7.** $677,200 **8.** 1, 2, 4, 5, 10, 20
9. $2^2 \cdot 5 \cdot 11$ **10.** 600, 20 **11.** 4 **12.** > **13.** −13
14. adding **15.** 83°F increase **16.** −270
17. −1 **18.** −2,100 ft **19.** $3(-5) = -15$ **20.** 60
21. 8, −3, 36, −6, 6 **22.** 35 **23.** −5,000 **24.** $\dfrac{6}{13}$

25. equivalent fractions **26.** $\dfrac{5}{7}$ **27.** $\dfrac{21}{128}$ **28.** $-\dfrac{3}{16}$

29. $\dfrac{17}{18}$ **30.** $19\dfrac{1}{8}$ **31.** $26\dfrac{7}{24}$ **32.** $-\dfrac{1}{3}$ **33.** $\dfrac{7}{64}$

34. $11\dfrac{1}{8}$ in. **35.** $\dfrac{3}{4}$ **36.** 0.001 in. **37.** <

38.

39. 130.198 **40.** 1.101 **41.** −8.136 **42.** 0.056012
43. 5.6 **44.** 0.0000897 **45.** 33.6 hr **46.** 157.5 in.²
47. 232.8 **48.** $0.41\overline{6}$ **49.** −2.325 **50.** 8, 8
51. 7 **52.** $\dfrac{15}{4}$ **53.** −6 **54.** 39

THINK IT THROUGH (page 418)

1. c **2.** f **3.** b **4.** e **5.** g **6.** d **7.** a **8.** h

STUDY SET SECTION 5.1 (page 419)

1. ratio **3.** unit **5.** 3 **7.** 10 **9.** $\dfrac{11 \text{ minutes}}{60 \text{ minutes}} = \dfrac{11}{60}$

11. $\dfrac{13}{9}$, 13 to 9, 13:9 **13.** $\dfrac{5}{8}$ **15.** $\dfrac{11}{16}$ **17.** $\dfrac{5}{3}$ **19.** $\dfrac{7}{4}$ **21.** $\dfrac{2}{3}$

23. $\dfrac{1}{2}$ **25.** $\dfrac{1}{3}$ **27.** $\dfrac{3}{4}$ **29.** $\dfrac{1}{3}$ **31.** $\dfrac{13}{3}$ **33.** $\dfrac{19}{39}$ **35.** $\dfrac{2}{7}$

37. $\dfrac{1}{2}$ **39.** $\dfrac{6}{1}$ **41.** $\dfrac{1}{5}$ **43.** $\dfrac{3}{7}$ **45.** $\dfrac{3}{4}$ **47.** $\dfrac{7}{12}$ **49.** $\dfrac{32 \text{ ft}}{3 \text{ sec}}$

51. $\dfrac{15 \text{ days}}{4 \text{ gal}}$ **53.** $\dfrac{21 \text{ made}}{25 \text{ attempts}}$ **55.** $\dfrac{3 \text{ beats}}{2 \text{ measures}}$

57. 12 revolutions per min **59.** $5,000 per year
61. 1.5 errors per hr **63.** 320.6 people per square mi
65. $4 per min **67.** $68 per person **69.** 1.2 cents per ounce

71. $0.07 per ft **73. a.** $\dfrac{2}{5}$ **b.** $\dfrac{5}{2}$ **75. a.** $\dfrac{1}{8}$ **b.** $\dfrac{8}{1}$

77. **a.** $\dfrac{2}{3}$ **b.** $\dfrac{3}{2}$ **79.** $\dfrac{3}{1}$ **81.** $\dfrac{1}{55}$ **83.** $\dfrac{1}{20}$

85. $\dfrac{33 \text{ compressions}}{2 \text{ breaths}}$ **87. a.** $1,800 **b.** $\dfrac{4}{9}$ **c.** $\dfrac{1}{3}$ **d.** $\dfrac{1}{18}$

89. $\dfrac{1}{1}$ **91. a.** ii **b.** iii **c.** i **93.** $\dfrac{329 \text{ complaints}}{100,000 \text{ passengers}}$

95. **a.** 108,000 **b.** 24 browsers per buyer
97. 0.03 ounce gold/cubic foot gravel **99.** 7¢ per oz
101. 1.25¢ per min **103.** $24,182 per lb **105.** 440 gal per min
107. **a.** 325 mi **b.** 65 mph **109.** the 6-oz can
111. the 50-tablet boxes **113.** the truck **115.** the second car
117. **a.** 25,000 **b.** 24,870 **123.** $\dfrac{8}{9}$ **125.** 12

STUDY SET SECTION 5.2 (page 434)

1. proportion **3.** cross **5.** variable **7.** isolated
9. true, false **11.** 9, 90, 45, 90 **13.** Children, Teacher's aides

15. $3 \cdot x$, 18, 3, 3, 6, 6 **17.** $\dfrac{20}{30} = \dfrac{2}{3}$ **19.** $\dfrac{400 \text{ sheets}}{100 \text{ beds}} = \dfrac{4 \text{ sheets}}{1 \text{ bed}}$

21. false **23.** true **25.** true **27.** false **29.** false
31. true **33.** true **35.** false **37.** yes **39.** no **41.** 6
43. 4 **45.** 0.3 **47.** 2.2 **49.** $3\dfrac{1}{2}$ **51.** $\dfrac{7}{8}$ **53.** 3,500 **55.** $\dfrac{1}{2}$

57. 36 **59.** 1 **61.** 2 **63.** $8\dfrac{1}{5}$ **65.** 180 **67.** 18 **69.** 3.1

71. $\dfrac{1}{6}$ **73. a.** 0.18 **b.** 0.18 **c.** 0.18 **75. a.** 6 **b.** $\dfrac{7}{9}$

77. $218.75 **79.** $77.32 **81.** yes **83.** 24 drops
85. 975 red blood cells **87.** 80 ft **89.** 65.25 ft = 65 ft 3 in.
91. 2.625 in. = $2\dfrac{5}{8}$ in. **93.** $4\dfrac{2}{7}$, which is about $4\dfrac{1}{4}$ **95.** 19 sec

97. 31.25 in. = $31\dfrac{1}{4}$ in. **99.** $309 **101.** 19.2 in.

107. 49.188 **109.** 31.428 **111.** 4.1 **113.** −49.09

STUDY SET SECTION 5.3 (page 447)

1. length **3.** unit **5.** capacity **7. a.** 1 **b.** 3 **c.** 36 **d.** 5,280 **9. a.** 8 **b.** 2 **c.** 2 **d.** 4 **11.** 1 **13. a.** oz

b. lb **15. a.** $\dfrac{1\ \text{ton}}{2{,}000\ \text{lb}}$ **b.** $\dfrac{2\ \text{pt}}{1\ \text{qt}}$ **17. a.** iv **b.** i **c.** ii **d.** iii **19. a.** iii **b.** iv **c.** i **d.** ii **21. a.** pound

b. ounce **c.** fluid ounce **23.** 36, in., 72 **25.** 2,000, 16, oz, 32,000 **27. a.** 8 **b.** $\dfrac{5}{8}$ in., $1\dfrac{1}{4}$ in., $2\dfrac{7}{8}$ in. **29. a.** 16

b. $\dfrac{9}{16}$ in., $1\dfrac{3}{4}$ in., $2\dfrac{3}{16}$ in. **31.** $2\dfrac{9}{16}$ in. **33.** $10\dfrac{7}{8}$ in. **35.** 12 ft

37. 105 ft **39.** 42 in. **41.** 63 in. **43.** $\dfrac{21}{352}$ mi ≈ 0.06 mi

45. $\dfrac{7}{8}$ mi = 0.875 mi **47.** $2\dfrac{3}{4}$ lb = 2.75 lb **49.** $4\dfrac{1}{2}$ lb = 4.5 lb

51. 800 oz **53.** 1,392 oz **55.** 128 fl oz **57.** 336 fl oz

59. $2\dfrac{3}{4}$ hr **61.** $5\dfrac{1}{2}$ hr **63.** 6 pt **65.** 5 days **67.** $4\dfrac{2}{3}$ ft

69. 48 in. **71.** 2 gal **73.** 5 lb **75.** 4 hr **77.** 288 in.

79. $2\dfrac{1}{2}$ yd = 2.5 yd **81.** 15 ft **83.** 24,800 lb **85.** $2\dfrac{1}{3}$ yd

87. 3 mi **89.** 2,640 ft **91.** $3\dfrac{1}{2}$ tons = 3.5 tons **93.** 2 pt

95. a. 864 in. **b.** $\dfrac{2}{3}$ yd **97. a.** 768 fl oz **b.** $\dfrac{3}{4}$ qt

99. 150 yd **101.** 2,880 in. **103.** 0.34 mi **105.** 100 ft

107. 128 oz **109.** $4\dfrac{19}{20}$ tons = 4.95 tons **111.** 68 quart cans

113. $71\dfrac{7}{8}$ gal = 71.875 gal **115.** 320 oz

117. $6\dfrac{1}{8}$ days = 6.125 days **121.** 209 **123.** 0.00283

STUDY SET SECTION 5.4 (page 462)

1. metric **3. a.** tens **b.** hundreds **c.** thousands **5.** unit, chart **7.** weight **9. a.** 1,000 **b.** 100 **c.** 1,000

11. a. 1,000 **b.** 10 **13. a.** $\dfrac{1\ \text{km}}{1{,}000\ \text{m}}$ **b.** $\dfrac{100\ \text{cg}}{1\ \text{g}}$ **c.** $\dfrac{1{,}000\ \text{mL}}{1\ \text{L}}$

15. a. iii **b.** i **c.** ii **17. a.** ii **b.** iii **c.** i **19.** 1, 100, 0.2 **21.** 1,000, 1, mg, 200,000 **23.** 1 cm, 3 cm, 5 cm **25. a.** 10, 1 millimeter **b.** 27 mm, 41 mm, 55 mm **27.** 156 mm **29.** 280 mm **31.** 3.8 m **33.** 1.2 m

35. 8,700 mm **37.** 2,890 mm **39.** 0.000045 km **41.** 0.000003 km **43.** 1,930 g **45.** 4,531 g **47.** 6 g **49.** 3.5 g **51.** 3,000 mL **53.** 26,300 mL **55.** 3.1 cm **57.** 0.5 L **59.** 2,000 g **61.** 0.74 mm **63.** 1,000,000 g **65.** 0.65823 kL **67.** 0.472 dm **69.** 10 **71.** 1.25 mg **73.** 5.689 kg **75.** 4.532 m **77.** 0.0325 L **79.** 675,000 **81.** 0.0000077 **83.** 1.34 hm **85.** 6,578 dam **87. a.** 0.045 m **b.** 450 cm **89. a.** 0.0045 L **b.** 4,500 mL **91.** 0.5 km, 1 km, 1.5 km, 5 km, 10 km **93.** 0.31 m, 0.91 m **95.** 12 cm, 8 cm **97.** 0.00005 L **99.** 3 g **101.** 3,000 mL **103.** 4 **105.** 1.5 L **111.** $0.\overline{8}$ **113.** $0.0\overline{7}$

THINK IT THROUGH (page 469)

1. 216 mm × 279 mm **2.** 9 kilograms **3.** 22.2 milliliters

STUDY SET SECTION 5.5 (page 472)

1. Fahrenheit, Celsius **3. a.** meter **b.** meter **c.** inch **d.** mile **5. a.** liter **b.** liter **c.** gallon **7. a.** $\dfrac{0.30\ \text{m}}{1\ \text{ft}}$

b. $\dfrac{0.45\ \text{kg}}{1\ \text{lb}}$ **c.** $\dfrac{3.79\ \text{L}}{1\ \text{gal}}$ **9.** 0.30 m, m **11.** 0.035, 1,000, oz

13. 10 in. **15.** 34 in. **17.** 2,520 m **19.** 7,534.5 m **21.** 9,072 g **23.** 34,020 g **25.** 14.3 lb **27.** 660 lb **29.** 0.7 qt **31.** 1.3 qt **33.** 48.9°C **35.** 1.7°C **37.** 167°F **39.** 50°F **41.** 11,340 g **43.** 122°F **45.** 712.5 mL **47.** 17.6 oz **49.** 147.6 in. **51.** 0.1 L **53.** 39,283 ft **55.** 1.0 kg **57.** 14°F **59.** 0.6 oz **61.** 243.4 fl oz **63.** 91.4 cm **65.** 0.5 qt **67.** 10°C **69.** 127 m **71.** −20.6°C **73. a.** about 0.14 lb **b.** about 28,123 g **75. a.** about 16.4 gal **b.** 235 L **77.** 5 mi **79.** 69 mph **81.** 4,511 km, 3,953 km **83.** 1.9 cm **85.** 578 lb, 1003 lb **87. a.** 226.8 g **b.** 0.24 L **89.** no **91.** about 62°C **93.** 38°C **95.** −5°C and 0°C **97.** the 3 quarts **99.** About 10 ft

101. 2.44 in. **107.** $\dfrac{29}{15}$ **109.** $\dfrac{4}{5}$ **111.** 7.8 **113.** 15.1

CHAPTER 5 REVIEW (page 475)

1. $\dfrac{7}{25}$ **2.** $\dfrac{15}{16}$ **3.** $\dfrac{2}{3}$ **4.** $\dfrac{3}{2}$ **5.** $\dfrac{1}{3}$ **6.** $\dfrac{7}{8}$ **7.** $\dfrac{4}{5}$ **8.** $\dfrac{3}{1}$

9. $\dfrac{7}{8}$ **10.** $\dfrac{5}{4}$ **11.** $\dfrac{1}{12}$ **12.** $\dfrac{1}{4}$ **13.** $\dfrac{16\ \text{cm}}{3\ \text{yr}}$ **14.** $\dfrac{\$3}{5\ \text{min}}$

15. 30 tickets per min **16.** 15 inches per turn **17.** 32.5 feet per roll **18.** 3.2 calories per piece **19.** \$2.29 per pair **20.** \$0.25 billion per month **21.** $\dfrac{37}{32}$ **22.** \$7.75 **23.** 1,125 people per min

24. the 8-oz can **25. a.** $\dfrac{20}{30}=\dfrac{2}{3}$ **b.** $\dfrac{6\ \text{buses}}{100\ \text{cars}}=\dfrac{36\ \text{buses}}{600\ \text{cars}}$ **26.** 2, 54, 6, 54 **27.** false **28.** true **29.** true **30.** true **31.** false **32.** false **33.** yes **34.** no **35.** 4.5 **36.** 16

37. 7.2 **38.** 0.12 **39.** $1\dfrac{1}{2}$ **40.** $3\dfrac{1}{2}$ **41.** $\dfrac{1}{3}$ **42.** 1,000

43. 192.5 mi **44.** 300 **45.** 12 ft **46.** 30 in. **47. a.** 16

b. $\dfrac{7}{16}$ in., $1\dfrac{1}{2}$ in., $1\dfrac{3}{4}$ in., $2\dfrac{5}{8}$ in. **48.** $1\dfrac{1}{2}$ in. **49.** $\dfrac{1\ \text{mi}}{5{,}280\ \text{ft}}=1$,

$\dfrac{5{,}280\ \text{ft}}{1\ \text{mi}}=1$ **50. a.** min **b.** sec **51.** 15 ft **52.** 216 in.

53. $5\dfrac{1}{2}$ ft = 5.5 ft **54.** $1\dfrac{3}{4}$ mi = 1.75 mi **55.** 54 in.

56. 1,760 yd **57.** 2 lb **58.** 275.2 oz **59.** 96,000 oz

60. $2\dfrac{1}{4}$ tons = 2.25 tons **61.** 80 fl oz **62.** $\dfrac{1}{2}$ gal = 0.5 gal

63. 68 c **64.** 5.5 qt **65.** 40 pt **66.** 56 c **67.** 1,200 sec

68. 15 min **69.** $8\dfrac{1}{3}$ days **70.** 360 min **71.** 108 hr

72. 86,400 sec **73.** $\dfrac{21}{176}$ mi ≈ 0.12 mi **74.** $20\dfrac{1}{4}$ tons = 20.25 tons

75. $484\frac{2}{3}$ yd   **76.** 100   **77. a.** 10, 1 millimeter

b. 19 mm, 3 cm, 45 mm, 62 mm   **78.** 4 cm

79. a. $\dfrac{1\ km}{1,000\ m} = 1$, $\dfrac{1,000\ m}{1\ km} = 1$ **b.** $\dfrac{1\ g}{100\ cg} = 1$, $\dfrac{100\ cg}{1\ g} = 1$

80. 5 places to the left   **81.** 4.75 m   **82.** 8,000 mm
83. 165,700 m   **84.** 678.9 dm   **85.** 0.025 mg   **86.** 8 g
87. 5.425 kg   **88.** 5,425,000 mg   **89.** 1.5 L   **90.** 3.25 kL
91. 40 cL   **92.** 1,000 dL   **93.** 1.35 kg   **94.** 0.24 L
95. 50 g   **96.** 1,000 mL   **97.** 164 ft   **98.** Willis Tower
99. 3,107 km   **100.** 193 cm   **101.** 850.5 g   **102.** 33 lb
103. 22,680 g   **104.** about 909 kg   **105.** about 2.0 lb
106. LaCroix   **107.** about 159.2 L   **108.** 221°F
109. 25°C   **110.** 30°C

CHAPTER 5 TEST   (page 489)

1. a. ratio   **b.** rate   **c.** proportion   **d.** cross   **e.** tenths, hundredths, thousandths   **f.** metric   **g.** Fahrenheit, Celsius   **2.** $\dfrac{9}{13}$, 9:13, 9 to 13   **3.** $\dfrac{3}{4}$ **4.** $\dfrac{1}{6}$ **5.** $\dfrac{2}{5}$ **6.** $\dfrac{6}{7}$

7. $\dfrac{3\ feet}{2\ seconds}$ **8.** the 2-pound can   **9.** 22.5 kwh per day

10. $\dfrac{15\ billboards}{50\ miles} = \dfrac{3\ billboards}{10\ miles}$ **11. a.** no   **b.** yes

12. yes   **13.** 15   **14.** 63.24   **15.** $2\dfrac{1}{2}$ **16.** 0.2   **17.** \$3.43

18. 2 c   **19. a.** 16   **b.** $\dfrac{5}{16}$ in., $1\dfrac{3}{8}$ in., $2\dfrac{3}{4}$ in.   **20.** introduce,

eliminate   **21.** 15 ft   **22.** $8\dfrac{1}{3}$ yd   **23.** 172 oz   **24.** 3,200 lb

25. 128 fl oz   **26.** 115,200 min   **27. a.** the larger one
b. the longer one   **c.** the right side   **28.** 12 mm, 5 cm,
65 mm   **29.** 0.5 km   **30.** 500 cm   **31.** 5 mg
32. 70,000 mL   **33.** 7.5 g   **34.** the 100-yard race   **35.** Jim
36. 0.9 qt   **37.** 114 cm   **38.** 182°F   **39.** A scale is a ratio
(or rate) comparing the size of a drawing and the size of an
actual object. For example, 1 inch to 6 feet (1 in.:6 ft).
40. It is easier to convert from one unit to another in the metric
system because it is based on the number 10.

CHAPTERS 1–5 CUMULATIVE REVIEW   (page 491)

1. a. five million, seven hundred sixty-four thousand, five hundred
two   **b.** $5,000,000 + 700,000 + 60,000 + 4,000 + 500 + 2$
2. a. 186 to 184   **b.** Detroit   **c.** 370 points   **3.** 69,658
4. 367,416   **5.** 20 R3   **6.** \$560   **7.** 1, 2, 3, 5, 6, 10, 15, 30
8. $2^3 \cdot 3^2 \cdot 5$ **9.** 140, 4   **10.** 81   **11.** <   **12.** −4
13. 15 shots   **14.** −9, 9   **15. a.** −8   **b.** undefined
c. −8   **d.** 0   **e.** 8   **f.** 0   **16.** 30   **17.** −5,000
18. $-\dfrac{4}{5}$ **19.** $\dfrac{54}{60}$ **20.** 59,100,000 sq mi   **21.** $A = \dfrac{1}{2}bh$

22. −1   **23.** $\dfrac{9}{20}$ **24.** $\dfrac{19}{15}$ **25.** $\dfrac{31}{32}$ in.   **26.** $6\dfrac{9}{10}$ **27.** $\dfrac{3}{4}$ hp

28. $-\dfrac{26}{15} = -1\dfrac{11}{15}$ **29.** >

30.

$$-3.2 \quad -1\frac{3}{4} \quad -0.5 \qquad 2.25\ \sqrt{9} \qquad \frac{11}{8} = 1\frac{3}{8}$$

(number line from −5 to 5)

31. −17.64   **32.** −23.38   **33.** 250   **34.** 458.15 lb
35. 0.025   **36.** 12.7   **37.** $0.08\overline{3}$ **38.** \$9.95   **39.** 23

40. $\dfrac{1}{5}$ **41.** the 94-pound bag   **42.** false

43. 202 mg   **44.** 15   **45. a.** 960 hr   **b.** 4,320 min
c. 480 sec   **46.** 2.5 lb   **47.** 2,400 mm   **48.** 0.32 kg
49. a. 1 gal   **b.** a meterstick   **50.** 36 in.

STUDY SET SECTION 6.1   (page 503)

1. Percent   **3.** 100, simplify   **5.** right   **7.** percent
9. 84%, 16%   **11.** 107%   **13.** 99%   **15. a.** 15%   **b.** 85%

17. $\dfrac{17}{100}$ **19.** $\dfrac{91}{100}$ **21.** $\dfrac{1}{25}$ **23.** $\dfrac{3}{5}$ **25.** $\dfrac{19}{1,000}$ **27.** $\dfrac{547}{1,000}$

29. $\dfrac{1}{8}$ **31.** $\dfrac{17}{250}$ **33.** $\dfrac{1}{75}$ **35.** $\dfrac{17}{120}$ **37.** $\dfrac{13}{10}$ **39.** $\dfrac{11}{5}$

41. $\dfrac{7}{2,000}$ **43.** $\dfrac{1}{400}$ **45.** 0.16   **47.** 0.81   **49.** 0.3412

51. 0.50033   **53.** 0.0699   **55.** 0.013   **57.** 0.0725   **59.** 0.185
61. 4.6   **63.** 3.16   **65.** 0.005   **67.** 0.0003   **69.** 36.2%
71. 98%   **73.** 171%   **75.** 400%   **77.** 40%   **79.** 16%
81. 62.5%   **83.** 43.75%   **85.** 225%   **87.** 105%

89. $16\dfrac{2}{3}\% \approx 16.7\%$ **91.** $166\dfrac{2}{3}\% \approx 166.7\%$

93. $\dfrac{157}{5,000}$, 3.14%   **95.** $\dfrac{51}{125}$, 0.408   **97.** $\dfrac{21}{400}$, 0.0525

99. 2.33, $233\dfrac{1}{3}\% \approx 233.3\%$ **101. a.** 0.0307   **b.** 0.307

103. a. $\dfrac{691}{10,000}$ **b.** $\dfrac{691}{1,000}$ **105. a.** 82%   **b.** 18%

107. a. 12%   **b.** 8%   **c.** 4% (Alaska, Hawaii)
109. a. 0.0475   **b.** 0.01   **c.** 0.1425
111. torso: 27.5%

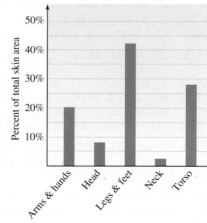

113. a. $\dfrac{15}{193}$ **b.** 0.0777   **c.** 7.77%   **115.** $33\dfrac{1}{3}\%$, $\dfrac{1}{3}$, $0.\overline{3}$

117. a. $\dfrac{13}{15}$ **b.** $86\dfrac{2}{3}\% \approx 86.7\%$ **119. a.** $\dfrac{1}{4}\%$ **b.** $\dfrac{1}{400}$

c. 0.0025   **121. a.** 0.27%   **b.** 0.0000003%   **123.** 12.43 volts;
recharge   **125.** 50%   **131 a.** 34 cm   **b.** 68.25 cm²

THINK IT THROUGH   (page 523)

45% are enrolled in college full time, 53% of the students work
less than 20 hours per week, 9% never

STUDY SET SECTION 6.2 (page 523)

1. sentence, equation 3. solved 5. part, whole 7. cross
9. Amount, base, percent, whole 11. 12% 13. a. 0.12
b. 0.056 c. 1.25 d. 0.0025

15. a. $x = 7\% \cdot 16, \dfrac{x}{16} = \dfrac{7}{100}$ b. $125 = x \cdot 800, \dfrac{125}{800} = \dfrac{x}{100}$

c. $1 = 94\% \cdot x, \dfrac{1}{x} = \dfrac{94}{100}$ 17. a. $5.4\% \cdot 99 = x, \dfrac{x}{99} = \dfrac{5.4}{100}$

b. $75.1\% \cdot x = 15, \dfrac{15}{x} = \dfrac{75.1}{100}$ c. $x \cdot 33.8 = 3.8, \dfrac{3.8}{33.8} = \dfrac{x}{100}$

19. 68 21. 132 23. 17.696 25. 24.36 27. 25%
29. 85% 31. 62.5% 33. 43.75% 35. 110% 37. 350%
39. 30 41. 150 43. 57.6 45. 72.6 47. 1.25% 49. 65
51. 99 53. 90 55. 80% 57. 0.096 59. 44 61. 2,500%
63. 107.1 65. 60 67. 31.25% 69. 43.5
71. a. 12.8 b. 20% c. 500 73. a. 2 b. 2% c. 1.125
75. 12M bytes = 12,000,000 bytes 77. $42.15
79. $658 81. yes 83. 5% 85. 120 87. 51 million
prescriptions are likely not appropriate 89. $2.342 trillion
91. 24 oz 93. 30, 12 95. 40,000% 97. 37,200 megawatt hours
99. a. 80,000 b. 4%

101. **Paying the college bill**

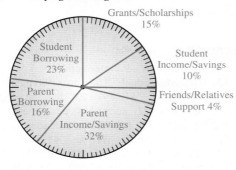

Grants/Scholarships 15%
Student Borrowing 23%
Student Income/Savings 10%
Parent Borrowing 16%
Friends/Relatives Support 4%
Parent Income/Savings 32%

103. 29%, 42%, 9%, 8%, 12%

2015 Federal Income Sources

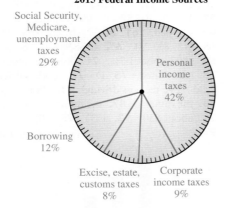

Social Security, Medicare, unemployment taxes 29%
Personal income taxes 42%
Borrowing 12%
Excise, estate, customs taxes 8%
Corporate income taxes 9%

111. 18.17 113. 5.001 115. 0.008

STUDY SET SECTION 6.3 (page 540)

1. commission 3. a. increase b. original 5. purchase
price 7. sales 9. a. $64.07 b. $135.00 11. subtract,
original 13. $3.71 15. $4.20 17. $70.83 19. $64.03
21. 5.2% 23. 15.3% 25. $11.40 27. $168 29. 2%
31. 4% 33. 10% 35. 15% 37. 20% 39. 10%
41. $29.70, $60.30 43. $8.70, $49.30 45. 19% 47. 14%
49. $56.25 51. a. $47.34, $2.84, $50.18
b. $5.22 + $22.32 = $27.54 c. 1.45% + 6.2% = 27.54

53. 8% 55. 0.25% 57. $249.50 59. a. 8%, 3.75%, 1.45%,
6.2% b. $5.22 + $22.32 = $27.54 c. 1.45% + 6.2% = 7.65%
61. 5% 63. 31% 65. 152% 67. 36% 69. 12.5%
71. a. 25% b. 36% 73. $5,955 75. 1.5% 77. 90%
79. $12,000 81. a. $7.99 b. $31.96 83. 41%
85. $349.97, 13% 87. 23%, $11.88 89. $76.50
91. $187.49 93. a. $450 b. 43% 99. − 50
101. 3 103. 13

STUDY SET SECTION 6.4 (page 550)

1. Estimation 3. two 5. 2 7. 4 9. 10, 5 11. 2.751, 3
13. 0.1267, 0.1 15. 405.9 lb, 400 lb 17. 69.14 min, 70 min
19. 70 21. 14 23. 2,100,000 25. 200,000 27. 4 29. 12
31. 820 33. 20 35. $9 37. $4.50 39. $18 41. $1.50
43. 8 45. 72 47. 12 49. 5.4 51. 180 53. 230 55. 6
57. 18 59. 7 61. 70 63. 12,000 65. 1.8 67. 0.49
69. 12 71. a. 0.01 b. 0.1 c. 1 73. a. 1 b. 2 c. 3
75. 164 students 77. $60 79. $6 81. $7.50
83. $30,000 85. 320 lb 87. 210 motorists 89. 220 people
91. 18,000 people 93. 3,100 volunteers

99. a. $\dfrac{4}{3} = 1\dfrac{1}{3}$ b. $\dfrac{1}{3}$ c. $\dfrac{5}{12}$ d. $\dfrac{5}{3} = 1\dfrac{2}{3}$

STUDY SET SECTION 6.5 (page 560)

1. interest 3. rate 5. total 7. a. $125,000 b. 5%
c. 30 years 9. a. 0.07 b. 0.098 c. 0.0625 11. $1,800
13. a. compound interest b. $1,000 c. 4 d. $50
e. 1 year 15. $I = Prt$ 17. $100 19. $252 21. $525
23. $1,590 25. $16.50 27. $30.80 29. $13,159.23
31. $40,493.15 33. $2,060.68 35. $5,619.26
37. $10,011.96 39. $77,775.64 41. $5,300 43. $198
45. $5,580 47. $46.88 49. $4,262.14 51. $10,000,
$7\dfrac{1}{4}\% = 0.0725$, 2 yr, $1,450 53. $192, $1,392, $58

55. $19.449 million 57. $755.83 59. $1,271.22
61. $570.65 63. $30,915.66 65. $159,569.75

67. 390% 73. $\dfrac{1}{2}$ 75. $\dfrac{29}{35}$ 77. $8\dfrac{1}{3}$ 79. −36

CHAPTER 6 REVIEW (page 564)

1. 39%, 0.39, $\dfrac{39}{100}$ 2. 111%, 1.11, $\dfrac{111}{100}$ 3. 61% 4. a. 54%

b. 46% 5. $\dfrac{3}{20}$ 6. $\dfrac{6}{5}$ 7. $\dfrac{37}{400}$ 8. $\dfrac{1}{500}$ 9. 0.27 10. 0.08

11. 6.55 12. 0.018 13. 0.0075 14. 0.0023 15. 83%
16. 162.5% 17. 5.1% 18. 600% 19. 50% 20. 80%

21. 87.5% 22. 6.25% 23. $33\dfrac{1}{3}\% \approx 33.3\%$ 24. $83\dfrac{1}{3}\% \approx 83.3\%$

25. $91\dfrac{2}{3}\% \approx 91.7\%$ 26. $166\dfrac{2}{3}\% \approx 166.7\%$ 27. a. 0.972

b. $\dfrac{243}{250}$ 28. 63% 29. a. 0.0025 b. $\dfrac{1}{400}$ 30. $6\dfrac{2}{3}\% \approx 6.7\%$

31. a. amount: 15, base: 45, percent: $33\dfrac{1}{3}\%$ b. Amount, base,

percent 32. a. 0.13 b. 0.071 c. 1.95 d. 0.0025

e. $\dfrac{1}{3}$ f. $\dfrac{2}{3}$ g. $\dfrac{1}{6}$ 33. a. $x = 32\% \cdot 96$ b. $64 = x \cdot 135$

c. $9 = 47.2\% \cdot x$ 34. a. $\dfrac{x}{96} = \dfrac{32}{100}$ b. $\dfrac{64}{135} = \dfrac{x}{100}$

c. $\dfrac{9}{x} = \dfrac{47.2}{100}$ 35. 200 36. 125 37. 1.75% 38. 2,100

39. 121 **40.** 30 **41.** 155.2 **42.** 5,300% **43.** 0.6 gal methane
44. 68 **45.** 87% **46.** $10.86
47. Oils and Fats Used in the Production of Biodiesel

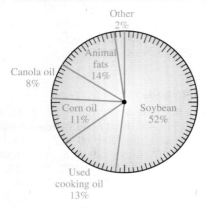

48. 139,531,200 mi² **49.** $3.30, $63.29 **50.** 4% **51.** $40.20
52. 4.25% **53.** $100,000 **54.** original **55.** 31%
56. 9.6% **57. a.** purchase price **b.** sales tax
c. commission rate **58. a.** sale price **b.** original price
c. discount **59.** $180, $2,500, 7.2% **60.** 5% **61.** 3.4203, 3
62. 86.87, 90 **63.** 4.34 sec, 4 sec **64.** 1,090 L, 1,000 L
65. 12 **66.** 120 **67.** 140,000 **68.** 150 **69.** 3 **70.** 10
71. 350 **72.** 1,000 **73.** 60 **74.** 2 **75.** $36 **76.** $7.50
77. about 12 fluid oz **78.** about 120 people **79.** 200
80. $30,000 **81.** $6,000, 8%, 2 years, $960 **82.** $27,240
83. $75.63 **84.** $10,308.22 **85. a.** $116.25 **b.** 1,616.25
c. $134.69 **86.** $2,142.45 **87.** $6,076.45 **88.** $43,265.78

CHAPTER 6 TEST   (page 580)

1. a. Percent **b.** is, of, what, what **c.** amount, base
d. increase **e.** Simple, Compound **2. a.** 61%, $\frac{61}{100}$, 0.61
b. 39% **3.** 199%, $\frac{199}{100}$, 1.99 **4. a.** 0.67 **b.** 0.123
c. 0.0975 **5. a.** 0.0006 **b.** 2.1 **c.** 0.55375 **6. a.** 25%
b. 62.5% **c.** 112% **7. a.** 19% **b.** 347% **c.** 0.5%
8. a. 66.7% **b.** 200% **c.** 90%
9. a. $\frac{11}{20}$ **b.** $\frac{1}{10,000}$ **c.** $\frac{5}{4}$ **10. a.** $\frac{1}{15}$ **b.** $\frac{3}{8}$ **c.** $\frac{2}{25}$
11. a. $3\frac{1}{3}\% = 3.3\%$ **b.** $177\frac{7}{9}\% = 177.8\%$ **12.** 6.5%
13. 250% **14.** 93.95% **15.** 90 **16.** 21 **17.** 134.4 **18.** 7.8
19. a. 1.02 in. **b.** 32.98 in. **20.** $26.24 **21.** 3% **22.** 23%
23. $35.92 **24.** 11% **25.** $41,440 **26.** $9, $66, 12%
27. $6.60, $13.40 **28. a.** two, left **b.** one, left **29. a.** 80
b. 3,000,000 **c.** 42 **30.** 18.3%
31.

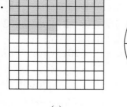

(a)   (b)

---

32. 100 trucks **33.** $4.50 **34.** 16,000 females
35. $150 **36.** $28,175 **37.** $39.45 **38.** $5,079.60

CHAPTERS 1–6 CUMULATIVE REVIEW   (page 583)

1. a. six million, fifty-four thousand, three hundred
forty-six **b.** 6,000,000 + 50,000 + 4,000 + 300 + 40 + 6
2. 279 days **3.** 42,156 **4.** 23,100 **5.** 15 R6 **6.** 80
7. 1, 2, 4, 5, 8, 10, 20, 40 **8.** $2 \cdot 3 \cdot 7^2$ **9.** 120, 6
10. 15 **11.** > **12.** 0 **13.** −$135 **14.** −36, 36
15. a. undefined **b.** 0 **c.** 0 **d.** 14 **16.** −30
17. −1,900 **18.** $\frac{9}{10}$ **19.** $\frac{36}{45}$ **20.** −60 **21.** 650 in.²
22. $-\frac{3}{4}$ **23.** $\frac{24}{35}$ **24.** $\frac{7}{6}$ **25.** $\frac{1}{12}$ lb **26.** −30
27. $35\frac{3}{4}$ in. **28.** $-\frac{5}{6}$ **29. a.** 452.03 **b.** 452.030
30. −5.5 **31.** $731.40 **32.** 0.27 **33.** $0.7\overline{3}$ **34.** −29
35. $\frac{5}{6}$ **36.** 4 **37.** 40 days **38.** 2.4 m **39.** 14.3 lb
40. 29%, $\frac{29}{100}$; 0.473, $\frac{473}{1,000}$; 87.5%, 0.875 **41.** 125
42. 0.0018% **43.** 78% **44.** $428, $321, $107, 25%
45. a. $12 **b.** $90.18 **46.** $1,450

STUDY SET SECTION 7.1   (page 594)

1. (a) **3.** (c) **5.** (d) **7.** axis **9.** intersection
11. pictures **13.** bars, edge, equal **15.** about 500 buses
17. $9.00 **19.** $5.93 ($24.82 − $18.89) **21.** fish, cat, dog
23. dogs and cats **25.** yes **27.** about 11,000,000 metric
tons **29.** 1990, 2000, 2015 **31.** 6,500,000 metric tons
33. seniors **35.** $50 **37.** Chinese **39.** yes **41.** 47%
43. 1,368,000,000 **45.** 494 **47.** 2002 to 2003; 2004 to 2005;
2005 to 2006; 2007 to 2008; 2008 to 2009; 2009 to 2010; 2011
to 2012; 2013 to 2014; 2015 to 2016 **49.** 2001 and 2003; 2014
and 2015 **51.** 2005 to 2006; a decrease of 14 resorts
53. 1 **55.** B **57.** 1 **59.** Runner 1 was running; runner 2
was stopped. **61. a.** 27 **b.** 22 **63.** $15,551.25
65. $129,833,231 **67. a.** $8,988.75 **b.** $6,432.50 **c.** $2,556.25
69. 2007; about 1.7% **71.** decrease; about 0.4%
73. it increased **75.** D **77.** reckless driving and failure to
yield **79.** reckless driving **81.** about $440 **83.** no
85. the miner's **87.** the miners **89.** about $42
91. about $30 **93.** 17% **95.** 10%
97.

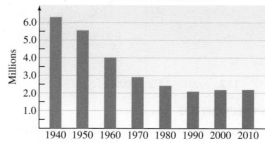

Number of U.S. Farms
Source: U.S. Dept. of Agriculture

99.

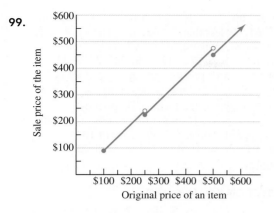

103. 11, 13, 17, 19, 23, 29   **105.** 0, 4

THINK IT THROUGH   (page 608)

**Median Weekly Earnings of Full-Time Workers
(25 years and older) by Education**

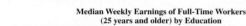
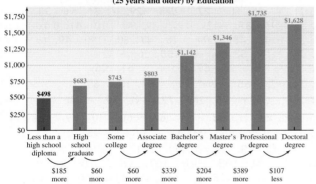

Source: Bureau of Labor Statistics, Current Population Survey

STUDY SET SECTION 7.2   (page 608)

1. mean   **3.** mode   **5.** the number of values   **7. a.** an even number   **b.** 6 and 8   **c.** 6, 8, 14, 7   **d.** 15, 4, 11   **9.** 8

11. 35   **13.** 19   **15.** 5.8   **17.** 9   **19.** 5   **21.** 17.2   **23.** $\frac{5}{8}$

25. 9   **27.** 44   **29.** 2.05   **31.** 1   **33.** 3   **35.** −6   **37.** 22.7

39. bimodal: $\frac{1}{3}, \frac{1}{2}$ **41.** mean: 5, median: 5; mode: 5; range: 5

43. a. mean: 3; median: 3; mode: 3   **b.** mean: 3; median: 3; mode: 3
45. a. 82.5   **b.** 83   **47. a.** 2,670 mi   **b.** 89 mi
49. a. $11,875   **b.** 125   **c.** $95   **51. a.** 65¢   **b.** 60¢
c. 50¢   **d.** 55¢   **53.** 61°   **55.** 2.23 GPA   **57.** 2.5 GPA
59. median and mode are 85   **61.** same average (56); sister's scores are more consistent   **63.** 22.525 oz, 25 oz, 17.3 oz
65. mean: 53.9, median: 24; mode: 10; range: 202.5
67. 5 lb, 4 lb, 10 lb   **73.** 65%   **75.** 42   **77.** 62.5%   **79.** 43.5

STUDY SET 7.3   (page 617)

1. experiment   **3.** sample space   **5.** Favorable   **7.** $\frac{s}{n}$

9. six: {1, 2, 3, 4, 5, 6}

11.

13. braces   **15.** {(1, H), (2, H), (3, H), (4, H), (5, H), (6, H), (1, T), (2, T), (3, T), (4, T), (5, T), (6, T)}   **17.** {a, e, i, o, u}
19. E = {2, 4, 6}   **21.** E = {(H, H), (H, T), (T, H)}

23. $\frac{1}{2}$ **25.** $\frac{3}{13}$ **27.** $\frac{5}{36}$ **29.** 0   **31.** $\frac{50}{90} = \frac{5}{9}$ **33.** $\frac{70}{90} = \frac{7}{9}$

35. $\frac{1}{6}$ **37.** $\frac{2}{3}$ **39.** $\frac{19}{42}$ **41.** $\frac{13}{42}$ **43.** $\frac{5}{12}$ **45.** $\frac{3}{12} = \frac{1}{4}$

47. $\frac{3}{8}$ **49.** 0   **51. a.** $\frac{1}{6}, \frac{5}{6}$ **b.** $\frac{1}{6} + \frac{5}{6} = \frac{6}{6} = 1$

53. a. $\frac{1}{36}, \frac{35}{36}$ **b.** $\frac{1}{36} + \frac{35}{36} = 1$ **55. a.** {(S, S, S, S), (S, S, S, F), (S, S, F, S), (S, S, F, F), (S, F, S, S), (S, F, S, F), (S, F, F, S), (S, F, F, F), (F, S, S, S), (F, S, F, F), (F, S, F, S), (F, F, S, F), (F, F, S, S), (F, S, S, F), (F, F, F, S), (F, F, F, F)}

b. $\frac{1}{16}$ **c.** $\frac{1}{4}$ **d.** $\frac{3}{8}$ **e.** $\frac{1}{4}$ **f.** $\frac{1}{16}$

57. a. $\frac{88}{141}$ **b.** $\frac{32}{119}$ **c.** $\frac{15}{71}$ **d.** $\frac{35}{282}$ **59.** $\frac{2}{24} = \frac{1}{12}$

65. −16   **67.** 39

CHAPTER 7 REVIEW   (page 620)

1. a. −18°   **b.** −71°   **2. a.** 30 mph   **b.** 15 mph   **3.** 25
4. 58   **5.** Ukraine, United Kingdom, and Canada
6. 24   **7.** 29%   **8.** men; 15% more   **9.** women
10. No, I would not date a coworker (31% to 29%)
11. about 3,700 animals   **12.** the Columbus Zoo; about 10,000 animals   **13.** about 2,300 animals   **14.** about 15,000 animals   **15.** oxygen   **16.** 4%   **17.** 13.5 lb
18. 166 lb   **19.** about 3,120 million eggs   **20.** about 3,060 million eggs   **21.** 2007; about 2,950 million eggs
22. about 5,750 million eggs   **23.** between 2006 and 2007
24. between 2014 and 2015   **25.** about 400 million more eggs   **26.** about 500 million more eggs   **27.** 60   **28.** 180
29. 160

30.

31. yes   **32.** median   **33.** 1.2 oz   **34.** 1.138 oz, 0.4 oz
35. 7.3 microns, 7.2 microns, 6.9 microns, 1.3 microns
36. 32 pages per day   **37.** $20   **38.** 2.62 GPA   **39.** $\frac{3}{8}$

40. $\frac{1}{2}$ **41.** $\frac{7}{8}$ **42.** $\frac{1}{18}$ **43.** 0   **44.** $\frac{1}{13}$ **45.** $\frac{1}{2}$ **46.** 1

CHAPTER 7 TEST   (page 629)

1. a. axis   **b.** mean   **c.** median   **d.** mode   **e.** central
2. a. 563 calories   **b.** 129 calories   **c.** about 8 mph
3. a. love seat; 150 ft   **b.** 50 feet more   **c.** 340 ft   **4. a.** 75%

b. about 15% **c.** lung cancer **d.** prostate cancer; about 32%
5. a. about 38 g **b.** about 15 g **6. a.** 14% **b.** 511,290
7. a. about 39,000 police officers **b.** 2011; about 33,800
police officers **c.** 2000; about 40,300 police officers
d. about 4,300 police officers **8. a.** bicyclist 1
b. Bicyclist 1 is stopped, but is ahead in the race. Bicyclist 2 is
beginning to catch up. **c.** time C **d.** Bicyclist 2 never led.
e. bicyclist 1 **9. a.** 22 employees **b.** 30 employees
c. 57 employees **10. a.** 7.5 hr **b.** 7.5 hr **c.** 5 hr **d.** 17 hr
11. 3 stars **12.** 3.36 GPA **13.** mean: about 2.28 million;
median: 1.85 million; mode: none; range: 2.84 million
14. Of all the existing single-family homes sold in December
2016, half of them sold for less than $233,800 and half sold for
more than $233,800. **15.** $\frac{1}{6}$ **16.** $\frac{1}{4}$ **17.** $\frac{1}{6}$ **18.** $\frac{1}{4}$
19. shade 7 out of 16 sections **20. a.** 0 **b.** 1

CHAPTERS 1–7 CUMULATIVE REVIEW (page 632)

1. seventy-two million, one hundred five thousand, four
hundred thirty-five; $70,000,000 + 2,000,000 + 100,000 + 5,000$
$+ 400 + 30 + 5$ **2.** 60,000 **3.** 54,604 **4.** 3,209
5. 27,336 **6.** 23 **7.** $1,683 + 459 = 2,142$ **8.** 40 in., 84 in.²
9. 2023 **10. a.** 1, 2, 3, 4, 6, 9, 12, 18, 36 **b.** $2^2 \cdot 3^2$
11. 2, 3, 5, 7, 11, 13, 17, 19, 23, 29 **12. a.** 24 **b.** 4
13. 35 **14.** 9 **15.**

16. a. 6 **b.** 5 **c.** false **17. a.** −30 **b.** −30 **c.** −175
d. 7 **18.** 1,100°F **19.** −5 **20.** 429 **21.** −4 **22.** −200
23. a. $\frac{5}{9}$ **b.** $\frac{3}{2}$ **24. a.** 0 **b.** undefined **25.** $\frac{8}{35}$
26. $\frac{4}{9}$ **27.** $-\frac{1}{6}$ **28.** $\frac{19}{20}$ **29.** 160 minutes are spent in
lecture each week. **30.** $-\frac{21}{20} = -1\frac{1}{20}$ **31.** $6\frac{3}{4}$ in. **32.** $10\frac{5}{8}$
33. $-\frac{3}{8}$ **34.** 428.91 **35.** $1,815.19 **36. a.** 345
b. 0.000345 **37.** 145.5 **38.** −0.744 **39.** 745
40. 0.01825 **41.** $0.\overline{72}$ **42.** 75 **43.** $\frac{2}{3}$ **44.** $59.95
45. $\frac{1}{7}$ **46.** 128 fl oz **47.** 6.4 m
48. 19.8°C **49.** $\frac{3}{100}$, 0.03; 2.25, 225%; $\frac{41}{1,000}$, 4.1%
50. 17% **51.** 24.36 **52.** 57.6 **53.** $225,000 **54.** 16%
55. $12 **56.** $3,312 **57.** $13,159.23 **58. a.** 8%
b. 6,460 cases **59. a.** 2008 and 2010; 36 **b.** 2015 to 2016;
an increase of 18 deaths **c.** 2014 to 2015; a decrease of
24 deaths **60.** mean: 3.02; median: 3.00; mode: 2.75;
range: 1.79 **61.** $\frac{3}{13}$ **62.** $\frac{1}{3}$

STUDY SET SECTION 8.1 (page 642)

1. Variables **3.** expressions **5.** terms **7.** coefficient
9. $(12 - h)$ in. **11. a.** $(x + 20)$ ounces **b.** $(100 - p)$ lb
13. 5, 25, 45 **15.** $4x$ **17.** $2w$ **19. a.** $x + y = y + x$
b. $(r + s) + t = r + (s + t)$ **21.** $0 \cdot s = 0$ and $s \cdot 0 = 0$
23. a. 4 **b.** 3, 11, −1, 9 **25.** 6, −75, 1, $\frac{1}{2}$, $\frac{1}{5}$, 1 **27.** term
29. factor **31.** $l + 15$ **33.** $50x$ **35.** $\frac{w}{l}$ **37.** $P + \frac{2}{3}p$
39. $k^2 - 2,005$ **41.** $2a - 1$ **43.** $\frac{1,000}{n}$ **45.** $2p + 90$
47. $3(35 + h + 300)$ **49.** $p - 680$ **51.** $4d - 15$

53. $2(200 + t)$ **55.** $|a - 2|$ **57.** $0.1d$ or $\frac{1}{10}d$ **59.** three-
fourths of r **61.** 50 less than t **63.** the product of x, y,
and z **65.** twice m, increased by 5 **67.** $(x + 2)$ in.
69. $(36 - x)$ in. **71.** 2 **73.** 13 **75.** 20 **77.** −12
79. −5 **81.** $-\frac{1}{5}$ **83.** 17 **85.** 36 **87.** 255 **89.** 8
91. a. $x - 23$ **b.** $x < 23$ **93. a.** $6(x + 8)$ **b.** $6x + 8$
95. a. Let x = weight of the Element (in pounds);
$2x - 340$ = weight of the Hummer (in pounds) **b.** 6,400 lb
97. a. let x = age of Apple; $x + 80$ = age of IBM;
$x - 9$ = age of Dell **b.** IBM: 112 years; Dell: 23 years
103. 60 **105.** $\frac{8}{27}$

STUDY SET SECTION 8.2 (page 652)

1. simplify **3.** distributive **5.** like **7. a.** 4, 9, 36
b. associative property of multiplication **9. a.** + **b.** −
c. − **d.** + **11. a.** $10x$ **b.** can't be simplified **c.** $-42x$
d. can't be simplified **e.** $18x$ **f.** $3x + 5$ **13. a.** $6(h - 4)$
b. $-(z + 16)$ **15.** $12t$ **17.** $63m$ **19.** $-35q$ **21.** $300t$
23. $11.2x$ **25.** $60c$ **27.** $-96m$ **29.** g **31.** $5x$ **33.** $6y$
35. $5x + 15$ **37.** $-12x - 27$ **39.** $9x + 10$ **41.** $0.4x + 1.6$
43. $36c - 42$ **45.** $-78c + 18$ **47.** $30t + 90$ **49.** $4a - 1$
51. $24t + 16$ **53.** $2w - 4$ **55.** $56y + 32$
57. $50a - 75b + 25$ **59.** $-x + 7$ **61.** $5.6y - 7$
63. $3x, -2x$ **65.** $-3m^3, -m^3$ **67.** $10x$ **69.** 0
71. $20b^2$ **73.** r **75.** $28y$ **77.** $-s^3$ **79.** $-3.6c$
81. $0.4r$ **83.** $\frac{4}{5}t$ **85.** $-\frac{5}{8}x$ **87.** $-6y - 10$ **89.** $-2x + 5$
91. does not simplify **93.** $4x^2 - 3x + 9$ **95.** $7z - 15$
97. $s^2 - 12$ **99.** $-41r + 130$ **101.** $8x - 9$
103. $12c + 34$ **105.** $-10r$ **107.** $-20r$ **109.** $3a$
111. $9r - 16$ **113.** $-6x$ **115.** $c - 13$ **117.** $a^3 - 8$
119. a. $60x$ **b.** $15x + 12$ **121. a.** $-4s + 8$ **b.** $4s + 8$
123. $12x$ **125.** $(4x + 8)$ ft **129.** 2

STUDY SET SECTION 8.3 (page 662)

1. equation **3.** solve **5.** equivalent **7. a.** $x + 6$
b. neither **c.** no **d.** yes **9. a.** c, c **b.** c, c
11. a. x **b.** y **c.** t **d.** h **13.** 5, 5, 50, 50, $\frac{?}{=}$, 45, 50
15. a. is possibly equal to **b.** yes **17.** no **19.** no
21. no **23.** no **25.** yes **27.** no **29.** no **31.** yes
33. yes **35.** yes **37.** 71 **39.** 18 **41.** −0.9 **43.** 3 **45.** $\frac{8}{9}$
47. 3 **49.** $-\frac{1}{25}$ **51.** −2.3 **53.** 45 **55.** 0 **57.** 21
59. −2.64 **61.** 20 **63.** 15 **65.** −6 **67.** 4 **69.** 4 **71.** 7
73. 1 **75.** −6 **77.** 20 **79.** 0.5 **81.** −18 **83.** $-\frac{4}{21}$
85. 13 **87.** 2.5 **89.** $-\frac{8}{3}$ **91.** $\frac{6}{7}$ **93.** 20 **95.** −5
97. −200 **99.** 21 **101. a.** 16 **b.** 5 **103. a.** 36 **b.** 4
105. 65° **107.** $474 million **113.** 0 **115.** $45 - x$

STUDY SET SECTION 8.4 (page 671)

1. solve **3.** simplify **5.** 4, 9 **7.** subtraction, multiplication
9. a. $-2x - 8 = -24$ **b.** $-20 = 3x - 16$ **11. a.** no
b. yes **13.** 7, 7, 2, 2, 14, $\frac{?}{=}$, 28, 21, 14 **15.** 6 **17.** 5
19. −7 **21.** 0.25 **23.** $-\frac{5}{2}$ **25.** −3 **27.** $\frac{10}{3}$ **29.** 6
31. 18 **33.** 16 **35.** 2.9 **37.** −4 **39.** $\frac{11}{5}$ **41.** −41
43. −6 **45.** 0.04 **47.** −6 **49.** −11 **51.** 7 **53.** −11
55. 1 **57.** $\frac{9}{2}$ **59.** −4 **61.** 3 **63.** $\frac{1}{4}$ **65.** $\frac{5}{6}$ **67.** 45

69. -49 **71.** 1 **73.** -12 **75.** -6 **77.** -5 **79.** 3.5

81. a. 2 **b.** 4 **83. a.** 6 **b.** 30 **87.** commutative property of multiplication **89.** associative property of addition

STUDY SET SECTION 8.5 (page 680)

1. Analyze, assign, equation, Solve, conclusion, Check
3. division **5.** addition **7.** borrow, add **9.** equal-size discussion groups, division **11.** $s + 6$ **13.** $g - 100$
15. 1,700, 425, jar, age, addition, 1,700, x, 1,700, 425, 425, 1,275, 1,275, 1,275, 1,700 **17.** 88, 10, first class, economy, first class, 10, $10x$, $10x$, 88, $11x$, 11, 11, 8, 8, 80, 10, 80, 88 **19.** She will need to borrow $248,000. **21.** Alicia could read 133 words per minute before taking the course. **23.** It would take 17 months for him to reach his goal. **25.** Last year 7 scholarships were awarded. This year 13 scholarships were awarded. **27.** She has made 6 payments. **29.** Dr. Dre earned $110 million in 2011. **31.** The length of the room is 20 feet and the width is 10 feet. **33.** The scale would register 55 pounds.
35. The first act has 5 scenes.
37. The value of the benefit package is $7,000.
39. His score for the first game was 28,460 points.
41. There were 6 minutes of commercials and 24 minutes of the program. **43.** They spend 150 minutes in lecture and 100 minutes in lab each week. **45.** The shelter received 32 calls each day after being featured on the news.
47. Three days ago, he waited for 35 minutes.
49. The initial cost estimate was $54 million.
51. The monthly rent for the apartment was $975.
53. She must complete 4 more sessions to get the certificate.
61. 600; 20 **63.** 140; 14 **65.** 3,528; 1 **67.** 2,178; 22

STUDY SET SECTION 8.6 (page 691)

1. exponential **3. a.** $3x, 3x, 3x, 3x$ **b.** $(-5y)^3$ **5. a.** add
b. multiply **c.** multiply **7. a.** $2x^2$ **b.** x^4 **9. a.** doesn't simplify **b.** x^5 **11.** x^6, 18 **13.** base 4, exponent 3
15. base x, exponent 5 **17.** base $-3x$, exponent 2
19. base y, exponent 6 **21.** base m, exponent 12
23. base $y + 9$, exponent 4 **25.** m^5 **27.** $(4t)^4$ **29.** $4t^5$
31. a^2b^3 **33.** 5^7 **35.** a^6 **37.** b^6 **39.** c^{13} **41.** a^5b^6
43. c^2d^5 **45.** x^3y^{11} **47.** m^{200} **49.** 3^8 **51.** $(-4.3)^{24}$
53. m^{500} **55.** y^{15} **57.** x^{25} **59.** p^{27} **61.** t^{18} **63.** u^{14}
65. $36a^2$ **67.** $625y^4$ **69.** $27a^{12}b^{21}$ **71.** $-8r^6s^9$ **73.** $72c^{17}$
75. $6,400d^{41}$ **77.** $49a^{18}$ **79.** t^{10} **81.** y^9 **83.** $-216a^9b^6$
85. n^{33} **87.** b^{60} **89.** $288b^{27}$ **91.** c^{14} **93.** $432s^{16}t^{13}$
95. x^{15} **97. a.** x^6 **b.** x^9 **c.** $2x^3$ **99. a.** n^{36} **b.** n^{36}

c. n^{12} **101.** $25x^2 \text{ ft}^2$ **105.** $\dfrac{3}{4}$ **107.** 5 **109.** 7 **111.** 12

CHAPTER 8 REVIEW (page 693)

1. a. $6b$ **b.** xyz **c.** $2t$ **2. a.** $c + d = d + c$
b. $(r \cdot s) \cdot t = r \cdot (s \cdot t)$ **3. a.** factor **b.** term

4. a. 3 **b.** 1 **5.** 16, -1, 25 **b.** $\dfrac{1}{2}$, 1 **6.** five hundred

less than m (answers may vary) **7. a.** $h + 25$ **b.** $100 - 2s$

c. $\dfrac{1}{2}t - 6$ **d.** $|2 - a^2|$ **8. a.** $(n + 4)$ in. **b.** $(b - 4)$ in.

9. a. $(x + 1)$ in. **b.** $\dfrac{p}{8}$ pounds **10. a.** Let $x =$ weight of the volleyball (in ounces), $2x + 2 =$ weight of the NBA basketball (in ounces) **b.** 22 oz **11.** 72 **12.** 64 **13.** 40

14. -36 **15.** $28w$ **16.** $24x$ **17.** $2.08f$ **18.** r
19. $5x + 15$ **20.** $-2x - 3 + y$ **21.** $3c - 6$
22. $12.6c + 29.4$ **23.** $7a, 9a$ **24.** $2x^2, 3x^2; 2x, -x$
25. $9p$ **26.** $-7m$ **27.** $4n$ **28.** $-p - 18$ **29.** $0.1k^2$
30. $8a^3 - 1$ **31.** does not simplify **32.** does not simplify
33. w **34.** $4h - 15$ **35. a.** x **b.** $-x$ **c.** $4x + 1$
d. $4x - 1$ **36.** $(4x + 4)$ ft **37.** yes **38.** no **39.** no
40. no **41.** yes **42.** yes **43.** equation **44.** true **45.** 21
46. 32 **47.** -20.6 **48.** 107 **49.** 24 **50.** 2 **51.** -9

52. -7.8 **53.** 0 **54.** $-\dfrac{16}{5}$ **55.** 2 **56.** -30.6 **57.** 30

58. -28 **59.** 3 **60.** -1.2 **61.** 4 **62.** 1 **63.** 20

64. 0.06 **65.** They needed to borrow $97,750. **66.** He originally had 725 patients. **67.** The original cost estimate was $27 million. **68.** There are 3,600 clients served by 45 social workers. **69.** It will take 6 hours for the hamburger to go from 71°F to 29°F. **70.** It cost $32 to rent the trailer. **71.** She runs 9 miles and she walks 6 miles. **72.** The attendance on the first day was 2,200 people. The attendance on the second day was 4,400 people. **73.** The width of the parking lot is 25 feet, and the length is 100 feet. **74.** The lunar module was 54 feet tall. **75. a.** base n, exponent 12 **b.** base $2x$, exponent 6 **c.** base r, exponent 4 **d.** base $y - 7$, exponent 3
76. a. m^5 **b.** $-3x^4$ **c.** a^2b^4 **d.** $(pq)^3$ **77. a.** x^4
b. $2x^2$ **c.** x^3 **d.** does not simplify **78. a.** keep the base 3, don't multiply the bases. **b.** multiply the exponents, don't add them. **79.** 7^{12} **80.** m^2n^3 **81.** y^{21} **82.** $81x^4$ **83.** 6^{36}
84. $-b^{12}$ **85.** $256s^{10}$ **86.** $4.41x^4y^2$ **87.** $(-9)^{15}$ **88.** a^{23}
89. $8x^{15}$ **90.** $m^{10}n^{18}$ **91.** $72a^{17}$ **92.** x^{200} **93.** $256m^{13}$
94. $108t^{22}$

CHAPTER 8 TEST (page 702)

1. a. Variables **b.** distributive **c.** like **d.** combined
e. coefficient **f.** substitute **g.** expressions **h.** equation
i. solve **j.** check **2. a.** $(b + c) + d = b + (c + d)$
b. $1 \cdot t = t$ and $t \cdot 1 = t$ **3.** $s - 10 =$ the length of the trout

(in inches) **4. a.** $r - 2$ **b.** $3xy$ **c.** $\dfrac{c}{3}$ **d.** $2w + 7$

5. three-fourths of t **6. a.** $h - 5 =$ the length of the upper base (in feet) **b.** $2h - 3 =$ the length of the lower base (in feet) **7. a.** factor **b.** term **8. a.** 4 terms **b.** 1, 8, -1, -6
9. -3 **10.** 36 **11. a.** $36s$ **b.** $-120t$ **c.** $12x$ **d.** $-72m$
12. a. $25x + 5$ **b.** $-42 + 6x$ **c.** $-6y - 4$ **d.** $0.6a + 0.9b - 2.1$
e. $m - 4$ **f.** $18r + 9$ **13.** $12m^2$ and $2m^2$ **14. a.** $12y$
b. $40a$ **c.** $21b^2$ **d.** $11z + 13$ **15.** $3y - 3$

16. It is not a solution. **17.** 4 **18.** 3.1 **19.** 11 **20.** -81

21. $\dfrac{1}{2}$ **22.** 24 **23.** 2 **24.** $\dfrac{1}{5}$ **25.** -6.2 **26.** 1 **27.** 16

28. -15 **29.** The sound intensity of a jet engine is 110 decibels. **30.** At this time, the college has 2,080 parking spaces. **31.** The string section is made up of 54 musicians. **32.** The developer donated 44 acres of land to the city. **33.** The smaller number is 23 and the larger number is 40. **34.** The width of the frame is 24 inches and the length is 48 inches. **35. a.** base: 6, exponent: 5 **b.** base: b, exponent: 4 **36. a.** $2x^2$ **b.** x^4 **c.** does not simplify
d. x^3 **37. a.** h^6 **b.** m^{20} **c.** b^8 **d.** x^{18} **e.** a^6b^{10}
f. $144a^{18}b^2$ **g.** $216x^{15}$ **h.** t^{15} **38.** Keep the common base 5, and add the exponents. Do not multiply the common bases to get 25.

CHAPTERS 1–8 CUMULATIVE REVIEW (page 704)

1. a. 7,535,700 **b.** 7,540,000 **2.** 3,194,754,000 barrels
3. 9,314 **4.** 3,322 **5.** 245,870 **6.** 875 **7. a.** 260 ft
b. 4,000 ft² **8.** $170 **9. a.** 1, 2, 4, 5, 10, 20 **b.** 2² · 5
10. a. 42 **b.** 7 **11.** 56 **12.** 2
13.

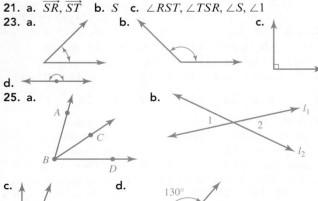

14. a. 11 **b.** 11 **c.** false **15. a.** −5 **b.** 38
c. −240 **d.** 8 **16.** 125°F **17.** −5 **18.** −200 **19.** $\frac{3}{8}$
20. $\frac{45}{54}$ **21.** $\frac{1}{7}$ **22.** $\frac{2}{5}$ **23.** $\frac{17}{18}$ **24.** 12 **25.** $42\frac{19}{22}$
26. $\frac{13}{3} = 4\frac{1}{3}$ **27. a.** He will have read $\frac{5}{6}$ of the book.
b. He will have $\frac{1}{6}$ of the book left to read. **28. a.** 1 hundredth
b. 7 **c.** 3 **d.** 7 thousandths **e.** 304.82 **29.** 658.04007
30. 182.894 **31.** −2,262 **32.** 3.16 **33.** 453.1 **34.** 13.60
35. 270 ÷ 90 = 27 ÷ 9 = 3 **36.** 67.5 mm **37. a.** 0.76
b. 0.01$\overline{5}$ **38.** −7 **39.** $\frac{9}{7}$ **40.** $93.75 **41.** 18.9 **42.** $1\frac{1}{3}$ hr
43. 7.5 g **44.** about 16 lb **45.** $\frac{1}{4}$, 25%, 0.$\overline{3}$, $\frac{21}{500}$, 0.042
46. 52% **47.** 65 **48.** $37.20, $210.80 **49.** 820 **50.** $556
51. a. the 18–49 age group **b.** 328 people **52.** mean: 6,
median: 5, mode: 10 **53.** $\frac{1}{2}$ **54.** $\frac{1}{12}$ **55.** −52
56. a. $x − 4$ **b.** $2w + 50$ **57. a.** $−15x$ **b.** $28x^2$
58. a. $−6x + 8$ **b.** $15x − 10y + 20$ **59. a.** $5x$ **b.** $12a^2$
c. $−x − y$ **d.** $29x − 36$ **60.** It is not a solution. **61.** −5
62. −16 **63.** 4 **64.** 18 **65.** She must observe 21 more shifts.
66. The length is 84 feet and the width is 21 feet. **67. a.** base: 8,
exponent: 9 **b.** base: a, exponent: 3 **68. a.** p^9 **b.** t^{15}
c. x^5y^7 **d.** $81a^8$ **e.** $108p^{12}$ **f.** $(−2.6)^{16}$

STUDY SET SECTION 9.1 (page 716)

1. point, line, plane **3.** midpoint **5.** angle **7.** protractor
9. right **11.** 180° **13.** Adjacent **15.** congruent
17. 90° **19. a.** one **b.** line
21. a. $\overrightarrow{SR}$, $\overrightarrow{ST}$ **b.** S **c.** $\angle RST$, $\angle TSR$, $\angle S$, $\angle 1$
23. a. b. c.
d.
25. a. b.

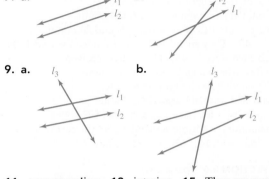

c. d.

27. congruent **29. a.** false **b.** false **c.** false **d.** true
31. true **33.** false **35.** line **37.** ray **39.** angle

41. degree **43.** congruent
45. a. b. c.

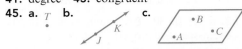

47. a. 2 **b.** 3 **c.** 1 **d.** 6 **49.** 50° **51.** 25° **53.** 75°
55. 130° **57.** right **59.** acute **61.** straight **63.** obtuse
65. 10° **67.** 27.5° **69.** 70° **71.** 65° **73.** 30°, 60°, 120°
75. 25°, 115°, 65° **77.** 60° **79.** 75° **81. a.** true
b. false, a segment has two endpoints **c.** false, a line does
not have an endpoint **d.** false, point G is the vertex of the
angle **e.** true **f.** true **83.** 40° **85.** 135° **87. a.** 50°
b. 130° **c.** 230° **d.** 260° **89. a.** 66° **b.** 156° **91.** 141°
93. 1° **95.** yes **97.** no **99. a.** about 80° **b.** about 30°
c. about 65° **101. a.** 27° **b.** 30° **103. a.** obtuse angle
b. vertical angles **c.** right angle **d.** vertical angles
e. acute angle **f.** vertical angles **109.** 40 **111.** 15

STUDY SET SECTION 9.2 (page 727)

1. coplanar, noncoplanar **3.** Perpendicular **5.** alternate
7. a. b.

9. a. b.

11. corresponding **13.** interior **15.** They are perpendicular.
17. right **19.** perpendicular **21. a.** $\angle 1$ and $\angle 5$, $\angle 4$ and
$\angle 8$, $\angle 2$ and $\angle 6$, $\angle 3$ and $\angle 7$ **b.** $\angle 3$, $\angle 4$, $\angle 5$, and $\angle 6$
c. $\angle 3$ and $\angle 5$, $\angle 4$ and $\angle 6$ **23.** m($\angle 1$) = 130°, m($\angle 2$) = 50°,
m($\angle 3$) = 50°, m($\angle 5$) = 130°, m($\angle 6$) = 50°, m($\angle 7$) = 50°,
m($\angle 8$) = 130° **25.** $\angle 1 \cong \angle X$, $\angle 2 \cong \angle N$ **27.** 12°, 40°, 40°
29. 10°, 50°, 130° **31. a.** 50°, 135°, 45°, 85° **b.** 180°
c. 180° **33.** vertical angles: $\angle 1 \cong \angle 2$; alternate interior angles:
$\angle B \cong \angle D$, $\angle E \cong \angle A$ **35.** 40°, 40°, 140° **37.** 12°, 70°, 70°
39. a. $x = 105$°; 115°, 65° **b.** $x = 25$°; 65°, 65°
41. a. $x = 15$°; 55°, 55° **b.** $x = 5$°; 55°, 125°
43. The plummet string should hang perpendicular to the top
of the stones. **45.** 50° **47.** The strips of wallpaper should
be hung on the wall parallel to each other, and they should be
perpendicular to the floor and the ceiling. **49.** 75°, 105°, 75°
51. parallel lines **53.** The angle of depression and the angle
of elevation are alternate interior angles of parallel lines. If two
parallel lines are cut by a transversal, alternate interior angles
are congruent. **61.** 72 **63.** 45% **65.** yes **67.** $\frac{1}{3}$

STUDY SET SECTION 9.3 (page 738)

1. polygon **3.** vertex **5.** equilateral, isosceles, scalene
7. hypotenuse, legs **9.** addition
11. a. b. c. d.

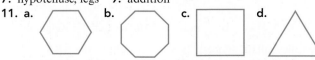

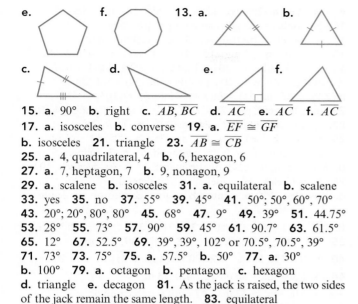

e. **f.** **13. a.** **b.**

c. **d.** **e.** **f.**

15. a. 90° **b.** right **c.** $\overline{AB}, \overline{BC}$ **d.** $\overline{AC}$ **e.** $\overline{AC}$ **f.** $\overline{AC}$
17. a. isosceles **b.** converse **19. a.** $\overline{EF} \cong \overline{GF}$
b. isosceles **21.** triangle **23.** $\overline{AB} \cong \overline{CB}$
25. a. 4, quadrilateral, 4 **b.** 6, hexagon, 6
27. a. 7, heptagon, 7 **b.** 9, nonagon, 9
29. a. scalene **b.** isosceles **31. a.** equilateral **b.** scalene
33. yes **35.** no **37.** 55° **39.** 45° **41.** 50°; 50°, 60°, 70°
43. 20°; 20°, 80°, 80° **45.** 68° **47.** 9° **49.** 39° **51.** 44.75°
53. 28° **55.** 73° **57.** 90° **59.** 45° **61.** 90.7° **63.** 61.5°
65. 12° **67.** 52.5° **69.** 39°, 39°, 102° or 70.5°, 70.5°, 39°
71. 73° **73.** 75° **75. a.** 57.5° **b.** 50° **77. a.** 30°
b. 100° **79. a.** octagon **b.** pentagon **c.** hexagon
d. triangle **e.** decagon **81.** As the jack is raised, the two sides
of the jack remain the same length. **83.** equilateral
85. a = 32°, b = 63°, c = 42°, d = 48° **87. a.** hexagon,
pentagon **b.** hexagon: 20, pentagon: 12 **93.** 22
95. 40% **97.** 0.10625

STUDY SET SECTION 9.4 (page 747)

1. hypotenuse, legs **3.** Pythagorean **5.** a^2, b^2, c^2
7. right **9. a.** $\overline{BC}$ **b.** $\overline{AB}$ **c.** $\overline{AC}$ **11.** 64, 100, 100
13. 10 ft **15.** 13 m **17.** 73 mi **19.** 137 cm
21. 24 cm **23.** 80 m **25.** 20 m **27.** 19 m
29. $\sqrt{11}$ cm ≈ 3.32 cm **31.** $\sqrt{208}$ m ≈ 14.42 m
33. $\sqrt{90}$ in. ≈ 9.49 in. **35.** $\sqrt{20}$ in. ≈ 4.47 in. **37.** no
39. yes **41.** $\sqrt{2}$ **43.** $\sqrt{10}$ **45.** 12 ft **47.** 25 in.
49. $\sqrt{16,200}$ ft ≈ 127.28 ft **51.** yes, $\sqrt{1,288}$ ft ≈ 35.89 ft
57. no **59.** no **61.** no **63.** no

STUDY SET SECTION 9.5 (page 757)

1. Congruent **3.** congruent **5.** similar **7. a.** No, they are
different sizes. **b.** Yes, they appear to have the same shape.
9. *PRQ* **11.** *MNO* **13.** $\angle A \cong \angle B, \angle Y \cong \angle T$,
$\angle Z \cong \angle R, \overline{YZ} \cong \overline{TR}, \overline{AZ} \cong \overline{BR}, \overline{AY} \cong \overline{BT}$
15. congruent **17.** angle, angle **19.** 100 **21.** 5.4
23. proportional **25.** congruent **27.** is congruent to

29.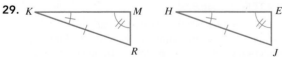

31. $\overline{DF}, \overline{AB}, \overline{EF}, \angle D, \angle B, \angle C$ **33. a.** $\angle B \cong \angle M$,
$\angle C \cong \angle N, \angle D \cong \angle O, \overline{BC} \cong \overline{MN}, \overline{CD} \cong \overline{NO}, \overline{BD} \cong \overline{MO}$
b. 72° **c.** 10 ft **d.** 9 ft **35.** yes, SSS **37.** not necessarily
39. a. $\angle L \cong \angle H, \angle M \cong \angle J, \angle R \cong \angle E$ **b.** *MR, LR, LM*
c. *HJ, JE, LR* **41.** yes **43.** not necessarily **45.** yes
47. not necessarily **49.** yes **51.** not necessarily **53.** 8, 35
55. 60, 38 **57.** true **59.** false: the angles must be between
congruent sides **61.** yes, SSS **63.** yes, SAS **65.** yes, ASA
67. not necessarily **69.** 80°, 2 yd **71.** 19°, 14 m **73.** 6 mm
75. 50° **77.** $\frac{25}{6} = 4\frac{1}{6}$ **79.** 16 **81.** 17.5 cm **83.** 59.2 ft
85. 36 ft **87.** 34.8 ft **89.** 1,056 ft **93.** 189 **95.** 21

STUDY SET SECTION 9.6 (page 769)

1. quadrilateral **3.** rectangle **5.** rhombus **7.** trapezoid,
bases, isosceles **9. a.** four; *A, B, C, D* **b.** four;
$\overline{AB}, \overline{BC}, \overline{CD}, \overline{DA}$ **c.** two; $\overline{AC}, \overline{BD}$ **d.** yes, no, no, yes
11. a. $\overline{VU}$ **b.** ∥ **13. a.** right **b.** parallel **c.** length
d. length **e.** midpoint **15.** rectangle **17. a.** no **b.** yes
c. no **d.** yes **e.** no **f.** yes **19. a.** isosceles
b. $\angle J, \angle M$ **c.** $\angle K, \angle L$ **d.** $\angle M, \angle L, \overline{ML}$
21. The four sides of the quadrilateral are the same length.
23. the sum of the measures of the angles of a polygon; the
number of sides of the polygon **25. a.** square **b.** rhombus
c. trapezoid **d.** square **27. a.** 90° **b.** 9 **c.** 18 **d.** 18
29. a. 42° **b.** 95° **31. a.** 9 **b.** 70° **c.** 110° **d.** 110°
33. 2,160° **35.** 3,240° **37.** 1,080° **39.** 1,800° **41.** 5
43. 7 **45.** 13 **47.** 14 **49. a.** 30° **b.** 30° **c.** 60°
d. 8 cm **e.** 4 cm **51.** 40°; m(∠A) = 90°, m(∠B) = 150°,
m(∠C) = 40°, m(∠D) = 80° **53. a.** trapezoid **b.** square
c. rectangle **d.** trapezoid **e.** parallelogram **55.** 540°
57. NBA: rectangle; International: trapezoid
63. two hundred fifty-four thousand, three hundred nine
65. eighty-two million, four hundred fifteen

THINK IT THROUGH (page 778)

about 108 ft²

STUDY SET SECTION 9.7 (page 783)

1. perimeter **3.** area **5.** area **7.** 8 ft · 16 ft = 128 ft²
9. a. $P = 4s$ **b.** $P = 2l + 2w$
11. a. **b.**

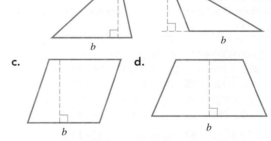

c. **d.**

13. a rectangle and a triangle **15. a.** square inch **b.** 1 m²
17. 32 in. **19.** 23 mi **21.** 62 in. **23.** 94 in. **25.** 15 ft
27. 5 m **29.** 16 cm² **31.** 6.25 m² **33.** 144 in.²
35. 1,000,000 mm² **37.** 27,878,400 ft² **39.** 1,000,000 m²
41. 135 ft² **43.** 11,160 ft² **45.** 25 in.² **47.** 27 cm²
49. 7.5 in.² **51.** 10.5 mi² **53.** 40 ft² **55.** 91 cm² **57.** 4 m
59. 12 cm **61.** 36 m **63.** 11 mi **65.** 102 in.² **67.** 360 ft²
69. 75 m² **71.** 75 yd² **73.** $1,200 **75.** $4,875 **77.** length
15 in. and width 5 in.; length 16 in. and width 4 in. (answers
may vary) **79.** sides of length 5 m **81.** base 5 yd and height
3 yd (answers may vary) **83.** length 5 ft and width 4 ft; length
20 ft and width 3 ft (answers may vary) **85.** 60 cm²
87. 36 m **89.** $28\frac{1}{3}$ ft **91.** 36 m **93.** x = 3.7 ft, y = 10.1 ft;
50.8 ft **95.** 80 + 1 = 81 trees **97.** vinyl **99.** $361.20
101. $192 **103.** 111,825 mi² **105.** 51 sheets **111.** 6t
113. $-2w + 4$ **115.** $-\frac{5}{8}x$ **117.** $9r - 16$

STUDY SET SECTION 9.8 (page 793)

1. radius **3.** diameter **5.** circumference **7.** twice
9. $\overline{OA}, \overline{OC}, \overline{OB}$ **11.** $\overline{DA}, \overline{DC}, \overline{AC}$ **13.** $\overset{\frown}{ABC}, \overset{\frown}{ADC}$
15. a. Multiply the radius by 2. **b.** Divide the diameter by 2.
17. π **19.** square 6 **21.** arc AB **23. a.** multiplication:
$2 \cdot \pi \cdot r$ **b.** raising to a power and multiplication: $\pi \cdot r^2$
25. 8π ft ≈ 25.1 ft **27.** 12π m ≈ 37.7 m **29.** 50.85 cm
31. 31.42 m **33.** 9π in.$^2 \approx 28.3$ in.^{2} **35.** 81π in.$^2 \approx 254.5$ in.^{2}
37. 128.5 cm^2 **39.** 57.1 cm^2 **41.** 27.4 in.^{2} **43.** 66.7 in.^{2}
45. 50π yd ≈ 157.08 yd **47.** 6π in. ≈ 18.8 in.
49. 20.25π mm$^2 \approx 63.6$ mm^2 **51.** 40 in., 31.4 in.
53. 96 ft, 50.3 ft **55. a.** 1 in. **b.** 2 in.
c. 2π in. ≈ 6.28 in. **d.** π in.$^2 \approx 3.14$ in.^{2}
57. π mi$^2 \approx 3.14$ mi^2 **59.** 32.66π ft ≈ 102.60 ft
61. 13 times **63.** 4π ft$^2 \approx 12.57$ ft^2; 0.25π ft$^2 \approx 0.79$ ft^2; 6.25%
65. 64π ft$^2 \approx 201.1$ ft^2 **71.** 90% **73.** 82.7%
75. 5.375¢ per oz **77.** five

STUDY SET SECTION 9.9 (page 802)

1. volume **3.** cone **5.** cylinder **7.** pyramid

9. 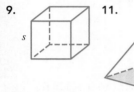 **11.** 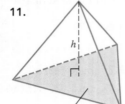 **13.**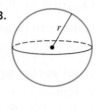
Base

15. cubic inches, mi^3, m^3 **17. a.** perimeter **b.** volume
c. area **d.** volume **e.** area **f.** circumference **19. a.** 50π
b. $\dfrac{500}{3}\pi$ **21. a.** cubic inch **b.** 1 cm^3 **23.** a right angle
25. 27 **27.** 1,000,000,000 **29.** 56 ft^3 **31.** 125 in.^{3}
33. 120 cm^3 **35.** 1,296 in.^{3} **37.** 700 yd^3 **39.** 32 ft^3
41. 69.72 ft^3 **43.** 6 yd^3 **45.** 192π ft$^3 \approx 603.19$ ft^3
47. $3,150\pi$ cm$^3 \approx 9,896.02$ cm^3 **49.** 39π m$^3 \approx 122.52$ m^3
51. 189π yd$^3 \approx 593.76$ yd^3 **53.** 288π in.$^3 \approx 904.8$ in.^{3}
55. $\dfrac{32}{3}\pi$ cm$^3 \approx 33.5$ cm^3 **57.** 486π in.$^3 \approx 1,526.81$ in.^{3}
59. 423π m$^3 \approx 1,357.17$ m^3 **61.** 60 cm^3
63. 100π cm$^3 \approx 314.16$ cm^3 **65.** 400 m^3 **67.** 48 m^3
69. 576 cm^3 **71.** 180π cm$^3 \approx 565.49$ cm^3
73. $\dfrac{1}{8}$ in.$^3 = 0.125$ in.^{3} **75.** 167 ft^3 **77.** 63π ft$^3 \approx 197.92$ ft^3
79. $\dfrac{32,000}{3}\pi$ ft$^3 \approx 33,510.32$ ft^3 **81.** 8:1
83. a. $2,250\pi$ in.$^3 \approx 7,068.58$ in.^{3} **b.** 30.6 gal **85.** $\dfrac{3}{64}$ in.$^3 =$
0.046875 in.^{3} **91.** -42 **93.** -4 **95.** $\dfrac{1}{5}$ or 1:5 **97.** 2,400 mm

CHAPTER 9 REVIEW (page 806)

1. points C and D, line CD, plane GHI **2. a.** 6 units
b. E **c.** yes **3.** $\angle ABC, \angle CBA, \angle B, \angle 1$ **4. a.** acute
b. B **c.** $\overrightarrow{BA}$ and $\overrightarrow{BC}$ **d.** 48° **5.** $\angle 1$ and $\angle 2$ are acute,
$\angle ABD$ and $\angle CBD$ are right angles, $\angle CBE$ is obtuse,
and $\angle ABC$ is a straight angle. **6.** yes **7.** yes
8. a. obtuse angle **b.** right angle **c.** straight angle
d. acute angle **9.** 15° **10.** 150° **11. a.** m($\angle 1$) = 65°
b. m($\angle 2$) = 115° **12. a.** 39° **b.** 90° **c.** 51° **d.** 51°

e. yes **13. a.** 20° **b.** 125° **c.** 55° **14.** 19° **15.** 37°
16. No, only two angles can be supplementary.
17. a. parallel **b.** transversal **c.** perpendicular
18. $\angle 4$ and $\angle 6$, $\angle 3$ and $\angle 5$ **19.** $\angle 1$ and $\angle 5$, $\angle 4$ and $\angle 8$, $\angle 2$
and $\angle 6$, $\angle 3$ and $\angle 7$ **20.** $\angle 1$ and $\angle 3$, $\angle 2$ and $\angle 4$, $\angle 5$ and
$\angle 7$, and $\angle 6$ and $\angle 8$ **21.** m($\angle 1$) = m($\angle 3$) = m($\angle 5$) =
m($\angle 7$) = 70°; m($\angle 2$) = m($\angle 4$) = m($\angle 6$) = 110°
22. m($\angle 1$) = 60°, m($\angle 2$) = 120°, m($\angle 3$) = 130°, m($\angle 4$) = 50°
23. a. 40° **b.** 50°, 50° **24. a.** 20° **b.** 110°, 70°
25. a. 11° **b.** 31°, 31° **26. a.** 23° **b.** 82°, 82°
27. a. 8, octagon, 8 **b.** 5, pentagon, 5 **c.** 3, triangle, 3
d. 6, hexagon, 6 **e.** 4, quadrilateral, 4 **f.** 10, decagon, 10
28. a. isosceles **b.** scalene **c.** equilateral
d. isosceles **29. a.** acute **b.** right **c.** obtuse **d.** acute
30. a. 90° **b.** right **c.** $\overline{XY}, \overline{XZ}$ **d.** $\overline{YZ}$ **e.** $\overline{YZ}$
f. $\overline{YZ}$ **31.** 90° **32.** 50° **33.** 71° **34.** 18°; 36°, 28°, 116°
35. 50° **36.** 56° **37.** 67° **38.** 83° **39.** 13 cm
40. 17 ft **41.** 36 in. **42.** 20 ft **43.** $\sqrt{231}$ m ≈ 15.20 m
44. $\sqrt{1,300}$ in. ≈ 36.06 in. **45.** 73 in.
46. $\sqrt{1,023}$ in. ≈ 32 in. **47.** not a right triangle
48. not a right triangle **49. a.** $\angle D$ **b.** $\angle E$ **c.** $\angle F$
d. $\overline{DF}$ **e.** $\overline{DE}$ **f.** $\overline{EF}$ **50. a.** 32° **b.** 61° **c.** 6 in.
d. 9 in. **51.** congruent, SSS **52.** congruent, SAS
53. not necessarily congruent **54.** congruent, ASA
55. yes **56.** yes **57.** 4, 28 **58.** 65 ft **59. a.** trapezoid
b. square **c.** parallelogram **d.** rectangle **e.** rhombus
f. rectangle **60. a.** 15 cm **b.** 40° **c.** 100° **d.** 7.5 cm
e. 14 cm **61. a.** true **b.** true **c.** true **d.** false
62. a. 65° **b.** 115° **c.** 4 yd **63.** 1,080° **64.** 20 sides
65. 72 in. **66.** 86 in. **67.** 30 m **68.** 36 m **69.** 39 ft
70. a. 9 ft^2 **b.** 144 in.^{2} **71.** 9.61 cm^2 **72.** 7,500 ft^2
73. 450 ft^2 **74.** 200 in.^{2} **75.** 120 cm^2 **76.** 232 ft^2
77. 152 ft^2 **78.** 120 m^2 **79.** 8 ft **80.** 18 mm **81.** $3,281
82. $4,608 **83. a.** $\overline{CD}, \overline{AB}$ **b.** $\overline{AB}$ **c.** $\overline{OA}, \overline{OC}, \overline{OD}, \overline{OB}$
d. O **84.** 21π ft ≈ 65.97 ft **85.** 45.1 cm
86. 81π in.$^2 \approx 254.47$ in.^{2} **87.** 130.3 cm^2 **88.** 6,073.0 in.^{2}
89. 125 cm^3 **90.** 480 m^3 **91.** 1,728 mm^3
92. $\dfrac{500}{3}\pi$ in.$^3 \approx 523.60$ in.^{3} **93.** 250π in.$^3 \approx 785.40$ in.^{3}
94. 2,000 yd^3 **95.** 2,940 m^3 **96.** $\dfrac{1,024}{3}\pi$ in.$^3 \approx 1,072.33$ in.^{3}
97. 1,518 ft^3 **98.** 3.125π in.$^3 \approx 9.8$ in.^{3} **99.** 1,728 in.^{3}
100. 54 ft^3

CHAPTER 9 TEST (page 828)

1. a. 135°, obtuse **b.** 90°, right **c.** 40°, acute
d. 180°, straight **2. a.** measure **b.** length **c.** line
d. complementary **3.** D **4. a.** false **b.** true **c.** true
d. true **e.** false **5.** 20°; 60°, 60° **6.** 133°
7. a. transversal **b.** $\angle 6$ **c.** $\angle 7$ **8.** m($\angle 1$) = 155°,
m($\angle 3$) = 155°, m($\angle 4$) = 25°, m($\angle 5$) = 25°, m($\angle 6$) = 155°,
m($\angle 7$) = 25°, m($\angle 8$) = 155° **9.** 50°; 110°, 70°
10. a. 8, octagon, 8 **b.** 5, pentagon, 5 **c.** 6, hexagon, 6
d. 4, quadrilateral, 4 **11. a.** isosceles **b.** scalene
c. equilateral **d.** isosceles **12.** 70° **13.** 84° **14. a.** 12
b. 13 **c.** 90° **d.** 5 **15. a.** 10 **b.** 65° **c.** 115° **d.** 115°
16. 1,440° **17.** 188 in. **18.** 15.2 m **19.** 360 cm^2
20. $864 **21.** 144 in.^{2} **22.** 120 in.^{2} **23. a.** $\overline{RS}, \overline{XY}$
b. $\overline{XY}$ **c.** $\overline{OX}, \overline{OR}, \overline{OS}, \overline{OY}$ **24.** π **25.** 21π ft ≈ 66.0 ft
26. $(40 + 12\pi)$ ft ≈ 77.7 ft **27.** 225π m$^2 \approx 706.9$ m^2
28. $\angle R, \angle S, \angle T$; $\overline{RT}, \overline{RS}, \overline{ST}$ **29. a.** congruent, SSS

b. congruent, ASA   **c.** not necessarily congruent
d. congruent, SAS   **30. a.** 8 in.   **b.** 50°
31. a. yes   **b.** yes   **32. a.** 6 m   **b.** 12 m   **33.** 21 ft
34. a. 26 cm   **b.** $\sqrt{28}$ in. $\approx$ 5.3 in.   **35.** $\sqrt{986}$ in. $\approx$ 31.4 in.
36. 1,728 in.^{3} **37.** 216 m^3 **38.** 480 m^3
39. 1,296π in.^{3} $\approx$ 4,071.50 in.^{3} **40.** 600 in.^{3} **41.** 1,890 ft^3
42. 63π yd^3 $\approx$ 197.92 yd^3 **43.** 400 mi^3
44. $\frac{256}{3}\pi$ in.^{3} $\approx$ 268.08 in.^{3} **45.** 11,250π ft^3 $\approx$ 35,343 ft^3

CHAPTERS 1–9 CUMULATIVE REVIEW   (page 832)

1. $8,995   **2.** 2,110,000   **3.** 32,034   **4.** 11,022
5. a. 602 ft   **b.** 19,788 ft^2 **6.** 33 R 10   **7.** 48 gal
8. a. $2^2 \cdot 5 \cdot 11$ **b.** 1, 2, 3, 4, 6, 12   **9. a.** 48   **b.** 8
10. 11   **11. a.** $\{\ldots, -3, -2, -1, 0, 1, 2, 3, \ldots\}$ **b.** 3
12. a. -12 **b.** 20   **c.** -64 **d.** 4   **e.** -16 **f.** 16
13. $-$$140   **14.** -2 **15. a.** $\frac{5}{4}$ **b.** $\frac{18}{48}$ **c.** $\frac{8}{9}$ **d.** $\frac{15}{2}$
16. 9 oz   **17.** $-\frac{3}{70}$ **18.** $\frac{1}{6}$ **19.** $\frac{3}{20}$ **20.** $-\frac{11}{20}$ **21.** $142\frac{7}{15}$
22. $\frac{38}{29} = 1\frac{9}{29}$ **23.** $\frac{3}{32}$ fluid oz   **24.** $13\frac{3}{4}$ cups   **25.** $\frac{8}{9}$
26. a. 3.1416   **b.** >
c. six million, five hundred ten thousand, three hundred forty-five and seven hundred ninety-eight thousandths
d. $7,000 + 400 + 90 + 8 + \frac{6}{10} + \frac{4}{100} + \frac{6}{1,000} + \frac{1}{10,000}$
27. 145.188   **28.** 3,803.61   **29.** -25.6 **30.** 17.05
31. 0.053   **32.** 22.3125   **33.** $2,712.50   **34. a.** 899,708
b. 0.899708   **35.** $18,000 \div 9 = 2,000$ **36.** -9.32
37. $0.1\overline{3}$ **38. a.** -2 **b.** $\frac{7}{9}$
39.

40. a. $\frac{3}{7}$ **b.** $\frac{1}{4}$ **41.** the smaller board   **42.** $6\frac{1}{2} = 6.5$
43. 125,000   **44.** 75 ft   **45. a.** 14 ft   **b.** 13.25 lb $= 13\frac{1}{4}$ lb
c. 120 quarts   **d.** 750 min   **46. a.** 1,538 g   **b.** 0.5 L
c. 0.000003 km   **47.** 240 km   **48. a.** about 4 m/gal
b. 11,370,000 L   **49.** about 4.5 kg   **50.** 167°F
51. $0.57, \frac{57}{100}, 0.1\%, \frac{1}{1,000}, 33\frac{1}{3}\%, 0.\overline{3}$ **52. a.** 93%   **b.** 7%
53. 67.5   **54.** 120   **55.** 85%   **56.** $205, $615
57. $1,159.38   **58.** 500%   **59.** $21   **60.** $1,567.50
61. a. 375,000 vehicles   **b.** 250,000 vehicles
c. 25,000 vehicles   **62. a.** 18%   **b.** 2,920,000
63. a. food: about $20.5 billion   **b.** about $2.2 billion
c. about $9.5 billion   **64.** mean: 0.86 oz, median: 0.855 oz,
mode: 0.85oz   **65. a.** $\frac{3}{13}$ **b.** $\frac{1}{3}$ **c.** $\frac{5}{36}$ **66.** 5
67. a. $2x - 16$ **b.** $75s + 6$ **68. a.** $48a$ **b.** $42b$
c. $27t - 90$ **d.** $32x - 40y + 8$ **69. a.** $3x$ **b.** $6c^2$
c. $-8m + 6n$ **d.** $-12x + 8$ **70.** It is not a solution.
71. -24 **72.** 5   **73.** 89   **74.** -11
75. She must make 7 more 6-hour classroom visits.
76. a. base: 4, exponent: 8   **b.** base: s, exponent: 4
77. a. s^{10} **b.** a^{35} **c.** r^5t^9 **d.** $8b^9c^{18}$ **e.** y^{22} **f.** $(-5.5)^{36}$
78. a. acute   **b.** right   **c.** obtuse   **d.** 180°   **79. a.** 75°
b. 15°   **80. a.** 50°   **b.** 50°   **c.** 130°   **d.** 50°   **81. a.** 75°
b. 30°   **c.** 105°   **d.** 105°   **82.** 46°, 134°   **83.** 73°
84. 26 m   **85.** yes   **86.** 42 ft   **87.** 540°   **88.** 48 m, 144 m^2
89. 126 ft^2 **90.** 91 in.^{2} **91.** 144 in.^{2} **92.** circumference:
14π cm $\approx$ 43.98 cm, area: 49π cm^2 $\approx$ 153.94 cm^2
93. 98.31 yd^2 **94.** 6,480 in.^{3} **95.** 972π in.^{3} $\approx$ 3,053.63 in^3
96. 48π m^3 $\approx$ 150.80 m^3 **97.** 20π ft^3 $\approx$ 62.83 ft^3
98. 1,728 in.^{3}

INDEX

APPLICATIONS INDEX

Examples that are applications are shown with **boldface** page numbers.
Exercises that are applications are shown with lightface page numbers.

Units of Measurement

American Units of Length

12 inches (in.) = 1 foot (ft)

3 ft = 1 yard (yd)

36 in. = 1 yd

5,280 ft = 1 mile (mi)

Metric Units of Length

1 kilometer (km) = 1,000 meters (m)

1 hectometer (hm) = 100 m

1 dekameter (dam) = 10 m

1 decimeter (dm) = $\frac{1}{10}$ m

1 centimeter (cm) = $\frac{1}{100}$ m

1 millimeter (mm) = $\frac{1}{1,000}$ m

Equivalent Lengths

1 in. = 2.54 cm	1 cm ≈ 0.39 in.
1 ft ≈ 0.30 m	1 m ≈ 3.28 ft
1 yd ≈ 0.91 m	1 m ≈ 1.09 yd
1 mi ≈ 1.61 km	1 km ≈ 0.62 mi

American Units of Weight

16 ounces (oz) = 1 pound (lb)

2,000 lb = 1 ton

Metric Units of Mass

1 kilogram (kg) = 1,000 grams (g)

1 hectogram (hg) = 100 g

1 dekagram (dag) = 10 g

1 decigram (dg) = $\frac{1}{10}$ g

1 centigram (cg) = $\frac{1}{100}$ g

1 milligram (mg) = $\frac{1}{1,000}$ g

Equivalent Weights and Masses

1 oz ≈ 28.35 g	1 g ≈ 0.035 oz
1 lb ≈ 0.45 kg	1 kg ≈ 2.20 lb

American Units of Capacity

1 cup (c) = 8 fluid ounces (fl oz)

1 quart (qt) = 2 pints (pt)

1 pt = 2 c

1 gallon (gal) = 4 qts

Metric Units of Capacity

1 kiloliter (kL) = 1,000 liters (L)

1 hectoliter (hL) = 100 L

1 dekaliter (daL) = 10 L

1 deciliter (dL) = $\frac{1}{10}$ L

1 centiliter (cL) = $\frac{1}{100}$ L

1 milliliter (mL) = $\frac{1}{1,000}$ L

Equivalent Capacities

1 fl oz ≈ 29.57 mL	1 L ≈ 33.81 fl oz
1 pt ≈ 0.47 L	1 L ≈ 2.11 pt
1 qt ≈ 0.95 L	1 L ≈ 1.06 qt
1 gal ≈ 3.79 L	1 L ≈ 0.264 gal

Geometric Formulas

Pythagorean Theorem: If the length of the hypotenuse of a right triangle is c and the lengths of its legs are a and b, then $a^2 + b^2 = c^2$.

Area Formulas

square	$A = s^2$
rectangle	$A = lw$
parallelogram	$A = bh$
triangle	$A = \frac{1}{2}bh$
trapezoid	$A = \frac{1}{2}h(b_1 + b_2)$

Circumference of a Circle: $C = \pi D$ or $C = 2\pi r$

$\pi = 3.14159\ldots$

Volume Formulas

cube	$V = s^3$
rectangular solid	$V = lwh$
prism	$V = Bh$
sphere	$V = \frac{4}{3}\pi r^3$
cylinder	$V = \pi r^2 h$
cone	$V = \frac{1}{3}\pi r^2 h$
pyramid	$V = \frac{1}{3}Bh$

B represents the area of the base.